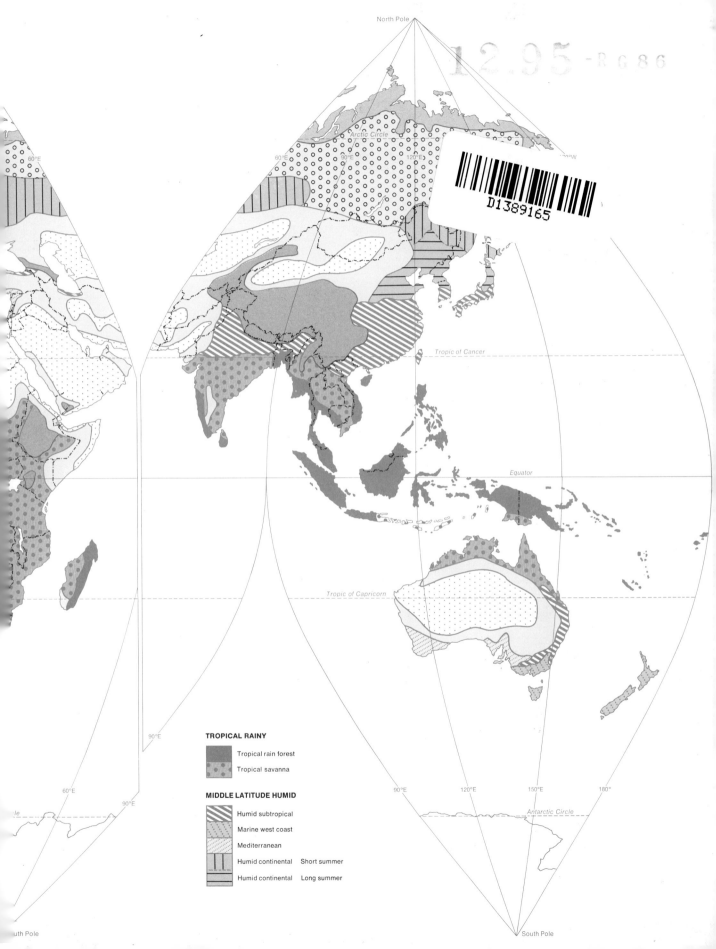

North Pole

Arctic Circle

60°E

90°E

120°E

90°W

Tropic of Cancer

Equator

Tropic of Capricorn

60°E

90°E

90°E

120°E

150°E

180°

Antarctic Circle

South Pole

South Pole

TROPICAL RAINY

Tropical rain forest

Tropical savanna

MIDDLE LATITUDE HUMID

Humid subtropical

Marine west coast

Mediterranean

Humid continental   Short summer

Humid continental   Long summer

Regional Geography of the World

Jesse H. Wheeler, Jr. University of Missouri
J. Trenton Kostbade University of Missouri
Richard S. Thoman Queen's University

Regional Geography of the World

Third Edition

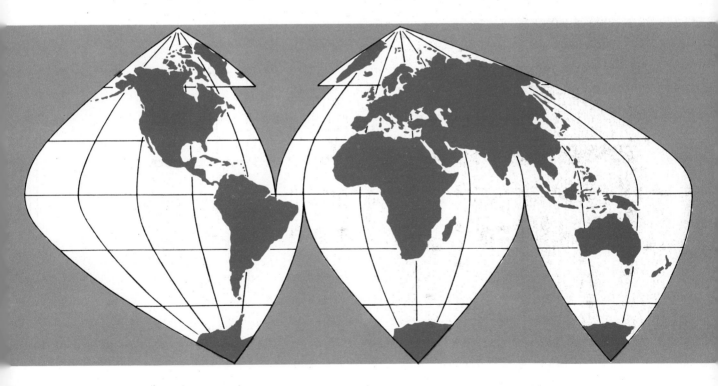

Holt, Rinehart and Winston, Inc.

New York   Chicago   San Francisco   Atlanta   Dallas
Montreal   Toronto   London   Sydney

PREFACE

This third edition, like its predecessors, seeks to assist college and university students in acquiring certain basic ideas and supporting facts about contemporary world geography which a person with a college education might reasonably be expected to know. Its aim, in short, is general education in world geography. It surveys the world importance, geographical characteristics, and major problems of eight world regions—Europe, the Soviet Union, the Middle East, the Orient, the Pacific World, Africa, Latin America, and Anglo-America. Attention is also given to important individual countries and regional groups of countries within each world region. Pertinent background material on systematic physical and cultural geography is presented at appropriate places. A general introduction to geography as a field of study is presented in Chapters 1–3. The book is designed as a text for courses in geography on an introductory college level.

In the third edition the basic structure of the book remains unchanged, although the material has been updated, many sections have been rewritten, a new chapter on geographic processes has been incorporated, and the chapter on the British Isles has been expanded to include general concepts of urban geography as applied to the internal structure of London. A list of selected readings and references, comprised primarily of books and articles published since 1960, has been prepared for each chapter. Extensive changes and improvements have been made in the maps. All maps carried over from earlier editions have been redrawn for presentation in color and many new maps or diagrams have been added. In addition, the number of photographs has been significantly increased.

The text is adaptable to a variety of teaching situations and the authors have tried to make it sufficiently flexible in organization to fit a multiplicity of course structures, intensities, and sequences. Some teachers may wish to follow the present order of chapters, either in a single course or in a sequence of courses, while other teachers will vary the order of chapters. A major theme woven into the various sections of the book is the worldwide impact of Westernization, beginning in Europe and spreading from there to other regions. In conformity with this theme, Europe comes first in the sequence of regional chapters, but the book has been organized in such a way that it is perfectly feasible to commence a course with any of the other regions.

If the time available does not permit the entire book to be covered, selected parts of the book may easily be combined to provide text material of the appropriate length. For example, teachers may wish to assign the general introductory chapters on the various regions, together with selected other chapters or parts of chapters, to provide a very general survey of the world and intensive case studies of certain countries or regions. Teachers may wish to assign certain chapters to be discussed in class, may assign other chapters to be read without class discussion, and may wish to lecture on the subject matter of some chapters without assigning them. General introductory material in Chapters 1 to 3 may or may not be assigned, according to the amount of previous training in geography that students have received. A workbook and study guide is available to aid students in systematic note-taking and self-testing on text material.

Instructors may not find it necessary to require students to purchase an atlas. Nearly all of the place names mentioned in the text are shown on maps. A considerable number of maps showing the world or regional distribution of population, types of climate, major crops, minerals, industrial concentrations, and other elements of geographical significance are included.

The photographs in the volume are not intended to give a comprehensive coverage, although it is hoped that each photo will convey at least one significant idea about the country or region portrayed. Teachers are urged to make such use as time permits of movies, slides, and film strips to illustrate, amplify, and supplement the ideas contained in the text and to give the students the "feel" of the areas considered.

Statistics have been employed in various places to reinforce important concepts. It was not felt necessary in an introductory textbook to give full documentation for all statistics. They have been drawn from a wide variety of standard sources.

Although most of the text has been specially written, a variety of readings, mostly short, are included. In editing the readings, minor changes in spelling, capitalization, and punctuation have been made in order to give uniformity to the book as a whole. Many of the sectional headings within the body of the readings represent editorial additions.

The basic core of ideas in *Regional Geography of the World* was to a large extent a product of many hours of discussion among the three authors during a fruitful association at the University of Missouri extending over a period of several years. The senior author has carried the overall editorial responsibility for the book and the chief responsibility for the planning and supervision of the cartography, the selection and captioning of photographs, and the preparation of chapter bibliographies. Chapters 1, 3, 4, 12, 13, 14, 15, 16, 17, 24, 25, and 26 were prepared by the senior author. Mr. Kostbade prepared Chapters 8, 9, 10, 11, 18, 19, 23, 29, and 30; in addition he was mainly responsible for the following chapters on which the senior author also did some work: 2, 5, 6, 20, 21, and 22. Chapters 27 and 28 were prepared by Mr. Thoman, while Chapter 7 was prepared jointly by Mr. Thoman and the senior author. Mr. Thoman also assisted in revising Chapters 6 and 8 for the third edition.

Special thanks are due Mr. Philip True, who revised Chapter 20, "The Chinese Realm," and Chapter 21, "Japan and Korea," for this edition. The authors are fortunate to have had the benefit of his knowledge, experience, and careful work. Mr. Herbert M. Eder revised the material on the Mexican *ejido* for Chapters 27 and 28. Mr. Walter A. Schroeder prepared sketches for the diagrams in Chapter 2; and Mr. John A. Soulsby contributed a number of photographs of France and Germany. Preliminary versions of the map panels on London and New York were executed by Mrs. Dorothy Woodson and Mr. Max Gilland, respectively. Mrs. Evelyn Thoman did library research for the updating of certain chapters. The authors gratefully acknowledge these various contributions.

The third edition owes much to the sustained effort and skill of Miss Rose McCall and Mrs. Nancy Ware Whitt. Miss McCall updated the tables and much of the statistical material in the body of the text; and Mrs. Whitt supplied accurate and tireless secretarial assistance.

The authors extend sincere thanks to the many geographers who have been kind enough to offer their comments on portions of the manuscript for the respective editions, or who have otherwise rendered assistance. Critical appraisal of the manuscript for the third edition was particularly searching and thorough. Suggestions of critics for all three editions have proved exceedingly useful and were heeded in the great majority of cases. The critics are, of course, absolved of any responsibility for errors of fact or interpretation in the book. Those to whom acknowledgment is due include the following:

Third edition: William E. Dooley, Herbert M. Eder, Paul Ward English, Howard F. Hirt, George W. Hoffman, William Horbaly, William E. Martin, Alexander Melamid, Aloys A. Michel, F. J. Monkhouse, Harry J. Schaleman, Jr., Walter A. Schroeder, John A. Soulsby, R. W. Steel, Fredric R. Steinhauser, Philip True, and Irene S. van Dongen.

Second edition: David H. K. Amiran, James E. Collier, Kenneth B. Cumberland, Andreas Grotewold, Lois Grotewold, Howard F. Hirt, Alexander Melamid.

First edition: John E. Brush, Wesley Calef, the late Charles C. Colby, Jerome D. Fellmann, Gerard Foster, Norton S. Ginsburg, William A. Hance, Robert A. Harper, Chauncy D. Harris, Howard F. Hirt, William Horbaly, Rayburn W. Johnson, James B. Kenyon, H. Louis Kostanick, Trevor Lloyd, James S. Matthews, Edwin S. Munger, Howard J. Nelson, Jerome P. Pickard, the late Robert S. Platt, Paul H. Sisco, Joseph E. Spencer, Edward J. Taaffe, Philip True, Philip L. Wagner, R. Kenton Wibking, and James R. Wray.

Thanks are expressed to the individuals, firms, and government agencies or other organizations who permitted the use of copyrighted textual matter, photographs, and maps. Such permissions are separately acknowledged.

The authors sincerely hope the third edition will be of service to college and university students of the 1970s as they attempt to gain perspective on the turbulent and varied world in which they find themselves.

Columbia, Missouri
March 31, 1968

J.H.W.,Jr.
J.T.K.
R.S.T.

CONTENTS

PART I

INTRODUCTORY
CONCEPTS

MAJOR WORLD REGIONS

Some
Introductory Concepts
of Geography

1 The main reason for studying geography is to gain a better understanding and appreciation of the world in which we live. Of course, many subjects other than geography contribute to this end. In fact, most school subjects are concerned in one way or another with enhancing our understanding and appreciation of the world about us. Geography, however, has one characteristic that tends to distinguish it from most other subjects, namely, that it centers attention on the study and interpretation of particular *areas* in the world. A well-known geographer, the late George B. Cressey, stated this idea as follows: "It is the task of geography . . . to draw information from widely scattered sources, and to give it a new significance as applied to the understanding of specific areas."[1]

The term "area," as used by geographers, may refer to any portion of the earth's surface. The largest of all areas is, of course, the entire world. However, in most

---

[1] By permission from *Asia's Lands and Peoples,* by George B. Cressey. Copyright, 1944, 1951, 1963 (New York: McGraw-Hill Book Company, Inc.), p. 61.

respects the world is much too large and complicated to be readily comprehended all at once. For purposes of study it generally must be subdivided into areas of a smaller size. In this book primary attention is given to two kinds of areas: the *countries* as outlined on a political map and the *major world regions* in which the individual countries are located.

INTRODUCTORY IDEAS ABOUT REGIONS

As customarily used by geographers, the term "region" refers to an area of considerable size which has a substantial degree of internal unity or homogeneity and which differs in significant respects from adjoining areas. A region may be wholly contained within a single country, or it may include several different countries or parts of countries. Many kinds of regions are employed in geographical study. They may or may not be defined and delimited with mathematical precision. Some regions are distinguished on a physical basis; others have a cultural, economic, social, or political basis; and still others are composites. Regions may be defined on the basis of a single characteristic or of multiple characteristics. In defining and interpreting regions, geographers make use of three fundamental ideas: the idea of *generalized areal homogeneity*, the idea of *coherent functional organization*, and the idea of *regional distinctiveness*. Each of these ideas is associated with a particular category of regions.

1. The *homogeneous region,* also known as the *uniform* or *formal region,* is an area having a unitary quality "because of the relative homogeneity of a certain characteristic, or characteristics, throughout. Put another way, it is an area within which a certain characteristic or combination of characteristics falls within a specified range of variation. Examples are a lowland region, containing only land below a certain elevation; a lowland forest region, containing only land below a certain elevation and forested to a specified degree; an industrial region, containing only statistical units with more than a specified percent of their labor force in manufacturing."[2]

2. The *functional region,* also called the *nodal region,* is a coherent structure of areal units organized into a functioning system by lines of movement or influence that converge on a central node or trunk. An example is the "milkshed" (milk supply area) of a city, exhibiting a pattern of dairy farms reached by transportation lines over which fresh milk moves to processing and distribution points, and ultimately to consumers in the city. A more complex example is the territory served by a commercial metropolis. It is comprised of a hierarchy of local and regional commercial centers (crossroads stores, hamlets, villages, small towns, small cities, and so on) which are bound to each other and to the central metropolis by the flow of people, goods, and information over an organized network of transportation and communication lines. Still a third instance is the territory served by a railway system, with its ramified pattern of feeder lines branching from a central trunk.

3. Many regions, sometimes called *general regions,* are too varied and loosely knit to qualify in any strict sense as "homogeneous regions" or "functional regions," but are recognized simply on the basis of overall distinctiveness. The "major world regions" that form the basic framework of this text are best thought of as "general regions."

The concept of regions has a utility that extends beyond formal course work in geography.

Regions are areal categories, and categories are essential to thought. Regionalization is either a type of classification or a process very similar to classification. Man cannot think without grouping his infinitely varying observations and experiences into categories; and he cannot orient himself with respect to the areally differentiated surface of the earth, and think about that differentiation, without grouping the infinite variety of places on the surface of the earth into regions. Thus regional concepts are not merely the tools of geographers. They are necessities for thinking man, as is indicated by their widespread use outside the field of geography and in everyday life and discourse.[3]

THE MAJOR WORLD REGIONS

This book is organized in terms of eight major world regions: Europe, the Soviet Union, the Middle East, the Orient, the Pacific World, Africa, Latin America, and Anglo-America (map, p. 1). Although these regions are commonly recognized as grand divisions of the world, geographers often differ as to the precise area which

---

[2] J. Trenton Kostbade, "The Regional Concept and Geographic Education," *Journal of Geography,* 67 (1968), 8. Used by permission of the *Journal of Geography.*
[3] *Ibid.,* p. 7. Used by permission. One footnote omitted in editing. Wording follows author's original manuscript.

should be included in each. Here the eight world regions are delimited as groups of countries, and thus their outer limits follow political boundary lines. This method of regional division has been adopted for convenience in presenting regional concepts on an introductory level. However, the student should not assume that the political lines employed as regional boundaries necessarily represent sharp lines of cleavage in other respects. In physical, cultural, and economic terms the major world regions tend to be separated by zones of transition where the characteristics of one region change gradually to those of the next. This tendency to merge gradually rather than to be separated by sharp lines is also true of regions of a smaller order of size. Since regions are merely convenient devices useful in generalizing about the world, it is more important to grasp the particular set of features which characterizes the core of a region than it is to search for an exact line where the characteristics of one region end and the characteristics of another region begin.

The scheme of eight major world regions employed in this book is only one among various alternative methods of subdividing the world for purposes of study. However, it is believed to be an especially suitable organizing device for accomplishing the major objective of the book: assisting the student to acquire a fund of concepts about the geography of the contemporary world that any educated person might reasonably be expected to know. The regional names employed already have some familiarity for the student and are in common use. Thus the student can proceed from the known to the unknown in acquiring the new perspectives which a sound knowledge of world geography can provide.

Each major world region has certain well-defined characteristics which tend to give it a unity or personality of its own.

Europe is a diversified industrial, commercial, and agricultural region in middle latitudes.[4] It includes a larger number of individual countries than any other area of similar size. These countries are small in area, being generally comparable to individual American states. However, many of them have relatively large and/or dense populations. Some European countries have overseas possessions acquired during the age of colonial expansion which extended from the fifteenth century to the twentieth. During this period Europe became the domi-

nant region in world affairs. Today this dominance has been lost as a result of profound world changes associated in part with World Wars I and II. Since World War II most European dependencies have gained independence (map, p. 65), and some of the remaining ones may become independent in the near future. The economic and military power which was once centered in Europe is now shared by other areas, notably Anglo-America and the Soviet Union. But despite the loss of its commanding position, Europe continues to be an important and influential region. Its impact on the rest of the world in recent centuries has been so great that a knowledge of Europe is a necessity for the understanding of any other major region.

[EDITORIAL COMMENT. The term "latitude" denotes position with respect to the equator and the poles. Latitude is measured in degrees, minutes, and seconds. The equator, which circles the globe midway between the poles, has a latitude of 0°. All other latitudinal lines are parallel to the equator and to each other, and therefore are called "parallels." Places north of the equator are said to be in north latitude; places south of the equator, in south latitude. The highest latitude a place can have is 90°. Thus the latitude of the North Pole is 90°N., and that of the South Pole is 90°S. Places near the equator are said to be in *low latitudes;* places near the poles, in *high latitudes* (map, p. 1). The Tropics of Cancer and Capricorn, at 23½°N. and 23½°S., and the Arctic and Antarctic Circles, at 66½°N. and 66½°S., form convenient and generally realistic boundaries for the low latitudes and the high latitudes, respectively (although a more symmetrical arrangement would place the boundaries at 30°N. and S., and 60°N. and S.). Places occupying an intermediate position with respect to the poles and the equator are said to be in *middle latitudes.* The middle latitudes have a marked seasonal quality in their yearly temperature regimes which contrasts strongly with the relatively constant and monotonous heat of low latitude lowlands and with the extremes of cold which prevail for much of the year in the high latitudes. (For a definition of longitude, see pages 291–292.)]

The Soviet Union, though a single country, is continental in size and resources. So vast, influential, and distinctive is this Communist state that one seems justified in regarding it as a major region comparable in

---

[4] For the guidance of students who have had no previous course work in college or high school geography, a limited number of important technical terms are explained in bracketed sections of the text following the paragraph in which they first appear.

world importance to regions containing many different countries. It is by far the largest country in area on the globe, being more than twice as large as Canada, China, or the United States. However, its size as seen on a map is somewhat deceptive. Much of the country is handicapped by a harsh physical environment that restricts the possibilities for human settlement and use, and large sections have few people. Most of the people live in a triangular coreland (map, p. 310) about half the size of the United States. One of the most significant distinguishing characteristics of the Soviet Union is its type of planned economy, instituted by the Communist Party since the Russian Revolution of 1917. Prior to the 1930s, the Soviet Union was predominantly an agricultural country. Today it has become an industrial power ranking second only to the United States. Since World War II, Communist forms of society patterned in varying degrees after that of the Soviet Union have been established in a sizable number of other countries, most of them located along or near the Soviet borders in Europe or Asia. During the late 1950s and the early and middle 1960s the Communist system in the Soviet Union and various other Communist countries seemed to lose some of its former rigidity, and the Soviet grip on the Communist bloc of nations appeared to loosen somewhat. But the continuation of such tendencies was brought into question in 1968 by armed Soviet suppression of liberalizing and westernizing trends in the Communist "client state" of Czechoslovakia.

The *Middle East* is predominantly a region of deserts and semiarid grasslands. Authorities differ regarding the extent of this region; some would disavow the term "Middle East" altogether as being too vague and thus tending to create confusion. However, it now seems firmly established as a regional name. In this book the Middle East comprises an area stretching across northern Africa and southwestern Asia from Morocco and the Spanish Sahara to West Pakistan and Kashmir. In large sections of this region human settlement is discouraged by lack of moisture and other natural resources. Modern types of manufacturing are poorly developed, though they are gradually being expanded in some places and have reached a relatively advanced state in Israel and limited areas elsewhere. Despite the rapid growth of scattered urban populations in recent years, the Middle East remains primarily a region of farmers and herdsmen. Crop growing is largely confined to irrigated oases along the major rivers and to some sections, particularly around the Mediterranean Sea, where winter rains bring more moisture than is normal in most other parts of the

region. Although the Middle East is not well supplied with most types of natural resources, it has an abundance of one vital commodity, oil. The world's largest deposits are found in the vicinity of the Persian Gulf, and other sizable deposits, more recently exploited, underlie the Sahara Desert of northern Africa. One of the most significant characteristics of the Middle East is the dominance in most countries of the Islamic (Moslem or Muslim) religion. This region forms the heart of the Islamic world. Until recently a colonial region in large part, this area is today comprised almost entirely of independent states, many of which are seriously underdeveloped and poverty-stricken.

The *Orient,* a region extending around the southern and eastern margins of Asia from India and Pakistan to Korea and Japan, is the home of about half of the human race. In most of its countries well over half, and in some cases four-fifths, of the inhabitants are farmers. But manufacturing and other sources of employment associated with growing urban areas are increasingly important; indeed, in Japan, which has become one of the world's most important industrial nations, less than a fourth of the labor force is agricultural. The majority of the region's inhabitants gain little more than subsistence from plots of land that seldom average larger than 3 to 5 acres per family and are often much smaller. In many parts of the Orient the pressure of population on resources is very severe. Most of the region's people live in tropical or subtropical climates, though large areas extend outside these climatic zones. Agriculture in many areas depends upon fluctuating monsoonal rains of the summer half-year. Many millions of people are vitally dependent on these rains; when the summer monsoon fails, or is weak, famine may result. Like the Middle East, the Orient is primarily composed of independent countries that have only recently emerged from a colonial status. Today this region is gripped by changes and conflicts associated with the decline of Western colonialism, the emergence of communism, with its powerful stronghold in mainland China, and the drive of many impoverished and underdeveloped nations to modernize and improve their economic and political status.

The *Pacific World* embraces the island continent of Australia, together with a host of islands, large and small, scattered about in the vast reaches of the Pacific Ocean. This region is composed of dependent territories—except for Australia, New Zealand, Western Samoa, and Nauru, which are independent countries—and the American state of Hawaii.[5] The Pacific World lies in the tropics, with the exception of New Zealand and southern Australia, which are in middle latitudes. Its island peoples long led a self-sufficient existence based upon simple forms of agriculture, the utilization of the coconut palm, and fishing. But the coming of the white man and the disturbances wrought by the two world wars have brought great changes, not always propitious, to the trade-wind isles of the "South Seas."

The massive continent of *Africa* is the largest in area of the major world regions considered in this book, though many of its vast expanses of desert, grassland, and forest are so sparsely inhabited that the continent's total population is only about three-fourths that of Europe. It will be noted that "Africa" overlaps the "Middle East," since North Africa is included in both regions. This should occasion no difficulty so long as one conceives of regions simply as useful tools designed to further geographic study. For various reasons it seems desirable to consider North Africa in the context of both "Africa" and the "Middle East." This procedure appears clearly to enhance our understanding of both world regions, and thus to be consistent with the aims of geographic study. Most of Africa is an underdeveloped tropical region inhabited by Negroid peoples, though the peoples of North Africa are basically Caucasian. White men of European descent, primarily inhabitants of the rapidly growing cities, are found principally in the southern and northwestern reaches, where Africa extends into subtropical latitudes, and in some parts of the tropical highlands of the continent. Possibly two-thirds of the European population is in the continent's most highly developed country, South Africa. In the world's economy, Africa is principally important as an exporter of minerals and tropical and subtropical agricultural commodities, as a market for European manufactured goods, and as a field for capital investment by European and Anglo-American nationals. Politically, the region has been in rapid transition from European colonialism to independence. In the immediate past Africa has exerted little direct influence on world affairs, its destinies being mainly controlled by forces outside the region. But in recent years a host of newly independent countries have taken their place in the world's councils. These new governments are faced with massive problems as they attempt to organize their countries effectively,

often in the face of strongly felt tribal loyalties, and to improve the lot of the impoverished farmers and herdsmen who comprise the mass of their populations.

Aside from giant Brazil, with its 3.3 million square miles and its 90 million people, and the sizable republics of Argentina and Mexico, *Latin America* is predominantly a region of small or medium-sized independent nations, together with a scattering of small dependencies of overseas nations in the West Indian isles or on the nearby mainland. The independent countries of Latin America were formerly colonies of Spain, Portugal, France, or Great Britain. Most of them achieved independence in the nineteenth century, though a number of former British dependencies have become independent only since World War II. Today the widespread predominance of Roman Catholicism and of European languages, primarily Spanish and (in Brazil) Portuguese, reflects the former importance of European colonial powers throughout this region. Like Africa, Latin America is mainly tropical or subtropical in climate, though in the south it extends far into the middle latitudes. Economically, this region is most important to the outside world as an exporter of minerals such as oil, copper, iron ore, bauxite, tin, and silver, and tropical agricultural commodities such as coffee, cane sugar, and bananas, and as a market for manufactured goods and investment capital from the United States and Europe. Its people—who are mostly of American Indian, European, and Negro stock and their various intermixtures—support themselves primarily by agriculture, often on a subsistence or near-subsistence level. Only in scattered areas, such as southeastern Brazil, Buenos Aires, central Mexico, central Chile, and coastal Venezuela, has much manufacturing developed. Rapid population increases and urbanization are posing difficult problems for a region that is, for the most part, seriously underdeveloped and poor.

Anglo-America consists of the United States, Canada, and, as a marginal appendage, Greenland. Like Europe and the Soviet Union, this region lies in middle or high latitudes in the Northern Hemisphere. The two nations that compose it, drawn together by cultural similarities, and closely interlocked both economically and militarily, are firm friends and allies despite occasional differences of opinion and viewpoint. The territory

---

[5] The independent island nations of Japan and the Philippines are considered in this text to lie within the Orient. The same is true of Indonesia except for Indonesian New Guinea, which is treated with the Pacific World as well as with the Orient.

they occupy is probably the richest of all the major world regions in the natural resources needed by advanced nations. On this resource base the United States and Canada have developed an extraordinarily productive economy. This world region leads all others in productivity per person and general material prosperity, although Canada falls somewhat behind the United States and is matched by a few European countries in these respects.

AREAL DIFFERENCES AND SIMILARITIES: THE BASIS FOR GEOGRAPHICAL STUDY

From the foregoing descriptions of major world regions, as well as from common knowledge, it is apparent that one of the most striking characteristics of our world is its diversity from place to place. Countries and regions differ widely in location, in world importance, in types of people and combinations of resources, in ways of living and levels of living. They differ in physical character and political character, in economic mainstays, in internal arrangement and organization, and in external connections and relationships. Since early times, geography as a field of study has found its essential rationale in these variations that exist from place to place over the world. In the ancient world geography first developed as a normal outgrowth of the curiosity of people regarding lands which were different from their own. As knowledge of the world has increased through discovery and exploration, an almost endless variety of landscapes and modes of life has been found to exist. Scholars have been intrigued by these differences in areas, and they have sought to describe and account for them. Thus through observation, description, and analysis has grown the modern science of geography. It is, of course, self-evident that if each area of the world were just like all the others, there would be no need to study geography. But each area has its own distinctive character, and so there *is* need for a field that tries to ascertain and teach what the different parts of the world are like, how and why they differ from or resemble each other, and how they fit together into a meaningful whole.

Differences between America and Other Areas: A Geographer's View of America's Queerness

The impact of modern technology and modern forms of economic, social, and political organization is bringing about far-reaching changes in the world's geography. It is tending to make various parts of the world more similar, at least outwardly. Yet striking differences in areas continue to exist, and the appreciation of these differences is an important aspect of one's education in geography. Four decades ago an American geographer, the late Ellsworth Huntington, illustrated such differences by showing some of the ways in which the United States was different from other parts of the world. Today his observations, as quoted in the following selection, are essentially as applicable as they were some forty years ago:[6]

One of the greatest troubles with Americans is that they cannot recognize their own queerness. . . . The average American fails to appreciate the fact that in most respects it is we who are queer and not the people of China, India, Japan, or even Russia. We realize, to be sure, that we are queer in the sense of being extremely prosperous. Outside of the United States not one-tenth of the world's . . . people live as comfortably as does the average skilled laborer in the United States. . . . Since not one person in ten among all the peoples of the world [enjoys such prosperity], it must be we who are queer and not the others.

Another way in which the uniqueness of America is apparent is illustrated by fences. In most parts of the world fences are built around houses, but there are no fences around the fields. With us the opposite is the case. There are few fences around houses, but practically all our fields are fenced. England and some of the neighboring regions of northwestern Europe have many hedges, but that condition prevails in only a relatively small area compared with the world as a whole. In most parts of the world the normal thing is to see field after field in unbroken succession covering all the land between the villages. The only thing to separate one field from another is a change in the crops or a little line, a foot or so wide, where an unsown strip separates one field of wheat, beets, or some other crop from another.

This lack of fences is symbolic of two far greater differences between America and the rest of the world. The first of these differences is that the size of land holdings

[6] Ellsworth Huntington, "A Geographer's View of America's Queerness," *Education,* 52 (1932), 254–257. Used by permission of The Bobbs-Merrill Company, Inc., publishers of *Education.*

is far smaller in most parts of the world than in the new lands in the temperate parts of North and South America and Australia. . . . [In most areas of the Old World] fences are a very expensive luxury. This is not only because the farms are so small, but also because [in many cases] the farmers of the Old World do not have all of their land in one place. Each man . . . owns several small separate parcels. If fences were built around every individual field, the cost would be prohibitive. Because our fields are so large it is relatively cheap to fence them.

The other thing illustrated by the absence of fences is that in the world as a whole materials are relatively more expensive than labor, but with us the reverse is the case. In [large] parts of the world . . . the place of fences is taken by people. A small boy or an old man or woman can be hired to herd the sheep or cattle far more cheaply than a fence can be made. . . .

Another of our queer habits is that even among our urban populations a vast number of us live in detached houses set among grass and green trees. It is hard for us to realize that there is anything strange about this, but the facts are clear. Even in our larger cities, like New York or Chicago, there are large areas where this type of detached dwelling prevails. Such houses sometimes contain two or three families but usually only one. In smaller cities, especially in those with less than 100,000 people, a large fraction of the population lives in such houses. And of course this is also true of our millions of suburban people. This seems to us so much a matter of course that we fail to appreciate the fact that it is very uncommon when the world as a whole is considered. In Italy, India, China, and most other countries, such houses are practically unknown. In Germany, and the neighboring parts of Europe, only a very few people live in single houses surrounded by lawns. . . . In cities like Madrid, Rome, Vienna, and . . . Berlin, big apartment houses built up solidly on regular city streets often adjoin the fields. Almost the only important parts of the Old World where

Village settlement in an intensively farmed section of the Old World. Farmers go out to the fields each day from their homes in the two villages shown in the view. Note the intricate patchwork of long, narrow strips; a farm may consist of numerous scattered strips, often separated from each other by considerable distances. Carefully tended woodlands occupy the higher and steeper slopes. The photo was taken in southern Luxembourg near Luxembourg city. (U.S. Air Force Photo.)

there are large numbers of detached suburban houses are Great Britain, Holland, and Scandinavia. Yet, even in England, the industrial worker who lives outside the central part of a town is usually housed in a little two-story house of brick or stone which forms part of a long monotonous row. He may have a bit of lawn as big as a tablecloth in front of his house, but the chances are that his house is part of a solid row. Even where such houses are separate from one another, they usually have merely a narrow paved walk between them.

An even more striking evidence of the queerness of the people of the United States is the fact that the vast majority of our farmers live on farms. This may seem absurd. It seems to us so natural and convenient for a farmer to live on his farm that we call anything else stupid. . . . In Canada, Australia, and New Zealand the same habits and opinions prevail. Yet, strange to say, in all the rest of the world there are probably not much more than [a few tens of millions of] farming people who live in this isolated way. In Scandinavia, to be sure, and in mountainous regions in many parts of the world, the farmers do live by themselves on their own land. In England this is true to a [considerable] degree, but even there [many] farmers live in villages and go out to their work each day. In other countries, such as Spain and Italy . . . the wealthier land owners may live in isolated houses on their own land, but the bulk of the agricultural population lives in compact villages. The hundreds of millions of farming people in India, China, Japan, and most other parts of the world practically all live in compact villages. . . .

In many ways such agricultural villages are the most characteristic feature of human geography. This is because they retain a stronger individuality than does either the land on which there are no houses or the land that is covered with houses in cities. Grain fields, pastures, forests, and gardens have a strong family resemblance all over the world. Of course, a field of cabbages is very different from a field of potatoes, and an apple orchard is not at all like a banana patch. Nevertheless, cabbages look much the same everywhere no matter who raises them, and so do potatoes, apples, and bananas. Then, too, all over the world the cities tend gradually to become more and more alike. When Yokohama and Tokyo suffered from earthquake and fire, large sections were rebuilt in a purely Western style. . . . Electric lights, steel frames, modern processes of using concrete, and a thousand other new methods tend constantly to become standardized all over the world. Cotton factories and hotels in Bombay are almost like those in North Carolina. A railroad station or moving picture house is much the same thing in Jerusalem and in Montevideo. Cities like [Istanbul], Cairo, Madras, and Hankow still possess a high degree of individuality, to be sure. Nevertheless, even in these cities some sections are so thoroughly modernized that the mere appearance of the streets scarcely tells a stranger whether he is in America, Europe, Asia, or Africa. On the other hand, the farming villages all over the world still display a persistent individuality. The flat-roofed, mud villages near [Izmir] in Turkey could never be confused with the palm-thatched huts near Madras or with the thatched houses of a wholly different type near Kyoto. Everywhere the farming villages, far more than almost any other feature of the surroundings, reflect the characteristics of the local geography. Since [the majority of] the world's people live in such villages, we may well say that they are one of the most fundamental features of human geography. Yet in this respect as well as in wealth, transportation, use of machinery, and the use of fences with all that they imply as to the size of land holdings and the value of labor, we are so unique that we well deserve to be called queer.

Similarities among Areas

While it is the *differences* among areas that provide the essential basis for the existence of a field called geography, a good knowledge and understanding of world geography requires that areas also be examined with regard to their *similarities*. Such resemblances are often rather superficial, but in most cases it is still very useful to know that areas located in different parts of the world have some geographic features in common, and that the global distribution of such areas and features often forms a pattern which is regular, orderly, and rational. A good example is the occurrence of areas with similar climates in analogous locations on the earth. For instance, the "marine west coast" type of climate, characterized by winters that are abnormally mild for such a relatively high latitude, occurs not only in a strip along the northwestern coast of Eurasia, but in analogous places along the western coasts of North and South America (see front endpaper climatic map).

Comparisons of areas with respect to their similarities and differences are basic to the study of geography. Such comparisons may be explicit, or they may be implied. When we say that the Rockies are higher than the Appalachians, the comparison is explicit; but even to say that the Rockies are "high" mountains is to make an implicit comparison with other mountain areas. Comparison of areas enables us to see each area in a broader perspective, and thus helps us discern those characteristics which are of primary importance in giving an area its distinctive quality. It helps us to better understand the relationships between different characteristics of given areas, and it suggests interesting geographic problems as to why certain things occur in some areas, but occur to a lesser degree, or not at all, in other areas. Much scholarly research in geography is concerned with problems of this type.

REFERENCES and READINGS

The Nature of Geography as a Field of Learning

Bobbs-Merrill Reprint Series in Geography. Over 225 articles available as individual reprints (Indianapolis, Ind.: The Bobbs-Merrill Company).

BROEK, JAN O. M., *Geography: Its Scope and Spirit* (Columbus, Ohio: Charles E. Merrill Books, 1965; abridged as *Compass of Geography,* 1966).

CHORLEY, RICHARD J., and PETER HAGGETT, eds., *Frontiers in Geographical Teaching* (London: Methuen and Co., 1965).

CLARK, ANDREW H., *"Praemia Geographiae:* The Incidental Rewards of a Professional Career," *Annals of the Association of American Geographers,* **52,** no. 3 (September 1962), 229–241.

COHEN, SAUL B., ed., *Problems and Trends in American Geography* (New York: Basic Books, 1967).

DOHRS, FRED E., and LAWRENCE M. SOMMERS, eds., *Introduction to Geography: Selected Readings* (New York: Thomas Y. Crowell Company, 1967).

Geography in Undergraduate Liberal Education (Commission on College Geography, Publication No. 1; Washington, D.C.: Association of American Geographers, 1965).

HARTSHORNE, RICHARD, *The Nature of Geography: A Critical Survey of Current Thought in the Light of the Past* (Washington, D.C.: Association of American Geographers, 1939, 1946, and subsequent reprintings); also, *Perspective on the Nature of Geography* (Chicago: Published for the Association of American Geographers by Rand McNally & Co., 1959); and "Why Study Geography?" *Journal of Geography,* **65,** no. 3 (March 1966), 100–102.

JAMES, PRESTON E., "Geography," in *The Social Studies and the Social Sciences* (New York: Harcourt, Brace & World, 1962), pp. 42–87; also, ed., *Geography as a Professional Field* (U.S. Department of Health, Education, and Welfare Bulletin 1966, No. 10; Washington, D.C.: Government Printing Office, 1966); also, "On the Origin and Persistence of Error in Geography," *Annals of the Association of American Geographers,* **57,** no. 1 (March 1967), 1–24; and, with CLARENCE F. JONES, eds., *American Geography: Inventory and Prospect* (New York: Published for the Association of American Geographers by Syracuse University Press, 1954).

KARIEL, HERBERT G., "Scope of Geographic Study," *Journal of Geography,* **66,** no. 4 (April 1967), 150–154.

KOSTBADE, J. TRENTON, "A Brief for Regional Geography," *Journal of Geography,* **64,** no. 8 (November 1965), 362–366; and "The Regional Concept and Geographic Education," *Journal of Geography,* **67,** no. 1 (January 1968), 6–12.

LUKERMANN, FRED A., "Geography as a Formal Intellectual Discipline and the Way in Which It Contributes to Human Knowledge," *Canadian Geographer,* **8,** no. 4 (1964), 167–172.

McDONALD, JAMES R., "The Region: Its Conception, Design, and Limitations," *Annals of the Association of American Geographers,* **56,** no. 3 (September 1966), 516–528.

MINSHULL, ROGER, *Regional Geography: Theory and Practice* (Chicago: Aldine Publishing Company, 1967).

"National Character in the Perspective of the Social Sciences," *Annals of the American Academy of Political and Social Science,* **370** (March 1967), 1–163.

New Approaches in Introductory College Geography Courses (Commission on College Geography, Publication No. 4; Washington, D.C.: Association of American Geographers, 1967).

PATTISON, WILLIAM D., "The Four Traditions of Geography," *Journal of Geography,* **63,** no. 5 (May 1964), 211–216.

SAUER, CARL O., "The Education of a Geographer," *Annals of the Association of American Geographers,* **46,** no. 3 (September 1956), 287–299.

SPATE, O. H. K., *Let Me Enjoy: Essays, Partly Geographical* (London: Methuen and Co., 1966).

STEINHAUSER, FREDRIC R., "The Study of Geography," *Journal of Geography,* **66,** no. 7 (October 1967), 350–359.

TURNOCK, DAVID, "The Region in Modern Geography," *Geography,* **52,** pt. 4 (November 1967), 374–383.

WHITE, GILBERT F., "The Changing Dimensions of the World Community," *Journal of Geography,* **59,** no. 4 (April 1960), 165–170.

WRIGHT, JOHN KIRTLAND, *Human Nature in Geography: Fourteen Papers, 1925–1965* (Cambridge, Mass.: Harvard University Press, 1966).

Atlases, Dictionaries, Gazetteers

Columbia Lippincott Gazetteer of the World, with 1961 Supplement (New York: Columbia University Press, 1962). This large, expensive volume is a basic source of information about places. Much of the information is repeated in *The Columbia Encyclopedia.*

Goode's World Atlas (12th ed.; Chicago: Rand McNally & Co., 1964). A college atlas.

MONKHOUSE, F. J., ed., *A Dictionary of Geography* (Chicago: Aldine Publishing Company, 1965).

Odyssey World Atlas (New York: The Odyssey Press, 1967). A college-level atlas. Some maps adapted for publication in a shorter paperbound atlas, *Man's Domain: A Thematic Atlas of the World* (New York: McGraw-Hill Book Company, 1968).

Oxford Economic Atlas of the World (3d ed.; New York: Oxford University Press, 1965). Abridged as *The Shorter Oxford Economic Atlas of the World* (3d ed., 1965). A college atlas.

Prentice-Hall World Atlas (2d ed.; Englewood Cliffs, N.J.: Prentice-Hall, 1963). A college atlas. Excellent cartography.

STAMP, L. DUDLEY, ed., *A Glossary of Geographical Terms* (New York: John Wiley & Sons, 1961).

Times Atlas of the World, 5 vols. (Boston: Houghton Mifflin Company, 1955–1959). Superb reference atlas produced in Great Britain.

Webster's Geographical Dictionary (rev. ed.; Springfield, Mass.: G. & C. Merriam Co., 1963).

Bibliographical Sources

Bibliographie Géographique International (Paris: Centre National de la Recherche Scientifique). Annual one-volume bibliography of books and articles in geography and related fields. Available in the United States from the American Geographical Society of New York.

CHURCH, MARTHA, ROBERT E. HUKE, and WILBUR ZELINSKY, comps., *A Basic Geographical Library: A Selected and Annotated Book List for American Colleges* (Commission on College Geography, Publication No. 2; Washington, D.C.: Association of American Geographers, 1966).

Current Geographical Publications. Mimeographed bibliography of books and articles, published ten times yearly by the American Geographical Society of New York.

Geographical Abstracts, in four series: A. *Geomorphology;* B. *Biogeography, Climatology, and Cartography;* C. *Economic Geography;* D. *Social Geography.* Each series published in six issues annually by the Department of Geography, London School of Economics, London, England. Each book or article cited is abstracted in a paragraph.

New Geographical Literature and Maps. Published twice yearly by the Royal Geographical Society, London, England, at a very modest price. Available in the form of booklets or as gummed sheets for pasting individual references on cards.

Research Catalog of the American Geographical Society, 15 vols. (Boston: G. K. Hall & Co., 1962). Massive bibliography of geographical books and articles. Available in large libraries.

Sources of Statistical Information

Encyclopedia Britannica Book of the Year. A highly valuable annual source of statistics and other information relating to geographical changes.

FULLARD, HAROLD, ed., *The Geographical Digest,* annual (London: George Philip and Son). Mainly statistical, but includes brief summaries of recent geographical changes.

"Statistics for Geography Teachers," *Journal of Geography,* 66, no. 1 (January 1967), 6–40.

United Nations statistical publications: *Statistical Yearbook, Demographic Yearbook, Yearbook of International Trade Statistics, F.A.O. Production Yearbook,* and others. Available from the United Nations, New York, N.Y.

U.S. Department of Agriculture, *Agricultural Statistics,* annual.

U.S. Bureau of Mines, *Minerals Yearbook,* annual.

Additional statistical sources for particular regions and countries are cited in the appropriate chapter bibliographies.

Periodicals

College students will find the following English-language geographical journals to be the most generally useful: *Geographical Review, Annals of the Association of American Geographers, Professional Geographer, Economic Geography, Journal of Geography, Focus, Geography* (British), *Geographical Journal* (British), and *Tijdschrift voor Economische en Sociale Geografie* (Dutch: carries many articles in English). The *National Geographic Magazine* is unsurpassed for pictures and place maps. *Foreign Affairs, Fortune, Current History, The New York Times Magazine, The Scientific American,* and *Foreign Agriculture* frequently carry articles of geographical interest. Other periodicals are cited in the appropriate chapter bibliographies.

Processes
That Shape the
Geography of Areas

2 The differentiation of the world into distinctive areas is the result of a multitude of processes through which nature and man determine the contents of segments of earth space. Natural processes endow areas with minerals and rocks, landforms, climates, water features, plant and animal communities, and soils—the interlocking assemblage of features that we call the natural environment. Groups of people have entered areas, have explored them, and have occupied, organized, and developed them. As a consequence of human use (in addition to the ongoing processes of nature itself), the natural environment of an occupied area is continually altered, while the culture of the occupying group also is modified as the group wrestles with the day-to-day problems of organizing and developing its habitat. Today the empty spaces of the world are far less extensive or inviting than they were a few centuries ago, and pioneer activity by migrating populations is increasingly restricted in scope. Though interest in possible pioneer areas continues, a far more meaningful activity at present is the reinvestigation and reorganization of areas and cultures in the parts of the world that are already settled. This process is under way on a global scale as men try to find ways to provide a better livelihood for increasing multitudes of people who have no rich new lands to occupy and develop.

Thus the geography of an area is the product of successive modifications by processes that have been ceaselessly active in the past and will continue in the future. Every area is constantly in flux. Natural processes continue to modify it, while men continue to investigate, reevaluate, and change both the area and their own use of it in accordance with their changing needs and aspirations, their changing social, economic, and political circumstances, and their changing technological competence.

[EDITORIAL COMMENT. The concept of *culture* is much used in geography, as in other disciplines which study man and his works. It is a very useful concept, but so inclusive that scholars do not agree on any exact definition. A group's culture includes the values, beliefs, and aspirations, modes of behavior, social institutions, knowledge, and skills which are transmitted and learned within the group. It also includes a material aspect—the group's material possessions and products. Important for geography is the fact that man sees his environment, and relates himself to it, in terms of his culture. Thus two human groups with different cultures may see the same environment as offering quite different opportunities and difficulties, and may relate themselves to it in quite different ways. In so doing, each group will alter the environment in its own way, thereby creating a distinctive *cultural landscape* (the portion of the total landscape attributable to man). Cultures, of course, are constantly changing. They change by the adoption of inventions (material and nonmaterial) made by their own members, and by the learning and adapting of traits from the cultures of other groups with whom they come in contact. Many of the characteristics of regions of the earth are better understood when one keeps in mind the concept of cultures as distinctive group ways of life, which include distinctive ways of looking at and using the environment.]

NATURAL PROCESSES AND THEIR GEOGRAPHIC RESULTS

Five major groups of natural processes are of prime importance in terms of their geographic results: processes that create rocks; processes that create landforms; processes that create climates; processes that create wild plant and animal communities; and processes that create soils. We may refer to them more succinctly as rock-forming, land-shaping, climatic, biotic, and soil-forming processes.

Rock-forming Processes and the Major Types of Rocks

The study of rocks is carried on mainly by geologists, but the student of geography is concerned with them in various ways. His interest lies in the fact that the rocks which are under, or in some areas are at, the surface of the earth vary widely in physical character from one place to another, and that these variations often are related to such other characteristics of places as landforms, soils, and mining.

Rocks are composed of minerals, which may be chemical elements or combinations of elements. Most frequently, minerals are chemical compounds such as silicon dioxide (which occurs in one form as quartz, the principal constituent of sand), hydrous aluminum silicates (the clay minerals), or calcium carbonate, commonly known as lime. The ultimate origins of minerals need not concern us here. It is important, however, to note the fact that the properties of a rock derive from its particular combination of minerals. Minerals that are economically useful to man may or may not occur in rocks in sufficient concentrations to be minable at a profit. A mineral concentration which can be profitably mined is known as an *ore*.

Sedimentary, Igneous, and Metamorphic Rocks

The rocks of the earth form three major groups according to mode of origin: sedimentary rocks, igneous rocks, and metamorphic rocks.

Sedimentary rocks have been formed from sediments (sand, gravel, clay, or lime)—deposited by running water, wind, or wave action either in bodies of water or on land—which have consolidated into rock over a span of geologic time by pressure of the accumulated deposits or the cementing action of chemicals contained in waters percolating through the rock materials. The main classes of sedimentary rocks are sandstone, shale (formed principally of clay), and limestone. Sedimentary rocks generally are comprised of multiple layers or *strata,* and the term "stratified rocks" often is used in reference to them. In a coal-bearing area, the layers of coal commonly are interspersed ("interbedded") with layers of sandstone and/or shale. *Igneous,* or *volcanic, rocks* are formed by the cooling and solidifying of molten materials, either by slow cooling within the earth (as in the case of granite), or by more rapid cooling when poured out on the surface by volcanic activity (as in the case of basalt). *Metamorphic rocks* are formed from igneous or sedimentary rocks through changes occurring in the rock structure as a result of heat, pressure, or the chemical action of infiltrating waters. Marble, formed from pure limestone, is a common example of a metamorphic rock.

These major types of rocks have certain general, although not invariable, associations with other geographic characteristics and features of areas. Igneous and metamorphic rocks, for instance, being generally harder and more closely knit together than sedimentary rocks, are usually more resistant to wearing down by weathering and erosion. Thus they tend to form uplands in areas where the sedimentary rocks have been weathered into lowlands. (But in purely sedimentary areas the more resistant sedimentaries tend to form higher lands than the less resistant ones.) In general, sedimentary rocks weather into more fertile types of soil than igneous or metamorphic rocks. However, certain types of volcanic rock break down into extremely fertile soils, and within the sedimentaries themselves there are great variations in this respect, with limestones usually forming soils of greater natural fertility than do shales or sandstones. Sedimentary rocks are also significantly different from igneous and metamorphic rocks with respect to associated minerals of economic value. Coal and petroleum, for example, are customarily found in sedimentary areas, whereas metal-bearing ores are found in igneous or metamorphic rocks, except for certain deposits (particularly of iron ore) formed of materials removed from these rocks by percolating waters and redeposited in sedimentary areas.

Land-shaping Processes and Their Results

The main processes that shape the earth's surface into landforms are (1) *warping, folding,* and *faulting* of the rock layers of the earth's crust (known collectively as diastrophic processes or "diastrophism"), (2) *vulcanism,* and (3) *weathering, erosion,* and *deposition.* Warping, folding, faulting, and vulcanism are called the *tectonic processes;* their basic long-term effect is to bring about unevenness in the land surface. Weathering, erosion, and deposition are the *gradational processes;* their long-term tendency is to wear down the higher parts of the surface and fill in the lower parts until the surface is reduced to a uniform plain.

Warping, Folding, and Faulting

Forces in the earth whose nature is not well understood subject the solid rocks of the crust to enormous stresses, and over an immense time slowly warp and bend, or break, them. Masses of rock may be warped upward to form domelike or arched structures, or downward to form structural basins. Such structures may be purely local, or they may be subcontinental in size. Often the uplift or depression is accomplished without much structural change in the rocks themselves. For example, in the Grand Canyon the Colorado River has cut its way through sedimentary rocks that initially lay at a much lower elevation but that have been subjected to warping processes which lifted the entire mass of rocks thousands of feet while leaving the horizontally bedded rock layers essentially undisturbed. On the other hand, intense bending of rock layers ("folding") may produce an accordionlike series of upfolds (*anticlines*) and downfolds (*synclines*) (see diagram). Such structures are very common in mountainous areas of the world. In some instances the anticlines form the present ridges and the synclines form valleys, but in other instances a complex erosional history has produced synclinal ridges or anticlinal valleys.

When they are crowded together or pulled apart by irresistible forces in the crust, rocks may break and the separate portions move past each other along the line of fracture. This process is known as faulting, and the break in the rocks is called a fault. A break due to rock masses being pulled apart is a *tensional* or *"normal" fault,* while a break due to rocks being pushed together until one mass rides up over the other is a *compressional fault* (see diagram). Horizontal displacement of rocks along faults may damage fences, transportation lines, or buildings, but the most pronounced landform effects occur when rocks are displaced vertically. Such displacements of rock masses upward or downward along faults have often been great enough to create mountains of a type called "block mountains." In the most common instance, the rock mass comprising the mountain is tilted in such a way as to create a steep face ("fault scarp") on the side where the fault occurs, and a longer, gentler "back slope" behind the scarp. The Sierra Nevada and the Wasatch Range of western United States (map, p. 754) are "tilted block mountains" of this kind. In the Sierra Nevada the fault scarp faces eastward, while in the Wasatch it faces westward. Some block mountains have been formed between roughly parallel faults and have a scarp face on two sides. A landform of this type is called a *horst* (see diagram). A steep-sided trough known as a *rift valley* or *graben* is formed when a segment of the crust is displaced downward between parallel tensional faults, or when segments of the crust which border it ride upward along parallel compressional faults (see diagram). The Great Rift Valley of eastern Africa and southwestern Asia (p. 558), Death Valley in California, and the Rhine Rift Valley (p. 144) on the border between West Ger-

FOLDED STRUCTURES

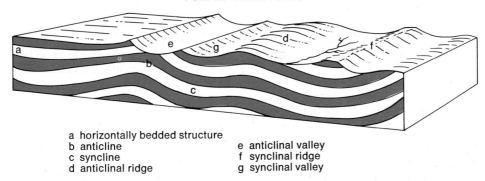

a horizontally bedded structure
b anticline
c syncline
d anticlinal ridge
e anticlinal valley
f synclinal ridge
g synclinal valley

The blue tint shows the more resistant rock strata.

FAULTED STRUCTURES

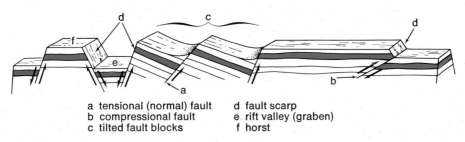

a tensional (normal) fault
b compressional fault
c tilted fault blocks
d fault scarp
e rift valley (graben)
f horst

The blue tint indicates a rock stratum that has been broken and shifted into various positions by faulting. Arrows indicate direction of rock movement along faults.

many and France are perhaps the best known of the world's rift valleys. Faulting is very widespread in mountainous regions. In the American West a whole succession of mountain ranges separated by flat, detritus-filled basins ("basin and range topography") (p. 759) has been produced by faulting.

Vulcanism

In some parts of the world the present terrain was shaped initially by outpourings of volcanic material from vents or cracks in the earth. A volcanic cone such as Mount Fuji or Mount Vesuvius is a familiar instance. In places, enormous amounts of molten material issuing in repeated flows through large fissures in the crust have buried the previous topography to a depth of hundreds or thousands of feet. Prominent examples are found in Ethiopia, interior India, and the Columbia-Snake Plateau of northwestern United States. This type of volcanic activity is known as *extrusive vulcanism.* The cooling and solidifying of the extruded molten matter often has produced the dark-colored rock called basalt. *Intrusive vulcanism* takes place when molten material rising through the crust penetrates the existing rocks and cools and solidifies without reaching the surface. Granite and other "intrusive" igneous rocks are

formed in this way. Uparching followed by erosion may cause the overlying rocks to be stripped away, revealing the intrusive igneous formations. These may then be sculptured into hills or mountains by erosive processes. For example, many of the present mountain ranges in the Colorado and Wyoming Rockies originated in this manner.

The Global System of High Mountains and Associated Tectonic Phenomena

Warping, folding, faulting, and vulcanism are interrelated processes, though the precise connections among them are by no means entirely clear. In recent geologic time tectonic activity has reached its maximum intensity in a branching worldwide zone of crustal instability, a good part of which is under the sea. The land portions of the zone comprise a system of high mountains which may be thought of as radiating in several arms from a central node, the "Pamir Knot," in Asia (back endpaper map of landforms). Most of the truly high mountains of the world—the Alps, Pyrenees, Atlas, Himalayas, Rockies, Andes, and others—lie within this system. Most of the mountains in the system are very young, geologically speaking, and the tectonic processes associated with mountain building are still active.

This fact often has disastrous consequences for people who live in or near the mountains, as the global system of young mountains is the main zone within which earthquakes and active volcanoes occur. The earthquakes result from rock movement along fault lines. Most of the devastating earthquakes of human history, including many recent ones, have occurred in or near the young-mountain belt. Active volcanoes occur in lines or clusters in some parts of the belt—for example, in Central America, in the Alaska Peninsula and Aleutian Islands, and in Java and other islands in Indonesia. These groupings of volcanoes are irregularly distributed, and are, indeed, by no means typical of the young-mountain system as a whole, most parts of which show no evidence of recent volcanic activity.

Weathering, Erosion, and Deposition

Landforms that have been shaped initially by tectonic processes are then sculptured by gradational processes; indeed the two types of processes go on simultaneously. One of the gradational processes is *weathering*. When rocks are subjected to weathering, they gradually crumble, and the fragments move downslope by gravity or are transported away by moving water, ice, or wind and deposited elsewhere. *Mechanical weathering* involves such processes as (1) the formation of cracks or joints in rocks as a result of stresses; (2) the wedging action of ice when water freezes in rock crevices; or (3) the expansive force exerted by plant roots growing in crevices. *Chemical weathering* is more complex than mechanical weathering. It involves interaction between surface or underground water and certain chemicals in rocks. Generally the water with which rocks come in contact contains weak acids acquired when falling raindrops interact with carbon dioxide in the atmosphere (forming carbonic acid), or when water comes in contact with decomposing plant life and absorbs acids in this manner. Especially when fortified by acids, water dissolves some of the minerals in rocks and carries them away, thus weakening the rocks and facilitating mechanical weathering (which in turn exposes still more rock surfaces to chemical action). In other instances rock minerals may swell when they interact with water and thus break the bonds that hold individual mineral grains together. In still other cases, minerals may be changed in form or structure, becoming softer or more easily dissolved.

A second gradational process is *erosion,* the wearing away and shaping of the earth's surface by running water, ice, or wind. Erosion by the cutting action of streams is extremely important in shaping the form of the land. In this process the force of running water is augmented by fragments of rock which are carried along by the stream and act as a kind of file, abrading the stream channel and helping the stream to cut its valley. As the valley is deepened, its edges tend to crumble, erode, and slump. Thus the valley grows wider as it is cut deeper. In arid areas deepening tends to occur more rapidly relative to widening than is the case in humid areas. Thus in high areas of arid regions streams that have risen in more humid areas carve deep, steep-walled canyons ("inverted mountains"). In humid areas, while streams are deepening valleys, weathering and erosion of the sides of the valleys occur more actively than they do in arid lands. A plateau may thus be carved into a mass of hills or mountains consisting of stream valleys separated by ridges which may be capped by remnants of the original surface. The hills and low mountains of the "Appalachian Plateau" in eastern United States (map, p. 725) were formed in this way. Long-continued erosion will reduce a highland to a lowland plain, but at any stage the process may be interrupted by upwarping. Many land surfaces as they exist today are primarily a reflection of some stage in the destruction of a raised surface by stream erosion.

Ice and wind are active agents of the erosional process in some areas. The grinding, scouring movement of glaciers over some parts of the land surface during the Great Ice Age left many characteristic features behind when the ice melted (see Editorial Comment, p. 79), and glaciation is still active today in restricted areas. The wind is particularly active as an erosive force in arid and semiarid areas where the protective cover of vegetation on the surface of the land is thin or absent.

Material transported by water, ice, or wind eventually is deposited by the agent that removed it; actually it may be picked up and redeposited many times. Large areas of the world are plains underlain by sedimentary rocks that were formed by the deposition of vast quantities of sediments in shallow seas. Prominent examples are the Interior Plains of North America (map, p. 707), the North European Plain (map, p. 78), and the interior plains of South America drained by the Amazon, Orinoco, and Paraná rivers (map, p. 664, and back endpaper map). Recently deposited sediments not yet consolidated into rocks form alluvial plains in river valleys; major areas are found along the lower Mississippi (map, p. 725), the Amazon, the Hwang Ho (Yellow River) of North China (map, p. 405), and many other rivers. Often a distinction must be made between older alluvium and newer alluvium, as in northern India. Here the main plain of the Ganges and Indus rivers

An escarpment, in this instance the "great escarpment" marking the edge of Brazil's mountainous eastern plateau, the Brazilian Highlands (see Chapter 28). A narrow coastal plain along the Atlantic shore of Brazil lies at the foot of the escarpment. The view shows the highway connecting the seaport of Santos with the inland metropolis of São Paulo. (Brazilian Government Trade Bureau.)

(map, p. 405), formed of older alluvium, is hundreds of miles wide, but with relatively narrow strips of newer alluvium forming the present flood plains of the rivers, which flow in valleys eroded into the main plain. Some of the flattest areas in the world are comprised of recent alluvium in river flood plains and deltas, in the beds of vanished lakes, or in stretches of former sea bottom that have been upwarped slightly to form dry land.

The Major Classes of Landforms

The results of these land-shaping processes are commonly classified into four very general types of landforms: plains, plateaus, hill lands, and mountains. Each type is susceptible to further subdivision, and the distinctions between the types are not always clear and objective.

In general, a *plain* is a relatively level area of slight elevation (although some parts of the "High

Plains" in the western United States reach elevations of 5000 feet above sea level). Some plains are flat, but most of them exhibit gentle to moderate slopes. However, it is the horizontal rather than the vertical dimension which predominates in plains country. Most of the world's people live on plains, so that consideration of plains and their significance becomes a highly important facet of geographic study.

To a degree, a *plateau* is simply an elevated plain. Most areas recognized as plateaus lie at elevations of 2000 feet or more, although some are considerably lower. To qualify as plateaus in the strictest sense, such areas should be terminated on at least one side by a steep edge, or *escarpment,* marking an abrupt transition from the plateau surface to areas at a lower elevation. However, the word "plateau" is often used loosely in referring to relatively level areas lying at considerable heights, whether terminated by a definite escarpment or not. Well-defined plateaus are often spoken of as "table-lands." To complicate matters, many plateaus in areas

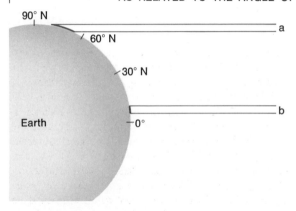

Solar ray *b* strikes the earth more nearly vertically at lower latitudes than solar ray *a* does in higher latitudes. Thus a given amount of solar energy is concentrated on a smaller extent of surface at lower latitudes than at higher latitudes.

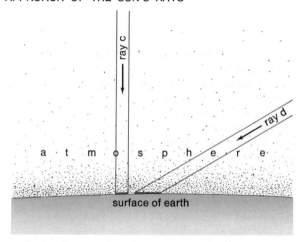

Solar ray *c* passes through a smaller thickness of absorbing and reflecting atmosphere before reaching the surface of the earth than solar ray *d* does.

with considerable rainfall have been cut into predominantly hilly or even mountainous country by their many streams, yet are still commonly referred to as plateaus.

In *hill lands* and *mountains* the vertical dimension predominates. No sure way has been devised to distinguish between these two classes of landforms. In general, mountains are higher and more rugged than hill lands, are more of a barrier to movement, and generally offer fewer possibilities for human settlement and use. Whereas many areas of hill land support moderate to dense populations, most mountain areas are sparsely populated. Local usage of the term "mountain" varies greatly from place to place over the earth. In plains country, areas of a few hundred feet elevation which stand out conspicuously above their surroundings may be called "mountains" by the local inhabitants. However, the term is most commonly used in reference to comparatively rugged areas lying at least 2000 to 3000 feet above sea level.

Climatic Processes and Their Geographical Results

The great climatic variety of the earth results primarily from areal differences in the processes by which the earth's atmosphere is heated and cooled. The fundamental processes are the radiation and absorption of heat energy. Solar energy (insolation) reaches the earth in the form of short-wave radiation, some of which is visible as light. A considerable part of the total radiation never reaches the ground, but is reflected back into space or absorbed by water vapor, carbon dioxide, other gases, or dust particles in the atmosphere. Most of the solar energy that reaches the land or ocean surface is absorbed, and some of it is returned to the atmosphere in the form of long-wave radiation, which generates heat and is the principal agent that warms the atmosphere. Thus the air is heated mainly from the underlying land or water surface, although ultimately and indirectly, of course, the source of the heat is the sun.

Air Temperature

Of primary geographical consequence is the fact that the earth varies greatly from place to place in the total amount of energy received annually from the sun and in the resulting air temperatures. If we omit the effects of cloud cover, we may say that the lower the latitude of a place (the nearer the equator it lies), the more solar energy it receives annually. One reason is that on the average through the year, the sun's rays strike the earth more nearly vertically at lower latitudes, and thus a given amount of solar energy is concentrated on a smaller extent of surface than at higher latitudes (see diagram). A second reason is that when they approach the earth at a more nearly vertical angle in lower latitudes, the solar rays must pass through a smaller thickness of absorbing and reflecting atmosphere before reaching the surface than is the case at higher latitudes (see diagram). However, many areas near the equator have a relatively abundant cloud cover, which reflects or absorbs much of the incoming solar radiation and reduces the amount reaching the surface. The result is that the areas of highest annual solar energy actually received at the surface appear to be in the vicinity of 20 to 30 degrees north and south of the equator. In these zones the average angle of receipt of the sun's rays is still close to vertical, while the cloud cover tends to be much less.

Processes transferring heat from one part of the earth to another are essential in maintaining its habitability. The differences in energy received from the sun are such that places near the equator receive more energy annually than they lose by radiation to outer space, while places near the poles radiate more energy than they receive. If heat were not transferred after its original receipt, low-latitude areas would become steadily hotter and high-latitude areas steadily colder, until both were uninhabitable. But warm air masses and ocean currents originating in lower latitudes transport heat to higher latitudes, and thus maintain an equilibrium ("heat balance") on the earth.

Extremes of heat and cold are tempered by the equalizing effects of moving air and ocean waters, but great differences in annual and seasonal temperatures from place to place still remain. In lowlands near the equator temperatures remain high throughout the year, and in areas near the poles temperatures remain low for most of the year. In the intermediate (middle) latitudes, well-marked seasonal changes of temperature occur, with warmer temperatures associated generally with the period of high sun (sun's rays striking the surface more nearly vertically) and cooler temperatures with the period of low sun (angle of impact more oblique). In addition, variability of temperature in these latitudes is increased by periodic incursions of polar or tropical air masses, bringing unseasonably cold or warm weather.

Absorption of earth radiation by air that is denser and that contains more water vapor and dust is greater than absorption by air that is thinner and contains less

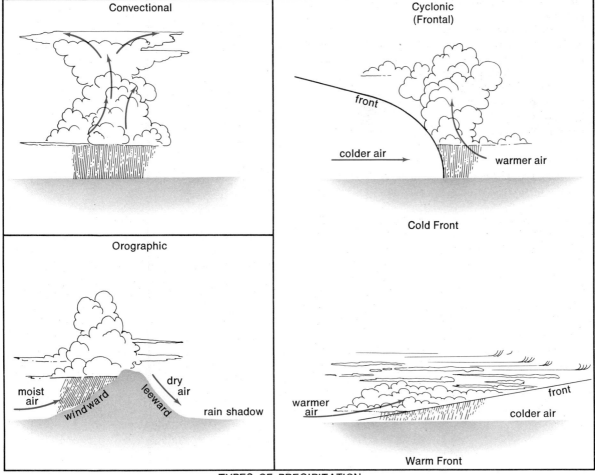

TYPES OF PRECIPITATION

of these materials. Such differences are characteristic between air at lower elevations and that at higher elevations. Consequently, temperatures at all latitudes are affected by altitude. On the average, air temperature decreases about 3.6°F for each increase of 1000 feet in altitude. Thus a place on the equator at an elevation of several thousand feet may have an average annual temperature comparable to that of a place in a middle-latitude lowland, though the tropical place will not experience the seasonal temperature changes of the middle latitudes.

Precipitation

Processes which cool the air are responsible for precipitation. Precipitation results when water vapor in the atmosphere is cooled to the point that it condenses and thus changes from a gaseous to a liquid or solid form. How much cooling is necessary depends on the original temperature of the air and the amount of water vapor in it, as warm air can hold more water vapor than cool air. Differences from place to place in type of air and occurrence of cooling processes sufficient to give precipitation are great enough that average annual precipitation ranges from practically none in many places to 400 inches or more annually in a few places.

For cooling and precipitation to occur, it is generally necessary for air to rise. This may be brought about in several different ways. In equatorial latitudes, or in the high-sun period elsewhere, air heated by intense surface radiation may rise rapidly, cool rapidly, and produce a heavy downpour of rain. Precipitation that originates in this way is called *convectional precipitation* (diagram p. 21). *Orographic* (mountain-associated) *precipitation* (see diagram) results when moving air strikes a topographic barrier and is forced upward. Some of the heaviest rainfall and snowfall in the world is produced in this way. Most of the moisture falls on the windward side of the barrier, and the lee side is apt to be excessively dry. Such dry areas are said to be areas of *rain shadow.* In some parts of the world the existence of arid or semiarid lands is due primarily to rain shadow. *Cyclonic* or *frontal precipitation* (see diagram) is generated in traveling low-pressure cells (*cyclones*) which bring different air masses into contact. A cyclone normally comprises a very extensive segment of the atmosphere, and may overlie hundreds of thousands of square miles of the earth's surface. In the atmosphere air moves from high pressure to low, just as water runs downhill, and a cyclone is a portion of the atmosphere into which different air masses are drawn. Generally one air mass will be cooler, drier, and more stable than the other. Such masses do not mix readily, but each tends to retain its separate identity. They are in contact within a zone which may be three to fifty miles wide, and is called a *front.* A front is named according to the air mass which is advancing. In a *cold front,* the cold air pushes in under the warm air and forces it upward and back. In a *warm front* the warm air rides up over the cold air and gradually pushes it back. But whether the front is a warm front or a cold front, precipitation is apt to result when the warmer air mass rises and condensation takes place. The middle latitudes, especially in winter, experience a succession of traveling cyclones moving from west to east in an airstream called the *westerly winds* (see diagram). These cyclones rotate slowly, and may be thought of as somewhat analogous to whirls and eddies that are carried along by a running stream. Normally two fronts are experienced as a cyclone passes. First comes the warm front, and then the cold front. But the cold front moves faster and eventually overtakes the warm front. When this occurs, the cyclone is said to be occluded, and it disappears from the atmospheric pressure map.

Large areas of the earth receive very low amounts of precipitation (see back endpaper map). Often such areas are primarily the result of high atmospheric pressure. Some parts of the atmosphere generally exhibit high pressures; these are known as semipermanent high-pressure cells or *anticyclones* (see diagram). In such a high, the air is descending, and thus becomes warmer as it comes under the increased pressure (weight) of the air above it. As it warms, its capacity to hold water vapor increases, and the result is minimal precipitation.

Cyclone (Northern Hemisphere): The entire system is moving west to east in the airstream of the westerly winds.

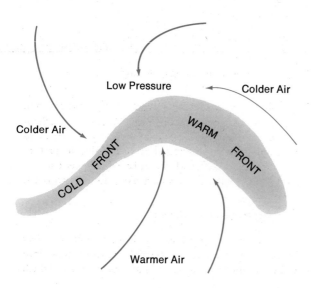

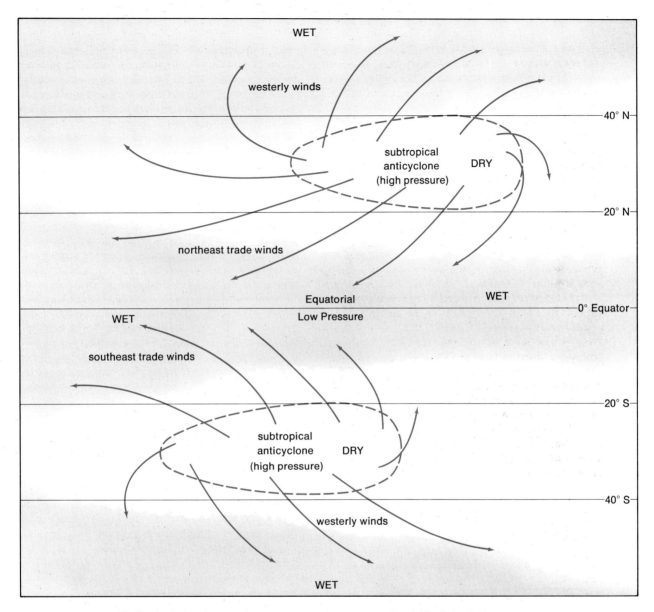

Idealized wind and pressure systems. Wetter areas are indicated by light shading.

The streams of dry, stable air that move outward from the anticyclones are apt to bring prolonged drought to the areas that lie in their path. Such is often the effect of the *trade winds*—streams of air that originate in semipermanent anticyclones on the margins of the tropics and are attracted equatorward by a semipermanent low-pressure cell, the *equatorial low* (see diagram). In some parts of the world, cold ocean waters appear to be mainly responsible for the existence of coastal deserts: air moving from sea to land is warmed, and thus instead of yielding precipitation, its capacity to hold water vapor is increased. But many, perhaps most, areas of excessively low precipitation stem from a combination of the foregoing influences. The Sahara of northern Africa, for example, seems to be primarily the result of high atmospheric pressures, but is also due in part to the rain-shadow effect of the Atlas Mountains and the presence of cold Atlantic waters along its western coast.

The Major Types of Climate

The foregoing processes produce an infinite variety of local climates. But it is possible to group these into a limited number of major *types of climate,* each of which occurs in more than one part of the world (see front endpaper climatic map). In the *ice-cap climate, tundra climate,* and *subarctic climate,* the dominant feature is the long, severely cold winter. In *desert* and *steppe climates* the dominant feature is aridity or semiaridity. Deserts and steppes occur both in low latitudes and in middle latitudes; the principal area, sometimes called the "Dry World," extends in a broad band across northern Africa and southwestern and central Asia. Rainy low-latitude climates include the *tropical rain forest* type and the *tropical savanna* type, the critical difference between them being that the tropical savanna type has a pronounced dry season whereas in the tropical rain forest climate the dry season is short or absent. The humid middle-latitude climates have mild to hot summers, and winters ranging from mild to cold. There are several relatively distinct types of climate in this group. The *marine west coast climate,* occupying the western sides of continents in the higher middle latitudes, has winters that are greatly moderated by the effects of warm ocean currents, and summers that tend to be on the cool side, though still warm enough for a great variety of crops. The *mediterranean climate* customarily occupies an intermediate location between a marine west coast climate on the poleward side and a steppe or desert climate on the equatorward side; in the high-sun period it lies under high atmospheric pressures and is rainless; in the low-sun period it lies in the westerly wind belt and receives precipitation of cyclonic or orographic origin. The *humid subtropical climate* occupies locations on the southeastern margins of continents and is characterized by hot summers, mild to cool winters, and ample precipitation for agriculture. The *humid continental climate* lies poleward of the humid subtropical type; it has cold winters, warm to hot summers, and sufficient rainfall for agriculture, with the greater part of the precipitation in the summer half-year. *Highland climates* exhibit a range of conditions according to altitude and exposure to wind and sun.

Biotic Processes and Natural Vegetation

Biotic processes are those which clothe the earth with living things, including man himself. Geographic interest in these processes and their results centers on those aspects that differ from one area to another. It does not focus primarily on aspects of biology which are universal or have little effect in rendering one area distinct from another.

One of the most striking and significant features of most areas is their natural vegetation—the part of their vegetation cover that is not cultivated or domesticated. The natural vegetation of an area is striking because it contributes so much to the appearance of an area's landscapes. It is significant because it provides resources or hindrances for man's activities and because it is very closely related to many other aspects of the area's natural environment. The particular type of natural vegetation in an area is influenced strongly, through various processes, by climatic conditions, light level and duration, slope and drainage, rock and soil type, and the area's wild fauna, as well as the evolutionary history of the flora itself. In addition, despite being called "natural" vegetation, it is often greatly influenced by such actions of man as burning, clearing, and the grazing of livestock. In turn, the natural vegetation exercises a powerful influence on certain other characteristics of an area, such as the quality of the soil, the speed of erosion, the nature of wild animal life, and the presence or absence of some economic potentials.

The geographic importance of different processes related to natural vegetation varies with the degree of generalization one employs. On a broadly generalized worldwide scale of observation and comparison, climatic conditions related primarily to intake of water by plants through their root systems and loss of it to the atmosphere by transpiration are the most important. Major types of natural vegetation are related to major types of climate closely enough that the vegetative results of the climatic conditions are often used to name the climate types; for example, the tropical rain forest type of climate or the tundra type of climate. However, when observation becomes more detailed, other factors assume considerable importance, and variations in natural vegetation are found to be related to various aspects of the natural environment other than climate.

Forests

When one looks at the broadly generalized world pattern of vegetation types (see back endpaper map), it is apparent that broadleaf forest tends to occupy the areas which are climatically most favorable to vegetative growth. Where heat and moisture are continuously, or almost continuously, available, the *tropical rain forest* is found. It is composed of broadleaf trees and is evergreen because the climate imposes no dormant season. The trees are comprised of more species per unit area than in any other forest, due to the ability of many species to compete in this optimum environment. Many trees grow straight and very tall in their competition with each other for sun, although smaller species requiring less sun flourish under the shade of the forest

The tropical rain forest. Many different species of broadleaf evergreen trees comprise the vegetation in this view. The photo shows a portion of the new highway, 1300 miles long, from Brazil's capital city of Brasília to the seaport of Belém at the mouth of the Amazon River. The Amazonian rain forest is the largest continuous expanse of such vegetation in the world. (Brazilian Government Trade Bureau.)

giants. At ground level the forest floor is so well shaded that there is little undergrowth until some circumstance, such as the occurrence of a river, breaks the dense leaf canopy overhead. In tropical areas with a noticeable dry season, but still enough moisture available for tree growth, the tropical rain forest is replaced by *tropical deciduous forest*. The broadleaf trees lose their leaves and cease to grow during the dry season, and then put on foliage and resume their growth during the wet season. The tropical deciduous forest approaches the luxuriance of rain forest in wetter areas, but declines to low, sparse scrub in drier areas. In middle-latitude areas with hot,

wet summers a *mid-latitude deciduous forest* is found, composed of broadleaf. species (oaks, maples, and the like) that are different from those in the tropical areas. Here the cold temperatures of winter freeze the water within reach of plant roots and render it seasonally unavailable. The trees shed their leaves and cease to grow, thus reducing water loss, and then resume their foliage and grow vigorously during the hot, wet summer when climatic conditions resemble those of rainy tropical lands.

Where water availability is high, needleleaf trees such as pine, spruce, and fir cannot, in general, compete successfully with the broadleaf trees. But in middle-

The pine forest of southeastern United States. Pine trees, often found in pure stands like the one in this view, grow rapidly in the humid subtropical climate of the lowland South and are an important source of pulpwood and lumber. The trees in the photo are longleaf pines, but the Southern pine forest also includes other species such as shortleaf pine or loblolly pine. In many parts of the South, particularly the uplands, pines are intermixed with deciduous hardwoods. A much smaller area of pine forest somewhat comparable in character to that of southeastern United States is found in extreme southern Brazil. (See Chapter 28 and back endpaper vegetation map). (U.S. Forest Service.)

However, needleleaf evergreen forests occupy some locations where temperature conditions are less severe than those of the taiga; for example:

1. In certain situations hot and wet enough for broadleaf trees, a porous sandy soil allows much water to percolate downward out of reach of tree roots, thus decreasing the available moisture and giving an advantage to needleleaf over broadleaf trees. An outstanding instance of this is the pine forests of the coastal plain along the Gulf of Mexico and Atlantic coasts of the American South. Many experts think, however, that repeated burning of this area helped to drive out the broadleaf trees and establish the dominance of pine. (Needleleaf trees are generally less vulnerable to fire than the broadleaf trees.)

2. Another type of situation where needleleaf conifers are sometimes dominant is to be found in areas that lie along the west coasts of continents and have the marine west coast type of climate. These areas are very wet, with mild winters, and summers that are warm but seldom hot. They appear to offer good conditions for broadleaf trees, and stands of these are found; but in some of these areas, particularly in North America, the forest is predominantly needleleaf. The reasons for this are not fully known, but the circumstance is fortunate for the United States and Canada in that the conifers, such as the redwoods of northern California and southern Oregon or the Douglas fir of Washington, Oregon, and southern British Columbia, grow to unusual size in these favorably moist conditions and are highly valuable commercially.

3. Finally, needleleaf forests are also found in certain tropical areas, principally in uplands or on porous soils.

As a rule, broad types of vegetation do not change abruptly along a line of division, although such a change

latitude forested areas where water conditions are somewhat less favorable, the needleleaf evergreen trees tend to become dominant. These trees, which are often called coniferous because most of them bear seed cones, are adaptable to a wide range of climatic conditions. In particular, they can stand long periods of freezing temperatures with attendant lack of water. Thus the most extensive needleleaf evergreen forests lie north of the broadleaf forests in the regions of subarctic climate which cover great areas of North America and Eurasia. They are generally called the northern coniferous or boreal forests, or, after Russian terminology, the *taiga* (see p. 296). The trees in these forests reach a considerable size and density in more favorable southerly locations, but become progressively smaller and sparser toward the north. Such forests are absent in the southern hemisphere because land masses in that hemisphere do not reach far enough south to provide the long, cold winters and short, warm summers in which the taiga forests develop.

may be represented symbolically by a line on a map. Rather, one type tends to grade into another through a transition zone. Thus there are sizable areas in which broadleaf trees and needleleaf trees compete with each other and are intermingled. Such transitional areas are designated as areas of *mixed forest*. They occur both north and south of the broadleaf forest zone of eastern United States, cover much of Europe and Russia, and occur in various other parts of the world. Within them stands of broadleaf trees often occupy places which are more favorable to tree growth, and needleleaf stands occupy less favorable spots.

Grasslands

Where still less moisture is available, trees cannot flourish, and grass becomes the dominant vegetation. If there were no trees, grass would be dominant over most of the world, as it is extremely adaptable to a very wide range of climatic conditions. But in areas wet enough to support forests it tends to be shaded out, and so it is ordinarily dominant only in areas with environments hostile to trees. One reason for its hardiness is that so much of the plant is underground in an extensive root system.

Grasslands that have short grass are called *steppes*. They are the semiarid transition zones between humid areas and truly arid deserts. They occur in areas with widely differing temperature conditions, both in the tropics and in middle latitudes. In general, their grasses grow higher and denser toward their wetter edges, and lower and sparser toward their drier edges. The amount of precipitation under which a steppe develops depends largely on temperature. Where temperatures are higher, evaporation is faster and an area can have a considerable amount of precipitation and still be too dry for forest; while areas with cooler temperatures have less evaporation and thus require a lower amount of precipitation to rule out forest and leave the land to grass. The major area of steppe in the United States occupies a broad band of country in the Great Plains along the eastern foot of the Rocky Mountains.

However, very large areas of grassland occur in some parts of the world where conditions do not appear to rule out a forest cover. These are the *prairies* of the middle latitudes and the *tropical savannas* of the low latitudes. The world's greatest expanse of prairie, now almost completely destroyed in favor of agriculture, was a triangular area in North America with its corners in Alberta, Texas, and Indiana. Before the area was settled, tall grasses ranging from knee-high to horse-high were dominant, although heat and moisture are sufficient for forest growth. The reasons for this anomaly are in dispute. There is a strong presumption that one

Savanna grassland in the New World tropics. The locale is the remote interior of Colombia to the east of the Andes. Here such grasslands, stretching into adjacent Venezuela, are known as the *Llanos*. The photo shows a ranch operated by a family of *mestizo* origin (a mixture of Indian and European). Cattle on the ranch are poor in quality, but hardy; they graze on the coarse and rather unnutritious natural grasses that surround the house. In the distance a band of trees follows the Rio Meta, a tributary of the Orinoco. (Standard Oil Company, N.J.)

Desert vegetation in a sandy part of the province of Hasa, eastern Saudi Arabia. Note the wide spacing of individual shrubs. In the almost rainless desert, plants cannot survive if they congregate too closely, but generous spacing maximizes the amount of soil moisture available to each one. The truck convoy in the photo belongs to an oil company. Immense oil deposits underlie deserts in the eastern part of the Arabian Peninsula, and these subsurface riches are bringing much wealth to lands that have very unpromising exteriors. (Standard Oil Company, N.J.)

important factor was repeated burning, which helped grass to replace forest, the latter being more vulnerable to fire.

Tropical savannas occur, like tropical deciduous forests, in low-latitude areas with marked wet and dry seasons. The savannas have scattered trees, and the deciduous forests are likely to have some grass on the forest floor. Thus the boundary between them is often indistinct and debatable, with forest gradually giving way to savanna as the dry season becomes longer and grass predominates more and more over trees. As in the case of prairies, the origin of the vast savannas of the tropics is a matter of scientific debate, with the presumption that repeated burning has probably been at least one important factor favoring grasses over trees. According to local conditions, savanna grasses vary between about one foot and twelve feet in height (see photo, p. 27).

Vegetation of Deserts, "Mediterranean" Climate, and Tundra

Three extreme climatic environments in which aridity plays a large role—desert, "mediterranean" climate, and tundra—are characterized by distinctive types of vegetation. The *deserts* of middle and low latitudes are too dry to support a cover of either trees or grass. For the most part they have a scattered *desert shrub* vegetation, although exceptional areas may have no vegetation

at all. Desert shrub consists of bushy plants which are *xerophytic;* that is, possess special adaptations for withstanding drought. Among these adaptations are (1) small leaves, providing a minimal leaf surface for transpiration (loss of water to the atmosphere); (2) special qualities of leaf and bark to reduce transpiration; and (3) extensive or deep root systems to gather moisture.

A rather similar, though more luxuriant, vegetation characterizes areas of the "mediterranean" type of climate. This climate exhibits an extreme quality in the desertlike dryness of its hot summers, which are combined with mild, wet winters. Due to the summers the vegetation must be xerophytic. It consists primarily of many species and sizes of shrubs. These are predominantly evergreen and broadleaf, with adaptations such as thick bark and waxy or leathery leaves to retard transpiration and thus conserve moisture during the dry summers. Such vegetation is called by a variety of names, including *maquis* and *chaparral.*

Still another extreme environment is the *tundra.* Here vegetation must be adapted to survive a long, cold winter when moisture is unobtainable because it is frozen, and a very short and very cool summer. Tundra vegetation includes mosses, lichens, shrubs, dwarfed trees, and some grass in favored spots. Over large areas there is practically no vegetation. As in deserts after a rare rainfall, ephemeral wild flowers will bloom during the summer.

Summary

Thus a process of competition among plants for moisture appears to explain in broad outline the distribution of major types of natural vegetation in the world. Broadleaf forests generally require more moisture than other types of natural vegetation and, where they can flourish, tend to exclude other types. Needleleaf forests can exist within a wider range of moisture conditions. They can and do exist in very moist conditions, but are usually excluded from such areas by the competition of broadleaf trees. However, they can also flourish in areas that have a smaller total of available moisture, and in areas where a long dry season must be endured each year. Hence they often, but not always, occupy areas less favorable in moisture supply than those occupied by the broadleaf type of forest. Grass can grow in a still wider range of conditions, from areas wet enough for rain forest to areas of near-desert conditions. It tends to be dominated or excluded by forest in wetter areas, however, and to be left as the dominant natural vegetation only in areas too dry for forest growth. Where even grass is largely excluded, shrubs and tundra plants are dominant.

Soil-forming Processes and the Major Groups of Soils

Variations from place to place in the thin earth mantle of decomposed rock and decayed organic material which is called soil are among the most significant of geographic patterns. The quantity and quality of foodstuffs and other vegetable products which man can wrest from various parts of the earth are intimately related to the characteristics of the soil and to man's ability to take advantage of these characteristics or to overcome them. And most other development of areas is severely handicapped if the soil cannot be made to yield an adequate agricultural production.

Soils vary enormously and in extremely complex ways in response to a wide variety of factors. These are often quite intricate in their operation. Some soils, even with little care, will yield abundant and varied agricul-

A summer scene in Arctic tundra. The photo shows the base camp of a Canadian government surveying party on Ellesmere Island, Northwest Territories. Here in the extreme environment of the high Arctic not a tree can be seen, but a low carpet of grass, moss, and lichens manages to survive. (National Film Board of Canada, Ottawa.)

tural products for many years. Some will yield very small quantities or only a small variety of useful products, or will deteriorate rapidly with use. Almost every intermediate point on the spectrum of soil quality between these extremes occurs in various places on the earth, presenting man with innumerable opportunities and problems in the exploitation of this vital resource.

Despite this complexity, however, it is possible to call attention to a few major circumstances and processes which help to make the geography of soils comprehensible in broad outline. This possibility is due to the fact that the characteristics of soils are closely related to other major aspects of nature, notably to climate, vegetation, and rock types. Many of the important characteristics of a soil are often due to the type of rock whose weathering has provided the initial soil material. Other soil characteristics are due to the effects of temperature and moisture, while still others are due to the effects of the natural vegetation whose decomposed remains, called *humus,* are incorporated into the soil.

Leaching and Its Effects

Leaching is a very important process contributing to broad and significant differences in the soil from one area to another. Plants are nourished by taking into their roots, in solution, a variety of minerals from the soil. The more adequate the mineral content of the soil, the more fertile the soil will be, other things being equal. Leaching is the process by which mineral plant foods are removed from the soil in solution. Since plants cannot utilize minerals unless the latter are dissolved in water, adequate water content within the soil is absolutely essential. Excess water, however, will dissolve the important minerals and carry them by underground percolation into strata beyond the reach of plant roots. Eventually the minerals will be carried to rivers and the sea. Thus leaching is a process of impoverishment of the soil, lowering its useful mineral content. As this process depends on moisture supply, there is a strong general tendency for areas which receive more precipitation to have soils that are more leached. Because they are more lacking in minerals required by plants, these soils are less fertile than many soils in areas of lower precipitation. Such areas are generally wet enough to have a natural vegetation of forest. Thus we can say as a broad generalization that humid, naturally forested areas tend to have soils impoverished by leaching. On the other hand, the soils of drier areas in which the natural vegetation is grassland or desert vegetation tend to be leached relatively little or not at all. Unfortunately, the high content of mineral nutrients in such soils may be of limited utility or no utility because of inadequate

moisture for most plants. In addition, too high a concentration of mineral salts, as in many dry-area soils, may itself be toxic to useful plants.

The amount of leaching a soil is subjected to is not entirely a function of the amount of precipitation in the area. The type of vegetation and temperature conditions can also influence it strongly. Thus areas with a natural vegetation of needleleaf forest tend to have heavily leached soils, partly because the mat of needles which accumulates on top of the soil imparts an acid character to water seeping through it. The resulting acidity of the water in the soil tends to speed the dissolution of mineral nutrients. Acidity of soil water under broadleaf forests tends to be less strong, and soils in such areas, particularly in the middle latitudes, tend to be somewhat better. Warm temperatures also speed leaching. Where winters are not cold enough for the soil to be frozen, the leaching process can proceed the year around instead of only part of each year. In addition, under high temperatures there is a general speeding up of chemical reactions, such as the dissolution of mineral plant foods. Thus very heavily leached soils are characteristic both of subarctic areas, with their vegetation of needleleaf forest, and of many tropical areas that have heavy precipitation and high temperatures. Such areas stand at the opposite end of the scale of soil leaching from deserts and dry grasslands. Heavily leached soils in the tropics, like those of subarctic areas, tend to be highly acid in reaction, due to the fact that bases are leached out. A highly acid chemical environment is unfavorable for most crop plants.

Decay and Absorption of Organic Material

A second important process contributing to broad differentials of soil fertility in different parts of the world is the decay and absorption into the soil of organic material. Such material is derived principally from the vegetation which grows on the land: natural vegetation before use of the soil or in intervals of nonuse, and crop remains in soils being farmed. This decomposed organic soil material, called humus, is a very important element in soil productivity. It adds to the store of mineral plant foods in the soil, especially the very critical element nitrogen; contributes to a physical structure of the soil which favors air circulation and root development; adds to the ability of a soil to store water; and retards leaching. Rock particles without organic material do not form a true soil.

Humus accumulation in the soil under natural conditions varies greatly from one part of the earth to another. It is generally highest in grassland soils because dryness slows the action of microorganisms which destroy organic material. It is somewhat lower in mid-latitude broadleaf forest areas, which are wetter. It is very low in three types of areas: deserts, which have little plant cover to provide humus; wet tropical areas, where heat, moisture, and the activity of microorganisms are so intense as to destroy and remove most dead organic material before it can be incorporated in the soil; and areas with a needleleaf forest vegetation, where the mat of needles on the surface of the soil is resistant to decomposition—an effect especially marked in cold environments where the action of microorganisms is greatly retarded.

Other Factors Affecting Soil Quality

The amount of leaching and of humus are important determinants of the quality of soils but there are, in addition, a number of other factors which display less regular patterns of relationship with climate and vegetation.

1. Some areas have stony soils where the rock has not been completely weathered. Such soils are likely to be difficult to work.

2. The soils of poorly drained areas tend to be waterlogged to such an extent that air in the soil, necessary for most plants, is deficient. However, they have a very high organic content and often become quite fertile when drained.

3. Soils which are too sandy are usually poor for several reasons. Sand is composed of particles of relatively large size compared to those that compose silt or clay. It offers a relatively small particle-surface area within a given volume compared with soils made up of finer particles. This handicaps it in supporting plant life, since plant roots feed only from the surface area of particles. In addition, sand holds water less well than soils composed of finer particles and therefore often provides drought conditions for plants in areas where moisture might otherwise be adequate.

4. Claypan or hardpan soils often develop in very level areas in both middle and low latitudes. These have a relatively impermeable layer not far below the surface, due to a high concentration of small particles at certain depths as a result of water movements within the soil. Sometimes these particles are actually cemented. The result is retardation of downward and upward movement of water within the soil. After a rain the soil tends to waterlog, as water cannot permeate downward through the pan; in a dry period the soil becomes excessively dry, as water from below cannot move upward through the pan by capillary action to replace that lost by evaporation and transpiration. The pan may also physically inhibit the downward penetration of roots. However, it may be valuable for some kinds of agriculture, for instance in ponding water in a rice paddy.

Certain widespread conditions and processes often result in soils superior to those of surrounding areas with similar climatic and vegetative conditions.

1. In many places soils resulting from the weathering of underlying limestone have superior character. A high lime content in a soil contributes to giving it an internal structure with abundant pore space for circulation of water and air; and counteracts acidity. In humid areas lime is often applied to cultivated land for these purposes.

2. Soils formed from alluvium deposited on the flood plains of rivers are often, though not always, unusually productive. They are level and easily cultivated, and may be subject to natural renewal of fertility by flooding. They are likely to have a rich humus content derived from swamp and marsh vegetation; they are also likely, of course, to require extensive flood control and drainage operations.

3. A third type of soil which is often unusually productive is soil derived from loess. Loess is material which has been picked up, transported, and deposited in its present location by wind. This means that it consists of very fine particles. Such particles, as mentioned above, afford a large amount of feeding surface for plants, along with adequate pore space for the necessary air and water in the soil.

4. Finally, certain volcanic materials weather into exceptionally fertile soil. This is especially noteworthy in a number of tropical areas where generally poor soils may be interspersed with much better ones derived from volcanic deposits. In some locations soil fertility is renewed intermittently by volcanic ash settling out of the air, or new lavas being eroded, transported by streams, and deposited as alluvium.

Man's Role in Determining Soil Quality

Man himself is an agent of the greatest importance in determining the quality of soils. In cultivating the land, or grazing animals on it, he can either increase or decrease its fertility. His activities in decreasing soil fertility generally receive more notice. They include (1) destruction of the plant cover by overgrazing or clean-tilled cultivation, leading to excessive erosion; (2) repeated burning of the vegetation cover, thus destroying humus content; and (3) cropping the land too continuously, thus depleting reserves of mineral nutrients. However, man also can do many things—given proper knowledge, incentive, and means—to improve the quality of soils. Bogs can be drained; mineral content built up by fertilizers; nitrogen added by growing legumes, whose roots serve as hosts to nitrogen-fixing bacteria; acidity corrected by liming; erosion checked by proper cultivation and crop selection; and organic content increased by manuring. There are at least small parts of the world where the soil is probably more productive today than when cultivation began centuries ago. Thus man must be added to such natural factors as leaching, humus supply, and material of origin as a major reason why soils exhibit great differences from one region to another.

MAN'S ROLE IN SHAPING THE GEOGRAPHY OF AREAS

The major processes of geographical significance that originate with human beings are so intertwined that they are very difficult to separate into categories. For the purpose here, we may list them as exploration, migration, cultural diffusion, and area organization and development. These processes have made their contribution to the shaping of distinctive areas when groups of people have investigated and occupied these areas, and have organized and developed them. This has involved the movement of groups possessing distinctive cultures into areas possessing distinctive natural environments, and in many instances already occupied by previous migrants. Sometimes the previous inhabitants have invited the newcomers to settle in the area and sometimes not; sometimes the entry of the new group has been peaceful, and sometimes it has been warlike. The previous inhabitants have on occasion been eliminated or displaced, or they have been absorbed, or simply left alone while the newcomers settled among them. Every new group settling in an area has attempted to organize and develop the area in terms of its particular needs and wants, its culture, its technology, and the opportunities provided by the area within the specific context of time and place. At the heart of the process has been the selection and development of individual sites or areas for particular enterprises and institutions: homes, farms, mines, factories, stores, warehouses, offices, banks, schools, churches, hospitals, hamlets, villages, towns, cities, roads, railways, airports, governmental units at various levels, and others. In the course of time the area has become an organized web of political, economic, and social units interconnected with each other and with the outside world by lines of transportation and communication. From time to time the inhabitants have devised new ways of using and organizing the area, or new ideas have reached them from other parts of the world through the entry of new settlers or through diffusion of ideas along established channels of communication. Both the area itself and the culture of its inhabitants have been subject to modifications, due partly to experiences of the occupying group in its new environment and partly to the contacts of the group with the outside world. In most parts of the world this activity has been going on for a long time, and in many places the geographical results are very complex. Elsewhere there are still some areas that are being effectively occupied and organized by human beings for the first time.

Geographical Results of European Expansion

Since the Age of Discovery, the most far-reaching human influence on the geography of the world has been exerted by people of European culture who initially fanned out over the globe by sea from the European homeland. This movement, the "Expansion of Europe," is described in some detail on pages 61–64. Throughout this text, reference is made constantly to the impact of Europeans on the geography of areas they have colonized, controlled, or influenced. The activities of Europeans revolutionized the geography of large sections of the earth. Most areas outside of Europe have been controlled politically by a European colonial power at one time or another, and most of the international boundaries that appear on the world political map were established by Europeans. The system of sovereign national states that now extends over the earth originated in Europe, and European ideas about government and politics greatly influence the political life of most countries. Europeans have migrated in large numbers to overseas areas and have carried with them the languages and cultures of their homelands. Today most people in the Americas, Australia, and New Zealand, as well as minorities in South Africa and various other countries, speak European languages. Even in newly independent Asian and African countries where there are few people of European descent, educated persons are generally conversant with the language of the European colonial power that was formerly in control. This is a decided asset to developing countries like India or Zambia in which many languages are used. In these countries European languages (for example, English in India and Zambia) afford a common means of communication for educated persons from all sections of the country. Many newly independent countries actually employ European tongues as official languages. It may be noted that the widespread use of these languages greatly simplifies the work of the United Nations and other international organizations. The Industrial Revolution which began in Europe has transformed large sections of the earth, and is still spreading. Europeans organized, and in large measure still operate, the worldwide system of transportation, communication, and trade. Even the world's commercial agriculture has been developed mainly by Europeans or their descendants.

The Present Tempo of Geographical Change

Today the worldwide process of migration by individuals or groups of many different cultures and nationalities is still going on, though at a slower pace than formerly. The process of cultural diffusion also continues, and at an accelerating rate, as a result of constantly improving networks of communication. The rapid spread of new ideas, coupled with the continuing search for greater security and more acceptable living arrangements, is stimulating a reinvestigation and reorganization of areas and cultures, is bringing new resources into play, and is creating new uses for old ones. To keep abreast of geographical change in this extremely fluid situation is a very difficult assignment. But even in the midst of rapid change, there are certain major characteristics of areas that tend to endure, and a thorough knowledge of these can help us keep perspective and avoid being overwhelmed by the flood of new details. To portray these long-term characteristics in an orderly way is a major purpose of this text.

THE ROLE OF MAPS IN GEOGRAPHICAL STUDY

In studying the geographical results of natural and human processes, it is essential that the student make constant use of maps. By and large the field of geography is concerned with things that can be mapped. Maps are indispensable records of man's exploration, occupation, use, and organization of the earth. By means of symbols they portray systematically the geographical aspects of man's unfolding knowledge about the world he inhabits. Essentially what maps do is to show places and things on the earth in their correct spatial relationship to each other. They are of great assistance in enabling us to grasp the pattern of geographical features which confers distinctiveness on each part of the world.

Maps make it possible for us to extract certain items from the totality of things, to see the patterns of distribution they form, and to compare these patterns with each other. They help us to grasp the internal layout and organization of areas and the connections which exist among areas or among different geographical components of areas. Maps can be used, at least in theory, to portray almost anything that shows a variation from one place to another. Some maps show the distribution and interrelations of things which are fixed in place—man-made structures, terrain features, and so on—while other maps portray the flow of people, goods, and ideas from place to place.

It is essential to keep in mind that a map is not a complete record of an area. It incorporates a selection of certain details that the map maker records to accomplish his particular purpose. These details are shown by symbols. Among the more common map symbols are those which represent fixed objects like railroads or towns, or quantities such as 1000 people or 1000 tons of coal, or averages such as 20 inches average annual rainfall. Dots are commonly used symbols for portraying quantities, and the *dot map* is one of the commonest types of maps showing the distribution of people or things on the earth. In a dot map of population, for example, each dot represents a stated number of people, and the dot is placed as near as possible to the center of the geographical space occupied by these people. In interpreting such a map, we are not greatly concerned with the individual dots, but rather with the way the dots are arranged; that is, with the *pattern* that they form. The best means of portraying almost any distribution on the earth's surface is to show it on a map. One can readily imagine what a hopeless task it would be to describe the distribution of the world's people in words. But by making a map, we are able to see the broad pattern at a glance (map, p. 42).

Maps showing different distributions can be compared with each other to see what correlations exist; for example, between rainfall and crop yields. Geographers are much concerned with the idea that the various elements in the geographical makeup of areas exist together in interlocking combinations, and they try to discern and interpret the specific interrelationships among these features. In this endeavor, the map is directly useful as a tool of analysis only on a rather simple level. But maps are an essential record of data, and the study of maps reveals endless problems that require geographical analysis. Much geographical research is concerned with examining the interrelationships among distributions shown on maps. Such research often involves elaborate statistical procedures and general theories as to why certain types of observed relationships exist.

Map Projections

All maps introduce a certain amount of distortion, because it is not possible to represent the curved surface of the earth with complete accuracy on a flat sheet of paper. If the area being represented is very small, the amount of distortion may be slight enough to be disregarded, but maps that represent larger areas may introduce very serious distortions. A classic instance of this is the well-known Mercator projection (p. 631). This projection was designed as an aid to navigation, and both its parallels of latitude and its meridians of longitude are shown as straight lines. On the globe, the lines of longitude draw closer together from the equator to the poles, where they converge, but on a Mercator map these lines remain parallel. Thus a Mercator map greatly exaggerates the east-west dimension of areas near the poles. In addition it exaggerates the north-south dimension, since the parallels are not spaced evenly between the equator and the poles, as they are on the globe, but gradually draw farther and farther apart with increasing distance from the equator. On a Mercator map the *mathematical location* (latitude and longitude) of every place is accurate, but areas near the poles are stretched out of shape and are greatly distorted in size, with the familiar result that Greenland appears bigger than South America, though in reality it is only one-eighth as large. Thus we cannot use the Mercator projection for visual comparison of the sizes of continents or countries; for that purpose we must use an *equal-area projection* (p. 1). Almost endless ways have been devised to project the curved surface of the globe onto a plane. One must take care not to use a particular projection for some purpose for which it was not designed. It must be remembered that no flat map can simultaneously represent true direction, true distance, true shape, and true area for a particular segment of earth space. For this a globe is required.

Map Scale

Students should keep in mind the importance of map *scale.* The scale of a map is the actual distance on the earth that is represented by a given linear unit on the map. This relationship may be shown on a *graphic scale,* with tick marks representing units of distance, or it may be stated in words, for instance, one inch [on the map] equals one mile [on the ground], or it may be expressed as a *representative fraction,* for example 1/62,500, wherein one unit on the map is equivalent to 62,500 of the same units on the ground. A large-scale map is one with a relatively large representative fraction (for example, 1/5000); a small-scale map is one with a relatively small representative fraction (for example, 1/5,000,000). A large-scale map permits more detail to be shown for a given area, whereas small-scale maps permit fewer details and require an increasing degree of generalization as the scale of the map grows less. On a map with a scale of 1/5000, a house can be represented in its true dimensions by a black square on the map, but on a map with a scale of 1/5,000,000 it would obviously be quite impossible to represent a single house at true scale. To convey relative magnitudes effectively, symbols on a small-scale map (for instance, symbols denoting cities of different sizes) often occupy disproportionately large portions of the map sheet.

Value of Maps in Spatial Planning

Maps are exceedingly valuable tools, not only for helping us understand the human record, but also for helping us plan the future. No agency concerned with planning things which involve location can accomplish its purpose properly without the use of maps. One can readily imagine the difficulties of a city or regional planning board, or almost any governmental agency, if it had to operate without maps.

REFERENCES and READINGS

General Physical Geography and Landform Geography

BACHMANN, H. G., "The Origin of Ores," *Scientific American,* **202,** no. 6 (June 1960), 146–156.

BARNETT, LINCOLN, and the Editorial Staff of *Life, The World We Live In* (New York: Time Inc., 1955).

BERTIN, LEON, *Larousse Encyclopedia of the Earth* (2d ed.; London: Paul Hamlyn, 1965). Illustrated with great numbers of photographs, many in color.

BULLARD, FRED M., *Volcanoes: In History, in Theory, in Eruption* (Austin: University of Texas Press, 1962).

BUNNETT, R. B., *Physical Geography in Diagrams* (New York: Frederick A. Praeger, 1968).

DOHRS, FRED E., and LAWRENCE M. SOMMERS, eds., *Physical Geography: Selected Readings* (New York: Thomas Y. Crowell Company, 1967).

DURY, G. H., *The Face of the Earth* (Baltimore, Md.: Penguin Books, 1959).

DYSON, JAMES L., *The World of Ice* (New York: Alfred A. Knopf, 1962).

KENDALL, HENRY M., ROBERT M. GLENDINNING, and CLIFFORD H. MacFADDEN, *Introduction to Physical Geography* (rev. ed.; New York: Harcourt, Brace & World, 1967).

KUENEN, P. H., *Realms of Water: Some Aspects of Its Cycle in Nature* (New York: John Wiley & Sons, 1956).

Life Nature Library, by various authors and the Editors of *Life* (New York: Time Inc.), particularly A. Starker Leopold. *The Desert* (1961), Peter Farb, *The Forest* (1961), and Louis J. Milne and Margery Milne, *The Mountains* (1962).

LOBECK, A. K., *Geomorphology: An Introduction to the Study of Landscapes* (New York: McGraw-Hill Book Company, 1939). Fine diagrams and photographs; succinct text.

McDIVITT, JAMES, *Minerals and Men: An Exploration of the World of Minerals and Its Effect on the World We Live In* (Baltimore, Md.: Johns Hopkins Press, for Resources for the Future, 1965).

McINTYRE, MICHAEL P., *Physical Geography* (New York: The Ronald Press Company, 1966).

MONKHOUSE, F. J., *Principles of Physical Geography* (6th ed.; New York: Philosophical Library, 1964); and *Landscape from the Air: A Physical Geography in Oblique Air Photographs* (Cambridge: At the University Press, 1959).

MOORE, W. G., *A Dictionary of Geography: Definitions and Explanations of Terms Used in Physical Geography* (rev. and enl. ed.; New York: Frederick A. Praeger, 1967).

MURPHY, RICHARD E., "Landforms of the World," Map Supplement No. 9, *Annals of the Association of American Geographers,* **58**, no. 1 (March 1968), with accompanying text by Norman J. W. Thrower, pp. 198–200. A large map sheet in color.

PEARSON, ROSS NORTON, *Physical Geography* (College Outline Series; New York: Barnes & Noble, 1968).

SHELTON, JOHN S., *Geology Illustrated* (San Francisco: W. H. Freeman and Co., 1966). Numerous diagrams and air photographs.

STRAHLER, ARTHUR N., *Introduction to Physical Geography* (New York: John Wiley & Sons, 1965); also, *Physical Geography* (3d ed.; New York: John Wiley & Sons, 1969); and *The Earth Sciences* (New York: Harper & Row, 1963).

TREWARTHA, GLENN T., ARTHUR H. ROBINSON, and EDWIN H. HAMMOND, *Physical Elements of Geography* (5th ed.; New York: McGraw-Hill Book Company, 1967). The same material is incorporated in the larger *Elements of Geography,* cited below (Trewartha *et al.*).

VAN RIPER, JOSEPH E., *Man's Physical World* (New York: McGraw-Hill Book Company, 1962).

WALTON, KENNETH, "The Unity of the Physical Environment," *Scottish Geographical Magazine,* **84**, no. 1 (Apr. 1968), 5–14.

WYCKOFF, JEROME, *Rock, Time, and Landforms* (New York: Harper & Row, 1966).

Climate

BAILEY, HARRY P., "Toward a Unified Concept of the Temperate Climate," *Geographical Review,* 54, no. 4 (October 1964), 516–545.

BLUMENSTOCK, DAVID I., *The Ocean of Air* (New Brunswick, N.J.: Rutgers University Press, 1959).

BROOKS, C. E. P., *Climate in Everyday Life* (New York: Philosophical Library, 1951).

CHANG, JEN-HU, *Climate and Agriculture: An Ecological Survey* (Chicago: Aldine Publishing Company, 1967).

FAIRBRIDGE, RHODES W., ed., *The Encyclopedia of Atmospheric Sciences and Astrogeology* (Encyclopedia of Earth Sciences, Vol. 2; New York: Reinhold Publishing Corp., 1967).

HARE, F. KENNETH, *The Restless Atmosphere* (4th ed. London: Hutchinson University Library, 1966).

JAMES, PRESTON E., "A New Concept of Atmospheric Circulation," *Journal of Geography,* 63, no. 6 (September 1964), 245–250.

KENDREW, W. G., *The Climates of the Continents* (5th ed.; Oxford: Clarendon Press, 1961). A major reference, with many useful tables of climatic data.

NELSON, H. L., *Climatic Data for Representative Stations of the World* (Lincoln: University of Nebraska Press, 1968).

RIEHL, HERBERT, *Introduction to the Atmosphere* (New York: McGraw-Hill Book Company, 1965).

RUMNEY, GEORGE R., *Climatology and the World's Climates* (New York: The Macmillan Company, 1968).

SEWELL, W. R. DERRICK, ROBERT W. KATES, and LEE E. PHILLIPS, "Human Response to Weather and Climate: Geographical Contributions," *Geographical Review,* **58,** no. 2 (April 1968), 262–280.

STEWART, GEORGE R., *Storm* (Modern Library; New York: Random House, 1947). A highly readable novel describing a middle-latitude cyclonic storm and its effects on the lives of people.

TREWARTHA, GLENN T., *The Earth's Problem Climates* (Madison: University of Wisconsin Press, 1961).

U. S. DEPARTMENT OF AGRICULTURE, *Climate and Man: Yearbook of Agriculture, 1941* (Washington, D.C.: Government Printing Office, 1941).

Biogeography and Soils

BUNTING, BRIAN T., *The Geography of Soil* (Chicago: Aldine Publishing Company, 1965).

CRITCHFIELD, WILLIAM B., and ELBERT L. LITTLE, *Geographic Distribution of the Pines of the World* (U.S. Department of Agriculture, Forest Service, Miscellaneous Publication No. 991; Washington, D.C.: Government Printing Office, 1966).

DANSEREAU, PIERRE, *Biogeography: An Ecological Perspective* (New York: The Ronald Press Company, 1957).

DARLINGTON, PHILIP J., Jr., *Zoogeography: The Geographical Distribution of Animals* (New York: John Wiley & Sons, 1957), particularly Chap. 1, "Introduction," pp. 1–38, and Chap. 11, "Principles of Zoogeography; Geographical History of Man," pp. 617–646.

de LAUBENFELS, DAVID J., "The Variation of Vegetation from Place to Place," *Professional Geographer,* **20,** no. 2 (March 1968), 107–111; and "The Status of 'Conifers' in Vegetation Classification," *Annals of the Association of American Geographers,* **47,** no. 2 (June 1957), 145–149.

EYRE, S. R., *Vegetation and Soils: A World Picture* (2d ed.; Chicago: Aldine Publishing Company, 1968).

HADEN-GUEST, STEPHEN, JOHN K. WRIGHT, and EILEEN M. TECLAFF, eds., *World Geography of Forest Resources* (American Geographical Society Special Publication No. 33; New York: The Ronald Press Company, 1956).

MATHER, JOHN R., and GARY A. YOSHIOKA. "The Role of Climate in the Distribution of Vegetation," *Annals of the Association of American Geographers,* **58,** no. 1 (March 1968), 29–41.

OOSTING, HENRY JOHN, *The Study of Plant Communities: An Introduction to Plant Ecology* (2d ed.; San Francisco: W. H. Freeman and Co., 1956).

RILEY, DENNIS, and ANTHONY YOUNG, *World Vegetation* (Cambridge: At the University Press, 1966).

U.S. Department of Agriculture, *Soils and Men: Yearbook of Agriculture, 1938,* and *Soil: The Yearbook of Agriculture,* 1957 (Washington, D.C.: Government Printing Office, 1938; 1957).

General Cultural and Human Geography

BROEK, JAN O. M., and JOHN W. WEBB, *A Geography of Mankind* (New York: McGraw-Hill Book Company, 1968).

BURTON, IAN, and ROBERT W. KATES, eds., *Readings in Resource Management and Conservation* (Chicago: University of Chicago Press, 1965).

DICKEN, SAMUEL N., and FORREST R. PITTS, *Introduction to Human Geography* (Waltham, Mass.: Blaisdell Publishing Co., 1963).

DOHRS, FRED E., and LAWRENCE M. SOMMERS, eds., *Cultural Geography: Selected Readings* (New York: Thomas Y. Crowell Company, 1967).

ENGLISH, PAUL WARD, "Landscape, Ecosystem, and Environmental Perception: Concepts in Cultural Geography," *Journal of Geography*, 67, no. 4 (April 1968), 198–205.

HOYT, JOSEPH BIXBY, *Man and the Earth* (2d ed.; Englewood Cliffs, N.J.: Prentice-Hall, 1967).

JAMES, PRESTON E., "A Conceptual Structure for Geography," *Journal of Geography*, 54, no. 7 (October 1965), 292–298; also, *One World Perspective* (Waltham, Mass.: Blaisdell Publishing Co., 1965); and, with the collaboration of Hibberd V. D. Kline, *A Geography of Man* (3d ed.; Waltham, Mass.: Blaisdell Publishing Co., 1966).

JONES, EMRYS, *Human Geography: An Introduction to Man and His World* (New York: Frederick A. Praeger, 1965).

LEIGHLY, JOHN, ed., *Land and Life: A Selection from the Writings of Carl Ortwin Sauer* (Berkeley and Los Angeles: University of California Press, 1963).

MARSH, GEORGE PERKINS, *Man and Nature*, edited by David Lowenthal (Cambridge, Mass.: The Belknap Press of Harvard University Press, 1965). A famous work originally published in 1864 and subtitled "Physical Geography as Modified by Human Action."

SPENCER, J. E., and WILLIAM L. THOMAS, Jr., *Cultural Geography: An Evolutionary Introduction to Our Humanized Earth* (New York: John Wiley & Sons, 1969).

THOMAS, WILLIAM L., JR., ed., *Man's Role in Changing the Face of the Earth* (Chicago: University of Chicago Press, 1956). A monumental collection of articles by well-known authorities.

TREWARTHA, GLENN T., ARTHUR H. ROBINSON, and EDWIN H. HAMMOND, *Elements of Geography* (5th ed.; New York: McGraw-Hill Book Company, 1967), pt. 3, "Cultural Elements of Geography," pp. 553–645.

WAGNER, PHILIP L., *The Human Use of the Earth* (New York: The Free Press of Glencoe, 1960); and, with Marvin W. Mikesell, *Readings in Cultural Geography* (Chicago: University of Chicago Press, 1962).

Maps

BAGROW, LEO, *History of Cartography* (rev. ed., trans. by D. L. Paisey; Cambridge, Mass.: Harvard University Press, 1964).

BROWN, LLOYD A., *The Story of Maps* (Boston: Little, Brown, and Company, 1949; New York: Bonanza Books, 1949).

DICKINSON, G. C., *Statistical Mapping and the Presentation of Statistics* (London: Edward Arnold, 1963).

FISHER, IRVING, and O. M. MILLER, *World Maps and Globes* (New York: Oxford University Press, Essential Books, 1944). Illustrates some of the distortions associated with different types of map projections.

GREENHOOD, DAVID, *Mapping* (Chicago: University of Chicago Press, 1964).

GREGORY, S., *Statistical Methods and the Geographer* (London: Longmans, Green and Co., 1963).

LOBECK, A. K., *Things Maps Don't Tell Us: An Adventure into Map Interpretation* (New York: The Macmillan Company, 1956).

PORTER, PHILIP W., *A Bibliography of Statistical Cartography* (Minneapolis: Department of Geography, University of Minnesota, 1964).

RAISZ, ERWIN, *General Cartography* (2d ed.; New York: McGraw-Hill Book Company, 1948); and *Principles of Cartography* (New York: McGraw-Hill Book Company, 1962).

ROBINSON, ARTHUR H., *Elements of Cartography* (2d ed.; New York: John Wiley & Sons, 1960).

Key Topics
in the Geographical
Interpretation
of Countries and Regions

3   To assess the geography of a country or region is a complex task. The chief problem is to avoid being overwhelmed by a heterogeneous mass of detail. If a geographic treatment of an area is not to become a mere compendium of assorted facts, a high degree of selectivity is essential. One must marshall the available data in such a way that an orderly, coherent, and meaningful picture emerges. No two geographers will approach this task in quite the same way. Procedures in the geographical interpretation of areas are not so formalized as they are in the exact sciences, and the field of geography often has been characterized as both an art and a science. A noted geographer, H. C. Darby, has summarized this situation as follows:[1]

In historical thinking there has been great debate about whether history is a science or an art. It was perhaps an unnecessary debate because history, like geography, is both. Geography is a science in the sense that what facts we perceive must be examined, and perhaps measured, with care and accuracy. It is an art in that any presentation (let alone any perception) of those facts must be selective and so involve choice, and taste, and judgment.

[1] H. C. Darby, "The Problem of Geographical Description," Institute of British Geographers, Publication No. 30, *Transactions and Papers, 1962,* p. 6. Used by permission of the author and the Institute of British Geographers.

In attempting to gain an understanding of the geography of countries and regions, one is apt to have the most satisfying results if the "catalog approach" is avoided. In other words, it is essential not to follow a set catalog of topics slavishly from one area to another. One needs to consider each area on its own merits and to single out concepts and themes which will bring out the distinctive character of the area and convey clear ideas about its geographical structure and functioning, the conditions under which its people live, and the role it plays in world affairs. We may say that the prime task of the geographer is to illuminate the geographical character of areas; he does this by spotlighting certain critical features of areas and giving little or no attention to others.

Despite the need for selectivity and avoidance of a set catalog of topics, the observant student will note certain key topics that almost invariably are touched on as different countries and regions are studied. These include (1) Location, (2) World Importance, (3) Population, (4) Political Status, (5) Natural Environment, (6) Type of Economy, (7) Internal Arrangement and Organization, (8) External Connections and Relationships, (9) Characteristic Landscapes and Their Origins, (10) Potentialities, and (11) Problems. In the following pages each of these topics will be introduced and briefly discussed.

LOCATION

One of the reasons for studying geography is to learn the location of important features on the earth's surface—in other words, to learn where things are in the world. One tries to acquire a mental map, on which countries, important cities, rivers, mountain ranges, climatic zones, agricultural and industrial areas, and other features are plotted in their correct relation to each other. No person can claim to be truly educated who does not carry this kind of map in his head. However, it is not enough simply to know the *facts* of location; one also needs to develop an understanding and appreciation of the *significance* of location. To a considerable degree the geographic characteristics of any area

Great Britain and New Zealand compared in location. Most of the main industrial concentrations and seaports on the mainland of Europe (excluding the Soviet Union) are found in the area shown in solid color. The broad bands of color tint indicate the middle latitudes.

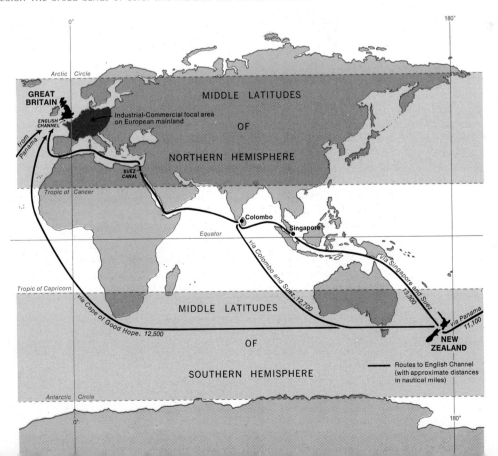

are due, directly or indirectly, to its location. For the most part, location is a concept which must be discussed with respect to particular areas. There is relatively little of real significance that can be said about location in general. Each area has a different location and must be separately evaluated.

Perhaps an illustration will point up the foregoing remarks. Let us compare the location of Great Britain with that of New Zealand (see map). Each is an island area, Great Britain being a single large island, while New Zealand has two main islands about equal in size. Both come within the influence of westerly winds which, blowing off the surrounding seas, bring abundant rain and moderate temperatures throughout the year. The climates of Great Britain and New Zealand are remarkably similar, in spite of the fact that these areas are located in opposite hemispheres and are about as far from each other as it is possible for two places on the earth to be.

Thus in certain ways the locations of Great Britain and New Zealand are similar. In other respects, however, their locations are vastly different. Great Britain is located in the Northern Hemisphere, which contains the bulk of the world's land and most of the principal centers of population and industry; New Zealand is on the other side of the equator, in the Southern Hemisphere. Great Britain is located near the center of the world's land masses (map, p. 101) and is separated by only a narrow channel from the densely populated industrial areas of western continental Europe; New Zealand is surrounded by vast expanses of ocean. Great Britain is located in the western seaboard area of Europe where many major ocean routes of the world converge; New Zealand is far away from the centers of world commerce. For more than four centuries Great Britain has shared in the development of northwestern Europe as a great organizing center for the world's economic and political life; New Zealand, meanwhile, has lived in comparative isolation. Great Britain, in other words, has had a *central* location within the existing frame of human activity on the earth, whereas New Zealand has had a *peripheral* location. These differences of location help to explain why Great Britain has become a densely populated industrial area and an important center of political and economic power, while New Zealand has remained a sparsely populated pastoral country of much less significance in world affairs. Centrality of location is a highly important factor to consider in assessing the geography of any country, region, or other place.

The student should realize that factors of location are not constant, but are relative to the circumstances

of a particular time. In other words, the *significance of location changes as circumstances change*. During the early centuries when the borderlands of the Mediterranean Sea contained the principal centers of European culture and political power, Great Britain was an unimportant area situated on the very edge of the known world, its location being scarcely more advantageous than that of New Zealand. Not until much later, when the European center of gravity had shifted from the Mediterranean region to the shores of the North Sea and the age of oceanic expansion had commenced, did the location of Great Britain become highly advantageous.

WORLD IMPORTANCE

The people of any country or locality are very apt to think of their home area as the most important place on the earth. But if one considers the affairs of the world as a whole, it becomes apparent that areas are very different in their importance to the world at large. Countries with large populations, for example, are apt to play a more important role than countries with smaller populations. This fact is recognized in the present text, which allots far more space to China or India or the Soviet Union than it does to Paraguay or Togo or Upper Volta. All six of these countries are sovereign nations, but it is quite clear that the larger countries are vastly more influential than the smaller ones. Such comparisons, however, can easily be pushed too far. Among the larger countries, variations in world importance depend far more on wealth, industrial strength, levels of education, military preparedness, and national cohesion and morale than they do on sheer numbers. Brazil, for instance, has a larger population than Great Britain or France, but one could scarcely say that it exceeds them in current world importance. Some of the smaller nations of the world have an importance that is out of proportion to their size. Among other reasons, this may be because they possess a natural resource of international consequence (for example, the petroleum of Kuwait or the bauxite of Jamaica or Guyana), or because they contain a section of an important international trade route (for example, Panama), or possess a highly developed industrial, commercial, and financial economy with worldwide links (for example, Belgium, the Netherlands, or Switzerland), or lie at the focus of an important international controversy (for example, Israel).

Within a given country, some areas play outstandingly important roles in the life of the country as a whole. These include such areas as (1) major metropolitan areas which provide unusually abundant and diversified opportunities for employment, and exert much control over the national economy; (2) major manufacturing areas, especially those that specialize in products that are of basic importance to the country's economy as a whole; (3) major seaports and other centers of commerce; (4) unusually productive or distinctive agricultural areas, particularly those that produce an important surplus for export; (5) areas that produce minerals on which the country's economy is very dependent; (6) areas that differ markedly in culture from the rest of the country, especially if this results in political disaffection which threatens the country's cohesion and stability; (7) areas with a pronounced strategic importance in wartime; or (8) areas of unusual poverty that consume a disproportionate share of the nation's revenues. In this text such areas often are treated at greater length than other areas that have a larger extent or population but are seen as less critically important for the country or the world as a whole. For example, London is treated at greater length than Scotland or Wales or New Zealand; the German industrial district of the Ruhr is allotted more space than Alaska or Nepal.

POPULATION

Population is a topic of major importance in geographic study. More than any other element, it is human life that gives character and geographic significance to areas. The numbers, density, distribution, and qualities of the population, together with population changes or trends, supply an essential background or focus in studying the geography of any area. These aspects of population will be illustrated in an introductory way through a brief consideration of world population as a whole.

Population Numbers

In 1968 the total population of the world was estimated at 3.5 billion, with an annual rate of increase of 2.0 percent. No very accurate total is yet possible, due to a lack of census data for some countries, as well as to great variations in the accuracy of census taking from one country to the next. Estimated populations of the eight major world regions are given in Table 1. A few important points to be noted from this table are (1) the tremendous population total of the Orient,

TABLE 1 THE MAJOR WORLD REGIONS: AREA AND POPULATION DATA [a]

REGION	AREA (MILLION SQUARE MILES)	ESTIMATED POPULATION (IN MILLIONS, 1968)	POPULATION DENSITY (PER SQUARE MILE)
Europe	1.8	455	253
Soviet Union	8.6	239	28
Middle East[b]	6.8	279	41
Orient[b]	7.7	1842	239
Pacific World[b]	3.3	19	6
Africa[b]	11.6	336	29
Latin America	7.9	268	34
Anglo-America[b]	8.3	222	27

[a] Unless otherwise noted, tables of area and population data in this text are estimates based on figures in standard sources such as the *United Nations Statistical Yearbook,* the *United Nations Demographic Yearbook,* the *Britannica Book of the Year,* the Population Reference Bureau, "World Population Data Sheet–1968," and statistical tables in Kurt Witthauer, "Bevölkerungszahlen der Erde 1967 und Gewichtsverschiebungen 1960/67," *Petermanns Geographische Mitteilungen,* vol. 111, no. 2 (1967), pp. 139–144. The student is cautioned that many figures in these tables are rough approximations. Many parts of the world have not been accurately surveyed, and many countries do not have adequate census data on which to base population estimates. Wide discrepancies often exist in area and/or population figures cited for a particular country in different sources of information. Data in this text are, in general, projections from official figures that have been reported by governments of the countries concerned.

[b] Areas and populations of certain units are included in two different regional totals. For example, Egypt is included under both Africa and Middle East.

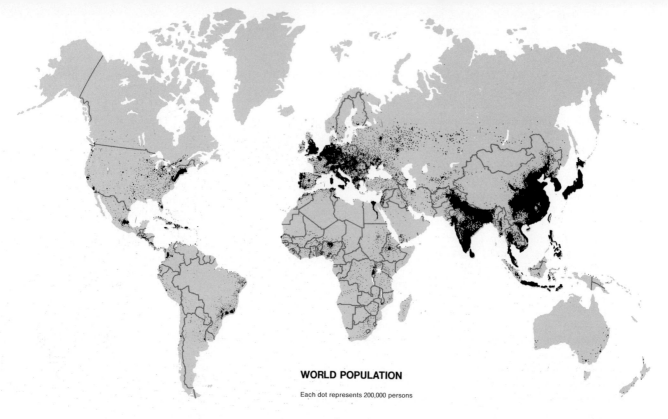

WORLD POPULATION

Each dot represents 200,000 persons

A French geographer, the late Jean Brunhes, maintained that the two most significant maps were the map of men, or population, and the map of rainfall. Much geography can be read or inferred from the map of men shown above. Among other things the map shows that the areas where large numbers of men have found it desirable to live are rather limited in extent. There are striking concentrations in the intensively cultivated farming regions of southern and eastern Asia and in the industrialized regions of Europe, western Russia, northeastern United States, and Japan. Note the tendency of men to congre-

which has over four times as many people as Europe, the region ranking second in total numbers; (2) the extremely small number of people in the Pacific World as compared with other world regions; and (3) the comparative similarity in population totals among the remaining five world regions.

Population Density

The total area of the earth's surface is approximately 197 million square miles. The land surface, however, including inland waters, comprises only about 57.4 million square miles, or slightly more than 29 percent of the total. A population total of 3.5 billion gives a figure of approximately 61 per square mile as the average density of population for the entire world. This figure is a little greater than the estimated average density for the United States in 1968 (56 per square mile).

Figures on population density for the world as a whole are practically meaningless, due to the extremely uneven distribution of population, resources, and productive facilities over the earth. However, such figures

do furnish a rough yardstick with which to measure the densities of various countries and regions.

Population Distribution

The distribution of population over the earth is extraordinarily uneven. Most of the world's people are concentrated in three major clusters (map, above). The largest of these occurs in the Orient. It includes portions of China, Japan, India, Pakistan, Indonesia, and several other countries. The people of this Oriental cluster comprise about half of mankind. A second major cluster, including Europe and portions of the Soviet Union, contains over a fifth of the world's people. The third major cluster is considerably smaller. It includes the eastern half of the United States and adjoining parts of Canada. Elsewhere in the world smaller concentrations of population are scattered about here and there, separated from each other by vast stretches of sparsely settled or unoccupied land.

Areas in which population is extremely sparse or absent make up more than three-fourths of the earth's

42

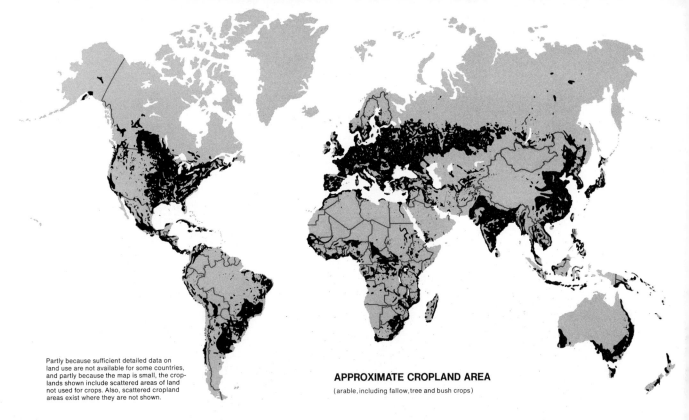

Partly because sufficient detailed data on land use are not available for some countries, and partly because the map is small, the croplands shown include scattered areas of land not used for crops. Also, scattered cropland areas exist where they are not shown.

APPROXIMATE CROPLAND AREA

(arable, including fallow, tree and bush crops)

gate on the margins of continents, near the sea. Geographers are students of *regions,* but they are also students of *distributions*—not only of men, but of crops, livestock, rainfall, minerals, industries, and other features on the earth. They are particularly interested in the various ways that different distributions *are related* to each other. Compare the map of population with the map of cultivated land for similarities and differences in the *pattern* of distribution. (Source: U.S. Department of Agriculture. The data are for 1961.)

land surface. Such areas fall into four principal categories, as follows: (1) arctic and subarctic areas, where settlement is hampered by excessive cold; (2) areas of desert and dry grassland, which are handicapped by lack of moisture; (3) areas of rugged highland, where settlement is restricted by steep slopes and high altitudes; and (4) areas of tropical rain forest and tropical grassland or scrub, where excessive heat and moisture, dense forest growth or rank grasses, and infertile soils discourage fixed settlements. Of the four main types of "negative areas," the tropical forests and grasslands seem to offer the best possibilities for large future increases in population. Indeed, some areas of this type, particularly in the Orient, already support large and dense populations.

A winter scene in the Canadian subarctic west of Hudson Bay. Aside from the possibility of valuable minerals, this lake-strewn area, with its harsh winters, short summers, and poor soils, offers few attractions for human settlement. The sparse coniferous forest in the photo has little commercial utility. The black dots in the foreground are a herd of caribou on a frozen lake. (Canadian Wildlife Service.)

However, areas as large or larger in Latin America, Africa, and the East Indies are still quite sparsely populated.

Population Trends

Various periods of history have witnessed great changes in the numbers, density, and distribution of people over the earth and in different countries and regions. Some of the most significant changes have occurred during the past three or four centuries. This period has seen an unparalleled increase in total population numbers. It is thought that since 1650 the world's population has increased six or seven times over, from an estimated 470 to 545 million to 3.5 billion. At the present time this increase is continuing at an estimated rate of 70 million or more a year. Fears are frequently expressed that mankind will outrun its food supply unless the increase in numbers is checked. However, some optimistic students of this problem believe that the earth could support several times its present population if all the available resources were more intensively and scientifically managed.

The tremendous increases in population during the past three centuries appear to be mainly the result of a declining death rate. The latter has resulted partly from improved medical and sanitary facilities and partly from the more abundant, varied, and dependable food supply made possible by improved agricultural techniques, better transportation facilities, and the opening of new lands for cultivation in the Americas and elsewhere.

At the present time population is increasing in nearly all of the world's countries, but the rate of increase is much greater in some countries than in others. Sharp upward trends of population in such countries as India (2.5 percent increase per year) and Indonesia (2.3 percent annual increase) contrast strikingly with the slow population growth of the United Kingdom (0.5 percent) or Sweden (0.8 percent).

During the past three or four centuries the world pattern of population distribution has been altered considerably. This was brought about principally by two great movements of population: the migration of Europeans to new lands overseas, and the migration of rural dwellers to cities.

Other Aspects of Population

The general topic of population encompasses not only numbers, density, distribution, and movements or trends but also the *qualities* of the population—in other words, such characteristics as skin color, language, religion, food habits, educational levels, health, and general cultural heritage. The peoples of various areas are vastly different in these respects. Such differences are of fundamental concern and interest to the student of geography.

POLITICAL STATUS

We live in a world which has been divided into a multitude of political units—sovereign states, subdivisions of these states (provinces, states, departments, counties, and so on), and various types of dependent political units (colonies, trusteeships, and the like) and their subdivisions. Separated by boundaries drawn on maps and in many cases shown by markers on the ground, these political divisions form a complicated patchwork enclosing most of the land surface of the earth.

The political status of any area is an important feature of its geography. In a broad sense, political

Excessive crowding in the lowlands of heavily populated agricultural regions often leads to the tilling of steep slopes such as the Puerto Rican tobacco field shown in the view, left. Erosion by heavy tropical downpours will soon destroy the soil and render this land worthless. In some parts of the world steep slopes have been made productive on a long-range basis by systems of terracing. (Government of Puerto Rico.)

status includes not only the political organization of the area in terms of functional political units, but also the distinctive role of the area in the political world. Although the latter is sometimes difficult to define precisely, it should not be ignored in geographic study.

The key unit in the worldwide political structure is the sovereign state. Each state occupies territory and administers the people and resources of that territory. In most matters the state, whether large or small, is its own master within its own boundaries. Despite the many attempts at supranational organization, the system of sovereign states shows a remarkable persistence, and seems likely to be for a long time the basic international frame within which citizens and their nations will operate. This fact is a major reason for the emphasis on the geography of individual sovereign states in this text.

An important geographical circumstance in any state is the degree to which it has attained political cohesion throughout its extent. Some states, such as Denmark or New Zealand, are very cohesive politically, but other states exhibit separatist tendencies in certain areas. India, Indonesia, the Sudan, and Canada are notable examples of states in which strong *centrifugal tendencies* exist.

NATURAL ENVIRONMENT

By natural environment is meant the total complex of natural conditions and resources occurring in an area. The principal elements of the natural environment are landforms, climate, natural vegetation, soils, native animal life, underground and surface waters, and mineral resources. These elements vary considerably in relative significance from one area to another and from one period of history to another. To a considerable degree the significance of a particular environmental feature depends on the type of culture prevailing at a given time in the area where the feature is found. For example, the rich coal deposits of western Pennsylvania were of little or no use to the Indian tribes who originally inhabited that area, but to the present inhabitants, armed with an adequate technology and living within the framework of a highly developed industrial civilization, these same coal deposits are of great significance.

One should beware of easy generalizations about the "influence" of the natural environment on human life.

The connections or relationships which exist between people and their environment are often extremely complicated. Any particular set of environmental features offers various possibilities for human use. Social, cultural, economic, technological, psychological, and historical factors will condition the actual employment to which such features will be put by a human group.

An American geographer, the late Robert S. Platt, summed up the foregoing ideas effectively in the section quoted below:[2]

The importance of our natural environment is obvious. People have been, and still are, conscious of the significance to them in everyday life of the weather, the soil, minerals, mountains, and plains. Travelers long ago observed, and still do, that differences from place to place in natural environment are associated with differences in the lives of people: that mountaineers live differently from plainsmen, jungle savages differently from Arctic Eskimos. . . .

[However] it is misleading to advance the hypothesis of an active influence of natural environment tending to shape human life in the natural and proper way and to look for coincidences between environment and life as evidence confirming this hypothesis. Increasing evidence shows that the hypothesis of a simple and direct relationship is not thus confirmed, that there is no proper natural way of shaping life but innumerable ways, not sorted out by nature but reduced by man's choices past and present. *People live differently in similar environments and differently at different times in the same environment,* without feeling any environmental pressure to lessen these differences.

In equatorial regions savage life is natural, but all the other forms of life there are natural also. There are not only jungle savages living in many different ways but also . . . many people who are not savages and are under no inducement to become so. On the contrary, some of them are engaged in horticultural enterprises as highly developed as any in the world. In addition, there are innumerable untried possibilities for other ways of living, some of which would require thousands of years for full realization through the development of plant or animal resources. *In any given type of regional environment, people have alternative ways of living, apparently many in some places and few in others.* Probably everywhere the conceivable number of possibilities is far greater than people can imagine and far beyond the range of choice now open to them.

Actually, people are limited by things other than natural environment, though set within the confines of that environment. Particularly are they limited by habits they have learned and facilities available to them, ac-

---

[2] Robert S. Platt, "Environmentalism Versus Geography," *American Journal of Sociology,* 53 (1948), 351–352, Reprinted by permission of The University of Chicago Press. Some portions italicized in editing.

cumulated through an unbroken series of choices and rejections in the entire course of their history—in other words, by the cultural heritage of the group to which they belong, by their culture defined in the broadest sense. The choices made in the past which now limit people in their activities have been impelled not by natural environment but by the play of history thereafter embodied in their culture.

For example, in the Great Lakes region farmers have been limited agriculturally to a certain range of possibilities, mainly involving cereals and livestock. But if some of our ancestors at the dawn of civilization had not chosen to domesticate certain grasses as cereals and certain four-footed animals as livestock but, instead, had chosen to develop fungus growths or edible insects as a basis of productive culture, our mode of life might be now utterly different in ways which the natural environment might support as well as it does our present agriculture, or conceivably better.

TYPE OF ECONOMY

One of the most significant characteristics of an area is its type of economy—in other words, the kind of mechanism which human beings have developed in the area as a means of satisfying their needs and wants for goods and services. A particularly important aspect of an area's economy is its *occupational structure,* or the numbers and proportion of people employed in each of the principal means of livelihood. In most countries a larger proportion are employed in *agriculture* than in any other occupation. Heavy dependence on agriculture is especially characteristic of less developed countries. In countries with more advanced technologies a large share of the people are engaged in *manufacturing, transportation, trade,* and *personal* and *professional services.* But even in highly industrialized countries agriculture is a significant means of livelihood. However, the proportion of the population engaged in this occupation is generally much smaller than in developing countries, although agricultural production per worker and per unit of land is often much higher.

The more sparsely settled regions of the world are characterized by such occupations as *grazing (nomadic herding* or *livestock ranching), hunting, fishing, trapping,* or the *forest industries.* In scattered areas *mining* serves as a source of livelihood. Mining areas are distributed irregularly over the earth wherever commercial deposits of minerals are found.

Besides the occupational structure, it is also important to consider the characteristic *units of economic organization* in an area (farms, factories, stores, and so on), together with the lines of transportation and communication which connect these units. Other pertinent questions regarding the economy of the area would be

1. What is the general status of technology—advanced, moderately developed, primitive, or a mixture of these?

2. What are the principal commodities produced, and in what quantities?

3. Is production primarily of a commercial or a subsistence character?

4. Is the economy largely self-contained, or is there a significant dependence on trade with other areas?

5. What are the principal imports and exports?

6. Have the productive facilities of the area been largely financed by domestic capital, or by outside capital?

7. What is the adequacy of livelihood provided by the economy?

8. What are the major economic trends in the area?

INTERNAL ARRANGEMENT AND ORGANIZATION

In studying an area geographically, one is constantly concerned with the ways in which the features that give it character are arranged and organized spatially—that is, how they are distributed within the area, how different distributions are related to each other, and how the geographical features of the area are linked together by natural processes or human enterprise into coordinated structures and functioning systems. To think geographically about the internal makeup of an area is to think in terms of such spatial patterns, associations, and interconnections. Almost endless examples of this type of analysis are to be found in the present text. For an illustration, it is suggested that the student turn to the discussion of the Ruhr industrial district on pages 172–175, or the discussion of London's internal geography on pages 112–122.

Students of geography should make a conscious effort at all times to relate the geographical features and distributions of an area to each other and to think in terms of overall spatial patterns into which details can be fitted. This inevitably means to think in terms of maps, for maps portray spatial patterns and associations with a clarity that is generally beyond the reach of words. Students should constantly refer to maps as they read geographical material, should use outline maps for note taking, should make sketch maps to show important spatial relationships, and should try to visualize map relationships when these are stated in words in text material. By using maps and striving to build up a coherent picture of the internal arrangement and organization of each area, students can keep geographical aims in focus, avoid the danger of mere accumulation of encyclopedic and unrelated details, and gain an appreciation of the internal arrangement and organization of geographical features in each area as an element conferring distinctiveness on the area and conditioning the lives of its inhabitants.

EXTERNAL CONNECTIONS AND RELATIONSHIPS

The overall pattern of an area's economic, cultural, social, and political interaction with other parts of the world is an important component of its total geography. Such connections and relationships involve trade flows, population movements, cultural, technical, and informational exchanges, political, strategic, and military influences and alignments, and the day-to-day flow of personal and group decisions which affect man's use and organization of the earth. This exceedingly broad topic cuts across all of the other topics considered in this chapter, including the topic of internal arrangement and organization. A little reflection will show how intimately the internal makeup of most areas is related to their interaction with the outside world. The agricultural economy of São Paulo State in Brazil, for example, is built around coffee, and the landscape betrays abundant evidence of this fact; but the geographical pattern of the state would be very different were it not for the extremely large market for imported coffee afforded by the United States.

CHARACTERISTIC LAND-SCAPES AND THEIR ORIGINS

One of the things that lends color and interest to geography as a field of study is the opportunity to observe and interpret the extraordinary variety of landscapes which the world affords. The study of landscapes carries intellectual as well as esthetic rewards, for a great deal of the human record, and nature's works also, are inscribed in the earth's surface and call for careful mapping and rigorous analysis if one is to truly understand and appreciate what is written there. From the

Although the rectangular survey system visible along some fence lines gives a trace of geometry to the Missouri Ozarks landscape shown above, the dominant motif is one of extreme irregularity in landscape patterns. Cleared land, found both on the relatively level ridge tops and on the narrow floors of stream valleys, forms an interrupted patchwork of fields that are varied in size, generally irregular in shape, and disposed in various directions. Farmsteads are scattered irregularly over the landscape. Farmers have laid out their crop fields in such a way as to take full advantage of the patches of land that are sufficiently level and fertile to justify cultivation. Steeper slopes between the ridge tops and the valley floors are left in forest. Close examination of the photo reveals that some cleared land is now reverting to forest. The stream at the bottom left is the Osage River, a tributary of the Missouri. The village at the top left is St. Thomas, a German settlement about 20 miles south of Jefferson City. (U.S. Department of Agriculture photo.)

earliest times the landscape has been a basic source of data in geographical work, and the marked differences in landscapes over the earth have stimulated much geographical inquiry. Some landscapes are rather irregular, nondescript, and hard to characterize, while others are

A landscape of straight lines. The geometric pattern of this wheat-farming area in southern Manitoba offers a remarkable contrast to the Ozark landscape seen in the preceding view. The photo shows a broad expanse of flat alluvium near the town of Portage La Prairie approximately 50 miles to the west of Winnipeg. A rectangular survey system has been employed here, as it has in the Missouri Ozarks, but in this uniform stretch of rich earth man has not found it necessary to depart from rectangles and straight lines in laying out his fields and roads. Farmsteads, marked in the photo by clumps of trees, are distributed in a regular pattern. The large rectangular fields make possible maximum efficiency in using large machines. (National Film Board of Canada, Ottawa.)

regular, sharply defined, and conducive to precise geographical description and characterization. One may, for instance, contrast the rambling patchwork of fields, woods, hills, valleys, and winding roads in the Missouri Ozarks with the rectangular pattern of large fields and straight roads on the flat Red River plain of southern Manitoba (see photos). In the study of landscapes, the geographical mind is concerned with the features that are typical; in other words, with landscape units that repeat themselves and are subject to mapping, classification, and orderly analysis. In general, the interest of the geographer in a landscape is different from that of the tourist. The tourist tends to be attracted by features that are quaint, unique, exotic, or bizarre; the geographer is excited by the things that are *representative:* his concern is to discover what the landscapes of an area are like *on the average.*

POTENTIALITIES

As a field of study, geography is primarily concerned with knowledge and understanding of areas as they are. But geography also gives attention to the *potentialities* of areas—that is, to areas as they may be, or could be, in the future. In this connection it needs to be reemphasized that every area offers alternative possibilities and limitations for human use. For example, the interior grasslands of North America were originally the habitat of Indian tribes pursuing a mode of life based on the

hunting of buffalo and other herbivorous animals. Later, however, the Indians were displaced by ranchers who utilized for the grazing of domesticated livestock the grasses that previously had supported the native wild animals. The ranchers, in their turn, have now given place in many parts of the grasslands to farmers who are growing crops on the deep, black, fertile grassland soils that previously lay undisturbed and unused. Each successive group of occupants has used this area in a way consistent with its particular cultural heritage and with the circumstances of a particular time. In all probability the potentialities of the area have by no means been exhausted, and centuries, or even decades, hence its occupants may use it in ways that are entirely unforeseen at present.

By and large, it seems idle to discuss the potentialities of an area in purely general terms. Potentialities have meaning only with reference to a particular human group that has a certain set of objectives, attitudes, and abilities, and only in the light of a specific set of future historical circumstances. Perhaps the two most fundamental questions that can be asked concerning the potentialities of an area are

1. How many people can the area support at specified levels of living?

2. What possibilities does the area afford for improving its present level of living?

Geographers, as well as scholars in many other fields, are constantly looking for answers to these questions, both for the world as a whole and for particular countries and regions.

PROBLEMS

The people of almost any area are confronted with serious problems which must be solved if the potentialities of the area are to be realized. Most of these fall under the heading of political problems, economic problems, social problems, or a mixture of these. Space will not permit an extended discussion of the endless variety which they assume. The following are a few specific categories of problems that are important on a world basis:

1. *Problems of severe population pressure* are found in many of the world's less developed areas. In such areas, resources are insufficient to support the population adequately under present systems of resource use, and rapid population increases often add to the difficulties.

2. *Problems of landownership* are likewise most serious in less developed regions, where the bulk of the productive land is often held by a relatively few individuals and the mass of the people are either tenants or owners of very modest plots of land. These problems are being alleviated in many countries by government-sponsored programs of land reform.

3. *Problems of racial and religious antagonism* are found in many parts of the world, a few prominent examples being South Africa (whites versus nonwhites, and Africans versus Asians), India-Pakistan (Hindus versus Moslems), and Israel and neighboring countries (Jews versus Moslem or Christian Arabs).

4. *Problems of unrest in colonial areas,* arising from local dissatisfaction over outside rule, are present to some degree in nearly all parts of the rapidly shrinking colonial world.

The four categories of problems listed above are not entirely separate from each other, but are interconnected. For example, unrest in the world's remaining colonial areas is partly due to population pressure, unequal distribution of land, and racial and religious antagonisms, as well as to other circumstances.

DIFFICULTY OF EXPLAINING AN AREA

To account for the complex web of features and relationships that make an area geographically distinctive is a challenging task. One must accept the fact that to explain fully the geography of an area may be very difficult or impossible due to the presence of historical or other factors whose effects cannot be known with certainty. However, this should not be allowed to deter a search for such understanding as the available data will permit. Much that is useful and enlightening can be learned about the geography of areas, even if all the facts cannot be fully explained.

REFERENCES and READINGS

Population and Food Supply

BEAUJEU-GARNIER, J., *Geography of Population,* trans. by S. H. Beaver (London: Longmans, Green and Co., 1966).

BERG, ALAN D., "Malnutrition and National Development," *Foreign Affairs,* **46**, no. 1 (October 1967), 126–136.

BROWN, HARRISON S., *The Challenge of Man's Future: An Inquiry Concerning the Condition of Man During the Years That Lie Ahead* (New York: The Viking Press, 1954; also Compass Books).

BURKE, TERENCE, "Food and Population, Time and Space: Formulating the Problem," *Journal of Geography,* **65**, no. 2 (February 1966), 58–66.

BURTON, IAN, and ROBERT W. KATES, "Slaying the Malthusian Dragon: A Review", *Economic Geography,* **40**, no. 1 (January 1964), 82–89; and *Readings in Resource Management and Conservation* (Chicago: University of Chicago Press, 1965).

CALEF, WESLEY, "World Food Production," *Geographical Review,* **58**, no. 2 (April 1968), 306–307. A brief review of recent publications on food production and world food needs.

CLARKE, JOHN I., *Population Geography* (New York: Pergamon Press, 1965).

GLASS, DAVID VICTOR, and DAVID EDWARD CHARLES EVERSLEY, eds., *Population in History: Essays in Historical Demography* (Chicago: Aldine Publishing Company, 1965).

HOOSON, DAVID J. M., "The Distribution of Population as the Essential Geographical Expression," *Canadian Geographer,* 17, no. 1 (1960), 10–20.

JONES, CLARENCE FIELDEN, and GORDON GERALD DARKENWALD, *Economic Geography* (3d ed.; New York: The Macmillan Company, 1965), Chap. 2, "The Human Factor in Economic Geography," pp. 9–19.

KINDLEBERGER, C. P., "Mass Migration, Then and Now," *Foreign Affairs,* 43, no. 4 (July 1965), 647–658.

LOWENTHAL, DAVID, and LAMBROS COMITAS, "Emigration and Depopulation: Some Neglected Aspects of Population Geography," *Geographical Review,* 52, no. 2 (April 1962), 195–210.

NICOL, HUGH, *The Limits of Man: An Inquiry into the Scientific Bases of Human Population* (London: Constable and Co., 1967).

NORTON-TAYLOR, DUNCAN, "What the U.S. Can Do About World Hunger," *Fortune,* 74, no. 6 (June 1966), 110–114 ff.

NOTESTEIN, FRANK W., "The Population Crisis: Reasons for Hope," *Foreign Affairs,* 46, no. 1 (October 1967), 167–180.

PIRIE, N. W., "Orthodox and Unorthodox Methods of Meeting World Food Needs," *Scientific American,* 216, no. 2 (February 1967), 27–35.

Population Bulletin (six issues a year) and other publications of the Population Reference Bureau, Washington, D.C.

SCOTT, FRANKLIN D., ed., *World Migration in Modern Times* (Englewood Cliffs, N.J.: Prentice-Hall, 1968).

SIMPSON, DAVID, "The Dimensions of World Poverty," *Scientific American,* 219, no. 5 (November 1968), 27–35.

SOPHER, DAVID E., *Geography of Religions* (Englewood Cliffs, N.J.: Prentice-Hall, 1967).

STAMP, L. DUDLEY, *Our Developing World* (London: Faber and Faber, 1960); and *The Geography of Life and Death* (London: Collins, 1964).

TROLL, CARL, "Plural Societies of Developing Countries: Aspects of Social Geography," in International Geographical Congress, 20th, London, 1964, *Congress Proceedings,* ed. by J. Wreford Watson (London: Thomas Nelson and Sons, 1967), pp. 9–33.

TREWARTHA, GLENN T., ARTHUR H. ROBINSON, and EDWIN H. HAMMOND, *Elements of Geography* (5th ed.; New York: McGraw-Hill Book Company, 1967), Chap. 25, "Population," pp. 519–552, and Chap. 26, "Settlements," pp. 557–575.

U.S. DEPARTMENT OF AGRICULTURE, ECONOMIC RESEARCH SERVICE, *Increasing World Food Output: Problems and Prospects* (Foreign Agricultural Economic Report No. 25; Washington, D.C.: Government Printing Office, 1965); and *Man, Land and Food: Looking Ahead at World Food Needs* (Foreign Agricultural Economic Report No. 11, Washington, D.C.: Government Printing Office, 1963).

U.S. PRESIDENT'S SCIENCE ADVISORY COMMITTEE, *The World Food Problem: Report of the Panel on the World Food Supply,* 2 vols. (Washington, D.C.: Government Printing Office, 1967).

"World Population," *Annals of the American Academy of Political and Social Science,* 369 (January 1967), 1–140. Articles by many different authors in a variety of fields.

ZELINSKY, WILBUR, *A Prologue to Population Geography* (Englewood Cliffs, N.J.: Prentice-Hall, 1966); and *A Bibliographic Guide to Population Geography* (University of Chicago, Department of Geography Research Paper No. 80, Chicago, 1962).

Political Geography

ALEXANDER, LEWIS M., *World Political Patterns* (2d ed.; Chicago: Rand McNally & Co., 1963). The basic organization of this text is regional.

DE BLIJ, HARM J., *Systematic Political Geography* (New York: John Wiley & Sons, 1967). Text matter interspersed with selected articles.

EAST, W. GORDON, and A. E. MOODIE, eds., *The Changing World: Studies in Political Geography* (New York: Harcourt, Brace & World, 1956). Organized by countries and regions.

FISHER, CHARLES A., ed., *Essays in Political Geography* (London: Methuen and Company, 1968).

HARTSHORNE, RICHARD, "The Functional Approach in Political Geography," *Annals of the Association of American Geographers,* 40, no. 2 (June 1950), 95–130.

JACKSON, W. A. DOUGLAS, ed., *Politics and Geographic Relationships: Readings on the Nature of Political Geography* (Englewood Cliffs, N.J.: Prentice-Hall, 1964).

POUNDS, NORMAN J. G., *Political Geography* (New York: McGraw-Hill Book Company, 1963).

SPROUT, HAROLD, and MARGARET SPROUT, *The Ecological Perspective on Human Affairs with Special Reference to International Politics* (Princeton, N.J.: Princeton University Press, 1965).

Economic Geography

ALEXANDER, JOHN W., *Economic Geography* (Englewood Cliffs, N.J.: Prentice-Hall, 1963).

ALEXANDERSSON, GUNNAR, *Geography of Manufacturing* (Englewood Cliffs, N.J.: Prentice-Hall, 1967); and, with Göran Norström, *World Shipping: An Economic Geography of Ports and Seaborne Trade* (New York: John Wiley & Sons, 1963).

BOESCH, HANS H., *A Geography of World Economy* (Princeton, N.J.: D. Van Nostrand Co., 1964).

CHISHOLM, MICHAEL, *Geography and Economics* (London: G. Bell and Sons, 1966); and *Rural Settlement and Land Use: An Essay in Location* (London: Hutchinson University Library, 1961).

DURAND, LOYAL, JR., *Economic Geography* (New York: Thomas Y. Crowell Company, 1961).

ESTALL, R. C., and R. O. BUCHANAN, *Industrial Activity and Economic Geography: A Study of the Forces behind the Geographical Location of Productive Activity in Manufacturing Industry* (rev. ed.; London: Hutchinson University Library, 1966).

FRYER, D. W., *World Economic Development* (New York: McGraw-Hill Book Company, 1965).

GINSBURG, NORTON, *Atlas of Economic Development* (Chicago: University of Chicago Press, 1961); and "On Geography and Economic Development," in Saul B. Cohen, ed., *Problems and Trends in American Geography* (New York: Basic Books, 1967), pp. 174–186.

GREGOR, HOWARD F., *Environment and Economic Life: An Economic and Social Geography* (Princeton, N.J.: D. Van Nostrand Co., 1963).

GROTEWOLD, ANDREAS, "Some Aspects of the Geography of International Trade," *Economic Geography,* 37, no. 4 (October 1961), 309–319; and, with Lois Grotewold, "Some Geographic Aspects of International Trade," *Economic Geography,* 33, no. 3 (July 1957), 257–266.

GUIDRY, NELSON P., *A Graphic Summary of World Agriculture,* rev. ed. (U.S. Department of Agriculture, Economic Research Service, Regional Analysis Division, Miscellaneous Publication No. 705; Washington, D.C.: Government Printing Office, 1964). Numerous maps and graphs.

HIGHSMITH, RICHARD M., JR., "How Types of Farming Divide World Agriculture into Regions," *Foreign Agriculture,* 4, no. 11 (March 14, 1966), 3–5; and, ed., *Case Studies in World Geography: Occupance and Economy Types* (Englewood Cliffs, N.J.: Prentice-Hall, 1961).

HOFFMAN, LAWRENCE A., *Economic Geography* (New York: The Ronald Press Company, 1965).

HOOVER, EDGAR M., *The Location of Economic Activity* (New York: McGraw-Hill Book Company, 1948; paperback ed., 1963).

HULL, OSWALD, *A Geography of Production* (New York: St. Martin's Press, 1968).

HUNKER, HENRY L., ed., *Erich W. Zimmermann's Introduction to World Resources* (New York: Harper & Row, 1964).

JONES, CLARENCE FIELDEN, and GORDON GERALD DARKENWALD, *Economic Geography* (3d ed.; New York: The Macmillan Company, 1965).

KLAGES, KARL H. W., *Ecological Crop Geography* (New York: The Macmillan Company, 1942).

McCARTY, HAROLD H., and JAMES B. LINDBERG, *A Preface to Economic Geography* (Englewood Cliffs, N.J.: Prentice-Hall, 1966).

McNEE, ROBERT B., "The Changing Relationships of Economics and Economic Geography." *Economic Geography,* **35**, no. 3 (July 1959), 189–198.

MILLER, E. WILLARD, *A Geography of Manufacturing* (Englewood Cliffs, N.J.: Prentice-Hall, 1962).

PARKER, GEOFFREY, *The Geography of Economics: A World Survey* (London: Longmans, Green and Co., 1965).

ROEPKE, HOWARD G., ed., with the assistance of Thomas J. Maresh, *Readings in Economic Geography* (New York: John Wiley & Sons, 1967).

RUTHERFORD, J., M. I. LOGAN, and G. J. MISSEN, *New Viewpoints in Economic Geography* (Sydney, Australia: Martindale Press, 1966).

SMITH, ROBERT H. T., EDWARD J. TAAFFE, and LESLIE J. KING, eds., *Readings in Economic Geography: The Location of Economic Activity* (Chicago: Rand McNally & Co., 1968).

SYMONS, LESLIE, *Agricultural Geography* (New York: Frederick A. Praeger, 1967).

THOMAN, RICHARD S., EDGAR C. CONKLING, and MAURICE H. YEATES, *The Geography of Economic Activity* (2d ed.; New York: McGraw-Hill Book Company, 1968).

THOMAN, RICHARD S., and DONALD J. PATTON, eds., *Focus on Geographic Activity: A Collection of Original Studies* (New York: McGraw-Hill Book Company, 1964).

TREWARTHA, GLENN T., ARTHUR H. ROBINSON, and EDWIN H. HAMMOND, *Elements of Geography* (5th ed.; New York: McGraw-Hill Book Company, 1967), Chap. 27, "Agriculture," pp. 576–602; Chap. 28, "Manufacturing," pp. 603–626; and Chap. 29, "Transportation and Trade," pp. 627–645,

WHITE, C. LANGDON, PAUL F. GRIFFIN, and TOM L. McKNIGHT, *World Economic Geography* (Belmont, Calif.: Wadsworth Publishing Co., 1964).

"World Fuel and Power Resources and Needs," *Advancement of Science,* **22**, no. 99 (September 1965), 281–341. A collection of articles from a symposium.

Miscellaneous

EAST, W. GORDON, *The Geography Behind History* (rev. ed.; New York: W. W. Norton & Company, 1967).

FREEMAN, T. W., *A Hundred Years of Geography* (Chicago: Aldine Publishing Company, 1961).

GUTKIND, E. A., *Our World from the Air* (Garden City, N.Y.: Doubleday & Co., 1952). A book of photographs with accompanying text.

HAGGETT, PETER, *Locational Analysis in Human Geography* (New York: St. Martin's Press, 1966).

JAMES, PRESTON E., "Toward a Further Understanding of the Regional Concept," *Annals of the Association of American Geographers,* **42**, no. 3 (September 1952), 195–222.

PLATT, ROBERT S., "Environmentalism Versus Geography," *American Journal of Sociology,* **53**, no. 5 (March 1948), 351–358.

TAYLOR, GRIFFITH, ed., *Geography in the Twentieth Century* (3d ed.; New York: Philosophical Library, 1957).

See also the reference lists for Chapters 1, 2, and 4.

PART II

EUROPE

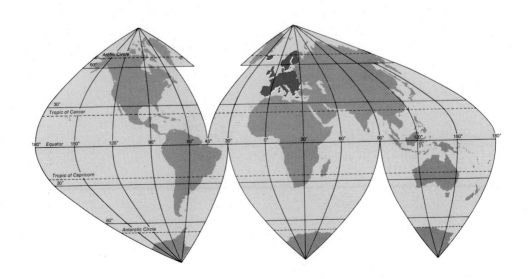

Introduction
to Europe

4 For the purposes of this book, "Europe" is defined as the group of countries in Eurasia lying west of the Soviet Union and Turkey. This definition is at variance with traditional ideas. Europe has customarily been regarded as a continent, separated from Asia by the Ural Mountains, Caspian Sea, Caucasus Mountains, Black Sea, Turkish Straits (Bosporus, Sea of Marmara, and Dardanelles), and Aegean Sea. Thus Europe has been considered to include parts of the Soviet Union and Turkey, often referred to as "European Russia" and "European Turkey," respectively. For various reasons, including important differences in culture and environment, it has been thought desirable in this book to exclude the Soviet Union and Turkey from "Europe." Essentially a great peninsula fringed by lesser peninsulas and islands, Europe is bounded on the west and north by the Atlantic and Arctic oceans and on the south by the Mediterranean Sea. Eastward, Europe merges with the main continental mass of Eurasia. (See maps, pp. 57 and 70.)

REGIONAL GROUPS OF COUNTRIES IN EUROPE

The countries of Europe can be divided into a number of fairly distinct regional groups.

At the northwest, separated from the mainland by the English Channel and the North Sea, are the *countries of the British Isles.* They include the United Kingdom of Great Britain and Northern Ireland, and the Republic of Ireland. These countries are now entirely separate in a political sense. However, they have been closely associated historically and are still closely linked by trade.

The *countries of West Central Europe* lie on the mainland opposite the British Isles. West Germany and France are the major countries in this group. Other countries include East Germany; the small but highly important Benelux nations of Belgium, the Netherlands, and Luxembourg; the mountainous countries of Switzerland and Austria; and the tiny semi-independent political units of Liechtenstein, Andorra, and Monaco. The countries of West Central Europe, considered as a group, are dominantly industrial and commercial in their activities and outlook. Together with the United Kingdom, they constitute the industrial and commercial core of Europe. In these productive countries agriculture is also of great importance; in some, indeed, large sections are dominantly agricultural. Generally speaking, the farmers of West Central Europe pursue a highly commercialized agriculture characterized by intensive and scientific methods and high yields. The countries of this area carry on a complex interchange of goods and services. They form a group to which the phrase "highly developed" may properly be applied.

The *countries of Northern Europe,* often referred to as the Scandinavian or Fennoscandic countries, include Denmark, Sweden, Norway, Finland, and Iceland. Relatively small in numbers, but industrious, thrifty, and progressive, the peoples of these lands have made skillful use of a restricted resource base in achieving general levels of health, education, and welfare which are exceeded by few nations. The different countries of this regional group have many environmental and cultural similarities and have had close historical ties with each other.

Very different from the cool, northerly lands just described are the warm and sunny *countries of Southern Europe,* often called the Mediterranean countries. This group includes Italy, Spain, Portugal, and Greece, together with the microstates of San Marino and Vatican City within Italy, the former British island colony of Malta, independent since 1964, and the British colony of Gibraltar, commanding the eastern entrance to the famous strait that connects the Mediterranean Sea with the Atlantic Ocean.[1] These countries considered as a group are mountainous, generally poor in natural resources, and less industrialized than Great Britain or the countries of West Central Europe. In spite of a historic past and ancient cultures, they lag considerably behind the lands of northwestern Europe in technological advancement, productivity per person, and general level of living. Some sections of Southern Europe, however, are notably better off than the average, the best example being northern Italy, where important industries exist and conditions for agriculture are more favorable than in most parts of the Mediterranean world. Italy as a whole is in a different economic class from its Southern European neighbors, being far more industrialized, urbanized, and productive. Indeed, in total output of goods and services, Italy is exceeded within Europe only by the United Kingdom, West Germany, and France.

Along or near the borders of the Soviet Union are the *countries of East Central Europe*—Poland, Czechoslovakia, Hungary, Romania, Bulgaria, Albania, and Yugoslavia. Following World War II these countries, together with East Germany, came under the control of the Soviet Union. Communist forms of political, economic, and social organization were instituted, and each country was brought within the economic and strategic orbit of the USSR. In 1948 one of the countries, Yugoslavia, broke away from Soviet direction and began to pursue a separate path of Communist development. A second country, Albania, staunchly Stalinist in outlook, diverged from the Soviet Union over doctrinal questions after Stalin's death, and aligned itself with the USSR's arch-rival, Communist China. After 1956, when a revolt in Hungary was put down by Soviet troops and serious unrest developed in Poland, manifestations of nationalism and separatism in the Soviet "client states" appeared to make some headway. But

---

[1] Malta is an independent parliamentary state within the Commonwealth of Nations (see p. 250). The independent Mediterranean island republic of Cyprus, also a former British colony and a member of the Commonwealth, is discussed with the Middle East in this text.

EUROPE

INTRODUCTORY LOCATION MAP

★ Capital of Country
● Largest city of Country
✪ Capital and largest city

Europe, showing political units as of January 1, 1969. Note the fact that in most European countries the political capital is also the largest city (metropolitan area).

future prospects for such trends were rendered uncertain in 1968, when the Soviets occupied Czechoslovakia militarily after that country's Communist government introduced liberal political reforms. Such uncertainty has long been characteristic of this region. Through the centuries political boundaries have fluctuated violently, and the various national groups have been repeatedly submerged by great powers bordering the region. The pattern of peoples, languages, religions, and cultures is more complicated than anywhere else in Europe, though it was notably simplified in many areas by the slaughter or deportation of minorities during or immediately after World War II. Slavic languages are dominant in the greater part of the region, although Hungarian and Albanian are significant exceptions, and the Romanian language—which has many Slavic words and expressions—is basically a Romance language. In large sections of East Central Europe, life still has a deeply rural and agricultural cast, but this is being eroded by a powerful drive to industrialize. Industrialism and urbanization have advanced farthest in Czechoslovakia and Poland. Levels of living in most parts of the region remain substantially lower than those of northwestern Europe, but East Central Europe, trodden under conquerors and landlordism for centuries, and enormously devastated in some areas by World War II, seems now to be achieving, economically at least, a life for its impoverished masses which is superior to anything they have known in the past.

Combining countries into regional groups in the manner of the preceding paragraphs is a useful device for securing an introductory general picture of an area as complicated as Europe. However, it needs to be emphasized that each regional group is composed of countries having definite "personalities" and worthy of study as individual units despite broad physical and cultural similarities.

Most European countries are different from their neighbors in language, and often in religion. In the following section, Europe's complex pattern of languages and religions is discussed, particularly with respect to its origins.

LANGUAGES AND RELIGIONS

Europe emerged from prehistory as the homeland of many different tribal peoples. In classical and medieval times, certain peoples experienced periods of vigorous expansion, and their languages and cultures became widely diffused. First came the expansion of the Greeks and the Celtic peoples; later that of the Romans, the Germanic (Teutonic) peoples, and the Slavic peoples. As each expansion occurred, traditional languages persisted in some areas but were displaced in others. Each of the important languages eventually developed many local dialects. With the rise of centralized national states in early modern times, particular dialects became the bases for standard national languages. Thus the dialect of Paris evolved as standard French, and the dialect of London as standard English. But peasants in many localities continued to use their accustomed dialects and do so today. Europe's language pattern, then, is complex: with certain major exceptions, every country has its own national language, and many local dialects and minor languages persist in isolated or deeply rural areas or in zones along international frontiers.

Expansion and Decline of the Greek and Celtic Languages

The first millennium B.C. witnessed a great expansion of the Greek and Celtic peoples. In peninsulas and islands bordering the Aegean and Ionian seas, the early Greeks evolved a civilization that reached unsurpassed heights of philosophical inquiry and literary and artistic expression. Greek traders and colonists spread the language and culture of their homeland to many parts of the Mediterranean world. In Roman times Latin became the principal language of administration and commerce in the Roman dominions, but Greek continued to be widely used, especially in areas bordering the eastern Mediterranean. The culture of ancient Greece forms one of the major foundations of Western civilization. But classical Greek has long since vanished as a vehicle of everyday speech. The Greek language is still predominant in Greece, but the various dialects of the modern spoken language differ greatly from the Greek of Sophocles, Herodotus, and Plato. However, the "pure" form of modern Greek taught in schools diverges less from the ancient language.

During the period from about 1200 to 400 B.C. a massive expansion of Celtic-speaking peoples took place from a nuclear area in the southwestern part of present-day Germany. This expansion affected a large part of continental Europe and the British Isles. But today Celtic languages survive in only a modest way on some far western peninsulas and islands: Breton in Brittany, Welsh in Wales, Irish Gaelic in Ireland, and Scottish Gaelic in northwestern Scotland. Few people speak a Celtic language exclusively; in nearly all cases English or (in Brittany) French is spoken as a second language.

Romance, Germanic, and Slavic Languages

In present-day Europe the overwhelming majority of the people speak Romance, Germanic, or Slavic languages. The Romance languages are descended from Latin, originally the language of ancient Rome and a small district around it. As the Roman Empire expanded, Latin was spread to a vast area in southern and western Europe, southwestern Asia, and northern Africa. It did not necessarily displace the languages of conquered peoples (though often it did), but the upper classes in each Roman province were encouraged to learn it, and it became increasingly common as the language of administration, commerce, and education. Over a long period, regional dialects of Latin evolved into the Romance languages of today: Italian, French, Spanish, Portuguese, and others.

Today each Romance language is centered in a particular national state, but often the language frontier does not coincide with the political frontier. For example, Italian is the principal language not only of Italy itself, but also of Corsica and the city of Nice in France, and of some parts of southern Switzerland. The French language extends beyond the borders of France to western Switzerland and also to southern Belgium, where it is known as Walloon. The Romance dialect of northwestern Spain called Galician is closely related to the neighboring Portuguese language.

In northeastern Spain, Andorra, and the Balearic Islands, the Romance language known as Catalan is spoken. It is distinctly different from the Castilian Spanish that is the standard language in most of Spain. Catalan is very similar to the language of southern France called Provençal. The latter has largely been superseded by standard French, though it is still spoken by many peasants, and some attempt is made to keep it alive as a literary language. Minor Romance dialects survive in some mountain valleys of southeastern Switzerland. The Romanian language is generally considered a Romance language, though it has acquired many Slavic words and expressions.

In the middle centuries of the first millennium A.D., the power of Rome declined and a prolonged expansion by Germanic and Slavic peoples began. Germanic peoples first appear in history as a group of tribes inhabiting the coasts of Germany and much of Scandinavia. Their incursions overthrew the Western Roman Empire in the fifth century, but left little permanent impress on the languages of many areas that were overrun. However, German became established as the language of present-day Germany, Austria, Luxembourg, Liechtenstein, and the greater part of Switzerland, and it is widely spoken in Alsace and eastern Lorraine in France. In the Netherlands, Dutch developed as a language closely related to the Low German dialects of northern Germany, while Flemish, almost identical to Dutch, became the language of northern Belgium. The Frisian language, once widespread along the North Sea coast, is today confined to the Frisian Islands and small districts on the adjoining mainland.

Except for Finnish, an Asiatic language belonging to an entirely different language family, the present languages of Northern Europe are descended from the same ancient Germanic tongue. Modern Icelandic has evolved least from the parent language and would be largely unintelligible to a speaker of Danish or Swedish. The dialect of the Faeroe Islands is very similar to Icelandic. From the fourteenth century to the nineteenth, Norway was ruled by the Danish crown, and Danish, in a slightly modified form, became its standard language. Many peasants, however, continue to speak the older, more purely Norwegian dialects known as *Landsmål,* and there is a desire in some quarters to use them as the basis of a national language to replace the Danish-inspired *Riksmål.* The Danish and Swedish languages are, in general, mutually intelligible. Swedish is the main language in parts of western Finland, though most of that country's population speaks Finnish. Danish is spoken in a small section of northern Germany adjoining Denmark.

English is basically a Germanic language, though it has many words and expressions derived from French, Latin, Greek, and other languages. Originally it was the language of the Angles and Saxons who invaded England in the fifth and sixth centuries A.D. The Norman conquest of the eleventh century established French as the language of the court and the upper classes. Modern English retains the Anglo-Saxon grammatical structure, but it has borrowed great numbers of words from French, as well as smaller numbers from many other languages. English is the principal language in most parts of the British Isles.

Slavic languages are dominant in most parts of East Central Europe. They were originally spread by migrations of Slavic peoples in the Middle Ages. The major languages today include Russian and Ukrainian, Polish, Czech, Slovak, Serbian, Croatian, and Bulgarian. As a result of postwar territorial changes and population transfers, few speakers of Russian and Ukrainian are found outside the Soviet Union. Czech and Slovak, spoken respectively, in western and eastern Czechoslovakia, are closely related to each other. The same is true of Serbian and Croatian, the principal languages of Yugo-

Europe's religious heritage is symbolized in this view of the famous medieval cathedral at Chartres, southwest of Paris. The cathedral towers above the otherwise even skyline of the French provincial town. Its spires can be seen for miles across the gently rolling countryside of the Île de France. (Jesse H. Wheeler, Jr.)

slavia. Though they are nearly identical in spoken form, Serbian is written in the Cyrillic alphabet, while Croatian utilizes the Latin alphabet.

In the Middle Ages, Europe was invaded periodically by nomadic peoples from the grasslands of Asia and southern Russia. Most of their conquests were transitory. But the Magyar invasion of the ninth century established modern Hungary as an enclave of non-Slavic speech in the midst of a predominantly Slavic region. A sizable Hungarian minority exists in Romania. The Hungarian (Magyar) language is distantly related to Finnish. Originally the Bulgarians were Asian nomads, but after settling in their present homeland they adopted a Slavic language.

Survivals of very ancient languages persist in mountainous or isolated corners of Europe. Basque, a language apparently unrelated to any known tongue (though it has borrowed great numbers of words from other languages), is spoken in the western Pyrenees Mountains of France and Spain. The Albanian language is a survival of an ancient group of languages that once was widespread in the Balkan Peninsula.

Religious Divisions

Since Roman times Europe has been a major stronghold of Christianity, and modern European civilization owes much to Christian influence. In the Middle Ages most Europeans belonged to the Roman Catholic Church, except for the Greeks, Serbians, Bulgarians, and Romanians, who adhered to the Orthodox Eastern Church, as they do today. But the pattern of religions was complicated by two developments that occurred in early modern times: the Turkish invasions of southeastern Europe and the Protestant Reformation. In the period from the fourteenth century to the seventeenth, the Ottoman Turks spread the Islamic (Moslem or Muslim) faith widely in the Balkan Peninsula, where it endures today as the principal religion of Albania and as a minor faith in Yugoslavia and Bulgaria. As a result of the sixteenth-century Reformation, Protestant Christian sects became dominant in northern Germany, Denmark, Norway, Sweden, Finland, Iceland, the Netherlands, and Great Britain. All of these areas have remained dominantly Protestant to the present day. Important Protestant groups are found in Czechoslovakia and Switzerland, and smaller groups in various other European countries. In total number of adherents, however, the Roman Catholic Church is Europe's largest religious division, as it has been since the Christian church was first established. Areas that are predominantly Roman Catholic include Italy, Spain, Portugal, France, rural Switzerland, Belgium, the Republic of Ireland, southern Germany and the German Rhineland, Austria, Poland, Hungary, much of Czechoslovakia, and the Croatian and Slovenian areas in northwestern Yugoslavia. Large Roman Catholic minorities are found in the United Kingdom and other European countries. Jewish minorities exist in many countries, although Jewish populations were decimated by persecution in Nazi Germany, Poland, and other areas formerly controlled by the Nazis. Hundreds of thousands of European Jews have emigrated to Israel. The effects of the official Communist philosophy of atheism in East Central Europe and East Germany are difficult to determine with any degree of precision.

THE IMPORTANCE OF EUROPE

In recent centuries the peoples of Europe have exerted a profound influence on the rest of mankind. In fact, these centuries have been in many ways a "European Age," characterized by great migrations of Europeans to overseas areas, establishment of European political control over vast territories peopled by non-Europeans, and penetration of European trade and cultural influence into all quarters of the globe. "For centuries Europe was at the root of every important world trend: modern civilization stemmed from it; science, art, trade, and migrations fanned out of it and imposed European supremacy on the other parts of the world; standards were set according to European normality."[2]

The interaction of European and non-European peoples and cultures resulting from European expansion has been one of the most significant developments, not only of modern times, but of all human history. Multitudes of people outside of Europe have had their ways of living profoundly altered by the spread of European goods, institutions, languages, customs, beliefs, ideas, and ideals, and the Europeans themselves have been deeply affected by their contacts with other lands.

Modern European expansion had its principal beginnings in the fifteenth, sixteenth, and seventeenth centuries, when a series of great voyages of discovery and exploration opened the way to vast new areas previously unknown or only dimly realized. The historian J. H. Parry has termed this period the "Age of Reconnaissance," and has summarized some of its salient features and results as follows:[3]

Between the middle of the fifteenth century and the late seventeenth, Europeans learned to think of the world as a whole and of all seas as one. Their lessons were those of experience and eye-witness report. During those two and a half centuries European explorers actually visited most of the habitable regions of the globe; nearly all those, in fact, which were accessible by sea. They found vast territories formerly unknown to them, and drew the rough outlines of the world which we know. The period, especially the earlier half of it, is commonly called the Age of Discovery, and with reason. Geographical exploration, however, is only one of many kinds of discovery. The age saw not only the most rapid extension of geographical knowledge in the whole of European history; it saw also the first major victories of empirical inquiry over authority, the beginnings of that close association of pure science, technology, and everyday work which is an essential characteristic of the modern western world. During this period, especially the latter half of it, European scientists sketched the outline of the physical universe which, broadly speaking, is that accepted by the ordinary educated man today, and formulated the laws they deduced from the movement and interaction of its parts. All forms of discovery, all forms of original thought, are connected in some way, however distant: and it is natural to see a connection between these particular forms. The seaman, exploring uncharted seas, needed the help of learned men, especially men learned in mathematics, astronomy, and physical science; also, though this came later, in medical science. The student of science, seeing the achievements of geographical exploration (most empirical of all forms of inquiry, and most destructive of purely *a priori* reasoning) was naturally stimulated to further exploration in his chosen field. Both kinds of discovery further stimulated, and were stimulated by, the work of philosophers, poets, and pamphleteers.

. . . . .

The initial steps in expansion were modest indeed: the rash seizure by a Portuguese force of a fortress in

[2] Jean Gottmann, *A Geography of Europe* (4th ed.; New York: Holt, Rinehart and Winston, 1969), p. 1. Used by permission of the author and the publisher.
[3] J. H. Parry, *The Age of Reconnaissance* (London: Weidenfeld and Nicolson; New York: The World Publishing Company, 1963), pp. 1 and 14–16. Copyright © 1963 by J. H. Parry. Reprinted by permission of the author and the publishers.

Morocco; the tentative extension of fishing and, a little later, trading, along the Atlantic coast of North Africa; the prosaic settlement by vine and sugar cultivators, by log-cutters and sheep-farmers, of certain islands in the eastern Atlantic. There was little, in these early- and mid-fifteenth-century ventures, to suggest world-wide expansion. In the later fifteenth century, however, new advances in the arts of navigation and cartography, made by a new combination of academic knowledge and nautical experience, enabled the explorers for the first time to observe and record the position—or at least the latitude—of a point on an unknown coast; and even, in favourable circumstances, of a ship at sea. New methods in the design of ships, consequent upon a marriage of European with oriental traditions, made it possible for sailors not only to make long voyages of discovery, but to repeat them, and so to establish regular communication with newly discovered lands. New developments in gunnery and the making of guns, particularly in ship-borne artillery, gave European explorers a great advantage over the inhabitants of even the most civilized countries to which they sailed; enabled them to defend themselves, upon arrival, sometimes against overwhelming numbers; and encouraged them to establish trading posts even in places where they were clearly unwelcome. This vital technical superiority in ships and guns ensured the continuous development of the Reconnaissance and the permanence of its results. By the end of our period European explorers had not only sketched the rough outlines of most of the continents of the world; they had established, in every continent except Australasia and Antarctica, European outposts—trading factories, settlements, or lordships, according to the nature of the area—small, scattered, diverse, but permanent. In so doing, they not only derived help from physical scientists and stimulated the further development of physical science—though tentatively and indirectly at first . . .; they also called attention to new and far-reaching problems in the social sciences, in economics, in anthropology, and in the arts of government. In these fields also there was a tentative but widespread Reconnaissance, a wide but uneven series of additions to knowledge, with momentous consequences for Europe and the world as a whole. In all branches of science, as the Reconnaissance proceeded and became less tentative, as the European picture of the world became fuller and more detailed, so the idea of continually expanding knowledge became more familiar and the links between science and practical life became closer. A technological attitude to knowledge, an extreme readiness to apply science in immediately practical ways, eventually became one of the principal characteristics which distinguish western civilization, the civilization originally of Europe, from other great civilized societies. The unprecedented power which it produced eventually led Europe from Reconnaissance to world-wide conquest, and so created the world of yesterday, much which was governed by Europeans, and the world of today, almost all of which has accepted European technology and European techniques of government, even if only to escape from actual European rule.

The latter theme has been amplified as follows by Marshall Hodgson:[4]

It was not merely, or perhaps even primarily, that the Europeans (and their overseas settlers) found themselves in a position to defeat militarily any powers they came in contact with. Their merchants were able to outproduce, outtravel, and outsell anyone, their physicians were able to heal better than others, their scientists were able to put all others to shame. Only a limited part of the world's surface was actually occupied by European troops, at least at first. European hegemony did not mean direct European world rule. What mattered was that both occupied ("colonial" or "settled") areas and unoccupied ("independent") areas were fairly rapidly caught up in a world-wide political and commercial system the rules of which were made by, and for the advantage of, Europeans and their overseas settlers. Even "independent" areas could retain their local autonomy, in the end, only to the extent that they provided European merchants, European missionaries, even European tourists, with a certain minimum of that type of international "law and order" to which they had become accustomed in Europe, so that the Europeans remained free to vaunt a privileged position and to display among all peoples the unexampled new physical and intellectual luxuries of Europe. (For otherwise, the European powers would feel forced to intervene; and wherever one of them concentrated its efforts it would almost infallibly succeed.) Thus all peoples had to adjust their governments to a modern European international political order; but also to adjust their economies—a harder task—to the competition of technically industrialized Europe; and finally to adjust their mental outlook to the challenge of modern science as studied in Europe. The mere presence of the Europeans was enough for their new power to be felt.

We may summarize this by saying that the Europeans (including, of course, their overseas descendants) had by 1800 reached a decisively higher level of *social power* than was to be found elsewhere. . . . Individual Europeans might still be less intelligent, less courageous, less loyal than individuals elsewhere; but when educated and organized in society the Europeans were able to think and to act far more effectively, as members of a group, than could members of any other societies. European enterprises, such as firms or churches or, of course, governments, could muster a degree of power, intellectual, economic, and social, which

---

[4] Marshall Hodgson, "The Great Western Transmutation," *Chicago Today,* 4, no. 3 (Autumn 1967), 40. Adapted from the author's forthcoming book, *The Venture of Islam,* to be published by Aldine Publishing Company, Chicago. Used by permission of the author, the Aldine Publishing Company, and *Chicago Today.*

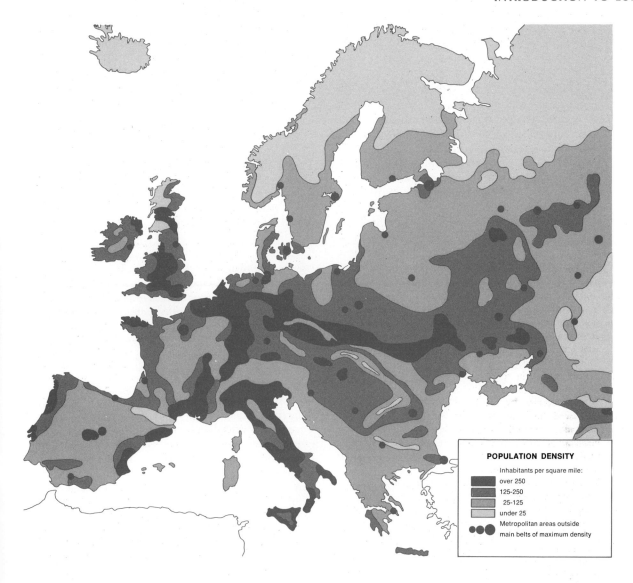

POPULATION DENSITY

Inhabitants per square mile:

over 250

125-250

25-125

under 25

Metropolitan areas outside
main belts of maximum density

Generalized pattern of population densities in Europe and the western Soviet Union. Black dots represent major urban agglomerations outside the main belts of maximum density. Largest dots represent metropolitan areas of 3,000,000 or over; second largest dots mark areas of 1,000,000 to 3,000,000, smallest dots mark areas of 500,000 to 1,000,000. Most European urban areas of 500,000 or over lie within the belts of maximum density shown as continuous strips of black. (Redrawn from a map in Jean Gottmann, *A Geography of Europe,* 4th ed., Holt, Rinehart and Winston, 1969.)

was of a different order from what could be mustered among even the most wealthy or vigorous peoples in the rest of the world.

By the nineteenth century Europe stood unchallenged as the dominant region in world affairs. As noted above, most areas outside of Europe became dependent on it in either a political or an economic sense, or both. Europe served as a great focus, or center of organization, for the world's economic and political life. From the European core, lines of trade and political influence extended to outlying territories in all parts of the world.

This worldwide economic and political system was mainly designed to facilitate the exchange of European manufactured goods for foodstuffs and industrial raw materials (as well as certain manufactures of a luxury character) produced in other parts of the world. Modern techniques of manufacturing were first developed in Europe, and for many decades the manufacturers of Europe had little effective competition in world markets. Most of the world's people depended primarily on Europe for those factory-made items they could afford. Europe, in turn, became dependent on other areas for a considerable part of its food supply and for minerals, fibers, and other materials not produced in Europe, or produced in insufficient quantities to meet the needs of European industry.

The impact of Europe and its civilization on all parts of the world has been so great that one cannot achieve a proper understanding of any other major region unless he first has a good knowledge of Europe. It is for this reason that the present survey of major world regions commences with Europe. Throughout the book, a major theme is found in each region's relationships with Europe and their geographical results.

In the twentieth century, Europe has lost its former predominance. The reasons for this are complicated, but certain relevant factors can be cited. Among these are (1) the weakening effects of two world wars, both of which were initiated and mainly fought in Europe; (2) the emergence of the United States and the Soviet Union as world powers; (3) the rise of nationalist movements in the colonial world; and (4) an increasing decentralization in the world of modern types of manufacturing; hence a lessening of the former dependence on Europe for manufactured goods.

Whatever the causes, there can be no doubt that the relative importance of Europe has declined. However, this should not be taken to mean that Europe is now unimportant, for this region is still tremendously sig-

nificant. Essentially, the continuing importance of Europe seems mainly connected with four factors: (1) its large, dense, and highly skilled population; (2) its large annual production of goods and services; (3) the political, economic, and cultural ties which it still retains with various overseas areas; and (4) its great potentialities for future development.

The Population of Europe

Although Europe is the smallest in area of the major world regions considered in this book, it is the second largest in population, being exceeded only by the Orient (Table 1, p. 41). In a space little more than half as large as the United States is concentrated a population more than double that of the United States and Canada combined. Despite the reverses of recent decades, Europe's 455 million people (1968 estimate), often highly skilled and long accustomed to political, economic, and cultural leadership, must still be reckoned with as a highly significant factor in world affairs.

With conspicuous exceptions in the case of certain northerly and/or rugged or infertile areas, all large sections of Europe have an overall population density that exceeds the world average. However, the greatest congestion is found along two major population axes—a west-east axis extending from Great Britain across northeastern France, Belgium, the Netherlands, West and East Germany, and western Czechoslovakia to southern Poland and continuing into the Soviet Union, and a north-south axis—broken by the Alps—that extends from Great Britain to Italy (map, p. 63). These axes of population correspond to the major axes of European industry. In Europe the densest populations are found in or near the principal industrial districts and depend mainly on manufacturing and associated activities for a livelihood.

The period 1940–1968 saw great changes in the political map of the world, resulting in considerable part from the emancipation of overseas areas held by European nations. Dotted bands in the oceans surround general areas within which the included islands were still held by European nations (nearly all by the United Kingdom or France) at the end of 1968. The "overseas possessions, dependencies, and quasi-dependencies of European nations" include some areas (for example, Greenland) that are officially regarded as integral parts of the European nations concerned. This category also includes several units in the eastern part of the Arabian Peninsula that have been under British protection and influence by virtue of special treaties, but Britain's commitments under these treaties are being phased out (1968). Several small units of former French India and former Portuguese India are not indicated on the map. Some uninhabited island possessions of European nations, as well as Antarctic claims, are not shown. (Briesemeister Elliptical Equal Area Projection, courtesy of the American Geographical Society.)

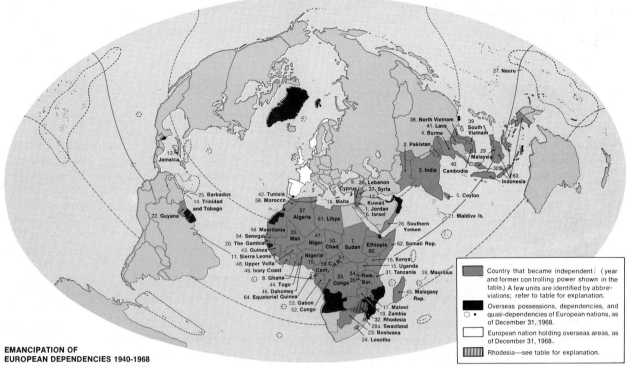

**EMANCIPATION OF
EUROPEAN DEPENDENCIES 1940-1968**

Map legend:

- Country that became independent; (year and former controlling power shown in the table.) A few units are identified by abbreviations; refer to table for explanation.
- Overseas possessions, dependencies, and quasi-dependencies of European nations, as of December 31, 1968.
- European nation holding overseas areas, as of December 31, 1968.
- Rhodesia—see table for explanation.

Great Britain

1. **Jordan** 1946
2. **Pakistan** 1947
3. **India** 1947
4. **Burma** 1948
5. **Ceylon** 1948
6. **Israel** 1948
7. **Sudan** 1956
8. **Ghana** 1957
9. **Cyprus** 1960
10. **Nigeria** 1960
    No. Br. Cameroons 1961
11. **Sierra Leone** 1961
12. **Kuwait** 1961
13. **Jamaica** 1962
14. **Trinidad and Tobago** 1962
15. **Uganda** 1962
16. **Kenya** 1963
17. **Malawi** 1964
18. **Malta** 1964
19. **Zambia** 1964
20. **The Gambia** 1965
21. **Maldive Is.** 1965
22. **Guyana** 1966
23. **Bostwana** 1966
24. **Lesotho** 1966
25. **Barbados** 1966
26. **Southern Yemen** 1967
27. **Nauru** 1968
    Prior to independence a joint trusteeship of Gt. Britain, Australia, New Zealand, with Australia the principal administrator.
28. **Mauritius** 1968
28a. **Swaziland** 1968

Great Britain (cont.)

29. **Malaysia**
    Malaya 1957
    Sarawak and Sabah 1963
30. **Singapore**
    part of Malaysia 1963
    independent state 1965
31. **Tanzania**
    Tanganyika 1961
    Zanzibar 1963; joined Tanganyika to form Tanzania ..1964
32. **Rhodesia** 1965
    Unilateral declaration of independence, not recognized by Gt. Britain or any other sovereign state.

Belgium

33. **Congo** 1960
34. **Rwanda** 1962
35. **Burundi** 1962

France

36. **Lebanon** 1944
37. **Syria** 1944
38. **North Vietnam** 1954
39. **South Vietnam** 1954
40. **Cambodia** 1954
41. **Laos** 1954
42. **Tunisia** 1956
43. **Guinea** 1958
44. **Togo** 1960
45. **Malagasy Republic** 1960
46. **Dahomey** 1960
47. **Niger** 1960
48. **Upper Volta** 1960

France (cont.)

49. **Ivory Coast** 1960
50. **Chad** 1960
51. **Central African Republic** 1960
52. **Congo** 1960
53. **Gabon** 1960
54. **Senegal** 1960
55. **Mali** 1960
56. **Mauritania** 1960
57. **Algeria** 1962

France, Spain

58. **Morocco** 1956

France, Gt. Britain

59. **Cameroon**
    East Cameroon independent from France 1960
    West Cameroon, independent from Gt. Britain; Federal Rep. of Cameroon formed 1961

Italy

60. **Ethiopia** 1942
    Eritrea, federated with Ethiopia 1952
61. **Libya** 1951

Italy, Gt. Britain

62. **Somali Rep.** 1960

Netherlands

63. **Indonesia** 1949
    Western New Guinea (West Irian) passed from Dutch control in 1962 and subsequently was incorporated into Indonesia

Spain

64. **Equatorial Guinea** 1968

Continuing Ties with Overseas Areas

Although their influence in overseas areas has declined, many European countries continue to have strong political, economic, and cultural ties with countries outside of Europe. For example, the United Kingdom, which has been the most important of the colonizing and imperial powers, still forms the center of a Commonwealth of Nations embracing about a fourth of the world's land and people (see pp. 126–128). In recent years many colonial possessions of European nations have secured independence (map, p. 65). However, some of them, such as the nations of the Commonwealth, have retained at least a nominal political tie with the home country.

Many countries outside of Europe, whether associated politically with European nations or not, still look to Europe as a market for their surplus foodstuffs and industrial raw materials and as a supplier of manufactured goods which they do not produce in sufficient quantities to meet their needs. The bulk of the world's seaborne commerce still moves in European ships, and the worldwide European network of communication, insurance, and banking facilities is still vitally important to the transaction of the world's business.

Less tangible than political and economic ties, but nevertheless important, are cultural ties between European nations and overseas countries. Such ties are especially well exemplified by the close cultural bonds between Great Britain and its former colonies of Australia, New Zealand, and Canada. Although the latter nations are entirely self-governing, their cultures reflect many affiliations with the home country.

Productivity of Europe

One of the most significant reasons for the continuing importance of Europe is its immense yearly production of goods and services. Over the centuries the peoples of Europe have developed a complex array of farms, fisheries, mines, and factories, capable of producing a tremendous volume of foodstuffs, raw materials, and manufactures. At present Europe, exclusive of the Soviet Union, produces about one-fourth of the world's wheat, two-fifths of its barley and oats and one-half of its rye, two-fifths of its coal, one-fourth of its iron ore and one-third of its steel, and large amounts of many other important commodities.

Especially significant on a world scale is Europe's production of manufactured goods. Although the relative importance of European manufacturing has declined, this region still represents one of the three greatest concentrations of industry on the globe. Only Anglo-America and the Soviet Union are in a class with Europe as manufacturing regions. Since World War II, most European countries have greatly expanded their total output of manufactured goods over prewar levels.

European production is not confined merely to production of commodities. In a broad sense the term "production" also includes services of various kinds, such as transport and communication services, banking, insurance, and investment services, and others. Europe has long been the center of a worldwide network of service facilities, and still retains great importance in this respect, although the United States has offered increasing competition in recent decades, especially in financial services.

Potentialities of Europe

The potentialities of Europe, like those of almost any region, are difficult to assess. Much will depend on events and trends in the rest of the world. But there is no doubt that Europe, given continued peacetime conditions and favorable trading relationships with other regions, has the necessary manpower, skills, experience, resources, productive plant, and worldwide commercial and political connections to achieve higher levels of productivity and a better life for its citizens, as well as maintain an influential place in world affairs. Realization of these goals will depend partly on the ultimate success of current efforts to bring about a greater degree of economic and political unity, or at least cooperation, among the European nations. This movement, essentially a post-World War II phenomenon, is described in the following section.

INTERNATIONAL COOPERATION IN POSTWAR EUROPE

In the years immediately following World War II, the peoples and countries of Europe were in difficult straits. Communist expansion, backed by the military power of the Soviet Union, threatened their liberties and way of life. Communist parties took control in the countries of East Central Europe and in East Germany, and seemed on the verge of doing so in Italy and France. In the overseas possessions of the European colonial powers there was increasing pressure for independence. Great Britain and France were faced with revolts and widespread disorder in their Middle Eastern, African, and Asian

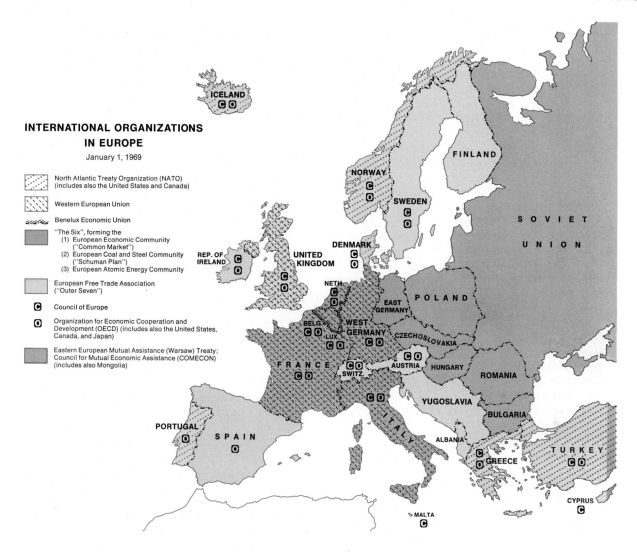

**INTERNATIONAL ORGANIZATIONS
IN EUROPE**

January 1, 1969

North Atlantic Treaty Organization (NATO)
(includes also the United States and Canada)

Western European Union

Benelux Economic Union

"The Six", forming the
(1) European Economic Community
("Common Market")
(2) European Coal and Steel Community
("Schuman Plan")
(3) European Atomic Energy Community

European Free Trade Association
("Outer Seven")

C Council of Europe

O Organization for Economic Cooperation and
Development (OECD) (includes also the United States,
Canada, and Japan)

Eastern European Mutual Assistance (Warsaw) Treaty;
Council for Mutual Economic Assistance (COMECON)
(includes also Mongolia)

possessions. The Netherlands, driven from its East Indian empire by Japan in World War II, was unable to regain effective control after Japan's defeat. Meanwhile the economies of most European nations labored under a severe strain. Financial reserves had been drained by six years of fighting, and a normal peacetime pattern of production and trade was slow to emerge after the war. Serious economic crises were experienced by several countries in 1946 and 1947.

A saving factor, American economic and military aid, now entered the situation. In 1948 the United States, faced with an urgent need to bolster the nations of western and southern Europe against communism, began sending them financial aid under the Marshall Plan to stimulate economic recovery. In the military sphere the

North Atlantic Treaty Organization (NATO), a defensive alliance, was formed in 1949. Comprised of the United States, Canada, and the majority of European nations outside the "iron curtain" (map, above), it provided a vehicle for funneling American military aid to Europe.

The European countries receiving American aid responded vigorously, and after 1948 the trend of economic recovery and military preparedness was steadily upward. Meanwhile a significant movement toward greater cooperation and unity, especially in the economic sphere, was developing. It grew out of a feeling on the part of many Europeans, including important leaders in several countries, that the security and development of their countries required their combination into larger economic and political units, though opinions varied as to the proper degree and form of combination. The result was a series of new "communities" and other intra-European organizations designed to foster unity. These efforts were actively encouraged by the United States, but the principal initiative came from Europe itself. One of the earliest organizations to be formed was the Council of Europe, a deliberative "parliament" composed of political and intellectual leaders from the participating states. It held its first session in 1949 at Strasbourg near the French-German boundary, and it has continued to meet in that city. Some idealists viewed it as the beginning of a federated "United States of Europe," but so far it has been only a useful forum to promote the general cause of European unity and social progress.

It is in the economic sphere that the most notable advances have been made. Many European leaders, feeling that schemes for political federation were impracticable or premature, have bent their primary efforts toward establishing economic unification, often with the hope that political unification would follow. In 1948 the Organization for European Economic Cooperation (OEEC) was formed to plan and coordinate the use of Marshall Plan funds and otherwise to promote recovery and economic advance. The Marshall Plan was terminated in 1952, but OEEC continued to function. In 1960 the name of the organization was changed to Organization for Economic Cooperation and Development (OECD), and the membership was broadened to include the United States, Canada, and, later, Japan.

A much publicized and highly significant aspect of the movement for greater economic unity has been the effort of six OECD members—France, West Germany, Italy, Belgium, the Netherlands, and Luxembourg—to fuse their economies through progressive removal of trade barriers among themselves and establishment of a common tariff against outside nations. In 1948 Belgium, the Netherlands, and Luxembourg inaugurated the Benelux Economic Union. Five years later the European Coal and Steel Community, formed in 1951 under the Schuman Plan, began to operate. Comprised of "the Six," its purpose was to expand and integrate the production and lower the unit costs of coal, iron ore, steel, and scrap, thus laying a foundation for general economic advance. The six nations pledged themselves to remove restrictions on the flow of these commodities across their respective frontiers. Hindrances to the flow of labor and investment capital in the coal, iron, and steel industries were also to be removed. A supranational High Authority was created as the Community's administrative arm, and legislative and judicial organs also were provided. In effect, each member state surrendered a little of its sovereignty to the supranational government of the Community, which was charged with regulating and developing the coal, iron, and steel industries as if no frontiers between them existed.

The Common Market and the "Outer Seven"

In 1958, after several years of negotiation, the six Schuman Plan nations inaugurated the European Economic Community or "Common Market." Its scope was much broader than that of the Coal and Steel Community. Tariffs and other trade restrictions were to be progressively removed on a wide range of industrial and agricultural commodities over a period of 12 to 15 years. Restrictions on the movement of workers and capital from one country to another were also to be removed. Thus the six nations would become, in most economic respects, one unit, with a standardized set of tariffs, quotas, and other regulations governing their trade with the outside world.

In creating the European Economic Community, the six countries have been trying to secure the benefits of large-scale production by pooling their resources and furnishing a larger assured market for efficient producers. It was anticipated that investment in mass-production enterprises would be stimulated and that increased specialization would develop, with each part of the Community expanding lines of goods it was best fitted to produce. Trusts and cartels, which had long restricted competition within and among the six nations, were to be eliminated or discouraged. By merging their economies, the six countries hoped to achieve greater production, larger exports, lower costs to consumers, higher wages, and a higher level of living than any one of them could achieve on its own.

The Common Market has moved very far toward attainment of its ambitious goals. In fact, it has achieved a degree of success that would hardly have been thought possible in the years immediately after World War II. By the end of 1968 both the elimination of customs duties within the EEC and the development of common tariffs on imports had reached an advanced stage. A particular achievement has been the gradual abolition of internal customs duties and the development of common external tariffs on products of agriculture. This, however, has caused hardships among marginal farmers throughout the six-nation area, although it has favored the larger, more efficient farms in areas of fertile land such as the district around Paris (p. 149). To enable the resulting dislocations and problems to be dealt with on a uniform basis, the six nations have instituted an overall system of agricultural subsidies to replace the varying subsidies administered by individual member nations in the past.

Collectively, the six members of the European Economic Community comprise a unit that is impressively large in total population and that is equipped with substantial resources and great numbers of highly skilled workers. Its population (estimated at 186 million in 1968) is smaller by only 15 million than that of the United States. Its international trade is of the first order of importance, amounting to slightly more than one-fourth of the world's total trade turnover (imports plus exports). Its total production of goods and services is a little less than half that of the United States (1967), but this gap has narrowed markedly since the Common Market was established. The six-nation Community has become a major force in the economic world, and bids fair to steadily increase its production, trade, and world influence in coming years.

The negotiations that led to the signing of the Common Market treaties in 1957 also produced the European Atomic Energy Community (Euratom). In forming this organization, the six nations agreed to pool their atomic resources and to proceed cooperatively in the development of atomic energy for peaceful purposes. The Common Market and Euratom were placed under supranational administrative bodies similar to the High Authority of the European Coal and Steel Community. Subsequently, proposals were advanced to merge the governing bodies of the three economic communities into one structure, and this came to fruition in 1967 with the establishment of a single European Commission to replace the separate administrative organs of the three communities. The communities already were being served by common legislative and judicial organs. The European Commission has its headquarters in Brussels, Belgium.

The United Kingdom, though it is ranged with "the Six" in a military alliance called the Western European Union (organized within the framework of the North Atlantic Treaty Organization to permit the rearmament of West Germany), did not initially apply for membership in the European Coal and Steel Community, Common Market, and Atomic Energy Community. Among other things, it feared that membership in these organizations would jeopardize its economic and political relationships with the Commonwealth of Nations and the United States. In 1959 the United Kingdom was the leader in a movement to create a seven-nation "outer" free-trade area in which the members would eliminate trade restrictions among themselves but would maintain separate tariff schedules against outside nations. This arrangement would enable the United Kingdom to continue operating within the system of tariff preferences in the Commonwealth of Nations. An agreement to establish the free-trade area was signed in the summer of 1959, and the new organization, officially the European Free Trade Association (EFTA), commenced to operate in 1960. Its members, often called the "Outer Seven," include the United Kingdom, Sweden, Denmark, Norway, Switzerland, Austria, and Portugal. The seven countries have a total population of 95 million (1968 estimate) and account for about one-sixth of the world's total trade turnover. In 1966 the goal of complete abolition of internal tariffs within the seven-nation area was achieved.

The European Free Trade Association offers less scope for economic expansion than the Common Market, and various EFTA members, particularly the United Kingdom, have been exploring the possibilities of membership in the larger organization. The United Kingdom has increasingly felt a need to tie its sluggish economy to the economic dynamism of the Common Market (which has been absorbing a growing percentage of British trade); and the general trend of world affairs, including liquidation of colonial empires, the rise of strong nationalism in former colonial areas, and the increasing importance of giant states and multinational economic blocs, has led many Britons to believe that they must now deemphasize some of their overseas commitments and seek their future primarily within the frame of an economically integrated Europe. The United Kingdon has applied formally for membership in the Common Market, but as this edition went to press its application had not been approved. Opposition has centered in France's President De Gaulle, who is reported to have feared that Britain's entry would dilute the French influence on the Market, would tend to bring about an undesirable increase in supranational control over the affairs of member states, and, in view of Britain's close relationships with the United States, would tend to increase the uncomfortably large American "presence" in continental Europe. Dissatisfaction with the American role in Europe has led France to reduce sharply its participation in the North Atlantic Treaty Organization, and led in 1967 to a shift of NATO headquarters from Paris to a town near Brussels, Belgium. Considerable uneasiness has been generated in many west European quarters by a vast inflow of American capital and business influence since World War II, leading to the establishment of numerous branches and subsidiaries of American corporations and an increasingly active voice by American business interests in the management of the west European economy.

Many other nations besides the EFTA group have shown great interest in a closer affiliation with the Common Market. The Market has made special trade treaties with some nations such as Israel, and associate membership in the Market has been granted to Greece, Turkey, and eighteen African nations which are former colonies of France, Belgium, or Italy. In this connection it should be noted that Finland, which has so many close

The main islands, seas, peninsulas, and rivers of Europe.

relationships with the Scandinavian countries, is an associate member of the European Free Trade Association. At present (1968), full membership for Finland is precluded by that nation's delicate situation vis-à-vis the USSR (see pp. 227–228).

MAJOR FEATURES OF THE EUROPEAN ENVIRONMENT

The diversity of Europe is not limited to cultural matters. The natural environment also is diverse. Probably no area of equal size in the entire world exhibits such a variety of natural conditions and resources. On the whole, this environment has proved favorable for human settlement and use. No other world region contains such a small proportion of unproductive land. In the following paragraphs certain significant features of the European environment are briefly discussed.

Irregular Outline

One of the most noticeable characteristics of Europe as seen on a map is its extremely irregular outline.

The main peninsula of Europe is fringed by numerous smaller peninsulas, including the Scandinavian, Iberian, Italian, and Balkan peninsulas of the second order of size, the still smaller peninsulas of Jutland, Brittany, and Cornwall, and many others (map, left). Offshore are a multitude of islands, including such large and well-known islands as Great Britain, Ireland, Iceland, Sicily, Sardinia, Corsica, and Crete. Around the indented shores of Europe, arms of the sea penetrate the land and countless harbors offer a protection for shipping. The complex mingling of land and water has created an environment which provides many opportunities for maritime activity, and every European nation possessing a stretch of coast has turned to the sea for a part of its livelihood. Excluding a few microstates such as Andorra, only five European countries—Switzerland, Austria, Luxembourg, Czechoslovakia, and Hungary—lack direct access to the sea. Even these countries are connected with the sea by rivers navigable for barges.

Northerly Location

A circumstance of European geography which is not always appreciated is the northerly location of this world region. Much of Europe, including some of the most densely populated areas, lies north of conterminous

Europe and the Mediterranean Sea compared in latitude and area with the United States and Canada. Most islands have been omitted. (Goode's Homolosine Equal Area Projection, copyright University of Chicago Press.)

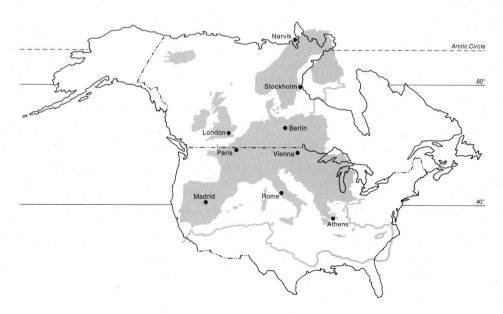

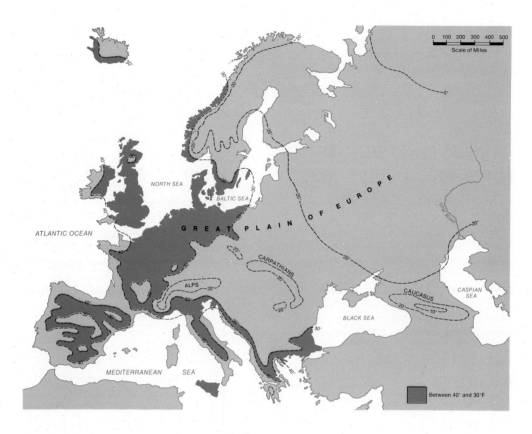

Actual temperatures in Europe in January in degrees Fahrenheit. The dashed lines, known as isotherms, connect places having the same average January temperature. (Redrawn from Jean Gottmann, *A Geography of Europe,* 4th ed., New York: Holt, Rinehart and Winston, 1969.)

United States. We are so accustomed to thinking of these northern lands as bleak and forbidding that it comes as a surprise to realize that Scotland lies in the same general latitude as Hudson Bay and that Norway has many communities located as far north as the northern mainland of Canada. The British Isles, the Scandinavian countries, Belgium, the Netherlands, and most of Germany and Poland lie north of conterminous United States (map, p. 71). Only in Spain, Italy, Greece, Malta, and Gibraltar do European latitudes reach as far south as North Carolina or Tennessee.

Temperate Climate

As might be inferred from the presence of some of the world's most densely populated areas in latitudes corresponding to those of Canada, the climate of Europe is more temperate than the northerly location would suggest. Winter temperatures, in particular, are very mild for the latitude. For example, London, England, has approximately the same average temperature in January as Richmond, Virginia, which is 950 miles farther south. Reykjavik, the capital of Iceland, is nearly as warm in January as St. Louis, 1750 miles to the south, and Tromsö, located on the coast of Norway 3° north of the Arctic Circle, has a slightly higher January average than Chicago, which is 1900 miles farther south. (In the preceding comparisons, distances given are approximate and the places named lie at generally comparable elevations above sea level.) Such anomalies of temperature are largely due to the influence of relatively warm currents of ocean water which wash the western shores of Europe during the winter. These currents, originating in tropical parts of the Atlantic Ocean, drift to the north and east, and make the waters around Europe in winter much warmer than the latitude would warrant. Westerly winds, blowing across these waters, bring considerable amounts of warmth to the land, making temperatures abnormally high, even in midwinter. Such effects are naturally most striking in places

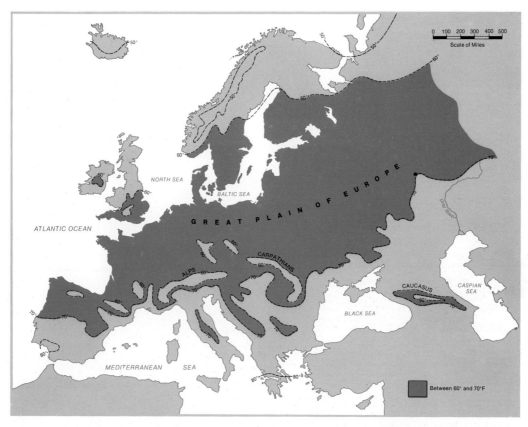

Actual temperatures in Europe in July. (Redrawn from Jean Gottmann, *A Geography of Europe*, 4th ed., New York: Holt, Rinehart and Winston, 1969.)

on or near the sea, although the moderating influence of the westerly winds and warm ocean waters is felt to some degree in all parts of Europe. In summer the ocean tends to have a cooling effect so that the seaward parts of Europe seldom experience excessively high summer temperatures.

Most large bodies of water temper the climate of adjoining land areas to some degree. Water gains and loses heat more slowly than land, and thus the oceans, seas, gulfs, and even large lakes have a moderating or stabilizing influence on the climate of lands which they border. The climate of Europe is affected not only by the open Atlantic, but also by the many arms of the Atlantic, such as the North Sea, the Baltic Sea, and the Mediterranean Sea, which penetrate deeply into the land. However, it is the Atlantic itself which has the most profound influence.

The same winds which bring warmth in winter and coolness in summer also bring abundant moisture to the land. Most of this falls in the form of rain, al-

though the higher mountains and more northerly areas have considerable snow. Most European lowlands receive 20 inches or more of precipitation a year (map, p. 74), and a few highland areas receive 100 inches or more. The general average of precipitation in the lowlands is 20 to 35 inches; few areas receive more than 40 inches a year. Although in some parts of the world 20 to 30 inches of precipitation would be distinctly marginal for agriculture, in western and northern Europe this amount is ample for a wide range of crops; this is essentially due to the effect of mild temperatures and high atmospheric humidity, which lessen the rate of evaporation and thus increase the effectiveness of the precipitation for crop growth. However, in Mediterranean Europe conditions are more unfavorable. In this area high summer temperatures cause excessive evaporation, and the yearly regime of rainfall concentrates most of the rain in the winter half-year. Thus moisture is most deficient during the summer season when it is needed most by growing crops.

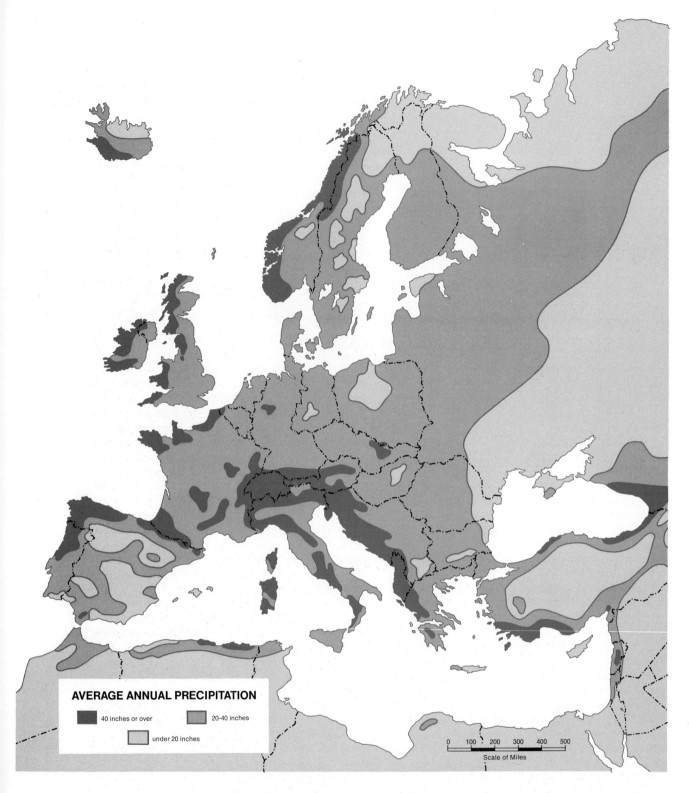

AVERAGE ANNUAL PRECIPITATION

- 40 inches or over
- 20-40 inches
- under 20 inches

Scale of Miles

0 100 200 300 400 500

Most parts of Europe receive sufficient total precipitation for crop production, although the utility of the rainfall in areas of mediterranean climate is lessened by its concentration in the colder half of the year. Highlands along windward coasts often receive excessively heavy precipitation. (After a map of the Office of Foreign Agricultural Relations, U.S. Department of Agriculture.)

Types of Climate

Geographers and climatologists have observed, primarily on the basis of records kept by weather stations, that climate tends to be fairly uniform over considerable areas. Various systems have been devised for classifying climates into "types," to which descriptive names have often been applied, varying with the particular climatic system. Each climate type has certain characteristics of temperature, precipitation, wind behavior, and so forth, which tend to distinguish it from other types. In addition, each type has certain associated features of vegetation, soil, and, to some degree, landforms, which tend to occur wherever the type is found. An area characterized by a particular climate type is known as a climatic region. Maps have been made showing the distribution of climatic types and regions on a world basis. Although such maps are somewhat inaccurate, due to lack of data for certain areas, they are extremely useful in geographic study. (See front endpaper climatic map and the introductory discussion of climate on pp. 20–24.)

MARINE WEST COAST CLIMATE. Climatic effects of the ocean are most pronounced in areas along the northwestern seaboard of Europe. The climate of these areas is often referred to as the *marine west coast* or *humid marine* climate. This type of climate is characterized by relatively mild winters and relatively cool summers, with ample precipitation for crops at all seasons, a high proportion of cloudy weather, especially in winter, and considerable fog. Throughout the year, changes of weather follow each other in rapid succession. Rainfall in the marine west coast climate tends to come in slow drizzles, often lasting for days. London has experienced as many as 72 successive rainy days. There are a large proportion of rainy days during the year. On the average London has 164 rainy days and Paris 188; parts of the Shetland Islands, north of Scotland, have 260 rainy days a year.[5] Winter days are characteristically short, cloudy, dull, and humid. In the lowlands winter snowfall is light, and the ground is seldom covered for more than a few days at a time. Summer days are longer, brighter, and more pleasant, but even in summer a good many chilly and overcast days occur. The frost-free season of 175 to 250 days is sufficiently long for most crops grown in the middle latitudes to mature. However, excessively cool and/or rainy summers in some areas are a handicap to certain crops such as corn and wheat. Regions of marine west coast climate are found in parts of the world other than northwestern Europe, a notable example being the Pacific coast of Anglo-America from San Francisco northward to southern Alaska.

HUMID CONTINENTAL CLIMATE. Inland from the coast in western and central Europe, the marine climate gradually changes. Winters become colder, summers hotter; the annual precipitation becomes somewhat less, with more precipitation in summer than in winter; and the percentage of cloudiness and fog decreases. At a considerable distance inland the climate becomes sufficiently different to be considered a separate type—the *humid continental* climate. This type of climate, which prevails in most of East Central Europe, is similar in many ways to the climate of east central United States.

MEDITERRANEAN CLIMATE. Southern Europe has an unusually distinctive type of climate—the *dry-summer subtropical* or *mediterranean climate*. The most characteristic feature of this climate type is the occurrence of a pronounced autumn and winter maximum of precipitation. The total yearly precipitation is less, on the average, than in the marine west coast and humid continental climates, and very little precipitation occurs during the summer months, when temperatures are most advantageous for crop growth. Mediterranean summers are warm to hot, and winters are mild. In the lowlands snow is rare, although it may accumulate to considerable depths on adjacent mountains. The frost-free season is very long, lasting practically the entire year in some of the more southerly lowlands. Winters in this climatic region are famous for their mild, bright, sunny weather—a great attraction to tourists from the damper, cloudier, cooler regions to the north. In the United States a mediterranean type of climate occurs in parts of central and southern California.

SUBARCTIC, TUNDRA, AND HIGHLAND CLIMATES. The greater part of Northern Europe has a *subarctic climate* characterized by long, severe winters and short, rather cool summers. Due to the shortness of the frost-free season, only the hardier crops can be grown. Most of the land in the subarctic areas is covered by coniferous forest.

The *tundra climate* is found in northerly areas where conditions are too severe for a normal stand of trees. In such areas the vegetation is composed of moss, lichens, grass, low bushes, and a scattering of stunted trees. Conditions are even less favorable for agriculture than in the subarctic zone. Frost is apt to occur at any time during the year. The principal areas of tundra

---

[5] Glenn T. Trewartha, *An Introduction to Climate* (New York: McGraw-Hill Book Company, 1954), p. 319.

climate in Europe are found along the Arctic shore of the European mainland and on the island of Iceland. Both the subarctic climate and the tundra climate of Europe are duplicated, on a much larger scale, in the Soviet Union and Canada. These climatic zones are among the most sparsely settled parts of Europe.

The higher mountains of Europe, like high mountains in other parts of the world, have a *highland climate*

TABLE 2 CLIMATIC DATA FOR SELECTED EUROPEAN STATIONS

LATITUDE (TO NEAREST WHOLE DEGREE)	STATION	ELEVATION ABOVE SEA LEVEL (FEET)	TYPE OF CLIMATE	AVERAGE TEMPERATURE (DEGREES F TO NEAREST WHOLE DEGREE)[a]			PRECIPITATION	
				ANNUAL	JANUARY	JULY	ANNUAL AVERAGE (TO NEAREST INCH)	PERCENT OCCURRING APRIL–SEPTEMBER (TO NEAREST WHOLE PERCENT)
53°N.	Dublin (Ireland)	266′	Marine west coast	49°	40°	59°	30″	51%
56°N.	Glasgow (Scotland)	29′	Marine west coast	48°	39°	59°	41″	44%
52°N.	London (England)	16′	Marine west coast	51°	40°	64°	23″	50%
49°N.	Paris (France)	174′	Marine west coast	52°	38°	66°	23″	53%
51°N.	Brussels (Belgium)	328′	Marine west coast	50°	36°	64°	31″	50%
56°N.	Copenhagen (Denmark)	72′	Marine west coast	47°	32°	64°	24″	54%
60°N.	Bergen (Norway)	144′	Marine west coast	46°	35°	59°	77″	45%
53°N.	Berlin (Germany)	180′	Humid continental	49°	31°	67°	22″	58%
52°N.	Warsaw (Poland)	351′	Humid continental	46°	28°	66°	19″	63%
45°N.	Belgrade (Yugoslavia)	456′	Humid continental	53°	32°	73°	28″	56%
43°N.	Marseilles (France)	10′	Mediterranean	58°	42°	74°	21″	40%
42°N.	Rome (Italy)	430′	Mediterranean	60°	44°	77°	26″	32%
38°N.	Athens (Greece)	351′	Mediterranean	64°	49°	82°	16″	21%
66°N.	Haparanda (Sweden)	23′	Subarctic	35°	13°	61°	22″	53%
70°N.	Vardö (Norway)	49′	Tundra	35°	24°	48°	21″	49%
47°N.	Santis (Switzerland)	8187′	Highland	29°	16°	42°	98″	56%

[a] The revised climatic data in this edition are taken from a variety of sources, primarily the publications of the U.S. Weather Bureau, Washington, D.C., and the Meteorological Office, Air Ministry, London; also W. G. Kendrew, *The Climates of the Continents* (5th ed.; Oxford: At the Clarendon Press, 1961), and publications by M. Y. Nuttonson, American Institute of Crop Ecology, Washington, D.C.

The French Alps. Snowclad summits rise above a foothill zone of pastures and woodlands. In the foreground is the rapidly growing industrial, educational, and scientific center of Grenoble, located on a tributary of the Rhone River. (French Embassy Press and Information Division.)

varying in character according to altitude. In parts of the Alps, for instance, a range of climates is found which varies from subtropical at the base of the mountains to tundra and ice-cap climates at the highest elevations.

Temperature and precipitation data for selected European climatic stations are given in Table 2.

Taken as a whole, the climates of Europe are temperate, humid, seasonal climates which have proved favorable for human settlement. The general range of climates in Europe is duplicated in other parts of the world, most notably in Anglo-America. It is to overseas areas having climates generally similar to those of Europe that Europeans have migrated in the largest numbers.

Varied Topography

The surface features of Europe are extremely diversified (map, p. 78). It is safe to say that no other area of equal size exhibits so much variety. Plains, plateaus, hill lands, and mountains form a complex pattern, and together with associated vegetation, water bodies, and

works of man, produce a variegated and often highly scenic landscape.

One of the most prominent surface features of Europe is a vast plain which extends from the Pyrenees Mountains across western and northern France, central and northern Belgium, the Netherlands, northern Germany, Denmark, and Poland, and stretches without a break far into the Soviet Union (map, p. 78). Outliers of this plain are found in Great Britain, in the southern part of the Scandinavian Peninsula, and in southern Finland. On this North European Plain are found the largest expanses of arable land in Europe. In addition, there are valuable deposits of coal, natural gas, potash, salt, iron ore, and other minerals. For many centuries the plain has been a major avenue of movement between east and west in Europe. Today it forms the principal focus of European industrial, commercial, and agricultural activity. For the most part the North European Plain is not flat, except in certain areas, but contains great stretches of undulating or rolling land, with many gentle hills and low escarpments.

South of the northern plain, Europe is predominantly mountainous or hilly, although the mountains and hill masses enclose many relatively level areas. The

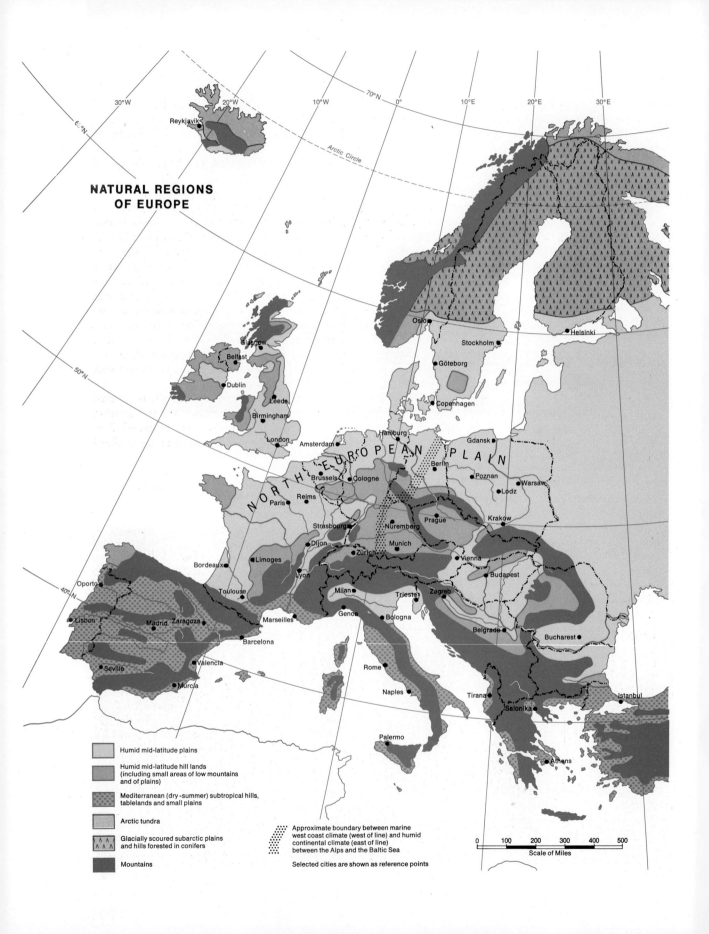

NATURAL REGIONS
OF EUROPE

NORTH EUROPEAN PLAIN

Humid mid-latitude plains

Humid mid-latitude hill lands
(including small areas of low mountains
and of plains)

Mediterranean (dry-summer) subtropical hills,
tablelands and small plains

Arctic tundra

Glacially scoured subarctic plains
and hills forested in conifers

Mountains

Approximate boundary between marine
west coast climate (west of line) and humid
continental climate (east of line)
between the Alps and the Baltic Sea

Selected cities are shown as reference points

0   100   200   300   400   500
Scale of Miles

Reykjavik

Arctic Circle

70°N

60°N

50°N

40°N

30°W   20°W   10°W   0°   10°E   20°E   30°E

Oslo   Stockholm   Helsinki
Göteborg
Copenhagen
Glasgow
Belfast
Dublin   Leeds   Hamburg   Gdansk
Birmingham   Amsterdam   Berlin   Poznan   Warsaw
London   Brussels   Cologne   Lodz
Paris   Reims   Prague   Krakow
Strasbourg   Nuremberg
Dijon   Munich   Vienna
Limoges   Zurich   Budapest
Bordeaux   Lyon   Milan   Triest   Zagreb
Toulouse   Genoa   Bologna   Belgrade   Bucharest
Oporto   Marseilles
Lisbon   Madrid   Zaragoza   Rome
Seville   Barcelona   Naples   Tirana   Istanbul
Valencia   Salonika
Murcia   Palermo   Athens

mountains of southern Europe are mostly rugged alpine mountains, with many jagged peaks and snowcapped summits. The highest peaks are found in the Alps, a massive snowclad range containing many summits of 12,000 feet or higher and culminating in Mont Blanc at 15,771 feet. These mountains constitute a considerable barrier to traffic, though on the whole a less formidable barrier than their appearance on a map would suggest, since they are cut through by many passes, river valleys, and tunnels. To the countries in which they lie, the mountains are valuable as sources of hydroelectric power, as tourist attractions, and as defensive ramparts in wartime. In some areas, particularly the Balkan Peninsula, they have tended to isolate small groups of people, and thus have aided in the development of many small, distinctive cultural regions.

The hill lands of Europe are lower and less rugged than the mountains. Generally speaking, the hill lands are older than the mountains in point of geologic time. Having been formed earlier, they have been exposed to weathering and erosion for a longer period and thus have a smoother and more rounded aspect. Many of the important deposits of metals in Europe are found in the hill lands.

To the north the North European Plain is fringed by glaciated lowlands and hill lands in eastern Sweden and Finland, and by rugged, ice-scoured mountains in western Sweden and Norway. In the British Isles a few sizable areas of lowland are fringed by hill country and low mountains.

[EDITORIAL COMMENT. During the Great Ice Age, massive continental ice sheets formed over the Scandinavian Peninsula and Scotland. The ice moved outward into Russia, Germany, Poland, the Netherlands, and England. Over a period of perhaps a million years ice sheets alternately advanced and retreated as glacial and interglacial ages succeeded each other. The latest retreat of the ice is thought to have begun at least 35,000 years ago. Continental ice sheets still cover about 10 percent of the earth's land surface in Greenland and Antarctica, as compared with an estimated 28 percent of the lands when the glaciers were at their maximum extent.

In some glaciated areas the ice sheets scoured deeply, removing most of the surface soil and gouging hollows which became lakes when the ice melted. Elsewhere the glaciers deposited the materials accumulated

A glaciated landscape in Scandinavia. The town in the view is Strängnäs, situated on the shore of Lake Mälaren in central Sweden. (Swedish National Travel Office.)

on their under side, either in the form of outwash carried by sheets of water issuing from under the ice, or in the form of morainal materials dropped in place as the ice melted. In most of Norway and Finland, much of Sweden, parts of the British Isles, and Iceland (which had a local ice sheet of its own) scouring predominated over deposition. The present landscape of these areas is dominated by thin soil, bare rock, and a multitude of lakes. The ice sheets interfered with the preexisting drainage pattern so that the present streams tend to wander about aimlessly, with many rapids and waterfalls. Thus many natural sites for the development of hydroelectric power have been created. This is an important circumstance in Norway, Sweden, Finland, and Iceland, which are poor in mineral fuels and take care of a large share of their power needs by the use of hydroelectric stations.

Glacial deposits of varying thickness were laid down on the North European Plain. Unfortunately,

many of the deposits were excessively sandy so that the present soils are not very fertile on the whole, although they have often been made productive by careful handling and large additions of fertilizer. Immediately south of the glaciated areas, deposits of a windblown material known as *loess* appear. This material appears to have been transported from the glaciated areas by winds after the surface was dry. A relatively continuous band of loess extends along the southern edge of the North European Plain, from northeastern France to southern Poland and southern Russia. Loess tends to weather into good loamy soils, and the loess belt represents one of the most fertile and productive farming areas in Europe.]

River Systems

[EDITORIAL COMMENT. Some terminology regarding rivers may be in order at this point. A *river*

This view of Rotterdam harbor in the Netherlands reflects both the highly commercialized character of Europe's economy and the region's worldwide economic connections. Wheat from overseas producing areas is being transferred from the ocean vessels to barges that will transport it to inland consuming areas. This traffic is facilitated by Europe's fine network of navigable rivers and canals. (*Foreign Agriculture.*)

system is a river together with its tributaries. A *river basin* is the area drained by a river system. The *source* or a river is its place of origin; the *mouth* is the point where it empties into another body of water. As they near the sea, many rivers become rather sluggish, depositing great quantities of sediments to form *deltas,* and often dividing into a number of separate channels, known as *distributaries.* The Rhine divides in this manner, its two principal distributaries being the Lek and the Waal.]

In the mountains and hill lands of Europe many important rivers rise. Most of the rivers that are the chief transportation arteries empty into the seas which border Europe on the north and west. Among the most important of these commercial waterways are the Rhine, Scheldt, Seine, Thames, Mersey, Clyde, Weser, and Elbe (map, p. 70). Along the lower courses of many streams deep *estuaries* have been formed by the submergence of the river mouths. Such estuaries are often capable of admitting ocean vessels for considerable distances. The seaward portions of many rivers experience tides of considerable height. The daily ebb and flow of tidal waters helps to keep navigation channels scoured free of sediments, although the action of the tides must often be supplemented by dredging. Were it not for the tides, it would be very difficult for many of the smaller ports located some distance upstream to keep open. However, in certain rivers the range of tides is so great that it constitutes a distinct handicap for the ports, requiring expensive installations to offset the rise and fall of the water level and thus permit continuous use of the wharves. The foregoing statements do not apply to the Mediterranean, Baltic, and Black seas, which are nearly tideless.

Many of the major seaports of Europe are located along the lower courses of rivers. Among the largest and most important of these ports are London, on the Thames; Rotterdam, in the delta of the Rhine; Antwerp, on the Scheldt; and Hamburg, on the Elbe.

The most important inland waterway of Europe is the Rhine. This river, together with tributary rivers and canals, is equaled or surpassed in economic significance only by the Great Lakes–St. Lawrence waterway of North America. The longest river of Europe, however, is the Danube. Although it is used considerably for navigation, it carries a much smaller volume of traffic than the Rhine. Flowing eastward to the Black Sea through regions which are less industrialized than those along the Rhine, the Danube carries a much smaller tonnage of the coal, iron ore, steel, and other bulky industrial materials which constitute the major

Industrial traffic on the Rhine. A barge loaded with coal moves upstream. In the distance are the Rhine Uplands, through which the river flows in the narrow, twisting valley called the Rhine Gorge. Terraced vineyards that produce the famous Rhine wines cling to the steep slopes of the valley wall. (Jesse H. Wheeler, Jr.)

traffic on the Rhine. It should be noted, however, that the volume of traffic on the Danube has increased greatly since World War II. This has been due primarily to expanded industrial development in Communist countries that the river touches.

The rivers of Mediterranean Europe, in Italy, Iberia, extreme southern France, and the Balkans, are used very little for navigation. Flowing mostly through hilly or mountainous country, they are typically steep and swift with many rapids and waterfalls (the sluggish Po River in northern Italy is a conspicuous exception). Due to the unbalanced yearly distribution of rainfall in the mediterranean climate, these streams are less regular in flow than the rivers of northwestern Europe. They are often choked with water during periods of heavy rain in the autumn and winter, but may become mere trickles in the drought of midsummer.

The rivers of the Scandinavian Peninsula, Finland, and Iceland, like those of Mediterranean Europe, are mostly short, rock-strewn, and swift. However, they are more even in flow, partly as a consequence of the more balanced distribution of precipitation and partly as a result of the fact that many of them originate in lakes which serve as natural reservoirs and release water gradually throughout the year. These rivers, with their many rapids and waterfalls, are not very useful for navigation, except for the transportation of logs. However, many of them are harnessed for the production of hydroelectric power.

The usefulness of rivers emptying into the Baltic Sea is somewhat lessened by the fact that they are ordinarily frozen for a considerable period in midwinter. Parts of the Baltic Sea itself freeze in winter; indeed the Gulfs of Bothnia and Finland are normally frozen for 3 months or more. In some winters the extreme north of the Gulf of Bothnia is frozen for 7 months. The Baltic and its gulfs, being almost completely enclosed by land and continuously supplied with fresh water by a large number of rivers, are less saline than the ocean and so are especially susceptible to freezing. However, extensive use of icebreakers lessens the handicap to transportation.

In most other parts of Europe sea ice is uncommon. The shores of the North and Irish seas, the Bay of Biscay, and the Mediterranean are normally ice free throughout the winter.

THE DIVERSIFIED ECONOMY OF EUROPE

Diversity is the keynote of the European economy, as it is of the European environment. Agriculture, fishing, manufacturing, mining, the forest industries, transportation, trade, and other forms of economic activity are highly developed. Much variety exists from one country and region to another, both in the relative emphasis given to each of the major pursuits and in the relative intensity of development.

Agriculture and Fisheries

Agriculture finds its highest development in the countries bordering the North Sea. Scientific methods of farming are employed, and yields per acre and per man are high. However, the different countries vary considerably among themselves. The Netherlands, for example, consistently obtains higher yields than France, mainly as a result of more intensive methods. In the North Sea countries, agriculture is more highly commercialized than in other parts of Europe, a fact principally due to the close proximity of large and dependable urban markets for surplus foodstuffs.

Types of Farming

Almost everywhere in Europe livestock are important in the farm economy. However, the emphasis on livestock raising is proportionately greatest in northwestern Europe. Two principal types or systems of farming are especially characteristic of the latter area—*dairy farming* and *commercial crop and livestock farming*. In dairy farming, cows are kept for the production of milk, butter, and cheese; calves not needed for the dairy herd are sold for veal; and the income from dairying is often supplemented by production of hogs and poultry. A high proportion of the available land is occupied by temporary or permanent pasture, and crop land is devoted principally to hay, feed grains, and root crops suitable for feed. In commercial crop and livestock farming, cattle and other animals, especially hogs, sheep, and poultry, are raised primarily for meat. The farm income results principally from the sale of livestock, although crop surpluses may be sold. The ratio of crop land to pasture is higher than in dairy farming, with a good balance being maintained among grains, root crops, and hay. In East Central Europe agriculture has been characterized in the past by poorer methods and lower yields than in the North Sea lands. Much of the farming has been of a semisubsistence nature, with a high proportion of the products being used at home rather than sold. Collective farming on the Russian model has been instituted in many parts of East Central Europe since the end of World War II (see pp. 266–267).

In Southern Europe agriculture is both favored and hindered by the climate. Mild winters and a long growing season make possible the production of specialized subtropical crops, such as citrus fruits and olives, together with early vegetables. However, lack of rain severely restricts the growing of crops in the summer, except for deep-rooted, drought-resistant perennials such as olives, grapes, dates, and figs, or crops grown by irrigation. Summer is the difficult season for most farmers in this part of Europe. Fall-sown grains (wheat or barley) are widely grown with the aid of fall and winter rains. In Southern Europe, as in East Central Europe, much of the agriculture is of a semisubsistence character. Farmers in this region are handicapped not only by a lack of summer rain but also by a dearth of land sufficiently level to grow crops. The most important exception to these generalizations is the Po Valley, which has summer rain coupled with wide stretches of level land.

Generally speaking, agricultural methods in Europe are most backward, and yields are lowest in areas where the largest proportion of the people depend di-

The productive agriculture of western Europe reaches a peak of intensity in glasshouse districts where vegetables and flowers can be grown during the cool season. This produce is marketed in large metropolitan areas near by. Many acres of land, especially in southeastern England and the Netherlands, are enclosed by glasshouses similar to those in the above view, taken in the London metropolitan area. (British Information Services.)

rectly on agriculture for a livelihood. Such areas, found mostly in Southern and East Central Europe, are frequently characterized by a serious degree of rural overpopulation.

The Major Crops

Approximately half of the crop land in Europe is devoted to cereals. Wheat, the leading cereal, both in acreage and yield, is grown to some degree in most farming areas in Europe, and tends to occupy the best soils. A map of wheat acreage probably comes closer to a map of general crop distribution than a map of any other single crop (maps, pp. 84–85).

On the sandy plains of northern Germany and Poland, rye is an important crop, though in both West and East Germany it has shown a marked decline in acreage and production since the early 1950s. It is the principal bread grain for the majority of rural people on the North European Plain east of the Elbe. Most of the world's rye is grown in Europe and the Soviet Union.

Growing of corn (maize) for grain is mainly restricted to areas with long, hot summers and a summer maximum of rainfall. Such conditions are found in the Danubian plains of Hungary, Romania, and Yugoslavia, in the Po Valley of northern Italy, and in the Southwestern Lowland of France. These five countries grow about four-fifths of Europe's corn.

Considerable quantities of oats and barley are grown in various parts of Europe. Both are used mainly for livestock feed. Irrigated rice is grown in a few districts in Italy, Spain, and a number of other countries. The Po Valley of Italy and the Ebro delta of Spain are the largest areas of production.

Root crops are far more important in European agriculture than in the agriculture of the United States. Potatoes are grown in enormous quantities, particularly in West and East Germany and Poland; they are used for human food, stock feed, and the production of alcohol. Sugar beets are extensively grown on the better loamy soils; the beet tops are fed to stock, as are the residues after the juice has been extracted for the production of sugar. Turnips, mangels, rutabagas, carrots, and other root crops are very important for stock feed, especially in the North Sea countries.

Many important areas of market gardening and specialized horticulture are found in Europe, particularly in the zone of mediterranean climate and in the vicinity of the larger cities.

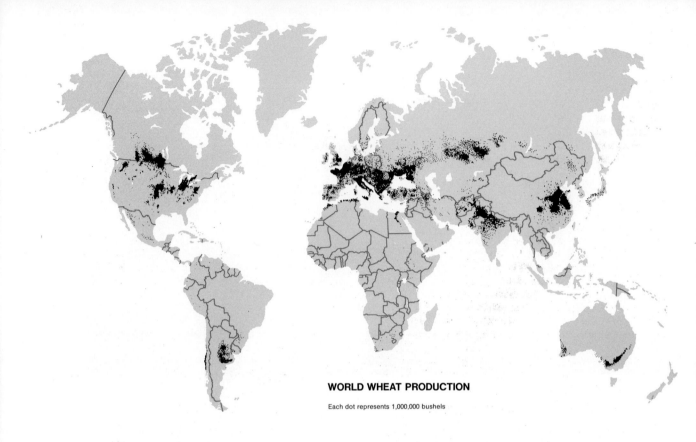

WORLD WHEAT PRODUCTION

Each dot represents 1,000,000 bushels

Source: U.S. Department of Agriculture. The data are averages for 1957–1961.

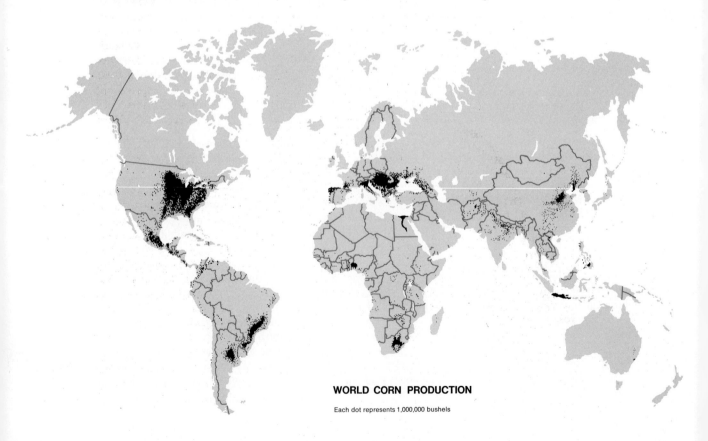

WORLD CORN PRODUCTION

Each dot represents 1,000,000 bushels

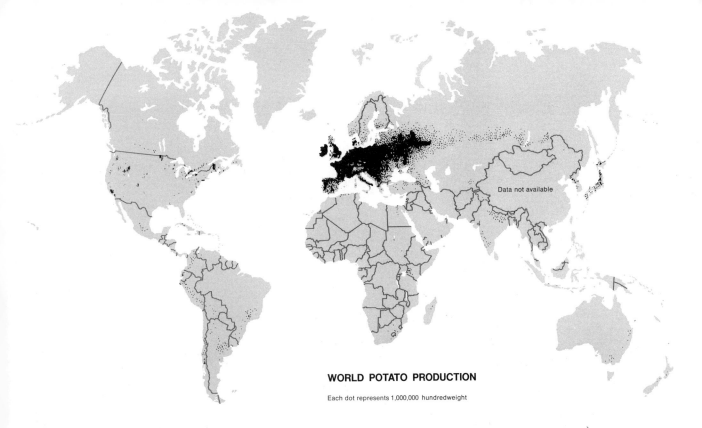

WORLD POTATO PRODUCTION

Data not available

Each dot represents 1,000,000 hundredweight

Although Europe's agriculture is overshadowed by the region's enormous development of manufacturing and commerce, this region is one of the world's most important agricultural areas—a fact brought out clearly by the maps on pages 84–86. Its agriculture is geared almost entirely to the production of foodstuffs for sale or use within the region itself, and while it remains the greatest food-importing region in the world, it should not be forgotten that Europe's farms supply by far the greater part of the total food requirements for nearly half a billion Europeans.

European Fisheries

The products of European agriculture are supplemented by products of the fisheries. Europeans eat more fish per capita than do Americans, and fishing has been intensively developed by a number of countries. The most important fishing nations are the North Sea countries and Iceland. The shallow seas which border Europe on the northwest are rich in small marine organisms known collectively as *plankton*. Such organisms are the principal food for schools of herring, cod, and other fish of economic value. Many thousands of fishermen gain their principal livelihood from these waters. The Dogger Bank in the North Sea is famous as a fishing ground. Norway is the leading European nation in total quantity of fish caught, although fishing is of greatest relative importance in the economy of Iceland, a country

that is notoriously poor in land resources and has depended mainly on the sea for support. In recent years an increasingly large proportion of the European (and world) fish catch has been converted into protein-rich fish meal, principally for use in fertilizers and livestock and poultry feeds. Much research, in European countries and elsewhere, is currently devoted to the development of fish-protein concentrates, which offer a promising avenue for increasing the amount of animal proteins in the diet of undernourished peoples in the less developed parts of the world.

Dependence on Food Imports

In spite of its great production from agriculture and sea fisheries, Europe is not self-sufficient in most foodstuffs. In fact, among the principal foodstuffs it is 99 to 100 percent self-sufficient only in whole milk, potatoes, and rye. Domestic production of the remaining foods must be supplemented by imports from other parts of the world. Among the more prominent food imports are wheat, corn, oil cake (for stock feed and fertilizer), animal and vegetable fats and oils, chilled and frozen meats, cane sugar, cocoa, tea, coffee, and tobacco. Europe also imports large quantities of fibers (cotton and wool predominantly) and natural rubber. As for many decades past, this region still remains the greatest market for the world's surplus agricultural commodities. However, the individual countries of Europe

85

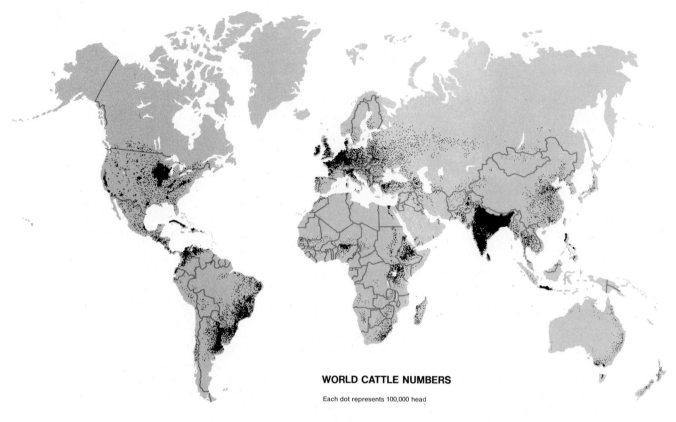

WORLD CATTLE NUMBERS

Each dot represents 100,000 head

Source: U.S. Department of Agriculture. The data are averages for 1957–1961.

WORLD SHEEP NUMBERS

Each dot represents 100,000 head

vary a great deal in their degree of dependence on agricultural imports. The United Kingdom is dependent on such imports for half of its food supply, but for such countries as Hungary, Romania, and Bulgaria they are relatively unimportant. There is a great deal of intra-European trade in foodstuffs, particularly in livestock products and fresh fruits and vegetables. Among the European nations, the Netherlands, France, and Denmark are the largest exporters of food surpluses, while Great Britain and West Germany are the largest importers, both of surpluses originating within Europe and those originating outside of it.

Manufacturing and Mining

The products of European factories run the gamut of industrial commodities, from pig iron and steel to the finest textiles, ceramics, optical goods, and metalwares.

Metalworking

European industries produce three main types of commodities: metal goods, chemicals, and textiles. The most important of the metalworking industries is the

The importance of western Europe as a major center of heavy industry and market for raw materials is prominently indicated in this diagrammatic map of ocean flows of iron ore. The arrows on the map do not necessarily emanate from the exact mining areas or show the exact routes of flow. Note the great changes that took place in world patterns of ore supply and demand during a 15-year period. The iron and steel industries of western Europe, the United States, and Japan are skimming the best resources of high-grade iron ore (and other minerals as well) from many "developing" countries in the world. This may in time create problems for such countries as their own industries develop, but for the moment they are finding the revenues derived from overseas ore sales a welcome addition to their national treasuries. (The source of the map was a study of the world market for iron ore by Gerald Manners for Resources for the Future. Map redrafted and reproduced by permission of the author and Resources for the Future.)

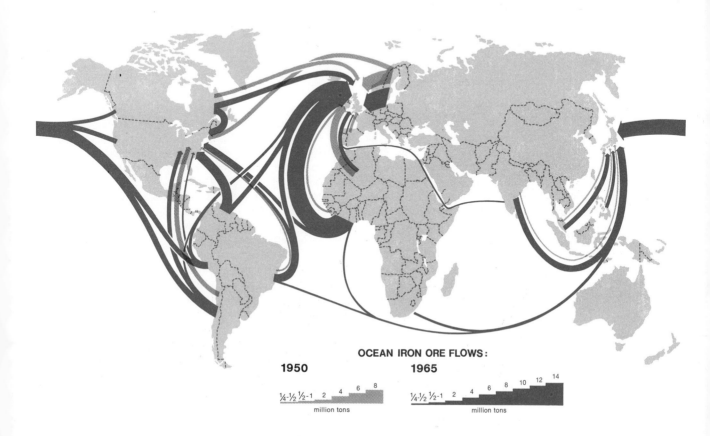

OCEAN IRON ORE FLOWS:

1950   **1965**

¼-½ ½-1 2 4 6 8
million tons

¼-½ ½-1 2 4 6 8 10 12 14
million tons

iron and steel industry, which is mainly concentrated in a few major districts in West Germany, the United Kingdom, France, Belgium, Luxembourg, Poland, and Czechoslovakia. These districts are located in close proximity to deposits of coal, iron ore, or both; coal is the more abundant and widespread. Coal is tremendously significant, not only to the iron and steel industry, but also to many other European industries which depend on it as an important source of fuel and power. However, in recent years it has been subject to increasing competition from other power sources, including (1) petroleum, most of which is imported; (2) hydroelectricity; (3) natural gas, from deposits in the Netherlands, Italy, the British-controlled sector of the North Sea (see p. 105), and a number of other areas; and (4) atomic energy, particularly in Great Britain. The chief coal fields are found in a west-east line extending from the United Kingdom to southern Poland. This axis of coal deposits corresponds generally with the main axis of population and industry described earlier. In Europe, as in other industrialized parts of the world, there has been a pronounced tendency in past times for industrial plants and workers to gravitate to coal.

Iron ore is less plentiful than coal, and many of the smaller deposits have been exhausted. However, two major reserves of ore still remain—the deposits of Lorraine in eastern France and of the Kiruna–Gällivare district of northern Sweden. The French ores are larger in amount, but the Swedish ores are of better quality. Smaller reserves of ore are found in West Germany, the United Kingdom, Austria, Luxembourg, Czechoslovakia, northern Spain, and elsewhere.

An increasingly large share of Europe's iron-ore needs is being met by imports from other parts of the world, particularly Africa. The map on page 87 shows the dramatic rise in oceanborne imports reaching western Europe in recent years. A considerable number of the newest European iron and steel plants—for example, at Dunkirk in northern France—are being built in coastal locations accessible to ore ships as well as ships bringing coal from the United States. American coal, mined at a relatively low cost by large-scale machine methods, has been outselling higher-cost European coal in many European markets since World War II. Italy, which has become Europe's fourth largest steel producer (after West Germany, the United Kingdom, and France), is overwhelmingly dependent on imported ore and coal, and has most of its iron and steel capacity in coastal situations. The countries of Communist Europe depend heavily on the Krivoy Rog deposits of the Ukrainian SSR in the Soviet Union as a source of iron-ore supply (see pp. 317 and 318).

The known deposits of nonferrous metals in Europe have been worked for a long period, some of them for thousands of years. Today these deposits can no longer supply the demand, though some countries have a sizable production of at least one or two metals. Large quantities of copper, zinc, lead, aluminum, and tin must be imported. In addition, Europe requires large imports of the ferroalloys, including manganese, nickel, chromium, tungsten, cobalt, molybdenum, and vanadium. The latter are used to impart special qualities of hardness, toughness, resistance to heat, or other desirable properties to steel.

The most valuable products manufactured from metals in Europe are machinery and transportation equipment. Such items form the largest single category of exports from the four largest iron and steel producers (outranking chemicals, textiles, and unmanufactured metals by a wide margin), and they are very important in the export trade of many other European countries. Actually there is an enormous amount of intra-European trade in machinery and transportation equipment, as exemplified by France, which sells great quantities of these items to its western European neighbors and also buys great quantities from them. In France, machinery and transportation equipment comprise both the largest category of exports and the largest category of imports.

The Manufacture of Chemicals

All industrial nations use large quantities of chemicals—acids, alkalis, dyes, and a host of others—because many industrial processes require chemicals of some type. Fortunately, the chemicals needed in the largest amounts by modern industry can be manufactured from raw materials which are relatively abundant. Among the latter, salt, limestone, sulfur, and coal are extremely important. Europe is well supplied with these materials, and it was in this region that the modern chemical industry was born, Great Britain and Germany being the early leaders in the industry.

The United States has now become the leading nation in chemical manufacture, but the European chemical industry remains significant. One of the most important branches of this industry is the manufacture of chemical fertilizers. Because most European soils are not particularly fertile by nature, it has taken large applications of commercial fertilizer (along with animal manures) to make them productive. The principal raw materials for fertilizer manufacture are available in Europe in large quantities: potash from huge natural deposits in West and East Germany and France; phosphates from blast furnace wastes and from deposits of phosphatic rock in North and West Africa; and nitrates secured by the fixation of atmospheric nitrogen and also as a by-product of coke manufacture. Europe uses more chemical fertilizer than any other world region.

In recent years Europe has seen a great upsurge in petroleum refining and the manufacture of petrochemicals from oil and natural gas. Though several European countries, most notably Romania, have oil fields, all but a very small percentage of the crude oil used in the region is imported, primarily from the Persian Gulf area, North Africa, and Venezuela; or, in Communist Europe, from the Soviet Union. Refineries and petrochemical plants tend to cluster around seaports or in major inland industrial cities or districts, particularly along the Rhine River. Among the outstanding concentrations of refineries are those located in or near Rotterdam, London, Southampton, Le Havre, Marseilles, Hamburg, the Ruhr, Cologne, Karlsruhe, and Ingolstadt. Natural gas, used primarily as an industrial and household fuel, comes mainly from huge deposits in the northern part of the Netherlands or from smaller, but sizable, deposits in northern Italy and the Transylvanian Basin of Romania. The floor of the shallow North Sea is being prospected actively for natural gas and petroleum, and a number of promising strikes of natural gas have been made. Gas is being piped to Great Britain from several wells. No oil in quantity has been reported (1968).

Textile Manufacture

The manufacture of textiles is a very widespread branch of European industry. Although in many areas textile milling is the main industry, there is also an extensive development of textile manufacture employing chiefly women and girls in some areas where the male workers are principally employed in other enterprises—coal mining, steel milling, shipbuilding, and the like. Woolen and cotton textiles are mainly produced from imported fibers. Linen manufacture, particularly important in Northern Ireland, relies on European (primarily Belgian and Baltic) flax, and the synthetic fiber industry (rayon, nylon, and the like) depends mainly on European materials. The manufacture of silk is less important than formerly. Some raw silk is produced in the Po Valley and a few other areas.

Europe continues to lead other manufacturing regions in the production of highly finished goods requiring painstaking care and a high degree of skill in manufacture. Swiss watches, English woolens, Scotch whiskies, German porcelain and cameras, Bohemian glassware, Irish linens, and the best French wines, brandies, and liqueurs are famous, and unsurpassed in quality. Such products find a ready sale in export markets.

The Role of Electric Power

Today the long-distance transmission of electric power is facilitating the development of lighter types of industry in areas which are less congested than the old coal-field industrial districts. In Northern Europe and Southern Europe, electricity is provided primarily by falling water, but between these two hydropower zones lies a broad zone, mainly on the North European Plain, where thermal power predominates (map, p. 90). The majority of thermal-electric stations are fueled by coal, and important clusters of these stations are found in the coal-field industrial districts. Increasingly, the utility of Europe's electric power supply is being enhanced by interconnections between the power systems of different countries, or between different power grids within the same country. The principal advantages of such interconnections have been summarized as follows by W. R. Derrick Sewell:[6]

There are five principal technical and economic advantages of interconnection. These stem from differences in the demand conditions or in the supply patterns of neighboring regions, and from differences in the stage of economic development in adjacent regions.

First, interconnection may make it possible to take advantage of the fact that peak loads in neighboring regions seldom coincide. The peak load of one system may occur one hour ahead of that of a neighboring system. The seasonal peak of one system may be in the summer, whereas, in another system it may be in the winter. Such differences in the time of occurrence of peak loads on adjacent systems are described as "diversity" in the electric power utility business. The amount of peaking capacity required to serve the interconnected systems is less than the sum of the peaking capacities required to serve the systems operated in isolation. Therefore, considerable savings can be derived through interconnections; for example,

[6] W. R. Derrick Sewell, "The Role of Regional Interties in Postwar Energy Resource Development," *Annals of the Association of American Geographers,* 54 (1964), 569–570. Used by permission of the author and the *Annals.* Two footnotes renumbered in editing.

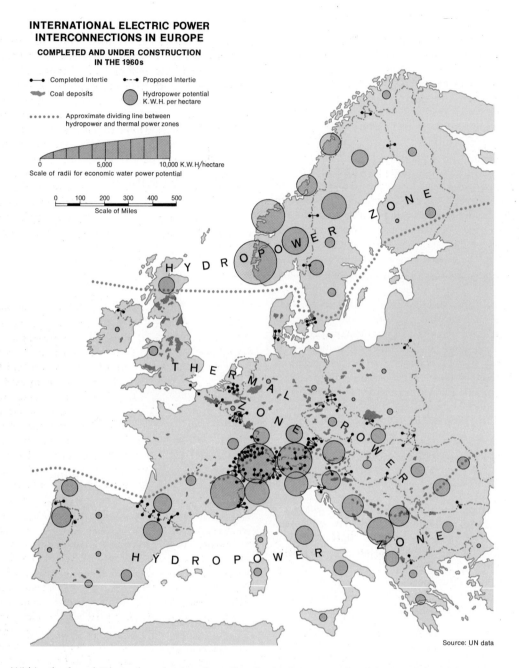

**INTERNATIONAL ELECTRIC POWER
INTERCONNECTIONS IN EUROPE**

**COMPLETED AND UNDER CONSTRUCTION
IN THE 1960s**

●—● Completed Intertie   ●--● Proposed Intertie

〰〰 Coal deposits   ⬤ Hydropower potential
                          K.W.H. per hectare

••••• Approximate dividing line between
      hydropower and thermal power zones

0   5,000   10,000 K.W.H/hectare
Scale of radii for economic water power potential

0   100   200   300   400   500
Scale of Miles

Source: UN data

Within the broad "thermal power zone" on the above map, most electricity is produced by thermal stations powered by coal (various ranks and grades), oil, gas (natural or manufactured), or atomic energy. Within the two "hydroelectric power zones" electricity is produced mostly in hydroelectric stations powered by falling water. The "interties" are connections permitting the transfer of electric power from one country to another. Interties are shown as they existed in the early 1960s. Note that the hydropower symbols represent potential power rather than developed power. Note also the heavy concentration of potential hydroelectricity in mountains or glacially-scoured areas (compare with the map of natural regions on p. 78). (Redrafted and reproduced from a map by W. R. Derrick Sewell, *Annals of the Association of American Geographers*, vol. 54, 1964, p. 573. By permission of the author and the Association of American Geographers.)

diversity in the peak loads of the United Kingdom and France encouraged the construction of the cross-Channel cable.[7] . . .

Second, interconnections make it possible to take advantage of differences in streamflow patterns in adjacent regions. Streams in one region may have a summer peak flow, whereas the streams in a neighboring region may have a winter peak flow. By interconnecting the two regions, therefore, the firm power capability of both regions can be increased.[8] France, for example, takes advantage of the fact that the Atlantic region (Massif Central and Pyrenees) has a winter peak, whereas the Alpine region's streams have a spring peak.

Third, interconnection makes it possible to complement thermal power production with hydropower production, and vice versa. Often it is possible to use thermal power capacity in the winter when streamflows are low and hydropower capacity in the spring when streamflows are high. In France the thermal power plants of the north carry most of the load in the winter, and the hydropower plants of the south service most of the load in spring and summer.

Fourth, interconnection makes it possible to reduce reserve capacity. All power systems have to carry a certain margin of spare capacity to cover such emergencies as breakdowns in generating equipment, and to make capacity available during routine overhauls and maintenance. By spreading the risk over a larger number of plants through interconnection, reserve capacity in the interconnected systems can be reduced. Considerable savings in costs of carrying spare capacity can be achieved as a result.

A fifth advantage of interconnection makes possible the enjoyment of the economies of large-scale production much earlier than would have been possible with isolated operation. Recently a number of electric power utilities in the eastern United States agreed to construct a huge thermal power plant as a cooperative venture. By connecting their systems to this station and by sharing its capacity, the various utilities will be able to share the resulting economies of large-scale production.

The Forest Industries

Europe's principal forest stands of commercial value are found in the Scandinavian Peninsula and Finland. Smaller commercial forest tracts are scattered throughout continental Europe, particularly in the mountains and hill lands. Despite an extensive development of lumbering in certain areas, the existing forests cannot fully supply the demand for timber products so that sizable imports from other world regions are required. A number of the European countries have developed scientific forestry to a high degree in an effort to conserve their remaining stands of timber. The present commercial forests are principally composed of softwood trees (spruce, fir, pine, and so on), although sizable stands of oak, beech, and other hardwoods occur in some areas. Among the European nations, Sweden and Finland lead as exporters of lumber, wood pulp, and paper, the main types of forest exports. Most of these exports go to other European countries.

Internal Transportation

Europe is equipped with an elaborate network of rail lines, water routes, highways, and airways over which flows a vast amount of freight and passenger traffic. Established long before the "iron curtain" separated Europe into rather distinct non-Communist and Communist realms of activity, this network is focused upon the industrial complexes and population concentrations to the northwest (map, p. 138). Most of the individual routes radiate out from the northwestern core of manufacturing and diversified agriculture to less populous and more agrarian places in the south, east, and northeast.

Europe's irregular coastline, together with numerous navigable rivers and man-made canals, is conducive to the movement by water of a sizable traffic. Like most coastwise and inland waterway traffic of technically advanced countries, this commerce consists largely of bulk goods—petroleum and its products, iron ore and concentrates, coal and coke, grain, cement, and even sand and gravel. Some general cargo and passengers also are carried. Large seagoing vessels tend to ply between major ports on regular schedules, and small coastwise shipping and river craft connect the large ports with smaller ports and receiving depots within their respective trading territories. In some places, coastwise or river transport is not ancillary to ocean shipping; movement of coal from the German

---

[7] The Cross-Channel Cable between the United Kingdom and France was completed in 1962. Its basic objectives are to take advantage of the diversity between the United Kingdom peak load and the French peak load in the winter; to facilitate the use of the most efficient capacity in either country; and to facilitate transmission of emergency supplies of energy between the two countries. For discussions of the Cross-Channel Cable see H. Boudrant, "Frontiers Go As Power Link-Up Grows," *Financial Times* (September 1962); "DC Cable Fits Channel Exigencies," *Electrical World* (August 13, 1962), p. 45; and George Kish, "A British-French Electricity Grid?" *The Professional Geographer*, 15 (July 1963), 35.

[8] Firm power is the amount of power a plant can be expected to deliver 100 percent of the time. In a hydropower system it corresponds to the amount of water that can be made available to a power plant under minimum water conditions.

Ruhr to France via the Rhine River and connecting waterways is a good example.

Overland transportation, faster but more expensive than water transport, carries a higher percentage of passengers, mail, and general cargo. But it is also highly important in the movement of bulk goods, accounting for far more freight tonnage in Europe as a whole than that carried by internal waterways.

Since World War II the motortruck has become an important long-distance freight carrier. As in the United States, the airplane is a major carrier of passengers and mail, but is comparatively minor in freight movements.

Europe lacks the vast and intricate pipeline network of the United States, but a large and expanding net of pipelines carrying natural or manufactured gas exists in some areas, and recent years have seen the construction of several long-distance lines carrying crude oil from Mediterranean or North Sea ports to inland refineries (mostly West German), or from oil fields in the Soviet Union to refineries in several countries of Communist Europe.

REFERENCES and READINGS

Standard Geography Texts and General References

CHURCH, R. J. HARRISON, PETER HALL, G. R. P. LAWRENCE, W. R. MEAD, and ALICE F. A. MUTTON, *An Advanced Geography of Northern and Western Europe* (London: Hulton Educational Publications, 1967).

CURRY-LINDAHL, KAI, *Europe: A Natural History* (New York: Random House, 1964).

DEFFONTAINES, PIERRE, ed., *Larousse Encyclopedia of Geography: Europe* (New York: G. P. Putnam's Sons, 1961). A large work by multiple authors, translated from the French. Lavishly illustrated with photos and maps.

DOLLFUS, JEAN, *Atlas of Western Europe* (Chicago: Rand McNally & Co., 1963).

EGLI, EMIL, and HANS MÜLLER, eds., *Europe from the Air* (New York: Wilfred Funk, Inc., 1960).

GOTTMANN, JEAN, *A Geography of Europe* (4th ed.; New York: Holt, Rinehart and Winston, 1969).

HOFFMAN, GEORGE W., *A Geography of Europe* (3d ed.; New York: The Ronald Press Company, 1969).

MONKHOUSE, F. J., *A Regional Geography of Western Europe* (3d ed.; New York: Frederick A. Praeger, 1967); and *The Geography of Northwestern Europe* (New York: Frederick A. Praeger, 1966).

OGILVIE, ALAN G., *Europe and Its Borderlands* (Edinburgh and London: Thomas Nelson and Sons, 1957).

PARKER, W. H., "Europe: How Far?" *Geographical Journal,* 126, pt. 3 (September 1960), 278–297.

PEARCY, G. ETZEL, "Geographic Terminology of Europe," *U.S. Department of State Bulletin,* 48, no. 1236 (March 4, 1963), 330–338.

POUNDS, NORMAN J. G., *Europe and the Soviet Union* (2d ed.; New York: McGraw-Hill Book Company, 1966).

SHANAHAN, E. W., *Western and Central Europe: A Regional Geography* (New York: St. Martin's Press, 1965).

VAN VALKENBURG, SAMUEL, and COLBERT C. HELD, *Europe* (2d ed.; New York: John Wiley & Sons, 1952).

Political, Cultural, and Historical Geography

ALEXANDER, LEWIS M., *Offshore Geography of Northwestern Europe: The Political and Economic Problems of Delimitation and Control* (Chicago: Rand McNally & Co., for the Association of American Geographers, 1963).

An Atlas of European Affairs. Text by Norman J. G. Pounds; maps by Robert C. Kingsbury (New York: Frederick A. Praeger, 1964).

BARKER, ERNEST, AND OTHERS, eds., *The European Inheritance,* 3 vols. (Oxford: At the Clarendon Press, 1954).

"Britain and the Common Market," *Geography,* 48, pt. 2 (April 1963), two articles: (1) Charles A. Fisher, "The Changing Significance of the Commonwealth in the Political Geography of Great Britain," pp. 113–129; and (2) M. J. Wise, "The Common Market and the Changing Geography of Europe," pp. 129–138.

The Cambridge Economic History of Europe (Cambridge: At the University Press); especially vol. 4, E. E. Rich and C. H. Wilson, eds., *The Economy of Expanding Europe in the Sixteenth and Seventeenth Centuries* (1967), and vol. 6, H. J. Habakkuk and M. M. Postan, eds., *The Industrial Revolutions and After: Incomes, Population and Technological Change,* 2 vols. (1965).

CHADWICK, H. MUNRO, *The Nationalities of Europe and the Growth of National Ideologies* (Cambridge: At the University Press, 1945).

CLEVELAND, HAROLD van B., and JOAN B. CLEVELAND, *The Atlantic Alliance: Problems and Prospects* (Headline Series, No. 177; New York: Foreign Policy Association, 1966).

DARBY, H. C., "The Face of Europe on the Eve of the Great Discoveries," in *The New Cambridge Modern History* (Cambridge: At the University Press, 1957), Vol. 1, Chap. 2.

DICKINSON, ROBERT E., *The West European City: A Geographical Interpretation* (2d ed. rev.; London: Routledge and Kegan Paul, 1961).

EAST, W. GORDON, *An Historical Geography of Europe* (5th ed.; New York: E. P. Dutton & Co., 1966).

"European Unity," *Britannica Book of the Year* (Chicago: Encyclopedia Britannica, annual.)

FIELDHOUSE, D. K., *The Colonial Empires* (New York: Dell Publishing Company, 1966).

HALE, JOHN R., and the Editors of Time-Life Books, *The Age of Exploration* (Great Ages of Man; New York: Time Inc., 1966).

LEVINE, MORTON H., "The Basques," *Natural History,* 76, no. 4 (April 1967), 44–51.

McNEILL, WILLIAM H., *The Rise of the West: A History of the Human Community* (Chicago: University of Chicago Press, 1963); and *A World History* (New York: Oxford University Press, 1967).

MIKESELL, MARVIN W., "Comparative Studies in Frontier History," *Annals of the Association of American Geographers,* 50, no. 1 (March 1960), 62–74. A review article.

MOLLER, HERBERT, ed., *Population Movements in Modern European History* (New York: The Macmillan Company, 1964).

PALMER, R. R., ed., *Atlas of World History* (Chicago: Rand McNally & Co., 1957).

PARRY, J. H., *The Age of Reconnaissance* (Cleveland, Ohio: The World Publishing Company, 1963; also available in New American Library, Mentor Books).

PEI, MARIO, "The Hidden Politics of Words," *Saturday Review* (January 15, 1966), pp. 22–24.

POUNDS, NORMAN J. G., *An Historical and Political Geography of Europe* (London: George G. Harrap and Co.; New York: Chanticleer Press, 1947); and, with Sue Simons Ball, "Core-Areas and the Development of the European States System," *Annals of the Association of American Geographers,* 54, no. 1 (March 1964), 24–40.

SCHECHTMAN, JOSEPH B., *European Population Transfers, 1939–1945* (New York: Oxford University Press, 1946); and *Postwar Population Transfers in Europe, 1945–1955* (Philadelphia: University of Pennsylvania Press, 1962).

SHEPHERD, WILLIAM R., *Historical Atlas* (9th ed.; New York: Barnes & Noble, 1964).

SMITH, C, T., *An Historical Geography of Western Europe before 1800* (London: Longmans, Green and Co., 1967).

WAYS, MAX, "Europe's New Nationalism," *Fortune,* 74, no. 4 (September 1966), 109–111 ff.

WEBB, WALTER PRESCOTT, *The Great Frontier* (Boston: Houghton Mifflin Company, 1952).

WHITTLESEY, DERWENT S., *Environmental Foundations of European History* (New York: Appleton-Century-Crofts, 1949).

WILLIAMSON. J. A., "The Expansion of Europe," in John Bowle, ed., *The Concise Encyclopedia of World History* (New York: Hawthorn Books, 1958), Chap. 11, pp. 263–278.

WOODRUFF, WILLIAM, *Impact of Western Man: A Study of Europe's Role in the World Economy* (London: The Macmillan Co., 1966).

Economic Geography

ALEXANDER, TOM, "All That Gas in the North Sea," *Fortune,* 74, no. 3 (August 1966), 110–115 ff.

AXELRAD, MARGARET A., "Petroleum Pipelines in Western Europe," *Professional Geographer,* 16, no. 4 (July 1964), 1–5.

BARACH, ARNOLD B., *The New Europe and Its Economic Future* (a Twentieth Century Fund Survey; New York: The Macmillan Company, 1964). Based on J. Frederic Dewhurst and Associates, *Europe's Needs and Resources: Trends and Prospects in Eighteen Countries* (New York: Twentieth Century Fund, 1961).

BENOIT, EMILE, *Europe at Sixes and Sevens: The Common Market, the Free Trade Association, and the United States* (New York: Columbia University Press, 1962).

BIRD, JAMES, "Seaports and the European Economic Community," *Geographical Journal,* 133, 3 (September 1967), 302–327.

COLBY, CHARLES C., *North Atlantic Arena: Water Transport in the World Order* (Carbondale: Southern Illinois University Press, 1966).

"The Common Market," *East Lakes Geographer,* vol. 1 (November 1964). A group of seven articles by different authors.

DURAND, LOYAL, JR., *Economic Geography* (New York: Thomas Y. Crowell Company, 1961), "The Western European Manufacturing Region," pp. 488–503.

FLEMING, DOUGLAS K., "Coastal Steelworks in the Common Market Countries," *Geographical Review,* 57, no. 1 (January 1967), 48–72; and, with Günter Krumme, "The 'Royal Hoesch Union': Case Analysis and Adjustment Patterns in the European Steel Industry," *Tijdschrift voor Economische en Sociale Geografie,* 59, no. 4 (July–August 1968), 177–199.

HARPER, ROBERT A., "The Geography of World Energy Consumption," *Journal of Geography,* 65, no. 7 (October 1966), 302–315.

HOFFMAN, GEORGE W., "The Role of Nuclear Power in Europe's Future Energy Balance," *Annals of the Association of American Geographers,* 47, no. 1 (March 1957), 15–40.

JONES, CLARENCE F., and GORDON G. DARKENWALD, *Economic Geography* (3d ed.; New York: The Macmillan Company, 1965), Chap. 19, "Mixed-Farming Economy of Western Europe," pp. 304–325, and Chap. 20, "Dairy Farming," pp. 326–340.

KENT, P. E., "North Sea Exploration—A Case History," *Geographical Journal,* 133, pt. 3 (September 1967), 289–301.

KISH, GEORGE, "European Economic Community," *Focus,* 13, no. 4 (December 1962), 6 pp.

LAMBERT, AUDREY M., "Farm Consolidation in Western Europe," *Geography,* 48, pt. 1 (January 1963), 31–48.

LINTON, D. L., "The Geography of Energy," *Geography,* 50, pt. 3 (July 1965), 197–228.

LISTER, LOUIS, *Europe's Coal and Steel Community: An Experiment in Economic Union* (New York: Twentieth Century Fund, 1960).

MANNERS, GERALD, "The Gas Industry in Europe," *Canadian Geographer,* 5, no. 4 (Winter 1961), 30–36; "The Pipeline Revolution," *Geography,* 46, pt. 2 (April 1962), 154–163; and "Transport Costs, Freight Rates and the Changing Geography of Iron Ore," *Geography,* 52, pt. 3 (July 1967), 260–279.

MATHER, EUGENE, and JOHN FRASER HART, "The Geography of Manure," *Land Economics,* 32, no. 1 (February 1956), 25–38.

NYSTROM, J. WARREN, and PETER MALOF, *The Common Market: The European Community in Action* (Searchlight Books; Princeton, N. J.: D. Van Nostrand Co., 1962).

POUNDS, NORMAN J. G., *The Geography of Iron and Steel* (rev. ed.; London: Hutchinson University Library, 1966); and, with William N. Parker, *Coal and Steel in Western Europe: The Influence of Resources and Techniques on Production* (Bloomington: Indiana University Press, 1957).

SEWELL, W. R. DERRICK, "The Role of Regional Interties in Postwar Energy Resource Development," *Annals of the Association of American Geographers,* 54, no. 4 (December 1964), 566–581.

SINCLAIR, ROBERT, "Cohesive and Inhibiting Factors Affecting International Exchange of Electric Power in Western Europe," *Tijdschrift voor Economische en Sociale Geografie,* 56, no. 2 (March–April 1965), 42–53.

STAMP, L. DUDLEY, *Land Use Statistics of the Countries of Europe* (World Land Use Survey, Occasional Papers, no. 3; Bude, Cornwall: Geographical Publications Ltd., 1965).

TRACY, MICHAEL, *Agriculture in Western Europe: Crisis and Adaptation Since 1880* (London: Jonathan Cape, 1964).

UNITED NATIONS, ECONOMIC COMMISSION FOR EUROPE, *Economic Bulletin for Europe,* and other publications.

VAN VALKENBURG, SAMUEL, "Land Use within the European Common Market," *Economic Geography,* 35, no. 1 (January 1959), 1–24.

THOMAN, RICHARD S., and EDGAR C. CONKLING, *Geography of International Trade* (Englewood Cliffs, N. J., Prentice-Hall, 1967).

THOMAS, TREVOR M. See articles on North Sea gas and oil in the list of references and readings for Chapter 5.

WARREN, KENNETH, "The Changing Steel Industry of the European Common Market," *Economic Geography,* 43, no. 4 (October 1967), 314–332.

See also the lists of references and readings for Chapters 1–3 and 5–12.

The
British Isles

5 The British Isles lie off the northwest coast of Europe. There are approximately 5500 islands in the group, but most of them are small, and only two islands, Great Britain and Ireland, are of major consequence. The largest island, Great Britain, lies only 21 miles across the Strait of Dover from France, and the whole island group is generally considered part of Europe.

Two countries occupy the islands: the Republic of Ireland, with its capital at Dublin, and the United Kingdom of Great Britain and Northern Ireland, with its capital at London. The latter country is usually referred to by shorter names such as "United Kingdom," "UK," "Britain," or "Great Britain." It incorporates the island of Great Britain, plus the northeastern corner of Ireland and most of the smaller islands, including the Isle of Wight, Isle of Man, Hebrides, Orkneys, Shetlands, and Channel Islands (map, p. 99). Altogether the United Kingdom comprises about four-fifths of the area and has about 95 percent of the population of the British Isles (Table 3).

POLITICAL SUBDIVISIONS OF THE UNITED KINGDOM

The United Kingdom is sometimes incorrectly referred to as "England." However, England is merely the largest of four main subdivisions of the country, the others being Scotland, Wales, and Northern Ireland (map, right). These were originally independent territories. Wales was conquered by England in the Middle Ages, but preserves some cultural distinctiveness associated with the Welsh language, still spoken by about a quarter of its population. Northern Ireland, together with the rest of Ireland, was twice conquered by England. The earlier conquest in the Middle Ages was followed by a lapse of English control during the Wars of the Roses in the second half of the fifteenth century. One of its outcomes was the settlement of Scottish and English Protestants in the North, where they became numerically and politically dominant. In 1921, when the Irish Free State (later the Republic of Ireland) was established in the Catholic part of the island, the Protestant North, for economic as well as religious reasons, elected to remain with the United Kingdom. Today Northern Ireland retains a certain amount of political autonomy within the United Kingdom. For example, it has its own parliament, which meets at Belfast and legislates in matters of local concern. However, like the

The two main islands and the major political divisions of the British Isles.

TABLE 3 BRITISH ISLES: AREA AND POPULATION DATA

POLITICAL UNIT	AREA (THOUSAND SQUARE MILES)	POPULATION (MILLIONS: 1968 ESTIMATES)	DENSITY (PER SQUARE MILE: TO NEAREST MULTIPLE OF 5)
United Kingdom	94.5	55.8	590
England	50.3	45.8	920
Wales	8.0	2.8	350
Scotland	30.4	5.6	185
Northern Ireland	5.5	1.5	275
Isle of Man[a]	0.225	0.055	240
Channel Islands[a]	0.075	0.101	1350
Republic of Ireland	27.1	2.85	105
British Isles[b]	121.7	58.7	480

[a] In a strict sense the Isle of Man and Channel Islands are not included in the United Kingdom, both being dependencies of the British Crown. But in practical effect they are part of the United Kingdom and are so regarded in this chapter.
[b] Discrepancies in totals are due to rounding of figures.

other subdivisions of the United Kingdom it sends representatives to the Parliament at London. Scotland was first joined to England when a Scottish king inherited the English throne in 1603, and the two became one country under the Act of Union passed in 1707. Scotland has no separate parliament, but it does have special administrative agencies in Edinburgh which deal with Scottish affairs. It also has its own system of courts and laws.

SIZE AND WORLD IMPORTANCE OF THE UNITED KINGDOM

In area and, to a lesser extent, population, the United Kingdom is a relatively small country. Its area of slightly less than 95,000 square miles is about the same as the combined total for New York and Pennsylvania, and its population of about 56 million (1968) is only slightly more than a quarter of that of the United States. In these respects it fits in a general class with the larger nations of continental Europe, but not with the larger nations of the world.

The international importance once attained by this small country and the strong imprint its influence has made on the world are among the more amazing aspects of modern history. When British power and prestige were at their peak in the century between 1815 and 1914, the United Kingdom was generally considered the world's greatest power. Its overseas empire eventually covered a quarter of the earth, and the influence of its language and culture spread still farther. Until the late nineteenth century it was the world's foremost manufacturing nation. Its navy dominated the seas, and its merchant ships carried half or more of the world's ocean trade. It was the world's greatest trading nation, controlling about a fourth of all international trade in the middle of the nineteenth century and about a fifth at the turn of the twentieth century. Invested profits from its industries and commerce helped finance the development of much of the rest of the world. Its investments were especially large in the United States, Canada, Australia, India, South Africa, and Argentina, but they played a prominent role in dozens of other countries.

A decline in the United Kingdom's relative importance in the world began in the later nineteenth century and has continued, with an acceleration since World War II. British naval dominance is now a thing

of the past. The country's merchant fleet is still the world's largest, but now includes only a little more than one-tenth of world shipping. In total trade the United Kingdom is now surpassed by two other countries (the United States and West Germany), and in industrial production by the United States, the Soviet Union, and West Germany. The large countries once included in the British Empire have all gained independence, and Britain's remaining overseas possessions, principally islands, are scattered and small.

However, relative decline still leaves this small country in a position of great consequence in the world: first among the nations in merchant shipping, third in trade, fourth in manufacturing. It is also one of the five countries possessing nuclear armaments (1968), is associated with many of its ex-colonies in a large Commonwealth of independent countries, and benefits from the enormous cultural impact it had on many parts of the world during its period of ascendancy as a political and economic power.

THE ISLAND OF GREAT BRITAIN

Aside from minor islands and a small part of Ireland, the physical base of the United Kingdom is Great Britain, an island about 600 miles long by 50 to 300 miles wide. For its small size this island provides a homeland with a notable variety of landscapes and modes of life. Some of this variety can be attributed to a broad fundamental division of the island between two contrasting parts of the British Isles, *Highland Britain,* and *Lowland Britain* (map, right).

Highland Britain

Highland Britain embraces Scotland, Wales, Ireland, and parts of northern and southwestern England. It is predominantly an area of treeless hills, uplands, and low mountains, formed of rocks that are generally older and harder than those of Lowland Britain. All three of the major classes of rocks—igneous, metamorphic, and sedimentary—are well represented. Moors,[1] tenanted by sheep farmers or uninhabited, cover broad areas. The principal highlands, together with the maximum elevation of each, include (1) the Scottish Highlands (Ben

---

[1] In its most common usage the term "moor" means a deforested upland, usually covered with grass or heather.

SHETLAND
ISLANDS

ORKNEY
ISLANDS

NORTH
SEA

HEBRIDES ISLANDS

SCOTTISH

HIGHLANDS

Loch Ness

GREAT

BRITAIN

B R I T A I N

SCOTTISH
Firth of Forth
Clyde Edinburgh
Glasgow
LOWLANDS

SOUTHERN

UPLANDS

IRELAND

Belfast

Newcastle
Tyne R.

Tees R.

LAKE
DISTRICT
Middlesbrough

PENNINES

ISLE
OF
MAN

IRISH
CENTRAL

PLAIN

Dublin

Shannon R.

IRISH SEA

LANCASHIRE

YORKSHIRE

Leeds   Hull

Liverpool

Mersey R.

Manchester
Scunthorpe
Sheffield

Humber R.

Nottingham

H I G H L A N D

MTS.

OF

WALES

Trent R.

THE

MIDLANDS

EAST
ANGLIA

Birmingham

B R I T A I N

Swansea

Severn R.
Avon R.

London

Cardiff
Bristol

Thames River

Dover

STRAIT OF DOVER

Southampton

L O W L A N D

Exe R.

ISLE OF
WIGHT

CORNWALL
Plymouth

ENGLISH CHANNEL

FRANCE

CHANNEL
ISLANDS

Predominantly mountains or uplands

0   50   100

Scale of Miles

Highland and Lowland Britain. (Based in part on a map by L. Dudley Stamp.)

This scene from northeastern Wales shows in close juxtaposition two major categories of landscape in the British Isles. The valley in the center of the view (Llangynog Valley) protrudes into the eastern edge of the main belt of highlands in Wales. A tongue of ice moved down it from the adjoining highlands in the Ice Age and gave it the U-shaped profile of glacially scoured valleys. Today the valley floor exhibits the field-and-hedgerow landscape characteristic of lowlands and valleys in many parts of Great Britain and Ireland. Note the lines of trees along the hedgerows. When seen from ground level these give an impression of extensive woody growth, but in an air view the small proportion of trees and woodlands is apparent. In this valley, located in a relatively high, cool, and wet area, most of the fields are kept in permanent grass and are used as pasture for cattle and sheep. But in many parts of the British Isles such fields, characterized in most places by the varied shapes and sizes seen here, are devoted to complex patterns of use involving permanent grass, rotational grass, grains, roots, and other crops. In highland areas such as the one shown, fields climb the walls of valleys for a distance (often to a remarkably uniform height coinciding in many instances to the upper limit of glacial deposition) and then give place to the treeless moorland that is the most widespread type of landscape in Highland Britain. Planting of trees is giving some valley sides and upland surfaces a wooded aspect, but treelessness is still the dominant motif in the highlands as a whole. Sections of moorland used for livestock grazing often are fenced with stone walls. The use of stone for fences, homes, and outbuildings lends a picturesque touch to the distinctive and scenic highland landscape. (Aerofilms Limited.)

Nevis, 4406 feet); (2) the Southern Uplands of Scotland (2764); (3) the Pennines (2930) and Lake District (3210) of northern England; (4) the mountains and uplands of Wales (3560); (5) the uplands of the Cornwall peninsula in southwestern England (2309); and (6) the mountainous and hilly rim of Ireland (3414).

Highland Britain includes two important lowlands: (1) the boggy, agricultural Central Plain of Ireland, and (2) the Scottish Lowlands, a densely populated industrialized valley separating the rugged Scottish Highlands from the gentler and more fertile Southern Uplands of Scotland. The Scottish Lowlands comprise only a fifth of Scotland's area, but they incorporate more than four-fifths of its population and its two main cities, Glasgow (1.9 million)[2] on the west, and Edinburgh (625,000), Scotland's political capital, on the east.

---

[2] Estimated population of metropolitan area (the total metropolitan aggregate, including Glasgow city plus suburban and satellite districts near by). Unless otherwise noted, population figures for cities in this text are metropolitan area estimates for 1968, extrapolated from data in standard sources such as the *United Nations Demographic Yearbook,* the *Rand McNally Commercial Atlas,* the *Britannica Book of the Year,* yearbooks published by various national governments, and International Urban Research, *The World's Metropolitan Areas* (Berkeley and Los Angeles: University of California Press, 1959). It must be emphasized that these estimates are rough approximations and are not comparable from one country to another, since individual countries define their metropolitan areas differently. Nor are the estimates directly comparable to figures in the second edition of this text, which drew extensively on *The World's Metropolitan Areas,* cited above. The latter publication has not been updated and was used only incidentally in preparation of the present edition of *Regional Geography of the World.*

Lowland Britain

Practically all of Lowland Britain lies in England; it is often called the English Lowland. In contrast to the old hard rocks of Highland Britain, the Lowland section is mainly an area of younger, softer sedimentary rocks that have produced better soils than those of the highlands and a gentler, though for the most part rolling, topography. The Lowland supports an extremely productive agriculture, based primarily on tilled crops in the eastern margins of the Lowland and on grass and livestock in most other areas. Most of Lowland Britain is overspread with a continuous mosaic of well-kept pastures, meadows, and crop fields, punctuated frequently by substantial villages, market towns, and industrial cities. The largest industrial and urban districts, aside from London, lie in the midlands and the north, around the margins of the Pennines. Here clusters of manufacturing cities, among the greatest in the world, have risen in the midst of the coal fields that have provided most of the power for English factories in the industrial age.

INSULARITY AS A FACTOR IN THE EARLY RISE OF BRITAIN

The rise of Great Britain to world importance began after Europe established contact by sea with the Americas and the Orient. Following the voyages of Columbus and Vasco da Gama at the end of the fifteenth century, the nations along the western seaboard of Europe launched an intense competition for trade and colonies in overseas areas. By the eighteenth century Britain had outdistanced her continental rivals in this struggle. In doing so the island was able to exploit certain geographic advantages as against its continental competitors.

Perhaps the greatest of these advantages was the fact that Britain had no land frontiers to defend. A strong navy was sufficient to protect the island and thus to allow Britain to choose the time and place in which to engage in land fighting. In contrast, the continental powers had to be constantly on the alert against possible military moves by their neighbors. Of necessity their attention was strongly divided between Europe itself and the colonial areas, whereas Britain's island security made possible a greater concentration of effort in the colonial field. Whenever one of the continental powers became a threat to Britain, the latter sought allies among the other continental nations. Thus in a series

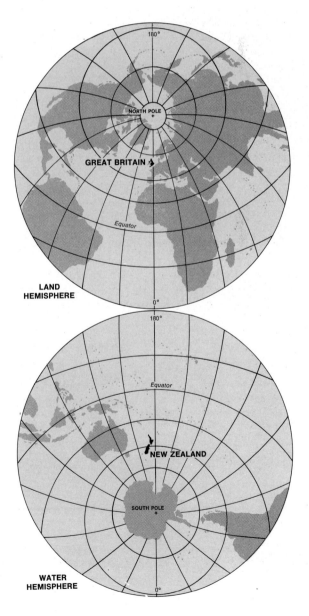

Note in the top map how the major land masses are grouped around the margins of the Atlantic and Arctic oceans. The British Isles and the northwestern coast of Europe lie in the center of the "land hemisphere," which comprises 80 percent of the world's total land area and has some 95 percent of the world's population. New Zealand lies near the center of the opposite hemisphere, or "water hemisphere," which has only 20 percent of the land and 5 percent of the population. (Base map © Hammond Incorporated No. 10336.)

of wars extending over most of two centuries, Spanish ambitions were thwarted by a coalition of Britain, France, and the Netherlands; the Netherlands was checkmated by Britain and France; and France was checkmated by Britain and a series of continental allies in shifting combinations. The rebellious American colonies secured their independence at a time when Britain lacked a continental ally against France, Spain and the Netherlands.

The Age of Discovery changed immensely the significance of Britain's location. Previously Great Britain had been an island outpost on the edge of the known world. During the Middle Ages the island was mainly a pastoral area supplying raw and semifinished wool to continental industries. Its peripheral location and type of economy have caused Britain to be described as the Australia of the time. Following the great discoveries in the fifteenth and sixteenth centuries, however, Great Britain found itself at the front door of Europe, opening toward the new lands overseas. Britain's position between the coast of northwestern Europe and the open Atlantic meant that continental shipping would have to pass by at close range, especially since most ships preferred the narrow English Channel to the stormy ocean outlet of the North Sea between Scotland and Norway. Thus Britain lay across or near the lines of communication connecting the continent with overseas areas, while the continental countries had no similar advantage with respect to Britain. This advantage of position could be exploited either peaceably to expand trade, or militarily to intercept and harry the overseas shipping of continental rivals.

and security of the people and their spirit of economic enterprise, the unusually large influence of commercial classes on the government, and the availability of such key resources as coal and iron ore were all probably favorable and stimulating circumstances.

At any rate, with the rapid development of the new techniques, after about 1750 a long period of industrial expansion began. Population increased rapidly and concentrated on the coal fields (map, right), where mining and industrial towns and adjacent seaports experienced a mushroom growth. Britain became the world's first essentially industrial and urban nation. The changes in production and in accompanying social and economic conditions were so great that they are often referred to as the Industrial Revolution. This revolution occurred first in Britain, transformed a few other countries in the nineteenth century, is convulsing others today, and has left some relatively untouched as yet.

Five industries were of outstanding prominence in Britain's industrial rise: coal mining, iron and steel, cotton textiles, woolen textiles, and shipbuilding. The early start of these industries as compared with those of other countries gave the United Kingdom industrial preeminence through most of the nineteenth century. Their exported surpluses paid for massive imports of food and raw materials and made Britain the hub of world trade. They made Britain wealthy. But their difficulties in the face of growing international competition and changing patterns of demand have been fundamental to many of the problems that have faced Britain in the twentieth century.

BRITAIN AND THE INDUSTRIAL REVOLUTION

In the eighteenth and nineteenth centuries the United Kingdom led the world into the modern age of large-scale, mechanized industrial production. Invention of a practical steam engine in Britain made coal a major resource and greatly increased the amount of power available. New processes and equipment made iron relatively abundant and cheap. The invention of new industrial machinery, soon made of iron and driven by steam engines, multiplied the output of manufactures, principally textiles at first.

Why most of the key inventions were first made or effectively applied in Britain cannot be answered with absolute assurance. The country's commercial supremacy and the resulting opportunities, the comparative freedom

The British Coal Industry

Coal became a major industrial resource between 1700 and 1800. It had been mined and used for domestic heating for some centuries before 1700, but in the eighteenth century it became fuel for the blast furnace and the steam engine. For use in blast furnaces it was coked, and coke replaced the charcoal previously used for smelting iron ore. With the development of an economically practical steam engine, coal supplanted human muscles and running water as the principal source of industrial energy. The steam engine bore an intimate relation to the expanding coal industry in a variety of ways. Not only did its increasing use to drive machinery greatly stimulate the demand for coal; but the steam engine added to the availability of coal by making it possible to pump water from the mines more efficiently and thus to utilize coal seams, particularly those at deep levels, that previously could not be worked because of drainage

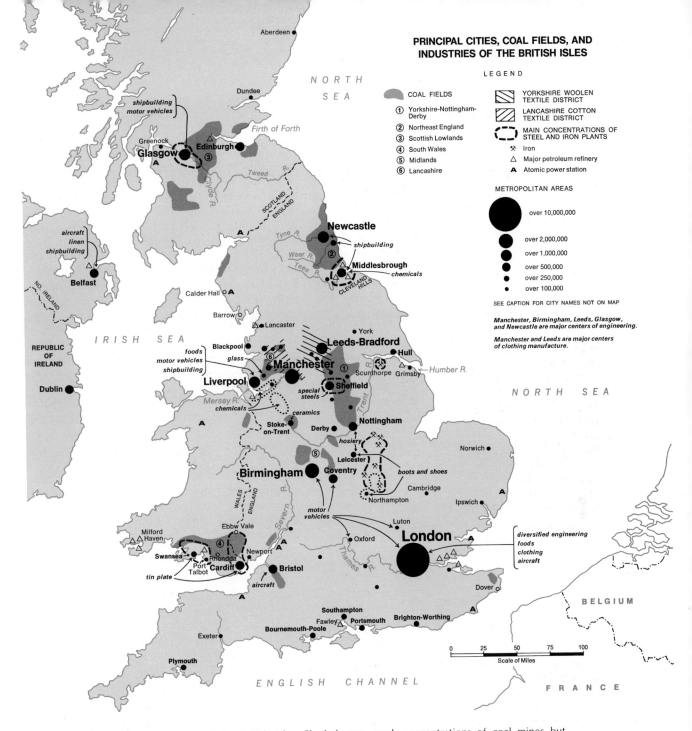

PRINCIPAL CITIES, COAL FIELDS, AND INDUSTRIES OF THE BRITISH ISLES

LEGEND

- COAL FIELDS
- ① Yorkshire-Nottingham-Derby
- ② Northeast England
- ③ Scottish Lowlands
- ④ South Wales
- ⑤ Midlands
- ⑥ Lancashire

- YORKSHIRE WOOLEN TEXTILE DISTRICT
- LANCASHIRE COTTON TEXTILE DISTRICT
- MAIN CONCENTRATIONS OF STEEL AND IRON PLANTS
- ⚒ Iron
- △ Major petroleum refinery
- ▲ Atomic power station

METROPOLITAN AREAS

- over 10,000,000
- over 2,000,000
- over 1,000,000
- over 500,000
- over 250,000
- over 100,000

SEE CAPTION FOR CITY NAMES NOT ON MAP

Manchester, Birmingham, Leeds, Glasgow, and Newcastle are major centers of engineering.

Manchester and Leeds are major centers of clothing manufacture.

Industrial and urban map of the British Isles. Shaded areas mark concentrations of coal mines but do not portray the full extent of known coal deposits. Some of the more important industrial emphases of different cities and areas are indicated by italicized words and arrows. Dotted lines in the vicinity of Manchester and Northampton enclose, respectively, major areas of chemical production and boot-and-shoe production. Each iron-mining symbol (pick and hammer) marks represent of several mines. The canal symbol from the Mersey estuary to Manchester marks the Manchester Ship Canal. Small metropolitan areas shown by symbol but not by name are, in northeastern England, Sunderland (near Newcastle) and Darlington-Auckland (near Middlesbrough); in Yorkshire (north to south), Halifax, Huddersfield, Barnsley, Doncaster, Chesterfield, and Mansfield-Sutton; in Lancashire (northeast to southwest), Burnley-Nelson, Blackburn-Accrington, Preston, Wigan-Leigh, St. Helens, Warrington; in southern England (east to west), Chatham-Rochester-Gillingham (on the Thames estuary), Aldershot-Farnborough, Reading (on the Thames), Swindon, Gloucester (on the Severn). The Middlesbrough metropolitan area, officially "Teesside," includes Stockton and West Hartlepool; metropolitan area Edinburgh includes its seaport, Leith.

problems. Furthermore, the steam engine facilitated the use of coal in the iron industry, since it was able to create the stronger air blast that was necessary when coke became fuel for smelting iron.

[EDITORIAL COMMENT. Coal is the residue from organic matter that has been compressed for a long period under overlying layers of the earth. Different original materials and different degrees and lengths of compression have given a variety of coals rather than a single uniform substance. The common ranks of coal as classified on the basis of hardness and heating value (caloric contents) are anthracite, bituminous, and lignite (sometimes called brown coal). Peat, which is coal in the earliest stages of formation, may be added at the lower end of the scale, since it can be burned if dried. Lignite is mostly used in areas where adequate deposits of bituminous and anthracite coal are lacking, being principally employed as household fuel (in the form of briquettes) and for raising steam—which may be used in generating electricity. Bituminous coals are used in larger quantities than any other rank. Gas is manufactured from them, and they are used for raising steam. Some bituminous coals can be used for making coke, which is coal with the volatile elements baked out, leaving a high content of carbon. Coke is essential in manufacturing iron on a large scale in blast furnaces. Access to a good supply of "coking coal" is a major necessity for most industrial nations. Anthracite coal burns with a hot flame and almost no smoke. It is excellent for domestic heating and can be used to produce gas and raise steam. It is not ordinarily used for coking. Often coals are referred to according to the uses for which their special properties fit them, as flame coals, gas coals, or coking coals. Not only the amount of coal, but the *kind of coal* available is very important to a country's industries. In recent times coal has become very important as a raw material in the modern chemical industry.]

Since the development of the coke-fired blast furnace and the steam engine in the eighteenth century, coal has been a key factor in the economy of the United Kingdom. Its use has changed the face of Britain. During the early period of industrialization electricity was still in the future and coal was expensive to transport; consequently the new factories were generally built on or near the coal fields. The working population, in turn, clustered near the factories and mines. Thus a group of industrial districts developed which incorporated mines, factories, and urban areas. These districts contained a sizable share of the British population, and do so today.

Coal production rose until World War I: 6 million tons (later estimate) in 1770; 65 million in 1854; 287

million, the all-time high, in 1913. Coal powered domestic industry and became a major export. By 1913 one-third of the production was for export, and British coal was in use by railways, industries, and shipping over much of the world. The principal markets were in Europe, but sizable quantities were sold in South America and at coaling stations along all major ocean routes. Britain's massive exports of coal were bulky in proportion to their value, but they were favored temporarily by the country's early start in mining and were facilitated by the coastal location of some fields.

Since 1913 the British coal industry has followed a generally declining path in production, exports, and employment, interrupted briefly at times by temporary expansions. No expansion, however, has been sufficient to bring the industry back to the level attained before the preceding decline. Over this long period the principal reason for decline appears to have been loss of export markets, a process which began after World War I and by the mid-1960s had practically reached completion. Prominent among the factors accounting for this loss have been (1) depletion of much of the British coal that was easiest and cheapest to mine, although very large reserves of less easily mined coal are still present; (2) increasing competition of coal fields in continental Europe and in other parts of the world—fields which were often brought into large-scale production by the spread of the industrial revolution at a later date than exploitation on a similar scale began in the British fields; and (3) increasing substitution of petroleum and, in some countries, gas or hydroelectricity, for coal.

In recent years, however, in Britain itself substitution of other fuels for coal has played a rapidly increasing role in the decline of the British coal industry. Britain's total power output has risen sharply since World War II, but coal has not maintained its share of power production. A much expanded British economy in the mid-1960s was using for its own consumption somewhat less coal than it had used in 1913, and total coal production was continuing to decline. Between 1951 and 1964 coal's share in total British fuel consumption declined from 89 percent to 65 percent, while the share of oil, practically all of which was imported, increased from 11 percent to 33 percent. During this period one of the world's largest petroleum-refining industries was built in Britain at various crude-oil receiving ports, especially on the lower Thames River near London, at Milford Haven, Wales, and at Fawley, near Southampton on the Channel Coast (map, p. 103).

Two other power sources—nuclear energy and natural gas—are of fairly minor consequence in Britain as yet, but are expected to make major contributions in the future.

Britain's program for development of nuclear power is one of the world's most advanced, with ten nuclear power stations either in operation or under construction in the mid-1960s and with much further expansion planned. Recent technical advances in oil and gas exploration have facilitated prospecting for these fuels in the shallow North Sea and adjacent inland areas—a search greatly stimulated by discoveries of huge natural gas deposits in the Netherlands, commencing in 1959. In 1965 and 1966 important discoveries of natural gas were made off the east coast of England, as well as some smaller discoveries in inland areas near the coast. These gas discoveries were considered to be a factor of much potential significance in Britain's power situation. Meanwhile, exploration for oil continues to be pressed in the North Sea area.

Emphasis on the past, present, and probable future decline of coal mining, however, should not be allowed to obscure the fundamental importance of coal in Britain's industrial rise and in its present economy. Well over half of the heat and energy generated in Britain still comes from coal, which is one of the relatively few important mineral resources that the United Kingdom possesses in anything like abundance. Furthermore, the localizing effect of coal on industrial development and population in former times was such that there is still a considerable correspondence between densely populated urban-industrial areas and the coal fields, even though local coal production is not always of major importance to the economy of these industrial areas today.

The Major Coal Fields

The United Kingdom's coal fields, which still supplied almost two-thirds of the nation's mechanical power in 1965, lie mostly around the edges of the highlands on the island of Great Britain. In England the Pennines are ringed on three sides by a series of fields. The *Lancashire fields* to the west and the smaller, scattered *Midlands fields* to the south of the Pennines (map, p. 103) provided fuel for the development of great industrial districts centering on Manchester and Birmingham. But more important than these fields today is the *Yorkshire-Nottingham-Derby field* in the lowland along the eastern flank of the Pennines. This field produces approximately half of the total British output, has the largest and most cheaply mined reserves of any field in the United Kingdom, and is the only field where coal mining is tending to expand rather than contract. Another major

field, the Northumberland-Durham or *Northeast England field,* lies on the North Sea coast between the Tees River and the Scottish border. It ranks second in production among British fields. Coking coals from the southern part of this field are reputed to be the finest in Britain.

Coal, mostly destined for markets in Britain itself, makes up the largest tonnages of outbound cargo at ports on the Humber and Tyne estuaries and at most other ports along the northeast coast of England. This group of ports accounts for the bulk of the United Kingdom's waterborne coal shipments. Most shipments move coastwise to southern England, particularly to electric generating stations, industrial plants, and gasworks in the London metropolitan area. The huge exports to foreign markets that made Newcastle-upon-Tyne a famous name in the world's coal trade have dwindled to a tiny fraction of their former size.

The *South Wales field,* which ranks second among British fields in reserves and third in production, has been described as follows:

The South Wales coal field has a distinct "personality." It is the only coal field in this country set amid moorlands. Many parallel, deep, and steep-sided valleys have been carved into the sandstones of central Glamorganshire and western Monmouthshire, and shafts sunk in these valley floors tap the rich seams, which are present at some depth, and send out their coal by rail. Today many of the mines employ more than 1000 men, and some more than 2000.

This coal field is different from other British fields in another respect, for it includes the entire range of coals, from soft house-coals in the east, through steam, gas, and coking coals in the center, to hard, shiny anthracite in the west. While any one valley will usually contain coal of several types, the reputations of Newport and Cardiff have been built up on their almost smokeless steam coals; while the chief outlet for the anthracite coals is Swansea.[3]

In former times steam coals exported from Cardiff were highly prized as ship fuel, but today the once enormous trade of the port in this "bunker coal" has almost disappeared.

The *Scottish Lowlands fields* stretch from sea to sea across the narrow waist of Scotland. The central field near Glasgow has contributed most to the industrial development of the Lowlands, but today the center of production is moving gradually eastward toward fields on either side of the Firth of Forth.

---

[3] Henry Rees, *British Ports and Shipping* (London: George G. Harrap and Co., 1958), p. 176. Copyright by Henry Rees. Used by permission.

Small coal fields exist in the vicinity of Dover and the important southwestern port of Bristol (625,000), as well as in northern Wales and along the coast of northwestern England, but their combined production is less than 3 percent of the United Kingdom total.

Iron and Steel

Before the Industrial Revolution steel was a scarce and expensive metal. Iron, none too cheap or plentiful itself, was more commonly used. It was made in small blast furnaces by heating iron ore over a charcoal fire, with temperatures raised by an air blast from a primitive bellows. Supplies of iron ore in Great Britain were ample for the needs of the time, but the island was largely deforested by the eighteenth century, and by 1740 it was importing nearly two-thirds of its iron from countries with better charcoal supplies, notably Sweden and Russia.

During the eighteenth century British ironmasters made revolutionary changes in iron production: replacement of charcoal by coke; improvement of the blast mechanism; invention and use of a refining process called puddling, to make the iron more malleable; and adoption of the rolling mill in place of the hammer for final processing. Thus the fuel shortage was solved, the scale and pace of production were expanded, and the island's general changeover to machine industry was promoted. Between 1720 and 1855 the output of iron in the United Kingdom expanded by something like 190 times; by the latter date the country was producing half the world's iron, much of which was exported.

Then in the 1850s British inventors developed the Bessemer converter and the open-hearth furnace. These allowed steel, a metal superior in strength and versatility, to be made on a large scale and cheaply for the first time. They are still the principal means of steel production today, although a new oxygen converter, in which an oxygen blast is employed on molten pig iron, has become increasingly important since the 1950s. In the Bessemer converter, steel is made rapidly from molten iron by passing a blast of air through it to burn out impurities. In the open-hearth furnace it is made by heating iron with gas flames over a period of many hours and drawing off the impurities. In both processes quality is controlled by the addition of carbon and alloys, but it is controlled better in the slower open-hearth process. After these inventions iron became primarily a raw material for steel, though great quantities of iron are still consumed directly by fabricating industries.

Britain now had a head start in steel production; but British resources and markets could not support a position of leadership in the industry once better-endowed countries began to industrialize on a large scale. Steel production in both the United States and Germany surpassed that of the United Kingdom before 1900. Then in the 1930s the Soviet Union surpassed both Britain and Germany; and in the 1960s Japan took a lead over Britain and West Germany. Thus in a century, though multiplying its own production, Britain has fallen from first to fifth rank among the world's steel-producing nations, behind the United States, the Soviet Union, Japan and West Germany. Britain still exports large quantities of steel, however, besides supplying her own metal-using industries.

Originally the iron industry in Great Britain relied entirely on domestic ores. A major problem in iron making is to bring together economically the industry's bulky raw materials, but in Britain suitable ores were often present in the coal fields. For a century after 1750 nearly all the ore used was "coal-measures" ore. Depletion of this ore led in 1850 to the start of large-scale iron mining outside the coal fields and in the 1870s to the first large imports of foreign ore. Since then the industry has depended mainly on low-grade domestic ores from the English Lowland (map, p. 103) and on high-grade imported ores. (In 1963 imported ore accounted for two thirds of the total iron content of all ore used.)

Major Iron and Steel Districts

In harmony with the widespread distribution of coal and ore deposits, British iron and steel plants have developed in widely scattered locations. With a few conspicuous exceptions in eastern England, most plants today are located in or near the major coal fields, and all of the coal-field industrial areas have a share of the industry. However, most of the industry is concentrated into four general districts. These may be summarized, in order of importance, as follows:

1. EASTERN ENGLAND. This district has concentrations of iron and steel plants at Scunthorpe (city proper, 70,000), just south of the Humber estuary; in and near Sheffield (750,000; Sheffield itself has steel plants only); and in the Northamptonshire iron-mining area to the south. The plants at Scunthorpe and in Northamptonshire are based on the major areas of domestic iron mining in Britain, and are supplied with coal from the country's leading field Yorkshire-Nottingham-Derby near by. The plants in the Sheffield area are based on the coal field, and draw a portion of their ores from the iron mines near by. In the case of Sheffield it may also be said that the plants are based on a centuries-old tradition of high-quality steel manufacture, expressed

today in Sheffield's leadership in stainless steel and other special alloys. Together the plants of eastern England account for approximately two-fifths of the United Kingdom's steel industry.

2. SOUTH WALES. This district accounts for a quarter of the British steel industry. The industry developed here on the basis of local coal and iron ore. Since the depletion of the coal-measures ore it has continued to expand, using steadily increasing quantities of imported ore. As this has occurred, production has concentrated more and more at a few large plants in coastal locations. Most of the present capacity is in or near three seaport cities—Swansea (275,000), Cardiff (625,000), and Newport (city proper, 115,000)—and at a smaller interior city, Ebbw Vale. The latter plant would probably not exist had it not been for the influence of the British government in getting it established in a community which was undergoing great economic distress. In addition to mining coal and manufacturing iron and steel, the South Wales industrial district is the United Kingdom's largest producer of tin plate and is important in smelting or refining imported ores and concentrates, particularly copper, zinc, aluminum, and nickel. Since 1945 the British government has attempted to bring about greater diversification of the South Wales economy and lessen its dependence on coal and metals production. Considerable success has been attained in bringing new industries, but mining, steel, and nonferrous metals are still dominant, and South Wales still has one of the highest regional unemployment rates in the United Kingdom.

3. NORTHEAST ENGLAND. Along and near the lower Tees River are located plants accounting for about a fifth of Britain's steel industry. They use coal from the Northeast England field and originally relied on low-grade ore from the Cleveland Hills just south of the river. Now, however, the local ore is depleted, and imported ore is mainly used. Sweden is the largest source, but ore is brought from areas as widely dispersed as North Africa, West Africa, Venezuela, and Canada. The principal city in this steel district is Middlesbrough (570,000), the leading iron-ore port in the United Kingdom. In addition to steel, the district is a major producer of chemicals, based on local deposits of salt and anhydrite.

4. THE SCOTTISH LOWLANDS IN THE VICINITY OF GLASGOW. This area produces about a tenth of Britain's steel. Development of the iron and steel industry has been broadly similar to that of South Wales: early iron manufacture based on local coal and local coal-measures ores; then, after the depletion of the local ores, a heavy reliance on imported ores and on scrap.

However, the Scottish industry is more poorly provided with local coking coal as compared with South Wales, and is farther from the major United Kingdom markets for steel in the Midlands and southeastern England. Consequently, it has been growing relatively slowly and has not relocated on tidewater.

A historically famous district lies around Birmingham (2.7 million), especially to the west and northwest of the city. This was the leading center for the British iron industry until the middle of the nineteenth century, and Birmingham was the first city to become world famous for large-scale iron production. But depletion of both iron ore and coking coal in the vicinity has reduced it to a center of only secondary importance today. Birmingham has developed a variety of engineering industries making secondary metal products. Its brass industry is the largest in Britain. Diversified engineering is the characteristic form of manufacturing in the English Midlands, though conspicuous exceptions are found in Stoke-on-Trent (450,000)—in which is concentrated nearly the whole of the British chinaware and earthenware industries—and in the hosiery and boot-and-shoe industries of Leicester (450,000) and other cities in the east Midlands (map, p. 103).

Cotton Milling in Lancashire

During the period of Britain's industrial and commercial ascendancy, cotton textiles were its leading export by a wide margin, amounting to almost a quarter of all exports by value in 1913. Though for a time it was rather scattered, the enormous cotton industry eventually became concentrated in Lancashire, in northwestern England, with Manchester (2.9 million) the principal commercial center (map, p. 103). Raw cotton was imported, and manufactured textiles were exported, through the Mersey River port of Liverpool (1.7 million), though Manchester itself became a supplementary port after completion of a ship canal to the Mersey estuary in 1894. Today Manchester remains the organizing, technological, and commercial center of the cotton industry, and a center of finishing industries (primarily bleaching, dyeing, and printing) and clothing manufacture, but most of the spinning of cotton yarns and weaving of cloth is carried on in other cities and towns of Lancashire.

A textile industry using imported materials, including cotton, and marketing a share of the product abroad long antedated the Industrial Revolution in Lancashire. In the Middle Ages handworkers spun and wove local wool and flax. Then in the seventeenth and eighteenth centuries the industry imported additional flax from Ireland, originated new fabrics using linen in combination with cotton, which was imported from the eastern Mediterranean, and sold part of the product in tropical markets in connection with the flourishing slave trade of Liverpool.

As it affected the Lancashire textile industry, the Industrial Revolution involved several lines of development. First a series of inventors, in good part Lancashire men, mechanized and thus speeded the spinning and weaving processes. Then power was applied to drive the machines, first falling water from Pennine streams, then steam raised by coal from the underlying Lancashire fields. Finally, in 1793, the American invention of the cotton gin—which separated the seeds from the raw cotton fibers economically—made cotton a relatively cheap material from which inexpensive cloth could be manufactured.

The stage was now set for the spectacular rise of the Lancashire cotton industry and, incidentally, for the spread of cotton farming in the American South, which supplied most of the raw material. Liverpool merchants financed much of the industrial development of Lancashire, but the cotton factories were built inland on the streams and the coal. In addition to sources of power, the inland locations gave access to supplies of soft water from the Pennines for use in the finishing processes and in boilers. For over a century, until World War I, Lancashire dominated world trade in cotton and cotton textiles.

But 1913 saw the peak of British production, and since that time the decline of the industry has been about as spectacular as its former rise. This has presented the United Kingdom with one of its more serious economic problems of the recent difficult decades. The root of the trouble has been increasing foreign competition. In 1913 Lancashire exported almost 80 percent of its production, but most of the markets it served have since been lost. The cotton industry is peculiarly adaptable to areas that are commencing to industrialize, and it is often the first large-scale industry to be established in such areas. Among the reasons for this are (1) the relatively low level of skill required of the labor force in cotton mills, especially those manufacturing the cheaper grades of cloth, (2) the almost universal importance of the product, and (3) the transportability of the raw material. Once the industry has been started in such an area, it often has the important advantages of (1) cheap labor, connected with the nonindustrial country's lower level of living, (2) newer equipment, and (3) a protective tariff to keep out competition from older producers—such as Britain.

By World War I the United States cotton industry, serving mainly the domestic market behind a protective tariff, was a larger producer than Lancashire, and important cotton industries had been established in continental Europe, Russia, India, and Japan. During World War I, markets were lost to Japan and other competitors who were not so involved in the war and thus not so commercially handicapped as the United Kingdom, and after the war these markets could not be regained. Between 1913 and the late 1930s, British cotton exports declined by approximately two-thirds, and economic distress was prevalent in Lancashire during most of the interwar period. Since World War II the downward trend of the Lancashire industry has continued, though at a somewhat slower pace. Production is now more concentrated on quality fabrics, which competitors find more difficult to duplicate, while large quantities of cheap cloth are actually imported by the United Kingdom. Production is also more geared to the home market, rather than export markets, and about half the cloth produced is woven from synthetic fibers such as nylon and rayon. However, Lancashire's future as an industrial area seems to lie mainly with other industries, which have been growing as cotton production has declined. Important among these are engineering industries producing textile machinery and electrical and electronic equipment; the clothing industry; chemical industries based on massive salt deposits south of the Mersey River; and, at Liverpool, the processing of imported foodstuffs, particularly wheat, cane sugar, and oilseeds.

The Woolen Textile Industry of Yorkshire

The United Kingdom has one of the largest woolen textile industries in the world[4] and is the world's greatest exporter of woolens. Most of the British woolen industry is concentrated in western Yorkshire in Leeds-Bradford (1.4 million) and a large number of smaller textile towns and cities. This area on the east slope of the Pennines contains the largest single concentration of

---

[4] The woolen and cotton textile industries consist of two major divisions and several lesser ones. The major divisions are spinning, which turns the fiber into thread and yarn, and weaving, which makes cloth from the products of the spinning mills. The United Kingdom produces more woolen yarn than any other country, but less woolen cloth than the USSR or Japan.

woolen and worsted textile plants in the world. Bradford is the largest producing center; Leeds, the main city in size, is a diversified manufacturing center in which engineering industries and clothing manufacture are more important than the woolen industry.

In the Middle Ages, Great Britain, an island well supplied with grazing land, exported raw wool and then woolen cloth to continental Europe. Thus the Industrial Revolution, as it applied to woolen manufacture, involved the mechanization of an industry that had long been important, rather than, as in the case of cotton milling, the development of a new industry based on a fiber that had never been of prime importance. With the mechanization of the woolen industry, production expanded rapidly and soon outran Britain's output of raw wool. Thus the country became not only the leading exporter of woolen goods but also the leading importer of raw wool, a position it still occupies.

The west Yorkshire area was the leader among several important woolen-manufacturing districts in Great Britain before the Industrial Revolution. It was well suited for the industry through its resources of water from the Pennine streams, which supplied both power and abundant soft water for cleaning wool and finishing cloth. With the coming of the Industrial Revolution, this was the quickest of the woolen-manufacturing areas to adopt the new machinery, first used in adjacent Lancashire, and it was the only such area that happened to be on a coal field. The industry largely died out in competing areas—except for some upland sections of Scotland, such as the Tweed Valley, producing distinctive tweeds and tartans—and as it grew it became concentrated more and more in west Yorkshire.

The British woolen industry was less spectacular in its rise and has undergone a less drastic decline than the cotton industry. Although it did gain a position of world leadership, its growth was somewhat checked by the rapid rise of the cotton industry after the invention of the cotton gin, and by the fact that many countries already had woolen industries by the time of the Industrial Revolution. It neither exported as much nor was as dependent on the export market as the cotton industry, since Britain's home consumption of woolens has been large. But it, too, reached an export peak before World War I and has since declined. However, the decline has been greatly checked by the importance of the home market as compared with the market for cotton goods; by the greater labor skills required in its processes, which makes it more difficult to initiate in new industrial areas; and by its greater emphasis on quality, for which the British industry has long had a worldwide reputation.

Nevertheless, Yorkshire now relies increasingly on industries that formerly were of secondary importance, including the production of textile and other machinery, and clothing.

Yorkshire is served by a group of seaports along the Humber estuary, of which the most important is Hull, Britain's third port in value of foreign trade after London and Liverpool. Imports of raw wool, especially from Australia and New Zealand, are the leading item in Hull's foreign commerce.

Shipbuilding

When wood was the principal material in building ships, Great Britain, as a major seafaring and shipbuilding nation, was presented with increasingly serious difficulties due to deforestation. Large timber imports were brought from the Baltic area and from America; and the importation of completed ships from the American colonies, principally New England, was so great that an estimated one-third of the British merchant marine was American-built by the time of the American Revolution. During the first half of the nineteenth century American wooden ships posed a serious threat to Britain's commercial dominance on the seas.

But in the later nineteenth century ocean transportation was transformed by iron and steel ships propelled by steam. Britain led the way, and by the 1890s was building four-fifths of the world's seagoing tonnage. Two districts developed as the world's greatest centers of shipbuilding. These were (1) the Northeast England district, with a great concentration of shipyards along the lower Tyne River between Newcastle (1.2 million) and the North Sea, and with other yards along the rivers Wear and Tees; and (2) the western part of the Scottish Lowlands, with shipyards lining the River Clyde for miles downstream from Glasgow. Both districts had a seafaring tradition—Newcastle in the early coal trade and in fishing, and Glasgow in trade with America; both were immediately adjacent to centers of the iron and steel industry; and both had suitable waterways for location of the yards. Their ships were built mainly for the United Kingdom's own merchant marine—as they still are—but even the minority built for foreign shipowners were an important item in Britain's exports.

During the twentieth century Britain's share in world shipbuilding has declined steadily, except for short-term recoveries after both world wars. Its output represented about three-fifths of the world total in 1913, one-third just before World War II, and is only about one-tenth today. Competition has come principally from

Japan and from continental Europe, especially Sweden and West Germany. Japan displaced the United Kingdom as the world's leading shipbuilder in the 1950s and now produces three times the British tonnage, while the West German and Swedish outputs closely approach and threaten to overtake the British. Japan has had cheaper skilled labor than Britain, a factor which is very important in shipbuilding and which has been a long-time British advantage over the United States. All three competitors possess the heavy industry and engineering skills upon which modern shipbuilding must be based, and they have appeared recently to adopt improved techniques more readily than the older British industry. The latter is especially true of Sweden, whose rapidly growing shipbuilding industry has made substantial progress toward converting peacetime shipbuilding from a custom-building to an assembly-line operation. However, the total British output of ships has fallen much less since the years before World War I than has the country's share in an expanding world industry. The "Clydeside" and "Tyneside" yards are still among the world's great centers of the industry, and a multitude of activities originally subsidiary to shipbuilding make both districts important centers of metal and machinery manufacture.

A LAND OF CITIES

The rise of the industries discussed in the preceding sections gave a powerful impetus to the growth of cities at the same time that the country's total population was increasing rapidly. The United Kingdom became the world's first predominantly urban nation, and by almost any measurement it is still more highly urbanized than any other nation. Only 2 percent of the employed population is supported by agriculture. The 1961 census showed that approximately 60 percent of the population lived in cities of over 50,000 inhabitants. The comparable figure for the United States, itself a highly urbanized country, was 36 percent in 1960. Approximately nine-tenths of the United Kingdom's people are classed as city or town dwellers.

Most of the large cities, with the notable exception of London, are on or near the coal fields (map, p. 103). A tremendous cluster is found on or near the fields that flank the Pennines on the east, south, and west; that is, in the industrial districts of Yorkshire, the Midlands, and Lancashire, respectively. Here the metropolitan areas of two cities, Birmingham and Manchester, have over 2 million people each; two others, Leeds-Bradford and Liverpool, have between 1 and 2

million; three others—Sheffield, Nottingham (650,000), and Coventry (625,000)—have over half a million; and another seventeen have between 100,000 and 500,000 people each. In this area around the Pennines, measuring only about 80 miles from east to west and 100 miles from north to south, are twenty-four metropolitan areas with a combined population of nearly 15 million people. Lesser urban clusters are apparent on and near the other major coal fields in the Scottish Lowlands, Northeast England, and South Wales.

In striking contrast, the largest metropolitan areas that lie at a considerable distance from the coal fields—aside from London—have less than half a million people. These are the English Channel cities of Portsmouth (475,000), a major naval base; Southampton (375,000), the main British terminal for large passenger liners; and Brighton-Worthing (425,000), a major resort center. For several decades, however, there has been a notable tendency for these and other cities away from the older industrial districts to grow especially fast. Electricity has freed many factories from the necessity of locating close to coal, and much of Britain's more recent industrial development has sought the relatively small, uncongested, and pleasant towns and cities of southern England.

Major Elements in the Economy of London

Britain's largest city is relatively far from the major coal fields and other big cities. The London metropolitan area sprawls over a considerable part of southeastern England and can be considered to include 11½ to 12 million people, which makes it the third largest urban agglomeration of the world, after New York and Tokyo.

London is not only much larger than the other British cities, but it differs from them economically and in historical background. It was the country's leading city, and already a large city, before the Industrial Revolution, though its population has since multiplied many times over. In the late seventeenth century, when most of the present cities were little better than hamlets, it already had nearly 700,000 people. This was roughly a tenth of the population of Great Britain at that time, while metropolitan London today has about a fifth of the total population of the United Kingdom. So London has more than held its own in population growth during the industrial age, despite its relative distance from the booming coal-field areas. But most of the major elements in its economy developed originally in the preindustrial period and have only been expanded and supplemented during the past two centuries.

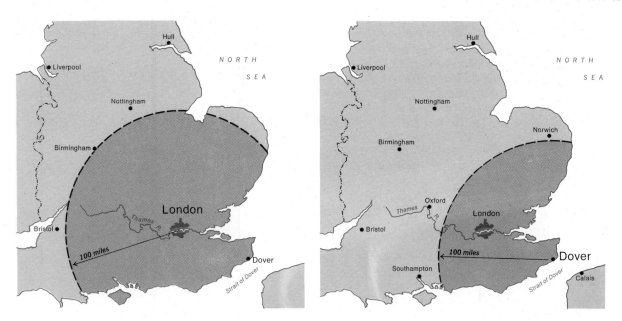

Colored areas lie within a 100-mile radius of the center of London and Dover, respectively.

The first and most fundamental of these elements is *commerce*. The city became an important seaport as early as the Roman occupation of England in the first century A.D., and has continued so to the present. Several aspects of its location help to account for its early development as a trading center. (1) It is located in the corner of Great Britain which is nearest the continent. This section of the island was almost the only part of England which did much trading overseas in the long centuries before the age of overseas expansion, and is still quite important in British trade. (2) It is located in the English Lowland, which was the most productive and populous part of Great Britain in the preindustrial period, supplying most of the wool which England exported and most of the market for imports. (3) It is located on the Thames River, which allows comparatively deep penetration of ocean ships into the Lowland. A port set inland in such a manner can serve more trading territory in the surrounding countryside, within a given radius, than can a port on a comparatively straight stretch of coast or on a promontory. This principle can be easily illustrated by drawing an arc of 100 miles radius from the Channel port of Dover, which is closer to the continent, and another arc of 100 miles radius from London, and comparing the respective sizes of the areas falling within the two arcs (map, above). Inland location of a port tends to maximize the distance goods can be carried toward many destinations by relatively

cheap water transportation and to minimize the more expensive land transport. This tends—other things being equal, which is a most important qualification—to give the inland port the advantage over the seaward port. (4) London is located where early conditions facilitated transport connections with much of the surrounding plain. One route, with branches, was provided by the Thames River and its tributaries. Before the age of the railroad and the motortruck, water transport had greater advantages over land transport than it has today; and these natural inland waterways were extensively used for transporting goods to and from London. Furthermore, London became a major junction of Roman roads, better than any Britain was to see for well over a thousand years. Again the location had advantages: the lower Thames was originally fringed by tidal marshes which were a major barrier to transportation, but at the original site of London firm ground penetrated these marshes and approached the river bank on both sides. Highways could avoid the marshes by crossing the river at London.

In view of these advantages with respect to transportation and trade, it is not surprising that London became very early the principal port and metropolis of Great Britain, and handled as much as three-quarters of the island's sea trade around 1700. In the industrial age it has faced strong competition from other British ports located closer to the rising industrial centers and some better located for access to the open Atlantic. But by the time this occurred London was too firmly established to lose its preeminence. It had a good harbor, which could be expanded by extending port facilities down the Thames. It had probably the world's greatest concentration of wealthy and well-established commercial and financial institutions for handling the enormously intricate operations and transactions of worldwide trade. The city itself was the country's single greatest market for many imported goods. It was well placed for trade with Europe and developed a large entrepôt trade, a trade in which London merchants imported goods from overseas and then resold them to continental importers, thus acting as middlemen between foreign producers on the one hand and foreign consumers on the other. Distances between the main population centers of Great Britain are not really great; for instance, London is only about 100 miles from the Midlands industrial area, and about 200 miles from Leeds. So London was not too badly handicapped by distance alone in its competition with ports such as Liverpool or Hull for the trade generated by the industrial areas.

The result has been a proportionate decrease but a great absolute increase in London's trade during the industrial age. It now handles only about one-third of the United Kingdom's total overseas trade, and is usually exceeded in value of exports by the second-ranking port of Liverpool; but as Britain's total trade has multiplied, so has that of London, despite the proportionate loss to other ports. One-third of the trade of the world's second greatest trading nation places London among the top three ports in the world, along with New York and Rotterdam. In 1967 the value of London's trade exceeded that of such *countries* as India, China, Australia, Sweden, Pakistan, and Brazil, as well as all but ten of the rest of the world's countries. This trade is extremely diversified, but includes an especially large proportion of many of the foodstuffs Britain imports and of its exports of automobiles and electrical equipment.

A second major element in London's economy is *administration*. During the Middle Ages the kings of England came more and more to reside in or near their principal city, then fixed parts of their growing administrative apparatus there, and London gradually evolved into the national capital. This English capital eventually came to administer the British Isles and then an empire stretching around the globe. At the same time the government became more centralized and dependent upon an ever-expanding permanent bureaucracy. To political administration and bureaucracy were added business administration and bureaucracy. The wealth of London merchants and of the landed nobility who came often to reside at the capital financed trading companies and,

later, manufacturing companies to take advantage of expanding world opportunities; and there was advantage to the management of these companies in being near the government which authorized the companies' existence, often gave them monopolistic privileges, and regulated them. Thus London became the administrative headquarters of large corporate enterprises with widespread interests in Britain and over the world. Finally, the city's pyramiding wealth made it a financial center, the greatest market for capital in the world until New York displaced it after World War I, and furnished the market for artistic and cultural goods and activities which made it the nation's cultural center.

A third major element in London's economy is *manufacturing*. It is not basically a manufacturing city. That is, its industries are not based directly on any local natural resource, but on its huge market and labor force and its advantages for transportation and commerce; thus they are in a sense subsidiary to the other elements in its economy. Nevertheless, Greater London is the largest single industrial center in the United Kingdom. Production is highly diversified and especially emphasizes consumer goods, though not exclusively. Major lines include clothing, related to the city's influence as a style center; processing of imported foods; refining of imported oil along the lower Thames; automobile production by English Ford on the Lower Thames; plus elements of the British aircraft, electrical, and electronic industries.

London's Internal Geography

The Internal Regionalization of Cities

It is well known and easily observed that cities possess distinctive internal geographies. As a city grows, various activities (functions) carried on within it become more or less segregated into particular sections of the city. Thus the city comes to consist of a collection of recognizable "regions," such as commercial sections, industrial sections, and residential sections of different types. Actually each section of any size is generally distinguished, not by the complete monopoly of any one function in it, but by a distinctive combination of functions.

The primary force behind this differentiation of the city into sections appears to be the economic competition of different functions—that is, of people and institutions desiring to carry on different activities—for locations within the city. This is a very complicated competition, as both the kinds of locations prized and the amount which can be paid for a given location vary from one function to another. For example, it may well be that the kind of location desired by an industrial firm will not be the same as the kind desired by a householder looking for a place to live, or by a company proposing to operate a department store, and these possible competitors will vary as to the amounts they feel able to pay for any given location. Theoretically, the competitor who stands to gain most from occupation of a given location can bid highest for it and is likely to occupy it. The location's value for various potential users is likely to be affected by such factors as topography, government policies, the types of development already present in the area and in adjacent areas, and, especially, accessibility. The factor of accessibility is generally important enough that the parts of the city most easily accessible to the greatest number of people—the central business district and to a lesser extent areas along major thoroughfares—tend to have the highest land and rental values. Since these various factors are subject to change (even topography can be changed extensively with modern equipment), the character of sections of the city may also change. In fact, the kind and rate of change going on within it is often one of the most significant characteristics of a section of the city.

Several generalizations concerning the typical arrangement of sections which cities develop have been proposed by scholars. Two of these, the concentric zone generalization and the sector generalization, are employed in the present description of London.[5] The *concentric zone generalization* holds that cities become articulated into contrasting zones arranged as concentric rings around the central business district. Thus as one progresses outward in any direction from the center of a city he should pass in sequence from (1) the central business district to (2) a zone of poor housing and light industry to (3) a zone of workingmen's housing to (4) a zone of middle-class housing to (5) a suburban zone of commuters' towns. The *sector generalization,* on the other hand, holds that the contrasting sections of a city tend to be arranged like wedge-shaped slices of a pie, with the inner points of the slices converging on the central business district. London's internal arrangement does not fit either generalization with any great exactness,

and it may well be that no city does so. Nevertheless, it is apparent that the metropolis is arranged in some respects and to some degree concentrically, and also in some respects and to some degree in sectors. Thus the two generalizations offer useful viewpoints from which to survey its historical development and present structure.

The Inner Core of London

However the structure of London is seen, that structure focuses, as in most cities, on an inner core which is the city's central business district. In the case of London, as not uncommonly in other cities, the inner core is not only the main business center of the present city but also the historical nucleus around which the present metropolis has developed. The core consisted initially of two small cities on the north bank of the Thames. One was the walled port and commercial city of London, developed originally by the Romans. The other, a bit upstream to the west, was Westminster, which became the seat of the British monarchs during the Middle Ages. (See maps and photos, pp. 114–117.)

These two settlements grew together physically during the latter seventeenth century, and then—slowly at first, more rapidly with the advent of railroads in the nineteenth century, and still more rapidly in the twentieth century—the present huge metropolis grew around them. As this occurred, they themselves developed as the heart of the metropolis and of the whole country. Together they occupy an area which extends for about 5 miles from east to west along the Thames and is much narrower from north to south. One author assigns them, along with some immediately adjacent sections which have become part of the core, an area of about 11 square miles, and says of their development:

Concentrated in these few square miles are the functions of the Crown and Government, Church, Law, Press, Finance, Banking and Insurance, the major commodity and wholesale markets, most of the institutions of the University of London, the teaching hospitals, the headquarters of the majority of Britain's leading business enterprises, nationalized industries and corporations, and the professional organizations and trade unions. Central London is also the nation's main cultural, entertainment, shopping and tourist

[5] For a discussion of these generalizations see Raymond E. Murphy, *The American City: An Urban Geography* (New York: McGraw-Hill Book Company, 1966), Chap. 12. The sector generalization has been developed especially with reference to investigations of the expansion and migration of urban residential areas, but growth and arrangement of other functions in a linear or sectoral pattern have often been observed. Another well-known generalization of urban growth and structure, not used here, is called the *multiple nuclei concept.* It was first proposed in Chauncy D. Harris and Edward L. Ullman, "The Nature of Cities," *Annals of the American Academy of Political and Social Science,* 242 (November 1945), 7–17.

1. LONDON IN ITS REGIONAL SETTING

Built-up areas

Approved Green Belt, 1962 (generalized)

Formally submitted and/or generally accepted Green Belt (generalized)

e   Expanded Towns

New Towns are in lighter type and underlined

ASPECTS OF LONDON'S GEOGRAPHY

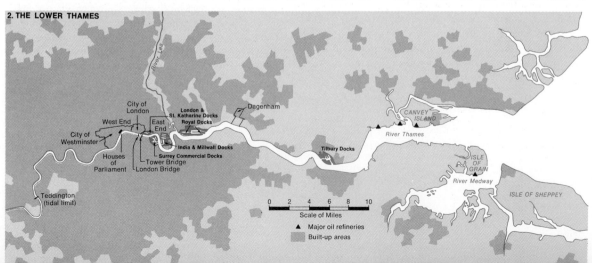

2. THE LOWER THAMES

▲ Major oil refineries

Built-up areas

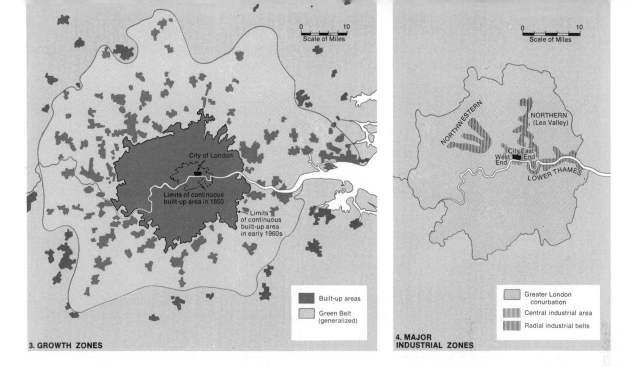

3. GROWTH ZONES

Built-up areas
Green Belt (generalized)

4. MAJOR INDUSTRIAL ZONES

Greater London conurbation
Central industrial area
Radial industrial belts

On the map "London in Its Regional Setting" small areas of parks and other open space within the main built-up area of London appear in the tint for "Approved Green Belt" but are not, strictly speaking, part of the green belt. The counties (shires) in the immediate vicinity of London are commonly referred to as the "Home Counties"; they include Middlesex, Essex, Kent, Surrey, and Hertfordshire. The historic area of Westminster as outlined on the map "The Lower Thames" was enlarged northward in 1963 to incorporate more of the West End. Built-up areas shown on the maps are generalized from the *Atlas of Britain and Northern Ireland* (1963), by permission of Clarendon Press, Oxford. Green belt after Ministry of Housing and Local Government, shown on Ordnance Survey base, Crown copyright. Various details of maps 2, 3, and 4 after J. T. Coppock and Hugh C. Prince, eds., *Greater London* (Faber and Faber, 1964).

centre and the focus of communication by road, rail, cable, and wireless.[6]

While they function jointly as the central core of London, the two original cities have maintained well-recognized separate identities. The resident population of the ancient City of London has been declining and its business population has been increasing, until now less than 5000 people live in it while 500,000 people work in it. The City "is almost entirely occupied by the headquarters and offices of banks, insurance companies, and other business concerns, packed tightly within the block outlines provided by a largely medieval street plan."[7] Until recently building regulations kept the heights of buildings to a modest level, precluding the skyscraper landscape that Americans might expect

in such an area. However, relaxation of these regulations has allowed a number of new office towers to arise since World War II, when bombing devastated many older buildings.

The part of the central core which has developed in and around Westminster continues to have, as it had from the beginning, a somewhat different character. This section, generally called "The West End," continues to be the seat of the British government, and shares some of the office development of The City. But it is much more diversified than The City, accounting for most of the other functions of the core mentioned in the quotation above, plus sizable though declining residential areas, some of which are very fashionable, and even a considerable amount of small-shop industry, especially the manufacture of clothing. A tremendous

[6] D. F. Stevens, "The Central Area," Chap. 7 in J. T. Coppock and Hugh C. Prince, eds., *Greater London* (London: Faber and Faber Limited, 1964), p. 167.

[7] A. E. Smailes, "Greater London—The Structure of a Metropolis," *Geographische Zeitschrift*, 52, no. 3 (August 1964), p. 169. The description of London in this text owes much to Smailes' excellent article.

The heart of London. In the foreground are government buildings in Westminster. Fronting the Thames at the lower right are the Houses of Parliament, with Westminster Abbey (the building with the twin white square towers) behind them in the center foreground. To the immediate left of the world's most famous clock, Big Ben, is a block of grayish buildings housing government offices, including the former headquarters of London's Metropolitan Police. Directly in rear of this block some of the most important government departments occupy portions of two immense rectangular buildings. The one with the rectangular central well houses the Foreign Office and the Home Office. The long side of this building away from the camera fronts on Downing Street, where the Prime Minister resides at Number Ten. The huge white building with the three rectangular wells contains the Air Ministry and the Board of Trade. Facing it on the opposite bank of the Thames is Shell Center, the white high-rise building containing the headquarters of one of the world's largest oil companies, Royal Dutch/Shell. Directly in rear of Shell Center, the large sooty structure with tracks leading in from the right is Waterloo Station. Westminster Bridge spans the Thames at the right on the view, with Waterloo Bridge next downstream beyond the railway bridge. At the upper right center on the north side of the Thames the dome of St. Paul's Cathedral can be discerned in the midst of the City of London. The City extends westward about as far as the midpoint of the line of four ships anchored on the north side of the river, and it extends eastward to a point well beyond London Bridge, a portion of which can be seen at the upper right edge of the view. A tree-lined boulevard on the Victoria Embankment follows the curve of the river in the left center of the photo. (Aerofilms Limited.)

variety of activities, architectural styles, and even ethnic groups is the keynote of the West End.

Concentric Aspects of London

Seen from the point of view of the concentric-zone generalization, London is comprised of a series of rings around the inner core. As one progresses outward from the core each ring is progressively newer as a part of the London agglomeration. In addition, population density reaches a peak in the ring immediately around the core and successive rings outward are less densely populated. As shown on the maps on pages 114–115, these rings may be summarized as follows:

A view looking westward across a portion of the City of London. Approximately the southwestern quarter of the City is visible. High-rise structures in the West End punctuate the distant skyline. The dome of St. Paul's rises prominently at the upper right of the photo, and the Thames is visible at the upper left. The twin brick walls in the foreground beside the Thames mark the Cannon Street Station. Some of the more famous buildings of the City, such as the Bank of England, the Royal Exchange, and Lloyd's, lie outside the photo to the right. The close-packed buildings of the City, dating from various periods, are a jumble of architectural styles. Damage from bombing in World War II was very heavy in this area, but most of the scars have now been erased. (Jesse H. Wheeler, Jr.)

1. *An area built up before 1850 and, along with the core, comprising the city as of that date.* Although the first railway reached London in 1836, this part of the city developed essentially before the railway age. Actually, the inner parts of this belt developed in medieval and early modern times, with the pace of development quickening in the eighteenth and nineteenth centuries. Buildings and tracts of many periods are now intermingled in the zone, with large modern urban renewal projects set amidst housing first built in the nineteenth century and earlier. Population density is the highest in London, although decreasing rapidly at present due to migration to more outlying zones. Some of this zone was originally developed with mansions for the well-to-do and

some of it with row housing for workers. Onetime mansions have been transmuted into apartment buildings, hotels, shops, or offices. Mixed with this dense residential development are many warehousing and wholesaling activities, as well as railroad lines, which are rather closely spaced here as they converge toward a ring of terminals in and around the edge of the inner core. In addition, a sizable portion of London's manufacturing is in this zone, carried on most typically at a small scale, with factories replacing dwelling units or occupying converted dwellings. Various districts within the zone specialize in such products as clothing, precision instruments, jewelry, shoes, furniture, and printing.

The part of this zone lying immediately east of

117

The City is often referred to as the East End. Its character contrasts very greatly with that of both The City and the West End. The East End developed very early as a densely populated industrial quarter, and has become especially associated with the British clothing industry. It was heavily bombed in World War II, and has witnessed very extensive urban redevelopment. Sharp reductions in resident population and population densities have occurred as workers have moved toward London's outskirts.

2. *A zone built up in the past century, as railroads and then, to some extent, buses and automobiles extended commuting ranges, and as outlying industrial and business centers grew.* The map shows that the continuously built-up area of the city, roughly circular in form, was approximately 10 miles in diameter in 1850, but is now approximately 25 miles across. Thus the great majority of the present continuously built-up area, called the Greater London Conurbation, is a product of the past century; in fact, about half of it is a product of the period since World War I, as the built-up area approximately doubled during the interwar period.

This increase in the built-up area was not synchronized with the increase in the city's population. Population growth in the conurbation during the past century took place mainly in the earlier part of the period. From 1851 to 1911 the population of the conurbation increased from 2.7 million to 7.3 million. From 1911 to 1939 it increased only from 7.3 million to 8.7 million, while the built-up area doubled; and from 1939 to 1961 the conurbation's population fell to 8.2 million. Continued building on the periphery of the conurbation was accompanied by progressive depopulation at the center. The new building on the periphery was done to standards of much lower density than in older residential sections, with fewer dwellings per unit of area and more open spaces retained. Population decline began in The City in the 1850s, and has spread outward until between 1951 and 1961 it characterized most of the conurbation. The zone of increasing population has moved outward ahead of the zone of decrease, and now does not lie within the conurbation at all, but in surrounding suburbs and satellites which are part of a London Region, but are not part of the continuously built-up conurbation.

The term conurbation, implying cities grown together, is appropriate for London in that the spread of the metropolis has engulfed numerous communities that formerly were separate. Often these communities are still identifiable by their names, now used for their sections of London, and often by the distinctive character of their streets and buildings, now forming parts of the London maze. Since they were in existence prior to the spread of metropolitan London, they may now appear as islands of older buildings in the zone of postrailway development.

3. *The Metropolitan Green Belt, a ring of territory beyond the limits of the Greater London Conurbation, but very much a part of the metropolitan complex.* The ring varies in width, but is generally on the order of 15 to 25 miles wide. Within it strong restrictions are enforced against new development, in order to check the further spread of the conurbation and to preserve large areas of open countryside adjacent to it. Green belt proposals began to be made in Britain in the later nineteenth century, and implementation by London authorities began in the 1930s. Then after World War II the idea of green belts around British cities became part of national planning policy. Under this policy local authorities are encouraged to submit plans for green belt areas, and if these plans are approved by the government the local authorities may then control land uses and development and prevent further building in the designated areas. This arrangement around London became fully operational in the 1950s. As a government policy it is subject to modification at any time, and the extent and nature of London's green belt have changed somewhat since its establishment, and continue to change. There has been a strong tendency to expand the belt, but also not to be completely rigid about shutting off further development.[8]

There is, of course, much urban development within the green belt, as it was far from empty when planning controls were first applied to it. Many suburbs and satellite towns already existed within it. Some of these were new and some were old, some were mainly dormitory settlements and some were industrial or commercial centers. Consequently, it is today an area of extremely mixed development. A recent study of a section of it northwest of the conurbation showed that somewhat over half the land in that section was used for agriculture, about one-fifth was in residential and com-

---

[8] The fully approved and operating Green Belt in the early 1960s averaged about 10 miles wide, though it was very irregular in outline. A much larger area, lying outside this approved area, was proposed and operative, though still subject to final approval, or possibly rejection, by the central government. Under the law, restrictions are enforced before final government approval.

mercial development, 9 percent was woodland, 6 percent was in recreational facilities, and the rest embraced a wide variety of other uses.[9]

Apparently the green belt has effectively checked the spread of the continuously built-up area of Greater London, which has grown little since 1939. In addition, the green belt itself has remained much more open than would apparently have been the case without planning controls. However, population increase within the belt continued to be rapid in the 1950s and early 1960s. Some of this increase was deliberately planned, in order to accommodate some of the continued migration out of London, as well as migration to the London area from other parts of Britain. To further this objective the British government, in the years immediately after World War II, acquired land and began building eight cities (all of which now lie within or very near the green belt), which it designated as New Towns.[10] By 1961 these towns, scattered in different directions from London, had a combined population of over 300,000. Their planned populations are to aggregate over half a million, supported partly by industries being developed in the towns themselves and partly by commuting. More New Towns may be developed in the future.

4. *The surrounding territory outside the green belt.* This area will have to absorb much of the London region's expected population increase in the future if current plans with respect to the conurbation and the green belt are carried out. Already in the 1950s some areas outside the green belt were experiencing rapid population increase, and most areas within a radius of 50 miles of central London were commuting areas. Thus far, only a small fraction of the workers of this outlying zone actually commute to London, as a number of sizable cities within the area are themselves centers of employment (for example, Oxford and Cambridge). However, commuting to London is tending to increase rapidly, as it would seem that it must if employment in the conurbation continues to increase while its population decreases and population growth in the green belt is restricted. To direct further population increase in the region into this outer zone a number of towns within it have been authorized and encouraged by the government to enter into contractual arrangements with London authorities to absorb population shifted from London. These places are called Expanded Towns. Those estab-

lished so far are located at distances of 30 to 110 miles from the center of London, but mostly outside the green belt and within 70 miles of London. To restrict the growth of long-distance commuting into London from such outlying centers, an attempt is being made to get industries as well as people to migrate to them, thus supplying adequate local employment.

Sectoral Aspects of London

To visualize London and its region as a structure of concentric rings points up some aspects of the metropolis but obscures others. Mainly what is obscured is that various sections of a given ring tend to differ somewhat from one another, and that in some respects adjacent sections of different rings are likely to have more similarity than different sections of the same ring. In other words, some facets of the London complex are arranged more in a pattern of radiating sectors than in rings.

Particularly prominent among these sectors are three belts of industrial development that radiate outward from the center of the metropolis. One of these runs eastward along the lower Thames, another runs generally northwestward, and the third approximately northward. These three industrial belts, along with industrial areas in and immediately adjacent to the core, contain most of London's manufacturing, although smaller industrial districts are scattered here and there in other sections.

1. *The lower Thames industrial belt* is intertwined physically with the port of London, and has close functional relationships with it. The port consisted originally of the river and its banks immediately downstream from the obstruction of London Bridge at The City, but has now grown downstream on both sides of the river to include "140 miles . . . of tidewater frontage"[11] besides its five great systems of docks. The docks are immense enclosed basins, rectangular in form, which have been excavated in the alluvium along the lower Thames. Ships enter the docks from the river by means of locks, which isolate the docks from the river's tidal fluctuations. The docks are lined with berthing space and specialized cargo-handling facilities.

The manufacturing industries of the lower Thames industrial belt process imports, such as oil, timber, grain,

---

[9] David Thomas, "The Green Belt," Chap. 12 in Coppock and Prince, p. 301.
[10] Some New Towns are enormous expansions of relatively small older settlements. Some have sprung from open countryside.
[11] James Bird, "The Growth of the Port of London," Chap. 8 in Coppock and Prince, p. 218.

The Port of London, looking downstream. The third bridge from the bottom of the photo is London Bridge, rebuilt since this view was taken. It marks the head of navigation for ocean shipping, aside from vessels that have superstructures low enough to permit them to pass under London Bridge and other bridges upstream. The banks of the river are lined for many miles with warehouses, factories, power stations, storage yards, and wharves. Stretches of river bottom visible near the banks indicate that the photograph was taken at low tide. The twin towers of Tower Bridge, a drawbridge, rise in the stretch of river known as the Pool of London. The famous Docks, comprising five distinct systems, commence just downstream from Tower Bridge. The oldest system (London and St. Katharine Docks) lies nearest Tower Bridge. The London Docks appear prominently as two rectangular sheets of water with ships tied up at wharves; the smaller, more irregular St. Katharine Docks lie closer to the bridge. Downstream on the right bank of the Thames, and just inside the point of the first large meander, are the Surrey Commercial Docks, devoted mainly to handling imported timber. Directly across the river, three rectangles of water in the neck of the narrow peninsula called the Isle of Dogs mark the West India Docks, the main component of the India and Millwall Docks. Still farther downstream (just to the right of the twin chimneys of a thermal-electric power station) are the three Royal Docks, comprising the largest of the five systems. Much farther away on the left bank are the Tilbury Docks, not discernible in this view. Each system of docks is connected to the river by locks which offset the rise and fall of the tide and maintain a constant depth of water at the wharves. This is an older photograph, and wartime devastation is still visible, especially in the City of London (lower left). (Aerofilms Limited.)

A close-up of the Royal Docks. The system is comprised of three interconnected docks: the Royal Albert Dock (on the right), the King George V Dock (on the left), and the Royal Victoria Dock (in the distance beyond the Royal Albert Dock). Like the other dock systems the Royal Docks have been excavated in the alluvium of the Thames flood plain, underlain here by an impervious clay that retards the escape of water. The docks, which are surrounded by walls, were originally begun in the early nineteenth century to give cargoes greater security from theft and to increase the amount of berthing space for ships using the port. Ships, tugs, and towed barges enter and leave by means of locks (bottom of photo). Inside the docks, ships tie up at wharves and unload or load cargo. The low buildings next to the wharves are transit sheds affording temporary storage. Large cranes, four of which are clearly visible to the right of the lock at the bottom left, move along the wharves on railway tracks and assist in the transfer of cargo between ship and shore. Some cargo is transferred between the ships at the wharves and barges that come alongside. Note the many clusters of barges in the docks and the river. They provide the principal means for transferring goods between different sections of the port (London is the only port in Britain that uses this lightering system). Some goods arriving at the docks are warehoused or processed inside the walls, but most items of cargo are carried to other destinations within a short time by ships, barges, trucks, or railway cars. (Aerofilms Limited.)

sugar, and wood pulp; or are engineering industries which have grown out of former shipbuilding activity along the river; or are nuisance industries, such as certain chemical industries which foul the air and have been compelled to locate on relatively open and isolated sites in marshy areas near the river. The largest plant in the belt is the Ford automobile factory at the city of Dagenham. This large industrial plant includes blast furnaces and a steel mill, and is supplied with iron ore and coal by sea transport. Residential areas occur within the lower Thames belt, but tend to lie inland from the marshy flood plain along the river.

2. *The northwestern industrial belt* has also developed in a strategic position with respect to transportation. It extends outward through the side of London which lies closest to the industrial Midlands, thus combining immediate access to London's market and labor with superior access to Britain's other economic center of gravity around the Pennines. Clusters of factories are strung along the numerous railroads and major highways which connect London with the Midlands and industrial areas beyond, and residential development is interspersed between the factory clusters. Small and medium-sized plants producing consumer goods are the most common, although these plants vary widely in scale of operations and type of product. The northwestern industrial belt has developed especially rapidly in the twentieth century and is now the leading industrial area of London except for the old inner area around the core.

3. *The northern industrial belt* follows the valley of the Lea River, which enters London from the north and flows into the Thames about 2 miles east of The City. Poorly drained, marshy, and subject to flooding, this valley was avoided by housing development as London grew, leaving it available for industrial sites with little competition. It has canal transportation as well as roads and railroads, and the lower end of the valley merges with the port area. Industrial development is especially intense in the lower end, but extends far up the valley to the north, where it is interspersed with large reservoirs and recreational areas such as parks and playing fields. Workingmen's housing lines both sides of the valley on slightly higher, better-drained land. As in the northwestern belt, the industries of the northern belt are extremely varied in scale and products.

Between these three industrial sectors are more predominantly residential sectors—a sea of housing, thinning and of generally newer construction toward its outer edges, dotted with shopping areas and cut through by commercial streets. This housing tends to be poorer in quality, more crowded and more densely occupied near the industrial sectors. In contrast, exceptionally prized residential areas occupied by more well-to-do people tend to occur on heights where the Thames Valley, in which most of the Greater London Conurbation lies, gives way to somewhat higher ground to the north and south. Two interruptions in this residential pattern are noteworthy. One is the sector or belt along the Thames upstream from central London in the direction of the royal residence at Windsor. This area is marked by unusual expanses of parkland, numerous large reservoirs that store Thames water for the city, and intensive recreational use of the river. The other is the

borough of Croydon, in the southern section of the conurbation, where the first considerable center of office employment outside of central London is developing. This section, with its high-rise buildings, represents the only notable decentralization of the administrative functions of London's inner core to occur thus far.

FOOD AND AGRICULTURE IN BRITAIN

One of the major characteristics of the United Kingdom is its dependence on imported food. It is estimated that Britain now produces about half of its total food requirements or two-thirds of all that can be grown in temperate climates. Importation of the remainder makes the United Kingdom the greatest importer of foodstuffs in the world. It is the leading importer of grain, meat, dairy products (except fluid milk), edible fats and oils, and tea; is second in fruit; and third in sugar and cocoa. For many food-exporting countries, both in middle latitudes and in the tropics, it is the most important foreign market.

In most of the country, crop production is unrewarding or impossible. Poor soils, steep slopes, cool summers, and excessive cloudiness and moisture are the principal handicaps. Highland Britain suffers the most on all these counts, but various parts of Lowland Britain are also affected. Comparatively slight rises in elevation bring sharp decreases in temperature and increases in precipitation. In Highland Britain the warmest month generally averages below 60°F, and precipitation is nearly always over 40 inches and in many places over 60 inches a year. Moorland, often waterlogged and boggy, is estimated to cover 25,000 of the 88,000 square miles of land in the island of Great Britain. In terms of economic use, such land is mainly unimproved natural pasture ("rough grazing"), generally with a low density of both animals and men.

Land and climate are more favorable to agriculture in Lowland Britain, and especially in the drier and warmer east. Even in Lowland Britain, however, much of the land is in pastures and meadows, though these are usually sown and tended and have a high carrying capacity. About one-third of the land in the United Kingdom is regularly plowed and planted in tilled crops. Eastern England is the largest section in which such arable land is predominant over grass (map, p. 124).

Britain's dependence on imported food arose with the growth of its population during the nineteenth century, from about 11 million to about 38 million. In earlier times the land had been adequate to feed the smaller population and it had afforded an export of wool and sometimes of grain. During the early part of the Industrial Revolution improvements were made in agriculture which amounted to an accompanying Agricultural Revolution. Among the advances were (1) new implements, (2) new crops such as potatoes and turnips, (3) new crop rotations which eliminated the frequent fallow periods of previous times, (4) increased application of fertilizers, (5) scientific stock breeding, and (6) the extension of crop production into marginal areas. Food production increased so much that for many decades it almost kept pace with population growth. But through the nineteenth century food imports increased steadily, especially imports of wheat, which was basic in the diet and relatively easy to transport. A decision to rely on the cheapest wheat available, imported or British, was embodied in the repeal of the Corn Laws in 1846. The Corn Laws had forbidden the importation of wheat until the price of British wheat rose to a specified point, thus protecting the British producer. Repeal of these laws was an important step in placing the country on a general free-trade basis.

Although imports continued to rise, the full effect of the free importation of grain was not felt until the 1870s and subsequent decades. By that time transportation facilities had become adequate to pour floods of grain into Britain cheaply from new producing areas in the Americas and Australia. The price of wheat fell to levels at which many British farmers, handicapped by relatively high land and production costs, could not survive. Widespread depopulation of rural areas and a drastic decline in British wheat acreage and production followed. Most other crops were also affected, though in lesser degrees.

This disaster to British cultivators accentuated a major characteristic of the United Kingdom's agriculture: its high concentration on grass farming for milk and meat production. Of course much of Britain had always been primarily pastoral, but now much of the land on which other crops were no longer profitable reverted to grass, for which the climate was relatively favorable. Meat and, to an even greater degree, fresh milk, are more difficult and expensive to transport than grains and many other vegetable products; consequently they afford local producers a degree of advantage over competitors located a greater distance from the market. Thus grasslands, dairy and beef cattle, and sheep have become the dominant elements in British agriculture. Fully two-thirds of farmers' income in Britain is derived from the sale of animals or their products. Most of the crop land is

used for producing supplementary animal feeds such as barley, oats, and turnips. Only about 7 percent of the United Kingdom, an area which if compacted would almost fit into a square 80 miles to a side, is used for producing vegetable products directly consumed by man—mainly wheat, potatoes, sugar beets, vegetables, and fruits—and substantial portions of some of these are fed to animals. Even so, the United Kingdom imports about one-fourth of the beef that moves in international trade, and more dairy products, other than fluid milk, than any other country.

This dependence on imported food is not due to lack of efficiency in British agriculture, but to the high density of population and the relatively high plane of living. Crop yields and the output of animal products per unit of land are among the highest in the world, being exceeded only by a few extremely productive countries in western continental Europe. Output per man is probably the highest in Europe, as farms are larger on the average in Britain than in the rest of Europe, and mechanization has now reached a point that is matched by very few countries anywhere. Total food production has almost doubled in the last quarter-century, mainly an outcome of various government policies aimed at reducing imports. Results have been achieved principally by inducing farmers to more than double the acreage in feed crops, mainly barley, at the expense of grassland, since a given acreage of such crops can generally feed more animals than the same acreage of grass—though the feed crops may well be more expensive to produce. Production of potatoes, which give a very high caloric yield per acre, and of wheat has also been sharply increased. But these programs have required various subsidies to farmers and thus have been expensive. In the final analysis, the amounts and kinds of food Britain produces would appear to depend on the relative costs of home production as against imports, on considerations as to the actual availability of imports in the future, and on the level of diet—now quite high—that can be afforded: if most of the country were planted in potatoes it could perhaps feed the whole population—at the levels of nutrition and enjoyment provided by a diet of potatoes.

Distribution of Crops and Livestock

All but a small fraction of the agricultural production of the United Kingdom comes from the island of Great Britain. Barley, oats, and wheat occupy greater acreages than other tilled crops on the island, with

VEGETABLES
harvested for sale
Main areas
Moderate to low production; scattered areas
Little or no production

BARLEY
Greatest density
Lighter density
Very minor or absent

OATS
Greatest density
Relatively high to moderate density
Minor or absent

WHEAT
Greatest density
Lighter density
Very minor or absent

SHEEP
Greatest density
Medium density
Low density
Few or absent

DAIRY CATTLE
Greatest density
Relatively high to moderate density
Relatively light density
Few or absent

BEEF CATTLE
Greatest density
Somewhat lighter density
Smaller numbers or absent

0   100
Scale of Miles

SCOTTISH HIGHLANDS

Firth of Forth

SCOTTISH LOWLANDS

Glasgow

SOUTHERN UPLANDS

NORTH SEA

Newcastle

PENNINES

LAKE DISTRICT

NORTHERN IRELAND

IRISH SEA

Leeds-Bradford

LANCASHIRE

YORKSHIRE

Liverpool

Humber Estuary

Manchester   Sheffield

THE WASH

W A L E S

THE MIDLANDS

Birmingham

THE FENLAND

EAST ANGLIA

London

Thames Estuary

K E N T

CORNWALL

ENGLISH CHANNEL

UNITED KINGDOM
LAND UTILIZATION
PRIMARY USE (Highly generalized)

Main urban areas

Cultivated crops and temporary grass; some permanent grass; some livestock

Permanent grass and livestock; some crops and temporary grass

Main areas of orchards and small fruits

Rough pasture and moorland; sheep; some permanent grass; some cattle

All maps in the panel opposite are broadly generalized from very detailed reference maps in *The Atlas of Britain and Northern Ireland* (Oxford: Clarendon Press, 1963), by permission. A great many small tracts of forest and woodland are not shown. The area shown on the map as The Fenland, bordering the broad but shallow east-coast bay called The Wash, is a tract of low, flat land that once was marsh or sea bottom but has been reclaimed for agriculture by diking and artificial drainage. The resulting farm land is the most fertile and productive area of its size in the British Isles. The Fenland, devoted very largely to cultivated crops rather than grass and livestock, is especially notable for vegetables, potatoes, fruits, wheat, and sugar beets.

barley accounting for about 5 million acres, wheat 2 million to 2½ million acres, and oats about 1 million acres. Oats, more tolerant of dampness than the other grains, is the most widespread, being grown in tilled areas throughout the island (map, left). Wheat and barley, and potatoes and sugar beets as well, are rather markedly localized in the sunnier, drier, more level, and more fertile eastern sections of England, though all are grown to some extent in other areas. Potatoes normally occupy between 600,000 and 700,000 acres and sugar beets around 450,000 acres. Fodder crops (turnips, swedes, mangels, rape, cabbage, and others), like oats, are extremely widespread and occupy, in the aggregate, somewhat more than a million acres. Intensive market gardening and fruit growing are carried on in a number of scattered districts in the English Lowland, the Scottish Lowlands, and Cornwall; the most important districts lie within 100 miles of the huge London market.

Cattle are by far the most prominent type of livestock in Great Britain; if the island's livestock are converted to standard animal units (broadly: 1 cow, bull, or horse is the equivalent of 2 calves or heifers, 7 sheep, 14 lambs, 7 goats, 5 hogs, 10 pigs, or 100 fowl), cattle represent about two-thirds of the total. The main area of dairy farming occupies most of Lowland Britain, except for the predominantly arable eastern sections, with a northward extension along the west side of the island to the Glasgow area, and westward extensions into parts of Wales also into Cornwall (map, left). Raising of beef cattle in the lowlands is so widespread that it is difficult to generalize; central and western England and northeastern Scotland are the most prominent large areas, while densities are comparatively light in eastern England and in most highland areas. Sheep are an element in the farm economy of most areas, both highland and lowland, but the largest numbers and densest concentrations are found in two areas: (1) the uplands of Wales, and (2) the Southern Uplands of Scotland, northern Pennines and Lake District. Since the 1930s they have declined sharply on farms in southeastern England. Sheep are raised primarily for meat, with wool an important secondary source of income.

TWENTIETH-CENTURY DIFFICULTIES AND TRENDS

Since World War I the United Kingdom has experienced an almost constant succession of economic difficulties, punctuated by military and political struggles which have tended to accelerate the country's declining power and influence. The economic difficulties have been due largely—though by no means wholly—to declining exports in the principal industries upon which Britain's export trade was based before 1914. By the time these difficulties appeared there was no turning back from an economy heavily dependent on trade, since the swollen population required large food imports, and manufactures to pay for the food required imported raw materials.

Although the foundations for trouble had been laid previously, World War I was the turning point for the British economy. Depressed economic conditions characterized many areas of the country—especially those depending greatly on cotton textiles, coal, shipbuilding, and woolen textiles—during most of the period between the world wars. However, the rest of the world was also in a depression, at least during the 1930s, and Britain's position did not yet appear crucial despite the fall in commodity exports. A major reason was that exports of goods had not been sufficient to pay for imports, even before World War I. A very important role had always been played by "invisible exports." These included (1) capital from the formerly immense profits of British industry, which was invested on a large scale in the development of many other countries, (2) shipping services, which the huge British merchant marine supplied for much of the world, and (3) various other services, such as banking and insurance, which London provided on an international scale. Essentially, Britain's accumulated wealth was made to do various jobs in and for other countries; and the profits from these activities covered the deficit in the United Kingdom's trade in goods. In the late 1930s they paid for as much as 30 percent of the imports, but no longer quite made up for the excess of imports over exports.

World War II was disastrous to this system. Exports could not be maintained at substantial levels and the war fought at the same time. By 1944 exports stood at less than a third of the 1938 level. Consequently, to pay for imports overseas investments amounting to more than a billion pounds sterling were sold, and debts amounting to over three billion pounds were contracted to overseas suppliers. At the same time much of Britain's industrial equipment was overworked and not replaced when it normally would have been. So Britain emerged in 1945 a much impoverished, debt-ridden, and needy victor—though a justly proud one.

Crucially important was the fact that invisible exports were no longer available in quantities sufficient to pay for an excess of goods imported over goods exported. The export industries would now have to earn the nation's way back to solvency. Despite a very large expansion of exports after World War II, they have not quite succeeded in doing so. Expanded production and export has been accompanied by expanded imports, and a rising standard of living has increased imports still further. Consequently, Britain has found itself subject to recurrent trade and payments crises, and this despite two separate devaluations of the pound designed to make British goods cheaper relative to foreign goods in the world market. The situation has been made more difficult by the growing strength of competing industrialized exporters such as West Germany and Japan and by rejection of Britain's attempts to join the European Common Market. Exclusion from the Common Market has meant that British goods are at a tariff disadvantage against those of member countries in the very important markets of these western European neighbors. Attempts, more or less desperate, to deal with the country's deteriorating trading position have occupied all British governments for several decades. Such attempts at times have seemed on the brink of success and at other times have seemed to be barely avoiding the brink of disaster.

Major Export Industries

The post-World War II struggle has not been carried on primarily by the old export industries, however. Even in the interwar period the growth of a number of newer industries provided bright spots in the generally gloomy picture, and since World War II some of these have burgeoned into dominant places in the economy. At present the major export industries are (1) *general engineering,* the production of machinery in great variety, of which textile machinery and tractors are examples prominent in the export trade; (2) *the automobile industry,* third largest in the world (1966) after that of the United States and West Germany; (3) *the iron and steel industry,* which still produces large direct exports as well as metal for other industries; (4) *the chemical industry,* which exports products ranging from cosmetics and medicines to fertilizers; and (5) *electrical engineering,* exporting generating equipment and many other products. Of these five, only iron and steel and general engineering were leading export industries in 1913, and the latter was much less important than today.

These newer industries—the aircraft industry is another that has been much publicized and should be mentioned—are generally types in which advanced scientific and production skills are highly important; and these are British assets, stemming from long experience and educational attainment, which cannot be so easily matched by many countries as can Britain's resources of power, materials, or cheap labor. Thus Britain has a greater competitive advantage in them than in the older industries, though an advantage that steadily decreases.

Although many of the new plants have been established in the coal-field industrial districts, many others have located in London or other cities and towns in southern England. An example is the automobile industry (map, p. 103), which has new plants in Liverpool and Glasgow, but is centered especially in the Midlands at Birmingham and Coventry and, to a lesser extent, Derby (280,000) and Nottingham, and also is quite important in London and in the smaller cities of Oxford (200,000) and Luton (195,000), which lie within a 50-mile radius of London to the north. Thus there is a modern tendency for the old distinction between industrial highland margins and nonindustrial lowland to diminish. It is more and more replaced by a distinction between a relatively dynamic and prosperous South, with its newer "growth" industries, and a population that is growing relatively rapidly; and a more stagnant and less prosperous North, with its coal-field industrial areas based on old and struggling industries, and its population drained by southward emigration.

The End of Empire

One result, and evidence, of Britain's declining power has been the rapid dissolution of the British Empire, which in the early twentieth century covered a quarter of the earth. Although independence was granted to some units of the empire earlier, its real dissolution

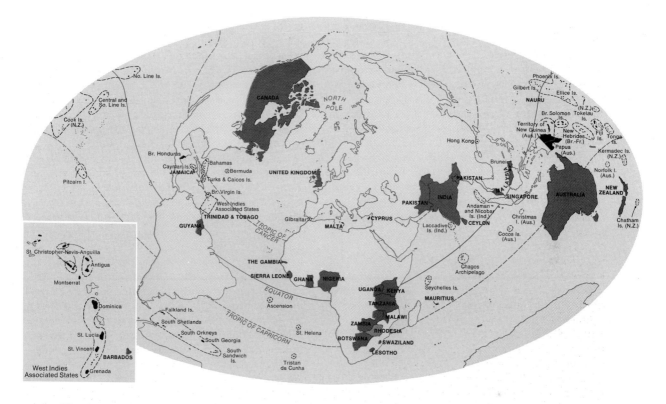

The Commonwealth of Nations as of December 1, 1968. Names in capitals indicate sovereign states. Rhodesia, shown in a separate pattern, has declared its independence but continues to be regarded officially by the United Kingdom as a self-governing colony (late 1968). Dependent units are indicated by capital and small letters; those not otherwise identified as to ownership are administered or protected by the United Kingdom. Small island dependencies are encircled by dashes; those encircled by one set of dashes do not in all cases form a single governmental unit. Some minor islands, often uninhabited, are not indicated on the map, and Antarctic claims are not shown. Persian Gulf dependencies from which British protection will soon be withdrawn are not indicated. The West Indies Associated States (inset), consisting of six island colonies encircled by dashes, do not comprise a sovereign state. Montserrat and the independent state of Barbados, also shown in the inset, do not belong to the Associated States. (Briesemeister Elliptical Equal Area Projection, courtesy of the American Geographical Society.)

has occurred rapidly in the decades since World War II, as one colony after another has been granted independence (map, p. 65). However, newly independent colonies have desired more often than not to maintain a special bond with Britain and with other ex-British colonies. Thus an association of independent countries, called the Commonwealth of Nations, has replaced much of the former empire. In 1968 the Commonwealth, still rapidly evolving, included twenty-eight independent countries besides the United Kingdom itself: thirteen in Africa; five in southern Asia; four in the West Indies and South America;

two island states in the Mediterranean Sea; one island state in the tropical Pacific; and Canada, Australia, and New Zealand (map, above). These states ranged from India and other giants to tiny states such as Malta. Their Commonwealth membership gave them some trading privileges with Britain, some claim to military protection, and some advantage in obtaining development capital from Britain or other Commonwealth sources. In the largely Anglo-Saxon countries (United Kingdom, Australia, New Zealand, Canada) sentiment plays a large role in maintaining Commonwealth ties. However, the declining political and economic power of Britain, combined with the diversity of interests among (and in some cases enmities between) Commonwealth states, has been tending to loosen Commonwealth ties and decrease the importance of the organization. In 1961 an important member, South Africa, left the Commonwealth. By 1968 all that remained of the empire itself (aside from the self-governing African colony of Rhodesia, which declared itself independent, without British sanction, in 1965) were a number of small and widely scattered island dependencies, along with a few patches of mainland in Latin America (British Honduras), Europe (Gibraltar), and the Orient (Hong Kong).

IRELAND

Ireland is a land of hills and lakes, marshes and peat bogs, cool dampness and verdant grassland. The island consists of a central plain surrounded on the north, south, and west by hills and low, rounded mountains (map, p. 99). The climate is marine to an extreme, with average temperatures ranging from the low 40s in January to about 60 degrees in July, and rainfall from over 100 inches in western uplands to about 30 inches in eastern lowlands. Glaciation blocked drainage lines and has left many lakes connected by sluggish rivers, of which the most famous is the River Shannon. Marshes and waterlogged moorland are common, and peat bogs cover one-seventh of the island. Nearly all of the natural forest is gone, but in the cool dampness luxuriant grass grows, giving the countryside its prevailing and proverbial green. The island's open, varied, and verdant rural landscapes are a major asset to a growing tourist industry.

Industrial Development

The island is almost completely lacking in important natural resources for industry. There are no mineral deposits or forests of much consequence, though government-sponsored programs of reforestation have had a limited success in small areas, as they have in Great Britain. However, in recent years a government program of electrification has had considerable success. This has greatly reduced a major and long-standing handicap to industrialization, although Ireland is still not a power-rich country. The main elements in this program have been a number of small hydroelectric installations, plus technical advances, beginning about 1950, allowing the use of peat in thermal generating stations. Peat is a very low-grade fuel, but is abundant in Ireland and has long been a major household fuel there. Supplementary power continues to be generated from imported fuels to meet Ireland's requirements, which are still modest.

But Ireland's comparative lack of industrial development is partly a consequence of its long subordinate relationship to England. After effective conquest of the whole island by the English in the seventeenth century, it was oppressed for two centuries under probably the harshest rule imposed on any British colonial territory. Most of the land was appropriated and divided among large estates held by English landlords, who were often absentees living in Britain; the Irish peasantry was reduced in many instances virtually to serfdom, and the proceeds of its labor were drained off—and often out of Ireland—in extortionate rents and taxes. Irish trade and industry were penalized, restricted, and at times nearly destroyed by governmental measures designed to favor British competitors. All parts of Ireland were affected, but the Catholic South more than the Protestant North. (There was a strong element of religious persecution in British rule, Catholics being regarded as potentially disloyal to the Protestant crown.) Ireland, especially the Catholic part, became a land of deep poverty, sullen hostility, and periodic violence.

The North, somewhat less handicapped, did develop important industries despite the poverty of natural resources. Northern Ireland today has essentially an industrial economy, with slightly more industrial employment than the whole of the larger Republic. Its major industries are textiles, especially the famous Irish linen, shipbuilding, clothing, and aircraft production. Other industries manufacture cigarettes, electrical goods, furniture, and various other products. About 130 new factories were built between 1945 and 1966. Belfast (595,000) is the principal industrial center as well as the political capital and chief seaport. It is by far the leading center for all of the main industries. The only material Northern Ireland provides for its major industries is a minor portion of the flax used in making linen. Perhaps it is a reflection in part of Northern Ireland's precarious resource base that, in spite of the new plants mentioned above, this has remained the part of the United Kingdom most afflicted by unemployment since World War II, and its plane of living remains below that of Great Britain.

The industries of the Republic of Ireland are devoted very largely to the manufacture of light consumer goods for the domestic market behind tariff protection. A considerable number of new factories have been built in recent years, and the annual value of the

Republic's manufactures now exceeds that of Northern Ireland. Domestic demand is far from satisfied, however, and well over half of the Republic's imports are manufactures in great variety. Dublin (760,000), situated on the Irish Sea at the eastern edge of the Central Plain, is the main industrial center, the leading seaport, and the political and cultural capital of the Republic. It is also the only large city. The southeastern port of Cork, next in size, has perhaps 150,000 people in its metropolitan area. The Republic's only steel mill is located at Cork.

Agricultural Production and Trade of the Irish Republic

In the comparative absence of large-scale industry, the Republic of Ireland is still very predominantly an agricultural country. The same physical and climatic handicaps to cultivation that affect Great Britain are present, but in even greater degree. Only about one-fifth of the Republic is classed as arable land, one-half is pasture, and most of the rest is wasteland. Grazing is the main element in the agricultural economy (as it is in Northern Ireland), and much of the arable land is devoted to fodder crops. The country's largest export is live cattle for fattening and slaughter in Britain. Attempts are being made to expand home production of food, as large imports of grain are now necessary, and also to fatten and slaughter more animals at home and export the more valuable meat instead of the live animals.

There is a strong tendency in the modern world for agricultural areas and societies to be poorer, at least by quantitative measurements such as per capita income, than those with more industrial employment. The farmer's net output, measured by value, tends to be less than that of the worker in mechanized industries and his return smaller. The tendency is exemplified in Ireland, where the average per capita income of the Republic is only slightly more than half that of the United Kingdom, and noticeably below that of most of the industrialized countries of western and northern Europe. The estates of the period of English control are gone, and today the Irish farmer is usually a small owner rather than a "rack-rented" tenant; but farms are often neither large nor highly mechanized. There is marked rural poverty, especially in western Ireland, where a particularly rugged and poor agricultural environment is inhabited by an unusually dense peasant population, stemming originally from flight to the area to escape English exactions.

Another result of the Republic's agricultural economy is its very close economic relationship to the United Kingdom, contrasting strongly with continued political hostility toward that country. The United Kingdom is both the logical and historical market for Irish agricultural produce, and more than two-thirds of the Republic's exports are sold there, essentially in return for British manufactures. Meanwhile the old political antagonism has been fired since Irish independence by the desire to unite Northern Ireland to the Republic and by terrorist activities of the underground Irish Republican Army in that cause. In World War II this antagonism was expressed in Ireland's cool official neutrality—though large numbers of Irish joined the British armed forces or worked in British war industries as individuals—and it has also been expressed in the attempt—thus far not very successful—to foster Irish Gaelic as a national language in place of English. In 1949 the Republic withdrew from the Commonwealth of Nations, and thus severed its last remaining political tie with the United Kingdom.

The Problem of Depopulation

An almost unique distinction of Ireland is depopulation. Few sizable areas in the world have smaller populations today than a century ago, but the island of Ireland has just over half as many people as the 8.2 million it had in 1841. In the late eighteenth and early nineteenth centuries its population increased very rapidly on the precarious basis of one main subsistence crop, the potato. In the late 1840s this crop failed for several years in succession, and the potato famine which resulted claimed the lives of nearly a million people in five years, while another million and a half fled the island, mainly to the United States, in the decade following 1846. A pattern of emigration was established that has continued to the present day, though most emigrants now go to the United Kingdom. Between 1841 and 1961 the total population of the island declined in ten of twelve intercensal periods, as emigration outweighed natural increase. In 1967, however, the Republic was estimated to have 74,000 more people than it had in the census of 1961, an increase of between 2 and 3 percent in six years. This was very low compared to most countries, and did not mean that heavy emigration had ceased. But it did mean that emigration was no longer greater than natural increase, in contrast to normal conditions previously. It gave hope that Ireland was at last providing new jobs and better conditions at home rapidly enough to avoid "national suicide."

REFERENCES and READINGS

General and Miscellaneous

The Aerofilms Book of Aerial Photographs (rev., enl., and retitled; London: Aerofilms Ltd., 1965). Primarily photos of Britain, with some from other parts of the world.

Atlas of Britain and Northern Ireland (Oxford: Clarendon Press, 1963). A superb reference atlas with large maps in color; perhaps the finest national atlas in the world.

BOWEN, E. G., ed., *Wales: A Physical, Historical and Regional Geography* (London: Methuen and Co., 1957).

Britain's Foreign Trade (London: Port of London Authority, 1966).

CAESAR, A. A. L., "Planning and the Geography of Great Britain," *Advancement of Science, 21,* no. 91 (September 1964), 230–240; and, with D. E. Keeble, "Regional Planning Problems in Great Britain," *Advancement of Science, 22,* no. 97 (July 1965), 177–185.

CHISHOLM, MICHAEL, "Must We All Live in Southeast England? The Location of New Employment," *Geography, 49,* pt. 1 (January 1964), 1–14.

COPPOCK, J. T., "Great Britain," *Focus, 14,* no. 10 (June 1964), 6 pp.

DARBY, H. C., ed., *An Historical Geography of England before 1800: Fourteen Studies* (Cambridge: At the University Press, 1961).

DURY, G. H., *The British Isles: A Systematic and Regional Geography* (2d ed [1st American ed.]; New York: W. W. Norton & Company, 1964).

East Midland Geographer (semiannual).

EAST, W. G., gen. ed., *Regions of the British Isles* (London: Thomas Nelson and Sons). Individual volumes thus far include A. E. Smailes, *North England* (1960), A. C. O'Dell and K. Walton, *The Highlands and Islands of Scotland* (1962), and T. W. Freeman, H. B. Rodgers, and R. H. Kinvig, *Lancashire, Cheshire, and the Isle of Man* (1966).

FISHER, CHARLES A., "The Changing Significance of the Commonwealth in the Political Geography of Great Britain," *Geography, 48,* pt. 2 (April 1963), 113–129.

THE GEOGRAPHER, U.S. DEPARTMENT OF STATE, *Commonwealth of Nations* (Geographic Bulletin No. 6), (Washington, D.C.: Government Printing Office, May 1968).

HOSKINS, W. G., *The Making of the English Landscape* (London: Hodder and Stoughton, 1955); and, with L. Dudley Stamp, *The Common Lands of England and Wales* (London: Collins, 1963).

HOUSE, JOHN W., "North-east England: Personality and Prospect," *Geographical Magazine, 36,* no. 11 (March 1964), 621–632.

HUGHES, MARGARET E., *Wales: A Physical, Economic and Social Geography* (London: University of London Press, 1961).

JONES, H. R., "Migration within Scotland," *Scottish Geographical Magazine, 83,* no. 3 (December 1967), 151–160.

KENDALL, HENRY MADISON, "Contributions to the Geography of the British Isles," *Geographical Review, 55,* no. 4 (October 1965), 581–586. A review article.

LAMB, HUBERT HORACE, *The English Climate* (London: English Universities Press, 1964).

LESLIE, EDWARD A., *Basic Data on the Economy of the United Kingdom,* U.S. Bureau of International Commerce, Overseas Business Reports, No. OBR–66–50 (July 1966). (Washington, D.C.: Government Printing Office, 1966.)

LOWENTHAL, DAVID, and HUGH C. PRINCE, "The English Landscape," *Geographical Review, 54,* no. 3 (July 1964), 309–346; and "English Landscape Tastes," *Geographical Review, 55,* no. 2 (April 1965), 186–222.

MACLEOD, IAN, "Reports of Britain's Death . . .," *Foreign Affairs, 45,* no. 1 (October 1966), 88–97.

MANLEY, GORDON, *Climate and the British Scene* (London: Collins, 1952); and "Climate in Britain," in International Geographical Congress, 20th, London, 1964, *Congress Proceedings,* ed. by J. Wreford Watson (London: Thomas Nelson and Sons, 1967), pp. 34–45.

MITCHELL, J. B., ed., *Great Britain: Geographical Essays* (Cambridge: At the University Press, 1962). Contributions by leading British geographers, mainly on the regions of Great Britain.

PEARSALL, W. H., *Mountains and Moorlands* (London: Collins, 1950).

RAWSTRON, E. M., and B. E. COATES, "Opportunity and Affluence," *Geography* **51**, pt. 1 (January 1966), 1–15.

REES, HENRY, *The British Isles: A Regional Geography* (London: George G. Harrap and Co., 1966).

Scottish Geographical Magazine (3 issues a year). See especially the complete issue on Scotland, **80**, no. 2 (September 1964), incorporating articles as follows: J. B. Caird, "The Making of the Scottish Rural Landscape," 72–80; W. H. K. Turner, "Wool Textile Manufacture in Scotland," 81–89; A. R. Wannop, "Scottish Agriculture," 90–98; A. MacPherson, "Scotch Whisky," 99–106; R. H. Campbell, "Scottish Shipbuilding: Its Rise and Progress," 107–113; C. J. Robertson, "New Industries and New Towns in Scotland's Industrial Growth," 114–123; and J. B. Caird and D. R. Diamond, "The Geography of Scotland: A Select Bibliography," 124–126.

SMITH, WILFRED, *An Economic Geography of Great Britain* (London: Methuen and Co., 1949).

STAMP, L. DUDLEY, and STANLEY H. BEAVER, *The British Isles: A Geographic and Economic Survey* (5th ed.; London: Longmans, Green and Co., 1963). A standard text.

STEEL, ROBERT W., and RICHARD LAWTON, eds., *Liverpool Essays in Geography: A Jubilee Collection* (London: Longmans, Green and Co., 1967).

STEERS, JAMES ALFRED, ed., *Field Studies in the British Isles,* published on the occasion of the 20th International Geographical Congress, United Kingdom, 1964 (London: Thomas Nelson and Sons, 1964).

THOMAS, DAVID, "England and Wales: Recent Changes in the Social and Economic Structure," *Geographisches Taschenbuch und Jahrweiser für Landeskunde, 1964–65,* pp. 115–135. Article in English.

THOMAS, J. GARETH, "The Geographical Distribution of the Welsh Language," *Geographical Journal,* **122,** pt. 1 (March 1956), 71–79.

WATSON, J. WREFORD, "Geography and Growth in Scotland," *Journal of Geography,* **64,** no. 9 (December 1965), 398–414; and, with J. B. Sissons, eds., *The British Isles: A Systematic Geography* (London: Thomas Nelson and Sons, 1964). Published on the occasion of the 20th International Geographical Congress. Contributions by well-known British geographers.

General References on Urban Geography

BEAUJEU-GARNIER, J., and G. CHABOT, *Urban Geography,* trans. by G. M. Yglesias and S. H. Beaver (New York: John Wiley & Sons, 1967).

CREESE, WALTER L., *The Search for Environment: The Garden City, Before and After* (New Haven, Conn.: Yale University Press, 1966).

DICKINSON, ROBERT E., *City and Region* (London: Routledge and Kegan Paul, 1964).

GETIS, ARTHUR, JUDITH GETIS, and Others, "To Understand Urban Environments," *Journal of Geography,* **65,** no. 5 (May 1966), 204–240.

HAUSER, PHILIP M., and LEO F. SCHNORE, eds. *The Study of Urbanization* (New York: John Wiley & Sons, 1965).

JOHNSON, JAMES H., *Urban Geography: An Introductory Analysis* (New York: Pergamon Press, 1967).

JONES, EMRYS, *Towns and Cities* (New York: Oxford University Press, 1966).

MAYER, HAROLD M., "Cities and Urban Geography," *Journal of Geography,* **68,** no. 1 (January 1969), 6–19; and, with Clyde F. Kohn, eds., *Readings in Urban Geography* (Chicago: University of Chicago Press, 1959).

MUMFORD, LEWIS, *The City in History: Its Origins, Its Transformations and Its Prospects* (New York: Harcourt, Brace & World, 1961).

London, the Green Belt, and the New Towns

BIRD, JAMES, *The Geography of the Port of London* (London: Hutchinson University Library, 1957).

BROWN, COLIN M., "The Structure of Manufacturing Industry in London's New Towns," *Tijdschrift voor Economische en Sociale Geografie,* 57, no. 3 (May–June 1966), 121–124.

CLAYTON, R., ed., *The Geography of Greater London* (London: George Philip and Son, 1964). See especially A. E. Smailes, "The Site, Growth, and Changing Face of London," pp. 1–52; James Bird, "Contrasted Scenes in the Port of London," pp. 53–73; J. E. Martin, "The Industrial Geography of Greater London," pp. 111–141; and M. J. Wise, "The Population of London," pp. 143–170; extensive bibliography of books and articles; and, ed., *Guide to London Excursions* (London: 20th International Geographical Congress, 1964).

HALL, PETER, *The World Cities* (London: Weidenfeld and Nicolson, World University Library, 1966), Chap. 2, "London," pp. 30–58; also *London 2000* (London: Faber and Faber, 1963); and *The Industries of London Since 1861* (London: Hutchinson University Library, 1962).

JEANS, D. N., "Competition, Momentum and Inertia in the Location of Commercial Institutions: Case Studies in Some London Commodity Markets," *Tijdschrift voor Economische en Sociale Geografie,* 58, no. 1 (January–February 1967), 11–19.

JOHNSON-MARSHALL, PERCY, *Rebuilding Cities* (Chicago: Aldine Publishing Company, 1966), Chap. 5, "Comprehensive Development in London," pp. 177–290.

London: Aspects of Change (Centre for Urban Studies, Report No. 3; London: Macgibbon and Kee, 1964).

MARTIN, J. E., *Greater London: An Industrial Geography* (London: G. Bell and Sons, 1966).

OSBORN, FREDERIC J., and ARNOLD WHITTICK, *The New Towns: The Answer to Megalopolis* (New York: McGraw-Hill Book Company, 1963).

SMAILES, A. E., "Greater London—The Structure of a Metropolis," *Geographische Zeitschrift,* 52, no. 3 (August 1964), 163–189.

THOMAS, DAVID, "London's Green Belt: The Evolution of an Idea," *Geographical Journal,* 129, pt. 1 (March 1963), 14–24.

THROWER, NORMAN J. W., "Crawley, Sussex: An English New Town," *Yearbook of the Association of Pacific Coast Geographers,* 25 (1963), 21–29.

WHITE, H. P., "London's Rail Terminals and Their Suburban Traffic," *Geographical Review,* 54, no. 3 (July 1964), 347–365.

WHITEHAND, J. W. R., "The Settlement Morphology of London's Cocktail Belt," *Tijdschrift voor Economische en Sociale Geografie,* 58, no. 1 (January–February 1967), 20–27.

Other British Cities

BRITISH ASSOCIATION FOR THE ADVANCEMENT OF SCIENCE. Surveys of British cities in their regional settings, particularly M. W. Beresford and G. R. J. Jones, eds., *Leeds and Its Region* (1967), K. C. Edwards, ed., *Nottingham and Its Region* (1966), F. J. Monkhouse, ed., *A Survey of Southampton and Its Region* (1964), C. F. Carter, ed., *Manchester and Its Region* (1962), Ronald Miller and Joy Tivy, eds., *The Glasgow Region* (1958), D. L. Linton, ed., *Sheffield and Its Region* (1956), and M. J. Wise, ed., *Birmingham and Its Regional Setting* (1950).

DICKS, BRIAN, "Sheffield: City of Steel," *Geographical Magazine,* 37, no. 11 (March 1965), 854–866.

EDWARDS, K. C., "Nottingham: Queen of the Midlands," *Geographical Magazine,* 38, no. 5 (September 1965), 329–347.

FREEMAN, T. W., *The Conurbations of Great Britain* (2d ed.; Manchester: Manchester University Press, 1966).

JOHNS, EWART, *British Townscapes* (London: Edward Arnold, 1965).

LAWTON, R., "Liverpool and the Tropics," in Robert W. Steel and R. Mansell Prothero, eds., *Geographers and the Tropics: Liverpool Essays* (London: Longmans, Green and Co., 1965), pp. 349–375; and "Vital Liverpool," *Geographical Magazine,* 40, no. 4 (August 1967), 299–312.

McGOVERN, P. D., "The New Towns of Scotland," *Scottish Geographical Magazine,* 84, no. 1 (April 1968), 29–44.

SCARGILL, D. I., "Metropolitan Influences in the Oxford Region," *Geography,* 52, no. 235 (April 1967), 157–165.

WALKER F., "Economic Growth on Severnside," Institute of British Geographers, *Transactions No. 37* (December 1965), pp. 1–13.

WILKINSON, H. R., "Humberside," *Geographical Magazine,* 38, no. 2 (June 1965), 120–131.

Minerals and Industries

BAILEY, RICHARD, "Coal in Britain and Europe: Problems and Solutions," *International Affairs,* 42, no. 3 (July 1966), 432–443.

CONKLING, EDGAR C., "South Wales: A Case Study in Industrial Diversification," *Economic Geography,* 39, no. 3 (July 1963), 258–272.

ELLIOT, N. R., "A Geographical Analysis of the Tyne Coal Trade," *Tijdschrift voor Economische en Sociale Geografie,* 59, no. 2 (March–April 1968), 71–93; and "Tyneside: A Study in the Development of an Industrial Seaport," *Tijdschrift voor Economische en Sociale Geografie,* 53, no. 11 (November 1962), 225–237, and no. 12 (December 1962), 263–272.

ESTALL, R. C., "Industrial Change in Lancashire and Merseyside," *Geography,* 46, pt. 1 (January 1961), 56–59.

GOODWIN, WILLIAM, "The Structure and Position of the British Motor Vehicle Industry," *Tijdschrift voor Economische en Sociale Geografie,* 56, no. 4 (July–August 1965), 145–156.

GRIFFITHS, IEUAN L., "The New Welsh Anthracite Industry," *Geography,* 47, pt. 4 (November 1962), 389–400.

KEELING, B. S., and A. E. G. WRIGHT, *The Development of the Modern British Steel Industry* (London: Longmans, Green and Co., 1964).

LINTON, D. L., "The Geography of Energy," *Geography,* 50, pt. 3 (July 1965), 197–228.

LUCKAS, M. R., "Britain's Crude Oil: Changing Patterns in Supplies and Destinations, and Natural Gas Implications," *Tijdschrift voor Economische en Sociale Geografie,* 57, no. 4 (July–August 1966), 141–148; and "Recent Developments in the United Kingdom Oil Industry," *Geography,* 50, pt. 2 (April 1965), 152–160.

MANNERS, GERALD, ed., *South Wales in the Sixties: Studies in Industrial Geography* (Oxford: Pergamon Press, 1964); also "The Tinplate and Steel Industries in West South Wales," *Geography,* 44, pt. 1 (January 1959), 38–40; "Transport Costs, Freight Rates and the Changing Economic Geography of Iron Ore," *Geography,* 52, pt. 3 (July 1967), 260–279; and *The Geography of Energy* (New York: Hillary House Publishers Ltd., 1964).

MOISLEY, H. A., "Harris Tweed: A Growing Highland Industry," *Economic Geography,* 37, no. 4 (October 1961), 353–370.

MORGAN, MICHAEL, "Natural Gas and Britain," *Geographical Magazine,* 39, no. 9 (January 1967), 736–743.

MOUNFIELD, P. R., "Nuclear Power in the United Kingdom: A New Phase," *Geography,* 52, pt. 3 (July 1967), 310–316; and "The Location of Nuclear Power Stations in the United Kingdom," *Geography,* 46, pt. 2 (April 1961), 139–155.

POCOCK, D. C. D., "Britain's Post-war Iron-ore Industry," *Geography,* 51, pt. 1 (January 1966), 52–55.

RODGERS, H. B., "The Changing Geography of the Lancashire Cotton Industry," *Economic Geography,* 38, no. 4 (October 1962), 299–314.

SALT, JOHN, "The Motor Industry on Merseyside," *Geography,* 53, pt. 3 (July 1968), 320–322.

SIMPSON, E. S., *Coal and the Power Industries in Postwar Britain* (London: Longmans, Green and Co., 1966).

SMITH, D. M., "Recent Changes in the Regional Pattern of British Industry," *Tijdschrift voor Economische en Sociale Geografie,* 56, no. 4 (July–August 1965), 133–144.

STORRIE, MARGARET C., "The Scotch Whisky Industry," Institute of British Geographers, Publication No. 31, *Transactions and Papers, December, 1962,* pp. 97–114.

STRINGER, E. *A Geography of Resources: World Survey and British Isles* (London: Cassell, 1966).

THOMAS, TREVOR M., "The North Sea Gas Bonanza," *Tijdschrift voor Economische en Sociale Geografie,* 59, no. 2 (March–April 1968), 57–70; "The North Sea and Its Environs: Future Reservoir of Fuel?" *Geographical Review,* 56, no. 1 (January 1966), 12–39; "Oil and Natural Gas: Discoveries and Exploration in the North Sea and Adjacent Areas," *Geography,* 49, pt. 1, no. 202 (January 1964), 50–55; "Recent Developments in the South Wales Coal Mining Industry," *Tijdschrift voor Economische en Sociale Geografie,* 55, no. 8–9 (August–September 1964), 185–196; "Geographic and Economic Factors in the Siting of a Major Integrated Steelworks: The Spencer Works, Newport, South Wales," *Tijdschrift voor Economische en Sociale Geografie,* 54, no. 8–9 (August–September 1963), 173–180; and *The Mineral Wealth of Wales and Its Exploitation* (London: Oliver and Boyd, 1961).

TURTON, B. J., "The British Railway Engineering Industry: A Study in Economic Geography," *Tijdschrift voor Economische en Sociale Geografie,* 58, no. 4 (July–August 1967), 193–202.

WARREN, KENNETH, "Locational Problems of the Scottish Iron and Steel Industry Since 1760," *Scottish Geographical Magazine,* 81, no. 1 (April 1965), 18–37; and no. 2 (September 1965), 87–103.

Transportation

APPLETON, J., "Transport and the Landscape of Northern England," in J. W. House, ed., *Northern Geographical Essays in Honour of G. H. J. Daysh* (Newcastle upon Tyne: Oriel Press, 1966), pp. 178–195.

BEAVER, S. H., "Ships and Shipping: The Geographical Consequences of Technological Progress," *Geography,* 52, no. 235 (April 1967), 133–156.

BIRD, JAMES, *The Major Seaports of the United Kingdom* (London: Hutchinson and Co., 1963).

BUCHANAN, COLIN D., "Britain's Road Problem," *Geographical Journal,* 130, pt. 4 (December 1964), 470–483.

DAVIDSON, FRANK P., "Chunnel," *Foreign Affairs,* 44, no. 2 (January 1966), 314–319. An article on the proposed English Channel Tunnel.

Digest of Port Statistics [for Great Britain] (London: National Ports Council).

FAIRHALL, DAVID, "The Manchester Ship Canal," *Geographical Magazine,* 35, no. 1 (May 1962), 32–46.

KINNIBURGH, IAN A. G., "New Developments in Clydeport," *Scottish Geographical Magazine,* 82, no. 3 (December 1966), 144–153.

PATMORE, J. A., "The Contraction of the Network of Railway Passenger Services in England and Wales, 1836–1962," Institute of British Geographers, *Transactions No. 38* (June 1966), pp. 105–118.

REES, HENRY, *British Ports and Shipping* (London: George G. Harrap and Co., 1958).

SEALY, KENNETH R., "The Siting and Development of British Airports," *Geographical Journal,* 133, pt. 2 (June 1967), 148–177.

THOMAS, BRINLEY, "A Transport Revolution in Wales," *Geographical Magazine,* 37, no. 9 (January 1965), 649–663.

WATTS, H. D., "The Inland Waterways of the United Kingdom in the 1960s," *Economic Geography,* 43, no. 4 (October 1967), 303–313.

WISE, M. J., "The Impact of a Channel Tunnel on the Planning of South-eastern England," *Geographical Journal,* 131, pt. 2 (June 1965), 167–185.

Rural Geography, Land Use, Agriculture, and Fisheries

BEST, ROBIN H., "Recent Changes and Future Prospects of Land Use in England and Wales," *Geographical Journal,* 131, pt. 1 (March 1965), 1–12; and, with J. T. Coppock, *The Changing Use of Land in Britain* (London: Faber and Faber, 1962).

BRAMLEY, MARGARET, *Farming and Food Supplies: The Case for Expansion of British Agriculture* (London: George Allen and Unwin, 1965).

BUCHANAN, KEITH, and D. J. SINCLAIR, *Types of Farming in Britain* (rev. ed.; London: Association of Agriculture, 1966).

CHISHOLM, MICHAEL, *Rural Settlement and Land Use: An Essay in Location* (London: Hutchinson University Library, 1962); and "Have English Villages a Future?" *Geographical Magazine,* 35, no. 5 (September 1962), 243–252.

COLEMAN, ALICE, "The Use of Britain's Land," *Geographical Magazine,* 36, no. 12 (April 1964), 687–697. A discussion of the Second Land Use Survey of Britain.

COPPOCK, J. T., "Agricultural Changes in Britain," *Geography,* 49, pt. 3 (July 1964), 322–327; *An Agricultural Atlas of England and Wales* (London: Faber and Faber, 1964); "Crop, Livestock, and Enterprise Combinations in England and Wales," *Economic Geography,* 40, no. 1 (January 1964), 65–81; "Postwar Studies in the Geography of British Agriculture," *Geographical Review,* 54, no. 3 (July 1964), 409–426 (a review article); and "The Recreational Use of Land and Water in Rural Britain," *Tijdschrift voor Economische en Sociale Geografie,* 57, no. 3 (May–June 1966), 81–96.

COULL, JAMES R., "Modern Trends in Scottish Fisheries," *Scottish Geographical Magazine,* 84, no. 1 (April 1968), 15–28.

DARBY, HENRY CLIFFORD, "British National Parks," *Advancement of Science,* 20, no. 86 (November, 1963), 307–318; "National Parks in England and Wales," in Henry Jarrett, ed., *Comparisons in Resource Management* (Baltimore: Johns Hopkins Press for Resources for the Future; Lincoln: University of Nebraska Press, Bison Books, 1961), pp. 8–34; and "The Regional Geography of Thomas Hardy's Wessex," *Geographical Review,* 38, no. 4 (July 1948), 426–443.

ELLISON, W., "Agriculture and Changes in Land Use," *Advancement of Science,* 23, no. 112 (October 1966), 287–296.

GASSON, RUTH, "The Changing Location of Intensive Crops in England and Wales," *Geography,* 51, pt. 1 (January 1966), 16–28.

GREAT BRITAIN, CENTRAL OFFICE OF INFORMATION, REFERENCE DIVISION, *Agriculture in Britain* (London: H.M.S.O., 1965).

GRIGG, D. B., "An Index of Regional Change in English Farming," Institute of British Geographers, *Transactions No. 36* (June 1965), pp. 55–67; and "Small and Large Farms in England and Wales: Their Size and Distribution," *Geography,* 48, pt. 3 (July 1963), 268–279.

GROTEWOLD, ANDREAS, and MICHAEL D. SUBLETT, "The Effect of Import Restrictions on Land Use: The United Kingdom Compared with West Germany," *Economic Geography,* 43, no. 1 (January 1967), 64–70.

HART, JOHN FRASER, *The British Moorlands: A Problem in Land Utilization* (Athens: University of Georgia Press, 1955).

O'DELL, A. C., "Highlands and Islands Developments," *Scottish Geographical Magazine,* 82, no. 1 (January 1966), 8–16.

PATERSON, J., "New Life for the Highlands," *Geographical Magazine,* 38, no. 1 (May 1965), 20–33.

ROBINS, D. L. J., "The Emergence of British Land-use Planning," in J. B. Whittow and P. D. Wood, eds., *Essays in Geography for Austin Miller* (Reading, England: Reading University, 1965), pp. 306–316.

SHIRLAW, D. W. GILCHRIST, *An Agricultural Geography of Great Britain* (New York: Pergamon Press, 1966).

STAMP, L. DUDLEY, *The Land of Britain: Its Use and Misuse* (3d ed.; London: Longmans, Green and Co., in conjunction with Geographical Publications, 1962); and "Land Use in the Scottish Highlands, *Advancement of Science,* 31, no. 90 (July 1964), 141–190.

Ireland

BARRITT, DENIS P., and CHARLES F. CARTER, *The Northern Ireland Problem* (New York: Oxford University Press, 1962).

DOOLEY, WILLIAM E., "Changing Attitudes in Irish Agriculture," *East Lakes Geographer,* 1 (November 1964), 53–58.

DWYER, D. J., "The Peat Bogs of the Irish Republic: A Problem in Land Use," *Geographical Journal,* 128, pt. 2 (June 1962), 184–193; and "Electricity and the Industrial Development of the Irish Republic," *Geography,* 45, pts. 1–2 (January–April 1960), 116–119.

EVANS, E. ESTYN, *Mourne Country: Landscape and Life in South Down* (rev. ed.; Dundalk, Ireland: Dundalgan, 1967).

FREEMAN, T. W., *Ireland: A General and Regional Geography* (3d. ed.; New York: E. P. Dutton & Co., 1965).

Irish Geography (one issue a year).

JOHNSON, JAMES H., "The Political Distinctiveness of Northern Ireland," *Geographical Review,* 52, no. 1 (January 1962), 78–91.

McCARTHY, JOE, and the Editors of *Life, Ireland* (Life World Library; New York: Time, Inc., 1964).

McGOVERN, P. D., "Northern Ireland: People with a Purpose," *Geographical Magazine,* 38, no. 7 (November 1965), 505–520.

SALAMAN, REDCLIFFE N., *The History and Social Influence of the Potato* (Cambridge: At the University Press, 1949). Discusses the causes of the nineteenth-century Irish emigration.

SOULSBY, J. A., "The Shannon Free Airport Scheme: A New Approach to Industrial Development," *Scottish Geographical Magazine,* 81, no. 2 (September 1965), 104–114.

VAN STEEN, MARCUS, "Northern Ireland Plans for Economic Growth," *Canadian Geographical Journal,* 69, no. 4 (October 1964), 130–139.

WOODHAM-SMITH, CECIL, *The Great Hunger: Ireland 1845–1849* (New York: Harper & Row, 1963). A graphic account of the Irish famine of the 1840s, in the perspective of general relationships between Ireland and England.

See also the lists of references and readings for Chapters 1–4, particularly the relevant chapters in the geographies of Europe listed on page 92.

France

6

The west central portions of the European mainland across the English Channel and North Sea from the British Isles are occupied by a group of highly developed industrial nations which have long had close relations with each other. Germany[1] and France are the largest nations in the group and the only ones which have ranked as great powers in the twentieth century. Four of the remaining countries—the Netherlands, Belgium, Luxembourg, and Switzerland—are found in a historic buffer zone separating the two major nations. Austria, though it is not in the buffer zone, may be included in the group because of its close proximity to both West and East Germany, its extensive trade with West Germany, and its close relationships with Germany both culturally and historically. Monaco and Andorra, adjoining France, and Liechtenstein, between Switzerland and Austria, are insignificant microstates with a

---

[1] In this chapter the name "Germany" will be applied to the German nation prior to the end of World War II, or to the two German republics of today taken together, whereas "West Germany" (the German Federal Republic) and "East Germany" (the German Democratic Republic) will be applied to the political units into which Germany was separated after World War II (see pp. 165–166).

Cities, industrial concentrations, waterways, and highlands in West Central Europe and adjoining areas. Most sections of the internal waterways shown are navigable for barge traffic. The "major industrial concentrations" are relatively continuous industrialized areas. Most of the industrial concentrations shown contain large deposits of bituminous coal or lignite, the exceptions are the Middle Rhine and the Po Basin. Note the clustering of industrial concentrations in border zones between highlands and plains. The highlands and rough uplands (in the darker color) offer some hindrance to transportation and are sufficiently steep, rugged, or infertile to discourage dense agricultural settlement.

quasi-independent status; only proximity entitles them to be numbered with the countries of this group.

Topographic variety, economic interdependence, historical antagonism, and burgeoning cooperation are key phrases describing the countries of West Central Europe as a group. In physical terms the area they occupy may be thought of as consisting of three concentric and widely different arcs of land (maps, p. 78 and above): an outer arc of lowland plains bordering the Bay of Biscay, English Channel, and North and Baltic seas;

a central arc of hills, low mountains, and small plains; and an inner arc consisting of the high Alps mountains, with another high range, the Pyrenees, offset to the west. France and Germany include portions of all three arcs, the Netherlands and northern Belgium lie within the outer arc of plains, southern Belgium and Luxembourg are in the central hilly zone, and Switzerland and Austria lie principally within the inner arc of Alpine mountains, though their main populated areas are in the southern fringes of the central arc.

The largest deposits of coal and iron ore in Europe (map, below) provide basic resources for an industrial development of the first magnitude. A series of great industrial concentrations stretches across this area from northern France to southwestern Poland, following the main axis of coal deposits (map, left). Other industrialized areas exist by the score. Coal and iron ore furnish an important basis for trade and cooperation among the countries of the group, most of which are deficient in one or both of these vital commodities. For example, West Germany exports coal and coke to France, while France, in turn, supplies much of the iron ore

for the iron and steel mills of West Germany's Saar district. Many other bases for trade exist, and all of the countries concerned have close economic relations with other countries of the group.

Political relations have been troubled, however, by antagonisms of long standing. The countries of this group have fought each other repeatedly in past wars. In the twentieth century the principal antagonists have been Germany and France, but most of the other countries have been drawn into their contention and strife. Today France and West Germany are making a strong effort to cooperate with each other, and they have joined with Belgium, the Netherlands, Luxembourg, and Italy in forming a number of important new supranational economic "communities" (pp. 68 and 147). These include the European Economic Community or Common Market, the European Coal and Steel Community, and the European Atomic Energy Community (Euratom). Meanwhile Switzerland and Austria have become linked with the United Kingdom, three Scandinavian countries, and Portugal in yet another economic grouping, the European Free Trade Association (p. 69). But both

Major coal fields of West Central Europe. Note the line of fields stretching across western Germany, the Low Countries, and northeastern France. Stars show national capitals. (After a map by the U.S. Geological Survey.)

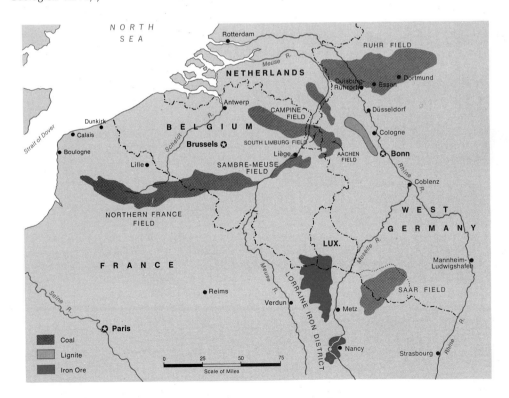

Switzerland and Austria remain closely connected in a multitude of ways with their neighbors in West Central Europe and enjoy generally harmonious relations with them.

THE INTERNATIONAL IMPORTANCE OF FRANCE

Of the major countries in West Central Europe, West Germany is the largest in population, with an estimated 60.3 million people in 1968 (including 2.2 million in West Berlin), as compared with an estimated 50.4 million for France. France, however, is much larger in area than West Germany: approximately 211,000 square miles (including the island of Corsica), as against 96,000 square miles for West Germany. In fact, excluding the Soviet Union, France is the largest European country in area.

The importance of France, however, does not rest primarily on its size, though its population of over 50 million is an element of very considerable significance. It is one of the world's more important industrial and trading nations and is an atomic power by virtue of independently manufacturing its own atomic weapons—one of only five powers to have done so (1968). In cultural prestige France stands extremely high among the nations of the world. Still other factors that add to the importance of France are its strategic geographical position in Europe and its special relationships with its overseas departments, dependencies, and former colonial possessions. France's industrial and commercial importance is discussed at some length in the latter part of this chapter. In the following paragraphs the country's cultural achievements, its strategic geographical position, and its relationships with overseas affiliates are briefly surveyed.

Achievements of French Culture

France was conquered for Rome by Julius Caesar in the first century B.C. and thereafter became thoroughly Romanized. It has thus been a highly civilized land longer than any other European country except the peninsular countries of Southern Europe. During the Middle Ages, Paris, the French capital, became the main center of Roman Catholic scholarship and culture. In modern times the French have made eminent contributions in all fields of scholarship and art, and Paris has generally been regarded as the world's foremost artistic and cultural center. Such names as Pasteur and Curie in science; Descartes, Pascal, Voltaire, Rousseau, and Bergson in philosophy; Molière, Racine, Hugo, Balzac, Zola, and Proust in literature; Matisse and Gauguin in art come readily to mind as evidence of France's contributions to civilization. Even to make such a short list is almost to falsify by selection. In addition, many illustrious foreigners have drawn their inspiration from and pursued their studies in France. The position of French as an international language, particularly as a major language of diplomacy, and the continued attraction of Paris for the world's artists, students, and tourists, as well as for important international conferences, are further evidences of the cultural prestige of France.

Strategic Importance of the French Beachhead

During the twentieth century France's strategic position has been that of a beachhead. In two world wars a major path to the conquest of Germany has lain through France. Location at the western end of Europe (except for Spain and Portugal, which have been militarily weak and relatively isolated behind the Pyrenees), sea frontage on three sides, proximity to Great Britain, and good communications with the continent to the east have combined to make France a critical area in the wars and hence a critical area in the politics of the twentieth century.

Special Overseas Relationships: The Legacy of Empire

The world significance of France has been heightened by its importance as an imperial and colonial power. Before it began to dissolve in 1954 the French overseas empire totaled about 4½ million square miles (21 times the size of France itself) and had a population of over 70 million—a colonial aggregate second only to that of Great Britain in population and second to none in area. Most of this empire was in Africa, though there were some important possessions elsewhere (map, right). Actual settlement overseas by Frenchmen was seldom on a large scale—the principal exception being Algeria—but political control was often accompanied by a strong penetration of French culture. Today many of the leading figures in France's former possessions speak French, have received the same type of education that Frenchmen receive, often including higher education in

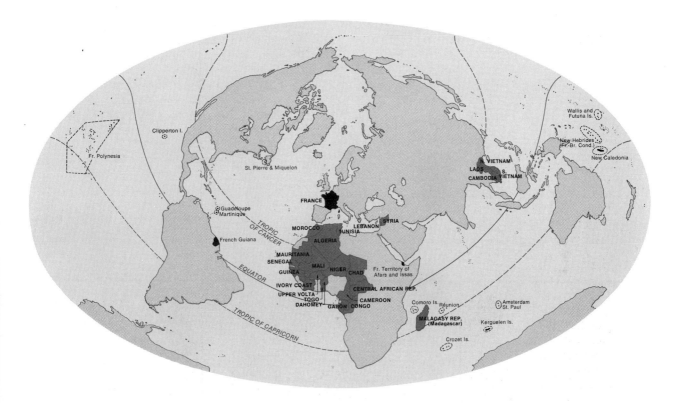

France and its overseas departments and territories (shown in black) as of December 1, 1968, and former French possessions that have received independence since 1940 (shown in blue). Guadeloupe, Martinique, French Guiana, and Réunion are overseas departments. St. Pierre and Miquelon, Comoro Islands, New Caledonia, French Polynesia, Wallis and Futuna Islands, and the French Territory of Afars and Issas are overseas territories. "French Southern and Antarctic Lands" (not named on map) is an overseas territory comprised of islands between the Tropic of Capricorn and Antarctica, plus France's claims in Antarctica. A few tiny coastal areas that were ceded to India by France are not shown. (Briesemeister Elliptical Equal Area Projection, courtesy of the American Geographical Society.)

French universities, and have had political or military training and experience in the service of the French government. In the Americas, areas of French-descended population and/or culture exist as survivals from an earlier empire that was lost before most of the more recent possessions were acquired.

France's colonial empire as such survives now in only a minor way, having practically disappeared in a few tumultuous years (map, p. 65). In 1954 military defeat by a Communist-led rebellion resulted in loss of control over Indochina, the major east Asian part of the empire; and in the same year the last French enclaves on the coast of India were ceded to that country. France's defeat in Indochina was a spur to nationalist movements, already strong, in her North African possessions—Morocco, Algeria, and Tunisia—and in 1956 independence was yielded to Morocco and Tunisia. Then in 1958 the units of the empire classed as "overseas territories"[2] were allowed to vote on the question of their continued political association with France.

[2] In French West Africa the territories of Senegal, French Sudan, Mauritania, French Guinea, the Ivory Coast, Dahomey, Upper Volta, and Niger; in French Equatorial Africa the territories of Gabon, Middle Congo, Ubangi-Shari, and Chad; in or bordering the Indian Ocean the territories of Madagascar, the Comoro Islands, and French Somaliland; in the Pacific Ocean the territories of New Caledonia and French Polynesia; in the North Atlantic Ocean off Newfoundland, the territory of St. Pierre and Miquelon.

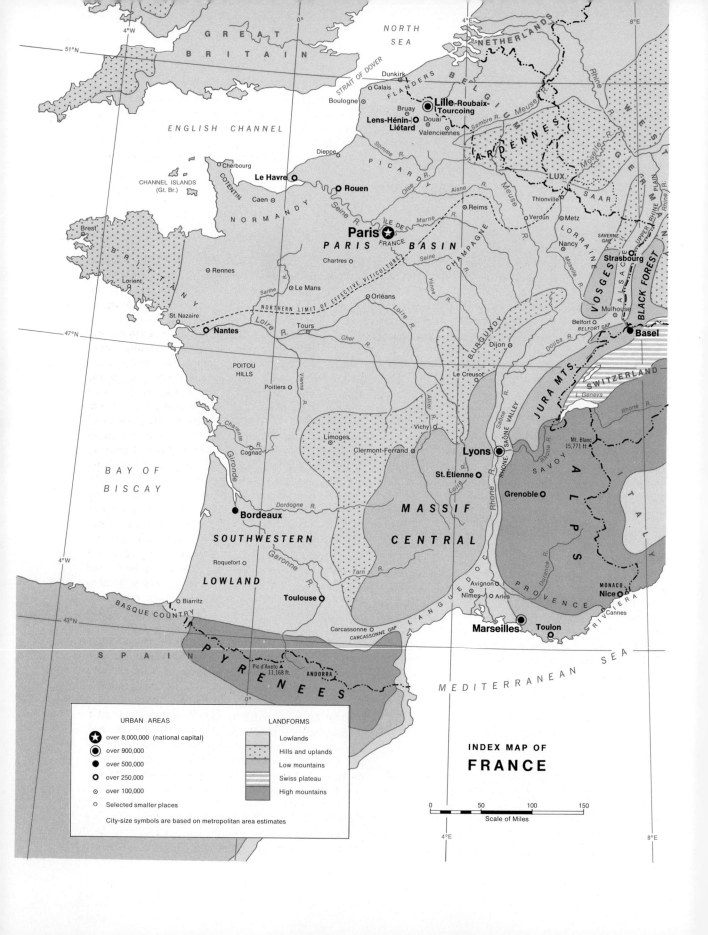

GREAT BRITAIN

NORTH SEA

51°N

0°

4°W

8°E

NETHERLANDS

ENGLISH CHANNEL

STRAIT OF DOVER

Dunkirk
Calais
Boulogne

BELGIUM
FLANDERS
Bruay
Lens-Hénin-
Liétard
Douai
Valenciennes

Lille-Roubaix-
Tourcoing

ARDENNES

Sambre R.
Meuse R.

LUX.

Rhine R.

SAAR

GERMANY

CHANNEL ISLANDS
(Gt. Br.)

Cherbourg
COTENTIN
Caen

Le Havre
Dieppe
Rouen

PICARDY
Somme R.
Oise R.

NORMANDY
Seine R.

Aisne

Reims
Marne
CHAMPAGNE

Thionville
Verdun
Metz

Moselle R.

LORRAINE

Nancy

SAVERNE GAP

Strasbourg

VOSGES
ALSACE
BLACK FOREST

UPPER RHINE PLAIN
Rhine R.

Brest

BRITTANY

Rennes

Lorient
St. Nazaire

Nantes

Paris
ÎLE DE FRANCE
PARIS BASIN

Chartres

Seine
Sarthe
Le Mans
Loire R.
Orléans
Loire R.
Yonne

Burgundy
Dijon
Le Creusot

BURGUNDY

Belfort
BELFORT GAP

Mulhouse

Basel

SWITZERLAND

47°N

Tours
Cher R.

POITOU HILLS

Poitiers
Vienne R.

Limoges

Clermont-Ferrand

Allier R.

Vichy

Saône R.
SAÔNE VALLEY

Lyons

Loire R.
St. Étienne

Grenoble

Rhône R.
SAVOY

Doubs R.

JURA MTS.

L. Geneva

Rhone R.

Mt. Blanc
15,771 ft.

ALPS

ITALY

BAY OF BISCAY

Charente R.
Cognac

Bordeaux

SOUTHWESTERN

Dordogne R.

MASSIF CENTRAL

Roquefort

LOWLAND

Garonne R.

Tarn R.

Toulouse

Biarritz

BASQUE COUNTRY

SPAIN

PYRENEES

Pic d'Aneto 11,168 ft.

ANDORRA

Carcassonne
CARCASSONNE GAP

LANGUEDOC

Avignon
Nîmes
Arles

PROVENCE

Durance R.

MONACO
Nice
Cannes
RIVIERA

Marseilles
Toulon

MEDITERRANEAN SEA

NORTHERN LIMIT OF EFFECTIVE VITICULTURE

4°W

43°N

0°

4°E

8°E

URBAN AREAS

⬤★ over 8,000,000 (national capital)
◉ over 900,000
● over 500,000
◯ over 250,000
⊙ over 100,000
○ Selected smaller places

City-size symbols are based on metropolitan area estimates

LANDFORMS

Lowlands
Hills and uplands
Low mountains
Swiss plateau
High mountains

INDEX MAP OF
FRANCE

0   50   100   150
Scale of Miles

The African territory of French Guinea voted against further association and was granted immediate independence, but twelve units in Africa voted to remain associated with France as semi-independent republics (the remaining overseas territories voted to retain that status). In 1960 full independence was granted to the twelve associated republics and to France's African trust territories of French Cameroons and French Togoland as well (see map, p. 141 and, for further discussion, pp. 575–582).

While the foregoing events were occurring, another important French possession, Algeria, was racked by bitter warfare between nationalist rebels and the bulk of the French army. The Algerian situation involved problems that were more difficult to solve than those of most colonial areas, mainly because of the fact that large numbers of Frenchmen had settled in Algeria and had become a favored and powerful minority in the country's predominantly Moslem population. It was complicated also by the discovery, in the late 1950s, of large deposits of commercially exploitable oil and natural gas in the Algerian Sahara. In 1962 Algeria became independent. An almost immediate result was the return to France of more than 850,000 French citizens who had been living in Algeria, some of them for generations. An important sociological tie with Algeria thus was severed—or nearly so. Economically, however, Algeria has remained oriented to France.

Despite the loss of political jurisdiction, France continues to maintain special relationships, closer and more formalized in some cases than in others, with most of its former possessions. They receive economic, technical, and military aid, and have special trading privileges in the French market. Many of them have been accorded associate membership in the European Common Market (see p. 69). Meanwhile France continues its affiliation with a variety of small overseas departments and overseas territories in widely separated parts of the world (map, p. 141).

TOPOGRAPHY AND FRONTIERS

France was once a very great power. In the seventeenth and eighteenth centuries its power was a primary factor in European politics, and in the early nineteenth century, under Napoleon I, it came near to subjugating the entire continent before it was checked by a continental coalition in alliance with Great Britain. France's power was based on the fact that the French kings unified a large and populous national territory at a relatively early date. Before Germany and Italy achieved political unification in the later nineteenth century, France was a larger country relative to its European neighbors than it is today.

Physical and Cultural Unity of the French Hexagon

Modern France lies mostly within an irregular hexagon framed on five sides by seas and mountains (map, left): in the south the Mediterranean Sea and Pyrenees Mountains form two sides of the hexagon; in the west and northwest the Bay of Biscay and the English Channel form two sides; in the southeast and east a fifth side is formed by the Alps and Jura mountains and, farther north, the Vosges Mountains with the Rhine River a short distance to the east. Part of the sixth or northeastern side of the hexagon is formed by the low Ardennes Upland, which lies mostly in Belgium and Luxembourg; but broad lowland passageways lead into France from West Germany both north and south of the Ardennes. Aside from the large Massif Central in south central France, all of the country's major highlands are peripheral. Even the Massif Central, a hilly upland with low mountains in the east, is skirted by lowland corridors which connect the extensive plains of the north and west with the Mediterranean coast. These corridors include the Rhone-Saône Valley between the Massif to the west and the Alps and Jura to the east; and the Carcassonne Gap between the Massif and the Pyrenees. Thus France is a country whose frontiers coincide, except in the northeast, with seas, mountains, or the Rhine, and which has no serious internal barriers to movement.

At a very early time the French hexagon began to form in some degree a recognizable political unit. It entered history as Roman Gaul, conquered mostly by Julius Caesar in the first century B.C. The Romans found a degree of cultural unity already established within it among the Gallic tribes. The Rhine was the frontier between these tribes and the Germans to the east, and, failing to conquer the Germans, the Romans established the Rhine as the frontier of the Empire. Five hundred years of Roman occupation and Latinization increased the unity of Gaul, and also increased the distinction between it and unoccupied Germany. During the Middle Ages the land within the hexagon was politically fragmented, like most of Europe, but Germanic invaders were absorbed and a considerable degree of unity was maintained in a culture now becoming French. An exception was the area nearest the Rhine, where Germanic languages and culture became established for some distance west of the river. In the later Middle Ages the physical character and underlying cultural unity of the territory within the hexagon were important factors aiding the French kings in their successful attempt to build a unified and centralized state extending over a broad area. This state emerged in the fifteenth, sixteenth, and seventeenth centuries, and attained in general the boundaries of ancient Gaul, except near the Rhine.

Mountain Frontiers of the South and East

The state which emerged as modern France was to a considerable degree a natural fortress, as two of its three land frontiers lay in rugged mountain areas. The Pyrenees, along the border with Spain, are a formidable barrier with two peaks over 11,000 feet and only one pass lower than 5000 feet. Even today only a handful of highways and railroads connect France and Spain. The main routes skirt the ends of the range. Most roads through the mountains are blocked by snow for a considerable period in winter. The microstate of Andorra lies on one of these routes in the eastern Pyrenees. In general, the lowest elevations in the Pyrenees are found in the west near the Bay of Biscay. Here the small linguistic group called the Basques—neither French nor Spanish in language—occupies an area extending into both France and Spain. Though mountain pastures in the Pyrenees are utilized by livestock graziers from both sides of the range, these mountains form a fairly effective zone of separation between the French and Spanish populations.

In southeastern France, the Alps and Jura mountains are followed by France's boundaries with Italy and Switzerland. The sparsely populated Alpine frontier between France and Italy is even higher than the Pyrenees: Mont Blanc, the highest mountain in Europe, reaches 15,771 feet, and several other summits exceed 12,000 feet. The mountains are cut by several deep passes which have long been important as routes; and a new highway tunnel beneath the northern shoulder of Mont Blanc has provided an important new connection between France and Italy since it was opened in 1963. But the number of important routes through the mountains is limited, and the Alps have been an important defensive rampart throughout France's history. The Jura Mountains, on the French-Swiss frontier, are a very different kind of range, but also difficult to cross. Their rounded crests generally reach only 3000 to 5500 feet, but are arranged in a series of long ridges separated by deep valleys. Easy passes through the ridges are comparatively few.

Territorial Components of France's Northeastern Frontier

France's northeastern frontier, extending from Switzerland to the North Sea, is less defensible than the mountain frontiers in the south, and, unlike them, extends for long distances through well-populated areas where different cultures—the French and the Germanic—meet and interlock. It was France's misfortune that a stronger industrial and military power coalesced across this frontier during the nineteenth century. Though France's eventual decline from its position as a truly great power was probably inevitable when the United States and the Soviet Union began to emerge as giant, effectively organized powers, it has been most closely and directly associated with the rise of Germany.

Belgium and Luxembourg, although they lie between France and Germany, have not effectively separated them in times of conflict. Hence the French-German frontier may be thought of as a zone reaching to the North Sea and including these smaller countries. The border zone comprises six well-defined physical sections: the Belfort Gap, the Upper Rhine Plain, the Vosges Mountains, Lorraine, the Ardennes Upland, and the northern plains.

1. THE BELFORT GAP OR GATE (also known as the Burgundian Gate). This is a lowland corridor between the northern end of the Jura Mountains and the southern end of the Vosges Mountains. It forms an easy, though narrow, passageway between the Rhone-Saône Valley and interior France to the west and the Rhine Valley and West Germany to the east.

2. THE UPPER RHINE PLAIN. From the Rhine port of Basel, in Switzerland, to a point north of the French Rhine port of Strasbourg, the boundary between France and West Germany follows the Rhine River. Here the river flows through a level-floored alluvial valley formed originally by the foundering of a section of the earth's crust. The valley, known technically as a rift valley or *graben,* lies between ranges of low mountains which are rather similar in physical character: the Vosges Mountains in France and the Black Forest in West Germany. These mountains are remnants of a continuous structure which existed before the central section slipped downward to form the rift valley. The floor of the valley is often called the Upper Rhine Plain. The French section

of the plain, about 10 to 20 miles wide, lies in the old province of Alsace.[3] The plain is outside France's strongest defense lines, and has long been an important pawn, and prize, in French-German struggles.

3. THE VOSGES MOUNTAINS. For about 70 miles this low, wooded range overlooks the flat Upper Rhine Plain. Lower uplands continue northward into West Germany beyond the Saverne Gap at the north end of the mountains proper. Although the highest summit is below 4700 feet, the Vosges range presents a steep face to the east and in the past has constituted a military barrier of considerable value.

4. LORRAINE. Between the Vosges Mountains and the Ardennes Upland, a rolling lowland connects France and West Germany across the old French province of Lorraine. It is sometimes called the Lorraine Gate. To the west it opens into France's largest lowland, which is centered on Paris and is often referred to as the Paris Basin. Between the West German border and Paris, the terrain is interrupted at intervals by long escarpments facing eastward. These escarpments, formed by the outcropping of resistant layers among the sedimentary rocks which underlie the Paris Basin, have proved useful military barriers in past wars when properly fortified. But the lowland gateway of Lorraine has nonetheless been an easier military route into France than the upland to the north or the mountains to the south.

5. THE ARDENNES UPLAND. This low plateau centers in southern Belgium and extends into adjacent Luxembourg and West Germany, and for a very short distance into France. It is one of many remnants in western and southern Europe of ancient mountains often called the Hercynian mountains. Over a vast span of geologic time, they were planed off by erosion, and then the remnants were submerged for long ages beneath the sea, where great thicknesses of sediments accumulated and hardened into rock. Disturbances in the earth's crust uplifted some sections, and they appear today as comparatively low, discontinuous uplands or highlands. Long-continued erosion has removed much of the sedimentary covering, and the original igneous or metamorphic rocks underlie the surface in many areas. Some areas, such as parts of the Massif Central of France, show evidence of comparatively recent volcanic activity. Around the margins of these old highlands are found sedimentary formations that contain Europe's principal coal deposits.

Besides the Ardennes and the Massif Central, the Hercynian remnants include the uplands of Brittany, Cornwall, southern Wales, and southern Ireland; the Pennines of England; the Vosges Mountains and the Black Forest; the Rhine Uplands of West Germany; the uplands and highlands of central Germany (West and East) and western Czechoslovakia; and sizable areas in the Iberian and Balkan peninsulas.

Stream valleys deeply entrenched below a rolling, forested surface make the Ardennes a considerable barrier to transportation, especially to east-west transportation, as the major valleys trend north and south. This area has generally been considered unfavorable terrain for military movements, although Germany launched two major offensives across it in World War II.

6. THE NORTHERN PLAINS. For about 80 miles, from the northern edge of the Ardennes to the North Sea, the French-Belgian frontier runs through low plains, which are a segment of the North European Plain and thus are continuous across Belgium and the Netherlands into northern Germany. In both world wars the plains of Belgium, forming a broad corridor equipped with excellent transportation facilities, were an important avenue of invasion from Germany into France. The seaward portion of these plains is the region of Flanders (map, p. 142), famous for the bitterly contested battles fought there between the German armies and the British and French in World War I.

Major French-German Conflicts and Their Consequences

Thus the Franco-German border zone consists of three lowland passageways into France—the northern plains, Lorraine, and the Belfort Gap—separated by two upland barriers, the Ardennes and the Vosges, plus Alsace lying outside one of the barriers. This border zone is the arena in which France and Germany fought three major conflicts between 1870 and 1945. These conflicts were disastrous for France and, while hardly the basic causes of France's difficulties and relative decline, were certainly major milestones along the path.

The first conflict was the Franco-Prussian War of 1870–1871. In this war German forces broke into France through Lorraine, destroying French armies around Metz

---

[3] At the time of the French Revolution, France's ancient provinces were abolished as administrative territorial units, and the present structure of small, less autonomous units called "departments" was instituted. But many of the provincial names from pre-Revolutionary France (Alsace, Lorraine, Burgundy, Champagne, Brittany, Normandy, and so on) are still well-recognized, significant, and useful regional designations.

and farther north along the edge of the Ardennes. Paris was invested and France was compelled to surrender. This stunning German victory demonstrated that France was no longer the dominant military power in continental Europe; and one of its immediate outcomes, the establishment of a unified German state under Prussian leadership, augured ill for France's future.

At the end of the Franco-Prussian War, France was compelled to cede to Germany Alsace and northeastern Lorraine, where a frontier was established which left Nancy and Verdun in France but placed Metz and Strasbourg in Germany. Germany's claim to the ceded areas was based on the fact that they were, as they still are, predominantly German in language, having been acquired piecemeal by France in the seventeenth and eighteenth centuries. But there is considerable evidence that the majority of their population would have preferred to remain in France, and Germany refused to allow a plebiscite to determine their preference. Their annexation gave Germany strategic military advantages vis-à-vis France, and also gave it control of very considerable mineral resources. These included major potash deposits in southern Alsace and salt deposits in Lorraine. But as it eventually turned out, the most valuable resource was the iron ore of Lorraine. The Lorraine deposits, extending from the vicinity of Nancy northward into southern Belgium and Luxembourg (map, p. 139) are the greatest source of iron ore in Europe. Control of them was probably not a major motive for annexation at the time, as they are low-grade ores with a high phosphorus content and hence were not highly valuable in 1871. (Phosphoric iron ores produced steel which broke easily under pressure or stress.) But in 1878 two British inventors, Thomas and Gilchrist, introduced a process to make good steel from such ores by lining the steel furnaces with refractory bricks made from crushed limestone to neutralize the phosphorus. Thus Germany found itself in possession of a most valuable resource, while only a minor part of the deposits were left in adjacent French Lorraine.

The second Franco-German conflict was the World War of 1914–1918. In 1914 German armies invaded France by the northern plains route. In a month Belgium was largely overrun, and northeastern France was penetrated to a point about 15 miles from Paris, but there the Germans were checked in the Battle of the Marne and fell back some miles. Four years of trench warfare followed, fought along various lines in northeastern France. During most of this time about a tenth of France was in German hands, and when the German armies were finally driven out in 1918, they added systematic destruction to the havoc wrought in this area

by the fighting itself. Thus large and important parts of France were devastated by 1918. But even this disaster was overshadowed by the enormous toll of human lives. The French armies had borne the main brunt of the Allied effort on the western front, and no less than 1.3 million men had died in service; close to half a million had perished in one battle—the six-month defense of Verdun in 1916. These losses amounted to over 3 percent of France's *total* population, and compared with a loss of just over 100,000 for the United States—a country with more than twice the population of France.

As a result of her "victory," France regained Alsace and the portion of Lorraine which had been lost in 1871. She also acquired control over the economy of the German industrial district called the Saar. This area, an important producer of coal and steel, lies just east of Lorraine and is closely tied to it economically. The Saar was not ceded to France, but was placed under an international political administration and attached to France economically, with the provision that the Saarlanders should eventually decide their own status in a plebiscite. When this plebiscite was held in 1935, the population voted overwhelmingly to be reunited with Germany.

The third French-German conflict came as part of World War II (1939–1945). Again France was invaded, in 1940, by the northern plains route. This time the Netherlands as well as Belgium was engulfed in the German assault. But the movement through Belgium proved to be only a powerful feint, and when large French and British forces had been drawn into northern Belgium to meet it, the main German thrust was delivered across the Ardennes Upland. This section had always been considered too difficult for major military operations, and was only lightly defended. But it proved to be no longer a sufficient barrier in a new age of mechanized and motorized warfare. German armies broke across and out of the Ardennes, and then turned north to reach the English Channel, thus trapping the Allied armies in the north in a huge pocket against the sea. These trapped armies were destroyed or captured, except for some 338,000 men who were evacuated to Britain through the Channel port of Dunkirk. The Germans were then able to brush aside the remaining French forces as they rapidly overran much of the country, sending it down to stunning defeat.

Four years later, in 1944, liberation from German occupation again made France a battlefield. After intensive preparatory bombing, which reached far into France, Allied armies landed in Normandy and swept the German forces back across northern France. Their advance was stubbornly contested, especially in Normandy, and many French communities were reduced to ruins, although Paris was spared. A secondary invasion force landed on the Mediterranean coast and moved into Germany via the Rhone-Saône Valley and the Belfort Gap. Liberated France emerged from the war with most of its transportation equipment and one-eighth of its housing destroyed, with wheat production at well under half of the prewar level, with most of its machine tools removed to Germany, and industrial production at about one-fifth of the prewar output. About 600,000 men had died in service, compared to about 400,000 for the United States with a population over three times that of France.

The New Era of Franco-German Cooperation

Thus for a century much of Europe's destructive military history has been enacted in the French-German frontier zone, and especially on the French side of the border. But since World War II this old zone of conflict has become the scene of constructive efforts to substitute peaceful cooperation for strife and antagonism. France, West Germany, Belgium, the Netherlands, and Luxembourg were leaders in the movement to create the deliberative assembly called the Council of Europe; and, with Italy, they have moved toward unification of their economies within the European Coal and Steel Community, the European Economic Community or Common Market, and the European Atomic Energy Community. These developments have been surveyed in some detail in Chapter 4. Most of the mines and factories within the jurisdiction of the Coal and Steel Community are within or near the old border zone of French-German conflict: in the German Ruhr and Saar; in Lorraine and northeastern France; and in central Belgium, Luxembourg, and the southern panhandle of the Netherlands. In a wider arena, the Organization for Economic Cooperation and Development (OECD) to which the six nations belong, continues to function actively (see p. 68).

Further results of French-German cooperation are to be seen in the amicable settlement of a dispute concerning the Saar which arose after World War II, and, as a part of the Saar settlement, in an agreement between the two nations to canalize the Moselle River. Following the collapse of Germany in World War II, control over the Saar's economy passed once more to France, and the area was given a quasi-independent political status. The resulting dispute between France and West Germany was ended, however, in 1957 by return of the Saar to West Germany following a plebiscite in which the Saar's population voted, as it had in 1935, to be reunited with the German state. Agreements concerning the Saar that were signed between France and West Germany in 1956 guaranteed France's right to purchase one-third of the Saar's annual coal production, and West Germany agreed to assist France in canalizing the Moselle River. This stream, which flows through the Lorraine iron-mining and steel-milling district and connects it with the Rhine (map, p. 139), has long been canalized for barge traffic in Lorraine itself, but from Lorraine to the Rhine the Moselle was practically useless for through navigation by large barges due to frequent rapids and stretches of shallow water. Canalization of the lower river, from Metz to Coblenz, was completed in 1964, providing Lorraine with good water transport to German industrial areas along and beyond the Rhine River. The necessary engineering involved the construction of 13 dams, which not only deepened the Moselle, but made possible the generation of large amounts of power at 10 hydroelectric stations attached to the dams.

CHARACTERISTICS OF FRANCE'S POPULATION

Until recently, the rate of French population growth was comparatively slow. Indeed, some authorities have held that this slow growth was partially responsible, during the century preceding World War II, for France's relative decline in military power and security. However, the rate of increase in the French population is now approximately 1 percent per year, which is still well below the world average of about 2 percent per year, but is slightly above the average in western Europe. Among the reasons for this increase have been government policies encouraging larger families. These policies have included bonuses for the birth of children, plus family allowances that provide partial support up to stipulated ages.

Another reason for the recent growth of France's population has been policies regarding immigration, whether from the European Economic Community, the overseas areas affiliated with France, or elsewhere. Absorption of refugees from Algeria increased the French population during and after the Algerian crisis. A key provision of the European Economic Community allows citizens of any of the six member nations to migrate to other member countries without penalty. The economic opportunities in France are numerous, and workers from other Common Market countries are attracted. People are also attracted, to a degree, from European countries outside the EEC and the overseas areas affiliated with France. By and large, French policies on immigration encourage movement of newcomers to France, but maintain a rather strict control by the granting of work permits. Thus a stonemason may be able to move freely to France, but he will not be able to obtain a work permit in his trade if there is a surplus of stonemasons. Such policies regarding immigration have resulted in a substantial in-migration of workers. Approximately 4 percent of the total population of France is now comprised of foreigners (1968), excluding tourists. In a recent year, nearly 45 percent of all population increase in the country was due to immigration.

Another important aspect of France's population involves the urban-rural ratio. By definition, France considers as urban all communities with 2000 or more inhabitants in the chief town. In 1946, 53.2 percent of the French population was urban according to this definition. But by the late 1960s nearly two-thirds of the population was classed as urban. The process of urbanization is obviously moving forward in France, although much of the country still retains a rural flavor in comparison with other highly industrialized and urbanized nations of western Europe and North America.

The French economy is trending toward increased concentration of employment in tertiary activities (trade, services, the professions, and the like), a plateau in secondary activities (manufacturing), and a definite decline in the primary activities (agriculture, fishing, mining, lumbering, and so on.). Estimates for the late 1960s (projections from the census of 1962) allocated nearly half of the total labor force to tertiary employment, approximately a fourth to manufacturing, a fifth to the primary activities, and the remainder to construction.

In this summary of some basic characteristics of France's population, we may note that 90 percent or more of the population is at least nominally Roman Catholic in religion. However, much of France's population has been termed "anticlerical" or even non-religious. Some writers have asserted that the weakening of religious ties has had much to do with the birth control measures credited with helping to slow the rate of population growth in France until recently.

FRENCH AGRICULTURE AND FISHERIES

France emerged from the French Revolution as a nation of small peasant farmers, most of whom owned their farms. The proverbial industry and thrift of the French peasant have been a great asset to the country, but his often narrow and unreasoning individualism and conservatism have been on occasion a provoking source of national weakness. Although agriculture is no longer the leading source of employment in France, it is still a very important activity there, employing nearly a fifth of the French labor force and contributing very significantly to the country's economy and social character.

Agricultural Advantages and Production

Compared to most European countries France has distinct agricultural advantages. Much of the country is composed of plains with soils that are at least reasonably fertile. Most of France lies within the mild and moist marine west coast type of climate, but the country lacks the large areas of excessive precipitation that restrict cultivation in the British Isles. Southern France's strip of mediterranean climate affords environmental variety and makes it possible to produce subtropical crops, such as early vegetables and rice, that are not duplicated in other countries facing the North Sea.

The ratio between population and good agricultural land is relatively favorable in France compared to most of western Europe. Over one-third of France's area is generally used for production of crops, a higher proportion than in any other country along the western seaboard of Europe except Denmark. As France's population density is relatively low for a west European country, its supply of arable land per capita is also greater than in any of the other highly developed North Sea countries except Denmark. The consequence is that France is able to approach self-sufficiency in food. Food imports are nearly balanced in value by food exports, mainly of cereals, animal products, and wine. Although there is regularly some net import of foods, France contrasts strongly in this respect with countries as heavily dependent on food imports as the United Kingdom and West Germany.

Considering the comparatively small area of the country (it is smaller by a fifth than the state of Texas), France's agricultural output is marked by surprising variety and volume. Wheat and other cereals, potatoes, sugar beets, and grapes can be distinguished as the crops receiving special emphasis, although many others are grown and are important in various localities. Wheat is the most important crop and is grown in every part of the country, occupying one-fifth of the total cultivated acreage. It is especially prominent, however, on the best lands of France, which occur in a belt extending from French Flanders to districts lying south of Paris. This

A street in an Alsatian agricultural village. Farm villages in different parts of France exhibit varied architectural styles; this one, located in the low uplands of northern Alsace between the Rhine-Marne Canal and the High Vosges, is closer in appearance to villages in nearby sections of the West German Rhineland than to those in most parts of France. The prosperous villages in this area, known as the Kochersberg Plateau, concentrate on the growing of hops, grains (especially barley for brewing), and tobacco on fertile loess deposits. Holdings are small family units but are often worked on a cooperative basis. Farmsteads are located in the villages; beyond the tractor on the right is a gateway leading to the large square courtyard enclosed by a farmhouse and other farm buildings. (J. A. Soulsby.)

area, located mainly within the Paris Basin, has exceptionally fertile soils, formed largely from limestone and loess (see p. 31). Large acreages and unusually high yields of wheat make the region the principal "breadbasket" of France. France is ordinarily the world's fifth-ranking wheat producer, after the Soviet Union, the United States, China, and Canada. In terms of acreage, wheat is followed by barley, oats, and corn (maize). Another cereal, rye, is of minor importance. Barley and oats, like wheat, are grown in many parts of France, but especially in the northwest; maize is concentrated particularly in the basins of the Garonne and lower Loire rivers in southwestern France.

In addition to the cereals, outstanding farm products in France include two root crops—potatoes and sugar beets—and the grapes which yield France's enormous production of wine.

The tolerance of the potato for poor soils and a cool, damp, climate, plus the fact that it gives a very high yield of calories per unit of land, makes this crop particularly adapted to the circumstances of much of Europe and adjacent parts of the Soviet Union. The countries of this area account for the great bulk of the world's potato production. With better soils and more agricultural land per person, France does not rely so heavily on the potato as do several other European countries, but the crop is cultivated in most parts of the country, and France is ordinarily the fourth-ranking producer in the world, after the Soviet Union, Poland, and West Germany. Although less land is devoted to sugar beets than to potatoes, more cash income is derived from beets than from potatoes. On the world scene, France ranks third in sugar beet output, exceeded only by the Soviet Union and the United States. Most beets are produced in the far north between Paris and the Belgian border.

The grape is a very special element in French agriculture, as wine is in the country's diet and culture. France produces 15 to 25 percent of the world's wine and consumes an even larger percentage. High-grade French wines, often identified with the name of the district or city from which they come, are exported in considerable quantities. The vineyards and wineries producing them are dominant economic elements in some districts as far north as the Loire Valley, Champagne, and Lorraine. Especially important concentrations are found in the valleys of the Loire, Garonne, Rhone, and Saône rivers. But most of the wine consumed in France itself is lower grade "ordinary wine," the production of which is especially concentrated in Languedoc, along the Mediterranean coast west of the Rhone. Domestic production has been supplemented by large imports from Algeria, although such imports have frequently glutted the market and resulted in the conversion of large quantities of wine to industrial alcohol. The exceptionally high rate of wine production and consumption in France can be attributed not only to cultural preferences and traditions of the population, but also to the influence of powerful economic interests on the government. The only country of the world which approaches France in its emphasis on wine is Italy.

Variety is a keynote in French agriculture, however, and the major crops named above are supplemented, replaced, or overshadowed by others in many localities. Among the lesser crops apples deserve special mention, as France is the world's leading producer. The orchards are principally localized in the cool, wet, and hilly lands of Normandy and Brittany, and most of the production is for cider.

Still to be discussed among major products of French agriculture are meat and milk. In meat production, with beef foremost but heavily supplemented by pork, France exceeds all other European countries and is actually one of the four or five leading countries in the world. In milk production France is definitely surpassed only by the Soviet Union and the United States. Cattle are an important element in the agriculture of most parts of France, but especially in northern coastal areas and in the Massif Central. The northern coastal region, from Brittany to Flanders, is wetter and cooler than most of France; and Brittany and Normandy are

very hilly. These conditions are favorable to hay and pasture and, in conjunction with the proximity of the Paris market, have led to the development of the principal dairy region of France. In the Massif Central poor soils and rugged topography discourage crop production, and this area depends mainly on a rather unproductive type of grazing agriculture. The same is true of the even more rugged Alps and Pyrenees.

Weaknesses of French Agriculture

The remarkable total output of agricultural products in France is achieved despite certain apparent and oft-noted agricultural weaknesses. One of these weaknesses is a maldistribution of land. About half of France's farmers have farms that are less than 25 acres in size; and one-third of them have farms of under 12 acres. Thus France's farm population includes a great many small farmers who make a relatively poor living and lack capital for the employment of modern methods. Such farms contribute strongly to the fact that French agriculture is less efficient in its use of the land, on the average, than the agriculture of several neighboring countries. Yields per acre are generally lower than in any neighboring countries except Italy and Spain, which are much less favored by nature and much poorer in general than France.

These weaknesses are the result of various circumstances which affected French agriculture until recently. Farming was valued not just as an element in the economy, but as a way of life, and a large peasant class was regarded as one of the foundations of a healthy society. Consequently, the agricultural policy of the French government was long directed toward maintaining the small farmer on the land, using tariff and quota protection from foreign competition and various subsidies as its instruments. At the same time, the relatively slow growth of France's urban and industrial economy did not drain labor from the countryside rapidly. Consequently, the farmer did not have strong incentives for increased efficiency, while the country's agricultural needs could be met without rapid increases in production. In recent decades, however, all of these circumstances have been reversed or weakened. It has become government policy to foster an agriculture with fewer, larger, and more efficient farms, and rapid trends in this direction are now apparent. For instance, between 1955 and 1963 the number of farms in the country decreased 17 percent, and over roughly the same period the farming population declined by 25 percent. At the same time the amount of land that was farmed remained relatively stable and overall production and yields increased substantially. Thus French agriculture, like some other aspects of France's economy, appears to be in a state of revolutionary change. But this change has come relatively recently and late, and many marks of the old order are still apparent.

Broadly speaking, France's agricultural weaknesses are least apparent in the north, and become progressively more marked in the part of the country that lies south of the Loire Valley. The north is in general a land of larger farms, more modern methods, and high productivity; the south a land of small and poor farms, more traditional methods, and low productivity, which pulls down national averages. The north may be somewhat more favored in the nature of its land and climate, but the contrast is probably more closely related to the slower and smaller development of urban markets in the south. In the country as a whole the more dynamic growth of urban population and industry since World War II has been accompanied by a new rise in agricultural efficiency and output after decades of comparative stagnation.

Fisheries

France's food supply from agriculture is augmented by the products of a sizable fishing fleet. Boulogne (110,000), located at the southern entrance to the Strait of Dover, is the most important fishing port, accounting for nearly a third of the total French catch. Its fishing fleet is dominated by large trawlers which fish primarily for cod in North Sea and North Atlantic waters. Some vessels operate as far afield as the banks off Newfoundland. Lorient (85,000) in Britanny and Cherbourg (75,000) in Normandy compete for second place in total fish landings. Many fishing ports are scattered around the rugged and indented coasts of the Brittany peninsula. Much of the fishing here is concentrated in the coastal waters of Brittany itself, fishing craft are of many sizes, and comparatively few vessels range much farther than the English Channel or Bay of Biscay. The catch includes a great variety of fish and shellfish. Brittany has long been famed as a nursery of sailors for the French navy and merchant marine, and many Bretons are employed in the fishing trawlers which operate from Boulogne. France's main Atlantic naval base, Brest (160,000), is located near the western end of the peninsula.

CITIES AND INDUSTRIES

The Primacy of Paris

Paris is the greatest urban and industrial center of France, completely overshadowing all other cities in both population and manufacturing development. With an estimated metropolitan population of 8½ to 9 million, it is by far the largest city on the mainland of Europe (excluding Moscow in the Soviet Union). Paris is located at a strategic point relative to natural lines of transportation, but it is especially the product of the growth and centralization of the French government and of the transportation system created by that government. Like London, it has no major natural resources for industry in its immediate vicinity, yet is the greatest industrial center of its country.

The city began on an island in the Seine River (photo, p. 152), which offered a defensible site and facilitated crossing of the river. The early growth of Paris, which dates from Roman times if not before, must have been furthered by its location in a highly productive agricultural area and at a focus of navigable streams as well as of land routes crossing the Seine. The rivers which come together in the vicinity of Paris include the Seine itself, which comes from the southeast and flows northwestward from Paris to the English Channel; the Marne, which comes from the east to join the Seine; and the Oise, which joins the Seine from the northeast (map, p. 142). In recent centuries these rivers have been improved and canals provided where necessary to give Paris waterway connections with seaports on the lower Seine, with the coal field of northeastern France, with Lorraine, and with the Saône and Loire rivers.

In the Middle Ages, Paris became the capital of the kings who gradually extended their effective control over all of France. As the rule of the French monarchs became progressively more absolute and centralized, their capital, housing the administrative bureaucracy and the court, grew in size and came to dominate the cultural as well as the political life of France. When, in relatively recent times, national road and rail systems were built, their trunk lines were laid out to connect Paris with the various outlying sections of the country. The result was a radial pattern, with Paris at the hub where all major lines met. As the city grew in population and wealth, it became increasingly a large and rich market for goods.

The local market, plus transportational advantages and proximity to the government, provided founda-

Paris: The Seine and Notre Dame (left) on the Île de la Cité. (Standard Oil Company, N. J.)

tions upon which a huge industrial complex has developed. Speaking broadly, this development comprises two major classes of industries. On the one hand, Paris is the principal producer of the high-quality luxury items—fashions, perfumes, cosmetics, jewelry, and so on—for which France has long been famous. The trades that produce these items are very old. Their growth before the Revolution of 1789 was based in considerable part on the market provided by the royal court, and since the Revolution it has been favored by the continued concentration of wealth in the city. On the other hand, and more important now, Paris has become the country's leading center of engineering industries and secondary metal manufacturing. These industries are concentrated in a ring of industrial suburbs that has sprung up in the nineteenth and twentieth centuries. Automobile manufacturing is the most important single industry in this group, but a great variety of metal goods is produced. Other industries produce chemicals, rubber goods, printed and published materials, foods and beverages, leather, glass, and many other goods. In addition, the city's highly diversified economic base includes an endless variety of services for a local, regional, national, and worldwide clientele. Recently there has been an increasing development of scientific research centers, including some that are concerned with atomic energy.

The Seine Ports

Two ports near the mouth of the Seine handle much of the overseas trade of Paris. Rouen (400,000) is the ancient capital of Normandy. Today it is a major center of France's cotton textile industry and the leading center of its paper industry, using imported raw materials in both activities. The city is located at the head of navigation for ocean vessels using the Seine. However, with the increasing size of ships in recent centuries more and more of Rouen's port functions have been assumed by Le Havre (250,000), located at the entrance to the Seine estuary. Various port industries are located at Le Havre, and the lower Seine area in general is a major district for refining the oil which France imports in large quantities. Le Havre is the principal transatlantic passenger port of France, by a wide margin over Cherbourg, on the Cotentin peninsula, and it is second to Marseilles as a freight port.

Urban and Industrial Districts Adjoining Belgium

France's second-ranking urban and industrial area is composed of cities clustered on or near the country's principal coal field, located in the north near the Belgian border (maps, pp. 139 and 142). A bit north of the coal field, several cities, of which Lille is the largest, form a metropolitan agglomeration of around 900,000 people. On the coal field itself a number of mining and industrial centers in the vicinity of Lens and Douai aggregate over 500,000 people, and smaller centers are scattered over the field. The Lille metropolitan area is the leading textile center of France, with a linen industry dating from medieval times but overshadowed now by cotton and woolen production. Cities on the coal field manufacture iron and steel, chemicals, cement, heavy machinery, and miscellaneous metal goods. They also constitute France's leading center for the production of ceramics, glassware, and pottery.

The mines of the northern coal field account for approximately half of France's total coal production. But the coal is deep underground and lies in thin and broken seams, so that it has been difficult and expensive to mine; and it includes relatively little coal of coking quality. It is increasingly used as a raw material for chemical plants and to fuel a cluster of giant thermal-electric power stations in the coal-field area. France's coal resources are much inferior to those of

Germany or Great Britain, and her total production has generally been seriously inadequate for the country's needs. Since World War II the mining industry has been nationalized and extensively modernized and re-equipped, but France remains a sizable importer of coal and coke.

Urban and Industrial Development in Southern France

France's third largest urban-industrial cluster centers on the metropolis of Lyons (metropolitan population about 1 million), in southeastern France. Here the valleys of the Rhone and Saône rivers join, providing routes through the mountains to the Mediterranean, northern France, and Switzerland. Lyons, located at this junction, has been an important city since Roman times. For several centuries silk production was the basic manufacturing industry of the city and many smaller towns round about, and the district is still the leading silk-manufacturing area of Europe. In the nineteenth and twentieth centuries, however, nylon and rayon textiles, chemical products, and engineering industries have

In western Europe carefully tended fields often lie in close juxtaposition to industrial areas. This view, taken in the coal field of northern France not far from Lille, shows shocks of harvested wheat, potatoes (the low, dark-colored crop), a coal-powered thermal-electric generating station (center of view), and piles of waste material from coal mines in the distance. (Standard Oil Company, N. J.)

A characteristic view of the French Riviera. Resort towns along this coast are tucked into indentations between rocky, maquis-covered headlands. The town along the bay in the center of the photo is Menton, located between Monaco and the Italian border. The view looks eastward toward Italy's Riviera. Flowers, possibly grown for sale to nearby perfume factories, occupy the foreground. (French Government Tourist Office.)

eclipsed silk in the district's industrial structure. The synthetic fiber and chemical industries are largely outgrowths of the silk industry. Electrical equipment related to the development of southeastern France's power resources is especially important in the engineering industries. Much of the area's industry is in smaller centers near Lyons, which functions partly as the business, administrative, and financial center of its region.

Abundant power resources are found in the surrounding area. Not far to the southwest the city of St. Étienne (315,000) mines coal from the most important of several small fields in the eastern margin of the Massif Central, and produces steel, machinery, and textiles. But the great power resource of the area is waterpower. To compensate for its inadequate deposits of mineral fuels, France has stressed the development of the hydroelectric potential of its mountain streams, and only five countries in the world—the United States, the Soviet Union, Canada, Japan, and Italy—surpass it in developed hydropower. Several large new projects have been completed since World War II on Rhone tributaries, and four on the Rhone itself south of Lyons. Uranium also is mined in the Massif Central for use in

France's rapidly expanding development in the field of atomic energy.

Not only is the increased power stimulating industrial development in Lyons and smaller cities of southeastern France, but additional dams on the Rhone will eventually bring the river under control and make it highly useful for navigation. At present the Rhone is of limited utility as a through waterway, due to the swift current, obstruction of the channel by alluvium deposited in spring floods, and water levels that fluctuate widely from season to season. It is a great contrast to its tributary, the Saône, which is a deeper and more placid river with a relatively even flow throughout the year. A considerable amount of barge traffic moves between Lyons and Paris via the Saône and connecting canals to the Seine (map, p. 138), but relatively little traffic of this kind moves from Lyons to the Mediterranean. However, the main rail and highway routes from Paris to the Mediterranean pass through Lyons and down the Rhone Valley to Marseilles.

France's principal Mediterranean cities—Marseilles, Toulon, and Nice—are strung along the coast east of the Rhone delta. Marseilles (950,000 to 1 million) is the

A view in foothills of the French Alps some distance to the north of the port of Toulon. The foreground is occupied by grapevines, and piles of bauxite appear in the center of the photo. Trucks bring the bauxite to this spot from mines in the nearby mountains, and the ore is then transferred to railway cars for further shipment. Some bauxite from French mines is processed at plants in valleys of the French Alps, and some is exported through Toulon. (Jesse H. Wheeler, Jr.)

country's third largest metropolitan city and its leading seaport as measured by tonnage of seaborne freight handled. Marseilles is located on a natural harbor far enough from the mouth of the Rhone to be free of the silt deposits which have clogged and closed the harbors of ports closer to the river which were once more important than Marseilles. Although it is a very old city, perhaps the oldest in France, Marseilles' modern growth was closely associated with the acquisition of France's trans-Mediterranean empire in North Africa and with the building of the Suez Canal, both in the nineteenth century. The French imperial system emphasized trade with the empire, especially with Algeria, and the Suez Canal made the Mediterranean a major world thoroughfare, while the Rhone-Saône trench provided an easy route between Marseilles and northern France. The city's industries process imported tropical foods, refine petroleum, build and repair ships, and manufacture various kinds of machinery.

Rapid growth of port traffic and manufacturing at Marseilles in recent years has been particularly associated with increasingly large oil imports and development of the petrochemical industry. Some experts speculate that Marseilles may soon rank second only to Rotterdam, in tonnage handled, among the ports of continental Europe. Lavéra, some distance to the west of Marseilles, is the chief oil terminal in the Marseilles area.

Toulon (250,000), a short distance southeast of Marseilles, is France's principal Mediterranean naval base, while Nice (350,000), near the Italian border, is the principal city of the French Riviera, probably the most famous resort district in Europe. Along this easternmost section of the French coast the Alps come down to the sea to provide a spectacular shoreline dotted with beaches. The mountains also provide a slightly drier and warmer variety of mediterranean climate than is usual in these latitudes by affording some protection from northerly and westerly winds. A string of resort towns and cities along the coast includes, near Nice, the tiny principality of Monaco, which is nominally independent but is closely related to France economically and administratively.

Deposits of bauxite scattered along the Mediterranean littoral are among France's more important mineral resources. The largest deposit is a few miles inland from Toulon. Economical manufacture of aluminum from bauxite requires large quantities of low-cost electric power, and France's aluminum works are mostly located in Alpine valleys near hydroelectric installations. Surplus ore is exported through Toulon. France ranks first among the countries of Europe, and fourth or fifth in the world, as a producer of both bauxite and aluminum.

Natural gas is the most important mineral resource in southwestern France; several fields are found in foothills along the front of the Pyrenees. The chemical plant in the foreground processes gas from deposits near by, with propane the main product. The plant is located at Boussens, on the Garonne River about 40 miles southwest of Toulouse. The railway bridge in the center of the photo crosses the Garonne; the stream visible at the upper right is a tributary. The view looks southward toward the Pyrenees foothills. Gentle slopes alongside the rivers are cultivated or provide sites for industrial plants, while the steeper hillslopes beyond are forested or cleared in patches for upland pastures. (French Government Tourist Office.)

Southwestern France has less urban and industrial development than the north or the southeast. Its two main cities are on the Garonne River. The principal seaport of the region, Bordeaux (500,000), is located at the head of the Garonne estuary, which is called the Gironde. The port has given its name to the fine wines which are a major product of the surrounding region and have been exported through Bordeaux for centuries. Toulouse (380,000), a very old and historic city with modern machinery, chemical, and aircraft plants, developed on the Garonne at the point where the river comes closest to the Carcassonne Gap, leading between the Pyrenees and the Massif Central to the Mediterranean coast.

In central and south-central France a considerable number of small to medium-sized cities are spaced at fairly regular intervals through the drainage basin of the Loire River. Besides St. Étienne, previously mentioned, the largest cities are the seaport of Nantes (415,000), located about 40 miles upstream from the mouth of the Loire, and the rubber-manufacturing center of Clermont-Ferrand (180,000) in the Massif Central.

Mining and Heavy Industry in Alsace-Lorraine

In eastern France, Alsace and Lorraine are important industrial areas, although their urban development is on a comparatively small scale. Lorraine is one of the world's more important iron-mining areas, but its importance in this regard is overshadowed by its large production of iron and steel, amounting to about 60 percent of France's output. The mines, blast furnaces, and steelworks extend in a line northward from Nancy (235,000) past Metz (185,000), and into southern Luxembourg and Belgium (maps, pp. 138 and 139). Part of West Germany's Saar coal field extends across the border into Lorraine, and coal is also brought from the Saar itself, from France's northern field, and from the West German Ruhr. Coke must be brought in large quantities, principally from the Ruhr. Access to adequate and low-cost supplies of coal, and especially of coke, has always been a problem to the Lorraine iron and steel industry, but may now be on the way to solution. Production of coal from the Lorraine field itself has been rising, and French industrialists have found an economical method of coking this coal, which had previously been of noncoking quality. The industries of Lorraine have benefited from the elimination of trade barriers under the European Coal and Steel Community and, in lesser measure, from the canalization of the Moselle River. Even without these improved conditions, however, Lorraine has made France the world's sixth-ranking producer of steel, after the United States, the Soviet Union, Japan, West Germany, and the United Kingdom.

Nevertheless, there is evidence that both coal and steel output in this area may decline relatively if not absolutely. Competing fuels are lowering the total demand for coal (for all uses and not iron and steel output alone). In the late 1960s coal supplied nearly three-fourths of France's fuel requirements, but by the late 1970s, it is expected to supply only about one-fourth. Petroleum, natural gas, and atomic energy are expected to replace coal in many phases of energy supply. Still other competition to Lorraine, in this case to the iron and steel industry, is appearing in the form of a number of large new iron and steel plants in coastal parts of EEC countries (including a plant at the port of Dunkirk [135,000] in northern France) (see p. 88). Because they can obtain overseas iron ore cheaply due to the increasing efficiency of progressively larger ocean carriers, and coal either from European sources or the United States, these coastal installations are offering serious competition to Lorraine and other inland producing regions of western Europe, despite such aids to the latter as the Moselle Canal.

Other mineral and power resources have also furthered the industrial development of Lorraine and Alsace. Large deposits of salt near Nancy and of potash near Mulhouse (225,000) in southern Alsace have provided raw materials for chemical plants located near the deposits. Streams of the Vosges Mountains have furnished water power, and Mulhouse as well as smaller cities around it have developed cotton textile and engineering industries. By 1968 a major navigation and hydroelectric power project was near completion along the Rhine in Alsace. It involves five low dams on the Rhine, plus several diversion canals paralleling the river. Eight power stations are located on the canals. The longest canal, known as the Grand Canal of Alsace, leaves the Rhine downstream from Basel, Switzerland, and reenters it about midway between Basel and Strasbourg (335,000), France's major Rhine River port. Barges using the Rhine leave the river at intervals to traverse the canals. The project has greatly improved the waterway connections southward from Strasbourg. Other significant water connections of Strasbourg include the Rhine-Marne Canal to Lorraine and the Seine River system, and a small canal through the Belfort Gap to the Saône and Rhone.

Some General Characteristics of French Urban and Industrial Development

France is a country which shows very striking contrasts in the nature of its development before and since World War II. From the industrial revolution until recent decades the country tended to fall behind its neighbors, except for Spain and Italy, in industrialization and modernization. It changed enough to become a considerable industrial power, but more slowly than most of its neighbors, and it remained more rural and less urban and industrial. Differences in this respect between France and the United Kingdom or Germany were particularly marked. Then after the shock of World War II the country acquired a new dynamism. New development of the most modern types has taken place at an impressive pace and scale. The old patterns often continue alongside the new, and the result is a situation of striking contrasts. There is, for instance, the contrast between the fact that the census of 1962 reported over 60 percent of French dwelling units to have been built before 1914 and almost a third of them before 1871, and the equally striking fact that new high-rise apartments break the skylines of many French cities. Or the contrast between the small artisan's shop manufacturing luxury goods by hand, and the massive factories and power plants of a new industrial age.

Despite the rapid development in recent years, the effects of the past are still manifest in many ways. For example, they are apparent in the relatively small size of French cities. Paris is a giant, but it is the only city with a metropolitan population of much more than a million, and only three others—Lyons, Marseilles, and Lille—are near the million mark. The marks of the past are also evident in France's employment structure and types of industrial development. In France a smaller proportion of the labor force is engaged in manufacturing than in any neighboring country except Spain, while agricultural employment still accounts for a greater proportion of the labor force than in any nearby countries except Spain, Italy, and Luxembourg. Small family-owned factories and shops were particularly characteristic of French industry in the past, and to some extent still are. Such establishments are often well adapted to light consumer goods of a luxury character which emphasize elegance, taste, and individual craftsmanship. But at the same time they are often undermechanized, with a low output per operative at a high production cost. They cannot take full advantage of the economies of mass production. Nor can they compensate for the retarded growth of the larger and more basic metal and machine-building industries—not generally so well adapted to small-scale enterprise—in the strenuous international economic and military competition of the twentieth century.

Factors in the Slow Growth of Large-Scale Industries in the Past

No definitive explanation can be given for the long retardation of large-scale industrial development in France, but a number of factors are worthy of note:

1. For over a century before 1946 population growth was slow. Between World Wars I and II there would actually have been a loss of population except for immigration. Thus French industry did not have the stimulation of a rapidly expanding domestic market, nor was there a rapidly expanding labor force to man new industries as they arose. Of course, slow industrial growth may also have retarded population growth in a "vicious circle."

2. The French government centered the railway and road networks of the country so strongly in Paris that, while the capital had excellent trunk-line connections with the various outlying regions of the country, these regions often had inadequate connections with each other. It has been alleged that they were handicapped with respect to efficient assembly of materials and distribution of products.

3. France's coal resources, in comparison with those of major industrial competitors, were inadequate in quality and quantity, and this during a time when alternative sources of power were much less important than today.

4. Adequate capital for investment in new industry was often unavailable. France's banking industry has frequently been criticized for its reluctance to lend venture capital; and the corporate form of business organization operated under certain legal and governmental handicaps. In addition, the small family-owned French company seems often to have been reluctant to sell stock widely or borrow heavily for fear of loosening family control over the enterprise, and consequently remained small and underequipped. A final circumstance lessening the amount of capital available for investment in new enterprises was the necessity of directing large sums into the replacement of industrial facilities destroyed in war.

5. Until after World War II, French governments generally did little to promote large-scale industrialization. They condoned and even helped to create a situation in which small and inefficient producers were protected from the threat of competition by tariffs and quotas on imports and by subsidies, cartel arrangements, and other practices in restraint of competition. The spur toward greater efficiency which competition provides was at least partially removed.

6. As a result of France's near self-sufficiency in food and its control over a valuable colonial empire whose trade was directed toward the home country, it was unnecessary for France to make an intensive effort to capture foreign markets for industrial goods.

7. Some observers maintain that various French attitudes were inimical to large-scale industry. Attitudes of this type often attributed to the French include a pronounced individualism and, in consequence, an antagonism toward the social regimentation and standardization which frequently accompany large-scale industrialization. In addition, there has been a tendency to accord the businessman a smaller degree of prestige and social status than the large landowner, the professional man, or the higher civil servant. As a result, the family-owned business seems often to have been regarded primarily as a stepping stone toward better things: if the business prospered enough that the family could acquire land and gives its sons a professional education, that was sufficient—expansion and improvement of the business itself was not an urgent matter. Generally speaking, French culture has tended to emphasize such values as a modest security and independence, leisure, and style in living somewhat more, and steadily expanding production and consumption of material goods somewhat less, than have the cultures of many Western countries.

Probably there was good as well as bad in the slow modernization of France, with its marked degree of social stability. Certainly foreigners have been much attracted to French life and culture. But a price was paid in declining national power and in the notable poverty of some sections of the population—especially industrial workers, small farmers, and small merchants—and of some sections of the country.

Now France's new development is rapidly eroding the old patterns. Population growth, encouraged by government welfare measures, has speeded up. Inadequate coal resources are becoming less of a handicap as hydroelectric power, natural gas, cheap imported oil, and atomic reactors steadily replace coal as sources of energy. Tariff protection of inefficient French producers from European competition is weakened or abolished by France's participation in the Common Market; and the French government is promoting combinations of companies into large-scale units capable of high efficiency and successful international competition.

NATIONAL AND REGIONAL PLANNING

The new dynamism of France has been achieved under a system of deliberate planning. Since 1947 the country has engaged in both national and regional planning with the stated objectives of improving national output and living standards, reducing the very considerable differences in prosperity of different sections of the country, and making France a pleasant place to live in a technological age. Unlike the all-encompassing state plans of Communist countries, this type of planning is envisaged as complementary to the activities of private enterprise, with the state using various means to influence the decisions of private companies while at the same time cooperating with them. A few key industries in France are government-owned, but government efforts are directed principally toward the encouragement of efficiency on the part of private enterprise and providing an adequate "infrastructure" (good transportation facilities, water supply and sewage facilities, cheap electrical power, a skilled and educated populace, and so forth) to support productive activities.

In addition to fostering overall national growth and improvement, French planning is directed toward reducing imbalances of long standing among the regions of France. A recent report on French planning[4] divided France into four regions of unequal size: the Paris Region, the Northeast, the Southeast, and the West (map, right). The Paris Region had 2 percent of the country's area, 19 percent of its population, and 29 percent of its industrial jobs. The West had 55 percent of the area, 37 percent of the population, and 24 percent of the industrial jobs. The Northeast and Paris together had 20 percent of the area, 38 percent of the population, and accounted for 46 percent of the total national product. The picture which emerges from such comparisons is one of Paris at the top of the regional scale, a highly industrialized and productive northeast, a southeast with a medium level of development, and an underprivileged western half of the country.

If France's sharp regional differences in development and prosperity are to be ameliorated, the Southeast and the West must be encouraged to grow more rapidly. Modern economic growth is closely associated with urbanization. Hence the planners of France, in their attempt to encourage regional development, are stimulating economic growth in selected urban places

France: Four regions with contrasting levels of development; in order, highest to lowest: (1) Paris Region, (2) Northeast, (3) Southeast, and (4) West. Why? (Source: *France Town and Country Environment Planning*, p. 10.)

which they intend to develop into major "regional metropolises," less subordinate to Paris than they are now. Selected smaller cities in each region are being developed as "relay centers" whose role is to transmit the influence of the regional metropolis into the various local sections of its region. Thus the plans for regional development focus sharply, though not completely, on expanded roles for cities other than Paris.

Among several urban centers receiving particular attention are Toulouse in the southwest and Grenoble (375,000) in the Alps. Deliberately encouraged transfers from Paris plus new enterprises are making Toulouse a major center of aircraft and aerospace industries. Educational institutions related to these activities are being transferred from Paris and pushed forward at Toulouse, and outside the old city a new planned community for 100,000 people is being constructed. At Grenoble, long a small industrial center, expansion is being sparked by hydropower, expansion of the university, and development of a Center of Nuclear Studies located there by the government. These developments are facilitating

[4] *France Town and Country Environment Planning* (New York: French Embassy Press and Information Service, 1965).

various engineering, electronics, and chemical industries. In the 16 years preceding 1968 the city's metropolitan population grew from 160,000 to an estimated 375,000.

More general evidence of the success of France's planned development lies in the country's rate of economic growth, which averaged 4.5 percent a year between 1950 and 1960, and 5.1 percent a year from 1960 to 1966. This rate of growth made France one of the world's more dynamic countries economically; between 1960 and 1966, for instance, France surpassed the growth rates of such countries as the United States, the Soviet Union, and West Germany. An impressively new and relatively affluent France appears to be replacing the old imperial power with its lagging economy.

REFERENCES and READINGS

English-Language References

ABRAHAM, JOHN, "Changing Face of Paris," *Geographical Magazine,* 41, no. 2 (November 1968), 129–143.

BAKER, ALAN R. H., "A Modern French Revolution," *Geographical Magazine,* 40, no. 10 (February 1968), 833–841. Discusses the consolidation of fragmented farm holdings.

CLARKE, JOHN I., "Demographic Revival in France," *Geography,* 48, pt. 3 (July 1963), 309–311.

Current History. Special issues on France: 50, no. 296 (April 1966), and 54, no. 307 (March 1968).

DIEM, AUBREY, "The Mont Blanc Tunnel: A New Route thru the Alps," *Journal of Geography,* 58, no. 7 (October 1959), 344–349.

EVANS, E. ESTYN, *France: An Introductory Geography* (New York: Frederick A. Praeger, 1966).

FARNSWORTH, HELEN C., "Determinants of French Grain Production, Past and Prospective," Stanford University, *Food Research Institute Studies,* 4, no. 3 (1964), 225–272.

FIELDING, A. J., "Internal Migration and Regional Economic Growth: A Case Study of France," *Urban Studies,* 3, no. 3 (November 1966), 200–214.

France Town and Country Environment Planning (New York: French Embassy Press and Information Service, 1965).

GALLOIS, LUCIEN, "The Origin and Growth of Paris," *Geographical Review,* 13, no. 3 (July 1923), 345–367.

THE GEOGRAPHER, U.S. DEPARTMENT OF STATE, *Survey of the French Republic* (Geographic Bulletin No. 4). (Washington, D.C.: Government Printing Office, 1965).

GOLDBERG, HARVEY, "The Myth of the French Peasant," *American Journal of Economics and Sociology,* 13, no. 4 (July 1954), 363–378.

GUERARD, ALBERT, *France: A Modern History* (Ann Arbor: University of Michigan Press, 1959).

HALL, PETER, *The World Cities* (London: Weidenfeld and Nicolson, World University Library, 1966), Chap. 3, "Paris," pp. 59–94. The main emphasis is on urban and regional planning of the Paris region. Excellent maps.

HARTSHORNE, RICHARD, "The Franco-German Boundary of 1871," *World Politics,* 2, no. 2 (January 1950), 209–250.

HAYTER, TERESA, "French Aid to Africa: Its Scope and Achievements," *International Affairs,* 41, no. 2 (April 1965), 236–251.

HOUSE, J. W., "The Franco-Italian Boundary in the Alpes Maritimes," Institute of British Geographers, Publication No. 26, *Transactions and Papers, 1959,* pp. 107–131.

KISH, GEORGE, "Hydroelectric Power in France: Plans and Projects," *Geographical Review,* 45, no. 1 (January 1955), 81–98.

LAMOUR, PHILIPPE, "Land and Water Development in Southern France," in Henry Jarrett, ed., *Comparisons in Resource Management* (Baltimore, Md.: Johns Hopkins Press for Resources for the Future; Lincoln: University of Nebraska Press, Bison Books, 1961), pp. 227–250. Deals with irrigation of the Languedoc vineyard area to make diversified agriculture possible, using water from the Rhone River.

LEVINE, MORTON H., "The Basques," *Natural History,* 76, no. 4 (April 1967), 44–51.

MARTIN, J. E., "Location Factors in the Lorraine Iron and Steel Industry," Institute of British Geographers, Publication No. 23, *Transactions and Papers, 1957,* pp. 191–212; and "New Trends in the Lorraine Iron Regions," *Geography,* 53, pt. 4 (November 1968), 375–380.

MARTONNE, EMMANUEL DE, *Geographical Regions of France* (2d ed.; London: Heinemann Educational Books, 1948; reprinted, 1962).

MAUROIS, ANDRÉ, *An Illustrated History of France,* trans. by Henry L. Binsse and Gerard Hopkins (New York: The Viking Press, 1960).

McGURRIN, A. JEAN, and ALEXANDER DAUMAN, *Basic Data on the Economy of France.* U.S. Bureau of International Commerce, Overseas Business Reports, OBR 67-56 (August 1967). (Washington, D.C.: Government Printing Office, 1967.)

"Menton, Riviera Resort," in Richard M. Highsmith, Jr., ed., *Case Studies in World Geography: Occupance and Economy Types* (Englewood Cliffs, N.J.: Prentice-Hall, 1961), pp. 155–161.

MICHEL, ALOYS A., "The Canalization of the Moselle and West European Integration," *Geographical Review,* 52, no. 4 (October 1962), 475–491.

ORMSBY, HILDA, *France, A Regional and Economic Geography* (2d ed.; New York: E. P. Dutton and Co., 1950; reprinted, 1964).

PILKINGTON, ROGER, "Moselle Waterway," *Geographical Magazine,* 40, no. 3 (July 1967), 190–200; and "Joining the Rhine and Rhone," *Geographical Magazine,* 39, no. 3 (July 1966), 214–228.

PLATT, ROBERT S., "The Saarland, an International Borderland: Social Geography from Field Study of Nine Border Villages," *Erdkunde,* 15, no. 1 (March 1961), 54–68.

POUNDS, NORMAN J. G., "Historical Geography of the Iron and Steel Industry in France," *Annals of the Association of American Geographers,* 47, no. 1 (March 1957), 3–14; and "Lorraine and the Ruhr," *Economic Geography,* 33, no. 2 (April 1957), 149–162.

SCARGILL, D. IAN, *Economic Geography of France* (New York: St. Martin's Press, 1968).

SCHOFIELD, G., "The Canalization of the Moselle," *Geography,* 50, pt. 2 (April 1965), 161–163.

SIEGFRIED, ANDRÉ, *France: A Study in Nationality* (New Haven, Conn.: Yale University Press, 1930).

SIMON, ANDRÉ L., *Wines of the World* (New York: McGraw-Hill Book Company, 1967).

THOMPSON, IAN B., "A Review of Problems of Economic and Urban Development in the Northern Coalfield of France," *Southampton Research Series in Geography* (Department of Geography, Southampton University, England), no. 1 (March 1965), pp. 31–60; "A Geographical Appraisal of Recent Trends in the Coal Basin of Northern France," *Geography,* 50, pt. 3 (July 1965), 252–260; and "Some Problems of Regional Planning in Predominantly Rural Environments: The French Experience in Corsica," *Scottish Geographical Magazine,* 82, no. 2 (September 1966), 119–129. The latter article includes some useful maps of mainland France.

WEIGEND, GUIDO G., "The Basis and Significance of Viticulture in Southwest France," *Annals of the Association of American Geographers,* 44, no. 1 (March 1954), 75–101; and "Bordeaux: An Example of Changing Port Functions," *Geographical Review,* 45, no. 2 (April 1955), 217–243.

See also the reference lists for Chapters 1–4, particularly the list of standard geographies of Europe, p. 92.

French-Language References

Annales de Géographie. One of the major geographical journals of the world. Includes many articles on the geography of France. Six issues a year. All articles are in French.

Atlas de France (2d ed.; Paris: Éditions Géographiques de France, 1951–   ). One of the world's great atlases.

CHABOT, GEORGES, *Géographie Régionale de la France* (Paris: Masson et Cie, 1966). Much physical and economic detail is shown on numerous maps.

CLOZIER, RENÉ, *Géographie de la France* (Paris: Presses Universitaires de France, 1967). A short geography.

DEFFONTAINES, PIERRE, and MARIEL JEAN-BRUNHES DELAMARRE, *Atlas Aérien: France,* 5 vols. (Paris: Éditions Gallimard, 1955–1964). A massive collection of air photographs, organized regionally, with accompanying text in French.

France de Demain. Paris: Presses Universitaires de France. A series of volumes on the geographical regions of France, for example, Pierre George and Others, *La Région Parisienne,* 2d ed., 1964. All volumes are in French.

LE LANNOU, MAURICE, *Les Régions Géographiques de la France* (2d ed., 2 vols.; Paris: Société d'Édition d'Enseignement Supérieur, 1964).

L'Information Géographique. A highly useful geographical journal in the French language, devoted very largely to articles, maps, and statistical materials on the contemporary economic geography of France. Five issues a year.

Germany

7   For many centuries German-speaking peoples have played a prominent role in the affairs of Europe. But a unified German national state did not emerge until after the middle of the nineteenth century, and up to the present it has failed to achieve territorial stability. Its boundaries have fluctuated a great deal in the twentieth century as a result of two world wars, and today it remains divided into Western and Eastern sections under separate governments. Serious questions raised by its partitioning are yet to be resolved, and its future remains one of the most critical issues of international politics.

EMERGENCE OF MODERN GERMANY

Prior to the year 1866, "Germany" was little more than a territorial expression referring to a loosely connected group of independent or semi-independent states, mostly small, in central Europe. This group of states, a holdover from the fragmented political order of medieval times, possessed a certain amount of economic unity resulting from the

The division of Germany between Western and Soviet spheres is evinced by this photo of a checkpoint in the outskirts of Berlin. The line of cars at the left has been halted so that visas of their occupants can be checked. The travelers, bound for West Berlin, have driven across East German territory by autobahn from Hanover and Brunswick in West Germany proper. As one traverses the highway into the city, it passes briefly within the city limits and then lies for 2 miles within East Germany before it again enters West Berlin. (German Information Center.)

gradual abolition of customs barriers after the Napoleonic wars. But political unity was not achieved until the period 1866–1871, when all of the states except Austria and Luxembourg were combined into a single country which shortly emerged as the most powerful nation on the mainland of Europe.

The historical background of German unification has been briefly sketched as follows:

In ancient and medieval times Germanic peoples overran large parts of Europe. Having invaded the ancient Roman Empire, they settled down in various parts of the conquered lands and were eventually absorbed by the native populace. Later conquests led in the ninth century to the creation of the Holy Roman Empire of the German Nation. This loose congeries of petty German states finally expired at the time of Napoleon (1806) because of internal weaknesses. Voltaire said it was "neither holy, nor

Roman, nor an Empire," and despite its name it was certainly not a national state in the modern sense.

Except for Italy, Germany was the last [important European power] to achieve national unity. During the centuries from the close of the Middle Ages down to 1870, while Great Britain, France, and Russia were strong and united, Germany remained subdivided among petty states, and its people as a whole felt little national consciousness. In political loyalty they were Prussians or Bavarians or Saxons, and Germans only in speech and culture.

Unification was a slow process, accomplished under the leadership of Prussia, which from about 1740 on gradually extended its own territories to include the whole of northern Germany. In 1871 Bismarck's statesmanship brought South Germany also into a new state, over which the King of Prussia was crowned as Emperor William I. Thus was created the Second German Empire, a firmly knit federation under Prussian domination.[1]

[1] *Geographical Foundations of National Power* (Army Service Forces Manual M 103–1; Washington, D.C.: Government Printing Office, 1944), p. 63.

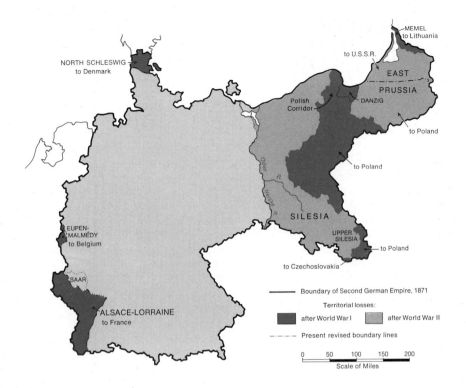

Territorial losses of Germany resulting from World War I and World War II. Minor frontier cessions of territory to the Netherlands, Belgium, and Luxembourg following World War II are not shown. (See Lewis M. Alexander, "Recent Changes in the Benelux-German Boundary," *Geographical Review*, v. 43 [1953], pp. 69–76.) (Map compiled from Isaiah Bowman, *The New World: Problems in Political Geography*, 4th ed., New York: Harcourt, Brace & World, 1928, and other sources.)

TERRITORIAL GAINS AND LOSSES

The Second German Empire was much larger in area than all of Germany today (map, above). It stretched eastward to include East Prussia and a sizable part of Poland, which had been partitioned among Prussia, Russia, and Austria in the eighteenth century. On the west it included Alsace-Lorraine, taken from France following the Franco-Prussian War of 1870–1871. In the north it included the former duchies of Schleswig, Holstein, and Lauenburg, taken over by Prussia and Austria after a war with Denmark in 1864, and annexed by Prussia after the Austro-Prussian War of 1866.

The new nation grew rapidly in economic and military power. Large coal-mining, iron and steel, and chemical industries were developed as a basis for general economic expansion. Germany built a large navy and merchant marine and developed a sizable foreign trade, centered in Europe, but extending to many other parts of the world. Between 1884 and 1900 a colonial empire was acquired, mainly in Africa but also including island territories in the Pacific Ocean and certain holdings in China.

Tensions and rivalries among the great powers of Europe led to World War I (1914–1918), in which Germany and its allies (Austria-Hungary, Turkey, and Bulgaria) were defeated by a worldwide coalition of nations, eventually including the United States. As a result of the Versailles Treaty following the war, Germany was obliged to cede much of its territory to other nations. Alsace-Lorraine was returned to France, and northern Schleswig was ceded to Denmark following a plebiscite. Poland was organized as an independent nation and was given a Baltic Sea outlet—the "Polish Corridor"—between East Prussia and the rest of Germany. Two small areas of German-speaking population bordering East Prussia were lost: (1) Danzig, organized as a Free City under League of Nations supervision, and (2) Memel, annexed by Lithuania in 1924 after several years of French administration. Small cessions of territory were made to Belgium and to the new state of Czechoslovakia. The cession to Czechoslovakia involved a segment of the important Upper Silesian-Moravian coal field, most of which was held by Germany before the war. A far larger share of the field went to Poland after a plebiscite, though Germany itself retained a small but very productive share until after World War II, when its remaining portion was taken over by Poland. Under the Versailles Treaty, the Saar was placed under international administration, with control of its coal mines vested in France. All German areas west of a line running 50 kilometers east of the Rhine were demilitarized. Germany's African and Pacific island possessions were taken away and distributed among the victorious powers as mandated territories under the League of Nations.[2]

After World War I, Germany experienced a period of severe inflation, and subsequently was plunged into the worldwide economic depression of the 1930s. In 1933 the Weimar Republic was succeeded by the Nazi dictatorship under Hitler, and Germany commenced a series of expansionist moves that led to World War II. In 1935 the Saar was recovered as a result of a plebiscite, and in 1936 the Rhineland was remilitarized. Annexation of Austria in 1938 was quickly followed by the dismemberment and partial annexation of Czechoslovakia (1938–1939). The European phase of World War II began with a German attack on Poland in September 1939. Once again, after six years of warfare and extensive conquests of territory, Germany was defeated by a massive coalition of allies, led by the United States, the Soviet Union, and Great Britain. As a result of the war it was deprived of the territories it had annexed or conquered since 1936, and extensive eastern sections of its pre-1936 territory were taken over by Poland and the Soviet Union.

Postwar Territorial Arrangements

Following the German surrender in 1945, the part of the country immediately east of the Oder and Neisse rivers was placed under Polish administration, along with a narrow strip lying west of the Oder estuary and including the port of Stettin (now Szczecin). East Prussia was partitioned between Polish and Russian administration. The Saar was made quasi-independent politically, but was meshed into the French economy. (In 1957 the Saar rejoined Germany as a state of the German Federal Republic—see pp. 146 and 147.) The rest of Germany was divided into four zones of occupation administered, respectively, by the United States, the United Kingdom, France, and the Soviet Union. The capital city of Berlin, deep within the Soviet Zone, was placed under joint administration of the four powers. As a result of postwar tensions between the Soviet Union and the three Western powers, the western boundary of the Soviet Zone crystallized in 1949 into a dividing line between two separate republics (map, right). To the west of the boundary is the German Federal Republic, or West Germany, with approximately 96,000 square miles and 60.3 million people (1968 estimate; the figure includes some 2.2 million people in West Berlin, which is still under Western occupation and is not technically a part of the German Federal Republic, though it is represented in its legislature). To the east is the Soviet-inspired German Democratic Republic, or East Germany, with about 42,000 square miles and an estimated 17.1 million people, including about 1.1 million in its political capital, East Berlin.

Thus the question still arises, what is meant by "Germany"? The present political divisions were reached on the basis of expediency and may or may not endure. In German thinking they are not necessarily permanent: present-day German-compiled statistics almost invariably include all of the territory occupied by the nation in 1936, although such figures may be classified under the headings of "Federal Republic," "Democratic Republic," "West Berlin," "the section administered by Poland," and "East Prussia."

It now appears, however, that the divisions of Germany after World War II will continue into the immediate future, if not longer. Hence the discussion in this chapter will be couched primarily in terms of "West Germany" and "East Germany". The term "Germany," will be used as a general regional designation for the whole of Germany in the past or the two German republics in combination today. "Prewar Germany" will be used to designate the country as delimited in 1936.

THE DIVERSIFIED GERMAN ENVIRONMENT
The North German Plain

Broadly conceived, the terrain of Germany can be divided into a low-lying, undulating plain in the north

---

[2] Subsequently they became United Nations trust territories except for South West Africa, which South Africa has repeatedly refused to place under the trusteeship system. Some of them have gained independence.

INDEX MAP OF
GERMANY

URBAN AREAS

● over 4,000,000

◉ over 1,000,000

● 500,000—1,000,000

✪ 300,000-500,000

○ Selected smaller places

Stars show capitals

City-size symbols are based on
metropolitan area estimates

LANDFORMS

Lowlands

Hills and uplands

Low mountains

High mountains

Canals

0   50   100
Scale of Miles

A pleasant countryside in southern Germany. The view, which looks southward, shows the valley of the Neckar River about midway between Mannheim and Stuttgart. The highway in the foreground crosses the bridge into the town of Bad Wimpfen. (German Information Center.)

before use and resembles the diked lands of the adjoining Netherlands and Belgium.

The Varied Terrain of Central and Southern Germany

The central and southern countryside is much less uniform than the northern plain. To the extreme south are the moderately high German (Bavarian) Alps (maximum elevation 9721 feet), fringed by a rolling piedmont or foreland that slopes gradually toward the east-flowing upper Danube River. Between the Danube and the northern plain is a complex series of uplands, highlands, and depressions. The higher lands are predominantly composed of rounded, forested hills or low mountains. Prominent individual highlands or uplands include (1) the Bohemian Forest, Ore Mountains (Erzgebirge), and Uplands of Saxony at the east, bordering Czechoslovakia; (2) the Black Forest, Oden Forest (Odenwald), and Rhine Uplands at the west, bordering the Rhine Valley; (3) the Harz Mountains (photo, right) and Thuringian Forest at the north center; and (4) the low Jura Upland at the south, bordering the Danube Valley on the northern side. Interspersed with these lands are agricultural depressions or basins draining to the Rhine, Danube, Weser, or Elbe.

The Major Rivers

The latter streams, together with the Oder, are the principal German rivers. Three of them, however—the Rhine, Elbe, and Oder—rise outside the borders of Germany, and three—the Rhine, Danube, and Oder—enter the sea via a foreign country. Only the Weser is a purely German stream. The Rhine, which is by far the most important internal waterway of either West or East Germany, rises in Switzerland. From the Swiss border to its junction with the Main River, the Rhine flows in a flat-floored valley bordered on the east by the Black and Oden Forests and on the west by the Vosges Mountains of France and lower uplands extending northward from these mountains into West Germany (see p. 144). Downstream, the river follows a narrow gorge through the Rhine Uplands before it emerges upon the North German Plain in the vicinity of the industrial Ruhr district (photo, p. 81). Most of the lower course of the Rhine from the Ruhr to the North Sea lies in the Netherlands. The Weser also flows to the North Sea, as does the Elbe, which rises in Czechoslovakia. The Oder, formerly German for almost its entire length, now lies mostly in Poland,

and higher country to the south. The lowland of northern Germany is a part of the much larger North European Plain. The German segment of the plain seldom rises much above 500 feet in elevation and is mostly below 300 feet. It is predominantly an area of former glaciation, and the present topography bears much evidence of glacial action. Among the more prominent glacial features are the low ridges or *terminal moraines,* formed by long-continued deposition at the front of an ice sheet during times when melting balanced the ice movement to such an extent that the ice front remained stationary. These moraines, which are especially numerous east of the Elbe River, rise intermittently above a surface that is fairly level but seldom flat and is principally composed of infertile sandy or gravelly land punctuated here and there by stretches of heavy clay, a considerable number of bogs, and, in the northeast, many lakes. Along the coasts and the river estuaries bordering the North Sea are found expanses of flat, somewhat more fertile land that in most instances had to be drained artificially although it forms the boundary for a distance between Poland and East Germany. The Oder enters the Baltic Sea through Polish territory. The Danube River rises in the Black Forest of West Germany, although several important tributaries enter the upper river from Austria. The Danube flows eastward to the Black Sea. Most of its course lies outside Germany.

Climate, Soils, and Vegetation

The climate of Germany is maritime in the northwest and becomes increasingly continental toward the east and south. On the whole the climate, like that of most parts of northwestern Europe, is characterized by moderate temperatures. The January average temperature is 34°F at Emden, 31° at Berlin, and 28° at Munich. July average temperatures are 62° at Emden, 67° at Berlin, and 64° at Munich. The annual precipitation in both West and East Germany is adequate but not excessive, with most places in the lowlands receiving an average of 20 to 30 inches per year. Distribution of the rainfall is fairly uniform throughout the year, although there does exist a tendency for a summer maximum toward the south in both republics.

The soils of the two Germanies, while extremely varied in type, are rather acid and infertile by nature. Soils significantly higher than average in fertility are found mainly in a discontinuous belt along the southern margins of the North German Plain. Here a cover of loess (windblown silt) has provided parent material for the most productive large group of soils in both countries. Another region possessing soils of comparatively high natural fertility is the alluvial Upper Rhine Plain between the city of Mainz and the Swiss border. Elsewhere, soils tend to be, at best, of moderate quality; many, especially those of the hilly or mountainous sections, are poor.

The original virgin forest is thought to have been a mixture of coniferous and deciduous trees; today conifers predominate, although deciduous species are relatively abundant. Tracts of forest are distributed irregularly throughout the two republics on the rougher lands and the better-drained areas of poor soil. Prewar Germany was famous for its state-administered and rigidly controlled program of forest management, and good management practices are being continued in both West and East Germany. Both republics have a large output of lumber, wood pulp, paper, and other forest-derived products, but the demand for these products is so great that supple-

A winter scene in the Harz Mountains. Coniferous forests occupy the rounded slopes behind the small resort town (Braunlage, located about 40 miles south of Brunswick and very near the border between West and East Germany). Note the ski jump at the upper right. (German Tourist Information Office.)

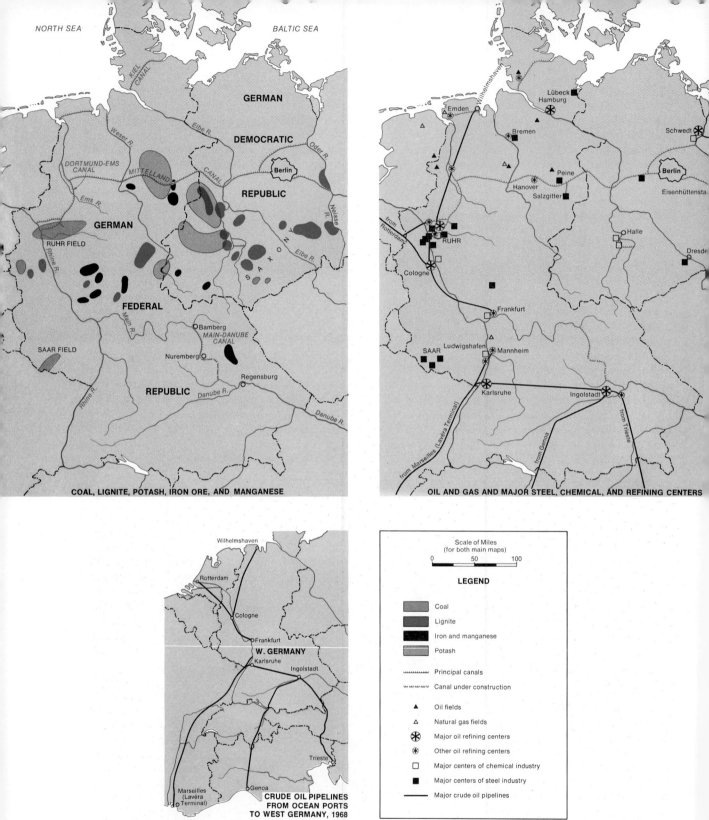

Major mineral resources, and centers of iron and steel manufacturing, chemical manufacturing, and oil refining in the two Germanies.

mental imports are required. This is particularly true of West Germany. Wood products entering that country come primarily from Austria, Sweden, Finland, or other European sources.

Coal, Potash, and Other Minerals

In both West and East Germany, coal is an abundant and vital resource. The reserves amount to much less than those of the United States or the Soviet Union, but are nevertheless extremely large. The largest and best deposits of bituminous coal, together with some anthracite, are found in the Ruhr coal basin (maps, pp. 139, 170, and 172). Other deposits of consequence exist in the Saar coal basin (bituminous) and the much smaller Aachen field (anthracite and bituminous). Lignite is mined extensively in the southern half of East Germany and to the west of the lower Rhine River in West Germany. The deposits lie close to the surface and are worked in huge open pits by large machines. In East Germany, which has very little bituminous coal, lignite is the leading source of energy. Both republics use lignite primarily in thermal generating plants to produce electricity, though in West Germany its importance in this regard is far less than that of bituminous coal. Most of the electricity used in the two Germanies is produced in plants powered by coal. Such plants are often located in or very near the coal fields. For example, about half of the electricity generated in West Germany comes from the Ruhr or other areas in the lower Rhineland. West Germany has some hydroelectric production from various sections of the Rhine and Danube drainage basins, but the annual output amounts to less than a fifth of the country's total production of electricity.

Potash, a vital material in the chemical industry, is another mineral which Germany has in outstanding quantities. The deposits, situated mainly between the city of Hanover and the upper Elbe River, are divided about equally between West and East Germany. They coincide in location rather closely with the principal area of fertile loess soils. West Germany's estimated production of potash in 1966 was slightly greater than that of East Germany.

Aside from coal and potash, and substantial deposits of common salt worked in southern East Germany and in the lower Rhineland, few minerals are exploited commercially in large amounts in either West or East Germany. Iron ore is mined at a number of places in West Germany, particularly in the vicinity of Salzgitter (city proper, 120,000) near the East German border. But the production, though substantial, is low grade, and must be supplemented by imported high-grade ores. Sweden is the largest source of such ores (1968), closely followed by the West African country of Liberia. Brazil and numerous other countries, primarily in Latin America or Africa, make contributions of varying size. The much smaller iron and steel industry of East Germany also depends on a combination of low-grade domestic ores with higher grade imported ores, primarily from the Soviet Union. Small deposits of a number of other metals exist in both West and East Germany, but the reserves and production are, in general, far from adequate. The output from several small oil fields in widely separated parts of the North German Plain makes West Germany the second largest oil producer in Europe (after Romania), but it supplied less than a fifth of the country's requirements in 1966. It may be noted, however, that this resource has been valuable to West Germany as a means of conserving foreign exchange. West Germany is also a small producer of natural gas. Recently the country's proved reserves have been enlarged by discoveries in areas immediately to the east of the immense new gas fields in the northern Netherlands.

POPULATION AND ECONOMIES OF THE TWO GERMANIES

Considered together, West and East Germany constitute one of the most densely populated parts of the world. Excluding a few of the microstates, their estimated overall density of approximately 562 per square mile in 1968 was exceeded in Europe only by the Netherlands (987), Belgium (814), and the United Kingdom (590). The German Federal Republic, with approximately 628 per square mile (including West Berlin), was more densely populated than the German Democratic Republic, with about 409 per square mile.

German is the household language of practically the entire population, though marked variations in dialect exist from place to place. With respect to religion, the population of West Germany is fairly evenly divided between Protestants (50 percent according to the 1964 census) and Roman Catholics (46 percent), but East Germany is overwhelmingly Protestant (82 percent, as opposed to 11 percent Roman Catholic). These figures reflect a traditional division between the North German Plain, which has been heavily Protestant, especially in the east and north, and the Rhineland and southern West Germany, which have been the principal strongholds of the Roman Catholic faith.

Distribution of Population and Cities

The distribution of rural population is fairly uniform throughout the country. The lightest densities are found in the more rugged parts of the central and southern uplands and highlands, and in some of the more infertile areas of the northern plain. The heaviest densities tend to occur in the zone of loess soils and in the southwestern depressions draining to the Rhine.

Both West and East Germany are highly urbanized, though less so than the United Kingdom. The primary axis of city development tends to follow the border zone between the North German Plain and the southern uplands. This axis is anchored on the west by a great cluster of industrial cities in the Ruhr and on the east by a smaller, less concentrated cluster of manufacturing cities in Saxony. The important cities of Hanover (790,000), Brunswick (280,000), and Magdeburg (370,000) are spaced along the central part of the urban axis. Berlin (4 million), the largest city of Germany, is situated slightly to the north of the axis at its eastern end.

Other major cities outside the main urban axis include (1) the North Sea ports of Hamburg (2.4 million) and Bremen (700,000); (2) the widely spaced cities along the Rhine and its tributaries, including (besides the Ruhr cities) Cologne (1.7 million), Bonn, the capital of West Germany (320,000), Mannheim-Ludwigshafen-Heidelberg (1.3 million), Frankfurt-on-the-Main (1.6 million), Stuttgart (1.5 million), and Wiesbaden-Mainz (555,000); and (3) the important, widely separated Bavarian cities of Munich (1.7 million) and Nuremberg (715,000).

Germany's Premier Industrial District—The Ruhr

The urban dwellers of Germany, and many of the rural inhabitants as well, are primarily dependent upon manufacturing for a livelihood. As in all great manufacturing countries, industrial development rests on three basic lines of goods: (1) metals, particularly iron and steel and products made from them, (2) textiles, and (3) chemicals. The largest single concentration of industry

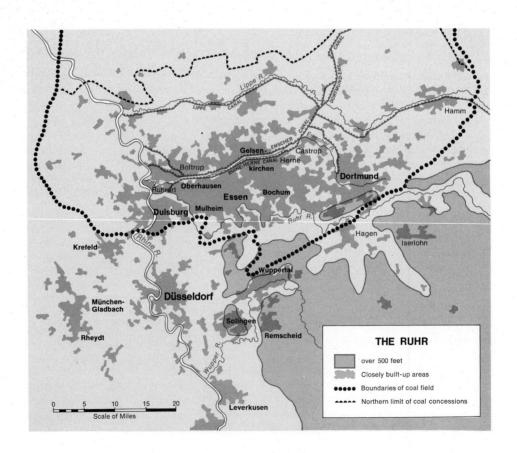

The industrial might of the Ruhr is reflected in the immense barge traffic of Duisburg-Ruhrort. The city's harbor, seen in the above view, provides the main point of contact for Rhine shipping serving the Ruhr. About a third of West Germany's raw steel production is handled here. The waterway at the extreme right in the photo is the Ruhr River, which joins the Rhine at the bottom of the view. Waterways parallel to the Ruhr River are canals leading into the barge harbor from the Rhine. A steel mill is visible at the upper left, and a crowded residential area lies between it and the harbor. (German Information Center.)

is found in the Ruhr, an industrial district so important as to have no real counterpart in all of Europe. The Ruhr is the principal seat of the German iron and steel and coal-mining industries, and has important metal-fabricating, chemical, and textile industries as well.

The Ruhr district takes its name from the Ruhr River, a small east-bank tributary of the Rhine. Heavy industry has been attracted to the district by Europe's most important coal field, and particularly by the great re-

serves of coking coal. Some salient features of the Ruhr are sketched by Norman J. G. Pounds in the following selection:[3]

[The Ruhr is located] in northwest Germany, close to the Belgian and Dutch frontiers. In the south the Ruhr is hilly, and large areas are forested. In the north it is low and rolling; movement is easy, and much of the land that has not been built upon is agricultural. Most of the cities and towns lie on the northern plain. . . .

---

[3] Norman J. G. Pounds, "The Ruhr," *Focus,* **1,** no. 3 (December 1950), 1–2. A subhead and populations of cities have been added. Used by permission of the author and *Focus,* published by the American Geographical Society of New York.

Across the western edge of the Ruhr flows the Rhine, a great highway of trade between South Germany and the Netherlands. The industrial area is drained by the Ruhr, Emscher, and Lippe rivers. . . . Canals have been built along the Emscher and Lippe valleys and at their east ends are joined to the Dortmund-Ems Canal, by which they are linked with the German North Sea ports of Emden and Bremen.

The Ruhr industrial area has grown up within the last 100 years. Its origins lay in the primitive iron refining carried on in the hills, where there were small deposits of iron ore, as well as charcoal for the furnaces and running streams to power the machines and the hammers. It was not until after 1850 that industry spread on any considerable scale from this early center into the area that we now call the Ruhr. It was attracted by the beds of coal, much of which was suitable for coking and thus for the iron furnace. In the years 1850 to 1870 many steelworks were built. Local supplies of ore were soon exhausted and the Ruhr came to depend more and more on imports from foreign countries.

As industry developed in this area, so the cities grew in size and importance. There are today three cities each with more than half a million inhabitants: Essen [720,000], Dortmund [660,000], and Düsseldorf [695,000].[4] There are [20] cities with more than 100,000 inhabitants.

[The Coal-mining Industry]

But without the coal field there would be no Ruhr as we know it today. The coal seams reach the surface along the valley of the Ruhr . . . and dip northward beneath the North German Plain. Mining began along the southern edge, but for more than a hundred years it has been spreading slowly northward, with the opening of mines of ever-increasing depth. The oldest mines are in the Ruhr Valley, which has thus given its name to the region; the newer are in the valleys of the Emscher and the Lippe. There seems to be no limit to this northward expansion except that set by the increasing depth and cost of mining.

The coal varies in quality. The lowermost seams, which come to the surface in the south, are anthracitic. Above them are, in order, coking, gas, and flame coals. Of these the coking coals, which are mined in the Emscher Valley, are of the greatest industrial importance.

Between the two wars the Germans concentrated production in a reduced number of larger mines. At many of these, coking furnaces were erected. Here the coal is pulverized, blended, and coked. The coke is sent to the blast furnaces, and the gas from the retorts is used for various heating purposes in the steelworks; the exhaust gases from the furnaces go to fire the retorts in the cokeries. Coal mining and steelworking have thus been integrated geo-

graphically, not merely in the balance sheets of the operating companies.

The Inner Ruhr

A distinction often is made between the "inner Ruhr," the area of concentrated coal mining, iron and steel manufacture, chemical manufacture, and urban development, and the "outer Ruhr," in which cities are more widely spaced. Extending for about 40 miles from the Rhine in the west to Dortmund in the east, and for about 15 miles from the Ruhr River northward (map, p. 172), the inner Ruhr incorporates many individual cities and comprises an estimated 5½ million people—the largest urban agglomeration in Europe except the London and Paris metropolitan areas. Though steelmaking and steel-using industries are spread through the area, most of the actual smelting of iron ore is done in large, integrated iron and steel mills located in or near Duisburg (city proper, 475,000) or Dortmund. The Ruhr's largest individual city, Essen, home of the famous Krupp works, is a major producer of high-grade steels, but does little smelting. Essen is the the principal administrative center of the Ruhr.

Chemical manufacturing in the inner Ruhr is based on a variety of raw materials, including coal, manufactured gas, waste gases from cokeries, phosphate-rich wastes from the iron and steel industry, imported petroleum, and natural gas piped from deposits near the lower Ems River. Chemical industries exist in many parts of the inner Ruhr, with especially noteworthy concentrations in or near Duisburg and in the Emscher Valley.

The Outer Ruhr

In the "outer Ruhr" cities are more scattered, though still numerous, and their manufactures consist primarily of a great variety of secondary metal products, chemicals, and textiles. The agglomeration of Krefeld-München Gladbach-Rheydt (690,000), west of the Rhine, is predominantly a textile-milling area, though it has numerous other industries. South of the Ruhr Valley, the cities of Wuppertal, Solingen, and Remscheid form an urban agglomeration of 925,000. Wuppertal is best known as a textile center, but Solingen and Remscheid make a variety of highly finished metalwares, including hardware, tools, machine tools, and the high-grade cutlery

---

[4] City populations in brackets are 1968 estimates for the city proper. All other city populations cited in this chapter and other chapters, unless otherwise noted, are 1968 estimates for the metropolitan area (see n. 2, p. 100).

An industrial landscape in the inner Ruhr. The plant in the foreground, located at Oberhausen, mines coal and processes part of the output to make coke, gas, and chemicals. Note the two hoists of the coal mine at the left of the view. They are used to raise and lower elevators in the vertical shafts of the mine. Innumerable chimneys of industrial plants punctuate the horizon. At the upper right are blast furnaces of an iron and steel plant. (German Information Center.)

for which Solingen is world famous. The large Düsseldorf metropolitan area (1.1 million), located on the Rhine some distance south of Duisburg, has a variety of engineering industries and a substantial production of iron and steel. In addition it is an important banking, commercial, and administrative center, with imposing new office buildings housing the headquarters of many Ruhr firms. Cologne, still farther south, is not generally considered a part of the Ruhr.

Industrial Districts outside the Ruhr

The Saxon Triangle

Some 200 miles east of the Ruhr, a second industrial area of high importance lies in Saxony, immediately to the northwest of Czechoslovakia. This district is located in East Germany. Its industrial production is extremely diversified. The district has long had a high reputation for types of goods requiring much skill and precision in manufacture—precision machinery, machine tools,

electronic equipment, cameras and other optical equipment, musical instruments, fine porcelain, and so on. Metal and engineering industries, chemicals, and textiles employ the largest share of the labor force, but food processing and various other kinds of manufacturing are important. The main industrialized area of Saxony is shaped like a triangle (map, p. 138). At the northwestern apex is the city of Halle (430,000), the principal urban center of an important potash, salt, and lignite-mining and chemical-manufacturing area. Halle is East Germany's most important center of chemical industries, and in addition has engineering industries that make mining machinery and many other products. It also has sugar refineries that process beets from areas of fertile loess soil near by. A short distance to the southeast is the largest city of Saxony, the diversified industrial center of Leipzig. At the southeastern corner of the triangle is another diversified industrial city, Dresden (635,000) on the Elbe River. Clusters of lignite mines that supply East Germany's basic energy source as well as an important raw material for chemical manufacture are grouped around Halle and Leipzig; while other mines cluster some distance north of Dresden. Westward from Dresden a line of

Assembly lines in the Volkswagen plant at Wolfsburg, West Germany. Automobiles are one of West Germany's major exports. (Volkswagenwerk GMBH.)

manufacturing towns and cities extends along the northern foothills of the Ore Mountains. The largest city is Karl-Marx-Stadt, formerly Chemnitz (420,000), an important textile-milling center which also manufactures machinery, including textile machinery. To the southwest are two other textile centers, Zwickau (175,000) and Plauen (city proper, 80,000), the latter being located at the southwestern corner of the triangle. With its varied range of production and skilled labor force, the Saxon Triangle represents one of the richest prizes to fall within the orbit of the Soviet Union as a result of World War II.

No brief résumé can do justice to the great variety and complexity of industrial production in West and East Germany. The range of production extends from the basic lines of metals, chemicals, and textiles (including an increasing output of synthetic textiles) through many types of machinery, motor vehicles, ships, electrical equipment, optical equipment, processed foods, and products made from wood, paper, leather, and tobacco, to such specialty items as toys, high-grade porcelain, and musical

instruments. Manufacturing is by no means confined to the Ruhr and the Saxon Triangle, but is spread through a multitude of other cities and towns outside these major industrial foci. Relatively small cities may house manufacturing plants of major size. Examples are the giant Volkswagen plant at Wolfsburg (city proper, 85,000), West Germany; and East Germany's large iron and steel plant at Eisenhüttenstadt (city proper, 37,000) ("Steel Mill City") on the Oder River to the southeast of Berlin.

Berlin and Hamburg

Berlin was the most important single manufacturing city before World War II, with Hamburg probably ranking second. Like Paris, Berlin is situated in a locality that lacks significant natural resources for manufacturing. But like Paris, the city was able to develop important industries as an indirect result of political centralization. Most of this development began after 1871, and Berlin is thus much younger as a major manufacturing city than Paris or London. As the political capital of a large, new, and strongly centralized state, Berlin quickly attracted great numbers of people, and its population grew from half a million in 1865 to over 2 million at the turn of the century and to 4 million by 1925. Thus its developing industries could rely on a large market in the immediate vicinity, and they had good connections with other areas via the impressive network of transportation lines which was focused on the capital by the government. Berlin became one of the world's more important centers for the manufacture of electrical equipment, and this remains the city's leading industry in both employment and value of production. But a wide variety of other industries exist, mainly secondary consumer-type industries such as food processing, clothing manufacture, light engineering, and the manufacture of pharmaceuticals.

Hamburg also has extremely diversified industries, including electrical and machine-building industries, shipbuilding, oil refining and chemical manufacture, rubber processing, printing and allied industries, food processing (grain milling, processing of imported oilseeds, and the like), and many others. Many of its industries are "port industries," such as shipbuilding and the processing of raw materials and foods brought in by sea. The Hamburg metropolitan area is the largest center of West Germany's shipbuilding industry, the third largest in the world (after Japan and the United Kingdom), though the shipyards of Bremen-Bremerhaven and Kiel have an important share of the industry.

Other Industrial Areas

Special mention should also be made of the important group of diversified industrial cities located along the Rhine and its tributaries south of the Ruhr—particularly Cologne and Mannheim-Ludwigshafen on the Rhine, Frankfurt-on-the-Main, and Stuttgart on the Neckar. Each of these cities has a metropolitan population of over a million, and in each case the metropolitan area is a major center of engineering industries. The first three are prominent centers of chemical manufacture.

The main city of West Germany's far south is Munich. Like the country's other "million cities," it has a highly diversified economic base.

The Saar district contains West Germany's second most important concentration of coal mines and iron and steel plants, though its output is far overshadowed by the enormous production of the Ruhr. For iron and steel manufacture the Saar depends primarily on Lorraine iron ore. Some coke and high-grade coking coal are brought from the Ruhr to supplement the Saar's own lower-grade production. However, improvements in technology enabling the Saar to make better coke from its own coal have considerably reduced its dependence on the Ruhr. The largest urban center of the district, Saarbrücken, is a city of 140,000, but its metropolitan area comprises an estimated 335,000 people.

The New Development of Oil Refining and Petrochemicals

A significant new industrial development in both West and East Germany is large-scale oil refining in inland locations. The refineries operate almost entirely on imported crude oil brought by recently constructed pipelines from North Sea and Mediterranean ports or, in the case of East Germany, from the Soviet Union. West Germany's refining formerly was done in or near the seaports, particularly Hamburg, or in a few "field refineries" adjacent to the country's own oil wells. But though Hamburg and the lower Elbe area are still the locale for a major cluster of refineries based primarily on imported oil, the main focus now has shifted to large refineries along or near the Rhine and Danube rivers. The largest capacity is found in the area of the Ruhr and Cologne. In this district refineries aggregating over a third of West German capacity are served by two pipelines, one from Rotterdam and the other from Wilhelmshaven (city proper, 100,000). The harbor of Wilhelmshaven is able to accommodate tankers of 100,000 tons or over, and new facilities under construction at Rotterdam's Europoort will be able to accommodate even larger ships. Pipelines from the two ports converge at Cologne, and an extension carries crude oil farther south to a new refinery at Frankfurt. In the Upper Rhine Plain several large refineries have developed around Karlsruhe (370,000), which is served by the South European Pipeline from Lavéra near Marseilles, France. A northward extension carries oil to a new refinery at Mannheim. An eastward extension connects Karlsruhe with a cluster of large refineries at Ingolstadt, Bavaria, on the Danube River. But oil also flows to Ingolstadt (and Karlsruhe) via the Transalpine Pipeline from the port of Genoa, Italy; and Ingolstadt will soon have a pipeline from the port of Trieste, Italy, on the Adriatic Sea. Ingolstadt is admirably located for distribution of refined products, having access to good road and rail facilities (and perhaps eventually to barge transport on the Danube), and being about midway between Munich and Nuremberg, and also midway between Augsburg (320,000) and Regensburg (city proper, 125,000). In East Germany the main oil-refining center is the town of Schwedt, located near the Oder River to the northeast

The market square at Tübingen, a university town just south of Stuttgart. The town, seat of a medieval university, was unable to live on the business brought by the university alone, and the market function has long been equally important. The open stalls around the periphery of the square sell produce from the surrounding region, which varies from the traditional vegetables and cut flowers to baskets, glassware, pottery, and clothing. The motor truck has replaced the ox-drawn or horse-drawn wagon. The market square is in the medieval nucleus of the town. Note the narrow winding streets leading off the square and the old half-timbered buildings. The latter have modern shops at street level and offices and apartments above. At the left of the view is a small hotel. (J. A. Soulsby.)

of Berlin. The town is the terminus of the northern branch of the "Friendship" Pipeline from the Volga-Urals oil fields of the Soviet Union. (The northern branch carries oil to Poland and East Germany, while the southern branch serves Hungary and Czechoslovakia.) In both West and East Germany important petrochemical industries have developed in the major oil-refining areas.[5]

The Scientific Agriculture of the German Republics

The fact that both West and East Germany are predominantly manufacturing countries should not be allowed to obscure their agricultural productivity. Though it is not so self-sufficient agriculturally as France, West Germany is estimated to supply from its own production a little more than two-thirds of its requirements for agricultural products. East Germany attains an even higher degree of self-sufficiency, though its present standard of living is considerably lower than that of the Federal Republic. The high output from soils which are, in the main, only moderately fertile, is a tribute to German enterprise and skill. It also derives in part from tariff protection and government subsidies of various kinds, as well as the market opportunities afforded by large urban populations with a rising standard of living. The art of soil improvement through careful use of natural and artificial fertilizers, crop rotation, and other scientific land management practices has been developed to a high degree, and the average crop yields in both West and East Germany are among the highest in the world. Only the much smaller, intensively cultivated neighboring countries of Belgium, the Netherlands, and Denmark

[5] This section has been based on a fine article by Peter P. Waller and Harry S. Swain, "Changing Patterns of Oil Transportation and Refining in West Germany," *Economic Geography*, 43, no. 2 (April 1967), 143–156. Students are urged to read the entire article.

definitely surpass West Germany in all-around per-acre yields. East Germany lags somewhat behind the Federal Republic, but its yields still are higher than those of the majority of European countries. The highly developed chemical industries of the two German republics have provided agriculture with large quantities of chemical fertilizers. West Germany ranks third or fourth in the world in consumption of commercial plant food per arable acre (after Belgium, the Netherlands, and possibly Japan), and East Germany probably ranks fifth. But despite the productivity of domestic agriculture, both West and East Germany must import sizable quantities of foodstuffs, fibers, and other agricultural products. In 1965 such items represented about 33 percent by value of all imports entering the German Federal Republic. It should be noted that German agricultural productivity is achieved at a relatively high cost. Were agriculture not subsidized and protected by government, it seems probable that a much higher percentage of agricultural requirements would be imported.

Rye, oats, and potatoes have long been staple crops of the North German Plain, although wheat and sugar beets have been more important in many localities in the zone of fertile loess soils along the southern margin of the plain. In recent years the acreage and production of oats have declined sharply as an accompaniment to a decreased number of horses. Rye acreage and production exceeds that of wheat in East Germany, but in West Germany both wheat and barley now exceed rye. The depressions in the central and southern upland country grow rye, wheat, barley, and oats in a variety of rotations. Potatoes and sugar beets also are important, though somewhat less so than on the northern plain. In the two Germanies rye and wheat are grown both as breadstuffs and for animal feed, barley is used for animal feed and in the beer industry, and oats is grown for animal feed. Potatoes are used for human food, for the manufacture of alcohol, starch, and other products, and for animal feed. The great importance of potatoes in German agriculture is shown by the fact that in West Germany potatoes occupy an acreage over half as large as the acreage occupied by wheat, and in East Germany an acreage nearly twice as large as that occupied by wheat. In both republics domestic sugar beet production supplies the great bulk of sugar requirements.

In the southwestern part of West Germany, orchards and vineyards are major specialties in the sheltered valleys of the Rhine and its tributaries, particularly the lower Main, Neckar, and Moselle.

Livestock production is an important branch of agriculture in all parts of Germany. In West Germany, livestock and livestock products account for an estimated

three-fourths of the value of all products marketed from farms. Major emphasis is placed on the growing of cattle and hogs: West Germany raises more hogs than any other European country (excluding the USSR) and more cattle than any European country except France. In the cooler, more infertile, and poorly drained portions of the North German Plain, large sections are given over principally to the grazing of livestock, and the same is true of the rougher areas in central and southern sections of West Germany. Dairying is especially prominent in two areas: (1) the German Alpine Foreland and German Alps, south of the Danube, and (2) the reclaimed coastal marshes bordering the North Sea. It should be noted, however, that cattle in Germany consist primarily of dairy varieties. There is no well-developed beef animal, and thus, although certain districts may specialize in fattening steers for slaughter, there is not the sharp

A traditional type of farmstead in many of the older settled areas of southern Germany consists of buildings around three sides of an open courtyard. The photo below was taken a short distance southwest of Stuttgart. In the building at the left, animals are housed at a lower level, below the living accommodations. The barn and cart shed with roof-high doors occupies the closed end of the courtyard, while on the right is a small hand implement shed and vegetable patch. Animal dung is piled over an open grate in the center of the courtyard, and rainwater percolates through the dung and grate into a storage tank below ground. The liquid manure is then lifted as required by means of the hand pump in the left foreground. (J. A. Soulsby.)

difference between the beef and dairy industries that one finds in the United States. Some hogs, cattle, and of course, poultry, are present on nearly all German farms. These livestock provide cash products which are readily salable in nearby urban markets; supply quantities of manure for enriching the soil; consume kitchen wastes and crop residues such as the tops and pulp left over from sugar-beet processing; and generally fit well in the intensive, careful agriculture of such a densely populated, urbanized area as Germany.

Transportation Facilities

The complex economies of West and East Germany are served by highly developed transportation facilities. The largest seaports are Hamburg on the Elbe and Bremen on the Weser, both in West Germany. These ports are located well upstream on the estuaries of their respective rivers, but are served by outports. Bremen's outport, Bremerhaven (155,000), handles much of Bremen's passenger traffic, as well as considerable amounts of freight; but in the case of Hamburg (which has a much smaller passenger traffic), most vessels make the journey up the Elbe to the main port instead of discharging their passengers at the outport of Cuxhaven (city proper, 45,000). Formerly a port of call for large liners, Cuxhaven has now become primarily a fishing port. The great Dutch seaport of Rotterdam on the lower Rhine handles large tonnages of West German trade (including a portion of the Ruhr's iron-ore imports) and in a functional sense may be counted as one of the major German ports. In the harbors of Rotterdam and Ham-

The port of Bremen opens off the River Weser, seen at the right in the view below. Three distinct harbors are shown: the Timber and Factories Harbor at the left, the Overseas Harbor in the center, and the European Harbor at the right. Note the extensive railway yards in the midst of these harbors. A large share of the port's traffic consists of the transfer of goods between oceangoing vessels and railway freight cars. (German Information Center.)

burg a major specialty is the direct transfer of goods between ocean ships and river or canal barges. The Rhine River, navigable for barges from the Swiss port of Basel to the sea, is the largest German carrier of barge traffic, and the city of Duisburg, located on the Rhine near the western end of the Ruhr industrial district, is one of the most important river ports in the world (most of the traffic is actually handled by the suburb of Ruhrort) (photo, p. 173). Canalization of the Moselle from its junction with the Rhine to the Lorraine iron-mining and steel-milling district was completed in 1964. Though completion of a planned canal connection from the Main tributary to the Danube is not anticipated in the immediate future, the lower and middle Main has been canalized for large barges, and the northern part of the Main-Danube Canal has been scheduled for completion as far as Nuremberg by 1969 or 1970. Outstanding among the other inland waterways of West and East Germany are (1) the Elbe River, navigable for barge traffic from Czechoslovakia to the North Sea (though underused because it is cut by the iron curtain); (2) the Mittelland Canal System connecting the Rhine and the Ruhr district with the Weser, Elbe, and, via Berlin, the Oder; (3) the Dortmund-Ems Canal (recently enlarged to accommodate 1500-ton barges) connecting the Ruhr with the North Sea via the Ems River and the port of Emden (city proper, 48,000), which is the most important receiving port for Swedish iron ore en route to the Ruhr; and (4) the North Sea-Baltic Canal or Kiel Canal, a ship canal which crosses the base of the Danish peninsula and thus eliminates the sea journey around Denmark for many vessels plying between the Baltic and the German North Sea ports.

Fast inland transportation is provided by an effective system of railways and superhighways (*autobahnen*), connecting the larger cities. An important postwar development has been the emergence of the truck as a major freight carrier. As in the United States, many trucking fleets are owned by manufacturing companies, while others serve as public carriers. Originally designed to focus on Berlin, the *autobahn* system has developed as a network joining all major cities and industrial districts. For West Germans, travel through East Germany to West Berlin sometimes is hindered by political conditions.

With the emergence of a dynamic trucking industry has come increased competition between truck and rail. Regulating bodies are considering a ban on heavy trucks for distances exceeding 30 miles, and on shipment of certain goods by truck. This would revitalize a lagging rail industry and provide relief to the overworked highways.

A major transport bottleneck in both West and East Germany, and indeed in western Europe generally, is the geographic layout of today's urban places. Characterized largely by small winding streets inherited from the past, they cannot easily accommodate large modern trucks. Yet such trucks are necessary for efficient movement of goods. The answer would appear to be freeways that bridge the congestion.

Large-scale transfer of passengers and freight by air is growing rapidly. One of the most significant contributions of air transportation is the quick and sure access it provides from the German Federal Republic to West Berlin. A spectacular demonstration of the latter was the famous Berlin airlift of 1948–1949, in which the Western occupying powers, through ferrying necessary supplies to West Berlin by air over a period of more than a year, succeeded in overcoming a Russian blockade of surface transportation into the city.

Pipelines are being introduced actively into the economies of both West Germany and East Germany, as previously discussed.

AFTERMATH OF PARTITION

Economic Dissection: The Example of Hamburg

As a consequence of defeat in World War II, Germany underwent not only a political but also a severe economic dissection. The country lost control of important coal-mining and steel-milling districts in Silesia (map, p. 165) and (temporarily) the Saar, as well as agricultural lands east of the Oder-Neisse line which normally had produced a considerable surplus of foodstuffs in prewar times. Probably the most damaging aspect of the postwar situation, however, was the disruption of normal trading relationships between West and East Germany. Some of the effects of this disruption are illustrated in the following concrete example.

During the late 1930s the port of Hamburg was one of the leading seaports of Europe and of the world. World War II brought great destruction to the city and particularly to the harbor. When peace came, the overall capacity of the port had been reduced to about one-fifth of its prewar level. Embarking upon an energetic rebuilding program, the port quickly restored its facilities and by 1955 had surpassed its 1936 tonnage of seaborne commerce. Today, Hamburg is the leading seaport in West Germany and third or fourth (depending on the reporting year) in continental western Europe, ranking behind Rotterdam and Antwerp, and, in some years, Marseilles. However, much of its prewar hinterland, notably the portion upriver on the Elbe, is now behind the iron curtain, and traffic to and from this territory stands at less than half of the prewar volume. At present the port's primary hinterland is the northern section of the German Federal Republic, especially the city-state of Hamburg itself, though government subsidies allow the port to compete with Bremen and Rotterdam in serving the southern section of West Germany. The composition of Hamburg's commerce has changed since the war. It is now made up more predominantly of bulk goods, particularly coal, petroleum, and their products, than it was in the 1930s.

The postwar severance of the German Democratic Republic has required extensive readjustments not only in Hamburg but in many other centers of industry and trade in both republics. Each republic needs the trade of the other in order to function most effectively. There appears to be little evidence, however, that the political and economic cleavage represented by the Russian-imposed iron curtain will be fundamentally altered within the foreseeable future.

Absorption of Political and Ethnic Expellees

German recovery from World War II was also hindered initially by the necessity of caring for a great number of expellees and other displaced persons. By 1950 the German Federal Republic had accepted nearly 8 million and the German Democratic Republic nearly 4½ million newcomers, most of them either political or ethnic expellees from Communist-held areas. The political expellees were former occupants of German national territory transferred to Polish or Russian administration following the war. The ethnic expellees were people of German descent who lived in countries outside of Germany before the war, but were alleged to have had strong pro-German leanings. The latter group came mainly from Czechoslovakia, Poland, Yugoslavia, the former Free City of Danzig (now Gdansk), Romania, and Hungary. Nearly a sixth of Germany's total population is now comprised of the two categories of expellees and their families. Further complexity was added by the voluntary migration of over 3 million persons from Communist to non-Communist Germany and by a much smaller movement in the opposite direction.

It was feared initially that the task of absorbing so many millions of homeless people into the economy of a defeated nation would prove insuperable. Yet West Germany had largely accomplished the task by 1954. Additional immigrants now are being absorbed under Common Market arrangements that permit unimpeded movement of people searching for work in the six member countries. Relocation is no longer a serious problem in East Germany. The success of the two republics in coping with it reflects their resurgence from the debacle of World War II.

Resurgence of the Two Germanies

For over three years following defeat in 1945, Germany was in ruins. The war had cost her an estimated 5.6 million lives, including 3.8 million in the armed forces. Much of her machinery had been carried off as reparations. Her merchant fleet was gone. She had practically no credit. Industrial production at the end of the war stood at less than one-tenth of the prewar level. Approximately one third of all housing had been bombed out; in the major cities the proportion was nearly twice as high. Unemployment, inflation, and hunger existed everywhere.

Beginning in 1949, however, the Western zones began rebuilding at a frenzied pace, and since that time the German Federal Republic has amazed the world with its rapid economic recovery. Aided initially by the United States and other nations, its economy soon acquired its own momentum, and it has emerged as one of the most dynamic in Europe. A measure of the achievement is seen in the fact that in 1967 West Germany was Europe's largest producer of coal and steel (excluding the Soviet Union). In its rebuilding program, attention was initially focused on the restoration of capital equipment, but consumer necessities were not overlooked. Despite an expanding population, there is no longer a pressing housing shortage in the country. The factories, commercial establishments, and other components of the country's large cities also have been essentially rebuilt, though in some cities, notably Berlin, little attempt has been made to reconstruct the devastated "dead heart" of the city, and new development has taken place primarily in the suburbs. Today the German Federal Republic figures prominently in any economic, political, or military program involving western Europe, and it is increasingly a factor of significance in the affairs of the world at large.

The German Democratic Republic, smaller in area and population, less well endowed naturally, linked politically and economically with war-ravaged nations, handicapped by rigid initial adherence to Marxist dogma, and burdened with heavy reparations demands by the Soviet Union, recovered more slowly than its western neighbor. But it, too, has experienced a marked resurgence, especially since reparations ceased in 1957. Indeed, it now ranks among the top ten nations of the world in total industrial output—perhaps as high as eighth. Its population, however, declined by about 2 million between 1953 and 1960 and has remained at about 17 million since that date.

Prospect of Reunification

Although prediction is hazardous in the present era of rapid political changes, there seems to be little prospect that Germany can recover its territories lost to Poland and the Soviet Union, or that the country can look forward to a reunification of the two German republics in the immediate future. Nor does the problem of Berlin seem near a solution, though a change for the better in relations between the Soviet Union and the Western powers might produce a formula that would end the Western occupation of West Berlin without surrendering its inhabitants to the Communist state which surrounds them. Poland and the Soviet Union have established firm control over the former German territories east of the Oder-Neisse line. Most

German residents were evacuated to either West or East Germany, their places were taken by Poles or Russians, and the territories were reorganized functionally to mesh into the economies of the governing nations. Meanwhile the two German republics have drawn apart in their economic and political relationships. The German Federal Republic, recognized as a fully sovereign entity by the Western nations since 1955, has become more and more interlocked with them in political, economic, and military affairs. But the German Democratic Republic looks largely to other Communist nations in political and economic relationships, though it is slowly forging a certain number of economic links with the West. The Soviet Union is by far its most important trading partner, being both the largest supplier of its imports and the largest market for its exports. Well over a third of its total trade (1967) is with the USSR. Meanwhile its internal political and economic organization, including the organization of its agriculture, has been remodeled along Communist lines.

Thus Germany is fragmented in political and economic fact, as well as in name. Probably most Germans would like to see their country reunified. But as months become years and years become decades, the possibility of reunification continues to recede into the indefinite future.

REFERENCES and READINGS

BARRINGTON, R., "The Hamburg 'Outer-Harbour' Project and Related Developments," *Tijdschrift voor Economische en Sociale Geografie,* 59, no. 2 (March–April 1968), 106–108.

BRADLEY, PETER, "Bremen: A City-State," *Geographical Magazine,* 37, no. 9 (January 1965), 696–708.

BURTENSHAW, D., "Recent Changes in Saar Coalmining," *Geography,* 53, pt. 4 (November 1968), 404–406.

CAHNMAN, WERNER J., "Frontiers between East and West in Europe," *Geographical Review,* 39, no. 4 (October 1949), 605–624.

CHILDS, D., "Recent East German Economic Progress," *Geography,* 51, pt. 4 (November 1966), 367–369.

DICKINSON, ROBERT E., *Germany: A General and Regional Geography* (2d ed.; London: Methuen and Co., 1961).

DILL, MARSHALL, JR., *Germany: A Modern History* (Ann Arbor: University of Michigan Press, 1961).

FRANKLIN, S. H., "Gosheim, Baden-Württemberg: A 'Mercedes Dorf,'" *Pacific Viewpoint,* 5, no. 2 (September 1964), 127–158.

FRIEDENSBURG, F., "The Geographical Elements in the Berlin Situation," *Geographical Journal,* 133, pt. 2 (June 1967), 137–147.

Geographische Rundschau (monthly). An informative, scholarly geographical journal intended primarily for teachers. The articles are in German, with summaries in English. Has world-wide coverage, but with a considerable focus on Germany.

GROTEWOLD, ANDREAS, and MICHAEL D. SUBLETT, "The Effect of Import Restrictions on Land Use: The United Kingdom Compared with West Germany," *Economic Geography,* 43, no. 1 (January 1967), 64–70.

GUTKIND, E. A., *Urban Development in Central Europe,* Vol. 1 in his *International History of City Development* (New York: The Free Press of Glencoe, 1964). Many interesting photos of German cities accompany the text of this large and handsome volume.

HALL, PETER, *The World Cities* (London: Weidenfeld and Nicolson, World University Library, 1966), Chap. 5, "Rhine-Ruhr," pp. 122–157.

HERMANN, RICHARD D., "Some Observations on Forestry in Present Day Germany," *Journal of Forestry,* 66, no. 3 (March 1968), 174–177.

HIRST, STEPHEN M., and LEON LEWINS, *Basic Data on the Soviet Zone of Germany,* U.S. Bureau of International Commerce, Overseas Business Reports, OBR 67–5 (February 1967). (Washington, D.C.: Government Printing Office, 1967.)

HOLZNER, LUTZ, "The Rhine-Main-Danube Waterhighway," *Journal of Geography,* 65, no. 6 (September 1966), 270–281.

MAURER, ALAN O., *Basic Data on the Economy of the Federal Republic of Germany,* U.S. Bureau of International Commerce, Overseas Business Reports, OBR 67–24 (May 1967). (Washington, D.C.: Government Printing Office, 1967.)

MUTTON, ALICE F. A., *Central Europe: A Regional and Human Geography* (2d ed.; New York: Frederick A. Praeger, 1968).

PERRY, NORMAN, "Recent Developments in the West German Oil Industry," *Geography,* 52, pt. 4 (November 1967), 408–411.

POUNDS, NORMAN J. G., *Divided Germany and Berlin* (Princeton, N.J.: D. Van Nostrand Co., Searchlight Books, 1962); also *The Economic Pattern of Modern Germany* (London: John Murray, 1963); and "East Germany," *Focus,* 17, no. 2 (October 1966), 6 pp.

U.S. Army Area Handbook for Germany (Department of the Army, Pamphlet No. 550–29); (2d ed.; Washington, D.C.: 1964).

WALLER, PETER P., and HARRY S. SWAIN, "Changing Patterns of Oil Transportation and Refining in West Germany," *Economic Geography,* 43, no. 2 (April 1967), 143–157.

"West Germany, 1966", *Current History,* 50, no. 297 (May 1966), 257–308. A series of articles by various authors.

See also the lists of references and readings for Chapters 1–4, particularly the relevant chapters in geographies of Europe listed on p. 92.

The Benelux Countries

8   The three small countries of Belgium, the Netherlands, and Luxembourg have been closely associated with each other throughout their long histories. During some periods they have been included in a single political unit, although they have been politically separated during the greater part of the past four centuries. During the past quarter century, they have been attempting to strengthen their mutual relations and to weld themselves into an economic union, while maintaining their respective political sovereignties. The economic union has been designated "Benelux" from the first syllable of each country's name, and the three nations are now often referred to collectively as "the Benelux Countries." Although the gradual absorption of these three countries into the European Economic Community has deemphasized the initial concept of a Benelux customs union, the three countries are still commonly designated as "Benelux". An older name often applied to the three is "the Low Countries." In its strictest sense, however, this term is properly applied only to the two larger countries, Belgium and the Netherlands, and will be so used in this chapter.

EUROPE
B E N E L U X

INDEX MAP

URBAN AREAS
Stars show national capitals

◉ ⬤	over 1,000,000
✪ ⬤	500,000-1,000,000
○	250,000-500,000
★ ○	Selected smaller cities

City-size symbols are based on metropolitan area estimates.

Zuider Zee dike and Delta Plan dams:

〰〰〰 Completed as of 1969.

▭▭▭ Under construction

⊢⊣ Adjustable flood barrier

⊨ Bridge

Boundaries of Ardennes Highland in Belgium-Luxembourg.

Polderlands

Coastal sand dunes

—— Selected rivers

∧∧∧ Selected canals

— — Flemish-Walloon language boundary

Scale of Miles
0 10 20 30 40 50

WEST FRISIAN ISLANDS

Groningen

WIERINGER POLDER

IJSSEL-MEER
(ZUIDER ZEE)

NORTHEAST POLDER

MARKERWAARD POLDER (under construction)

EAST FLEVOLAND POLDER

SOUTH FLEVOLAND POLDER

Zwolle

Ems R.

Enschede

Ijmuiden

NORTH SEA CANAL

Haarlem

Amsterdam

Leiden

Utrecht

AMSTERDAM-RHINE CANAL

Arnhem

Ijssel R.

Rhine R.

DORTMUND EMS CANAL

NETHERLANDS

The Hague

NORTH SEA

NEW WATERWAY

Rotterdam

Lek R.

Waal R.

Maas R.

Breda

Tilburg

Eindhoven

Maas R.

Rhine R.

Duisburg-Ruhrort

Dortmund

THE RUHR

Flushing

Terneuzen

Antwerp

CAMPINE CANAL

ALBERT CANAL

C A M P I N E

JULIANA CANAL

SOUTH LIMBURG

Maastricht

WEST GERMANY

Ostend

Bruges

GHENT-TERNEUZEN CANAL

Ghent

Scheldt R.

B E L G I U M

Brussels

Aachen

F L A N D E R S

Lys R.

(FLEMISH)
(WALLOON)

Liège

Verviers

Tournai

L O E S S

Namur

Meuse R.

Charleroi

Mons

Sambre R.

Escaut R. (Scheldt R.)

C O N D R O Z

Rhine R.

F R A N C E

Somme R.

Bastogne

A R D E N N E S

Meuse R.

LUXEMBOURG

Luxembourg

Oise R.

Aisne R.

LANDS OF LOW RELIEF

"Low Countries" is very descriptive of Belgium and the Netherlands, since approximately the northern two-thirds of the former and practically all of the latter consists of a very low plain facing the North Sea. This plain is the narrowest section of the great plain of northern Europe which extends from France into the Soviet Union. In the Low Countries the plain seldom reaches as much as 300 feet above sea level. Large sections near the coast, especially in the Netherlands, are actually below sea level. They are protected from flooding only by a coastal belt of sand dunes, by man-made dikes, and by constant artificial drainage.

The only land in the Low Countries with even a moderate elevation lies mainly in Belgium, south of the line formed by the Sambre and Meuse rivers (map, left). Here the Ardennes Upland rises in some places above 2000 feet, although its surface is more commonly around 1200 feet above sea level. Much of the Ardennes is an area of little relief, though in places intrenched rivers produce a more rugged terrain. The term "Ardennes," as used in this chapter, includes the lower northern foreland or Condroz, located between the Ardennes proper and the Sambre-Meuse Valley. The edges of the Ardennes overlap Belgium's frontiers, extending without a break into the Rhine Uplands of the German Federal Republic on the east and for a short distance into France on the west. To the south the Ardennes includes the northern half of the tiny country of Luxembourg. Southern Luxembourg and a small adjoining tip of Belgium have a lower, rolling terrain similar to that of neighboring French Lorraine.

POPULATIONS AND STANDARDS OF LIVING

High population densities are an outstanding characteristic of the Benelux nations, and especially of the

Netherlands and Belgium. In fact, no sovereign nation of comparable or larger area in the entire world can match either of the latter two countries in overall population density. The density of population in the Benelux area as a whole is greater than that of the most densely populated American state, New Jersey. California, which has the largest population of any American State, is about six times as large in area as Benelux, but has a slightly smaller population (1968). Areas, populations, and population densities of the Benelux countries are given in Table 4.

In many countries of the world an unusually high density of population is associated with a low standard of living. In the Benelux countries, however, the reverse is true. By effective utilization of such opportunities and resources as their small and crowded national territories afford, the peoples of these countries are able to maintain standards of living which are equaled or surpassed by few nations. Estimates of per capita income credit the Benelux countries with a figure at least three times the world average. Belgium and Luxembourg rank substantially above the Netherlands in this respect. While income level is not exactly equivalent to standard of living, the relationship is sufficiently close to indicate the high position of the Benelux units among the world's countries.

THE SIGNIFICANCE OF TRADE

The economic life of the Benelux countries is characterized by an intensive development of three interrelated activities—industry, agriculture, and trade. An especially distinctive and significant feature of their economies is an unusually high development of, and dependence on, international trade. In total value of foreign trade (imports plus exports), the Netherlands and Belgium–Luxembourg ranked fifth and sixth, respectively, among the countries of Europe in 1967,

TABLE 4 BENELUX: AREA AND POPULATION DATA

COUNTRY	AREA (THOUSAND SQUARE MILES)	POPULATION (MILLIONS: 1968 ESTIMATES)	DENSITY (PER SQUARE MILE: TO NEAREST WHOLE NUMBER)
Netherlands	13.0 [a]	12.8	987
Belgium	11.8	9.6	814
Luxembourg	1.0	0.34	338
Totals	25.8	22.7	883

[a] Excluding inland waters.

being exceeded only by four much larger nations, the United Kingdom, West Germany, France, and Italy. (Since 1922 Belgium and Luxembourg have maintained a customs union, and trade figures for the two countries are reported as a unit.) In per capita trade, Belgium-Luxembourg usually ranks second, after Iceland, and the Netherlands usually ranks third.

Several factors help to explain the unusual development of foreign commerce in the Benelux countries:

1. Trade is essential to such small countries if they are to make maximum use of a limited variety of internal resources. Such resources as Luxembourg's iron ore or Belgium's coal must be either left in the ground or exported for comparatively small returns unless complementary materials needed for manufacturing can be imported.

2. The need of these countries to trade is matched by their ability to trade. Specializing in activities which offer the greatest possibilities for effective use of limited resources, the Benelux nations are able to export large surpluses of certain manufactures and, in the case of the Netherlands, livestock products, vegetables, and flower bulbs. Entry into the European Common Market has removed the tariff and associated barriers to larger markets in West Germany, France, and Italy.

3. The position of the Benelux countries is highly favorable for trade. These countries lie in the heart of the most highly developed part of Europe. Their nearest neighbors, West Germany, France, and the United Kingdom, are among the world's foremost producing, consuming, and trading nations. The resulting commercial opportunities are reflected in the fact that between 50 and 60 percent of Benelux foreign trade is accounted for by trade between Belgium-Luxembourg and the Netherlands or between these countries and their three larger neighbors. A second significant aspect of position lies in the location of the Netherlands and Belgium at or near the mouth of the Rhine. This river is the greatest inland waterway of Europe, and one of the greatest in the world. Location where the Rhine meets the sea permits the Low Countries to handle in transit much of the foreign trade of Switzerland, eastern France, and, especially, West Germany with its Ruhr industrial district.

4. The trade of the Low Countries has profited somewhat from the fact that both the Netherlands and Belgium have held large colonial empires in the tropics. Before they became independent, the Netherlands East Indies and the Belgian Congo offered assured markets for goods and capital, and allowed the home countries to act as European entrepôts for tropical agricultural products and certain minerals, particularly nonferrous metals. Today, industrial and trading specialties connected originally with this former colonial trade continue to function in the Low Countries, though only a tiny fraction of their present foreign trade is with their former colonies.

IMPORTANCE OF THE DUTCH AND BELGIAN PORTS

Exploitation of their commercial opportunities by the Low Countries is reflected in the presence of three of the world's major port cities within a distance of about 80 miles: Rotterdam and Amsterdam in the Netherlands, and Antwerp in Belgium. During the twentieth century Rotterdam has normally ranked second only to New York among the world's seaports on the basis of total tonnage of goods handled. Recently it has surpassed New York in tonnage (though not in *value* of goods handled), and now ranks first in the world. By the same criterion Antwerp has ordinarily been exceeded only by Rotterdam, London, and Hamburg among European ports; since World War II the slow recovery of Hamburg has enabled Antwerp to surpass this German port and rank second only to Rotterdam on the continent. The larger city of Amsterdam is a considerably less important port than Rotterdam or Antwerp, though its trade is large, nonetheless. The commercial activities of the three major ports are supplemented by those of numerous smaller ports in the Low Countries, particularly Ghent in Belgium.

Amsterdam—The "Colonial" Port of the Netherlands

Amsterdam (1.8 million) is the largest city and the constitutional capital of the Netherlands, although the government is actually located at The Hague (840,000). As a port, Amsterdam handles mainly the trade of the Netherlands itself, plus a considerable entrepôt traffic, especially in tropical products. The latter specialty stems from the fact that Amsterdam was the main port of the Netherlands during the centuries when the country's colonial empire was being acquired and organized, before the rapid rise of Rotterdam in the later nineteenth and twentieth centuries. Although the East Indian empire has been lost, Amsterdam has retained business connections and specialized marketing facilities built up during the colonial period, and so it has been able to keep a considerable amount of its "colonial" trade. In addition, the profits accumulated during several centuries of the East Indian trade have been used to finance industrial development in Amsterdam, and, by supplying capital for the large foreign investments of the Netherlands, have helped to make Amsterdam a center of international finance. Although the Eastern trade no longer appears to be a really major element in the economy of the Netherlands, it is still far from negligible, and has been of great importance in the historical development of the country, and particularly of Amsterdam.

In modern times Amsterdam has experienced considerable difficulties as a port. The original approach from the sea by way of the Zuider Zee eventually proved too shallow for modern shipping. This problem was solved in 1876 with the opening of the North Sea Canal, through which ships now enter the harbor. More recently, location away from the mouth of the Rhine proved a disadvantage, but the city has been attracting a larger share of Rhine traffic since the opening of the Amsterdam-Rhine Canal in 1952.

Rotterdam—The Major Port for Rhine Traffic

Rotterdam (1.1 million) is better situated with respect to Rhine shipping than either Amsterdam or Antwerp, being located directly on one of the navigable distributaries of the river rather than to one side. Accordingly, Rotterdam controls and profits from the major portion of the river's transit trade, receiving goods by sea and dispatching them upstream by barge, and receiving goods downstream by barge and dispatching them by sea. Two developments have been mainly responsible for the port's tremendous expansion in recent times. First was the opening in 1872 of the New Waterway, an artificial channel to the sea far superior to the shallow and treacherous natural mouths of the Rhine, and superior also to the sea connections of either Amsterdam or Antwerp. Second, and more fundamental, has been the increasing industrialization of areas near the Rhine, particularly the Ruhr district, for which Rotterdam has become the main sea outlet. The need of the inland industrial districts for imports of ores, crude oil, cereals, and other bulky commodities, coupled with their exports of such heavy goods as coal, steel, and chemicals, accounts for the huge tonnage of Rotterdam's trade. A large addition to Rotterdam's facilities known as Europoort has been under development since 1958. It lies along the New Waterway at the North Sea entrance to the Rotterdam port area and is specially designed to handle supertankers and other large carriers of bulk cargoes.

The Scheldt Port of Antwerp

Antwerp (1.1 million), located about 50 miles up the Scheldt River, is primarily a port for Belgium itself, handling some four-fifths of the foreign trade of Belgium and Luxembourg. However, the port also accounts for an important share of the Rhine transit trade. Belgium's coast is straight and its rivers shallow, so that the deep estuary of the Scheldt gives Antwerp the best harbor in the country, even though it must be reached through the Netherlands. The city has become a major focus in Belgium's dense railway net, and has river and canal connections to Ghent, Brussels, and Liège. Transit trade is facilitated by Antwerp's position in relation to the major inland industrial areas of northwestern continental Europe. It is slightly closer than either Rotterdam or Amsterdam to these areas (map, p. 138), and it has somewhat shorter and more direct rail connections. Thus Antwerp ordinarily handles a large share of the transit trade in goods of a type which can move more profitably by the relatively expensive but faster rail lines than by the cheaper but slower barge.

Competition between Antwerp and the Dutch ports has provided a fertile field for controversy between the Low Countries in the past. Throughout the seventeenth and eighteenth centuries the Dutch were able to keep the lower Scheldt, which runs through their territory, closed to traffic, and thus to throttle Antwerp almost completely. Later disputes have arisen over the maintenance of the river's shipping channel, over Belgium's desire for a canal through Dutch territory to connect Antwerp directly with the Rhine, and over Belgium's payment of special subsidies to attract Rhine barges through the southwestern part of the Netherlands to Antwerp in preference to the more convenient Rotterdam. One of the tasks confronting the European Economic Community is to resolve problems arising from the strenuous competition among the Belgian and Dutch ports (particularly Antwerp and Rotterdam) and other leading seaports within the EEC.

MANUFACTURING IN THE BENELUX COUNTRIES

All of the Benelux countries are highly industrialized. Before major discoveries of natural gas were made in the northeastern Netherlands at the end of the 1950s, Belgium and Luxembourg were clearly better provided with domestic mineral resources. This asset helped them become somewhat more industrialized than the Netherlands. (The total value of manufactured products from the Netherlands greatly exceeded that of Luxembourg, but Luxembourg's dependence on manufacturing and the proportion of the labor force employed in it were greater.) The differential is decreasing, though Belgium continues to have a far greater development of the heavier types of industry and Luxembourg continues to exhibit a far greater dependence on industries of this kind than does the Netherlands. Belgium, best endowed with coal, is the most important manufacturing nation of the three.

Heavy Industry in the Sambre-Meuse District

One of Europe's major coal fields crosses Belgium in a narrow east-west belt about a hundred miles long (map, p. 139). It follows roughly the valleys of the Sambre and Meuse rivers, and extends into France on the west and West Germany on the east. Liège (600,000), Charleroi (400,000), and smaller industrial cities strung along this Sambre-Meuse field account for most of Belgium's metallurgical, heavy chemical, and other heavy industrial production. The iron and steel industry is the most important. It originally developed using local resources. These included small iron ore deposits, with the ore being smelted first with local charcoal and later with coke made from local coal. Now, however, the industry must import its iron ore and even much of its coking coal and coke, which the Belgian fields can no longer supply in adequate quantities at low cost. A large part of the production is sold outside of the relatively small home market. Liège, the largest manufacturing center on the coal field, has metallurgical industries dating back to handicraft days, and was the first city in continental Europe to develop modern, large-scale iron and steel manufacture following the industrial revolution. It will be noted that most of the industries described above are more or less long-standing and traditional. The district has attracted only a moderate share of such fast-growth industries as electronics and motor vehicles. Indeed, a new iron and steel plant now functions at Ghent, which is near the coast and well beyond the Sambre-Meuse district. Here, again, we see an inland heavy-industrial district giving way, however slowly, to locations on and near the sea.

Small deposits of zinc, as well as iron ore, in the Ardennes originally furnished raw material for the Sambre-Meuse metallurgical industries. Although the zinc, like the iron ore, has long been exhausted, it started Belgian industry toward what has now become a remarkable specialty in nonferrous smelting. Belgium produces about 6 percent of the world's smelter zinc, a production exceeded only by that of the United States, the Soviet Union, Japan, and Canada. In addition, the little country has become Europe's third greatest smelter of tin (after the United Kingdom and the Netherlands) and fourth greatest of lead, after West Germany, France, and Yugoslavia. The nonferrous ores and concentrates for the smelting industry are now entirely imported, and the bulk of the production is exported to other European nations in one form or another. Belgium thus acts as a processing middleman between the overseas ore producers and the European industrial consumers of these metals. Certain other specialties have also developed in the Sambre-Meuse district, notably the production of cement, glass, and nitrogenous chemicals. These products, which require much coal in manufacture, are principally made from local raw materials.

The Light, Diversified Industries of Brussels

Throughout Belgium even the smaller towns are generally characterized by some industrial development. Away from the coal fields there is a tendency to specialize in lighter types of industry having smaller power requirements than the industries located near to coal. In these industries labor is often of great importance as a factor in the total cost of production. The Belgian capital, Brussels, is the outstanding center of such industries. Like most European political capitals, it is the largest city in the country, with a metropolitan population of 2.1 million, and is primarily a product of political and cultural centralization. The labor force, transport facilities, and local market it provides, however, have led to the development of a great variety of light industries, including textiles and clothing, skilled metalwork, printing, food processing, and luxury crafts. Antwerp, also, is an important center of diversified industries, particularly those that are generally found in a major seaport.

The Belgian Textile Industry

As it is in most countries, textile manufacturing is the leading branch of Belgium's lighter industries. As far back as the twelfth century, Flanders (roughly the part of Belgium north of the Scheldt and west of Antwerp) was a center of commercial cloth production, importing raw wool (mainly from England) and selling woolen cloth in various parts of Europe. The same area is now the center of Belgium's cotton, linen, synthetic textile and jute industries. It still imports its raw materials and exports a large part of the finished product. Flanders readily adapted its long textile tradition to machine production during the early nineteenth century simply by concentrating the former cottage workers into factory cities. The greatest of these cities today is the ancient commercial center of the industry, Ghent (370,000).

Flanders lost its leadership in Belgium's woolen industry to Verviers (80,000), in the edge of the Ardennes, by the eighteenth century at the latest. The pure and soft water flowing from the Ardennes gave this place a considerable advantage in washing the wool preparatory to spinning. Its value can still be seen in the fact that Verviers not only produces a major share of Belgium's woolen goods but also specializes in cleaning raw wool for distant mills, both in Belgium and in neighboring countries.

Belgium's Export Trade

Belgium's export trade closely reflects the industrial emphases brought out in the preceding discussion. Processed metals (iron and steel, copper, zinc, tin, aluminum, lead) are by far the largest single category of exports by value. Other leading categories include machinery and vehicles; textile yarns, fabrics, and fibers; chemicals; cut diamonds from Antwerp; petroleum products from refineries at Antwerp; and glass.

Manufacturing in Luxembourg

Luxembourg's simple industrial structure presents a picture of dependence on foreign trade proportionately equal to, or greater than, that of Belgium. About one-sixth of the country's entire working population is employed in the iron and steel industry, which is carried on in several small centers near the southern border, where the Lorraine iron-ore deposits of France overlap into Luxembourg. Annual production of both pig iron and steel approximates four million tons. Thus the tiny country ranks tenth among the steel producers of Europe, and about sixteenth in the world. The very small home market both necessitates and permits export of most of the production. In return, Luxembourg buys fuel (mainly coal and coke from West Germany), alloy metals, and many other products.

Industries of the Netherlands

The Netherlands differs from Belgium and Luxembourg in the smaller part which domestic mineral re-

sources have played in the rise of its industries. It resembles Belgium–Luxembourg, however, in the dependence of its industries on international trade. Basic factors in the rise of Dutch industrial production have been (1) the ability of Dutch industries to import both fuel and raw materials cheaply, and (2) the availability of an abundant, skilled, and relatively low-priced labor force. Generally speaking, in their location the industries of the Netherlands are transportation and labor oriented rather than materials or market oriented.

Rotterdam and Amsterdam are the main seats of industry as well as of commerce. Prominent among the diversified industries of both these ports are shipbuilding and associated engineering industries. For a brief time in the seventeenth century the Netherlands was probably the world's greatest sea power, and a strong maritime tradition has survived. A disproportionately large merchant marine is an important factor in the Dutch economy, and shipbuilding is a prominent industry. The ports of the Netherlands are also the locale for a variety of industries which process overseas materials for sale in European markets. Industries that process foodstuffs of tropical origin, notably oilseeds, sugar, and cocoa, play a substantial role. But in value of exports they are overshadowed by industries that process imported minerals. The most important of the latter is the oil-refining industry. A huge concentration of refineries, especially at Rotterdam, is partly an expression of the Netherlands' important financial and managerial interest in the worldwide operations of the Royal Dutch Shell Company. Among the nations of Europe the Netherlands ranks fifth in oil refining, after the United Kingdom, West Germany, Italy, and France. Immediately downstream from Rotterdam—indeed, a projection of the Rotterdam harbor—is Europoort (Gateway to Europe), begun in 1958 and still under construction. This ultramodern seaport is being built especially to accommodate deep-draft vessels carrying bulk cargoes like oil from the Middle East and iron ore from various sources. Petroleum products are by a wide margin the country's leading export among processed minerals. Iron and steel manufactured from imported iron ore (primarily Swedish ore) rank second, though the Netherlands' imports of iron and steel are somewhat larger than its exports, and its small steel industry produces primarily for the domestic market. The largest producing unit is an integrated iron and steel plant located at Ijmuiden (city proper, 70,000) on the North Sea Canal near its seaward entrance. Also notable among mineral-processing industries is the smelting of tin. Indonesian tin, channeled into Europe through the Netherlands, has made the country the world's fifth nation in tin smelting, after Malaysia, the Soviet Union, China, and the United Kingdom. This industry is centered at Arnhem (225,000) on the Rhine rather than in one of the major port areas. As in many countries, chemical industries largely involving by-products or derivatives of steel and petroleum tend to be located within or near the steel and petroleum districts.

Away from the ports, and especially in some of the poorer agricultural districts, surplus labor and good transportation have led to the development of numerous small industrial cities. Textiles, light machinery, and consumer goods of various kinds are their major products. The industries of these cities rely heavily on imported raw materials and export a large part of the finished product. The only really important manufacturing industry which is based primarily on domestic raw materials is the widespread food-processing industry, which prepares for the domestic market and for export the meat, milk, butter, cheese, and other products of the Netherlands' specialized agriculture.

THE INTENSIVE AGRICULTURE OF BENELUX

Through manufacturing and trade the Benelux nations are able to support considerably larger populations than their agriculture can feed under present conditions. Like the United Kingdom, Belgium and Luxembourg import large quantities of food, together with feed for livestock and fertilizers to help increase domestic agricultural production. The Netherlands, while a net exporter of food, achieves this position partially by processing and reexporting foodstuffs. Domestic production in the Benelux countries is particularly inadequate in the case of grains. Only limited areas are really well suited to the production of grains, particularly wheat, the grain consumed in the largest quantities for human food. Imported wheat has proved able to undersell the domestic product on the Benelux market, and thus farmers in the Benelux

The intricate canal network of the Netherlands polders. Windmills, such as those in this photo, have largely been replaced by motor-driven pumps. (Netherlands Information Service.)

Milking the dairy herd on a farm in the northern Netherlands. The milking machines draw their power from the motor of the truck. The farmhouse and barn at the right are housed in one structure. This type of farm dwelling, with its thatched roof, is characteristic of older farmhouses in large parts of the Netherlands and much of northern Germany as well. Note the windbreak of trees at the left of the view. (Netherlands Information Service.)

nations have usually concentrated their main efforts on farm commodities other than wheat. This is especially true of the Netherlands.

The extremely high density of population makes domestic food production insufficient despite the fact that crop yields in the Low Countries are among the highest in the world. Wheat, for example, ordinarily gives an average yield of 50 to 65 bushels per acre. This compares with a world average of perhaps 20 bushels per acre, and is exceeded by the average yield in only one other country, Denmark. However, such yields are obtained only through large expenditures of both labor and capital. So intensive is the labor expended on the cultivation of each acre that agriculture often resembles gardening. This type of farming is made both possible and necessary by the density of rural population and the small size of most farms. Nearly 40 percent of all Dutch farms and nine-tenths of all Belgian farms are smaller than 12½ acres. Most Belgian farms are actually closer to 4 acres. Along with labor, capital is applied to the soil abundantly, mainly in the form of fertilizer. The Netherlands uses more commercial fertilizer per arable acre than any other country in the world, and Belgium is not far behind.

A close adaptation of agriculture to climatic and soil conditions helps the Low Countries to obtain the maximum output from their restricted acreages of farm land. Since climatic conditions tend to be relatively uniform over both countries, differences in type of agriculture are mainly related to differences in type of land. Although in detail the picture is intricate, three main types of land, each with its associated types of agriculture, may be identified. These are the *polder lands,* the *loess lands,* and the *infertile lands.*

Dairy Farming in the Polder Lands

Polder lands are those which have been surrounded by dikes and artificially drained. The process of turning former swamps, lakes, and shallow seas into agricultural land has been going on for over seven centuries. An individual polder, of which there are a great many of various sizes, is an area enclosed within dikes and kept dry by constant pumping (photo, left) into the drainage canals which surround it. About 50 percent of the Netherlands now consists of an intricate patchwork of polders and canals, while Belgium has a narrow strip of such lands behind its coastal sand dunes. The polder lands of the Netherlands extend for about 180 miles between the Belgian and German borders in an irregularly shaped belt that is roughly parallel to the seacoast, and is separated from the sea by dunes, swamps, or dikes.

The polders are the best lands in the Netherlands, and their production is the heart of Dutch agriculture. The reclaimed soil is very rich, water supply is subject to considerable control, and the canals form a complete transport network. The drier polders are often used for crop farming, and produce huge harvests. Most polders, however, are kept in grass, and dairy farming is the main type of agriculture. It thrives so well that large quantities of dairy products are exported, mainly to nearby industrial areas in West Germany, Belgium, and the United Kingdom. Pork production, as well as production of beef from males and overage cows and veal from male calves, is ordinarily a part of dairy farming, since the necessary feed for the hogs can largely be supplied by skim milk left over from butter making. The production of vegetables, fruits, and horticultural specialties, such as the famous Dutch flower bulbs, is also of considerable importance in the polder zone. Dairying and other specialized types of agriculture supply about a fifth of the total exports of the Netherlands, and more than compensate, in value at least, for the necessary food imports. The main categories of exports provided by the Netherlands' agriculture include (1) meat, (2) evaporated, condensed, and dry milk, and cheese and butter, (3) vegetables, (4) flowers and bulbs, and (5) eggs.

The Fertile Loess Lands

A second section of outstanding fertility and agricultural production is found in Belgium. In central Belgium a gently rolling topography was mantled in glacial times by a blanket of loess, a fine dust picked up by the wind from glacial debris. This material has formed soils of exceptional fertility in Belgium, as in many other places. These soils are found in a belt across the country between the Sambre-Meuse Valley on the south and a line just beyond Brussels on the north.

Three characteristics distinguish the agriculture of the loess belt. One is an emphasis on wheat, which has been able to survive foreign competition on these good lands. Another is the production of sugar beets, a crop requiring an exceptionally fertile soil. Third is the production of fodder crops, which give very high yields and permit the loess area to specialize in the production of

beef cattle, often brought in for fattening from poorer areas. This combination of food, feed, and livestock production supports the largest and most prosperous farms in Belgium.

The Infertile Lands

The remaining lands of the Low Countries, and most of Luxembourg as well, can be described as relatively infertile, though small areas sometimes contradict this generalization. Sandy soils are dominant in northern Belgium and practically all of the Netherlands outside of the polder areas, though the sands are often interspersed with peat bogs and patches of clay, both of which can be made into good soil if properly managed. The soil of the sandy areas is generally lacking in humus, lime, and plant nutrients. That of the Ardennes is no better.

The agriculture of these areas is generally centered about livestock production, and crops are grown primarily for feed. The most common crops are rye, potatoes, oats, turnips, and hay. All are well adapted to the climate, and all except hay suffer relatively little, or actually grow better, in light and infertile soils. Rye and potatoes are used in these areas for human food as well as stock feed. Potatoes and turnips give especially good returns per acre, provided the necessary labor is available for harvesting. In addition, these root crops fit well with rye in that they give a crop during the summer on the same land that produced rye during the winter and spring. Livestock may be either beef cattle or dairy cattle, often both. Besides providing a cash income, they supply an important part of the large quantities of fertilizer necessary in farming such lands.

Although the density of population and intensity of agricultural production on the poorer lands are generally less than on the polders or the loess, they are often surprisingly great. In one predominantly sandy area, Flanders, the average production per acre is almost as large as on the naturally better lands, and the density of population is even greater. Here intensive agriculture and systematic improvement of the land have a continuous history dating back at least to the eleventh century. The precocious urban, commercial, and industrial development of Flanders stimulated Flemish agriculture during these early times. A garden type of agriculture has developed, especially centered on livestock, dairy, and vegetable production, but making some use of almost every crop that can be grown in the Low Countries and using soils almost completely transformed by centuries of improvement. Few of the world's farming areas are more skillfully handled.

MAJOR PROBLEMS OF THE BENELUX NATIONS

The Benelux countries share the general problems of the modern world. Like many other countries, they are concerned with achieving and maintaining international peace and cooperation, national security, and internal unity, freedom, and economic well-being. However, due to peculiarities of position, resources, population, and economic development, the Benelux countries are faced with certain types of problems which present themselves in distinctive form or with particular urgency.

The Belgian Problem of National Unity

Belgium has had the most difficulty among the Benelux nations in securing and maintaining internal unity and cooperation. Division of the population into two major language groups has been at the root of the difficulty. If one draws an east-west line across Belgium just south of Brussels, it will approximate a line of sharp linguistic division (map, p. 186). To the north Flemish, a slight variant of Dutch, is the dominant language except in Brussels. To the south live the French-speaking Belgians known as Walloons. Rivalries and antagonisms have sometimes arisen between the two groups. The Flemish people, until recently the minority group, have occasionally shown separatist tendencies. Differential birth rates have now given the Flemish group a majority, however, and Walloon uneasiness has appeared. There is always a danger in Belgium that any political issue will be fought primarily along language lines and erupt into communal violence. How explosive such a situation can be was shown in 1967 when violent disturbances threatened over the movement of a section of a university from its Flemish location to a Walloon area. In the face of this language division constant care must be exercised to maintain a precarious national unity.

Problems of National Security

At present, however, all three of the Benelux nations find problems of national security and economic welfare more pressing than others. The two types of problems are closely related due to the heavy costs involved in war and in the peacetime provision of armaments against possible attack.

The position of the Benelux countries between Germany, France, and the United Kingdom, and their inability to defend themselves against these more powerful neighbors, have tended to involve them in every general European war, usually as a battleground. Both world wars of the twentieth century have added to Belgium's long history as an international battlefield. The Netherlands has offered a valuable base for operations from the continent against England and, conversely, an approach to the industrial heart of Germany. Flooding of the polders is no longer the effective defense against invading armies that it has been in times past.

Trade Problems

The resulting concern of the Benelux countries for international peace has been paralleled by a concern with international trade which is proportional to their heavy dependence on importing and exporting. They have had to face particular problems in this field associated with (1) the loss of the East Indian empire by the Netherlands and of the Congo by Belgium, (2) restriction of trade with European areas within the Communist orbit, and (3) the necessity for trading more with the United States.

Most of these recent trade difficulties have been overcome, largely in connection with western Europe's general recovery from the devastation and dislocations of the war and its subsequent economic boom. Much has been achieved by the Benelux countries through their participation in various international projects (particularly the European Economic Community) to secure freer trade among the European nations. Since the Benelux countries are so very dependent on favorable political and economic circumstances for doing business with other countries, it is understandable that they have generally been ardent supporters of international organization. In the field of economic cooperation they have provided an example for the European Common Market in their own Benelux Economic Union, which was begun earlier and has moved a long way toward its goal of eradicating trade barriers between the Netherlands and Belgium-Luxembourg.

Population Problems

Internally, the main problem of both Belgium and the Netherlands seems to be the expansion of agriculture and industry, already intensively developed, to support still increasing populations. The situation is particularly acute in the Netherlands, which has one of the highest rates of natural increase in Europe and has witnessed the addition of over 5 million people to the population of

Reclamation of the Zuider Zee

The most important and celebrated reclamation work currently going on in the Netherlands, however, is the formation of new polders from the former Zuider Zee (map, p. 186). A massive 18-mile-long dike (photo, left) was completed across the entrance of the Zuider Zee in the early 1930s, and a sizable area of sea bottom has now been reclaimed. By 1968 three of the five large polders planned were under cultivation, a fourth polder had been drained, and the fifth was under construction. Meanwhile the remainder of the Zuider Zee has become a fresh water lake, the Ijsselmeer. The completed project will add about 7 percent, all excellent farm land, to the land area of the Netherlands. All in all, some 1.6 million acres—nearly one-fifth of the total area of the Netherlands—has been reclaimed in this way from the sea, and much more from swamps and bogs.

The Delta Plan

Another massive and much-publicized project of the Netherlands, called the Delta Plan, is aimed primarily at flood control and only incidentally at land reclamation. Its purpose is to prevent a recurrence of the devastating flood of February 1953, when the sea broke through the dikes in the southwestern part of the country during a storm, taking 1800 lives and causing property damage estimated at nearly 400 million dollars. It is intended also to combat salination, especially in the lower sections of rivers, and hence to increase agricultural output; to connect islands to the mainland by roads across the dikes; and to stimulate the tourist industry. The heart of the Delta Plan is the construction of dikes connecting the islands in the triple delta of the Rhine, Maas (Meuse), and Scheldt with each other and with the adjoining mainland. The New Waterway, giving Rotterdam access to the sea, and the Scheldt River outlet for Antwerp will remain unobstructed, but the other channels through the delta will be closed by huge dikes. It is estimated that it will take at least until 1980 to complete the entire scheme, which involves various engineering works besides the major dikes referred to above.

an already crowded country in the last four decades. In such a small, densely populated country as the Netherlands, such an increase may be disastrous if long continued—a fact that is reflected in the nation's very unusual policy of officially encouraging and assisting emigration. As in most other countries of western Europe, regional development programs have been active. We have described such a program for France, emphasizing a slowdown in population movement to Paris (p. 159), and have seen Britain's attempt to check the growth of London (p. 119). In the small Benelux countries, some efforts likewise are being made to slow the movement of populations to larger cities and towns. However, an even greater effort involves land reclamation, as explained below.

Reclamation of New Land for Agriculture

Land reclamation has been one response to increasing populations in both countries. Cultivation of the infertile lands has been greatly expanded in recent times, and the process continues. In Belgium such effort has centered in the Ardennes and, especially, in the Campine (Flemish: Kempen), the sandy area south and east of Antwerp. In the Netherlands special attention has been directed toward bringing into productivity some of the vast peat bogs which are interspersed with the sandy areas of the eastern part of the country, and also toward an extension of dairy farming into the poorer lands.

Industrial Expansion

Agricultural expansion alone cannot maintain the present high standard of living in the Low Countries. The governments of both Belgium and the Netherlands, especially the latter, regard increased industrialization as essential. This is in spite of the fact that both are already among the most highly industrialized countries in the world.

A reasonably favorable situation with regard to power has characterized past industrial development in the Low Countries and seems likely to continue. Belgium's exceptional degree of industrialization is associated in part with unusually early utilization of coal resources. The Sambre-Meuse field was the first major coal field on the continent to be intensively developed. Its history has paralleled that of most of the British fields, with output reaching a peak in 1913 and then declining as thin, broken, and inclined seams have had to be worked at increasing depths. In Belgium, however, the situation was alleviated by the discovery in 1898, and development after 1917, of a new field, located in the Campine (map, p. 139). The coal lies from 1500 to 3000 feet or more below the surface, but thick seams, large reserves, and high quality have resulted in relatively efficient production, and a certain amount of heavy industry has already been drawn to the Campine in preference to the older Sambre-Meuse district. The field that underlies the Campine extends across the southernmost part of the Netherlands, known as South Limburg, and on to the vicinity of Aachen, Germany. Here the Netherlands, long considered to be essentially barren of domestic power resources, has developed since World War I an unusually efficient mining industry which supplies all but a minor share of the country's coal requirements. The South Limburg field yields varied types of coal, including much high-grade coking coal and some anthracite. Both of the Low Countries, as well as Luxembourg, lie within relatively short distances of major foreign coal fields. They have long imported some coal and coke, especially from the German Ruhr, and are now favored in obtaining their requirements beyond domestic production by their membership in the European Coal and Steel Community.[1]

Nor are the Low Countries likely to suffer as other fuels continue to replace coal in importance. Both Rotterdam and Antwerp are major European oil-receiving ports, and the Netherlands has very large reserves and a rapidly increasing production of natural gas. In addition, both countries have the capital and the advanced technology for nuclear power development in the future.

An astonishing record of past accomplishment in these small countries suggests that insofar as solutions to their present difficulties can be achieved by vigorous action within their own borders, such solutions are likely to be forthcoming. Problems wider in scope, however, can be solved only in international action on a regional or a world scale, and on such action the Benelux countries can exercise, at best, only a modest influence. They lie at the opposite pole from national self-sufficiency—in an age when even large and powerful countries are far from self-sufficient—and may only press forward within their own boundaries and hope for the best from the world on which they are so dependent.

REFERENCES and READINGS

General

RACHLIS, EUGENE, and the Editors of *Life, The Low Countries* (Life World Library; New York: Time Inc., 1963).

Belgium

BRUYNE, HECTOR DE, "Antwerp on the Scheldt," *Geographical Magazine,* **40**, no. 6 (October 1967), 481–490.

[1] It should be noted that the Low Countries are both importers and exporters of coal and coke, though in both countries there is a net import. This situation is accounted for by differing requirements of particular industries with respect to types and grades of coal, and by transportation and price differentials that often make it more economical for a user to buy from a foreign supplier than from one in some other part of the home country.

ELKINS, T. H., "Liège and the Problems of Southern Belgium," *Geography*, 41, pt. 2 (April 1956), 83–98.

MONKHOUSE, F. J., "Albert and Juliana: Two Great Waterways," *Scottish Geographical Magazine*, 72, no. 3 (December 1956), 163–176.

NEESEN, V., "The Albert Canal As Shipping Artery and Site for the Location of Industry," *Tijdschrift voor Economische en Sociale Geografie*, 55, no. 6–7 (June–July 1964), 172–177.

RILEY, R. C., "Changes in the Supply of Coking Coal in Belgium Since 1945," *Economic Geography*, 43, no. 3 (July 1967), 261–270; and "Recent Developments in the Belgian Borinage: An Area of Declining Coal Production in the European Coal and Steel Community," *Geography*, 50, pt. 3 (July 1965), 261–273.

TAMSMA, R., "The Projected Antwerpen-Rhine Canal," *Tijdschrift voor Economische en Sociale Geografie*, 55, no. 6–7 (June–July 1964), 150–164.

VERBURG, M. C., "The Gent-Terneuzen Developmental Axis in the Perspective of the European Economic Community," *Tijdschrift voor Economische en Sociale Geografie*, 55, no. 6–7 (June–July 1964), 143–150.

WALKER, ROBERT H., *Basic Data on the Economy of Belgium*, U.S. Bureau of International Commerce, Overseas Business Reports, OBR 67–88 (December 1967). (Washington, D.C.: Government Printing Office, 1967.)

Luxembourg

EDWARDS, K. C., "Historical Geography of the Luxembourg Iron and Steel Industry," Institute of British Geographers, Publication No. 29, *Transactions and Papers, 1961*, pp. 1–16.

LAMBERT, JOHN, "Smallest of the Six" [Luxembourg], *Geographical Magazine*, 36, no. 10 (February 1964), 559–568.

WALKER, ROBERT H., *Basic Data on the Economy of Luxembourg*, U.S. Bureau of International Commerce, Overseas Business Reports, OBR 68–60 (July 1968). (Washington, D.C.: Government Printing Office, 1968.)

The Netherlands

BILLINGTON, DAVID P., "Holland–Between River and Sea," *Civil Engineering*, 37, no. 5 (May 1967), 40–45. An article about the Delta Plan.

BOXER, C. R., *The Dutch Seaborne Empire, 1600–1800* (New York: Alfred A. Knopf, 1965).

BULL, G. B. G., "The Netherlands Delta Plan," *Geography*, 47, pt. 1 (January 1962), 87–89.

BURKE, GERALD L., *Greenheart Metropolis: Planning the Western Netherlands* (New York: St. Martin's Press, 1966). Authoritative discussion of planning problems, accomplishments, and prospects; covers the Zuider Zee reclamation, the Delta Plan, and urban development and planning; numerous maps and photographs); also, *The Making of Dutch Towns: A Study in Urban Development from the Tenth to the Seventeenth Centuries* (New York: Simmons-Boardman Publishing Corporation, 1960).

"The Changing Human-Geographical Aspects of the Netherlands" (5 articles), *Tijdschrift voor Economische en Sociale Geografie*, 54, no. 2 (February 1963), 30–59.

DIEM, AUBREY, "Urban Development Problems of the Ports of Rotterdam and Amsterdam," *Cahiers de Géographie de Québec*, 11, no. 22 (April 1967), 5–25.

"The Dutch Turn the Tide," Fortune, 77, no. 3 (March 1968), 132–137. Photographs and brief discussion of the Delta Plan.

DUTT, ASHOK K., "Levels of Planning in the Netherlands, with Particular Reference to Regional Planning;" *Annals of the Association of American Geographers*, 58, no. 4 (December 1968), 670–685.

HALL, PETER, *The World Cities* (London: Weidenfeld and Nicolson, World University Library, 1966), Chap. 4, "Randstad Holland," pp. 95–121. A discussion of the geography and planning problems of the "Rim City" (Amsterdam-Hague-Rotterdam-Utrecht) in the southern Netherlands.

JOHNSON-MARSHALL, PERCY, *Rebuilding Cities* (Chicago: Aldine Publishing Company, 1966), Chap. 7, "Rotterdam," pp. 319–348.

KUIPERS, HENDRIK, "The Changing Landscape of the Island of Rozenburg (Rotterdam Port Area)," *Geographical Review,* 52, no. 3 (July 1962), 362–378.

"Land Reclamation in the Netherlands: The Northeast Polder," Chap. 8 in Richard M. Highsmith, Jr., ed., *Case Studies in World Geography: Occupance and Economy Types* (Englewood Cliffs, N.J.: Prentice-Hall, 1961), pp. 55–59.

LINGSMA, J. S., *Holland and the Delta Plan,* trans. by C. van Amerongen (2d ed.; Rotterdam and The Hague: Uitgeverij Nijgh and Van Ditmar, 1964).

MURRAY, JOHN J., *Amsterdam in the Age of Rembrandt* (Centers of Civilization, Vol. 21; Norman: University of Oklahoma Press, 1967), Chap. 3, "Market of the World," pp. 48–90.

NORTH, GEOFFREY, "Gargantuan Rotterdam," *Geographical Magazine,* 40, no. 17 (September 1968), 1465–1476.

ORME, A. R., "The Reclamation of the Zuiderzee," *Geographical Viewpoint* [Journal of the Association of Geography Teachers of Ireland], 1, no. 3 (1966), 101–123.

PILKINGTON, ROGER, "The Rhine Delta Project," *Geographical Magazine,* 39, no. 11 (March 1967), 926–936.

SEVERIENS, JACOBUS T., *Basic Data on the Economy of the Netherlands,* U.S. Bureau of International Commerce, Overseas Business Reports, OBR 65-9 (February 1965). (Washington, D.C.: Government Printing Office, 1965.)

SMIDT, M. de, "Foreign Industrial Establishments Located in the Netherlands," *Tijdschrift voor Economische en Sociale Geografie,* 57, no. 1 (January–February 1966), 1–19.

STEIGENGA-KOUWE, SUZANNE E., "The Delta Plan," *Tijdschrift voor Economische en Sociale Geografie,* 51, no. 7 (July 1960), 167–175.

"The Storm Floods of 1st February, 1953" (9 articles), *Geography,* 38, pt. 3 (July 1953), 132–189.

TAKES, CH. A. P., and A. J. VENSTRA, "Zuider Zee Reclamation Scheme: Post War Developments in Planning and Execution Especially As Regards Layout and Colonisation," *Tijdschrift voor Economische en Sociale Geografie,* 51, no. 7 (July 1960), 162–167.

VOORMOLEN, C., "Post-war Changes in the Economic Structure of the Northern Province of the Netherlands," in J. W. House, ed., *Northern Geographical Essays in Honor of G. H. J. Daysh* (Newcastle upon Tyne, England: Oriel Press, 1966), pp. 241–255.

van VEEN, JOH., *Dredge, Drain, Reclaim: The Art of a Nation* (3d ed.; The Hague: Martinus Nijhoff, 1952).

WEVER, E., "Pernis-Botlek-Europoort: un Complexe à Base de Pétrole," *Tijdschrift voor Economische en Sociale Geografie,* 57, no. 4 (July–August 1966), 131–140.

See also the lists of references and readings for Chapters 1–4, especially the relevant chapters in standard texts on Europe, p. 92.

Switzerland
and Austria

9   The two small countries of Switzerland and Austria, located in the heart of Europe, have often been contrasted and otherwise compared. Despite certain environmental and cultural similarities, these neighbors have been remarkably different in their historical development. Switzerland represents one of the world's foremost examples of the economic and political success of a small nation, whereas Austria has experienced great economic and political difficulties, although these have lessened markedly in recent years. The likenesses and differences of the two countries will form a major theme in the present chapter.

PHYSICAL SIMILARITIES OF SWITZERLAND AND AUSTRIA

With respect to physical environment Switzerland and Austria have much in common. Over half of each country is occupied by the high and rugged Alps (map, right).

A tiny village in the Swiss Alps. Note the characteristic church tower, jagged skyline, and forested slope at right behind the small inn. (Standard Oil Company, N. J.)

North of the Alps both countries include part of the rolling morainal foreland of the mountains. The Swiss section of the Alpine Foreland is often called the Swiss Plateau. It lies mostly between 1500 and 3000 feet in elevation and extends between Lake Geneva, on the French border, to the southwest, and Lake Constance, on the German border, to the northeast. The Austrian section of the foreland, which is slightly lower in elevation, lies between the Alps and the Danube River from Salzburg on the west to Vienna on the east. The Swiss and Austrian sections are separated from each other by a third portion of the foreland which lies in southern Germany. North of the foreland both Switzerland and Austria include mountains or hills which are much lower in elevation and smaller in areal extent than the Alps. These highlands differ in character. The Jura Mountains, on the border between Switzerland and France, consist largely of parallel ridges formed of sedimentary rocks, while the Bohemian Hills of Austria, on the border with Czechoslovakia, represent the irregular, eroded southern edge of the extremely old Bohemian massif, and are formed of igneous and metamorphic rocks. North of the Jura in Switzerland the area around Basel opens onto the Upper Rhine Plain of Germany and France, while in eastern Austria the Vienna Basin and a strip of lowland to the south (Burgenland) adjoin the Little Hungarian Plain on the east and the

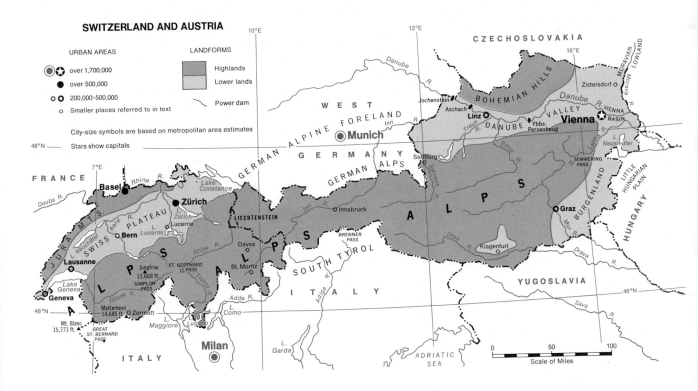

SWITZERLAND AND AUSTRIA

Farms and woodlands in the Swiss part of the Alpine Foreland (Swiss Plateau). The view was taken near Lake Geneva. Note the lack of fences and farm animals in the photo. For the most part, the dairy cattle that provide the principal farm income are kept in barns and fed with hay, commercial cattle feed, or green fodder that is gathered regularly from the fields. (Standard Oil Company, N.J.)

TABLE 5 SWITZERLAND AND AUSTRIA: AREA, POPULATION, AND LAND USE[a]

ITEM	AUSTRIA	SWITZERLAND	SWITZERLAND AS APPROXIMATE PERCENTAGE OF AUSTRIA
Area (thousand square miles)	32.4	15.9	49%
Population (millions; 1968 estimates)	7.4	6.2	84%
Approximate population density (per square mile)	227.0	391.0	172%
Cultivated land, including fallow and orchards (000 acres)	4268 (21%)	1042 (10%)	24%
Permanent meadows and pastures (000 acres)	5572 (27%)	4305 (42%)	77%
Forests and woodlands (000 acres)	7850 (38%)	2423 (24%)	31%
Other land use, mostly waste (000 acres)	3021 (15%)	2428 (24%)	80%

[a]Land-use data from *FAO Production Yearbook*, 1965.

A view of the Austrian section of the Alpine Foreland (Danube Valley). This photo of the deeply intrenched Danube and the rolling surface of the Foreland was taken about midway between Linz and the West German border. (Austrian Information Service.)

Moravian Lowland of Czechoslovakia on the north. The Alps extend from Switzerland southward into Italy and southwestward into France, and from Austria southward into Italy and Yugoslavia.

RELATIVE SUPERIORITY OF AUSTRIAN NATURAL RESOURCES

The physical similarities of Switzerland and Austria are not paralleled by an equal endowment of natural resources. The resources of Austria are, on the whole, superior to those of Switzerland, and in addition Austria is favored by a lower density of population. Some of these advantages of Austria appear in Table 5.

The more favorable situation of Austria with respect to land under cultivation is due primarily to its higher proportion of relatively level terrain. It reflects also to some degree Switzerland's greater specialization in dairy farming, which leads the Swiss to keep a larger percentage of potentially cultivable land in pasture. As might be expected from these circumstances, Switzerland is much more dependent than Austria on food imports. The greater proportion of forest land in Austria primarily reflects the fact that the Austrian Alps are in general lower and less rugged than the Swiss Alps, with a smaller percentage of their land above the tree line. Austria has important exports of forest products, whereas Switzerland is a considerable importer.

In addition, Austria has advantages with respect to mineral and power resources. Switzerland is almost devoid of important mineral resources, while Austria's mineral production is varied and valuable, although exhibiting on the whole more variety than quantity. Oil, primarily from deposits around Zistersdorf, northeast of Vienna, is extracted in sufficient quantities to place Austria fourth among European oil producers, after Romania, West Germany, and France. However, production has declined from a peak reached in 1955, and the annual output of oil is now much less than the country's consumption. Domestic iron ore produced for the most part in the Alps northwest of Graz supplies the greater part of Austria's needs, though substantial imports are necessary. There are scattered deposits of lignite, coal, bauxite, magnesite, and salt, as well as other minerals. Both Switzerland and Austria have large hydroelectric resources relative to their areas, but Austria's potential is greater.

THE SUPERIOR SWISS STANDARD OF LIVING

The advantages of Austria with respect to natural resources have been overweighed by other factors in the historical development of the two countries so that the economic situation of Austria is today more adverse than that of Switzerland and the Austrian standard of living is much lower. Switzerland, in fact, is generally regarded as having one of the world's soundest and most successful economies, while Austria has been saved from economic collapse several times since World War I only by the extension of large amounts of foreign aid. The Swiss standard of living is one of the highest in Europe, while Austria's is closer to that of the relatively poor countries that comprise southern and eastern Europe. Austria's per capita income is only one-half that of Switzerland. These contrasts between the two countries can be attributed primarily to a long period of peace enjoyed by Switzerland. During this time its citizens have been able to develop an economy finely adjusted to their country's potentialities and opportunities and supported by a stable and democratic government. Meanwhile the economy of Austria has been badly disoriented, warped, and hindered by a series of military and political calamities. Since the end of the post-World War II military occupation in 1955 (p. 207), Austria's economic and political situation has been relatively favorable and the country has made rapid economic progress. But it still has far to go to equal the prosperity of its more fortunate neighbor.

ROLE OF SWITZERLAND AS A NEUTRAL BUFFER STATE

Except for some minor internal disturbances in the nineteenth century, Switzerland has been at peace inside stable boundaries since 1815. The basic factors underlying this long period of peace seem to have been (1) Switzerland's position as a buffer between larger powers, (2) the comparative defensibility of much of the country's terrain, (3) the relatively small value of Swiss economic production to an aggressive state, (4) the country's value as an intermediary between belligerents in wartime, and (5) Switzerland's own policy of strict and heavily armed neutrality. The difficulties which a great power might encounter in attempting to conquer Switzerland are often popularly exaggerated, since the Swiss Plateau, the heart of the country, lies open to Germany and France, and even the Alps have frequently been traversed by strong military forces in past times. On the other hand, resistance in the mountains might well be hard to thoroughly extinguish. In World War II Switzerland was able to hold a club over the head of Germany by mining the tunnels through which Swiss rail lines avoid the crests of the Alpine passes. Destruction of these tunnels would have been very costly to Germany, as well as to its military partner, Italy, since the Swiss railways were depended on to carry much traffic between them.

THE PRODUCTIVE SWISS ECONOMY

The efficient and highly successful Swiss economy finds its greatest development in four lines of enterprise: (1) the production of highly finished, specialized manufactures for export, (2) dairy farming, (3) the world's most renowned tourist industry, and (4) a high development of international banking and other international financial services.

Industrial and Urban Development

In terms of employment and income Switzerland is primarily an industrial country, despite the fame of its picturesque resorts and Alpine dairy farms. Swiss industry is based on the country's two major industrial

A product of Swiss industry. This enormous turbine was manufactured by a Swiss firm for installation in one of Switzerland's many hydroelectric power stations. (Bell Engineering Works, Ltd., Kriens-Lucerne, Switzerland.)

categories in 1965 included machinery (excluding vehicles), 28 percent; chemicals, 20 percent; watches, clocks, and parts, 14 percent; and textiles, excluding apparel, 10 percent. Instruments and apparatus of various types comprised an additional 5 percent. Switzerland's export and import trade is worldwide, but in 1965, 40 percent of its exports went to the European Common Market countries (17 percent to West Germany alone). Its next biggest markets, the United States and the United Kingdom, took only 10 percent and 7 percent, respectively. This situation has made the question of future relations with the Common Market, or inclusion in it, a subject of some concern in Switzerland, as in most of the small, trade-dependent countries of Europe.

Reliance on hydroelectricity as the major source of power has facilitated the development of many small industrial centers throughout the country. The majority of these are located on the Swiss Plateau, but some are found in valleys of the Jura and Alps. Six Swiss cities have metropolitan populations of more than 100,000: Zürich (750,000), Basel (540,000), Geneva (330,000), Bern, the capital (265,000), Lausanne (210,000), and Lucerne (145,000). Five of the six cities form a roughly linear pattern across the Swiss Plateau, from Zürich on the northeast to Geneva on the southwest (map, p. 201). Basel, however, lies beyond the Jura at the point where the Rhine River turns northward between Germany and France. This city, located at the head of navigation for large barges on the Rhine, handles most of Switzerland's riverborne commerce. In addition, it is the largest railway center in the country.

resources—hydroelectric power and the skill of Swiss workers. Most raw materials as well as supplementary fuels are imported. The country's industries are specialized along lines that minimize Switzerland's lack of bulky raw materials, while making full use of its power resources and traditions of workmanship that date back to handicraft days before the industrial revolution. The major products are (1) metal goods, including machinery (much of it designed to order—see photo above) and, of course, the famous Swiss watches, (2) chemicals, especially pharmaceuticals, and (3) textiles, generally of very high quality and including silks, artificial silks, and embroideries, in addition to more ordinary materials. In recent years chemicals and aluminum products have been gaining in importance. Most of Switzerland's export trade is comprised of the foregoing lines of goods. Major

Swiss Agriculture

Switzerland is one of the world's least self-sufficient countries agriculturally, relying very heavily on food imports except for dairy products and potatoes. The salability of Swiss exports has made such food imports possible and has enabled Swiss agriculture to become adjusted to lands which on the whole are better suited to pasture and hay than to cultivated crops. The result has been a high degree of specialization on dairy farming. The dairy industry is centered in small, intensively worked farms on the rolling to hilly lands of the central plateau, although mountain pastures in the Alps are extensively used. The latter is an especially well-known aspect of Swiss agriculture. Dairy cattle and goats, accompanied by herdsmen or the entire farm family, are driven in the spring to pastures near the snow line, where they are kept throughout the summer. This seasonal migration of farm people and their livestock between the valley floors and high mountain pastures is known as *transhumance* and is practiced in many mountainous regions of the world. The high pastures of Switzerland, known as "alps," have given their name to the Alps Mountains. In recent times there has been an increasing tendency among Swiss farmers to pasture only the young cattle on the high meadows and to keep the main dairy herds at lower levels the year round. Switzerland has a limited export of cheeses, milk chocolate, and condensed milk. Government policies encourage greater self-sufficiency in foodstuffs, particularly wheat, in order to place the country less at the mercy of belligerents in time of war and to subsidize Swiss farm income.

The Importance of "Invisible" Exports

The total cost of Swiss imports is not ordinarily covered by exports of goods, but the deficit is more than made good by various "invisible" exports, of which exports of capital and services to tourists are the most important. Few countries have so much to offer the tourist, and probably no country has developed and organized the tourist trade so completely and successfully. Special training programs are made available to personnel in the tourist industry, and high standards are enforced. Alpine resorts such as Zermatt, Davos, and St. Moritz have become world famous, as has the name of a pioneer in the trade, Ritz. Switzerland, more than any other, is a country which exports its scenery. It entertains more visitors compared to its resident population then any other major tourist country. However, even more important than tourism as a source of international income are the returns from foreign loans and investments and charges for international banking and insurance services. As a relatively wealthy and stable country and a long-standing neutral, Switzerland is in a good position to supply financial services to a worldwide clientele. Zürich is one of the world's great centers of international banking.

THE NATIONAL UNITY OF SWITZERLAND

In their successful pursuit of economic goals, the Swiss have been aided by an effective national unity expressed in a stable, democratic, and competent government.

The unity of the Swiss is the more remarkable in that it embraces a population divided in both language and religion. At present approximately two-thirds of the Swiss speak German as a native tongue, about one-fifth speak French, about one-tenth Italian, and about 1 percent Romansch, the latter an almost extinct descendant of Latin which has been preserved in the mountains of southeastern Switzerland. As might be expected, German, French, and Italian are dominant, respectively, in the sections of Switzerland adjoining Germany, France, and Italy. A religious division also exists, since approximately 53 percent of the Swiss are listed as Protestants and about 46 percent as Roman Catholics.

The internal political organization of Switzerland expresses and makes allowance for the ethnic diversity of the population. Originally the country was a loose alliance of small sovereign units known as cantons. When a stronger central authority became desirable in the nineteenth century, not only were the customary civil rights of a democracy guaranteed, but governmental autonomy was retained by the cantons except for limited functions specifically assigned to the central government. Although the functions allotted the central government have tended to increase with the passing of time, each of the local units (now 25 in number, including 22 cantons of which 3 are divided into half-cantons) has preserved a large measure of authority. Local autonomy is supplemented by the extremely democratic nature of the central government. In no country are the initiative and the referendum more widely used. Through these devices most important legislation is submitted directly to the people for their decision. Thus guarantees of fundamental rights, local autonomy, and close governmental responsiveness to the will of the people have been successfully used to foster national unity despite the potential handicaps of ethnic diversity and local particularism.

The central government, in turn, has pushed the country's economic development vigorously. Two outstanding accomplishments have been the construction and operation of Switzerland's railroads and the development of the hydroelectric power system. Despite the difficult terrain the Swiss government has succeeded in building a rail network whose density on a countrywide basis is exceeded only by the railway systems of Belgium and Great Britain. It carries not only Switzerland's own traffic, but a volume of international transit traffic sufficiently large to be an important source of revenue. The highly developed hydroelectric power system, utilizing the many torrential streams of the mountains, places Switzerland second only to Norway among the world's nations in the amount of hydroelectricity available per capita. In total developed capacity Switzerland ranks sixth in Europe, behind Italy, France, Norway, Sweden, and Spain. The hydroelectric system has more than doubled in capacity since World War II and is still being expanded in order to relieve some of the continuing need for supplementary imports of fuel. Well under half of Switzerland's estimated potential capacity is now in use.

Thus internal unity and effective government have contributed greatly to Switzerland's economic success. On the other hand, economic success has undoubtedly reinforced the internal unity and political stability of the country. And a century and a half of peace has provided highly favorable conditions for both political and economic adjustment. Few modern nations have been so fortunate.

POLITICAL AND ECONOMIC PROBLEMS OF AUSTRIA

In contrast to Switzerland's happy circumstances, Austria has been the victim during the twentieth century of a series of military and political disasters which have required large and difficult readjustments of its economic life. These difficulties have been occasioned by the two world wars and the political and economic arrangements following them. Austria emerged in its present form as a defeated remnant of the Austro-Hungarian Empire when that empire disintegrated in 1918 under the stress of war and internal difficulties. Austria's population is essentially German in language and cultural background, and there is evidence that a majority wished to unite with Germany following World War I. This was forbidden by the victors, and Austria became an independent national state. Torn by internal strife and with an economy seriously disoriented by the loss of its empire, the country limped through the interwar period until absorbed by Nazi Germany, against the will of a majority, in 1938. In 1945 it was reconstituted a separate state, but as in the case of Germany, was divided into four occupation zones administered, respectively, by the United States, the United Kingdom, France, and the Soviet Union. Vienna, like Berlin, was placed under joint occupation by the four powers. During 10 years of occupation the Soviet Zone, already badly damaged in the war, was subjected to extensive removal of industrial equipment for reparations. In 1955 the occupation was ended by agreement among the four powers. At the insistence of the Soviet Union, it was stipulated that Austria maintain neutrality in the East-West political struggle and that she continue to make reparations payments, especially in oil, to the Soviet Union. These payments ended in 1964.

As the core area of an empire of 50 million people, Austria developed a diversified industrial economy during the years prior to 1914. Iron and steel were manufactured in a number of small centers, of which Graz was the most important. Production of a variety of secondary metal goods was centered largely in Vienna. Textiles were manufactured in the Alps and at Vienna. Wood industries made use of the Alpine forests. These industries developed within, and were dependent on, an empire which was an extraordinarily self-contained economic unit. Austrian ore was smelted mainly with coal drawn from Bohemia and Moravia, now in Czechoslovakia. The Austrian textile industry specialized to a considerable extent on spinning, leaving much of the weaving to be done in Bohemia. In general, Austrian industry was not outstandingly efficient, but it had the benefit of a protected market in the agricultural parts of the empire to the east—areas now included in Hungary, Romania, Czechoslovakia, Yugoslavia, Poland, and the Soviet Union. In turn, Austria drew foodstuffs and some industrial raw materials from these areas, while Austrian agriculture was relatively neglected.

When the empire disintegrated at the end of World War I, the areas which had formed Austria's protected markets were incorporated in the independent states referred to in the preceding paragraph (excepting the Soviet Union, which later gained a share of these lands only as a result of World War II). These states, motivated by a desire to develop industries of their own, began to erect tariff barriers. Other industrialized nations began to compete with Austria in their markets. The resultant decrease in Austria's ability to export to its former markets made it more difficult for the country to secure the imports of food and raw materials which its unbalanced economy required. Such difficulties were further increased after World War II by the absorption of East Central Europe into the Communist sphere. The necessary reorientation of Austrian industries toward new markets, now found principally in West Germany, Italy, and other countries of free Europe, has not been easy. In these new markets Austrian products must compete with the products of domestic industries (which often enjoy superior resources and tariff protection) and with the products of other industrialized exporting nations. As a consequence, Austrian exports of goods have been consistently inadequate to pay for necessary imports, which include foodstuffs, fuels, many raw materials, and certain types of manufactured goods. Since Austria has not had such extensive "invisible" exports as Switzerland, the result has been frequent deficits in international trade since 1918.

Needed readjustments to this situation have been rendered still more difficult by certain internal conditions and problems. One of these is the financial drain of a notoriously oversized bureaucracy, originally developed to meet the needs of a great empire and tending to persist despite the fact that it is far too large for the small Austrian state. Another inheritance from the empire is the rigidly noncompetitive structure of Austrian industry. Firms and cartels enjoying a high degree of monopoly control over their respective sectors of the internal market and having great influence in Austrian politics have not always been amenable to changes which would be in the best interests of the country and the people. In addition, Austria was subjected to violent class and party disputes during the period between World Wars I and II. Economic crisis deepened class antagonisms, and feelings were further embittered by disagreements over the role of the church. (Austria's population is 90 percent Roman Catholic, but like the populations of France and Belgium it has contained strong anticlerical elements.) In 1934 the anticlerical socialists who controlled Vienna were suppressed in open civil war by a clerically oriented authoritarian government drawing its main support from the peasants and the upper classes. Pro-German and anti-German sentiments and parties also divided the country and resulted in further violence. Such internal divisions contributed materially to the ease with which Austria was absorbed by Germany in 1938.

AUSTRIAN EFFORTS TOWARD ECONOMIC READJUSTMENT

Despite the difficulties and distractions described above, Austria has made notable progress since World War I, and especially rapid progress since World War II, in building an economy suited to its new status as a small independent state. This has involved a closer adaptation of the economy to domestic resources and a more intensive exploitation of them. Lines of development for which the country's resources are especially suitable have been emphasized. Some of Austria's resources closely parallel those of Switzerland, and some of the main lines of Austrian development are also very similar to those of the neighboring country. However, Austria has a greater variety of resources than Switzerland, and some resources and industries not present in the latter country play an important role. Thus an industrial structure of surprising diversity, considering the country's size, has developed. It consists of a blend of newer industries, usually closely related to the country's natural resources, with older industries which have persisted since imperial times and are often not so closely connected with domestic natural resources.

The Principal Austrian Industries

Forest Industries and Tourism

From the viewpoint of export production, Austria's forest resources have been of the most fundamental importance. Almost 40 percent of Austria consists of forested areas (a higher percentage than in any European countries except Finland and Sweden), while most surrounding countries have a shortage of wood. In this situation large and consistent increases in the production and export of a variety of wood products have been possible. Forest-derived products now form the leading category of Austrian exports, accounting for nearly one-fifth of the total value of all exports. Lumber and paper are the main items, but there is also a growing output and export of rayon.

In addition, the attraction of forested mountain areas has been a major factor in a very rapid recent increase in tourism. This finally reached such a point in the early 1960s that its receipts were able to cover the chronic Austrian deficit in international trade in goods. The country is generally visited by more foreign tourists per year than Switzerland, but not quite so many relative to its resident population. Nor are the visitors so affluent and free-spending on the average as those of Switzerland. Austria, in fact, has frequently been called the "poor man's Switzerland." The drawing power of its Alpine resorts and its scenery, only slightly less imposing than that of Switzerland, is supplemented by a number of carefully fostered cultural attractions, such as the world-famous music festivals at Salzburg and Vienna.

The Iron and Steel Industry

From an export standpoint, the iron and steel industry holds second rank in Austria to the wood-based industries, accounting generally for well over one-tenth of the country's total exports. Although by no means large compared to that of many countries, the Austrian steel industry is larger than it probably would be except for a series of historical accidents. After it was shorn of important markets and sources of coal by the breakup of the Empire in 1918 it never regained its pre-World

The hydroelectric power dam at Jochenstein, a joint project of Austria and West Germany. In this stretch the Danube forms the boundary between the two countries. (Austrian Information Service.)

War I production until the period of German domination. Nazi Germany, however, for reasons of security connected with Austria's interior European location, more than doubled the country's steel capacity. American economic aid provided further help after the war, and since then the economic growth of Europe has afforded generally favorable conditions for expanding steel production and exports. Recently the Austrian industry developed the newest major technological breakthrough in steelmaking—the Linz-Donawitz oxygen blast process, which has spread rapidly to many other countries. However, the industry has had to make progress in the face of increasingly inadequate natural resources. About three-quarters of its iron ore requirements are supplied by mines in the eastern Alps, but large supplemental imports, notably from the Soviet Union in recent years, are necessary. Austrian coal production, never large, rose to a peak during the 1930s and then declined until the last mine was closed, perhaps permanently, in 1965. Thus the steel industry must now import all its coal and coke requirements; must secure sizable amounts of ore from suppliers a considerable distance away; and then must sell much of its output in highly competitive foreign markets. The principal centers of the industry are located at or near Graz, where production began under the Empire, and at Linz, where it was developed by Nazi Germany.

The Rising Importance of Hydroelectric Power

Increased power development has been a basic aspect of Austria's recent progress. Hydroelectricity is the key element, although Austrian capacity is still only about half that of Switzerland. Most installations thus far are in the Alps, but recent development and future plans emphasize the Danube River. A series of fifteen large dams and power stations are planned to control completely the Austrian course of the river. The first three were completed at Jochenstein (jointly with West Germany), Ybbs-Persenbeug, and Aschach in 1956, 1960, and 1963, and work is proceeding on the project. The Aschach plant is the largest hydroelectric station in Europe.

Increasing supplies of cheap electric power have been important to Austrian development in a number of ways. Among other things, the country has been converted from an importer of electric power to Europe's principal exporter of electricity. At the same time, Austrian electricity consumption has been rising at the rate of approximately 7 percent a year, to the benefit of the standard of living and of many industries. Especially notable among the industries that have benefited from increased power supplies are the chemical industry, which also has available extensive salt deposits near Salzburg, and the aluminum industry, which requires very large amounts of cheap power.

Misfortunes of Austria's Oil Industry

Austria is fortunate to have such a favorable hydropower situation, as its other power resources are quite inadequate. Not only are the country's small coal reserves practically exhausted, but its oil production provides an excellent example of the handicaps under which the country has had to operate. A tiny pre-World War II output was multiplied almost 40 times by 1944 under the German occupation, but the enlarged production was consumed by the German war machine. Following a period of disruption in the immediate postwar years, the 1944 level of production was surpassed in 1950 and was approximately tripled by 1955. But the Zistersdorf deposits lay in the Soviet zone of occupation, and Soviet exactions left Austria in a deficit situation with respect to oil despite the fact that the country was Europe's second largest producer (after Romania). Between the end of occupation in 1955 and the early 1960s Soviet reparations took about a third of the production. It now appears that the Zistersdorf field has passed its peak, and production has never regained the levels of the mid-1950s. Thus Austria has not reaped anything like maximum benefit from its oil resources, which were originally among the largest possessed by any European country. The output from the Zistersdorf field is still valuable to Austria, but it cannot meet the country's requirements, and petroleum products are a major import, along with coal.

Magnesite and Graphite

In two mineral resources, magnesite and graphite, Austria holds a very outstanding position in the world. Deposits in the eastern Alps make the country second only to the Soviet Union in the production of magnesite. Austria accounts for about one-sixth of world production. Magnesite refractory brick for high-temperature furnace linings comprise the highest export value from this material. However, exports of magnesium metal, a very light alloy metal with a variety of uses, have increased in importance during recent years, and Austria's shipments of magnesite ore to foreign smelting centers have also become important. Austria is also a major world producer and exporter of natural graphite, a mineral with many uses in metallurgical, chemical, and electrical industries.

Clothing and Jewelry

Certain consumer goods industries which were natural to an imperial center such as Vienna have survived to play some role in modern Austria's economic revival. The most important of these from a modern export point of view are clothing and jewelry manufacture. In the case of clothing, recent progress has been connected with Austria's growing reputation as a winter sports center and the related possibilities for setting styles.

Engineering and Textile Industries

Although the above industries—wood industries, iron and steel, electric power, aluminum, magnesite, clothing, and jewelry—provide Austria's main lines of net exports, certain other industries are quite important within the country in employment and production largely for the home market. Outstanding among these are the engineering (machine-building) and textile industries. Various branches of the engineering industries contribute importantly to exports also, but development in these lines is small enough compared to demand that Austria is still a heavy net importer of machinery. As in many countries with a long industrial history, textiles, in contrast to engineering, have been a relatively stagnant industry except for those branches associated with artificial fibers.

Trends in Austrian Agriculture

In Austrian agriculture the main trend since the end of the Empire has been an increased development of dairy and livestock farming, representing an attempt to make more intensive use of the relatively large acreages that are best adapted to pasture and to produce more valuable products from the land in general. This trend has resulted in a significant increase in the acreage devoted to pasture and a decrease in cropland. The smaller acreage in crops has meant chiefly a decreased acreage in rye, formerly the main crop on the country's poorer soils.

Modest exports of dairy products and live animals have been achieved, but as yet they amount to only a small fraction of comparable exports from the Netherlands or Denmark, and are exceeded by exports from a number of other European countries. Physical conditions in Austria are very suitable for dairy and livestock farming, but attempts to specialize in this branch of agriculture have been limited by the difficulty of paying for imports of grain and other foods. Since a fair degree of self-sufficiency has had to be maintained, wheat acreage has been kept at levels comparable to those which prevailed under the Empire. Barley grown for animal feed has replaced rye as the country's second crop in acreage and production. A relatively large potato acreage has also been maintained, and the production and acreage of sugar beets have been greatly expanded. In addition, the country's overall crop yields have been steadily increased by more intensive management, except for the years when agriculture was disrupted by World War II and its aftermath. However, comparisons of crop yields with those of the North Sea countries reveal that Austrian agriculture still has substantial room for improvement.

Despite the policies and achievements summarized above, foodstuffs and supplementary animal feeds still represent approximately 12 percent of the country's imports, and in view of the limited amount of arable land and the poor quality of much of it, a higher degree of self-sufficiency in food promises to be increasingly difficult to attain. Actually, more self-sufficiency is probably not desirable if overall exports can be expanded to pay for larger food imports and thus free Austrian land for more specialized uses.

As the foregoing survey indicates, Austria has made impressive progress toward a viable economy since the end of the Empire. Loss of assured markets and sources of materials in territories that were formerly under Austrian political control or domination has led to new economic emphases based on a fuller utilization of Austria's domestic resources. The fact that the country's chronic trade deficit was not overcome until the early 1960s, and then perhaps not finally, points to the disastrous state in which the Austrian economy was left by World War I and the magnitude of the obstacles it has faced since that time.

The Problem of Trade with the Common Market

In the 1960s a new threat to Austria's stability and progress appeared. By the mid-1960s approximately half of the country's foreign trade was being carried on with the six nations of the European Common Market. As trade

A summer pasture above the timber line in the Austrian Alps. The photo was taken in extreme western Austria near the Swiss border. (Austrian Information Service.)

barriers within the Market were progressively reduced toward extinction, while barriers against non-Market countries were maintained, Austria found itself at an increasing disadvantage in maintaining its trading relationships. Consequently, protracted negotiations were begun with a view to Austria's inclusion in, or at least economic association with, the Common Market.

Two main obstacles continue to block Austrian inclusion in the trade benefits of the Common Market. One of these is the treaty under which Soviet troops evacuated eastern Austria in 1955. It specifies that Austria is to remain perpetually neutral, and the Soviet Union holds that Austrian association with the Common Market (which has ultimate aims of greater political unification among its members) would be a breach of this treaty. Austria, on the other hand, argues that it can associate with the Market countries economically while remaining politically neutral.

The second major obstacle lies in the fact that one of the Common Market countries—Italy—has been hostile

A view across Vienna. The spire of St. Stephen's Cathedral dominates the center of the photo; the Cathedral is in the heart of the city's Central Business District. In the foreground, the Burgtheater. (Austrian Information Service.)

to Austrian inclusion. This results from a long-standing political quarrel between the two countries, centering on the status of the part of Italy that lies adjacent to Austria in the Alps just south of the Brenner Pass. This area, known to Austrians as the South Tyrol and to Italians as the Alto Adige, had been part of the Austrian Empire, but was acquired by Italy in 1919 and has been held by Italy since that time. It is inhabited by approximately 200,000 German-speaking people. Many (perhaps most) of these would prefer reinclusion in Austria. Italy argues that the territory is necessary to its national security, as it places the frontier at the crest of the Alps. Terrorist activities have been used by elements of the Tyrolese population to enforce their demands for self-determination, and Italy has accused Austria, apparently with some truth, of allowing its territory to be used as a base for these activities. In retaliation it has opposed Austrian

association with the Common Market. During the 1960s this quarrel was the most active border dispute in Europe.

THE ROLE OF VIENNA

A significant aspect of Austrian readjustment since World War I has been the somewhat lessened importance of the famous Austrian capital, Vienna (2 million). Although Vienna is still by far the largest city and most important industrial center of Austria, there has been a pronounced tendency for population and industry to shift away from the capital and toward the smaller cities and mountain districts. In 1968 Vienna had an estimated population of 1.6 million in the city proper, as compared with about 2 million in 1910. But the smaller cities of Linz (285,000), Graz (255,000), Innsbruck (195,000), and Salzburg

(120,000) have all experienced rapid growth. From being the capital of a great empire, Vienna has regressed to a more modest status as the capital of a small country, and the resultant damage to its situation is reflected in the fact that its present metropolitan population is about the same as the population of the city proper half a century ago. Vienna's failure to match the growth of other Austrian cities also reflects the increased importance of forests and waterpower in the Austrian economy, as well as the dispersal of Nazi war industries into less vulnerable locations away from the capital.

Vienna's importance, however, has now persisted since Roman times and is based on more than purely Austrian circumstances. The city is located at the crossing of two of the European continent's major natural routes: the Danube Valley route through the highlands separating Germany from the Hungarian plains and southeastern Europe, and the route from Silesia and the North European Plain to the head of the Adriatic Sea. The latter route follows the lowland passageway of Moravia to the north and makes use of the passes of the eastern Alps, especially the Semmering Pass, to the south. These routes have been important corridors of movement throughout European history. Whenever political conditions have permitted, Vienna has been a major focus of transportation and trade, and is now regaining some of its old importance in this respect as restrictions imposed by the iron curtain are relaxed. But the city's position has also made it a major strategic objective in time of war—a fact that brought it great damage in World War II and that contributed to the long occupation of Austria after that war.

REFERENCES and READINGS

Across the Alps: Aerial Views between Nice and Vienna. 92 views by Swissair: texts by Hans Annaheim and others (Bern: Kümmerly and Frey, 1959).

Atlas der Republik Österreich [Atlas of the Austrian Republic] (Vienna: Freytag-Berndt und Artaria 1961). A large and elaborate national atlas of Austria with magnificent maps in color.

BONJOUR, EDGAR, *Swiss Neutrality: Its History and Meaning* (2d ed.; London: George Allen and Unwin, 1952).

BURGHARDT, ANDREW F., *Borderland: A Historical and Geographical Study of Burgenland, Austria* (Madison: University of Wisconsin Press, 1962).

CANTWELL, JOHN A., *Basic Data on the Economy of Austria,* U.S. Bureau of International Commerce, Overseas Business Reports, OBR 66–24 (May 1966), (Washington, D.C.: Government Printing Office, 1966.)

EGLI, EMIL, *Swiss Life and Landscapes* (London: P. Eleh, 1949).

GUTKIND, E. A., *Urban Development in the Alpine and Scandinavian Countries* (International History of City Development, Vol. 2; New York: The Free Press of Glencoe, 1965). Elaborately illustrated with photographs and maps of cities.

HEROLD, J. CHRISTOPHER, *The Swiss without Halos* (New York: Columbia University Press, 1948).

HOFFMAN, GEORGE W., "The Survival of an Independent Austria," *Geographical Review,* 41, no. 4 (October 1951), 606–621.

KUBLY, HERBERT, and the Editors of *Life, Switzerland* (Life World Library; New York: Time Inc., 1964).

MAYER, KURT B., *The Population of Switzerland* (New York: Columbia University Press, 1952); also "The Impact of Postwar Immigration on the Demographic and Social Structure of Switzerland," *Demography,* 3, no. 1 (1966), 68–89; and "Post-war Migration to Switzerland," *International Migration,* 3, no. 3 (1965), 122–134.

MINGHI, JULIAN V., "Boundary Studies and National Prejudices: The Case of the South Tyrol," *Professional Geographer,* 15, no. 1 (January 1963), 4–8.

MUTTON, ALICE F. A., *Central Europe: A Regional and Human Geography* (2d ed.; New York: Frederick A. Praeger, 1968).

SCHEIDL, LEOPOLD G., "The Development of Hydro-electricity in Austria," *Advancement of Science,* 23, no. 109 (July 1966), 133–145.

SINCLAIR, ROBERT, "Austria's Place in Europe's Electric Power Exchange," *Geographical Review,* 54, no. 1 (January 1964), 93–103; and "The Austrian Danube: A Central European Electric Power Bastion," *Professional Geographer,* 16, no. 5 (September 1964), 18–22.

YETT, WILLIAM E., *Basic Data on the Economy of Switzerland,* U.S. Bureau of International Commerce, Overseas Business Reports, OBR 66–42 (July 1966). (Washington, D.C.: Government Printing Office, 1966.)

See also the lists of references and readings for Chapters 1–4, especially the appropriate sections of standard textbooks on Europe listed on p. 92.

Countries

of

Northern Europe

The five countries of Denmark, Norway, Sweden, Finland, and Iceland may be defined as the countries of Northern Europe. This regional grouping accords with a geographic concept which is well established in the countries themselves. The peoples of these lands recognize their close relationships with each other and habitually group themselves and their countries geographically under the regional term *Norden,* or "The North."

A more common term used in referring to some or all of these countries is Scandinavia or the Scandinavian countries. But this regional name is somewhat ambiguous, being used sometimes to refer only to the two countries which occupy the Scandinavian Peninsula, Norway and Sweden; more often to include these countries plus Denmark; and sometimes being extended to include these three plus Finland, Iceland, and even Greenland. Sometimes when Finland is included in the group, the term "Fennoscandia," or Fennoscandian countries, is used, thus taking account of the greater difference of the Finns in ancestry, historical relationships, and language as compared with the other nations. These differences, however, are overshadowed by many cultural, economic, environmental, and other similarities between Finland and the other countries of the group.

215

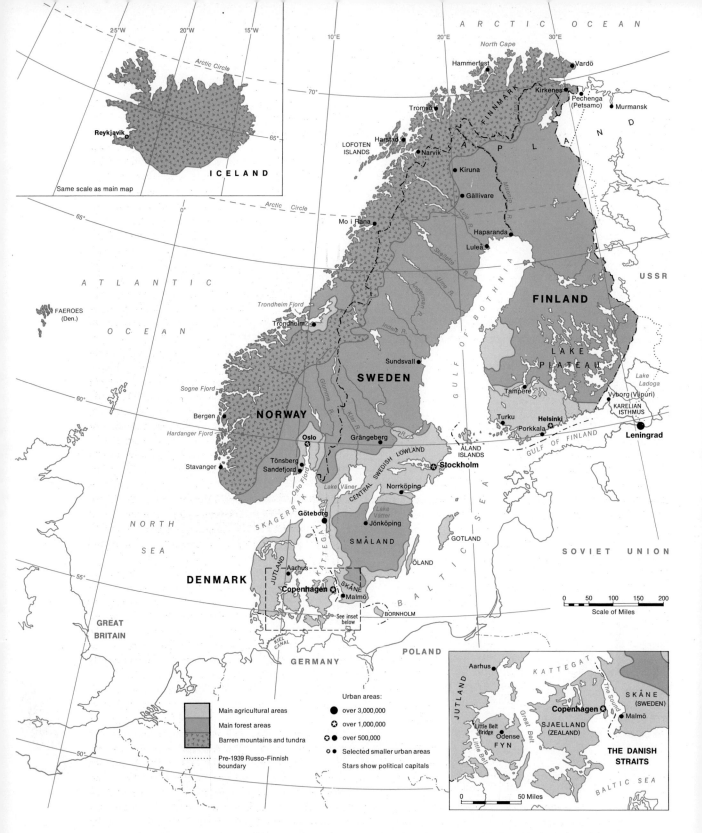

Index map of Northern Europe. Small tracts of agricultural land are scattered through the areas of forest, mountain, and tundra, and small forest tracts occur within the agricultural areas. City-size symbols are based on metropolitan area estimates.

NORTHERN EUROPE AS A REGION

"The North" is a good descriptive term for these lands. No other highly developed countries have their principal populated areas so near the pole. Located in the general latitude of Alaska and occupying a geographical position in Eurasia analogous to that of Alaska in North America, the countries of Northern Europe represent the northernmost seat of advanced culture in the world. The 280,000 people of Alaska inhabiting a slightly larger total area can hardly compare in numbers or in economic and cultural achievements with the 21.5 million people of Northern Europe. The Soviet Union has several cities and areas of importance in these northern latitudes, but its economy and population are based for the most part on areas farther south.

Climatic Effects of the Atlantic

West winds from the Atlantic, warmed here in winter by the North Atlantic Drift, moderate the climatic effects of northern location considerably. Temperatures average above freezing and harbors are ordinarily ice-free in winter over most of Denmark and along the coast of Norway. But away from the direct influence of the west winds, winter temperatures average below freezing and are particularly severe at elevated, interior, and northern locations. In summer the ocean tends to be a cooling rather than a warming influence, and most of Northern Europe has July temperatures averaging no higher than the fifties or low sixties. Highlands have temperatures sufficiently low that a number of glaciers exist, both on the Scandinavian Peninsula and in Iceland. Despite the overall moderation of the climate of Northern Europe as compared with what might be expected from the latitude, the populations of the various countries tend to cluster in the southern sections, and all of the countries except Denmark have considerable areas of sparsely populated terrain where the problems of development are largely those of overcoming a northern environment.

Historical and Cultural Unity

Close historical interconnections and cultural similarities are more important factors in the regional unity of Northern Europe than partial similarity of environmental problems, however. Historically, each of the countries of Northern Europe has been more closely related to others of the group than to any outside power. Finland was ruled by Sweden from the twelfth century until 1809. Denmark and Norway had a common sovereign from the fourteenth century until 1814, as did Sweden and Norway from 1814 to 1905. The southern part of Sweden (*Skåne*) passed from Danish to Swedish control as late as the seventeenth century. Iceland was a possession of Denmark from the fourteenth century until 1918; in the latter year it secured independence, but remained under the Danish sovereign until 1944. At times, warlike relations have prevailed among various countries in the group. For the past century and a half, however, internal relations in Northern Europe have been peaceful, and the feelings of relationship among these countries have come to be expressed in close international cooperation.

Cultural similarities among the countries of Northern Europe are many. Similarities of language, religion, and form of government are probably the most important in fostering regional unity. The languages of Denmark, Norway, and Sweden are descended from the same ancient tongue and are mutually intelligible, though not identical. Icelandic, though a branch of the same root, is more difficult for the other peoples only because it has evolved less and become less modernized. Only Finnish, which belongs to a different language family, is entirely distinct from the other languages of Northern Europe. Even in Finland, however, about 7 percent of the population is of Swedish descent and speaks Swedish as a native tongue. Swedish is recognized as a second official language in Finland.

Among these countries there are no exceptions to the cultural unity embodied in a common religion. The Evangelical Lutheran Church is the dominant religious organization in each country. Over 90 percent of the respective populations adhere to it, at least nominally. It is a state church, supported by taxes levied by the respective governments, and is probably the most all-embracing organization outside of the state.

The countries of Northern Europe also exhibit basic similarities with respect to law and political institutions. These countries have very old traditions of individual rights, broad political participation, limited governmental powers, and democratic control. Thus old foundations have been available for building modern democracies, and the countries of Northern Europe are recognized as outstanding strongholds of democratic institutions. Iceland claims to have the world's oldest legislature, founded in A.D. 930. Today Iceland is a republic, as is Finland, while the other three states are constitutional monarchies. In all these countries real power rests with an elected parliament. In the twentieth century the countries of Northern Europe have consciously and actively worked to increase their similarities respecting legal codes and political institutions by coordinating their laws wherever feasible.

One other similarity not yet mentioned is the small size of these countries, especially in population. Comparative figures on area, population, and population density for the five countries are given in Table 6.

Disadvantages of an "In-between" Position

Small populations and limited resources have forced the countries of Northern Europe to give up imperial ambitions during recent times, though their armies and fleets were the scourge of much of Europe in times past. A policy of neutrality plus a relatively isolated position in one corner of Europe allowed them a long period of peace between the Napoleonic Wars and World War II. In the twentieth century, however, the increasing strategic importance of North Atlantic air and sea routes has jeopardized their safety. In the present world situation, these countries occupy an "in-between" position. They lie on the most direct routes between the United States and the western coreland of the Soviet Union. The coast of Norway, which adjoins the Soviet Union in the far north, offers some of the world's best and most strategically located naval harbors in the famous fjords. This coast is especially suitable as a base for submarine operations against North Atlantic shipping and was so used by the German navy in World War II. Denmark, Norway's neighbor to the south, lies across the outlet from the Baltic Sea, on which some of the

main Soviet seaports and naval bases are located. Finland lies between the Soviet Union and the Scandinavian Peninsula, and Sweden, the largest and most powerful country of Northern Europe, lies in the midst of these various positions. The changed significance of their position has presented the countries of Northern Europe with a common problem of national security, evidenced by the fact that only Sweden escaped involvement in World War II. It has also focused on them an increased degree of world attention and concern.

Social and Cultural Achievements

Small size and resource limitations have made it necessary for each of the countries of Northern Europe to build a highly specialized economy in attempting to attain a high standard of living. Success in such endeavors has been so marked that these countries are probably known as much for high living standards as for any other characteristic. Except for Finland, which falls somewhat short of the other countries in the group, they are, along with Switzerland, the leading countries of Europe in per capita income. Their high standards of health, education, security for the individual, and creative achievement are evidenced by impressive health statistics, long life expectancy, almost nonexistent illiteracy, disproportionately great achievements in art and science, and the reputation, particularly in Sweden, of having "abolished poverty."

In their attack on economic problems the countries of Northern Europe have employed a moderate socialism, consciously seeking a "middle way" between uncontrolled capitalism and communism. They have attempted to put a floor under the living standard of every member of

TABLE 6 NORTHERN EUROPE: AREA AND POPULATION DATA

COUNTRY	AREA (THOUSAND SQUARE MILES)	POPULATION (MILLIONS: 1968 ESTIMATES)	DENSITY (PER SQUARE MILE: TO NEAREST WHOLE NUMBER)
Denmark[a]	16.6	4.8	291
Norway[b]	125.2	3.8	30
Sweden	173.7	7.9	45
Finland	130.1	4.7	36
Iceland	39.7	0.2	5
Totals	485.3	21.5[c]	44

[a] Excluding the Faeroe Islands and Greenland.
[b] Excluding Svalbard and Jan Mayen.
[c] Apparent discrepancy is due to rounding of figures.

Denmark. Looking across the Little Belt (see inset on the Northern Europe index map) from the island of Fyn to closely settled farmland on the Jutland peninsula. The bridge accommodates both highway and rail traffic. (Luftfoto Nowico, Copenhagen.)

the community, while closely limiting the accumulation of wealth. Great emphasis on conservation of resources, the exercise and general acceptance of economic control and initiative by the state, and often the development of resources cooperatively by the state and private enterprise are prominent features of economic life in these countries. At the same time private business and ownership are fostered by the state in many lines of activity, as is trade unionism.

Parallel with the development of this "middle way," the countries of Northern Europe have experienced the world's greatest development of the private cooperative type of economic enterprise, reaching into almost every phase of production, distribution, and consumption. Cooperatives are particularly widespread and important in two countries of the group, Finland and Denmark. They are important, however, in all five countries.

DENMARK

Denmark has the somewhat paradoxical distinction of possessing the largest city in Northern Europe and of being at the same time the most dependent on agriculture of any country in the region. The Danish capital of Copenhagen (1.4 million) has well over a fourth of

Denmark's population in its metropolitan area. Denmark has a much greater density of population than the other countries of Northern Europe, a fact accounted for by the presence of Copenhagen, the greater productivity of the land agriculturally, and the lack of any sparsely populated zone of frontier settlement.

Copenhagen and the Danish Straits

Copenhagen lies on the island of Sjaelland (Zealand) at the extreme eastern margin of Denmark (map, p. 216). Sweden lies only 12 miles away across The Sound, the main passage between the Baltic and the wide Kattegat and Skagerrak straits leading to the North Sea. Copenhagen grew beside a natural harbor well placed to control all traffic through The Sound, which is the most direct and most used channel for traffic in and out of the Baltic. For many years before the seventeenth century Denmark controlled adjacent southern Sweden, as well as the less favored alternative channels to The Sound—the Great Belt and Little Belt. Toll was levied on all shipping passing to and from the Baltic. Although the days of levying tolls are now long past, Copenhagen still benefits from its strategic location. The city does a large transit and entrepôt business in North Sea–Baltic trade, and has encouraged this business by setting up a free zone in its harbor, where goods destined for redistribution may be landed without paying customs duties. Increasing trade has led to the development of industry, and the city is the principal industrial center of Denmark as well as its chief port and capital. Within a diversified industrial structure shipbuilding and food processing are the most important branches. The latter industry prepares products of Denmark's specialized agriculture for export and processes the many foods that Denmark must import.

Agriculture

Denmark stands out sharply from the other countries of Northern Europe in the nature of its land and the place of agriculture in its economy. The topography of the country results principally from glacial deposition. The western part of the peninsula of Jutland consists mainly of sandy outwash plains and coastal dunes. Eastern Jutland and the Danish islands exhibit a rolling topography of ground and terminal moraine. The highest hill in the country is less than 600 feet in elevation. Although the sandy areas of the west are not very fertile in their natural condition they have mostly been reclaimed and are cultivated, while the clay soils which characterize the moraine areas, with greater natural fertility, support a very intensive and productive agriculture upon which the country's prosperity is largely based.

[EDITORIAL COMMENT. *Outwash plains* are formed of glacial materials deposited by meltwater flowing from an ice sheet. Sheets of material deposited directly by the ice sheet are known as *ground moraine*. A ridgelike accumulation of such materials at the edge of an ice sheet is a *terminal moraine*.]

Most of Denmark is available for farming, and about two-thirds of the entire country is normally sown to crops, the largest proportion of any European country. It is fortunate that so much of Denmark is arable because the country is practically without natural resources except for soil, climate, and its strategic position for trade. Danish agriculture is so efficient that less than a fifth of the working population is employed on the land. Agriculture, however, is basic to the country's economy. Many additional workers are engaged in processing and marketing agricultural products, and many others in supplying the needs of the farms. About half of Denmark's exports ordinarily come from its farms.

Few countries or areas which depend so heavily on agriculture are as materially successful as Denmark has been during the first half of the twentieth century.

Danish agriculture is based on a highly specialized and very consciously and carefully fostered development of animal husbandry, which began to be emphasized when the competition of cheap grain from overseas brought ruin to the previous Danish system of grain farming in the latter part of the nineteenth century. Until recently, Danish animal husbandry was primarily dairy farming, with supplementary meat production, but in recent years the production and sale of animals for meat has become dominant over dairying. The following factors help to explain the outstanding success of animal industries in Denmark:

1. The land and climate of the country are well suited to fodder crops. Barley and other cereals, together with fodder beets, potatoes, and other root crops, are the most emphasized. These crops generally take precedence over hay, since they yield a larger quantity of nutrients per acre. Hay tends to be relatively more important in the less fertile west than elsewhere in the country. Despite the emphasis on fodder crops, however, the total production of feedstuffs is inadequate, and Denmark is a substantial importer of livestock feeds. As such, the country has been likened to a large agricultural "factory" which imports raw materials and exports the finished products.

2. Nearby markets, primarily in Great Britain and Germany, have generally been adequate to accommodate the bulk of Danish export production. In 1966 almost three-fifths of all Danish exports were taken by three North Sea countries and the United States: United Kingdom, 23 percent; West Germany, 15 percent; Sweden, 12 percent; United States, 9 percent. The same four countries also provided almost three-fifths of all Danish imports: West Germany, 21 percent; United Kingdom, 17 percent; Sweden, 14 percent; United States, 7 percent.

3. The Danish government has ranged itself actively and intelligently behind the development of a prosperous agriculture (though until the early 1960s Danish agriculture, unlike that of most European countries, was conducted without direct government subsidies). Probably the most important government measures have been (a) encouragement of family farming, in contrast to the large estates which formerly dominated the countryside, by the provision of liberal and low-cost credit for farmers desirous of purchasing land, including loans covering up to nine-tenths of the total purchase price; (b) financial aid to agricultural education and research; and (c) financial encouragement of reclamation projects carried out by private societies, especially in the sandy west. Denmark has become a land of small to medium-sized farms, about half being less than 25 acres in size. Over 90 percent of all farms are owned by their

operators, and most of them are very efficiently and scientifically handled.

4. Cooperative societies have played a major role in the development of agricultural prosperity. Practically all Danish farmers are members of farm cooperatives. Cooperative dairies and other food-processing plants prepare the products of the farms for sale, cooperative export societies conduct a large share of the marketing in foreign countries, and still other cooperatives supply many of the farmers' needs. Thus the individual farmer derives the benefits of large-scale marketing and buying, and the profits of a middleman are largely eliminated. Through his membership in the cooperative the farmer has become to some extent his own middleman. A further important function largely performed by the cooperatives is rigid quality control and grading of farm products. Danish agricultural products enjoy an enviable reputation for dependable quality—which is an important factor in ensuring a steady market.

5. Probably underlying much of the success of Danish agriculture, and especially the success of the cooperative movement, is a very high level of education. The traditional school system, noted for high standards, is supplemented by various forms of adult education. An emphasis on the continuing education of adults is characteristic not only of Denmark, but of all the countries of Northern Europe.

Thus Danish agriculture has become noted for its efficiency, rationality, and prosperity. Meat and milk production from dairy cattle is combined with supplementary activities, particularly the raising of hogs, fed partly on skim milk, and the raising of poultry. Meat, butter, and milk are Denmark's main contributions to the European economy.

In certain respects the Danish economy is very fragile. If the few export markets which customarily absorb the major share of Danish production were closed, the country would find itself in a critical situation. British austerity and German depression and disorganization following World War II retarded Danish recovery seriously. But a pronounced lessening in specialization probably would mean a lessening in prosperity.

NORWAY

Norway stretches for well over a thousand miles along the west side and around the north end of the Scandinavian Peninsula. The peninsula, as well as Finland, occupies part of the Fennoscandian Shield, a block of very ancient and hard rocks, especially gneiss and granite,

similar geologically to the Canadian or Laurentian Shield of North America. The western margins of the block in Norway are extremely rugged, being composed of mountains, uplands, and steep-sided valleys that were heavily eroded by the continental glaciation which centered in Scandinavia. Most areas have little or no soil to cover the rock surface, scraped bare by glaciation. About three-quarters of Norway is classified as wasteland. Only about 3 percent is classed as arable or pasture land, and the remainder is forest land.

Not only agriculture but also transportation is hindered by the nature of the terrain. Glaciers deepened the valleys of streams flowing from the mountains into the sea, and when the ice melted, the sea invaded the valley floors. Thus were created the famous fjords—long, narrow, and deep extensions of the sea into the land, usually edged by steep valley walls. Some run 100 miles or more inland. The difficulties of building highways and railroads parallel to such a coast are obvious. Coastal steamers and planes must bear a major transport burden. Coastal shipping lanes are sheltered from Atlantic storms by a screen of small islands paralleling the coast. The fjords of Norway provide some of the finest harbors in the world, but most of them have practically no hinterland. Much of the Norwegian population is scattered in relatively small and isolated settlements on narrow ribbons of lowland along the various fjords. The basic economic activity of such settlements is ordinarily a combination of livestock farming and fishing.

Northern Norway

The northern sections of Norway, Sweden, and Finland and adjacent parts of the Soviet Union are the homeland of the Lapps, a relatively primitive people whose best-known occupation is reindeer herding on a nomadic or seminomadic basis. Today a few thousand Lapps continue to gain their livelihood in this manner, but a large majority are settled in permanent homes and support themselves primarily by farming, fishing, or a combination of these. About 22,000 Lapps are inhabitants of northern Norway, while Sweden has about 10,000, Finland about 3000, and the USSR about 2000.[1]

The Norwegian coast as far south as Trondheim is occupied mainly by small, scattered fishing villages inhabited by Norwegians or Lapps. Places sufficiently large and varied to be classed as commercial towns are few in number, and none exceeds 15,000 in size. Especially noteworthy among the latter are Tromsö, Narvik, Hammerfest, and Harstad, each an important fishing port as well as the commercial center for a large sparsely popu-

[1] Daniel M. Epstein, "The Lapps: Nomads in Transition," *Journal of Geography*, **68** (1969), 28.

One of Norway's largest aluminum plants is located on one of the arms of Sogne Fjord. In the photo, some of the factory buildings are visible at the lower right, while homes of workers appear in the center of the view near the water of the fjord. The plant was constructed by Germany during the occupation of Norway in World War II. Water to turn the turbines of the hydroelectric station at the site falls more than half a mile through a conduit hewn in rock. (Norwegian Information Service.)

lated area. One of these ports has an additional activity which gives it much international importance. This is Narvik, which is the Atlantic outlet, ice-free the year round, for the large iron ore exports of northern Sweden. The importance of Narvik was brought to the attention of the world in 1940, when Germany conquered Norway, and the British fought a bitter though unsuccessful campaign to keep possession of the port.

At Mo i Rana, located at the head of a fjord just south of the Arctic Circle, the Norwegian government has established Norway's first integrated iron and steel plant. It is supplied with iron ore from various sources in northern Norway, including deposits a few miles inland from the plant. Electric furnaces, supplied with current from a hydroelectric power station at the site, are used for smelting the ore. A large share of the product is sold abroad, although Norway as a whole is a net importer of steel.

The Southwest Coast

From Trondheim south the intensity and scale of settlement along the coast are somewhat greater. The major communities are Trondheim (115,000), which benefits from an unusually large area of agricultural lowland roundabout as well as from gaps in the highlands giving access to Sweden and Oslo; Bergen (225,000), the most important center for the Norwegian fishing industry and the commercial metropolis of western Norway; and Stavanger (110,000), another fishing and commercial center.

The southwestern coastal area has a number of advantages over the coast farther north. Agricultural conditions are slightly more favorable, some areas of forest occur, and the location is better for utilization of the hydroelectric power which is one of Norway's primary resources. Norway already uses far more electricity per capita than any other nation in the world, with less than two-fifths of its estimated potential hydropower developed. Along the southwest coast a considerable electrometallurgical and electrochemical industry has come into being. The raw materials are imported for the most part and the products exported. Among the more important products are aluminum, copper, ferroalloys (mainly nickel), zinc, carbide, and nitrogenous fertilizers.

The Southeastern Core Region

More than half of the population of Norway lives in the southeast, which centers on the capital, Oslo (690,000). This is the core region of modern Norway. Here, where valleys are wider and the land is less rugged, are found the most extensive agricultural lands and the largest forests in the country. Streams coming down from the mountains to the west and north furnish power for sawmilling, pulp and paper production, metallurgy, electrochemical industries, and industries which process imported materials, such as textile yarns, for Norwegian consumption. Oslo lies at the head of the Oslo Fjord where several valleys converge. It is the principal seaport and industrial, commercial, and cultural center of Norway as well as its capital and largest city. Along Oslo Fjord are Norway's principal oil refinery and the country's largest concentration of shipbuilding and ship repair yards.

Resources and National Economy

Norway's basic natural resources are land, waterpower, fish, and forests. Although a variety of metalliferous ores are present, they are mostly of such low metallic content that they cannot be worked economically under present conditions. Some iron, copper, and pyrites, and considerable quantities of titanium ore are mined. Kirkenes, in the far northeast, is the largest center of iron mining. Most of its product is exported in the form of concentrates, largely to West Germany or Great Britain, though some of it moves to Norway's iron and steel plant at Mo i Rana.

Despite the meager amount of land that is usable for farming, agriculture is an important, though declining, element in the Norwegian economy. It currently employs a little less than a fifth of the labor force. Most

agricultural land in Norway grows feedstuffs for livestock—primarily grass, hay, barley, and oats—or potatoes, used both for livestock feed and for human food. The country has a small net export of meat and dairy products, though sizable imports of grains (primarily wheat), fruits and vegetables, sugar, and coffee are necessary.

Many Norwegians combine farming with part-time fishing or forestry. Norway ranks first in Europe as a fishing nation. In 1965 the country's total catch—comprised largely of cod and herring—was exceeded in tonnage by only five nations: Peru, Japan, China, the Soviet Union, and the United States. The forests of Norway, mainly composed of coniferous trees, are much less extensive than those of Sweden or Finland. Nevertheless, pulp, paper, and lumber account for about 15 percent of all Norwegian exports.

Manufacturing employs a larger number of workers than any other occupation. It is based mainly on domestic resources of hydroelectric power, forests, and fish. Electrically processed metals and chemicals, pulp and paper, some lumber and other timber products, and fish in various forms make up the great bulk of Norway's exports. Industries besides those indicated (for example, textile industries) serve mainly the domestic market.

International earnings from the products of export industries are supplemented by those of the world's third largest merchant marine. Merchant shipping is probably the most important element in the country's economy except for manufacturing. Norway's merchant fleet is exceeded in size only by the merchant shipping of the United States and the United Kingdom,[2] and it is by far the largest in the world on a per capita basis. It is a tangible expression of the country's intimate relation with the sea and long and distinguished seafaring traditions.

SWEDEN

Sweden is the largest in area and population and the most diversified of the countries of Northern Europe. In the northwest it shares the mountains of the Scandinavian Peninsula with Norway; in the south it has rolling, fertile farmlands like those of Denmark; in the central area of the great lakes another relatively extensive area of good farmland occurs. To the north, between the mountains and the shores of the Baltic Sea and Gulf of

---

[2] Liberia's tonnage of merchant ships is also greater than that of Norway, but these ships are merely registered in Liberia (to escape taxes and labor regulations), and are not actually owned there.

Bothnia, Sweden consists mainly of ice-scoured, forested uplands similar to those which constitute the greater part of Finland. A smaller area of the latter type occurs south of Lake Vätter.

A large Swedish sawmill on the Gulf of Bothnia at the mouth of the Ångerman River (used for transporting logs from the interior). Logs from the mill-pond move up inclined trackways into the mill, where large circular saws cut them into the proper lengths and reciprocating band saws process them into boards. Note at the upper right the stacks of lumber (resembling small houses) for export. (K. W. Gullers Studio, Stockholm.)

Agriculture

Swedish agriculture normally supplies about 90 percent of the country's food requirements. In addition, it provides some small exports. Since Sweden has other important sources of exports, its agriculture has not had to be as specialized as that of Denmark. A major emphasis on fodder crops and livestock, representing a good adaptation to the soil and climate, is accompanied by a considerable production of bread grains. Though most grain production is for the home market, some wheat is regularly exported. Fruits and coffee are the largest food imports. About 8 percent of the country's land area is cultivated, most of this land being found in the two most favorable areas, *Skåne,* and the central lowland or lakes district. Skåne is a regional name for the southernmost tip of Sweden, an area with a surface largely morainal, like that of the adjoining Danish islands. It has always been the most productive agricultural area in Sweden, with soils of above average fertility and the country's mildest climate. Skåne is the most densely populared part of Sweden, having about 15 percent of the national population on about 2 percent of the total area. The main urban center of the region, Malmö (270,000), is the third largest city of Sweden. It lies slightly to the southeast of Copenhagen on the opposite shore of The Sound.

The Central Swedish Lowland, extending between the important port cities of Stockholm and Göteborg, ranks second to Skåne in agricultural importance. However, agriculture in the Central Lowland is subordinate to manufacturing in terms of employment and value of product. The average elevation of this part of Sweden is several hundred feet lower than that of the granitic uplands to the north and south, and a considerable amount of glacial and marine deposition has provided the basis for fairly fertile soils. Patches of farmland are scattered through a forested terrain with many lakes. Farming in this area is built around dairying and associated fodder crops, leaving most of Sweden's substantial wheat production for the better soils and milder climate of Skåne.

Some farming is found in other parts of Sweden, but it is severely restricted by bare rock surfaces, thin and infertile soils, and in the north, by a harsher climate. In these areas agricultural settlement is scattered, and farming is often combined with seasonal work in other occupations, especially logging.

Forest Industries

Practically the whole of Sweden is naturally forested, with spruce, Scots pine, and other conifers pre-

A trainload of iron ore en route from the Kiruna mines of northern Sweden to the ice-free Norwegian shipping port of Narvik. Note the electrified railway. (Swedish State Railways.)

dominating north of the Central Swedish Lowland, and mixed conifers and broadleaf deciduous hardwoods in the southern part of the country. In the extreme north the forest cover becomes sparser and the trees stunted.

As a group, the wood-products industries based on these forests are the most important export industries of Sweden. This group includes sawmilling, pulp milling, papermaking, the manufacture of wood chemicals and synthetic fabrics, and the production of fabricated articles such as plywood, window and door frames, furniture, and prefabricated houses. Pulp, paper, and lumber account for most of the value of production. While logging and wood industries are characteristic of most of Sweden, the main concentration is found in areas to the north of the Central Lowland. In these areas other economic opportunities are less abundant, large quantities of good timber are available, and logs can be transported with relative ease—in winter by sled, or in summer by floating them to mills on the numerous rivers. Most of the sawmills and pulp mills are located in industrial villages and towns which dot the coast of the Gulf of Bothnia at the mouths of the rivers (photo, left). From these mills huge quantities of wood products are exported to other countries by ship during the summer months when the Gulf of Bothnia is free of ice. Power for the milling operations is supplied by numerous hydroelectric stations, and much electricity is transmitted by high-tension systems to the central and southern parts of the country,

where growing demands cannot be met by streams which are now almost completely developed for power.

Mining and Manufacturing

The mineral wealth of Sweden and the associated high development of metallurgical and mechanical industries are the features which most distinguish the Swedish economy from that of the other individual countries of Northern Europe. Sweden is abundantly supplied with high-grade iron ore, and mines some nonferrous and alloy metals as well. However, the utility of these minerals is somewhat lessened by a shortage of coal, a deficiency shared with the other countries of the region. Very small amounts of low-grade coal are produced from mines in Skåne, but this production meets only a tiny fraction of the country's needs.

The largest and best known deposits of iron ore are located in the far north beyond the Arctic Circle. Here, at Kiruna and Gällivare, underground mining is supplemented by surface operations in which entire mountains of ore are being mined with power shovels and dumped into railway cars for export (photo, above). The iron ore of northern Sweden has a high content of phosphorus impurities but is very rich in metallic iron. Most of the production is marketed in West Germany, Great Britain, and other European countries. An electrified rail line connects the ore fields with the port of Luleå (city proper, 35,000) on the Gulf of Bothnia, through which ore shipments move from May until the first of December, and with the ice-free Norwegian port of Narvik, which ships ore the year round.

Sweden itself is an important steel producer, though more in terms of quality than quantity. The Swedish industry is mainly centered in the Bergslagen ore-producing region just to the north of the central lowland. The ores are extremely high-grade and some of them are very free of phosphorus. Such ores have been smelted since medieval times, originally with charcoal, and now mainly with imported coke. The steel made from them has long been renowned for quality. "Swedish steel" has become practically a synonym for fine steel. About half of the country's steel output today comes from electric furnaces rather than conventional open-hearth furnaces. The electric process is expensive, but gives extraordinary control of quality. However, in recent decades Sweden has greatly expanded its output of ordinary types of steel, and has built a number of new plants. Some of these are in interior central Sweden, but there has been a strong tendency (also notable elsewhere in Europe) for new steel plants to be located on the coast. Several of the largest Swedish plants have such locations, including one in the far north at Luleå and one approximately 50 miles south of Stockholm. While ore receipts by sea are of major concern in the location of most new coastal steel plants in Europe, the Swedish plants are more concerned with seaborne imports of coal and coke.

The Swedish steel industry's emphasis on skill and quality carries over into the finishing and fabricating industries which use the steel. Among Swedish specialties which have acquired a worldwide reputation are such items as cutlery, tools and machine tools, surgical instruments, antiaircraft artillery, ball bearings, home appliances, business machines, and electrical equipment. Emphasis on skill in design and execution is also basic to the success of some Swedish industries outside the metal-goods field. The most noted of these are the glassware and furniture industries of Småland, the infertile and rather sparsely populated plateau south of the Central Lowland.

Swedish manufacturing is principally carried on in numerous small industrial centers in the Central Lowland. These places, ranging in size from about 100,000 to mere villages, have grown up in an area favorably located with respect to minerals, forests, waterpower, labor, food supplies, and trading possibilities. The Central Lowland has been the historic core of Sweden and has long maintained an important agricultural development and a relatively dense population. At opposite ends of the lowland are Sweden's two major ports and major cities, Stockholm, on the Baltic Sea, and Göteborg on the Kattegat.

Stockholm and Göteborg

Stockholm (1.3 million) is the second largest city in Northern Europe. The location of the capital reflects the role of the Central Lowland as the early core of the Swedish state and the early orientation of that state toward the Baltic and trans-Baltic lands. Stockholm is the principal administrative, financial, and cultural center of the country and shares in many of the manufacturing activities typical of the Central Lowland. It is the country's most important single industrial center.

In the past century, as Sweden has come to do more and more trading via the North Sea, Stockholm has been displaced by Göteborg (550,000) as the leading port of the country and, in fact, of all Northern Europe. Besides its advantage of position, the latter city has a harbor which is ice-free the year round, whereas icebreakers are needed to keep open the harbor of Stockholm in midwinter. Göteborg combines its trade functions with numerous manufacturing activities. In recent decades it has enjoyed a spectacular rise as a shipbuilding center. Sweden has recently become the world's fourth-ranking builder of ships, and Göteborg, whose yards have pioneered significant technical advances, is the leading Swedish producer.

The Swedish Policy of Neutrality and Preparedness

In the Middle Ages and early modern times Sweden was a powerful and imperialistic country. Finland was conquered in the twelfth and thirteenth centuries, and in the seventeenth century the Baltic became almost a Swedish lake. During the eighteenth century, however, the rising power of Russia and to some extent of Prussia put an end to Swedish imperialism, aside from a brief campaign in 1814 through which Sweden won control of Norway from Denmark. Since 1814 Sweden has never been engaged in a war, and it has become known as one of Europe's most successful neutrals. A century and a half of peace has undoubtedly been partially responsible for the country's success in attaining a high level of economic welfare and a reputation for social advancement. At present, however, Sweden is carrying a heavy burden of armaments. As the strongest military power in Northern Europe, it represents an important obstacle to possible Soviet domination of the Baltic area. Although the country escaped involvement in World War II, successful neutrality in another general war seems less likely. The dangers inherent in the international situation have led the Swedish government to build up strong armed forces while maintaining, at least formally, the nation's traditional policy of neutrality.

FINLAND

Conquered and Christianized by the Swedes in the twelfth and thirteenth centuries, Finland was ceded in 1809 by Sweden to Russia, and it was controlled by the latter nation until 1917. Under both the Swedes and the Russians the country's status was that of a semiautonomous grand duchy, and its people developed their own culture and feelings of nationality to such an extent that the opportunity for independence provided by the collapse of Tsarist Russia toward the end of World War I was eagerly seized.

Importance of Forestry and Livestock

Most of Finland is a sparsely populated, glaciated, subarctic wilderness of coniferous forest, ancient igneous and metamorphic rocks, thousands upon thousands of lakes, and numerous swamps. Despite this somewhat hostile environment Finland was primarily an agricultural country until very recently, and the majority of the population is concentrated in relatively fertile and warmer lowland districts scattered through the southern half of the country. Hay, oats, and barley are the main crops of an agriculture which is predominantly directed toward livestock production, especially dairying. Food crops—principally wheat, potatoes, and rye—occupy a much smaller acreage. A pattern similar to that of Norway and Sweden results, with small export surpluses of dairy products and heavy imports of other foods—although neither Norway nor Finland can duplicate Sweden's wheat exports. Coffee, fruits, and cereals are the major food items on Finland's import list.

To pay for food and many other imports, the nation depends primarily on exports of forest products. About two-thirds of Finland is forested, mainly in pine or spruce mixed with birch, and ordinarily over two-thirds of its total exports consist of lumber, wood pulp, paper, and other timber products of a type largely duplicating the forest exports of Sweden. Forest production is especially concentrated in the south central part of the country, often referred to as the Lake Plateau (map, p. 216). A poor and rocky soil discourages agriculture here, but the timber is of good quality and a multitude of lakes, connected by streams in interlocking systems, provide ready transportation for the logs.

Problems of a Buffer State

For centuries Finland has been a buffer between Russia and Scandinavia. Since 1939 its position has been especially difficult. In that year, shortly after the outbreak of World War II, Finland refused to accede to Soviet demands for the cession of certain strategic frontier areas and, despite a valiant resistance, was overwhelmed by the Soviet Union in the "Winter War" of 1939–1940. Then, in an attempt to regain what had been lost, Finland fought with Germany against the Soviet Union from 1941 to 1944, and again was defeated as German power waned.

The peace settlement following the war left Finland still in existence as an independent country, although shorn of considerable areas in the east and north which were annexed by the USSR. In the southeast the Karelian Isthmus between the Gulf of Finland and Lake Ladoga passed into Soviet control, and with it went the city of Viipuri, now Vyborg, which had been Finland's main timber port. More territory was lost along the central part of the eastern frontier, and in the north the area around the small port of Petsamo, now Pechenga, was ceded. The latter cession cut Finland off from access to the Arctic Ocean, and gave the Soviet Union rich nickel mines and smelting facilities, as well as a land frontier with Norway.

Besides territorial concessions the Soviet Union demanded large reparations payments, which were completed in the early 1950s, and it has insisted that Finland exhibit a reasonably friendly and cooperative attitude toward the USSR. In view of these events and circumstances, plus the steady growth of her huge neighbor's power, Finland has felt compelled to follow a very careful line of neutrality with respect to big-power quarrels. At the same time, however, the Finns have increasingly participated in inter-Scandinavian cooperation, both as a matter of practical interest and as an expression of their basically Western preference and orientation. Following the war the economy of Finland was heavily burdened by (1) the necessity for rebuilding the northern third of the country, which was devastated by retreating German soldiers after Finland surrendered to Russia, (2) the necessity for resettling a tenth of the total population of the country after they fled as refugees from the areas ceded to Russia, (3) the loss of the ceded areas themselves, some portions of which were of disproportionate importance in the prewar economy of the country, and (4) the necessity for making large reparations payments to the Soviet Union. In spite of these various difficulties, however, the hardworking and thrifty Finns made a rapid recovery from the war period. In the course of this the economy of the country was drastically changed in some respects. One of the most striking changes was the rise of metalworking industries. This was made necessary by the fact that the Soviet Union required a large part of the reparations to be paid in metal goods. Several small steel plants and other metalworking establishments were built to meet this demand, and they have continued to operate since the end of reparations. The steel plants are too small to supply all of Finland's needs, and both the coal and a portion of the iron ore which the plants require must be imported. The metalworking establishments, which manufacture a variety of secondary metal products, serve mainly the domestic and Soviet markets. Finland has only a modest endowment of metal-bearing ores, but its mines do contribute somewhat to the country's metal industries and exports. The principal metals extracted include some iron, copper, nickel, zinc, vanadium, cobalt, and titanium.

The Major Cities

The most important industrial center of the country, as well as its capital, largest city, main seaport, and principal commercial and cultural center, is Helsinki (730,000), located about in the center of the southern coast. This city has diversified food-processing, shipbuilding, mechanical, chemical, and other industries. Smaller industrial centers include Tampere (190,000), located in the southwestern interior, and, farther south, the port of Turku (195,000) at the southwestern corner of the country. Each is especially important for textile production. The industries of all these cities depend heavily on hydroelectricity, the main source of power in Finland, as in all the other countries of Northern Europe except Denmark.

Few countries today occupy a more precarious strategic position than Finland. The Finns struggle to get along as best they can with their powerful neighbor to the east, meanwhile striving to maintain democratic institutions and hoping that nothing causes the Soviet Union to move militarily toward Scandinavia or to regard Finnish independence as a threat.

ICELAND

Iceland is a fairly large, mountainous island in the Atlantic Ocean just south of the Arctic Circle. Its rugged surface shows the effects of intense glaciation and vulcanism. Some upland glaciers and many active volcanoes and hot springs remain. The vegetation consists mostly of tundra, with considerable grass in some coastal areas and valleys. Trees are few, being discouraged by summer temperatures averaging in the low fifties or below as well as by the prevalence of strong winds. The cool summers are also a great handicap to agriculture. Mineral resources are almost nonexistent.

Despite the deficiencies of its environment, however, Iceland has been continuoulsy inhabited at least since the ninth century, and is now the home of a progressive and democratic republic which had an estimated population of 205,000 in 1968. Practically all of the population lives in coastal settlements, with the largest concentration in the vicinity of the capital, Reykjavik, which itself has about 100,000 people in its metropolitan area. Due to the proximity of the relatively warm North Atlantic Drift, the coasts of Iceland, and especially the southern coast where Reykjavik is located, have winter temperatures which are unusually mild for the latitude. Reykjavik has an average January temperature that is only 2° lower than the January average for New York City.

Agriculture, fishing, and the processing of their products are the main occupations of Iceland. Agriculture is centered around the raising of cattle and sheep. Farmland is used for hay and pasture, and a limited production of potatoes and hardy vegetables. Although some agricultural products are exported, the real backbone of the economy is fishing. Fish products supply about eight-tenths of all exports by value, and thus pay for the many kinds of goods which must be imported. Manufacturing is mostly, though not entirely, confined to food processing. It has been encouraged somewhat in recent years by the development of a small part of the island's considerable potential of hydroelectric power. This relatively simple economy, on a barren land, has attained one of the world's higher standards of living.

For centuries before 1918 Iceland was a colony of Denmark. In that year it became an independent country under the same king as Denmark, and in 1944 declared itself a republic. It attracted much notice at the beginning of World War II because of its strategic position along major sea and air routes across the North Atlantic. The island was used by British and American forces as an air and sea base during the war. As a member of the North Atlantic Treaty Organization, Iceland is host to a small

American military contingent which maintains base facilities at the important Keflavik international airport, 20 miles from Reykjavik. American expenditures connected with this base have played an important role in the country's increasing prosperity.

GREENLAND, THE FAEROES, AND SVALBARD

Two of the countries of Northern Europe, Denmark and Norway, possess outlying islands of some significance. Denmark holds Greenland, the world's largest island, off the coast of North America, and the Faeroe Islands between Norway and Iceland. Both of these areas are now considered integral, though self-governing, parts of Denmark, and their peoples have equal political rights with other Danish citizens. Although the area of Greenland, approximately 840,000 square miles, is nearly a fourth that of the United States, about 85 percent is covered by an ice cap, and the population, mainly distributed along the west coast, amounts to only about 40,000 persons. The latter figure does not include American forces stationed at the large air base maintained at Thule in the extreme northwest of Greenland under the auspices of the North Atlantic Treaty Organization, to which Denmark belongs. Aside from a few nonnative Danes, and a small number of pure Eskimos in the remote northwest and other isolated places, the population of the island is of mixed Eskimo and Scandinavian descent. The principal means of livelihood are fishing, hunting, trapping, a limited amount of sheep grazing, and the mining of cryolite, a mineral used in the aluminum industry and certain other industries. A deposit at Ivigtut in southwestern Greenland has been quarried for more than a century and represents the only large commercial deposit in the world. Some of the world's cryolite is made artificially, however.

The Faeroes are a group of treeless islands where some 40,000 people of Norwegian descent make a living by fishing and grazing sheep. They enjoy considerable autonomy under the Danish government.

Norway controls the island group of Svalbard, which is located in the Arctic Ocean and is commonly known as Spitsbergen. Although largely covered by ice, the main island of West Spitsbergen contains the only substantial deposits of high-grade coal which are known to exist in Northern Europe. Mining operations are carried on by Norwegian and Soviet Russian companies, with the coal being shipped to Norway and to the Soviet port of Murmansk. Approximately 1000 Norwegians and 2000 Russians are employed in the mines. Norway also holds the volcanic island of Jan Mayen in the Arctic plus two small islands in the far South Atlantic, and claims a share of the Antarctic Continent.

REFERENCES and READINGS

General and Miscellaneous

BIRKET-SMITH, KAJ, *Primitive Man and His Ways: Patterns of Life in Some Native Societies,* trans. from the Danish by Roy Duffell (London: Odhams Press; New York: New American Library, 1960), "The Lapps: Hunters and Nomads; Reindeer," pp. 103–140.

EPSTEIN, DANIEL M., "The Lapps: Nomads in Transition," *Journal of Geography,* 68, no. 1 (January 1969), 26–33.

GOODWIN, WILLIAM F., JR., "Scandinavia," *Focus,* 5, no. 5 (January 1955), 6 pp.

GUTKIND, E. A., *Urban Development in the Alpine and Scandinavian Countries* (International History of City Development, Vol. 2; New York: The Free Press of Glencoe, 1965).

HAVINDEN, M. A., "Rural Policies in Northern Europe," in John Higgs, ed., *People in the Countryside: Studies in Rural Social Development* (London: National Council of Social Service, 1966), pp. 52–80.

"The Lapps, Nomads of the North," Chap. 17 in Richard M. Highsmith, Jr., ed., *Case Studies in World Geography: Occupance and Economy Types* (Englewood Cliffs, N.J.: Prentice-Hall, 1961), pp. 117–121.

MALMSTRÖM, VINCENT H., *Norden: Crossroads of Destiny and Progress* (Princeton, N.J.: D. Van Nostrand Co., Searchlight Books, 1965).

MEAD, W. R., *An Economic Geography of the Scandinavian States and Finland* (London: University of London Press, 1958); and "Problems of Scandinavia and Finland," Chap. 5 in W. Gordon East and A. E. Moodie, eds., *The Changing World: Studies in Political Geography* (New York: Harcourt, Brace & World, 1956), pp. 138–161. Penetrating studies by an outstanding authority.

MILLWARD, ROY, *Scandinavian Lands* (New York: St. Martin's Press, 1964). A geography.

O'DELL, ANDREW C., *The Scandinavian World* (London: Longmans, Green and Co., 1957). A geography.

ORVIK, NILS, *Europe's Northern Cap and the Soviet Union* (Cambridge, Mass.: Harvard University Center for International Affairs, 1963); and "Scandinavia, NATO, and Northern Security," *International Organization, 20,* no. 3 (Summer 1966), 380–396.

OXENSTIERNA, ERIC, "The Vikings," *Scientific American,* **216,** no. 5 (May 1967), 67–78.

SÖMME, AXEL, ed., *A Geography of Norden: Denmark, Finland, Iceland, Norway, Sweden* (New York: John Wiley & Sons, 1962). A major work by Scandinavian and Finnish geographers, authoritative and well illustrated with photos and maps (some maps are in color).

Yearbook of Nordic Statistics (Stockholm: The Nordic Council, annual). Published in English.

Denmark

BUNTING, B. T., "The Present Re-organization of Agriculture in Denmark", *Geography,* **53,** pt. 2 (April 1968), 157–162; and "Vallundgaard: A Study of Land Use in a Pod-solized Area of Western Jutland," (Institute of British Geographers, Publication No. 26, *Transactions and Papers, 1959*), pp. 89–105.

"Copenhagen," *Danish Foreign Office Journal,* 1967. A special issue.

GOLDSTEIN, SIDNEY, "Rural-Suburban-Urban Population Redistribution in Denmark," *Rural Sociology,* **30,** no. 3 (September 1965), 267–277.

HART, JOHN FRASER, "Vestergaard: A Farm in Denmark," Chap. 8 in Richard S. Thoman and Donald J. Patton, eds., *Focus on Geographic Activity: A Collection of Original Studies* (New York: McGraw-Hill Book Company, 1964), pp. 45–48.

HEARE, GERTRUDE, *Basic Data on the Economy of Denmark,* U.S. Bureau of International Commerce, Overseas Business Reports, OBR 65–88 (December 1965). (Washington, D.C.: Government Printing Office, 1965.)

HINKSTON, EUGENE R., "Denmark—Where Reason Rules," *The Progressive,* **28,** no. 1 (January 1964), 30–33.

JACOBSEN, NIELS KINGO, ed., *Collected Papers: Denmark,* International Geographical Congress, 20th, London, 1964 (Copenhagen: Københavns Universitets Geografiske Institut, 1964). A series of articles by Danish geographers, including a number on Denmark's population, agriculture, and fisheries.

LASSEN, AKSEL, "The Population of Denmark, 1660–1960," *Scandinavian Economic History Review,* **14,** no. 2 (1966), 134–157.

NEWCOMB, ROBERT M., "Geographic Aspects of the Planned Preservation of Visible History in Denmark," *Annals of the Association of American Geographers,* **57,** no. 3 (September 1967), 462–480.

NIELSEN, NIELS, ed., *The Atlas of Denmark,* 2 vols. (Copenhagen: Hagerup, 1949–1961).

"A Sample of Danish Farming," Chap. 10 in Richard M. Highsmith, Jr., ed., *Case Studies in World Geography: Occupance and Economy Types* (Englewood Cliffs, N.J.: Prentice-Hall, 1961), pp. 65–69.

U.S. DEPARTMENT OF AGRICULTURE, ECONOMIC RESEARCH SERVICE, *The Agricultural Economy and Trade of Denmark* (ERS–Foreign 244; Washington, D.C.: Government Printing Office, 1968).

Norway

Basic Data on the Economy of Norway, U.S. Bureau of International Commerce, Overseas Business Reports, OBR 67–71 (November 1967). (Washington, D.C.: Government Printing Office, 1967.)

"Fjord Farming in Norway," Chap. 13 in Richard M. Highsmith, Jr., ed., *Case Studies in World Geography* (Englewood Cliffs, N.J.: Prentice-Hall, 1961), pp. 87–92.

GILPIN, MARGARET C., "Møre and Romsdal, Norway: A Study of Changes in the Rural Transport System of a Coastal Area, *Geography,* 53, pt. 2 (April 1968), 145–156.

LANDMARK, KAY, and Others, "Northern Norway: Nature and Livelihood," *Norsk Geografisk Tidsskrift,* no. 1–4 (1959), 138–167.

LLOYD, TREVOR, "Iron Ore Production at Kirkenes, Norway," *Economic Geography,* 31, no. 3 (July 1955), 211–233.

MALMSTRÖM, VINCENT, *Marine Orientation in Norway: An Assessment of the Role of the Sea in the Life of the Country* (Middlebury, Vt.: Middlebury College Library, 1963).

PICKARD, JEROME P., "The Manufacturing Regions of Norway," *Norsk Geografisk Tidsskrift,* 18, no. 1–2 (1961) 65–77.

SOMMERS, LAWRENCE M., "Commercial Fishing in Norway," *Tijdschrift voor Economische en Sociale Geografie,* 53, no. 11 (November 1962), 237–242; and "The Antarctic Pelagic Whaling Crisis: The Geographic Impact of a Declining International Resource," *Tijdschrift voor Economische en Sociale Geografie,* 58, no. 3 (May–June 1967), 126–134.

SYMES, D. G., "Changes in the Structure and Role of Farming in a West Norwegian Island (Radøy)," *Economic Geography,* 39, no. 4 (October 1963), 319–333.

Sweden

ANDERSSON, INGVAR, and Others, *Introduction to Sweden,* trans. by Nils G. Sahlin (5th ed.; Stockholm: Swedish Institute, 1961).

CARPENTER, AILEEN M., "Recent Development at Oxelösund, Sweden," *Geography,* 53, pt. 4 (November 1968), 407–409.

DIEM, AUBREY, "An Alternative to Unplanned Urban Growth: The Case of Stockholm," *Canadian Geographer,* 9, no. 4 (1965), 193–204.

JOHNSON, J. A., "Developments in the Swedish Iron Ore Industry," *Geography,* 52, pt. 4 (November 1967), 420–422.

JONASSON, OLOF, ed., with assistance by Bo Carlsund, *Economic Geografical Excursion to Middle Sweden.* Prepared for the International Geographical Congress, 19th, Sweden, 1960 (Göteborg: 1960). Concise accounts of industry, forests, hydroelectric power, the iron and steel industry, Göteborg, and other topics.

McNITT, HAROLD A., *Basic Data on the Economy of Sweden,* U.S. Bureau of International Commerce, Overseas Business Reports, OBR 66–26 (May 1966). (Washington, D.C.: Government Printing Office, 1966.)

NOTT, KATHLEEN, *A Clean, Well-Lighted Place: A Private View of Sweden* (London: Heinemann and Co., 1961).

SIDENBLADH, GÖRAN, "Stockholm: A Planned City," *Scientific American,* 213, no. 3 (September 1965), 107–118.

STONE, KIRK H., "Swedish Fringes of Settlement," *Annals of the Association of American Geographers,* 52, no. 4 (December 1962), 373–393.

STREYFFERT, THORSTEN, "Management of Small Forest Holdings in Sweden," in Henry Jarrett, ed., *Comparisons in Resource Management* (Baltimore, Md.: Johns Hopkins Press for Resources for the Future; Lincoln: University of Nebraska Press, Bison Books, 1961), pp. 52–73.

SWEDISH GEOGRAPHICAL SOCIETY, *Atlas över Sverige* [Atlas of Sweden] (Stockholm: Generalstabens Litografiska Anstalts Forlag, 1953–   ). One of the world's great reference atlases, with first-class cartography.

Finland

GEOGRAPHICAL SOCIETY OF FINLAND, *Atlas of Finland, 1960* (Helsinki: Kuslannusosakeyhtiö Otava, 1960). Excellent large reference atlas with multicolored maps. Text in Finnish and English.

HELIN, RONALD A., "Finland Regains an Outlet to the Sea: The Saimaa Canal," *Geographical Review,* 58, no. 2 (April 1968), 167–194.

McNITT, HAROLD A., *Basic Data on the Economy of Finland,* U.S. Bureau of International Commerce, Overseas Business Reports, OBR 66–74 (October 1966). (Washington, D.C.: Government Printing Office, 1966.)

MEAD, W. R., *Finland* (Nations of the World; New York: Frederick A. Praeger, 1968); also, *Farming in Finland* (New York: Oxford University Press, Essential Books, 1953); also, "Frontier Themes in Finland," *Geography,* 44, pt. 3 (July 1959), 145–156; and "The Adoption of Other Lands: Experiences in a Finnish Context," *Geography,* 48, pt. 3 (July 1963), 241–254; also, with Helmer Smeds, *Winter in Finland: A Study in Human Geography* (New York: Frederick A. Praeger, 1967); and, with S. Jaatinen, "The Intensification of Finnish Farming," Economic Geography, 33, no. 1 (January 1957), 31–40.

PLATT, RAYE R., ed., *Finland and Its Geography* (New York: Duell, Sloan and Pearce, 1955).

STONE, KIRK H., "Finnish Fringe of Settlement Zones," *Tijdschrift voor Economische en Sociale Geografie,* 57, no. 6 (November–December 1966), 222–232.

U.S. DEPARTMENT OF STATE, THE GEOGRAPHER, *Finland–USSR Boundary* (International Boundary Study, No. 74; Washington, D.C.: 1967).

Iceland

ALEXANDER, LEWIS M., "Iceland," *Focus,* 10, no. 2 (October 1959), 6 pp.

ASHWELL, I. Y., "Recent Changes in the Pattern of Farming in Iceland," *Canadian Geographer,* 7, no. 4 (1963), 174–181.

HEIEN, BOBETTE P., *Basic Data on the Economy of Iceland,* U.S. Bureau of International Commerce, Overseas Business Reports, OBR 67–11 (March 1967). (Washington, D.C.: Government Printing Office, 1967.)

"Land of Fire, Ice, and Grass," *Farm Quarterly,* Spring, 1967, pp. 86–89 ff.

MALMSTRÖM, VINCENT H., *A Regional Geography of Iceland* (National Research Council, Publication No. 584; Washington, D.C.: 1958).

NAWRATH, ALFRED, *Iceland: Impressions of a Heroic Landscape* (Bern, Switzerland: Kümmerly and Frey, 1959).

THORARINSSON, SIGURDUR, "Population Changes in Iceland," *Geographical Review,* 51, no. 4 (October 1961), 519–533.

Greenland, the Faeroes, and Svalbard

"Arctic Denmark," *Danish Foreign Office Journal,* no. 58 (1967), pp. 10–49. Articles by several authors.

HARRINGTON, RICHARD, "Spitsbergen: Norway's Arctic Outpost," *Canadian Geographical Journal,* 74, no. 6 (June 1967), 207–213.

KAMPP, AA. H., "Sheep Farming in Greenland," in Niels Kingo Jacobsen, ed., *Collected Papers: Denmark,* International Geographical Congress, 20th, London, 1964 (Copenhagen: København's Universitets Geografiske Institut, 1964), pp. 68–72; and "The Faeroes: Today's Problems," *Inter-Nord,* no. 9 (March 1967), pp. 83–97.

See also the lists of references and readings for Chapters 1–4, particularly the relevant chapters in standard geographies of Europe listed on p. 92.

Countries

of

Southern Europe

11

On the south the continent of Europe is separated from Africa by the Mediterranean Sea, into which three large peninsulas extend (map, p. 235). To the west, south of the Pyrenees Mountains, is the Iberian Peninsula, unequally divided between two countries, Spain and Portugal. In the center, south of the Alps, is the Italian Peninsula and its southern offshoot, the island of Sicily. To the east, between the Adriatic and Black seas, is the Balkan Peninsula, from which the Greek subpeninsula extends still farther south between the Ionian and Aegean seas. The four main countries which occupy the two western peninsulas and the Greek subpeninsula— Portugal, Spain, Italy, and Greece—along with the island microstate of Malta, the mainland microstates of Vatican City State (enclosed within the city of Rome) and San Marino, and the British colony of Gibraltar—may be conveniently grouped as the countries of Southern Europe. Three of the peninsular countries include islands in the Mediterranean, the largest of which are Sicily and Sardinia, held by Italy; Crete, held by Greece; and the Balearic Islands, held by Spain. These islands are governed as integral parts of their respective countries. Some Mediterranean islands are held by other countries, the most notable being Corsica, administered as a department of France; and Cyprus, which is an independent republic. The independent island state of Malta was a possession of the United Kingdom until 1964. Gibraltar, a peninsula on the south coast of Iberia, remains a British possession despite Spanish demands that it be returned to Spain.

The countries of Southern Europe exhibit many natural and cultural similarities as a group. Most of their natural characteristics, however, while tending to differentiate them from other parts of Europe, are shared with lands of northern Africa and southwestern Asia which front on the Mediterranean. Throughout the Mediterranean area the broad pattern of natural features tends to be much the same, despite differences in detail from place to place. There also tends to be a broad similarity in agricultural practices. However, the countries of Southern Europe are distinguished as a group from their Mediterranean neighbors by important cultural differences, including differences in religion and language. Most of the African and Asian lands fronting on the Mediterranean are predominantly Moslem in religion, whereas Roman Catholicism is the prevailing religious faith in Spain, Portugal, and Italy, and the Orthodox Eastern Church is dominant in Greece. The Spanish, Portuguese, Italian, and Greek languages are, of course,

quite distinct from Arabic, the principal language in most of the non-European Mediterranean countries. Moreover, the countries of Southern Europe have shared in the development of Western or Occidental culture, whereas most of the African and Asian countries of the Mediterranean realm have principally been influenced by the culture of Islam (see pp. 359–361).

The areas, populations, and population densities of the Southern European countries are given in Table 7.

THE DISTINCTIVE MEDITERRANEAN CLIMATE

A distinctive natural characteristic of Southern Europe is its climate, which typically combines mild, rainy winters with hot, dry summers. The Mediterranean area has given its name to this particular combination of climatic qualities. In systems of climatic classification a "mediterranean" type of climate is customarily recognized, although the designation "dry-summer subtropical," also in common use, is perhaps more descrip-

TABLE 7 SOUTHERN EUROPE: AREA AND POPULATION DATA

COUNTRY[a]	AREA (THOUSAND SQUARE MILES)	POPULATION (MILLIONS: 1968 ESTIMATES)	DENSITY (PER SQUARE MILE: TO NEAREST WHOLE NUMBER)
Italy	116.3	52.8	453
Spain[b]	194.9	32.4	166
Portugal[b]	35.3	9.5	267
Greece	50.5	8.8	170
Malta	0.122	0.320	2600
Totals	397.1	103.8	261

[a] Microstates not shown in the table include Vatican City State (area 108.7 acres, pop. 855 in 1967, density 5029 per sq. mi.) and San Marino (area 24 sq. mi., pop. 17,850 in 1967, density 744 per sq. mi.). Both units are enclosed within Italy, but technically are sovereign states. The internally self-governing British colony of Gibraltar, not shown, has an area of 2.25 sq. mi., a population of about 25,000, and an average population density of slightly over 11,000 per sq. mi.
[b] Figures for Spain include the Balearic and Canary Islands. Figures for Portugal include the Azores and the Madeira Islands. The island groups named are governed as integral parts of Spain and Portugal, respectively.

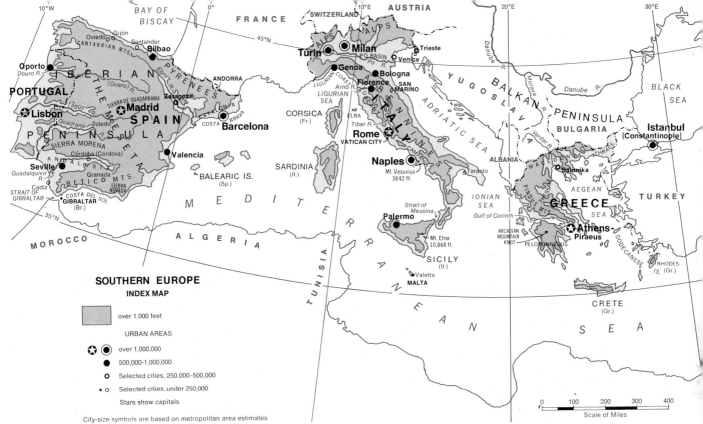

tive. Other areas having this type of climate occur in southern California, central Chile, southwestern South Africa, and southern Australia (see endpaper climatic map).

Generally speaking, the countries of Southern Europe experience temperatures averaging 40° to 50°F in the coldest month and 70° to 80°F in the warmest month. The total precipitation received during a year varies considerably from place to place, in response to differences in elevation and exposure to rain-bearing winds. The general average is between 15 and 35 inches per year, with most places falling in the lower half of this range. But regardless of total precipitation, most areas experience the characteristic seasonal regime of relatively moist winters and dry summers.

In Southern Europe the characteristics associated with the mediterranean climate become increasingly pronounced toward the south. The northern extremities of both Spain and Italy have atypical climatic characteristics. Except for a strip along the Mediterranean, northern Spain has a marine climate like that of northwestern Europe (see endpaper climatic map), cooler and wetter in summer than the typically mediterranean areas, while the basin of the Po River in northern Italy is distinguished by cold-month temperatures in the lowlands averaging just above freezing and a relatively wet summer. Much of the high interior plateau of Spain, the Meseta, cut off from rain-bearing winds by fringing mountains, has somewhat less precipitation and colder winters than is typical of the mediterranean climate, al-

though the seasonal regime of precipitation is characteristically mediterranean. Table 8 illustrates some of the climatic characteristics of Southern Europe.

MEDITERRANEAN AGRICULTURE

In the areas of mediterranean climate frosts are rare, the summers are hot and sunny, and thus the temperature regime is, in general, excellent for agriculture. But total precipitation is generally low, and a summer drought must be faced each year. Thus water is a critical factor. Where irrigation water is available during the summer months, a notable variety of crops can be produced. But in areas which are not irrigated—and they comprise the great majority—inadequate amounts of seasonal deficiencies of moisture limit the range or agricultural possibilities.

The Basic Pattern of Agriculture

Agriculture in the Mediterranean Basin is principally based on crops that are naturally adapted to the prevailing climatic regime of winter rainfall and summer drought. Winter wheat is the single most important crop. It occupies more land in each of the four main Southern European countries than any other crop. Barley,

TABLE 8 CLIMATIC DATA FOR SELECTED SOUTHERN EUROPEAN STATIONS

CLIMATIC TYPE OR AREA AND STATION	JANUARY AVERAGE TEMPERATURE (DEGREES F)	JULY AVERAGE TEMPERATURE (DEGREES F)	AVERAGE ANNUAL PRECIPITATION (INCHES)	AVERAGE PRECIPITATION JUNE-AUGUST (INCHES)
Typically mediterranean				
Athens	49°	82°	15.8″	1.1″
Rome	44°	77°	26.0″	2.3″
Palermo	52°	75°	30.7″	1.5″
Valencia	50°	76°	15.4″	2.2″
Seville	51°	79°	22.0″	1.6″
Lisbon	51°	72°	27.9″	2.2″
Spanish Meseta				
Madrid	41°	76°	17.2″	3.3″
Po Basin				
Milan	35°	76°	37.9″	12.1″
Northern Spain (marine west coast climate)				
Oviedo	44°	64°	36.8″	6.3″

a less prized but more adaptable grain, tends to supplant wheat in some particularly dry or infertile areas, such as the southern part of the interior plateau of Spain. Other typical crops are olives, grapes, and vegetables. The olive tree and the grapevine have extensive root systems and certain other adaptations which allow them to survive the summer droughts, and they yield for many years. Olive oil is the main source of fat in the typical Mediterranean diet, and virtually all of the world supply is produced in countries that touch or lie near the Mediterranean Sea. Italy, Spain, Greece, and Portugal together normally account for about four-fifths of the world total, and the Mediterranean countries of Turkey and Tunisia are also major producers. The principal use of grapes is for wine, a standard household beverage in Southern Europe and a major export product. The countries surrounding the Mediterranean Sea, including Portugal, produce almost three-fourths of the world's wine. Italy and France are the leaders by a wide margin, but Spain, Portugal, Algeria, Yugoslavia, Greece, Morocco, and Tunisia are significant producers. Where irrigation water is lacking, types of vegetables are grown which will mature during the wetter winter season or in the spring. Most important among the vegetables are several kinds of beans and peas. These

are a source of protein in an area where meat animals make only a limited contribution to the food supply. Feedstuffs are not available in sufficient quantities to fatten large numbers of animals, and parched summer pastures further inhibit the development of an adequate meat supply. Extensive areas which are too rough for cultivation are used for grazing, but their carrying capacity is generally low. Sheep, which can survive on a sparser pasturage than cattle, are the favored animals. They are kept only partially for meat, and the total amount they supply is relatively small. In many places grazing depends on a system of transhumance utilizing lowland pastures during the wetter winter and mountain pastures during the summer. In some areas nonfood crops supplement the basic Mediterranean products. An example is the tobacco of Greece, which is grown on farms in most parts of the country and forms Greece's largest export.

Areas in which the supply of available moisture is either considerably above or considerably below average tend to diverge from the normal pattern of agriculture described above. The drier areas depend more on barley than wheat, and the very driest areas depend mainly on grazing of sheep and goats. Some wet and rough areas which have remained in forest are also grazed, particularly oak forests where pigs can feed on the fallen acorns or mast. The bark of one type of oak, the cork oak, supplies a major export for Portugal and a minor export for Spain. Portugal is the world's largest exporter of cork.

The Intensive Agriculture of Irrigated Areas

Mediterranean agriculture comes to its peak of intensity and productivity in areas where the land is irrigated. In such areas relatively abundant and dependable supplies of moisture allow full exploitation of the subtropical temperatures, and the growing of a variety of fruits and vegetables, often with a large proportion destined for export, tends to supplement and sometimes largely to displace other types of production.

Although irrigation on a small scale is found in many parts of Southern Europe, a few irrigated areas stand out from the rest in size and importance. They are usually outstanding also for high population densities. Among these major areas of irrigation farming are northern Portugal; the Mediterranean coast of Spain; the northern coast of Sicily; and the Italian coastal areas near the city of Naples. The largest and most important irrigated area of all, the plain of the Po River in northern Italy, uses irrigation water to supplement year-round rainfall and is discussed separately.

In northern Portugal irrigation is used mainly to intensify a type of agriculture that does not differ radically from the normal Mediterranean type. In this area, however, irrigated corn replaces wheat as the major grain crop, and some cattle are raised on irrigated meadows. Grapes and sheep are the other agricultural mainstays.

In the coastal regions of Spain that front on the Mediterranean Sea, irrigation has made possible the development of extensive orchards. Oranges are the most important product, and Spain, though far from the greatest producer, is by a wide margin the world's largest exporter of this fruit. The preeminent orange-growing district of Spain, around the city of Valencia, has given its name to a type of orange. Small acreages of irrigated rice in the coastal areas provide an important element in Spanish cookery and a minor export. Vegetables, and in some places even tropical fruits like dates and bananas, also are grown to supplement wheat, vines, and olives.

As an agricultural area, northern Sicily is mainly differentiated from ordinary Mediterranean areas by its concentration of irrigated citrus groves. Lemons are particularly important. Sicily is the largest producer of lemons in Europe and vies with the United States for first place in the world. The island normally produces around nine-tenths of Italy's lemon crop and about three-quarters of its oranges. Italy is the world's principal exporter of lemons and is one of the leading exporters of oranges.

The district around Naples, known as Campania, is more intensively farmed, productive, and densely populated than any other agricultural area of comparable size in Southern Europe. Over considerable tracts rural population densities exceed 1200 per square mile. Relatively abundant irrigation water is applied to exceptionally fertile soils formed of volcanic debris from Mount Vesuvius. A remarkable variety of production includes almost every crop grown in Southern Europe. The rather typical Mediterranean practice called polyculture is highly developed, with a given field supporting as many as four layers of agriculture simultaneously—for instance, vegetables, grapevines or bush fruits, low fruit trees, and higher fruit trees or walnut trees. Despite the intensive cultivation and agricultural productivity of this area, the overcrowded rural population on its tiny farms is notably poor.

Agriculture in Areas with Summer Rainfall

In the northern parts of the countries of Southern Europe sizable areas are found which do not have truly mediterranean climates, and these areas exhibit corresponding differences in agriculture. In northern Spain wheat becomes subordinate to corn and rye, and the summer rainfall permits a greater development of cattle raising than is customary in Southern Europe. Another large area that diverges from the normal pattern of climate and agriculture is the Po Basin in northern Italy. This area is of outstanding importance in Italian agriculture. A considerable amount of rain falls during the summer months on the level plain of the Po River, and the surrounding mountains provide superior water supplies for irrigation. In this area corn, grown both in irrigated and unirrigated fields, and irrigated rice become important cereal crops along with wheat. Vineyards are supplemented by orchards of peaches and other temperate fruits. In addition, the plain of the Po is the center of Italian production for such industrial crops as sugar beets and hemp, and for cattle, nourished on fodder crops and irrigated meadows. Crop yields and the general welfare of the peasants stand at considerably higher levels than in the more typically Mediterranean areas of central and southern Italy. The parts of northern Greece known as Macedonia and Thrace are also atypical. They are distinguished by a strong tendency to substitute cotton for grapes and olives to supplement grains and tobacco, which are the agricultural mainstays.

RELIEF AND POPULATION DISTRIBUTION

In terrain and population distribution as well as in climate and agriculture the countries of southern Europe present various points of similarity. Rugged terrain predominates in all four major countries, and lowland plains occupy a relatively small part of the total land area. Individual plains tend to be small and to face the sea. They are separated from each other by the sea and by mountainous territory.

Population distribution corresponds in a general way with topography, with the lowland plains being densely populated and the mountainous areas much less so, although a number of comparatively rough areas attain surprisingly high densities. Thus on the whole the picture of population distribution is one of relatively isolated areas of dense population facing the sea and separated from one another by large areas of comparatively low population density.

In the Iberian Peninsula the greater part of the land consists of a plateau, the Meseta, with a surface lying at a general elevation of between 2000 and 3000 feet. The plateau surface is interrupted at intervals by deep river valleys and ranges of mountains rising above the general level (map, p. 78). Population density is restricted, mainly by lack of rainfall, to figures ranging generally between 25 and 100 per square mile. For the most part the edges of the plateau are steep and rugged. The Pyrenees and the Cantabrian Mountains border it on the north, and the Betic Mountains, culminating in the Sierra Nevada, on the southeast. Most of the population of Spain and Portugal is distributed peripherally on discontinuous coastal lowlands which ring the peninsula. In the part of southern Spain known as Andalusia, the depression followed by the Guadalquivir River extends the coastal plain inland for more than 100 miles. In southern and central Portugal, also, the coastal plains are broader than in other parts of Iberia.

In Italy the Alps and the Apennines are the principal mountain ranges. Northern Italy includes the greater part of the southern slopes of the Alps. The Apennines form the backbone of the peninsula, extending from their junction with the southwestern end of the Alps to the toe of the Italian boot, and appearing again across the Strait of Messina in Sicily. The Apennines vary considerably in height and appearance from place to place. East of Rome the mountains reach more than 9000 feet elevation. In Sicily, Mount Etna, a volcanic cone, reaches 10,868 feet. This is the highest elevation in the entire mountain chain. Other cones tower above the general summit level in various places. Near Naples, Mount Vesuvius, which rises to 3842 feet, is one of the world's most famous volcanos. West of the Apennines, most of the land between Florence on the north and Naples on the south is occupied by a tangled mass of lower hills and mountains, often of volcanic origin. Between these highlands and the sea there is often a narrow coastal plain. Both Sicily and Sardinia, the two largest Italian islands, are predominantly mountainous or hilly.

Parts of the Italian highlands have population densities of more than 200 per square mile. Yet even these areas are sparsely populated as compared with most Italian lowlands. The largest lowland, the Po Plain, contains over two-fifths of the entire Italian population, with densities in rural areas that are often over 500 per square mile. Other lowland areas with extremely high population densities are the narrow Ligurian Coast centering on Genoa, the plain of the Arno River as far inland as Florence, the Campania around Naples, much of the eastern coastal plain of the peninsula, and the northern and eastern coastal areas of Sicily.

In Greece most of the peninsula north of the Gulf of Corinth is occupied by the Pindus Range and the ranges which branch from it. Extensions of these ranges form islands in the Ionian and Aegean seas. Greece south of the Gulf of Corinth, commonly known as the Peloponnesus, is composed mainly of the Arcadian mountain knot. Along the coasts of Greece many small lowlands face the sea between mountain spurs and contain the majority of the people. Probably the best known of these lowlands, though far from the largest, is the Attic Plain, still dominated as in ancient times by Athens and its seaport, the Piraeus. Larger lowlands are found in Thessaly and to the north in Macedonia and Thrace.

HISTORICAL CONTRASTS IN WEALTH AND POWER

Despite recent economic gains, the countries of Southern Europe are among the poorest in Europe. In each of the four main countries present conditions offer a striking contrast to a past period of economic and political power and leadership. In Greece this period of past glory is the most remote, centering in the fifth and fourth centuries B.C., when Greek city-states were spreading the seeds of Western civilization through the Mediterranean area. To some degree there was a rebirth of Greek power and influence in the Middle Ages, when Constantinople was the capital of a Byzantine Empire which was largely Greek in population and control.

These carefully preserved ruins in Italy's capital bear witness to the antiquity of civilization in the Mediterranean Basin and the ancient power of Rome. The right of the view is occupied by remnants of the Forum Romanum, which reaches toward the Arch of Titus at the right center and the larger Arch of Constantine beyond (partly obscured by the Arch of Titus). To the left of the Arches, and partly demolished, is the Colosseum. (Pan American Airways.)

Italy's main period of former eminence was, of course, the centuries when the Roman Empire embraced the whole Mediterranean Basin and lands beyond. During the later Middle Ages, some centuries after the final collapse of the Roman Empire in the fifth century A.D., many of the Italian cities became independent centers of trade, wealth, and power. Venice became the center of a maritime empire within the Mediterranean area, as did Genoa to a lesser extent. Such inland cities as Milan, Bologna, and Florence also prospered and grew powerful on the basis of their trade with Europe north of the Alps. The growth of a hostile Turkish Empire astride routes to the East, the unification of larger states such as France and Spain, and the discovery of sea routes to the East which bypassed the Mediterranean were factors contributing to the end of this second period of Italian preeminence. Following its appearance in the nineteenth century, the modern unified state of Italy made an attempt to emulate ancient Rome. A colonial empire was gradually acquired which included Libya, the Dodecanese Islands, Eritrea, Italian Somaliland, and Ethiopia. All of these colonial territories were lost as a result of World War II.

Italy has been the only nation of Southern Europe to attempt the role of a great power during modern times. The futility and unreality of this attempt became manifest with the country's military collapse during World War II.

The main period of Spanish power and influence began in the Middle Ages when Spain stood as the bulwark of Christian Europe against Moslem civilization and, in a struggle lasting for centuries, eventually expelled the Moors from Europe. In the same year that this expulsion was finally accomplished, 1492, Christopher Columbus, an Italian navigator in the pay of the Spanish court, crossed the Atlantic and discovered the lands that were eventually to be called the Americas. For a century thereafter Spain stood at its greatest peak of power and prestige. It was the greatest power not only in Europe, but in the entire world, and built one of the largest empires ever known, in areas as diverse and widely separated as Italy, the Netherlands, North and South America, Africa, and the Philippine Islands. This empire shrank in size with the gradual decline in the relative power of Spain. All that remained of it in early 1969 were minor possessions along the coast of northwestern Africa. The

Spanish-held Canary Islands, off Morocco, are governed as an integral part of Spain.

Portugal also played a part in expelling the Moors from Iberia, and took the lead in the fifteenth century in seeking a sea route around Africa to the Orient. The first Portuguese expedition to succeed in the voyage, headed by Vasco da Gama, returned from India in 1499. For the better part of a century thereafter Portugal dominated European trade with the East, and built an empire there and across the Atlantic in Brazil. However, commencing in the latter part of the sixteenth century there was a rapid decline in Portuguese fortunes. This was associated in part with the conquest of the small Portuguese homeland by Spain, which held it from 1580 to 1640. Under Spanish rule, Portugal's interests were often neglected or subordinated to those of Spain. Meanwhile other European powers, particularly the Netherlands, offered increasingly successful competition in trade and colonization. By 1640 many of Portugal's possessions had fallen to the Dutch and could not subsequently be regained. However, the Portuguese held Brazil until the nineteenth century, when it gained independence, and Portugal still has a fairly large colonial empire today. By the late 1960s its possessions— located mainly in Africa and its offshore Atlantic islands, but also including some minor holdings in Asia and the East Indies (maps, pp. 401 and 550)—comprised the largest overseas empire remaining in the world. Retention of the two largest colonial units—Angola and Mozambique —in the face of nationalist revolts was requiring a major military effort by Portugal, with the ultimate issue in doubt (see p. 595). The Madeira Islands, located in the Atlantic Ocean to the north of the Canaries, are administered as a part of the Portuguese homeland.

RESOURCES, DEVELOPMENT, AND WEALTH

One of the striking characteristics of Southern Europe is its relatively low productivity and standard of living as compared with overall levels in northwestern Europe. As measured by per capita gross national product (approximately equivalent to per capita national income), Portugal is the most poverty-stricken country in Europe except for Albania. The other countries of Southern Europe are somewhat better off, but are still far below the general level of countries of northern and western Europe. Within Southern Europe there are wide variations among the different countries, with relatively prosperous Italy having a per capita income more than twice that of Portugal and

with the other countries falling between these two. But even Italy's per capita income is not much more than half that of such countries as Denmark, the United Kingdom, West Germany, and France. On the whole, per capita incomes in Southern Europe bear a greater resemblance to those of countries in a "developing" region such as Latin America than to those of the more prosperous countries of Europe. In all of the Southern European countries widespread poverty has been associated with a considerable amount of social unrest, with political instability, and with tendencies toward authoritarian government. Unrest in these countries has been aggravated by great internal contrasts in wealth, between regions as well as between social classes.

The relative poverty of Southern Europe is a characteristic that stubbornly persists in spite of very rapid recent progress. In the nine years between 1957 and 1966, Greece, Portugal, Spain, and Italy had the most rapidly growing economies in non-Communist Europe, and quite possibly in all of Europe—during a time when European growth rates as a whole were high. Economic advances in the two larger countries have attracted the most attention, and it has become common to hear of the Italian or Spanish "economic miracle." But the differentials between Southern Europe and northwestern Europe are so great that it will probably take many years to close the gap, even if present trends continue.

Retardation and Progress in Southern European Industry

Many of the difficulties of the countries of Southern Europe are related to the fact that until very recently their industrial development was greatly retarded. During the nineteenth century and the early part of the twentieth century, while northwestern Europe and the United States were forging ahead industrially, the Southern European countries remained predominantly agricultural. This is still true of Greece; and even Italy, the most industrialized country of Southern Europe, still has a larger proportion (over one-fifth) of its population in agricultural occupations than is common in northwestern Europe. Much more industrial development is still needed in Southern Europe to reduce the numbers of people depending on inadequate agricultural resources.

During the nineteenth and early twentieth centuries Southern Europe was characterized by a host of traits that have generally been inimical to industrialization.

1. The countries of the region were not major trading nations and did not develop an appreciable class of wealthy merchants. Thus they lacked a group which

has led the way in organizing and financing industrial development in many of the world's countries.

2. The people of the Southern European countries were generally poor and uneducated, offering little in the way of skilled labor or a promising market for products of industry.

3. Governments in the region were generally undemocratic, frequently unstable, often ineffective, and usually little oriented toward economic development.

4. Political, economic, and social affairs were controlled to a large extent by wealthy landowners whose interests lay in maintaining the agrarian societies which they dominated.

5. Internal transportation systems were poorly developed and could not easily be improved due to the rugged terrain.

6. Populations in the individual political units, especially before the unification of Italy in 1870, were small and hence offered only limited scope for the development of viable industries.

Deficiencies in Industrial Resources

7. In addition to the handicaps summarized in the foregoing points, some of the natural resources upon which early industrialization was frequently based in more fortunate countries have not been plentiful in Southern Europe. A prominent example is wood. In some areas outside of Southern Europe wood was important in early industrial growth, either as a raw material or as a fuel for smelting metal-bearing ores. But in most of Southern Europe usable timber has been scarce, due to dryness and centuries of deforestation. Many "forested" areas are covered only with scrub, a vegetation so typical as to have special names, such as *maquis* or *garrigue* in French and *macchia* in Italian. Waterpower is another resource that was important in early industrial growth in some of the world's countries, but has been of limited utility in Southern Europe. Although some parts of the region offered considerable scope for early industries powered by water wheels, many streams in the areas of mediterranean climate are dry channels during part of each year. But probably the most significant resource deficiencies in Southern Europe have been shortages of coal and iron ore. In the age of the early industrial revolution, before hydroelectricity and oil became major sources of power, coal was the dominant power source in the development of many of the world's industrial districts. Although each of the four main countries of Southern Europe has at least some minor coal resources and production, there is only one sizable coal field in the whole region. The same generalizations can be made with respect to iron ore. In both cases the exception is northern Spain, where both coal and iron ore occur in the vicinity of Oviedo (135,000) and Gijón (150,000), and higher grade iron ore near Bilbao (750,000) and Santander (135,000). Here an industrial area developed relatively early; but it has never become a major European district comparable to such areas as Lorraine, the Sambre-Meuse Valley, or South Wales. It has been handicapped not only by the general political, economic, and social conditions in Spain, but also by the fact that its coal includes only a limited proportion suitable for coking, so that imports of coal have been common. Today even Spain, to say nothing of the other countries of Southern Europe, continues to be an importer of coal.

Had it not been for deficiencies with respect to these critical natural resources, some of the mineral resources which are relatively abundant in Southern Europe might have been of much greater use. Small deposits of a great many metals and other minerals do exist in the region, and some are abundant enough to make the countries of Southern Europe important world suppliers. Outstanding among these is mercury, of which Spain and Italy together normally provide about half of the world's production. Other important minerals which are produced in appreciable quantities today include pyrites (for sulfur), lead, and zinc in all four of the major countries; potash in Spain and Italy; and bauxite, principally in Greece and Italy. In addition, a very substantial production of some lesser known but important industrial minerals—such as pumice, fluorspar, gypsum, and others—comes from Southern Europe. But these materials have offered only limited industrial advantages to a country when it lacked such fundamental resources as coal, iron ore, wood, and oil.

Major Industrial Areas Prior to World War II

Only three sizable industrial areas developed in Southern Europe before World War II. The most important of these by far was in the Po Plain of northern Italy. In this area imported coal and raw materials, hydroelectric power from stations in the surrounding mountains, and cheap Italian labor formed the basis for an industrial area specializing in textiles, but including some heavy industry and engineering. An area very similar in its foundations and specialties, but developed on a much smaller scale, came into existence in the northeastern corner of Spain, in and around the city of Barcelona (2.5 million). A third area, somewhat diversified but especially focused on heavy industry, developed in the coal and iron mining section of northern coastal Spain. It is noteworthy that two of the three areas—the Po Plain and northern Spain—are not typically Mediterranean in environment, and the third area, northeastern Spain, lies near the foot of mountains with humid conditions that provide the basis for hydroelectric power development. In addition to the industries in the three main concentrations, some local industries developed in other cities, especially in major seaports. But most of Southern Europe remained rural, agricultural, and poor.

Recent Economic Progress

In the years following 1950 the economies of the four main countries of Southern Europe began to show a new and unusual dynamism. By the late 1960s these countries had made some real progress in narrowing the economic gap between them and the countries of northwestern Europe, which were themselves continuing to develop rapidly. Industrial and service activities led the way, and in all countries the proportion of the population still employed in a generally poverty-stricken agriculture declined. In the 1960s agricultural employment represented about half of all employment in Greece, under half in Portugal, about one-third in Spain, and under one-quarter in Italy—figures still higher than in northwestern Europe, but decreasing rapidly.

A number of factors have played a part in the revitalization of the Southern European economies.

1. Improved and relatively cheaper transportation has made the import of fuel and raw materials for manufacturing more economical. Thus countries poor in these materials are less handicapped relative to richer countries than was formerly the case. Ocean transport has been especially important in this respect. Even today, internal land transport is not really adequate in Southern Europe except in much of Italy. In that country a tradition of road building that dates back to Roman times is expressed today in a relatively good railway system and a growing superhighway (*autostrada*) system. Due to the rugged terrain which must be overcome, these routes are often characterized by spectacular scenery and massive feats of engineering.

2. Coal has ceased to be the dominant source of power, its place being taken by oil or, in Italy, by natural gas and oil. All four of the main countries in Southern Europe have become major importers of oil, as have their industrial competitors in northwestern Europe. The latter countries have long had easy access to large coal resources, whereas now they must import oil over somewhat greater distances than is generally true of the Southern European countries. Thus Southern Europe is on more of an even footing with northwestern Europe

in securing power than it was when it had to depend on large imports of coal. In Italy the power situation has also been altered for the better by the exploitation of natural gas deposits. Spectacular gas discoveries were made after World War II, principally in the Po Plain. Production multiplied so rapidly that Italy was the world's sixth-ranking producer by the late 1960s and was deriving about half its power requirements from this source, which had played no part at all in its economy 20 years before. The other countries of Southern Europe, however, have no significant natural gas production as yet.

Accommodations for vacationers (a hotel and an apartment building) under construction at Playa d'Aro on Spain's Costa Brava. Spanish tourism has increased greatly with the development of charter flights and package-tour vacations. Guaranteed sunshine attracts many visitors from industrial northwestern Europe. A recent development in the smaller resorts is the construction of apartment houses with a basement restaurant and laundry. Real estate developers let or sell these accommodations to individuals or small groups, mainly from Belgium or West Germany. The new owners or lessees occupy the apartments during the height of the season and then rent them out cheaply for the rest of the year. (J. A. Soulsby.)

3. Market conditions for Southern European products have improved greatly in recent decades. The growth and prosperity of northwestern Europe, toward which Southern Europe's trade is largely oriented, has been basic to this improvement. In the case of Italy membership in the Common Market, and in the case of Greece associate membership in that organization since 1962, have given additional advantages not yet held by the other Southern European countries.

4. There have been massive infusions of foreign money into the Southern European countries since World War II. Each country has been the recipient of much American economic aid in one form or another, including military aid. Favorable economic circumstances and government policies have attracted a considerable amount of foreign investment in industrial and other facilities. And, not least, the countries of Southern Europe have been the beneficiaries of a tremendous boom in tourism. This picked up momentum in the 1950s and began to reach flood tide in the 1960s. To increasingly prosperous Europeans and Americans, Southern Europe has offered a combination of spectacular scenery, historical depth exceeding that of even the other parts of Europe, a sunny subtropical climate, the sea, and beaches. The ancient cities, often bustling and modern around their monuments of the past, were supplemented as attractions by new or greatly expanded resort areas along Spain's Costa Brava and Costa del Sol, in the Balearic Islands, on the Italian Riviera, and in various other places. Tourism has focused particularly on larger and more famous attractions such as Rome, Athens, or the Costa Brava, but innumerable areas from Oporto to the Aegean Islands have benefited from the massive infusion of tourist money.

THE PO BASIN: SOUTHERN EUROPE'S MAJOR INDUSTRIAL AREA

The Po Basin is the economic heart of modern Italy and the most highly developed and productive part of Southern Europe. Comprising just under a fifth of Italy's area, it contains almost half of the country's population (by far the more prosperous half) and accounts for half of its agricultural production and two-thirds of its industrial output. Economically, peninsular Italy is largely a poverty-stricken and overpopulated appendage to the Po area. These differences between north and south in Italy are recognized within the country in strong regional feelings embodying a degree of antagonism and mutual disdain.

However, many people of the relatively prosperous North have southern roots, as there is a steady flow of migration from south to north.

The modern industrial development of the Po area began in the nineteenth century on the foundations of the region's already outstanding position in agriculture and trade. A gradual development of the textile industries succeeded here, while industrial attempts further south generally failed. Imported coal and raw materials were necessary, but the northern factories had year-round streams for water power and were located in the richest market area in Italy. In addition, there was some accumulation of capital from trade, as some of the northern cities had been way stations for centuries on routes leading northward from the Mediterranean across the Alpine passes.

With the twentieth century came the age of hydroelectric power development in Italy. The best possibilities were in the Alps, and to a lesser degree the northern Apennines—the mountain areas framing the Po Plain on three sides and not afflicted by the dryness of the Mediterranean summer. By World War II the Po area was one of the major textile manufacturing regions of Europe. It still imported coal to supplement its hydroelectric power and still imported raw materials. The latter were processed with relatively cheap Italian labor, and much of the finished product was exported. By this time also a normal industrial evolution had added various engineering industries as a secondary element in the industrial economy of the area. Imported iron and steel were used to manufacture a growing variety of machines, particularly those types which emphasized the input of cheap skilled labor and minimized the amount of metal needed.

A very rapid further expansion of Po Basin industry has occurred since World War II. Italian industrial production, most of it in the Po Basin, more than doubled in the ten years from 1951 to 1961 and continued its rapid growth into the 1960s. Prominent components of this expansion were (1) the growth of engineering and chemical industries, (2) the development of a sizable iron and steel industry, mainly in the North, but partly in peninsular Italy, and (3) a shift to domestic natural gas and imported oil as the predominant sources of power but with hydroelectric energy continuing to play an important role. Textile industries remain important in the Po area, but have become secondary to engineering.

Many cities share the region's industry, but the leading ones are Milan (3.5 million), Turin (1.5 million), and Genoa (900,000). Milan is the region's industrial capital, the largest city of Italy, and the country's most important center of finance, business administration, and railway transportation. It is also a leading cultural capital, especially in the field of opera. The city is located between the port of Genoa and important passes, now shortened by railway tunnels, through the Alps to Switzerland and the North Sea countries. Turin is the leading center of Italy's automotive industry, though several other cities have a share. It is particularly associated with the Fiat company, which is by far the largest Italian auto manufacturer. The port of Genoa functions as a highly crucial part of the Po Basin industrial complex, although it is not actually in the basin. It is reached from the Milan-Turin area by passes across the narrow but rugged northwestern end of the Apennines. The city, which dominates the overseas trade of the Po area, is Italy's leading seaport and ranks with Marseilles as one of the two leading ports of the entire Mediterranean. Genoa is now much larger and economically more important than its old rival Venice (420,000), whose location in the eastern edge of the Po Plain at the head of the Adriatic Sea is relatively unfavorable for serving the main present-day industrial centers of northern Italy. But Venice's trade, actually handled by mainland suburbs rather than the old city on its numerous islands, is still sufficiently large to place the city third among Italian seaports after Genoa and Naples.

MAJOR URBAN CENTERS OUTSIDE THE PO BASIN

The generally low level of industrial development in Southern Europe, aside from northern Italy and the two smaller industrial areas of northeastern and northern Spain, is indicated by the basically nonindustrial character of the leading cities. The latter are mostly political capitals or ports serving especially productive agricultural areas, or both. They are, of course, not without industries appropriate to a port or to any large city, particularly one forming a reservoir of cheap labor, but their industrial functions are subordinate on the whole to government and trade.

In *Greece* the largest city by an overwhelming margin is Athens. With a metropolitan population of over 2 million (including the port of Piraeus), amounting to about one-fourth of the entire population of Greece and growing rapidly, Athens has no national rival. It is not only the national capital and the main port of the country but also the main industrial center. This development is logical enough in a situation where imported fuels and materials are used to manufacture products that are marketed almost entirely in a domestic market whose main center is Athens itself. The second city of Greece, Salonika (450,000), is the port for a particularly productive agricultural area, some of it on reclaimed and irrigated lands on the flood plains of the Vardar and other rivers. These lands produce most of the cotton which is one of Greece's major exports. Salonika is also the sea outlet for an important route through the mountains of southern Yugoslavia formed by the combined valleys of the Morava and Vardar rivers (see p. 264).

In *Italy* the largest cities outside of the northern industrial area are the capital, Rome (2.9 million), Naples (1.9 million), and Palermo (660,000). Rome has a location that is roughly central within Italy. It lies a few miles inland about halfway down the west coast of the peninsula, in a position to some extent intermediate between the country's contrasting northern and southern regions. It is almost entirely a governmental, religious, and tourist center in a region that is poorly developed both agriculturally and industrially. Its central location and enormous historical prestige made it almost inevitably the favored choice for national capital when Italy emerged as a unified state in the nineteenth century. Naples is the port for the populous and productive, but miserably poor, Campanian agricultural region described earlier in this chapter. It is also the main urban center of one major focus of Italy's tourist industry, with attractions such as Vesuvius, the ruins of Pompeii, and the island of Capri in the vicinity. Palermo is the main port for Sicily's citrus trade and the island's largest city. The largest city of Italy which is neither capital, port, nor Po Basin industrial center is the industrial, cultural, and tourist center of Florence (600,000), located at the head of the plain of the Arno River and at the southern end of an important route through the Apennines.

In *Spain* the three largest cities are Madrid (3.1 million), the capital, and the two Mediterranean ports of Barcelona (2.5 million) and Valencia (810,000). Madrid was deliberately chosen as the capital of Spain in the sixteenth century because of its location near the mathematical center of the Iberian Peninsula, approximately equidistant from the various peripheral areas of dense population and sometimes of separatist political tendencies. Located in a poor countryside, it has had little economic excuse for existence during most of its history. But its position as the capital has made it the center of the Spanish road and rail networks and thus has given it certain business advantages. In Spain's recent economic boom it has begun to attract industry on a considerable scale.

Barcelona is something of an exception in Southern Europe in that it is not only its country's major port but the center of an industrial district of some consequence, the main such district in Spain. Industrial development up to the present resembles that of the Po Basin at an earlier stage. The main elements have been (1) the early formation of a commercial class, (2) hydroelectric power, mainly from the Pyrenees, (3) cheap and sufficiently skilled labor, and (4) imported raw materials. Unlike the Po Basin, however, the Barcelona area is still primarily a textile producer, and is not an important exporter of manufactured goods. Its products still are sold primarily in the internal market of Spain.

Barcelona is the main urban and political center for a distinct people—the Catalans—who inhabit the part of northeastern Spain known as Catalonia and sometimes think of themselves as a nationality apart from the Spanish. The Catalan language is distinct from Castilian Spanish, the major tongue of Spain. Cultural, economic, and historical distinctiveness have fostered a jealous local patriotism in Catalonia and a distinct feeling of separateness from the rest of Spain. Close control by the central government in Madrid, such as exists under the present Spanish regime, has been resented, and repeated demands for a greater degree of local autonomy have occurred in the past. The situation of the Catalans is duplicated in some measure by that of the Basque people at the western end of the Pyrenees Mountains. Portugal is a larger expression of such separatism, and other parts of the Iberian perimeter have their own, lesser, cultural distinctiveness. Within Spain, however, all tendencies toward regional separatism have been rigidly suppressed by the present dictatorship since it came into power as a result of the Civil War of 1936 to 1939.

Valencia, Spain's third city in population, is the business center and port for an unusually productive section of coastal Spain. An extensive development of irrigation in the area originated with the Moors in the Middle Ages. Today the density of population on irrigated land resembles that of Italy's Campania. But here much of the agricultural effort goes into producing the oranges which are Spain's greatest single export.

The two large cities of *Portugal* are both seaports on the lower courses of rivers which cross the Meseta and reach the Atlantic through Portugal. Lisbon (1.4 million), on a magnificent natural harbor at the mouth of the Tagus River, is both the leading seaport of the country and the capital. The smaller city of Oporto (800,000), at the mouth of the Douro River, is the regional capital of northern Portugal and the commercial center for Portugal's trade in the famous port wine, which comes from terraced vineyards along the hills overlooking the Douro. Both cities have developed some industry, but they are fundamentally commercial centers and in addition Lisbon is a governmental center.

AGRICULTURAL PROBLEMS OF SOUTHERN EUROPE

Up to the present time, retarded development of industry has left to agriculture the burden of supporting too great a share of the people of Southern Europe. While agriculture has developed sufficiently to produce an export surplus of certain specialties—principally fruits and vegetables, but also tobacco and cotton from Greece and cork and wine from Portugal—it has not supplied adequate quantities of cereals and animal products or a good living for most of the people engaged in agricultural occupations. An indication of the unsatisfactory agricultural situation lies in the fact that all of the

A rural scene in the Mediterranean Basin. This villager on the island of Cyprus has collected a load of pine and cedar branches for household fuel. Securing an adequate supply of fuel is a constant problem in the deforested lands surrounding the Mediterranean. The widely spaced trees on the rocky hillside have extensive root systems and other special adaptations enabling them to survive the dry Mediterranean summers. (U.S. Department of Agriculture.)

Southern European countries must import some grain to supplement their home production. This is true despite the heavy emphasis on the production of grains in the agriculture of this region, and despite a somewhat lower caloric intake per person than in northwestern Europe.

Until a few decades ago, the countries of Southern Europe were characterized by high rates of population increase. Rapid growth of population, coupled with retarded industrial development, required the absorption of an increasing number of people into agriculture. The result was an excessive subdivision of land, hence a decrease in the amount of land available for the support of each family. Today the rates of population growth in these countries are much lower than formerly, and the number of people supported by agriculture is declining in most areas. But the total farm population is still very large, and most farms are very small. In addition, they are often fragmented into numerous separate parcels, which may be miles apart. Such farms are difficult or impossible to operate efficiently.

Farms which are excessively small or fragmented may in themselves supply an adequate reason for rural poverty. This difficulty, however, has been augmented in Southern Europe by low crop yields, at least by European standards. For example, the Spanish farmer, on the average, gets only about one-third as much wheat from each acre as does his neighbor in France, and one-fourth to one-fifth as much as the Danish, Dutch, or Belgian farmer.

Reasons for Low Agricultural Productivity

Many different factors underlie the generally low productivity of Southern European farms. Pressure of population on the available land has led to the farming of some rather infertile areas, especially rugged areas where the growing of cultivated crops soon leads to soil erosion with consequent damage not only to hillside farms, but also to farms in the valley below. The lack of rainfall in summer, and sometimes in winter, is a hindrance. Some irrigated areas produce good yields, but only limited amounts of land have been or can be irrigated. In many areas a lack of adequate markets near at hand, or a lack of transportation giving access to distant markets, has prevented farmers from specializing in types of production that would bring higher returns and has obliged them to operate on a semisubsistence basis. Farmers in Southern Europe are ordinarily too poor to afford sufficient amounts of fertilizer, proper seeds, or satisfactory farm implements to make their

fields productive. Comparatively few have access to adequate technical information on modern methods of farming, though such information is spreading. The difficulty of accumulating a little capital, of getting ahead a little, is increased by the variability of the rainfall from year to year. Very often the earnings from sale of products in a year of adequate rain will be absorbed in the losses of a succeeding year of drought.

In recent years some agricultural gains have been made in Southern Europe. The amount of land under irrigation has been increased, with a number of sizable projects being carried out by the various governments. Consumption of chemical fertilizers in the various countries of the region is now two to four times as great as it was in the 1940s. Production of some export crops, such as irrigated cotton in Greece, has greatly expanded, and overall crop yields have increased. Total agricultural output increased by one-quarter to one-third in the countries of the region between the early 1950s and the middle 1960s. But these increases were actually less than those that characterized northwestern Europe during this period. Thus the situation in Southern Europe improved, but the region's farmers and its land remained relatively unproductive, and agriculture continues to be a major locus of poverty in the region.

Problems of Landownership

In some areas of Southern Europe, rural poverty is partly the result of a system of landownership under which most of the land is held in large estates and the actual farmers are tenants or day laborers. Tenants on these estates are required to yield a considerable share of their meager produce to the landowner as rental for the land. Large estates are particularly characteristic of parts of Italy and of southern Spain and Portugal. Portions of such estates are often devoted to grazing or some type of commercial crop which is profitable to the landowner but is resented by the tenants, who would prefer to use the land for the production of food. Since World War II a number of peasant riots have occurred, resulting from the efforts of landowners to eject tenants from portions of large estates which they have simply occupied and begun to use without permission. Such happenings are a clear indication of intense land hunger among the peasants. The Italian government has indicated that it considers land reform in peninsular Italy to be of pressing importance for political stability in that area, and has made some progress in redistributing land. In Spain substantial reforms were instituted during a brief period of republican government in the early 1930s, but most of these were rescinded by the fascist regime which succeeded to power as a result of the Civil War of 1936 to 1939. About 40 percent of Spain's agricultural land is still owned by approximately 1200 families.

Rural Poverty—An Example from Macedonia

However, land tenure is not really at the root of Southern Europe's agricultural difficulties, though it is a contributing factor in some areas. This is clearly shown by the conditions of poverty which generally prevail even in those areas where the peasants customarily own their own land. Some idea of the situation and problems of the individual Southern European farmer can be gained from the following selection:[1]

A farmer in Macedonia who had the reputation of being one of the most progressive in his village said when interviewed that he had not bought fertilizer for his fields, even though he knew very well that it would improve the harvest, simply because he could not afford the risk. In a good year, he calculated, fertilizer would pay off, for the increased yield meant more grain to sell and enough money to pay back the cost of the fertilizer. In a bad year, however, fertilizer would not make his fields produce enough grain for him to feed his family and at the same time pay the fertilizer debt. Consequently, rather than risk indebtedness, and be forced to sell some part of his small capital— half a dozen sheep, a pig, a donkey, and a mule—he preferred to do without the fertilizer, even though he knew that in the long run his fields would be impoverished and the task of feeding his family would become all the harder.

This man was both intelligent and hardworking. He had just returned to his little two-room house after cultivating his vines with a spade all day in the fields. He was dressed in carefully patched trousers and a tattered shirt, and on his feet he wore homemade moccasins. He owned about 15 acres of land, of which 5 were devoted to vines and fruit trees and the rest produced wheat and beans in alternate years. Each morning the village shepherd came and collected his six sheep and drove them out to the hill slope behind the village where they were pastured with the others from the community; in the evening they were returned to a fold made out of brushwood, where the farmer milked them. His donkey and mule were needed for plowing, and served the year round as beasts of burden. They were fed on straw from the wheat through the winter, and picked up what they could along the paths and field margins of the village in summer. In all this, he was quite average.

He had four children, three sons and a daughter, between eleven and eighteen years of age. The problem which worried him most was not the hardship of day-to-day existence, real though it was, but rather the question of his children's future. With 15 acres to divide among four children, less than 4 acres could be given to each; and even though each child married someone who had an equivalent amount of land, there would be only 8 or 10 acres to support each of the new families. Life on his farm of 15 acres was a desperate struggle already; with less land, his children would face an impossible situation.

In this particular case, the peasant hoped to keep his farm intact for the eldest son. He was preparing to send his second son to the high school in Katerini, a town nearby, where he could learn to be a clerk and with luck find a job. But the problem of supporting the boy away from home was serious, for he would have to buy both food and lodging. The farmer expected to sell his pig, although it was not full grown, in order to get the cash he needed to keep his son at school until the harvest, when, with a good season, he hoped to have a surplus of grain to sell for the same purpose. Looking ahead, he wondered how he would be able to manage when his eldest son was called up to serve in the army and his help on the farm would no longer be available. For the boy of eleven he had made no plans. The girl, he expected, would marry in the village.

Life for this peasant family was a grim and laborious business, a constant struggle to escape from sinking into still deeper poverty. With luck, good harvests and enormous hard work the farmer could hope to pass his farm down unimpaired to his eldest son, and open a way to a tolerable living for his other children. But he realized all too well that illness, bad weather, or accident could bring the whole structure of his plans and effort down in ruins and impose an apparently irremediable poverty on his children.

One result of agricultural conditions in Southern Europe is a long tradition of emigration, which still continues. Some of this is permanent emigration, such as once brought southern Italians and Greeks to the United States in large numbers. Other prominent destinations have been Australia, Argentina (especially for Italians and Spaniards), and Brazil (especially for Italians and Portuguese). Much emigration today is temporary. Large numbers of workers from each South European country emigrate regularly to the more prosperous nations of northwestern Europe as temporary workers. To give one small example, over 3000 southern Italians are employed at the Volkswagen plant in West Germany. They send money home and return once a year on vacation to visit their families.

---

[1] Frank Smothers, William Hardy McNeill, and Elizabeth Darbishire McNeill, *Report on the Greeks* (New York: The Twentieth Century Fund, Inc., 1948), pp. 85–87. Used by permission of The Twentieth Century Fund, Inc.

Probably the largest stream of migration, however, is from countryside to cities within each country. Such a movement improves the situation only if the urban areas can provide the jobs needed. In many Southern European cities this is not the case, and unemployment levels in these places are high. This is particularly true of cities in the more agricultural regions—that is, in the ones closest to the source of emigration—such as Naples, Palermo, and other south Italian cities. Consequently, major efforts are being made in the Southern European countries to direct a sizable share of their industrial growth into cities and areas which heretofore have had, and in general still have, relatively little industry. As these are generally not the places most preferred by private investors, direct government investment and special financial inducements to private companies are used. For instance, in Italy a number of the largest companies are government owned, and they have been ordered to put at least a certain percentage of their new investments into facilities in the south. In both Italy and Spain such efforts are showing some results, without transforming as yet the areas they affect. Instances in point are the placing of two of Italy's new integrated iron and steel plants in the south, at Taranto (220,000) and Naples, respectively, and another on the coast of Tuscany. Only one of four large plants built relatively recently has been placed in the heavily industrialized part of the country. It is located in a suburb of Genoa. These plants, which have been highly successful, use imported coal, largely from the United States, and also depend mainly on iron ore from overseas sources. They are intended to attract further industrial growth to their areas. Recent discoveries and a rising production of oil in Sicily and of natural gas there and in the southern part of the peninsula, as well as the siting of Italy's first two atomic power stations just north of Naples, give further hope for the continued development of south Italy. The depressed condition of this region has for some time been regarded as a major national problem.

THE STRATEGIC POSITION OF SOUTHERN EUROPE

The fact that they have failed to attain great-power status in modern times has not made the countries of Southern Europe unimportant. Rather, they have retained a considerable political significance deriving in part from a strategic position. Their international political orientation has been a matter of grave concern to various outside nations, which have often vied with each other for power and influence in Southern Europe.

Greece—A Mediterranean Gatekeeper

The international importance of modern Greece stems largely from the fact that the Greek Peninsula lies between the Aegean Sea and the entrance to the Adriatic, and thus commands the routes of access to the Mediterranean from the Balkan Peninsula and Black Sea. In addition, the important route of sea trade running through the Mediterranean and Red seas can be effectively threatened from Greece. The strategic implications of its position has made Greece to some extent a pawn between land power and sea power in past times. After its liberation from Turkey in the first half of the nineteenth century, Greece was customarily under the influence and protection of Great Britain. As a major sea power vitally dependent on the Mediterranean–Red Sea route, the latter nation was naturally anxious to keep strong land powers away from Mediterranean shores. Following World War II, Greece became the scene of a civil war between Communist and anti-Communist forces, and it seemed for a time that the country might be added to the list of Russian satellite nations. However, this threat was averted by assistance furnished the anti-Communist elements by the United States, taking over this responsibility from a weakened Britain. The position of Greece made such action feasible, since the country was easily accessible from the sea. Since then Greece has remained closely allied with the United States.

The Italian Causeway to North Africa

Italy's position has made it a natural sea outlet for parts of central Europe, a potential threat to the security of the sea route through the Mediterranean, and a land bridge extending most of the way across the Mediterranean toward northern Africa. In World War II Italian participation on the side of Germany forced Britain to largely abandon the strategic Mediterranean-Red Sea route and to rely on the old pre-Suez route around Africa for traffic with areas surrounding the Indian Ocean. Separated from Africa only by the narrow stretch of water between Sicily and Tunisia, the Italian causeway also permitted German and Italian land forces to wage their African campaigns in relative security from Allied sea power.

The population of Italy, third largest in Europe, and the growing output of its industries add to the country's international importance. Several outside nations have been deeply involved or interested in Italian affairs during recent times. In the latter stages of World War II the country was the scene of a desperate and destructive struggle between Allied and German armies. After the war a struggle for influence in Italy was carried on between the Soviet Union and the Western bloc of nations headed by the United States. This contest was mainly fought in the arena of Italian internal politics, with Soviet influence being exerted through the powerful Italian Communist party and American support being given the anti-Communist elements of the country. The victory went to the West, but the Italian Communist party remains the largest in the world outside of Communist countries, normally polling between 20 and 25 percent of Italy's vote. Nevertheless, Italy is closely linked with the Western nations in the North Atlantic Treaty Organization, the European Economic Community, and other international organizations of this type.

The Spanish Base for Air Power

In Spain the Civil War of 1936–1939 reflected the general tendency of internal political struggles in Southern Europe to be directly influenced by the action of outside powers. In this war the fascist or Insurgent faction eventually triumphed with the aid of an entire army provided by Fascist Italy and an air force and other special units largely provided by Nazi Germany. The losing republican or Loyalist side profited somewhat from aid by the Soviet Union, at that time a major antagonist of Germany and Italy on the European stage. In addition, a considerable number of volunteers from several nations fought in the Loyalist armies. The Insurgent victory was regarded as a considerable defeat for Britain and France, neither of which had intervened directly in the war, since it placed a hostile Spain on the flank of important British sea routes and left France almost surrounded by fascist countries. As it turned out, the new Spanish government refrained from entering World War II, though it maintained a generally hostile attitude toward the Allied nations. Recently Spain has become more closely aligned with the Western nations, particularly the United States. In 1953 an agreement was signed to permit the establishment of American air and naval bases in Spain. The present Spanish government has been a bitter opponent of communism, and

the country's position behind the Pyrenees, well removed from Soviet-held territory, provides some protection for bases. Thus strategic considerations have brought about an accord between Spain and the United States, despite strained official attitudes which have generally prevailed since the fascist victory in the Spanish Civil War. The principal American bases established in Spain thus far include three air bases near Madrid, Seville (600,000), and Zaragoza (420,000), respectively, plus a supply base near Seville and a naval base near Cadiz (135,000).

The Gibraltar Question

In recent years Spain has been involved in a dispute with the United Kingdom over possession of Gibraltar. Britain took the Gibraltar peninsula from Spain in the eighteenth century and has held it since that time as a fortified base guarding the eastern entrance to the Strait of Gibraltar. It is now a self-governing British colony with an area of about 2 square miles and a population of 25,000. With the dissolution of the British Empire and the decline in British power, Spain's claims to Gibraltar were revived. They have been continued despite a referendum in 1967 in which the population voted almost unanimously to retain its ties with Britain.

The Smaller Strategic Significance of Portugal

Portugal has had a lesser degree of political and strategic significance during recent times than the other three nations of Southern Europe, and it has been less involved in international conflicts. The country's proximity to Atlantic sea routes passing near its coast has made it an object of British attention and support during most of its modern history, and the lack of any strong power in a position to challenge British influence has given Portugal a relatively tranquil existence in international politics. Since World War II it has also come into association with the United States, which maintains an air base in the Azores Islands southwest of mainland Portugal. These islands and another Atlantic group, Madeira, are governed as parts of metropolitan Portugal. However, Portugal's harmonious international relations were disturbed somewhat during the 1960s by widespread international criticism of its determined effort to maintain control of its African empire.

Political and Economic Readjustment in Malta

The island state of Malta, which was granted independence by Britain and became a member of the Commonwealth of Nations in 1964, has based its livelihood in the past primarily on its strategic position. Located in the narrow waist of the Mediterranean, 60 miles from Sicily and 180 from Africa, the two small islands of the present Maltese state, Malta and Gozo, have been a naval base for many centuries. The islands were long a stronghold of the crusading Knights of St. John. From 1800 to 1964 they were in British hands, and the income from British naval expenditures and personnel became their main source of support. In 1967, however, Britain began to phase out its operations, with a target date of 1972 for complete withdrawal. The state of Malta has an area of 122 square miles and a population of over 320,000, for an average density of about 2600 per square mile. Two-thirds of the population lives in a multi-city urban complex centering on the capital and port of Valletta (city proper, 17,000), but rural density is still extremely high. Overall levels of income are low, and it was estimated that withdrawal of British naval expenditures would push unemployment near the 20 percent mark. Emigration and a heightened development of tourism appear to offer the principal hope for improving the economic situation. A determined effort is being made to expand tourist facilities and increase the number of visitors, to whom Malta can offer both natural and historical attractions. Meanwhile Britain is helping to subsidize emigration, and financial inducements are attracting some new industries.

REFERENCES and READINGS

General and Miscellaneous

GIORDANO, GUGLIELMO, "The Mediterranean Region," Chap. 15 in Stephen Haden-Guest and Others, eds., *A World Geography of Forest Resources* (American Geographical Society, Special Publication No. 33; New York: The Ronald Press Company, 1956), pp. 317–352.

GLESINGER, EGON, "Mediterranean Project," *Scientific American,* 203, no. 1 (July 1960), 86–96. Discusses a joint plan of the Mediterranean countries and FAO to increase food output in the region.

GUTKIND, E. A., *Urban Development in Southern Europe: Spain and Portugal* (International History of City Development, Vol. 3; New York: The Free Press of Glencoe, 1967).

HAVINDER, M. A., and JOHN HIGGS, "Rural Policies in Southern Europe," in John Higgs, ed., *People in the Countryside* (London: National Council of Social Service, 1966), pp. 81–98.

HOUSTON, J. M., *The Western Mediterranean World: An Introduction to Its Regional Landscapes* (New York: Frederick A. Praeger, 1967; first published in England by Longmans, Green and Co., 1964).

Land Use in Semi-Arid Mediterranean Climates (Arid Zone Research, Vol. 26; Paris: UNESCO, 1964).

WALKER, D. S., *The Mediterranean Lands* (New York: John Wiley & Sons, 1960).

WILLIMOTT, S. G., "The Cork Oak as an Economic Tree Crop," *World Crops,* 15, no. 5 (May 1963), 172–179.

Greece

BOWRA, C. M., and the Editors of *Time-Life* Books, *Classical Greece* (Great Ages of Man; New York: Time Inc., 1965).

CARTER, F. W., "Population Migration to Greater Athens," *Tijdschrift voor Economische en Sociale Geografie,* 59, no. 2 (March–April 1968), 100–105.

DICKS, T. R. B., "Greater Athens and the Greek Planning Problem," *Tijdschrift voor Economische en Sociale Geografie,* **58,** no. 5 (September–October 1967), 271–275.

HOFFMAN, GEORGE W., "The Problem of the Underdeveloped Regions in Southeast Europe: A Comparative Analysis of Romania, Yugoslavia, and Greece, *Annals of the Association of American Geographers,* **57,** no. 4 (December 1967), 637–666.

HUERTLEY, W. A., and Others, *A Short History of Greece: From Early Times to 1964* (Cambridge: At the University Press, 1967).

KAYSER, BERNARD, and KENNETH THOMPSON, *Economic and Social Atlas of Greece* (Athens: National Statistical Service of Greece, Center of Economic Research, Social Sciences Center, 1964).

LAEMMERZAHL, ARTHUR J., and ROBERT FRANCIS, *Basic Data on the Economy of Greece,* U.S. Bureau of International Commerce, Overseas Business Reports, OBR 67–100 (December 1967). (Washington, D.C.: Government Printing Office, 1967.)

NUTTONSON, M. Y., *Ecological Crop Geography of Greece and Its Agro-Climatic Analogues in North America* (Washington, D.C.: American Institute of Crop Ecology, 1947).

SANDERS, IRWIN T., *Rainbow in the Rock: The People of Rural Greece* (Cambridge, Mass.: Harvard University Press, 1962).

SMITH, MICHAEL LLEWELLYN, *The Great Island: A Study of Crete* (London: Longmans, Green and Co., 1965).

THOMPSON, KENNETH, *Farm Fragmentation in Greece: The Problem and Its Setting, with 11 Case Studies* (Athens, Greece: Center of Economic Research, 1963).

WAGSTAFF, J. M., "Rural Migration in Greece," *Geography,* **53,** pt. 2 (April 1968), 175–179; and "Traditional Houses in Modern Greece," *Geography,* **50,** pt. 1 (January 1965), 58–64.

Italy

BANFIELD, EDWARD C., *The Moral Basis of a Backward Society* (New York: The Free Press of Glencoe, 1958).

BELOTTI, MARIO L., "The Development of the Italian South," *Land Economics,* **42,** no. 4 (November 1966), 495–502.

BURTON, A. N., "The Anguish of Calabria," *Geographical Magazine,* **38,** no. 3 (July 1965), 207–221.

CARLYLE, MARGARET, *The Awakening of Southern Italy* (London: Oxford University Press, 1962).

CLOUGH, SHEPARD B., and CARLO LIVI, "Economic Growth in Italy: An Analysis of the Uneven Development of North and South," *Journal of Economic History,* **16,** no. 3 (September 1956), 334–349.

COLE, J. P., *Italy: An Introductory Geography* (New York: Frederick A. Praeger, 1966).

DICKINSON, ROBERT E., "Geographical Aspects of Economic Development in Southern Italy," *Heidelberger Geographische Arbeiten,* **15** (1966), 340–359; and *The Population Problem of Southern Italy: An Essay in Social Geography* (Syracuse, N.Y.: Syracuse University Press, 1955).

DIEM, AUBREY, "An Evaluation of Land Reform and Reclamation in Sicily," *Canadian Geographer,* **7,** no. 4 (1963), 182–191; and "Land Reform and Reclamation in Sicily," *Canadian Geographical Journal,* **66,** no. 3 (March 1963), 88–91.

GUZZARDI, WALTER, JR., "Boom Italian Style," *Fortune,* **77,** no. 5 (May 1968), 136–139 ff.

KISH, GEORGE, *Land Reform in Italy: Observations on the Changing Face of the Mediterranean: Final Report* (Ann Arbor: Department of Geography, University of Michigan, 1966).

MAURER, ALAN O., *Basic Data on the Economy of Italy,* U.S. Bureau of International Commerce, Overseas Business Reports, OBR 67–57 (September 1967). (Washington, D.C.: Government Printing Office, 1967.)

MAZZUCCONI, VITTORIO, "The New City of Milan," *Ekistics,* **21,** no. 125 (April 1966), 281–287.

PRICE, EDWARD T., "Viterbo: Landscape of an Italian City," *Annals of the Association of American Geographers,* **54,** no. 2 (June 1964), 242–275.

RODGERS, ALLAN L., "The Port of Genova: External and Internal Relations," *Annals of the Association of American Geographers,* 48, no. 4 (December 1958), 319–351; *The Industrial Geography of the Port of Genova* (University of Chicago, Department of Geography Research Paper No. 66, 1960); "Regional Industrial Development with Reference to Southern Italy," in Norton Ginsburg, ed., *Essays on Geography and Economic Development* (University of Chicago, Department of Geography Research Paper No. 62, 1960), pp. 143–173; and "Naples: A Case Study of Government Subsidization of Industrial Development in an Underdeveloped Region," *Tijdschrift voor Economische en Sociale Geografie,* 57, no. 1 (January–February 1966), 20–32.

SHEARER, ERIC B., "Italian Land Reform Re-appraised," *Land Economics,* 44, no. 1 (February 1968), 100–106.

VON HAGEN, VICTOR W., *The Roads That Led to Rome* (New York: Harcourt, Brace & World, 1967).

WALKER, D. S., *A Geography of Italy* (2d ed.; London: Methuen and Co., 1967).

WHEELER, DAVID L., "Land Reclamation in the Po River Delta of Italy," *Land Economics,* 41, no. 4 (November 1965), 376–382; and "An Experiment in Cooperative Agriculture: The Rice Communities of the Po River Delta," *Journal of Geography,* 62, no. 7 (October 1963), 310–316.

Portugal

BENEDICT, JOHN, and RICHARD HUMBERT, *Basic Data on the Economy of Portugal,* U.S. Bureau of International Commerce, Overseas Business Reports, OBR 66–92 (December 1966). (Washington, D.C.: Government Printing Office, 1966.)

STANISLAWSKI, DAN, *The Individuality of Portugal: A Study in Historial-Political Geography* (1959), and *Portugal's Other Kingdom, the Algarve* (1963). (Austin: University of Texas Press.)

Spain

BENNETT, HUGH HAMMOND, "Soil Erosion in Spain," *Geographical Review,* 50, no. 1 (January 1960), 59–72.

BOLT, ANNE, "Water for Andalucia," *Geographical Magazine,* 40, no. 10 (February 1968), 850–859.

BRADFORD, SAX, *Spain in the World* (Princeton, N.J.: D. Van Nostrand Co., Searchlight Books, 1962).

BRENAN, GERALD, *The Spanish Labyrinth: An Account of the Social and Political Background of the Civil War* (2d ed.; Cambridge: At the University Press, 1950; reprinted, 1964); also, *The Face of Spain* (London: Turnstile Press, 1950; Penguin Books, 1965); and *South from Granada* (New York: Farrar, Straus & Giroux, 1957).

CHILCOTE, R. H., *Spain's Iron and Steel Industry* (Research Monograph No. 32; Austin: University of Texas, Bureau of Business Research, 1968); "The Spanish Iron and Steel Industry," *Geography,* 52, pt. 1 (January 1967), 60–64; "Spain's Iron and Steel: Renovations of an Old Industry," *Geographical Review,* 53, no. 2 (April 1963), 247–262; and "Spain and European Integration: Heavy Industry in Economic Development," *International Affairs,* 42, no. 3 (July 1966), 444–455.

COURTENAY, P. P., "Madrid: The Circumstances of Its Growth," *Geography,* 44, pt. 1 (January 1959), 22–34.

EDDY, JOHN H., *Basic Data on the Economy of Spain,* U.S. Bureau of International Commerce, Overseas Business Reports, OBR 63–157 (December, 1963). (Washington, D.C.: Government Printing Office, 1963.)

ENGGASS, PETER M., "Land Reclamation and Resettlement in the Guadalquivir Delta–Las Marismas," *Economic Geography,* 44, no. 2 (April 1968), 125–143.

FISHER, WILLIAM BAYNE, and HOWARD BOWEN-JONES, *Spain: An Introductory Geography* (New York: Frederick A. Praeger, 1966).

HEASMAN, D. J., "The Gibraltar Affair," *International Journal,* 22, no. 2 (Spring 1967), 265–277.

HOUSTON, J. M., "Land Use and Society in the Plain of Valencia," in R. Miller and J. Wreford Watson, eds., *Geographical Essays in Memory of Alan G. Ogilvie* (New York: Thomas Nelson and Sons, 1959), pp. 166–194.

INTERNATIONAL BANK FOR RECONSTRUCTION AND DEVELOPMENT, *The Economic Development of Spain* (Baltimore, Md.: Johns Hopkins Press, 1963).

KERNAN, HENRY S., *Reforestation in Spain* (World Forestry Publication, No. 3; Syracuse, N.Y.: State University College of Forestry at Syracuse University, 1966).

MYHILL, HENRY, *The Spanish Pyrenees* (London: Faber and Faber, 1966).

NAYLON, JOHN, "Tourism—Spain's Most Important Industry," *Geography,* 52, pt. 1 (January 1967), 23–40; "Land Consolidation in Spain," *Annals of the Association of American Geographers,* 49, no. 4 (December 1959), 361–373; "Progress in Land Consolidation in Spain," *Annals of the Association of American Geographers,* 51, no. 3 (September 1961), 335–338; "Irrigation and Internal Colonization in Spain," *Geographical Journal,* 133, pt. 2 (June 1967), 178–191; and "The Badajoz Plan: An Example of Land Settlement and Regional Development in Spain," *Erdkunde,* 20, no. 1 (January 1966), 45–60.

"Olive Culture in the Mediterranean Lands: The Hacienda de Córdoba," Chap. 5 in Richard M. Highsmith, Jr., ed., *Case Studies in World Geography: Occupance and Economy Types* (Englewood Cliffs, N.J.: Prentice-Hall, 1961), pp. 30–35.

OWEN, CHARLES, "Barcelona: City Transformed," *Geographical Magazine,* 39, no. 5 (September 1966), 351–363.

PARSONS, JAMES J., "The Acorn-Hog Economy of the Oak Woodlands of Southwestern Spain," *Geographical Review,* 52, no. 2 (April 1962), 211–235; and "The Cork Oak Forests and Evolution of the Cork Industry in Southern Spain and Portugal," *Economic Geography,* 38, no. 3 (July 1962), 195–214.

PRITCHETT, V. S., *The Spanish Temper* (London: Chatto and Windus, 1954).

WAY, RUTH, assisted by Margaret Simmons, *A Geography of Spain and Portugal* (London: Methuen and Co., 1962).

Malta

OWEN, CHARLES, "Island in Transition" [Malta], *Geographical Magazine,* 39, no. 3 (July 1966), 203–213.

See also the lists of references and readings for Chapters 1–4, especially the relevant chapters in the geographies of Europe listed on p. 92.

Countries
of East Central
Europe[1]

12

In the aftermath of World War II, Communist parties took control of eight countries occupying the eastern margins of Europe between the Baltic, Black, and Adriatic seas. Since then, the national life of these countries has been restructured along Communist lines. They exhibit the customary features of Communist states: government by one party; planning and direction of the economy by organs of the state; abolition of private ownership (with minor exceptions) in the fields of manufacturing, mining, transportation, finance, commerce, and services; and measures, more thoroughgoing in some countries than in others, to convert agriculture from a private basis to a socialized basis. But the countries of Communist Europe must not be thought of simply as replicas of each other. Each country has implemented the Communist system in a somewhat different way (Yugoslavia differing the most widely from the rest), each one has its own distinctive relationships within the Communist world and with non-Communist nations, and each one operates within its own frame of physical, cultural, economic, and historical circumstances. Few parts of the world, in fact, exhibit so much diversity within a comparable extent of land. One of the countries,

[1] A committee of specialists has recommended that "East Central and Southeast Europe" be employed as an overall name for this region (information from Professor George Hoffman, Department of Geography, University of Texas). Discussions of the region frequently include Greece, although that country is grouped with Italy, Spain, and Portugal in this text.

East Germany, has been considered in Chapter 7, but will be brought into the discussion here at appropriate points. The remaining seven—Poland, Czechoslovakia, Hungary, Romania, Bulgaria, Yugoslavia, and Albania—are grouped in the present chapter as the countries of East Central Europe.[1] They have a combined area of 450,000 square miles (the size of Texas plus California and Maine), and had an estimated population total of 107 million in 1968 (see Table 9). Their area is only one-nineteenth as great as that of their eastern neighbor, the Soviet Union, but their population amounts to 45 percent of the total for the Soviet Union or, combined with East Germany's, a little more than half of the USSR total.

RECENT RELATIONSHIPS WITH THE SOVIET UNION

Comparisons between Communist Europe and the Soviet Union are very appropriate, for six of the eight countries in Communist Europe lie within the economic and strategic orbit of the USSR, and their productive capacities and markets contribute in no small measure to the USSR's position as a major world power. This situation has persisted since the closing stages of World War II, when Soviet troops pursued the retreating Germans across eastern Europe, and in the process brought most of the region under Soviet control. During a transitional period from 1944 to 1948, local Communist parties, often led by men who had been exiles in the Soviet Union for many years, were given such backing by the USSR as was necessary for them to gain control

of the governments in their respective countries. Since then, the Soviet Union has made some concessions to the desire of these countries to run their own affairs, but it has never wavered from its insistence that the direction of affairs be in the hands of the Communist party and that party leaders in each country maintain a responsive attitude to the desires and policies of the USSR. In 1956 Soviet troops crushed an uprising against the Communist government of Hungary, and in 1968 the Soviets, in company with East Germany, Poland, Hungary, and Bulgaria, effected a military occupation of Czechoslovakia after the Communist leadership of that nation had introduced political reforms regarded as dangerously liberal by the USSR.

Two nations of Communist Europe—Yugoslavia and Albania—have had a very different postwar history from the others. In Yugoslavia, Marshal Tito's Communist party led the Partisan resistance against the Nazi occupation in World War II, and was able to take immediate control of the country when the war ended. Soviet troops entered Yugoslavia briefly but were withdrawn for service elsewhere as the fighting swirled on toward Germany. Tito's government then moved to establish a Communist order throughout Yugoslavia, but in the process showed an independence of action that roused suspicions and dislike on the part of the Soviet dictator, Stalin. In 1948, despite expressions of loyalty to Stalin and the USSR, the Yugoslav Communists were outlawed from the world Communist movement, and for a time the political and economic relations of Yugoslavia with other Communist states were almost completely severed. Tito thereupon sought and received economic and military aid from Western nations, principally the United States. Since then, Yugoslavia has continued to carry forward its program of separate

TABLE 9 EAST CENTRAL EUROPE: AREA AND POPULATION DATA

COUNTRY	AREA (THOUSAND SQUARE MILES)	POPULATION (MILLIONS: 1968 ESTIMATES)	DENSITY (PER SQUARE MILE: TO NEAREST WHOLE NUMBER)
Poland	120.6	32.3	268
Czechoslovakia	49.4	14.4	292
Hungary	35.9	10.2	284
Romania	91.7	19.4	212
Bulgaria	42.8	8.4	195
Albania	11.1	2.0	178
Yugoslavia	98.8	20.2	205
Totals	450.3	106.9	237

Communist development. The widespread thaw in the Communist world which developed after Stalin's death in 1953 saw the restoration of many political and commercial links between Yugoslavia and other Communist countries. But the special Yugoslav pattern of development still diverges more from Soviet theory and practice than is true of any other European Communist state, and Yugoslavia's economic relationships with the non-Communist world are stronger than those of any other Communist country.

In Yugoslavia's close neighbor, Albania, several partisan groups carried on resistance against the Axis, and at the end of the war the strongest group, thoroughly Stalinist in character, took control of the country. In 1948 Albania sided with Stalin in the dispute with Yugoslavia, and after 1953 the country continued to follow a hard Stalinist line. This led in 1961 to a complete break in economic and diplomatic relations with the USSR after the latter repudiated Stalinism and became embroiled in an ideological dispute with Communist China. Today Albania relies chiefly on the Communist Chinese for support, and it has generally maintained a severely isolationist and suspicious attitude in its dealings with the USSR, the nations of Communist Europe, and the nations of the West.

Thus neither Yugoslavia nor Albania can presently be termed a "client" or "satellite" of the USSR, though these terms still seem appropriate to the other European Communist states, including East Germany. But even these states are freer of outright Soviet dictation than they were in the Stalin era, though Czechoslovakia's experience in 1968 indicated that this freedom has its limits. They have been able in some measure to reassert their separate identities and to make more decisions on the basis of national self-interest. Meanwhile, they maintain close economic and military relationships with the USSR and with each other, and all of them adhere, in name if not always in spirit, to the Soviet-sponsored Council for Mutual Economic Assistance (COMECON or CEMA) and the Warsaw Pact military alliance (map, p. 67). For the USSR the European client states are a buffer zone providing defense in depth against a possible resurgence of German militarism; and the manpower, resources, production, and markets of these countries are important adjuncts to the USSR's own economic potential.

POLITICAL INSTABILITY IN THE "SHATTER BELT"

The extension of Soviet power into East Central Europe at the end of World War II was in keeping with the general history of the region throughout modern times. In the Middle Ages several peoples in the region—the Poles, Czechs, Magyars, Bulgarians, and Serbs—enjoyed political independence for long periods and at times had control of extensive territories outside of their homelands. Their situation then deteriorated as stronger powers—the Germans and Austrians on the west, and the Ottoman Turks and Russians on the east—pushed into East Central Europe, and carved out empires. In the process the empires came into collision, and the local peoples were caught in numerous wars which devastated great areas, often resulted in a change of masters, and sometimes brought about large transfers of population from one area to another. Frequently there were revolts against ruling powers, and these were often put down with great severity. Sizable sections of East Central Europe were virtually depopulated for long periods. Such conditions reached their zenith in the long history of wars and disorders that attended the rise and slow decline of the empire of the Ottoman Turks. In the fifteenth, sixteenth, and seventeenth centuries Turkish armies overran the Balkan Peninsula and then pushed northward through Hungary to the borders of Austria. Their advance was checked at Vienna in 1683 by forces of the Holy Roman Empire, abetted by the Poles and other Christian nations of Europe. Thereafter, the Turks were pushed back by armies of the Austrian and Hungarian empires or by Russia. As Turkish power weakened, the Ottoman possessions in Europe were taken over by Austria, Hungary, or Russia; or local peoples were able gradually to achieve self-government and independence. In some areas from which the Turks were expelled, the Austro-Hungarian Empire

Index map of East Central Europe. The "mountainous areas" are rather broadly generalized to bring out the major outlines of the topography. The Bakony Forest, separating the Great Hungarian and Little Hungarian Plains, and the narrow highland spur extending eastward from Slovenia in northern Yugoslavia are hilly rather than truly mountainous, and the same is true of certain other areas shown in the same shading. For identification of cities shown by letter in the major industrial concentrations, see the map on page 138.

BALTIC SEA

15°E

POMERANIA

EAST
GERMANY

Berlin ●★

Elbe R.

H
L **SAXONY**
INDUSTRIAL CONCENTRATION
D
K·M·S
SAXON
P GATE
ORE
MTS.

BOHEMIA
INDUSTRIAL CONCENTRATION
Prague ✪

BOHEMIA
Pilsen ○

BOHEMIAN
FOREST

WEST
GERMANY

ALPS

AUSTRIA

ITALY

Ljubljana ○

ISTRIAN
PEN. →

Rijeka-Susak ○

Trieste ●

45°N

ADRIATIC
SEA

ITALY

Kaliningrad
(Königsberg) ○

Gdynia ○
Gdansk
(Danzig) ○

Szczecin ○
(Stettin)

MASURIA

USSR

25°E

20°E

Bydgoszcz ○

Warta

Poznan
(Posen) ●

Plock ○

Narew

Bialystok ○

Bug

Vistula R.

Brest
(Brest-Litovsk) ○

Warsaw ✪

S I L E S I A

Lodz ●

Oder R.

Wroclaw
(Breslau) ●

Walbrzych
(Waldenburg) ○

Czestochowa ○

Lublin ○

UPPER SILESIA
INDUSTRIAL CONCENTRATION

SUDETEN MTS.

Z·C·B·S
G·C·C·B

Vistula R.

50°N

Neisse R.

Katowice ○

Nowa Huta ○

Krakow ●

Lvov ●

G A L I C I A

MORAVIAN
GATE

Ostrava ○

Brno ●

MORAVIA

TATRA
MTS.

C A R P A T H I A N

S L O V A K I A

Kosice ○

RUTHENIA

BUKOVINA

MOLDAVIA

Dniester R.

Southern Bug

30°E

USSR

Vienna ✪

Bratislava ✪

LITTLE
HUNGARIAN
PLAIN

Gyor ○

Ozd ○

Miskolc ○

Tisza R.

Budapest ✪

GREAT

HUNGARIAN

Debrecen ○

MTS.

Iasi
(Jassy) ○

Prut R.

B E S S A R A B I A

BAKONY FOREST

Lake
Balaton

Dunaujvaros ○

PLAIN

Cluj ○

TRANSYLVANIA

Drava R.

SLOVENIA

Pecs ○

Szeged ○

Subotica ○

Timisoara ○

BIHOR
MTS.

Hunedoara ○

Brasov ○

Galati ○

Mouths
of the
Danube

Zagreb ●

CROATIA

V O J V O D I N A

B A N A T

TRANSYLVANIAN
ALPS

Braila ○

Novi Sad ○

Resita ○

Ploesti ○

DINARIC ALPS

BOSNIA

Sava R.

SERBIA

Morava R.

Belgrade ✪

IRON
GATE

W A L A C H I A

Bucharest ✪

Constanta ○

DOBRUJA

BLACK

SEA

Sarajevo ○

Nis ○

DALMATIAN COAST

Split ○

HERZEGOVINA

MONTENEGRO

Dubrovnik ○

Titograd ○

Cetinje ○

Bar ○

Shkodër ○

KOSOVO-
METOHIJA

Skoplje ○

Danube R.

Varna ○

Maritsa R.

Mt. Musala
9,596 ft. +

Sofia ✪

BALKAN RANGE

Burgas ○

Plovdiv ○

RHODOPE MTS.

Vardar R.

Tirana
★

MACEDONIA

Salonika ●

Istanbul
(Constantinople) ✪

Bosporus

SEA OF
MARMARA

TURKEY

40°N

GREECE

IONIAN
SEA

AEGEAN
SEA

Dardanelles

**LITHUANIA, LATVIA,
ESTONIA** (off map)
absorbed by USSR

LITHUANIAN
SSR

LATVIAN
SSR

GERMANY

EAST
PRUSSIA

to USSR
from
Germany

Danzig

Free City
of DANZIG
to POLAND

to POLAND
from Germany

CZECHOSLOVAKIA

to USSR
from
Poland

POLAND

to USSR
from
Czecho-
slovakia

AUSTRIA

HUNGARY

to USSR
from
Romania

Trieste ○

to YUGOSLAVIA
from Italy

ROMANIA

ITALY

YUGOSLAVIA

to BULGARIA
from Romania

ALBANIA

GREECE

BULGARIA

main territorial changes since 1938

EAST CENTRAL EUROPE

INDEX MAP

mountainous areas

major industrial concentrations

URBAN AREAS
(by approximate population categories)

● over 3,000,000

✪ ◎ over 1,000,000

✪ ● 500,000–1,000,000

○ 250,000–500,000

★ ○ selected smaller places

stars show national capitals

0   50   100   150

Scale of Miles

brought in colonists of many nationalities to occupy the land, and the result was an intricate intermixture of peoples. The principal area where this occurred was the Banat, a fertile plains region that is now divided between the Vojvodina of Yugoslavia and a section of Romania.

In the eighteenth century the national independence of Poland was extinguished in a series of partitions which gave control of the east and center to Russia, the south to Austria, and the north and west to Germany (Prussia). These areas were integrated politically and economically into the respective empires. A differential economic development then took place, with the German sections advancing at a faster rate than those held by Austria or Russia.

Thus for several centuries the map of East Central Europe was in a state of flux as empires rose and fell, boundaries shifted, and populations were passed back and forth from one imperial master to another. The fragmented and unstable pattern of nationalities and political units has led some students of this region to call it "the Shatter Belt" or "Crush Zone" of Europe.

ORIGINS OF THE PRESENT PATTERN OF COUNTRIES

The present pattern of countries in East Central Europe resulted from the disintegration of the Ottoman Turkish, Austro-Hungarian, German, and Tsarist Russian empires in the nineteenth and early twentieth centuries. This process was hastened by World War I, in which Germany, Austria-Hungary, and Turkey were on the losing side. Russia, though on the side of the victors, withdrew from the war following the Bolshevik Revolution of 1917 and was not represented at the Paris Peace Conference of 1919. At this conference the political map of East Central Europe was rearranged in an attempt to satisfy the aspirations of the various nationalities which had been included in the old empires. Poland was reconstituted as an independent country. Czechoslovakia, the homeland of two closely related Slavic peoples, the Czechs and Slovaks, was carved out of the Austro-Hungarian Empire as an entirely new country. Hungary, greatly reduced in size, was severed from Austria. The Kingdom of Serbia, which had won independence from Turkey by stages in the nineteenth century, was joined with the small independent state of Montenegro and several regions taken from Austria-Hungary to form the new Kingdom of the Serbs, Croats, and Slovenes, later known as Yugoslavia. Ro-

mania, which had been independent of Turkish control since the mid-nineteenth century, was enlarged by territories taken from Austria-Hungary, Russia, and Bulgaria. Independence from Turkey had been achieved by Bulgaria in the second half of the nineteenth century and by Albania immediately before the outbreak of World War I. The sovereignty of these nations was confirmed by the peace conference.

Boundary Problems

The territorial settlement in East Central Europe following World War I left a number of important boundary issues unresolved. These arose from the fact that several of the new states, as constituted by the Paris conference, included frontier areas claimed by neighboring countries on historical or ethnic grounds. At various times in past centuries a number of nationalities in East Central Europe had ruled over extensive territories outside the boundaries of their new political units. This became a basis for territorial claims on neighboring states. More important, however, were claims based on ethnic considerations. Due to intermixture of ethnic groups as well as to complex economic and strategic factors, the delimitation of political boundaries following the war left large ethnic minorities in Poland, Czechoslovakia, Yugoslavia, and Romania, and smaller but substantial minorities in Bulgaria and Hungary. Such groups, often localized in frontier zones, provided a basis for territorial demands by adjacent units regarding themselves as mother states for the minorities concerned.

Between World Wars I and II, boundary questions resulted in much bickering among the nations of East Central Europe, as well as between these nations and neighboring countries such as Germany, the Soviet Union, or Italy. A few of the more important disputed areas, as they existed between the wars, may be summarized as follows (map, p. 257):

1. Transylvania, Bessarabia, and the southern Dobruja, acquired by Romania from Hungary, Russia, and Bulgaria, respectively, but still claimed by those nations.

2. Macedonia, divided between Yugoslavia and Greece, with parts of Yugoslav Macedonia claimed by Bulgaria and Albania, and Greek Macedonia claimed by Bulgaria and Yugoslavia.

3. The Banat, divided between Yugoslavia and Romania, with the Romanian Banat claimed by Hungary and the Yugoslav Banat claimed by Hungary and Romania.

4. Various border areas separating Yugoslavia from Italy and claimed by both nations.

5. Silesia, divided among Germany, Poland, and Czechoslovakia, and subject to a complicated series of claims by them.

6. Portions of the Sudeten Mountains and Ore Mountains incorporated in Czechoslovakia but claimed by Germany.

7. The part of Poland known as the "Polish Corridor" and the Free City of Danzig, separating East Prussia from the remainder of Germany and claimed by Germany.

Boundary Changes and Population Transfers Since 1939

As a result of World War II, several shifts of territory occurred involving some of the disputed areas listed above. Romania was forced to cede Bessarabia and the northern part of adjoining Bukovina to the Soviet Union, and the southern Dobruja was returned to Bulgaria. The Soviet Union annexed a wide strip of territory in eastern Poland amounting to some 46 percent of the prewar area of that country. However, Poland was compensated for the loss when it took over large areas from Germany—areas included in the prewar German provinces of Pomerania and Silesia immediately east of the Oder and Neisse rivers and the southern half of East Prussia. Poland also took control of Danzig. The sum of these areas amounts to only half as much in square miles as the area taken by the Soviet Union from Poland, but the industrial development, mineral wealth, and agricultural productivity of Poland's new territories (the "Regained Territories" or "Western Territories" in Polish terminology) more than counterbalance her losses to the USSR, at least in an economic sense. By far the most valuable area economically is the former German province of Silesia, with its coal and other mineral deposits, heavy industries, and productive farmlands. The northern half of former East Prussia was taken over by the Soviet Union as a result of the war. Thus the Polish Corridor no longer exists, since Germany has lost the whole of East Prussia. Ruthenia, or Carpathian Ukraine, a mountainous, underdeveloped area at the extreme eastern end of prewar Czechoslovakia, was ceded by that country to the Soviet Union. A number of boundary changes benefiting Yugoslavia occurred along the Yugoslav-Italian frontier after the war. In 1947 the seaport of Trieste (city proper, 280,000), near the head of the Adriatic Sea, was included, together with adjoining areas, in the Free Territory of Trieste, under United Nations supervision. This enclave existed as a sore spot until 1954, when the two countries signed an agreement which placed the city of Trieste under Italian administration but made it a free port. Most of the remaining area comprising the Free Territory was transferred to Yugoslavia.

In addition to boundary adjustments, and often as a consequence of them, extensive population transfers have taken place in East Central Europe since the beginning of World War II. Such transfers, involving millions of people—Germans, Poles, Hungarians, Italians, and others—have notably simplified the ethnic pattern. Nearly a third of the prewar population of Poland and of Czechoslovakia was composed of minorities, but today these have largely disappeared. Ethnic minorities now comprise less than 1 percent of Poland's population and less than a tenth of Czechoslovakia's. The two countries have transferred most of their German population to Germany (p. 182), including nearly all the Germans who lived in Poland's Western Territories. The Soviet Union has absorbed most of the Ukrainians and Belorussians (White Russians) of Poland and most of Czechoslovakia's Ruthenian minority; and Czechoslovakia has transferred part of its Hungarian minority to Hungary. The only countries in East Central Europe that still have truly large minorities are Yugoslavia (primarily Albanians and Hungarians, with smaller numbers of Turks, Romanians, and others; see map, p. 260) and Romania (mostly Hungarians). Even in these countries, minorities comprise a smaller proportion of the population than before the war.

DOMINANCE OF SLAVIC PEOPLES

The postwar expulsion of Germans from East Central Europe and the extermination or flight of more than 3 million Jews during the period of Nazi control intensified the Slavic character of the region. Peoples speaking Slavic languages form a large majority within the region as a whole. The original homeland of these peoples is thought to have lain in the area between the Vistula and Dnieper rivers. In the early Middle Ages groups of Slavs began migrating into other parts of East Central Europe, as well as eastward into Russia. The Elbe River seems to have marked the limit of their penetration toward the west. Later on there were eastward movements by groups of Germans who established themselves among the Slavic peoples of East Central Europe and Russia. Sometimes the way was cleared for such groups by warfare, but in other situations German colonists came by invitation of local rulers, as their skills in agriculture, mining, industry, and trade were valued highly. Persistent colonization by Germans over a period of centuries resulted in the large German minorities found in several countries of East Central Europe prior to World War II.

The Major Slavic Groups

The Slavic peoples are often grouped into three large divisions: (1) *East Slavs,* including Russians, Ukrainians, Belorussians (White Russians), and Ruthenians; (2) *West Slavs,* including Poles, Czechs, and Slovaks; and

SOURCE: Based on *Stanovnistvo Po Narodnosti* (*Population by Ethnic Nationality*), Volume IX, Belgrade, 1954. (Final results of the population census of March 15, 1948.) The map below has been redrawn from a map in George W. Hoffman and Fred Warner Neal, *Yugoslavia and the New Communism* (New York: Twentieth Century Fund, 1962), p. 30. By permission of the authors and the Twentieth Century Fund. Names in the largest type denote the six socialist republics that comprise the Socialist Federal Republic of Yugoslavia. The Vojvodina and Kosovo-Metohija (Kosmet) are "autonomous regions" attached to Serbia.

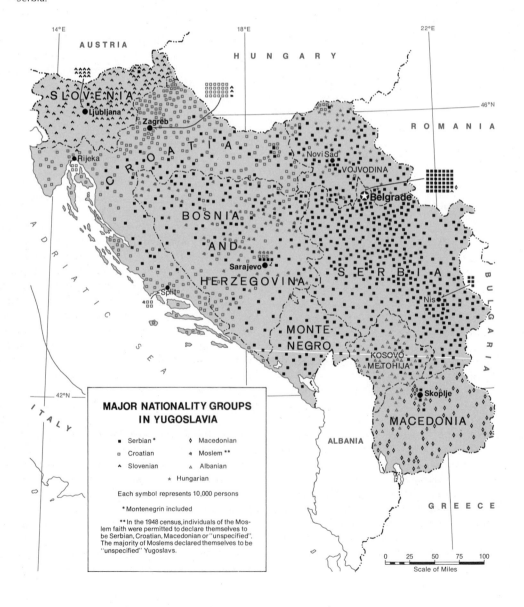

MAJOR NATIONALITY GROUPS IN YUGOSLAVIA

- ■ Serbian *
- ▫ Croatian
- ▲ Slovenian
- ◆ Macedonian
- ◀ Moslem **
- △ Albanian
- ★ Hungarian

Each symbol represents 10,000 persons

* Montenegrin included

** In the 1948 census, individuals of the Moslem faith were permitted to declare themselves to be Serbian, Croatian, Macedonian or "unspecified". The majority of Moslems declared themselves to be "unspecified" Yugoslavs.

Scale of Miles
0 25 50 75 100

(3) *South Slavs,* including Serbs, Croats, Slovenes, Bulgarians, and Macedonians. Another group, the Romanians, came under the influence of the later Roman Empire, and the Romanian language is classed with the Romance languages derived from Latin. However, the Romanians have been affected by Slavic influences, and their language contains many Slavic words and expressions.

Although the various Slavic peoples speak related languages, such languages may not be mutually intelligible. For example, the Polish language is not easily understood by a Czech, though Poles and Czechs are customarily grouped as West Slavs. Other significant differences also exist. Serbian and Croatian, for example, are essentially one language in spoken form, but the Serbs, like the Bulgarians and most of the East Slavs, use the Cyrillic alphabet, while the Croats use the Latin alphabet, as do the Slovenes and the West Slavs. A religious division exists between the West Slavs, Croats, and Slovenes, who are Roman Catholics for the most part, and the East Slavs, Romanians, Bulgarians, Serbs, and Macedonians. who adhere principally to various branches of the Orthodox Eastern Church. Religious differences have been important factors in conflicts among the peoples of the region.

Non-Slavic Peoples

The principal non-Slavic population elements in East Central Europe, excluding the Romanians, are the Hungarians and the Albanians. The Hungarians, also known as Magyars, are the descendants of Asian nomads who settled in Hungary in the ninth century. They are distantly related to the Finns and speak a language which is entirely distinct from the Slavic languages. Roman Catholicism has been the dominant religious faith in Hungary, though substantial Calvinist and Lutheran groups have long existed there. The Albanians speak an ancient Thraco-Illyrian language which is not related to the Slavic languages except in a very distant sense. Excluding Turkey, Albania is the only European country in which Moslems form a majority of the population. Substantial Orthodox and Roman Catholic minorities also exist there. Yugoslavia and Bulgaria have Moslem minorities of considerable size. In Yugoslavia, Moslems are estimated to comprise close to one-tenth of the population, and in Bulgaria perhaps one-twelfth. The Moslem element in the three countries is a legacy from the long period of Turkish rule in the Balkan Peninsula prior to World War I.

MAJOR PHYSICAL DIVISIONS

The peoples of East Central Europe inhabit a region in which cultural and political complexity is matched by, and is in some measure a product of, great physical diversity. The mountains, forests, and swamps of the region inhibited contacts between neighboring peoples in earlier times (and in some areas still do), and thus contributed to local and regional differences in culture and political allegiance. When viewed in detail, the mountains, hill lands, plains, and plateaus of East Central Europe form a pattern which is very complicated. Considered in broad outline, however, a certain order can be discerned in the arrangement of surface features. These tend to group themselves into four major physical belts or zones: (1) the northern plain, (2) the central mountain zone, (3) the Danubian plains, and (4) the southern mountain zone.

The Northern Plain

Most of Poland lies in the northern plain, between the Carpathian and Sudeten mountains on the south and the Baltic Sea on the north (map, p. 257). The plain is a segment of the North European Plain, which extends westward from Poland into Germany, Denmark, the Low Countries, and France, and eastward into the Soviet Union. Central and northern Poland are comprised of land that is rather sandy and infertile, with many swamps, marshes, and lakes, especially in the eastern Masurian Lakes district. The level expanses of plain are broken at intervals by terminal moraines, the low, regular, elongated hills created by the continental ice sheets which covered this area during the Ice Age. At the south the plain gradually rises to low uplands. Here a thick cover of loess provides the most fertile soils of Poland.

Poland's largest rivers, the Vistula and Oder (Polish: Odra), wind across the northern plain from Upper Silesia to the Baltic Sea. The Oder is the country's main internal waterway, carrying some barge traffic between the important Upper Silesian industrial area (with which it is connected by a canal) and the Baltic. The main seaports of Poland—Gdansk (formerly Danzig), Gdynia, and Szczecin (formerly Stettin)—are along or near the lower courses of these rivers. Danzig, located at the western edge of the Vistula delta, was created a Free City in 1919 because of its predominantly German population. Poland was given a privileged position in using the port, but decided to build Gdynia a short distance to the northwest as an additional all-Polish port. The two cities form a metropolitan area with an estimated population of 600,000. Szczecin (city proper, 320,000) is located about 50 miles inland on the west bank of the Oder. Formerly an all-German river except for its headwaters, the Oder is now controlled by Poland for most of its length. The lower river is shared with East Germany for a considerable distance, although the part from just above Szczecin to the Baltic Sea is entirely within Polish territory.

The Central Mountain Zone

The central mountain zone is formed by the Carpathian Mountains and lower ranges rimming the western part of Czechoslovakia. The Carpathians extend in a giant arc for a thousand miles from Slovakia and southern Poland to south-central Romania. Geologically, these mountains are a continuation of the Alps. They are lower than the Alps, however, and are cut by a greater number of easy passes. Elevations of 8000 feet or more are reached only in the Tatra Mountains of Slovakia and Poland and in the Transylvanian Alps of Romania.

West of the Carpathians, lower mountains enclose the hilly basin of Bohemia, the industrial core of Czechoslovakia. On the north the Sudeten Mountains and Ore Mountains separate Czechoslovakia from Poland and East Germany. Between these ranges, at the Saxon Gate, the valley of the Elbe River provides a lowland connection and a navigable waterway leading from the Bohemian Basin to the highly developed industrial region of Saxony in East Germany, and, farther north, to the West German seaport of Hamburg and the North Sea. However, the imposition of the iron curtain has largely severed traffic on the Elbe between Czechoslovakia and Hamburg (see pp. 180–181), and Czechoslovakia's ocean-borne trade now moves primarily through the Polish seaport of Szczecin. To the southwest the Bohemian Forest occupies the frontier zone between Czechoslovakia and West Germany. Lower highlands border Bohemia on the southeast. The mountainous rim of Bohemia has a general elevation of only 2000 to 5000 feet, while the floor of the Bohemian Basin lies generally at 1500 feet or less.

Between the mountain-rimmed upland basin of Bohemia and the Carpathians of Slovakia, a convenient and historic passageway is provided by the lowland corridor of Moravia. Through this corridor run major routes of transportation connecting Vienna and the Danube Valley with the plains of Poland. To the north, near the Polish frontier, the corridor narrows at the Moravian Gate between the Sudeten Mountains and the Carpathians. Just beyond this gateway, East Central Europe's most important concentration of coal mines and iron and steel plants has developed in the Upper Silesian-Moravian coal field of Poland and Czechoslovakia.

The Danubian Plains

Two major lowlands, bordered by mountains and drained by the Danube River and its tributaries, comprise the Danubian plains. One of these, the Great Hungarian Plain, occupies two-thirds of Hungary and smaller adjoining portions of Romania, Yugoslavia, and the Soviet Union. The Great Hungarian Plain is very level in most places, and contains much poorly drained land in the vicinity of its rivers. To the northwest an outlier, the Little Hungarian Plain, extends into the margins of Czechoslovakia and Austria. The second major lowland drained by the Danube is comprised of the plains of Walachia and Moldavia in Romania, together with the northern lowland fringe of Bulgaria. Walachia lies between the Carpathians and the Danube, while Moldavia is between the Carpathians and the Soviet Union. The Danubian plains represent the most fertile large agricultural region in East Central Europe. Broad expanses of level land with deep, rich soils, an average growing season of 190 days or more, and a summer maximum of rainfall provide good natural conditions for growing corn and wheat, which occupy the greater part of the arable land.

The Danube River, which supplies a navigable water connection between these lowlands and the outside world, is the longest river in Eurasia west of the Volga. It rises in the Black Forest of southwestern West Germany and follows a winding course of some 1750 miles to the Black Sea, which it enters through three main channels. The Danube is customarily divided into three principal sections: Upper, Middle, and Lower. The Upper Danube, above Vienna, is fed principally by tributaries from the Alps. This section of the river is swift and difficult to navigate, though large barges use it as far upstream as Regensburg, West Germany. Regensburg is to be the southern terminus of a canal now under construction from the Main River to provide a good water link between the Rhine and Danube river systems (see p. 181). Below Vienna the Middle Danube flows leisurely across the Little Hungarian and Great Hungarian plains past the Czechoslovak river port of Bratislava (city proper, 270,000) and the Hungarian and Yugoslav capital cities of Budapest and Belgrade. Three major tributaries enter the river between the latter cities: the Tisza from the

The Danube River at the Iron Gate. Boats have long been assisted upstream against the swift current by a railway on the bank, but this will be rendered unnecessary by a new power and navigation dam that Romania and Yugoslavia are constructing as a joint project in the Iron Gate gorge. (Yugoslav Information Center.)

Carpathians and the Drava and Sava from the Alps. In the border zone between Yugoslavia and Romania the Danube follows a winding series of gorges through a belt of mountains about 80 miles wide where the Carpathians reach southward to merge with the Balkan Range (Stara Planina). At the eastern end of these gorges is the famous Iron Gate. Beyond the Iron Gate the Lower Danube forms the boundary for a long distance between the level plains of southern Romania and the low plateaus of northern Bulgaria. The river then turns northward into Romania and enters the Black Sea through a low, marshy delta with many lakes and swamps. The inland river ports of Braila (city proper, 130,000) and Galati (city proper, 120,000) in Romania are accessible to seagoing vessels drawing 21 feet of water or less. Constanta (city proper, 130,000) is the principal Romanian seaport on the Black Sea itself. No river in the world touches so many different countries as the Danube—eight in all, including West Germany, Austria, Czechoslovakia, Hungary, Yugoslavia, Romania, Bulgaria, and the Soviet Union. Traffic on the Danube does not compare in volume with traffic on the Rhine. However, it has increased greatly in recent decades. Very prominent among the items of cargo carried by Danube shipping (barges and small seagoing ships) are coal, petroleum and its products, and iron ore. Various schemes to improve navigation on the river are under way. Perhaps the most important is a large navigation and power dam being constructed jointly by Yugoslavia and Romania in the Iron Gate gorge. The reservoir will eliminate a stretch of rapids in the gorge and enable larger vessels and a much greater volume of shipping to traverse this stretch of river. Locks will permit vessels to bypass the dam. Completion of the project by the early 1970s is anticipated.

The Southern Mountain Zone

The southern mountain zone occupies most of the Balkan Peninsula. Bulgaria, Yugoslavia, and Albania, the East Central European countries that share this zone, are very mountainous, though Yugoslavia and Bulgaria contain substantial areas of lowland, and Albania has lowlands along the coast.

In Bulgaria the principal mountains are the Balkan Range, extending east-west across the center of the country, and the more extensive Rhodope Mountains in the southwest. These are rugged mountains that attain heights of 9000 feet in a few places. Mt. Musala (9596 feet) in the western Rhodope of Bulgaria is the highest summit in East Central Europe. Between the Rhodope Mountains and the Balkan Range is the productive basin of the Maritsa River, constituting, together with the adjoining Sofia Basin, the economic core region of present-day Bulgaria. North of the Balkan Range a low plateau, covered with loess and cut by deep river valleys, slopes to the Danube.

In central and southern Yugoslavia a tangled mass of hills and mountains constitutes a major barrier to travel.

Through this difficult region a historic lowland passage connecting the Danube Valley with the Aegean Sea follows the trough of the Morava and Vardar rivers. At the Aegean end of the passage is the Greek seaport of Salonika (Thessaloniki), which handles some traffic for Yugoslavia. A free zone enables Yugoslav transit traffic to pass through the port without paying customs duties. The Morava-Vardar corridor is linked with the Maritsa Basin by an important east-west route which leads through the high basin in which Sofia, the capital of Bulgaria, is located.

Along the rugged, island-fringed Dalmatian Coast of southwestern Yugoslavia, mountains rise steeply from the Adriatic Sea. The western part of Yugoslavia is almost continuously mountainous from the Italian and Austrian Alps in the extreme north to the mountains of Albania in the south. The principal ranges run parallel to the Adriatic Coast and impose a succession of rocky heights crossed by only a few significant passes. The general range of elevations is 4000 to 6000 feet, but scattered summits reach 8000 to 9000 feet, and the highest summit, located in extreme northwestern Yugoslavia, reaches 9395 feet. Some extremely picturesque

Barren karst highlands of western Yugoslavia tower above Cetinje in southwestern Montenegro. (Yugoslav Information Center.)

scenery is found in the Dinaric Ranges (Dinaric Alps), bordering the Dalmatian Coast. Much of the mountainous region of western Yugoslavia is characterized by karst or sinkhole topography caused by the solvent action of underground waters in limestone bedrock. The dry, inhospitable karst plateaus are among the most desolate and sparsely settled parts of Europe. Railroads crossing the mountains at the north, where they are narrowest, give access to Yugoslavia's seaports of Rijeka (city proper, 120,000)–Susak, and Split (city proper, 120,000), and to the Italian seaport of Trieste. Farther south, the Yugoslavs have developed a new seaport at Bar and have connected it by rail with Titograd (city proper, 30,000), the capital of Montenegro. The rail connection will eventually be extended to Belgrade.

Most of Albania is composed of rugged mountains, the main exception being a flat coastal lowland. Perhaps best of all the Balkan countries Albania exemplifies Gottmann's assertion that "The Balkan highlands have lived as a tissue of isolated cells, preserving archaic forms of life and resisting to the best of their abilities the multitude of influences and external interferences that have swirled across the peninsula throughout history.[2]

THE CLIMATIC PATTERN

The climates of East Central Europe are transitional between the marine climate of northwestern Europe, the extreme continental climates of Russia, and the dry-summer subtropical or mediterranean climate of Southern Europe. In most areas winters are colder than in the British Isles, France, or the Low Countries, although not so cold as in Russia. The average January temperature is 28°F at Warsaw, 31° at Prague, 31° at Budapest, 32° at Belgrade, 27° at Bucharest, and 29° at Sofia. Only in sheltered valleys and coastal lowlands of the extreme south do midwinter average temperatures rise significantly above freezing. An example is provided by the small seaport and resort city of Dubrovnik (city proper, 25,000), on the Dalmatian Coast of Yugoslavia, which has a January average of 48°. Summer temperatures are higher than in northwestern Europe, the average for July being 66° at Warsaw, 67° at Prague, 71° at Budapest, 73° at Belgrade, 73° at Bucharest, 71° at Sofia, and 77° at Dubrovnik.

In most lowlands of East Central Europe the average annual precipitation is between 20 and 25 inches.

The heaviest precipitation occurs in Yugoslavia along the Adriatic coast (map, p. 74). Dubrovnik averages 50 inches a year and Rijeka 73 inches; limited areas on windward slopes in the coastal mountains receive 180 inches or more–the greatest precipitation in Europe. Most areas in East Central Europe have a summer rainfall maximum except for the extreme south of Yugoslavia, the Dalmatian Coast, Albania, and extreme southern Bulgaria, which have the strong winter maximum that is characteristic of the Mediterranean region. Only a tenth of the annual rainfall at Dubrovnik comes in June, July, and August.

In the Danubian plains hot summers decrease the efficiency of the rainfall, and periods of drought make agriculture somewhat more hazardous than in the more dependable climate of northwestern Europe. But in most years moisture is sufficient for a good harvest. The humid continental long-summer climate of the Danubian plains is comparable in many respects to the climate of the corn and soybean region (Corn Belt) in the United States Middle West, although Middle-Western summers are somewhat hotter on the average and the annual precipitation in the corn belt is greater. The plains of Poland have a humid continental short-summer climate comparable to the climate of the American Great Lakes region.

RURAL LIFE IN EAST CENTRAL EUROPE

In most parts of East Central Europe the nature of the climate has a more decisive impact on large masses of people than is true in the highly industrialized lands of northwestern Europe. Though industry and urbanism have made impressive strides since World War II, great areas in East Central Europe still have a strongly rural and agricultural cast. It is very difficult to make comparisons among the different countries of the region with respect to their rural-urban ratio, due to poor data and differing official definitions of "rural" and "urban." However, it seems safe to say that Czechoslovakia is the most urbanized country, that Albania is the least urbanized, and that Hungary and Poland are more urbanized than Yugoslavia, Bulgaria, or Romania.

Increasing numbers of workers in rural areas of East Central Europe commute to jobs in urban places, but most rural people still support themselves wholly or

---

[2] Jean Gottmann, *A Geography of Europe* (4th ed., New York: Holt, Rinehart and Winston, 1969). Used by permission.

primarily by agriculture. On the whole, their efforts do not provide a very good livelihood, despite significant gains in recent years. Great numbers of rural homes now have electricity, and overall levels of rural income, diet, health, education, and general welfare are much improved over what they were before World War II. But even this improvement still leaves rural levels of living in the respective countries far below those of urban places, and it leaves rural dwellers in the region as a whole well below the general level of their counterparts in the countries bordering the North Sea. Within East Central Europe there are sharp differences in rural livelihood from one region to another. For example, farmers in the more industrialized western sections of Czechoslovakia or Poland are better off than those in the less industrialized east, and are far better off than farmers in remote mountain areas of the Balkan Peninsula.

Land Reform
before the Communist Era

The discrepancy in rural levels of living between East Central Europe and the North Sea lands has endured throughout modern times. For centuries prior to World War I, and in some areas up to World War II, the countryside of Poland, Slovakia, Hungary, and Romania was dominated by estates worked by tenants or hired labor. Strong rulers might give workers a certain amount of protection against the landowning class, but the history of estates often was one of exploitation or neglect of the laborers or tenants who eked out a living there and had little hope of anything better because law or custom bound them to the land. Bulgaria, Serbia, Montenegro, and Albania mainly were lands of small landowning peasant farmers, but the latter were severely handicapped by ignorance, brigandage, feuds, and the indifference, corruption, and greed of rulers, as well as by inadequate and fragmented landholdings, poor transportation, and, in mountainous areas, the poor and scanty soil. Some peasants managed to obtain a reasonable living, but the great majority were among the most poverty-stricken in Europe.

In most East Central European lands held by Germany and Austria, estates were prominent, but many farmers owned the land they worked and some of them were well off. This was particularly true in certain areas where manufacturing developed, attracting surplus population off the land, and urban markets were available for large amounts of farm produce. The height of rural prosperity was reached in Bohemia and Moravia, the industrial heart of the Austro-Hungarian Empire. Here a diversified crop and livestock agriculture provided rural levels of living generally comparable to those of northwestern Europe. But this area was very exceptional. In most parts of East Central Europe population accumulated on the land, the amount of land per person became smaller, strips of the fragmented small farms were subdivided and resubdivided among heirs, and the general situation of farmers deteriorated.

In the nineteenth century some of the pressure was relieved by the emigration of millions of East Central Europeans to overseas lands, principally the United States. Those who remained clamored for land, and the result was a series of land reform programs in various countries of the region. Such programs commenced in the nineteenth century, but did not gain much headway until after World War I. The main intent was to break up large holdings and sell the land to small farmers or landless laborers. In many instances this took the form of expropriating and breaking up properties belonging to the aristocracy of defunct empires. After World War I, Hungarian landlords in Slovakia were dispossessed, as were Russians in eastern Poland. In Hungary, however, the politically powerful Hungarian landowners generally managed to retain their properties, as did many large native landowners in Poland and Romania.

On the whole, redistribution of land was most successful, and benefits to farmers were greatest, in Czechoslovakia. This was especially true in the more progressive and prosperous western provinces of Bohemia and Moravia. Elsewhere in East Central Europe, the transfer of land to peasant ownership did not generally result in notable improvement of the standard of living, though it did help to keep down discontent among the peasants, for whom ownership of land had an intense emotional significance. Many landless farm laborers were too poor to buy land, even under generous governmental credit arrangements, and farmers who did acquire land generally lacked the necessary capital, machinery, and knowledge to make effective use of it. Machinery would have been at a serious disadvantage in any case, due to the fragmented arrangement of most farms and the small size of individual strips of land. Some attempts were made to induce peasants to consolidate their scattered strip holdings into more compact and efficient farm units, but such programs generally were unsuccessful.

Agricultural Development
under Communism

After World War II the remaining large private holdings were liquidated by the new Communist governments and programs of collectivized agriculture on the Soviet model were introduced. Some farmland was placed in large state-owned farms on which workers were paid wages, but most of the land was organized into collective farms owned and worked jointly by peasant families who shared the proceeds after the operating expenses of the collective had been met. Farm families normally were allowed to retain garden plots of one to three acres and limited numbers of livestock for their individual use, but most privately owned land and livestock became the property of the collectives. Collectivization met with strong resistance in the countryside, and in Yugoslavia and Poland it was discontinued in the 1950s and the collectives were allowed to disband. Today about nine-tenths of the farmland in each of these countries is owned and worked by private farmers, the remaining land being held by large state farms. The remaining East Central European countries, and East Germany as well, pushed ahead with collectivization, though they back-tracked or slowed the pace from time to time. Today most of their farmland is in collective or state ownership.

Since World War II the agriculture of East Central Europe has increased significantly in productivity over prewar levels. Rates of increase have been the most rapid since the early 1960s. The output per farm worker, per unit of land, and per head of population is not growing as rapidly as in the United States or the North Sea countries, but is well ahead of growth rates in most "developing" lands of Africa, Asia, or Latin America. After collectivization was introduced, agriculture was handicapped for a time by direct or indirect taxation which drained wealth from the farms in order to support industrial development. In addition, the efficiency of agriculture suffered because of excessively rigid control by planners and collective-farm managers who often were distinguished more by political orthodoxy and loyalty to the regime than by agricultural expertise. The lack of a personal stake in the land, as well as the low government-controlled prices for farm products, created apathy and indifference on the part of collectivized farmers. In all countries of the region many of the younger and more energetic workers forsook the farms for work in industry. In recent years, however, there has been a definite turn for the better in the conditions affecting farm output. The control and planning of collectivized agriculture has, in general, become less rigid and doctrinaire, the collective farms have been allowed greater latitude in making production and marketing decisions, and systems of pricing and taxation in all countries of Communist Europe (and in the Soviet Union as well) have been changed in order to give greater financial incentives for increased output. Agri-

cultural investment in the form of machinery, artificial fertilizer, irrigation, drainage, and so on has been increasing, and the general level of agricultural education and technique has been rising. Results in the form of increased output have not been spectacular, with certain exceptions such as Bulgaria's greatly increased output of irrigated vegetables, but there has been an unmistakable upward trend in all countries of the region. It should be noted also that the character of agriculture has changed somewhat. It has become more mechanized, diversified, and intensive than it was before World War II, with more attention given to protective foods (vegetables, fruits, meat, milk, eggs) for sale in urban markets; "industrial crops" such as sugar beets, sunflowers, and tobacco; and green forage and silage crops to feed dairy cattle. In the countries where agriculture has been collectivized, a sizable share of the vegetables, poultry, eggs, and milk that reach the market come from the privately owned garden plots and animals of collective-farm members.

The increased importance of the products named above should not be allowed to obscure the predominant role of grains in the crop farming of the region. Grains occupy about three-fourths of all land in crops in East Central Europe as a whole. The grain acreage decreased slightly in the 1960s, but the overall output showed a moderate rise. Romania and Bulgaria are the only countries that regularly have a surplus of grain for export, though Yugoslavia and Hungary have small surpluses in some years. Most grain exports are marketed in Communist Europe, particularly in East Germany, Czechoslovakia, or Poland. All countries in Communist Europe earn a certain amount of foreign exchange by agricultural exports to Western nations. Some grain enters this trade, but most countries depend primarily on exports of specialized items like Polish hams, Bulgarian tobaccos, or Yugoslavian and Hungarian wines.

The Pattern of Crops and Livestock

The principal crop-growing areas of East Central Europe are the plains of Poland, the Danubian plains of Hungary, Romania, Yugoslavia, and Bulgaria, the uplands and valleys of western Czechoslovakia, and the Maritsa Basin of Bulgaria.

On the sandy plains of central and northern Poland rye and potatoes are the most important crops, and Poland is the world's second producer of both (after the Soviet Union). More rye and potatoes are grown in Poland than in all the other countries of

East Central Europe combined. They are important in all parts of the country, though wheat and sugar beets are predominant in some loess-covered upland sections of the south.

In the Danubian plains corn and wheat are the main crops. The long, warm, rainy summers and the large expanses of fertile soil, often formed of loess or alluvium, make these plains one of the few areas in Europe suitable for growing corn on a large scale.

Bringing in the corn harvest to central bins of a collective farm near Bucharest, Romania. Horse-drawn wagons are still the principal mode of transportation on Romanian farms. (Eastfoto.)

Romania, Yugoslavia, Hungary, and Bulgaria produce about two-thirds of Europe's corn. Most of it is used within the producing countries for livestock feed, though in Romania corn meal is made into a porridge that is an important part of the diet of rural people. Wheat is an important crop in all of the East Central European countries, being more extensively grown through the area as a whole than any other grain. It is the preferred breadstuff in most parts of the region, though rye bread is more important in rural Poland. Wheat and rye are important sources of animal feed as well as human food. The bulk of East Central Europe's wheat production comes from the Danubian plains, the loess zone of southern Poland, and Bohemia and Moravia in Czechoslovakia. Barley also is widely grown, mainly for animal feed or for use in the beer industry. The total production for the region as a whole is only about one-third that of wheat.

Poland and Czechoslovakia are the largest producers of oats (grown to feed horses or other farm animals), potatoes, and sugar beets. These crops are of consequence, however, in all countries of the region except Albania. Potatoes are grown mainly for human food, although in Poland they are an important livestock feed and a source of alcohol, some of which is made into vodka. Throughout East Central Europe the production of sugar beets is tending to increase rather rapidly in areas where the soil and climate are favorable. Another "industrial crop," sunflowers, is an important source of vegetable oil in the Danubian plains and the Maritsa Basin. Limited amounts of irrigated rice are grown on river flood plains and drained marshes in Hungary, Romania, and the countries to the south. Deciduous fruit orchards are widespread in all countries of the region. Perhaps their most notable product is plums, used for making a popular brandy in Bulgaria, Yugoslavia, Romania, and other countries. Citrus orchards and olive groves along the sheltered coasts of Albania and extreme southern Yugoslavia reflect the subtropical temperatures of the mediterranean climatic zone. Vineyards are important in many areas from southern Czechoslovakia southward. In Bulgaria, Hungary, and Romania there has been a large increase in irrigated vegetable acreage.

In general, the greatest variety of crops tends to be found in the southern margins of the region. This is well illustrated in the southern half of Bulgaria. Here the growing of crops is centered in lowlands and valleys drained by the Maritsa River or its tributaries, but also extends into adjacent uplands and lower mountain slopes. All of the crops named above are grown to some extent (except citrus fruits), and in addition the area

Apartment buildings in the rebuilt central part of Warsaw. The heart of the city was almost totally demolished during World War II by shells, bombs, and demolition. The tall building at the right (with spire) is the Palace of Culture and Science. It houses government ministries dealing with these fields. (Eastfoto.)

produces cotton, strawberries, and various other crops. This part of Bulgaria has developed sizable exports of fresh vegetables (especially tomatoes) and fruits to Czechoslovakia, East Germany, West Germany, and other European countries. Valleys along the southern slopes of the Balkan Range are famous for fields of roses from which attar of roses is secured for export to the world's perfume industry. However, rose scents can be made synthethically and Bulgaria's rose industry has declined in significance, though it has shown some resurgence in recent years. The unusually diversified pattern of crops in southern Bulgaria is made possible by the area's long growing season, summer warmth, varied soils, wide range of altitudes, and relatively abundant moisture. The area lies near the Mediterranean and there is some tendency for summer droughts to occur, but these are being counteracted, and agriculture is being intensified, by a rapidly increasing development of irrigation. A somewhat comparable agricultural pattern is found in extreme southern Yugoslavia and in drained and irrigated marshlands along the coast of Albania.

In mountainous areas throughout East Central Europe, cropland tends to be scanty, generally infertile, and devoted to a limited range of hardy crops such as rye, oats, potatoes, or hay.

The raising of cattle, hogs, horses, poultry, and associated feed crops is a prominent aspect of agriculture in all countries of the region. Horses are gradually declining in numbers as mechanization takes hold, but they are still far more important as a source of draft power than in northwestern Europe. Sheep are the most numerous in Romania, Yugoslavia, and Bulgaria. Throughout East Central Europe stock raising is the principal form of agriculture in the highlands. Transhumance, the seasonal movement of farm animals between the valleys and mountain pastures, is still practiced in some areas, but is much less important than formerly.

URBAN DEVELOPMENT

Although cities in East Central Europe are growing very rapidly, only two areas—the Upper Silesian-Moravian coal field of Poland and Czechoslovakia, and the Bohemian Basin of Czechoslovakia—exhibit a closely knit web of numerous urban places. Elsewhere, cities of much size tend to be few in number and very widely spaced. Aside from the Upper Silesian-Moravian agglomeration of medium-sized mining and metallurgical centers

—a Ruhr in miniature—all of the largest cities are national capitals. In every instance the capital is its country's leading center of diversified industries, and except in Poland and Yugoslavia it far overshadows all other urban areas in size. The overwhelming predominance of one city in several different countries of the region shows clearly in the following summation (population figures are conservative 1968 estimates for the city proper; in a few cases metropolitan-area estimates also are given):

COUNTRY	LARGEST CITY	SECOND CITY
Poland	Warsaw (1.3 million; metropolitan area, 1.8 million)	Lodz (760,000; metropolitan area, 900,000)
Czechoslovakia	Prague (1.1 million)	Brno (340,000)
Hungary	Budapest (2.0 million; metropolitan area, 2.4 million)	Miskolc (180,000)
Romania	Bucharest (1.3 million; metropolitan area, 1.5 million)	Cluj (170,000)
Yugoslavia	Belgrade (710,000)	Zagreb (510,000)
Bulgaria	Sofia (800,000)	Plovdiv (230,000)
Albania	Tirana (160,000)	Shkodër (50,000)

INDUSTRIAL DEVELOPMENT AND RESOURCES

Prior to World War II, large sections of East Central Europe had been little affected by the industrial and urban modes of life that swept over northwestern Europe after the industrial revolution. In most parts of the Balkan Peninsula, the Danubian plains, the Carpathians, and Poland, peasant life had a deeply rural quality that was matched only in the most isolated outlying sections of the countries around the North Sea. In all countries, industries grew considerably in number and size through the period between the two world wars, but they were concentrated very heavily in a handful of large urban districts or national capitals: Prague and other cities of the Bohemian Basin, the cities located in or near the Upper Silesian-Moravian coal field, and the Budapest, Warsaw, and Bucharest metropolitan areas. Many other cities scattered through East Central Europe had a certain amount of industry, but it was generally of a very simple type. The region was seriously lacking in the sophisticated and complex forms of manufacturing that characterized dozens of large industrial centers in Britain, France, Germany, the Low Countries, Switzerland, northern Italy, and Scandinavia. Such industries as existed often were financed and controlled from western Europe, and industrial life had a markedly "colonial" quality. The above description was least true of western Czechoslovakia. As the industrial heart of the Austro-Hungarian empire, the Czech lands in Bohemia and Moravia had developed diversified, advanced, and efficient industries that were able to adapt to new conditions and continue to operate successfully when the empire disintegrated at the end of World War I. But these lands were only an exception to the underdevelopment that characterized East Central Europe as a whole.

Industrial Trends in the Communist Era

The accession of communism in East Central Europe inaugurated a new industrial and urban era. With minor exceptions the existing industries were taken out of private hands, and national economic plans modeled on the Five-Year Plans of the Soviet Union (see pp. 283–291) were developed in the various countries. Under the Communists the older industrial districts have continued to grow in size and complexity, but Communist planners in all countries have placed much emphasis on the development of new industries in weakly industrialized areas. By this means they have sought (1) to raise incomes and standards of living in less-developed areas so as to make all regions more nearly uniform in socioeconomic levels, (2) to increase the contribution of less-developed areas to national economies, (3) to induce workers to remain in their home areas instead of migrating to the older, more congested industrial and urban districts, and (4) to keep down political discontent in less-developed areas by providing better economic and cultural opportunities. There is evidence of partial success in achieving these objectives, though it should be noted that the older industrial districts continue to be the predominant centers, and in most instances have suffered only a moderate decline, or have actually increased, in relative importance. Many of the newer industries in outlying areas have been criticized as uneconomic, but there have been undeniable social and political gains from their establishment. These have not, however, been sufficient to alter the imbalance between major regions in socioeconomic level. For instance, in Yugoslavia wide discrepancies continue to exist between the more developed and prosperous northern part of the country—with its larger cities and industries, gentler terrain, better soils, and closer historic relationships with the western world—and the mountainous, less developed south, with its history of stagnation or regression during the long period of Turkish control.[3]

Initially the central planning agencies of the Communist governments maintained a rigid control over individual industries. Plants were directed to produce certain goods in quantities determined by the planners, and the success of a plant was judged by its ability to meet production targets rather than its ability to sell its products at a profit in competition with other plants. Political reliability and conformity to the central plan were qualities much desired in plant managers. Recently, however, the system has been changing, though with more speed in some countries than in others. There is now a strong tendency to allow the managers and labor force of individual plants greater latitude in making their own decisions with regard to production and marketing, though such decisions must still be made within the general frame of national, regional, and local economic plans. Plants are judged not on their ability simply to produce predetermined quantities of specified goods, but on their ability actually to market their products on a profitable basis. Workers in plants that make a profit receive better wages than those in plants that do not, and plants that operate at a loss are apt to be merged with other enterprises or allowed to go bankrupt. This much publicized development, which parallels recent practice in the Soviet Union, began earliest and has gone farthest in Yugoslavia. It is a particularly desirable change in the present period when more stress is being placed on improving the output and quality of consumer goods.

In the early stages of industrialization, Communist planners did not aim at a balanced development of all types of industry. Instead, they stressed the types that were deemed most essential to industrial development as a whole. New investment was channeled heavily into a relatively few fields: mining, iron and steel, machinery, chemicals, construction materials, and electric power. Development of a strong base of heavy industries and sources of power was viewed as an essential prerequisite for overall industrial development, and investment in such industries was favored at the expense of consumer-type industries and agriculture. Today heavy industries are still pushed, but more attention is gradually being paid to diversified industries making consumer goods. In this respect, as in many others, East Central Europe is following the same general path of development as the Soviet Union. It should be noted that there has recently been some incursion of business enterprise from west of the iron curtain into both East Central Europe and the Soviet Union. Yugoslavia and Romania, for example, have certain cooperative business ventures involving their own nationals and businessmen from western Europe. An Italian company (Fiat) has been invited to produce automobiles in the USSR (commencing in the early 1970s) under a special arrangement with the Soviet government.

As a means of facilitating industrial development, great stress has been placed on the improvement of transportation. This effort has been devoted largely to the railroads, which carry most long-distance freight in East Central Europe. Water transportation is poorly developed as compared with northwestern Europe, though some efforts are being made to improve it. As in the Soviet Union, long-distance truck transportation is not well developed. Trucks are used a great deal, however, for short hauls within cities or to connect cities and towns with their immediate hinterlands. Long stretches of good highway are much less numerous than in northwestern Europe, but they do exist here and there, a prime example being the throughway in northern Yugoslavia which connects Belgrade with Zagreb, the capital of Croatia, and Ljubljana (190,000), the capital of Slovenia.

The planned expansion of mining and industry in Communist Europe has produced notable increases in the total output of minerals, manufactured goods, and power. The largest absolute increases have occurred in the countries that already were the most industrialized:

---

[3] This paragraph draws heavily on an excellent article by George W. Hoffman, "The Problem of the Underdeveloped Regions in Southeast Europe: A Comparative Analysis of Romania, Yugoslavia and Greece," *Annals of the Association of American Geographers*, 57, no. 4 (December 1967), 637–666. See also M. I. Logan, "Regional Economic Development in Yugoslavia, 1953–1964," *Tijdschrift voor Economische en Sociale Geografie*, 59, no. 1 (January–February 1968), 42–59.

Czechoslovakia, Poland, and, of course, East Germany. But the most spectacular percentage increases have occurred in the other countries, which began their postwar industrialization from a much smaller base.

Czechoslovakia and Poland are by far the largest producers of manufactured goods, minerals, and electric power among the countries of East Central Europe, although they share this lead with East Germany within Communist Europe as a whole. These countries lead in the production of iron, steel, machine tools, machinery, textiles, chemicals, and many other lines of manufactured goods. Hungary, Yugoslavia, Romania, and Bulgaria have all shown large percentage increases in their output of metals, fuels, machinery, chemicals, and electric power since World War II. Textiles, food processing, and other light consumer industries continue to play an important role, however, as they do in the more industrialized countries named above. In Albania, consumer industries account for most of the relatively meager industrial production.

The Expanding Iron and Steel Industry

One of the outstanding features of the new industrial age in East Central Europe has been the construction of a number of large new iron and steel plants. Most of these are located in or near the principal industrial districts. They are *integrated* plants, possessing coke ovens, blast furnaces, steel furnaces, and rolling mills which shape hot steel into bars, sheets, rods, or pipes.

1. POLAND. The largest iron and steel plant in East Central Europe is the Lenin Works at Nowa Huta ("New Furnace"), a suburb of Krakow in southwestern Poland. The plant operates with Polish coal and coke, iron ore drawn mainly from the Krivoy Rog deposits of the Soviet Union, and limestone from local sources. A second new plant in Poland is located at Czestochowa (city proper, 180,000), about 50 miles northwest of Krakow. Both of these plants lie relatively near Poland's main cluster of integrated iron and steel plants in the Upper Silesian industrial complex (see discussion below). Limestone escarpments in the Czestochowa-Krakow region contain deposits of low-grade iron ore that have been worked for many years, but supplies of ore from these and scattered other low-grade deposits in Poland are completely inadequate for the country's needs. Most of the imported ore is furnished by the Soviet Union, which also supplies manganese, a necessary alloy mineral in the manufacture of steel from pig iron.

2. CZECHOSLOVAKIA. In Czechoslovakia a new integrated plant has been built in the Ostrava area, near two other integrated plants in Czechoslovakia's segment of the Upper Silesian-Moravian coal field. These plants use coke made from local coal and some ore from deposits in Bohemia and eastern Slovakia, but the greater part of the ore has to be imported, mainly from the USSR. The Ostrava area is the country's leading center of iron and steel production. A much publicized new integrated plant in Czechoslovakia, the "East Slovakian" plant near Kosice (city proper, 100,000), has been under construction intermittently for a long period. Intended to be a key factor in the industrialization of Slovakia, this plant, located a long way from its raw materials, has often been branded as uneconomic. It lies between Ostrava coking coal and Krivoy Rog iron ore, and will operate with these materials.

3. ROMANIA. Romania's largest plant will be an integrated works at the seaport of Galati on the Danube. It was still under construction at the time of writing (1968). Two older integrated plants at Resita (city proper, 50,000) and Hunedoara (city proper, 50,000) in the western part of the country have been modernized and expanded. They are located near coal and ore deposits. Romania imports additional supplies of coking coal and iron ore from the USSR and other sources.

4. HUNGARY. The largest iron and steel mill in Hungary is a new integrated plant, the Danube Iron Works, located at Dunaujvaros ("Danube New Town"; 40,000) on the west bank of the river about 40 miles downstream from Budapest. Factories in the metropolis provide the largest market for iron and steel from the plant. A site this far from the capital was chosen in order to avoid further congestion in the Budapest metropolitan area. Hungary has two other integrated plants, located at Ozd (40,000) and Miskolc (170,000) in the northeast. The country has some coking coal in the southwest, and scattered deposits of iron ore, but it normally imports around half of its coking coal, mainly from the USSR and Czechoslovakia, and four-fifths of its iron ore. The Krivoy Rog field in the Soviet Union is the largest supplier of ore.

5. YUGOSLAVIA. In Yugoslavia, four widely separated, integrated plants exist: in Bosnia, Macedonia, Croatia, and Slovenia, respectively. They utilize iron ore from Yugoslavia itself but import most of their coking coal, primarily from the United States or the Soviet Union.

6. BULGARIA AND ALBANIA. Bulgaria's main iron and steelworks is a new integrated plant about 15 miles northeast of Sofia. It uses local ore, and coke that comes partly from Bulgarian sources and partly from

the USSR. No iron and steel plant yet exists in Albania (late 1960s). The country has some high-grade iron ore, though no coking coal.

Mineral Resources

On the whole, mineral resources to support industrial expansion in East Central Europe are distinguished more by diversity than by massive size. Every country has a miscellany of useful minerals, but very few of the deposits are outstandingly large. In most highlands of the region, metal-bearing ores have been exploited at one time or another. Many deposits have been exhausted, and others have proved uneconomical to work in an age of large-scale enterprise and strenuous international competition among mineral-producing areas. Since the beginning of the Communist era, however, production has been expanded in some preexisting mining areas, and an intensive search for new sources of minerals has been carried on, often with considerable success. Clearly the outlook for minerals in the region is improving. The Carpathian Mountains of Romania, for example, are a storehouse of nonferrous ores. Mineral deficiencies of individual countries generally are filled, at least in large part, by imports from other East Central European countries or the Soviet Union. Soviet iron ore and petroleum, for example, are crucially important to the present economy of the region. The major mineral resources of East Central Europe itself may be summarized as follows:

1. COAL. The Upper Silesian-Moravian field in Poland and Czechoslovakia is in the same broad class with West Germany's Ruhr in total reserves. However, a much smaller proportion of its coal can be used for coking, and much of the coke is lower in quality than that made from Ruhr coal. Nevertheless, substantial reserves of coking coal exist in this field (the best coal is in the Czechoslovak part) and in the much smaller Walbrzych (Waldenburg) field in the Sudeten Mountains of Lower Silesia. Gas, flame, and steam coals are abundant in the Upper Silesian-Moravian field, and the output from this field, along with some production from other areas, enables Poland and Czechoslovakia to rank sixth and twelfth, respectively, among the world's countries in bituminous coal production. All the remaining countries in East Central Europe have some bituminous coal, but the fields are small and supply only a fraction of each country's needs. Deficits are made up by imports from Poland, the Soviet Union, Czechoslovakia, or the outside world. Deposits of lignite are numerous and widespread in East Central Europe, and production is increasing rapidly in all countries. By far the largest producing area is the foothills of the Ore Mountains in western Czechoslovakia. Most of East Central Europe's production of these low-grade coals comes from open-pit mines and is burned in thermal plants to generate electricity. Lignite is critically important for this purpose in Czechoslovakia, Hungary, and Bulgaria, though in Poland bituminous coal is used more in thermal plants, and in Romania natural gas is the main fuel used for this purpose. Yugoslavia and Albania have some thermal plants, based primarily on lignite, but these countries stand apart from others in East Central Europe in their great reliance on hydroelectric power. Numerous mountain streams in both countries have been harnessed, and falling water now provides over two-thirds of Albania's electricity supply and over three-fifths of Yugoslavia's. Yugoslavia's total hydroelectric potential is second only to Norway's in Europe. Hydropower stations, the great majority of which have been constructed since World War II, are well distributed over the country. In the other East Central European countries, hydroelectric development is still in an early stage, but some stations exist in mountainous areas of every country and there are ambitious plans and projects for increasing the hydroelectric supply. A number of larger schemes envision greater use of the region's major rivers. A plan to create a series of dams and reservoirs on the Vistula (the "Vistula Cascades") is still in the discussion stage; but Yugoslavia and Romania are at work on a large dam and power station on the Danube in the Iron Gate gorge (see p. 263).

2. PETROLEUM AND NATURAL GAS. All countries in East Central Europe have some production of petroleum and natural gas, but one country, Romania, far overshadows all the others combined, being in fact the largest producer of both oil and natural gas in Europe (1968). Romania's oil production, amounting to a little under 1 percent of the world total, comes from a series of fields strung in a long arc through the southern and eastern foothills of the Carpathians in Walachia and Moldavia. Foreign companies, mainly British or American, developed the original fields around Ploesti (city proper, 170,000), commencing in the late nineteenth century. Their properties, severely damaged by Allied bombing in World War II (Romanian oil was highly important to the Nazi war machine), were nationalized by the new Communist government after the war. Since then, production in the older workings around Ploesti has been declining, but new areas have developed to the west of Ploesti, and a small field in Moldavia has been greatly expanded. The newer fields now account for most of Romania's production. The

Ploesti area, however, remains the country's principal refining center. Some of Romania's petroleum production is exported; much of the remainder is utilized in the country's growing petrochemical industries. Natural gas production comes partly from the petroleum fields (where pockets of gas often overlie petroleum), but the main producing area is a large gas field in the Transylvanian Basin. Most of the gas output is used industrially in thermal power stations or in chemical and metallurgical industries. Pipelines connect the gas fields with Romania's principal cities and also carry some gas westward to a large new petrochemical combine on the Tisza River in Hungary. All of the other countries in East Central Europe have small oil and gas fields, but only Albania has a surplus (petroleum from coastal fields) for export. The region as a whole has a sizable deficit in petroleum, which is made good primarily by imports from the huge Volga-Urals fields (p. 320) or the smaller Ukrainian fields of the Soviet Union. The "Friendship" Pipeline from the Volga-Urals region to East Central Europe divides into two sections, one serving Poland and East Germany, and the other serving Hungary and Czechoslovakia. In Poland a large refining and petrochemical center at Plock on the Vistula River utilizes crude oil from the Soviet pipeline.

3. BAUXITE. Reserves of bauxite in Hungary and Yugoslavia are very large. Both countries produce it on an export basis and also manufacture small amounts of aluminum. Hungary's bauxite exports move primarily to other countries in Communist Europe, while Yugoslavia's go mainly to western Europe.

4. LEAD AND ZINC. About a tenth of the world's lead production and a slightly smaller proportion of its zinc comes from East Central Europe. Poland is the main zinc producer and Yugoslavia, followed closely by Bulgaria, is the largest lead producer. Poland's zinc production, along with some lead, comes primarily from ores in limestone rocks that overlie coal beds in the northern part of the Upper Silesian coal field. Thus the ores are conveniently located for smelting, which is done in the vicinity, using coal as a fuel. Zinc and lead mining in Upper Silesia are extremely ancient, antedating the coal-mining industry by many centuries. Today the deposits are less productive than formerly, and Poland imports considerable quantities of zinc concentrates for processing.

5. OTHER MINERALS. Scattered through the mountains of Yugoslavia, Bulgaria, and Albania are a variety of metal-bearing ores. The main products not already mentioned include *copper* in all three countries (Yugoslavia is Europe's leading producer), *chromium* in Albania and Yugoslavia (the two countries rank first and second in Europe), and *nickel* in Albania (Europe's leading producer). Hungary and Romania produce *manganese* for their own needs and for export. Poland has commenced to exploit sizable reserves of copper ore and native *sulfur,* both essential minerals for a modern industrial nation. *Uranium* is produced in Czechoslovakia, Hungary, and Bulgaria. One of the most critical mineral deficiencies within the region as a whole is a shortage of *iron ore.* All of the countries have some deposits, but they are generally small and/or low-grade. Yugoslavia mines ore for its own needs from deposits in Bosnia and Macedonia, and Albania produces some high-grade ore for export. All of the Soviet "client states" in Communist Europe are dependent in varying degrees on ore imports from the USSR.

MAJOR INDUSTRIAL REGIONS

Before the Communist period in East Central Europe, most of the area's coal mining and associated iron and steel production was concentrated in the Upper Silesian-Moravian coal field or in the Bohemian Basin. Today these areas still comprise the major concentrations of mining and heavy industry. In addition, the Bohemian Basin is East Central Europe's leading area of diversified industries, particularly those based on highly skilled labor and long industrial experience.

Coal and Steel in Upper Silesia-Moravia[4]

The Upper Silesian-Moravian coal field, situated mainly in Poland, but extending for a short distance into northern Czechoslovakia, is one of the two largest coal fields of continental Europe (excluding the Soviet Union). Sometimes it is referred to as the "Second Ruhr" or "Eastern Ruhr," and its total reserve of coal is indeed of the same order of magnitude as that of the German field (an estimated 80 billion metric tons, as

---

[4] This section draws extensively on two articles by Norman J. G. Pounds, "The Spread of Mining in the Coal Basin of Upper Silesia and Northern Moravia, *Annals of the Association of American Geographers,* 48, no. 2 (June 1958), 149–163, and "The Industrial Geography of Modern Poland," *Economic Geography,* 36, no. 3 (July 1960), 231–253.

compared to 120 billion in the Ruhr). But in most respects it suffers by comparison with the Ruhr. It lacks the Ruhr's anthracite coal and, as previously noted, its reserves of coking coal are much smaller and generally of poorer quality than those of the Ruhr. Only in the relatively small Czech (Moravian) part of the field are coking coals found that compare favorably in quality with those of the Ruhr, though coke of lower quality is made in sizable quantities from deposits along the western margins of the Polish part of the field. This field also lacks the Ruhr's convenient access to water transportation. A canal links the western end of the Polish part to the Oder River, which is navigable for large barges to the Baltic Sea. But no network of connecting waterways extends through the mining and industrial area as in the Ruhr. Thus in Upper Silesia it is not possible to load coal onto barges directly from the mines, or to deliver iron ore by water directly to smelters.

In two respects at least, the Upper Silesian-Moravian field does have an advantage over the Ruhr. Its coal seams are thicker on the average than those of the Ruhr, and those currently mined lie nearer the surface. In addition, the strata have been less disturbed by warping and faulting. Thus mining is somewhat easier and cheaper than in the Ruhr.

During the nineteenth and early twentieth centuries this coal field was divided politically among the Prussian (later German), Austrian, and Russian empires. By far the most extensive development of coal mining and iron and steel manufacture, however, took place in the western segment controlled by Prussia. Political arrangements following World War I gave most of the German-held part of the field to Poland or Czechoslovakia, and after World War II Germany's remaining share went to Poland. Since then, mining, iron and steel production, chemical production, and electric power production, along with some machinery and textile manufacture, have greatly expanded, though the total development is much smaller than that of the Ruhr. In the Polish part of the field, a large urban and industrial concentration is formed by a cluster of small to medium-sized towns and cities, of which the largest is Katowice (city proper, 290,000). The entire agglomeration comprises about 2.2 million people and is Poland's most important industrial area. Ostrava (city proper, 260,000) is the largest urban and industrial center in the part of the field held by Czechoslovakia.

A notable share of present-day Poland's industrial development is concentrated in or relatively near the

Upper Silesian coal field. The industrial and university city of Krakow (city proper, 530,000), with its large steel mill in the suburb of Nowa Huta, lies on the Vistula River about 40 miles east of the main coal mining area, while another industrial center, Wroclaw (metropolitan area, over 500,000), formerly Breslau, is on the Oder about 90 miles to the northwest, and Lodz, Poland's main textile center, is about the same distance to the north. Lodz, sometimes called the "Polish Manchester," and its environs account for approximately half of Poland's total employment in textile industries. Cotton spinning and weaving industries lead, but woolens, linens, and synthetic fabrics also are manufactured. German capital and technicians played a leading role in establishing the industry at Lodz, as they did in the case of many other industries in Poland and Czechoslovakia. The city is still primarily a textile center, but it also has diversified industries of other types.

Industrial Development in Bohemia

Czechoslovakia is the most industrialized and generally advanced of the East Central European countries. Its factories manufacture an extremely wide range of goods. The Czech part of the Upper Silesian-Moravian coal field is the country's most important focus of coal mining and iron and steel production, but the main area of diversified industries, including many heavy industries, is found in the Bohemian Basin. Here Pilsen (city proper, 150,000) is a famous center of steel, armaments, machinery, and beer production, but the capital, Prague, has a much larger overall industrial development.

Prague (Praha) has a population of over one million. It is a great historical center of administration and learning, with many monuments and beautiful residential sections. . . . Until 1850 the city had less than 100,000 inhabitants, mostly foreigners, particularly Germans. The railroads and modern industry, using the coal of the Kladno field, made it a large city–300,000 people in 1878, 400,000 in 1910, 921,000 in 1947, and 1,025,000 in 1966. The industrial function, spread among the suburbs, is varied: iron and steel finishing industries and engineering make the core of it; anything that can be made out of steel is produced here, railroad rolling stock, automobiles, electrical machinery, and machine tools playing a great part. Textile and leather industries are well represented as are chemical manufacturing and, of course, glassware.[5]

[5] Gottmann, *op. cit.* Used by permission.

A host of smaller industrial cities and towns in the Bohemian Basin carry on diversified and often highly specialized types of manufacturing. Many of these places are near Prague and Pilsen, while others are found in the foothills of the Ore Mountains and Sudeten Mountains.

ORIENTATION OF TRADE TOWARD COMMUNIST COUNTRIES

In the period between World Wars I and II, the countries of East Central Europe traded primarily with Germany and other North Sea lands. German coal and manufactures, for example, were sent to the countries of the region in exchange for surplus agricultural products and industrial raw materials. Since 1945, however, the trade of these countries has been drastically reoriented. Except for Yugoslavia, all countries in Communist Europe (including East Germany) now trade primarily within the Communist world. Of the total trade turnover (imports plus exports) of Romania, Poland, and Hungary in 1965, 60 to 70 percent was accounted for by trade with other Communist nations, in East Germany, Czechoslovakia, and Bulgaria the figure was 70 to 80 percent; and in Albania it was over 90 percent. For all of these countries except Albania, the Soviet Union is by far the largest trading partner. It exports fuels, ores and metals, some machinery, wheat, raw cotton, and timber products to them, and receives in return a wide variety of manufactures plus a few raw materials. For Albania, Communist China is overwhelmingly the main trading partner. Yugoslavia trades widely within the Communist world, but well over half of its total trade is with non-Communist nations.

REFERENCES and READINGS

BLUNDEN, GODFREY, and the Editors of *Life, Eastern Europe: Czechoslovakia, Hungary, Poland* (Life World Library; New York: Time Inc., 1965).

BURCK, GILBERT, "East Europe's Struggle for Economic Freedom," *Fortune,* 75, no. 5 (May 1967), 124–127 ff.

BYRNES, ROBERT F., general ed., *East-Central Europe under the Communists.* Separate volumes include Robert F. Byrnes, ed., *Yugoslavia* (1957); Vratislav Bušek and Nicholas Spulber, eds., *Czechoslovakia* (1957); L. A. D. Dellin, ed., *Bulgaria* (1957); Stephen Fischer-Galati, ed., *Romania* (1957); Oscar Halecki, ed., *Poland* (1957); E. C. Helmreich, ed., *Hungary* (1957); and Stavro Skendi, ed., *Albania* (1956). (New York: Frederick A. Praeger.)

CAESAR, A. A. L., "Yugoslavia: Geography and Post-War Planning," Institute of British Geographers, Publication No. 30, *Transactions and Papers, 1962,* pp. 33–43.

COMPTON, P. A., "The New Socialist Town of Dunaujvaros," *Geography,* 50, pt. 3 (July 1965), 288–291.

COUSENS, S. H., "Changes in Bulgarian Agriculture," *Geography,* 52, pt. 1 (January 1967), 12–22.

CREIGH, J. C., "Montenegro's Changing Economy," *Scottish Geographical Magazine,* 83, no. 2 (September 1967), 131–138.

DARBY, H. C., and Others, *A Short History of Yugoslavia from Early Times to 1966,* ed. by Stephen Clissold (Cambridge: At the University Press, 1966). Many useful maps.

den HOLLANDER, A. J. N., "The Great Hungarian Plain: A European Frontier Area," I and II, *Comparative Studies in Society and History,* 3, no. 1 (October 1960), 74–88; no. 2 (January 1961), 155–169.

"East Central Europe: Continuity and Change," *International Affairs,* 20, no. 1 (Spring 1966). An entire issue with articles by various authors.

ENYEDI, GYÖRGY, "The Changing Face of Agriculture in Eastern Europe," *Geographical Review,* 57, no. 3 (July 1967), 358–372.

FISHER, JACK C., ed., *City and Regional Planning in Poland* (Ithaca, N.Y.: Cornell University Press, 1966; individual chapters by Polish geographers and planners; much historical detail on urbanism and city planning; many interesting maps and photos); also, *Yugoslavia—A Multinational State: Regional Difference and Administrative Response* (San Francisco: Chandler

Publishing Co., 1966; fine maps and photos); and "Urban Analysis: A Case Study of Zagreb, Yugoslavia," *Annals of the Association of American Geographers,* 53, no. 3 (September 1963), 266–284.

Geographia Polonica. From Vol. 1 (1964) to date. Innumerable articles on Poland by Polish geographers, published in English. Great numbers of maps. A source of the highest value for geographical detail on Poland.

HAMILTON, F. E. I., *Yugoslavia: Patterns of Economic Activity* (New York: Frederick A. Praeger, 1968); "The Changing Pattern of Yugoslavia's Manufacturing Industry 1938–1961," *Tijdschrift voor Economische en Sociale Geografie,* 54, no. 4 (April 1963), 96–106; "Location Factors in the Yugoslav Iron and Steel Industry," *Economic Geography,* 40, no. 1 (January 1964), 46–64; "Yugoslavia's Hydro-Electric Power Industry," *Geography,* 48, pt. 1 (January 1963), 70–73; and "Bulgaria: Land of Change," *Geographical Magazine,* 38, no. 4 (August 1965), 276–289.

HELIN, RONALD A., "The Volatile Administrative Map of Rumania," *Annals of the Association of American Geographers,* 57, no. 3 (September 1967), 481–502.

HOFFMAN, GEORGE W., "The Problem of the Underdeveloped Regions in Southeast Europe: A Comparative Analysis of Romania, Yugoslavia and Greece," *Annals of the Association of American Geographers,* 57, no. 4 (December 1967), 637–666; *The Balkans in Transition* (Princeton, N.J.: D. Van Nostrand Co., Searchlight Books, 1963; covers Yugoslavia, Bulgaria, Albania); "Thessaloniki: The Impact of a Changing Hinterland," *East European Quarterly,* 2, no. 1 (March 1968), 1–27; "Transformation of Rural Settlement in Bulgaria," *Geographical Review,* 54, no. 1 (January 1964), 45–64; "Eastern Europe," in George W. Hoffman, ed., *A Geography of Europe* (3d ed.; New York: Ronald Press Company, 1969); and, with Fred Warner Neal, *Yugoslavia and the New Communism* (New York: Twentieth Century Fund, 1962; Part I, "The Land of the South Slavs," pp. 9–65, is an excellent geographical and historical summary).

JOHNSTON, W. B., and I. CRKVENCIC, "Examples of Changing Peasant Agriculture in Croatia, Yugoslavia," *Economic Geography,* 33, no. 1 (January 1957), 50–71.

Journal of the Czechoslovak Geographical Society. Supplement for the International Geographical Congress, 20th, London, 1964. Numerous articles on the geography of Czechoslovakia, written in English.

KARCZ, JERRY F., ed., *Soviet and East European Agriculture* (Berkeley and Los Angeles: University of California Press, 1967).

LAEMMERZAHL, ARTHUR J., and ROBERT FRANCIS, *Basic Data on the Economy of Yugoslavia,* U.S. Bureau of International Commerce, Overseas Business Reports, OBR 67–80 (December 1967). (Washington, D.C.: Government Printing Office, 1967.)

LEDERER, IVO J., *Yugoslavia at the Paris Peace Conference: A Study in Frontiermaking* (New Haven, Conn.: Yale University Press, 1963).

LEWINS, LEON, and JERSCHKOWSKY, OLEG, *Basic Data on the Economy of Poland.* U.S. Bureau of International Commerce, Overseas Business Reports, OBR 65–65 (September 1965). (Washington, D.C.: Government Printing Office, 1965.)

LOGAN, M. I., "Regional Economic Development in Yugoslavia, 1953–1964," *Tijdschrift voor Economische en Sociale Geografie,* 59, no. 1 (January–February 1968), 42–52.

LONDON, KURT, ed., *Eastern Europe in Transition* (Baltimore, Md.: Johns Hopkins Press, 1966).

MATLEY, IAN M., "Transhumance in Bosnia and Herzegovina," *Geographical Review,* 58, no. 2 (April 1968), 231–261.

MAY, JACQUES M., *The Ecology of Malnutrition in Five Countries of Eastern and Central Europe* [East Germany, Poland, Yugoslavia, Albania, Greece] (1963), and *The Ecology of Malnutrition in Central and Southeastern Europe: Austria, Hungary, Rumania, Bulgaria, Czechoslovakia* (1966). (New York: Hafner Publishing Co.)

MONTIAS, JOHN MICHAEL, *Economic Development in Communist Rumania* (Cambridge, Mass.: M.I.T. Press, 1967).

OSBORNE, R. H., *East-Central Europe: An Introductory Geography* (New York: Frederick A. Praeger, 1967). The most useful geographical reference in English on the region as a whole. Admirable historical summaries, as well as remarkably full treatment of geographical detail and very full bibliography.

PÉCSI, MÁRTON, and BÉLA SÁRFALVI, *The Geography of Hungary* (London: Collet's, 1964).

PERHALLA, FERDINAND F., *Basic Data on the Economy of Czechoslovakia,* U.S. Bureau of International Commerce, Overseas Business Reports, OBR 66–30 (May 1966); and *Basic Data on the Economy of Bulgaria,* OBR 65–40 (June 1965). (Washington, D.C.: Government Printing Office.)

POUNDS, NORMAN J. G., *Geographical Essays on Eastern Europe* (Indiana University Russian and East European Series, No. 24; Bloomington: Indiana University Press, 1961); *Poland between East and West* (Princeton, N.J.: D. Van Nostrand Co., Searchlight Books, 1964); *The Upper Silesian Industrial Region* (Bloomington: Indiana University Publications, Graduate School, Slavic and East European Series, Vol. 11, 1958); "The Spread of Mining in the Coal Basin of Upper Silesia and Northern Moravia," *Annals of the Association of American Geographers,* 48, no. 2 (June 1958), 149–163; "Fabryka Im. Juliana Marchlewskiego: A Textile Plant in Lodz, Poland," Chap. 24 in Richard S. Thoman and Donald J. Patton, eds., *Focus on Geographic Activity: A Collection of Original Studies* (New York: McGraw-Hill Book Company, 1964), pp. 154–158; and, with Nicholas Spulber, *Resources and Planning in Eastern Europe* (Bloomington: Indiana University Publications, Graduate School, Slavic and East European Series, Vol. 4, 1957).

SINGLETON, F. B., *Background to Eastern Europe* (New York: Pergamon Press, 1965); and "Macedonia on the Move," *Geographical Magazine,* 40, no. 7 (November 1967), 537–547.

WANKLYN, HARRIET G., *Czechoslovakia: A Geographical and Historical Study* (New York: Frederick A. Praeger, 1954).

WILKINSON, H. R., "Perspective on Some Fundamental Regional Divisions in Yugoslav Illyria," in Robert W. Steel and Richard Lawton, eds., *Liverpool Essays in Geography: A Jubilee Collection* (London: Longmans, Green and Co., 1967), pp. 565–587.

WILLIAMSON, DAVID, "The New Warsaw," *Geographical Magazine,* 38, no. 8 (December 1965), 596–607.

See also the lists of readings and references for Chapters 1–4, particularly the relevant sections in standard geographies of Europe listed on page 92.

PART III

THE
SOVIET   UNION

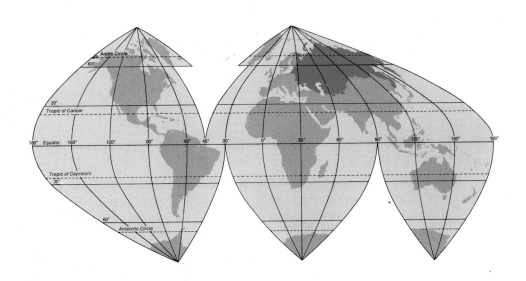

Introduction

to the

Soviet Union

13

The Union of Soviet Socialist Republics[1] emerged from World War II as a major center of world power. Such a development would have been difficult to foresee before the war. In the prewar period there was a pronounced tendency in the outside world to undervalue the political, economic, and military strength of the USSR. Russia was defeated by the Central Powers in World War I, after suffering possibly a greater toll of military casualties than any other participant. Between 1917 and 1921 the country experienced a destructive period of revolution, foreign intervention, and

---

[1] Union of Soviet Socialist Republics, frequently shortened to USSR or Soviet Union, is the official name of the Soviet state. The USSR is the successor to the Russian Empire of tsarist (czarist) times. It came into existence following the overthrow of the last of the Romanov tsars in 1917. Pre-Revolutionary Russia is often spoken of as Old Russia, Tsarist Russia, or Imperial Russia. Post-Revolutionary Russia is often referred to as Soviet Russia. The name Russia is used loosely to refer to the country either before or after the Revolution, though technically this name now applies only to the Russian Soviet Federated Socialist Republic (RSFSR), the largest of fifteen Union Republics that comprise the USSR. The part of the country west of the Ural Mountains and north of the Caucasus Mountains is often referred to as European Russia. The part east of the Urals and north of Soviet Middle Asia is called Siberia.

civil war. The October Revolution of 1917 and the success of the Red Army in the subsequent civil war gave control of the government to the radical Bolshevik faction of the Communist party. During the 1920s and 1930s a series of major internal crises occurred as the Communist leaders attempted to fasten a program of Marxian socialism on an exhausted and partly unwilling country, and in the meantime carried on a struggle for power among themselves.

EMERGENCE OF THE USSR AS A MAJOR WORLD POWER

When Germany attacked the Soviet Union in June 1941, it was freely predicted in outside countries that the USSR would prove too weak and disunited to withstand the German assault for more than a few weeks or months. The Russians, however, showed an unexpected ability to resist. After suffering tremendous casualties and losses of territory in the early part of the war, their forces eventually rallied. Late in 1942 the German invasion was checked in a decisive battle at Stalingrad (now Volgograd) on the Volga River. Thereafter the Germans were gradually pushed back, and by the end of the war in 1945 Soviet armies had taken Berlin and effected a junction with American forces at the Elbe River. As a consequence of victory, the Soviets annexed the northern part of East Prussia and took over the zone of occupation that subsequently became the German Democratic Republic (East Germany). They still have troops in East Germany today (1968).

Soviet success in World War II was due in part to economic and military assistance furnished by outside nations, especially the United States, United Kingdom, and Canada. These allies sent large shipments of munitions and other supplies to the USSR, and their military campaigns in North Africa, Italy, and northwestern Europe diverted sizable German forces from the Russian front. Mismanagement of the Russian campaigns by the German dictator, Adolf Hitler, may also have worked in favor of the Soviets. Nevertheless, the failure of Germany to conquer Russia was a clear indication that the strength of the USSR had been underrated. The end of World War II found the Soviet Union established as a world power ranking second in importance only to the United States.

Soviet Territorial Gains

Following the war the actions of the Soviet Union created a state of profound uneasiness among the non-Communist nations of the world. Beginning in 1939 the Soviet government had considerably enlarged the national territory of the USSR by successive annexations of areas taken from neighboring countries. Most of these annexations took place in the western frontier zone of the Soviet Union between the Black Sea and the Arctic Ocean. The three Baltic republics of Lithuania, Latvia, and Estonia were absorbed in 1940. Other European nations losing territory to the USSR included Poland, Finland, Romania, Czechoslovakia, and Germany (maps, pp. 216 and 257). The greater part of the annexed lands had been Russian-held territory prior to World War I, but some areas were acquired for the first time. In the Far East the Soviet Union took the southern half of Sakhalin Island and the Kuril Islands from Japan. Both of these had been ceded to Japan by Tsarist Russia. In addition, the small semi-independent Asian country of Tannu Tuva was absorbed. All of the foregoing annexations were complete by the end of 1945. They added to the national domain of the USSR a total area of approximately 287,000 square miles (20,000 square miles larger than Texas), with an estimated population of 20 to 25 million.

After 1945 the political influence of the Soviet Union was extended still further through its dominance over the European satellite nations of Poland, Czechoslovakia, Romania, Hungary, Bulgaria, Albania, and East Germany. (See Chapter 12 for recent changes in Soviet relationships with these countries.) In eastern Asia, Communist governments took power in North Korea, in mainland China, and in North Vietnam. (A Communist state already existed in Mongolia—see p. 485.) Thus was created a Communist political system embracing about a fourth of the world's lands (excluding Antarctica) and about a third of its people.

The Cold War

Although the United States and Britain had rapidly demobilized after World War II, the USSR was slow to follow suit. Large military forces continued to be maintained, and the production of armaments was kept at a high level. Meanwhile, agitation by local Communist groups was carried on in many countries outside of the Soviet orbit. Actual fighting between Communist and anti-Communist forces broke out in Greece, Indochina, Korea, and a number of other countries. Such events gradually created a state of alarm in the non-Communist world. Many nations outside the Soviet iron curtain began reluctantly to rearm and to form defensive alliances against possible Communist aggression. In the period of cold war which developed, it became increasingly clear that the world had divided into three groups of nations. One group was dominated by the Soviet Union and Communist China, while another looked to the United States for leadership and aid. A third group of countries, mainly in southern Asia and in Africa, maintained a precarious neutrality.

Thus the Soviet Union by success in war, economic and territorial expansion, military threat, and political intrigue has gained a position of immense importance in world affairs. Soviet technical and scientific progress has been highlighted repeatedly by its achievements in the fields of nuclear energy and the exploration of space.

THE PLANNED ECONOMY OF THE SOVIET UNION

One of the characteristics for which the Soviet Union is best known is its tightly controlled, socialized type of planned economy. Since 1928 the economy of the USSR has been operated under a series of Five-Year Plans touching every major sector of economic life. These plans prescribe the goals of production for the entire nation. They specify the types and quantities of minerals, manufactured goods, and agricultural commodities to be produced, the factories, rail lines, highways, canals, and dams to be built or improved, the locations of new residential areas to house industrial workers, and so on.

The Soviet state exercises a monopolistic control over the whole economic structure and resources of the country. It owns and operates large-scale industry, mines, power plants, railways, shipping, and other means of communication. It engages in farming on its own account through the institution of state farms, and it largely controls peasant agriculture through the organization of collective farming. It has an exclusive monopoly of banking, foreign trade, and exchange operations. It controls the domestic channels of distribution in its capacity as a manufacturer, farmer, merchant, shipper, and banker. . . .

All these branches of economic life are subject to the system of economic planning by the state; they are within the orbit of "planned economy" as it is understood and practiced in the Soviet Union.

It is true that the private market, however diminished or limited in scope, has never become entirely extinct, at any rate so far as petty trade is concerned. Although the Soviet policy toward private enterprise has generally been unmistakably restrictive, it has occasionally relaxed in the direction of greater liberality, at least in the realm of trade.

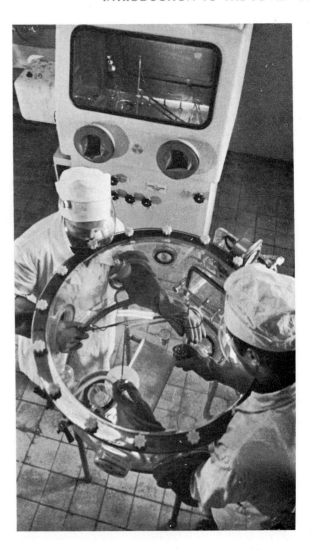

The upward push of the Soviet Union in technical and scientific fields is well exemplified by this photo of medical technicians handling radioactive isotopes for diagnostic purposes in a clinical hospital at Perm in the Ural region. (Novosti from Sovfoto.)

But whatever its concessions to private enterprise, the Soviet state has maintained, unaltered, its dominance in the economic sphere. From the fields of large-scale industry and foreign trade, over which the Soviet state early asserted a monopoly, it extended its dominance to domestic trade and finally to agriculture. Since the early 1930s collectivization of agriculture has been an achieved fact.[2]

---

[2] Lazar Volin, *A Survey of Soviet Russian Agriculture* (U.S. Department of Agriculture Monograph No. 5, 1951), p. 10.

THE SOVIET UNION

INDEX MAP

URBAN AREAS

★	over 6,000,000 (national capital)
●	over 3,000,000
◉	over 1,000,000
•	500,000—1,000,000
○	250,000—500,000 (selected places)
∘	Selected smaller places

0   500   1,000

Scale of Miles

Map labels:

SWEDEN, NORWAY, FRANZ JOSEF LAND, ARCTIC OCEA, BALTIC SEA, FINLAND, Pechenga (Petsamo), BARENTS SEA, NOVAYA ZEMLYA, KARA SE

Kaliningrad, POLAND, Vilnyus, Riga, Tallin, GULF OF FINLAND, Murmansk, KOLA PENINSULA, KARELIA, WHITE SEA–BALTIC CANAL, WHITE SEA

CZECHOSLOVAKIA, HUNGARY, Lvov, Minsk, Leningrad, Lake Ladoga, Novgorod, Lake Onega, MARIINSK CANAL, Arkhangelsk (Archangel)

ROMANIA, Kiev, Kishinev, Cherepovets, Rybinsk Res., Vologda, Northern Dvina R., Ukhta, Vorkuta

Odessa, Nikolayev, Krivoy Rog, Nikopol, MOSCOW CANAL, Moscow, Ivanovo, Yaroslavl, Tula, Kursk, Volga R., Gorkiy, Pechora R., Arctic Circle

Sevastopol, CRIMEA, Yalta, Kharkov, Dnepropetrovsk, Zaporozhye, Zhdanov, Lipetsk, Voronezh, Don R., Oka R., Kazan, Solikamsk, Berezniki, Ob R., Igark

Kerch, SEA OF AZOV, Donetsk, Makeyevka, Lugansk, Ulyanovsk, Izhevsk, Perm, Kama R.

Taganrog, Rostov, Krasnodar, Saratov, Syzran, Volga R., Kuybyshev, Nizhniy Tagil, Surgut

Maykop, Don R., Stavropol, VOLGA-DON CANAL, Volgograd, Ufa, Zlatoust, Sverdlovsk, Ural R.

Batumi, Chiatura, Astrakhan, Groznyy, Magnitogorsk, Chelyabinsk, Tobolsk, Ob R.

TURKEY, Tbilisi, Yerevan, CASPIAN SEA, Orsk, Tobol R., Kustanay, Ishim R., Petropavlovsk, Irtysh R.

Baku, Krasnovodsk, ARAL SEA, Tselinograd (Akmolinsk), Omsk, Yenisey, Tomsk, Novosibirsk, Kemerovo, Krasnoyars

IRAN, Ashkhabad, Amu Darya, Bukhara, Dzhezkazgan, Ekibastuz, Temir Tau, Karaganda, Barnaul, Prokopyevsk, Novokuznetsk, Tashtagol, Abakan, Leninogorsk, TANNU

Balkhash, LAKE BALKHASH, Ob R., Irtysh R., Yenisey

AFGHANISTAN, Amu Darya, Samarkand, Bekabad (Begovat), Tashkent, Almalyk, Fergana, Syr Darya, Frunze, Alma Ata, CHINA (SINKIANG), MONGOL PEOPLE REPUB

Dushanbe, Syr Darya

PAKISTAN, KASHMIR

ARCTIC OCEAN

90°E 110°E 130°E 150°E 80°N 170°E 70°N Cape Dezhnev (East Cape) U.S. (ALASKA) 170°W

WRANGEL ISLAND

BERING SEA

SEVERNAYA ZEMLYA (NORTH LAND)

NEW SIBERIAN ISLANDS

EAST SIBERIAN SEA

Cape Chelyuskin

LAPTEV SEA

Arctic Circle

60°N

TAIMYR PENINSULA

Khata R.

Nordvik

Omolon R.

KAMCHATKA

170°E

Kotuy R.

Lena R.

Olenek R.

Yana R.

Indigirka R.

Kolyma R.

Verkhoyansk

Arctic Circle

Aykhal

Magadan

Lower Tunguska R.

Vilyuy R.

Aldan R.

Okhotsk

Petropavlovsk

60°N

Yakutsk

Mirnyy

Lena R.

SEA OF OKHOTSK

50°N

Aldan

KURIL ISLANDS

PACIFIC

a River

Ust Ilim

Vitim R.

Nikolayevsk

SAKHALIN

GULF OF TATARY

170°E 150°E

Ust Kut

Lena R.

Bratsk

Komsomolsk

Sovetskaya Gavan (Soviet Haven)

Cheremkhovo

Amur R.

Bureya R.

Khabarovsk

Irkutsk

Ulan-Ude

Chita

Shilka R.

Argun R.

Amur R.

Ussuri R.

40°N

Selenga R.

Onon

110°E

CHINA (MANCHURIA)

OCEAN

RUSSIAN LATVIAN SFSR SSR

LITHUANIAN SSR ESTONIAN (Tallin) SSR

RUSSIAN SSR (RUSSIAN)

(Minsk) (Riga) (Vilnyus)

MOLDAVIAN SSR (Kiev)

(Kishinev) UKRAINIAN SSR

GEORGIAN SSR

N SSR (Tbilisi) (Yerevan) (Tiflis)

BAIJAN SSR (Baku)

(Ashkhabad) UZBEK SSR
TURKMEN SSR (Tashkent) (Alma Ata)

(Dushanbe) (Frunze)

TADZHIK SSR KIRGIZ SSR

CONSTITUENT UNION REPUBLICS
(Political capitals are in parentheses)

Moscow

RUSSIAN SOVIET FEDERATED SOCIALIST REPUBLIC

K A Z A K H   S S R

Ussuriysk

Nakhodka

Vladivostok

SEA OF JAPAN

JAPAN

KOREA

130°E

0   500   1,000 Miles

The economic system described above was initiated by stages following the Bolshevik Revolution of 1917. Essentially the aim of the revolutionary leaders was twofold. They meant to abolish the old aristocratic and capitalist institutions of Tsarist Russia and to develop a strong socialist state able to stand on an equal footing with the major industrial nations of the West. For the first decade they were mainly occupied in consolidating their hold on the country and in putting a limited part of their program into effect. Large-scale industry, banking, and foreign trade were nationalized, but a certain amount of private trading, together with private ownership of small industries and agricultural land, was permitted under the New Economic Policy (N.E.P.) announced in 1921. This compromise policy was the result of a near breakdown in the newly instituted Communist economy during the difficult period of civil war and foreign intervention following the Revolution. However, it was intended only as a temporary expedient.

The Five-Year Plans

In 1928 the first of the Five-Year Plans was announced, and succeeding plans have covered the period up to the present.[3] The two most prominent outcomes of these plans have been (1) the collectivization of agriculture and (2) a rapid expansion of heavy industry.

Collectivization of Agriculture

Between 1929 and 1933 about two-thirds of all peasant households in the Soviet Union were collectivized. By 1940 this figure had risen to 97 percent. Today the number of independent peasant cultivators is insignificant. Collectivization was fiercely resisted by the land-hungry peasants, and drastic measures were necessary to put the government decrees into effect. Millions of livestock were slaughtered by the peasants and nomads, and crops were burned to avoid turning them over to the socialized sector. Two main types of farm units now exist: the collective farm (*kolkhoz*) and the large factory-type state farm (*sovkhoz*). They are described in the following selection.

On the state-operated farms, the workers are paid cash wages [in the same manner as industrial workers], with bonus payments for high performance. In addition, they have small garden allotments and the privilege of keeping a cow, a pig or two, chickens, geese, and sometimes . . . sheep or goats. On the collective farms, the workers share in the net income from operations in accordance with the days worked [and types of work done]; and they . . . receive bonus rewards for special efforts. Workers on collective farms also have garden allotments [private plots] and the privilege of keeping individually owned livestock.

The farm people live in villages, with their individual allotments of land usually extending back of each house. Consequently, there are no individual farmsteads dotting the countryside as there are in the United States.

In the process of organizing the large collective and state farms, the land was consolidated into huge fields wherever the terrain would permit. The first impression one gets of the major grain areas is of a vast expanse of wheat, with some . . . corn, sunflowers, [sugar beets], and other . . . crops. In some areas, shelterbelts of trees have been planted to break the force of the wind and to hold the snow cover. In the naturally forested areas of the north, the fields are smaller, and the terrain is broken up by streams, ponds, and wooded hills; the farms also tend to be smaller. . . .

Almost always, several villages are included within the boundaries of one collective or state farm.[4]

As originally conceived, the system of collectivized agriculture was supposed to result in the following major advantages:

1. The old arrangement of small, fragmented individual holdings separated by uncultivated boundary strips would be replaced by a system of larger fields incorporating the boundary strips, thus increasing the amount of cultivated land and giving scope for mechanized farming.

---

[3] Though Soviet planning has generally been on a five-year basis, an exception occurred in the 1950s. The 1956–1960 plan was scrapped in 1957, and after two years of planning on an annual basis, a Seven-Year Plan for 1959–1965 was instituted.

[4] Sherman E. Johnson, "Impressions of Agriculture in the Soviet Union," *Foreign Agriculture*, 22, no. 10 (October 1958), 3–4. Material has been inserted in brackets in some places without compromising the sense of the original. As of the end of 1966, the state farms numbered over 12,000 and accounted for about 43 percent of the total sown area in the USSR. Collective farms, which are still the dominant form of agricultural organization, numbered over 37,000. The original structure of collective farms did not provide for payment of wages to farm members. Collective farm members shared in the distribution of the farm's income after all taxes and other payments to the state had been deducted and a certain percentage set aside for capital investment in the farm itself. As of July 1, 1966, guaranteed monthly incomes were established for farmers in the collectives on the same scale as state farm wages. The extent to which this reform had actually been put into practice was not clear at the time of writing (mid-1968).

2. Increased mechanization would release surplus farm labor for employment in factories and mines, thus facilitating industrialization and bringing into existence the large urban working class looked to as the principal support for the Communist system.

3. Mechanization, improved methods of farming, and reclamation of new land under state supervision would result in greater overall production.

4. Increased production, plus easier collection of surpluses from a greatly reduced number of farm units, would result in larger and more dependable food supplies for the growing urban populations.

5. Liquidation of individual peasant farming would remove the most important capitalist element still remaining in the USSR.

Changes in Soviet Farming

Soviet agriculture experienced many setbacks in the first quarter-century of collectivization, and overall production was slow to expand. Various factors have been cited as contributing to the stagnation of farm production; they include (1) rapid draining of agricultural workers (including many of the more skilled workers) for employment in industry, without an adequate supply of farm machines being provided to take their place; (2) peasant resistance to collectivization, including large-scale slaughter of draft animals and other livestock; (3) climatic handicaps (see pp. 294–298); (4) excessively rigid and inefficient management by the Soviet bureaucracy; and (5) devastation of many of the best farming districts in World War II. An extremely significant factor was (6) the government's deliberate policy of draining wealth from agriculture for the support of industry. Farm products were requisitioned from the collectives at low prices and were then sold in state-owned stores at inflated prices to skim off excess consumer purchasing power and to provide funds for industrial investment. Quotas set by the government for obligatory deliveries tended constantly to rise. By the time a collective had met its financial obligations for seed, fertilizer, other supplies, and services, only a meager surplus of cash and produce was left to be distributed among the members. Under such conditions there was a natural tendency for peasants to shirk their obligations to the collective in favor of work on their personal plots. Any surplus from these plots and from privately owned livestock could be sold on the collective-farm market in a nearby town or city, generally at much higher prices than those paid the collectives by the state. Meanwhile, much of the country's farmland suffered from a lack of fertilizer. The production of chemical fertilizer was given a low priority by government

planners and consequently it was not available for most crops. Animal manure was also in short supply, due to the slow recovery of livestock from the disastrous slaughter of the early collectivization period and the ravages of World War II.

After World War II, and particularly after the death of Stalin in 1953, important changes were made in Soviet farming. The government revamped its system of agricultural procurement, pricing, and taxation in the hope of stimulating the peasants to produce more on collectively farmed land. By merging adjoining units, the total number of collective farms was sharply reduced. Some collectives were absorbed by or converted to state farms.

Accompanying the trend to larger units was a government decision to permit individual collectives to own and operate tractors and farm machines. Heretofore, machines for planting, harvesting, and other farm tasks had been supplied on a custom-work basis by machine-tractor stations (MTS). These stations, employing salaried workers to operate and service the machines, were paid by the collectives partly in cash and partly in farm products. They served as an important channel for collecting farm surpluses to feed urban populations. Furthermore, through their control of tractors and machinery they were able to exercise supervision over the collectives and to acquaint government officials with the true size of the harvest, thus giving the government a basis for demanding that its quotas be delivered. But in 1958 the system of machine-tractor stations was abolished. With limited exceptions in special situations, the stations were directed to sell their tractors and machinery and to transfer their operators and mechanics to the collectives. A new system of repair-technical stations (RTS) was instituted to supply the collectives with fuels, lubricants, spare parts, seeds, fertilizers, insecticides, and so on, to make certain types of repairs to machines owned by collectives, and to make available to the collectives some types of highly specialized machines. Transfer of machinery to the newly enlarged collectives was one facet of a program to give greater powers of day-to-day decision making to the officials of each collective farm. It was contemplated, however, that such decisions would be made within the framework of overall policies laid down by Moscow.

After the removal of Nikita Khruschev from power in 1964, the new regime headed by Brezhnev and Kosygin endeavored to provide further incentives for increased farm production. Various restrictions on the use of personal plots and privately owned livestock were removed, guaranteed monthly incomes for farmers in the collectives were promised, farm income taxes were lowered, the system of procurement and pricing of farm products was

The heavy dashed line encloses most of the acreage of new lands brought under cultivation in the period 1954–1957. The light dotted line shows the boundaries of the Kazakh SSR. (After a map in N. N. Baransky, *Economic Geography of the USSR,* Moscow: Foreign Languages Publishing House, 1956, p. 71.)

altered still further to raise incomes of farmers, and prices of consumer goods, heretofore much higher in rural than in urban areas, were adjusted downward in rural areas as a means of raising living standards there. How thoroughgoing such reforms would prove and what their ultimate impact on agricultural production would be remained unclear at the time of writing (mid-1968). The history of Soviet agricultural policies and programs yields many instances of promises to farmers that were never fulfilled in any significant way. But the totality of government measures since 1953 undoubtedly has made it more worthwhile for farmers to exert themselves in the interest of greater agricultural output.

Another facet of the drive to increase the national supply of farm products has been a sizable enlargement of the cultivated area. In 1954 a vast program was instituted to increase the amount of grain (mainly spring wheat) by bringing tens of millions of acres of "virgin and idle lands" into production in the steppes of the northern Kazakh SSR in Soviet Middle Asia and adjoining sections of western Siberia and the Volga region. Hundreds of large new state farms were organized in the area and a network of narrow-gauge railroads was constructed to serve them. Between 1954 and 1960 the cultivated area of the USSR was enlarged by over 90

million acres, with most of the increase taking place in 1954–1956 in the new wheatlands of Kazakhstan, Siberia, and the Volga region. In the agrarian history of Russia and the Soviet Union, expansion of cultivation into "new lands" has been a recurrent theme, but no expansion comparable in scale to that of 1954–1956 has ever occurred in such a short time. Despite adverse weather in some years, with consequent crop failures or low production, the "new lands" program has added large increments to Soviet grain output and has moved the center of gravity of grain farming eastward. On the average, lands in the zone of spring wheat east of the Volga River now supply over half of the USSR's grain and over two-thirds of its wheat. The bulk of this production, coming as it does from very large farms worked primarily with machines, is available to feed urban populations. However, the Soviet government faces important problems in maintaining satisfactory production from the "new lands" on a long-term basis. Such lands are very marginal in precipitation, and they require extremely careful management to conserve moisture in the soil and to prevent serious wind erosion. A great deal of technical knowledge about these matters has accumulated, for example in the United States and Canada as a result of disastrous experiences with "dust bowls" along the dry margins of agriculture in the Great Plains. The application of such knowledge to the Soviet "new lands" will be necessary if these lands are to have a stable agriculture in coming years.

Doubtless of greater overall significance to the USSR in the long run, however, is the current attempt to make the agriculture of older farmlands more productive. These lands, located primarily west of the Volga, are favored over the "new lands" from the standpoint of annual moisture and temperature regimes. Their agricultural environment is distinctly inferior, however, to that of the best lands in the United States—a fact that became very evident in the unsuccessful attempt of the Khruschev regime to add large increments to Soviet grain supply by growing corn. In the highly productive agriculture of the United States corn plays a key role, being in fact the single most important crop by a very wide margin. It is grown primarily to feed livestock, and the Soviets attempted to foster its production on a large scale for the same purpose. On a visit to the United States, Khruschev had been greatly impressed with the productivity of corn and livestock farms in the Middle West. But climatic conditions in most farming areas of the Soviet Union are more comparable to those of Canada than they are to those of Iowa or Illinois, and the requirements of corn in regard to moisture, temperature, and length of growing season cannot be satisfactorily met. In the best year of the corn-for-grain program (1961) the tonnage of corn harvested for grain was one-fourth that of wheat. But in most years corn has played a much more subordinate role in Soviet grain supply. The bulk of the crop is cut for silage or green fodder before the grain is fully mature.

Despite their climatic limitations in regard to corn-for-grain, however, Soviet farmlands offer many possibilities for increased production of crops and livestock through greater use of chemical fertilizers and animal manures, the liming of acid soils, use of improved varieties of crops and animals, better control of plant and animal diseases, increased irrigation, mechanization, and drainage; construction of better storage facilities; use of more scientific methods of cultivation and animal husbandry, and so on. Such measures have increased since the Stalin era, and despite reverses such as the serious crop failures in 1963, there has been a definite upward trend in Soviet agricultural output. It is estimated that overall farm output increased by 70 percent from 1950 to 1966, and though a good part of the increase was due simply to the cultivation of additional land, some of it was due to intensification of farming on the older lands. Today Soviet agriculture provides the average citizen with a diet that is adequate in caloric content and is better balanced than formerly, though still not equal in quality to the average diet in the United States, Canada, or most of western Europe. It may be noted that the personal plots and privately owned livestock on Soviet farms play a vital role in maintaining and improving the overall quality of the Soviet diet. Only about 3 percent of the country's tilled land is in the personal plots, and they contribute very little to the supply of grain, sugar, or vegetable oil, but in the production of many other foods their role is outstanding. In the early 1960s the "private sector" was producing an estimated 40 to 50 percent of the USSR's vegetables, meat, and milk, over two-thirds of its potatoes and fruit, and over three-fourths of its eggs. These proportions have probably declined somewhat in recent years, though not enough to change the overall situation in any fundamental way. A large share of the produce from personal plots and livestock is consumed by the farm families themselves, but the residue that is sold in urban areas is an indispensable element in the country's food supply.

In overall output of many farm products (for example, wheat, rye, barley, potatoes, beet sugar, flax, sunflower seeds) the USSR is the world's leading nation. It does not achieve this rank, however, through high productivity of its agriculture per unit of land and labor. As compared with the United States, for example, the Soviet Union uses about one and a half times as much crop acreage and four times as many farm workers to produce a volume of farm commodities that is probably smaller than that of the United States and must meet the needs of a larger population. In mechanization, fertilization, and the technical proficiency of its agriculture, the USSR lags well behind the United States. For instance, most of the work on the personal plots is done by hand or with horses, and it is estimated that two-thirds of the country's cows are still milked by hand. Soviet croplands receive on the average only about one-quarter as much chemical fertilizer as American lands, and this factor, along with climatic and other handicaps, contributes to an average yield per acre that is only 40 to 60 percent as great as that of the United States for most of the important crops. The Eighth Five-Year Plan for 1965–1970 called for increased investment in the agricultural sector of the economy, including a stepped-up output of tractors, trucks, grain combines, and other farm machines, and a doubling of the supply of chemical fertilizer. Heretofore, chemical fertilizer has been applied primarily to "industrial crops" such as cotton and sugar beets, or to vegetables, but sizable amounts are commencing to be applied to wheat and other crops. It is apt to be a long time, however, before the amount of fertilizer applied per acre in the Soviet Union approaches that of the United

States or before there is a truly significant narrowing of the differential between the two countries in agricultural output per acre and per farm worker.

One of the most important long-term problems of Soviet agriculture has been the stagnation of animal industries for long periods. Since the Stalin era the Soviet leadership has placed much stress on the need to increase the supply of livestock products and associated forage crops. This program has had its ups and downs, but on the whole has achieved considerable success. For example, between 1953 and 1964 cattle numbers rose from 56.6 million to 85.4 million, and milk production increased from 36.5 million tons to 63.1 million tons. The average yield of milk per cow, however, remains very low when compared with that of nations with advanced dairy industries. It is only about half the average yield in the United States.

Industrial Emphases under the Five-Year Plans

The most important objective of the Five-Year Plans has been a large increase in industrial production, with emphasis on the production of heavy machinery and other capital goods, increased exploitation of minerals, increased development of electric power, and improvement of transportation facilities, especially railroads. On the whole this phase of Soviet planning has experienced far more success than the planned expansion of agriculture. Since 1928, and especially since 1945, the Soviet Union, which was largely an agricultural country in 1928, has industrialized at a rapid rate. Masses of peasants have been converted into factory workers, new industrial centers have been created and old ones enlarged, and overall industrial production has greatly increased. Various estimates credit the Soviet Union with a fivefold to eightfold increase in total industrial output between 1928 and the late 1950s. Since then production has continued to rise, but the annual rate of increase has slowed, and the Soviets have found it necessary to use increasing amounts of capital for the replacement of outdated or worn-out buildings and industrial equipment. The prolonged drive to industrialize has made the Soviet Union the second industrial power of the world, but its stated objective of surpassing the United States in overall production is still very far from realization. In the late 1960s its total output of manufactured goods was estimated to be no more than half that of the United States and only two-fifths on a per capita basis. The gap between the two countries is especially wide in the production of automobiles and many other consumer

items, which have been slighted by the USSR in favor of heavy industry. Soviet planners have consistently given the highest priority to iron and steel production, other forms of heavy metallurgy, and the manufacture of machinery, power-generating and transportation equipment, and industrial chemicals. Many of the consumer items that can readily be purchased in any town or city in the United States are in short supply in the Soviet Union, though greater attention has been given to consumer needs in recent years and the output of consumer goods is rising. In the middle 1960s it was estimated that by the end of 1970 a little over half of all Soviet families would own television sets, about two-fifths would own washing machines, and nearly a third would own mechanical refrigerators. Such advances are still confined almost entirely to urban areas, which have a much higher level of living in terms of conveniences, amenities, and consumer services than is true of the countryside.

One of the greatest contrasts between the United States and the Soviet Union lies in the lack of automobiles and good highways in the USSR. Up to the present, Soviet production of automobiles has been extremely low, considering the size of the country and its overall industrial strength. In the middle 1960s the output of passenger vehicles was only about 3 percent as great as in the United States. It is planned to increase the automobile output, partly by inviting foreign companies to open plants in the USSR; for example, production by Italy's Fiat company is scheduled to commence in the early 1970s. But it is likely to be many years before the Soviet Union has anything approaching an "automobile civilization."

An important recent development in Soviet industry has been the government's attempt to increase productivity by assigning greater powers of decision to managers of individual plants and then judging the performance of each plant on the basis of quantities of goods sold. This is a departure from the past practice of judging plants on the basis of quantities of specified goods produced in relation to predetermined targets set by government planners. It is hoped that the new scheme will bring about a closer correspondence between the country's actual needs and the types and quantities of goods produced. It is also hoped that goods of better quality will be produced. In the past, stocks of goods often have accumulated in warehouses because they were not needed or because their quality was so low that prospective purchasers refused to buy them. Under the new arrangement there will be more stress on market research, advertising, and modern techniques of plant management. Plants will be required to pay interest on investment capital supplied by the government, and the incomes of workers will be geared more closely to the plant's success in marketing its wares than has been true in the past. A smaller share of each plant's profits will be requisitioned by the government for the general budget than in the past, and larger funds will thus be available for the plant to use in reinvestment, payment of bonuses to reward efficient work, improvement of workers' housing, and so on. Thus the overall industrial milieu is being changed to something more approaching the norm in Western countries, but this does not mean that the USSR is reverting to private capitalism. Prices of goods and basic wage levels will continue to be set by the government, and overall planning will rest with national economic ministries in Moscow. It remains to be seen whether the new scheme will succeed in stimulating initiative, innovation, and higher performance at the plant level and overcoming the excessively rigid control by an intrenched bureaucracy that has handicapped Soviet industry in the past.

One of the less publicized but nonetheless vital industries in the USSR is residential construction. Housing, especially in the cities, has been critically short for many years. The Soviets have devoted much effort to mass construction of standardized housing units, and these are gradually relieving the worst overcrowding. Despite the extensive slums that characterize many American "inner cities," however, the Soviet Union is still very far behind the United States in the overall adequacy of its housing and residential services.

The USSR ranks far below the United States as a trading nation, though its foreign trade has been steadily expanding in recent years. About 70 percent of the Soviet Union's international trade in 1965 was carried on with other Communist countries. It is primarily an exporter of minerals and other industrial raw materials and an importer of manufactures (primarily capital equipment), along with some raw materials and foods.

SIZE, BOUNDARIES, AND RESOURCES

We may now proceed to consider some of the pertinent physical and locational characteristics and the territorial evolution of this immense country of 239 million people (1968 estimate) which has witnessed such revolutionary developments since 1917, and whose future still poses so much uncertainty for mankind.

Size

The territorial base from which Soviet ambitions proceed is the largest country in area on the globe. It is, in fact, a country which ranks with the continents in size. With a total area of approximately 8.6 million square miles, Soviet Russia is larger than either the North American mainland or South America. It is more than twice as large as China, Canada, the United States, or Brazil, 40 times as large as France, and 171 times as large as England. Out of every 7 square miles of the earth's land surface, 1 is Soviet territory (1 of every 6 if the ice caps of Antarctica and Greenland are excluded).

Even more impressive, perhaps, than its area is the tremendous east-west extent of the Soviet Union. The westernmost point of the country, near Kaliningrad (formerly Königsberg) on the Baltic Sea, is separated from Cape Dezhnev (East Cape) on Bering Strait by approximately 170° of longitude, representing a distance of nearly halfway around the world. In terms of sun time this means a difference of approximately 11 hours between the easternmost and westernmost points of the USSR. When it is 6:00 A.M. by sun time at Kaliningrad, it is already 9:00 A.M. in the Ural Mountains and is approaching 1:30 in the afternoon at Vladivostok. At Cape Dezhnev it is shortly after 5:00 P.M. The time difference between Kaliningrad and Cape Dezhnev is more than twice the difference between New York and London (5 hours) and is nearly four times the difference between New York and San Francisco (3 hours).

[EDITORIAL COMMENT. The concepts of longitude and time are closely related. The meridians of longitude are imaginary straight lines connecting the poles. Every meridian runs due north and south. The meridians converge at the poles and are widest apart at the equator. Longitude, like latitude, is measured in degrees, minutes, and seconds. The meridian which is customarily used as a base, or starting point, runs through the Royal Astronomical Observatory in Greenwich, England. It is known as the meridian of Greenwich, or prime meridian. Every point on a given meridian has the same longitude. The prime meridian has a longitude of 0°. Places east of the prime meridian are in east longitude; places west of it are in west longitude. The meridian of 180°, exactly halfway around the world from the prime meridian, is the dividing line between places east and west of Greenwich. The longitude of Kaliningrad is correctly stated as 20°31' East of Greenwich, or simply as 20°31' E. The longitude of Cape Dezhnev is 170° W. At the equator, 1° of longitude

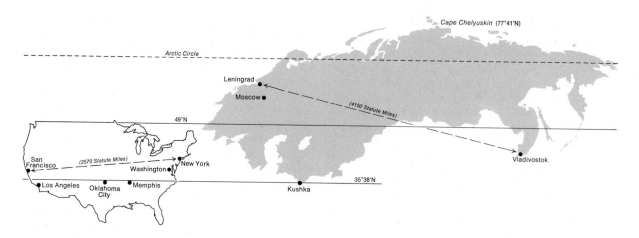

The Soviet Union compared in latitude and area with conterminous United States.

is equivalent to 69.15 statute miles. Since the circumference of the earth decreases toward the poles, however, a degree of longitude at 60°N., the latitude of Leningrad, is equivalent to only 34.6 miles.

The earth, rotating on its axis, turns completely around once every 24 hours. Thus it turns through 15° of longitude every hour (360° divided by 24) or 1° every 4 minutes. Since the earth turns toward the east, the day breaks and the sun comes into view on the Pacific side of the USSR while places farther west are still in darkness. It is noon at a place when the earth's turning has brought the meridian of that place directly under the sun, that is, when the sun has climbed highest above the horizon for that day. Noon occurs simultaneously at all points along a given meridian. When it is noon at Vladivostok by sun time, the earth must still turn for approximately 7½ hours before the meridian of Kaliningrad will come under the sun and it will be noon at that place. In other words, the apparent solar time (sun time) at Kaliningrad is about 4:30 A.M. when the time at Vladivostok is 12 noon.

Development of modern means of transportation has made it desirable to establish standard time zones so that the same time prevails over a considerable extent of longitude. Watches of travelers are set forward or back 1 hour when crossing the boundary of such a zone. The United States, excluding Alaska and Hawaii, has four standard time zones (Eastern, Central, Mountain, Pacific). The Soviet Union has eleven such zones.]

The east-west length of the Soviet Union is much greater than its north-south width, but even the latter is impressive. The maximum latitudinal extent, exclusive of islands in the Arctic Ocean, is about 42°, which is short by only 5° of the combined extent of the United States and the Canadian mainland. Cape Chelyuskin (77°41' N.) in the Soviet Arctic is about 6° farther north than any point on the mainland of Canada, while the town of Kushka (35°38' N.) in Soviet Middle Asia is in the general latitude of Memphis, Tennessee, and Oklahoma City, Oklahoma, and is only 1° farther north than Los Angeles, California (map, above).

Sea Frontiers

The land and water boundaries of the USSR have a total length of about 37,500 miles.[5] More than two-thirds of this distance is coastline, giving the Soviet Union a longer sea frontage than any other country. To only a limited degree, however, does the USSR have unrestricted access to the world ocean. Despite its great length of coast, it is essentially a landlocked country. Most of the main ports are located on seas that have entrances controlled by other countries. The water passages that give access to the Baltic Sea from the Atlantic Ocean (the Danish Straits and the Kiel Canal) pass through Danish, Swedish, or West German territory, while the entrance to the Black Sea is controlled by

[5] N. T. Mirov, *Geography of Russia* (New York: John Wiley & Sons; London: Chapman and Hall, 1951), p. 3.

Turkey. The entrances to the Sea of Japan are controlled or flanked by Japan and South Korea. Acquisition of the Kuril Islands from Japan has given unrestricted access to the Sea of Okhotsk, but that body of water borders remote and sparsely populated regions, is frozen for a considerable period in winter, has no rail connection with the rest of the country, and is used relatively little by shipping.

Most of the Arctic Ocean ports are small, and except in the extreme west are completely icebound for more than half of the year. Nearly all of the major Soviet ports are hampered to some degree by winter freezing, although most ports on the Black Sea, Baltic Sea, and Sea of Japan can be kept open all winter by the use of icebreakers. However, Leningrad, the foremost seaport of the USSR, is frozen in for a few weeks in midwinter. Only in the far northwest does a stretch of coast bordering the Barents Sea give unrestricted access to the open ocean. A tongue of water from the North Atlantic Drift warms this coast in winter and makes it possible for icebreakers to keep open the harbor of Murmansk. The latter city is connected with Leningrad and Moscow by rail. However, its value as a port is lessened by remoteness from the main centers of population and industry in the USSR.

Land Frontiers

The Soviet Union has a common land frontier with no less than twelve other countries in Europe and Asia. Along the 8000-mile boundary between the Black Sea and the Pacific Ocean, it borders Turkey, Iran, Afghanistan, China, the Mongolian People's Republic, and North Korea (map, pp. 284–285). The Soviet frontiers in Asia are mostly found in mountainous or arid regions which form sparsely populated zones of separation between the USSR and the core regions of the bordering countries. In Middle Asia only a narrow corridor of Afghan territory separates the Soviet Union from the northern boundary of the Indian subcontinent. However, the frontier zone lies in extremely rugged and difficult mountain country, so that the narrow political separation as seen on a map does not express the true reality of military and economic separation. In the Far East the Soviet island of Sakhalin is separated by a 25-mile-wide strait from the northernmost main island of Japan, and even narrower straits separate Japan from the Russian-held Kuril Islands. If India, Pakistan, and Japan are included, the Soviet Union in Asia is a neighbor of about half the world's people.

The Soviet frontiers on the west are very different from the frontiers in Asia. This is especially true of the 900-mile-long boundary zone between the Black Sea and the Baltic Sea. Here the USSR abuts on Romania, Hungary, Czechoslovakia, and Poland. The boundary line passes mostly through well-populated lowlands that have long been disputed territory between the Russian state and its western neighbors. The fluctuating boundaries in this area have contrasted with Soviet boundaries in Asia, which have changed relatively little during the twentieth century. After 1939 the USSR made important gains of territory along the Black Sea–Baltic frontier. The boundary was pushed westward 100 to 200 miles in Poland, although not so far west as the boundary of Russian-held territory in Poland prior to World War I. From Czechoslovakia the Soviet Union acquired a section of the Carpathian Mountains known as Ruthenia or Carpathian Ukraine, and thus gained a common frontier with Hungary. From Romania the USSR took Bessarabia and Northern Bukovina, thus pushing the Romanian boundary 50 to 125 miles farther west. Previous mention has been made of the absorption of Lithuania, Latvia, and Estonia in 1940. In addition, the northern half of the former German province of East Prussia was incorporated by the USSR (see map, p. 257).

The Black Sea–Baltic line is seriously lacking in natural defenses. During the course of history it has been crossed many times by invading armies, including the French under Napoleon in 1812 and more recently the German armies in 1941. The recent frontier changes shortened the boundary and gave the Soviet Union additional defense in depth. Other strategic advantages accruing to the USSR were (1) a better frontage on the Gulf of Finland and a new frontage on the Baltic Sea proper, (2) possession of a number of strategic junction points on major rail lines leading westward, including the important rail hub of Lvov (525,000), acquired from Poland, (3) control over important passes leading to the plains of Hungary through former Czechoslovak territory in the low central Carpathians, and (4) a frontage on the northernmost of the three main distributaries of the Danube River.

The USSR also made important strategic gains along the 900-mile frontier between the Gulf of Finland and the Barents Sea. It took from Finland the Karelian Isthmus between Lake Ladoga and the Gulf of Finland (map, p. 216). By this means the defensibility of the important city of Leningrad was increased. Prior to 1940 the city lay only about 12 or 15 miles from the nearest points on the Finnish border. Conversely, Soviet acquisition of the Karelian Isthmus lessened the ability of the

Finns to defend themselves against a military move by the USSR and deprived them of the important timber port of Viipuri (now Vyborg) (see p. 227). A further cession of Finnish territory to the USSR in the far north gave the latter nation a common frontier with Norway and important nickel mines and smelting facilities in the Pechenga (Petsamo) district.[6]

Natural Resources

Soviet economic and military strength rests in considerable measure on an outstanding endowment of natural resources. Within the vast confines of the USSR, nature has assembled a rich and varied complement of minerals and water resources, together with enormous areas of forest and grazing land, some of the world's most fertile soils, and sizable fish and fur resources. Although much of the country is handicapped by excessive cold, low rainfall, poor drainage, acid soils, or rugged terrain, the overall resource base of the Soviet Union is undoubtedly superior to that possessed by any other nation with the exception of the United States.

The Superior Mineral Position of the USSR

The resource position of the Soviet Union is particularly strong with regard to minerals. It seems adequately (or in many cases, abundantly) supplied for the foreseeable future with mineral fuels, most of the important metals and materials for chemical manufacture, and a miscellany of other useful minerals such as diamonds, asbestos, china clay, fire clay, and mercury. On the whole the Soviet Union is remarkably self-sufficient in minerals, though in the case of some minerals it meets its needs by mining low-grade deposits at a relatively high cost. Its position seems likely to be improved by future mineral discoveries in the varied geological formations with which the country is endowed.

Coal and lignite reserves in the Soviet Union are exceedingly widespread, but the known reserves of most other minerals tend to be rather heavily concentrated in a relatively few areas. The outstanding mineral-producing areas, together with the minerals for which each area is particularly noteworthy, are as follows:

1. The Ukraine (coal, iron, manganese, natural gas, salt).

2. The Ural Mountains (iron, ferroalloys, copper, aluminum, potash, asbestos).

3. The Volga-Urals oil fields (petroleum, natural gas).

4. The Caucasus-Caspian region (petroleum, natural gas, manganese, lead, zinc).

5. Soviet Middle Asia (coal, copper, iron, natural gas, petroleum, sulfur, lead, zinc, aluminum, uranium, ferroalloys).

6. The Kuznetsk Basin of south central Siberia (coal).

7. The Kola Peninsula, adjoining Finland and Norway (phosphate, nickel, copper, aluminum, iron).

These and various other mineralized areas are discussed in some detail in Chapter 14 (see map, p. 312).

THE SEVERE CLIMATE OF THE SOVIET UNION

Despite its immense area and resources, large and growing population, and increasing industrial production, the Soviet Union suffers from a number of important disadvantages which tend to lessen its effectiveness as a political, economic, and military power. Among these might be mentioned (1) the difficulty of providing adequate transportation and communication facilities in a country of such vast distances and containing so much marshy, arid, densely forested, and mountainous terrain, and (2) the additional burden on transportation and the problems of defense caused by the peripheral location of many important centers of population, mining, agriculture, and manufacturing. Not the least among Soviet disadvantages is (3) the severe climate.

General Nature of the Climate

Most parts of the Soviet Union have a continental climate characterized by long, cold winters, warm to hot summers, and low to moderate precipitation. The severe winters are the result of a northerly continental location, coupled with mountain barriers on the south and east. Four-fifths of the total area is farther north than any point in the conterminous United States (map, p. 292). The Soviet Union lies mainly in the higher middle and lower high latitudes of the Northern Hemisphere where land masses reach their greatest extent relative to the bordering oceans. In these latitudes the

---

[6] The foregoing discussion of Soviet frontier changes is mainly based on W. Gordon East, "The New Frontiers of the Soviet Union," *Foreign Affairs*, **29**, no. 4 (July 1951), 591–607.

climatic influence of the land is paramount over the influence of the sea in continental interiors, and the most extreme continental climates of the world prevail. The effects of continental location are heightened in the Soviet Union by the presence of the Arctic Ocean on the north, which is frozen for most of the year and thus in a sense forms an extension of the land. Continuous chains of mountains and high plateaus occupying the southern and eastern margins of the country act as a screen against moderating influences from the Indian and Pacific oceans. In addition, most of the USSR is a zone of high atmospheric pressure in winter, with resultant outflowing winds. Westerly winds from the Atlantic serve to moderate the winter temperatures somewhat, but their effects become steadily weaker toward the east. Thus Leningrad (60°N., 30°E.) has an average January temperature of 18°F, whereas Yakutsk (62°N., 130°E.) has a January average of –45° and Verkhoyansk (68°N., 133°E.) a January average of –52°.

Aside from the interior of eastern Siberia, most parts of the USSR experience winter temperatures which do not differ greatly from the temperatures of places located at comparable latitudes and altitudes in the interior of North America. Indeed, due to westerly winds from the Atlantic, many places in the west and south of the Soviet Union are actually warmer in midwinter than comparable interior places in North America. Kiev (50°N.) is 21° warmer in January on the average than Winnipeg, Odessa (46°N.) is 12° warmer than Montreal, and Tashkent (41°N.) is 10° warmer than Omaha. Moscow is about 1° warmer in January than Minneapolis, although located 11° farther north. Leningrad, located a thousand miles north of Minneapolis, is 4° warmer in January than the latter city.

Nevertheless, the Soviet Union as a whole is definitely handicapped by the winter climate. Huge areas are permanently frozen at a depth of a few feet. The average frost-free season of 150 days or less in most parts of the country except the extreme south and west is too short for a wide range of crops to mature.

Most places are relatively warm during the brief summer, and the southern steppes and deserts are hot. A factor partially offsetting the brevity of the summer is the length of summer days. Leningrad has 19 hours of daylight on June 22, and Moscow has 17½ hours. Throughout the summer long hours of sunlight facilitate the growth of plants and thus compensate somewhat for the shortness of the growing season.

From an agricultural standpoint, lack of moisture is probably an even greater handicap to the Soviet Union than low temperatures. The annual precipitation is less than 20 inches nearly everywhere except in the extreme west, along the eastern coast of the Black Sea and the Pacific littoral north of Vladivostok and in some of the higher mountain areas. North of the Arctic Circle and in the southern desert zone, the average precipitation is less than 10 inches. Since most of the precipitation is derived from the Atlantic Ocean, the total amount tends to decrease from west to east. Precipitation also decreases toward the south, being 10 to 20 inches in the fertile black-earth belt, but only 5 to 10 inches or even less in the dry steppes and deserts farther south. The country's most important grain-producing region, the black-earth belt (map, p. 310), is subject to severe droughts and to hot, desiccating winds (*sukhovey*) which may greatly damage or destroy a crop. Areas to the north of this belt in European Russia have a somewhat greater and more dependable rainfall, coupled with a lower rate of evaporation resulting from lower temperatures. Thus they are better supplied with moisture than the black-earth belt, but this advantage is offset by their poorer soils, cooler summers, shorter growing season, and larger proportion of poorly drained land.

Types of Climate

In the USSR five main east-west climatic belts—tundra, subarctic, humid continental, steppe, and desert—are customarily recognized in classifications of climate. These belts, each with its associated vegetation and soils, succeed each other from north to south in the order named (see front endpaper climatic map).

Tundra Climate

The zone of *tundra climate* occupies a continuous strip, 50 to several hundred miles in width, along the Arctic coast from the Norwegian frontier to Bering Strait. Tundra is both a climatic and a vegetational term. Climatically, it signifies a region which has at least 1 month averaging above 32°F but which has no month averaging higher than 50°. In most areas of tundra 2 to 4 months average above freezing. Generally speaking, the climatic conditions of the tundra are too severe for trees to grow, except in a few protected localities and toward the southern margins, where trees are found along the streams. The typical vegetation of this treeless region consists of mosses, lichens, sedges, and scrubby bushes. Some of the islands in the Arctic Ocean, including northern Novaya Zemlya, have temperatures averaging below freezing or only slightly above freezing even during the warmest month of the year. Here much of the land is covered with glaciers, and an ice-cap

A lumbering operation in the taiga. The photo was taken in the Karelian Autonomous SSR, adjoining Finland. (Sovfoto.)

climate similar to that of Antarctica or interior Greenland prevails.

Subarctic Climate

At the south, the tundra climate gradually merges with the zone of *subarctic climate*. In this climatic zone from 1 to 4 months average above 50°F. Thus the subarctic climate differs from the tundra climate in having a warmer summer. Winters, however, are extremely long and cold, most places averaging below freezing for 5 to 7 months. This climate type occupies a wedge-shaped area extending from the Finnish border to the Pacific Ocean. It is narrowest at the west and broadens eastward to the neighborhood of Lake Baikal, where it occupies the full width of the country except for a narrow strip of tundra at the north.

The zone of subarctic climate is essentially coextensive with the *taiga*, or northern coniferous forest. The taiga of the Soviet Union is the largest continuous area of forest land on earth. A similar belt of forest covers much of Canada (see back endpaper vegetation map). The prevailing tree species are spruce, fir, larch, and pine. These coniferous softwood trees, while useful for pulpwood and firewood, are frequently too small, twisted, or knotty to make good lumber. Nevertheless, vast reserves of timber suitable for lumber exist in parts of the taiga, and this part of the Soviet Union is one of the major areas of lumbering in the world. Intermixed with the conifers of the taiga are stands of certain broadleaf species such as birch and aspen. These deciduous trees are of little commercial utility. They are ordinarily second growth, replacing stands of conifers removed by fires or cutting.

Some agricultural settlement occurs in the taiga, especially toward the south. However, the farming is definitely of a marginal character. Agriculture in this forested region is handicapped by (1) the short summers, (2) the severe cold of winter, which makes it difficult for winter grains and fruit trees to survive, (3) the unseasonable frosts that often occur in late spring or early fall, (4) the marshy or swampy character of much of the land, and (5) the prevalence of poor soils. The dominant soils of the taiga are the *podzols* (see back endpaper soils map). This term, derived from Russian words translated as "ashes underneath," refers to a group of soils which characteristically have a grayish, bleached appearance when plowed, are lacking in well-decomposed organic matter, are poorly structured, and are very low in natural fertility. The acidity of these soils is unfavorable for most crops and also for bacteria, earthworms, and other soil-improving organisms.

The boundary between the taiga and the tundra is not a sharp line, although it appears so when generalized on small-scale maps. A transitional area of "wooded tundra" (small forest stands alternating with tundra) marks the contact between the tundra and taiga in most places.

Humid Continental Climate

South of the subarctic zone a triangular area of *humid continental climate* extends eastward from the western border of the USSR to the vicinity of Novosibirsk. At the west the triangle is more than 600 miles wide, but the Siberian portion averages only 100 to 200 miles in width. More than half the total population of the USSR is found within this climate zone. The humid continental climate of the USSR differs from the subarctic climate in having longer summers, milder winters, and more precipitation. However, these are only differences of degree. Even within the humid continental zone most places have an average frost-free period of 150 days or less, except at the extreme west (Leningrad, 160 days; Kiev, 172 days). Winters are long and cold, with average cold-month temperatures ranging from $-3°F$ (Novosibirsk) to $24°F$ (Riga). The average annual precipitation is relatively low, being only 18 to 25 inches at most places. However, the low rate of evaporation increases the effectiveness of the precipitation, so that the available moisture is generally adequate for the staple crops of the climate zone. The Soviet Union has the short-summer subtype of humid continental climate. A comparable climate is found in the Great Lakes region and the northern Great Plains of the United States and Canada. In this climate zone of the USSR mixed or deciduous forest supplants the evergreen taiga forest. Oak, ash, maple, elm, and other deciduous species alternate with coniferous trees in the vegetation cover. At the south is found the "wooded steppe," a transitional zone between the deciduous forest and the open steppe grassland.

Both climate and soil are more favorable for agriculture in the area of humid continental climate than in the subarctic zone. Soils developed under a cover of deciduous or mixed forest are normally more fertile by nature than the podzols of the taiga, although less fertile than grassland soils. The forest soils in the area of humid continental climate are of major importance in Soviet agriculture, supporting a varied development of crop and livestock farming. Among the climatic regions of the USSR, only the steppe region vies with the humid continental area in total farm production.

Steppe Climate

The grassy plains of Russia south of the forest zone are known as the *steppe,* or steppes. This term refers both to a type of climate and to the characteristic form of vegetation associated with it. On climatic maps the zone of steppe climate in the USSR is shown as an east-west band of varying width extending from the Romanian border to the Altai Mountains in Soviet Asia. As compared with the humid continental short-summer type of climate, the steppe climate is characterized by warmer summers, a somewhat longer frost-free season, and less precipitation. The average annual precipitation is 10 to 20 inches, an amount barely sufficient for growing crops without irrigation. Recurring periods of drought add to the hazards of farming in the steppe zone. Nevertheless, this part of the Soviet Union is a major area of crop and livestock production. The handicap of low and variable rainfall is partially offset by the fertility of the soils in the famous *black-earth belt*. This expanse of deep, black, exceedingly fertile soils extends for about 3000 miles in an east-west direction and 300 to 600 miles from north to south (map, p. 310). The characteristic soils of the belt are the *chernozems*—a term meaning "black earth." Chernozem soils are exceptionally thick, productive, and durable, being in fact among the best soils to be found anywhere. A similar belt of soils occurs in the eastern Great Plains of North America. The great fertility of the chernozems is largely due to an abundance of well-decomposed organic matter or *humus* in the topsoil and to the presence of sufficient lime to neutralize excessive acidity. Strictly speaking, the term humus refers to organic remains, both plant and animal, which have been decomposed by bacteria and have become part of the soil. Complete decomposition of humus makes available to plants the nutrients it contains, and its presence in the soil is essential to the development of a friable structure that permits easy cultivation and helps the soil to retain moisture. The accumulated remains of annual grasses in mid-latitude prairies and steppes are a better source of humus than the leaves and twigs of a forest. The steppe zone of the

USSR includes extensive areas of *chestnut soils* in the areas of lighter rainfall south of the chernozems. The chestnut soils are lighter in color than the chernozems and lack the superb fertility of the latter soils. Nevertheless, they are among the better soils of the world. In North America a belt of chestnut soils occurs to the west of the chernozems in the Great Plains.

Within the zone of steppe climate a considerable range of natural vegetation types is found, varying from tall grass and scattered forest stands in the "wooded steppe," where moisture is more abundant, to a sparse cover of low grasses and shrubs in the drier areas. The most characteristic form of natural vegetation is short grass, forming a carpet which is continuous, or nearly so. The treeless steppe grasslands, stretching monotonously over a vast area between the forest zone and the southern mountains and deserts, were tenanted from an early time by tribes of pastoral nomads with their flocks and herds. Today, however, much of the steppe is cultivated, with wheat as the main crop. In addition, it is a major producer of sugar beets, of sunflowers grown for vegetable oil, and of various other crops.

Some confusion is occasioned by the fact that the wooded steppe and a portion of the black-earth belt extend into the zone of humid continental climate as customarily defined on climatic maps. However, the treeless part of the steppe is largely outside of the humid continental zone, as is the larger part of the black-earth belt.

Desert Climate

To the east and immediate north of the Caspian Sea, the steppe zone trends gradually into an extensive area of desert. In the desert zone rainfall is even more scanty and erratic than in the steppe. Widely spaced shrubs and occasional tufts of grass afford only the sparsest pasturage for livestock. Agriculture is largely precluded by lack of moisture, except for oases watered by streams that originate in the high mountains which border this desert region on the south and east. The desert zone of the Soviet Union includes two extensive areas of sandy desert. Such deserts are a type made familiar by the movies, but are much less common on a world basis than deserts floored by combinations of sand with gravel, rock fragments, or the bare bedrock.

Subtropical, Monsoon, and
Highland Climates

A number of relatively small but distinctive climatic areas remain to be mentioned. The south coast of the Crimean Peninsula, sheltered by the Yaila Mountains, has the mild temperatures and summer rainfall minimum associated with the *mediterranean* or *dry-summer subtropical* climate. This picturesque area, with its orchards, vineyards, and resorts, of which Yalta is the most famous, is sometimes referred to as the "Russian Riviera." A subtropical climate also prevails in the coastal lowlands and valleys south of the high Caucasus Mountains. Mild winters and warm to hot summers are characteristic throughout the lowlands of Transcaucasia. However, the rainfall is very unevenly distributed. The lowlands bordering the Black Sea in western Transcaucasia receive the heaviest rainfall of the USSR (50 to 100 inches annually) and are classed as *humid subtropical* in climate, while the lowlands of eastern Transcaucasia, bordering the Caspian Sea, receive so little precipitation in most places that they are classed as steppe.

The coastal regions of the Soviet Far East, from Vladivostok northward to the mouth of the Amur River, have a humid continental climate characterized by a distinct *monsoon* tendency. As in nearby parts of Korea and of China, most of the annual rainfall results from moist onshore winds of the summer monsoon. In contrast, the cold outflowing winds of the winter monsoon produce little precipitation. The average annual precipitation is 20 to 30 inches, of which three-fourths or more falls from April through September.

The mountains of the USSR are characterized by climates varying according to altitude and exposure to wind and sun. However, the range of climates is smaller than in mountains located closer to the equator.

Climatic data for selected Soviet stations are given in Table 10.

TERRITORIAL EVOLUTION
OF MODERN RUSSIA

Territorial and economic expansion, so much in evidence during recent decades, is not a new theme in Russian history. The gigantic Russian state has reached its present limits through a long process of colonization, conquest, territorial annexation, and associated economic development extending back for more than a thousand years.

The modern phase of this process began toward the end of the European Middle Ages. In the second half of the fifteenth century Ivan III, the ruler of Muscovy, a small feudal state with its capital at Moscow, began to expand his holdings rapidly by military con-

TABLE 10 CLIMATIC DATA FOR SELECTED SOVIET STATIONS

STATION	LATITUDE AND LONGITUDE TO NEAREST WHOLE DEGREE	ELEVATION ABOVE SEA LEVEL (FEET)	TYPE OF CLIMATE	AVERAGE TEMPERATURE (DEGREES F TO NEAREST WHOLE DEGREE)			AVERAGE LENGTH OF FROST-FREE SEASON (DAYS)	AVERAGE ANNUAL PRECIPITATION TO NEAREST INCH
				AN-NUAL	JAN-UARY	JULY		
Arkhangelsk (Archangel)	65°N., 87°E.	50′	Subarctic	35°	11°	61°	120	21″
Igarka	67°N., 87°E.	115′	Subarctic	17°	−20°	59°	no data	16″
Nordvik	74°N., 111°E.	102′	Tundra	7°	−21°	41°	no data	5″
Verkhoyansk	68°N., 133°E.	400′	Subarctic	5°	−52°	60°	65	6″
Leningrad	60°N., 30°E.	30′	Humid continental	40°	18°	65°	160	22″
Moscow	56°N., 38°E.	480′	Humid continental	40°	14°	66°	130	23″
Sverdlovsk	57°N., 61°E.	925′	Humid continental	35°	6°	64°	no data	18″
Novosibirsk	55°N., 83°E.	436′	Humid continental	31°	−3°	66°	122	15″
Irkutsk	52°N., 104°E.	1532′	Subarctic	31°	−5°	64°	95	18″
Vladivostok	43°N., 132°E.	95′	Humid continental	40°	7°	68° (August)	152	31″
Odessa	46°N., 31°E.	210′	Steppe	50°	28°	72°	208	15″
Yalta	45°N., 34°E.	135′	Mediterranean	56°	39°	75°	245	20″
Batumi	42°N., 42°E.	20′	Humid subtropical	58°	43°	73°	308	93″
Baku	40°N., 50°E.	0′	Steppe	57°	38°	77°	296	10″
Akmolinsk	51°N., 71°E.	1148′	Steppe	35°	0°	70°	no data	11″
Tashkent	41°N., 69°E.	1568′.	Steppe	56°	32°	81°	206	16″
Krasnovodsk	40°N., 53°E.	−56′	Desert	58°	36°	82°	no data	4″

quest or the threat of conquest. To the north the rival principality of Novgorod was annexed and a domain thus secured which extended to the Arctic Ocean and eastward to the Ural Mountains. Ivan IV, the Terrible [or Dread], added large new territories by conquering the Tatar (Tartar) khanates of Kazan and Astrakhan in 1552 and 1554, respectively, thus giving the Russian state control over the entire course of the Volga River. Upon ascending the throne in 1547 Ivan was crowned as tsar (czar or Caesar), being the first Russian ruler to assume this title. Later tsars pushed the frontiers of Russia westward toward Poland and the Baltic Sea and southward toward the Black Sea. Peter the Great (reigned 1689–1725) gained a secure foothold on the Baltic Sea by defeating the Swedes under Charles XII, and Catherine the Great (reigned 1762–1796) secured a frontage on the Black Sea at the expense of Turkey.

Meanwhile the conquest of Siberia had proceeded rapidly. This vast, thinly populated wilderness between the Urals and the Pacific Ocean was already being penetrated by traders and Cossack military pioneers at the end of Ivan the Dread's reign. In 1639 a Cossack expedition reached the Pacific. Russian expansion toward the east did not stop at Bering Strait, but continued down the west coast of North America as far as northern California, where a Russian trading post existed between 1812 and 1841. In 1867, however, Alaska was sold to the United States and Russia withdrew from North America.

Role of the Rivers in Russian Expansion

Early Russian expansion in Eurasia followed the river lines. The Moscow region lies in a low upland from which a number of large rivers radiate like spokes of a wheel. The longest rivers lead southward (map, pp. 284–285): the Volga to the landlocked Caspian Sea, the Dnieper (Dnepr) to the Black Sea, and the Don to the Sea of Azov, which is connected with the Black Sea through a narrow strait. Shorter rivers lead north and northwest to the Arctic Ocean and the Baltic Sea. These river systems are accessible to each other by easy por-

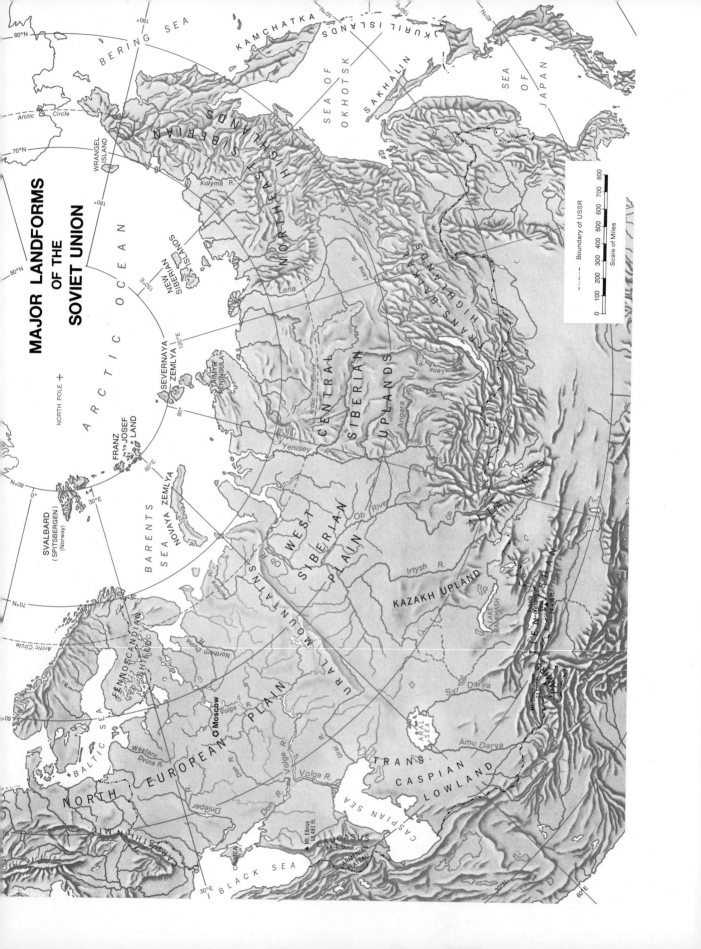

MAJOR LANDFORMS
OF THE
SOVIET UNION

------- Boundary of USSR

Scale of Miles

0 100 200 300 400 500 600 700 800

BERING SEA

KAMCHATKA

KURIL ISLANDS

SEA OF JAPAN

SEA OF OKHOTSK

SAKHALIN

NORTHEAST SIBERIAN HIGHLANDS

Arctic Circle

WRANGEL ISLAND

Kolyma R.

NEW SIBERIAN ISLANDS

ARCTIC OCEAN

NORTH POLE +

SEVERNAYA ZEMLYA

TAIMYR PENINSULA

FRANZ JOSEF LAND

Lena

Lena R.

Arctic Circle

Aldan

TRANS-BAIKAL HIGHLANDS

Amur

CENTRAL SIBERIAN UPLANDS

Angara

Vitim

Yenisey

SVALBARD (SPITSBERGEN) (Norway)

BARENTS SEA

NOVAYA ZEMLYA

Pechora R.

ALTAI MTS.

Ob R.

Ob River

Ob R.

WEST SIBERIAN PLAIN

FENNOSCANDIAN SHIELD

URAL MOUNTAINS

Irtysh R.

KAZAKH UPLAND

Pobeda (Victory) Peak 24,406 ft.

TIEN SHAN

Northern Dvina R.

BALTIC SEA

Moscow

Volga

Western Dvina R.

Don R.

NORTH EUROPEAN PLAIN

Dnieper

Don

Volga R.

Ural R.

Volga R.

Syr Darya

LAKE BALKHASH

ARAL SEA

Amu Darya

PAMIR

Mt. Communism 24,590 ft.

TRANS-CASPIAN LOWLAND

CASPIAN SEA

CARPATHIAN MTS.

CRIMEA

Mt. Elbrus 18,481 ft.

CAUCASUS MTS.

ARMENIAN PLATEAU

BLACK SEA

tages. In the early history of Russia the rivers formed natural passageways for trade, conquest, and colonization. The technique of expansion followed by the tsars was to dominate the river lines by *ostrogs* (blockhouses) built at portages and other strategic places. From these strong points political control could gradually be extended over the hinterland.

Expansion in Siberia

The rivers were especially valuable aids to expansion in the enormous reaches of Siberia. The latter region, half again as large as the United States, is drained by some of the greatest rivers on earth: the Ob, Yenisey, Lena, and Kolyma, flowing to the Arctic Ocean, and the Amur, flowing to the Pacific. By following these rivers and their lateral tributaries the Russians advanced from the Urals to the Pacific in less than a century. The latter movement has been summarized as follows by a specialist on Russian expansion, Robert J. Kerner:[7]

The earliest background of the eastward movement across the Urals is to be found in the fur-trading enterprise of Novgorod. Daring merchants and trappers from Novgorod exploited the lower reaches of the Ob from about the fourteenth century by portaging from the tributaries of the Pechora. They and the Muscovites, who carried out expeditions in 1465, 1483, and 1499, raided the inhabitants beyond the Urals for the purpose of obtaining tributes of furs, of which there was a diminishing supply in European Russia. The Russian raids were often followed by counterraids of Siberian natives, which endangered the security of the Ural frontier. it was the latter which especially concerned Tsar Ivan the Dread. He received news that the Volga pirate Yermak, who was wanted for offenses against the laws of tsardom, had, in the employ of the Novgorodian family of the Stroganovs, raided beyond the Urals. The tsar, in fact, ordered the Stroganovs to bring him back for trial. He feared the Ural frontier would be overrun by the tribesmen of the Tatar khan, Kuchum. Yermak's success in capturing Sibir, the capital of Siberia, caused the tsar to change his intentions in regard to Yermak; instead of beheading him he gave Yermak his blessing and a real coat of armor. Incidentally, it was this heavy accoutrement that caused Yermak to lose his life by drowning.

Moscow took over in 1538 and ended the practice of raids. It initiated a planned domination of rivers and portages through the building of blockhouses, called *ostrogs* in Russian. This was in line with centuries of Russian tradition in Europe. The original motive for the advance into Siberia was the acquisition of furs. Moscow sought to add to it the search for gold and silver. The conquest of the Tatar khanate gave security to the Ural frontier and created a base for further expansion. Thus from its origins to the present day, Russian rule in Asia was planned and regimented from Moscow.

The first Russians in Siberia were fur merchants and trappers, government officials, Cossacks, and Orthodox priests. The advance was rapid, once the khanate had been subdued. Within a decade and a half after Yermak's death the basin of the Ob had become a Russian possession with the ostrog of Tobolsk on the Irtysh, founded in 1587, as the key, but with ostrogs guarding the route to it from Russia and others built on the lower reaches of the Ob and the upper reaches of its tributaries. The pattern set in the Ob basin was followed, in the years 1607–1625, in that of the Yenisey, directly to the east with its center at Yeniseysk (1618). The basin of the Lena was occupied between 1630 and 1648, with its center at Yakutsk, an ostrog founded in 1632. The Russians reached the Pacific in 1639. Their occupation of the Lake Baikal and Amur region, with its center at Irkutsk (1652) followed in the 'fifties and 'sixties. Forced out of the Amur River region by the Chinese under the Manchu emperor Kang-hsi by the negotiations which resulted in the Treaty of Nerchinsk (1689), the Russians lost the base of an adequate supply of grain and vegetables. This was to make their hold on eastern Siberia precarious for nearly two centuries. Even more than that, they lost easy access to the Pacific, which they did not regain until 1858–1860.

Large areas in the Caucasus region and in Turkestan (the arid or semiarid area east of the Caspian Sea) were not secured by Russia until the nineteenth century. These regions, and Siberia as well, were administered as colonial areas by the tsars.

The Role of Topography

Most of the important rivers of Russia wind slowly for hundreds or thousands of miles across vast expanses of plain. Early expansion was facilitated not only by the long, continuous highways provided by the rivers themselves, but also by the easy overland connections between river systems. Only in the extreme south and east of Russia are the river basins separated by ranges of high mountains. Elsewhere the divides are ordinarily found in areas of low hills, or even in level plains where the

---

[7] Robert J. Kerner, "The Russian Eastward Movement: Some Observations on Its Historical Significance." © 1948 by The Pacific Coast Branch, American Historical Association. Reprinted from *Pacific Historical Review*, 17 (1948), 136–137, by permission of the Branch. Spellings of some proper names have been slightly modified.

gradients between headstreams are so gentle as to be scarcely perceptible.

Plains and low hills occupy nearly all of Russia from the Yenisey River to the western border of the country (map, p. 300). The only mountains which rise in the midst of this vast lowland are the Urals, a low range located about midway between the western frontier of Russia and the Yenisey. The Urals trend due north and south, but do not occupy the full width of the lowland. A wide gap between the southern end of the mountains and the Caspian Sea permits uninterrupted east-west movement. Actually the Urals themselves do not constitute a serious barrier to transportation, the main range being less than 100 miles wide in most places and cut by river valleys offering easy passageways through the mountains. Today the main rail lines connecting Moscow with Siberia pass directly through the Urals. In general form and elevation these mountains bear many resemblances to the Appalachian Mountains of the United States. Like the Appalachians, they are old, worn-down mountains with rounded contours. The average elevation is less than 2000 feet, and the highest summit, located toward the northern end of the range, is only 6214 feet above sea level. The central Urals are especially low, being little more than high hills. Rail lines connecting Moscow with the important Ural city of Sverdlovsk cross this part of the range over a divide lying at only 1350 feet above sea level. The southern and central Urals are forested, but at the extreme north the mountains extend into the zone of Arctic tundra.

The Russian lowlands west of the Urals are mostly undulating or rolling rather than flat. To the east, however, between the mountains and the Yenisey River, the great lowland of western Siberia is one of the flattest areas on earth. Much of it is covered by immense swamps and marshes through which the Ob River and its tributaries slowly wend their way. This waterlogged country, underlain by a permanently frozen subsoil that blocks drainage, is a major barrier to land transportation and is extremely uninviting to settlement. Large sections are almost devoid of people. In the spring tremendous floods occur when the breakup of ice in the upper basin of the Ob releases great quantities of water while the river channels farther north are still frozen.

The area between the Yenisey and Lena rivers is occupied by hilly uplands lying at a general elevation of 1000 to 1500 feet. Occasional summits reach 3000 feet. East of the Lena River and Lake Baikal the landscape is dominated by mountains. Extreme northeastern Siberia is especially wild and difficult mountain country, with a few peaks reaching elevations of 10,000 feet or more. This bleak region is one of the least inhabited and least known areas on earth.

High mountains border Russia on the south from the Black Sea to Lake Baikal, and lower mountains from Lake Baikal to the Pacific. Elevations of 15,000 feet or higher are reached in the Caucasus Mountains between the Black and Caspian Seas and in the Pamir, Tien Shan, and Altai mountains east of the Caspian Sea. From the foot of the latter ranges and lower ranges between the Pamirs and Caspian Sea, arid or semiarid plains and low uplands extend northward and gradually merge with the West Siberian Plain and the broad plains and low hills lying west of the Urals.

REFERENCES and READINGS

General Geographies, Atlases, Bibliographies, and Journals

An Atlas of Soviet Affairs. Text by Robert N. Taaffe; maps by Robert C. Kingsbury (New York: Frederick A. Praeger, 1965).

Atlas Mira [Atlas of the World] (Moscow: Glavnoye Upravleniye Geodezii i Kartografii, 1967). A huge, expensive reference atlas with magnificent cartography. Available in Russian-language edition or English-language edition.

COLE, J. P., *A Geography of the USSR* (Baltimore, Md.: Penguin Books, 1967); and, with F. C. German, *A Geography of the USSR: The Background to a Planned Economy* (Washington, D.C.: Butterworth, Inc., 1961).

CRESSEY, GEORGE B., *Soviet Potentials: A Geographic Appraisal* (Syracuse, N.Y.: Syracuse University Press, 1962).

DEWDNEY, JOHN C., *A Geography of the Soviet Union* (New York: Pergamon Press, 1965).

EAST, W. GORDON, *The Soviet Union* (Princeton, N.J.: D. Van Nostrand Co., Searchlight Books, 1963).

Geograficheskiy Atlas dlya Uchiteley Sredney Shkoly [Geographical Atlas for Teachers in Middle Schools] (rev. ed.; Moscow: Glavnoye Upravleniye Geodezii i Kartografii, 1959). Text in Russian. English *Guide* (mimeographed) by John P. Cole and Dennis R. Mills (Department of Geography, University of Nottingham, 1956). A world atlas, but with heavy emphasis on the Soviet Union. Fine cartography.

GREGORY, JAMES S., *Russian Land: Soviet People: A Geographical Approach to the USSR* (London: George G. Harrap and Co., 1968).

HOOSON, DAVID J. M., *The Soviet Union: People and Regions* (Belmont, Calif.: Wadsworth Publishing Co., 1966).

HORECKY, PAUL L., ed., *Basic Russian Publications: An Annotated Bibliography on Russia and the Soviet Union* (1962), especially Chap. 2, "The Land," by Chauncy D. Harris, pp. 25–48; and, ed., *Russia and the Soviet Union: A Bibliographic Guide to Western-Language Publications* (1965), especially Chap. 3, "The Land," by Chauncy D. Harris, pp. 46–54 (Chicago: University of Chicago Press).

JORRÉ, GEORGES, *The Soviet Union,* trans. by E. D. Laborde, rev. by C. A. Halstead (3d ed.; New York: John Wiley & Sons, 1967).

LYDOLPH, PAUL E., *Geography of the U.S.S.R.* (New York: John Wiley & Sons, 1964).

MELLOR, ROY E. H., *Geography of the U.S.S.R.* (New York: St. Martin's Press, 1964).

Oxford Regional Economic Atlas: The U.S.S.R. and Eastern Europe (New York: Oxford University Press, 1956).

SHABAD, THEODORE, *Geography of the USSR: A Regional Survey* (New York: Columbia University Press, 1951). An encyclopedic survey, outdated in many respects, but still a useful reference.

Soviet Geography: Review and Translation (ten issues a year). Innumerable articles on Soviet geography, translated from the Russian.

Soviet Studies (quarterly). Numerous articles on the economy and society of the USSR.

Other General References

FITZSIMMONS, THOMAS, ed., *USSR: Its People, Its Society, Its Culture* (New Haven, Conn.: Human Relations Area Files Press, 1960).

FLORINSKY, MICHAEL T., ed. *McGraw-Hill Encyclopedia of Russia and the Soviet Union* (New York: McGraw-Hill Book Company, 1961).

FRENCH, R. A., "The Making of the Russian Landscape," *Advancement of Science,* **20**, no. 83 (May 1963), 44–56; and "Contemporary Landscape Change in the U.S.S.R.," in Robert W. Steel and Richard Lawton, eds., *Liverpool Essays in Geography: A Jubilee Collection* (London: Longmans, Green and Co., 1967), pp. 547–563.

MAXWELL, ROBERT, ed. and comp., *Information U.S.S.R.: An Authoritative Encyclopedia about the Union of Soviet Socialist Republics* (New York: Pergamon Press, 1962). Translated from Soviet materials.

SALISBURY, HARRISON E., ed., *The Soviet Union: The Fifty Years* (A New York Times Book; New York: Harcourt, Brace & World, 1968). Chapters by different writers give an interesting and up-to-date view of many aspects of life in the Soviet Union today.

Soviet Geography: Accomplishments and Tasks (American Geographical Society, Occasional Publication No. 1, New York: American Geographical Society, 1962). A symposium prepared by prominent Soviet geographers, translated from the Russian by Lawrence Ecker; English edition edited by Chauncy D. Harris. A survey and appraisal of scholarly work in the field of geography in the Soviet Union.

SPATE, O. H. K., "Theory and Practice in Soviet Geography," *Australian Geographical Studies,* **1**, no. 1 (April 1963), 18–30.

THAYER, CHARLES W., and the Editors of *Life, Russia* (Life World Library; New York: Time Inc., 1965).

Historical Background and Political Relationships

ADAMS, ARTHUR E., IAN M. MATLEY, and WILLIAM O. McCAGG, *An Atlas of Russian and East European History* (New York: Frederick A. Praeger, 1966).

CARMICHAEL, JOEL, *An Illustrated History of Russia,* ed. by Georges and Rosamond Bernier (New York: Reynal & Company, 1960).

CHEW, A. F., *An Atlas of Russian History: Eleven Centuries of Changing Borders* (New Haven, Conn.: Yale University Press, 1967).

DEUTSCHER, ISAAC, *The Unfinished Revolution: Russia, 1917–1967* (New York: Oxford University Press, 1967).

EARLE, EDWARD MEAD, ed., *Makers of Modern Strategy: Military Thought from Machiavelli to Hitler* (Princeton, N.J.: Princeton University Press, 1943), Chap. 14, "Lenin, Trotsky, Stalin: Soviet Concepts of War," pp. 322–364.

EAST, W. GORDON, "The New Frontiers of the Soviet Union," *Foreign Affairs,* **29,** no. 4 (July 1951), 591–607; and, with A. E. Moodie, eds., *The Changing World: Studies in Political Geography* (New York: Harcourt, Brace & World, 1956), Chaps. by East as follows: 14, "The Soviet Union: Its Geographical Setting and Historical Background," pp. 347–364; 17, "Land Frontiers and Frontier Problems of the Soviet Union," pp. 410–431; 18, "The Soviet Union and the 'Heartland,'" pp. 432–450.

FITZGERALD, C. P., "Tension on the Sino-Soviet Border," *Foreign Affairs,* 45, no. 4 (July 1967), pp. 683–693.

FULLARD, HAROLD, ed., *Soviet Union in Maps: Its Origin and Development* (1965 ed.; distributed by Denoyer-Geppert Co., Chicago, Ill).

THE GEOGRAPHER, U.S. DEPARTMENT OF STATE, International Boundary Studies ("– – – -U.S.S.R. Boundary"), nos. 24, *Norway* (1963); 25, *Iran* (1963); 26, *Afghanistan* (1963); 29, *Turkey* (1964); 43, *Rumania* (1964); 59, *Korea* (1965); 64, *China* (1966); 74, *Finland* (1967); 76, *Hungary* (1967); 77, *Czechoslovakia* (1967).

JACKSON, W. A. DOUGLAS, *The Russo-Chinese Borderlands: Zone of Peaceful Contact or Potential Conflict?* (2d ed.; Princeton, N.J.: D. Van Nostrand Co., Searchlight Books, 1968); and "Mackinder and the Communist Orbit," *Canadian Geographer,* 6, no. 1 (Spring 1962), 12–21.

JONES, SHEILA, *A Student's History of Russia* (New York: Pergamon Press, 1966).

KARAN, P. P., "The Sino-Soviet Border Dispute," *Journal of Geography,* **63,** no. 5 (May 1964), 216–222.

KEEP, JOHN LESLIE HOWARD, "Fifty Years of Communism, 1917–1967," *Britannica Book of the Year, 1967* (Chicago: Encyclopaedia Britannica, Inc., 1967), pp. 770–772.

KERNER, ROBERT J., "The Russian Eastward Movement: Some Observations on Its Historical Significance," *Pacific Historical Review,* **17,** no. 2 (May 1948), 135–148.

SMITH, ROBERT E. F., "Russian History and the Soviet Union," *Comparative Studies in Society and History,* 4, no. 3 (April 1962), 373–387.

WOLFE, BERTRAM D., *Three Who Made a Revolution: A Biographical History* (2d ed.; New York: The Dial Press, 1960), Chap. 1, "The Heritage," pp. 11–38.

Planned Economy of the Soviet Union

BERNARD, PHILIPPE J., *Planning in the Soviet Union,* trans. from the French by I. Nove (New York: Pergamon Press, 1966).

BURCK, GILBERT, "The Toughest Management Job in the World," *Fortune,* 74, no. 1 (July 1 1966), 73–79 ff.; and "The Auspicious Rise of the Soviet Consumer," *Fortune* 74, no. 3 (August 1966), 130–168.

CHABE, ALEXANDER M., "Soviet Economic Life at Mid-Century," *Journal of Geography,* 67, no. 8 (November 1968), 490–498.

GOLDMAN, MARSHALL I., "Economic Revolution in the Soviet Union," *Foreign Affairs,* **45,** no. 2 (January 1967), 319–331; "The Soviet Standard of Living and Ours," *Foreign Affairs,* **38,** no. 4 (July 1960), 625–637; and *Soviet Foreign Aid* (New York: Frederick A. Praeger, 1967).

GRUNWALD, HENRY ANATOLE, "The Quality of Life Behind the Soviet Statistics," *Fortune,* 69, no. 3 (March 1964), 146–147 ff.

JERSCHOWSKY, OLEG, and LEON LEWINS, *Basic Data on the Economy of U.S.S.R.,* U.S. Bureau of International Commerce, Overseas Business Reports, OBR 66–9 (April 1966). (Washington, D.C.: Government Printing Office, 1966.)

KISH, GEORGE, with the assistance of IAN M. MATLEY and BETTY BELLAIRE, *Economic Atlas of the Soviet Union* (Ann Arbor: University of Michigan Press, 1960).

LIBERMAN, YEVSEI, "The Soviet Economic Reform," *Foreign Affairs,* 46, no. 1 (October 1967), 53–63.

New Directions in the Soviet Economy, published in 5 parts, 1966; and *Soviet Economic Performance 1966–67,* published in 1968 (studies prepared for the Subcommittee on Foreign Economic Policy of the Joint Economic Committee, U.S. Congress; Washington, D.C.: Government Printing Office).

POWELL, RAYMOND P., "Economic Growth in the U.S.S.R.," *Scientific American,* 219, no. 6 (December 1968), 17–23.

"The Russian Revolution–Fifty Years After," *Foreign Affairs,* 46, no. 1 (October 1967), 1–94. A series of articles.

SOSNOVY, TIMOTHY, "The New Soviet Plan: Guns Still Before Butter," *Foreign Affairs,* 44, no. 4 (July 1966), 620–632.

TURKEVICH, JOHN, "Soviet Science Appraised," *Foreign Affairs,* 44, no. 3 (April 1966), 489–500.

Soviet Agriculture

Atlas Sel'skogo Khozyaistya S.S.S.R. [Atlas of Agriculture of the U.S.S.R.] (Moscow: Glavnoye Upravleniye Geodezii i Kartografii, 1960). An outstanding reference atlas, with text in Russian.

BIRDSALL, STEPHEN S., "The Effect of Management on Crop Yields in Soviet Agriculture," *Journal of Geography,* 67, no. 2 (February 1968), 95–103.

BROWN, G. STANLEY, "Soviet Agriculture after Khrushchev: A Brief Survey," *Foreign Agriculture,* 5, no. 11 (March 13, 1967), 3–6.

DALRYMPLE, DANA G., "American Technology and Soviet Agricultural Development, 1924–1933," *Agricultural History,* 40, no. 3 (July 1966), 187–206.

DURGIN, FRANK A., "The Virgin Lands Programme 1954–1960," *Soviet Studies,* 13, no. 3 (January 1962), 255–280.

JACKSON, W. A. DOUGLAS, "The Virgin and Idle Lands Program Reappraised," *Annals of the Association of American Geographers,* 52, no. 1 (March 1962), 69–79.

KARCZ, JERZY F., ed., *Soviet and East European Agriculture* (Los Angeles: University of California Press, 1967).

LAIRD, ROY D., ed., *Soviet Agricultural and Peasant Affairs* (Lawrence: University of Kansas Press, 1963); and, ed., *Soviet Agriculture: The Permanent Crisis* (New York: Frederick A. Praeger, 1965).

NOVE, ALEC, "Soviet Agriculture Marks Time," *Foreign Affairs,* 40, no. 4 (July 1962), 576–594.

SYMONS, LESLIE, "Agricultural Production and the Changing Roles of State and Collective Farms in the U.S.S.R.," *Pacific Viewpoint,* 7, no. 1 (May 1966), 54–66.

U.S. DEPARTMENT OF AGRICULTURE, *Economic Aspects of Soviet Agriculture* (1959); *Livestock in the Soviet Union* (1961); *Soviet Agriculture Today* (1963); *Agriculture in the United States and the Soviet Union* (1963); and *The Europe and Soviet Union Agricultural Situation* (annual). (Washington, D.C.: Government Printing Office.)

Natural Environment and Resources

ARMSTRONG, TERENCE, "Farming on the Permafrost" [of northern Siberia], *Geographical Magazine,* 40, no. 11 (March 1968), 961–967.

BERG, L. S., *Natural Regions of the U.S.S.R.,* trans. from the Russian by Olga Adler Titelbaum, ed. by John A. Morrison and C. C. Nikiforoff (New York: The Macmillan Company, 1950).

BONE, ROBERT M., "The Soviet Forest Resource," *Canadian Geographer,* 10, no. 2 (1966), 94–116; see also Brenton M. Barr, "The Importance of Regions in Analyses of the Soviet Forest Resource: A Reply," *Canadian Geographer,* 10, no. 4 (1966), 234–237.

BORISOV, A. A., *Climates of the U.S.S.R.,* trans. by R. A. Ledward from the Russian of the 2d ed.; ed. by Cyril A. Halstead (Chicago: Aldine Publishing Company, 1965).

BROWN, R. J. E., "Comparison of Permafrost Conditions in Canada and the U.S.S.R.," *Polar Record,* 13, no. 87 (September 1967), 741–751.

FIELD, N. C., "Environmental Quality and Land Productivity: A Comparison of the Agricultural Land Base of the USSR and North America," *Canadian Geographer,* 12, no. 1 (1968), 1–14.

Fiziko-Geograficheskiy Atlas Mira [Physico-Geographical Atlas of the World] (Moscow: Akademiya Nauk SSSR i Glavnoye Upravleniye Geodezii i Kartografii GGK SSSR). A magnificent reference atlas with text in Russian. An English translation of legends is available in *Soviet Geography: Review and Translation.* 6, nos. 5–6 (May–June 1965), entire issue.

HARRIS, CHAUNCY D., "U.S.S.R. Resources: Agriculture," rev. ed., *Focus,* 13, no. 5 (January 1963), 6 pp.; and "U.S.S.R. Resources for Heavy Industry," *Focus,* 13, no. 7 (March 1963), 6 pp.

HODGKINS, JORDAN A., *Soviet Power: Energy Resources, Production and Potentials* (Englewood Cliffs, N.J.: Prentice-Hall, 1961.

LYDOLPH, PAUL E., "The Russian Sukhovey," *Annals of the Association of American Geographers,* 54, no. 3 (September 1964), 291–309. Deals with the hot, dry winds that periodically damage crops in the steppe regions of the Soviet Union.

MATLEY, IAN M., "The Marxist Approach to the Geographical Environment," *Annals of the Association of American Geographers,* 56, no. 1 (March 1966), 97–111.

MELLOR, R. E. H., "Some Influences of Physical Environment upon Transport Problems in the Soviet Union," *Advancement of Science,* 20, no. 88 (March 1964), 564–571.

MIROV, N. T., *Geography of Russia* (New York: John Wiley & Sons, 1951). Primarily a physical geography.

SHABAD, THEODORE, "The Resources of a Nation," in Harrison E. Salisbury, ed., *The Soviet Union: The Fifty Years* (New York: Harcourt, Brace & World, 1967), pp. 222–238.

SHIMKIN, DEMITRI B., "Resource Development and Utilization in the Soviet Economy," in Marion Clawson, ed., *Natural Resources and International Development* (Baltimore, Md.: Johns Hopkins Press for Resources for the Future, 1964), pp. 155–238.

TSEPLYAEV, V. P., *The Forests of the Soviet Union,* trans. from the Russian by A. Gourevitch (Jerusalem: Israel Program for Scientific Translations; New York: Daniel Davey and Co., 1965).

See also the reference lists for Chapters 1–3 and the lists of general geographies of Europe and Asia, pp. 92 and 415. Textbooks on the geography of Europe or Asia frequently include a section on the Soviet Union.

Population and
Major Territorial Divisions
of the USSR

14

In 1959 the Soviet Union held its first nationwide census since 1939. The results showed a population of 209 million, as compared with an estimated 1939 population of 193 million and an estimated total in 1913 of 159 million for the present national territory of the USSR. The increase in population from 1939 to 1959 would have been much higher had World War II not intervened. It is estimated that this war cost the Soviet Union 30 million people or more in deaths and unborn children. In the 1960s the overall rate of population increase declined, but absolute growth was sufficient to give the country an estimated total of 239 million people in mid-1968. During recent years the general pattern of population growth in the Soviet Union has been rather similar to that of the United States. In both countries the annual rate of population increase was around 1.6 to 1.8 percent in the 1950s, but by 1965 the estimated rate had dropped to 1.1 percent in the United States and 1.2 percent in the Soviet Union. The decrease was due to a declining birth rate in both cases. Limited housing and a desire for a better life have noticeably reduced the size of urban families in the USSR, particularly in the larger cities. In the Soviet Union birth rates much higher than the national average are found among some non-Slavic peoples, the most notable example being the Moslems of Soviet Middle Asia.

The Soviet Union's population is very heavily concentrated in the western parts of the country. In 1965 an estimated 68 percent of the population lived west of the main Ural Mountain divide and north of the Caucasus divide, and an additional 5 percent lived west of the Caspian Sea and south of the Caucasus crest. About 12½ percent were in the five republics of Soviet Middle Asia, and the remaining 14½ percent were in Siberia. Of those in Siberia, nearly a third lived along or very near the eastern slopes of the Urals. The remainder of Siberia, an area considerably larger than the United States, had only 25 million people and was increasing in population more slowly than the USSR as a whole. There was a pronounced tendency for outmigration of workers seeking jobs in urban and industrial areas of the western USSR. In Siberia, living conditions are generally less attractive and living costs are higher than in European Russia. The climate is harsher; housing, heating, and transportation are more expensive to provide and maintain; the food supply is less varied and more costly; and social services and amenities are less adequate. Incentives in the form of higher pay and fringe benefits have not been sufficient in many cases to counterbalance the disadvantages of life in the east, and many Siberian enterprises have found it very difficult to maintain a stable supply of labor. This tendency has been accentuated by trends in Soviet economic planning. Today there is more emphasis on purely economic considerations than was formerly the case, and this tends to channel new investment into the more developed western sections of the country rather than the less developed east. Certain lines, such as metals, fuels, hydroelectricity, and timber, are pushed vigorously in the east, but for many types of production preference is given to the west, with its greater market potential, its larger and more diversified pool of skilled labor, and its greater development of services. The resulting employment opportunities tend to drain workers, especially younger people, from the areas east of the Urals.

RECENT URBAN DEVELOPMENT

For several decades the Soviet Union, like the United States and many other countries, has been witnessing a large migration from rural districts into the cities. An American geographer, Chauncy D. Harris, has written as follows concerning urbanization in the period 1926–1939:[1]

The rapid growth of cities in the Soviet Union between the 1926 and 1939 censuses is unequaled either in the earlier history of Russia or in the urban development of other countries. The numerical increase in urban population in the Soviet Union in this 12-year period exceeded the total numerical increase in all the previous centuries of Russian history. During this time the urban population more than doubled (112.5 per cent increase), in contrast with the rural population, which decreased slightly. Of the urban increase of 29.6 million, 5.4 million came from the natural increase within the urban areas themselves and 24.2 million from rural areas. The proportion of the total population living in cities increased from 17.9 per cent to 32.8 per cent. A comparable increase in degree of urbanization required about 31 years in the United States (1856–1887) and probably about a century in most European countries, though exact comparisons are difficult because of the incompleteness of early statistics.

• • • • •

Urbanization has been part of the industrialization that has been planned and executed by the government in a series of 5-year plans. The long-delayed impact of the Industrial Revolution has struck in the Soviet Union with a suddenness unknown in Western Europe or the United States.

From 1939 to 1959 the number of urban dwellers continued to climb (from 60 million to 100 million within the present national territory—a much slower *rate* of increase than from 1926–1939), while the rural population decreased from 130 million to 109 million. The population of the USSR was listed as 48 percent urban by the 1959 census, and was probably 54 or 55 percent urban in 1968. It should be noted, however, that as late as the early 1960s more than two-fifths of all employed workers were still in agriculture, as against somewhat less than a third in manufacturing and mining. The Soviet Union still has far to go before it becomes a truly urban country on the order of the United States, Great Britain, or West Germany.

THE SLAVIC MAJORITY

Slavic peoples represent the dominant population element in the USSR, both in numbers and political and eco-

[1] Chauncy D. Harris, "The Cities of the Soviet Union," *Geographical Review,* 35 (1945), 104–112. Used by permission of the author and the American Geographical Society of New York. Three footnotes are omitted.

nomic power. The major Slavic groups are the Great Russians, often referred to simply as Russians, the Ukrainians, and the Belorussians or White Russians. Great Russians constitute about 55 percent, Ukrainians about 17 or 18 percent, and Belorussians about 4 percent of the total population of the USSR. Together the three groups comprise slightly more than three-quarters of the population, a proportion which has changed little since 1927.

Early Scandinavian and Byzantine Influences

Slavic peoples have inhabited Russia since the early centuries of the Christian era. During the Middle Ages Slavic tribes living in the forested regions of western Russia came under the influence of Viking adventurers from Scandinavia. These newcomers, known as Rus or Varangians, "fought and traded up and down the waterways between the Baltic and Black seas."[2] In the ninth century they were instrumental in organizing a number of principalities in the upper basin of the Dnieper River and in the region extending northward from the Dnieper headwaters to the Gulf of Finland. One of these units, the principality of Kiev, ruled by a nobility of mixed Scandinavian and Slavic descent, achieved mastery over the others and became a powerful state. The culture which it developed was the foundation on which the Great Russian, Ukrainian, and Belorussian cultures later arose.

Kievan Russia was much affected by its contacts with Constantinople (now Istanbul). This famous city, located on the straits connecting the Black Sea with the Mediterranean, was the capital of the Eastern Roman or Byzantine Empire which endured for nearly a thousand years after the collapse of the Western Roman Empire in the fifth century A.D. It became an important magnet for Russian trade and was the fount of much cultural borrowing by the Russians. "At the end of the tenth century, the Kievan Slavs accepted Christianity from Byzantium (A.D. 988), and Byzantine influence in religion, art, and political life became permanently established among the eastern Slavs."[3] After Constantinople fell to the Ottoman Turks, a Moslem people, in the fifteenth century, the Russian tsars carried on the traditions of the Eastern Empire. They came to regard Moscow as a "Third Rome," and the Russian branch

of the Orthodox Eastern Church was made the official church of the Russian Empire.

The Tatar Invasion

Another cultural influence reached the Russians from the heart of Asia. Since the earliest known times the steppe grasslands of southern Russia had been the haunt of nomadic horsemen of Asiatic origin. During the later days of the Roman Empire and in the Middle Ages these grassy plains, stretching far into Asia, provided a passageway for inroads into Europe by the Huns, Magyars, Bulgars, and other Asiatic peoples. In the thirteenth century the Tatars (Tartars) appeared. Many steppe and desert peoples were represented in their ranks, but they were led by Mongols. In 1240 Batu Khan, grandson of Genghis Khan, launched a devastating invasion which brought all of the Russian principalities except the northern one of Novgorod under his rule. The Khanate of Kipchak, established by Batu, endured for more than two centuries. It was one of the main divisions in a vast Mongol empire stretching eastward to China and southward into the Middle East. Eventually it broke apart into the separate khanates of Kazan, Astrakhan (Sarai), Crimea, and Sibir. Batu and his successors collected taxes and tributes from the Russian principalities, but generally allowed the rulers of these units considerable latitude in matters of local government. Even the princes of Novgorod, who were not technically under Mongol control, paid tribute in order to avoid trouble. The Tatars used the southern steppes to graze their livestock herds, and they established their main camps and political centers there. When their power declined in the fifteenth century, the rulers of the Moscow principality were able to begin the process of territorial expansion previously described, which resulted in the formation of present-day Russia (see pp. 298–301).

West European Influences

Thus four different cultures—Slavic, Scandinavian, Byzantine, and Tatar—furnished significant elements in the civilization of Old Russia. Important elements of west European culture were added in early modern times, especially during the reign of Peter the Great. Western

---

[2] *Aspects of Contemporary Ukraine* (New Haven, Conn.: Human Relations Area Files, 1955), p. 30.
[3] *Ibid.,* p. 31.

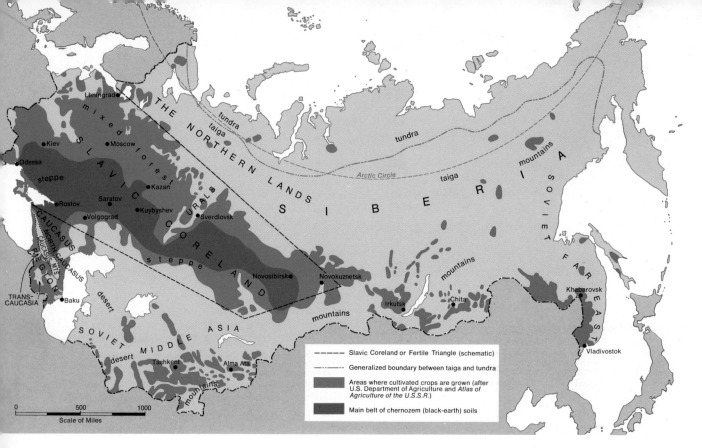

Major regional divisions of the Soviet Union as considered in the present chapter.

manners and languages were introduced into the Russian court, industry was stimulated to develop along Western lines, and the capital was moved in 1713 from Moscow to the new city of St. Petersburg (now Leningrad) on the Gulf of Finland, thus affording a closer contact with western Europe. Further cultural importations from the latter area have continued up to the present. Many of the basic doctrines of the Communist party were developed by west European thinkers, and the modern industrial techniques which the Soviet leaders have introduced on a large scale in the USSR were originally a west European development.

Minority Groups and the Political Structure

About a fourth of the population of the Soviet Union is composed of non-Slavic peoples. Over 100 ethnic groups can be distinguished in the USSR, most of which are very small. However, some 19 nationalities other than the three principal Slavic groups number over a million, and perhaps a dozen additional ones number more than 250,000. Soviet policy has permitted the different minorities to retain their own languages and other elements of their traditional cultures. Alphabets

have been created for nationalities that previously had no written language, and newspapers and books are published in most of the existing languages, though the amount of printed material in many of the minor languages is rather small. These procedures have facilitated mass education, regarded by the Communist party as a necessity for the success of the Communist system. However, a definite tendency toward Russification of minority areas has been apparent for many years. Instruction in the Russian language is compulsory in all schools, and non-Russians who aspire to rise above an ordinary station must learn to speak it fluently. Immigration of Russians and Ukrainians into non-Slavic areas has been a major theme in Russian history and continues today on a sizable scale.

The multinational composition of the Soviet Union is directly reflected in the political structure. From the standpoint of political control the USSR is, of course, a highly centralized authoritarian state. From the standpoint of administrative organization, however, it is a federation of fifteen Union Republics (inset map, p. 285), each organized primarily around a particular national group. The largest and most important is the Russian Soviet Federated Socialist Republic, or RSFSR. It includes more than three-fourths of the area and over half of the total population of the Soviet Union. The Great

Russian ethnic group constitutes an estimated 83 percent of its population. The fourteen smaller republics are located around the margins of the RSFSR. They fall into three distinct groups: a Middle Asian group of five republics separating the RSFSR from Iran, Afghanistan, and China; a Transcaucasian group of three republics bordering Turkey and Iran; and a western group of six republics bordering East Central Europe. Area and population figures for the fifteen republics are given in Table 11.

Peoples not considered sufficiently numerous or advanced to form Union Republics are organized into Autonomous Soviet Socialist Republics, Autonomous Oblasts (Regions), or lesser autonomous units. Each autonomous unit is politically subordinated to a Union Republic. The majority are under the jurisdiction of the Russian Soviet Federated Socialist Republic.

THE SLAVIC CORELAND

Most of the Slavic population of the Soviet Union is found within an irregular triangle extending from the Black and Baltic seas to the neighborhood of Novosibirsk and Novokuznetsk (formerly Stalinsk) in Siberia. This part of the USSR is customarily referred to as the Slavic Coreland, Fertile Triangle, or Agricultural Triangle (map, p. 310). It is the core region of the country in every sense, containing about 70 percent of the total population, all of the principal industrial concentrations, most major mining areas, four-fifths of the cities having estimated populations of 500,000 or more (1968), and by far the greater part of the total sown acreage. The coreland is about half as large as the United States in area and has about 85 percent as large a population. It is really the "Russia that matters," despite the fact that it occupies little more than a fifth of the total area of the USSR. Lowlands predominate within the coreland, the only mountains being the Urals, a small segment of the Carpathians, and a minor range in the south of the Crimean Peninsula. The original vegetation was mixed coniferous and deciduous forest in the northern part of the coreland and steppe grassland in the south. Moscow (6.6 million, or 8.6 million including suburbs)[4] and Leningrad (3.7 million, or 4.1 million with suburbs) are by far the largest cities. Moscow, the capital of

TABLE 11 UNION REPUBLICS: AREA AND POPULATION DATA

REPUBLIC	AREA (THOUSAND SQUARE MILES)	POPULATION 1968 ESTIMATES (MILLIONS)
Russian SFSR	6591.1	128.6
Ukrainian SSR	232.0	46.8
Belorussian SSR	80.1	8.9
Moldavian SSR	13.0	3.5
Lithuanian SSR	25.2	3.1
Latvian SSR	25.6	2.4
Estonian SSR	17.4	1.3
Azerbaijan SSR	33.4	5.1
Georgian SSR	26.9	4.7
Armenian SSR	11.5	2.5
Kazakh SSR	1048.0	12.7
Uzbek SSR	173.5	11.4
Kirgiz SSR	76.5	3.0
Tadshik SSR	55.2	2.9
Turkmen SSR	188.4	2.1
Totals	8597.8	239.0

SOURCE: Population figures are extrapolated from Harrison E. Salisbury, ed., *The Soviet Union: The Fifty Years* (New York: Harcourt, Brace & World, 1967), p. 239.

[4] Unless otherwise specified, population figures for cities in this chapter are 1968 estimates for the city proper. For most Soviet cities this figure represents at least 80 or 90 percent of the population in the total urban area. There is not the sharp difference between population totals for the central city and the entire metropolitan area that one often finds in American cities.

MINERAL RESOURCES OF THE USSR

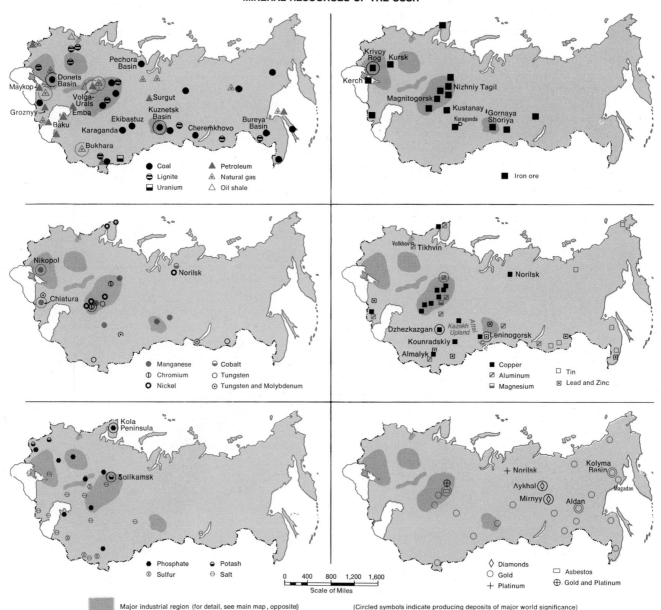

These maps are based in considerable part on qualitative data. Most mineral symbols show producing deposits. The basic symbols for industrial centers are arranged in three size groups according to magnitude of manufacturing as estimated by Richard E. Lonsdale and John H. Thompson, "A Map of the USSR's Manufacturing," *Economic Geography,* vol. 36, no. 1 (January 1960), pp. 36–52. The largest symbols, for Moscow and Leningrad, show cities with an estimated 8.2 percent and 4.9 percent, respectively, of the USSR's manufacturing. Cities in the second size group (Donetsk, Volgograd, Kiev, etc.) had an estimated .60 percent or more of all manufacturing. Cities in the third size group (Ivanovo, Zlatoust, etc.) are selected smaller centers. Shifts in the relative importance of manufacturing centers since the study by Lonsdale and Thompson are not believed to have been sufficiently great to alter significantly the rankings indicated on the map. The subsidiary symbols (petroleum refining, etc.) have no quantitative significance. On the map of railroads the line from Chita to Vladivostok via Harbin traverses Chinese territory for most of its length. Maps by Hooson, Lydolph, and others were used as a basis for indicating the railroad lines carrying the heaviest freight traffic.

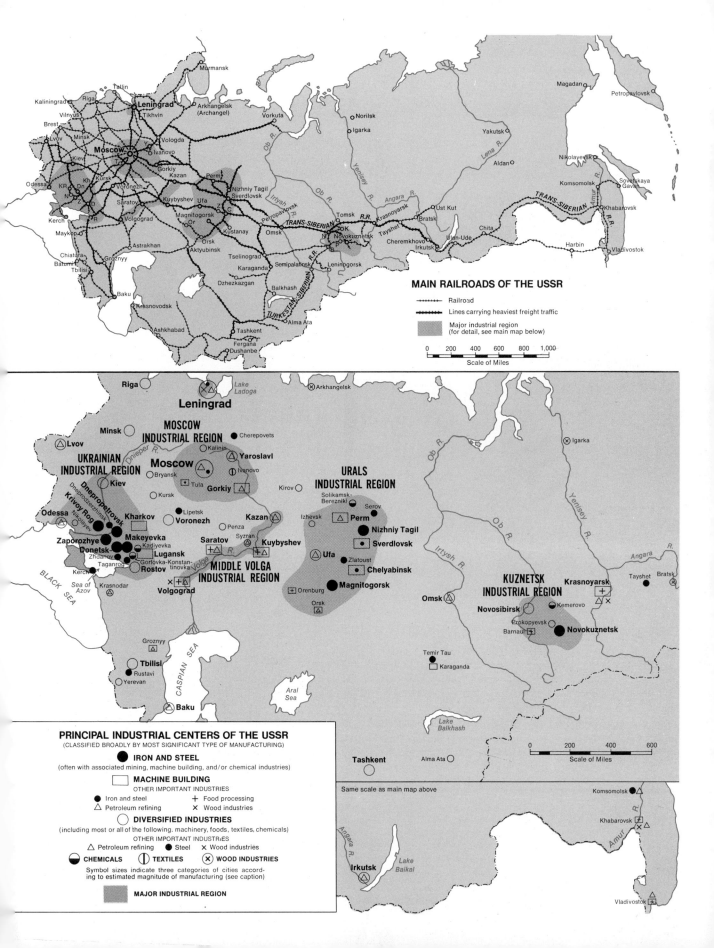

The main shrine, showplace, and nerve center of the USSR. Red Square and Lenin's Tomb are in the left and center foreground, with the outer wall of the Moscow Kremlin to the right. At the end of Red Square is St. Basil's Cathedral with its famous onion-shaped domes, and to its left, also fronting on the square, is the Soviet Union's largest department store, G. U. M. At the upper right the Moscow River is visible. The prominent building with the three towers rising high above the skyline at the upper left is the Ukraine Hotel. (Sovfoto.)

latter are found in the Ukrainian Soviet Socialist Republic. Although closely related to the Great Russians in language and culture, the Ukrainians are a distinct national group. The name Ukraine means "borderland." In this area the Russian tsars fought for centuries against nomadic steppe peoples and also against Poles, Lithuanians, and Turks before the Ukraine was finally brought under control in the eighteenth century.

Today the Ukraine is one of the most densely populated and productive areas in all of Russia. It is a major producer of coal, iron ore, manganese, natural gas, iron and steel, chemicals, machinery, wheat, corn, livestock, vegetable oil (mainly from sunflower seeds) and beet sugar. Probably no area of equal size contributes more to the total economy of the USSR.

The Ukraine lies partly in the forest zone and partly in the steppe. In the border between these vegetation realms is the historic city of Kiev (1.5 million), the capital of the Ukrainian SSR and a major industrial and transportation center.

Russia before 1713 and of the Soviet Union since 1918, is today the most important manufacturing city, transportation hub, and cultural, educational, and scientific center. Leningrad, the capital from 1713 to 1918, is second to Moscow in the above respects and is also a major seaport. Most of the other major cities within the coreland are found in the principal industrialized regions, or in other words, in the vicinity of Moscow, in the Ukraine, in the Urals, in the vicinity of the Kuznetsk coal basin in Siberia, or along the Volga River.

The Ukraine

Great Russians form the largest ethnic group in the coreland, followed by the Ukrainians. Most of the

Lesser Republics of the Coreland

The Ukrainian SSR is separately represented in the General Assembly of the United Nations. The same is true of the adjoining Belorussian (White Russian) Republic. By virtue of this arrangement the Soviet Union commands three votes in the General Assembly, as compared with one for the United States. The Belorussian SSR is much smaller in area and population than the Ukrainian SSR, and is far less productive. The political capital and main industrial center is Minsk (780,000), located on the direct rail line from Moscow to Warsaw and Berlin. This republic suffers from a lack of mineral resources (aside from a large potash deposit recently opened in the south) and a retarded industrial development. The absence of fuel resources, other than peat, has recently been offset by the construction of major oil and natural gas pipelines from the Volga region. The cool, damp climate and infertile soils of Belorussia are not favorable for wheat, though small amounts are grown, and most cropland grows rye, oats, barley, hay, potatoes, or flax. Dairy cattle and hogs are the principal livestock. Southern Belorussia incorporates the greater part of the Pripyat (Pripet) Marshes. Much of the marshland has been drained, but extensive reclamation work remains to be done.

The Lithuanian, Latvian, and Estonian SSRs lie between the Belorussian SSR and the Baltic Sea. These small republics were part of the Russian Empire before World War I, but successfully asserted their independence following the Russian Revolution. In 1940, however, they were reabsorbed by the USSR as Union Republics. The Latvians and Lithuanians have some Slavic affinities, but are not considered to be true Slavs. The Estonians, related to the Finns, are distinctly non-Slavic. In the Baltic republics dairy farms alternate with forested areas in a hilly, glaciated landscape. Lumbering is an important activity in all three republics. Mineral resources are generally lacking, although Estonia has extensive deposits of oil-bearing shale. The shale is processed in distillation plants to produce gas, much of which is transmitted by pipeline to electric-power stations in Leningrad. The principal industries are concentrated in a few medium-sized cities, of which the largest is the important seaport of Riga (680,000) in the Latvian SSR. A significant expansion of industry has occurred in all three republics since World War II. Adjoining the Lithuanian SSR on the southwest is former German East Prussia, the northern half of which is now a part of the Russian Soviet Federated Socialist Republic.

Also included within the coreland at the extreme southwest is the small Moldavian SSR, largely constituted of territory acquired from Romania in 1940. The Moldavians, a people with many Slavic characteristics, speak a dialect of the Romanian language. Mainly a fertile black-earth steppe upland, the Moldavian SSR is largely agricultural, though industry has shown significant gains since World War II. Agricultural-processing industries predominate. Corn, wheat, and other grains occupy around three-fourths of the republic's crop acreage. Dairy cattle and hogs are the principal livestock in the north, but sheep and goats predominate in the drier south. An important specialty is the growing of vegetables and fruits, especially grapes. This republic is one of the major grape-growing and wine-producing areas of the USSR, its natural environment for vineyards being somewhat similar to that of the famous French district of Champagne.

Industrial Development in the Coreland

Although pre-Revolutionary Russia was basically an agricultural country, a slow development of modern industry, partially financed by foreign capital, took place before World War I. At the time of the Revolution industrial development in the country was mainly confined to three areas: the Moscow region, Leningrad (then called Petrograd), and the Ukraine. Under the Soviets, these areas have continued to represent major concentrations of industry. In addition, industry has been greatly expanded in the Urals and in widely spaced cities along the Volga River, and an entirely new industrial concentration has been created in the Kuznetsk Basin. All of these major industrialized areas are within the coreland (see maps, pp. 310 and 313), along with many smaller industrial centers.

Industry in the Moscow Region

The industrialized area surrounding Moscow is often referred to as the Central Industrial Region, Old Industrial Region, or Moscow-Tula-Gorkiy Region. These names indicate important characteristics of the region. It has a central location physically within European Russia, and is functionally the major focus for the entire Soviet Union. It lies at the center of the Soviet rail and air networks, and is connected by river and canal transportation with the Baltic, White, Azov, Black, and Caspian seas. It is an old region in the sense of being the first major concentration of modern large-scale industry to have been developed in pre-Revolutionary Russia. From Moscow the region extends southward to the mining, metallurgical, and machine-building center of Tula (390,000) and eastward to the automobile and diversified industrial center of Gorkiy (1.1 million, or 1.5 million with suburbs)—the "Soviet Detroit." To the north the region includes the important textile center of Ivanovo (420,000)—the "Soviet Manchester"—and Yaroslavl (510,000), important in oil refining, synthetic rubber, tire production, linen textiles, and a variety of other industries. Textile milling, largely on the basis of imported American cotton and Russian flax, was the earliest form of large-scale manufacturing to be developed in the Moscow region. It gradually replaced the earlier handicraft industries in the nineteenth century. Although the area still has the USSR's main concentration of textile plants, with Moscow and Ivanovo the leading centers, a great variety of other light and heavy manufactures have developed. Metal-fabricating industries, emphasizing types of construction requiring a relatively high degree of skill and precision, are the most important in value of product. This industrial region accounts for perhaps one-fifth of Soviet manufacturing, as measured by employment and amount of fixed capital in manufacturing.[5] Moscow and its immediate suburbs and satellites probably account for about one-tenth of the USSR total.

The industrial eminence of the Moscow region has been achieved in spite of a notable lack of natural resources. The minerals that have been of greatest consequence in recent times are large deposits of lignite south and west of Moscow and peat deposits north and east of the city—both used in steam-generating plants to produce electricity. In addition, small iron ore deposits and fairly sizable deposits of phosphate rock occur in the vicinity of Tula. The iron ore provided the initial basis of an iron and steel industry at Tula (which now depends increasingly on ore from other sources), and the phosphate is used for the manufacture of chemical fertilizer. Other major industrial minerals are largely lacking in the region. However, the well-developed railway connections of the capital, partly a product of political centralization, have provided good facilities for a constant inflow of minerals, other materials, and foods, and a return outflow of finished products to all parts of the USSR. Relatively minor amounts of freight are handled by water, highway, or air transportation. A significant recent development has been the construction of long-distance pipelines to bring petroleum and its products or natural gas to the Moscow region from fields in several different areas. Piped oil and gas have become major fuels in the region and provide the basis for large petrochemical industries. Moscow and the upper Volga region are now interconnected with an electric grid extending to the Urals and to the hydroelectric plants along the middle and lower Volga.

Industrial Development in the Leningrad Area

Leningrad does not form the center of an industrial region comparable in area and population to the Moscow region, the Ukraine, or the Urals. Nevertheless, the city and its immediate environs account for perhaps 4 or 5 percent of the USSR's manufacturing. Leningrad is even more handicapped than Moscow from the standpoint of mineral resources. Local deposits of peat and deposits of oil shale and lignite at a somewhat greater distance are supplemented by hydroelectricity

from several generating stations to the north, west, and east of the city. High-grade coal, as well as petroleum and natural gas, must be transported from hundreds of miles away, as must all of the metals except aluminum, produced east of Leningrad at Volkhov from bauxite deposits near Tikhvin. Even the Volkhov aluminum works relies increasingly on aluminum-bearing raw material (nephelite) from the Kola Peninsula (p. 330). Nevertheless, the metal-fabricating industries are by far the most important branch of manufacturing in Leningrad's diversified industrial structure.

Supported by university and technological institute research workers, Leningrad's highly skilled labor force played an extremely significant role in early Soviet industrialization, pioneering the development of many complex industrial products such as power-generating equipment and synthetic rubber, and supplying groups of experienced workers and technicians to establish new industries in various parts of the USSR.

The Ukrainian Industrial Region

From the beginning of the Five-Year Plans, great stress has been laid on the heavier types of industry. Today heavy metallurgy, together with associated coal and iron mining and chemical manufacturing, is mainly concentrated in two regions: the Ukraine and the Urals, with a lesser, though important, concentration in the Kuznetsk Basin and smaller centers in various other places. Together the Ukraine and Urals account for over four-fifths of the pig iron and over three-quarters of the steel production in the USSR.

Heavy industry in the Ukraine is based essentially on major deposits of five minerals: coal, iron, manganese, salt, and natural gas. The coal is found in the Donets Basin coal field of the eastern Ukraine, often referred to as the Donbas field. Mining began in this area under the tsars, and as late as 1913 the Donets field accounted for nearly nine-tenths of the total tonnage of hard coal (bituminous and anthracite) mined in Russia. Today it accounts for less than half of the national output of hard coal, and little more than a third of the total output (caloric value) of hard coal and lignite together, but it is still the leading coal field, and its *absolute* production is at least eight times as great as it was in 1913. The Donbas field produces a variety of bituminous and anthracite coals, including much coal that is suitable for coking. However, its coking coals have a relatively high

---

[5] See Richard E. Lonsdale and John H. Thompson, "A Map of the USSR's Manufacturing," *Economic Geography*, 36 (1960), 36–52.

content of undesirable impurities (ash and sulfur) as compared with the best American coals.[6] Total coal reserves appear to be sufficient for many decades at present rates of production, though much of the best coking coal has been extracted and costs are rising due to increasing depths of mining. Between one-half and three-fifths of Soviet coking-coal production comes from the Donets Basin. A little over a quarter comes from the Kuznetsk Basin, which has better coal, larger reserves, and easier physical conditions for mining, but is less favorably located with respect to markets.

The Soviet Union's largest iron-mining district is found in the western part of the Ukraine in the vicinity of Krivoy Rog (520,000). Ore has been extracted here since tsarist days, and the Krivoy Rog field has been the leading Russian producer for many decades. The best ore is virtually exhausted, but large reserves of lower-grade ore remain. This ore, mined by large-scale open-pit methods (in contrast to the high-grade ore, which is deep-mined) is processed at concentrating plants prior to shipment. The Krivoy Rog deposits supply about half of the Soviet Union's total output of iron ore. Most of the production is utilized by iron and steel plants in the Ukraine, but some of it is shipped to the Moscow region. In addition, the deposits are a major source of ore for several countries in East Central Europe (see p. 272).

Manganese, the most important of the ferroalloys, is mined in the vicinity of Nikopol (110,000), about 60 miles southeast of Krivoy Rog. The Soviet Union is the world's largest producer of manganese and has very large reserves. More than nine-tenths of these are located in the vicinity either of Nikopol or of Chiatura in Transcaucasia. The bulk of the production is utilized in the USSR itself, but sizable amounts are exported, mainly to countries in East Central Europe.

The largest concentration of iron and steel plants in the Ukraine is located on the Donets Basin coal field, in the general vicinity of the important coal-mining and heavy industrial centers of Donetsk, formerly Stalino (850,000) and Makeyevka (420,000). Iron ore and manganese are brought by rail from Krivoy Rog and Niko-

pol. This area is also an important center of chemical manufacturing, based in part on blast-furnace wastes, coke-oven gases, and huge deposits of common salt. Nitrogenous fertilizers are an important product. Other important iron and steel plants are found along the Dnieper River at Dnepropetrovsk (820,000) and at Zaporozhye (600,000), the site of a large dam and hydroelectric power station. These cities of the Dnieper Bend carry on a variety of electrochemical and electrometallurgical activities, including the manufacture of special alloy steels in electric furnaces at Zaporozhye. An iron and steel plant at Krivoy Rog that was destroyed during World War II has been rebuilt on a greatly enlarged scale.

South of the Donbas at Kerch (120,000), on the Crimean Peninsula, immense deposits of low-grade, high-phosphorus iron ore are found. The ore is utilized locally for manufacturing iron and steel, and is shipped by water across the Sea of Azov to iron and steel plants at Zhdanov, formerly Mariupol (390,000), and Taganrog (250,000). Coal and manganese for these operations are brought from the producing fields to the north.

Surrounding the inner core of mining and heavy metallurgical districts in the Ukraine is an outer ring of large industrial cities which carry on metal-fabricating and various other types of manufacturing. These cities include the Ukrainian capital of Kiev (1.5 million) on the Dnieper, the machine-building and railway center of Kharkov (1.2 million), about 250 miles east of Kiev, and the seaports and diversified industrial centers of Odessa (780,000) on the Black Sea and Rostov (760,000) on the lower Don River just outside of the Ukrainian SSR. East of Odessa the smaller Black Sea port of Nikolayev (300,000) has become one of the major shipbuilding centers of the USSR.[7]

The Ukrainian industrial region accounts for around one-fifth of the Soviet Union's manufacturing. Heretofore, the industries of the region have depended largely on coal as a source of power, and it is still the most important single source, but they are drawing more and more on petroleum and its products brought by pipelines from the large Volga-Urals fields (see p. 320), hydro-

---

[6] High ash content lessens the heating power of coke and thus increases the volume of coke and limestone required in the blast furnace. It also adds to transportation costs due to the waste matter in the coke. A further difficulty is that coke containing excessive ash may be too brittle to stand transportation well. Too much sulfur in a coke necessitates larger amounts of coke and limestone in the blast furnace. It also lowers the tensile strength of pig iron, and makes it necessary to use additional manganese in making steel. See Allan Rodgers, "Coking Coal Supply: Its Role in the Expansion of the Soviet Steel Industry," *Economic Geography,* **40**, no. 2 (April 1964), 113–150. Inferior grades of coke will be less of a problem as the Soviet iron and steel industry shifts to new oxygen processes for smelting and makes greater use of natural gas to increase the efficiency of furnaces.

[7] The Soviet Union's expanding naval, merchant, and fishing fleets secure many of their ships from abroad. Nikolayev and Leningrad are probably the foremost shipbuilding centers in the USSR itself, though several other ports have a share of the industry.

electricity transmitted from chains of dams on the Dnieper and Volga rivers, and natural gas from large deposits in the eastern and western Ukraine and the northern Caucasus. Gas from the latter fields is piped to Moscow, Leningrad, and many other cities outside the Ukraine, as well as to the Ukrainian cities themselves.

Industrial Development in the Urals

The Ural Mountains contain an extraordinarily varied collection of useful minerals. Although deficient in coking coal, this highly mineralized area has valuable deposits of iron, copper, nickel, chromium, manganese, tungsten, zinc, lead, bauxite, platinum, gold, asbestos, magnesium, potash, industrial salt, and various other minerals. Much low-grade bituminous coal and lignite and some anthracite occur, though relatively little of the coal is suitable for coking. Oil and natural gas are produced in the western foothills of the mountains (see p. 320).

The Soviet regime has placed great emphasis on the development of the Urals as an industrial region well removed from the exposed western frontier of the Union. This emphasis was amply repaid in World War II, when the Ukraine was overrun by German armies. Today the region accounts for perhaps one-eighth of the USSR's manufacturing. The major industrial activities are as follows:

1. Heavy metallurgy, including the manufacture of iron and steel and the smelting of nonferrous ores.

2. The manufacture of heavy chemicals, especially concentrated in the area of Solikamsk (90,000) and Berezniki (140,000), where some of the world's largest deposits of potassium and magnesium salts are found.

3. The manufacture of machinery and other metal-fabricating activities, carried on in the important industrial and transportation centers of Sverdlovsk (970,000), Chelyabinsk (840,000), Perm (formerly Molotov, 800,000), Ufa (710,000), and dozens of smaller places.

Old pre-Soviet metallurgical and machine-building plants in the Urals have been modernized and expanded, and a number of immense new plants have been constructed. Probably the most famous and spectacular development has been the creation of an iron and steel center at Magnitogorsk (370,000) in the southern Urals. This place, located near a large reserve of exceptionally high-grade iron ore, was not even a village prior to 1931. In that year construction of a huge plant was begun and a city was built to house the workers. Today the iron and steel plant at Magnitogorsk is the largest in the Soviet Union and possibly in the world.

The years that saw the creation of Magnitogorsk also witnessed a large expansion of coal mining in the Kuznetsk Basin (Kuzbas), located in southern Siberia more than a thousand miles to the east of the Urals. A railway shuttle developed, with Kutznetsk coal and coke moving westward to Magnitogorsk and other industrial centers in the Urals, and Urals iron ore (primarily from Magnitogorsk) moving eastward to a new iron and steel plant in the Kuznetsk Basin. Thus was created the famous Urals-Kuznetsk Combine. Each end of the Combine, however, soon became partially independent of the other. A coal field was developed at Karaganda in Soviet Middle Asia to provide fuel for the Urals, while iron mining was developed south of the Kuznetsk Basin at Tashtagol and other places in the foothill area called the Gornaya Shoriya. After 1937 the Combine, as such, was deemphasized following a transportation crisis that led to a search for greater regional self-sufficiency. Nevertheless, due to qualitative shortcomings of Karaganda coal and Gornaya Shoriya iron ore, the Kuznetsk Basin continued to send large quantities of coal and coke to the Urals and received high-grade iron ore in return. Karaganda coal has too high a content of ash to make good metallurgical coke, and it is mixed with higher quality Kuznetsk Basin coal in the coke ovens at Magnitogorsk. (Many coke works in the Urals use little or no Karaganda coal, but rely overwhelmingly on Kuznetsk coal.) Similarly, Urals iron ore has been mixed with lower quality local ore in Kuznetsk furnaces. Today, the iron and steel industries of the Urals and Kuznetsk Basin, like those of the Ukraine, are having to adjust their operations to lower grades of ore and coal. During World War II the best grades of iron ore and coking coal in the major Soviet deposits were heavily exploited, and it is reported that serious shortages of both now exist. The best iron ores of the Krivoy Rog and Magnitogorsk deposits appear to be virtually exhausted; and shortages of the best coking coals have been reported from both the Donets and Kuznetsk basins. However, in 1960 the USSR reported a large discovery of high-grade iron ore north of the Donets Basin in a formation called the Kursk magnetic anomaly (from its disturbing effect on compass needles). Only low-grade ores had previously been found in the formation. Both high-grade and low-grade ores are now mined, though the high-grade ores are more difficult to extract, as they are buried under water-bearing rocks. Production is expected to increase, and will be used by expanded iron and steel plants at Tula and Lipetsk (250,000) or by the Ukrainian iron and steel industry. A little high-grade ore from the Kursk formation is reported to reach the Urals. However, that area will rely increasingly on low-grade ores from large deposits near Kustanay (120,000) in the northwestern part of the Kazakh SSR, or from even larger and lower-grade deposits in the Urals north of

Nizhniy Tagil. Large-scale open-pit mining in both of these areas has been under way for some time. The ore is processed in concentrating plants prior to shipment.

Though Magnitogorsk is the largest center of iron and steel production in the Urals, important plants are found at several other places, particularly Nizhniy Tagil (380,000) and Chelyabinsk. Zlatoust (180,000) is a metallurgical center that manufactures special alloy steels.

Sverdlovsk, located on the eastern flank of the mountains, is the largest city of the Urals and the preeminent economic, cultural, and transportation center. Direct trunk rail lines connect the city with Moscow and Leningrad. The second most important rail center is Chelyabinsk, about 120 miles south of Sverdlovsk. Chelyabinsk also has direct rail connections with Moscow. In western Siberia between the Urals and the large industrial and trading center of Omsk (780,000), rail lines from Sverdlovsk and Chelyabinsk join to form the Trans-Siberian Railroad, the main artery linking the Soviet Far East with the coreland (railroad map, p. 313).

The Kuznetsk Industrial Region

From Omsk the Trans-Siberian leads eastward to the Kuznetsk industrial region, which accounts for perhaps 3 or 4 percent of the Soviet Union's manufacturing. The principal localizing factor for industry in this region is an enormous deposit of coal, much of it suitable for coking. The total reserves are estimated to be about four or five times the size of the reserves in the Donbas field.

The manufacture of iron and steel is a major industrial activity of the Kuznetsk region. The main center of the industry is Novokuznetsk, formerly Stalinsk (500,000), where the first iron and steel plant was commenced in 1932, concurrent with the development of Magnitogorsk. A second large plant was under construction in the 1960s. The industry now draws its iron ore very largely from the Gornaya Shoriya area previously mentioned or from deposits farther east in the upper basin of the Yenisey River around Abakan (80,000). The Kuznetsk Basin produces steel primarily for use by fabricating industries in Siberia, but sizable quantities move to factories west of the Urals. North of Novokuznetsk coke is made and chemical manufacturing is carried on at Kemerovo (370,000), the most important center of the chemical industry in Siberia. The industry at Kemerovo is based in large measure on by-products of the coke ovens. The largest urban center of the Kuznetsk region, however, is Novosibirsk, a diversified industrial, trading, and transportation center located on the Ob River at the junction of the Trans-Siberian and Turkestan-Siberian (Turk-Sib) railroads. Sometimes called the

"Chicago of Siberia," Novosibirsk has developed from a town of a few thousand at the turn of the century to a city of 1.1 million people. Over 300 factories produce mining, power-generating, and agricultural machinery, tractors, machine tools, and a wide range of other products. Novosibirsk has the Soviet Union's only gold refinery and one of its two tin refineries. The city's much-publicized *Akademgorodok* ("Academy Town" or "Science Town"), Siberia's main center of scientific research, is an outlying suburb.

The Volga Cities

Numerous industrial cities outside of the five industrial concentrations discussed above are scattered through the coreland. The most notable of these are a group of cities spaced at fairly regular intervals along the Volga River between Gorkiy and the Caspian Sea. The largest are Kuybyshev (1.0 million), Kazan (830,000), Volgograd, formerly Stalingrad (750,000), and Saratov (730,000). These four cities in the "Middle Volga industrial region" (maps, p. 313) have diversified machinery, chemical, and food-processing plants. The smaller city of Astrakhan (360,000) in the Volga delta also has diversified industries and in addition is the main headquarters for the fishing industry of the Caspian Sea.

In recent years the cities and towns along the Volga have experienced a marked upsurge of industrial activity. One of the most publicized aspects has been the construction of immense dams and hydroelectric power stations in the vicinity of Kuybyshev and Volgograd. For its supply of electricity the Soviet Union relies primarily on thermal generating stations, often extremely large, that are fueled by bituminous coal, lignite, natural gas, or oil. Such stations frequently are located in coal or lignite fields where the necessary fuel can be obtained cheaply by open-pit mining. An increasing number of plants now burn natural gas or oil brought by pipeline. Since World War II, however, there has been a very extensive development of hydroelectricity, which now accounts for somewhat more than a fifth of the total electric-power supply. Numerous power dams, some of which are very large, have been built along rivers in the mountains of the Caucasus region or Soviet Middle Asia, and still others are under construction or planned in those regions. But the main element in the program has been the construction of large dams, or chains of dams, along such major rivers as the Volga, Dnieper, Don, Kama, Irtysh, Ob, Yenisey, and Angara. In the process the middle and upper Volga has largely been converted to a series of lakes, with consequent improvement in navigation. The Volga, the most important internal waterway of the USSR, is connected with Moscow by the Moscow-Volga Canal, with the Baltic Sea by the Mariinsk Canals, and with the Don River and the Black Sea by the Volga-Don Canal, opened in 1952. It is estimated that the main river, its tributaries, and connecting canals carry around two-thirds of the USSR's river and canal traffic. Timber from the taiga, petroleum and its products from the Volga-Urals and Caucasus fields, and mineral building materials such as cement, sand, and gravel, constitute the largest tonnages. Much of the timber is processed at Volgograd, which has become one of the largest sawmilling centers in the Soviet Union. The importance of Volga traffic, however, should not be overstressed. In the Soviet Union, only about 5 percent of intercity freight tonnages move by river and canal, despite strenuous attempts by the Soviet government to foster this type of transportation. The amount of freight carried on the Volga and other internal waterways has substantially increased since World War II, but the importance of these waterways relative to other forms of transportation has declined. They are at a competitive disadvantage in several ways. River transportation is more circuitous and much slower; the channels are blocked by ice for several months a year; and facilities for transshipment of goods at river ports often are inadequate. The new reservoirs behind power dams stay frozen longer, and windstorms on these broadened sheets of water cause increased hazards and delays in the operation of river craft. The flow of goods in the Soviet Union is prevailingly east-west, whereas the river courses are oriented generally north to south. For many types of freight movement, transshipment to or from river craft is inconvenient and costly. Thus in spite of deepened waterways, improved carriers, and rate structures designed to favor water transportation, the waterways are tending to lose ground in their competition with other means of transport. It is reported that railroads paralleling the Volga River often carry as much or more freight than the river does.[8]

The Volga-Urals Oil Fields

The Soviet Union's main area of oil production is found in the Volga-Urals fields, sometimes referred to as the "Second Baku." These fields, stretching from the Volga River to the western foothills of the Ural Mountains and containing many separate deposits of oil, account for more than two-thirds of the USSR's oil output.

They also are a considerable producer of natural gas, though the country's largest gas reserves are elsewhere. Among the world's countries the USSR is second only to the United States as a producer of petroleum and natural gas. Pipelines carry crude oil or petroleum products from the Volga-Urals fields to Moscow and many other cities in the western USSR; also to the Urals, Siberia, and several countries in East Central Europe (see pp. 273–274). Important petrochemical industries based on oil and natural gas from the Volga-Urals fields and other sources are developing very rapidly in many parts of the USSR. These industries manufacture synthetic rubber, artificial fibers, plastics, nitrogenous fertilizers, and various other products. In the Volga-Urals region Kuybyshev, Syzran (170,000), and Ufa are the most prominent oil-refining and petrochemical centers, though all of the larger cities along the Volga and Kama rivers from Volgograd northward to Gorkiy and Perm, as well as some smaller cities in the oil fields, have a share of these industries.

In the past the Soviet economy has been oriented to coal as its major fuel. But oil and natural gas, which supplied only 22 percent of the Soviet Union's fuel consumption (caloric value) in 1954, reached an estimated 55 percent by 1967 and were scheduled for further advances. Soviet railroads, which carried around 80 percent of all intercity freight traffic in the 1960s, have converted very largely from steam traction to diesel-electric and electric traction. Many thermal-electric power stations also are shifting to gas and oil for fuel, though coal will continue to be a prominent element in the electric-power industry. Thus with the new development of oil and natural gas resources and hydroelectricity, the Soviet economy has entered an age of diversified, abundant, and flexible sources of power.

Agriculture in the Coreland

Nearly all of the important crops of the middle latitudes are grown in the Soviet Union. Grain crops lead in acreage and production. Wheat and rye are the principal bread grains, while the major feed grains are oats, barley, and corn. The greater part of the corn is harvested in an immature stage for silage and green forage (see p. 289). Wheat normally accounts for somewhat more than half of the total tonnage of grain produced. Grasses grown for hay and silage have increased rapidly in acreage in

[8] For a fuller discussion, see Robert N. Taaffe, "Volga River Transportation: Problems and Prospects," in Richard S. Thoman and Donald J. Patton, eds., *Focus on Geographic Activity: A Collection of Original Studies* (New York: McGraw-Hill Book Company, 1964), pp. 185–193.

recent years and now occupy over a third as much land as grains. Millet, buckwheat, and rice occupy relatively small though substantial acreages. Millet has had some success as a substitute for corn. Its yields are lower, but it is a more reliable crop because of its resistance to drought. Potatoes and sugar beets are the most important root crops; flax, cotton, and hemp the main fiber crops; and sunflowers the major source of vegetable oil. Substantial quantities of tobacco and hardy fruits are produced, and various other crops, including some citrus fruits in a few areas, are grown. Cattle, hogs, sheep, and horses are the major types of livestock. Although tractors have largely replaced horses for plowing and harvesting, large numbers of horses are still used on farms for drawing wagons, and they often provide draft power for the cultivation of the small personal plots. In Soviet Middle Asia, particularly the Kazakh SSR, horse meat is a common item of diet.

The Fertile Triangle is the agricultural core of the Union. It is by far the leading area of production for all of the major crops except cotton and subtropical fruits and for all the major types of livestock. Within the Fertile Triangle two major crop zones have long been recognized: a black-soil zone in the southern steppes and a nonblack-soil zone roughly corresponding to the region of mixed forest. Although some of the same crops and types of livestock are raised in each, the differences tend to

outweigh the similarities, and these zones are essentially complementary rather than competitive.

The Black-Soil Zone

The black-soil zone includes not only the chernozem soils proper, but also associated areas of chestnut and other grassland soils. These are excellent wheat soils, and this crop zone is one of the major wheat-growing areas of the world (map, p. 84). Wheat is ideally suited to the large-scale, mechanized agriculture stressed in Soviet planning. Winter wheat is grown west of the Dnieper and spring wheat east of the Volga; areas between the two rivers grow both winter and spring wheat. The black-soil zone is also the principal producing area for sugar beets, sunflowers, hemp, barley, and corn.

The Nonblack-Soil Zone

In the nonblack-soil zone, with its cooler and more humid climate and poorer soils, rye replaces wheat as the major grain crop and well over half of the country's rye is grown in this zone. However, increased liming and fertilization of soils, plus the development of frost-resistant and quick-ripening varieties of wheat, have made possible some gain in wheat acreage and production during recent

Dairy farming in the nonblack-soil zone of the coreland. This view was taken on a collective farm in the Minsk region. (Sovfoto.)

Rugged, snow-capped heights in the Georgian Caucasus. (Sovfoto.)

years. The nonblack-soil zone is the major producing region for potatoes, oats, and flax. Dairy farming is a major branch of agriculture in this zone.

THE OUTLYING REGIONS

The majority of the non-Slavic population of the USSR is found outside the coreland. Four major outlying regions can be distinguished: the Caucasus region, Soviet Middle Asia, the Soviet Far East, and the Northern Lands (map, p. 310).

The Caucasus Region

The Caucasus region occupies the mountainous southern borderland of the USSR between the Black and Caspian seas. It includes the rugged Caucasus Mountains, a fringe of foothills and level steppes north of the mountains, and the area to the south known as Transcaucasia.

The Greater Caucasus Range forms practically a solid wall from the Black Sea to the Caspian. It is similar in age and general character to the Alps, but is considerably higher: Mt. Elbrus, the highest summit, stands at 18,481

feet, as compared with 15,771 feet for Mt. Blanc, the highest peak of the Alps. Railroads to Transcaucasia follow narrow coastal lowlands at either end of the range. In the south of Transcaucasia is the Armenian Plateau, a mountainous, volcanic highland reaching 13,435 feet. Between the Greater Caucasus Range and Armenian Plateau are subtropical valleys and coastal plains where the majority of the people in Transcaucasia are found. Russians and Ukrainians predominate in the North Caucasus, but non-Slavic groups form a large majority in Transcaucasia. The Caucasian isthmus between the Black and Caspian seas has been an important north-south passageway for thousands of years, and the present population is composed of many different peoples who have migrated into this region at various times. At least 25 or 30 nationalities can be distinguished, most of which are small in numbers and largely confined to mountain areas that have served as places of refuge in past times. Besides Russians and Ukrainians, mostly found north of the Greater Caucasus, the nationalities of greatest importance are the Georgians, Armenians, and Azerbaijanians, each represented by a separate Union Republic.[9]

Subtropical crops and minerals are the main contributions of the Caucasus region to the economy of the USSR. The lowlands bordering the Black Sea, with the heaviest rainfall and warmest winters of the USSR, produce such specialty crops as tea, tung oil, tobacco, almonds, camphor, silk, and wine, together with some citrus fruits that are grown on a rather marginal basis (there is too much freezing weather for the citrus industry to really flourish). This area, the most densely populated part of Transcaucasia, is located in the Georgian SSR. The Caspian lowlands of eastern Transcaucasia, located in the Azerbaijan SSR, have colder winters and less rain; here irrigated cotton, grown in rotation with alfalfa, is the main crop except in the narrow coastal lowlands at the extreme south where warmer temperatures and greater precipitation permit the culture of rice, tea, citrus, and other subtropical specialty crops. Temperate fruits and nuts, including apples, pears, apricots, peaches, plums, chestnuts, and walnuts, are grown in many parts of Transcaucasia, and viticulture and sericulture are widely practiced, not only in the lowlands but also on the warmer slopes of the mountains.

[9] These nationalities have racial characteristics and cultural traditions that are primarily Asian or Mediterranean in origin. They differ in religion, with the Azerbaijanians (also known as Azeri Turks) being Moslems, while the Georgians and Armenians belong to separate branches of the Orthodox Eastern Church. There has been a considerable history of animosity between the Armenians and Azeri Turks, growing out of the persecution of Armenians in the Ottoman Empire prior to and during World War I.

Livestock, principally sheep and cattle, are grazed on mountain pastures during the summer months and wintered in the lowlands. The dry steppes of eastern Transcaucasia are especially prominent as a winter grazing ground.

Petroleum and natural gas are the most important minerals of the Caucasus region. The main petroleum field is on the Caspian Sea at Baku (1.2 million), and lesser fields occur north of the mountains in the vicinity of Groznyy (340,000) and Maykop (110,000). Baku, the largest city of Transcaucasia and capital of the Azerbaijan SSR, was the leading center of petroleum production in Old Russia (in fact, it was for a time at the turn of the century the leading center in the world), and under the Soviets it maintained its position as the oil capital of the country for several decades. But during World War II its production declined, and since the war it has been decisively surpassed by the large new Volga-Urals fields (see p. 320). The main natural gas fields in the region are found in the North Caucasus around Stavropol (170,000). Most of the production is piped to the Moscow region.

The Caucasus region also has deposits of coal and is an important producer of metals. At Chiatura in the Georgian SSR is one of the largest known deposits of manganese. Other metals include zinc and lead, mined on the northern slopes of the Greater Caucasus, copper in Armenia, and iron in Azerbaijan. The region has large reserves of hydroelectric power, and a large number of hydroelectric installations.

Besides Baku, only two cities of the Caucasus region exceed half a million in population: Tbilisi or Tiflis (850,000), the capital of the Georgian SSR, and Yerevan (670,000), the capital of the Armenian SSR. Both are diversified industrial centers.

The Caucasus region vies with the Crimea as a center of resort development. The principal resorts are found along the Black Sea coast and in the mountains of the Georgian SSR.

Soviet Middle Asia

Across the Caspian Sea from the Caucasus region lies a vast expanse of deserts and dry grasslands, bounded on the south and east by high mountains. Like Transcaucasia this region, Soviet Middle Asia, has a preponderance of non-Slavic nationalities. The most numerous are four peoples speaking closely related Turkic languages— the Uzbeks, Kazakhs, Kirgiz, and Turkmen—and a people of Iranian origins, the Tadzhiks. Each has its own Union Republic. By far the largest in area is the Kazakh Republic (Kazakhstan), which covers about a million square miles and is more than twice as large as the remaining four republics combined (Table 11).[10]

Soviet Middle Asia is predominantly composed of plains and low uplands except for the Tadzhik and Kirgiz SSRs, which are extremely mountainous. In these republics are found the highest summits of the USSR: Mount Communism (formerly Mt. Stalin), 24,590 feet, in the Pamirs (Tadzhik SSR), and Pobeda (Victory) Peak, 24,406 feet, in the Tien Shan (Kirgiz SSR) (map, p. 300). The Kazakh Upland in east central Kazakhstan is a hilly area with occasional ranges of low mountains. Soviet Middle Asia is remote from the world ocean, and is almost entirely a region of interior drainage. Only the waters of the Irtysh, a tributary of the Ob, reach the open sea; all the other streams either drain to enclosed lakes and seas or gradually dry up and disappear in the arid wastes of the Middle Asian deserts.

The native peoples of Soviet Central Asia and Kazakhstan are Moslems (Muslims). But the twentieth century has seen a large incursion of Russians and Ukrainians, who now comprise well over a third of the population in the five republics combined. Indeed, in the Kazakh SSR they outnumber the native Kazakhs, and in many cities in the five republics they make up half or more of the population. They include administrative and managerial personnel, engineers, technicians, factory workers, and, in the north of Kazakhstan, a sizable number of farmers in the "new lands" wheat region. Most of the Slavic newcomers to Soviet Middle Asia live apart from the Moslems, either in separate quarters of the cities or in separate agricultural villages. A specialist on Russian-Moslem relationships, Richard Pipes, has described the situation in the cities of Soviet Central Asia as follows:[11]

In pre-Revolutionary times the Russians moving into the cities of Turkestan followed the custom of Europeans in other colonial areas of the world, and rather than make

---

[10] The four smaller republics are often referred to as Soviet Central Asia or Turkestan. But some writers apply these names to the region as a whole, including Kazakhstan. In the present chapter the name Soviet Middle Asia will be employed when the five republics are referred to as a group.

[11] Richard Pipes, "Muslims of Soviet Central Asia: Trends and Prospects (Part II)," *Middle East Journal,* 9 (1955), p. 299. Reprinted by permission of the author and the *Middle East Journal.*

Soviet-made mechanical cotton pickers in operation on irrigated land in the Uzbek SSR near Tashkent. (Sovfoto.)

their homes in the Oriental quarters, insanitary and over-crowded by Western standards, or build new cities from scratch, they constructed separate European sections adjoining the native towns. . . . After the advent of the Communists this basic pattern remained, with one important change: leaving the Old and the New sections of the towns intact, the Soviet authorities constructed in most towns new quarters to house the officials, employees, and other privileged elements of Soviet society, who are predominantly, but not exclusively, European. Thus most Central Asian towns today are divided into three parts, one predominantly native, another predominantly European, and a third one mixed, with Europeans dominant.

Irrigated Agriculture

The majority of the people in Soviet Middle Asia live in irrigated valleys at the base of the southern mountains (map, p. 325). Here most soils (often formed of loess) are fertile, the growing season is long, and rivers issuing from mountains provide large supplies of water for irrigation. The irrigated area has been substantially increased in size and productivity under the Soviets, and further increases are planned. The principal rivers in the heart of the region are the Amu Darya (Oxus) and Syr Darya, both of which empty into the enclosed Aral Sea. But a large share of the irrigation water is not obtained from these streams directly, but from their tributaries or from other rivers. Numerous irrigation canals tap the rivers and distribute water to oases. Expansion of irrigation is being facilitated by new dams and reservoirs in the mountains to store water and produce hydroelectricity, some of which is used to operate pumps which lift water from one conduit to another or from rivers and canals to the fields. The most important future source of water is the Amu Darya, which has two to three times the volume of the Syr Darya. Elaborate and expensive control works will be needed before it can be fully utilized.

Most of the larger irrigated districts are found in the Uzbek SSR. Especially important are the fertile Fergana Valley and the oases surrounding the cities of Tashkent (1.2 million), Samarkand (250,000), and Bukhara (110,000). The latter cities are historic trading centers on ancient caravan routes connecting southwestern Asia and the Mediterranean basin with eastern Asia. Today Tashkent, the capital of the Uzbek SSR, is the largest city and principal center of diversified industry in Soviet Middle

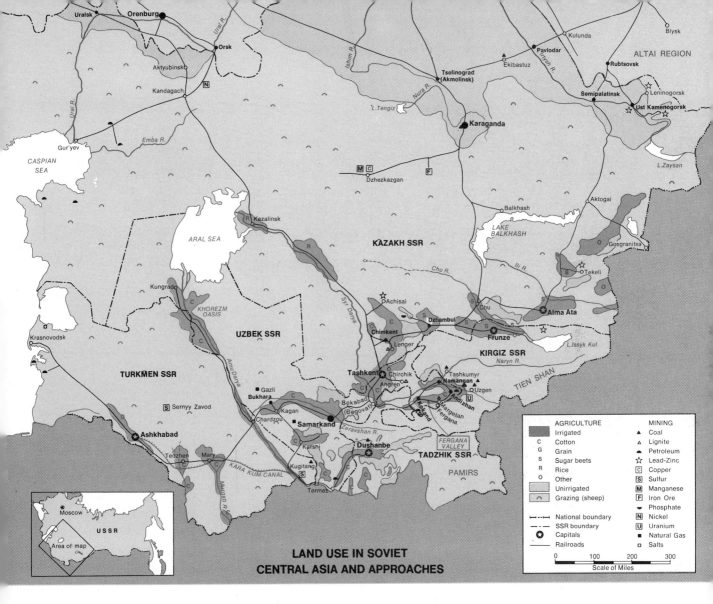

**LAND USE IN SOVIET
CENTRAL ASIA AND APPROACHES**

AGRICULTURE		MINING	
▨ Irrigated		▲ Coal	
C Cotton		△ Lignite	
G Grain		⬤ Petroleum	
S Sugar beets		☆ Lead-Zinc	
R Rice		C Copper	
O Other		S Sulfur	
▨ Unirrigated		M Manganese	
⌒ Grazing (sheep)		F Iron Ore	
		⌐ Phosphate	
⊢–·–⊣ National boundary		N Nickel	
–·–·– SSR boundary		U Uranium	
✪ Capitals		■ Natural Gas	
―― Railroads		◻ Salts	

Scale of Miles: 0 100 200 300

Only selected mining areas, confined generally to those of greatest importance, are shown on this map. Agricultural land is shown only for areas within the boundaries of the five Soviet Middle Asian republics. Modified and redrafted from a map by Robert N. Taaffe, "Transportation and Regional Specialization: the Example of Soviet Central Asia," *Annals of the Association of American Geographers*, 52, no. 1 (March, 1962), 81. Used by permission of the author and the *Annals*.

Asia. Among the prominent factories in the Tashkent area are an agricultural machinery plant, several cotton mills, and a plant utilizing hydroelectric power to manufacture nitrate fertilizers from atmospheric nitrogen.

The major crop of the oases is cotton (photo, p. 324, and maps, pp. 325 and 326). This area grows nine-tenths of the USSR's output. Early-maturing American-upland varieties account for the bulk of the crop, though

some long-staple Egyptian-type or American-type cotton is grown in the Turkmen and Tadzhik SSRs. A visitor described as follows the importance of cotton in the most concentrated areas of production:[12]

The concentration on cotton growing is so heavy . . . that it resembles the one-crop system existent in the U.S. Cotton Belt 30 years ago. Alfalfa needed for winter feed is

[12] Foreign Agricultural Service, U.S. Department of Agriculture, *Cotton in the Soviet Union* (Washington, D.C.: Government Printing Office, 1959), pp. 1–2.

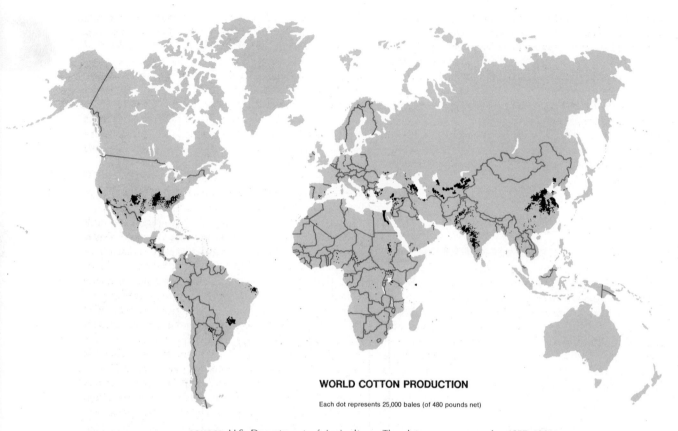

WORLD COTTON PRODUCTION

Each dot represents 25,000 bales (of 480 pounds net)

SOURCE: U.S. Department of Agriculture. The data are averages for 1957–1961.

the only commercial crop used for partial rotation with cotton except for small fields of melons and rice. Cultivable areas are seldom used as pasture even for rotation. Livestock consist mainly of sheep, goats, burros, and cattle (small numbers), tethered individually along the banks of irrigation ditches or tended by shepherds in rocky areas not suitable for cultivation. There are no fences along cotton fields or pasture areas.

The areas nearest the sources of water were developed first, resulting in a heavy concentration of cotton cultivation in the eastern portion of Uzbekistan. This republic accounts for two-thirds of the Soviet crop.

Mulberry trees to feed silkworms are grown around the margins of irrigated fields in some areas. Sugar beets are increasingly important as an irrigated crop, while irrigated rice is grown in sufficient quantities in the Uzbek SSR to make that republic the Soviet Union's leading producer (though the USSR's rice acreage is very minor compared to that of cotton). Vineyards and fruit orchards (apples, pears, peaches, apricots) are found in many oases; commercial orchard farming is especially prominent around

Alma Ata ("father of apples"; 660,000), the capital of the Kazakh SSR.

Nonirrigated Grain Farming and Livestock Herding

Most of the area outside of the oases is too dry for cultivation, although a considerable amount of non-irrigated grain farming is carried on by Russians and Ukrainians in the north of Kazakhstan, which extends into the black-earth belt. Huge acreages, primarily on chestnut soils, were planted in wheat under the "virgin and idle lands" program of 1954–1956 (see pp. 288–289). Nonirrigated grain culture is also found on the rainier slopes in mountain foothills of the south. The latter areas often have good soils formed of loess. In former times nomadic herding provided the livelihood of many Middle Asian tribes. It was based primarily on the natural forage of the steppes and on mountain pastures in the Altai and Tien Shan ranges. But over the centuries there was a slow drift away from nomadism, and this process was accelerated by the Soviet government. It collectivized the remaining nomads, often in the face of strong resistance, and settled them in permanent villages. Frequently such villages were made possible by the drilling of deep wells to provide a dependable year-round supply of water and by the planting of forage crops for supplemental feed. Livestock raising in Soviet Middle Asia has now become a form of ranching, although herdsmen must often accompany the grazing animals for considerable distances as the latter are moved from one area of range to another. Sheep, cattle, and horses are the principal types of livestock.

Mining and Manufacturing

During recent decades Soviet Middle Asia has become increasingly important as a producer of minerals, particularly coal, iron ore, nonferrous metals, natural gas, petroleum, chemical materials, and uranium (maps, pp. 312 and 325). Large reserves of bituminous coal are found at Karaganda in central Kazakhstan. The Karaganda coal basin has become the third largest producer of bituminous coal in the USSR (after the Donets and Kuznetsk basins), and Karaganda itself has experienced spectacular growth from a tiny village in 1926 to a city of over 500,000 today. Much of the coal mined here is shipped to metallurgical works in the Urals. However, Karaganda coal has an excessively high ash content, which limits its usefulness for coking (see p. 318). Large open-pit coal mines have recently been opened in the vicinity of Ekibastuz, about 150 miles northeast of Karaganda. Some of this coal can be coked, though its ash content is high. Among other uses, it provides fuel for thermal-electric power production near the mines.

Karaganda, Ekibastuz, and Kuznetsk Basin coals, as well as coal and lignite from smaller deposits in the Uzbek and Kirgiz SSRs, provide fuel for large-scale metallurgy in Soviet Middle Asia. Of primary importance is the smelting of nonferrous metals, especially copper, lead, and zinc. The largest copper reserves and the most important copper mines of the USSR are found in central Kazakhstan at Dzhezkazgan (60,000), about midway between Karaganda and the Aral Sea. Other copper mines are found at Kounradskiy, just north of Lake Balkhash, and at Almalyk (70,000) in Uzbekistan. Complex ores yielding zinc and lead, plus silver, gold, and copper, are mined in the Altai Mountain foothills of eastern Kazakhstan near Leninogorsk (70,000). The Leninogorsk area is the Soviet Union's chief source of zinc and lead. Smelting works are found in close proximity to all of the nonferrous deposits named. Impressive reserves of chrome and

nickel in the Urals foothills of northern Kazakhstan, petroleum in the fields bordering the Caspian Sea, large iron-ore deposits in the vicinity of Kustanay (120,000) in northern Kazakhstan, and deposits of these and many other minerals scattered throughout the region add further to the picture of Soviet Middle Asia as a rich and diversified mineralized area. Mines in the Tien Shan range are thought to provide the Soviet Union's main supplies of uranium, though some uranium is secured as a by-product from the Krivoy Rog iron mines and the Estonian oil shales.

An important recent development in Soviet Middle Asian mining has been the discovery and exploitation of large natural gas deposits in several different parts of the region. The main producing field at present is near Bukhara, though other fields are being developed. Gas is shipped to the Urals through pipelines 40 inches in diameter, and is supplied to manufacturing plants, electric-power stations, and urban households in Soviet Middle Asia itself. Over three-quarters of the fuel requirements in the Uzbek SSR, for example, are now supplied by natural gas. Expanding chemical industries, particularly in the Uzbek SSR, use gas as a fuel or raw material. The most important product of these industries is chemical fertilizer for Soviet Middle Asia's irrigated agriculture.

Manufacturing in Soviet Middle Asia is steadily increasing in importance. Except for handicrafts, manufacturing scarcely existed at the time of the 1917 revolution. But under the Soviets textile-milling and agricultural-processing industries have been developed on a sizable scale and these have been supplemented, especially since 1940, by machine-building and chemical industries and by an integrated iron and steel plant at Temir Tau (160,000) near Karaganda and a small steel mill at Bekabad (formerly Begovat; 60,000) in the Uzbek SSR. A considerable number of electric power stations have been built, both thermal stations and hydroelectric stations.

The Soviet Far East

The Soviet Far East is the mountainous Pacific margin of the USSR. Most of it is a thinly populated, underdeveloped wilderness in which the only settlements are fishing ports, mining or lumber camps, or the villages or encampments of aboriginal tribes. It is remote from the main centers of Soviet life, and its contribution to the national economy is relatively meager. Most of the Russians and Ukrainians who compose the great majority of its population are found in a narrow strip of lowland behind the coastal mountains in the southern part of the region. This lowland, drained by the Amur River and its tributary, the Ussuri, is the main axis of industry, agriculture, transportation, and urban development in the region. Six main cities form a north-south line along two important arteries of transportation: the Trans-Siberian Railroad and the lower Amur River. At the south on the Sea of Japan is the port of Vladivostok (390,000). It is the main terminus of rail, sea, and air routes connecting the Far East with the rest of the Union. Vladivostok has modern port facilities and its harbor is kept open throughout the winter by icebreakers. About 50 miles east of the city, an additional port, nearly ice free, is being developed at Nakhodka (100,000). Some 50 miles to the north of Vladivostok is Ussuriysk (formerly Voroshilov; 130,000), a food-processing center located in the district that is the most important center of the Soviet Far East's meager agriculture. Mills in the city prepare cereals, soybeans, sugar beets, and milk from the surrounding farms for Far Eastern consumption. The Soviet Far East, however, is far from self-sufficient in food, or, for that matter, in consumer goods of other types, and local production is supplemented by large shipments from the coreland or overseas sources. The diversified industrial and transportation center of Khabarovsk (440,000) is located at the confluence of the Amur and Ussuri rivers about 400 miles northeast of Ussuriysk. Here the main line of the Trans-Siberian Railroad turns south to Vladivostok and Nakhodka, and the Amur River turns north to the Sea of Okhotsk. To the northeast in the Amur Valley is Komsomolsk (220,000), which has the only steel mill in the region. The mill has no blast furnaces, and normally has imported its pig iron supply from Manchuria. A branch line of the Trans-Siberian reaches Komsomolsk and the port of Sovetskaya Gavan on the Gulf of Tatary (railroad map, p. 313). Farther downstream on the Amur near its mouth is Nikolayevsk, a fishing port and point of transshipment between sea and river transportation.

The Pacific littoral north of Nikolayevsk is very mountainous and sparsely populated. The largest settlement is Magadan (80,000), the supply port for an important gold-mining district in the basin of the Kolyma River.

Prior to World War II the Soviet Union and Japan held the northern and southern halves, respectively, of the large island of Sakhalin. But at the end of the war the USSR annexed southern Sakhalin and the Kuril Islands, and repatriated the Japanese population of these areas. (In recent years Japan has been endeavoring to regain the two southernmost islands in the Kurils.) Sakhalin has considerable mineral wealth, including coal, petroleum, and natural gas, and important forest and fishing industries. The Kurils are a sparsely populated chain of small volcanic islands that screen the Sea of Okhotsk from the Pacific. Fishing is the main economic activity. At the

north the islands approach the large peninsula of Kamchatka. Like the Kurils, Kamchatka is a region of active vulcanism, having many hot springs in addition to nineteen active volcanoes, one of which reaches a height of 15,912 feet. Petropavlovsk (130,000) is a fishing port in southern Kamchatka.

Fisheries and Forest Industries

The Soviet Far East is of minor economic significance to the Soviet Union. This region is generally poorer in known mineral resources than the Urals, the Ukraine, or Kazakhstan, although the gold of the Kolyma Basin and the petroleum of Sakhalin (refined at Komsomolsk, Khabarovsk, and Vladivostok) have considerable economic importance, and deposits of coal, lead and zinc, tin, and other minerals are worked to some extent (maps, p. 312). Fish, wood products, gold, and furs are the major commodities produced for sale outside the region. Salmon, herring, sardines, and crabs are abundant in the waters bordering the region and provide the basis for a sizable fishing industry. Herring and sardines caught in coastal waters of the Sea of Japan are processed by canneries at Vladivostok. The Sea of Okhotsk is also a source of herring, but is more important for coastal salmon and crab fisheries. In recent times Kamchatka, particularly the west coast, has come to the fore as a leading center of fishing. The crab fisheries of Kamchatka are reputed to be the world's largest, and salmon are caught in large numbers as they ascend the streams of the peninsula to spawn. The estuary of the Amur River is also an important center of salmon fishing.

Lumber exports from the region consist primarily of softwoods, but include some oak and other hardwoods from the basin of the Ussuri River. Sawmilling is centered at Khabarovsk, at Vladivostok, at points between these cities where the Trans-Siberian Railroad crosses tributaries of the Ussuri River, and at Sovetskaya Gavan. Sizable quantities of wood, however, are carried westward by rail in the form of sawn logs. Southern Sakhalin is the main area of pulp and paper milling. Several mills developed by Japan were taken over by the USSR when it annexed the area in 1945. On the mainland a large complex of wood-chemical plants is being developed on the Amur River upstream from Komsomolsk. Increasing quantities of forest products from the Soviet Far East are being sold in Pacific markets, particularly in Japan and Australia.

The Northern Lands

To the north and east of the Agricultural Triangle and west of, and partially including, the Pacific littoral,

A fur warehouse in the far northeastern arctic of the USSR. The view shows bear and fox pelts being aired and processed. (Sovfoto.)

are enormous stretches of coniferous forest (taiga) and tundra extending completely across the USSR from the Finnish and Norwegian borders to the Pacific. These outlying areas of wilderness may be designated for convenience as the Northern Lands, although parts of the Siberian taiga extend to the southern border of the USSR. The Northern Lands include more than half of the Soviet mainland and also a number of island groups in the Arctic Ocean, among which the two mountainous, fjorded islands of Novaya Zemlya are especially well known.

These difficult lands comprise one of the most sparsely populated large regions of the world. Ordinary types of agriculture are largely precluded by the climate, although hardy vegetables, potatoes, hay, and barley are grown in scattered localities and some dairy farming is carried on. Most of the people are supported by a few primary activities, including lumbering, mining, reindeer herding, hunting, fishing, and trapping. Towns and cities of any size are limited to a small number of sawmilling, mining, transportation, and industrial centers, found in most cases along the Arctic coast to the west of the Ural Mountains, along the major rivers, or along the

Trans-Siberian Railroad between the Agricultural Triangle and Khabarovsk.

The largest city along the Arctic coast is Arkhangelsk (Archangel; 320,000), located on the Northern Dvina River about 25 miles from the White Sea. This city is the most important sawmilling and lumber-shipping center of the USSR. Softwood lumber from Arkhangelsk moves primarily to the coreland, though some of it is marketed in western Europe. The city has direct rail connections with Moscow, and branching lines from Vologda (170,000), north of Moscow, reach Leningrad and the Urals (railroad map, p. 313). Arkhangelsk also has water connections with Leningrad and Moscow via the White Sea-Baltic Canal and the connecting Mariinsk Canals system. Despite its brief season of navigation (June to October), the city is one of the more important seaports of the USSR.

Another Arctic port, Murmansk, located on the northern shore of the Kola Peninsula west of Arkhangelsk, is the headquarters for important trawler fleets which operate in the Barents Sea or range widely into North Atlantic waters. Murmansk is the largest center of the Soviet fishing industry, which has increased greatly since

World War II. Cod, herring, haddock, and sea perch are the main species landed at Murmansk. Fish processing is the city's most important type of industry. The harbor of Murmansk is normally ice free, thanks to the warming influence of the North Atlantic Drift, and thin sheets of ice that occasionally form can easily be broken with icebreakers. The city has experienced a rapid growth from a population of less than 9000 in 1926 to an estimated 290,000 in 1968.

Murmansk is a western terminus (along with Arkhangelsk) of the Northern Sea Route, a waterway developed by the Soviet government to provide a connection with the Soviet Far East via the Arctic Ocean and to link the Arctic lands and seas more effectively with the rest of the country. The economic importance of the route is relatively slight. Despite the use of icebreakers to lead convoys of ships, and radio broadcasts concerning weather and ice conditions, it is navigable throughout its length for only a few weeks in the year. Except for the timber of Igarka, and the nickel and other metals of Norilsk, the areas along the route provide few cargoes. Most ships operating from Murmansk and Arkhangelsk go no farther east than the Yenisey River. A large proportion of the supplies carried by coasting vessels to settlements along the Arctic shore of Siberia are shipped northward by river transportation from cities along the Trans-Siberian Railroad.

A 900-mile rail line connecting Murmansk with Leningrad serves important mining districts in the interior of the Kola Peninsula and areas of lumbering in Karelia, adjoining Finland.[13] The Kola Peninsula is one of the more important mineralized areas of the USSR, producing nickel, copper, iron, aluminum-bearing ore (nephelite), and other metals, as well as phosphates from large reserves of the mineral apatite. The Kola Peninsula and Karelia are physically a prolongation of Scandinavia, being located on the same ancient, glaciated granitic shield which underlies most of the Scandinavian Peninsula and Finland. Karelia, with its thousands of lakes, short, swift streams, extensive softwood forests, timber industries (photo, p. 296), and hydroelectric power stations, bears a close resemblance to adjoining areas of Finland.

Important coal deposits, including good coking coals, are found in the basin of the Pechora River east of Arkhangelsk. The principal mining center is Vorkuta (70,000). Coal from the Pechora fields moves by rail to Leningrad and to an iron and steel plant at Cherepovets (180,000) which processes iron-ore concentrates from the Kola Peninsula. The Pechora deposits may ultimately prove of critical importance as a source of coking coal for metallurgical works in the Urals.

East of the Urals the swampy, roadless plain of western Siberia north of the Trans-Siberian Railroad is witnessing an impressive new development of petroleum and natural gas production. It is thought that this region between the Urals and the Yenisey River may have as much as a third of the USSR's oil and gas reserves. The principal area of production lies in the vicinity of Surgut, located on the Ob River about 420 air miles north of Omsk. Oil is shipped by pipeline and barge to a refinery and synthetic rubber plant at Omsk. Siberia will be the main market area for production from the new fields, but some of the oil will flow to European Russia. The main natural gas fields are located hundreds of miles to the northeast or northwest of Surgut. A pipeline from the westernmost fields supplies gas to Nizhniy Tagil and other industrial centers in the Urals.

Still farther east, the town of Igarka has become a sawmilling center of major importance. Igarka lies on the Yenisey River about 425 miles inland, but can be reached by ocean shipping during the few weeks of summer when the river is free of ice. About 125 miles from Igarka to the north-northeast is the mining center of Norilsk (130,000), producing nickel, copper, platinum, and cobalt. Few settlements of special note occur north of the Arctic Circle between Norilsk and the Lena River, though east of the Lena a handful of scattered mining towns produce tin or gold.

In central and eastern Siberia the taiga forest extends southward beyond the Trans-Siberian Railroad. A vast but sparsely settled hinterland is served by four cities spaced at wide intervals along the railroad: Krasnoyarsk (580,000) on the Yenisey River, Irkutsk (420,000) on the Angara tributary of the Yenisey near Lake Baikal, and Ulan-Ude (230,000) and Chita (210,000) east of Lake Baikal. From Tayshet, east of Krasnoyarsk, a branch line of the Trans Siberian leads eastward to the railhead of Ust-Kut on the Lena River (map, p. 313). The line passes through Bratsk (165,000), the site of a huge dam and hydroelectric power station on the Angara River. Lake Baikal, which the Angara drains, is the deepest body of fresh water in the world, being more than a mile deep in places. Fed by hundreds of streams, it supplies an enormous and relatively constant volume of flow to the Angara. In the development of Siberia, great emphasis is being placed on hydroelectric power installa-

---

[13] The area designated as Karelia was formerly a Union Republic, the Karelo-Finnish SSR, but in 1956 it was downgraded to the status of an Autonomous SSR.

tions along the major rivers. The largest ones thus far are the Bratsk installation, which has a generating capacity over twice that of Grand Coulee Dam in the United States, and a dam and power station on the Yenisey River at Krasnoyarsk which will have an ultimate generating capacity considerably greater than that of Bratsk. A dam almost as large as the Bratsk dam is under construction on the Angara at Ust Ilim about 160 river miles downstream from Bratsk, and a third large dam is planned still farther downstream. Upstream from Bratsk a smaller, though sizable, dam and power station has been constructed at Irkutsk. It is planned to construct a huge installation on the Yenisey around 250 miles upstream from Krasnoyarsk. The power stations at these dams supply large quantities of cheap electricity, for consumption primarily by large, highly mechanized industries that are voracious users of power. Cellulose plants and aluminum plants at Irkutsk, Bratsk, and Krasnoyarsk are prominent examples. These plants process Siberian timber and bauxite. Surplus electricity from Siberia may ultimately be transmitted by high-voltage lines to European Russia. In the meantime Siberia is benefiting from the eastward flow of another power resource—petroleum and its products—from the Volga-Urals and West Siberian fields. Pipelines paralleling the Trans-Siberian Railroad reach as far east as Irkutsk and may ultimately be extended to the Pacific coast. A refining and petrochemical complex stretches along the Angara just downstream from Irkutsk.

Important gold-mining centers are scattered throughout central and eastern Siberia (map, p. 312). Although production figures are not definitely known, this part of the Soviet Union is considered to be one of the major gold-mining areas of the world. The USSR almost certainly is the second gold producer of the world, after South Africa. Mining activities are mainly centered in the basins of the Kolyma and Indigirka rivers in northeastern Siberia, and in the basin of the Aldan River, a tributary of the Lena. Large coal deposits are mined at Cheremkhovo (110,000) on the Trans-Siberian Railroad west of Irkutsk (part of the production, along with better quality coal from the Kuznetsk Basin, will be used in a new iron and steel plant under construction at Tayshet), and a variety of metals, including aluminum, tin, tungsten, molybdenum, manganese, lead, and zinc are secured in small to medium quantities from widely scattered mining areas tributary to the railroad. Large coal reserves in the basins of the Lena River, the Lower Tunguska tributary of the Yenisey, and the Bureya tributary of the Amur have scarcely begun to be exploited. In the period 1954–1956, major discoveries of diamonds were made in the eastern part of the Central Siberian Uplands. Soviet sources report that the deposits, found in kimberlite "pipes" similar to those of South Africa, are among the largest in the world and yield gem stones as well as industrial diamonds. The principal mining centers at present are Mirnyy on the Vilyuy River, a Lena tributary, and Aykhal, located about 250 miles farther north near the Arctic Circle. Oil and natural gas have been found along the upper Lena east of Bratsk, and a newly developed natural gas field on the middle Lena supplies gas by pipeline to the main city on the river, Yakutsk (100,000).

Aside from the irregularly distributed centers of lumbering, mining, and transportation, the Northern Lands are mainly occupied by non-Slavic peoples who eke out a living by reindeer herding, trapping, fishing, and in the more favored areas of the taiga, by cattle raising and precarious forms of cultivation. The domesticated reindeer is an especially valuable source of livelihood for the tundra peoples, providing meat, milk, hides for clothing and tents, and serving as a draft animal. The Yakuts, a Mongoloid people inhabiting the basin of the Lena River, are among the most prominent of the non-Slavic ethnic groups in the Northern Lands. Their political unit, the Yakut Autonomous Soviet Socialist Republic, is larger in area than any of the Union Republics except the Russian Soviet Federated Socialist Republic, within which it is contained. However, in 1965 it had only an estimated 612,000 people inhabiting an area of 1,182,000 square miles. Actually the Yakuts do not form a majority in this republic, being outnumbered by immigrant Slavs. The republic's capital city of Yakutsk on the middle Lena River has highway connections with the Trans-Siberian Railroad between Chita and Khabarovsk and with Magadan on the Sea of Okhotsk. River boats on the broad and deep Lena River provide a connection with the Northern Sea Route and the Ust Kut railhead (p. 330) during the warm season.

Although the efforts of the Soviet government to develop the Northern Lands have received much publicity and have achieved successes in the fields of mining, lumbering, hydroelectricity, and transportation, it seems probable that these lands, especially the more northerly portions, will continue to be a thinly settled frontier region for a long time to come. A considerable amount of experimentation has been carried on with quick-growing and frost-resistant crops and with various forms of hothouse culture and forced plant growth, but it seems doubtful that these can form the basis of a significant agriculture except in very limited areas. At present there is no disposition on the part of Soviet planners to stimulate any large flow of population to these lands, though selected enterprises will continue to be developed. The situation of the Soviet Union in the foregoing respects is similar to that of Canada, and the experiences of the two nations in developing their sparsely inhabited northern territories will afford interesting comparisons in the coming decades.

REFERENCES and READINGS

General and Miscellaneous

ANDERSON, JEREMY, "Fodder and Livestock Production in the Ukraine: A Case Study of Soviet Agricultural Policy," *East Lakes Geographer,* 3 (October 1967), 29–46.

ARMSTRONG, TERENCE, "Soviet Sea Fisheries Since the Second World War," *Polar Record,* 13, no. 83 (May 1966), 155–186.

BONE, ROBERT M., "The Fertile Triangle of Soviet Russia," *Canadian Geographical Journal,* 67, no. 5 (November 1963), 167–171; and "Regional Planning and Economic Regionalization in the Soviet Union," *Land Economics,* 43, no. 3 (August 1967), 347–354.

CONOLLY, VIOLET, *Beyond the Urals: Economic Development in Soviet Asia* (New York: Oxford University Press, 1967). An excellent book, with much recent information and good maps.

DUNN, STEPHEN P., and ETHEL DUNN, "The Great Russian Peasant: Culture Change or Cultural Development?" *Ethnology,* 2, no. 3 (July 1963), 320–338.

ELLISON, HERBERT J., "Economic Modernization in Imperial Russia: Purposes and Achievements," *Journal of Economic History,* 25, no. 4 (December 1965), 523–540.

FIELD, N. C., "Land Hunger and the Rural Depopulation Problem in the U.S.S.R.," *Annals of the Association of American Geographers,* 53, no. 4 (December 1963), 465–478.

FOX, DAVID J., "Odessa," *Scottish Geographical Magazine,* 79, no. 1 (April 1963), 5–22.

FROLIC, B. MICHAEL, "The Soviet City," *Town Planning Review,* 34, no. 4 (January 1964), 285–306.

FUCHS, ROLAND J., "Moscow," *Focus,* 6, no. 5 (January 1966), 6 pp. An excellent brief survey.

GOLDHAGEN, ERICH, ed., *Ethnic Minorities in the Soviet Union* (New York: Frederick A. Praeger, 1968).

HALL, PETER, *The World Cities* (London: Weidenfeld and Nicolson, World University Library, 1966), Chap. 6, "Moscow," pp. 158–181.

HARRIS, CHAUNCY D., "City and Region in the Soviet Union," in R. P. Beckinsale and J. M. Houston, eds., *Urbanization and Its Problems* (New York: Barnes and Noble, 1968), pp. 277–296.

HELIN, RONALD A., "Soviet Fishing in the Barents Sea and the North Atlantic," *Geographical Review,* 54, no. 3 (July 1964), 386–408.

HOOSON, DAVID J. M., *A New Soviet Heartland?* (Princeton, N.J.: D. Van Nostrand Co., Searchlight Books, 1964); "A New Soviet Heartland," *Geographical Journal,* 128, pt. 1 (March 1962), 19–29; "The Middle Volga: An Emerging Focal Region in the Soviet Union," *Geographical Journal,* 126, pt. 2 (June 1960), 180–190; and "The Growth of Cities in Pre-Soviet Russia," in R. P. Beckinsale and J. M. Houston, eds., *Urbanization and Its Problems* (New York: Barnes and Noble, 1968), pp. 254–276.

JACKSON, W. A. DOUGLAS, "The Soviet Non-Chernozem Wheat Base," *Annals of the Association of American Geographers,* 49, no. 2 (June 1959), 97–109.

LE FLEMING, STEPHEN, "Novosibirsk: The City," *Geographical Magazine,* 39, no. 7 (November 1966), 495–503.

LIPSET, HARRY, "The Status of National Minority Languages in Soviet Education," *Soviet Studies,* 19, no. 2 (October 1967), 181–189.

MELEZIN, ABRAHAM, "Soviet Regionalization: An Attempt at the Delineation of Socio-economic Integrated Regions," *Geographical Review,* 58, no. 4 (October 1968), 593–621.

MIECZKOWSKI, Z., "The Economic Administrative Regions in the U.S.S.R.," *Tijdschrift voor Economische en Sociale Geografie,* 58, no. 4 (July–August 1967), 209–219; and "The Major Economic Regions of the U.S.S.R. in the Khrushchev Era," *Canadian Geographer,* 9, no. 1 (1965), 19–30.

MILNER-GULLAND, ROBIN, "Moscow's Countryside," *Geographical Magazine,* 35, no. 4 (August 1962), 215–227.

NEWTH, J. A., "The Soviet Population: Wartime Losses and the Postwar Recovery," *Soviet Studies,* 15, no. 3 (January 1964), 345–351.

PROCIUK, S. G., "The Manpower Problem in Siberia," *Soviet Studies,* 19, no. 2 (October 1967), 190–210.

SUSLOV, SERGEI PETROVICH, *Physical Geography of Asiatic Russia* (San Francisco: W. H. Freeman and Co., 1961).

TASKIN, GEORGE A., "The Soviet Northwest: Economic Regionalization," *Geographical Review,* 51, no. 2 (April 1961), 213–235.

THOMAS, C., "Population Trends in the Soviet Union, 1959–64," *Geography,* 52, pt. 2 (April 1967), 193–196.

U.S. DEPARTMENT OF STATE, THE GEOGRAPHER, *U.S.S.R.: Civil Divisions* (Geographic Report, No. 12) (Washington, D.C.: 1967).

VARDYS, V. STANLEY, "How the Baltic Republics Fare in the Soviet Union," *Foreign Affairs,* 44, no. 3 (April 1966), 512–517.

Industry, Transportation, and Trade

ADAMS, RUSSELL B., "U.S.S.R. Transportation," *Focus,* 18, no. 10 (June 1968), 1–8, and "Soviet Foreign Trade," 8–11.

CAMPBELL, ROBERT W., *The Economics of Soviet Oil and Gas* (Baltimore, Md.: Johns Hopkins Press for Resources for the Future, 1968).

CLARK, M. GARDNER, "Magnitogorsk: A Soviet Iron and Steel Plant in the Southern Urals," Chap. 21 in Richard S. Thoman and Donald J. Patton, eds., *Focus on Geographic Activity: A Collection of Original Studies* (New York: McGraw-Hill Book Company, 1964), pp. 128–134.

GARBUTT, P. E., "The Trans-Siberian Railway," *Journal of Transport History,* 6, no. 4 (November 1954), 238–249.

GUEST, B. ROSS, "Soviet Gas Pipeline Development during the Seven-Year Plan," *Professional Geographer,* 19, no. 4 (July 1967), 189–192; and "The Growth of Soviet Air Cargo," *Journal of Geography,* 65, no. 7 (October 1966), 323–327.

HUNTER, HOLLAND, "Transport in Soviet and Chinese Development," *Economic Development and Cultural Change,* 14, no. 1 (October 1965), 71–84.

KISH, GEORGE, "Railroad Passenger Transport in the Soviet Union," *Geographical Review,* 53, no. 3 (July 1963), 363–376.

LONSDALE, RICHARD E., "Siberian Industry before 1917: The Example of Tomsk Guberniya," *Annals of the Association of American Geographers,* 53, no. 4 (December 1963), 479–493; and, with John H. Thompson, "A Map of the USSR's Manufacturing," *Economic Geography,* 36, no.1 (January 1960), 36–52.

LYDOLPH, PAUL E., and THEODORE SHABAD, "The Oil and Gas Industries in the U.S.S.R.," *Annals of the Association of American Geographers,* 50, no. 4 (December 1960), 461–486; and "The Chemical Industries in the USSR," *Tijdschrift voor Economische en Sociale Geografie,* 53, nos. 8–9 (August–September 1962), 169–179.

MICHEL, ALOYS A., and STEPHAN A. KLAIN, "Current Problems of the Soviet Electric Power Industry," *Economic Geography,* 40, no. 3 (July 1964), 206–220.

MUCKLESTON, KEITH W., "Volga Transport System," *Yearbook of the Association of Pacific Coast Geographers,* 27 (1965), 67–76; and, with Fred E. Dohrs, "The Relative Importance of Transport on the Volga before and after the Communist Revolution," *Professional Geographer,* 17, no. 2 (March 1965), 22–25.

PADICK, CLEMENT, "Reorientation in Power Generation in the Volga Basin, U.S.S.R.," *Yearbook of the Association of Pacific Coast Geographers,* 27 (1965), 27–37.

PETROV, VICTOR P., "Soviet Canals," *United States Naval Institute Proceedings,* 93, no. 7 (July 1967), 32–44.

PRYDE, PHILIP R., "The Areal Deconcentration of the Soviet Cotton-Textile Industry," *Geographical Review,* 58, no. 4 (October 1968), 575–592.

RODGERS, ALLAN, "Coking Coal Supply: Its Role in the Expansion of the Soviet Steel Industry," *Economic Geography,* 40, no. 2 (April 1964), 113–150.

SPENCER, D. L., "The Role of Oil in Soviet Foreign Economic Policy," *American Journal of Economics and Sociology,* 25, no. 1 (January 1966), 91–107.

TAAFFE, ROBERT N., "Interregional Passenger Movement in the Soviet Union," *East Lakes Geographer,* 3 (October 1967), 47–79; and "Volga River Transportation: Problems and Prospects," Chap. 29 in Richard S. Thoman and Donald J. Patton, eds., *Focus on Geographic Activity: A Collection of Original Studies* (New York: McGraw-Hill Book Company, 1964), pp. 185–193.

WOOD, DONALD S., "The New Volga: A Soviet Transformation of Nature," *Journal of Geography,* 57, no. 2 (February 1963), 49–56.

Outlying Regions

ALLWORTH, EDWARD, ed., *Central Asia: A Century of Russian Rule* (New York: Columbia University Press, 1967).

ARALDSEN, O. P., "The Soviet Union and the Arctic," *United States Naval Institute Proceedings,* 93, no. 6 (June 1967), 49–57.

ARMSTRONG, TERENCE, *Russian Settlement in the North* (Cambridge: At the University Press, 1965); "Labour in Northern U.S.S.R.," *Polar Record,* 13, no. 87 (September 1967), 769–774; and "The Population of the North of the U.S.S.R.," *Polar Record,* 11, no. 71 (May 1962), 172–178.

BONE, ROBERT M., "Soviet Tea Cultivation," *Annals of the Association of American Geographers,* 53, no. 2 (June 1963), 161–173.

FIELD, NEIL C., "The Amu Darya: A Study in Resource Geography," *Geographical Review,* 44, no. 4 (October 1954), 528–542.

GIBSON, JAMES R., "Russia on the Pacific: The Role of the Amur," *Canadian Geographer,* 12, no. 1 (1968), 15–27.

JENSEN, ROBERT G., "Soviet Subtropical Agriculture: A Microcosm," *Geographical Review,* 54, no. 2 (April 1964), 185–202.

KOLARZ, WALTER, *The Peoples of the Soviet Far East* (New York: Frederick A. Praeger, 1954).

KOVDA, V. A., "Land Use Development in the Arid Regions of the Russian Plain, the Caucasus and Central Asia," in L. Dudley Stamp, ed., *A History of Land Use in Arid Regions* (Paris: UNESCO, 1961), pp. 175–218.

LEWIS, ROBERT A., "Early Irrigation in West Turkestan," *Annals of the Association of American Geographers,* 56, no. 3 (September 1966), 467–491; and "The Irrigation Potential of Soviet Central Asia," *Annals of the Association of American Geographers,* 52, no. 1 (March 1962), 99–114.

MIECZOWSKI, Z., "The Soviet Far East: Problem Region of the USSR," *Pacific Affairs,* 41, no. 2 (Summer 1968), 214–229.

NOVE, ALEC, and J. A. NEWTH, *The Soviet Middle East: A Model for Development?* (New York: Frederick A. Praeger, 1967).

PIPES, RICHARD, "Muslims of Soviet Central Asia: Trends and Prospects," *Middle East Journal,* 9, no. 2 (Spring 1955), 147–162; and no. 4 (Autumn 1955), 295–308.

TAAFFE, ROBERT N., *Rail Transportation and the Economic Development of Soviet Central Asia* (University of Chicago, Department of Geography Research Paper No. 64, Chicago: 1960); and "Transportation and Regional Specialization: The Example of Soviet Central Asia," *Annals of the Association of American Geographers,* 52, no. 1 (March 1962), 80–98.

THIEL, ERICH, *The Soviet Far East: A Survey of Its Physical and Economic Geography,* 1st English language ed. trans. from the German by Annelie and Ralph M. Rookwood (London: Methuen and Co., 1957).

WAYS, MAX, "The 'House of the Dead' [Siberia] Is Now the Liveliest Part of the U.S.S.R.," *Fortune*, **78**, no. 2 (August 1968), 112–115 ff.

WHEELER, GEOFFREY, *The Modern History of Soviet Central Asia* (London: Weidenfeld and Nicolson, 1964); *The Peoples of Soviet Central Asia: A Background Book* (London: Bodley Head Press, 1966); and *Racial Problems in Soviet Muslim Asia* (New York: Oxford University Press, 1960).

WILLIAMS, D. S. M., "The City of Tashkent, Past and Present," *Royal Central Asian Journal*, **54**, pt. 1 (February 1967), 33–43.

See also the list of references and readings for Chapter 13.

PART IV

THE MIDDLE EAST

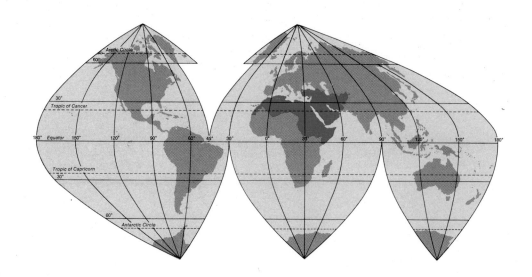

Introduction

to the

Middle East

15   The term "Middle East," as used in this book, refers to an elongated region stretching for 6000 miles across northern Africa and southwestern Asia from the Atlantic Ocean to the borders of India, China, and Soviet Asia. This region includes numerous countries (map, p. 340), but many of them are are small. Only nine countries have populations of 10 million or more: West Pakistan (50–60 million), Turkey and the United Arab Republic (Egypt) (30–40 million), Iran and Ethiopia (20–30 million); and Afghanistan, the Sudan, Morocco, and Algeria (10–20 million) (Table 12). But in the Middle East, as in other world regions, the importance of countries is not necessarily a function of size. The small state of Israel, for example, has carved out a niche for itself despite the opposition of larger neighbors; and the rich petroleum deposits of Kuwait, Qatar, and other minor territories along the Persian Gulf have magnified the importance of these small units.

The margins of the Middle East are mainly occupied by oceans, seas, high mountains, and deserts: to the west the Atlantic Ocean; to the south the Sahara Desert, the highlands of East Africa, and the Indian Ocean; to the north the Mediterranean, Black,

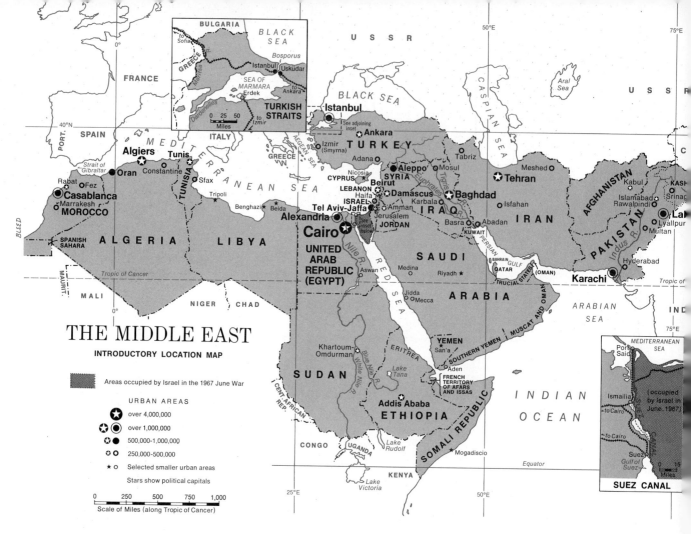

THE MIDDLE EAST

INTRODUCTORY LOCATION MAP

Areas occupied by Israel in the 1967 June War

URBAN AREAS

over 4,000,000

over 1,000,000

500,000-1,000,000

250,000-500,000

Selected smaller urban areas

Stars show political capitals

0   250   500   750   1,000
Scale of Miles (along Tropic of Cancer)

Introductory location map of the Middle East, showing political units as of January 5, 1969. The Somali Republic is comprised of former Italian Somaliland (later the Trust Territory of Somalia) and former British Somaliland.

and Caspian seas, together with mountains and deserts lining the southern land frontiers of the Soviet Union; to the east the Thar Desert of India and the great mountain knot of Inner Asia. The Middle East itself is mainly composed of arid or semiarid plains and plateaus, together with considerable areas of rugged mountains.

Although this region exhibits much variety from place to place, two important factors help to give it unity: (1) the dominance of dry climates and (2) the Islamic (Moslem or Muslim) religion. The latter is the principal religion in all of the Middle Eastern countries except four: Israel, where Judaism prevails; Lebanon and

TABLE 12 NOTES

[a] Area and population figures for Israel, the United Arab Republic, Jordan, and Syria are based on the extent of territory controlled by each of these countries prior to the 1967 Arab-Israeli War (see p. 364).

[b] This figure omits the large southern desert called the Empty Quarter, where political boundaries are undefined. Other sources estimate Saudi Arabia's area at 850,000 to 900,000 sq. mi. There is also much variation in estimates for other political units of the Arabian Peninsula.

[c] Pakistan is divided into two provinces, West Pakistan and East Pakistan, separated by the Republic of India. West Pakistan and the disputed area of Kashmir are considered in this text to be transitional between the Middle East and the Orient. East Pakistan has a very different type of natural environment and economy. For the purposes of this text it is not considered a part of the Middle East.

[d] This figure, based on official estimates by the Ethiopian government, may be several million too high (see p. 392).

[e] Comprised of five small enclaves along the Mediterranean coast of Morocco; the most important are the cities of Melilla (80,000) and Ceuta (75,000).

TABLE 12 MIDDLE EAST: AREA AND POPULATION DATA

POLITICAL UNIT	POLITICAL STATUS (EARLY 1969)	AREA (THOUSAND SQUARE MILES)	POPULATION (MILLIONS: 1968 ESTIMATES)	DENSITY (PER SQUARE MILE)
Arab League states		4280.2	114.2	27
United Arab Republic (Egypt)[a]	Independent republic	386.0	31.8	82
Lebanon	Independent republic	4.0	2.6	650
Syria[a]	Independent republic	71.5	5.8	81
Iraq	Independent republic	167.6	8.9	53
Jordan[a]	Independent kingdom	37.8	2.2	58
Saudi Arabia	Independent kingdom	617.8[b]	7.1	12
Southern Yemen	Independent republic	112.4	1.2	10
Kuwait	Independent sheikdom	6.2	0.5	81
Yemen	Independent republic	75.3	5.2	69
Sudan	Independent republic	967.2	14.8	15
Libya	Independent kingdom	679.2	1.8	3
Tunisia	Independent republic	63.4	4.7	74
Algeria	Independent republic	919.4	12.9	14
Morocco	Independent kingdom	172.4	14.7	85
Other independent states				
Turkey	Independent republic	301.4	33.8	112
Iran	Independent kingdom	636.3	26.5	41
Afghanistan	Independent kingdom	250.0	16.0	64
West Pakistan[c]	Province of independent Republic of Pakistan	310.2	51.0	164
Kashmir[c]	Area disputed between Pakistan and India	85.9	6.0	69
Israel[a]	Independent republic	8.0	2.8	349
Cyprus	Independent republic	3.6	0.6	168
Ethiopia	Independent empire	471.7	23.8[d]	50
Somali Republic	Independent republic	246.1	2.8	11
Totals: All independent states		6593.4	277.5	42
European affiliates				
Affiliated with Great Britain		123.0	1.19	10
Muscat and Oman	Sultanate in close treaty relationship with Great Britain	82.0	0.8	9
Bahrain	Protected sheikdom	0.23	0.20	905
Qatar	Protected sheikdom	8.5	0.07	9
Trucial States (Trucial Oman)	Group of protected sheikdoms	32.3	0.14	4
Affiliated with France		8.5	0.10	12
French Territory of Afars and Issas (former French Somaliland)	Overseas territory	8.5	0.10	12
Affiliated with Spain		102.7	0.21	2
Spanish Sahara	Spanish African province	102.7	0.05	0.5
Spanish North Africa[e]	Administered with Spain	0.012	0.16	13,333
Totals: European affiliates		234.2	1.50	6
Grand totals		6827.6	279.0	41

The widespread importance of the Islamic faith is a major distinguishing characteristic of the Middle East. In this view, camels wait patiently as their Saudi Arab Bedouin masters face meditatively toward the holy city of Mecca and pray during one of the five prayer periods of the day which Moslems observe. (Arabian American Oil Company.)

Ethiopia, where ancient forms of Christianity are of major importance; and Cyprus, where the majority of the population adheres to the Greek Orthodox Church. Even these countries have Moslem minorities; indeed in Lebanon Moslems are estimated to comprise at least a third of the population, and may possibly constitute a majority.[1]

WORLD IMPORTANCE OF THE MIDDLE EAST

Throughout history the sparsely populated deserts and mountains of the Middle East, separating the humid lands of Europe, Africa, and Asia, have been a hindrance

[1] The authors have elected not to discuss in the "Middle East" chapters certain areas that lie along the edges of the region and have many Middle Eastern characteristics. The principal examples are Soviet Middle Asia, the province of Sinkiang in western China, and northern sections of western tropical Africa. For the purposes of this text the foregoing areas are conveniently discussed in the context of the Soviet Union, the Orient, and Tropical Africa, respectively. It may be noted that the island of Cyprus is discussed with the Middle East, but it could with equal justification have been considered with Europe. The student is again reminded that the great cultural realms of the world grade into each other at the margins and that any attempt to separate them by sharp lines for convenience in teaching is bound to be somewhat arbitrary.

to overland travel between those regions. Yet circulation of people, goods, and ideas has taken place along certain favorable routes, and the scattered population centers of the Middle East have had a history of vigorous interaction with the outside world and with each other. No areas of the world have been invaded so often, have seen so many empires rise and fall, or have been subjected to such a variety of cultural influences and political pressures. None have made greater contributions to mankind. The earliest foundations of Western civilization were laid in the river valleys of ancient Egypt and Mesopotamia. In the Middle East the earliest cities arose, and the great monotheistic religions of Judaism, Christianity, and Islam were developed. Many of the plants and animals upon which the world's agriculture is based were first domesticated in this region.

Most of the vegetable foods we eat originated in this area. Wheat, barley, and rye, of the small grains, and possibly millet. Broad beans, chickpeas, lentils, and vetches, of the legumes. Onions, leeks, garlic, of the lilies. Figs and the vine, and all the delicious varieties of muskmelon that cool and slake the thirsty traveler in the heat of summer. Pomegranates . . .; olives and sesame for oil; apples, quinces, and pears, of the roses . . .; peaches, plums, apricots, and almonds; walnuts, saffron. Dates in the desert oases and along the exotic rivers.[2]

The list of common animals domesticated in the region includes oxen, sheep, and goats.

We who eat roast beef [or lamb chops] . . . seldom wonder whom to thank for these gifts, other than the ultimate and divine Source of all bounty. It was the ancient hunters and earliest farmers of the Middle East who first rounded up these animals and tamed them for their use. Try to imagine yourself on foot, armed with a bow and arrow, a length of cordage, and a stone ax, either alone or accompanied by a dozen of your fellows, setting out to catch a wild bull in the forest or a wild sheep on the mountain crags.[3]

Most of the world's great empires have included portions of the Middle East. Some of these empires were indigenous, while others, such as the Roman Empire, were imposed from outside. Commencing in the seventh century a powerful Islamic empire, organized by Arabs,

arose in the Middle East and evolved the most brilliant civilization of its day (see p. 360). Later it decayed. In the sixteenth century most of the territories it had controlled were conquered by the Ottoman Turks. Still later, when the Ottoman Empire was declining, Great Britain, France, and Italy extended their control over sizable areas. For centuries, while the great powers of the Western world were industrializing and achieving high levels of living, most of the Middle East remained a poor and backward area dominated by foreigners. Today, however, the peoples of the region are reasserting their independence and importance. The Ottoman Empire was liquidated at the end of World War I, and the Italian Empire following World War II. Since World War I the Middle Eastern possessions of Great Britain and France have gained independence, aside from the small unit known as the French Territory of Afars and Issas (formerly French Somaliland), and some minor sheikdoms on the Persian Gulf that were still under British protection (though not for long) at the time of writing (see p. 384). It may be added that Spain still holds a few minor Middle Eastern possessions along the borders of Morocco (footnote, p. 376), though its main Middle Eastern territory, Spanish Morocco, gained independence in 1956 as a part of the Kingdom of Morocco.

As European colonialism has declined in the Middle East, many political trouble spots have developed. The region lacks political stability, a fact partly associated with the illiteracy and poverty of many of its people. Several international crises have been precipitated by Middle Eastern questions since World War II. These reflect the growing importance of the region in world affairs. One of the main factors is the rich oil deposits of the Persian Gulf area and the Sahara Desert. Great Britain, France, and several other European countries import large quantities of oil from the Middle East, and European companies, especially British, French, and Dutch, play a prominent role in the region's oil industry. American companies are prominent also, though the United States is less dependent than Europe upon Middle Eastern oil. It should be noted that the relationship between the Middle East and Europe with respect to oil is symbiotic. The bulk of Middle Eastern production could at present be marketed only in Europe, and Europe, on the other hand, would find it difficult or impossible to secure adequate oil supplies elsewhere, at least on a short-term basis.

[2] Carleton S. Coon, *Caravan: The Story of the Middle East* (rev. ed.; New York: Holt, Rinehart and Winston, 1958), p. 23. Used by permission of the author and the publisher.

[3] *Ibid.,* p. 24. Used by permission of the author and the publisher. Supplementary material has been inserted in brackets.

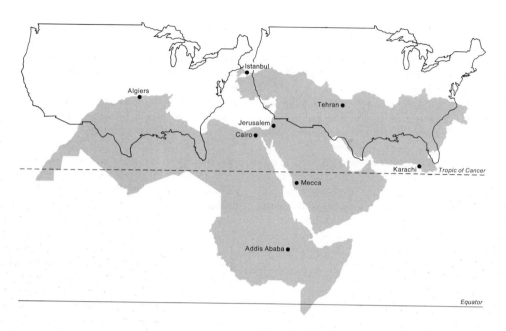

The Middle East compared in latitude and area with conterminous United States.

An important route of sea transportation, the Mediterranean-Asiatic route, crosses the heart of the Middle East by way of the Suez Canal and the Red Sea. It connects the nations of the Atlantic community with the Persian Gulf oil fields, the eastern coasts of Africa, the Indian subcontinent, eastern Asia, and Oceania. At the time of writing (early 1969) the future of this route was uncertain. The Suez Canal had been closed since the short Arab-Israeli war of 1967 (see pp. 361–365), and the oil and other cargo that formerly passed through this waterway was being handled increasingly by tankers and cargo vessels too large for the canal to accommodate. Important air routes span the Middle East, focusing on Cairo, Istanbul, Beirut, Baghdad, Karachi, Khartoum, Algiers, Tel Aviv, and many smaller centers. Long-distance rail and highway transportation are poorly developed, however. Today the importance of the region as a transit land lies principally in its traffic by sea and air, although in earlier times its caravan routes carried small but valuable amounts of intercontinental land traffic.

Though the age of European colonialism in the Middle East is closing, the great powers of the world continue to take a very active interest in the region, with its abundant oil, its strategic "crossroads" location, and its host of young nations whose power and influence will increase as the forces of modernization take hold.

Programs of economic, social, technical, and military assistance to Middle Eastern nations are carried on by the United States, the Soviet Union, and several other powers. Ground and air forces of outside nations have largely been withdrawn from the region, aside from advisers and certain technical personnel, but the major sea powers continue to maintain a strong naval presence there. Meanwhile the nations of the Middle East, jealous of their new liberties, and very conscious of their long histories as civilized lands, are determined to exert increasing control over their own destinies and to push forward with programs of modernization which they themselves direct. In this enterprise some nations are aided by large oil royalties, but in general the region is handicapped by a natural resource base that is seriously inadequate for the needs of modern societies. Some salient features of the Middle Eastern environment are discussed in the following section.

DOMINANT FEATURES OF THE ENVIRONMENT

On the whole, nature has not been kind to the Middle East. With a few conspicuous exceptions, this region is notably deficient in natural resources.

Dominance of Dry Climates

For the Middle East as a whole, the most critical resource deficiency is lack of moisture. Most of this region is part of the Dry World—a vast belt of deserts and dry grasslands extending across Africa and Asia from the Atlantic Ocean nearly to the Pacific. At least three-fourths of the Middle East has an average yearly rainfall of less than 10 inches—an amount too small for most types of nonirrigated agriculture under the prevailing temperature conditions. Some fair-sized areas bordering the Mediterranean Sea have 20 to 40 inches of rain, most of which falls during the cool season and thus is not available for growing the many crops that require the higher temperatures of the summer months. Rainfall sufficient for unirrigated summer cropping is confined to areas along the southern or northern margins of the region: the highlands of Ethiopia and Yemen, the southern Sudan, strips of territory bordering the Black and Caspian seas, and limited areas in Kashmir and northern West Pakistan.

Temperature Regimes

Middle Eastern climates exhibit the comparatively large seasonal and diurnal ranges of temperature which are characteristic of dry lands. Summers in the lowlands are very hot almost everywhere. Many places regularly experience daily maxima of 100°F for weeks at a time. Shade temperatures of 130° or higher have been recorded in parts of the Sahara, Arabia, Iran, and Pakistan. Day after day a baking sun assails the parched land from a cloudless sky, and hot, dusty winds add to the discomfort of the inhabitants. Only in the mountainous sections or in some places near the sea do higher elevations or sea breezes temper the intense heat of midsummer.

Lower temperatures of winter bring relief from the summer heat, and the more favored places receive enough precipitation to grow winter wheat or barley and a limited number of other cool-season crops. In general, Middle Eastern winters may be characterized as cool to mild. However, cold winters are experienced in the high interior basins and plateaus of Iran, Afghanistan, West Pakistan, and Turkey. Only in the southernmost reaches of the region, such as the upper Nile Basin, do temperatures remain consistently high throughout the year.

The Middle Eastern deserts have exceptionally wide daily ranges of temperature—one of the most characteristic features of desert climates. Clear skies, the relatively low humidity of the air, and the lack of a vegetation cover permit the sun's rays to heat the earth rapidly by day, but also promote a rapid escape of heat from the earth at night. Extreme cases are recorded of places in

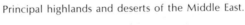

Principal highlands and deserts of the Middle East.

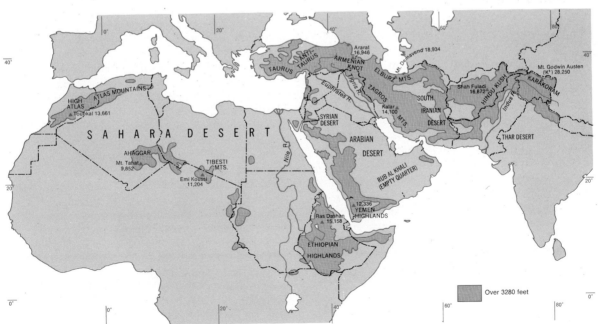

Many ingenious devices are employed to secure water in the deserts and steppes of the Middle East. The two photos, taken in east central Saudi Arabia near the capital city of Riyadh, show one means of raising water from a well. The donkeys and the horse walk back and forth on the inclined runway to pull up to the surface camel skins filled with water, which are dumped into an irrigation ditch. Traditional water-lifting devices such as this donkey well are being replaced in many instances by motor-driven pumps. (Standard Oil Company, N.J.)

the Sahara that have witnessed a maximum of over 90° and a minimum of below 32° within a 24-period period.

Types of Climate

Maps showing climates classified by type reveal the Middle East to be predominantly an area of desert or steppe climate, with smaller fringing areas of mediterranean (dry-summer subtropical) climate, highland climate, and, in the extreme south, atypical tropical savanna climate. (See front endpaper climatic map.)

The areas of *desert climate* include the great Sahara Desert of northern Africa, largest in the world; the immense desert which covers most of the Arabian Peninsula and extends northward into Iraq, Jordan, and Syria; and smaller deserts in Iran and Afghanistan and along the lower Indus River in West Pakistan. These areas have low and erratic rainfall averaging 5 to 10 inches a year or less. Occasional violent downpours alternate with rainless periods lasting for months or years at a time. Scattered areas of desert exhibit the familiar dunes of shifting sand, but rocky or gravelly surfaces are far more common. The characteristic vegetation is composed of widely spaced woody shrubs, with occasional tufts of grass. Short-lived flowering annuals lend a touch of color to the landscape following the infrequent rains.

Areas of *steppe climate,* while dry, are better supplied with moisture than the deserts because of greater rainfall, less evaporation, or a combination of these. Such areas are classed as semiarid rather than arid. An annual average of 10 to 20 inches of precipitation is typical, though deviations occur at both the upper and lower ends of this scale. The natural vegetation is more closely spaced than in the deserts and includes a larger proportion of grasses. Thus better forage is available for livestock, and the steppes are far superior to the deserts as grazing lands. However, the undependable nature of the rainfall renders crop growing precarious except where irrigation water can be brought to the land.

The Middle Eastern areas of *mediterranean* or *dry-summer subtropical climate* are principally confined to the borderlands of the Mediterranean Sea in northwestern Africa and southwestern Asia. This climate zone encloses Lebanon and Cyprus, and the most productive farming areas of Morocco, Algeria, Tunisia, Turkey, Syria, Israel, and Jordan are found within it. Rainfall averaging 15 to 40 inches annually provides more moisture for crop growth than is available in the steppe climate, but the characteristic regime of rainy winters and dry summers precludes nonirrigated production of crops that require a combination of adequate moisture and high temperatures at the same season. Both the climate and associated agricultural activities conform generally to the pattern described for Southern Europe in Chapter 11.

Areas of *tropical savanna climate* are found near the equator in parts of the Sudan, Ethiopia, and the Somali Republic. This humid climatic type, occupying the outer margins of the rainy tropics, is not typical of the Middle East. It is characterized by high temperatures the year round, a complete absence of frost, an annual rainfall averaging 20 to 60 inches or more, and a dry season lasting most or all of the winter half-year. The tropical savanna climate is described more fully in Chapters 17 and 24.

Highland climates, varying in character according to altitude and distance from the equator, are found in the higher mountain areas of the Middle East.

Table 13 gives climatic data for some representative Middle Eastern stations. Note the unbalanced rainfall regime at all stations listed.

Role of the Mountains in Water Supply

Mountains play a vital role in the economy of the Middle East. Indeed, by furnishing the principal supplies

TABLE 13 CLIMATIC DATA FOR SELECTED MIDDLE EASTERN STATIONS

STATION	LATITUDE TO NEAREST WHOLE DEGREE	ELEVATION ABOVE SEA LEVEL (FEET)	TYPE OF CLIMATE	AVERAGE TEMPERATURE (DEGREES F TO NEAREST WHOLE DEGREE)			PRECIPITATION	
				ANNUAL	JANUARY (OR COOLEST MONTH)	JULY (OR WARMEST MONTH)	ANNUAL AVERAGE (TO NEAREST INCH)	PERCENT OCTOBER-MARCH
Tehran (Iran)	36°N.	4002'	Steppe	62°	38°	85°	8"	75%
Baghdad (Iraq)	33°N.	110'	Desert	73°	50°	95°	6"	85%
Karachi (Pakistan)	25°N.	13'	Desert	79°	64°	89° (June)	8"	16%
Cairo (Egypt)	30°N.	67'	Desert	69°	54°	82°	1"	86%
Khartoum (Sudan)	16°N.	1247'	Desert	84°	73°	92° (May, June)	6"	2%
Istanbul (Turkey)	41°N.	164'	Mediterranean	57°	42°	74° (Aug.)	26"	71%
Algiers (Algeria)	37°N.	194'	Mediterranean	63°	51°	77° (Aug.)	27"	79%
Jerusalem (Israel-Jordan)	32°N.	2485'	Mediterranean	64°	48°	76° (July, Aug.)	19"	96%
Juba (Sudan)	5°N.	1509'	Tropical savanna	79°	75° (Aug.)	83° (Mar.)	39"	21%
Addis Ababa (Ethiopia)	9°N.	8038'	Tropical highland	61°	59° (Nov., Dec.)	65° (May)	49"	13%

Millions of goats and sheep graze on the sparse pasturage of Middle Eastern lands that are unsuited to crops because of steep slopes, thin and stony soils, or lack of moisture. This photo of an Arab goatherd and his flock was taken in the West Bank section of Jordan that was overrun by Israeli forces in the summer of 1967. (Trans World Airlines.)

West Pakistan, and includes the highest peaks in the region. The loftiest and best-known mountain ranges in Turkey and Iran are the Taurus, Anti-Taurus, Elburz, and Zagros mountains, which radiate outward from the rugged Armenian Knot in the tangled border country where Turkey, Iran, and the Soviet Union meet. The higher summits attain elevations of 10,000 to nearly 19,000 feet. The mountains in this quarter of the Middle East culminate in the great Hindu Kush mountain system of Afghanistan and West Pakistan, which reaches 25,426 feet, and the Karakoram Range, in Kashmir, where Mount Godwin Austen (K^2), second highest in the world, rises to 28,250 feet. (3) The third principal area of mountains is found in Ethiopia and Yemen, bordering the southern end of the Red Sea. The extensive highland of Ethiopia has a number of scattered summits higher than 13,000 feet, and one of 15,158 feet. The smaller highland of Yemen reaches 12,336 feet. Lower, discontinuous mountain ranges extend northward from Ethiopia and Yemen along either side of the Red Sea.

Scattered mountain areas occur elsewhere in the Middle East, for example, the Tibesti Mountains in the central Sahara, which rise to 11,204 feet.

Water originating in mountain rainfall or snowfall often percolates for long distances underground and reaches the surface in springs or is drawn upon by wells or qanats (p. 379) in populated areas at the base of the mountains or beyond. Water supplies of this kind are ordinarily sufficient for only a limited local development of irrigation farming. Most of the larger irrigated districts depend on water carried from the mountains by surface streams, a few of which are among the great rivers of the world. The Nile, Tigris-Euphrates, and Indus rivers provide water for many millions of farmers, and urban dwellers as well, in Egypt and the Sudan, Iraq, and West Pakistan. The earliest known civilizations arose in the valleys of these rivers, each of which has supported agriculture for several thousand years.

of water for irrigation and household use, they make life possible for most of the inhabitants of this dry region.

The mountains of the Middle East are found in three principal areas (map, p. 345): (1) In northwestern Africa the Atlas Mountains of Morocco, Algeria, and Tunisia lie between the Mediterranean Sea and Atlantic Ocean and the Sahara Desert. The High Atlas of Morocco reaches 13,667 feet. (2) A larger area of mountains occupies the northeastern quarter of the Middle East. It stretches across Turkey, Iran, Afghanistan, and

The Scarcity of Wood

Extensive forests existed in early times in the Middle East, but overcutting and overgrazing have almost wiped them out. Timber has been cut faster than nature could grow it, and the young seedlings have been grazed off by sheep, goats, and camels, with the result that the forests have been unable to reproduce themselves. Lumber in commercial quantities can still be obtained from a few mountain areas, such as the Atlas region of Morocco and Algeria and the Elburz Mountains of Iran, but the total supply falls far short of the need.

A cedar forest in the Atlas Mountains of Morocco. Forests of this quality are found in only a few parts of the Middle East. (French Embassy Press and Information Division.)

Carleton S. Coon, a leading authority on the Middle East, has written as follows concerning the scarcity of wood:[4]

The fact is that except for China no part of the earth's surface seems to have been so denuded as the Middle East. We read of the orchard that was Morocco in Roman times and see barren hillsides. The Egyptians imported thousands and thousands of trunks of the cedars of Lebanon; and in every museum in the world which contains Egyptological specimens, pieces of that wood appear in the form of coffins or other types of furniture. Solomon used the cedar of Lebanon in building his temple. If you go to Lebanon today, you will see the cedar depicted on the flag of the republic and on the hats of its customs inspectors. But on Mount Lebanon itself, despite recent attempts at conservation and reforestation, not enough remains to furnish the timber for one Gloucester schooner.

. . . In Iran one is struck by the difference between the complete barrenness of the landscape which is either under cultivation or within walking distance of a village, and that of the uninhabited deserts. In the inhabited land children go out every morning with long-handled hoes and sacks to remove every spear of inflammable vegetation, including camel thorn, that the sheep and goats have left behind them; their work is thorough. In the desert, in places too distant from villages to warrant this attention, the traveler is impressed by the abundance of natural flora, in the form of sagebrush and dry stalks of aromatic annuals. Up on the eastern flank of the Elburz . . . I once saw some juniper trees which had been lopped for firewood. "Aren't these protected by law?" I asked a soldier who was with me. "Yes," he replied, "but the government cannot afford to station one of us beside each tree." . . .

In Iran one of the most striking sights is the half-acre patch of dense poplar forest that grows up-ditch beside each village. These trees grow quickly, and their limbs cling close to the trunks; trees with 4-inch stems grow no more than 18 inches apart. No more efficient way of producing 20-foot poles quickly and on a minimum space could be found. These poplar poles are the standard units of material for rafters, door jambs, window frames, and the like. The forest has moved from the mountain to the valley.

It is easy for us, newly aroused to an interest in conservation, to decry the deforestation of the Middle East. We have lived in a rich land for periods ranging from one to three hundred years, depending on the part of the country, and we have wasted probably no more than half our natural resources in forest and topsoil during that time. In a far poorer environment, over a period of 5000 years and more, the peoples of the Middle East have destroyed only 90 percent of it. What if they had proceeded at our pace!

Shortage of Minerals

The Middle East is further handicapped, especially with respect to industrialization, by a widespread shortage of mineral resources. Good deposits of coal are rare; in fact only Turkey has deposits of major size. The region is

[4] *Ibid.*, pp. 20–23. Used by permission of the author and the publisher. It may be noted that the problem of deforestation is becoming worse, due to the fact that trucks and jeeps have extended the range of fuel collectors.

rich in petroleum and natural gas, but the largest deposits are confined to a few countries bordering the Persian Gulf, plus Algeria, Libya, and Egypt. Although scattered deposits of metals occur, only a few are of much importance on a world scale. The principal metals mined are iron ore (1 percent of the world total), lead (4 percent), zinc (2 percent), manganese (2 percent), and chromite (10 percent). Most of the value of production of lead, zinc, and manganese is accounted for by Morocco, while Turkey, Algeria, and Morocco lead as iron-ore producers, and Turkey produces practically all of the region's chromite. Good-sized deposits of salt are fairly common in the Middle East, and phosphate rock, useful as a chemical and fertilizer material, is mined on a large scale in Morocco and Tunisia, and on a lesser scale in Algeria, Israel, the United Arab Republic, and Jordan. Israel extracts potash, another chemical and fertilizer material, from the briny waters of the Dead Sea. But the general outlook for extraction of minerals in the Middle East is rather poor. Among the Middle Eastern countries, Turkey and Morocco have the most varied endowment of minerals, including metals, fuels, and chemical materials, but their overall output and reserves are not outstanding in world terms. The total value of their output is much less than that of the main Middle Eastern oil producers.

THE MIDDLE EASTERN LANDSCAPE

The landscape of this dry region alternates between green and brown, according to the presence or absence of water. The late George B. Cressey described the land surface as one might view it from a plane flying by either of two alternate routes across the heart of the Middle East:[5]

What does one see on a flight between Cairo and the Persian Gulf? For 5 hours, at 300 miles an hour, there is an almost total absence of life. As soon as one leaves the last irrigation ditch outside Cairo, the green of the Nile oasis instantly changes to the brown of the desert. One hour eastward there is a glimpse of the Red Sea. Then for 4 hours or more there is the desolation of Arabia. One looks down from the plane at sand dunes and bare rock. If there is an occasional stunted tree or water hole, it is not visible from the air. Not a trace of green is to be seen. Once or twice a keen eye will catch a group of a dozen mud houses, nothing more. This may be a place to study geomorphology, but it is not an attractive spot to make a living.

Then comes the green blue of the Persian Gulf, and one flies over the fabulous oil fields of Dhahran and Bahrain Island, with oil wells, refineries, docks, and air-conditioned comfort, American style.

Or fly from Cairo northeast to Tehran, 7 hours away. Here there is plenty of desert, but more of interest. Such a route provides a better glimpse of the Suez Canal, and, in peace time, of Jerusalem and the Dead Sea. It is a commentary on the size of Palestine that while one is over the Dead Sea, the Mediterranean is still in sight. . . . Near Baghdad one crosses the thin green strips of cultivation along the Tigris and Euphrates. In places old irrigation works, long abandoned, stand out in the desert, and ruined cities recall the dramatic history of these valleys. Eastward across the mountains, snowcapped in winter, lies the barren plateau of Iran and the capital city of Tehran. Radiating down the broad alluvial fans one may notice heaps of earth at intervals of a hundred yards. These locate the shafts spaced along the underground tunnels called qanats or karez which bring irrigation water from water-bearing gravels at the base of the mountain (photo, p. 380). This route from Cairo crosses the Dead Sea depression, with a minus elevation of 1286 feet, and comes within sight of Mount Demavend at [18,934] feet. This is an area of contrasts.

PEOPLES AND WAYS OF LIFE

The Middle East is principally inhabited by dark-haired peoples of the Caucasoid racial group. Especially important among the many languages spoken in the region are Arabic and other Semitic languages (including Hebrew), Berber and other Hamitic languages, Turkish, Persian, Urdu (spoken in West Pakistan), and Pushtu (the main language of Afghanistan). Most of the inhabitants of the Middle East are farmers. Towns and cities of 5000 or over contain only an estimated 20 to 40 percent of the population in most countries. Nomadic herdsmen are a well-known element in the region, but they are dwindling in numbers and today represent well under a tenth of the total population. The principal areas of dense population in the Middle East are the irrigated valleys of the

---

[5] George B. Cressey, "The Land of the Five Seas," *Journal of Geography,* 51 (1952), 222–223. Used by permission of the *Journal of Geography.*

major rivers and the coast lands of the Mediterranean Sea. The few large cities are mostly political capitals or seaports; those with estimated metropolitan populations of 900,000 or more (1968) include Cairo, Karachi, Tehran, Alexandria, Istanbul, Lahore, Casablanca, Baghdad, Algiers, and Ankara. The cities of the region, particularly the larger ones, have been increasing in population very rapidly due to declining death rates in the cities themselves, and emigration from the countryside, where death rates also have been falling and the population has been growing much faster than the supply of arable land. For unemployed and poverty-stricken villagers the city has offered the possibility of employment, education, some medical care, and excitement and ferment to replace the monotony of village life and make poverty easier to bear. Unfortunately, the influx of villagers has generally outrun the ability of the city to provide jobs, housing, and services, and the result often has been unsightly and insanitary shantytowns (*bidonvilles*) at the edge of the city.

Increased employment opportunities and a rising level of living in the Middle East will have to be provided primarily by manufacturing industries, despite the general shortage of raw materials and labor skills, and the governments of the region are making strenuous efforts to increase the number and output of modern factories. Most large factories are located in the larger cities, though many smaller factories and traditional handicraft industries are present in lesser cities and towns. Textiles, foods, and other relatively simple types of industries are the principal ones in most countries of the region. More intricate manufactures such as automobiles, typewriters, and electrical machinery must, with rare exceptions, be imported. Only in Israel and a handful of cities elsewhere has much of a start been made in the more sophisticated forms of industry.

The Middle East, like other world regions, is sufficiently varied in its ways of living that generalizations concerning the culture of the region as a whole nearly always contain a certain amount of distortion. Nevertheless, a surprising amount of similarity, at least on a superficial level, can often be discerned in the modes of life pursued in places as far apart as Morocco and Afghanistan. Because of the fact that Islam is not simply a religion, but a way of life, it is possible to distinguish a "Middle Eastern culture" which is characteristic of the entire region in greater or less degree. This culture in its original form has been much affected by influences from Europe and America, particularly in the cities. However, much of the old culture still remains in rural areas and in the older sections of urban places. Some of its characteristic features have been described as follows by Carleton S. Coon:[6]

Although late to take over the fruits of the industrial revolution, these countries originated two earlier and equally important crises without which the third could not have arisen: the neolithic revolution which brought in agriculture and animal husbandry, and the urban revolution which introduced city life, metalworking, and writing. For over 5000 years the peoples of these countries . . . have been experimenting in the arts of getting a living out of the landscape and getting along with each other. . . .

City Life

Three major ways of life have arisen in concert with one another. One is city life, with thousands of skilled workers, specialists in all kinds of hand industries; with merchants prepared to handle both local and international trade; with palaces and courts in which the political institution functions; with mosques, churches, and temples housing the religious hierarchies, whose prime duty is to maintain the equilibrium of the whole society; and with universities and other educational institutions preparing students for both religious and secular leadership. . . .

Some of the cities, notably Mecca, Kerbela [Karbala], and Meshed [and also Jerusalem, a holy place for Islam as well as Judaism and Christianity], are centers of pilgrimage, and in them the mechanism of caring for religious visitors is well worked out. Pilgrims, coming from all parts of the Islamic world to the holy places, from the Philippines and Morocco, India and Bukhara and China, do much to fortify the unity of the countries with which we are concerned. On the pilgrimage fellow devotees from different countries meet and converse and exchange ideas. Every year a number of pilgrims decide to stay in one of the countries en route, rather than going home. The uniformity of skills, of architecture, of music, and of literature in these countries reflects this annual intercourse. . . .

Village Life

From the life of cities one can move without pain to that of villages, for the city is a big village as well as a center of industry. The land on the outskirts of the

---

[6] Carleton S. Coon, "Point Four and the Middle East," *Annals of the American Academy of Political and Social Science,* **270** (1950), 88–92. One heading has been changed, and some additional material has been placed in brackets without compromising the sense of the original. Used by permission of the author and the *Annals.*

A Saudi Arabian oasis (Qatif in Hasa Province) on the shore of the Persian Gulf. The close-set, flat-roofed buildings of the town lie in the midst of date palms watered by springs. (Standard Oil Company, N.J.)

city is tilled by farmers who commute in reverse, and they have done this since the days of Hammurabi. A village is a group of houses located near water and surrounded by agricultural land. It contains a dozen to several hundred families, usually closely related. If the village is on the flatlands, it is usually owned by a sheik or a landlord. While the sheik may be in residence, the landlord usually lives in the city, governing through a local, appointed agent who is in effect the mayor or headman. Rarely, such a village is free, in that the farmers own their land and select their own leader.

If the village is in the mountains, its inhabitants usually form part of a tribal system, and their allegiance is to their tribal chiefs. Since the plains are accessible and defenseless, lowland villagers are seldom noted as fighters,

and are usually docile subjects of the central government, whatever it may be. The mountain villagers are . . . more warlike, since they have some chance of defending their rocky hideouts, and are used to intervillage and intertribal raids and feuds arising from population pressure. The mountaineers often preserve some archaic form of speech, foreign to that of the dwellers on the plain, wear distinctive clothing, sing distinctive songs, nurse their children on distinctive tribal lore, and have long been a problem to shahs and sultans.

Some of these tribes are seasonally migratory. They pasture their animals in the mountain meadows in summer, and in the fall pack all their belongings onto horses, donkeys, and cattle, and move in a wild rush of dust and sheep dung down to the warmer plain, where they spend the winter pasturing their animals in the new grass which grows with the rain. In the spring they climb back to the mountains. This annual migration may involve the movement of several thousands of human beings and hundreds of thousands of animals over a distance of several hundreds of miles. The country through which they move is inhabited by sedentary people, with whom all kinds of conflicts could easily arise. Hence the tribe needs leadership and a competent private police force to keep all hands in line. This policing of the migration is the practical basis of the tribal organization, which follows patriarchal family lines. Since international boundaries often follow watersheds, and since mountain watersheds make prime summer pasture, many of these tribes are international; hence the worries of shahs and sultans.

Desert Life

Another kind of nomadism exists in the deserts. The Baluchi, the Bedouin, the Tuareg, and other camel people of the South Iranian, Arabian, and Sahara deserts are probably the best known of all Middle Easterners to the American public, but the American public little realizes what complex lives the nomads lead. It is as true as it is trite to liken the desert to a sea and the camel to a ship. The agricultural communities along the shore serve as ports and havens to which the fleets of Bedouin come for victualing and re-equipping. But the ships carry their carpenters and machinists, and supply ships come out to service them when they are too busy to leave the banks.

Speaking less metaphorically, the camel nomads are not simple strangers at all, but Arabs or Berbers like their sedentary kinsmen, just as cowboys are Americans or Canadians, and not some special breed or nation. In years of good rainfall the desert provides bountiful pasture during the cool weather; in dry years there must be a scramble for greenery and for water, and in their scramble the strong may well push out the weak. We have had similar troubles over water holes and range in the West. In the good season, traders come out to live with the Bedouin and sell them the goods they need against livestock. At all seasons, a blacksmith will accompany each camp. He is no kin to the men he serves, and hence if they are raided, he, like the merchants, will not be harmed. Thus automatically the supply of metal tools and implements is assured.

If one walks down the streets and through the markets of any large [Middle Eastern] city, one is likely to see not only city men, but also villagers, mountain men, and camel men off the desert. . . . These people have come to town to see the sights and to trade. . . .

Some Details of Middle Eastern Material Culture

Many of the picturesque details of Middle Eastern material culture are based on excellent common sense and a maximum utilization of scarce raw materials. Old-fashioned Moslems have little furniture in their houses. They take off their shoes when they enter a room, and sit on clean carpets. They have enough wool to weave carpets, but not enough wood to build bulky tables and chairs. Their food includes much butter, cheese, curds, and other milk products, but little meat. Milking animals in a land of little abundance is less extravagant than butchering them. Pigs have long been forbidden, and one must remember that in a barren land where all vegetable foods grown are needed to feed men, it would be antisocial to fatten a pig.

Old-fashioned Middle Eastern clothing is designed to give the wearer the maximum of comfort, privacy, and dignity. The jellaba of the North African keeps him warm in winter and cool in the hot sun; he can sleep in it, and shield himself from the public gaze when performing natural acts on a barren landscape. The chuddar of the Persian lady covers her decently, and conceals the state of her clothes underneath. On the street, the widow and the judge's wife look alike. Middle Eastern street clothing, by its uniformity and voluminousness, is an instrument of democracy. The shift to Western garb can hardly be in every sense an improvement.

Middle Eastern architecture, with its adobe walls and clever domes, creates out of local mud and a few sticks what we cannot erect here, from exotic materials, for a dollar a cubic foot—a comfortable home. Fuel is a huge bottleneck in the Middle East, and hence the scarcity of burnt brick. . . . We can teach the Middle Easterners little about noninstitutional building.

CURRENTS OF CHANGE IN THE "ECOLOGICAL TRILOGY"

The impact of modern life is being felt more and more in the villages and among the nomadic tribes of the Middle East as well as in the cities, and traditional relationships among the three are being altered. In a fine article, Paul W. English[7] has characterized these basic elements of Middle Eastern society as an "ecological trilogy," and his summary of their interdependence and recent changes in relationships appears on pages 354-356.

---

[7] Paul Ward English, "Urbanites, Peasants, and Nomads: The Middle Eastern Ecological Trilogy," *Journal of Geography,* 66 (1967), 55-59. Used by permission of the author and the *Journal of Geography.*

Most Middle Eastern cities act as administrative, commercial, and cultural centers for large rural and nomadic populations for it is in the cities that the elite of Middle Eastern society, the wealthy, powerful, and literate decision-makers reside. These city-dwellers are principally engaged in collecting and processing raw materials from the hinterland—wool for carpets and shawls, vegetables and grain to feed the urban population, and nuts, dried fruit, hides, and spices for export. In return, the urbanites supply peasants and nomads with basic economic necessities such as sugar, tea, cloth, and metal goods as well as cultural imperatives such as religious leadership, entertainment, and a variety of services. This concept of urban dominance is basic to the idea of an interdependent ecological trilogy; it replaces earlier notions which tended to stress the isolation of peasant and nomad from the "civilizing" influences of urban life.

The patterning of these Middle Eastern cities until very recently differed markedly from that of the industrial centers of the Western world. The heart of the city was dominated by two Islamic institutions, the Friday-prayer mosque and the bazaar. The mosque was often the most beautiful and impressive building in the city, in contrast with our commercial skyscrapers and government buildings.

A downtown view of one of the larger Middle Eastern cities. The photo shows Ramleh Station in Alexandria, United Arab Republic. Note the tall minaret of the Moslem mosque in the background. (Arab Information Center.)

The bazaar was a welter of crowded lanes, covered stalls, and associated caravanseries, far noisier and more aromatic than our Wall Streets. Within the bazaars, a concentric hierarchy of trades and crafts existed. The shops of prestigious tradesmen such as booksellers, carpet merchants, jewelers, and silversmiths were located at the center of the bazaar, the lesser crafts progressively nearer the margins. This same principle, stressing central location, operated in the residential quarters of the city as well. Wealthy and influential people lived near the center of town to maintain contact with government and religious leaders and for reasons of security. The beggars, the prostitutes, the poor, and those tainted by religion, race, or occupation lived on the margins of the city, a striking contrast with the central slums and neat suburbs of Western cities. The streets in these residential quarters formed a maze of narrow alleys, twisting lanes, and *culs-de-sac* surrounded by high compound walls. A few major avenues ran from the outskirts of the city to the mosque-bazaar complex and they were usually the only thoroughfares suitable for modern transportation.

But this traditional Middle Eastern urban setting is being reshaped by the introduction of Western transportation and communication systems, by the development of new attitudes, values, and ideologies, and by the spreading availability of Western technology. Wide avenues suitable for motor transport have been built in most cities, and commercial activity has been liberated from the bazaar. Modern buildings have sprung up on these avenues—drugstores, gasoline stations, photography shops, hotels, and even department stores. Newly-built suburbs have spread beyond the confines of ancient city walls; the cemeteries and shrines which previously had acted as a collar on urban expansion have been levelled, providing striking evidence of the rising importance of secular considerations in a traditionally religious society. The urban elite are moving to these suburbs, where houses are less crowded and street patterns are less tortuous. Unfortunately, another modern institution, the shanty-town or *bidonville,* has also emerged on the outskirts of large cities pointing up one of the major movements that is occurring in the Middle East today, urbanization.

. . . . .

Most Middle Eastern villages are small, compact clusters of mud-walled dwellings with few facilities other than a mosque, a shrine, a few shops, and possibly a bath. Within the village, there is a strong sense of group solidarity. Its people are of one race and one faith, bound by the dictates of custom and convention, and the rhythm of the agricultural cycle. Personal relationships are face-to-face in the village, technology is simple, and division of labor is slight. A man's status is largely fixed at birth, for this is a stable society. When modern writers speak of social conditions in the Middle East—of poor health, of high infant mortality rates, of illiteracy, tenancy, and numbing poverty—they are describing conditions in the Middle Eastern village. Yet despite the stark contrast between these mud-walled communities and Middletown, U.S.A., terms such as "primitive," "backward," and "underdeveloped" must be used with great caution, because the Middle Eastern villager is very often an urbane, sophisticated individual who has drunk deeply from the mainstream of a great urban civilization.

The primary relationship between cities and villages in the Middle East is of course economic; cities rely on peasants for their food supply and village farmers must market their crops. Commonly, this economic bond was formalized by a land tenure contract between an urban landlord and a village sharecropper. The landlord usually supplied most of the elements of production—land, water, seed, and even plow animals—and received upwards of 70 percent of the crop. The peasant supplied his labor and received about 30 percent of the harvest. While equitable in theory, this system was, in practice, a mechanism of urban dominance which annually drained economic surpluses from the village leaving the mass of the rural population poverty-stricken. Other urban representatives, tax collectors, money lenders, and peddlers also penetrated into the village and young men were called away for military service. In the course of these various intermittent contacts, the culture and technology of the city diffused to most rural areas of the Middle East, deepening and broadening the interdependence of these two sectors of the ecological trilogy.

In modern times, these bonds between city and village have intensified dramatically as political regimes in the new nation states of the Middle East have attempted to bring their countries into modern twentieth-century life. A revolution in transportation and communication systems is gradually carrying social and economic ferment from the bazaars of the city into the countryside. To the village, these forces of change have brought a new cast of characters—the land reform agent who supervises the distribution of land to former sharecroppers, the government teacher convinced of the value of literacy and the equality of women, and the doctor who preaches sanitation and health. Modern technology in the form of tractors, bicycles, radios, sewing machines, and hurricane lamps is already available in many rural areas. Essentially, the "revolution of rising expectations" has begun to arrive in the Middle Eastern village.

That these forces of change are disrupting many characteristics of the Middle Eastern village is to be expected. New crops, improved agricultural technology, and new techniques of cultivation are modifying the agricultural basis of village life. The traditional reverence for age and religion is decreasing as power flows into the hands of young, educated representatives of modern governments and as religious leaders are shorn of their control of communication, education, and law. The young of the village are beginning to view their family, community, religion, and nation in a new way—and if this new vision prevails it may alter the very roots of the Middle Eastern village,

A classic view of Bedouin with their black tents and camels. The animals are gathered at a watering trough supplied from a deep well that originally was drilled in connection with an oil prospecting venture. The photo was taken near Jauf in northern Saudi Arabia. (Standard Oil Company, N.J.)

which until recently has been the most stable, unchanging component of the Middle Eastern ecological trilogy.

. . . . . .

The nomads . . . are the stockbreeders of the Middle Eastern agriculture. Utilizing the desert and mountain regions of the Middle East, they play a vital role in supplying beasts of burden, meat, milk, and other animal products to the sedentary population, whose cultivated land is too precious to devote to growing fodder crops. In return, the villagers provide tribesmen with grain, dates, and fruit and the townsmen supply sugar, tea, cloth, and weapons. In most areas of the Middle East, this economic interdependence led to peaceful, if disdainful, relations between the nomadic and sedentary populations. In some cases, the arrangement was sweetened by a "tribute of friendship" paid to the nomads by villagers and merchants since the nomads were feared and respected warriors. But in times of stress, or when the central government was weak, nomads often engaged in the ancient and exciting custom of "raiding." As long as both the government and the nomads had the same technology, that is horses, camels, and rifles, equilibrium was maintained. But now that this balance has shifted, the memory of wild-riding, mounted nomads repeated by townsmen and cherished by writers of this region is leading to the destruction of nomadic life.

There is no political place for tribal nomads in the new nation states of the Middle East and they are rapidly becoming alienated from an impossible world.

So the tribal peoples of the Middle East have most disastrously felt the impact of modern forces of change and many of their previous functions within the ecological trilogy have disappeared. Airplanes and trucks have conquered the deserts and mountains of the Middle East. These vehicles need little protection and their use has disrupted the camel market. In addition, planes, tanks, and machine guns have given modern governments in the Middle East the capacity of destroying raiding tribesmen who in earlier times were able to melt away into the deserts and mountains. In some cases, tribesmen attempted to adjust to changing conditions. A hundred years ago, for example, the *amirs* of the Baluch sent emissaries as far as Istanbul to buy sword blades and matchlock rifle barrels. More recently, occasional mention was made of tribal raiders in Fords and Chevrolets. Only in Yemen and Kurdistan, however, have tribesmen successfully resisted this antinomad technology and retained local autonomy. For the most part, the lot of the Middle Eastern nomad has been disarmament, destruction of herds, and enforced sedentarization. The economic losses of meat, milk, and animals have not been sufficiently expensive to influence the political decisions of modern Middle Eastern leaders.

REFERENCES and READINGS

Standard Geography Texts

BRICE, WILLIAM C., *South-West Asia.* A Systematic Regional Geography, Vol. 8 (London: University of London Press, 1966).

CRESSEY, GEORGE B., *Crossroads: Land and Life in Southwest Asia* (New York: J. B. Lippincott Co., 1960).

FISHER, W. B., *The Middle East: A Physical, Social, and Regional Geography* (5th ed.; New York: E. P. Dutton & Co., 1963).

LONGRIGG, STEPHEN H., *The Middle East: A Social Geography* (Chicago: Aldine Publishing Company, 1963).

See also the lists of texts on Asia and Africa in the readings and references for Chapters 17 and 24.

Other References

ADAMS, ROBERT, "The Origin of Cities," *Scientific American,* **203,** no. 3 (September 1960), 153–168.

AHMAD, M. S., "Middle East International Highways from Caravan Routes to Modern Roads," *Middle East Journal,* **21,** no. 1 (Winter 1967), 101–107.

AMIRAN, DAVID H. K., "Arid Zone Development: A Reappraisal under Modern Technological Conditions," *Economic Geography,* **41,** no. 3 (July 1965), 189–210.

Atlas of the Arab World and the Middle East (New York: St. Martin's Press, 1960).

BRAIDWOOD, ROBERT J., "The Agricultural Revolution," *Scientific American,* **203,** no. 3 (September 1960), 131–148.

BRIGGS, LLOYD CABOT, *Tribes of the Sahara* (Cambridge, Mass.: Harvard University Press, 1960). Especially Chap. 1, "Landscape and Natural Resources," pp. 1–33.

"Britain East of Suez—Special Issue," *International Affairs,* **42,** no. 2 (April 1966).

BULLARD, READER, ed., *The Middle East: A Political and Economic Survey* (3d ed.; London: Oxford University Press, 1958).

CAPOT-REY, ROBERT, "Problems of Nomadism in the Sahara," *International Labour Review,* **90,** no. 5 (November 1964), 472–487.

COON, CARLETON S., *Caravan: The Story of the Middle East* (rev. ed.; New York: Holt, Rinehart and Winston, 1958).

CRESSEY, GEORGE B., "Water in the Desert," *Annals of the Association of American Geographers,* **47,** no. 2 (June 1957), 105–124.

Current History, **52,** no. 306 (February 1967). An issue devoted to Middle Eastern subjects, mainly articles on individual countries.

DRESCH, JEAN, "Utilization and Human Geography of the Deserts," Institute of British Geographers, *Transactions No. 40* (December 1966), pp. 1–10.

ENGLISH, PAUL WARD, "Urbanites, Peasants, and Nomads: The Middle Eastern Ecological Trilogy," *Journal of Geography,* **66,** no. 2 (February 1967), 54–59.

FISHER, SYDNEY N., ed., *Social Forces in the Middle East* (Ithaca, N.Y.: Cornell University Press, 1955); especially "The Villager," by Douglas D. Crary, pp. 43–59. And *The Middle East: A History* (New York: Alfred A. Knopf, 1959).

GOLDMAN, MARSHALL I., "A Balance Sheet of Soviet Foreign Aid," *Foreign Affairs,* **43,** no. 2 (January 1965), 349–360.

GULICK, JOHN, ed., "Dimensions of Cultural Change in the Middle East," *Human Organization,* **24,** no. 1 (Spring 1965), 104 pp. A special issue, with contributions by various authors.

HILLS, E. S., ed., *Arid Lands: A Geographical Appraisal* (London: Methuen and Co., 1966).

HITTI, PHILIP K., *A Short History of the Near East* (Princeton, N.J.: D. Van Nostrand Co., 1966).

LEOPOLD, A. STARKER, and the Editors of *Life, The Desert* (Life Nature Library; New York: Time Inc., 1961).

LEWIS, BERNARD, *The Middle East and the West* (Bloomington: Indiana University Press, 1964).

McCONNELL, JAMES E., "The Middle East: Competitive or Complementary?" *Tijdschrift voor Economische en Sociale Geografie,* **58,** no. 2 (March–April 1967), 82–93.

McNEILL, WILLIAM H., *The Rise of the West: A History of the Human Community* (Chicago: University of Chicago Press, 1963); and *A World History* (New York: Oxford University Press, 1967). Numerous chapters and sections on the Middle East.

Middle East Journal. Multidisciplinary scholarly journal on the region; carries articles, chronology, book reviews, and comprehensive lists of books and articles on Middle Eastern subjects.

The Middle East and North Africa (London: Europa Publications), annual. Statistical and other data on the area as a whole and on individual countries.

MIKESELL, MARVIN W., "The Deforestation of Mount Lebanon," *Geographical Review,* **59,** no. 1 (January 1969), 1–28.

NOLTE, RICHARD H., ed., *The Modern Middle East* (New York: Atherton Press, 1963). A collection of articles by a variety of specialists on the region.

"Nomads and Nomadism in the Arid Zone," *International Social Science Journal,* **11,** no. 4 (1959), 479–585. Nine articles by various authors.

Oxford Regional Economic Atlas: The Middle East and North Africa (New York: Oxford University Press, 1960).

PATAI, RAPHAEL, "The Middle East as a Culture Area," *Middle East Journal,* **6,** no. 1 (Winter 1952), 1–21.

PERETZ, DON, *The Middle East Today* (New York: Holt, Rinehart and Winston, 1963). A college text on the region, stressing history, culture, and politics.

POLK, WILLIAM R., "The Nature of Modernization: The Middle East and North Africa," *Foreign Affairs,* **44,** no. 1 (October 1965), 100–110.

QUALE, G. ROBINA, *Eastern Civilizations* (New York: Appleton-Century-Crofts, 1966), Part II, "The Middle East," pp. 33–149.

SIMOONS, FREDERICK J., *Eat Not This Flesh: Food Avoidances in the Old World* (Madison: University of Wisconsin Press, 1961). Especially Chap. 3, "Pigs and Pork," pp. 13–43.

UNESCO, *Arid Zone Research,* especially the following titles: Vol. 17, L. Dudley Stamp, ed., *A History of Land Use in Arid Regions* (1961); Vol. 18, *The Problems of the Arid Zone* (1962); Vol. 26, *Land Use in Semi-arid Mediterranean Climates* (1964); Vol. 28, Peveril Meigs, *Geography of Coastal Deserts* (1966). (Paris: UNESCO.)

U.S. DEPARTMENT OF THE ARMY, *Middle East: Tricontinental Hub: A Strategic Survey.* Pamphlet No. 550–2 (Washington, D.C.: Government Printing Office, 1965). A comprehensive bibliography.

VON WISSMANN, HERMANN, and Others, "On the Role of Nature and Man in Changing the Face of the Dry Belt of Asia," in William L. Thomas, Jr., ed., *Man's Role in Changing the Face of the Earth* (Chicago: University of Chicago Press, 1956), pp. 278–303.

WHITE, GILBERT F., ed., *The Future of Arid Lands* (Washington, D.C.: American Association for the Advancement of Science, 1956).

See also the lists of readings and references for Chapters 1, 2, 3, 16, 17, and 24. General references on the Arab world and Islam are listed in Chapter 16.

Individual
Middle Eastern Countries
and Their
Characteristics

16

In this chapter the individual countries of the Middle East are surveyed under the following headings: (1) The Arab World and Israel, (2) Persian Gulf Oil and the States That Produce It, (3) Turkey, Cyprus, and Afghanistan, (4) Ethiopia and Somaliland. West Pakistan and Kashmir are discussed with the Indian Subcontinent in Chapter 18.

THE ARAB WORLD AND ISRAEL

The Arab world, stretching east-west from the Indian Ocean to Morocco and southward from the Mediterranean to the fringes of Negro Africa, is a major component of the Middle East. The peoples who inhabit it are racially and culturally diverse, but the majority speak some form of Arabic and are thought of broadly as Arabs. Originally the Arabs were inhabitants of the Arabian Peninsula. Here arose, in the seventh century A.D., the Moslem (Muslim) religion which most Arabs profess. It was founded on the teachings of the Prophet Mohammed. The rise and beliefs of the religion are briefly described below.[1]

The Prophet was born in A.D. 570 at Mecca and died at Medina some sixty-two years later. His contribution was not so much a new religious concept as a new pattern of life. At the time of his birth, Arab society was virtually without moral precept or guidance. Mohammed conscientiously set out to correct what he thought were the worst features of that society. For example, he ended the practice of female infanticide, which before that time had been left to the judgment of the father. Through his teachings, the position of women greatly improved, as the number of wives was limited to four at a time and divorce was regularized. Mohammed laid down laws for what he considered the proper treatment of orphans, slaves, prisoners, and animals. He forbade bearing false witness, worshiping idols, or speaking ill of chaste women. Because the main drink of the Arabs then was a potent beverage made from the heart of palms, which quickly reduced the user to a state of belligerent stupidity and led to many quarrels and sometimes to bloodshed, Mohammed forbade the use of intoxicating liquors. He also felt that the practice of gambling had weakened Arab society in his day and prohibited it too.

In the eyes of true Moslems, the Koran is the word of God, or Allah, which was transmitted through the Angel Gabriel to Mohammed and then recited by him to the faithful. . . . The Koran is not only the Holy Book of Islam, . . . but it is the textbook used by most Moslems in learning to read; thus it standardizes written Arabic throughout the entire world. . . .

The duties of a Moslem are built around the five Pillars of Islam. The first of these is the profession of faith: "There is but one God, and Mohammed is his apostle." The second is frequent daily prayer, which is largely supplication. . . . The most important prayer of the week occurs at noon on Friday, a public ceremony which must be observed by all Moslems. Mohammed emphasized the necessity for this gathering, and the need for the pilgrimage to Mecca, in order to bring Moslems together and weld them into a single community.

The third Pillar of the Moslem faith is the giving of alms. This started as a voluntary act, became obligatory for a while, and in most parts of the Arab world today has again become voluntary. The fourth Pillar is fasting in the month of Ramadan, the month during which the Koran was first revealed to Mohammed. . . . During this month the true believer can eat no food and swallow no drink from dawn until sunset, unless he is ill or on a journey. . . .

Making the pilgrimage to Mecca is the fifth Pillar of Islam. Every true Moslem tries to do this once during his lifetime. . . . Special ceremonies peculiar to the pilgrimage include striking down the devil with stones in the Valley of Mina, sacrificing a sheep or camel there, entering into the prescribed holy places wearing a seamless garment, walking seven times around the Kaaba, and devoting oneself to prayer.

After the death of the Prophet in 632, his Arab followers spread the Islamic faith through a vast area in Asia, Africa, and Europe. Before a century had passed, Arab armies had subjugated an empire reaching to Morocco and Spain on the west, to present-day West Pakistan on the east and to Soviet Middle Asia and Transcaucasia on the north, and southward into the Sahara. Within the new empire a high civilization developed. Great architectural works were created, significant advances were made in agriculture, medicine, mathematics, and science; and a vigorous intellectual life flourished in such university centers as Cairo, Fez, Baghdad, Damascus, Cordova, Toledo, Seville, and Granada. At a time when western Europe was relatively stagnant, the Arab lands contained many of the foremost centers of culture and learning in the world.

But these lands did not enjoy an overall political unity for very long. The original empire, ruled from Medina, then from Damascus, and finally from Baghdad, commenced to disintegrate soon after it was established. Separate Islamic states arose in Spain, northwestern Africa, Egypt, and elsewhere. Beginning in the eleventh century there were conquests of portions of the empire by Christian Crusaders and somewhat later by Turkic and Mongolian peoples. In the fifteenth century Christian armies of the Spanish kingdom, after centuries of intermittent warfare by Spaniards and Portuguese against Islam, completed the reconquest of the Iberian Peninsula. By the middle of the sixteenth century most parts of the original Arab empire, weakened and disunited, had been conquered by the Ottoman Turks.

Today most Arabs are found in countries that belong to the League of Arab States. This organization, formed in 1945, is comprised (1968) of the United Arab Republic (Egypt), Syria, Iraq, Jordan, Lebanon, Saudi Arabia, Kuwait, Yemen, Southern Yemen, Sudan, Libya, Tunisia, Algeria, and Morocco. Quarrels and jealousies among the Arab states have often hindered their functioning as a group. But these divisive tendencies are being counteracted in some measure by a consciousness of an overall Arab unity or solidarity that is developing among the peoples

---

[1] Richard H. Sanger, *The Arabian Peninsula* (Ithaca, N.Y.: Cornell University Press, 1954), pp. 95–97. Used by permission of the publisher. The word Islam means "submission" (to the will of God or Allah; a Moslem is "one who submits"). The Kaaba, a building at Mecca housing the venerated Black Stone, is the chief shrine of the Islamic world. It was a holy place long before Mohammed's time.

and governments of these lands. This feeling has several important roots. Partly it is the product of a common language, religion, and cultural heritage, coupled with pride in past achievements and a desire to regain past glories. Partly it is due to the realization of a need for united effort to improve social and economic conditions throughout the Arab lands. And partly it grows out of opposition to foreign economic and political domination and opposition to the state of Israel, which was founded under foreign auspices and is looked upon as usurping territory that is rightfully Arab.

Israel and Its Neighbors of the Fertile Crescent

Since 1948, when Israel was formed from the British mandated territory of Palestine, antipathy to the new state has been a major rallying point for the League of Arab States, particularly the members that are Israel's immediate neighbors. To the north and east these neighbors include Lebanon, Syria, and Jordan (formerly Transjordan). All three, and Palestine, were held by Turkey prior to World War I. When the Ottoman Empire was liquidated following the war, Syria and Lebanon were mandated to France by the League of Nations, while Great Britain received a mandate for Palestine and Transjordan. Today Syria and Lebanon are republics, having secured independence by stages in the period 1943–1946. Transjordan received independence in 1946 as the Hashemite Kingdom of Jordan. In 1958 Syria formed a political union with Egypt called the United Arab Republic, but withdrew from it in 1961. Egypt still calls itself the United Arab Republic, however, while Syria is officially the Syrian Arab Republic.

Palestine was divided among Israel, Jordan, and Egypt when Israel was constituted in 1948 as an independent homeland for the world's Jewish population. This action, the result of warfare after British and United Nations' efforts at peaceful partitioning had failed, gave Israel up to 1967 some 77 percent of the territory of the Palestine mandate, including the coastal areas (except the Gaza Strip in the south, which was occupied by Egypt), the northern hill country of Galilee, the dry, thinly populated southern triangle called the Negev, and a portion of the city of Jerusalem in the Judean hills, plus a corridor leading to the city from the coastal plain (map, p. 362). Prior to the independence of the Israeli state, the Zionist movement had brought Jews back to Palestine since about 1880, although some had returned there even earlier. After independence the inflow greatly increased. A total of 687,000 Jewish immigrants, mostly from Eastern Europe or the Middle East, entered Israel between May 14, 1948, when the new state was officially proclaimed, and the end of 1951. This influx more than doubled the country's Jewish population. Since 1951 the annual immigration has been much smaller, and it has fluctuated considerably from year to year. There has also been a limited amount of out-migration. The population reached an estimated 2.8 million by mid-1968, exclusive of more than 1 million in territory conquered by Israeli forces in the six-day Arab-Israeli war of June 1967. Almost nine-tenths of the population of pre-1967 Israel was Jewish, the remainder being comprised mainly of Moslem or Christian Arabs.

Aided by large amounts of outside capital, primarily from the United States, the Israelis have been developing their small republic as a modern Westernized state. Money from outside the country has come from both private and government sources, and has included loans, investments, and gifts, as well as reparations paid by the West German government (the reparations ceased in 1965). Israel is still far from self-sufficient economically, but is making strenuous efforts to correct its adverse trade balance and lessen its dependence on foreign aid.

One of the country's greatest achievements since 1948 has been the expansion and intensification of agriculture. The cultivated acreage in Israel proper has more than doubled, and the acreage under irrigation has quadrupled. The country produces the bulk of its food supply, though sizable amounts of grain and some meat, sugar, beverages, and oilseeds are imported. In addition, irrigated citrus groves provide exports of oranges and orange juice amounting to a little less than a fifth of Israel's total exports by value (1965). Food production for the domestic market concentrates on dairying, beef production, poultry raising for eggs and meat, and vegetable growing. A large share of the crops are grown as feed for animals. Pond fisheries play a significant role in the agricultural economy. The intensive, mechanized agriculture of Israel, oriented to nearby urban markets, resembles in many ways the agriculture of densely populated areas in western Europe. But in Israel a distinctive touch is added by the collectivized settlements called *kibbutzim* (singular, *kibbutz*) in which the land is held and worked by all families in common and meals are served in a central dining hall. Such settlements exist side by side with more numerous cooperative villages called *moshavim* (singular, *moshav*) and villages of private farmers.

Existing agricultural development is concentrated in the northern half of the country, with its heavier rainfall (largely concentrated in the winter half-year) and ampler supplies of surface and underground water for irrigation. Annual precipitation averages 20 inches or more in most sections north of Tel Aviv and in the interior hills where

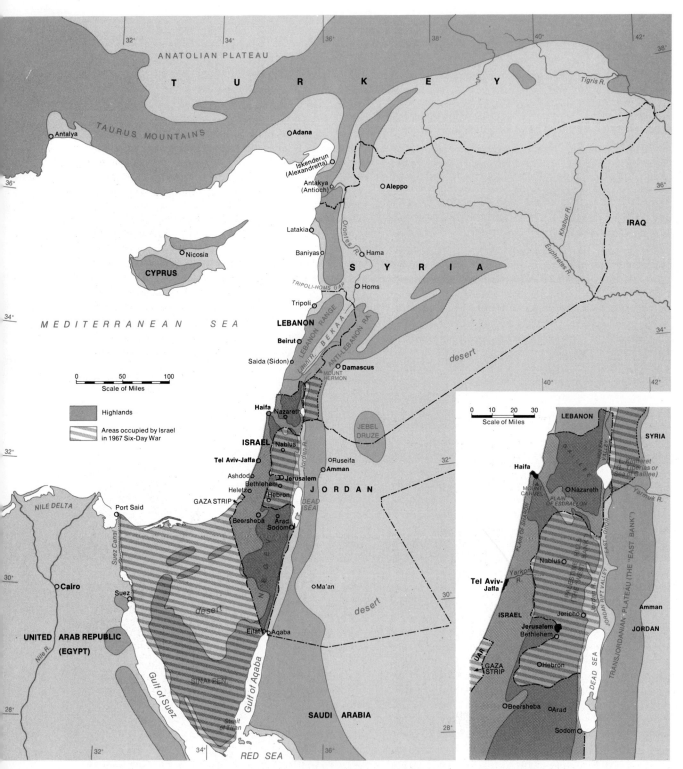

Israel and its neighbors. Most of the highland areas consist predominantly of mountains except in Israel and Jordan, where hill lands predominate. International boundaries are shown as they existed prior to the 1967 Arab-Israeli War.

Jerusalem lies, though it drops below that figure in the deep rift valley of the Jordan River. Tel Aviv averages 20.4 inches, Jerusalem 21.7, Jericho (in the rift valley section of Israeli-occupied Jordan) 5.6, Haifa 24.9, and Nazareth 25.2. But from Tel Aviv southward to Beersheba, in the northern Negev, the rainfall decreases to 7.9 inches, and in the central and southern Negev it drops to 4 inches or less (Eilat, 1.2). In the northern half of the country irrigation water from a variety of sources (streams, lakes, springs, wells, artificial reservoirs to catch runoff, and reclaimed sewage water) is used to intensify agriculture and overcome the summer drought. In the semiarid or arid south, on the other hand, surface water is scanty, and underground water, where it has been found at all, is apt to be too saline for most crops. To bring agriculture to the Negev, Israel has undertaken the transfer of large quantities of water from the north by pipelines. The main sources are the Yarkon River, which enters the Mediterranean just north of Tel Aviv, and Lake Kinneret (Lake Tiberias or Sea of Galilee), which is fed by the upper Jordan River. Existing irrigation development in the Negev is concentrated on fertile loess soils in the region around Beersheba. Local irrigation schemes exist in several parts of Israel, and these are being linked into a coordinated network by a national water planning agency.

But Israel's main hope for future support of its growing population, most of which lives by occupations other than agriculture, lies in expanded industry, trade, and tourism. The country has been steadily industrializing, though it is hampered somewhat by a lack of mineral resources. Metal-bearing ores are scanty, coal and natural gas are almost nonexistent, and Israel's oil deposits supply only a tiny part of the country's requirements. Imported oil is the main source of energy. Much emphasis is being placed on expansion of chemical manufacturing (including chemical fertilizers), based in part on potash and other chemical salts extracted from the brines of the Dead Sea and on rock phosphate mined in the Negev. Many skilled craftsmen have immigrated to Israel, and industries have been developed to capitalize on their skills, for example the cutting and polishing of imported diamonds. Diamonds are the country's largest export by value. In manufacturing as in agriculture, Israel is the most technically advanced country in the Middle East. Israeli engineers and other

Laying a conduit for irrigation water in the northern Negev of Israel. (Israel Office of Information.)

technically trained personnel have served as advisers to various "developing" countries in Africa and Asia.

Tel Aviv–Jaffa (950,000) on the Mediterranean coast is Israel's largest city and industrial center. (Unless otherwise noted, city populations in this chapter, as in other chapters of this text, are metropolitan area approximations for 1968.) Its industries produce textiles and a variety of other consumer goods. A factory in one of its suburbs is the Middle East's largest producer of heavy electrical equipment. The leading seaport and center of heavy industry is Haifa (350,000) in northern Israel. Among the many factories in its metropolitan area are an oil refinery, a large chemical plant manufacturing superphosphate fertilizers, a plant smelting nonferrous metals, and a steel mill. Approximately three-quarters of the population of pre-1967 Israel was concentrated in Tel Aviv, Haifa, or the narrow coastal strip around and between them. The ancient town of Beersheba (city proper, 70,000) is being developed as the northern gateway, administrative center, and main industrial center of the Negev frontier region. A rail line from Tel Aviv reaches Beersheba, and highways fan out from the city to Tel Aviv, the Dead Sea potash works, the chemical-manufacturing center of Arad on the edge of the Judean plateau overlooking the Dead Sea, Jerusalem, the Gaza Strip, Sinai, and Israel's small Red Sea port of Eilat (Elat, Elath, or Eilath) on the Gulf of Aqaba. Eilat is the southern terminus of a pipeline which carries imports of crude petroleum to the refinery at Haifa. (Small amounts of oil from Israel's own field at Heletz also move through the pipeline.) Aside from oil, Eilat handles much smaller amounts of traffic than does the port of Ashdod, south of Tel Aviv, which has been developed as the main sea outlet for central and southern Israel.

Israel's relations with its Arab neighbors continue to cloud the country's future. Immediately after the proclamation of independence in 1948, armed forces of Jordan, Egypt, Iraq, Syria, and Lebanon invaded Israel, but they were defeated or stalemated by the Israeli army. Armistice agreements were signed in 1949, though sporadic border fighting and raids continued. Prior to and during the fighting of 1948, over 500,000 Palestinian Arabs fled to the neighboring Arab countries. Refugee camps for these displaced persons were established by the United Nations in Jordan, Egypt, Lebanon, and Syria. Subsequently, the number of refugees swelled through natural increase and illegal registration to over 1 million, of whom more than half were in Jordan and another quarter in the Gaza Strip administered by Egypt. The refugees quickly became an important focus of bad feeling between Israel and the Arab states. Little was done to resettle them in permanent homes, and both the Arab governments and the refugees themselves continued to insist on the return of the refugees to Israel and the restoration of their properties there. New complexities were added as a result of the June war of 1967, in which Israel took over the areas where most of the camps were located. Many persons in the camps again took flight, and an entirely new group of refugees was created by the departure of a sizable number of permanent residents from areas overrun by Israeli forces. Some of the latter subsequently returned to their homes, but many others either did not attempt to return or were denied permission by Israeli authorities.

In addition to the Arab refugee question, other difficulties between the new state of Israel and the Arab states, particularly Jordan, arose over the city of Jerusalem (300,000). The problem of Jerusalem originates in the fact that the city is a holy place of Islam as well as Judaism and Christianity, being regarded as the third holiest city in the Islamic world (after Mecca and Medina). As a result of the first Arab-Israeli war in 1948–1949, Jerusalem was partitioned between Israel and Jordan. Despite its perilous situation on the boundary with a hostile country, the Israeli-held section—incorporating most of the "New City"—was made the capital of Israel. Most of the sacred shrines, however, were in the "Old City" in the Jordanian sector. In the 1967 war, the Israelis took the Jordanian part of the city and then commenced to integrate it economically and politically with Israeli Jerusalem. Israel's government has proclaimed that Jerusalem is "not negotiable" in any peace settlement with the Arab states. Should Israel retain permanently the shrines in Old Jerusalem, and perhaps those in Bethlehem and other places taken from Jordan, its tourist potential, already large, will be greatly enhanced, while Jordan's economy will have been dealt a crippling blow.

Still another major dispute between Israel and its Arab neighbors concerns the use of water from the Jordan River. This famous (though comparatively small) stream flowed southward in a deep rift valley and empties into the Dead Sea. After independence, Israel's government desired to use water from the river for irrigation and hydroelectric power development, and expressed willingness to share the water with Jordan, across whose territory the lower river flows. However, Jordan, Syria, and Lebanon, which controlled the river's headstreams, objected strenuously to schemes for developing the Jordan which Israel, the United States, and the United Nations proposed. Israel nonetheless proceeded to implement its water diversion plans. In the 1967 war the Israeli army took over a section of Syria in which an important headstream of the Jordan rises, and it occupied the parts of northwestern Jordan lying west of the river.

Since independence, Israel has fought three wars, two of them very short, against Arab states. The 1948–1949 war has been described. The 1956 war against Egypt (see p. 372) resulted in a reopening of the Gulf of Aqaba to Israeli traffic. The gulf had been closed to Israel from 1950 to 1956 because of an Egyptian-Saudi Arabian blockade of the entrance. In the 1967 war, precipitated in part by a further Egyptian closure of the gulf, Israel's armed forces routed the Egyptians, Jordanians, and Syrians in six days. Israel took over the Gaza Strip and all Egyptian territory as far west as the Suez Canal and Gulf of Suez. All territory in northwestern Jordan lying west of the Jordan River and Dead Sea was occupied, and also the Golan Heights section of Syria overlooking Israel's Huleh Valley in which the upper Jordan flows. All of these lands were still in Israel's possession in early 1969. Their acquisition, which quadrupled the area controlled by Israel's government, gave new defense in depth for the country's core region along the Mediterranean coast, possibly created more defensible frontiers, and gave Israel control of oil wells and manganese mines in Sinai. Conversely, the administration of Arab populations in these territories provided troublesome problems. How much of the conquered territory Israel would ultimately keep, and how and when some long-term settlement could be worked out were questions of great import for the entire world. Israel was created under the sponsorship of the Western powers, especially the United States, and it depends on them for financial aid and military equipment. On the other hand, Egypt, and in smaller measure other adversaries of Israel, have received a flow of arms from the Soviet Union. Thus the Arab-Israeli conflict, involving as it does the Suez Canal, has become a vortex of the Cold War and a matter of concern to many of the world's nations. Complexity is added by the fact that Western nations have from time to time supplied arms to Arab states as well as to Israel.

JORDAN, bordering Israel on the east, is mostly desert or semidesert. The main agricultural areas and centers of settlement, including the principal city and capital, Amman (350,000), are in the northwest. As it existed prior to the 1967 war with Israel, northwestern Jordan included two upland areas, the Palestine Hills on the west and the Transjordanian Plateau on the east, separated from each other by the deep rift valley occupied by the Jordan River, Lake Kinneret, and the Dead Sea (map, p. 362). The valley is the deepest depression on the earth's land surface, being about 600 feet below sea level at Lake Kinneret and nearly 1300 feet below sea level at the Dead Sea (the bottom of the Dead Sea reaches –2600 feet). Though the rainfall gradually increases northward to an average of more than 12 inches in the vicinity of Lake Kinneret, most of the valley bottom lies in deep rain shadow and receives less than 4 inches of precipitation a year. But the uplands that border it, rising to heights of 2000 to 3500 feet or more above sea level, and lying in the path of moisture-bearing winds from the Mediterranean, receive precipitation during the cool season that averages 20 inches or more annually within two north-south strips about 80 miles long by 15 to 20 miles wide. It was within these belts of maximum precipitation, one in the Palestine Hills (the "West Bank") and the other in the hilly Transjordanian Plateau (the "East Bank"), that most of Jordan's pre-1967 population was found. The largest percentage were villagers supporting themselves primarily by the cultivation of winter grains (wheat and barley), vegetables (grown both winter and summer on irrigated and nonirrigated land), and fruits, especially olives and grapes. The Transjordanian Plateau was the major area of grain production; olive production was heavily localized in the Palestine Hills.

Less than 5 percent of pre-1967 Jordan was cultivated, and less than 10 percent of the cultivated land was irrigated. The irrigated land produced a disproportionately large share of the country's crop production and exports, however, tomatoes being the largest agricultural export. Part of the irrigated land was occupied by Israel in the 1967 war; of the remainder, a large share is located on the east side of the rift valley floor. The latter district, known as the East Ghor, is irrigated from a canal that brings water from a Jordan River tributary, the Yarmuk.

Jordan's manufacturing establishment is extremely modest, consisting chiefly of flour mills, vegetable canneries, soap factories, cement mills, and other uncomplicated enterprises, mostly small. Amman is the leading industrial center. The country is poorly endowed with minerals, though phosphate rock, quarried at Ruseifa, northeast of Amman, provides Jordan's largest export (1966).

Prior to the establishment of Israel, Jordan's trading relationships were primarily with coastal Palestine, and most of the country's seaborne trade moved through Haifa. But these relationships were severed as a result of Israeli independence and the subsequent warfare between the two countries. Today Jordan's overseas trade moves through Beirut, Lebanon, or the small Jordanian port of Aqaba, located a few miles east of the Israeli port of Eilat. Aqaba handles the country's phosphate exports.

Jordan's economic and political situation has always been precarious. The country's creation under British auspices following World War I was largely a matter of political expediency. From the outset its trade balance was heavily adverse, and sizable grants and loans by foreign nations (primarily Great Britain and the United States) were required to keep its economy operative. In 1946 its population was approximately 400,000, of whom about one-quarter were desert nomads or seminomads and the remainder villagers or urban dwellers. But by 1950 the population had more than tripled, due to the annexation of 400,000 "West Bank" Palestinians and the entry of 572,000 refugees or exiles from areas incorporated by

Lebanon is famous for the flights of rock-walled terraces that climb steep hillsides and valley walls in the Lebanon Range. The narrow strips of flat land created in this manner grow olives, grapes, other fruits, vegetables, and grain. Olive trees dominate the left foreground of the view, which was taken at Beit el Dine about 15 miles southeast of Beirut. See Norman N. Lewis, "Lebanon—The Mountain and Its Terraces," *Geographical Review,* **43,** no. 1 (January 1953), 1–14. (Photo: U.S. Department of Agriculture.)

Israel. On the whole the Palestinian newcomers were more literate, skilled, and Westernized than the native "East Bank" Jordanians, and their presence created internal stresses that added to the country's political instability. It is among the Palestinians that the main anti-Israel and anti-Western elements have been found.

SYRIA and LEBANON, Israel's neighbors to the north, exhibit a pattern of physical and climatic features broadly similar to that of Israel and Jordan. As in Israel, narrow coastal plains backed by highlands front the sea. But the highlands of Lebanon and Syria are loftier than those of Israel and Jordan; indeed the Lebanon and Anti-Lebanon ranges and Mount Hermon are high mountains, rising to slightly more than 10,000, 7000, and 9200 feet respectively. The coastal mountains of Syria rise to about 4000 feet. Thus to cyclonic precipitation brought by air masses off the Mediterranean is added a considerable fall of orographic rain and snow created by the mountain chains that parallel the coast and block the path of the prevailing winds. Precipitation averages 20 to 40 inches annually in most areas that lie within 50 miles of the sea. West-facing slopes of the mountains often receive 40 to 50 inches.

Lebanon and Syria conform to the customary mediterranean pattern of winter rain and summer drought. In both countries the coastal plains and seaward-facing mountain slopes that border them are settled by a rather dense population of villagers. These people depend for a livelihood primarily on the cultivation of characteristic mediterranean crops: wheat, barley, vegetables, legumes, grapes, melons, olives, figs, citrus (primarily in Lebanon), and nuts. Tobacco and cotton are prominent export crops in coastal Syria. In Lebanon, fruit exports from the coastal region are augmented by apples from the highlands. Much of Lebanon's agriculture is found on spectacular flights of terraces that climb the lower mountain slopes (photo, above). Livestock play a relatively minor role in the seaward parts of Lebanon and Syria; in most areas cattle tend to be less prominent than goats or sheep. In Lebanon there has recently been a large increase in poultry and egg production.

In Syria and Lebanon, as in Jordan, precipitation decreases toward the east and the importance of pastoral activities becomes greater. The Anti-Lebanon range and Mount Hermon are drier than the Lebanon range, which screens out moisture that they would otherwise receive.

These mountains, and the eastern slopes of the Lebanon range as well, are used primarily for grazing and the cultivation of limited amounts of grain. An interior valley in Lebanon, the Bekaa, lies between the Lebanon and Anti-Lebanon ranges and is a northward continuation of the Jordan rift valley. It has fertile soils and is a rather productive agricultural area.

East of the mountains a zone of semiarid climate with an annual precipitation of 8 to 20 inches covers central and northern Syria. In southeastern Syria it trends gradually into the Syrian Desert, most of which lies in Iraq and Jordan. The semiarid region, much of which has good soils, is Syria's main producer of cotton, wheat, and barley. Cotton, the country's largest export, is grown as an irrigated summer crop, while wheat and barley, primarily nonirrigated, are winter crops. The cultivated areas of interior Syria are not continuous, but are separated by much land that is vacant or used for grazing. Cotton and wheat are grown in two main districts. One of these, between Homs (city proper, 200,000) and Aleppo, secures water from the Orontes River. The other district stretches eastward from Aleppo across the northern edge of the country. It has farms that are larger, more mechanized, and more purely commercial than most farms in Syria and Lebanon. Small irrigated districts growing cotton and grain are found along the valley of Syria's largest river, the Euphrates.

Both Lebanon and Syria are attempting to increase the amount of irrigated land. In Lebanon, plans for development center around the Litani River, while in Syria—where the irrigation potential is much greater—major reliance to date has been on the Orontes, but with future plans centering on the much larger Euphrates and its tributary the Khabur. Both countries have improved their agriculture considerably in recent years, and the rural level of living is higher than in most parts of the Middle East. In Syria this has been achieved in part by extension of cultivation into unused tracts of land that were initially fertile, but that will require increasing amounts of chemical fertilizer if their productivity is to be maintained on a long-term basis. Up to the present the use of fertilizers has not been great, and periodic fallowing has been the principal device for restoring some of the land's fertility.

Syria is somewhat more industrialized than Lebanon, but neither of them compares with Israel as an industrial country. Textile milling, agricultural processing, and other light industries predominate. Both countries are very deficient in mineral resources, though Syria has an oil field of moderate importance in the east. Both Syria and Lebanon have refineries that operate with crude oil brought by pipeline from Persian Gulf countries (Syria also refines a portion of its own oil), and both countries benefit from transit fees paid by the oil companies.

Western and northern Syria, Lebanon, northern Israel, and northwestern Jordan form the western half of the Fertile Crescent, a famous strip of cultivable land between the Syrian Desert and the high mountains of southeastern Turkey and southwestern Iran. The eastern half of the crescent lies mostly in the plain of the Tigris and Euphrates rivers (plain of Mesopotamia) in Iraq. Through the centuries, from the beginning of recorded history, this semicircle of land, leading from the Persian Gulf and the Iranian plateaus to the Mediterranean coast and southward toward the Nile Delta, was followed by the main caravan tracks from interior Asia to the Mediterranean and northern Africa. It also served as a pathway for marching armies—Assyrians, Hittites, Persians, Egyptians, Macedonians, Romans, Crusaders, Arabs, Mongols, Turks, and others have come this way at various times. The countries of the Fertile Crescent have probably been overrun by foreign armies more times than any other group of countries in the world. These are historic lands, with a record of civilized human settlement dating back for thousands of years.

Three of the five largest cities in the Fertile Crescent are located in Syria or Lebanon. All of them have been important cities for many centuries. Damascus (600,000), the capital of Syria, is situated in a large irrigated district in the southwestern part of the country. The main city of northern Syria is Aleppo (600,000), an ancient caravan center located in the "Syrian Saddle" between the Euphrates Valley and the Mediterranean. Aleppo and Damascus are the leading industrial and commercial centers of present-day Syria. Aleppo is the country's principal grain and cotton market.

The old Phoenician city of Beirut (550,000), on the Mediterranean, is the capital, largest city, and main seaport of Lebanon. Equipped with modern harbor facilities and an important international airport, Beirut is one of the busiest centers of transportation, commerce, and tourism in the Middle East. This thriving port symbolizes the importance of Lebanon as a trading nation. Income from transit and entrepôt trade and financial transactions helps redress the country's unfavorable balance of trade, as do remittances to their families from the many Lebanese who have emigrated to the Americas and other overseas areas. In recent times Beirut has become an important banking and investment center for oil revenues accruing to the Arab states. Lebanon handles an important share of the overseas trade of Syria and Jordan. But Syria has been expanding its port of Latakia (city proper, 85,000), which now handles most exports except for the international transit trade in oil (which moves through

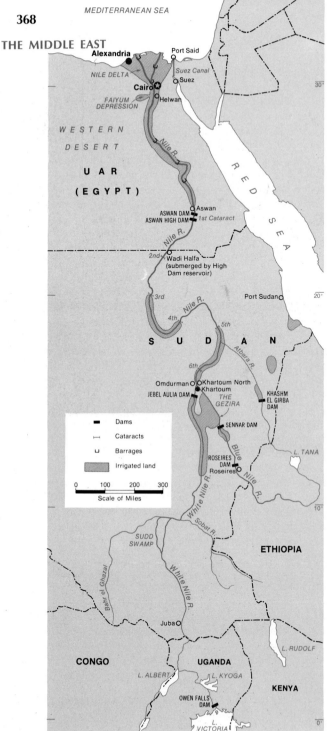

MEDITERRANEAN SEA

The Nile Basin. The width of the irrigated strips along the Nile is somewhat exaggerated. (Based partly on a map in *Focus,* published by the American Geographical Society.)

Baniyas), and Jordan is attempting to orient its trade to its small Red Sea port of Aqaba as much as possible. Lebanon has a tourist industry of some size, based on the summer coolness and winter skiing of the Lebanon range, plus Mediterranean beaches, the night life of Beirut, and historical monuments covering a span of thousands of years.

A complex pattern of minorities exists in both Syria and Lebanon and in Iraq and Jordan as well. However, in Syria Moslems of the Sunni (Orthodox) persuasion[2] comprise an estimated two-thirds or more of the population, the rest being divided among some seventeen significant religious minorities. In Lebanon, Christians—somewhat more than half of whom belong to the Maronite sect—are probably a majority. Most of the remaining population is Moslem, about equally divided between the Sunni and Shia sects. The presence of a large and highly influential Christian element is a major characteristic distinguishing Lebanon from the other Arab states.

Egypt—The "Gift of the Nile"

At the southwest, Israel borders Egypt (officially the United Arab Republic or UAR), the largest in population and the most influential of the Arab countries. This ancient land, the home of one of the oldest known civilizations, is described in the following selection:[3]

As long ago as 430 B.C., Herodotus observed in his *History* that "Egypt is an acquired country, the gift of the river," This characterization is just as valid today; for, with the exception of the Faiyum depression and some smaller oases in the Western Desert, the only part of Egypt that is regularly in cultivation [is found in] the 3½ percent of the land that [lies on the irrigated alluvial plain along] the Nile. . . . In this small area, made up of the triangular delta north of Cairo and the narrow, ribbonlike valley to the south [map, left], live more than 95 percent of Egypt's [32] million people. The rest live either in the oases or

---

[2] The Islamic community of believers is divided into two main branches, the Sunni (Sunnite) Moslems and the Shia (Shii or Shiite) Moslems. Sunni Moslems are in the majority in most Moslem countries, though Shia Moslems predominate in Iran and are very influential in Iraq and in Yemen.

[3] George H. T. Kimble and Dorothy L. Weitz, "Egypt," *Focus,* 2, no. 4 (December 15, 1961), 1–2. Used by permission of *Focus,* published by the American Geographical Society of New York. Bracketed sections contain updated or supplementary material.

An Egyptian fellah breaks the rich alluvial earth of the Nile Delta in preparation for a fall-sown crop. Note the date palms and field of ripening corn in the background. Corn bread is a staple in the Egyptian diet. (Peter Schmid—Pix, Inc.)

on the surrounding deserts in the wake of their wandering flocks.

Nowhere else in the world is the contrast between the desert and the sown so dramatic, or the transition from solitary waste to teeming valley so sharp. And nowhere else does the well-being of man hang by so tenuous a thread. Fortunately, the "thread" never breaks. Formerly, it is true, the annual fluctuations of the Nile gave rise to constant anxiety, but the construction of the dam at Aswan in 1902, and the six supplementary barrages[4] downstream, [as well as dams in Sudan and Uganda, and now the Aswan High Dam] have removed the grim specter of famine. Even so, the rises and falls of the water level are still the subject of as much comment among the *fellahin* [peasants] as the oscillations of the barometer are among the farmers of the prairies. Indeed, the Nile is to the Egyptian's conversation what the weather is to the American's.

The ancient Egyptians, who believed that the swelling of the Nile represented the tears of the goddess Isis

shed in compassion for her rain-starved people, early learned to raise crops on the flooded margins of their river and to supplement the floods by means of simple water-lifting devices. In most parts of the Nile Valley the ancient *shaduf* (consisting of little more than a pole and a bucket), the *saqia* (in which the water is raised by a rotary wheel furnished with buckets, powered by a beast of burden), and the Archimedean screw are still employed to fill the irrigation ditches. Thanks to these and more modern machinery, including . . . motor-driven pumps . . . [about 50–55 percent] of the total population of Egypt contrive to live off the soil. In few other places on earth is land cropped so intensively or so continuously. Two, and even three, crops are often raised off the same piece of soil in the course of a twelvemonth, and yields are generally high; for this the well-regulated water supply, the long-lasting fertility of the valley silts, and the growing employment of artificial manures are primarily responsible. Given a more lavish use of such manures, . . . even greater yields could be obtained.

---

[4] *Editorial note:* A *barrage* is a low dam designed to raise the level of a stream high enough that the water will flow by gravity into an irrigation canal. Barrages permit a constant flow of water to irrigated land even in periods of low water. They are not designed for storage of large amounts of water in the manner of the Aswan Dam or Aswan High Dam.

Cotton, the long-staple Egyptian variety, is the leading commercial crop; although occupying only [about a fifth] of the total cultivated area, it [accounted for about 55 percent of the country's exports in the middle 1960s]. . . .

The usual food crops are corn (a staple in the Egyptian diet), wheat, barley, rice, fruit, grain sorghum, and sugar cane.[5]

For the fellah, life is hard. His tools are those of his ancestors—the hoe, the wooden plow, the hand sickle, and the threshing board. His dwelling is a crude, fly-infested, two-room mud hut sheltering family, water buffalo, and chickens alike.

Diet, Sanitation, and Disease

Like rural people in many other "developing" regions, Egypt's *fellahin* are beset by serious health problems. In large measure these originate in dietary deficiencies and poor sanitation, as well as from the necessity for standing many hours in the water of irrigation canals and ditches.

Lacking both variety and abundance, the diet of the majority of Egyptians is dominated by a single staple, *bettai* (maize bread). . . . In the south, millet and wheat are more widely used as a maize substitute. Vegetables and fruits are available in limited quantities. These include onions, eggplants (the potatoes of the Middle East), tomatoes, turnips, cucumbers, green peppers, lettuce, green beans, lentils, marrows, citrus fruits, and dates. At most, 10 percent of the average Egyptian's caloric intake is obtained from these products; to them may be added an occasional egg and a coarse sour variety of cream cheese made of goat or buffalo milk preserved in salt water. . . .

Meat is an expensive delicacy rarely enjoyed by the fellah. A prosperous peasant family might have meat once a week; the poorer families see it only at festival times or when an animal has been killed by accident. . . .

. . . . .

The general sanitation of all but a few "model" villages is extremely poor. Dust beclouds the air of narrow alleys strewn with offal and rubbish. . . . Sewage disposal facilities in the village are few and primitive. Pits dug near the houses serve as latrines, as do the irrigation ditches in the field.

Weakened by malnutrition and living in unsanitary and crowded surroundings, the Egyptian is prey to an impressive array of debilitating or fatal diseases.[6]

Among the most widespread of the many serious diseases in Egypt are bilharzia (more properly bilharziasis or schistosomiasis) and trachoma. Bilharzia, an extremely common disease in Africa, is caused by parasitic worms. The immature worm, microscopic in size, spends part of its life cycle in the body of a fresh-water snail. Upon leaving the snail the organism swims about in river or canal water. The disease is contracted by drinking contaminated water or simply by wading, washing, or swimming in it, since the microscopic form of the bilharzia parasite is able to burrow through human skin. Bilharzia damages internal organs, causes many disagreeable symptoms, saps the vitality, and if untreated, often causes death. Trachoma is a contagious disease caused by a virus that inflames the eyes and may cause blindness. A common means of transmission is for infected and noninfected persons to use the same towel.

In recent decades, public health programs in Egypt have made considerable progress in alleviating the worst of the conditions described above. Supplies of pure water from chlorination plants or deep drilled wells are far more widespread than formerly and have lessened the use of polluted water from the Nile, the irrigation canals, and village ponds.[7] Treatment and prevention of disease is carried on by clinics that have been established throughout the country. Such measures have contributed greatly to a recent decline in death rates. Birth rates, however, have continued high, and Egypt is confronted with rapid population increases that threaten to outrun the food supply and bring to naught all attempts to raise the average level of living. Alarmed by this pros-

---

[5] *Editorial note:* Four Egyptian crops—cotton, corn, wheat, and berseem (clover)—stand out from the others in total acreage. Wheat, barley, and berseem are winter crops, whereas cotton and corn are grown in summer, as are sugar cane, grain sorghums, and rice. Berseem, a legume, is grown for livestock feed and to enrich the soil with nitrogen. An important development in recent times has been the rise of irrigated rice production on an export basis. However, large wheat imports make Egypt a net importer of grain.

[6] Donald N. Wilber, ed., *UAR-Egypt* (New Haven, Conn.: Human Relations Area Files Press, 1968), pp. 267–272. Reprinted by permission.

[7] *Editorial note:* Merely to provide a source of uncontaminated water does not necessarily ensure that villagers will actually use it. "Even where pure water has been provided, the fellah is apt to go on drinking the untreated waters of the Nile—his wife may find it inconvenient to go to the pump, the smallest fee for the pure water will seem too much, or, if he is not one of the many who believe that raw Nile water will enhance his fertility, he is convinced that it is pure by reason of the river's flow. Even more polluted than the Nile are the irrigation canals, which are laden with human and animal waste—and continue to be a . . . source of water supply." (Wilber, *op. cit.,* p. 268. Reprinted by permission.)

pect, Egypt's government has actively sponsored programs of birth control since the early 1960s.

Land Reform and Industrialization

In 1952 the corrupt and inefficient government of King Farouk was overthrown by a revolution sponsored by certain elements of the Egyptian army. An army lieutenant colonel, Gamal Abdel Nasser, soon rose to leadership in the new regime. In June 1953 the revolutionary government proclaimed Egypt a republic. One of its earliest important acts was to institute a nationwide program of land reform. Much of the Egyptian countryside had been dominated by large estates worked by tenants (generally sharecroppers) or wage laborers. Often such estates were absentee-owned, generally by urban merchants, professional men, bankers, or industrialists, and their actual operation was delegated to hired managers and overseers. Rents were often excessively high. Rising land values and the dire poverty of most peasants made it exceedingly difficult for landless peasants to acquire land or for small owners to increase their holdings. Under a land-reform decree of September 1952, the revolutionary government expropriated several hundred thousand acres from large landholders and redistributed this land among the peasants. To enable the latter to pay for it, long-term government loans at a low rate of interest were extended.

Land reform alone cannot solve the problem of rural poverty in Egypt, though the decree of 1952 and subsequent measures have had beneficial results. The basic difficulty is one of rapid population increase in rural areas that are already overcrowded. Improved living standards resulting from land reform, reclamation of new land, or greater output from the existing land will quickly be nullified unless the increasing pressure of population on the land can be checked. Birth control offers one answer to the problem. Another answer is an increase in manufacturing to draw surplus population off the farms; and the regime has thrown its weight behind a program to industrialize. Many additional considerations have motivated this program, among them the desire to overcome the economic and military weakness that made Egypt a pawn of foreign powers for many centuries and that recently led to humiliating defeats by Israel in the Palestine war of 1948–1949, the Sinai invasion of 1956, and the six-day June war of 1967.

Actually Egypt has been industrializing in a modest way for a long period and holds undisputed industrial leadership among the Arab countries. Cotton textiles and food processing are the largest industries. Clothing manufacture and cement production are important, and there is a growing output of chemicals, particularly nitrate and phosphate fertilizers. The manufacture of automobiles, refrigerators, and other durable consumer goods involves primarily the assembling of imported components. Egypt, in other words, has a fairly extensive range of industries, but is far from being a highly developed industrial nation. It should be noted that a large share of the country's manufacturing plants, as well as many other types of business enterprises such as banks and insurance companies, are government owned. Under the Nasser regime Egypt has become a socialistic country. But, despite the fact that Egypt receives much technical, economic, and military aid from the Soviet Union, the Communist party has been outlawed.

The two main industrial centers in Egypt by an overwhelming margin are Cairo (4.9 million), the capital, and Alexandria (1.9 million), the principal seaport, located, respectively, at the head and western seaward edge of the Nile Delta (map, p. 368).

Egypt shares the customary handicaps of underdeveloped countries that attempt to industrialize: an impoverished home market, a shortage of investment capital and skilled managerial and technical personnel, and a dearth of skilled industrial labor. A rapid development of public education in recent years will help remedy deficiences in technically trained personnel. Mineral resources in Egypt are not abundant, but some minerals of considerable value are present. They include oil, phosphate, iron ore, salt, some manganese and coal, and ample supplies of stone, clay, gravel, and sand. Israel's capture of the Sinai Peninsula in 1967 gave it control of some of Egypt's oil wells located along the Gulf of Suez; in addition, Israel gained control of manganese mines and small deposits of coal. However, Egypt has alternate, and probably much larger, sources of oil west of the Gulf of Suez and in the Western Desert near the Mediterranean Sea. Promising strikes have been made in the general area of the World War II battlefield of El Alamein, and production began there in 1968. Iron ore from deposits in Upper Egypt[8] a short distance east of Aswan is utilized in an iron and steel plant opened in 1958 at Helwan on the Nile River about 20 miles south

---

[8] Since ancient times the triangular Nile Delta has been referred to as Lower Egypt; the narrow Nile Valley between Cairo and the Sudan is Upper Egypt.

of Cairo (map, p. 368). The ore is smelted with imported coke. The Western Desert contains an iron-ore reserve that is closer to Helwan than the ore near Aswan, but it is not accessible at present due to lack of transportation. Egypt's phosphate deposits, mined for export as well as for use in Egypt itself, lie near the upper Nile some distance north of Aswan, and along the middle section of Egypt's Red Sea coast. Phosphate and nitrate fertilizers have become increasingly vital to Egypt as perennial irrigation has replaced basin irrigation.[9] Formerly the land's fertility was renewed during the annual flood by the deposit of rich silt brought by the Nile from volcanic highlands in Ethiopia, but in the future most of this will settle behind the new High Dam at Aswan.

For power with which to operate its industries, Egypt will rely increasingly on electricity produced at two hydroelectric stations on the Nile. The first of these was opened at the Aswan Dam in 1960. A plant near the dam uses part of the electricity to produce nitrate fertilizer by fixation of atmospheric nitrogen. The main future source of electricity, however, will be the station at the Aswan High Dam. This enormous rock-filled dam, some 360 feet high and 11,500 feet wide at the top, was nearing completion in 1968. It is located about 5 miles upstream from the older Aswan Dam. The reservoir, when filled, will be more than 300 miles long and will store at least 25 times as much water as the older dam. The first three of twelve large generators to be installed were transmitting power to Cairo by high-tension lines in 1968, and the power station was scheduled to reach full production by 1976. By storing water in years of high flood for use in years of low, the High Dam will regularize the flow of the river in Egypt to a much greater extent than has been possible heretofore. It will provide a larger and more constant water supply for the existing 6 million irrigated acres, thus enhancing the productivity of the land, will make it possible to convert all of the remaining basin irrigation to perennial irrigation, and will make possible the reclamation of an additional 1 to 2 million acres of farm land. National hopes for future economic progress center heavily in this billion-dollar dam.

The Suez Canal

In 1956 Egypt's High Dam project helped precipitate a major international crisis. In 1955 the United States and the United Kingdom had indicated their willingness to help Egypt finance the costly dam. But in July 1956 they withdrew their offers, and cancellation of a promised loan by the World Bank followed. President Nasser of Egypt then nationalized the Suez Canal, declaring that canal tolls would be used to build the High Dam. The canal had been constructed and operated by the Universal Suez Canal Company, a semiprivate Egyptian corporation in which the controlling financial interest was British but the actual management largely French. Great Britain and France protested Egypt's action and froze Egyptian financial assets in their countries. In October 1956, Israel invaded the Sinai Peninsula of Egypt for the avowed purpose of ending Egyptian border raids into Israeli territory. Great Britain and France promptly intervened, ostensibly to protect the canal. Port Said and Egyptian airfields were bombed, and troops were landed in the Canal Zone. These events, coupled with a revolt in Hungary against Soviet control, created a tense international situation. Diplomatic pressure by the United Nations, led by the United States and the USSR, plus protests by opposition parties in Great Britain and France, brought about a withdrawal of the invading forces, though Israeli forces lingered for several months in the Gaza Strip and along the Gulf of Aqaba. Commencing in 1950, Egypt barred Israeli ships from the Suez Canal and frequently interfered with ships of other nations carrying cargoes to or from Israel. The June War of 1967, however, put the Canal out of service, and it became a frontier between hostile armies. This situation was still in effect at the time of writing (late 1968).

The Canal connects the Mediterranean Sea with the Gulf of Suez and Red Sea across the low Isthmus of Suez (refer to inset map, p. 340). Unlike the Panama Canal, in which ships are raised or lowered to successive water levels by massive locks, the Suez Canal is a sea-level canal with no locks of any kind. It was completed in 1869 under the direction of a noted Frenchman, Ferdinand de Lesseps. The total length of the waterway, including dredged approach channels and the Bitter Lakes, is 107 miles. At opposite ends of the canal are the seaports of Port Said (230,000) on the Mediterranean and Suez (160,000) on the Gulf of Suez. Before the 1967 war the main item of traffic passing through the canal was petroleum, most of it carried by northbound tankers moving from the Persian Gulf to destinations in the United Kingdom, continental Europe, or the United

---

[9] In *basin irrigation*, the traditional type, floodwaters are led into fields surrounded by low embankments; after the water has stood for several weeks, the excess is drained back into the Nile and crops are sown in the muddy fields. For the period of growth they are dependent on the moisture stored in the soil. But in *perennial irrigation* water stored behind dams or raised in level by barrages is released to the fields whenever it is needed by the crops.

States. However, the rise of huge supertankers and other bulk carriers too large to negotiate the present canal makes the future of the waterway problematical. An extensive and costly program to deepen and widen the canal will be necessary if it is to resume its former role. Egypt has been considering the possibility of a pipeline across the isthmus to transfer oil from tankers in the Gulf of Suez to tankers in the Mediterranean.

The Sudan

Egypt is bordered on the south by the vast, sparsely populated tropical republic of the Sudan. Formerly an Anglo-Egyptian condominium, the Sudan received independence in 1956. To Egypt its southern neighbor is vitally important, for it is from the Sudan that the Nile River brings the irrigation and drinking water that sustains the Egyptian people. The Nile receives all of its major tributaries in the Sudan (map, p. 368), and storage reservoirs exist there (at Sennar Dam and Roseires Dam on the Blue Nile, Jebel Aulia Dam on the White Nile, and the Khashm el Girba Dam on the Atbara River) which benefit Egypt as well as the Sudan itself.

The Nile's two main branches, the White Nile and Blue Nile, originate, respectively, at the outlets of Lake Victoria (Uganda) and Lake Tana (Ethiopia). It is the Blue Nile, Atbara, and Sobat rivers, flowing from seasonally rainy highlands in Ethiopia, which are primarily responsible for the annual summer floods of the main river. In contrast to the Ethiopian rivers, however, the White Nile, fed by one of the world's largest lakes, maintains a fairly constant flow throughout the year.[10] Completion of the Owen Falls Dam in Uganda has increased the storage capacity of Lake Victoria by raising its level 3 feet, and thus has facilitated control of the White Nile. In the southern Sudan the river passes through a region of swamps and marshes, the Sudd, where much water is lost by evaporation. It has been proposed that this wastage be eliminated by construction of an artificial channel to straighten the river's course and increase the velocity of its flow through the Sudd.

Approximately 3 percent of the Sudan is cultivated, about the same proportion that is true of Egypt. But only a minor share—perhaps 10 to 20 percent—of the cultivated land is irrigated. The Sudan's largest block of irrigated land is found in the Gezira (Arabic, "island")

between the Blue and White Niles (map, p. 368). Water flows by gravity to the fields through canals from Sennar Dam. The newest large canal (Manaqil Canal), opened in 1959, added some 800,000 acres to the 1 million acres already under cultivation. A dam under construction in 1968 on the Blue Nile at Roseires, upstream from Sennar, will make possible a further large expansion of irrigated acreage and additional water supplies for the existing acreage. Substantial irrigated areas exist in the Sudan at many points along the main Nile and its tributaries. Though some basin irrigation is practiced and extensive use is made of traditional water-lifting devices, an increasing amount of water is being supplied by diesel-powered pumps.

The most important cash crop of the Sudan is irrigated long-staple Egyptian-type cotton. In the Gezira project it is grown in a systematic rotation with *dura* (a general name for grain sorghums that form the basic food crop in most parts of the Sudan) and *lubia* (a legume that yields fodder for cattle and beans for human or animal food). The Sudan Gezira Board, a government agency, administers the project. The land is farmed by tenants, who receive 42 percent of the proceeds from the cotton crop and are allowed to retain crops other than cotton. Nonirrigated short-staple cotton is grown with the aid of summer rains in several parts of the central and southern Sudan, but it accounts for only a minor part of the total cotton output. Cotton provided about half of the country's exports by value in 1965. Fluctuating yields, demand, and prices from year to year have made cotton a rather uncertain prop for the Sudanese economy, but in good years the exports have paid excellent returns to the government and have given the Gezira tenants a higher living standard than that of most farmers in the Middle East.

In the Sudan, as in most tropical countries, a variety of crops attain local importance (for example, dates, wheat, barley, corn, peanuts, tobacco, sugar cane, sesame, coffee, cassava, bananas, and rice), but for the country as a whole cotton and dura are the outstanding crops. An interesting export product from the Sudan is gum arabic. The bulk of the world supply is gathered, primarily in Kordofan Province, from resinous "tears" that form on certain varieties of acacia trees at the beginning of the dry season.

For the majority of the Sudan's population, livestock raising, largely on a subsistence basis, is the prin-

---

[10] The White Nile receives floodwater brought from Ethiopia in the summer rainy season by a tributary, the Sobat, and consequently there is a much greater seasonal variation in flow below the Sobat junction than there is above it.

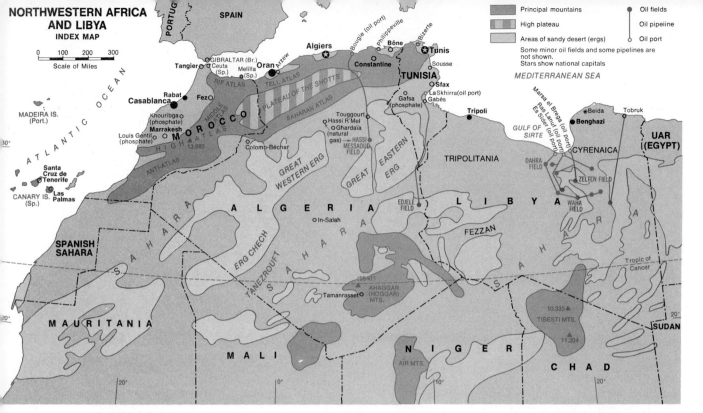

cipal means of support. In the middle 1960s the country was estimated to have some 7 million cattle (predominantly long-horned varieties in the southern savannas), 2 million camels, 8½ million sheep, and 7 million goats. The Sudan has few minerals of economic value and its industries, confined mainly to agricultural processing, cotton textiles, cement manufacture, and a miscellany of simple consumer goods, are relatively meager.

Most of the country is an immense plain, broken frequently by hills and in a few places by mountains. Climatically and culturally, the Sudan is transitional between the Middle East and Negro Africa. A great contrast exists between the arid Saharan North, peopled mainly by Arabic-speaking Moslems of mixed Negroid-Caucasoid ancestry, and the seasonally rainy equatorial South, with its grassy savannas, papyrus swamps, and pagan or Christian Negroid tribes. The population of the North is about twice that of the South, and the Sudan's government is dominated by the Northerners. There has been a considerable amount of political friction between the two sections, manifested by uprisings, guerrilla warfare, and a great deal of bloodshed in the southern Sudan since independence. The central government is making strong efforts to propagate the Islamic faith and the Arabic language in the south and to bring the region under firm control.

The Sudan is overwhelmingly a rural nation. The largest urban district, formed by the capital, Khartoum, and the adjoining cities of Omdurman and Khartoum North, is located toward the center of the country at the junction of the Blue and White Niles. Its total population was estimated at 450,000 in 1968. Transportation within the Sudan as a whole is poorly developed. Through navigation on the Nile north of Khartoum is barred by the famous series of cataracts (stretches of rapids) (map, p. 368), though the White Nile is continuously navigable at all seasons from Khartoum to Juba in the deep south. Most of the country is devoid of railways or highways. Khartoum, however, has rail connections with several widely-separated sections (map, p. 386). The main trunk routes for rail freight connect Khartoum and the Gezira with Port Sudan (city proper, 60,000) on the Red Sea. Port Sudan, a modern, well-equipped port, handles most of the Sudan's seaborne trade, including its vital exports of cotton. Sudan Airways operates a widespread air net within the country, and, along with foreign airlines, offers international services.

Libya

Egypt's neighbor to the west, Libya, was formerly an Italian colony, but became an independent kingdom in 1951 following a postwar period of transition under United Nations supervision. It is populated for the most part by Moslem Arabs and Berbers with a small remnant of Italians; racially the population is Caucasoid, with some Negroid elements and admixtures. Three major areal divisions (former provinces)—Tripolitania in the northwest, Cyrenaica in the east, and the Fezzan in the south-west—are commonly recognized. Libya's principal cities are Tripoli (230,000) in Tripolitania and Benghazi (150,-000) in Cyrenaica. Both cities function as national capitals, Benghazi being the principal place of residence of the king and Tripoli the meeting place of the national legislature. Still a third capital, Beida, is being developed in Cyrenaica. Located in low highlands directly behind the coast, it has a more pleasant summer climate than Benghazi or Tripoli and thus far it has functioned primarily as a "summer capital."

Most of Libya lies in the Sahara and thus is too dry to support cultivation except at scattered oases watered by wells or springs. The average density of population is less than 3 per square mile. This figure has little meaning, since most of Libya has, for practical purposes, no population at all. Libya's 1.8 million people (mid-1968 estimate) are heavily concentrated in coastal lowlands and low highlands along the Mediterranean border, where winter rains produce an annual precipitation averaging 8 to 17 inches in coastal Tripolitania and 8 to 24 inches in the northern "hump" of Cyrenaica. Libya's scanty agriculture, irrigated only in small and scattered areas, is devoted to the growing of winter wheat and barley, tree crops such as olives, dates, figs, citrus, and almonds; vegetables, especially tomatoes, and a scattering of other crops. Most of the production is used at home or marketed within the country. An east-west strip of semi-arid country between the desert proper and the coastal areas is devoted to the raising of livestock (sheep, goats, and camels) on a nomadic or seminomadic basis; along the southern shore of the Gulf of Sirte this strip reaches the Mediterranean. Livestock, including some cattle, also are raised in the better-watered coastal sections, especially in Cyrenaica.

For many years the economy and government of Libya operated at a deficit made good by foreign subsidies. Following World War II the principal contributors were the United States and the United Kingdom, each of which has maintained some armed forces in Libya since the war. The largest installation is an American air base (Wheelus Field) outside Tripoli. Although this American base and a small British base in Cyrenaica were still functioning in 1968, prominent elements in Libya's political life were asking that they be closed and all foreign military forces removed from the country. Since 1961, when the first Libyan oil was exported, the country's financial situation has taken a sharp turn for the better. Intensive prospecting by numerous Western oil companies has revealed deposits in several widely separated parts of Libya, including offshore waters in the Gulf of Sirte. The main producing fields to be devel-

oped thus far (the largest is the Zelten field) are clustered in Cyrenaica between 100 and 200 miles inland from the Gulf. Pipelines connect the fields with shipping points on the coast (map, left). By 1966 Libya had become the world's seventh-ranking producer of oil (after the United States, the USSR, Venezuela, Saudi Arabia, Iran, and Kuwait). Petroleum and its products, including liquefied natural gas, make up 99 percent of the country's exports by value, and royalties and taxes paid by the oil companies supply nearly all of the government's revenue. Though the highly mechanized oil industry does not require a large labor force, it has opened up direct or indirect employment opportunities for a great many Libyans at good rates of pay. A growing "middle" element of businessmen and skilled workers has developed, and farmers have found a new demand and higher prices for their products. A large share of the government's oil revenue is being spent on public works, education, housing, aid to agriculture, and other projects for social and economic betterment. All in all, Libya's economic and social prospects have improved immeasurably since the advent of oil. The country is apt to be a major oil producer for a long time, and it has a relatively small population to support. An important new phase of the Libyan oil industry is liquefaction of natural gas for shipment to markets in Italy and Spain. The gas, obtained by separation from crude oil, is prepared for export at a liquefaction plant reputed to be the world's largest. The plant is located at one of the tanker terminals (Marsa el Brega) on the Gulf of Sirte. Most of Libya's oil, as well as the liquefied gas, is marketed in western Europe.

Northwestern Africa— The Arab "Maghrib"

For many centuries Arabs have spoken of the northwestern fringes of Africa as the "Maghrib" ("West") of the Arab world. Despite its marginal location, this area is Middle Eastern in many essential characteristics. Today it includes three main units: Morocco, Algeria, and Tunisia. (Libya is sometimes included in the Maghrib concept, but for convenience is excluded here.) In the nineteenth and early twentieth centuries France became the dominant power in the area. French occupation of Algeria commenced in 1830; Tunisia was made a French protectorate in 1881; and a French protectorate was established in Morocco in 1912 (though parts of Morocco came under Spanish control, and a zone under international jurisdiction was established at Tangier on the Strait of Gibraltar). Though France invested large sums to develop mines, industries, irrigation works, electric-power stations, railroads, highways, and port facilities, her rule was not popular. In 1956, after a long period of agitation by local nationalists, Tunisia and Morocco secured independence.[11] In Algeria, independence in 1962 came only after warfare lasting for 8 years (see p. 143).

Moslem Arabs form the largest population element in all three countries of the Maghrib; Berbers, who antedate the Arabs in the area, are numerous in Morocco and Algeria. Many Berbers now speak Arabic and many have adopted Arab customs. All are Moslems. More than 1¾ million Europeans, primarily of French but also of Spanish or Italian origin, formerly lived as permanent settlers in the Maghrib. About two-thirds of them were in Algeria. Most Europeans were concentrated in Algiers (estimated population of metropolitan area in 1968, a little over 1 million), Oran (500,000), Casablanca (1.1 million), Tunis (530,000), and other cities, where they formed a class of business and professional people, administrators, office workers, and skilled artisans. In Algeria possibly 25,000 Europeans were supported directly by agriculture. Their farms, worked by Moslem laborers or tenants, were generally much larger and more mechanized than farms owned by Moslems, and they included a large share of Algeria's best agricultural land. European farmers also were found in Morocco and Tunisia, but their numbers were much smaller than in Algeria. Since these countries became independent, there has been a large exodus of Europeans, mainly to France. Less than a quarter of the original number in the Maghrib as a whole still remain. In Algeria European farmland has been nationalized and a cooperative system of agriculture introduced. In Morocco some European farms have been nationalized and most others sold to Moroccans. In Tunisia a comparable movement, while making headway, has left a good share of the best farmland in European hands, due to government recognition that a complete departure of European farmers would cripple the country's commercial agriculture and export trade. It is still too early to make accurate judgments concerning long-term effects of the transfer of ownership and management from European to Moslem in the "modern sector" of the Maghrib's agricultural economy. All three governments have ambitious plans to maintain or increase the output of the "modern sector" and to modernize the "traditional sector."

Most of the people in the Maghrib live in the coastal belt of mediterranean climate (called in Algeria "The Tell"), with its cool-season rains and hot, rainless summers. Here, in a landscape of hills, low mountains, small plains, and scrubby mediterranean vegetation, are grown the familiar products of mediterranean agriculture —wheat, barley, legumes, olives, grapes, citrus fruits, and vegetables. Some corn, small grains other than wheat and barley, and a miscellany of other crops are grown. Export production, largely from the "modern sector," concentrates mainly on grapes for wine, citrus fruits (largely oranges), olives, and vegetables. In Algeria wine and citrus fruits are the largest agricultural exports by a wide margin; in Morocco, vegetables (mainly tomatoes and dried beans) and citrus; and in Tunisia, olive oil and citrus. It may be noted that France is still by far the leading trade partner for each of the Maghrib countries, being the largest market for exports (though not necessarily for each individual category) and the largest supplier of imports in each case. Each country, however, is making efforts to lessen its dependence on economic relationships with France.

The great majority of cultivators and livestock raisers in the Maghrib are in the "traditional sector"; they eke out a living from small farms or nomadic herding, or drift off to *bidonvilles* (shantytowns) in the cities. In most cases tools are primitive, machinery essentially absent, and methods outmoded. Most livestock (primarily sheep and goats) are raised in the traditional sector. Probably two-thirds or more of the population in the Maghrib as a whole still can be classed as rural. The average level of living is low and the rate of population increase is high. All three governments are faced with massive problems as they attempt to bring their poverty-stricken populations up to acceptable levels of livelihood and welfare.

Inland from the coastal belt of the Maghrib, the Atlas Mountains extend in relatively continuous chains from southern Morocco to northwestern Tunisia. These mountains block the path of moisture-bearing winds from the Atlantic and the Mediterranean, and some mountain areas receive 40 to 50 inches or more of precipitation annually. In places the precipitation nourishes good forests of cork oak or cedar (photo, p. 349). The highest peaks are found in the High Atlas of Morocco, where one summit reaches 13,661 feet. In Algeria the mountains form two east-west chains, the Tell Atlas nearer the coast and the Saharan Atlas (map, p. 374); between them lies the semiarid Shotts Plateau (average elevation about 2700 feet), with its basins of interior drainage ("shotts"), subsistence

---

[11] Moroccan sovereignty over the Spanish zones, one in the north and one in the south, was recognized by Spain, but the latter continued to hold the adjacent units of Ifni, Spanish Sahara, and a number of localities along the Strait of Gibraltar. The Tangier International Zone was abolished and its territory became a part of the Kingdom of Morocco. Ifni was ceded by Spain to Morocco on January 4, 1969.

grain fields, patches of esparto (alfa) grass (gathered for export to the world's paper industry), and herds of sheep and goats. In Tunisia the mountains are lower than in Morocco or Algeria, but they contain many deep valleys.

At the south all three countries reach the Sahara; indeed the greater part of Algeria is Saharan. A prominent line of oases fed by springs, wells, mountain streams, and foggaras (photo, p. 380) is found along or near the southern base of the Atlas Mountains. A sparse population of nomads roams the Sahara, and clusters of oases exist in a few places where mountains rise high enough to catch moisture from the passing winds. Several large areas of sandy desert (*ergs*) and a barren, lifeless gravel plain, the *Tanezrouft,* occupy portions of the Algerian Sahara (map, p. 374).

The economic future of the Maghrib is tied in considerable measure to a valuable endowment of minerals. Morocco and Tunisia rank third and fourth, respectively, among world producers of phosphate rock (after the United States and the Soviet Union), and phosphate is the largest single export from each. Algeria also has phosphate deposits and a small production. All three countries of the Maghrib have a moderate export production of high-grade iron ore, and large unexploited reserves of iron ore and manganese exist in western Algeria near the Moroccan border. All three countries produce lead, with Morocco the leader by a wide margin. The most diversified collection of ferroalloys and nonferrous metals is found in Morocco, which produces manganese, cobalt, lead, zinc, and small quantities of several other metals. Morocco has small oil deposits, and both Morocco and Algeria have small deposits of coal. But the minerals of greatest value are the oil and natural gas of the Algerian Sahara. In 1956 two major oil fields—the Hassi Messaoud and Edjelé fields—were discovered in the sand sea called the Great Eastern Erg (map, p. 374). A 400-mile pipeline connects the Hassi Messaoud field and smaller fields to the south with the oil port of Bougie on the coast of Algeria about 200 miles east of Algiers. The Edjelé field and smaller neighboring fields are connected with the oil port of La Skhirra on the Tunisian coast by a 450-mile pipeline. One of the world's largest deposits of natural gas was discovered in 1957 at Hassi R'Mel, located in the northern edge of the Sahara about 50 miles north of the oasis city of Ghardaïa (map, p. 374). Gas is piped from Hassi R'Mel to industrial plants in Algiers and Oran, and to the port of Arzew near Oran, where a plant converts gas into liquid form for export. Algeria's oil production has not grown as rapidly as Libya's, amounting, in fact, to a little less

than half of Libyan production in 1966. Nevertheless, oil and gas are major economic assets, and large new discoveries of both may yet be made. French companies have played a major role in developing these resources, and the bulk of the oil production is marketed in France or is shipped to West Germany via pipeline from the Marseilles area.

In the Maghrib modern factories are not present in large numbers, though this area is more industrialized than the Sudan, Saudi Arabia, or Jordan. Textiles, processed foods, and miscellaneous other types of simple consumer goods are the main products; in addition there is some production of cement and of chemicals, particularly superphosphate fertilizers. All three countries of the Maghrib have oil refineries, and Tunisia produces iron and steel at a plant completed near Bizerte, northwest of Tunis, in 1966.

PERSIAN GULF OIL AND THE STATES THAT PRODUCE IT

No mineral resource is more vital to present-day civilization than petroleum. Modern transportation is mainly powered by it, and it provides the essential lubricants that enable power-driven vehicles and machines to function. The world demand for petroleum is constantly rising. In 1947 world production amounted to about 8 million barrels per day; by 1957 this had risen to 16 million; and by 1967 it reached 33 million barrels per day.[12]

Although the United States produces more petroleum than any other country (an estimated 24 percent of the world total in 1967), its production is surpassed by the total output from the Middle East, and the latter region is by far the leading area of *surplus* production. The Middle East (including North Africa) produces 35 percent of the world's oil (1967) and sells most of its output in foreign markets. The United States, by contrast, uses most of its own production for domestic needs and imports additional quantities of oil, partly from the Middle East.

Though Algeria and, more recently, Libya, have become important suppliers (pp. 377 and 375), most of the known reserves and production of oil in the Middle East are found in a group of states that ring the Persian Gulf. All but one of these states belong to the Arab world; the remaining state, Iran, has had close historical and cultural relations with the Arab countries and shares their Islamic religion, but most of its people are not

---

[12] A barrel of oil is equivalent to 42 United States gallons.

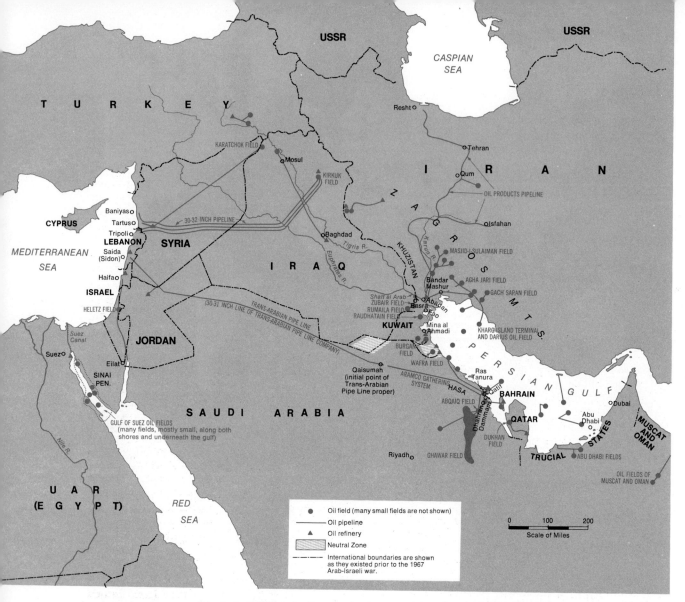

Principal oil fields, pipelines, and refineries in the heart of the Middle East.

Arabs. The oil deposits of the Persian Gulf states are by far the largest in the world. In 1967 their proved reserves were reported to be approximately 249 billion barrels, or about 70 percent of the estimated world total of 355 billion barrels. Most of the deposits lie along, under, or relatively near the Persian Gulf. This shallow, desert-rimmed arm of the Indian Ocean, three times the size of Lake Superior, ranks with the Gulf of Mexico and Caribbean Sea among the world's great centers of oil production and transportation. In 1947 the production of oil per day in the Persian Gulf states averaged 839,000 barrels; by 1959 this had risen to an estimated 4.6 million barrels, and by 1967 to an estimated 10.3 million barrels.

Western Europe is the principal market for Persian Gulf oil, though some oil is shipped to the Orient (principally Japan), Australia, the Western Hemisphere, and Africa. In 1967 oil from the Persian Gulf states supplied around 54 percent of western Europe's requirements, with an additional 26 percent coming from Libya and Algeria. Most of the oil is shipped to market in crude form, though some of it is processed at refineries located along the Persian Gulf, at pipeline termini on the Mediterranean coast, or at the northern and southern entrances to the Red Sea (map, above).

Large corporations, principally British or American in ownership, dominate the oil business. Foremost among them are the British Petroleum Company (formerly the Anglo-Iranian Oil Company); the Arabian American Oil Company (Aramco), owned by four United States concerns; the Kuwait Oil Company, part British and part American in ownership; and the Iraq Petroleum Corporation, owned by British, American, French, and Dutch interests, with a 5 percent interest being held by the C. S. Gulbenkian estate. The British government is the majority stockholder in the British Petroleum Company, which holds a half interest in the Kuwait Oil Company, slightly less than a quarter interest in the Iraq Petroleum Corporation, and substantial interests in Iran, Qatar, and Abu Dhabi. Persian Gulf oil is extremely important to Great Britain, not only because it supplies the bulk of the country's annual requirements, but also because sales of oil and refined products in world markets are a much needed source of foreign exchange.

Royalty and tax payments by the companies that produce and market the oil supply the bulk of the governmental revenues of the Persian Gulf states and are making possible a variety of projects to raise the level of living in these lands. Several Arab countries that contain only modest oil deposits or none at all have derived profit from oil operations through their control of transportation routes; they include the United Arab Republic, with its Suez Canal (not operating in 1968; see p. 372), and Syria, Lebanon, and Jordan, whose territories are crossed by pipelines (map, left) from which they derive transit fees and taxes.

Oil Wealth, Rural Poverty, and National Development in Iran

Iran (Persia) was the earliest of the Middle Eastern countries to produce oil in large quantities. The initial discovery at Masjid-i-Sulaiman dates from 1908, and commercial production commenced in 1912. The country's main oil fields form a line, oriented northwest-southeast, along the foothills of the Zagros Mountains in the southwest (map, left). Two fields, the Agha Jari and Gach Saran fields, account for well over two-thirds of Iran's production of crude oil. The oil facilities were originally developed by the Anglo-Iranian Oil Company, now the British Petroleum Company. This concern built a huge refinery on tidewater at Abadan (370,000) and pipelines to the inland oil fields. Abadan lies on the Shatt al Arab (the "River of the Arabs"), formed by the junction of the Tigris and Euphrates rivers. Located about 50 miles from the Persian Gulf, it is accessible to tankers up to 20,000 tons in size. Such vessels haul refined products economically, but they are not large enough for economical long-distance transportation of crude oil. A loading terminal for tankers up to 40,000 tons was developed at Bandar Mashur (map, left), while a newer terminal at Kharg Island allows the largest tankers afloat (100,000 tons and over) to load in deep water in the Persian Gulf. Most of Iran's exports of crude oil are shipped from Kharg Island. Prior to 1951, Iran's oil pro-

duction was the greatest in the Middle East. However, in 1951, as a result of anti-British agitation and a strong tide of nationalist feeling, the oil facilities were nationalized, despite strong British protests. Most of the British managers and technicians left the country, and Iranian oil was boycotted by the major oil interests, which refused to buy it. Production soon declined to a very low figure. Not until the summer of 1954 was a compromise settlement of the dispute finally reached. Since then, production has been carried on by the National Iranian Oil Company in cooperation with a consortium of foreign companies. The British Petroleum Company holds a 40 percent interest in the consortium. Iran's total oil production in 1967 was an estimated 2.6 million barrels a day, the second highest in the Middle East (after Saudi Arabia, 2.8 million; Kuwait, the third-ranking country, produced 2.5 million).

The greater part of Iran is comprised of arid plateaus and basins, bordered by high, rugged mountains. Only an estimated 10 to 15 percent of the country is cultivated; the land actually in crops at any one time may amount to as little as 3 percent. (The reason for the latter figure is that much cropland is allowed to lie fallow at intervals in order to accumulate moisture and regain some of its fertility.) Enough rain falls during the winter half-year in the northwest to permit unirrigated cropping. Rainfall sufficient for intensive agriculture, however, is largely confined to a densely populated strip of lowland between the Elburz Mountains and the Caspian Sea. The western half of this lowland receives more than 40 inches of precipitation annually, including both summer and winter rainfall, and produces a variety of subtropical crops, such as wet rice, cotton, oranges, tobacco, silk, and tea. In the other regions of Iran, agriculture depends mainly or exclusively on irrigation. Many of the irrigated districts are located on alluvial fans at the foot of the Elburz and Zagros mountains; the capital, Tehran (Teheran; 2.8 million), is found in one of these areas south of the Elburz. Qanats (tunnels) furnish the water supply for perhaps three-fifths of the country's irrigated acreage (see photo and caption, p. 380). Approximately two-thirds of the crop acreage in Iran is devoted to either wheat or barley, while an estimated 5.5 million cattle, 33 million sheep, and 14.5 million goats comprise the bulk of the livestock. A considerable share of the livestock is raised by seminomadic mountain peoples—the Qashqai, Baktiari, Lurs, Kurds, and others—whose independent ways have long been a source of friction between these tribesmen and the central government. The area inhabited by the Kurds extends into Iraq, Turkey, Syria, and the Soviet Union. Agitation for an independent Kurdistan has been prominent in past times, but seems quiescent today.

The Iranian people, possibly two-thirds of whom are peasant farmers or seminomadic tribesmen, have long been beset by the customary Middle Eastern ills of drought, dusty winds, illiteracy, disease, poverty, and landlordism. In the villages modern conveniences are largely lacking. Farming is mostly on a subsistence or near-subsistence basis, by traditional methods with crude tools. Justice William O. Douglas, a frequent and perceptive visitor to Iran, has given an effective picture of the customary methods of threshing grain:

There is not much farm machinery in Iran. Most of the wheat is cut by hand sickles, and the wheat is tied in sheaves or bundles and stacked in the field. These are loaded on donkeys and taken to a central thrashing floor. The floor is usually a flat place in the field with no covering. The sheaves are untied here and the wheat stalks scattered on the floor. Some farmers hitch cattle or donkeys, or cattle and donkeys together shoulder to shoulder and drive them around and around the floor, depending on the beat of the hoofs to dislodge the wheat from the head. More often, they use a heavy sled to help thrash the wheat.

These shafts in Saudi Arabia mark the course of a tunnel used to carry irrigation and drinking water from an underground source by gravity flow. Such devices, called "qanats" (kanats), "karez," or "foggaras," are very common in some Middle Eastern countries, notably Iran, Afghanistan, Algeria, and Morocco. "The essential idea is that of a gently sloping tunnel, often along the radius of an alluvial fan, which extends upslope until the water table is tapped and emerges at the downslope end to supply an oasis. To give access to the tunnel, vertical shafts are dug at closely spaced intervals. The length of a qanat ranges from a few hundred yards to tens of miles, and the upper end may be several hundred feet below the surface. . . . By this means thousands of acres are irrigated and hundreds of villages receive their sole water supply." (George B. Cressey, "Qanats, Karez, and Foggaras," *Geographical Review*, **48,** 1958, 27. Used by permission of the author and the American Geographical Society of New York. Students are urged to read the entire article, pp. 27–44.) (Photo: Standard Oil Company, N.J.)

Frequently the animals draw a crude drag that has one or more large drumlike wheels that pound the heads of the wheat. Women and children usually drive the animals. Round and round they go for days until the wheat is separated. Then men use a fork (usually a wooden one) to toss the residue into the air. The wind catches the chaff and blows it to one side. Sometimes a large round wooden traylike vessel is used to pick up the residue. A man shakes it until the kernels are at the bottom. Then, tipping the tray, he blows the chaff away. After many backbreaking days, the golden kernels lie in piles and are transferred by hand into burlap sacks.[13]

Iran has exhibited the common Middle Eastern picture of an illiterate peasantry, living under conditions of grinding poverty, in sharp contrast to a relatively small well-to-do upper class which has owned most of the land, held most of the money, and exercised considerable control over political affairs. The Shah of Iran and his government, aided by financial grants and loans from the United States (this aid was terminated by mutual agreement in 1967) and by increasingly large revenues from oil, have inaugurated programs designed to improve the condition of rural people and to set Iran on the road to modernization. One conspicuous achievement has been a sweeping program of land reform. Large amounts of land owned by the Iranian crown or government, or acquired by compulsory purchase from large private landowners, have been turned over to peasant families under long-term mortgages at low rates of interest. It is claimed that some 14 million rural dwellers have benefited from this program. Thus the stranglehold of the landowning upper class on the peasantry has been at least partially broken, and peasants who now own their land are able to use more of the crops they raise instead of turning over to the landlord anywhere from a fifth to four-fifths of the crop as rent.[14] Other facets of the drive to modernize include a widespread attack on illiteracy, the emancipation of women, public health measures, the development of new industries, the building of highways, some extension of the rail system, the introduction of better agricultural methods, provision of more chemical fertilizer and farm machinery, and, not least, the building of storage dams and hydropower stations to increase the output of electricity and permit more land to be irrigated. The most important river-control scheme involves the Karun River and smaller rivers in oil-rich Khuzistan Province adjoining Iraq (map, p. 378). Sugar cane is to become a major irrigated crop on the alluvial plain of the Karun. Iran hopes by this means to lessen its dependence on imported sugar (the country's largest agricultural import) and thus to conserve foreign exchange. Efforts to modernize Iran and correct the country's social and economic ills have been hindered in the early years by insufficient capital, the inertia and suspicion of the conservative Moslem peasantry, and the active or passive opposition of intrenched interests, including many of the larger landowners. But the movement, led by the energetic and forward-looking Shah, has gained momentum and will be aided not only by revenues from oil, but by the development of large natural gas reserves in the oil fields. An agreement has been made with the Soviet Union to export natural gas to that country, and the necessary pipeline facilities are expected to be completed in the early 1970s.

Iran's small but growing industrial establishment is dominated by the petroleum, food-processing, and textile industries. A beginning has been made in various other types of modern factory industries, such as petrochemicals and aluminum in Khuzistan, an iron and steel plant near Isfahan, built with Soviet aid, and some engineering plants to produce automobiles, tractors and household appliances from Iranian or foreign-made components. The country is still heavily dependent on imports of manufactured goods, however, and is likely to remain so for many years. Some traditional handicrafts survive, including the manufacture of the famous Persian rugs. Modern factories are confined almost exclusively to Tehran, Abadan, Tabriz (400,000), Isfahan (350,000), Meshed (320,000), Ahwaz (160,000), and a few other cities. The national government has stimulated, closely controlled, and partially financed the development of modern industries. Mineral resources in Iran other than oil and natural gas are not outstanding, but do include some coal and a miscellany of metal-bearing ores, including iron ore. The new iron and steel plant near Isfahan will operate with Iranian coal and ore. The country has considerable hydroelectric reserves and a fair amount of timber, both localized rather heavily in the rainier west. Iran's transportation system is inadequate for a modern state, but is gradually improving. Rail lines connect Tehran with the Shatt al Arab and Persian Gulf, with the Caspian Sea, with Tabriz and the Soviet border in the northwest, and with Meshed

---

[13] William O. Douglas, *West of the Indus* (Garden City, N.Y.: Doubleday & Company, 1958), p. 295. Copyright © 1958 by William O. Douglas. Copyright © 1958 by the National Geographic Society. Reprinted by permission of William Morris Agency, Inc., on behalf of William O. Douglas.

[14] Under sharing arrangements, it has been customary for the landlord to receive 20 percent of the crop as land rental, another 20 percent if he supplies irrigation water, a third 20 percent if he furnishes draft animals, and a fourth 20 percent if he furnishes seed.

in the northeast (map, p. 386). A line partially completed across central Iran eventually will connect with the rail system of Pakistan, and a link is under construction in the northwest to tie the Iranian system to that of Turkey. A loose web of highways covers the country; somewhat more than a fourth of it is asphalt surfaced and about half of the remainder is surfaced with gravel.

Iraq

Prior to 1958, all of the Persian Gulf states were monarchies. But in July of that year a short revolution precipitated by army officers made Iraq a republic. The revolution stemmed partly from resentment against Great Britain, though other factors were involved. Britain had governed Iraq for a decade under a League of Nations mandate when the country was brought into existence following World War I (previously it had been a part of the Ottoman Empire), and British influence continued to be strong after independence was granted in 1932. In 1955 the government of the monarchy had aligned itself militarily with Great Britain, Turkey, Iran, and Pakistan in the Baghdad Pact.[15] But despite their anti-British leanings, the revolutionists did not interfere with the operations of the Iraq Petroleum Corporation, which is partly British in ownership. Revenues from oil are the main source of income for the government, which sets aside an important share for a Development Board that is engaged in a long-term program of public works to raise the country's low standard of living.

Iraq's principal oil field, first opened in 1927, is located in the vicinity of Kirkuk (230,000) in northern Iraq (map, p. 378). The Kirkuk field, approximately 60 miles long by 2 miles wide, is one of the largest oil-bearing formations in the world. It is connected by pipelines with shipping points on the Mediterranean coast at Tripoli, Lebanon, and Baniyas, Syria. The pipelines also carry oil from wells near Mosul (390,000). The newer Zubair and Rumaila fields, located in southern Iraq near the country's main seaport, Basra (430,000), are connected by pipeline with export loading facilities built on an artificial island in the Persian Gulf off the old port of Fao.

Iraq, formerly known as Mesopotamia, occupies a broad plain drained by the Tigris and Euphrates rivers, together with fringing highlands (in the north) and deserts (in the west). In ancient times the country had an elaborate system of irrigation works on the Tigris-Euphrates plain, and its population at certain periods is thought to have been four or five times as great as it is today. But in the Middle Ages much of the irrigation system was destroyed or fell into disrepair during periods of political disorder and foreign invasion, and it has never been reconstructed. Today Iraq is underpopulated in relation to the amount of land that is irrigable. The Iraq Development Board has been attempting to expand irrigation, partly through the construction of multipurpose dams (for flood control and hydroelectric development as well as irrigation storage) on the Euphrates and on the Tigris and its tributaries. The board is also attempting to reclaim land by draining portions of large swamps in southern Iraq and by flushing salt from fields that were formerly irrigated but had to be abandoned because of salification due to improper drainage.

Well over half of Iraq's people are villagers depending directly on agriculture for a living. Small groups of Bedouin raise camels, sheep, and goats in the western deserts. Some of the Kurdish tribesmen in the mountain foothills of the north are seminomadic livestock graziers, though most Iraqi Kurds are sedentary agriculturalists. The Kurds are the country's largest minority. They have revolted against the central authority of Iraq on numerous occasions.

Agriculture is dependent on irrigation in most parts of Iraq, though the northern highlands receive 10 to 20 inches or more of precipitation in the winter half-year and thus are able to support nonirrigated fall-sown crops. Three grains—wheat, barley, and rice—occupy most of Iraq's cultivated acreage. Dates are the main export crop, though they contribute only about 2 percent of all exports (crude petroleum accounted for 94 percent in 1965). Iraq normally supplies more than three-fourths of the dates that enter international trade. Most of them are produced in irrigated gardens that stretch for a hundred miles along the Shatt al Arab in lower Iraq. Irrigation is facilitated by the changing water level of the Shatt al Arab, which fluctuates with the tides in the Persian Gulf. At high tide, water from the river is led through gates to the plantations. The water is brackish but can be utilized by the date palms, which are tolerant of salt.

The limited manufacturing of Iraq, confined largely to agricultural processing, textiles, and other relatively simple types of industry, is centered in the historic capital city of Baghdad (1.1 million) on the Tigris River and in Mosul.

---

[15] Iraq withdrew from the pact in 1959. The remaining members then changed the name of their alliance to Central Treaty Organization (CENTO).

This pipeline across the desert Trans-Arabian Pipe Line) carries oil for more than a thousand miles from fields in Saudi Arabia to a shipping point on the Mediterranean Sea in Lebanon. (Trans-Arabian Pipe Line Co.)

The Desert Kingdom of Saudi Arabia

Oil production in Saudi Arabia has been developed by the Arabian American Oil Company (Aramco). The company is officially a Saudi corporation but is owned by four United States companies. The first oil discovery in commercial quantities was made in the vicinity of Dammam in 1938, but large-scale production was delayed until after World War II. The oil fields, the largest of which are the 150-mile-long Ghawar field and the Abqaiq field, are located in the province of Hasa in the eastern part of the country (map, p. 378). The company has its headquarters and homes for its employees and their families at the town of Dhahran. Though some oil is refined at Ras Tanura or on nearby Bahrain Island, most of the production leaves Saudi Arabia in crude form via the Trans-Arabian Pipeline (photo, above) to Saida (Sidon), Lebanon, or via tanker from Ras Tanura.

Saudi Arabia occupies the greater part of the Arabian Peninsula, the original homeland of Arab civilization

and the Moslem religion. The Saudi state, a feudalistic monarchy which now is undergoing changes to modernize its administrative structure, was gradually consolidated by the late King Ibn Saud between 1902 and 1926. Prior to the discovery of oil, the country had felt the impact of modern civilization very little. Aside from oil, its resources are meager. The interior is an arid plateau, fronted on the west by mountains that rise steeply to heights of 3000 to 9000 feet from a narrow coastal plain bordering the Red Sea. The mountains are highest at the south, where they reach more than 12,000 feet in the neighboring state of Yemen. Most of the plateau lies at an elevation of 2000 to 4000 feet, with the lowest elevations near the Persian Gulf and in the eastern part of the desolate, sandy, southern desert called the Empty Quarter (Rub al Khali). Mountains rise locally from the plateau, and its surface is trenched by many *wadis* (stream valleys) that carry water whenever it happens to rain. Most interior sections average less than 4 inches of precipitation annually; often a year or more will pass with no precipitation at all. Such precipitation as there is comes generally in winter, and is heaviest along the mountainous western margin of the peninsula.

Although the best-known inhabitants of Saudi Arabia are the nomadic Bedouin tribesmen with their herds of camels, sheep, and goats, many of the people are farmers in oases. Recent times have seen a considerable movement of rural people seeking jobs in oil fields and cities. The country's principal crops are grain sorghums, dates, wheat, and barley. The total production is insufficient to feed the population, and sizable food imports, including such items as wheat, flour, rice, tea, coffee, and sugar, are required. Nearly all the manufactured goods used in Saudi Arabia are imported. The country has little to export besides petroleum, and most of the government's revenues come from that source.

Much of Saudi Arabia's income from oil has been spent on ostentatious living by the ruling class, but the present regime under King Faisal is carrying on many programs for economic and social betterment. For example, the country's educational system is rapidly expanding, deep wells are being drilled to tap sizable underground stores of water, and a network of asphalt-surfaced highways is well along. Until lately an almost medieval land, Saudi Arabia is commencing to move with some speed into the modern world.

The best-known city is Mecca (170,000), the birthplace of the Prophet Mohammed and the principal holy city of the Islamic world. Jidda (160,000), on the Red Sea some 40 miles from Mecca, is the main west-coast seaport and point of entry for pilgrims to Mecca, Medina, and other Arabian holy places. The first transpeninsular asphalt-surfaced highway, completed in 1967, links Jidda and Mecca with the capital city of Riyadh (180,000) and the port of Dammam on the Persian Gulf. A rail line completed in 1951 connects Riyadh with Dammam.

Lesser States of the Arabian Peninsula

Saudi Arabia is rimmed to the east and south by a series of smaller Arab countries that lie along the Persian Gulf, Indian Ocean, or Red Sea. They include (1) several small sheikdoms on the Persian Gulf, the most notable of which is Kuwait; (2) Muscat and Oman; (3) Southern Yemen; and (4) Yemen. Except for Yemen, all of these units have been under British influence and protection or actual British control in past times, though Britain today (late 1960s) is phasing out its political and military commitments in the area.

The Persian Gulf Sheikdoms

A number of small units bordering the Persian Gulf—Kuwait, Bahrain, Qatar, and the Trucial States (map, p. 378)—have had internal autonomy, but have had special treaty relations with Great Britain under which the latter has acted for them in foreign affairs and has guaranteed their military security. This situation has also been true of Muscat and Oman, though the latter state has had more latitude in foreign relations than the others. The Sheikdom of Kuwait achieved full independence in 1961, and Britain later announced its intention of withdrawing its remaining military forces from the other units by 1971.

The arid sheikdoms along the Gulf contain notable resources of oil, some of which is in offshore deposits. KUWAIT, indeed, has proved reserves larger than those of any country on the earth except Saudi Arabia. In 1967 Kuwait's proved reserves were estimated at 76.7 billion barrels, as compared with 83.7 billion for Saudi Arabia, 43.8 for Iran, 23.5 for Iraq, 15.0 for Abu Dhabi (one of the Trucial States), 3.7 for Qatar, and 2.5 for Muscat and Oman.[16] Kuwait was the Middle East's leading producer for several years, but in 1966 it was surpassed slightly by Saudi Arabia and in 1967 by Saudi Arabia and Iran.

Production in Kuwait centers in the Burgan field south of Kuwait city, though other fields exist, including the large Raudhatain field in northern Kuwait (map, p. 378). There is no pipeline connection to the Mediterranean coast, and the oil moves out by tanker from immense loading docks in the Persian Gulf at Mina al Ahmadi. Some oil is processed in a refinery at the latter point. The character and situation of the Burgan field greatly facilitate large-scale operations. It is possibly the largest single deposit of oil in the world. Virtually every well drilled produces oil, generally in quantities of 4000 to 5000 barrels or more per day. The enormous output of each well, many times as great as the flow of the average well in the United States, is due to the size of the underground pool of oil and the great pressure under which it lies. Such conditions are often found in the southwest Asian and North African fields, which produce over a third of the world's petroleum from a remarkably small number of wells. In the Burgan field the oil lies at comparatively shallow depths (within 5000 feet of the surface), which adds to the ease of extraction. It is pumped to storage tanks on a low ridge at the oil company's headquarters and residential town of Ahmadi. From here it flows by gravity through pipelines to the loading point a few miles away. Associated with the oil are abundant reserves of natural gas, some of which is used as fuel for thermal-electric power production.

Royalties and taxes paid by the Kuwait Oil Company and smaller companies have changed Kuwait from an impoverished country of boatbuilders, sailors, small traders, and pearl fishermen to a prosperous welfare state supported almost entirely by the oil industry. Kuwait city (150,000) boasts an impressive array of public schools, free hospitals and other medical facilities, a modern sewage and water system, asphalt-surfaced streets, public housing units, and electric-generation facilities. Not the least of the benefits from oil are large plants which distill salt water from the Persian Gulf and provide the country's principal water supply.

Oil reserves and production in Bahrain, Qatar, and Abu Dhabi are smaller than in Kuwait, but oil revenues nonetheless have conferred great benefits on these tiny sheikdoms. The BAHRAIN sheikdom occupies a small archipelago in the Persian Gulf between the Qatar peninsula and Saudi Arabia. The Saudi Arabian oil towns of Dhahran and Ras Tanura are about 30 miles away. In addition to a very modest production of oil, Bahrain

---

[16] Most statistical data on Middle Eastern oil in this chapter are from the *Oil and Gas Journal,* as cited in *Aramco Handbook: Oil and the Middle East* (rev. ed.; Dhahran, Saudi Arabia: Arabian American Oil Company, 1968), or the 1968 *Encyclopedia Britannica Book of the Year.*

is a center of oil refining, seaborne trade, and air transportation. Springs on Bahrain Island, the largest island of the group, nourish productive date groves. The capital, Manama (city proper, 80,000), also on Bahrain Island, is a city with modern facilities financed by oil revenues. QATAR and the TRUCIAL STATES are extremely arid, and prior to the oil era supported only a meager subsistence economy, but Qatar (population about 75,000) and the adjoining Trucial sheikdom of Abu Dhabi (18,000) now have oil revenues that are extraordinarily large in relation to the size of their populations. Oil also has been found in the Trucial sheikdom of Dubai (80,000), adjoining Abu Dhabi on the east.

The small NEUTRAL ZONE between Saudi Arabia and Kuwait, used for seasonal grazing by tribesmen of both countries, but not politically a part of either, is an oil producer of considerable consequence. Oil revenues from the Neutral Zone are divided about equally between the two countries that border it. Two American companies own and operate the onshore facilities, while a Japanese-owned company produces oil offshore.

The Southern Margins of Arabia

YEMEN and MUSCAT and OMAN occupy, respectively, the mountainous southwestern and southeastern corners of the Arabian Peninsula. These states receive the heaviest rainfall of the peninsula, particularly Yemen, where the annual precipitation in the highlands is estimated to be 15 to 40 inches. Elsewhere in the peninsula the scanty rainfall is confined mostly to the cool season, but Yemen has a summer rainfall maximum, and its main subsistence crops are spring-sown grains: millet, corn, and sorghums. A sparsely populated coastal zone lies between the Red Sea and the highlands. Cotton growing has been introduced on the coastal plain, but the country's most famous products and major exports, coffee and qat, a drug, are grown in the highlands in a zone between 4000 and 8000 feet. The highlands of Yemen are the most densely populated part of the Arabian Peninsula. An autocratic, old-style Moslem state, whose rulers have frequently been assassinated by rival claimants, Yemen has long been one of the world's most isolated, backward, and little-known countries. In the middle 1960s a republican government supported by Egypt took control of large sections of the country, including the capital city of San'a (100,000), though the royalist regime continued its resistance from bases in the north. Egyptian troops were withdrawn in 1967 following the six-day Arab-Israeli war. Yemen's civil war was still in progress in late 1968.

The Sultanate of Muscat and Oman, commonly known as Oman, lies along the Indian Ocean immediately west of the entrance to the Persian Gulf. The coastal region is very arid, and most of the population is found in interior highlands where mountains rising to 9900 feet induce a fair amount of rain and furnish water for irrigation. The country once controlled a sizable empire and was a major center of trading in African slaves, as attested by the Negroid character of much of its present coastal population. Most of the present inhabitants are farmers in interior oases, though a few camel-herding nomads exist and many of the coast dwellers support themselves as fishermen, sailors, or traders. Oil has been discovered in the interior, a pipeline has been laid to the coast, and exports of oil have begun.

The arid and mountainous coastlands of Arabia to the immediate south and west of Oman were formerly controlled by Great Britain. Britain held the small Crown Colony of Aden and supervised the numerous minor sheikdoms, sultanates, and other units which comprised the Aden Protectorate. In 1967 the area became independent as the People's Republic of Southern Yemen. Aden city (a general name for several municipalities; population, over 220,000) is the main urban place. Located on an excellent harbor, it is a center of oil bunkering, transshipment, and entrepôt trade. Crude oil from the Persian Gulf states and Egypt is processed at a refinery operated by the British Petroleum Company. The main product of the refinery is fuel oil for ships.

TURKEY, CYPRUS, AND AFGHANISTAN

Turkey and Afghanistan contain some oil, though the deposits do not compare in size with those of the countries bordering the Persian Gulf. These mountainous countries are Middle Eastern in many respects, but in other ways are transitional to other world regions. Like their joint neighbor, Iran, they lie outside the Arab world. But both are Moslem countries; indeed, Afghanistan is a major stronghold of traditional Islam. The island republic of Cyprus, a predominantly Christian (Greek Orthodox) country with a substantial Moslem (Turkish) minority in its population, is described briefly in this section following the discussion of Turkey.

Turkey: Past and Present

In world terms Turkey is considerably more important than Afghanistan. Once a reactionary Moslem state at the center of a disintegrating empire, Turkey

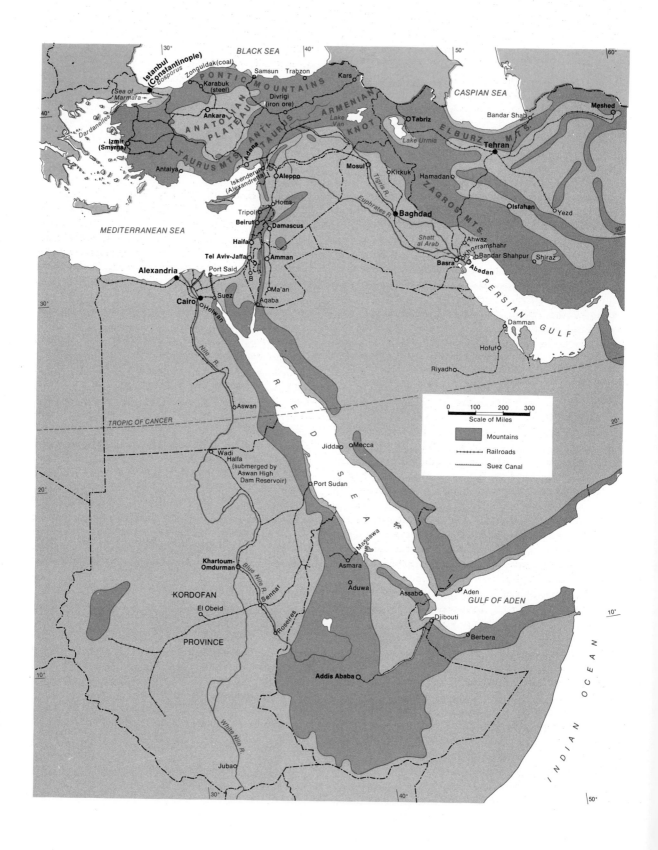

has been transformed in the twentieth century and has become a cohesive national state oriented to western Europe and America. Today it is militarily the strongest of the Middle Eastern countries. Its standard of living, though low as compared with that of western Europe, is one of the highest in the Middle East.

Ataturk and the Transformation of Turkey

The Turks who organized the Ottoman Empire in the fourteenth and subsequent centuries were originally pastoral nomads from Central Asia, where important Turkish elements still exist in the Soviet Union and China. From the sixteenth century to the nineteenth their empire was an important power. At its height in the sixteenth century its territories included North Africa as far west as Algeria and south to the Sudan; most of southeastern Europe; the Fertile Crescent and parts of the Arabian Peninsula; and, of course, the Anatolian Peninsula between the Mediterranean, Aegean, and Black seas which forms most of the national territory of the present Republic of Turkey. That republic was created from the wreckage of the old empire after World War I. Its founder, Kemal Ataturk, was determined to westernize the country, raise its plane of living, and make it a strong and respected national state. He inaugurated a series of social and political reforms that were designed to break the hold of traditional Islam and open the way for modernization and the growth of Turkish nationalism. Under the Ottoman Empire Islam had been the state religion, but Ataturk disestablished it. The Caliphate that had been held by the Ottoman Sultan was abolished. Wearing of the fez, an important symbolic act under the Ottoman caliphs, was prohibited, and the religious schools that had monopolized education were replaced by state-supported secular schools. To facilitate public education and remove further traces of Moslem dominance, the Arabic script of the Koran was supplanted by Latin characters. Polygamy was outlawed (not retroactively), along with slavery, and women were lifted from seclusion and were given full citizenship. Legal codes based on those of various Western nations replaced Moslem law, and the forms of democratic representative government were instituted, though Ataturk actually governed as a dictator.

The transformation that Ataturk envisioned could not be accomplished overnight, though his reforms made remarkable changes in Turkish life. The peasantry, mostly illiterate and devoutly Moslem, submitted reluctantly to many elements in his program, though they supported his objective of creating a strong Turkey. Since Ataturk's death the government has made certain concessions in religious matters, including the introduction of religious instruction in the public schools.

Characteristics of Turkish Agriculture

Though a little over one-third of the population is now classed as urban, Turkey is basically a country of small landowning farmers, most of whom live in villages. Turkish agriculture is often on a near-subsistence level, but very few farmers are sharecroppers without work animals of their own, and the general level of living in the rural areas is somewhat higher than in most parts of the Middle East. Cultivated land, including tree crops and vineyards, occupies perhaps one-third of the country. Grains, primarily wheat and barley, are grown in all parts of the country, and are the main crops in total acreage. Tree crops (of which olives are the most important) and vineyards are confined largely to coastal sections. Livestock, principally sheep, goats, and cattle, are an important component of agriculture in most areas. Most of Turkey's exports consist of agricultural and pastoral products, cotton and tobacco being the largest items by value. The country's farm production has greatly expanded under the republic, but yields are still low compared to those of western Europe. Expanded irrigation and greater use of commercial fertilizers are pressing needs, and some progress along these lines is being made. But irrigation is hindered by the difficulty of raising water to the level of the fields from streams that are often deeply intrenched; and the poverty of most peasants precludes large purchases of fertilizer. Animal manures that might be used to fertilize the fields are burned as fuel in many peasant households.

The Anatolian Plateau

From a physical standpoint Turkey is comprised of (1) the Anatolian Plateau and associated mountains, occupying the interior of the country (map, left), and (2) the coastal regions of hills, mountains, valleys, and small plains bordering the Black, Aegean, and Mediterranean seas. The Anatolian Plateau is a country of

---

Mountains, railroads, and selected rivers and cities in the central part of the Middle East. The areas of mountains shown have been broadly generalized and often incorporate important agricultural valleys (as in western Turkey) or agricultural uplands (as in Ethiopia).

Threshing wheat with oxen and sleds in eastern Turkey. The grains are separated from the wheat heads by the hooves of the oxen and the weight of the sleds. Several different threshing floors are visible in the photo. After the grain is threshed, it will be winnowed to remove the chaff. (*Foreign Agriculture.*)

wheat and barley fields and grazing lands. The annual precipitation, averaging 8 to 16 inches and concentrated in the winter half-year, is barely sufficient for grain. The plateau lies at an elevation of 2000 to 6000 feet; it is highest in the east where it adjoins the high mountains of the Armenian Knot. Its surface is rolling, treeless, windswept country, hot and dry in summer and cold and snowy in winter, with a natural vegetation of short steppe grasses and shrubs. The farming system on the plateau combines cereals and the raising of cattle, sheep, goats, and donkeys. There has been a substantial increase in farm mechanization since World War II, but the bulk of the farming is still done by traditional methods. Oxen are a major source of draft power.

Along its northern side, the Anatolian Plateau is bordered by the Pontic Mountains. These ranges, lying generally at 3000 to 6000 feet in the west but rising to a maximum of over 12,000 feet in the east, are clothed with Turkey's best forests. The trees, mainly hardwoods, are nourished by the heavy precipitation that these mountains receive the year round. The Taurus Mountains, bordering the plateau on the south, are in general somewhat higher than the Pontic Mountains, lying mostly at 6000 feet or higher. But their annual precipitation is smaller and their summers are dry; some good forests exist, but mediterranean woodland of little commercial value is the predominant vegetation. Turkey's forests, which are government-owned, were long subjected to overcutting and destructive grazing practices. However, under government conservation programs in recent years their productivity has increased and Turkey is able to meet most of its timber needs from domestic sources.

Aegean, Black Sea, and Mediterranean Coastlands

The coastal plains and valleys along or near the Aegean Sea, Sea of Marmara, and Black Sea are, in general, the most densely populated and productive parts of Turkey. Here are grown several of the country's principal export crops, including tobacco, hazelnuts (filberts), grown primarily in the Black Sea section, and grapes for sultana raisins, and figs, raised in the central Aegean section around Izmir (Smyrna; 430,000). Izmir, located on an excellent natural harbor, is Turkey's second-ranking seaport (after Istanbul). It exports a varied list of characteristic mediterranean products: tobacco, raisins, figs, olive oil, wheat, and others. The Black Sea coastlands differ from the rest of Turkey in having summer as well as winter rain (though even here the maximum rainfall comes in the winter half-year) and in having a much greater total precipitation, 35 to 100 inches annually in most sections, than other parts of the country. The easternmost section has the heaviest rainfall and, like the adjoining Georgian SSR in the Soviet Union, grows tea and some citrus fruit. Most of Turkey's corn crop is grown in the Black Sea coastlands. Corn is the country's third-ranking grain in total production, but the output is far less than that of wheat or barley. In the interior of Thrace (European Turkey), sugar beets have become the leading commercial crop. Turkey was formerly a large importer of sugar, but in recent decades has achieved self-sufficiency, or even small exports, through the widespread introduction of sugar beets.

The southern (Mediterranean) coast of Turkey is more sparsely populated than the Aegean or Black Sea coasts. Most of the coast is mountainous, and only a few sizable lowlands or valleys occur. The most important is the plain that centers on the city of Adana (300,000) in southeastern Turkey. Irrigated cotton, grown primarily on the Adana Plain, was Turkey's largest export in the middle 1960s.

Industry, Mining, and Transportation

Except for handicrafts there was almost no industrial development in Turkey under the Ottoman Empire. During the latter part of the Ottoman period various European powers secured an increasing hold over Turkish finances by virtue of loans made to the Ottoman sultans. When the Turkish Republic was formed in 1923, its leaders were anxious to avoid further financial involvement with foreign nations, and they imposed severe restrictions on foreign investment in Turkey. This policy handicapped Turkish industry, because Turkey itself had insufficient capital to finance industrial development. Private capital was so lacking that the state, which was anxious to stimulate industry, was forced to play a large role in financing it. Many of the present industries are state-owned, and the state exerts a close control over many private industries.

Turkey's predominant industries are those customarily found in underdeveloped nations: textiles, agricultural processing, cement manufacture, simple metal industries, assembly of vehicles from imported components, and so on. The country is very dependent on imports for much of its machinery as well as many other types of manufactured goods. An integrated iron and steel mill at Karabuk, located in the northern interior about 100 miles from Ankara (map, p. 386), is the principal heavy industrial establishment. Coke is manufactured at Karabuk from coal supplied by the Zonguldak coal basin (map, p. 386). The coal is brought by rail a distance of about 50 miles through mountainous terrain. High-grade iron ore comes from Divrigi, about 500 miles away. Turkey's steel mill at Karabuk, the largest in the Middle East, was located with strategic rather than purely economic considerations in mind, and its operations have never been particularly efficient or economical.

In addition to having the largest coal field in the Middle East (the Zonguldak field) and substantial deposits of iron ore, Turkey has important deposits of chrome and is a fairly important exporter of that ferroalloy. Other mineral production includes salt and sulfur for chemical manufacturing, copper, and a variety of other metallic and nonmetallic minerals produced in small quantities. Production of oil from small deposits in southeastern Turkey has been increasing. Turkey is not outstanding as a mineral producer in world terms. But it is one of the larger Middle Eastern producers of minerals other than oil, and its resources offer scope for much future expansion.

Turkey's transportation system is more adequate than that of most Middle Eastern countries, though it falls far short of western European standards. A coarse mesh of rail lines extends into all major sections of the country (map, p. 386), and a network of all-weather automobile roads has developed to the point that it is considered satisfactory for present needs. Financial and technical assistance by the United States has been a major factor in the improvement of the road system.

Istanbul and the Straits

Istanbul (1.8 million), formerly Constantinople or Byzantium, is Turkey's main metropolis, industrial center, and port. One of the world's most historic and cosmopolitan cities, Istanbul was for many centuries the capital of the Eastern Roman (Byzantine) Empire. It became the capital of the Ottoman Empire when it fell to the Turks, after withstanding many sieges, in 1453. However, the capital of the Turkish Republic was established in 1923 at the more centrally located and more purely Turkish city of Ankara (910,000) on the Anatolian Plateau.

Istanbul is located at the southern entrance to the Bosporus, the northernmost of the three water passages (Dardanelles, Sea of Marmara, Bosporus) that connect the Mediterranean and Black seas and are known as the Turkish Straits (inset map, p. 340). The Straits have long been a focus of contention between Turkey and its powerful northern neighbor, Russia. Several of the main Russian seaports are located on the Black Sea or Sea of Azov, and Russia has often demanded a voice in the control of the straits outlet to the Mediterranean. But her chances of achieving this goal by any means except force have been poor since World War II, as Turkey has gravitated toward the side of the Western nations and has become one of the strongest and most active members of the North Atlantic Treaty Organization. Turkey's position vis-à-vis the Soviet Union has been bolstered by extensive economic and military aid from the United States.

The Problem of Cyprus

Most of the islands that border the Aegean and Mediterranean coasts of Turkey are predominantly Greek in population and are controlled by Greece. But the large Mediterranean island of Cyprus, located about 40

Afghanistan in its regional setting.

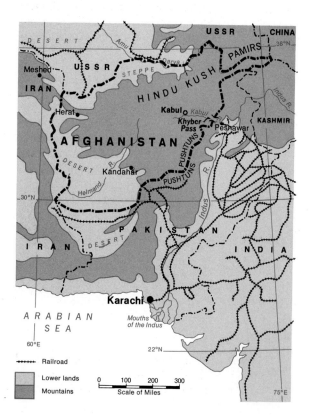

Railroad

Lower lands

Mountains

0   100   200   300

Scale of Miles

miles off the coast of southwestern Turkey, came under British control in 1878 after centuries of Ottoman rule. About 77 percent of the island's population of somewhat over 600,000 is Greek, and about 18 percent is Turkish. Agitation by the Greek majority for union with Greece (*enosis*) was prominent after World War II and led in 1954 to serious rioting and disorder. Both Great Britain and Turkey opposed *enosis* on the ground that the island's Turkish minority could not be safely entrusted to political control by Greece. In addition, Great Britain was anxious to retain control of military bases in Cyprus. An agreement to make Cyprus an independent republic was signed in 1959 by the United Kingdom, Greece, and Turkey. Approved by the leaders of both the Greek and the Turkish Cypriotes after lengthy negotiations, it guaranteed the United Kingdom's right to the continued use and control of its bases on the island. The Republic of Cyprus was proclaimed on August 16, 1960. It has remained within the Commonwealth of Nations. Renewed communal violence in Cyprus led to the establishment of a United Nations peacekeeping force in the island in 1964 and threatened to cause war between Greece and Turkey in 1966. Great Britain's military presence in Cyprus is gradually diminishing, in line with Britain's policy of phasing out most of its military commitments outside of western Europe. Nicosia (60,000) is the main city and political capital of Cyprus.

The Highland Kingdom of Afghanistan

In the twentieth century the kingdom of Afghanistan has been remote from the main currents of world affairs. A landlocked and mountainous country with limited resources, poor internal transportation, and relatively little foreign trade, it is today very underdeveloped and little known. But the territory it occupies played a strategic role in empire building of the past by virtue of important routeways and passes leading across it from the steppes and oases of Central Asia and the plateaus of Iran to the plains of northern India that have been a goal of Asian conquerors for thousands of years.

The present Afghan state arose in the eighteenth century. Throughout much of the nineteenth century its independence was jeopardized by its position between the expanding British and Russian empires. But in 1907 Great Britain and Russia agreed to maintain it as an independent buffer state between their respective domains. British influence continued to be strong, however, until after World War I.

Even today no rail line across Afghanistan connects Soviet Russia with the Indian subcontinent, though the Soviet and Pakistani rail systems extend to its borders (map, left). The limited overseas commerce of the country has customarily been routed through the port of Karachi, now in West Pakistan. But strained relations between Afghanistan and Pakistan caused the latter nation to halt Afghan traffic for several months in 1955. Recently Afghanistan has gravitated toward the economic orbit of the Soviet Union. The Soviet government has extended loans, gifts, and technical assistance to help the country build up its economy, road system, and social services, and it has made energetic and successful efforts to attract Afghan trade. The United States also has extended a sizable amount of economic and technical aid, particularly for road improvement and for irrigation and power development in the Helmand Valley.

Afghanistan's Population and Economy

No nationwide census has ever been taken in Afghanistan. Estimates that may be seriously in error place the population at about 16 million (1968). All but a small proportion of the people are illiterate. Most of the population is found in irrigated valleys around the fringes of a mass of high mountains occupying a large part of the country. The most heavily populated section is the southeast, particularly the fertile valley of the Kabul River in which is located the capital and largest city, Kabul (460,000; elevation, 6200 feet). Most of the inhabitants of the southeast are Pushtuns, also known as Pashtuns, Pakhtuns, or Pathans. The Pushtuns are the largest and most influential of the numerous ethnic groups that comprise the Afghan state. Sometimes they are referred to as Afghans, although this name is being applied more and more to all inhabitants of Afghanistan regardless of ethnic affiliation. Their language, Pushtu, is related to Persian. It is the principal language of Afghanistan in number of speakers, though Persian is a strong second and is the main language of administration and commerce.

Afghanistan's second most populous area occupies a belt of foothills and steppes on the northern side of the central mountains. Most of its inhabitants are found in a series of oases that form an east-west chain along the base of the mountains. But the area's most valuable commercial product, karakul lambskins, is provided by the millions of sheep that graze on the adjoining steppes. The skins, taken from lambs soon after birth (or even before birth in the case of the most expensive skins), are a major export from Afghanistan. They are marketed primarily in the United States or in western Europe. Northern Afghanistan faces the Soviet Union, and millions of people on the Afghan side are related to

peoples of Soviet Central Asia. Tadzhiks are the most numerous, followed by Uzbeks, Turkmen, and Kirgiz.

Southwestern Afghanistan is comprised of a large desert basin, part of which extends into Iran. An area of interior drainage, it receives the waters of the Helmand River (map, p. 390) and several others originating in Afghanistan's central mountain region. Extensive irrigation works supported a large population in this region in ancient times, but the water system was largely destroyed by Mongolian invaders in the thirteenth and fourteenth centuries and the damage has never been repaired. Today the government of Afghanistan, through its Helmand Valley irrigation and power project, is reclaiming and resettling large acreages of land by providing irrigation water from storage and flood-control reservoirs on the Helmand and other rivers. Most of the present inhabitants of southwestern and western Afghanistan are found in oases that lie in the zone of contact between the central mountains and the desert. The largest oases surround Afghanistan's second and third cities, Kandahar (150,000) and Herat (100,000). Persians form the largest population element in this part of the country.

Afghanistan is overwhelmingly a rural and agricultural or pastoral country. There is only a handful of modern factories, though craft and household industries are fairly numerous. The country is so mountainous and arid that only an estimated 12 percent is cultivated. Enough rain falls in the main populated areas during the winter half-year to permit nonirrigated growing of winter grains. Fall-sown wheat is the country's staple food grain. In most parts of Afghanistan irrigation is necessary for summer cropping. Livestock raising on a seminomadic or nomadic basis is a widespread and important part of the country's economy. Most livestock graziers raise some crops. Afghanistan's agriculture bears many of the customary Middle Eastern earmarks: excessive reliance on traditional methods, crude tools, inadequate fertilization of the soil, poor varieties of crops and animals, low yields. Large acreages are held by a comparatively small number of big landowners. However, the country's total food supply is probably as adequate as that of most Middle Eastern areas. Afghanistan exhibits a wide range of temperature conditions, corresponding in general to differences in altitude. A variety of cultivated crops, fruits, and nuts are locally important. In 1965–1966 the main items in the country's small export trade were fruits and nuts, karakul skins, and cotton.

The Pushtunistan Issue

Present-day Afghanistan is an amalgam of many ethnic and tribal groups. Traditional modes of thought and behavior based on family, tribal, or religious custom and allegiance are important influences in Afghan life. The conservative Moslem clerics (*mullahs*) are more powerful than in most Middle Eastern countries. But a leaven of modernity is stirring, and a united, detribalized Afghan nation is emerging. The most important region that is still essentially tribal lies along the eastern frontier with Pakistan. A variety of tribes and clans, mostly Pushtuns, inhabit this mountainous borderland on both the Afghanistan and Pakistan sides. Many migrate seasonally across the international boundary. These peoples, among the most intractable and militant in the world, have been slow to recognize the authority of the central governments under whose jurisdiction they theoretically reside. The northwestern frontier region was a major trouble spot in the old Indian Empire of Great Britain, and Peshawar, at the foot of the Khyber Pass on the Indian side, became a famous British garrison town. More recently, this region has been an important source of friction between Afghanistan and Pakistan. Partially the friction has been due to border incidents, but mainly it has grown out of proposals that the Pushtun-inhabited areas of Pakistan be incorporated in a separate state ("Pushtunistan"), either independent or affiliated with Afghanistan. Such proposals, sponsored or at least encouraged by Afghanistan, have been vigorously and firmly opposed by Pakistan.

ETHIOPIA AND SOMALILAND

East of the Sudan section of the Nile Basin a great volcanic mountain mass rises steeply from the desert. Much of this mountainous area, the heartland of the empire of Ethiopia, lies above 10,000 feet and one peak reaches 15,158 feet. The highland receives heavy rains during the summer half of the year rather than the winter rain characteristic of mediterranean climatic areas. Important Nile tributaries such as the Blue Nile, the Sobat, and the Atbara rise here (map, p. 368). Temperature conditions vary from tropical to temperate as elevation increases. Thus crops varying from bananas, coffee, and dates through oranges, figs, and temperate fruits to cereals can be produced without irrigation. Large expanses of upland pasture a variety of livestock, primarily cattle and sheep.

Ethiopia, together with Eritrea, is inhabited by perhaps 24 million people of very diverse racial and cultural characteristics. The country has not had a census, and estimates of total population vary widely. Although dark-skinned, most of the population is descended from the Hamitic or Semitic branches of the Caucasian race.

Possibly 55 to 65 percent, including the politically dominant Amhara peoples, adheres to the Coptic Christian faith, a very ancient branch of Christianity which penetrated Ethiopia from Egypt, where there is still a sizable Coptic minority. The rest of the people are divided between the Moslem faith and a variety of pagan beliefs. This diverse population is held in relatively loose political union under the Emperor of Ethiopia, whose seat is in Addis Ababa (570,000), the only sizable city in the country. Tribal forms of life remain important, and disputes are still often settled on a private basis with the rifle. Probably 95 percent of the empire's population is illiterate. Energetic efforts are being made to expand the educational system, but widespread results will not be apparent for a long time.

East of the mountain mass of Ethiopia, and partly included within the empire, lower plateaus and coastal plains descend to the Red Sea, the Gulf of Aden, and the Indian Ocean. Extreme heat and aridity assert themselves at these lower levels, and nomadic or seminomadic Moslem tribesmen make a poor living from camels, goats, and sheep. Precarious forms of irrigated agriculture are carried on in scattered oases.

Ethiopia and its borderlands constitute a rather marginal part of the Middle East, especially the rainy tropical Ethiopian highlands. However, the arid lowland sections possess many typically Middle Eastern characteristics, and the entire area has had important cultural and historical links with the core of the Middle East in Egypt, the Fertile Crescent, and Arabia.

European Imperialism

European powers seized coastal strips of this territory in the latter nineteenth century. Britain was first with British Somaliland in 1882, then France annexed French Somaliland in 1884, and lastly Italy asserted dominance over Italian Somaliland and Eritrea in 1889. The strategic importance of these holdings along the Suez–Red Sea route and adjacent to the independent native state of Ethiopia is obvious. Their economic importance is negligible.

Italy attempted to extend its domain from Eritrea over the much more attractive and potentially valuable land of Ethiopia in 1896, but the Italian forces were annihilated by the Ethiopian tribesmen at Aduwa. Forty years later, in 1936, a second attempt was successful. Hopes of developing the country's potential wealth and of using it as an outlet for surplus Italian population were frustrated in World War II, when the Italian empire in eastern Africa was conquered by British forces in 1941. Ethiopia was restored to independence and Eritrea was federated with it in 1952. Italian Somaliland was returned to Italian control to be administered as the Trust Territory of Somalia under the United Nations, pending independence in 1960. The Somali Republic was proclaimed on July 1, 1960. It includes not only the former Italian territory, but former British Somaliland as well. The British territory was granted independence at the same time and chose to unite with Somalia. However, the people of French Somaliland, in a 1958 referendum, indicated their desire to remain in a dependent relationship to France for the present. In 1968 the name of French Somaliland was changed to "French Territory of Afars and Issas." Officially an "overseas territory," it is France's only remaining dependency on the mainland of Africa. Influential elements among the Somali peoples advocate the formation of a "Greater Somalia" which would include the Somaliland units plus substantial parts of Ethiopia and possibly some of Kenya. Claims to Ethiopian areas rest primarily on seasonal use of these lands by Somali graziers who migrate with their livestock across the international boundary. Needless to say, the Ethiopian government takes a poor view of these proposals to annex portions of its territory, and it has advanced counterclaims to sizable parts of the Somali Republic.

Problems of Ethiopian Development

Ethiopia has substantial resources, possibly including a fair amount of mineral wealth. But its resource base remains largely undeveloped, though the emperor's government, with foreign assistance, is doing what it can. A start has been made in developing factory industries, and a few power dams have been built to harness some of the country's large hydroelectric potential. A primitive transportation system—whose best parts tend to be the 3000 miles of good road built by the Italians during their occupation—is a major factor in keeping the country isolated and undeveloped. The French-built railroad from the port of Djibouti (city proper, 40,000) in the French Territory of Afars and Issas to Addis Ababa (map, p. 386) lacks feeder lines and has never proved as successful as hoped, although a substantial part of the coffee which comprises Ethiopia's major export does move through the port. An estimated 50 to 60 percent of the empire's foreign trade now moves through the small Eritrean ports of Massawa and Assab. Both have highway connections with the interior, and a rail line extends inland from Massawa to the northern highlands via the main city of Eritrea, Asmara (140,000) (map, p. 386). Most routes into the Ethiopian heartland are mere trails, though better road connections are gradually developing, and scheduled air services link Addis Ababa to the outside world. An internal air net helps somewhat in overcoming the deficiencies in ground transportation. But all in all Ethiopia is badly handicapped by inaccessibility. Under such circumstances one of the Middle East's and the world's areas of potential wealth remains isolated and underdeveloped, though it should perhaps be added that the isolation probably had something to do with its remaining in independence during most of the colonial age.

REFERENCES and READINGS

Arab World and Islam: General

BAER, GABRIEL, *Population and Society in the Arab East,* trans. from the Hebrew by Hanna Szöke (New York: Frederick A. Praeger, 1964).

CRESSEY, GEORGE B., *Crossroads: Land and Life in Southwest Asia* (Philadelphia: J. B. Lippincott Co., 1960). A standard geography. Extensive bibliographies.

CRIST, RAYMOND E., "Land for the Fellahin: Land Tenure and Land Use in the Near East," *American Journal of Economics and Sociology,* one installment in each issue, 17, no. 1 (October 1957), through 20, no. 2 (January 1961).

HOLT, P. M., "The Arabs and the Expansion of Islam," Chap. 7 in John Bowle, ed., *The Concise Encyclopedia of World History* (New York: Hawthorn Books, 1958), pp. 146–174.

HOURANI, A. H., *Minorities in the Arab World* (New York: Oxford University Press, 1947).

ISSAWI, CHARLES, "The Arab World's Heavy Legacy," *Foreign Affairs,* 43, no. 3 (April 1965), 501–512.

KANOVSKY, E., "Arab Economic Unity," *Middle East Journal,* 21, no. 2 (Spring 1967), 213–235.

LANDAU, ROM, *Arab Contribution to Civilization* (San Francisco: American Academy of Asian Studies, College of the Pacific, 1958).

MORRISON, S. A., "Arab Nationalism and Islam," *Middle East Journal,* **2,** no. 2 (April 1948), 147–159.

PLANHOL, XAVIER de, *The World of Islam* (Ithaca, N.Y.: Cornell University Press, 1959).

RIAD el GHONEMY, M., "Land Reform and Economic Development in the Near East," *Land Economics,* 44, no. 1 (February 1968), 36–49.

SHOUBY, E., "The Influence of the Arabic Language on the Psychology of the Arabs," *Middle East Journal,* **5,** no. 3 (Summer 1951), 284–302.

SIMOONS, FREDERICK J., *Eat Not This Flesh: Food Avoidances in the Old World* (Madison: University of Wisconsin Press, 1961).

SMITH, C. G., "Arab Nationalism: A Study in Political Geography," *Geography,* **43,** no. 4 (November 1958), 229–242.

SMITH, REUBEN W., ed., *Islamic Civilization in the Middle East: Course Syllabus* (Chicago: University of Chicago, Committee on Near Eastern Studies, 1965).

Israel and the Jordan Valley

AMIRAN, DAVID H. K., "The Pattern of Settlement in Palestine," *Israel Exploration Journal,* 3, nos. 2, 3, 4 (1953), 65–78, 192–209, 250–260; also, "Land Use in Israel," in *Land Use in Semi-Arid Mediterranean Climates* (Arid Zone Research, No. 26; Paris: UNESCO, 1962), pp. 101–112; with Y. Ben-Arieh, "Sedentarization of Beduin in Israel," *Israel Exploration Journal,* 13, no. 3 (1963), 161–181; with A. Shahar, "The Towns of Israel: The Principles of Their Urban Geography," *Geographical Review,* 51, no. 3 (July 1961), 348–369; and, with Y. Karmon, "The Expansion of the Dead Sea Works," *Tijdschrift voor Economische en Sociale Geografie,* 55, no. 10–11 (October-November 1964), 210–223.

"The Arab-Israeli War" [of 1967], *Foreign Affairs,* vol. 46, no. 2 (January 1968), three articles: Charles W. Yost, "How It Began," pp. 304–320; Bernard Lewis, "The Consequences of Defeat," pp. 321–335; and Don Peretz, "Israel's Administration and Arab Refugees," pp. 336–346.

BALY, DENIS, *Geographical Companion to the Bible* (New York: McGraw-Hill Book Company, 1963).

BEN-DAVID, JOSEPH, ed., *Agricultural Planning and Village Community in Israel* (Arid Zone Research, No. 23; Paris: UNESCO, 1964).

BERMAN, MILDRED, "The Evolution of Beersheba as an Urban Center," *Annals of the Association of American Geographers,* 55, no. 2 (June 1965), 308–326; and "Israel's Dynamic Frontier Region—the Negev," *Journal of Geography,* 63, no. 9 (December 1964), 393–402.

BRAWER, MOSHE, "The Geographical Background of the Jordan Water Dispute," in Charles A. Fisher, ed., *Essays in Political Geography* (London: Methuen and Co., 1968), Chap. 12, pp. 225–242.

EFRAT, ELISHA, "Patterns in the Urban Development of Modern Jerusalem," *Tijdschrift voor Economische en Sociale Geografie,* 55, no. 10–11 (October–November 1964), 223–229; and "The Hinterland of the New City of Jerusalem and Its Economic Significance," *Economic Geography,* 40, no. 3 (July 1964), 254–260.

GARBELL, MAURICE A., "The Jordan Valley Plan," *Scientific American,* 212, no. 3 (March 1965), 23–31.

HALPERN, BEN, *The Idea of the Jewish State* (Cambridge, Mass.: Harvard University Press, 1961).

HALPRIN, LAWRENCE, "Israel, the Man-Made Landscape," *Landscape,* 9, no. 2 (Winter 1959–1960), 19–23.

HAUPERT, JOHN S., "Israel," *Focus,* 14, no. 7 (March 1964), 6 pp.; and "Development of Israel's Frontier Port of Elat," *Professional Geographer,* 16, no. 2 (March 1964), 13–16.

HOROWITZ, DAVID, *The Economics of Israel* (New York: Pergamon Press, 1967).

ISAAC, ERICH, "A Deteriorating Urban Core, Ideology and Economics in the Landscape of Tel Aviv," *Tijdschrift voor Economische en Sociale Geografie,* 52, no. 5 (May 1961), 113–119.

Israel Surveys Department, *Atlas of Israel* (Tel Aviv: Shemer Ltd., 1956–).

KANOVSKY, ELIYAHU, *The Economy of the Israeli Kibbutz* (Cambridge, Mass.: Distributed for the Center for Middle Eastern Studies of Harvard University by Harvard University Press, 1966).

KARMON, Y., "Ashdod: A New Mediterranean Port in Israel," *Geography*, 51, pt. 3 (July 1966), 254–258; also, "Eilath, Israel's Red Sea Port," *Tijdschrift voor Economische en Sociale Geografie*, 54, no. 5 (May 1963), 117–126; and "The Drainage of the Huleh Swamps," *Geographical Review*, 50, no. 2 (April 1960), 169–193.

LAUFER, LEOPOLD, *Israel and the Developing Countries: New Approaches to Cooperation* (New York: Twentieth Century Fund, 1967).

LOWDERMILK, WALTER C., "The Reclamation of a Man-Made Desert," *Scientific American*, 202, no. 3 (March 1960), 55–63.

ORNI, EFRAIM, and ELISHA EFRAT, *Geography of Israel* (2d ed.; New York: Daniel Davey, 1966).

PERETZ, DON, "Israel's New Arab Dilemma," *Middle East Journal*, 22, no. 1 (Winter 1968), 45–57.

RAPHAELI, NIMROD, "Israel's Water Economy," *Land Economics*, 41, no. 4 (November 1965), 361–364.

RITTER, WIGAND, "Some Geographical Aspects of Tourism and Recreation in Israel," *Tijdschrift voor Economische en Sociale Geografie*, 58, no. 4 (July–August 1967), 169–182.

ROKACH, AVSHALOM, "New Pattern of Rural Development in Israel," in John Higgs, ed., *People in the Countryside: Studies in Rural Social Development* (London: National Council of Social Service, 1966), pp. 146–159.

SMITH, C. G., "The Disputed Waters of the Jordan," Institute of British Geographers, *Transactions No. 40* (December 1966), pp. 111–128; and "Israel after the June War," *Geography*, 53, pt. 3 (July 1968), 315–319.

U.S. Bureau of International Commerce, *Basic Data on the Economy of Israel*, Overseas Business Reports, OBR 66–79 (December 1966). (Washington, D.C.: Government Printing Office, 1966.)

Jordan, Syria, and Lebanon

CHAUVEL, J. F., "Counting House of the Middle East" [Beirut], *Geographical Magazine*, 40, no. 7 (November 1967), 579–587.

HAUPERT, JOHN S., "Recent Progress of Jordan's East Ghor Canal Project," *Professional Geographer*, 18, no. 1 (January 1966), 9–13.

HINDLE, PETER, "Aqaba: An Old Port Revived," *Geographical Journal*, 132, pt. 1 (March 1966), 64–68.

HUDSON, JAMES, "Syria," *Focus*, 18, no. 8 (April 1968), 1–7, and "The Role of Irrigation," 8–11.

PERETZ, DON, "River Schemes and Their Effect on Economic Development in Jordan, Syria, and Lebanon," *Middle East Journal*, 18, no. 3 (Summer 1964), 293–305.

TALAL, H., "Growth and Stability in the Jordan Economy," *Middle East Journal*, 21, no. 1 (Winter 1967), 92–100.

U.S. Army Area Handbook for Syria (Development of the Army Pamphlet No. 550–47; Washington, D.C., 1965).

VOURAS, PAUL P., "Jordan," *Focus*, 17, no. 6 (February 1967), 6 pp.; and, with Alice Taylor, "Lebanon," *Focus*, 15, no. 10 (June 1965), 6 pp.

WARD, RICHARD J., "Focus in Jordan Agriculture," *Land Economics*, 42, no. 2 (May 1966), 189–194.

Egypt, the Sudan, and the Nile

ABU-LUGHOD, JANET, "Tale of Two Cities: The Origins of Modern Cairo," *Comparative Studies in Society and History*, 7, no. 4 (July 1965), 429–457.

BARBOUR, K. M., *The Republic of the Sudan: A Regional Geography* (London: University of London Press, 1961); and "Irrigation in the Sudan: Its Growth, Distribution and Potential Extension," Institute of British Geographers, Publication No. 26, *Transactions and Papers, 1959*, pp. 243–263.

BEDDIS, R. A., "The Aswan High Dam and the Resettlement of the Nubian People," *Geography*, 48, pt. 1 (January 1963), 77–80.

BERRY, L., and A. J. WHITEMAN, "The Nile in the Sudan," *Geographical Journal,* 134, pt. 1 (March 1968), 1–37.

DAVIES, H. R. J., "Nomadism in the Sudan: Aspects of the Problem and Suggested Lines for Its Solution," *Tijdschrift voor Economische en Sociale Geografie,* 57, no. 5 (September–October 1966), 193–202; and "The West African in the Economic Geography of Sudan," *Geography,* 49, pt. 3 (July 1964), 222–235.

FROOD, A. McKIE, "The Aswan High Dam and the Egyptian Economy," in Robert W. Steel and Richard Lawton, eds., *Liverpool Essays in Geography: A Jubilee Collection* (London: Longmans, Green and Co., 1967), pp. 363–379.

GRAY, ALBERT L., Jr., "The Egyptian Economy: Prospects for Economic Development," *Journal of Geography,* 66, no. 9 (December 1967), 510–518.

HAMDAN, G., "The Growth and Functional Structure of Khartoum," *Geographical Review,* 50, no. 1 (January 1960), 21–40.

HOLZ, ROBERT K., "The Aswan High Dam," *Professional Geographer,* 20, no. 4 (July 1968), 230–237.

HURST, H. E., *The Nile: A General Account of the River and the Utilization of Its Waters* (rev. ed.; London: Constable and Co., 1957).

NUTTONSON, M. Y., *The Physical Environment and Agriculture of Libya and Egypt with Special Reference to Their Regions Containing Areas Climatically and Latitudinally Analogous to Israel* (Washington, D.C.: American Institute of Crop Ecology, 1961).

OLIVER, J., "Port Sudan: The Study of its Growth and Functions," *Tijdschrift voor Economische en Sociale Geografie,* 57, no. 2 (March–April, 1966), 54–61.

RANDELL, JOHN R., "The Sudan," *Focus,* 17, no. 1 (September 1966), 6 pp.

SAAB, GABRIEL, *The Egyptian Agrarian Reform, 1952–1962* (New York: Oxford University Press, 1967).

WARREN, CLINE J., *Agricultural Development and Expansion in the Nile Basin,* U.S. Department of Agriculture, Economic Research Service, Foreign Agricultural Economic Report No. 48 (Washington, D.C.: October 1968).

YAUKEY, RAYMOND S., *Basic Data on the Economy of the United Arab Republic* [*(Egypt)*]. U.S. Bureau of International Commerce, Overseas Business Reports, OBR 67–25 (May 1967). (Washington, D.C.: Government Printing Office, 1967.)

Libya, Northwestern Africa, and the Sahara

BARBOUR, NEVILL, ed., *A Survey of North West Africa* (*the Maghrib*) (2d ed.; London: Oxford University Press, 1962).

BIRKET-SMITH, KAJ, *Primitive Man and His Ways* (London: Odhams Press, 1960), "The Tuareg: Desert and Oasis; The Camel," pp. 141–173.

BRIGGS, LLOYD CABOT, *Tribes of the Sahara* (Cambridge, Mass.: Harvard University Press, 1960).

CAPOT-REY, ROBERT, "The Present State of Nomadism in the Sahara," in *The Problems of the Arid Zone: Proceedings of the Paris Symposium* [1960] (Arid Zone Research, No. 18; Paris: UNESCO, 1962), pp. 301–310.

CLARKE, JOHN I., "Economic and Political Changes in the Sahara," *Geography,* 46, pt. 2 (April 1961), 102–119; and "Oil in Libya: Some Implications," *Economic Geography,* 39, no. 1 (January 1963), 40–59.

COULTER, JOHN WESLEY, "Libya's Black Gold," *Journal of Geography,* 66, no. 6 (September 1967), 294–305.

FARRELL, J. D., "Libya Strikes It Rich" [in Oil], *Africa Report,* 12, no. 4 (April 1967), 8–15.

GALLAGHER, CHARLES F., "Tunisia Modernizes," *Africa Report,* 13, no. 3 (March 1968), 7–15.

GULICK, JOHN, *Tripoli: A Modern Arab City* (Cambridge: Harvard University Press, 1967).

HARRISON, ROBERT S., "Libya," *Focus.* 17, no. 3 (November 1966), 6 pp.; and "Tunisia," *Focus,* 19, no. 5 (January 1969), 1–7.

International Bank for Reconstruction and Development, *The Economic Development of Morocco* (Baltimore, Md.: Johns Hopkins Press, 1966).

LEWIS, WILLIAM H., "Algeria Against Itself," *Africa Report,* 12, no. 9 (December 1967), 9–15.

MAY, JACQUES M., *The Ecology of Malnutrition in Northern Africa: Libya, Tunisia, Algeria, Morocco, Spanish Sahara, and Ifni* (New York: Hafner Publishing Co., 1967).

MIKESELL, MARVIN W., "Algeria," *Focus,* 11, no. 6 (February 1961), 6 pp.; also, *Northern Morocco: A Cultural Geography* (University of California Publications in Geography, Vol. 14; Berkeley and Los Angeles: University of California Press, 1961); "Deforestation in Northern Morocco: Burning, Cutting, and Browsing Are Changing a Naturally Wooded Area into a Land of Scrub," *Science,* 132, no. 3425 (August 19, 1960), 441–448; and "The Role of Tribal Markets in Morocco: Examples from the 'Northern Zone,'" *Geographical Review,* 48, no. 4 (October 1958), 494–511.

MURPHEY, RHOADS, "The Decline of North Africa Since Roman Times: Climatic or Human?" *Annals of the Association of American Geographers,* 52, no. 3 (September 1962), 242–254.

NUTTONSON, M. Y., *An Introduction to Northern Africa and a Survey of the Physical Environment and Agriculture of Morocco, Algeria, and Tunisia, with Special Reference to Their Regions Containing Areas Climatically and Latitudinally Analogous to Israel* (Washington, D.C.: American Institute of Crop Ecology, 1961).

STECHSCHULTE, ROGER F., *Basic Data on the Economy of Morocco,* U.S. Bureau of International Commerce, Overseas Business Reports, OBR 66–48 (July 1966). (Washington, D.C.: Government Printing Office, 1966.)

U.S. Army Area Handbook for Algeria (Department of the Army Pamphlet No. 550–44; Washington, D.C.: 1965).

U.S. Army Area Handbook for Morocco (rev. ed.; Department of the Army Pamphlet No. 550–49; Washington, D.C.: 1965).

ZARTMAN, I. WILLIAM, "Morocco," *Focus,* 15, no. 6 (February 1965), 6 pp.

Persian Gulf Countries (except Iran); Arabian Peninsula

Aramco Handbook: Oil and the Middle East (2d ed.; Dhahran, Saudi Arabia: Arabian American Oil Company, 1968). A handsome and very useful volume of historical and economic information on the Persian Gulf oil industry, emphasizing Saudi Arabia. Many maps and charts in color.

BRAUN, AMELIA J., *Basic Data on the Economy of Saudi Arabia,* U.S. Bureau of International Commerce, Overseas Business Reports, OBR 67–78 (December 1967). (Washington, D.C.: Government Printing Office, 1967.)

EBERT, CHARLES HORST VINCENT, "Water Resources and Land Use in the Qatif Oasis of Saudi Arabia," *Geographical Review,* 55, no. 4 (October 1965), 496–509.

EL MALLAKH, RAGAEI, "Economic Development through Cooperation: The Kuwait Fund," *Middle East Journal,* 18, no. 4 (Autumn 1964), 405–420; and "Planning in a Capital Surplus Economy: Kuwait," *Land Economics,* 42, no. 4 (November 1966), 425–440.

GULICK, JOHN, "Baghdad: Portrait of a City in Physical and Cultural Change," *Journal of the American Institute of Planners,* 33, no. 4 (July 1967), 246–255.

HAUPERT, JOHN S., "Saudi Arabia," *Focus,* 16, no. 9 (May 1966), 6 pp.

INGRAMS, HAROLD, *Arabia and the Isles* (3d ed.; London: John Murray, 1966).

International Bank for Reconstruction and Development, *The Economic Development of Kuwait* (Baltimore, Md.: Johns Hopkins Press, 1965).

KEIM, KATHLEEN, *Basic Data on the Economy of Kuwait,* U.S. Bureau of International Commerce, Overseas Business Reports, OBR 68–55 (July 1968). (Washington, D.C.: Government Printing Office, 1968.)

"Kuwait, Cinderella of Petroleum," Chap. 20 in Richard M. Highsmith, ed., *Case Studies in World Geography: Occupance and Economy Types* (Englewood Cliffs, N.J.: Prentice-Hall, 1961), pp. 139–145.

LONGRIGG, STEPHEN H., *Oil in the Middle East: Its Discovery and Development* (3d ed.; New York: Oxford University Press, 1968).

MELAMID, ALEXANDER, "Eastern Arabia," *Focus,* 18, no. 3 (November 1967), 6 pp.; and "South Yemen," *Focus* 18, No. 5 (January 1968), 6 pp.

PEPPELENBOSCH, P. G. N., "Nomadism on the Arabian Peninsula, *Tijdschrift voor Economische en Sociale Geografie,* **59,** no. 6 (November–December 1968), 335–346.

SHEHAB, FAKHRI, "Kuwait: A Super-Affluent Society," *Foreign Affairs,* **42,** no. 3 (April 1964), 461–474.

SIMMONS, JOHN L., "Agricultural Development in Iraq: Planning and Management Failures," *Middle East Journal,* **19,** no. 2 (Spring 1965), 129–140.

THESIGER, WILFRED, *Arabian Sands* (New York: E. P. Dutton & Co., 1959). A fascinating account of travels in the Empty Quarter, with much colorful detail on Arabian life and landscapes.

U.S. Army Area Handbook for Saudi Arabia (Department of the Army Pamphlet No. 550–51; Washington, D.C.: 1966).

U.S. Bureau of International Commerce, *Basic Data on the Economy of Bahrain, Qatar, Muscat and Oman, and the Trucial States,* Overseas Business Reports, OBR 68–22 (March 1968). (Washington, D.C.: Government Printing Office, 1968.)

Iran, Afghanistan, Turkey, and Cyprus

AKTAN, RESAT, "Problems of Land Reform in Turkey," *Middle East Journal,* **20,** no. 3 (Summer 1966), 317–334.

BALDWIN, GEORGE B., *Planning and Development in Iran* (Baltimore, Md.: Johns Hopkins Press, 1967).

BARTH, FREDRIK, "Nomadism in the Mountain and Plateau Areas of South West Asia," in *The Problems of the Arid Zone* (Arid Zone Research, No. 18; Paris: UNESCO, 1962), pp. 341–355.

BECKETT, P. H. T., and E. D. GORDON, "Land Use and Settlement Round Kirman in Southern Iran," *Geographical Journal,* **132,** pt. 4 (December 1966), 476–490.

BILL, JAMES A., "The Social and Economic Foundations of Power in Contemporary Iran," *Middle East Journal,* **17,** no. 4 (Autumn 1963), 400–418.

BROOKE, CLARKE, "Khat [Qat] (*Catha Edulis*): Its Production and Trade in the Middle East," *Geographical Journal,* **126,** pt. 1 (March 1960), 52–59.

CAROE, SIR OLAF, "Afghan Highways," *Geographical Magazine,* **3,** no. 7 (November 1965), 537–550.

CRESSEY, GEORGE B., "Qanats, Karez and Foggaras," *Geographical Review,* **48,** no. 1 (January 1958), 27–44.

DOBELL, W. M., "Division over Cyprus," *International Journal,* **22,** no. 2 (Spring 1967), 278–292.

DOUGLAS, WILLIAM O., *West of the Indus* (New York: Doubleday & Company, 1958). A perceptive travel account.

DUPREE, LOUIS, "A Suggested Pakistan-Afghanistan-Iran Federation," *Middle East Journal,* **17,** no. 4 (Autumn 1963), 383–399.

ELDRIDGE, ROBERT H., "Emigration and the Turkish Balance of Payments," *Middle East Journal,* **20,** no. 3 (Summer 1966), 296–316.

ENGLISH, PAUL WARD, *City and Village in Iran: Settlement and Economy in the Kirman Basin* (Madison: University of Wisconsin Press, 1966).

FISHER, W. B., ed., *The Land of Iran* (*The Cambridge History of Iran,* Vol. 1; Cambridge: at the University Press, 1968).

FLETCHER, ARNOLD, *Afghanistan, Highway of Conquest* (Ithaca: Cornell University Press, 1965).

HALE, PETER B., *Basic Data on the Economy of Turkey,* U.S. Bureau of International Commerce, Overseas Business Reports, OBR 66–91 (December 1966). (Washington, D.C.: Government Printing Office, 1966).

HEARN, JACKSON B., *Basic Data on the Economy of Afghanistan,* U.S. Bureau of International Commerce, Overseas Business Reports, OBR 66–34 (June 1966). (Washington, D.C.: Government Printing Office, 1966.)

HILTNER, JOHN, "The Distribution of Turkish Manufacturing," *Journal of Geography,* **61,** no. 6 (September 1962), 251–258.

HUREWITZ, J. C., "Russia and the Turkish Straits: A Revaluation of the Origins of the Problem," *World Politics,* 14, no. 4 (July 1962), 605–632.

ISSAWI, CHARLES, "Iran's Economic Upsurge," *Middle East Journal,* 21, no. 4 (Autumn 1967), 447–461.

KOLARS, JOHN F., *Tradition, Season, and Change in a Turkish Village* (University of Chicago, Department of Geography Research Paper No. 82; Chicago: 1963); also, "Bahtili: A Turkish Village in Transition from Subsistence to Commercial Agriculture," Chap. 3 in Richard S. Thoman and Donald J. Patton, eds., *Focus on Geographic Activity: A Collection of Original Studies* (New York: McGraw-Hill Book Company, 1964), pp. 12–18; and "Locational Aspects of Cultural Ecology: The Case of the Goat in Non-Western Agriculture," *Geographical Review,* 56, no. 4 (October 1966), 577–584.

MELAMID, ALEXANDER, "The Geographical Distribution of Communities in Cyprus," *Geographical Review,* 46, no. 3 (July 1956), 355–374; and "The Geographical Pattern of Iranian Oil Development," *Economic Geography,* 35, no. 3 (July 1959), 199–218.

NOWLAND, JOHN L., "The Port of Istanbul," *Scottish Geographical Magazine,* 77, no. 2 (September 1961), 67–74.

PLANHOL, XAVIER de, "Aspects of Mountain Life in Anatolia and Iran," Chap. 11 in S. R. Eyre and G. R. J. Jones, eds., *Geography as Human Ecology* (New York: St. Martin's Press, 1966), pp. 291–308.

"The Qanat in Iran, A Desert Water Supply System," Chap. 4 in Richard M. Highsmith, Jr., ed., *Case Studies in World Geography: Occupance and Economy Types* (Englewood Cliffs, N.J.: Prentice-Hall, 1961), pp. 23–29.

QURESHI, S. M. M., "Pakhtunistan: The Frontier Dispute Between Afghanistan and Pakistan," *Pacific Affairs,* 39, nos. 1 and 2 (Spring–Summer 1966), 99–114.

RAMAZANI, ROUHOLLAH K., *The Northern Tier: Afghanistan, Iran, and Turkey* (Princeton, N.J.: D. Van Nostrand Company, Searchlight Books, 1966).

REINER, ERNST, and ALICE TAYLOR, "Afghanistan," *Focus,* 15, no. 5 (January 1965), 6 pp.

SEVERANCE, ROGER D., *Basic Data on the Economy of Iran,* U.S. Bureau of International Commerce, Overseas Business Reports, OBR 68–64, (July 1968). (Washington, D.C.: Government Printing Office, 1968.)

STAUFFER, THOMAS R., "The Economics of Nomadism in Iran," *Middle East Journal,* 19, no. 3 (Summer 1965), 284–302.

STEWART, DESMOND, and the Editors of *Life, Turkey* (Life World Library; New York: Time Inc., 1965).

STIRLING, PAUL, *Turkish Village* (New York: John Wiley & Sons, 1966).

U.S. Army Area Handbook for Cyprus (Department of the Army Pamphlet No. 550–22; Washington, D.C.: 1964).

WILBUR, DONALD N., *Afghanistan: Its People, Its Society, Its Culture* (rev. ed.; New Haven: Human Relations Area Files Press, 1962); and *Iran Past and Present* (6th ed.; Princeton, N.J.: Princeton University Press, 1967).

WULFF, H. E., "The Qanats of Iran," *Scientific American,* 218, no. 4 (April 1968), 94–105.

Ethiopia and Somaliland

DRYSDALE, JOHN, "The Problem of French Somaliland," *Africa Report,* 11, no. 8 (November 1966), 10–17.

Ethiopian Geographical Journal (semiannual).

LEWIS, I. M., "Recent Developments in the Somali Dispute," *African Affairs,* 66, no. 263 (April 1967), 104–112.

SIMOONS, FREDERICK J., *Northwest Ethiopia: Peoples and Economy* (Madison: University of Wisconsin Press, 1960).

SMITH, MAUREEN R., *Basic Data on the Economy of Ethiopia,* U.S. Bureau of International Commerce, Overseas Business Reports, OBR 67–14 (February 1967). (Washington, D.C.: Government Printing Office, 1967.)

SOMMER, JOHN W., "Ethiopia," *Focus,* 15, no. 8 (April 1965), 6 pp.

TERNES, ALAN, "Ethiopia," *Natural History,* 77, no. 2 (February 1968), 28–37, and no. 3 (March 1968), pp. 46–53.

See also the lists of references and readings for Chapters 15, 17, and 24, particularly the relevant sections of general geographies of the Middle East, Asia, and Africa listed on pages 357, 415, and 570.

PART V

THE ORIENT

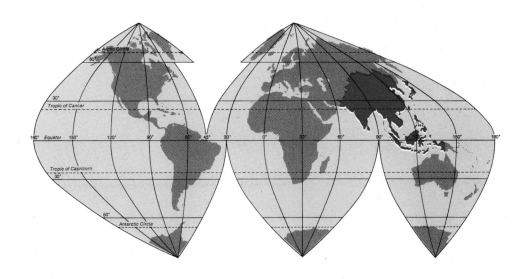

Introduction

to the

Orient

17   "The Orient" (East) is used in this book as a regional name for the countries occupying the southeastern quarter of Eurasia. Besides mainland countries extending from India and Pakistan to Korea, the Orient, as herein defined, also includes an arc of island countries stretching for thousands of miles between Ceylon and Japan (map, p. 404).

REGIONAL GROUPS OF COUNTRIES

For convenience in study in the chapters that follow, the countries of the Orient are divided into a number of regional groups on the basis of such factors as (a) proximity, (b) environmental, economic, and cultural similarities, (c) political relationships, and (d) historical ties.

    1. The *countries of the Indian subcontinent* include India, Pakistan, the disputed area called Kashmir (map, p. 404), and the small Himalayan states of Nepal, Bhutan,

THE ORIENT
INTRODUCTORY LOCATION MAP

✪ Political capitals, over 1,000,000

✪ Political capitals, less than 1,000,000

◉ Other urban areas, over 1,000,000

| 0 | 250 | 500 | 750 | 1,000 |

Scale of Miles

City-size symbols are based on metropolitan-area estimates

and Sikkim. Ceylon is often included with this group, although it is considered with Southeast Asia in the present treatment.

2. The *countries of Southeast Asia* include (a) Burma; (b) Thailand (Siam); (c) the units that once comprised French Indochina—South Vietnam, Cambodia, Laos, and the Communist state of North Vietnam; (d) Malaysia, comprised of the mainland unit called Malaya, together with Sarawak and Sabah (North Borneo) occupying contiguous portions of the large island of Borneo; (e) the small but densely populated and commercially important island country of Singapore, essentially a single seaport; (f) the large island country of Indonesia; (g) the smaller island countries of Ceylon and the Philippines; and (h) two minor island dependencies: the British-protected sultanate of Brunei on the island of Borneo, and Portugal's colony of Portuguese Timor, which shares the island of Timor with Indonesia.

3. The *countries of the Chinese realm,* as considered herein, include Communist China, Nationalist China

**MAJOR LANDFORMS
OF THE ORIENT**

-··-··- International boundary

(Taiwan [Formosa] and smaller islands), the Mongolian People's Republic or Mongolia (placed in this group for convenience and because of its historical associations with China), and the European coastal possessions of Hong Kong (British) and Macao (Portuguese), which are largely Chinese in population.

4. Japan, for the sake of convenience, is discussed here with its former colony of Korea, now divided between North Korea (Communist) and South Korea (non-Communist).

THE COLONIAL BACK-GROUND OF THE ORIENT

Modern European penetration of the Orient began at the end of the fifteenth century. The early comers established trading posts and gradually extended political control over limited areas near the coast. In the eighteenth and nineteenth centuries the pace of annexation quickened, and large areas came under European sway. By the end of

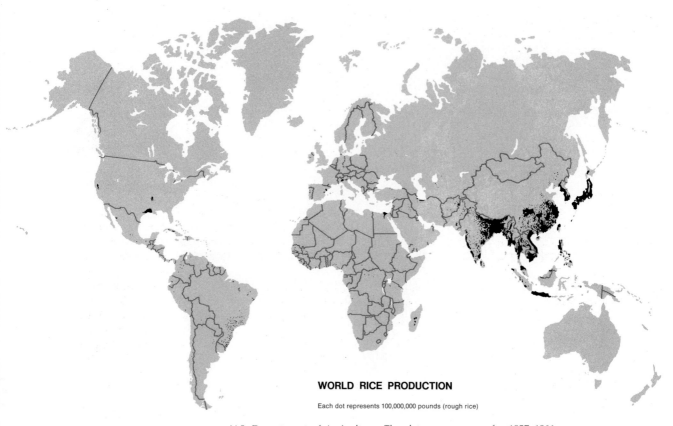

WORLD RICE PRODUCTION

Each dot represents 100,000,000 pounds (rough rice)

SOURCE: U.S. Department of Agriculture. The data are averages for 1957–1961.

the nineteenth century Great Britain was supreme in India, Burma, Ceylon, Malaya, and northern Borneo; the Netherlands possessed most of the East Indies; and France had acquired Indochina and a number of small holdings around the coasts of India. Meanwhile Portugal, the supreme colonial power of the Orient in the sixteenth century, had been displaced from her early holdings with the exception of Macao, a part of the island of Timor, and Goa and two other small holdings on the west coast of India. China, though retaining a semblance of territorial integrity, was forced to yield possession of strategic Hong Kong to Britain in the mid-nineteenth century and to grant special trading concessions and extraterritorial rights to various European nations and the United States. At the end of the century the Philippines, dominated by Spain after the mid-sixteenth century, passed into American control. The only Oriental countries of any importance to escape domination by the Western powers during the colonial age were (1) Thailand, which formed a buffer between British and French colonial spheres in Southeast Asia; (2) Japan, which withdrew into almost complete seclusion in the seventeenth century, but emerged in the latter nineteenth century as the first modern,

industrialized Oriental nation, and soon acquired a colonial empire of its own; and (3) Korea, which also followed a policy of isolation from foreign influences from the early seventeenth century until 1876, when a trade treaty was forced on it by Japan.

During the age of European expansion the Orient constituted an extraordinarily rich colonial area from which Western nations extracted vast quantities of such valuable commodities as rubber, sugar, tea, copra, palm oil, spices, and tin, and in which Western manufacturers found large markets for cheap textiles, metalwares, and other inexpensive types of goods. Westerners also found the Orient a fertile field for investment in plantations, factories, mines, transportation, communication, and electric power facilities.

Decline of Colonialism in the Twentieth Century

Western dominance of the Orient was ended in the twentieth century by a complex chain of circumstances, including (1) the weakening effects of conflicts among the

Western nations in the two world wars, (2) the rise of Japan to great-power status and its successful, though temporary, military challenge to the West in the early stages of World War II, and (3) the rise of anticolonial movements in areas subject to European control. World War II ended Western colonialism in China, and after the war all colonial possessions in the Orient, whether European (map, p. 65), American, or Japanese, gained independence, except for a few minor units listed earlier.

In the Orient the twentieth century has been an age of revolution, war, and general turmoil. All Oriental nations have been affected by these stresses, and the spotlight of world attention has focused first on one area and then another. At the time of this writing (1968), the most publicized upheaval was the long-drawn-out war in Vietnam, but it seemed most unlikely that cessation of this war would bring general tranquillity to the Orient. The active or potential trouble spots are too numerous and widespread to recount here. The major ones will be touched on in the chapters that follow.

THE VARIED PHYSIOGRAPHIC SETTING

The stage on which the Oriental drama is being enacted is a complex intermingling of many types of topography— high, rugged mountains, arid plateaus and basins, humid hill lands and river plains, and a vast number of offshore islands rising from the floor of shallow seas. Although the picture is extremely complicated in detail, a certain order appears if the surface features of the Orient are conceived of as three concentric arcs or crescents of land—an inner arc of high mountains, plateaus, and basins; a middle arc of lower mountains, hill lands, and river plains; and an outer arc of islands and seas (map, p. 405).

The Inner Highland

The inner highland of the Orient is composed of the highest mountain ranges on earth, interspersed with plateaus and basins. At the south the great wall of the Himalaya, Karakoram, and Hindu Kush mountains overlooks the north of the Indian subcontinent. At the north, the Altai, Tien Shan, Pamirs, and other mountains separate the Orient from the Soviet Union. Between these mountain walls lie the sparsely inhabited Tibetan Plateau, over 15,000 feet in average elevation, and the dry, thinly populated basins and plateaus of Sinkiang and Mongolia.

River Plains and Hill Lands

The area between the inner highland and the sea is principally occupied by river flood plains and deltas, bordered and separated by hills and relatively low mountains. Major components of the topography are (1) the immense alluvial plain of northern India, built up through countless ages by the Indus, Ganges, and Brahmaputra rivers; (2) the hill lands of peninsular India, geologically an ancient plateau, but largely hilly in aspect; (3) the plains of the Irrawaddy, Chao Praya (Menam), Mekong, and Red rivers in peninsular Southeast Asia, together with bordering hills and mountains; (4) the hill lands and small alluvial plains of southern China; (5) the broad alluvial plains along the middle and lower Yangtze River in central China and the mountain-girt Red Basin on the upper Yangtze; (6) the large delta plain of the Yellow River and its tributaries in North China, backed by loess-covered hilly uplands; and (7) the broad central plain of Manchuria, almost completely enclosed by mountains with low to moderate elevations. The floor of the Red Basin and the central plain of Manchuria are structural rather than river-made plains, and are rolling or hilly in aspect rather than flat.

Offshore Islands and Seas

Offshore, a fringe of thousands upon thousands of islands, mostly grouped in great archipelagoes, borders the mainland. On these islands high interior mountains with many volcanic peaks are flanked by broad or narrow coastal plains where most of the inhabitants live. Three major archipelagoes include most of the islands—the East Indies, the Philippines, and the Japanese Archipelago. Ceylon, Taiwan, and Hainan are large islands not included in an archipelago. Between the archipelagoes and the mainland lie the China Seas, and, to the north, the Sea of Japan. At the southwest the Indian peninsula projects southward between two immense arms of the Indian Ocean—the Bay of Bengal and the Arabian Sea.

THE PATTERN OF CLIMATES AND VEGETATION

In detail the climatic, like the physiographic, pattern is one of almost endless variety. However, two unifying elements are present throughout most parts of the Orient inhabited by any considerable number of people. These are

(1) the dominance of warm climates and (2) a characteristic monsoonal regime of precipitation.

Temperature and Precipitation

In the parts of the Orient where most of the population is found, temperatures are tropical or subtropical and the frost-free season ranges from around 200 to 365 days. The principal exceptions exist in northern sections of China, Korea, and Japan. Here summers are warm to hot in the lowlands, but the growing season is generally less than 200 days and winters are often severely cold. The arid, sparsely populated basins and plateaus of Sinkiang and Mongolia also have a continental type of climate with warm summers and cold winters. The higher mountain areas and Tibetan Plateau have highland climates varying with the altitude and latitude. Permanent snowfields and glaciers occur at the higher elevations.

Annual precipitation varies from near zero in the Tarim Basin of Sinkiang to an average of more than 400 inches in parts of the Khasi Hills of northeastern India. A monsoon climate, or at least a climate with monsoon tendencies, prevails nearly everywhere in the populous middle arc of plains and hills, and in many parts of the islands as well. Technically a monsoon is not, as some people imagine, a violent downpour of rain with accompanying winds and lightning (though such are often its effects), but is simply a current of air blowing fairly steadily from a given direction for several weeks or months at a time. Although conditions vary from place to place, the Orient is broadly characterized by two monsoons: a summer monsoon blowing from the sea to the land and bringing high humidity and rain, and a winter monsoon blowing seaward and bringing little or no rain and cool or cold, clear weather. The characteristic features of the monsoonal type of climate then, are (1) the seasonal reversal of wind direction, (2) the strong summer maximum of rainfall, and (3) the long dry season, typically lasting for most or all of the winter half-year.

Types of Climate

Seven main types of climate in the Orient are customarily recognized in climatic classifications: tropical rain forest, tropical savanna, humid subtropical, humid continental, steppe, desert, and highland (front endpaper climatic map).

Tropical Rain Forest

The rainy tropical climates of the Orient are found along or relatively near the equator. Consequently, high temperatures are experienced throughout the year in the lowlands. There is a complete absence of frost, and the year-round growing season offers the maximum possibilities for agriculture from the standpoint of temperature. However, the high temperatures and heavy rain promote rapid leaching of mineral nutrients and destruction of organic matter, with the result that most soils in the rainy tropics are relatively infertile despite the thick cover of deep-rooted trees and grasses they often support. Two main types of humid tropical climate, each associated with characteristic forms of vegetation, are recognized. These are (1) tropical rain forest climate and (2) tropical savanna climate.

The *tropical rain forest climate* is typically found in lowlands within 5 or 10 degrees of the equator. A rainfall of at least 30 to 40 inches, often 100 inches or more, is spread throughout the year so that every month has considerable rain. Average temperatures vary only slightly from month to month; Singapore, for example, exhibits a difference of only 2° between the warmest and coolest months. Monotonous heat prevails the year round, although excessively high temperatures of 95° or 100°F or more are seldom or never experienced. Some relief is afforded by a drop of 10° to 25° in the temperature at night, and conditions are more pleasant along coasts subject to periodic sea breezes.

The tropical rain forest produced by the climatic conditions described above is characteristically a thick forest of large broadleaf evergreen trees, mostly hardwoods, from 50 to 200 feet in height, and forming an almost continuous canopy of foliage. The trees are often entangled in a mass of vines, and dense undergrowth occurs wherever sufficient light can penetrate to the ground.

Tropical rain forest climate and vegetation are characteristic of most parts of the East Indies, the Philippines, and the Malay Peninsula. Rain forest vegetation is also found in certain other areas which experience a dry season, but in which the precipitation of the rainy season is sufficiently heavy to promote a thick growth of trees. Such areas exist (1) along the west coast of India south of Bombay and along the south coast of Ceylon, (2) along the coasts of Cambodia, Thailand, and Burma, extending northward to the Ganges-Brahmaputra delta in India and East Pakistan, (3) along part of the east coast of Vietnam, and (4) in parts of the northern Philippines.

TABLE 14 CLIMATIC DATA FOR SELECTED ORIENTAL STATIONS

STATION	COUNTRY	LATITUDE TO NEAREST WHOLE DEGREE	ELEVATION ABOVE SEA LEVEL (FEET)	TYPE OF CLIMATE	AVERAGE TEMPERATURE (DEGREES F TO NEAREST WHOLE DEGREE)			AVERAGE PRECIPITATION		
					ANNUAL	JANUARY	JULY (OR WARMEST MONTH)	ANNUAL (TO NEAREST INCH)	PERCENT OCCURRING APRIL–SEPTEMBER (TO NEAREST WHOLE PERCENT)	PERCENT OCCURRING JUNE–SEPTEMBER (TO NEAREST WHOLE PERCENT)
Karachi	West Pakistan	25°N.	13'	Desert	79°	64°	89° (June)	9"	84%	83%
Bombay	India	19°N.	37'	Tropical savanna	81°	75°	86° (May)	86"	95%	94%
Madras	India	13°N.	51'	Tropical savanna	83°	76°	91° (May, June)	49"	37%	30%
Calcutta	India	24°N.	21'	Tropical savanna	80°	68°	88° (May)	62"	83%	73%
Mandalay	Burma	22°N.	252'	Tropical savanna	81°	68°	89° (April)	31"	74%	53%
Singapore	Singapore	1°N.	16'	Tropical rain forest	80°	79°	81° (May)	95"	45%	30%
Jakarta	Indonesia	6°S.	26'	Tropical rain forest	80°	79°	81° (May)	71"	29%	15%
Delhi	India	29°N.	718'	Humid subtropical	78°	58°	94° (May, June)	28"	86%	84%
Canton	China	23°N.	29'	Humid subtropical	71°	56°	83° (August)	67"	81%	56%
Shanghai	China	31°N.	23'	Humid subtropical	59°	38°	81°	44"	69%	52%
Wuhan	China	31°N.	121'	Humid subtropical	63°	40°	86°	50"	73%	47%
Chengtu	China	31°N.	1611'	Humid subtropical	62°	42°	79°	52"	87%	74%
Peking	China	40°N.	125'	Humid continental (long summer)	54°	25°	80°	24"	93%	84%
Shenyang (Mukden)	China (Manchuria)	42°N.	141'	Humid continental (long summer)	46°	11°	78°	28"	84%	71%
Harbin	China (Manchuria)	46°N.	526'	Humid continental (short summer)	38°	−1°	72°	19"	86%	73%
Tokyo	Japan	36°N.	19'	Humid subtropical	58°	39°	80° (August)	62"	61%	44%
Hakodate	Japan	42°N.	13'	Humid continental (short summer)	47°	25°	71° (August)	46"	59%	45%
Kashgar	China (Sinkiang)	40°N.	4296'	Desert	55°	22°	80°	3"	47%	31%
Ulan Bator	Mongolia	48°N.	4347'	Steppe	26°	−14°	61°	8"	93%	85%
Gyantse	China (Tibet)	29°N.	13,110'	Highland	42°	24°	58°	12"	97%	90%

Tropical Savanna

Although the *tropical savanna climate,* like the tropical rain forest climate, is characterized by high temperatures the year round, this climate type is customarily found in areas farther from the equator, and the average temperatures vary somewhat more from month to month. However, the most striking and important difference between the two climate types lies in the fact that the savanna climate has a well-defined dry season, lasting in some areas for as much as 6 or 8 months of the year. The annual precipitation is less, on the average, than in the rain forest climate, but it is the seasonal distribution of the rain rather than inadequate total precipitation which represents the principal handicap for agriculture. The main areas of tropical savanna climate in the Orient occur in southern and central India, the Indochinese Peninsula except for most coastal areas and some northerly areas, and eastern Java and smaller islands to the east. Much larger areas of this climate type occur in Africa and Latin America and a smaller area in northern Australia.

In the Orient the characteristic natural vegetation associated with the savanna climate is a deciduous forest of smaller trees than those found in the tropical rain forest. The forest growth deteriorates to scrub in the drier areas. Tall, coarse tropical grasses, a very common vegetation form in the African and Latin American savannas, are found in only limited areas, and even there are thought to have been produced by repeated burning of forest growth during the dry season. In parts of Southeast Asia long-continued burning has fostered pure

Very large interior sections of the Orient consist of dry, mountainous terrain. The view shows sheep and lambs grazing on sparse pasturage at the base of arid mountains in Sinkiang province, western China. (Eastfoto.)

stands of certain tree species peculiarly resistant to extinction by fire. Among these are a number of economically valuable types, especially teak, which is exploited on a considerable scale in Burma and Thailand. In the more densely populated and long-settled parts of the Oriental savanna lands the natural vegetation has been so modified by centuries of human occupation that the original conditions are difficult or impossible to determine.

Climates outside the Humid Tropics

The *humid subtropical climate* occurs in southern China, the southern half of Japan, much of northern India, and a number of other Oriental countries. This climate type is characterized by warm to hot summers, mild or cool winters with some frost, and a frost-free season lasting 200 days or longer. The annual rainfall of 30 to 50 inches or more is fairly well distributed throughout the year, although monsoonal tendencies produce a dry season in some areas. The natural vegetation, now largely removed over extensive areas (especially in China), is a mixture of evergreen hardwoods, deciduous hardwoods, and conifers. A generally comparable climate occurs in the southeastern part of the United States, although rainfall is more evenly distributed through the year in the latter area.

The northern part of China proper, most of Korea and Manchuria, and northern Japan have a *humid continental* type of climate marked by warm to hot summers, cold winters with considerable snow, a frost-free season of 100 to 200 days, and less precipitation than the humid subtropical areas. Most areas experience a definite dry season in winter. The predominant natural vegetation is a mixture of broadleaf deciduous trees and conifers, although prairie grasses are thought to have formed the original cover in parts of North China and Manchuria. Many aspects of the humid continental climate of the Orient are duplicated in comparable latitudes of eastern North America, though the American areas lack the dry season in winter. The Orient has the long-summer subtype of humid continental climate except for northern Manchuria and extreme northern Japan, which have the short-summer subtype.

Steppe and desert climates, whose characteristics have been previously described in Chapters 13 and 15, are found in Sinkiang and Mongolia and in parts of western India and West Pakistan. A severe *highland climate* characterizes Tibet and adjoining mountain areas. Some of the higher mountains in the East Indies are also best classified as having highland climates.

Climatic data for representative stations in the Orient are presented in Table 14.

POPULATION AND ECONOMY OF THE ORIENT

Approximately half of the world's people live in the Orient. They range from some of the most primitive tribes on earth (in the remoter highlands) to peoples such as the Indians and Chinese, who have been civilized for thousands of years and are possessors of rich and varied cultures. Mongoloid peoples form a majority in China, Japan, Korea, Burma, Thailand, Cambodia, North and South Vietnam, and Laos, but the majority of the people in India, though darker skinned than Europeans, are considered to belong to the Caucasian race; and brown-skinned Malays or kindred peoples form a majority among the native inhabitants of the Malay Peninsula, the East Indies, and the Philippines. This cursory survey does little justice to the tremendous variety of racial stocks found in the region as a whole.

The picture with respect to religion is also complicated. Hinduism is dominant in India, although many other religions are practiced, while the Moslem faith is dominant in Pakistan, most parts of the East Indies, parts of the southern Philippines, parts of outer China, and among the native Malays of the Malay Peninsula. Various forms of Buddhism are dominant in Burma, Thailand, Cambodia, Laos, Tibet, and Mongolia. Ceylon and Nepal divide between Buddhism and Hinduism. The Chinese are hard to categorize, even if effects of Communist rule in mainland China are disregarded. Among them, Buddhism, Confucianism, and Taoism have all exerted an important influence, often in the same household. The same general situation has prevailed in Korea and Vietnam. In Japan religious affiliations have been divided between Buddhism and the strongly nationalistic religion of Shintoism. The Philippines, with a large Roman Catholic majority, is the only Christian nation of the Orient, though Christian groups are found in various other areas such as Korea, India, and Indonesia. Oriental religions (excepting the Moslem religion and Christianity, which are not indigenous to the region) may be broadly described as contemplative in nature, emphasizing meditation rather than active work for social betterment, and seeking converts with much less vigor than Christianity or Islam. Veneration of ancestors is a prominent feature, especially among the Chinese, Vietnamese, and Koreans. Often an elaborate ritual for everyday living is followed, this being particularly characteristic of the Hindu religion. The more primitive hill tribes of the Orient are largely Animists.

Buddhism in its various forms is a major religion in many different parts of the Orient. This scene shows the Buddhist "Temple of Dawn" in Bangkok, Thailand. (Trans World Airlines, Inc.)

Distribution of Population

The densest populations of the Orient are found, generally speaking, on river and coastal plains, although surprisingly high densities occur in some hilly or mountainous areas. The higher mountains and the steppes, deserts, and some areas of tropical rain forest are very sparsely inhabited.

Most countries in the Orient have experienced large increases in population during recent centuries, especially since the beginning of the nineteenth century. Rural densities of 500 to 1000 persons per square mile are fairly common, and some of the more fertile irrigated areas support 2000 persons, or even more, per square mile. At the present high rates of population increase in many areas of the Orient, a number of Oriental countries—for example, India and Pakistan—are faced with serious problems of population support. Should present rates of population growth continue for a considerable period, the problem of providing even a minimum diet for so many millions will become increasingly critical.

Means of Livelihood

Japan was the first country of the Orient to develop modern types of manufacturing on a really large scale. Since World War II, it has become one of the world's most important industrial powers. But Communist China and India also have important and expanding industries. These giant countries, by far the largest in the world in population, are much better supplied with mineral resources than Japan, and they have cheaper labor. But Japan's labor force is on the average more skilled, and Japan has been shifting more and more to types of industry requiring skilled labor. The remaining countries of the Orient lag far behind the three leaders in industrial development, though at least one of them, Indonesia, has sufficient natural resources to support a large expansion of industry.

For the Orient as a whole, agriculture remains the dominant source of livelihood. In most Oriental countries —Japan being a conspicuous exception—the majority of the population is supported directly by agricultural activities. A number of major types of agriculture may be distinguished in the Orient. Two types, plantation agriculture and shifting cultivation, are discussed in some detail in Chapter 19 on Southeast Asia, the part of the Orient in which these forms of agriculture are the most prominent. In the steppes and deserts of outer China, the Mongolian People's Republic, and West Pakistan, nomadic or seminomadic herding and oasis farming are practiced. Over large sections of the Orient, most farmers make a living by cultivation of small rain-fed or irrigated plots worked by family labor. (In Communist-held areas, collectivized agriculture has been replacing the traditional family farms.) Large amounts of hand labor are applied to the land (photo, right), and production is often of a semisubsistence character. This type of agriculture, which is often referred to as intensive subsistence agriculture, is built around the growing of cereals, although other types of crops are raised. Where natural conditions are not suitable for irrigated rice, such grains as wheat, barley, millet, grain sorghums, or corn are raised. However, irrigated rice is the grain which yields the largest amount of food per unit of area where conditions are favorable for its growth, and this crop is the agricultural mainstay in the areas inhabited by a large majority of the Orient's people.

Importance of Rice in the Oriental Economy

The nature and significance of rice growing in the Orient is briefly presented in the following selection by Smith and Phillips:[1]

The old adage that bread is the staff of life is a striking example of the ease with which a half-truth is perpetuated as a universal verity. The fact is that hundreds of millions of healthy and industrious men have never seen bread as we in the Occident know it, but that is no sign that these men are savage, barbarian, or heathen. Throughout the Orient from India to Japan, teeming millions obtain their carbohydrate from rice, which is low in gluten and will not make light bread. . . .

Since rice does not make light bread because it lacks the gluten, the Oriental boils the grain and eats it in that form. He flavors it with a bit of meat or fish if he can afford it; or uses curry, a hot seasoning preparation made in endless varieties. With peas and beans and some greens, rice furnishes almost the entire nourishment for

The arduous hand labor of hundreds of millions of farmers in the Orient is reflected in this view of a Japanese farmer harvesting a rice crop near Tokyo. To avoid waste, the rice stalks are cut close to the ground. In Japan, the most technically advanced nation of the Orient, an increasing proportion of the farm work is being done with small power-driven machines. (Foreign Agricultural Service, U.S. Department of Agriculture.)

hundreds of millions of people. Peas and beans are widely grown by almost all Eastern peoples who raise rice, and they are the substitutes for meat, milk, and cheese of the West, while the starch of rice is the substitute for bread, potatoes, and many puddings as well. . . .

The Rice Environment

Among the environmental factors affecting rice production, water supply is most important, for the great bulk of all rice is grown under irrigation, the rice fields being submerged under approximately 6 inches of fresh, slowly moving water for at least 75 days. While the amount of water needed in a given area varies with such factors as rate of evaporation, relative humidity, and soil conditions, a total of 45 to 65 inches of water is generally required for rice production. . . . Furthermore, rice requires a mean

---

[1] J. Russell Smith and M. Ogden Phillips, *Industrial and Commercial Geography* (3d ed.; New York: Holt, Rinehart and Winston, 1946), pp. 428–434. Authors' footnotes and all headings except one have been omitted. One heading has been added, one figure updated, and one word italicized in editing. Used by permission.

Intensive cultivation of irrigated rice on flights of terraces in the British colony of Hong Kong. Co-operation among the owners of these small fields is essential if the intricate and fragile system of land use is to be maintained. (British Information Services.)

The Technique of Rice Growing

Lowland rice must be grown by irrigation, and the devices used in fitting and keeping the land for this service are among the greatest monuments of human diligence in the world. . . . In Ceylon, for example, the railway that goes from the seacoast to the highlands goes through an irrigated plain divided by low banks into ponds of small area—rice fields, each of which has by great labor been leveled so that the water may be of uniform and proper depth for rice growing. As the railroad climbs the slopes of the hills the rice patches continue, with smaller area and higher banks, turning at last into a giant flight of gentle water steps, one of the most beautiful landscapes that the world possesses. . . .

The common treatment of . . . lowland rice is alternately to flood it and draw off the water during the early periods of its growth. It is kept under water during a large part of its development, the water being entirely drawn off as it ripens. The water must not become stagnant, and to keep it in motion it is the common practice on the hillsides to lead a stream to the top terrace, and let the water pass from terrace to terrace down the slopes. . . .

The labor of rice growing often involves the raising of plants in small seed beds and transplanting them in little bunches to the rice field itself. This work, as most of the other work in connection with terrace-grown rice, can be done only by hand. The small fields make it impossible to use such machinery as reapers and at times even the ox and water buffalo. . . .

When the Asiatic rice field is finally drained, the ripened grain is usually cut by hand, tied up in bundles, and allowed to dry. To accomplish this in moist places, it is often necessary to put the sheaves upon bamboo frames. It is usually threshed by hand with the aid of some very simple devices. One of these is a board with a slit in it. Drawing the rice through the slit pulls the grains from the heads and allows them to fall into a receptacle. The grain at this stage is called *paddy* because of a closefitting husk not unlike that which protects the oat kernel. As with oats, these husks cause the grain to keep much better than when the husk is removed and the final husking of rice for home use is usually deferred until the time of use approaches. Among the Oriental people the husking of the paddy to prepare it for food is a daily occurrence, commonly done by hand. One of the commonest sounds throughout the East . . . is the pounding of a heavy mallet or pestle as it falls into a vessel full of paddy in the process of pounding the grain and loosening the husk.

temperature of more than 70°F during the growing season of 4 to 6 months. Hence, rice is a product of the tropics and subtropics, most of it being grown in regions of reeking humidity with frequent, almost daily rains. . . .

Level land is obviously essential for irrigation, and where it is lacking man must create it artificially as he has done by laboriously building terraces on the steep hillsides in many parts of Japan, China, the Philippines, and the East Indies. Although rice is grown on a variety of soils, there must be an impervious subsoil to prevent the loss of valuable irrigation water by seepage. Ideal soil conditions are found on many an alluvial plain, where a topsoil of fertile and friable silt has been deposited above a layer of impervious clay.

Although rice is now produced in the tropic and subtropic lands of every continent, [nine-tenths] of the world's crop is grown each year in southern and eastern Asia. . . .

REFERENCES and READINGS

Standard Geography Texts

CRESSEY, GEORGE B., *Asia's Lands and Peoples* (3d ed.; New York: McGraw-Hill Book Company, 1963).

DOBBY, E. H. G., *Monsoon Asia* (3d ed.; London: University of London Press, 1966).

EAST, W. GORDON, and O. H. K. SPATE, eds., *The Changing Map of Asia* (4th ed.; New York: E. P. Dutton & Co., 1961).

GINSBURG, NORTON S., ed., *The Pattern of Asia* (Englewood Cliffs, N.J.: Prentice-Hall, 1958).

RAWSON, R. R., *The Monsoon Lands of Asia* (Chicago: Aldine Publishing Company, 1963).

ROBINSON, HARRY, *Monsoon Asia: A Geographical Survey* (New York: Frederick A. Praeger, 1967).

SPENCER, J. E., *Asia, East by South: A Cultural Geography* (New York: John Wiley & Sons, 1954).

STAMP, L. DUDLEY, *Asia: A Regional and Economic Geography* (12th ed.; London: Methuen and Co., 1967).

Other References

ADAMS, INEZ, "Rice Cultivation in Asia," *American Anthropologist,* **50**, no. 2 (April–June 1948), 256–278.

BARTLETT, H. H., "Fire, Primitive Agriculture, and Grazing in the Tropics," in William L. Thomas, Jr., ed., *Man's Role in Changing the Face of the Earth* (Chicago: University of Chicago Press, 1956), pp. 692–720.

BLACK, C. E., *The Dynamics of Modernization: A Study in Comparative History* (New York: Harper & Row, 1966).

BUCHANAN, KEITH, "Profiles of the Third World," *Pacific Viewpoint,* **5**, no. 2 (September 1964), 97–126.

CHANG, JEN-HU, "The Agricultural Potential of the Humid Tropics," *Geographical Review,* **58**, no. 3 (July 1968), 333–361.

COCKRILL, ROSS, "The Water Buffalo," *Scientific American,* **217**, no. 6 (December 1967), 118–125.

COURTENAY, P. P., *Plantation Agriculture* (London: G. Bell and Sons, 1965).

EMBREE, AINSLIE T., and Others, comps., *Asia: A Guide to Basic Books* (New York: The Asia Society, 1966).

GOUROU, PIERRE, *The Tropical World: Its Social and Economic Conditions and Its Future Status* (4th ed.; New York: John Wiley & Sons, 1966).

GREGOR, HOWARD F., "The Changing Plantation," *Annals of the Association of American Geographers,* **55**, no. 2 (June 1965), 221–238.

GRIST, D. H., *Rice* (4th ed.; London: Longmans, Green and Co., 1965).

JONES, CLARENCE F., and GORDON G. DARKENWALD, *Economic Geography* (3d ed.; New York: The Macmillan Company, 1965), Chap. 12, "Primitive Subsistence Agriculture," pp. 145–157; Chap. 13, "Commercial Plantation Farming in the Tropics," pp. 158–208; and Chap. 15, "Intensive Subsistence Farming in Monsoon Lands," pp. 242–258.

LACH, DONALD F., *Asia in the Making of Europe,* Vol. I, *The Century of Discovery* (Chicago: University of Chicago Press, 1964); other volumes to follow.

LEE, DOUGLAS H. K., *Climate and Economic Development in the Tropics* (New York: Harper & Row, 1957).

McNEIL, MARY, "Lateritic Soils," *Scientific American,* **211**, no. 5 (November 1964), 97–102.

MURPHEY, RHOADS, *An Introduction to Geography* (2d ed.; Chicago: Rand McNally & Co., 1966), Chap. 23, "Man and the Land in Monsoon Asia," pp. 357–381; and "New Capitals of Asia," *Economic Development and Cultural Change,* **5**, no. 3 (April 1957), 216–243.

MYRDAL, GUNNAR, *Asian Drama: An Inquiry into the Poverty of Nations* (3 vols.; New York: Pantheon Books, 1968).

Pacific Viewpoint (New Zealand). A geographical journal of high quality. Many articles on Asia.

PEDALABORDE, PIERRE, *The Monsoon,* trans. by M. J. Clegg (London: Methuen and Co., 1963; distributed in the United States by Barnes & Noble, New York).

PHILLIPS, JOHN, *The Development of Agriculture and Forestry in the Tropics: Patterns, Problems, and Promise* (rev. ed.; New York: Frederick A. Praeger, Inc., 1967).

RICHARDS, P. W., *The Tropical Rain Forest: An Ecological Study* (Cambridge: At the University Press, 1952; reprinted, 1964).

Scientific Problems of the Humid Tropical Zone Deltas and Their Implications (New York: UNESCO, 1966).

SPENCER, J. E., and G. A. HALE, "The Origin, Nature, and Distribution of Agricultural Terracing," *Pacific Viewpoint,* **2**, no. 1 (March 1961), 1–40.

UNITED NATIONS, ECONOMIC COMMISSION FOR ASIA AND THE FAR EAST, *Economic Survey of Asia and the Far East.* Annual.

WINT, GUY, *Asia: A Handbook* (New York: Frederick A. Praeger, 1966).

See also the reference lists for Chapters 1–3.

The Indian Subcontinent

18 The northern reaches of the Indian Ocean are split into two enormous bays—the Bay of Bengal and the Arabian Sea—by a triangular peninsula which thrusts southward for a thousand miles from the main mass of Asia. To the north the peninsula is bordered by the alluvial plain of the Indus and Ganges rivers, beyond which lie ranges of high mountains (map, p. 405). The entire unit—peninsula, alluvial plain, and fringing mountains—is often called the Indian subcontinent. Today it is occupied by the five countries of India, Pakistan, Nepal, Bhutan, and Sikkim (map, p. 418). India and Pakistan are by far the largest countries in area and population. Nepal, Bhutan, and Sikkim are small states[1] on the southern flank of the Himalaya Mountains. They have strong economic and political ties with India, and Bhutan and Sikkim are under formal Indian protection. The three Himalayan states are mountainous and difficult of access. Their chief political significance lies in their location between India and China, which have been engaged in a bitter and sporadically violent border dispute

[1] Area estimates in square miles: Nepal, 54,362; Bhutan, 19,305; Sikkim, 2744. Population estimates (1968): Nepal, 10,700,000; Bhutan, 765,000; Sikkim, 185,000.

Political units of the Indian subcontinent. The future status of Kashmir, disputed between India and Pakistan, remains undecided at the time of writing.

nation of India and the Moslem nation of Pakistan. Both countries have remained within the Commonwealth of Nations as republics which recognize the British monarch as the head of the Commonwealth and symbol of Commonwealth unity, but not as the titular head of their respective governments.

IMPORTANCE OF INDIA AND PAKISTAN

Population and Area

India and Pakistan are among the most influential of the many nations that have gained independence since World War II. While many factors are involved, one important element contributing to their influence in world affairs is their large size. Both are among the largest countries of the world in population, and both occupy sizable areas. Of the two countries, however, India is much the larger, being in fact the second country of the world in population and the seventh country in area. India's estimated population in 1968 was around 525 million, as opposed to about 110 million for Pakistan; her area of about 1,174,000 square miles compares with approximately 365,000 square miles for Pakistan (area figures exclude the disputed state of Kashmir). But Pakistan, though it is much smaller than India, is sixth among the world's countries in total population, and ranks with Indonesia as one of the two largest Moslem countries. Pakistan's population is divided between the two widely separated provinces of East Pakistan, with about 59 million people inhabiting 54,500 square miles, and West Pakistan, with approximately 51 million people occupying slightly more than 310,000 square miles.

since 1959. The large island of Ceylon off the southeast coast of the Indian peninsula is often included with India and Pakistan in regional groupings. However, in many respects Ceylon is more similar to the countries of Southeast Asia, and is considered with them in Chapter 19.

The Indian subcontinent is entirely enclosed on its landward borders by mountains. Its northern boundary lies in the highest ranges in the world, the Himalaya and the Karakoram. From each end of this massive wall, lower flanking ranges trend southward, to the Arabian Sea on the west and the Bay of Bengal on the east. Until 1947 it was customary to refer to the entire area now included in India and Pakistan as "India." It was for well over a century the most important unit in the British colonial empire. Then in 1947 it gained freedom, but in the process became divided along religious lines into two sovereign nations, the predominantly Hindu

Economic Importance and Potentialities

The subcontinent's production of goods and services is not impressive if it is measured on a per capita

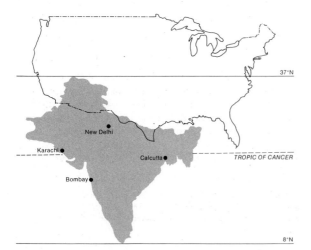

The Indian subcontinent compared in latitude and area with conterminous United States.

basis. But the total output of certain commodities is very great, and a considerable potential exists for economic development. India is a more important country economically than Pakistan and her resources are much greater. She is one of the world's largest producers of textiles (mostly cotton or jute) and ranks third in the Orient, after Japan and Communist China, as a producer of iron and steel. Her proven reserves of high-grade iron ore are larger than those of any other country. She also has abundant coal reserves, though comparatively little of her coal is suitable for coking. She ranks third in the world as a producer of manganese (after the Soviet Union and South Africa), and is a minor producer of various other minerals. Pakistan's mineral resources are relatively meager, but do include substantial reserves of natural gas, salt, and chromite, and small deposits of other minerals, including oil and low-grade coal. Both countries have sizable reserves and a modest but growing production of hydroelectric power. India's reserves and production are much larger than Pakistan's. The annual production of agricultural commodities in the two countries bulks large in sum total, but is dangerously small in terms of the enormous number of people to be supported. There are few important agricultural exports, though India is the world's second largest exporter of tea and Pakistan is the largest exporter of raw jute and a sizable exporter of raw cotton.

Some further characteristics and aspects of the subcontinent are summarized in the following selection from an important book on India and Pakistan by a regional specialist on South Asia, W. Norman Brown:[2]

The population figures show that the subcontinent has a large labor supply. Not only is the source practically inexhaustible; Indian labor also had a tradition of skill celebrated in Europe from the times of classical Greece and Rome. Five centuries ago India was producing wares which induced Europe to seek a water route to that land. Hence the discovery of America! . . .

Intellectual Achievements and Potentialities

Besides economic potentiality, India and Pakistan have the prospect of significant intellectual accomplishment. In the Indian subcontinent flourish today two of the world's greatest historic civilizations—the native Indian and the imported Islamic. Of these the first has existed for more than four thousand years. Islam, though younger in India, is the heir to a thousand years more of civilization in the Near East. Both of these have been supplemented and turned in new directions in the subcontinent by Europeanism since it established itself militarily and politically in the latter half of the eighteenth century. Thought in India has never been static, nor has social custom. As Jawaharlal Nehru repeatedly points out in his *Discovery of India* (1946), it is just the adaptability of Indian civilization to new ideas and its conjunction of intellectualism with imagination which enabled it to modify or abandon outworn institutions and maintain its vitality and strength by developing new ones. It is this same adaptability which makes India and Pakistan potentially strong for intellectual development in the modern world. It is reasonable to expect that in the future, as in the past, the Indian subcontinent will be intellectually creative and may some day come to rank with Europe and America in scientific, social, and humanistic accomplishment. It may rival, though not duplicate, the great Western nations.

•   •   •   •   •

Elements of Weakness

But accompanying these potential sources of strength, India and Pakistan have serious elements of weakness. One is the conflict between the Islamic and Hindu communities which separated the subcontinent into two political entities in 1947 and has since led to further strife in both nations. There are perilous divisions within each country among social groups, linguistic groups, and geographical regions. The economy of each remains weak. Poverty is general, and sudden disaster can produce a famine, local or widespread according to the circumstances. The population problem is critical. Before these two countries can realize their potentialities, they must get these various problems under control.

A LAND OF VILLAGES

India and Pakistan live today, as they have done in the past, primarily by means of agriculture, with relatively little supplement from industry. The imbalance is shown by the small number of industrial workers. . . . [Modern factories in the two nations employ only around 4 million persons. . . . This is less than 1 percent of the total population.] Whether you travel in the subcontinent by rail, automobile,

---

[2] Reprinted by permission of the publishers from W. Norman Brown, *The United States and India and Pakistan.* Cambridge, Mass.: Harvard University Press, Copyright, 1953, 1963, by The President and Fellows of Harvard College. Pp. 3–15 in the 1963 Revised and Enlarged Edition. All headings and one editorial footnote have been added, and some material has been inserted in brackets without compromising the sense of the original.

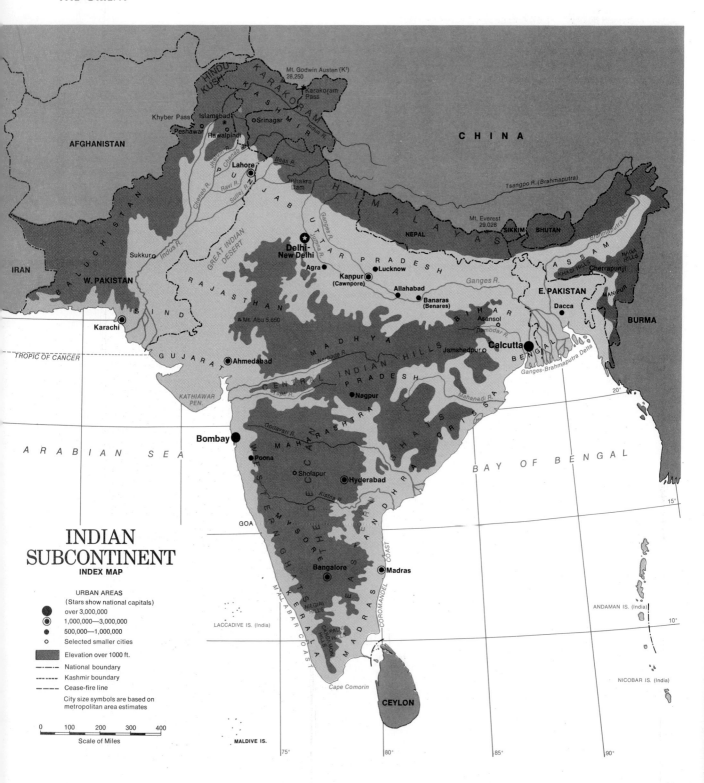

INDIAN
SUBCONTINENT

INDEX MAP

URBAN AREAS
(Stars show national capitals)

● over 3,000,000
◎ 1,000,000—3,000,000
● 500,000—1,000,000
○ Selected smaller cities

Elevation over 1000 ft.

—·—·— National boundary

········· Kashmir boundary

— — — Cease-fire line

City size symbols are based on
metropolitan area estimates

0   100   200   300   400

Scale of Miles

or air, you emerge suddenly from one of the few large cities to proceed for hundreds of miles across open country, dotted with drab little villages, lightened by only an occasional white-washed Hindu shrine or Muslim [Moslem] mosque. Here and there you come to a small town. There is nothing to compare with the Atlantic coastal stretch of almost continuous urbanization from Portland, Maine, to Washington, D.C., or with some other American industrial areas, as around Pittsburgh and Chicago.[3] . . . Most of the people . . . live in the more than 650,000 villages, which are [mostly between 500 and 5000 in population]. Of the villagers much the greater part (about 70 percent of the total population) are cultivators of the adjacent land; the others supply services or pursue handicrafts (weaving, pottery, metal work, oil pressing, or other). The country, as one sees it, consists of clusters or even long stretches of tiny fields, streaked with inarable land and jungle. Except at a few centers like Calcutta, Bombay, Madras, Jamshedpur, Asansol, Ahmedabad, Kanpur (Cawnpore), Sholapur, Karachi [map, left], factory chimneys are solitary or absent. . . . In the hills and mountains, where agriculture is more difficult, villages are fewer and smaller. In the deserts . . . they are still scarcer. . . .

THE PERENNIAL PROBLEM OF WATER SUPPLY

Agriculture in almost every part of India and Pakistan has to depend upon a scanty or fickle water supply, and the possession of water is a perennial and never fully solved problem. Nowhere else do so many people put so much labor into getting water; and nowhere else is the problem of getting it more dramatic. In a region where great rivers can be used for irrigation, such as the Punjab or Sind, the water is impounded by huge dams or barrages and led off through canals. The major outlets are large; from these run off smaller canals, and from these still smaller ones, and so by graduated decrease until the water finally reaches the fields through minute capillary-like distribution. For part of the year many of these channels must be cut off, to save the water until it is needed later. At the latter time the peasants are busy throughout the day opening the sluices leading to their land so as to get the maximum flow during the hours for which they have contracted and to direct the water first to one part of their farms and then to another. This is the easiest of India's ways of using human labor to water fields, and the most certain. The rivers of north India are largely fed by the melting

snows of the Himalayas and the rain that falls upon their southern ranges during the annual summer monsoon. At that time the clouds, after crossing the plains, have to rise, are cooled, and so are made to condense and precipitate their moisture. River-based irrigation is also practiced in other parts of the subcontinent, and since independence new schemes of irrigation have been inaugurated in all parts of the subcontinent, many drawing their water from rivers which are not fed by the Himalayan snows. Still other schemes are under construction or are planned. The flow of water in the rivers fluctuates from year to year, but in no year does it cease entirely. Hence, peasants who live in regions where there are river-fed irrigation systems are usually better off and less subject to crop failure and famine than those who must rely upon other sources.

In many parts of India and Pakistan the farmer gets water for his fields from a well. In North India he may use a great Persian wheel, filling the air with an interminable creaking, which dips an endless chain of earthenware pots into the water, raises them above ground level, and empties them into a trough from which the water flows through runlets to the land. A blindfolded camel or a pair of bullocks operate the mechanism, treading a tiny circle all day long while a man or boy stands by to keep them moving. Elsewhere a peasant may use a buffalo or a bullock or a pair of bullocks, which alternately raise and lower a leather sack into the water. There is a ramp built up to the well, rising above its mouth some six feet or more. Along it the bullocks tread forward and backward, relaxing the rope to which the sack is tied, and so letting it fall, then, when they reverse, raising it to the wooden crossbeam at the high end of the incline, where it is emptied into a channel to flow away. It takes one man to empty the sack, another to drive the bullocks. Very poor peasants who own no bullocks and cannot arrange to use those of a neighbor may operate such a well with only human labor, the subcontinent's cheapest commodity, drawn from the ranks of the family.

In still other areas peasants raise the water with a sweep, like an Egyptian *shaduf*. This is a seesaw-like apparatus with a leather sack or an earthenware pot at one end of the beam and a counterpoise balanced at the other. All day long a man may walk back and forth along the beam, first lowering the container into the well and then raising it to the top, where another man empties it into a runway. Or at the extremity of the sweep there may be a bamboo pole with the vessel fastened to its end. A man stands beside the well forcing the pole down until the vessel is submerged, then letting it go so that the counterpoise can raise it to the surface. In South India one may see a channel or ditch full of water, into which two peasants, often women, jointly dip a shallow scoop, rhythmi-

---

[3] *Editorial note:* Only an estimated 8½ percent of India's population in 1965 and perhaps 7½ percent of Pakistan's in 1961 lived in metropolitan areas of 100,000 or over. In the United States an estimated 63 to 64 percent of the population in 1965 lived in Standard Metropolitan Statistical Areas of 100,000 or over (see footnote 2, Chapter 30).

cally swinging it back and forth between them, lifting the water and emptying it into a higher channel through which it flows to the fields.

Throughout the land there are hundreds of thousands of "tanks," or artificial ponds, which are filled by the annual rains and serve in the dry season for irrigation, washing, even drinking. It is important to keep these tanks in good working order. Silt must not be left to accumulate; plants, such as the rapidly spreading water chestnut, must be prevented from choking them and absorbing the water supply. A tank has to be cleaned out regularly. But when it is cleaned, it is necessary to do the work carefully, so as not to dig out the bottom and let the water seep away.

.   .   .   .   .

PHYSICAL GEOGRAPHY OF THE SUBCONTINENT

What are the facts about the subcontinent's topography, its fertile plains, its river systems, its mountainous areas, its deserts, its basic water supply? [Maps, pp. 405 and 420.]

The Outer Wall of High Mountains

The subcontinent is shaped roughly like a quadrangle—or, more precisely, like a pentagon, though the fifth side is disproportionately short. One long point of the figure is the Deccan peninsula projecting sharply southward into the Indian Ocean, with the Arabian Sea on its west and the Bay of Bengal on its east. The rest is mountain-bound. Along the northwestern side, now held by Pakistan, is a protective barrier of [mountains] and desert, generally difficult for armies or peoples on the trek, yet penetrable at various points, and providing the chief means of ingress to India by land throughout recorded history. The northern side is a concave arc of lofty mountains, containing the world's highest peaks. . . . These ranges are geologically young and they and the nearby plains are disturbed by frequent tremors and occasional disastrous earthquakes. . . . On the east the short fifth side is the Burma frontier, whose jungle-covered mountains are all but impassable.

The Northern Plain

Inside the northwestern and northern walls is the great plain of the subcontinent. It extends from the peninsula of Kathiawar in the present Republic of India beside the Arabian Sea, in a direction slightly west of north to Sind in Pakistan; thence a little east of north for about 700 miles to the Himalayas, including all of West Pakistan except some mountain areas. From here it turns eastward into India below the curve of the Himalayan chain, across the breadth of the country and coming to an end against the hills of Assam and East Pakistan and the coast of the Bay of Bengal.

This plain varies from 80 to 200 miles in width. In its sweeping course it contains the greatest of the subcontinent's river systems. In the west is the Indus, which rises on the northern side of the Himalayas, flows westward behind them to round their end, separating them from the Karakoram and Hindu Kush ranges, and then drives inside the northwestern frontier down to the sea. The Indus is fed by the five rivers of the Punjab ("Land of Five Streams"), one of which (the Sutlej) also rises north of the Himalayas not far from the source of the Indus, but, unlike it, finds a route directly through them. The four others (Jhelum, Chenab, Ravi, Beas) rise in the range and flow directly to the lowlands. In the upper Punjab the moderate rainfall and the irrigation based upon its rivers support an extensive agriculture. Farther south, in the lower Punjab and Sind, rainfall is scanty, varying from ten to twenty inches annually and in many places being even less. Here agriculture is impossible except with the aid of irrigation. This has been practiced for millennia, but never on a scale to support a large population. Today there are great systems of irrigation in this region, and others are under construction or planned. East of the long course of the Indus is the Great Indian [Thar] Desert. . . .

The northern part of the Great Indian Plain, often called Hindustan, contains the Ganges-Jumna (Jamna) river system, which only a low rise of ground separates from the Indus system. The soil of this region is a deep alluvial deposit. Its two main rivers, the Ganges and the Jumna, rise on the lower side of the Himalayas and curve southeast in gradually converging arcs until they unite at Allahabad. From there the Ganges continues eastward to Bengal, absorbing many tributaries, to join the Brahmaputra. This last, like the Indus and the Sutlej, rises on the northern side of the Himalayas, but, as though to polarize the Indus, flows eastward to turn and circle the eastern end of the mountain chain, where it finds an opening, reverses itself, and flows southwest. It and the Ganges unite and form an immense delta. Their waters reach the Bay of Bengal through many mouths, steadily depositing silt, and today as for millennia in the past, continually projecting the land area into the bay.

The part of the northern plain east of the Punjab, already favored by its rivers, is in normal years also well watered by seasonal rains. It is agriculturally the most desirable part of India and has always been the goal of invaders. Its population density is around 800 per square mile and, though it comprises only about one-sixth of the subcontinent's total area, it contains about two-fifths of its total population.

Peninsular India

Below [south of] the northern plain is a complex highland, the upper end of which is embraced by the two extremes of the plain. Along its northern part are various ranges of low hills, of which the highest peak, Mount Abu, is 5650 feet in elevation. The terrain of central India makes it a difficult area to traverse, and permits it to support only a moderately dense population. Indian literature has for 2500 years spoken fearfully of the wild jungles and the primitive peoples in this area. . . .

Still farther south is the part of the plateau known specifically as the Deccan ("South"), which comprises most of the triangle of peninsular India. This tableland (varying from about 1000 to 2500 feet in elevation) tilts gently from west to east. Its great rivers rise on the western side, flow eastward across it, and empty in the Bay of Bengal. It is not well watered, either by streams or by rainfall, and much of its area is rocky or has soil of only inferior quality. Nevertheless it supports [between 200 and 300] persons to the square mile.

The Deccan is bordered on east and west by low ranges of mountains known as Ghats ("Steps"). The Western Ghats, a kind of seaboard scarp, which have a few peaks of approximately 5000 feet but average around 3000 feet, descend in thickly forested, bold declivities to the alluvial seaboard. The southern part of this shoreline, known as the Malabar coast, is one of the best-watered, most fertile, and most thickly populated parts of India, having well over 1000 persons to the square mile. On the other side of the peninsula the discontinuous Eastern Ghats, averaging about 1500 feet in altitude, are less picturesque. They lead down to another well-watered, productive, and thickly inhabited alluvial plain, wider than that on the west and known as the Coromandel coast. The central plateau terminates southward in clusters of hills called the Nilgiris ("Blue Mountains") and Palni, which respectively have peaks as high as 8640 and 8841 feet. Finally, . . . at the extreme south, are the Cardamon Hills. Beyond these last is Cape Comorin, the southernmost point of India, east and south of which lies the fragrant island of Ceylon.

. . . . .

ROLE OF THE MONSOON RAINS

The most important climatic feature of the subcontinent is the annual southwest monsoon, which brings "the rains" and gives India 90 percent of its heaven-dispensed water. So impressive has this phenomenon been upon India's consciousness that in her languages the commonest words for year primarily mean "rain" or "rainy season."

. . . . .

If "the rains," that "annual gamble" from the southwest, are "normal" and widespread, the subcontinent is prosperous. That is, people do not actually suffer starvation; the government can collect the land revenue; the peasantry do not have to borrow from the village moneylenders at a ruinous rate of interest and may even do something toward reducing the principal of their debts. But if the rains are scanty in any area or fail, not only do the fields get no direct water from heaven; the sources of irrigation dwindle too. Rivers fall; the village tanks are not replenished; the water table is lowered; wells dry up. So, too, if in northern India, the rains are too full and make the rivers flood . . ., seed may be washed out, cattle carried off, villages destroyed, and ruin come upon the peasantry and their land. Where there is irrigation from snow-fed streams, as in the Punjab, the case is not so desperate, for the mountain slopes always get a share of rain, which ultimately collects in the rivers. Elsewhere the inevitable result is poor crops or none at all. Agriculture stops; food is exhausted; there follows "distress," "scarcity," or "famine"; and relief must be brought in from outside. Such conditions have been reported since the third century B.C., just after India's historical records start; the case has not been different since.

The Wet Monsoon

The southwest or wet monsoon is at its height during the months from June to September, and most parts of the subcontinent receive the bulk of their annual rainfall during those months (Table 14). Two main arms of this monsoon can be discerned. One arm, approaching from the west off the Arabian Sea, strikes the Western Ghats and precipitates heavy rainfall on these mountains and the coastal plain at their base. But the amount of rain diminishes sharply in the interior Deccan to the east of the mountains. Here the annual precipitation over a sizable area is barely sufficient for unirrigated agriculture and in some years is so low as to result in serious crop failures.

The second major arm of the monsoon, approaching from the Bay of Bengal, brings moderate amounts of rain to the eastern coastal areas of the peninsula and heavy precipitation to the Bengal delta region (Ganges-Brahmaputra delta) and the northeastern Indian state of Assam. Exceptionally heavy rainfall occurs on the forward slopes of low mountains that rise behind the delta. One climatic station, Cherrapunji in the Khasi Hills on the border between Assam and East Pakistan, has an average annual rainfall of 425 inches (over nine-tenths of which occurs in the summer half-year), and thus ranks with a station on the island of Kauai in the Hawaiian group as one of the two spots with the greatest annual rainfall yet recorded on the globe. Part of the moving air in the Bay of Bengal arm of the wet monsoon passes up the Ganges Valley and precipitates moisture that diminishes with some regularity in total amount from east to west. Both major arms of the monsoon bring some rainfall to West Pakistan, but the total is so small as to result in semiarid or desert conditions in most areas.

The Dry Monsoon

The monsoon of the winter half-year is often called the northeast monsoon, though it blows generally from the west over most of the northern plain. The wind is often northeasterly, however, over the peninsula, the Bay of Bengal, and the Arabian Sea. To most parts of the subcontinent this monsoon brings dry weather, with occasional light rains in areas outside the tropics. An exception is found in the far south of the peninsula, where the heaviest rainfall of the year occurs along the eastern coast and in adjacent interior uplands during the period October-January. This precipitation is brought by retreating maritime air of the summer monsoon and by continental air of the winter monsoon which has accumulated moisture in its passage over the Bay of Bengal. In addition to widespread drought, the winter monsoon brings cooler weather to the subcontinent, especially in the north. But a period of stifling heat is experienced in the spring before the onset of the rainy season.

THE EXTENT AND EFFECTS OF POVERTY

Standards of living in the Indian subcontinent are among the lowest in the world. The extent and effects of the subcontinent's poverty have been summarized by Brown:[4]

The bulk of the village population gets only the most meager living in terms of food, clothing, and shelter. Urban factory labor lives no better, possibly worse. Without seeing Indians in their villages, towns, and cities, it is difficult for a Westerner to visualize the extent and effect of their poverty. . . . If the average American visitor wants to remain sensitive to the conditions in which the masses of the people live, it is well for him not to stay in the country long. Very quickly the want, the disease, the discomfort, the misery, become only accepted facts.

[In the middle 1960s the United Nations estimated the *average* individual diet in India to be 1990 calories a day and in Pakistan 2220—against a needed *minimum* of 2300 calories. The United States average, by no means the world's highest, was 3140 calories. During the middle 1960s famine occurred in some parts of India, though not on the massive scale of some past famines. Even when the diet is comprised of sufficient calories, it is apt to be poorly balanced, with a great deficiency of health-protecting foods.]

Housing is equally inadequate. In the villages most dwellings are made of mud and wattle or sun-dried brick, crowded together in an irregular huddle, affording little protection from the winter cold, the burning heat of summer, and the torrents of the rainy season. . . . [A United Nations estimate for 1960 put the average number of people per room of dwelling space in India at 2.6 and in Pakistan at 3.1 (United States, 0.7), both figures being among the highest in the world. Often urban quarters are] without a chimney or a window, with no lights or water supply, and no sanitary arrangements. . . . Every city, at least during the present century, has had a large number of people with no housing at all, who sleep each night in the open.

In typical village and urban dwellings furniture scarcely exists. A house, or hut, has a fireplace consisting of a few bricks or stones or molded clay set to form three sides of a rectangle over which a pot or pan can be placed; it may also contain a few metal cooking vessels and some primitive implements for farming or the pursuit of a handicraft. That is likely to be all. Scavenging is a function of the village dogs. With these basic handicaps to health has gone heavy incidence of disease and paucity of preventive and curative medicine. . . .

---

[4] Reprinted by permission of the publishers from Brown, *The United States and India and Pakistan,* pp. 16–19. Material has been inserted in brackets in two places without compromising the sense of the original.

[5] For the early 1960s, estimates which are probably not highly reliable place the annual crude death rate in India at about 13 deaths per thousand population, and in Pakistan at 16 to 17 deaths per thousand. These estimates compare with a rate of between 9 and 10 for the United States in the same period. The Indian rate has been greatly lowered in recent years, notably by the control of malaria, but is still exceeded by very few countries outside of Africa. Average life expectancy at birth in India was 45 years, a drastic improvement from life expectancy of 32 years in 1941. This compared with a figure of 62 years for the United States in the late 1950s, increased to 70 by the middle 60s. Life expectancy in India and Pakistan is, of course, scaled sharply downward by high infant mortality. This was estimated for India during the 1950s as 139 infant deaths per 1000 live births. In the mid-1960s, by which time the Indian rate had probably dropped somewhat, the United States infant mortality rate was 25. Although comparisons are made with the United States here, this is not meant to imply that American figures in these respects were the best in the world. They are not as good as those of a few European countries.

The combined effect of poor diet, insufficient clothing, substandard housing, lack of medical resources, is a high mortality rate.[5]. . . The brevity and ills of life in India have often been held responsible for her preoccupation with religion, emphasis upon family organization, and intense desire to have progeny and have it early in life, thus conducing to early marriage. In our time they look like an invitation to extremist remedies, such as Communism or any other that claims to have a quick cure for social ills.

For some five millennia man in the Indian subcontinent has not merely held his own against the disadvantages which nature puts upon him, but has searched out and utilized means to maintain a life of high achievement in the arts of civilization. If on the spiral of history South Asia once was more accomplished in those arts than the West but now is less so, it may again reach a position of equality. That, at least, is the hope of many citizens of India and Pakistan. But the two young nations started life after a wearying struggle to achieve independence, and a destructive conflict between Hindus and Muslims, the ill effects of which still continue. With no time granted for recuperation, they have had to attack their critical basic living problems, build new sources of national strength, and assume international responsibilities. Their resources and energy have not yet been equal to the demands, though there has been progress.

The "basic living problems" which India and Pakistan must attack are among the most difficult and pressing to be found anywhere in the world. Though the various problems are closely interrelated, they may be classified for convenient discussion under the headings of agricultural problems, industrial problems, problems of social and political relations, and population problems. In the following discussion the major problems are viewed against the general economic and social order and regional articulation of the subcontinent.

AGRICULTURAL CHARACTERISTICS AND PROBLEMS

From the standpoint of total production of agricultural commodities the Indian subcontinent is clearly one of the world's outstanding agricultural areas, as indeed it has to be to support its immense population even at the present low level of subsistence. It accounts for one-tenth or more—in some cases much more—of the annual production of at least seven or eight of the world's major crops and it has almost one-fifth of the world's cattle. Its enormous cattle population is of limited benefit, however, as will be seen.

Major Crops of the Subcontinent

Major Food Grains

Somewhat more than half of the cropland in India and Pakistan is devoted to a group of major food grains which furnish most of the calories in the meager diet of the people. In order of acreage these include (1) rice, (2) the combined acreage of various millets and sorghums, and (3) wheat, a poor third in acreage but a much stronger third in total yield. Rice is the leading crop of the subcontinent in both acreage and production. It occupies over 20 percent of the cropland of India and over 40 percent of the much smaller cropped area of Pakistan. The total rice production of the two countries combined amounts to more than one-fourth of the world total. Rice is generally preferred as a food over other grains by the subcontinent's people, and under favorable conditions it gives a markedly higher food yield per unit of land than other cereals. Thus it generally dominates agriculture where there is sufficient water and level land. The main rice-producing areas include (1) the delta area of Bengal and the lower Ganges Valley—areas in which unusually heavy monsoonal rainfall is supplemented by floods—and (2) the coastal lowlands fringing both the eastern and western sides of peninsular India. These rice-growing areas are the most productive agricultural sections of the subcontinent, but are not necessarily the best fed, as they are also the most densely populated areas. (See pp. 406 and 84 for maps of world rice and world wheat production.)

Millets and sorghums are grown principally in areas of marginal water supply, where a low rainfall cannot be supplemented by irrigation. The average yield of these grains per unit of land is relatively low—perhaps one-third that of rice. Some millets appear to be more nutritive than rice, but rice is generally preferred over them as a food. In the Republic of India the annual production of millets and sorghums combined amounts to almost one-fourth of the world total, and the total acreage devoted to these crops is greater than the acreage in rice. But they are relatively unimportant in Pakistan, and rice exceeds them in acreage within the subcontinent as a whole. Millets and sorghums dominate the agriculture of large areas in the interior of peninsular India where rainfall and irrigation water are too limited for other grains to thrive. They are supplemented, however, by patches of irrigated rice in local areas where water is more abundant. Millets and sorghums are associated in general with less productive and less populous areas than those that depend primarily on rice. But even these areas often have an average population density of 200 or more per square mile. In such areas these crops provide a basis for life that other grains could not.

Wheat, the third great food crop of the subcontinent, is grown principally on the northern plain in a belt extending from the Punjab to the middle Ganges Valley. It is the most important food crop of West Pakistan and adjoining parts of India, occupying a larger acreage in those areas than any other crop. Precipitation decreases markedly along the northern plain from east to west until steppe and semidesert conditions are reached in the Punjab, which produces more wheat than any other region of equal size in the subcontinent. (It may be noted that the Punjab formed a single province in the Indian Empire but is now politically divided between India and Pakistan, with much the greater part of the total area and wheat acreage being in Pakistan.) Thus wheat is of greatest importance in some of the driest farming areas of the subcontinent. It is produced, however, not by extensive machine methods, as in semiarid parts of Anglo-America or Australia, but by intensive cultivation of small irrigated holdings. Water is supplied by a system of irrigation dams and canals that may well be the world's largest. Originally a product of British governmental and financial enterprise, this irrigation system is still being extended. During the last 70 years or so, a rapid colonization of the new irrigated lands has taken place. Thus wheat-growing areas which were sparsely populated before the irrigation system was built have now become intermediate in population between the heavily populated rice lands and the millet and sorghum lands of the subcontinent, and the wheat surpluses that the new irrigated sections were able to provide at first have practically vanished. However, the average yield of wheat per acre in the Punjab is nearly twice the average yield of millets and sorghums for the subcontinent as a whole, and the average population density of the Punjab is still well below that of the rice lands. Thus the Punjab is a relatively well-fed and prosperous part of the subcontinent.

Minor Food Crops

Minor food crops in the subcontinent are legion. Outstanding among these are various kinds of peas and beans and a variety of oilseed crops. Peas and beans are widely grown over the subcontinent and supply vegetable

Immense quantities of low-grade cane sugar are produced for household use in the Indian subcontinent. The bullocks in the photo are supplying motive power for a small sugar-cane press. The crushed cane pulp in the foreground will be used for fuel. (Foreign Agricultural Service, U.S. Department of Agriculture.)

protein in an area where meat is not plentiful. These legumes grow reasonably well under unfavorable conditions of soil and climate, and they benefit the soil by adding to its nitrogen content. Oilseeds are also widely produced, although the main crop in this group, peanuts, is concentrated largely in peninsular India. Oilseed crops supply cooking oil for the households of the subcontinent and are a considerable export item. Corn, barley, sugar cane, coconuts, bananas, spices, and a long list of other minor crops are grown. The overall magnitude of the subcontinent's agricultural production is reflected in the fact that the total output of one of these "minor" crops, sugar cane (photo, left), amounts to over one-quarter of world production and ranks India second only to Cuba as a cane producer. The middle Ganges plain is the most important producing area.

Industrial and Export Crops

Cotton, jute, and tea are the principal crops grown in the subcontinent for industrial use or export. India has the largest cotton acreage of any country in the world, although her average yield per acre is so low that she ranks only fourth in total cotton production (see Chapter 14 for map of world production). Indian cotton is a small-farm crop, grown mainly in the interior Deccan on certain soils that are unusually retentive of moisture and produce a crop without irrigation. India's large textile industry absorbs most of the country's production of raw cotton and requires supplementary imports, part of which come from Pakistan. Small farmers growing cotton by irrigation in the Punjab and the lower Indus Valley make Pakistan a considerable producer, though its total output is much smaller than India's. With a much smaller textile industry, however, Pakistan is an important exporter of raw cotton.

Approximately three-quarters of the world's jute, the principal raw material for burlap and gunnysacking, is produced annually in the Ganges-Brahmaputra delta region. The canelike jute plant is grown in several feet of water and requires a long growing season with high temperatures. Its production also requires much hand labor. Thus it is well suited to the delta region, which is tropical, practically amphibious, and settled by a dense agricultural population. The plant became the commercial mainstay of the area in the nineteenth century, after British entrepreneurs had developed the necessary technology. The Ganges-Brahmaputra delta lies in the former British province of Bengal, which was partitioned between India and Pakistan at the end of British rule. Most of the jute-growing areas went to East Pakistan, while the jute-manufacturing plants, which had developed in the

Calcutta area, went to India. Since partition, India has markedly increased the production of raw jute on its side of the border. It now produces more than Pakistan, from which it continues to import sizable quantities of raw fiber. Pakistan supplies around nine-tenths of the world's total exports of raw jute, and one-third of its export revenues come from this source, while jute manufactures provide almost one-fourth of India's exports.

Of the three major commercial crops in the subcontinent, only tea is exported from the Republic of India on a sizable scale. But in tea growing and exporting India leads the world, accounting for one-third of world production and exports. Tea is India's second most important export after jute products. Unlike cotton, jute, and most other crops of the subcontinent, tea is principally a plantation crop. The plantations were developed and are still mainly owned by British interests. Production centers in the northeastern state of Assam, with a secondary center in the mountainous Nilgiri "Hills" and adjacent mountainous sections in the far south of peninsular India. Development of tea plantations in both Assam and south India was favored by the availability of lands which were lightly occupied, reasonably fertile and well drained, and supplied with ample moisture by a heavy rainfall.

General Aspects of the Agricultural Situation

The huge and varied agricultural production of the subcontinent must, unfortunately, be balanced against the fact that approximately one-sixth of the world's people live there and support themselves primarily by farming. The average production per man and per acre is abysmally low, despite the pressing need for food that is indicated by chronic undernourishment of large segments of the population (probably the majority) in India and Pakistan, coupled with periodic famines in various areas.

Trends in population and food production in the subcontinent offer only small grounds for optimism. It has been estimated that in 1941 the production of food grains in the Indian Empire was still at the same general level that it had been in 1921, despite the fact that population had increased during this period from 306 to 389 million. An estimate for the Republic of India indicates that total food production in the area which India now occupies increased about 23 percent between 1936–1938 and 1956–1957, but that during the same period population increased by about 29 percent. Thus the trend of *per capita* food production in both of these periods was downward. Some hope arises from the fact that this latter trend appears to have been reversed, though just barely, in more recent years. Between 1956 and 1965 per capita food production increased 2 percent in India and 7 percent in Pakistan. It is heartening that this involved increases in *total* food output of 22 percent and 27 percent, respectively, but frightening that these considerable accomplishments gave such small gains when set against population growth. In recent years very large grain imports have been necessary to fend off starvation. Paying for such imports is itself a major problem, especially when imports of capital goods for industrialization are so badly needed.

Low Productivity of the Individual Farmer

The shortage of food and the extreme degree of rural poverty over the subcontinent are closely related to

Cows, held sacred by Hindus, may frequently be seen on the sidewalks and streets of Indian cities. Those in the photo are on the sidewalk of a business street in Calcutta. (John Morgan of the Charleston, W. Va., *Gazette.*)

Making cow-dung cakes, India's most common cooking fuel. The number of cakes in the photo indicates a commercial rather than a household operation. (Foreign Agricultural Service, U.S. Department of Agriculture.)

the individual farmer's lack of productivity. One of the reasons for the latter is the small size of farms. In the years between 1891 and 1941 the average size of farm in that part of the subcontinent for which statistics are available decreased from 2.23 acres to 1.90 acres, and this trend has apparently continued. Many "farms" total less than a single acre. Many peasants have been detached completely from the land and have come to form a growing landless rural "proletariat."

Even if the individual farmer were able to make the most of his small holding his total production could not be large, but he suffers from additional handicaps. His time is often wasted in traveling between parts of his holding which are widely separated, for fragmentation of farms into numerous small, separated plots·has been the rule. Cases are recorded where one acre has been split among the holdings of sixteen different cultivators. The efficiency of his labor is decreased by lack of adequate equipment. It has often been noted that the plows of many farmers in the subcontinent are no more than ironshod sticks which merely scratch the soil rather than

turn it over, a practice which is adequate for some crops but far from optimum for others. Tillage is retarded in some areas by lack of adequate draft power, despite the huge number of cattle, principally oxen and water buffaloes, whose primary function is to serve as draft animals. In these circumstances the Hindu prohibition on the killing of cattle may well have a rational basis, considering the contributions of draft power, milk, and dung for fuel and fertilizer made by the enormous cattle population, and also considering the fact that many of these cattle are starving scavengers that do not compete with humans or work animals for food, but occasionally produce calves.[6] However, feed and forage are so scarce, and scientific animal husbandry so unknown or difficult to practice, that the cattle population is inadequate for its functions despite its impressive numbers and "protected" status.

On his small holding the farmer of the Republic of India obtains some of the world's lower crop yields. Rice yields per acre in India are approximately one-third the average yield in Japan, which has the Orient's most

<hr />

[6] See the argument to this effect in Marvin Harris, "The Myth of the Sacred Cow," *Natural History,* 76, no. 3 (March 1967), 6–12A.

productive agriculture. Wheat yields averaging 7 to 9 bushels per acre are the lowest for any major producer and among the lowest anywhere. Yields in Pakistan are generally a little better than in India, but are still well below the world average. Such low yields are the result of poor techniques of cultivation, lack of scientific seed selection, moisture deficiencies, and the poverty of the soil. Most of the soils now in use have been intensively worked for many centuries without adequate provision for maintaining their fertility. They have been leached by the concentrated rains of the monsoon season, and they receive little or no fertilizer. Most farmers cannot afford artificial fertilizers and, unlike Japanese or Chinese farmers, are prejudiced against the use of human wastes for fertilizer. Animal manures must often be used as fuel for cooking (photo, p. 429), due to the lack of wood over extensive areas long since deforested.

That the low average yields of the subcontinent are not absolutely necessary is shown by the success that has attended experiments aimed at increasing them. Village projects set up by the Indian and Pakistani governments, often with American financial aid, have succeeded in raising yields more than 50 percent through improved farming techniques, including the provision of better, though simple, equipment, scientifically improved seeds, and a modest amount of fertilizer.

However, before much in the way of permanent improvement can be accomplished, the peasant must have some incentive to improve and must be able to obtain credit on reasonable terms. Incentive for many peasants has been decreased by a landlord system whereby the peasant holds his land only as a tenant and must pay for its use with anywhere from one-sixteenth to over one-half of his produce. In addition, the economic insecurity of the peasants has led to a very large proportion, and in some areas the majority, being permanently in debt to the village moneylender. Studies of different areas have shown interest rates varying from 9 to 300 percent. It is generally recognized in India and Pakistan that these conditions must be altered in order to offer at least a minimum of hope to the cultivator if he is to exert himself toward improvement. Without such exertion on the part of individual farmers, any program for solving the agricultural problems of the subcontinent seems doomed from the start. In both India and Pakistan governmental programs are under way to eliminate large land holdings and control interest rates.

Possibilities of Land Reclamation

One obvious path toward solving the problem of greater food production is the reclamation of additional land to be farmed. The actual possibilities seem quite uncertain and are a matter of great debate, both among native inhabitants of the subcontinent and outsiders. For many years the cultivated acreage has expanded very slowly despite great and increasing need. Present farming methods seem incapable of coping with the aridity, roughness, and lack of fertility which characterize large areas not now farmed. Modern mechanized methods, especially in connection with increased irrigation, hold considerable promise, but are expensive to install. Nevertheless, several major irrigation and reclamation projects, and many minor ones, are under way in the subcontinent. Most experts seem to feel that reclamation alone is far from being an adequate solution to the problem, and that higher crop yields from the existing acreage offer a more promising approach. Improved yields will have to be obtained largely by application of fertilizer to the exhausted land, though improved water supply facilities and other applications of scientific and engineering skill can make an important contribution.

INDUSTRY AND ITS PROBLEMS

In the partitioning of the Indian Empire, the Republic of India received almost all of the developed capacity in modern industry which existed in the subcontinent at that time. It also received the great bulk of the area's mineral and power potential. Pakistan was left with a much smaller industrial plant and a poorer endowment of natural resources.

India, however, inherited a sizable industrial structure, which it has since expanded. In the Orient India stands third to Japan and Communist China in total industrial development. The labor force employed in factories in India can be placed at around 3 million. In addition, something like 16 million workers are still employed in handicraft production.

The Major Industries

Cotton Milling

The chief branches of Indian factory industry are cotton textiles, jute, and iron and steel. Modern cotton mills employ approximately 1 million workers. The industry is mainly concentrated in the Bombay and Ahmedabad metropolitan areas and neighboring areas along or near the west coast, and in smaller industrial cities in or close to the mountains at the southern end of the peninsula. However, over one-third of the total production comes from more widespread handicraft-scale producers. Of basic importance to cotton manufacturing are the large home market, abundant and cheap labor, hydroelectric power available from stations in the Western Ghats and the Nilgiri Hills, and the large domestic production of cotton. Not only does India's cotton industry supply most of the needs of the domestic market, but its output is sufficient to make India the second largest cotton textile exporter of the world (after Japan) in most years.

Jute Milling

The jute industry is closely concentrated in and near the largest city of the subcontinent, Calcutta. Almost 250,000 workers are employed in the industry, which lost most of its raw material base, and consequently suffered a large loss of employment, when the jute-growing areas of East Pakistan were partitioned off from India. Supplies of raw jute subsequently imported from Pakistan have not always been adequate and reliable. The two nations are highly competitive in the jute industry, with India attempting to increase the acreage in jute in West Bengal and Pakistan attempting to increase the number of jute-manufacturing plants in East Bengal. However, despite some success in both attempts, Calcutta continues to be the world's leading center of the manufacturing side of the industry, while East Pakistan remains the leading fiber-producing area. Calcutta's jute manufactures are India's leading export, somewhat more valuable than tea and more than twice as valuable as cotton goods.

Iron and Steel

India's iron and steel industry is concentrated in the northeastern corner of the peninsular uplands and in immediately adjacent parts of the Bengal lowland. The leading center is Jamshedpur (400,000), approximately 150 miles due west of Calcutta. Iron and steel production is also carried on in two towns near the coal-mining center of Asansol (250,000), north of Jamshedpur, and in two other towns to the southwest of Jamshedpur. The southernmost of the five iron and steel centers is located on the upper Mahanadi River. The most valuable collection of mineral resources in the subcontinent is found in or near this belt of heavy industry. It includes deposits of iron ore, coal, manganese, and tungsten.

These resources have drawn modern industry into a part of India that in other respects is one of the least developed parts of the country, its inhabitants being largely comprised of primitive tribesmen. Steel production in the area was pioneered by the Tata Iron and Steel Works at Jamshedpur, developed by the famous Tata family of Indian industrialists. This plant became the largest single steel-producing unit in the British Empire between World Wars I and II.

But the Indian iron and steel industry has never been large by world standards. As late as 1966 it produced just 2 percent of the world's pig iron and about 1.4 percent of its steel. Production on this scale is insufficient to supply India's needs for steel under its plans for industrial development, and expansion of the iron and steel industry continues to be a major goal of Indian planning. Three of the six main steel plants in the heavy industrial belt were constructed during the 1950s and only began production in 1959 and after. Their output, plus expanded production from an enlarged Tata Works, had more than tripled India's 1958 output of steel by 1966. This has brought the country to a production level above that of such countries as Australia and Sweden, but still below such smaller countries as Belgium, Czechoslovakia, and Poland. Except for coking coal, the resources of the area are adequate for much greater increases in the future. Iron, manganese, and noncoking coal reserves are very large. Coking coal is adequate for the near future, but presents an important long-term problem for Indian metallurgy.

Industrial Diversification: Engineering and Chemical Industries

As India's industrial expansion continues, its output of manufactured goods gradually becomes more diversified and sophisticated. Thus such smaller industries as engineering and chemicals are growing more rapidly than the cotton-textile and jute industries. The machine-building industries are still quite inadequate for the country's needs, necessitating large imports, but they employ over half a million workers and are growing at a good, though not spectacular, pace. Today there are very few items among the great variety of machines used in a modern society which are not produced somewhere in India. "In the Calcutta area . . . new industries engaged in the manufacture of textile machinery, instruments, machine tools, cutlery, screws, bolts, electrical machinery, and transportation equipment have developed during the 1950's."[7] There are several automobile and truck assembly

---

[7] Pradyumna P. Karan, "Changes in Indian Industrial Location," *Annals of the Association of American Geographers,* 54, no. 3 (September 1964), 349.

factories in the country, and recent additions to the array of manufactured products include tractors, bulldozers, data processing machines, and microscope slides. In similar fashion, a larger quantity and wider variety of chemical production, ranging from vitamins to fertilizers, are steadily being achieved, although large imports are still needed. The coal fields of the Damodar Valley and the major textile zones are the main areas in which chemicals are produced.

Industry in Pakistan

Pakistan's industrial inheritance from the Indian Empire was meager compared to that of India, and Pakistan is still far below India in industrial output and degree of industrialization. However, the country has made rapid progress in certain lines since independence. Its principal success has been the development of a cotton-textile industry. Cotton milling scarcely existed in Pakistan at independence, but as early as 1954 the industry had reached a point where it supplied the bulk of domestic needs and even a small export. New jute mills, fertilizer plants, and various consumer-goods factories have also been built. Pakistan appears to be poorly endowed with mineral resources, and this fact throws some doubt on the country's ability to develop a heavy industrial base. It does have some iron ore, low-grade coal, and sizable reserves of chromite. However, substantial reserves of natural gas, found in both West and East Pakistan, are its most important mineral resource. These gas deposits, together with considerable reserves of salt, a little oil, and other chemical raw materials, are providing the basis for an expanded chemical industry. Most of Pakistan's industries and known mineral resources are in West Pakistan.

Retarded Development of Factory-Type Industries and Urbanization

India's industrial potential is undoubtedly a large one, and in some respects the country's present industrial output is rather impressive. If employment and output of all the various factory industries are combined with those of handicraft industries, India probably ranks as one of the eight or ten leading industrial countries in the world. But cotton and jute milling are the only factory industries in which India ranks among the world's leading countries, and the total development of most basic lines of modern factory industry is still very small. The comparative lack of such industries is reflected in a low degree of urban development. Despite India's enormous population total, only eight cities had estimated metropolitan populations of more than 1 million in 1968. These include the major seaports and industrial centers of Calcutta (over 7 million) and Bombay (5.1 million); the political center of Delhi-New Delhi[8] (3.3 million); the principal seaport of south India, Madras (2.3 million); Hyderabad (1.4 million), a state capital in the interior Deccan; Bangalore (1.5 million), the capital of Mysore state and a prominent industrial center in south India's main hydroelectric power area; Ahmedabad (1.4 million), one of the two main centers of cotton milling in the subcontinent; and Kanpur (1.1 million), the largest of several sizable cities in the middle Ganges plain. Only six other metropolitan areas in India had more than 500,000 people according to 1968 estimates. These include Lucknow (760,000), Agra (600,000), Banaras (Varanasi: or Benares; 580,000), and Allahabad (500,000) in the Ganges Valley, and Poona (830,000) and Nagpur (820,000) in the Deccan. Pakistan's largest city, Karachi (2.3 million), is the chief seaport of West Pakistan. Its second largest city is Lahore (1.5 million), the economic capital of the Punjab. Dacca (800,000), the third largest city, is the main city of East Pakistan. Pakistan's political capital is the new and specially designed city of Islamabad (230,000) in the Punjab, built from the ground up during the 1960s for the specific purpose of housing the central government.

Governmental Promotion of Industrial Development

India and Pakistan would benefit greatly from increased industrialization, and strenuous efforts in that direction are being made in both countries. Specific benefits anticipated from industrial expansion include (1) a larger supply of goods for domestic consumption, (2) a greater number of jobs for the ever-increasing labor force, (3) a reduction in nonfood imports, which now impose a considerable drain on financial resources, and

---

[8] Delhi has been the most important political center during much of the subcontinent's history; the adjacent British-built city of New Delhi became the capital of Great Britain's Indian Empire and is the present capital of the Republic of India. Unless otherwise noted, city populations in this chapter, as in other chapters of the text, are metropolitan area estimates for 1968. They should be treated as rough approximations only.

The Bhakra Dam on the Sutlej River is one of the Republic of India's main power and irrigation dams. The dam was still under construction when this photograph was taken in the early 1960s. It has made possible a large output of hydroelectricity and provides a dependable water supply for large acreages in the Punjab and Rajasthan. (Government of India Tourist Office.)

(4) an increased supply of equipment and fertilizers for agricultural progress. Industrial development is being promoted by the governments of the two countries in a variety of ways, such as (1) direct governmental construction of plants, (2) extension of financial incentives to stimulate private investment in industry, and (3) improvement of education and services to provide a better social environment for industry. An important and continuing phase of government-sponsored development in India and Pakistan is the construction of large dams to supply hydroelectricity and impound water for expanded irrigation. Flood control and improvement of river navigation are supplementary objectives in some cases. A number of dams have been completed and others are under construction. Probably the most celebrated river-control scheme thus far is the multipurpose Damodar Valley project in India's main coal-mining area. Through a series of dams the Damodar River has been harnessed for irrigation, power production, and flood control. Hydroelectric development is especially needed in the many parts of the subcontinent that are remote from important coal fields. Both governments are also continuing to press

the search for oil and natural gas. Pakistan's discoveries of gas have lessened the emphasis given to waterpower development in that country.

PROBLEMS OF SOCIAL AND POLITICAL RELATIONS

On a physical map the Indian subcontinent looks like an obvious physical unit, marked off from the rest of the world by its mountain borders and seacoasts. Yet the social complexity of this area is so great that the apparent physical unity has never been paralleled by overall political unity except during a relatively brief period in the nineteenth century and the first half of the twentieth century. During this time political unity was imposed from outside by Great Britain. Even so, however, a great variety of small to fairly large political units under native rulers retained varying degrees of autonomy, although all of these native states were ultimately under British control. As soon as British power was withdrawn, the divisive force of conflicting social groups again asserted itself and the subcontinent split into two nations. At present the governments of both India and Pakistan have serious problems to overcome in reconciling the aims and aspirations of sharply divided groups within their respective countries, as well as in creating satisfactory relations between the two new countries themselves.

Religious Divisions and Conflicts

The major social divisions within the subcontinent are in religion and language. Of these, religious divisions have generally appeared the more important and have received more attention. The most serious division has been between the two major religious groups, the Hindus and the Moslems. Hinduism is a religion native to the subcontinent and was the dominant religion at the time Islam made its appearance. It has continued to have the largest number of adherents. In 1968 Hindus in the subcontinent are estimated to have numbered over 410 million, as compared with somewhat over 150 million Moslems.

Islam made its appearance in the subcontinent as a proselytizing and conquering religion in the eighth century A.D. Periodic later invasions penetrated the relatively weak northwestern frontier until at the peak of Islamic power in the sixteenth and seventeenth centuries the Mogul Empire dominated most of the subcontinent. British penetration of India was aided by the internal disintegration of this Moslem empire in the eighteenth century.

Seldom have two large groups with such differing beliefs lived in such close association with each other. To Islam's uncompromising monotheism and insistence on uniformity in religious beliefs and practices, polytheistic Hinduism opposes the view that a variety of religious observances is consistent with the differing natures and social roles of human beings. To Islam's essential intolerance of all other faiths, Hinduism opposes an essentially tolerant attitude. To Islam's belief in its divine mission to convert all men to the true religion, Hinduism opposes the belief that proselytizing is essentially useless and wrong. To Islam's democratic belief in the essential equality of all believers, Hinduism opposes a social system founded on the inequalities of caste. To Islam's use of the cow as food and for sacrifice, Hinduism opposes the view that the cow is an especially sacred animal and must under no circumstances be killed. The exuberant and noisy celebrations of the Hindu faith are a great contrast to the austere and silent ceremonials of Islam. Bloodshed has sometimes resulted when a Hindu parade with its jingling bells and firecrackers disturbed the solemnity of Moslems gathered for a particularly sacred religious rite.

The antagonism to be expected between such differing groups was intensified when the formerly subordinate Hindus came, under the British occupation, to dominate most Indian business as well as the civil service. Many Moslems feared the results of being incorporated into a single state with the Hindu majority, and their demands for political separation led to the creation of two independent states from the Indian Empire rather than one. The creation of Pakistan in two widely separated parts was due to the distribution of the main areas of predominantly Moslem population at the time of partition. Immediately preceding and following partition violence broke out between the two peoples on a huge scale, and hundreds of thousands of lives were lost in wholesale massacres before the new governments could establish control. Mass migrations between the two countries involved some 12 million people, most of whom became a burden on the country of their choice until they could be resettled and integrated into the life of their new communities, a task not yet entirely finished.

Partition of the subcontinent has left a minority of perhaps 10 million Hindus in East Pakistan and over 50 million Moslems in India. In addition, India has religious minorities of some importance in about 13 million Christians, living mainly in the south of the peninsula, and about 9 million Sikhs, concentrated mainly in the Punjab. Smaller numbers of Jains, Parsees, Jews, and Buddhists are present, mainly in India. Perhaps the most significant, though not the largest, of these groups is the Parsees. Though numbering only somewhat over a hundred thousand, mainly in Bombay, this group has attained wealth and economic power in India far out of proportion to its numbers. The Parsees derive originally from Persia, although the group has been in India for more than a thousand years. Their religion is the ancient Persian faith of Zoroastrianism.

Territorial Disputes between India and Pakistan

The traditional religious conflict of the subcontinent between Moslems and Hindus continues today largely in the form of strained relations between India and Pakistan. Partition created a number of problems, including territorial disputes, which have embittered subsequent relations. One dispute, involving the division of water resources in the Punjab, apparently was settled to the mutual satisfaction of the two nations by an agreement in 1960. The problem stemmed from the fact that all of the major rivers—the Indus and the rivers of the Punjab—on which the life of West Pakistan depends have their upper courses in Indian-controlled territory. How much of their water was India to be allowed to divert and how much could Pakistan count on? The agreement finally reached allocates the water of the three eastern rivers of the Punjab—the Beas, Sutlej, and Ravi—to India, which in return undertakes to allow unrestricted and undiminished flow of the Chenab, the Jhelum, and the Indus itself into Pakistan. A system of new canals is being built to bring water from the western rivers to areas in Pakistan previously irrigated from the eastern streams. Most of this project's cost is being borne by India and a group of Western nations, including the United States.

Certain territorial disputes have proved more intractable, however. The most strident and persistent of these is the dispute over Kashmir, in the mountainous northwestern corner of the subcontinent. Three-fourths of Kashmir's population of about 6 million is Moslem, which is the basis of Pakistan's claim to the territory. Before independence the state was administered, within British India, by a Hindu ruler. Under the partition arrangements, each native ruler was to have the right to join either India or Pakistan, as he chose. The choice of India by Kashmir's ruler is the legal basis of India's claim, although this is somewhat clouded by the fact that India forcibly absorbed certain areas of Hindu population whose Moslem rulers had opted for Pakistan. Fighting after partition led to a cease-fire line leaving eastern Kashmir, with most of the state's population, in India. After 1959, Chinese claims on remote northern mountain areas of Kashmir adjoining Tibet and Sinkiang were pressed and some border territories controlled by India or Pakistan were occupied. Pakistan ceded the territory claimed by China to that country and established friendly relations with the Chinese, while India rejected Chinese claims. Recurrent small-scale military actions have not dislodged the Chinese from the occupied section of Indian Kashmir, however. Raids and skirmishing also have continued along the Kashmir cease-fire line between India and Pakistan.

The situation reached its most serious phase thus far (1968) in 1965. In the spring of that year serious fighting erupted over a lesser disputed area, the Rann of Cutch, along the southern section of the India–West Pakistan boundary. Although this outbreak was brought to a cease-fire by British mediation, it was followed by still more serious fighting, amounting to a short but full-scale war involving tank and airborne forces and widespread air raids, in September of 1965. This conflict began in Kashmir but spread into the Punjab, where heavy fighting took place. The United Nations achieved a cease-fire and the conflict was eventually mediated by the Soviet Union. The basic territorial disputes have not been solved, however, and the situation remains potentially explosive.

The Problem of Caste

In the Republic of India, serious internal problems of religious division still exist, not only between the Hindus and minority groups, but within the body of Hinduism itself. One of the fundamental features of Hinduism has been the division of its adherents into the most elaborate caste system ever known. This system is described in the following paragraphs by Brown:[9]

Every Hindu is in traditional theory born to a caste, in which he must remain for life, and he is bound to live by its rules, subject to severe consequences for failure. A caste is a hereditary, endogamous group, which has a name of its own and some special traits of occupation, cult, or custom, giving it a separate place in the system. A man must take his wife from his caste—there are a few well-defined exceptions—usually can eat only with caste fellows, and is ranked in the social scale by the nature of the traditional customs of his caste. . . .

Caste stratifies Hindus into more than two thousand mutually exclusive groups, most of which, however, have limited geographical extent, so that no more than fifty to a couple of hundred may exist in any single locality. At the top of the caste hierarchy are the Brahmans, whose various castes included about 6.4 percent of the Hindus in 1941 (no figure available since then); at the bottom are the Untouchables . . . forming about 15.3 percent in 1951. . . . The Brahmans define social position, officiate in religious ceremonies, have custody of sacred lore, and enjoy marked privileges.

Modernization of India comes inevitably into conflict with this rigid social system. Brahman privileges are being increasingly threatened and in some areas have been restricted by law. Untouchables, so called because

---

[9] Reprinted by permission of the publishers from Brown, *The United States and India and Pakistan,* pp. 33–34.

their touch has been held to defile a high-caste Hindu, now number approximately 80 million and must be drawn into a modern economic system in ways which undermine restrictions on their caste. Their cause has been championed for both moral and "practical" reasons by Indian leaders of higher caste, and untouchability is now officially outlawed in India, though with doubtful effectiveness. Agitation from below and reaction from above combine to create friction as the rigidity of the caste system loosens under the impact of modern conditions and needs. Especially in the cities the close intermingling of large numbers of people in factories, rooming districts, public eating places, and public transportation has been a major factor in hastening the disintegration of the caste structure, although it is still strong in the villages.

Language Divisions

Language supplements religion as a divisive factor in India and Pakistan. The present languages of the subcontinent fall into two chief groups, the Aryan languages in the north and the Dravidian languages in roughly the southern half of peninsular India. The various languages within each of these groups are fairly closely related to one another. Languages of major importance, with many millions of speakers each, include four languages of the Dravidian group—Telugu, Tamil, Kanarese (Kannada), and Malayalam—and eight Aryan languages—Hindi, Bengali, Bihari, Marathi, Punjabi, Rajasthani, Gujarati, and Oriya. Dozens of other languages and dialects are spoken by lesser numbers of people. Although bi- and trilingualism are common, the difficulties of communication created by so many languages are obvious. Such difficulties are increased by the fact that less than one-quarter of the population can be regarded as literate in any language, even by the simplest of tests.

English has been the *lingua franca* of the subcontinent, or at least of the educated classes. However, only about 1 to 2 percent of the population is literate in it. This fact plus nationalist feeling led to a desire to adopt some native tongue as an official language (the business of both governments is largely carried on in English) to be propagated as a common medium of instruction and communication. Disputes naturally arose in each country as to which language should be chosen. India decided on Hindi, which is spoken by the largest linguistic group (well over 100 million), and made it the country's official language in 1965. But this decision on the part of the Indian government resulted in strong protests, especially from the Dravidian south, with the result that English

was designated the "associate language." In Pakistan, Urdu, a language similar to Hindi but written with Perso-Arabic script, is dominant in the West and Bengali in the East, although many other languages are spoken, especially in West Pakistan. Both Urdu and Bengali, and English as well, have been declared national languages. Bloody rioting between different language groups has occurred at times in areas of India where such groups are mixed, and demands by language groups have been major factors in reshaping the boundaries of a number of India's internal political units.

POPULATION PROBLEMS

In the 50 years between 1901 and 1951 the total population of the Indian subcontinent (excluding the small Himalayan states) increased by 153 million people or about 54 percent. Between 1951 and 1961 it increased by 96 million or about 22 percent, and between 1961 and 1968 it is estimated to have increased by another 102 million, or nearly 20 percent. At the present rate of increase the subcontinent would have over a billion people by the end of the century. With falling death rates the rate of population growth has been increasing, and the base from which the increase takes place is so large that the total increment is very great.

In essence this massive population growth is negating most of the gains being made by the economies of India and Pakistan. Over the period 1953 to 1965 the United Nations estimated that India's food production increased 29 percent; but its *per capita* food production increased during the same period less than 1 percent. Over the same period Pakistan's total food output was estimated to have increased 30 percent, and its per capita food production 3 percent. Agricultural results fluctuate widely with weather conditions in given years and periods. Thus in the shorter period 1960 to 1965 India achieved only a 7 percent rise in total food output and suffered a 3 percent *decline* in per capita output. Despite increasingly satisfactory agricultural improvements there is real question as to whether these can gain on, or even stay ahead of, population growth.

Thus few students of the matter would fail to agree that a sharp check on population increase is highly desirable. Few also would question that the basic cause of increase is a very high birth rate, since the death rate remains relatively high despite diminution in recent years. Thus the governments of India and Pakistan have begun the task of bringing to their large and mostly illiterate populations a knowledge of birth control, the means of birth control, and a desire to use it. In order to succeed, they must counteract sheer ignorance, traditional religious views placing a premium on large families, and the desire for children as a source of additional labor on the subcontinent's intensively worked farms.

REFERENCES and READINGS

The Subcontinent as a Whole

BASHAM, A. L., *The Wonder That Was India* (3d ed.; London: Sidgwick and Jackson; New York: Grove Press [paperbound], 1967). The ancient culture of India before the coming of the Moslems.

BHARADWAJ, O. P., "The Arid Zone of India and Pakistan," in L. Dudley Stamp, ed., *A History of Land Use in Arid Regions* (Arid Zone Research, No. 17; Paris: UNESCO, 1961), pp. 143–174.

BROWN, W. NORMAN, *The United States and India and Pakistan* (rev. ed.; Cambridge, Mass.: Harvard University Press, 1963); and, ed., *India, Pakistan, Ceylon* (rev. ed.; Philadelphia: University of Pennsylvania Press, 1965).

CHANG, JEN-HU, "The Indian Summer Monsoon," *Geographical Review,* 57, no. 3 (July 1967), 373–396.

DAVIES, CUTHBERT COLLIN, *An Historical Atlas of the Indian Peninsula* (2d ed.; New York: Oxford University Press, 1959).

HUDSON, G. F., "Civilizations of India and the Far East," Chap. 14 in John Bowle, ed., *The Concise Encyclopedia of World History* (New York, Hawthorn Books, 1958), pp. 319–349.

LAMB, ALASTAIR, *The Kashmir Problem: A Historical Survey* (New York: Frederick A. Praeger, 1967).

MANSERGH, NICHOLAS, "The Partition of India in Retrospect," *International Journal,* 21, no. 1 (Winter 1965–1966), 1–19.

MICHEL, ALOYS A., *The Indus River: A Study of the Effects of Partition* (New Haven, Conn.: Yale University Press, 1967).

NANDAN, RAM, "Jammu and Kashmir," *Focus,* 13, no. 1 (September 1962), 6 pp.

RANA, RATNA SHUMSER, "The Summer Monsoon in India," *Journal of Geography,* 67, no.5 (May 1968), 293–300.

SPATE, O. H. K., and A. T. A. LEARMONTH, with the collaboration of A. M. Learmonth, *India and Pakistan: A General and Regional Geography,* with a chapter on Ceylon by B. H. Farmer (3d ed.; London: Methuen and Co., 1967). The standard geography; very detailed and authoritative.

India

AHMAD, QAZI, *Indian Cities; Characteristics and Correlates* (University of Chicago, Department of Geography Research Paper No. 102; Chicago: 1965).

BERRY, BRIAN J. L., *Essays on Commodity Flows and the Spatial Structure of the Indian Economy* (University of Chicago, Department of Geography Research Paper No. 111; Chicago: 1966).

BETTELHEIM, CHARLES, "India's Third Five-Year Plan: Some Problems of Realization," *Pacific Viewpoint,* 4, no. 2 (September 1963), 139–154.

BHAT, L. S., and A. T. A. LEARMONTH, "Recent Contributions to the Economic Geography of India: Some Current Preoccupations," *Economic Geography,* 44, no. 3 (July 1968), 189–209. A review article.

BHATIA, SHYAM S., "Patterns of Crop Concentration and Diversification in India," *Economic Geography,* 41, no. 1 (January 1965), 39–56.

BOSE, NIRMAL KUMAR, "Calcutta: A Premature Metropolis," *Scientific American*, 213, no. 3 (September 1965), 91–102.

BREESE, GERALD, "Urban Development Problems in India," *Annals of the Association of American Geographers*, 53, no. 3 (September 1963), 253–265.

BRUSH, JOHN E., "The Distribution of Religious Communities in India," *Annals of the Association of American Geographers*, 39, no. 2 (June 1949), 81–98.

CHANDRASEKHAR, S., "How India Is Tackling Her Population Problem," *Foreign Affairs*, 47, no. 1 (October 1968), 138–150.

DAYAL, EDISON, "The Changing Patterns of India's International Trade," *Economic Geography*, 44, no. 3 (July 1968), 240–269; and "Crop Combination Regions: A Case Study of the Punjab Plains [India]," *Tijdschrift voor Economische en Sociale Geografie*, 58, no. 1 (January 1967), 39–47.

FONAROFF, L. SCHUYLER, and ARLENE FONAROFF, "The Cultural Environment of Medical Geography in Rural Hindu India," *Pacific Viewpoint*, 7, no. 1 (May 1966), 67–84.

Geographical Review of India (quarterly).

GREENWOOD, GORDON, ed., "Modern India," *Australian Journal of Politics and History*, 12, no. 2 (August 1966), 125–281. A special number, with 12 contributions.

HARRIS, MARVIN, "The Myth of the Sacred Cow," *Natural History*, 76, no. 3 (March 1967), 6–12A.

HARRISON, SELIG S., "Troubled India and Her Neighbors," *Foreign Affairs*, 43, no. 2 (January 1965), 312–330.

HIRT, HOWARD F., "Lock Making: A Handicraft Industry in Aligarh, India," Chap. 19 in Richard S. Thoman and Donald J. Patton, eds., *Focus on Geographic Activity: A Collection of Original Studies* (New York: McGraw-Hill Book Company, 1964), pp. 113–119.

HUTTON, J. H., *Caste in India: Its Nature, Function and Origins* (4th ed.; New York: Oxford University Press, 1963).

"India in Transition," *Foreign Affairs*, 46, no. 3 (April 1968), 519–561. Three articles by different authors.

JAYAWARDENA, CHANDRA, "Migration and Social Change: A Survey of Indian Communities Overseas," *Geographical Review*, 58, no. 3 (July 1968), 426–449.

KARAN, PRADYUMNA P., "Changes in Indian Industrial Location," *Annals of the Association of American Geographers*, 54, no. 3 (September 1964), 336–354; and, with William M. Jenkins, Jr., "Geography of Manufacturing in India," *Economic Geography*, 35, no. 3 (July 1959), 269–278.

KIRK, WILLIAM, "The Cotton and Jute Industries of India: A Study in Concentration and Dispersal," *Scottish Geographical Magazine*, 72, no. 1 (April 1956), 38–52; also, "The Inner Asian Frontier of India," Institute of British Geographers, Publication No. 31, *Transactions and Papers, December 1962*, pp. 131–168; and "The Sino-Indian Frontier Dispute: A Geographical Review," *Scottish Geographical Magazine*, 76, no. 1 (April 1960), 3–13.

KOSAMBI, D. D., "Living Prehistory in India," *Scientific American*, 216, no. 2 (February 1967), 105–114. An article about tribal groups pursuing ancient forms of livelihood.

LALL, AMRIT, and RANJIT TIRTHA, "India's Urbanization," *Focus*, 19, no. 1 (September 1968), 1–7, and Amrit Lall, "Delhi: A Capital City," 7–11 (same issue).

LAMB, ALASTAIR, *The China-India Border: The Origins of the Disputed Boundaries* (Chatham House Essays, No. 2; New York: Oxford University Press, 1964).

MAHESHWARI, P., "Botany and the Food Problem of India," *Science and Culture*, 32, no. 3 (March 1966), 104–114.

MASON, PHILIP, ed., *India and Ceylon: Unity and Diversity* (New York: Oxford University Press, 1967). A symposium.

MILNER, DONALD, "Calcutta—A City in Despair," *Geographical Magazine*, 41, no. 1 (October 1968), 35–49.

MURPHEY, RHOADS, "The City in the Swamp: Aspects of the Site and Early Growth of Calcutta," *Geographical Journal*, 130, pt. 2 (June 1964), 241–256.

NEALE, WALTER C., *India: The Search for Unity, Democracy, and Progress* (Princeton, N.J.: D. Van Nostrand Co., Searchlight Books, 1965).

PLATT, RAYE R., with the assistance of ROBERT C. KINGSBURY, JAMES L. McPHERSON, DAVID E. SOPHER, and Others, *India: A Compendium* (New York: American Geographical Society, 1962).

REED, WALLACE E., *Areal Interaction in India: Commodity Flows of the Bengal-Bihar Industrial Area* (University of Chicago, Department of Geography Research Paper No. 110; Chicago: 1967).

SAMUEL, T. J., "The Development of India's Policy of Population Control," *Milbank Memorial Fund Quarterly,* 44, no. 1, pt. 1 (January 1966), 49–67.

SCHWARTZBERG, JOSEPH E., "The Distribution of Selected Castes in the North Indian Plain," *Geographical Review,* 55, no. 4 (October 1965), 477–495; and "Three Approaches to the Mapping of Economic Development in India," *Annals of the Association of American Geographers,* 52, no. 4 (December 1962), 455–468.

SIMOONS, FREDERICK J., *Eat Not This Flesh: Food Avoidances in the Old World* (Madison: University of Wisconsin Press, 1961). Especially Chap. 4, "Beef," pp. 45–63.

SOPHER, DAVID E., "India's Languages and Religions," *Focus,* 6, no. 6 (February 1956), 6 pp.; also, "Landscapes and Seasons: Man and Nature in India," *Landscape,* 13, no. 3 (Spring 1964), 14–19; and "Pilgrim Circulation in Gujarat," *Geographical Review,* 58, no. 3 (July 1968), 392–425.

TAYLOR, ALICE, and S. REZA AHSAN, "India's Agricultural Problems," *Focus,* 14, no. 1 (September 1963), 6 pp.

TURNER, ROY, ed., *India's Urban Future* (Berkeley and Los Angeles: University of California Press, 1962). Especially Chap. 3, "The Morphology of Indian Cities," by John E. Brush, pp. 57–70.

VANDERVER, TIMOTHY A., *Basic Data on the Economy of India,* U.S. Bureau of International Commerce, Overseas Business Reports, OBR 65–81 (December 1965). (Washington, D.C.: Government Printing Office, 1965.)

WHYTE, R. O., *Land, Livestock and Human Nutrition in India* (New York: Frederick A. Praeger, 1968).

Pakistan

AHMAD, KAZI S., *A Geography of Pakistan* (Karachi: Oxford University Press, 1964).

AHMAD, NAFIS, *An Economic Geography of East Pakistan* (2d ed.; New York: Oxford University Press, 1965).

DICHTER, DAVID, with the Collaboration of NATHAN S. POPKIN, *The North-West Frontier of West Pakistan: A Study in Regional Geography* (New York: Oxford University Press, 1967).

GULICK, LUTHER H., "Irrigation Systems of the Former Sind Province, West Pakistan," *Geographical Review,* 53, no. 1 (January 1963), 79–99.

HELIN, RONALD A., "Uniting the Wings of Pakistan: A Matter of Circulation," *Professional Geographer,* 20, no. 4 (July 1968), 251–256.

JOHNSON, B. L. C., "Rural Population Densities in East Pakistan," *Pacific Viewpoint,* 3, no. 1 (March 1962), 51–62.

MACNAB, J. W., "The Pakistan Cotton Industry," *Pacific Viewpoint,* 2, no. 1 (March 1961), 85–97.

MICHEL, ALOYS A., "Pakistan," *Focus,* 17, no. 5 (January 1967), 6 pp.

NIKSCH, LARRY A., *Basic Data on the Economy of Pakistan,* U.S. Bureau of International Commerce, Overseas Business Reports, OBR 66–29 (May 1966). (Washington, D.C.: Government Printing Office, 1966.)

Pakistan Geographical Review (semiannual).

PLATT, RAYE R., with the assistance of ROBERT C. KINGSBURY, JAMES L. McPHERSON, and Others, *Pakistan: A Compendium* (New York: American Geographical Society, 1961).

SOPHER, DAVID E., "The Swidden/Wet-Rice Transition Zone in the Chittagong Hills," *Annals of the Association of American Geographers,* 54, no. 1 (March 1964), 107–126.

STEPHENSON, GLENN V., "Pakistan: Discontiguity and the Majority Problem," *Geographical Review,* 58, no. 2 (April 1968), 195–213.

TAYLOR, GEORGE C., Jr., "Water, History, and the Indus Plain," *Natural History,* 74, no. 5 (May 1965), 40–49.

TAYYEB, A., *Pakistan: A Political Geography* (New York: Oxford University Press, 1966).

U.S. Army Area Handbook for Pakistan (Department of the Army Pamphlet No. 550–48; Washington, D.C.: 1965).

WHITTINGTON, G., "The Irrigated Lands of the Indus Plains in West Pakistan," *Tijdschrift voor Economische en Sociale Geografie, 55,* no. 1 (January 1964), 13–18.

ZAIDI, IQTIDAR H., "Toward a Measure of the Functional Effectiveness of a State: The Case of West Pakistan," *Annals of the Association of American Geographers, 56,* no. 1 (March 1966), 52–67.

Nepal, Bhutan, and Sikkim

KARAN, PRADYUMNA P., *Bhutan: A Physical and Cultural Geography* (Lexington: University of Kentucky Press, 1967); also, "The Changing Geography of Bhutan," *Canadian Geographical Journal, 74,* no. 1 (January 1967), 33–39; "Sikkim and Bhutan: A Geographical Appraisal," *Journal of Geography, 60,* no. 2 (February 1961), 58–66; *Nepal: A Cultural and Physical Geography* (Lexington: University of Kentucky Press, 1960); and, with William M. Jenkins, Jr., "Population, Land Utilization, and Possible Expansion of Cultivated Land in Nepal," *Pacific Viewpoint, 2,* no. 1 (March 1961), 41–58.

See also the list of references and readings for Chapter 17, especially the relevant chapters in general geographies of Asia listed on page 415.

Southeast Asia

19

Southeast Asia, from Ceylon on the west to the Philippines on the east, is a region of peninsulas, islands, and intervening seas. East of India and south of China, between the Bay of Bengal and the South China Sea, the large Indochinese Peninsula, occupied by Burma, Thailand (Siam), Cambodia, Laos, South Vietnam, and North Vietnam, projects southward from the continental mass of Asia (map, p. 442). From it the long, narrow subpeninsula of Malaya extends another 900 miles toward the equator. Ringing the south and east of this continental projection are thousands of islands, among which Sumatra, Java, Borneo, Celebes, Mindanao, and Luzon are outstanding in size. Another large island, New Guinea, east of Celebes, is culturally a part of the Melanesian archipelagoes of the Pacific World. However, the western half of New Guinea, held by Indonesia, may be considered a marginal part of Southeast Asia in a political sense. To the west the island of Ceylon, while merely a detached part of the Indian subcontinent in a geologic sense, may be considered the western outpost of the Southeast Asian region from the standpoint of economic development, culture, and climate.

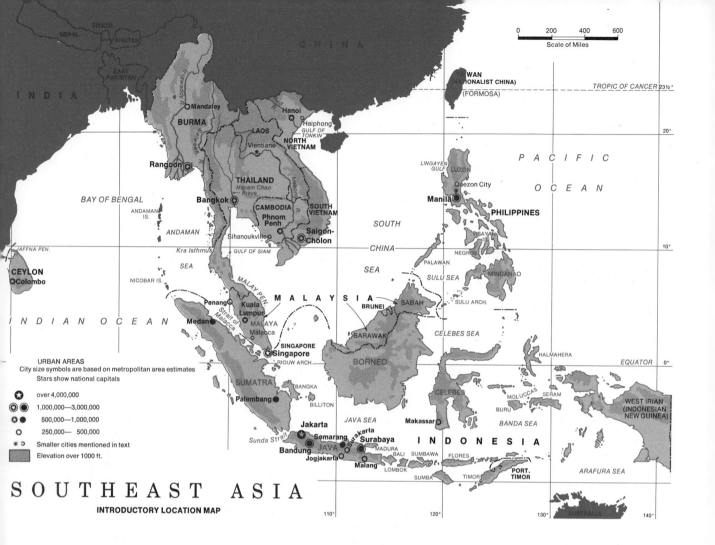

SOUTHEAST ASIA

INTRODUCTORY LOCATION MAP

Southeast Asia, thus conceived, is composed politically (1968) of eleven independent states and two colonial or semicolonial territories. The independent states are Ceylon, Burma, Thailand, Laos, Cambodia, North Vietnam (Communist), South Vietnam (non-Communist), Malaysia, Singapore, Indonesia, and the Philippines. With the exception of Thailand, none of these states dates its present independence farther back than 1946. The remnants of colonialism are the British-protected sultanate of Brunei, on the north coast of Borneo, and Portuguese Timor, which occupies the eastern half of the island of Timor plus a small exclave on the northwestern coast.

AREA, POPULATION, AND ENVIRONMENT

It is approximately 4000 miles from Ceylon to central New Guinea and 2500 miles from northern Burma to southern Indonesia, but the total land area of Southeast Asia is only about 1,762,000 square miles, or less than half the area of the United States. The population of the region is estimated at about 282 million, giving a density of 160 per square mile—an average which embraces very great extremes within the region. Area and population data for the individual political units are given in Table 15.

Relative Sparseness of Population

The average population density of Southeast Asia is high compared to that of much of the world but low compared to most other areas on the seaward margins of the Orient. For example, although Southeast Asia has almost three times the overall density of the United States, its density is well under half that of India-Pakistan or China proper and less than a fourth of the average density in Japan. Southeast Asia has apparently been less

TABLE 15 SOUTHEAST ASIA: AREA AND POPULATION DATA

POLITICAL UNIT	AREA (THOUSAND SQUARE MILES)	POPULATION (MILLIONS: 1968 ESTIMATES)	DENSITY (PER SQUARE MILE: TO NEAREST WHOLE NUMBER)
Ceylon	25.3	12.2	482
Burma	261.8	26.4	101
Thailand	198.5	33.7	170
North Vietnam	61.3	20.8	340
South Vietnam	66.9	17.4	260
Cambodia	69.9	6.6	94
Laos	91.4	2.8	31
Malaysia[a]	127.7	10.2	80
Singapore	0.224	2.0	8909
Indonesia	735.3	112.8	153
Philippines	115.8	35.9	310
Brunei	2.2	0.11	50
Portuguese Timor	5.8	0.57	100
Totals	1762.1	281.5	160

[a] Including Sabah and Sarawak.

populous than India and China proper throughout history. It has been estimated that the whole region contained only about 10 million people around the year 1800. Despite the fact that the region was known in early times by Chinese, Japanese, and Indians, that trade has moved through it for many centuries, and that some early immigrants reached it by sea (particularly from India), widespread dense populations did not develop.

It seems likely that difficulties imposed by the natural environment of Southeast Asia had much to do with the slowness of population growth there. By land the region is relatively isolated, as the Indochinese Peninsula abuts in the north on high, rugged, and malaria-infested mountains. Over a period of many centuries the forebears of most of the present inhabitants entered the area as recurrent thin trickles of population, crossing the mountain barrier under pressure, usually as refugees driven from previous homelands to the north.

Environmental difficulties within the region itself are formidable. The most significant of them derive from the fact that Southeast Asia is truly tropical, with continuous heat in the lowlands, torrential rains, a prolific vegetation difficult to clear and keep cleared, soils which are generally leached and poor, and a high incidence of disease. The unfavorable agricultural effects of a 4 to 6 months dry season in the Indochinese Peninsula and

scattered smaller areas in the islands are accentuated by a high rate of evaporation imposed by the tropic heat. The lush vegetation of Southeast Asia masks the infertility of most of its soils, an infertility that becomes apparent and generally increases rapidly when the land is cleared and cultivated. Heat and humidity speed bacterial and chemical action so that the humus in the topsoil is rapidly destroyed and the mineral plant foods become thoroughly depleted by leaching. In addition, much of Southeast Asia is mountainous, and erosion is

Southeast Asia compared in area with conterminous United States.

so rapid that the rivers generally carry enormous volumes of mud and silt, eventually to be deposited on flood plains and deltas at or near the sea. More spectacular, though probably less significant, environmental difficulties include frequent volcanic eruptions and earthquakes in the East Indies (Indonesia) and the Philippines, and the violent wind-and-rain storms known as typhoons which strike coastal Vietnam and the northern Philippines. Prescientific techniques were apparently unable to foster large populations under these conditions. Early civilizations did not expand widely or even maintain themselves, as is evidenced by the monumental ruins found in a few parts of the region. In other places climate and vegetation have almost destroyed the last vestiges of early cities and states.

Western scientific knowledge and technology have mitigated the effects of these environmental conditions somewhat and have stimulated a tremendous population increase in the last century and a half—an increase which has been accelerating in recent years. By 1959 the region's population was an estimated 212 million; but between that year and 1968 it climbed to 282 million, an increase of 70 million, or 33 percent in 9 years. The magnitude and consequences of this recent increase have varied from country to country. From the inadequate data available it appears that some countries have been able to keep their general economic development and food production ahead of their population increase, thus improving the average lot of their citizens. Other countries, however, have not, and in these lands the welfare and diet of the average citizen have deteriorated. All of the countries remain poor. Available estimates of average per capita incomes for 1966 ranged from $50 per year for Laos to $508 for Singapore, with no country except Singapore exceeding $250. In the late 1960s great hopes for adequate increases in food production were being placed in new strains of rice, developed in the Philippines, which were reported to give greatly increased yields.

THE ECONOMIC PATTERN

The overwhelming majority of the people of Southeast Asia are farmers. Industry has made much less headway than in Japan, China, or India.

Shifting Cultivation

A permanent form of agriculture has proved extremely difficult to establish in most parts of this mountainous tropical realm. Much of the land is used only for a shifting, primitive subsistence form of cultivation in which fields are cleared and used for a few years and then allowed to lapse back into jungle while the cultivator moves on to clear a new patch. The cultivators are generally in a primitive state with respect to both agricultural techniques and social organization. Clearing is ordinarily accomplished by fire, and the crops are often grown with little or no cultivation. Unirrigated rice is usually the principal crop, although corn, beans, and root crops (such as yams and cassava) are also grown. Agricultural activities are often supplemented by hunting and gathering in the forest. These migratory subsistence farmers ordinarily differ ethnically from adjacent settled populations, for they are the remnants of peoples driven to refuge in the back country by stronger invaders. Since it is estimated that 15 years are generally necessary for an abandoned clearing to regain its fertility, it is obvious that only a small proportion of the land can be cultivated at any one time; therefore, populations living under such conditions must necessarily be sparse.

Densely Populated Districts of Sedentary Agriculture

Most of Southeast Asia's people live, often in extremely dense clusters, in scattered areas of permanent sedentary agriculture. Such areas form the core regions of the various countries and stand in striking contrast to the relatively empty spaces of the adjoining districts. A superior degree of soil fertility appears to have been the main locational factor in most instances. In Southeast Asia soils of better than average fertility have ordinarily resulted from one or more of the following factors:

1. The presence of lava and volcanic ash of proper chemical composition to weather into fertile soils.

2. The accumulation and periodic renewal of plant nutrients washed from upstream areas and deposited on river flood plains and deltas.

3. A slowing down or seasonal reversal of the leaching process in areas having a distinct dry season and relatively low rainfall.

4. Occurrence in some areas of uplifted coral platforms which weather into a superior soil.

A few areas of Southeast Asia exhibit fairly dense populations without any corresponding soil superiority. Such areas are ordinarily characterized by a plantation type of agriculture based on crops adapted to poor soils, such as rubber.

Side by side with the present areas of dense settlement, some of which are seriously overcrowded, are sizable areas still undeveloped despite the fact that they apparently are capable of supporting large populations. Their development must wait until the need for their utilization becomes sufficiently great and the available resources sufficiently large to encourage large-scale clearing of forests, draining of swamps, and construction of irrigation systems. It also must await the initiative and resourcefulness of a sufficient number of people who are willing to leave established homes and accept the risks of pioneering in areas often plagued by relatively low and undependable rainfall and/or by malaria.

The typical inhabitant of the present areas of permanent sedentary agriculture is a subsistence farmer whose main crop is wet rice, grown with the aid of natural flooding or by irrigation. In some drier areas unirrigated millet or corn, the latter especially in Indonesia and the Philippines, takes the place of rice. The major crop is supplemented, to a degree tending to vary positively with the pressure of population, by secondary crops which can be grown on land not suited for rice or other grains. Prominent secondary food crops include coconuts, yams, cassava, beans, and garden vegetables. If the year is a good one, a portion of the produce may be sold, and some farmers grow a secondary crop such as tobacco, coffee, or rubber primarily for sale. Fish are an important component of the food supply and a considerable source of income, not only along the coasts but also inland along streams and lakes. Many fish are harvested from artificial fish ponds or from flooded rice fields in which they are grown as a supplementary "crop."

The Role of Commercial Agriculture

Subsistence production in Southeast Asia exists side by side with an important development of commercial agriculture, directed mainly toward supplying commodities for export. The region is one of the world's major supply areas for tropical plantation crops as well as a supplier of surplus rice to other parts of the Orient. These commercial elements in Southeast Asian agriculture are principally the result of Western colonial enterprise in the region.

European imperialism in Southeast Asia began in the sixteenth century. Until the nineteenth century, however, the principal interest of the newcomers lay in the region's strategic location on the route to China, and they were generally content with effective control over scattered patches of land along the coasts and with gleaning for trade such surpluses of value as the native economies happened to offer. But during the nineteenth century increasing populations and an expanding technology in Europe and North America greatly enlarged the demand for products of tropical agriculture. Consequently, effective European political control was forcibly extended over almost the whole of Southeast Asia, and European capital and knowledge were applied to bring about a rapid increase in production for export. Certain indigenous commodities, such as copra and spices, were given increased emphasis, and a number of entirely new crops were introduced, a case in point being the rubber seedlings smuggled out of Brazil in 1876 and introduced into Malaya in 1879.

Harvesting coconuts with a long pole on a plantation in Malaya. *(Foreign Agriculture.)*

**TABLE 16 EXPORTS OF MAJOR SOUTHEAST ASIAN PLANTATION-TYPE
AGRICULTURAL COMMODITIES**

COMMODITY	APPROXIMATE PERCENT OF WORLD EXPORTS ORIGINATING IN SOUTHEAST ASIA	PRINCIPAL EXPORTING COUNTRIES IN SOUTHEAST ASIA WITH APPROXIMATE PERCENTAGE OF WORLD EXPORTS	
Rubber	85–95	Malaysia	40–50
		Indonesia	25–35
Tea	35–45	Ceylon	30–40
		Indonesia	5–10
Copra	75–85	Philippines	55–65
		Indonesia	10–20
Coconut oil	80–90	Philippines	45–55
		Ceylon	20–30
		Malaysia	5–15
Palm oil	40–50	Indonesia	15–25
		Malaysia	15–25
Sugar	5–10	Philippines	6
Abaca	95–100	Philippines	95–100

The usual method of introducing commercial production was to establish large estates or plantations managed by Europeans but worked by indigenous or imported Oriental labor. Development of these large commercial farming enterprises was assisted not only by the favorable climate, but also by the availability of land in a relatively empty part of the Orient and by the amphibious nature of the topography, which made possible a close dependence on cheap, efficient, and easily established transportation by water. The major difficulty to be overcome was in most cases the recruitment of an adequate labor force from a population already fully engaged in food production and not sufficiently pressed by poverty and hunger to be easily attracted from the communal life of the villages into labor for wages on the plantations. The eventual solution to the latter problem in many areas was large-scale importation of contract labor from India and China.

Plantation activity came to have widespread repercussions on the economic life of the natives themselves, for many native farmers and ex-estate laborers learned by example and entered commercial production on their own account. Even shifting cultivators could plant a few rubber trees and return when the trees reached the producing stage, 7 to 10 years later, to tap them for latex if the price warranted. Small holders of the foregoing types now command an important share in the export production of most plantation commodities.

Many different plantation-type commodities have been produced in Southeast Asia on a medium to large scale at one time or another. Wide fluctuations have occurred from time to time in the crops grown, the

centers of production, the amounts exported, and the prosperity of the producers. These shifts have been occasioned by such factors as fluctuating world demands, intraregional and interregional competition, changing political conditions, and the occasional ravages of plant diseases. This general situation has prevailed not only in Southeast Asia but in plantation areas throughout the tropical world. Excluding rice, the major agricultural exports from Southeast Asia together with the main producing countries at the present time are indicated in Table 16. A smaller development of some of the major commodities listed in the table has occurred in various Southeast Asian countries other than those indicated. In addition, a variety of other products such as tobacco, kapok, spices, and sisal enter the total export picture. In the supply of some of these lesser commodities such as kapok and certain spices, Southeast Asia is the world's leading region.

Political convulsions since 1941 have effected great changes in the commercial agriculture of this region. Especially notable have been the great decline and only partial recovery of Indonesian export production as a consequence of war and revolution, governmental mismanagement, and emphasis on increased production of food for use within the country. Prior to World War II, for example, Indonesia was second only to Cuba as a sugar exporter and second only to the United States as a tobacco exporter, but postwar exports of both of these commodities from Indonesia have been relatively insignificant. Many other instances of Indonesia's worsened situation in plantation-type production could be cited.

Commercial Rice Farming

One very significant aspect of Western influence on the economy of Southeast Asia has been the stimulation of commercial rice farming in certain areas that were formerly unproductive. The development of plantation agriculture, mining, and trade provided a market for rice by bringing into existence a large class of people who worked for wages and had to buy their food. During the same period Western economic, medical, and sanitary innovations helped bring about an enormous increase in Oriental population generally and a growing demand for food. Western technology made possible the bulk processing and movement of rice, and aided in the development of drainage, irrigation, and flood control facilities needed to produce it on a large scale in areas previously undeveloped. As a consequence, an Asian pioneer movement took place into certain areas capable of greatly expanded rice production. Among these latter areas three gained outstanding importance in commercial rice growing: the deltas of the Irrawaddy, Chao Praya, and Mekong rivers, located in Burma, Thailand, and South Vietnam and Cambodia, respectively (map, p. 449).

Almost impenetrable swamps and uncontrollable floods had kept these deltas thinly settled previously, as other swampy areas in Southeast Asia are today, but adequate incentives and methods for pioneering have turned them into densely settled areas within the past century. Despite growing populations, the farms in these areas are larger than is common in the Orient, and the surplus of rice is such that half or more of the world's total rice exports generally originate in the three deltas. In the years before World War II Burma ranked first among the world's rice exporters, generally accounting for about a third of the world total, and was followed by Thailand and French Indochina (which then controlled the Mekong delta), each supplying approximately 15 percent of world exports. Unsettled conditions in Burma and Indochina during the war and postwar periods led to a serious decline in rice production and exports. By the 1960s political and military conditions in the Mekong delta of South Vietnam were so disruptive that exports temporarily almost ceased. Thailand, on the other hand, has experienced fewer political and military difficulties and has been able to greatly expand its output and exports of rice. In the middle 1960s Thailand accounted for 23 percent and Burma for 20 percent of the world's total rice exports, while Cambodia contributed another 5 percent from its share of the Mekong delta.

These exports of the basic food of the Orient go mainly to Indonesia, Malaya, and Ceylon within Southeast Asia, and to India, Japan, and Hong Kong outside of the immediate region. City populations, especially near the coast, and laborers in plantation and mining areas are the principal consumers of imported rice. The vital need for such imports to make up for deficiencies in domestic production has given Burma, Thailand, and the lower Mekong states great significance in the eyes of other Oriental nations, and indirectly in the eyes of all nations interested in Oriental affairs. And that significance is not lessened by the fact that full development has not yet been reached in any of the surplus rice-producing areas.

Production and Reserves of Minerals

Southeast Asia has considerable mineral wealth, much of which remains essentially unexploited. Tin and oil are the principal minerals exploited thus far. In 1966 this region produced and exported about half of the world's tin: 33 percent from Malaya, 11 percent from southern Thailand, and most of the rest from the islands of Bangka and Billiton in Indonesia. Oil is produced on the islands of Sumatra, Java, Borneo, and New Guinea in Indonesia, the region's main producing country, as well as in Malaysian Borneo (Sarawak), Brunei, and Burma, but the total regional production amounts to only about 2 percent of the world total. However, added significance is given to Southeast Asian oil by excellent possibilities of expansion and by its status as the most important source between the Persian Gulf and California.

About 5 percent of the world's bauxite is produced in Malaya, Sarawak, and the Riouw Archipelago of Indonesia, and about 10 percent of its chromite in the Philippines. Iron ore is fairly abundant in the region. It is exported from Malaya and the Philippines in relatively small quantities, and is worked on a very small scale in Thailand and North Vietnam. In addition, there is present in the region, and often worked on at least a minor scale, a great range of other mineral resources: gold and silver; the alloys tungsten, nickel, and manganese; the major nonferrous metals lead, zinc, and copper; and others.

Production of minerals thus far has been almost entirely for export, with Japan, western Europe, and the United States representing the principal markets. Use of Southeast Asian mineral resources for industrialization within the region itself would face the handicap of a shortage of coal suitable for large-scale metallurgy. Lignite is present, especially in Burma and Thailand, and low-grade bituminous coal is mined in Sumatra, Borneo, and the Philippines, but the only metallurgical coal known to exist in quantity in the region is the anthracite of North Vietnam. Electricity produced from low-grade coal and lignite and by utilization of the region's sizable resources of waterpower, although a necessity for future industrial development, could hardly replace metallurgical coal in heavy industry. Nevertheless, Southeast Asia appears to have the resources for a considerable expansion of both mining and manufacturing, in addition to its undoubted and perhaps more important potential for further expansion of agriculture. But like most underdeveloped areas, this region faces serious problems in securing the necessary skill, capital, and markets to make economic expansion a success.

DISTINCTIVE QUÀLITIES OF THE INDIVIDUAL COUNTRIES

While the countries of Southeast Asia have many broad similarities, each also has its own distinctive qualities. Each exhibits a different combination of environmental features, native and immigrant peoples, economic activities, and culture traits. Many of the dominant characteristics and problems of these countries are the outcome of 400 years of European imperialism, but in no two countries have the influences and results of imperialism been the same. The following brief portraits of the major countries in the region will perhaps be sufficient to give an idea of the distinctive characteristics of each.

Ceylon

The island of Ceylon consists of a coastal plain surrounding a knot of mountains and hill lands in the south central part. Most of the people live either in the wetter southwestern portion of the plain, in the hilly areas of the south center, in the drier Jaffna Peninsula of the north, or in a limited area on the east central coast. Coconuts and rice are the major crops of the low southwestern coast and Jaffna Peninsula, while tea and rubber plantations dominate the economy of the uplands. Colombo (900,000; unless otherwise noted, city populations in this chapter are metropolitan area estimates for 1968) in the southwest, is the capital, chief port, and only large city.

Centuries of recurrent invasion from India were followed by Portuguese domination in the early sixteenth century, Dutch in the seventeenth, and British from 1795

until 1948. In the latter year Ceylon was granted independence as a dominion in the British Commonwealth. This eventful history plus the island's long-standing commercial importance on the sea route around southern Asia has given Ceylon a polyglot population. The two major ethnic groups, distinguished from each other by language and religion, are the predominantly Buddhist Sinhalese, making up about 70 percent of the population, and the Hindu Tamils, constituting about 20 percent. The Tamils, whose main center of settlement is the Jaffna Peninsula, are descendants of early invaders and more recent imported laborers from southern India. In addition to the numerically superior Sinhalese and Tamils, there is a remnant of Arab population and the Burghers, descendants of Portuguese and Dutch settlers. About 10 percent of the people are Moslems and about 5 percent Roman Catholics. In the interior a few thousand tribesmen known as Veddas practice shifting cultivation. There is some antagonism among the various ethnic groups; in recent years the Sinhalese majority and the Tamils have often been at odds politically, and rioting between the two groups has occurred.

The economy of Ceylon is rather highly commercialized, with three export crops—coconuts (1,070,000 acres), rubber (665,000 acres), and tea (525,000 acres)—occupying two-thirds of the land under cultivation. Most of the remaining acreage is in rice, but the production is insufficient to feed the growing population of Ceylon, and half of the needed rice must be imported. Much potential agricultural land is currently unused, and the government is attempting to reduce food imports by expanding rice production through the establishment of modern irrigation works. Relatively prosperous export industries and stable political conditions seem to favor these and other forward-looking projects. However, an increase in the island's total population from 5.3 to 12.2 million since 1931 has considerably aggravated the food problem.

Burma

Burma centers in the basin of the Irrawaddy River and includes surrounding uplands and mountains. Within the basin are two distinct areas of dense population: the Dry Zone, around and south of Mandalay (city proper, perhaps 250,000), and the Delta, focusing on the capital and major seaport of Rangoon (1.6 million). The Dry Zone has been the historical nucleus of Burma. The annual rainfall of this area (31 inches at Mandalay) is exceptionally low for Southeast Asia and there is a dry season of about 6 months (Table 14). The people are

Index map of the Southeast Asian mainland. Stars show political capitals and the independent state of Singapore. Horizontal line pattern shows river deltas. Heavy white dot pattern shows the Dry Zone of Burma. Light stipple shows the Tin and Rubber Belt of Malaya. Malaya is the mainland part of the country of Malaysia.

country's total population. Prior to World War II around 1 million Indians (about equally divided between Hindus and Moslems) lived in the Delta as laborers, merchants, moneylenders, and owners of farmland, much of it rented to Burmese tenants. This Indian minority has been reduced to perhaps 500,000 by homeward migrations during World War II and during the 1960s. The later reverse migration was associated with thorough nationalization of the Burmese economy, including land, by the country's military government.

The Shan Plateau to the east is inhabited by two distinct groups, each accounting for between 5 and 10 percent of the country's population—the Shans to the northeast and the Karens to the southeast and in the delta of the Salween River. Some of these people eke out a meager livelihood by shifting cultivation in the hills, although the majority are plain and valley dwellers growing irrigated rice. A variety of hill-dwelling tribes inhabit the Arakan Mountains of the west and the northern highlands. Most of these people are shifting cultivators. Burma is predominantly a Buddhist country, with over 80 percent of its total population adhering to this religious faith.

Burma was conquered piecemeal by Great Britain in three wars between 1824 and 1885. In 1948 it abandoned all formal ties with Britain and became an independent republic. Japanese conquest and Allied reconquest in World War II were succeeded by a period of civil war following the establishment of an independent Burma. At one time no less than eight different rebellions were in progress. Communist agitation has been an important cause of disaffection, and also ethnic separatism, especially among the Karens. The country is being gradually reduced to order, but its economy has been badly damaged. Shipments of rice, principally to India, Japan, Malaya, and Indonesia, accounted for three-fifths of all exports by value in 1966.

Thailand

Thailand, formerly known as Siam, centers in the delta of the Chao Praya River, often referred to as the Menam (a Thai word meaning river; the full name as used in Thailand is Menam Chao Praya). The annual floods of this river irrigate the commercial rice production that forms the mainstay of the country's economy. More than 60 percent of all cultivated land in Thailand grows rice, and this product made up about one-third of all exports by value in 1965. Rubber (accounting for about one-fifth of total export value), tin, and jute are the other main exports. On the lower Chao Praya in the most

supported by mixed subsistence and commercial farming, with millet, rice, cotton, beans, peanuts, and sesame as major crops. During the past century the Dry Zone has been surpassed in population by the Delta, where a commercial rice-farming economy now supplies most of the country's exports. The Irrawaddy River forms a major artery of transportation uniting these two core areas.

The native Burmese, most of whom live in the Irrawaddy Basin, number only about 66 percent of the densely populated part of the country is the capital, main port, and only large city, Bangkok (2.5 million). Areas outside of the delta region are more sparsely populated, largely by sedentary farmers, although some shifting cultivators are present. These areas include mountainous territories in the west and north, inhabited mainly by

Karens and a variety of mountain tribes, and the dry Korat Plateau to the east, populated by the Thai and related Laotian and Cambodian peoples. To the south in the Kra Isthmus, a part of the Malay Peninsula, live some 700,000 Malays. An estimated 95 percent of the population of Thailand adheres to the Buddhist faith, the only significant exception being the Malays, who are Moslems.

The Thai monarchy enjoys the distinction of being the only Southeast Asian country to preserve its independence throughout the period of Western colonialism. Its success appears to have been largely due to its position as a buffer between British and French colonial spheres. A number of border territories were lost, however, mostly inhabited by peoples related to the Thai, and the government of the country has exhibited irredentist tendencies for many years. In return for Thailand's cooperation during World War II, the Japanese allowed the country to take over certain border territories from neighboring countries, but these areas were lost again at the end of the war. Recently there have been border disputes with Cambodia, Burma, and Laos, not explicitly centering, however, on irredentist claims.

The major ethnic minority of Thailand is Chinese. About 3 million Chinese are resident in the country and are claimed as citizens both by Thailand and by China. The Chinese control much of the country's rice trade and other business, and have extensive interests in the rubber plantations and tin mines of the Malay Peninsula section. This situation has roused a considerable amount of ill feeling on the part of the Thais, and makes the loyalty of the Chinese a matter of considerable importance. Thus far (1968) Thailand has escaped serious civil conflict, just as it escaped serious involvement in World War II. However, in the late 1960s Communist insurrection on a small but growing scale was beginning to affect northeastern and southern Thailand. Apparently this development was supported by China and North Vietnam; it may have been related to the fact that Thailand

River craft in a navigation lock at the Chainat Dam on the Menam Chao Praya about 300 air miles to the north of Bangkok. The dam, built to reduce flood damage and regularize the supply of irrigation water on the Menam Plain, was constructed with the assistance of the World Bank. Much of Thailand's internal transportation is handled by barges similar to those in this view. They utilize the Menam, its tributaries, and interconnecting canals. See David Love, "Controlling the Chao Phya," *Finance and Development*, **4,** no. 4 (December 1967), 237-249. (Photo: International Bank for Reconstruction and Development.)

Puddling a bed for rice seedlings in Thailand. Women reduce the soil to a creamlike consistency with their feet, while men perform the heavier task of breaking ground. (Foreign Agricultural Service, U.S. Department of Agriculture.)

had become a major United States ally in Southeast Asia, with air bases from which sorties were being flown to Vietnam.

Vietnam, Cambodia, and Laos

North and South Vietnam, Cambodia, and Laos are conveniently discussed together, as they are the states which emerged when the former French possession of Indochina disintegrated under Communist and nationalist pressure in 1954. The areas they occupy were conquered by France in stages between 1858 and 1907. In the process, the French extinguished an Annamite empire covering the approximate territory of present-day Vietnam, and defeated the Chinese, whom the Annamese called to their aid. Subsequently two sizable areas—Laos (in 1893) and western Cambodia (in 1907)—were taken from Thailand.

Three distinct regions—Tonkin, Annam, and Cochin China—have long been recognized in the territory now occupied by Vietnam. Tonkin, the northern region, includes the heavily populated delta of the Red River and a surrounding frame of sparsely populated mountains. Annam, in the center, includes the sparsely populated Annamite Cordillera and numerous small, densely populated pockets of lowland along its seaward edge. Cochin China, to the south, is mainly comprised of the Mekong River delta, which has a fairly high density of population. When France attempted to regain control of Indochina after the Japanese occupation of World War II, it was confronted with a strong and militant nationalist movement armed and led by Communists. Eight years of warfare resulted in French defeat in 1954. Vietnam was partitioned between the Communist North (the "Democratic Republic of Vietnam"), including Tonkin and northern Annam, and an independent non-Communist South (the Republic of Vietnam), occupying Cochin China and southern Annam.

Cambodia and Laos also became independent non-Communist states as French colonial control ended. Except for Cambodia's short and commercially little developed sea front on the Gulf of Siam (map, p. 449), these two states occupy interior, but very different, regions. Cambodia is comprised principally of plains along and to the west of the lower Mekong River, although it has mountainous fringes to the northeast and southwest. Laos is a very mountainous country which borders every other state on the Indochinese Peninsula and also Communist China. It is so sparsely populated, poorly organized, weak, and politically and ethnically divided that it hardly forms a functioning unit.

A number of different peoples inhabit the four states. However, each state is dominated by a particular people. About nine-tenths of the 38 million people in North and South Vietnam belong to the group traditionally known as Annamese, although Vietnamese is now the more common and preferred term. They are closely related to the Chinese in language and general culture. Their religious beliefs have been strongly influenced by Confucianism and ancestor worship, but with large elements of Buddhism and Catholic Christianity. Cambodians comprise a majority in their own state (estimated population about 6.6 million) and also form a small minority in adjacent Cochin China. More than 2 million Thais, speaking Laotian dialects of Thai, are distributed between Laos, where almost 2 million of them form the dominant group, and the mountains of Tonkin. Both Cambodians and Thais are predominantly Buddhist in religion. A variety of primitive tribes in the mountains of Vietnam and Laos aggregate perhaps 4 million people. A final element in the four states is comprised of approximately 2 million Chinese. Specializing in trade, they have an economic importance far out of proportion to their numbers. Most Chinese are found in the larger cities: Saigon-Cholon (2 million) in South Vietnam, Hanoi (over 650,000) and Haiphong (city proper, over 200,000) in North Vietnam, Phnom Penh (perhaps 500,000) in Cambodia, and Vientiane (city proper, perhaps 200,000) in Laos.

France's colonial administration, centered in Hanoi and Saigon, proved relatively ineffective from the standpoint of economic development, though it did succeed in opening the Mekong delta for commercial rice production. Economic functions connected with the rice industry, plus French administrative functions, were largely responsible for the development of Saigon-Cholon as French Indochina's only city of truly large size. The city is located at the northeastern edge of the Mekong delta. During the period of French control a number of other commercial crops besides rice were introduced in Indochina, both into native small-farm production and into plantation agriculture. None of these attained a really large export production. The area's substantial mineral resources, located mostly in present North Vietnam, were developed either on a small scale or not at all.

After independence the four states followed varying economic paths. South Vietnam and Cambodia, both of which received large amounts of American aid, concentrated largely on attempting to resuscitate and expand their agriculture. Cambodia also developed a new port city at Sihanoukville to avoid dependence on South Vietnam for a trade outlet through Saigon-Cholon. Laos received even more massive American support, at least on a per capita basis, but its domestic economy remains relatively primitive and its international trade very small. North Vietnam began to develop its economy along Communist lines with Chinese and Russian support, emphasizing a violent land reform and formation of peasant cooperatives in agriculture while pushing industrial development as rapidly as possible. Agriculturally it is the most crowded and hard pressed of the four Indochinese states, with rural densities on alluvial rice lands frequently exceeding 1500 per square mile. But it contains their largest reserves of known mineral wealth—including the best coal deposits in Southeast Asia plus some phosphate, iron ore, tin, zinc, and other minerals—so that its industrial resources may eventually compensate somewhat for its unusually overcrowded agricultural lands.

In Laos and Vietnam, however, economic development became a secondary consideration as new military and political conflicts followed the withdrawal of France. From 1954 to the time of writing (1968) Laos has been beset by a small-scale, intermittent, and desultory civil war. The government, officially neutral on the international level but leaning heavily on the United States, has been opposed by a Communist organization called the Pathet Lao, which is strongly supported by North Vietnam and controls considerable parts of Laos. Various attempts at compromise and coalition governments have had only temporary successes.

Vietnam was not supposed to be permanently partitioned according to the Geneva agreements which ended the French-Vietnamese struggle in 1954. Partition at the 17th parallel resulted from the fact that some Vietnamese were Communist-nationalist while some were anti-Communist, with both of these groups supported by outside powers. After partition a large-scale migration took place, mostly of anti-Communists, especially Catholic Christians, from North to South, with some migration in the opposite direction from South Vietnam to North Vietnam. The agreement was that all-Vietnam elections to unify the country were to be held within 2 years of partition. However, the non-Communist government established in the South maintained that truly free elections were impossible in the North, and they were not held. The division then hardened, with large-scale United States support for the government at Saigon, and support for the Hanoi government from the Communist powers.

During the 1950s a Communist-led insurrection gradually gained momentum in South Vietnam, and in the 1960s large-scale warfare engulfed the country. United States aid and advisers to the government of South Vietnam were involved against the insurrection almost from the first, and as it gained momentum in the 1960s American involvement rapidly expanded to a full-scale military commitment of half a million men. On the other side North Vietnam's support of the insurrection also rapidly expanded and North Vietnamese regular troops came into action in South Vietnam against American forces and those of the South Vietnamese government. North Vietnam itself came under intensive air attack by the United States air force and navy, although it was not invaded. Chinese and Russian help, including military equipment of steadily greater sophistication and destructive power, was supplied to North Vietnam and the Viet Cong (South Vietnamese insurgents). At the time of writing (1968) the conflict continued with unremitting ferocity, although American air attacks had been restricted to the southern part of North Vietnam and negotiations attempting to end the struggle had begun in Paris. Vietnam itself, both South and North, had been so badly damaged physically and disrupted socially and economically that, however the conflict eventually ended, a major effort of rehabilitation by the authorities then in power would be necessary.

Malaysia and Singapore

The small island of Singapore, narrowly separated from the southern tip of the mountainous Malay Peninsula, lies at the eastern end of the Strait of Malacca, which is the major passageway through which sea traffic is funneled between the Indian Ocean and the China Seas. For centuries European sea powers contested with each other for control of this passageway. Malacca city, on the west coast of the peninsula bordering the strait, was a Portuguese stronghold from 1511 and Dutch after 1641. British control over the strait was continuous from 1824. It was exercised from the port of Singapore, founded on the southern side of the island five years earlier. Under the British, Singapore, with its relatively central position among the islands and peninsulas of Southeast Asia, developed not only into a major naval base but also into the region's major entrepôt.

From Singapore, Britain gradually extended its political control over the adjacent southern end of the Malay Peninsula. Nine Malay sultanates occupying the area were reduced to the status of protected states between 1874 and 1909, the northern four being detached from

Thai suzerainty in the latter year. These units, plus the crown colonies of Malacca and Penang, gave Britain control over the part of the Malay Peninsula now included, along with northern Borneo, in the independent country of Malaysia. This southern end of the peninsula is customarily referred to as Malaya, though officially it is now West Malaysia.

During the period of British administration Malaya developed a highly commercialized and, for Southeast Asia, a prosperous economy. Tin and rubber became the

A plantation scene in the rubber belt of Malaya. Records are being made of the amount of latex (rubber sap) collected by each tapper. (British Information Services.)

Tin mining in Malaya. Nearly all Malayan tin ore is recovered by processing the alluvial earth with which it is intermixed. The ore is separated from unwanted earth by making a slurry and passing it through a sluice called a *palong*. Baffles in the sluice catch the ore, which is heavier, but allow the waste matter to be carried off by the moving water. The worker in the photo is raking the slurry to assist in the separation of the ore from the alluvium in which it is embedded. (British Information Services.)

Moslem Malays played only a minor role in this development. For the most part they have remained subsistence rice farmers in small coastal deltas along both sides of the peninsula. Chinese and Indian immigrants and their descendants became the principal commercial farmers, wage workers, and businessmen of the tin and rubber belt. So heavy was the immigration that Malays now comprise under 60 percent of Malaya's population, while Chinese constitute almost one-third and Indians (and Pakistanis) about 10 percent. The growth of Singapore involved a proportionately heavier immigration from overseas, so that Singapore's present population of 2 million is about 80 percent Chinese, 12 percent Malay, and 7 percent Indian.

Ethnic antagonisms between Malays and Chinese have been fundamental in shaping the political geography of Malaysia and Singapore. Malaya accepted independence in 1957 only on condition that Singapore not be included. This was done in order to prevent the Chinese population of Singapore from being added to that of Malaya itself, and by this means the majority position of the Malays within the new state was protected. In 1963, however, the Federation of Malaysia was formed, including Malaya, Singapore, and the former British possessions of Sarawak and Sabah in northern Borneo. The latter areas are sizable but relatively undeveloped and sparsely populated. They have Chinese minorities of considerable size, but their non-Chinese majorities were apparently counted on to counterbalance the admission to the Federation of Singapore's Chinese. This experiment in union lasted until 1965, when Singapore was expelled from the Federation and left to go its own way as an independent state. The remainder of the Federation, in Malaya (West Malaysia) and Borneo (East Malaysia), remains intact, though not without some political strain between the country's core area in Malaya—one of the most prosperous and productive parts of the Orient—and the outlying and relatively backward areas in northern Borneo.

major commercial products. The tin-mining industry was pioneered in the later nineteenth century primarily by Chinese interests, though these are now overshadowed by British concerns. A line of rich tin deposits along the western foothills of the mountains was made accessible by the building of railroads. These same rail lines provided transportation for rubber when new electrical and automotive industries stimulated a greatly increased demand for that product in the world's industrial countries during the early decades of the twentieth century. A densely populated belt of tin mines and rubber plantations developed in the foothills between Malacca and the hinterland of Penang (400,000). Within this belt the inland city of Kuala Lumpur (500,000) became the leading commercial center and capital (both before and after independence) of the country. From the relatively small area of the tin and rubber belt Malaya ordinarily supplies about one-third of the world's tin production and almost half of its rubber exports.

The development of a commercial economy gave Malaya an ethnically mixed population. The native

Since their separation, Malaysia and Singapore have followed their individual and somewhat divergent economic paths. Economic relations between the two are still important, as evidenced by the fact that about a quarter of Singapore's trade is with Malaya and perhaps a fifth of Malaya's trade is with Singapore, but this relationship is much less close than formerly. Malaya has been expanding its basic export industries, especially rubber, and is increasing irrigation in an attempt to become independent of the large rice imports now required. For Singapore an agricultural economy of much significance seems out of the question; its main efforts have been to maintain and develop its position as a general entrepôt in the midst of Southeast Asia and to achieve a rapid expansion of manufacturing, more or less on the pattern of the somewhat similar, though still colonial, state of Hong Kong. Its difficulties were compounded in the late 1960s by the rapid reduction of British strength at Singapore's military bases, which formerly employed 40,000 of the state's 2 million people.

Indonesia

Indonesia is by far the largest and most populous of the countries of Southeast Asia. Its several thousand islands comprise well over one-third of the region's land area and contain about two-fifths of its total population. The large population of Indonesia results from the enormous concentration of people which has developed on the island of Java. Here live at present (1968) about 73 million people, representing nearly two-thirds of Indonesia's population and more than a fourth of the population of all Southeast Asia. The island's population density is approximately 1400 per square mile; in contrast the remainder of Indonesia, which is about 13 times larger than Java in land area, has only about half as many people and an average population density of just under 60 per square mile.

The extraordinary concentration of population on this one island can be partly accounted for by the superior average fertility of its soils, the best of which have been derived from materials poured out by its many volcanic peaks. However, it also is the result in good measure the centering of Dutch colonial activities in Java. Other islands of Indonesia (with the notable exception of Borneo) have areas of volcanic soil, but such areas, while generally more densely populated than adjoining nonvolcanic areas, seldom attain the extremely high population densities found on Java. The economic development of some islands has undoubtedly been handicapped by the unfriendly nature of their coastlines, which present a front of coral reefs, steep cliffs, or extensive swamps, and thus create difficulties of access.

The Dutch East India Company, after long-continued hostilities with Portuguese and English rivals and with native states, secured effective control of most of Java in the eighteenth century. Large sections of the remaining islands, however, were not brought under control until the nineteenth century; and only in 1904 was the conquest of northern Sumatra completed. Strenuous efforts to develop and exploit the natural wealth of Java were undertaken by the government of the Netherlands with the introduction of the Culture System in 1830. Under this system forced contributions of land and labor were required of Javanese farmers for the intensive production of export crops under Dutch supervision. The system was undeniably harsh but, from the Dutch point of view, successful. Although the Culture System was abolished in 1870, native commercial agriculture continued to expand along with plantation production, and a means of support for increasing numbers of people was provided. Thus the introduction of the Culture System appears to have set off not only a great increase in the output of export commodities, but also the enormous rise in Java's population, which has multiplied more than 15 times in a century and a half.

Development of the other main islands began later and has been less intensive. Nevertheless the eastern coastal plain of Sumatra, inland from the great fringing swamp, has now surpassed Java in agricultural exports—especially of rubber, the single most important of Indonesia's numerous export crops. Java's export production is of declining importance; Borneo's is increasing, and the lesser islands, including Celebes, lead in export production of only one major crop—coconuts. Borneo, interior Celebes, and the mountainous western section of Sumatra are the habitat of primitive tribes who gain a meager subsistence from shifting cultivation and associated hunting, fishing, and gathering activities. Five areas outside of Java and Sumatra exhibit an uncommon density of agricultural settlement. These are the island of Madura, adjoining Java on the northeast and sharing its intensive commercial development although lacking its volcanic soils; the two volcanic islands of Bali and Lombok east of Java, given over largely to intensive subsistence cultivation; and the volcanic southwestern and northeastern arms of Celebes.

Irrigated rice, often grown on picturesque terraces in the mountainous areas, is the most important food crop of Indonesia. It is supplemented by such subsistence crops as sweet potatoes, cassava, corn, peanuts, and soybeans. As has already been indicated, the country is of some importance as a producer of certain minerals, notably tin (from the islands of Bangka and Billiton between Sumatra and Borneo), oil (from Sumatra, Java, and Borneo), and bauxite (from the Riouw Archipelago between Sumatra and Malaya). The leading Indonesian exports by value in 1965 were petroleum and its products (38 percent), rubber (30 percent), tin (5 percent), and coffee and palm oil (4 percent each), with coconut products, tobacco, and tea representing lesser values.

The intensive development of Java and the outstanding importance of Sumatra in export production are exemplified by the greater urban development of these two islands. The Indonesian capital of Jakarta, located near the western end of Java, has probably between 4 and 5 million people in its metropolitan area and is the largest city in Southeast Asia. Two other Javanese cities, Surabaya and Bandung, have metropolitan populations estimated at well over 1 million, while the following cities probably have metropolitan populations of more than 400,000 but less than 1 million: in Java the cities of Surakarta, Semarang, Malang, and Jogjakarta; in Sumatra, the cities of Medan and Palembang; and in Celebes the city of Makassar. An important share of the business of these cities is carried on by Chinese immigrants and their descendants, the only alien group of much significance. The number of Chinese in Indonesia is generally placed at 2 to 3 million, and they form for the most part an urban commercial class.

As of the late 1960s the Indonesian state appeared to be a political success and an economic failure. Increasing nationalism before and during the Japanese occupation of World War II led to a bitter struggle for independence from the Netherlands following the war. This struggle was eventually successful in 1949 after considerable fighting and political intercession by outside powers. Independent Indonesia did not initially include western New Guinea, which is Papuan rather than Indonesian in culture, but a confrontation with the Netherlands in 1962 brought this territory (West Irian in Indonesian terminology) into the Indonesian state. For some years it proved difficult to establish peace, order, and unity in the far-flung Indonesian island units. Ethnic diversity was at least partially responsible. Although about 90 percent of the population is at least nominally Moslem, there are great cultural differences in other respects, as reflected in the fact that over 200 languages and dialects are in use. Consequently, ethnic separatism played an important part in the affairs of the new state. A special degree of antagonism has existed in some localities against the numerically and politically dominant Javanese, and this has been reinforced by questions of economic relations between Java and the less developed islands. These conflicts have at times involved armed insurrection.

In addition, by 1965 the country seemed on the point of succumbing to steadily increasing Communist power in its internal politics, but an attempted Communist coup d'état in that year was unsuccessful. It resulted in the massacre by the Indonesian army and Islamic and nationalist groups of probably 300,000 Communists and Communist supporters and the establishment of a government in which army influence was predominant. Despite all of these circumstances and events, the new Indonesian state maintained itself as an entity in control of the vast archipelago. Economically, however, the first decades of independence during the 1950s and 1960s were years of almost consistent deterioration. War, military actions, political disorder, mismanagement, official preoccupation with political and military rather than economic matters, ill-advised nationalizations of foreign holdings, and extensive corruption might all be blamed. Whatever the causes, the results were an almost catastrophic failure to maintain exports and expand food production and to keep up an adequate level of general economic development. By the late 1960s the government seemed to be changing its priorities and laying greater stress on economic development. The size and natural wealth of the country gave some promise for the future.

The Philippines

The Philippine group includes over 7000 generally mountainous islands, but the two largest islands, Luzon and Mindanao, almost equal in size, account for two-thirds of the total area. Most of the mountainous districts are inhabited by a sparse population of relatively primitive shifting cultivators, although in northern Luzon the Igorot tribes have developed a spectacular and world-famous system of wet rice cultivation on terraced mountain sides.

Most of the population is concentrated in three areas:

1. The Visayan Islands in the center of the archipelago, where soils derived from volcanic materials and uplifted corals support an intensive subsistence agriculture based on rice and corn. Negros Island in this group is a major center of plantation sugar production.

2. The plains extending from south of Manila to Lingayen Gulf and thence north along the west coast of Luzon. Rice is the main food crop of these plains, and sugar, much of it produced on small native farms, is the main commercial product.

3. The southeastern peninsula of Luzon. Subsistence rice and commercial coconut production are basic to the economy here except in the extreme southern part, where large plantations utilize volcanic soils to produce a major share of the Philippines' output of abaca or Manila hemp.

The Philippines were a Spanish colonial possession from the later sixteenth century until 1898, and with the exception of the Japanese occupation of 1942–1944 were controlled by the United States from 1898 until 1946, when independence was granted. The Spanish legacy is still important in the country. Ever since its founding in 1571 the Spanish capital of Manila (2.5 million) has been the metropolis and only large city of the islands. The society created by Spain was composed of a relatively small upper class of Hispanicized Filipino landowners and a great mass of landless peasants. Problems created by the maldistribution of agricultural land have remained as a major source of difficulty for the new Republic of the Philippines. Discontented peasants gave much support to a Communist-led revolt after World War II. It was eventually suppressed, in large part through the granting of land (generally on sparsely populated Mindanao) to surrendered rebels, but flared up again during the 1960s. In the Philippines Spanish missionary activity succeeded in creating the only Christian nation in the Orient. Today over 80 percent of the population is listed as Roman Catholic, while almost 10 percent adheres to other Christian denominations. Spanish Catholicism failed to penetrate only the most remote mountain areas, still pagan, and southern Mindanao and the Sulu Archipelago, where the Moros, a Moslem people, successfully resisted Spanish control for 300 years.

The American period, although it began with the suppression of a Philippine independence movement, eventually brought independence, along with economic development and tutelage in democracy. The growth of major Philippine export industries—coconut products, cane sugar, and abaca—was stimulated by preferential treatment in the American market. In the 1930s a large degree of self-government was granted, full independence was promised for 1946, and legislation was passed in the American Congress to cushion the economic effects of independence by continuing the preferential treatment of Philippine exports to the United States. The latter is to be gradually reduced and will probably end in the 1970s. At the same time education was furthered so that even now English serves as the medium of instruction in Philippine schools and serves to some extent to bridge the gap between the diverse linguistic groups of the population. After independence one of the most widely used native languages, Tagalog, was selected as the principal base for an official national language and is now taught in all of the schools.

Since independence the Philippines have had considerable success in expanding their economy. Decline in the production of abaca, which is less in demand than it was before World War II, has been more than compensated for by expansion of other export commodities. Coconut products now take first place among the country's exports, while a spectacular expansion in timber production, principally of Philippine mahogany, has brought wood and wood products to second place on the export list. Sugar, which has long been important, comes third, and a variety of metal ores taken collectively rank fourth. During the 1960s Japanese and American capital was quite active in developing Philippine resources of copper and iron ore. In the late 1960s development of new strains of "miracle rice" by scientists in the Philippines offered the hope of ending by higher yields the country's growing dependence on grain imports—a hope which extended to many other hard-pressed countries into which the new rice might be introduced.

The official capital of the Philippine Republic is Quezon City, a suburb of Manila. Most government offices, however, are in Manila.

REFERENCES and READINGS

General and Miscellaneous

Atlas of South-East Asia, with an introduction by D. G. E. Hall (New York: St. Martin's Press, 1964).

BROEK, J. O. M., "Diversity and Unity in Southeast Asia," *Geographical Review,* 34, no. 2 (April 1944), 175–195.

BUCHANAN, KEITH, *The Southeast Asian World: An Introductory Essay* (New York: Taplinger Publishing Co., 1967).

CADY, JOHN F., *Southeast Asia: Its Historical Development* (New York: McGraw-Hill Book Company, 1964).

CHANG, SEN-DOU, "The Distribution and Occupations of Overseas Chinese," *Geographical Review*, **58**, no. 1 (January 1968), 89–107.

COE, MICHAEL D., "Social Typology and the Tropical Forest Civilizations," *Comparative Studies in Society and History*, **4**, no. 1 (November 1961), 65–85.

COURTENAY, P. P., "Changing Patterns in the Tin Mining and Smelting Industry of Southeast Asia," *Journal of Tropical Geography*, **25** (December 1967), 8–17; and *Plantation Agriculture* (New York: Frederick A. Praeger, 1966).

DOBBY, E. H. G., *Southeast Asia* (7th ed.; London: University of London Press, 1960). A geography text.

DWYER, D. J., "The City in the Developing World and the Example of Southeast Asia," *Geography*, **53**, pt. 4 (November 1968), 353–364.

FISHER, CHARLES A., *Southeast Asia: A Social, Economic, and Political Geography* (New York: E. P. Dutton & Co., 1964); the outstanding geography of the region: long, detailed, authoritative, and well written, with extensive bibliography. Also, "Southeast Asia: The Balkans of the Orient?" *Geography*, **47**, pt. 4 (November 1962), 347–367; and *The Reality of Place: An Inaugural Lecture* (London: University of London, School of Oriental and African Studies, 1965).

HEENAN, L. D. B., "The Petroleum Industry of Monsoon Asia," *Pacific Viewpoint*, **6**, no. 1 (May 1965), 65–95.

HUNTER, GUY, *South-east Asia: Race, Culture, and Nation* (New York: Oxford University Press, 1966).

KARNOW, STANLEY, and the Editors of *Life*, *Southeast Asia* (Life World Library; New York: Time Inc., 1962).

LEBAR, FRANK M., GERALD C. HICKEY, and JOHN K. MUSGRAVE, *Ethnic Groups of Mainland Southeast Asia* (New Haven, Conn.: Human Relations Area Files Press, 1964).

McGEE, T. G., *The Southeast Asian City: A Social Geography* (New York: Frederick A. Praeger, 1967); and "Aspects of the Political Geography of Southeast Asia," *Pacific Viewpoint*, **1**, no. 1 (March 1960), 39–58.

PELZER, KARL J., "Man's Role in Changing the Landscape of Southeast Asia," *Journal of Asian Studies*, **27**, no. 2 (February 1968), 269–279; and *Pioneer Settlement in the Asiatic Tropics: Studies in Land Utilization and Agricultural Colonization in Southeastern Asia* (New York: American Geographical Society, 1945).

POLHAMUS, LOREN GEORGE, *Rubber: Botany, Production, and Utilization* (New York: Interscience Publishers, 1962).

PURCELL, VICTOR, *The Chinese in Southeast Asia* (2d ed.; New York: Oxford University Press, 1965).

ROBEQUAIN, CHARLES, *Malaya, Indonesia, Borneo and the Philippines: A Geographical, Economic, and Political Description of Malaya, the East Indies, and the Philippines,* trans. from the French by E. D. Laborde (2d ed.; London: Longmans, Green and Co., 1958). A standard work.

SCHUSTER, W. H., "Fish Culture in Conjunction with Rice Cultivation," *World Crops*, **7**, nos. 1 and 2 (January and February 1955), 11–14 and 67–70.

SPENCER, J. E., *Shifting Cultivation in Southeastern Asia* (University of California Publications in Geography, Vol. 19; Berkeley and Los Angeles: University of California Press, 1966); and, with Ronald J. Horvath, "How Does an Agricultural Region Originate?" *Annals of the Association of American Geographers*, **53**, no. 1 (March 1963), 74–92.

VANDENBOSCH, AMRY, and RICHARD BUTWELL, *The Changing Face of Southeast Asia* (Lexington: University of Kentucky Press, 1966).

WATABE, TADAYO, and KEIZABURO KAWAGUCHI, "Increasing the Rice Yield in South and Southeast Asia," *Asian Survey*, **8**, no. 10 (October 1968), 820–828.

WILLIAMS, LEA E., *The Future of the Overseas Chinese in Southeast Asia* (New York: McGraw-Hill Book Company, 1966).

Ceylon

BAKER, SIMON, "Tea Production in Ceylon," *Journal of Geography*, **67**, no. 1 (January 1968), 49–55; and "Shifting Agriculture in Ceylon," *Journal of Geography*, **67**, no. 9 (December 1968), 564–568.

Basic Data on the Economy of Ceylon, U.S. Bureau of International Commerce, Overseas Business Reports, OBR 66–87 (December 1966). (Washington, D.C.: Government Printing Office, 1966.)

Ceylon Geographer (4 numbers a year).

FARMER, B. H., "Ceylon," in O. H. K. Spate and A. T. A. Learmonth, *India and Pakistan: A General and Regional Geography* (3d ed.; London: Methuen and Co., 1967), Chap. 26, pp. 786–824; "Agriculture in Ceylon," *Geographical Review,* 40, no. 1 (January 1950), 42–66; "The Ceylon Ten-Year Plan, 1959–1968," *Pacific Viewpoint,* 2, no. 3 (September 1961), 123–136; "Ceylon's Frontiersmen," *Geographical Magazine,* 39, no. 8 (December 1966), 611–620; "Peasant and Plantation in Ceylon," *Pacific Viewpoint,* 4, no. 1 (March 1963), 9–16; "Rainfall and Water-Supply in the Dry Zone of Ceylon," in R. W. Steel and C. A. Fisher, eds., *Geographical Essays on British Tropical Lands* (London: George Philip and Son, 1956), pp. 225–268; and "Ceylon: Some Problems of a Plural Society," in Charles A. Fisher, ed., *Essays in Political Geography* (London: Methuen and Co., 1968), pp. 147–159.

HUNG, FREDERICK, "The Tea Plantation System in Ceylon: Persistence and Change," *Professional Geographer,* 20, no. 5 (September 1965), 321–325.

JOSHI, TULASI R., "Ceylon," and "Planning for the Future" [of Ceylon], *Focus,* 18, no. 9 (May 1968), 1–8 and 8–11.

SWAN, S. B. ST. C., "Paddy Crop Failure and Variable Yields in Ceylon," *Pacific Viewpoint,* 8, no. 2 (September 1967), 159–174.

Burma

LEACH, E. R., "The Frontiers of 'Burma,'" *Comparative Studies in Society and History,* 3, no. 1 (October 1960), 49–73.

NASH, MANNING, *The Golden Road to Modernity: Village Life in Contemporary Burma* (New York: John Wiley & Sons, 1965). Focuses on two villages in central Burma.

SPATE, O. H. K., "The Burmese Village," *Geographical Review,* 35, no. 4 (October 1945), 523–543.

TRAGER, FRANK N., *Burma, from Kingdom to Republic: A Historical and Political Analysis* (New York: Frederick A. Praeger, 1966).

Thailand

GIRLING, J. L. S., "Northeast Thailand: Tomorrow's Viet Nam?" *Foreign Affairs,* 46, no. 2 (January 1968), 388–397.

PENDLETON, ROBERT L., *Thailand: Aspects of Landscape and Life* (An American Geographical Society Handbook; New York: Duell, Sloan & Pearce, 1962).

TRESCOTT, PAUL B., "Rice Production in Thailand," *World Crops,* 20, no. 4 (September 1968), 49–56.

VAN ROY, E., "Economic Dualism and Economic Change among the Hill Tribes of Thailand," *Pacific Viewpoint,* 7, no. 2 (September 1966), 151–168.

WEBBERT, M. VIRGINIA, *Basic Data on the Economy of Thailand,* U.S. Bureau of International Commerce, Overseas Business Reports, OBR 66–60 (September 1966). (Washington, D.C.: Government Printing Office, 1966.)

Vietnam, Cambodia, and Laos

BLACK, DOROTHY J., *Basic Data on the Economy of South Viet-Nam,* U.S. Bureau of International Commerce, Overseas Business Reports, OBR 67–10 (March 1967). (Washington, D.C.: Government Printing Office, 1967.)

FALL, BERNARD B., "Two Thousand Years of War in Viet-Nam," *Horizon,* 9, no. 2 (Spring 1967), 4–22. A concise summation of Vietnam's internal development through history, as related to foreign invaders and influences.

FISHER, CHARLES A., "The Vietnamese Problem in Its Geographical Context," *Geographical Journal,* 131, pt. 4 (December 1965), 502–515.

HAMMER, ELLEN, *Vietnam, Yesterday and Today* (Contemporary Civilizations Series; New York: Holt, Rinehart and Winston, 1966).

U.S. Army Area Handbook for South Vietnam (Department of the Army Pamphlet No. 550–55; Washington, D.C.: 1967).

WHITE, GILBERT F., "River Basin Planning and Peace: The Lower Mekong," in Saul B. Cohen, ed., *Problems and Trends in American Geography* (New York: Basic Books, 1967), pp. 187–199; and, with W. R. Derrick Sewell, "The Lower Mekong: An Experiment in International River Development," *International Conciliation,* no. 558 (May 1966), 63 pp.

Malaysia and Singapore

BOYCE, P. J., "Singapore as a Sovereign State," *Australian Outlook,* 19, no. 3 (December 1965), 259–271.

CLARKSON, JAMES D., *The Cultural Ecology of a Chinese Village: Cameron Highlands, Malaysia* (University of Chicago, Department of Geography Research Paper No. 114; Chicago: 1968).

FISHER, CHARLES A., ed., *Essays in Political Geography* (London: Methuen and Co., 1968), Chap. 7, "Malaysia: A Study in the Political Geography of Decolonization," pp. 75–145.

FRYER, D. W., and JAMES C. JACKSON, "Peasant Producers or Urban Planters? The Chinese Rubber Smallholders of Ulu Selangor," *Pacific Viewpoint,* 7, no. 2 (September 1966), 198–228.

GINSBURG, NORTON S., and CHESTER F. ROBERTS, *Malaya* (Seattle: University of Washington Press, 1958). A broad treatment of the country's environment, economy, political life, and society.

HILLSON, LOUISE H., *Basic Data on the Economy of Singapore,* U.S. Bureau of International Commerce, Overseas Business Reports, OBR 67–7 (February 1967); and *Basic Data on the Economy of Malaysia,* OBR 66–73 (October 1966). (Washington, D.C.: Government Printing Office.)

JONES, L. W., "Malaysia's Future Population," *Pacific Viewpoint,* 6, no. 1 (May 1965), 39–51.

KINLOCH, ROBERT F., "The Growth of Electric Power Production in Malaya," *Annals of the Association of American Geographers,* 56, no. 2 (June 1966), 220–235.

LEE, Y. L., "Historical Aspects of Settlement in British Borneo," *Pacific Viewpoint,* 2 (September 1961), 187–212; also, "The Population of Sarawak," *Geographical Journal,* 131, pt. 3 (September 1965), 344–356; and *North Borneo (Sabah): A Study in Settlement Geography* (Singapore: Eastern Universities Press, 1965).

MISSEN, G. J., "Manufacturing in Malaya: A Survey with Special Reference to the Impact of Political Changes," in J. Rutherford, M. I. Logan, and G. J. Missen, *New Viewpoints in Economic Geography* (Sydney, Australia: Martindale Press, 1966), pp. 296–313; and "The Big City in Malaya: A Study of Kuala Lumpur, Capital of a New Nation," *ibid.,* pp. 394–414.

NEVILLE, WARWICK, "Singapore: Ethnic Diversity and Its Implications," *Annals of the Association of American Geographers,* 56, no. 2 (June 1966), 236–253.

OOI, JIN-BEE, *Land, People, and Economy in Malaya* (London: Longmans, Green and Co., 1963); also, "The Nature and Distribution of the Natural Vegetation of Malaya, *Pacific Viewpoint,* 1, no. 1 (September 1960), 183–204; and "Some Aspects of Peasant Farming in Malaya," *Tijdschrift voor Economische en Sociale Geografie,* 56, no. 5 (September–October 1965), 170–185.

PARKINSON, BRIEN K., "Non-Economic Factors in the Economic Retardation of the Rural Malays," *Modern Asian Studies,* 1, pt. 1 (January 1967), 31–46.

RUTHERFORD, J., "Intensive Padi Farming in Malaya: Developmental Problems and General Effects of Government Planning," in J. Rutherford, M. I. Logan, and G. J. Missen, *New Viewpoints in Economic Geography* (Sydney, Australia: Martindale Press, 1965), pp. 71–92.

RUTHERFORD, JOHN, "Double Cropping of Wet Padi in Penang, Malaya," *Geographical Review,* 56, no. 2 (April 1966), 239–255.

SANDHU, K. S., "Indians in the Modern Malayan Economy," *India Quarterly,* 23, no. 2 (April–June 1967), 106–126.

U.S. Army Area Handbook for Malaysia and Singapore (Department of the Army Pamphlet No. 550-45; Washington, D.C.: 1965).

WARD, MARION W., "Major Port Hinterlands in Malaya," *Tijdschrift voor Economische en Sociale Geografie,* 57, no. 6 (November–December 1966), 242-251; and "A Review of Problems and Achievements in the Economic Development of Independent Malaya," *Economic Geography,* 44, no. 4 (October 1968), 326-342.

WHEATLEY, PAUL, *The Golden Khersonese: Studies in the Historical Geography of the Malay Peninsula before A.D. 1500* (Kuala Lumpur: University of Malaya Press, 1961).

WHEELWRIGHT, E. L., *Industrialization in Malaysia* (New York: Cambridge University Press, 1965).

Indonesia and Portuguese Timor

BENNETT, DON C., "The Basic Food Crops of Java and Madura," *Economic Geography,* 37, no. 1 (January 1961), 75-87.

GEERTZ, CLIFFORD, *Agricultural Involution: The Process of Ecological Change in Indonesia* (Berkeley and Los Angeles: University of California Press, 1963).

LUBAR, ROBERT, "Indonesia's Potholed Road Back," *Fortune,* 77, no. 6 (June 1968), 104-107 ff.

PAUKER, GUY J., "Toward a New Order in Indonesia," *Foreign Affairs,* 45, no. 3 (April 1967), 503-519.

WEATHERBEE, DONALD E., "Portuguese Timor: An Indonesian Dilemma," *Asian Survey,* 6, no. 12 (December 1966), 683-695.

WITHINGTON, WILLIAM A., "The Major Geographic Regions of Sumatra, Indonesia," *Annals of the Association of American Geographers,* 57, no. 3 (September 1967), 534-549; "The Kotapradja or 'King Cities' of Indonesia," *Pacific Viewpoint,* 4, no. 1 (March 1963), 75-86; "Migration and Economic Development: Some Recent Spatial Changes in the Population of Rural Sumatra, Indonesia," *Tijdschrift voor Economische en Sociale Geografie,* 58, no. 3 (May–June 1967), 153-163; and "Changes and Trends in Patterns of North Sumatra's Estate Agriculture, 1938-1959," *Tijdschrift voor Economische en Sociale Geografie,* 55, no. 1 (January 1964), 8-13.

The Philippines

HUKE, ROBERT E., *Shadows on the Land: An Economic Geography of the Philippines* (Manila: The Bookmark, Inc., 1963).

IMUS, RICHARD H., and MAURICE KOGON, *Basic Data on the Economy of the Philippines,* U.S. Bureau of International Commerce, Overseas Business Reports, OBR 68-74 (August 1968). (Washington, D.C.: Government Printing Office, 1968.)

Philippine Geographical Journal (quarterly).

RUTTAN, V. W., A. SOOTHIPAN, and E. C. VENEGAS, "Changes in Rice Growing in the Philippines and Thailand," *World Crops,* 18, no. 1 (March 1966), 18-33.

SCHUL, NORMAN W., "A Philippine Sugar Cane Plantation: Land Tenure and Sugar Cane Production," *Economic Geography,* 43, no. 2 (April 1967), 157-169.

STERNBERG, DAVID T., "The Philippines: Contour and Perspective," *Foreign Affairs,* 44, no. 3 (April 1966), 501-511.

WERNSTEDT, FREDERICK L., and J. E. SPENCER, *The Philippine Island World: A Physical, Cultural, and Regional Geography* (Berkeley and Los Angeles: University of California Press, 1967). The standard geography; comprehensive and authoritative.

See also the lists of references and readings for Chapters 1-3 and 17, particularly the relevant sections of works on Asia as a whole listed on p. 415.

The Chinese
Realm

20 China is the largest country of the world in population and is exceeded in area only by the Soviet Union and Canada.[1] It is mainly an agricultural country and very probably produces a greater total volume of crops than any other nation. Like several other countries of the Orient it has had a serious problem of rural overpopulation. The Chinese leadership has vacillated in its approach to the population problem, but at present the regime gives quiet sanction to birth control techniques, late marriage, and other control methods. Chinese leaders have long felt that industrialization was necessary to alleviate the country's poverty and to increase China's national power;

---

[1] The People's Republic of China (Communist China or mainland China) had an estimated population of at least 750 million in 1968 and many demographers would place this total significantly higher. The total was given as 583 million following the nationwide census of 1953, the first comprehensive census ever taken in mainland China. The lack of official statistics since 1958 makes predictions and estimates hazardous. Most experts, however, believe that the rate of annual increase is at least 2 percent or about 15 million people. Official Chinese Communist statistics give the area of mainland China as 3,691,000 square miles. The area of the Soviet Union is about 8,600,000 square miles, the figure for Canada is 3,852,000 square miles, and the United States 3,615,000 square miles.

and sufficient resources are present to support an industrial development of major size. However, modern types of manufacturing were little developed before the Communist accession to power in 1949, except in a few scattered areas; and they are still poorly developed in relation to the country's vast size and population despite impressive, though erratic, growth under the Communist regime.

China suffered more than a century of civil unrest, disorder, and costly warfare beginning with the Opium War of 1839–1842 and ending with the Korean Armistice in 1953. During most of this period, China's unity and even its national existence were threatened by the combined and largely successful efforts of other nations to extract trading rights and other economic and political concessions. The humiliations of this dark era in Chinese history have been bitterly resented by most Chinese—a resentment intensified by China's long history as the dominant power in Asia and its great achievements in numerous fields. Although these achievements of the Chinese past and the country's highly developed culture were the envy of the West in the eighteenth century, China's preoccupation with the past inhibited its adoption and assimilation of the rapidly spiraling scientific and technological advances made in the Western world during the nineteenth century. Pride in culture, love of the past, and disinterest in the outside world contributed powerfully to the downfall of the old order in China, and violation of Chinese pride, as well as territorial encroachment by outside powers, helped foster a powerful nationalism during the twentieth century.

The foregoing facts have gained added significance for the world at large with the emergence of mainland China as a military power under Communist rule. The full meaning of this development has yet to be unfolded, but with the explosion of its first nuclear device in October 1964, and several successful nuclear tests since then, there can be no doubt that Communist China has become an extremely important factor in world affairs. In the Korean War and later in the short-lived border war with India in 1962, the Chinese Communists demonstrated their military prowess. Thus under the Com-

China compared in latitude and area with conterminous United States.

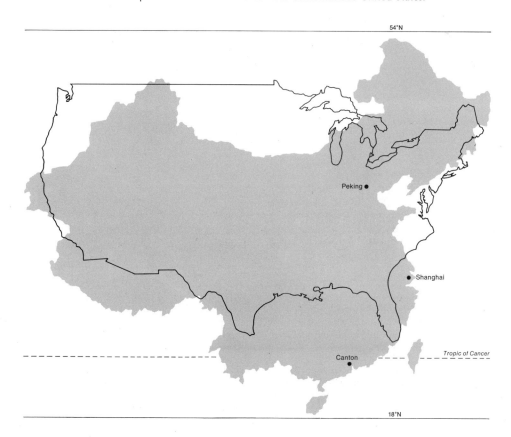

munist regime, China has thrust itself upon world attention by force of arms. The haltingly successful efforts of Chinese Communist planners to build an industrial society and the continued development of an advanced weapons program have added to China's power potential. The country's foreign policy, frequently belligerent, abusive, and aggressive in words, though not always in action, impinges directly on the many countries that border China (map, p. 467) and is a factor of importance in world affairs generally.

HISTORICAL BACKGROUND

In order to understand the significance of China's new role in the world, a certain amount of historical perspective is needed. This perspective, up to the beginning of war with Japan in 1937, is briefly supplied in the following selection:[2]

Early China

The foundations of Chinese history go back many centuries before the Christian era. The unification of the people into an empire came in the third century B.C. Since that time there have been many dynasties, some of them dynasties of conquering rulers leading nomadic peoples from the north, some of them native rulers. There were periods of weak central government characterized by wars between feudal states. The area known as China has shrunk and grown with the changing fortunes of these dynasties. Under the Mongol rulers of the Yuan dynasty (1280–1367) the boundaries of China were extended to the borders of Europe and encompassed Manchuria, Korea, Indochina, Burma, and Tibet. The Ming dynasty (1368–1644) that followed the Mongol rule was under native sovereigns and in this period some of the conquests of the Mongols were lost. Manchu rule replaced the Mings in 1644 in another period of aggressive expansion. Under this rule the boundaries again extended into Tibet, Central Asia, and other areas lost during the Ming dynasty. Much of this territory was very loosely held and allegiance amounted to sending tribute to the Imperial Court at stated times. The Republic of China dates from 1911. In the revolution that brought about this change of dynasty native control was re-established. However, China was subjected to almost continuous civil war from the Revolution [to the outbreak of war with Japan in 1937]. . . .

The civilization that grew within the areas known as China was of a very high order. Learning was held in great regard, and officials were scholars, appointed to their posts under a democratic examination system that was fairly well developed as early as the sixth century. Literature, philosophy, poetry, and the arts flourished and rose to high levels of excellence. An ingenious handicraft system produced objects of great beauty and delicacy as well as sturdy objects for daily use.

Transportation and communications were well organized and amazingly swift. Post offices were set up at short intervals along the main roads and letters were carried by relays of riders on horseback so that the Emperor in Peking was informed within a very few days of events occurring on the outer edges of his empire. This stage post system began in the Han period before the Christian era. Public inns also served the main highways and were open freely to all who had occasion to travel, and there were stores of fodder for the horses. Some of the more important roads were paved with stones. In the Sui dynasty in the sixth century the Grand Canal was built to connect the north and the south. It was rebuilt in the thirteenth century by the great Khublai Khan to improve communications between the capital and the Yangtze Valley and to facilitate the administration of the empire. Great use was made of the rivers as means of internal communication. The Yangtze and its tributaries were then, as they are now, important factors in the life of the nation.

The Chinese felt no need of the outside world. Their civilization was superior to that of neighboring lands and was drawn upon and borrowed from by those lands. It reached also into the outside world and influenced distant cultures. Japan took from China her written language and much of her art and literature and industry. Korea also drew heavily on Chinese culture. China gave the world silk and paper and many medicines, the art of printing, the magnetic needle, gunpowder, and many other things. Chinese philosophy reached far beyond the widest borders of the empire.

Contacts with the West

Influences from the outside world did come in and affect Chinese culture. Some of these influences came in by way of the conquering barbarians from the North; some came through peaceful trade; some were brought back by Chinese forces that had moved into the outer regions in the waves of empire expansion. Buddhism came from India and Christianity from the West, and they have had a wide and deep influence upon Chinese thinking and upon Chinese art. China also received many gifts from the

---

[2] *Geographical Foundations of National Power* (Army Service Forces Manual M 103–3; Washington, D.C.: Government Printing Office, 1944), pp. 1–3. One footnote has been omitted and two headings added. Some passages have been updated by words inserted in brackets.

outside, including cotton, which has become the primary fiber of her economy. By the eleventh or twelfth century, the use of the compass brought more sailing ships to China's shores, and by the sixteenth century contact with Europe by sea had become frequent.

China's self-sufficient economy had little interest in these first visitors from the West and the trade that they wanted. Even later, when the British East India Company tried to develop a trade, because they wanted China's tea and silk, pottery, lacquer, and cotton cloth, there was little besides silver that they could use in exchange —silver and a few luxuries and trinkets like watches and clocks. Their woolen cloth could find no market. The Chinese were not interested in the staple products of the West. America traded ginseng and [furs] for China's silk and teas in the early trade.

The Chinese were not eager for trade with the West. They erected a barrier to commerce comparable to the Great Wall they had built against the northern invaders in the first years of the Empire. This wall against commerce was a strict control of all trade by sea through the Co-hong, a guild system operated under strict supervision of a commissioner, or emperor's merchant, appointed by the throne. Under guild regulations foreign commerce was permitted only through the port of Canton. For many years this strict monopoly held the foreign traders under control. The advance of industrialization in the West, however, was a much stronger force than the Emperor's trade monopoly.

Early in the nineteenth century the invention of the steamship broke down the barrier of the Pacific Ocean. The Western nations had advanced their industrial development and were looking for markets, not for a luxury trade, but for the disposal of a large surplus production in exchange for China's tea and silk. For many decades their factory-made goods could find no market in China. They were inferior in quality to China's handmade products, and they were also sometimes more costly. The lighter weight cotton cloth for work clothes was not as elegant as the finer cloths still being spun by skilled native craftsmen. Gradually the factory goods improved in quality and cheapened in price, and the disintegration of China's great handicraft economy began. As the factory goods came into the coastal cities and displaced the handmade goods of the country, the economy was thrown out of adjustment. Following on the heels of the goods themselves, new industrial methods were imported into the coastal cities under foreign influence. The new methods established on the shores of China offered more serious competition to the old industries than the foreign factory goods, and the disintegration of the old order was hastened.

The superior ships and weapons of war of the Western nations forced political and economic concessions from isolationist China. Hong Kong was lost to the British and Macao to the Portuguese. The French secured a lease on Kwangchowan; the Germans, a lease on [Kiao-chow Bay (Tsingtao)]; the Russians, Liaotung in Manchuria.

Japan obtained the Nansei (Ryukyu) Islands, Taiwan (Formosa), and the Pescadores by arbitrary annexation or military conquest. Concessions were granted to foreign powers in Shanghai, Tientsin, Hankow, and many other so-called "treaty ports." Railway concessions were yielded to the Russians and Japanese in Manchuria, the British, French, Belgians, and Americans in China proper. Mining rights went to British, German, and French nationals. Legation quarters were set up in Peking with special privileges for the foreign powers. . . .

The Chinese Revolution and Subsequent Events

The Revolution that brought in the establishment of the Republic of China came about in 1911. The [four decades that] followed the Revolution [were] marked by both civil wars and foreign invasion. The old economy . . . continued to disintegrate, while a new economic structure [was] slow to build under the competition offered by the more industrialized nations.

The attempt to rebuild China politically following the overthrow of the Manchus in 1911 proved time-consuming and difficult. A period of "warlordism" ensued, with local rulers ("warlords") controlling provinces and other areas, until Chiang Kai-shek and the Kuomintang (Nationalist) Party managed, during military campaigns (1926–1928), to nominally unify most of the country. Although the new Nationalist government, with its capital at Nanking, was able to make noticeable progress in several fields, the strength of the regime was sapped by its efforts to defeat the Chinese Communists. Chiang Kai-shek finally drove the Communists from their base areas in the hills of South China in 1934, but in their "long march" the Communists managed to reestablish themselves and regroup in the loessial hills of northwest China in 1935.

In 1931 Manchuria was invaded and taken over by Japan, and in 1937 the Japanese provoked a full-scale war with China. Although the resistance of the Chinese Nationalist armies proved tougher than the Japanese had anticipated, the Nationalist forces and the government were forced to retreat to the interior. Here, with its wartime capital of Chungking protected by the mountains rimming the Szechwan (Red) Basin, the government was able to hold out, aided by a trickle of supplies from the outside, until final defeat of Japan by the Allies in 1945. Meanwhile, a kind of uneasy truce prevailed between the Nationalists and Communists. In the absence of pressure from the Nationalists, the Communists were able not only to engage in guerrilla warfare against the Japanese

Selling yak milk in Arid China to buyers from a nearby milk-powder processing plant. The photo was taken in Sinkiang. *(Foreign Agriculture.)*

but at the same time to build and expand their bases in the countryside. Thus by 1945 they were relatively much stronger than before the war. Despite efforts at negotiation, full-fledged civil war once again commenced in 1946. Although aided in this struggle by American equipment and supplies, the Nationalist government lost the support of large segments of the population, and the Nationalist armies gradually melted away. In 1949 the government, together with its remaining forces, fled to Taiwan (Formosa). Since then, it has continued to hold Taiwan, the neighboring Pescadores Islands, and some small islands along the mainland coast, the most important of which are the heavily fortified islands of Quemoy and Matsu (map, right). During the 1950s several crises developed in the Formosa Strait that threatened to involve other powers, specifically the United States and the Soviet Union, and the area remains a focus of tension.

MAJOR AREAL DIVISIONS

China is vast in size, but like all extremely large countries contains a great deal of unproductive land—mainly in the western and northwestern parts of the country in the Tibetan Highlands, Sinkiang, and Inner Mongolia. These outlying areas are principally comprised of high mountains and plateaus, together with great expanses of arid or semiarid plains where rainfall, generally speaking, is insufficient for agriculture. This dry, sparsely settled country probably contains only 5 percent or so of China's population, consisting mostly of various peoples other than the Han Chinese.[3] This half of China is in marked contrast to the better-watered, densely settled eastern half of the country where nearly 95 percent of the population lives. In a general way, then, China is divided into a western half, or Arid China, and an eastern core region, or Humid China, the latter containing the bulk of China's population and most of the country's resources and productive capacity. Arid China, however, is important in Chinese planning since it contains some important mineral resources and a little reclaimable land; additionally, it possesses large desolate areas necessary for the testing of various types of advanced weapons.

[3] The term Han Chinese is commonly used to designate the dominant ethno-linguistic group in China, with the term Han derived from the first great dynasty of China. Some 50 non-Han Chinese ethno-linguistic groups, nearly all of them racially of Mongoloid stock, number about 40 million and live primarily in the western and southwestern parts of the country.

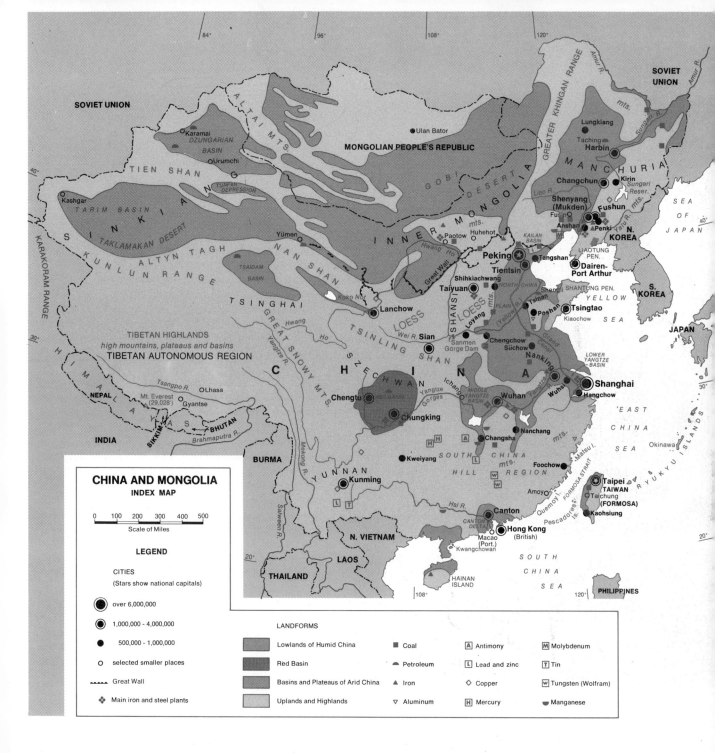

CHINA AND MONGOLIA
INDEX MAP

0   100   200   300   400   500
Scale of Miles

LEGEND

CITIES
(Stars show national capitals)

◉ over 6,000,000

◎ 1,000,000 - 4,000,000

● 500,000 - 1,000,000

○ selected smaller places

---- Great Wall

❖ Main iron and steel plants

LANDFORMS

- Lowlands of Humid China
- Red Basin
- Basins and Plateaus of Arid China
- Uplands and Highlands

■ Coal
⚓ Petroleum
▲ Iron
▽ Aluminum

Ⓐ Antimony
Ⓛ Lead and zinc
◇ Copper
Ⓗ Mercury

Ⓜ Molybdenum
Ⓣ Tin
Ⓦ Tungsten (Wolfram)
⬛ Manganese

ARID CHINA

A rough boundary between Humid and Arid China would be a line drawn from the northeastern corner of India to the northern tip of Manchuria. In a general way this line corresponds to the line of 20 inches average annual rainfall.

Except in parts of Manchuria, most of the country east of this line is densely populated and the people are predominantly Chinese; west of the line the population is nearly everywhere sparse, large tracts are uninhabited, and many of the people are non-Chinese, chiefly Mongols, Tibetans, or Turks. Nearly everywhere east of the line crops are raised without dry-land irrigation; west of it grazing largely takes the place of agriculture [photo, left], which is confined to scattered oases where water for irrigation is available along belts between mountains and dry plains.[4]

The principal regions west of the line include the Tibetan Highlands, Sinkiang, and Inner Mongolia.

Tibetan Highlands

The Tibetan Highlands comprise about one-quarter of China's land area, are made up politically of the Tibetan Autonomous Region, Tsinghai Province, the western portion of Szechwan, and small areas of other adjoining provinces, and culturally are unified by the dominance of about 3 million Tibetans who espouse Lamaism—the Tibetan variant of Buddhism. Most of this vast region is a high, mountainous plateau averaging nearly 3 miles in elevation. Still higher mountain ranges form its borders: the Himalayas on the south, containing the world's highest peak, Mt. Everest (29,028 feet) on the Nepal-Tibet border; the Karakoram and other ranges to the northwest; the Kunlun, Altyn Tagh, and Nan Shan ranges to the north; and the Great Snowy Range and other mountains on the east (map, p. 467). The Tibetan Highlands are further subdivided into two major physical regions: a very high northwestern or inner plateau region; and a somewhat lower and topographically more varied southeastern or outer plateau region. The inner plateau is characterized by internal drainage and a number of large salt lakes, low annual precipitation, scanty vegetation, and a very sparse, almost entirely nomadic population. Large areas of the inner plateau are uninhabited. The outer plateau region contains the Tsangpo River (the Brahmaputra of India) and the headwaters and upper courses of several other major rivers—the Salween, Mekong, Yangtze, and Hwang (Yellow). Marked contrasts in elevation, precipitation, temperatures, and vegetation result from the great dissection of the plateau surface by these rivers and their tributaries. The population, though comparatively small and dispersed, is considerably greater than that of the inner plateau and consists of settled agricultural communities as well as nomads. The northwest-to-southeast trend of the ridge and canyon country of eastern Tibet makes east to west communication difficult, and roads built from Szechwan to Lhasa and southern Tibet require much maintenance. The combination of lower elevations, warmer temperatures, and greater precipitation in parts of the outer plateau region results in some extensive grasslands and, in espe-

cially favored areas, stands of timber, mostly conifers. Included within the Tibetan Highlands is a third division or subregion, the Tsaidam Basin. Although rimmed by much higher plateau and mountain country, the basin itself is only 9000 to 10,000 feet in elevation. It is comprised of a barren sand-and-gravel western half that merges eastward into salt flats, marshy depressions, and bogs.

The area now designated by the Chinese regime as the Tibetan Autonomous Region—comprising about one-half of the Tibetan Highlands—coincides more or less with the traditional limits of effective Tibetan authority exercised from Lhasa (population, perhaps 70,000). The remaining parts of the Tibetan Highlands, though retaining spiritual ties with Lhasa, were fragmented by terrain and distance into virtually autonomous and often unruly nomadic groups and petty feudatories. These areas were influenced and at times loosely controlled by China, particularly the eastern territories lying nearest the Chinese provinces. Prior to the Communist era, Chinese control over the Tibetan Highlands had never been strong or long-lasting. The great distances from Chinese seats of power to southern Tibet and the lack of any significant agricultural land suitable for Chinese colonization contributed to the tenuous hold of China on the area. The last vestiges of traditional Chinese authority vanished with the overthrow of the Manchus, and from 1912 until the Chinese Communist military occupation of 1951 Tibet was in fact, if not legally, an independent state. After roads were completed to Lhasa and southern Tibet in late 1954, an increase in restrictive measures by the Chinese Communists contributed to the rise of guerrilla warfare. This culminated in the large-scale Tibetan revolt of 1959 and the flight of the Dalai Lama (the spiritual and political head of Lamaism) to India. The Chinese, freed from the necessity of dealing through the Dalai Lama and the Tibetan hierarchy, then moved more rapidly, driving most of the monks from their monasteries, expropriating the large monastic land holdings, completing their program of land reform, and making other institutional changes similar to those designed to implement socialism in the rest of China.

Agricultural land in the Tibetan Highlands is extremely limited and found only at elevations of 13,000 feet and under—primarily in the valleys of the Tsangpo and other major streams and their tributaries. A few hardy crops, especially barley and root crops, are grown. In grasslands at higher elevations, nomads still graze their flocks of yaks, sheep, and goats, and some of them combine herding with the growing of a little barley in the

---

[4] *Geographical Foundations of National Power*, p. 5.

valleys. Estimates suggest that the overall ratio of sedentary farmers to nomadic herdsmen in the Tibetan Highlands is 6 to 1 or greater. The Chinese did not directly enumerate the Tibetan population in 1953, but estimated the total to be about 2.8 million in the entire Tibetan Highlands; the total probably is now about 3 million. Aside from military personnel, relatively few Han Chinese, mostly administrators and technicians, live in the Tibetan Highlands.

Sinkiang

Sinkiang adjoins the Tibetan Highlands on the north. It has an area of roughly 635,000 square miles and had a population of some 5.6 million in 1957 and possibly 7 million or more in 1968. In physical terms, Sinkiang consists essentially of two great basins, the Tarim Basin to the south and the Dzungarian Basin to the north (map, p. 467). These basins are separated by the lofty Tien Shan range. The Tarim Basin is rimmed to the south by the mountains bordering Tibet, and the Dzungarian Basin is enclosed on the north by the Altai and other ranges along the southern border of the Soviet Union and Mongolia. Both basins are arid or semiarid. The Tarim Basin, particularly, is dry, being almost completely enclosed by high mountains which block off rain-bearing winds. The Tarim Basin is occupied by the Taklamakan Desert, perhaps the driest region in Asia. The basin varies in altitude from 2000 to 6000 feet above sea level; the smaller adjoining Turfan Depression drops to 928 feet below sea level. The Dzungarian Basin averages about 1000 feet in elevation, is more open than the Tarim Basin, and has somewhat more rain, though not enough for much agriculture.

The great majority of Sinkiang's population is concentrated in oases at points around the edges of the great basins where streams from the high mountains debouch on the basin floors. For many centuries these oases were stations on caravan routes crossing Central Asia from Humid China toward the Middle East and Europe. The most important routes followed the northern and southern piedmonts[5] of the Tien Shan. These ancient routes have now been superseded by new transportation lines. The Trans-Sinkiang railroad connects Lanchow (1.2 million in 1959),[6] a major industrial center and supply base in northwestern China proper, with Urumchi (320,000), the capital of Sinkiang; several major roads link the important oases centers in both the Tarim and Dzungarian basins; and several major airfields have been built in Sinkiang during the past two decades.

The new motor roads and the new railway are key elements in a general Chinese Communist drive to expand the economic significance of Sinkiang and bring this remote area under firmer political control. Demobilized soldiers and urban youths from eastern China have been organized into quasi-military production and construction divisions to spearhead expansion of irrigated land on state-operated farms. Cotton is a major crop, as it is in irrigated areas of adjacent Soviet Middle Asia, but wheat, corn, and rice, and such local specialties as fruit and melons are important. Sinkiang has deposits of coal, petroleum, iron ore, and other minerals. Expansion of mining and related manufacturing is planned for the region, though the available information (1968) suggests that little has been accomplished since 1960–1961.

Economic development in Sinkiang, especially the improvement of transportation, serves important Chinese political purposes by tying this remote area more closely to China's core area. The population is predominantly Moslem and of Turkic stock, and has often been restive under Chinese rule. In addition, some of the frontier regions have been contested between China and Russia, and Russian economic and political influence has been strong in the province—particularly during the 1930s and 1940s. Sinkiang is now formally ruled by China as an "autonomous region" (the Sinkiang Uighur Autonomous Region), but it is quite possibly under a tighter control by the Chinese central government than it has ever been in the past. The grip of the Communist regime on the region is being strengthened by better communications and an increasing influx of Han Chinese settlers.

Inner Mongolia

North and northwest of the Great Wall, rolling uplands, barren mountains, and lifeless basins stretch into

---

[5] A piedmont is a belt of country at an intermediate elevation along the base of a mountain range.

[6] In general, city population figures cited in this chapter are metropolitan-area approximations as of 1958, and represent extrapolations of the 1953 census data and some 1957 year-end estimates contained in the Chinese Communist publication *Ten Great Years*. These figures are reported in Morris B. Ullman, *Cities of Mainland China: 1953 and 1958*, U.S. Bureau of the Census, International Population Reports, Series P-95, No. 59 (August 1961). It is reasonable to assume that most cities have grown some since 1958.

the arid interior of Asia. Here lie slightly more than a million square miles customarily referred to as Mongolia, but traditionally subdivided into Inner Mongolia, the area nearest China, and Outer Mongolia. The latter region, now an independent political entity, is discussed in a separate section at the end of this chapter.

The Chinese Communists have taken more or less the traditional area of Inner Mongolia and with the inclusion of additional territory have created the Inner Mongolian Autonomous Region (IMAR). The IMAR, a province-level administrative unit with a population of about 10 million (of whom nearly 90 percent are Han Chinese), extends in a vast crescent from Sinkiang east and northeast to the headwaters of the Amur River. The northeastern section includes areas detached from the western part of what was formerly considered to be Manchuria. These areas are composed of grassy steppe country and, in addition, some sections of the heavily forested Greater Khingan range. The western part of Inner Mongolia is mostly a sand and gravel desert—a harsh and desolate country of widely spaced springs and meager pasturage where the annual precipitation is less than 5 inches. To the east and northeast the amounts of precipitation gradually increase, and produce fair to good pasture lands. This is particularly true of the area nearest the Great Wall and in the northeast, where cooler temperatures and less evaporation contribute to the formation of good grazing lands. The traditional picture of Mongols herding their flocks of sheep and goats and using camels and horses for riding and pack purposes is fast disappearing. In part this results from the long history of incursions by Chinese settlers into the better watered grasslands. For centuries Inner Mongolia, like other grassland areas along the dry margins of world agriculture, has been a zone of competition and conflict between grain farmers and herders. In addition, the comparatively few nomadic Mongol herdsmen who remain have been induced more and more by the Chinese to maintain fixed locations and use permanent pastures—an ancient device for maintaining political control over the nomads and easing the transition to sedentary occupations. A few areas, such as the northern loop of the Hwang Ho, have long contained a high-density sedentary agricultural population based upon the availability of irrigation water. Over all, however, only a small fraction of the IMAR is under cultivation, with millet, kaoliang, spring wheat, and oilseeds the predominant crops. The primary urban place in the region is Paotow

(490,000), on the Hwang Ho, a rapidly expanding industrial center with an iron and steel works. The capital of the IMAR is Huhehot (320,000), located about 100 miles east of Paotow.

HUMID CHINA

Humid China, sometimes referred to as Eastern or Monsoon China, the core region of the country, includes the densely settled parts of China south of the Great Wall, and also includes Manchuria. The 18 ancient provinces of China south of the Wall (increased to 19 in 1959) are often referred to by outsiders as "China proper," although the Chinese themselves do not employ the term. China proper includes two major divisions, North China and South China, which differ from each other in various physical, economic, and cultural respects. Although each area exhibits much variety from place to place, South China may be characterized in general as subtropical, humid, and hilly or mountainous, with irrigated rice as the main crop, while North China is continental, subhumid, has larger stretches of level land, and depends mainly on nonirrigated grain crops other than rice. Prior to collective agriculture, individual farms in the North averaged about twice the size of those in the South; in the latter region, however, double-cropping practices and the dominance of rice, with its higher yields per unit of land, compensated for the smaller size of farm. Oxen and other draft animals such as mules and donkeys are common in the north, but water buffaloes, with their ability to withstand heat and work in the muddy rice fields, predominate in the South. The North Chinese are typically taller, heavier, more purely Mongoloid, lighter complexioned, and mainly speak Mandarin (Chinese proper) or one of its variants. The South Chinese tend to be shorter, somewhat darker, and speak a variety of mutually unintelligible Chinese languages or dialects including Cantonese, Hakka, Min, Wu, and several others. The use of Mandarin, or *kuo-yü* ("the national language") is spreading. Historically, the system of written characters that was common to all literate Chinese and the tradition of governing through an educated elite helped the Chinese to administer a large country with a population speaking a number of distinctive languages.

The two major regional divisions of China proper are described in the selection on pp. 471–473:[7]

---

[7] *Geographical Foundations of National Power,* pp. 1 and 6–8. One footnote has been omitted and one heading added.

North China

North China is subhumid, average annual precipitation ranging from 17 inches in the Loess Upland to 21 inches in the North China Plain. But owing to the very high variability from year to year the area is subject to famine from droughts and floods. The winters are cold, the summers hot. . . . The people depend on wheat, millet, and kaoliang (a sorghum), as their primary food. The great plain of the unruly Yellow River (Hwang Ho) dominates the region and is densely populated. Chinese civilization originated and developed in this area. . . .

North China . . . comprises two very different subregions: (1) the uplands of the interior and (2) the great alluvial plain of the Hwang Ho, or North China Plain, with the included uplands of Shantung.

The Uplands

The Hwang Ho (Yellow River), which rises in Tibet, drains the western portion of the uplands; thence it makes a great swing far to the north into the steppes of Mongolia, where it turns back south through the uplands, only to make a right angle bend to the east, 200 miles before flowing out into the plain. The upland area as a whole was the historic frontier zone between China proper and the dry lands of inner Asia. Much of the region is covered with deep deposits of loess, which the Chinese descriptively call "Yellow Earth," an exceedingly fine-grained but fertile soil, laid down by the wind. The scarce and unreliable rainfall renders agriculture precarious, however. In periods of drought famine occurs. Conversely, when a series of years brings more than average rain, much of the steppe borderland can be tilled, and farmers push out into it. . . .

The North China Plain

The North China Plain [Yellow Plain] is an immense complex delta. The Hwang Ho shifts its course from time to time, reaching the sea first to the north and then to the south of the Shantung hills. Numerous [other streams] and ditches cross the Plain. These waterways are shallow and most of them are fordable and they have generally been

Flights of terraces for growing unirrigated crops climb the loess hills in this scene from Shansi Province in North China. Streams have cut miniature canyons into the loess, which has the property of standing in vertical walls when subjected to stream erosion. Many people in this part of China live in caves that have been excavated in the valley sides. Such dwellings are dry, and are insulated by the loess from summer heat and winter cold. (China Photo Service, from Eastfoto.)

diked in an effort to keep them within bounds during flood season. As the [Hwang] deposited silt and debris, the dikes were built higher and higher to keep above the rising water level until the bed of the stream in many sections is above the level of the surrounding country. This situation placed the residents of the area in a precarious position. Few rivers are subject to such violent floods as the Hwang. Hardly any of its waters, shallow and full of shifting sand bars, are navigable for large craft.

The country is so dry that it is a yellow and dusty land except during spring and early summer. There is generally sufficient rain, however, to bring crops to bearing without irrigation. . . .

South China

South China is a different world from that of the North. Its abundant rainfall and warmer climate make it lush and green at times of the year when North China is parched with drought or withered with cold. Much of South China is hilly or mountainous. The principal lowlands are the three large basins drained by the Yangtze River and the smaller Canton delta section.

The Yangtze Basins

The three Yangtze basins are set off from each other by narrows in the river valley. The Lower Basin is a delta merging with that of the Hwang Ho. The Middle Basin is likewise largely built up of river sediments and is flat, low, and dotted with lakes. The soil of these two basins is not quite as fertile as that of the North China Plain, but the more abundant rainfall ensures a good crop every year and provides water for the irrigation of rice, which cannot be extensively grown farther north because of the long, cold winters. The third or Red Basin lies farther upstream [in the province of Szechwan; it is] separated from the Middle Basin by a mountain wall through which the river has cut a deep gorge. The Red Basin stands several hundred feet above the Middle and Lower basins and is a land of hills, among which lies the highly productive Chengtu plain.[8]

The three Yangtze basins are set off from North China by a combination of barriers. On the east the many rivers and lakes of the delta flats form a considerable obstacle. Farther west, first a range of hills and then the Ch'in Ling (Tsinling) mountains, which rise to more than 10,000 feet, mark the border zone. These barriers are crossed by minor trails in many places, but the main highways . . . are few and have remained little changed since

the beginning of China's history. Even the construction of canals and railroads has not altered the pattern of the principal connecting routes. . . .

Throughout its course in China proper the Yangtze River is broad, deep, and, in places, swift. It is [an important element in] the Chinese communication system. The lower part of its course is paralleled by railroads, but the river itself constitutes the sole through connection between the three basins. It has always carried naval vessels as well as trading ships. Large ocean craft can ascend to Nanking at all seasons. Ships drawing ten feet of water can continue, at all seasons, to [Wuhan], the metropolis of the Middle Yangtze Basin, and river craft drawing seven feet can ascend to Ichang at the foot of the gorge. Chungking, 1300 miles from the river mouth, is [accessible by] small power vessels. The gorge between Chungking and Ichang is hazardous because of rocks, whirlpools, and the swift current, and specially constructed, powerful steamers are required for its navigation. . . .

The Canton Delta

The principal river of China's far south is the Hsi, or West, River, a shorter stream than either the Hwang or the Yangtze. Its course is through uplands, except where its delta merges with those of lesser streams to form the Canton delta, a densely populated lowland, smaller in area than those farther north. This lowland, which is separated from the Yangtze by a wide belt of hills and low mountains, was not finally incorporated into China until the time of the Tsin and Han dynasties, nearly 1000 years after the basins of the Hwang Ho and Middle and Lower Yangtze had been consolidated. . . .

The Hill Lands

The coastal zone between the deltas of the Yangtze and the Hsi is so difficult to penetrate that it was not annexed to China until the third century A.D. The range of relief is between 1500 and 2500 feet, although some peaks rise to 6000 feet. The rivers are short, swift, and unnavigable, and each basin constitutes a unit isolated except on the seaward side. The size of the towns at the river mouths is limited by the productivity of the basins they serve. Agriculture is restricted to small and scattered valley lands. The coast is dotted with fishing hamlets. This is the only section of China in which the people have taken much interest in seafaring.

Most of Yunnan, the southwesternmost province of China proper, is a deeply dissected plateau some 6000 feet high. In the long history of the southward advance of

---

[8] The Basin of Szechwan or Red Basin contains one of the largest concentrations of population in China, currently estimated at 60 to 65 million people. Nature has provided comparatively little level land in the basin, but rice and other crops are grown on enormous numbers of small fields in narrow ribbons of valley land and on easily terraced hillsides.

A view of the famous British colony of Hong Kong on the rugged coast of South China. The central business district of Victoria city on Hong Kong Island appears in the foreground. Most of the larger buildings—banks, hotels, office buildings, and apartment houses—have been constructed since World War II. The large residence of the British Governor-General appears to the right of center at the farther edge of the park in the center foreground. Apartment buildings climb the steep slopes behind the business section. Across the harbor lies the mainland suburb of Kowloon. Some of Hong Kong's cargo traffic is handled at piers, but most of it is carried by lighters that ply between the shore and ships anchored in the harbor. The colony's airstrip appears as a light-colored horizontal line in the upper right of the photo. Hong Kong Island is an outright possession of the United Kingdom, but the mainland portion is held under a 99-year lease granted in 1898 by the government of China. (Hong Kong Tourist Association.)

the Chinese, it was the area most recently penetrated. Many primitive groups still live there unassimilated to the Chinese way of life. . . .

Cities of China

The enormous rural population of China and the predominance of small rural villages and hamlets (estimated to number about 1 million) obscures the fact that there are a very sizable number of large cities in China.[9] No official population data have been released since 1958, and changes in municipal boundaries have made it very difficult to make comparisons with earlier totals and to compute trends. It appears, however, that as of 1958 there were 15 cities in China with populations in excess of 1 million. Of these all but Canton (1.9 million), a port city in the Hsi delta, are located along or north of the Yangtze River.[10] There were also an estimated 19 cities with populations between 500,000 and 1 million and some 32 with populations between 200,000 and 500,000. By far the largest city, and the second largest in Asia, is Shanghai (7 million; an estimate places the metropolitan-area total at 9.6 million in 1967), located on a navigable tributary

---

[9] Early western visitors to China, such as Marco Polo and later the Jesuit Fathers, were greatly impressed by the number, size, and wealth of Chinese cities.

[10] Another "million city" in the south, Chinese in population though not in political control, is the metropolitan area formed by the cities of Victoria (on Hong Kong island) and Kowloon (on the mainland) in the British colony of Hong Kong (photo, above). The colony lies a few tens of miles seaward from Canton. Most of its population of about 4 million (1968 estimate) is in Victoria, Kowloon, or suburbs. Near Hong Kong is the Portuguese colony of Macao (population about 170,000), also overwhelmingly Chinese in population.

of the Yangtze near the mouth of the latter stream. Three other cities with a million population or over are spaced upstream along the Yangtze. They include Nanking (1.5 million), the capital of Nationalist China before 1937 and again from 1945 to 1949; the conurbation of Wuhan (Hankow, Wuchang, and Hanyang) with a combined population of 2.2 million; and, in the Red Basin, Chungking (2.2 million), the capital from 1937 to 1945. Northwest of Chungking in the Red Basin is Chengtu (1.1 million). In North China are Peking (4.1 million, but annexations now boost the figure to 6 or 7 million), the capital before 1911 and reinstated by the Communists in 1949; the important seaport of Tiensin (3.3 million), 70 miles southeast of Peking; and Tsingtao (1.1 million) on the Shantung Peninsula. To the west Sian (1.4 million) and Taiyuan (1.1 million) are expanding industrial centers, both of which were relatively small cities of about 200,000 population in 1938. The remaining cities with over a million people in 1958 are in Manchuria. All of them are relatively modern cities that have experienced most of their growth within the present century. These cities, along the main north-south rail axis, include Shenyang (Mukden) with a population of 2.4 million; nearby Fushun (1 million); Harbin (1.6 million), the key transportation center and industrial city of northern Manchuria; and Dairen (1.6 million), at the tip of the Liaotung peninsula, Dairen is administratively termed the municipality of Luta, which includes Dairen, Port Arthur, and some surrounding territory. Other cities that are reported to have passed the million mark since 1958 include Changchun, located on the rail line between Harbin and Shenyang; Kunming, the major urban center of southwest China; and Lanchow in the northwest.

Manchuria

Manchuria, or the Northeast as it is commonly termed by the Chinese, is the part of Humid China north and northeast of the Great Wall. Manchuria, as now defined, has an area of about 308,000 square miles and a population of 50 million (1958). To the west are the Mongolian steppes and deserts; to the east Manchuria is separated from the Sea of Japan by Korea and Russia. In recent times this area has claimed an important share of world attention, chiefly through its role as a zone of rivalry and conflict among China, Russia, and Japan. Russian interest in Manchuria has been connected with the Chinese Eastern Railway, a shortcut across northern Manchuria between Vladivostok and the Trans-Siberian Railroad east of Lake Baikal, and with the naval and

port facilities of Dairen–Port Arthur, which the Russians have used extensively at various times. The Japanese were attracted to Manchuria as a source of supply for coal, iron, aluminum, timber, and other industrial resources, as a source of surplus foodstuffs, as an expanding market for Japanese manufactures, and as a promising area for capital investment. After the Russo-Japanese War of 1904–1905, mainly fought in Manchuria, Japan was increasingly active in the area, and took over actual control in 1931 with the establishment of the puppet state of Manchukuo. For the Chinese, Manchuria has served as an outlet for surplus population migrating from south of the Wall. Movement of Chinese colonists into Manchuria began centuries ago, but gained greatest momentum after 1900. Millions of Chinese farmers migrating to this area have made it one of the twentieth century's major zones of pioneer agricultural settlement. However, of great significance to Communist China was the considerable development of heavy industry left by the Japanese, and the best rail network in China, begun by the Russians and later taken over and expanded by Japan.

In physical terms Manchuria consists of a broad, rolling central plain surrounded by a frame of mountains which seldom rise higher than 6000 feet (map, p. 467). The mountains on the east and north are forested and contain much valuable timber, including oak and other hardwoods as well as conifers. The central plain, oriented northeast-southwest, is approximately 600 miles long by 200 to 400 miles wide. The northern portions of the plain are drained by the Sungari River, a tributary of the Amur, and the southern portions by the Liao River, which flows to the Yellow Sea. The soils are relatively deep, dark-colored, and very fertile, and the summer rainfall of 15 to 25 inches is generally sufficient for the crops that are grown, though spring drought is often troublesome. Conditions are not favorable for irrigated rice, except in limited areas, and the winters are generally too severe for winter grains. However, spring-sown wheat, millet, kaoliang, corn, and barley do well during the relatively short but warm frost-free season of 150 to 180 days. Large acreages are devoted to soybeans, a versatile leguminous crop which provides protein food for both men and livestock, is an important source of vegetable oil, and enriches the soil through the addition of nitrogen. Manchuria was the world's largest exporter of soybeans for considerable periods prior to World War II. Manchurian farms are larger, the density of the agricultural population is less, and much greater possibilities for future mechanization of agriculture are offered than in the parts of Humid China south of the Great Wall. Manchuria is also one of the few areas in

Heavy industry in Manchuria. The tall black structures to the right of center are the number 3 and 4 blast furnaces of the Penki Iron and Steel Company. Their silhouettes are characteristic of blast furnaces the world over. Local deposits of both coal and iron ore occur in the vicinity of Penki, located approximately 40 miles southeast of Manchuria's largest city, Shenyang, and about the same distance to the northeast of China's leading center of iron and steel production, Anshan. (China Photo Service, from Eastfoto.)

China where significant areas of land exist that are potentially cultivable but unoccupied or lightly used.

The industries of Manchuria are mainly centered in or relatively near the largest city, Shenyang (or Mukden) in south central Manchuria. Shenyang is a diversified center of metal-fabricating and agricultural-processing industries. The largest center of iron and steel production, however, is Anshan (830,000), about 50 miles south of Shenyang. The industry is based on deposits of iron ore in a belt which crosses southern Manchuria. The ore is mostly low-grade with an average iron content of around 35 percent. Some pockets of richer ore occur along or near the Korean border. Substantial deposits of bituminous coal also occur in Manchuria, although the total estimated reserves are far smaller than those of China south of the Great Wall, and only a small fraction of Manchurian coal is suitable for coking. In general, Manchuria appears to have the necessary coal and iron resources for substantial increases in heavy industry, although its reserves are not really outstanding on a world basis.

COMMUNIST CHINA'S POPULATION AND ECONOMY

Agricultural Emphases

Something like 550 million mainland Chinese, three-quarters of Communist China's population, are directly engaged in and supported by agriculture. This enormous mass of peasant humanity must wring food for itself and for the remaining fourth of the population from a country where roughly nine-tenths of the land is not in cultivation because of steep slopes, dryness, short growing seasons, and/or technological inadequacies. China's arable area, concentrated in large measure on the plains and river-valley lands of Humid China, is less than 60 percent as large as that of the United States, but through multiple-cropping practices (growing successive crops on a given field throughout the year) the sown acreage is about the same. It provides, on the average, only about half an acre of arable land per person to the agricultural population, and only two-fifths of an acre per person to the total population. If the farming unit were the family farm—which it is not under Communist rule—the average farm for a family of five would comprise about 2½ acres of arable land. These are, of course, crude averages, and conditions vary—for better or worse—from one part of the country to another. Production per acre also varies markedly from region to region and from one type of land to another. Thus farmers in North China and Manchuria, where the arable acreage per person is above average, are not necessarily more productive or better off than farmers in South China, where the acreage per person is less but the productivity of each acre is often greater than in the north. One fact, however, holds true throughout the country: nearly everywhere the population is increasing at a pace equal to or greater than the annual increases in agricultural production. Under such conditions it is not surprising that a desperate struggle for food is a hallmark of Communist China's economic character and that, consequently, food grains dominate the agricultural picture.

Humid China falls into two major regions of contrasting physical conditions and related contrasts in basic crops. From the southern borders of the country as far north as, and including, the three Yangtze basins, precipitation is relatively abundant (40 to 60 inches a year), temperatures are subtropical, and rice is the dominant crop. North of the Yangtze basins the climate is humid continental or steppe, with moisture that is much less abundant than in the south and highly erratic from year to year, and wheat and a variety of lesser grains are the dominant crops. Roughly two-thirds of the cropland in the rice region is irrigated, and long growing seasons allow double cropping (two successive crops a year from a given field) on most of this acreage. In the drainage basin of the Hsi, temperatures and water supplies normally are adequate for the production of two rice crops a year. In the northern part of the rice region cropping patterns tend to be more varied and include interplanting (growing of two or more crops simultaneously on a field), rice followed by winter wheat or an oilseed, and limited areas of two-crop rice where sufficient water is available. Some rice fields, especially those that are hard to drain, are left fallow in the winter or are sown with a green manure crop. Adjacent to and extending for about 150 miles north of the middle and lower Yangtze basins, a transitional cropping region (rice-wheat) exists where rice is gradually supplanted to the north by winter wheat as the dominant crop. In the wheat region there is much less irrigation; in fact not more than about a fifth of all cropland is irrigated. Winter wheat is the basic crop of

the North China Plain, although millets and sorghums, particularly the sorghum known as kaoliang, are also of major importance. These crops, which require even less moisture than does wheat, are grown in summer. They are well suited to the climate of the North China Plain, which is generally classed as humid, but is barely so. Peking, for instance, averages only 24 inches of precipitation a year, which is hardly abundant considering the high evaporation and transpiration rates that are associated with summers in which the warmest month averages around 80°F. Farther inland, in the loess uplands, distinctly semiarid conditions prevail, and drought has often brought famine conditions in past times. The cropping system of the uplands is built around winter wheat and millet, each of which is relatively low in moisture requirements. Farther north, in Manchuria and Inner Mongolia, spring wheat replaces winter wheat in response to winters that are distinctly colder than those of the North China Plain.

Secondary crops in great variety are grown, some for food and some for industrial uses. In various parts of the northern agricultural region, corn, potatoes, and sugar beets are important secondary food crops, as are soybeans and peanuts, both of which are important export commodities. Oddly enough, considering the type of climate, cotton is the major industrial crop of northern China proper, while hemp, flax, and tobacco are also grown in various districts of the north. In the southern agricultural region, sweet potatoes, corn, and a little wheat and barley are secondary food crops. The Yangtze lands account for a substantial share of the country's cotton production, and they also produce two other important fibers—silk and ramie. Still another fiber, jute, is produced farther to the south. Tea is an important crop in the hilly areas south of the Yangtze. Citrus fruits are grown in the warmer parts of the south, as is the tung tree, whose fruits yield a drying oil with many industrial uses. The growing of garden vegetables is a component of agriculture in all parts of the country.

As in the case of India, the total volume of agricultural products is impressive, for China is a major world producer of a number of secondary crops as well as of the major food crops. But as in the case of India, the total production is much less impressive when measured against the tremendous numbers of people engaged in production and the even larger numbers who must be fed. The relatively small production of foods other than cereals is highlighted by estimates that perhaps 80 percent of the calories in the average Chinese diet are derived from cereals. As in most lands that are hard pressed for food, the diet of the people is very deficient in animal proteins. Among the conditions that discourage the widespread use of meat are the following:

1. In the crowded areas of arable farming, little land can be spared for the feeding of livestock.

2. The very extensive natural pastures of less populous areas are not adequate to supply great quantities of meat; this would still be true even if a productive commercial grazing economy were developed on such lands to replace the present subsistence or semisubsistence type of grazing.

3. Most of China is so far from the sea and has such inadequate facilities for transportation and storage of perishables that sea fisheries cannot supply the protein element. Such fisheries are, however, locally important, especially along the coast of South China. In addition, river or pond fisheries make some contribution to the food supply in many parts of the country.

In most parts of Humid China, hogs and poultry are the principal animals raised as sources of food. Both can be fed in large part on kitchen scraps and other waste materials. In Arid China, sheep, goats, and yaks are locally important for milk and meat.

Recent Attempts to Increase Agricultural Production

Since it took control of the country in 1949, the Chinese Communist government has attempted to increase China's agricultural production. Continued increases in farm output are regarded as necessary, not only to provide food for a rapidly increasing population, but also to supply industrial raw materials, capital for industrialization, and agricultural exports to underwrite the costs of industrial equipment purchased in other countries.

The problem of increasing food production is many-sided, and the efforts and priorities given to it by the Chinese Communists have varied. The major lines of attack may be summarized under the headings of (1) water conservancy and increases in irrigation, (2) intensification of existing agricultural techniques, particularly an increase in the use of chemical fertilizers, (3) reclamation of land and increase in the cultivated acreage, (4) other noninstitutional measures, including mechanization, improved crop varieties, and elimination of insect pests, and (5) collectivization measures. These approaches should be viewed in the overall context of economic planning that initially relegated agriculture to a secondary role while giving priority in investment and resources to heavy industry. After the collapse, in 1960, of the program of forced economic development called the "Great Leap Forward," there was a reassessment of priorities and recognition of the importance of agriculture and of industries supplying agriculture.

Water Conservancy and Expansion of Irrigation

Projects to control and conserve water and to expand the irrigated acreage have been given a major emphasis. Floods and droughts have periodically afflicted Chinese crop production, especially in North China. On occasion they have caused millions of deaths from famine. Thus a major objective of water-control measures is to prevent water from running off in disastrous floods and to conserve it in reservoirs for times when it is needed. Three specific types of projects have been and are being carried out to achieve this objective:

(a) *Construction of small-scale dams, ponds, canals, and dikes by local communities.* These have been constructed in very large numbers. Usually such projects have been undertaken during the agricultural off-season when large inputs of manpower were available. Although considerable benefit undoubtedly has been gained, there is evidence that some of this work has been hasty and poorly planned, necessitating excessive maintenance or reconstruction.

(b) *Building of large-scale dams and associated structures to control larger streams.* A number of large-scale, multipurpose projects were initiated during the early years of the Communist regime, of which the most widely publicized was the Sanmen Gorge dam (map, p. 467), the key project of the Yellow River Plan.[11] This dam was scheduled for completion in the early 1960s, but apparently remains inoperative (1968) because of lack of electrical generating equipment and problems with siltation of its reservoir that were greater than anticipated. Although some of the other major dams in the Yellow River Plan were completed by 1960, little or no work seems to have been done on the remaining projects since that time.

(c) *Afforestation and reforestation.* These relate closely to water control. Only a tenth of China is forested,

---

[11] Envisioned as a seventy-year project, the Yellow River Plan would provide some 46 dams on the main river and many others on tributaries. The objectives are flood control, generation of hydroelectricity, provision of increased supplies of irrigation water, and conversion of the Hwang into a navigable stream.

and large areas are without significant vegetative cover. Most forest growth in China proper has long since been removed, primarily by peasants for use as fuel. The lack of a vegetative cover on nonfarm lands causes a rapid runoff of much of the rain that falls, thus raising stream levels and increasing flood hazards as well as effecting serious erosion. The Chinese Communists have stressed large-scale afforestation and reforestation projects, plus shelter belts and other protective measures.[12] Success has been spotty—a condition attributable to lack of investment in forestry projects, poor management, improper care, and failure to coordinate with other programs. Silting of reservoirs and crop fields, particularly in the loessial soils of North China, continues to be a massive problem and reflects the magnitude of the work yet to be done.

A major objective of water conservancy measures has been to increase the amount of land under irrigation and thereby increase the output of crops. The picture of achievement since 1949 is unclear because of the welter of conflicting statements and statistics. It appears that some modest increase in the total amount of land under irrigation has been achieved, many of the existing irrigation works have been improved, and use of power equipment has increased the efficiency of irrigation in some areas. On the other hand, various setbacks have occurred. For example, some ill-planned irrigation canals that were dug during the Great Leap Forward were actually harmful in that insufficient attention was paid to drainage; in some areas this resulted in an increase in the salt content of the soil and a drop in agricultural output.

Intensification of Existing Agricultural Techniques

A very important emphasis in Communist China's attack on the problem of agricultural production has been placed on further intensification of the existing agriculture in order to raise yields per acre. Irrigation and flood-control measures, where applied to land already farmed but not irrigated or protected from floods, may be seen as one phase of this effort. This is especially true of irrigation, which generally raises yields and may allow double cropping of a high-yielding but water-demanding crop such as rice. Another method of approach lies in the application of scientific knowledge and experimentation to develop better seed, equipment, and farming techniques, together with the development of adminis-

trative and propaganda machinery to spread their use. Still a third approach lies in increased fertilization. Chinese farmers have long fertilized their fields with animal and human manures, pond mud, plant residues and ashes, oilseed cakes, and green manure crops. Although the traditional fertilizers are still by far the most important source of plant nutrients, the Communist regime has now, if somewhat belatedly, begun to stress the use of chemical fertilizers. China's relatively small chemical fertilizer industry has received a high priority and has been expanded through purchase of whole plants from abroad. Despite sizable increases in the availability of chemical fertilizers, however, the amounts used are very low in comparison to current practice in Japan or Taiwan and to what is believed desirable by agronomists.

Land Reclamation

An obvious means for helping increase the food supply would be an expansion of the cultivated acreage. One thing that has often impressed observers in China is the small use that is made of second-rate land even in places where population is excessively crowded on adjacent good land. For instance, in South China, narrow valleys often teem with an overcrowded and struggling population, but broader adjacent uplands are little used. The explanation for this seems to lie largely in the fact that the Chinese farmer and his family, unequipped with machinery, cannot till a sufficient amount of second-rate land by hand labor to make a living. Only by concentrating this labor on a smaller acreage of good land can they feed themselves. The same principle seems to apply to much of the semiarid land in interior China, where it is probable that only large-scale mechanized methods could produce enough food per worker to make cultivation feasible.

In the early 1950s Chinese Communist statements placed considerable stress upon land reclamation and on the amount of potential land existing for future cultivation. Small gains in the amount of cultivated land were made through reclamation of poorly drained land, extension of irrigation into desert lands, and similar measures. Since 1957, however, the statistics suggest that there has been a slight decrease in the amount of land under cultivation. Some land may have been lost because of the added number of large and small reservoirs (which often submerge productive valley lands), the expansion of urban areas, and the development of numerous industrial sites. In recent years the Chinese have scaled down

---

[12] A first-hand account of forestry and land utilization practices is contained in S. D. Richardson, *Forestry in Communist China* (Baltimore, Md.: Johns Hopkins Press for Resources for the Future, 1965).

A production team at work in wheat fields of a commune near Shanghai. The team is examining the condition of the young wheat crop. Compost piles to fertilize the land dot the flat landscape of the Lower Yangtze Plain. (China Photo Service, from Eastfoto.)

their estimates of reclaimable land. Potential areas of significant size are claimed for Manchuria, Sinkiang, and scattered areas elsewhere. In general these areas are agriculturally marginal. They suffer from short growing seasons, lack of water, poor drainage, and other physical or economic handicaps that make them uneconomic to bring into production under present conditions. Some potential is claimed for the hill lands of South China, but reclamation would require mechanized equipment and coordination with overall water and soil conservancy schemes so as not to create problems for the intensively used and productive valley lands.

Mechanization, Crop Breeding, and Control of Diseases and Pests

Other methods to increase agricultural production probably have played a relatively minor role. Mechaniza-

tion has developed slowly, and progress throughout the nation has been uneven. The use of tractors is increasing, but primarily in Manchuria and in the west where state farms and extensive farming techniques are feasible. In recent years some priority has been given to improved farming techniques, implements, and electric pumps. Thus there has been progress compared to the situation existing in 1949, but by and large the rural scene still is dominated by human and animal power and simple hand tools. Progress also has been reported in crop breeding and in the control of insects and plant disease, but little definitive information is available concerning the efficiency and overall success of such programs. The lack of trained agricultural scientists and technicians is a hindering factor. Developmental work is being carried out with hybrid corn and possibly other crops. There is evidence that insecticides and herbicides are coming into wider use. Pest control was greatly publicized during the early years of the Communist government, with great campaigns to eradicate rats, mice, flies, sparrows, and other pests. Observers were amazed at the sight of Chinese peasants fanatically hunting down pests at the exhortation of the government. A major objective of these campaigns was to reduce food losses caused by pests, which have been credited with the destruction of many millions of tons of grain each year in the fields and in storage. Antipest campaigns against flies, mosquitoes, and rats have had, in addition, an obvious public health aspect. The evidence suggests that these programs have had at least a limited success and that progress may have been very considerable. Additional measures that have been undertaken to reduce food losses to pests and diseases include the construction of better storage facilities for crops, and programs of agricultural education and research.

Collectivization Measures

Although economic growth in China during the First Five-Year Plan (1953–1957) was impressive, the government was disappointed in the overall economic achievements, and in the performance of agriculture in particular, when measured against China's needs. With massive amounts of manpower available, much of which was underemployed during part of each year, the Chinese leadership reasoned that this asset could, if properly mobilized, be used to greatly accelerate economic performance. Essentially, production might be raised by making an already labor-intensive rural economy still more so. Thus the rural commune was conceived as the major institutional vehicle to provide the needed impetus and direction for the much publicized Great Leap Forward that characterized all facets of Chinese life from 1958 through 1960. With "politics in command," all things were deemed possible, from doubling food-grain production in a single year to overtaking Great Britain in steel production in 10 years.

Some 24,000 rural communes amalgamating about 750,000 collective farms were to fuse agriculture, industry, commerce, education, the militia, and local administration into self-sufficient units designed to meet all the needs of the people and the state. For a short time the outlook was sanguine: public mess halls were established; private property abolished; necessities provided "according to need"; backyard iron smelters built; and women liberated from household tasks to devote more time to production. But the task of organizing China's vast countryside proved much greater than anticipated; uneconomic production drives proved wasteful and harmful, particularly the backyard iron and steel drive; and the dislocations caused by the communes interrupted normal transportation, harvesting, and other activities to the overall detriment of the economy. By 1960 the rural populace, tired and apathetic from the physical and emotional demands of the Leap, could no longer engage in both agricultural production and mass-labor construction projects. Output fell, the economy was in disorder, and the leadership backtracked and tacitly admitted failure. The communes continue, now increased in number to about 76,000, but much has changed: small private plots have been returned and the individual may once again raise pigs and poultry; rural markets have reopened; domestic handicrafts are once again permitted; mess halls have been abolished; and large mass-labor projects have been largely discontinued. The basic units of rural organization now have shifted to the production brigade, which owns the collectivized land, and the production team. But the ultimate form of the commune and rural organization remains uncertain. With the production team, usually made up of 20 to 40 households, as the basic unit of farm management, rural organization is essentially that which existed prior to 1958. Exceptions occur, particularly near large urban centers where a variety of factors facilitate large units of management.

One observer has summarized in the following selection the impact of the Communist system on China's peasants. His article, written in 1959 at the height of the Leap Forward, remains an excellent analysis of the inherent problems of organizing the peasantry, and a prophetic forecast of changes in the communes:[13]

So thoroughly have the Communists shattered China's old family and social institutions that the peasant has lost his former lust for life and work. The concept of "face" which once was such a powerful social prop virtually has vanished. No longer can the peasant expend himself to provide for the future economic well-being of his family or for his own old age. His children are not taught to venerate elders and the familial security and respect from his grandsons that a peasant once dreamed of achieving in his latter years now is hardly even a mirage. He knows, too, that the prospect of his descendants sweeping his grave and paying homage at his tablet in the clan shrine are minimal since all such "feudal customs" are being abolished by the state. Even the gods who were his refuge

---

[13] Albert Ravenholt, "The Chinese Communes: Big Risks for Big Gains," *Foreign Affairs,* 37 (1959), 585–586. Reprinted by permission of the Council on Foreign Relations, publishers of *Foreign Affairs.*

and source of support in times of personal and natural disaster are now proscribed.

Though he is denied the satisfaction of values which have been imbedded in his culture for some 35 centuries, the peasant has not in general accepted those offered by the new order. The Communists moved too rapidly and ruthlessly. They insisted upon such total verbal conformity that prospects for genuine re-education of this generation of peasants have not yet been realized. The result is a mass of peasantry who "labor without heart." On major construction and industrial projects where cadre supervision is most effective, passive resistance and "unseen sabotage" among the peasants may not be critical. "Civil disobedience" is hard to gauge. But the price of anything like a genuine peasant enthusiasm for agricultural performance within the communes will be a revamping of present policies, particularly those affecting the family. Unless the Party becomes more responsive to the "human element," the entire system of people's communes might be a success organizationally and yet bog down in actual performance. China's demographic dilemma and the narrow margin on which Chinese agriculture must perform mean that the consequences of a short-fall could be critical.

It is difficult to assess the results of Communist China's efforts in agriculture, due to uncertainty concerning the reliability of official reports by the Communist government, as well as uncertainty growing out of the conflicting views of specialists. Clearly, some increases in food-grain production and in most industrial crops were recorded during the early years. But informed observers largely discount official reports of great increases in food-grain output during 1958 and 1959. During 1960–1961, the dislocations caused by the communes, coupled with poor crop years, considerably reduced farm output. Malnutrition was widespread during the hard winter of 1960–1961, forcing the import of 5 to 6 million tons of grain, an amount that has continued to be imported annually since then (1968). Lacking official statistics since 1959, experts disagree over agricultural performance, although all agree that some recovery has been made. In addition to imports, there may be some increase from use of better varieties and the effect of irrigation and other techniques. Undoubtedly, a greater share of the diet is now supplied by nongrain products as a result of the return of small private plots to farm households. Improvement in agricultural production after 1961 is attributed largely to a more pragmatic approach to agricul-

ture. Incentives and other aspects of capitalism have been permitted to emerge, with a corresponding retreat from dogma. The ideological battle continues, however, and has resurfaced in the unfolding of the Cultural Revolution[14] begun in 1966; consequently, the direction and form of future agricultural policies and programs cannot be predicted with any degree of certainty.

Industrial Development and Resources

Imperial China had a system of handicraft industries which produced goods of such high quality as to attract early Western traders in spite of China's official disinterest in such trade. The handicraft sector of Chinese industry continues to supply many of the consumer goods used in the country and is even increasing its output, though modern factory-type industries are growing at a much more rapid rate. As late as 1951 it was estimated that workers in handicraft industries outnumbered those in factories perhaps 9 or 10 to 1.

Modern industries producing by the factory method began to develop in China at about the turn of the present century. Such industries were largely foreign owned. Foreign capital was attracted to China by the huge Chinese market for cheap consumer goods, by the cheapness of Chinese labor, and by the presence of certain resources that were absent or inadequate in the investing countries. The major branch of modern industry in pre-Communist China was the manufacture of cotton textiles—a form of industry that is very characteristic of countries in an early stage of industrialization. Japanese and other foreign-owned firms, together with some native Chinese companies, developed a cotton industry that by the 1930s was exceeded in output only by the cotton textile industries of the United States, the Soviet Union, Japan, and India. Domestic cotton production in central and northern China supplied increasing quantities of raw materials and was able to meet most of the needs of Chinese mills. Most of China's textile production was sold in the domestic market, but some of it was exported. Shanghai and coastal cities farther north, notably Tientsin, were the principal textile centers. Other factory-made consumer goods such as tobacco products and processed foods were also produced for the most part in coastal

[14] The Cultural Revolution, launched in early 1966, ushered in a political struggle, often violent, throughout China. It involved the youth, organized into Red Guards; all ranks of the Communist party; and, eventually, the Army. The struggle highlighted basic differences of long standing among party leaders with respect to the shaping of policy decisions—from those who would rely almost exclusively on political ideology and the "revolutionary" tactics of guerrilla warfare days to those who would pursue less doctrinaire approaches to China's problems.

cities. In southern Manchuria Japan built an iron and steel industry at Anshan and engineering industries at other Manchurian centers. These industries made Manchuria the leading focus of heavy industry in China when it came back into Chinese control after World War II.[15]

The Chinese Communists inherited a damaged and poorly functioning industrial base. Initial emphasis (1949–1952) was placed upon rebuilding and restoring production to prewar levels—a goal reached and sometimes exceeded by the end of 1952. Determined to follow the example of the Soviet Union, the Chinese embarked upon their First Five-Year Plan (1953–1957), in which priority was given to expanding the output of basic industrial commodities—steel, coal, electric power, cement, and petroleum. A buildup of the machine-building and armaments industries also was stressed. Considerable, often impressive, gains in both capacity and output were recorded through 1960 with the help of Soviet aid projects and technology. But in mid-1960, as a concomitant of the deepening Sino-Soviet dispute and Soviet disenchantment with the Chinese Great Leap Forward, Soviet technicians were withdrawn and aid projects ceased. Output dropped dramatically, many projects were left half-finished, and some finished plants were crippled because of the lack of spare parts, Chinese inexperience, and a lack of technically trained personnel. Recovery was slow and emphasis was narrowed to a few key industries: advanced weapons, chemicals, petroleum refining, and industries supporting agricultural production. The withdrawal of the Soviet aid program also caused the Chinese Communists to turn to capitalist countries—Japan and western European nations—for imports of plants and industrial equipment.

Coal, Iron Ore, and Steel

A number of important mineral deposits were known to exist in China before the Communist era, and a substantial production of a few minerals was achieved. Since 1949, scientific exploration has added valuable new discoveries, and mineral production has increased. Coal and iron ore, the basic raw materials for steel manufacture, have naturally received major attention. The total reserves of each appear to be very extensive and widely distributed. The country's coal reserves consist largely of bituminous coal (including some coking coal), and some anthracite. From a peak pre-Communist coal production

of 62.5 million tons, achieved in 1944 when most mines were under Japanese control, output had mounted to a reported 270 million tons by 1958—an annual production surpassing that of any other country except the Soviet Union and the United States. In recent years production is believed to have averaged between 210 and 240 million tons a year. The main producing areas are in southern Manchuria, where China inherited mines developed by Japan, and in the North China Plain and the hills and uplands that border it. Fushun and Fusin (290,000) are the leading Manchurian centers, while the Kailan Basin, noted for its coking coal, is probably the leading producing area outside Manchuria. The largest reserves of coal are found in the loess uplands that lie inland from the North China Plain. Iron-ore production has varied considerably in recent years, but probably has ranged from 20 to 35 million metric tons, making China one of the major world producers. Southern Manchuria remains the single most important iron-mining area, but other output comes from deposits, often sizable, in North China, the middle and lower Yangtze basins, and Hainan Island.

Before 1949 the peak production of pig iron and steel achieved in any year in China was 1.9 million tons of pig iron and 0.9 million tons of steel. By 1958 crude steel production had risen to a reported 8 million tons, despite the admitted failure of a program of small-scale production by the communes. Since then production has slowly risen to an estimated 12 million tons in 1967, making China the world's eighth-ranking producer. The increase was accomplished by expansion of pre-Communist steel centers, especially Anshan (830,000) and Penki (500,000) in southern Manchuria, and by the development of new centers at Wuhan and Paotow. Both of these new plants use nearby coal and iron ore deposits. Older plants at Shanghai, Chungking, and Taiyuan also are major producers, while smaller iron and steel mills are scattered throughout the country, primarily in the Yangtze basin region and in North China.

Metals Other Than Iron

China has very large reserves and is an important producer of several metals other than iron. It is the world's leading producer of tungsten, and is a sizable producer of manganese, tin, mercury, molybdenum, and

---

[15] Large amounts of industrial equipment were removed from Manchurian factories by the Soviet Union following its occupation of Manchuria in the last weeks of World War II. Production of iron and steel at Anshan was brought nearly to a standstill by such dismantling. After the Communist victory in 1949, however, the Soviet Union gave much assistance to the Communist government in rebuilding Manchurian industries.

antimony. Manganese is mined primarily in the central Yangtze region, tungsten, mercury, and antimony in the hill country of South China, molybdenum in southern Manchuria, and tin principally in Yunnan (map, p. 467). Reserves of copper, lead, zinc, and aluminum-bearing ores are smaller but probably adequate for China's needs. Chrome, cobalt, and nickel, used in the production of certain special steels, are in short supply.

Power Resources

Coal is by far the major source of mechanical energy in Communist China. But oil reserves are being intensively searched out, and waterpower is being developed. The oldest oil field at Yümen, located in the lowland corridor leading from North China to Sinkiang, still is a major producer. Other fields discovered in western China and developed in the 1950s include those in the Tsaidam Basin and at Karamai in the Dzungarian Basin. Since 1950 important new fields have been opened at Taching in northern Manchuria and at Shengli in the North China Plain. These recent discoveries have made China virtually self-sufficient in petroleum products. China's level of consumption, however, is very low, and by world standards it is a minor producer.

Secondary and Light Industries

Although receiving lower priorities and fewer resources than does heavy industry, secondary and light industries have expanded significantly since 1949. Changchun in Manchuria produces trucks and railroad equipment; Loyang (500,000) makes tractors; and a wide variety of machine-building industries (transportation equipment, electrical equipment, agricultural and textile machinery, and machine tools) are scattered throughout China, primarily in manufacturing centers of North China and Manchuria. The cotton textile industry is the most important of the light industries; traditionally it has centered in the coastal cities, but there has been some expansion in recent years into the interior. Even during times of peak output, however, the amount of cotton cloth produced on a per capita basis is small, and cotton cloth is rationed.

Improvement of Transportation

In order that mainland China's widespread resources and widely scattered producing and consuming centers may be integrated into a functioning whole, the Chinese Communist government has placed great importance upon improvement in the transportation system. In spite of considerable progress, however, transportation facilities are still poorly developed and unevenly distributed throughout the country. Nearly half of China's internal commerce is still transported by human porters, pack animals, wheelbarrows, or carts drawn by animals or drawn and pushed by humans. Such transportation is slow, expensive, and of little utility for carrying the heavy, bulky commodities, low in value per unit of volume, that are the principal raw materials of modern industries. The country has a rudimentary but expanding highway system. By 1965 it was estimated that the road network was about seven times the length of the serviceable roads existing in 1949. However, about half of the current road net consists of earth roads subject to seasonal disruption. Important new roads have been constructed during the past two decades into western China, particularly to Tibet, where feeder routes extend to the Sino-Indian border and to Nepal. Roads constructed to the Tsaidam Basin and in parts of Sinkiang have permitted the Chinese to exploit the mineral resources of those areas. Although relatively more important than in previous times, roads largely supplement other forms of transport and, with some exceptions, are designed for short-haul usage. Most roads are poor in quality and require considerable maintenance.

Water transportation was the main long-distance mover of Chinese goods in past times. The Yangtze and its many tributaries provide a massive, branching system of navigable waterways reaching into many sections of the country, and the smaller Hsi River system in the south is also very useful. Historically, the Grand Canal was of great importance as it linked the agriculturally rich middle and lower Yangtze provinces with the locus of political power in North China. But only a third of the present inland waterway network is located in the industrially developed North and Northeast, and not more than a fourth of the total length of waterways considered navigable can be utilized by cargo vessels other than relatively small, flat-bottomed junks and sampans. The internal waterway system, like the country's highway system, is being improved. But the major effort of the Communist government in the field of transportation is being directed toward improvement and extension of railways.

Prior to 1949, Manchuria had a fairly extensive and reasonably efficient rail system, developed originally by Tsarist Russia and later taken over and expanded by Japan. Manchuria and North China contained about 75 percent of China's rail mileage. Most of China south of the Yangtze and all of the western provinces were without rail lines. Since 1949, the Chinese Communists have added about 8000 miles of main and branch lines for a current total of 23,000 miles. There has been some double-tracking, several new connections have been built between existing lines, and a two-deck railway and highway bridge across the Yangtze at Wuhan, completed in 1957, provided the first through connection between North China and South China. Other important rail lines built during the period up through 1960 included:

1. A line from the Szechwan (Red) Basin northward to connect with the rail system of North China.

2. A connection between the rapidly growing industrial centers of Lanchow and Paotow.

3. The Trans-Mongolian line initially built to shorten travel time between Moscow and Peking.

4. A line to the southeast coast opposite Taiwan to enhance Chinese Communist military capabilities in this region.

5. The Trans-Sinkiang line initially planned to connect with the Trans-Siberian Railroad. This line, however, has been extended only to Urumchi, the capital of Sinkiang, and is unlikely to be continued west to the Soviet border unless there is a change in Sino-Soviet relations.

Almost all rail construction was in abeyance during the economic slump of the early 1960s, but in recent years construction has been resumed on railways leading to the mountainous southwestern provinces and integrating these outlying areas into the national rail network.

The Role of Education

A modern industrial nation requires an educated and technically trained labor force, and this aspect of industrialization and modernization has been emphasized by the Chinese Communist government. A nationwide program has been pushed to eliminate the illiteracy and ignorance which have been major handicaps to the achievement of a modern society in China and to bring into existence a complement of scientists, engineers, technicians, and administrators equipped to develop, operate, and improve the large-scale, complex, and thoroughly modern industrial mechanism which the Communist government is attempting to create. The outcome of this program, and of the Communist effort generally, is awaited with much interest and some trepidation by the rest of the world, which often finds events in mainland China difficult to interpret, and watches uneasily as this giant Asian nation becomes more and more a force to be reckoned with in world affairs.

TAIWAN

Since 1949, interest in Taiwan or Formosa (Portuguese, "beautiful") has been more widespread than formerly, due to the presence there of the Chinese Nationalist government and its remaining armed forces. Taiwan is separated from the coast of South China by the Formosa Strait, about 100 miles wide. From 1895 to 1945 the island was governed by the Japanese, who developed its economy as an adjunct to that of Japan, and called the island by its ancient Chinese name of Taiwan.

The area of Taiwan, including the Pescadores and other small neighboring islands, is approximately 13,900 square miles. The population, largely Chinese in origin, was estimated at 13.5 million in 1968. High mountains rise steeply from the sea on the eastern side of the island, but slope on the west to a broad coastal plain bordering the Formosa Strait. Taiwan lies on the northern margin of the tropics, and its temperatures are further moderated by the warm, northward-flowing Kuroshio or Japanese Current. Irrigated rice, the main food crop, is favored by the hot, humid summers of the lowlands and by an abundance of irrigation water provided by streams originating in the mountains. Some of the higher mountain areas receive nearly 300 inches of rainfall a year. Cultivated land, mostly found on the western coastal plain, occupies about a fourth of the total area of the island. Rice accounts for more than half of the total crop acreage. Two rice harvests a year are commonly secured from the same field. Sweet potatoes are also a major food crop. Sugar cane, the most important cash crop, is grown primarily in the southern part of the island where hot, rainy summers and dry winters are ideal for its cultivation.

Taiwan has few mineral resources in significant quantities. The most important include coal, salt, sulfur, and copper. Recent discoveries of natural gas may be significant for the future. The island has many short, swift rivers, and hydroelectric power comprises about 50 percent of the national capacity. Lack of reserve capacity frequently reduces hydroelectric power generation during the dry winter and spring months, and this sometimes affects industrial production. Slightly more than half of Taiwan is forested. The forest cover includes both broadleaf evergreens and conifers, with the latter being the more important for timber production. A significant timber industry exists, though poor accessibility to the better stands hinders maximum exploitation of this resource. Overall, the most important group of manufacturing industries is the food processing and related industries. These include sugar refining, the canning of fruit, meat, vegetables, and fish, and the extraction of

Harvesting pineapples on the broad plain of western Taiwan. High mountains rimming the east side of the island rise steeply in the distance. (*Finance and Development,* a quarterly publication of the World Bank Group and the International Monetary Fund.)

oil from peanuts and soybeans. Other leading industries include textiles (dependent upon imported cotton), chemicals, cement, electronics, and various light consumer goods. Though small, the petroleum and iron and steel industries are expanding. Both are largely dependent on imports of raw materials.

The Taiwan economy showed remarkable growth during the 1960s. This was highlighted by the increased productivity of the agricultural sector. Rice output continues to climb, largely through improved yields reflecting greater use of chemical fertilizers and improved varieties. A greater variety of agricultural products are now grown than in past times, and this is permitting the export trade to be weaned away from the traditional dependence on sugar and rice. The industrial base inherited from the Japanese also has been expanded and is growing at a rapid pace, with diversification and a wider range of products constituting the most significant trends. The direction of the Taiwan economy is indicated by the relative decline of sugar, long the leading export. Cane sugar represented 52 percent of all exports by value in 1958, but in 1966 it constituted only 11 percent. Other important agricultural exports include bananas, processed foods (pineapples [photo, above], mushrooms, and asparagus), and rice. Textiles, wood and wood products, metals and machinery, and chemicals are major nonfood exports. Japan and the United States continue to be the major trading partners, with Japan taking about one-third of Taiwan's exports in 1966. Japanese products represented approximately 37 percent of Taiwan's imports in the same year.

The capital and largest city of Taiwan is Taipei (estimated metropolitan population 1.2 million in 1968), located near the northern end of the island.

MONGOLIA

The Mongolian People's Republic, referred to as Outer Mongolia in the past, was once part of the Chinese empire, but since 1924 it has been a separate country organized along Soviet lines and under strong Soviet influence. Mongolia includes an area of about 600,000 square miles and had an estimated population of about 1.2 million in 1968. About 90 percent of the population consists of various Mongol groups. The country contains large desert plains in the south and east, locally termed *gobis* (hence the "Gobi Desert"); mountain ranges in the west; and grassy valleys and wooded hills and mountains to the north. Vast herds of livestock, primarily sheep and cattle, are estimated at about 24 million head. Probably two-thirds of the population still derive their livelihood from animal husbandry. Collectivization of the nomadic population was accomplished (with some difficulties) in 1958–1959, the objective being to stabilize the population territorially and transform the rural economy to one based on crop tillage and dairying. Crops grown include wheat, barley, oats, and millet. Ulan Bator (220,000) is the capital, major urban center, and focus of the major transportation routes.

REFERENCES and READINGS

AIRD, JOHN S., "Estimating China's Population," *Annals of the American Academy of Political and Social Science,* 369 (January 1967), 61–72.

ANDERSON, EUGENE N., "Changing Patterns of Land Use in Rural Hong Kong," *Pacific Viewpoint,* 9, no. 3 (May 1968), 33–50.

BANKS, MIKE, "Europe's Side-Door to China" [Macao], *Geographical Magazine,* 39, no. 11 (March 1967), 887–898.

BOARDMAN, EUGENE P., "Chinese Mandarians and Western Traders: The Effect of the Frontier in Chinese History," in Walker D. Wyman and Clifton B. Kroeber, eds., *The Frontier in Perspective* (Madison: University of Wisconsin Press, 1958), pp. 95–110.

BOXER, BARUCH, *Ocean Shipping in the Evolution of Hong Kong* (University of Chicago, Department of Geography Research Paper No. 72; Chicago: 1961).

BUCHANAN, KEITH, *The Chinese People and the Chinese Earth* (London: G. Bell and Sons, 1966); also "The People's Communes after Six Years: Notes on Four Communes in Kwangtung Province," *Pacific Viewpoint,* 6, no. 1 (March 1965), 52–64; and "West Wind, East Wind," *Geography,* 47, pt. 4 (November 1962), 333–346.

BUCK, JOHN LOSSING, OWEN L. DAWSON, and YUAN-LI WU, *Food and Agriculture in Communist China* (New York: Frederick A. Praeger, 1966).

CHANG, KUEI-SHENG, "Geographical Bases for Industrial Development in Northwestern China," *Economic Geography,* 39, no. 4 (October 1963), 341–350; and "The Changing Railroad Pattern in Mainland China," *Geographical Review,* 51, no. 4 (October 1961), 534–548.

CHANG, SEN-DOU, "Peking: The Growing Metropolis of Communist China," *Geographical Review,* 55, no. 3 (July 1965), 313–327; also, "The Distribution and Occupations of Overseas Chinese," *Geographical Review,* 58, no. 1 (January 1968), 89–107; "The Historical Trend of Chinese Urbanization," *Annals of the Association of American Geographers,* 53, no. 2 (June 1963), 109–143; and "The Million City of Mainland China," *Pacific Viewpoint,* 9, no. 2 (September 1968), 128–153.

CHEN, CHENG-SIANG, "The Changing Economy of Taiwan," *Pacific Viewpoint,* 6, no. 2 (September 1965), 179–190; and "Taiwan," *Focus,* 17, no. 4 (December 1966), 6 pp.

CLUBB, O. EDMUND, *Twentieth Century China* (New York: Columbia University Press, 1964).

CRESSEY, GEORGE B., *Land of the 500 Million: A Geography of China* (New York: McGraw-Hill Book Company, 1955).

DAWSON, RAYMOND, *The Chinese Chameleon: An Analysis of European Conceptions of Chinese Civilization* (New York: Oxford University Press, 1967).

DONNITHORNE, AUDREY, *China's Economic System* (New York: Frederick A. Praeger, 1967).

DWYER, D. J., "The Coal Industry in Mainland China since 1949," *Geographical Journal,* 129, pt. 3 (September 1963), 329–338; also "The Development of China's Inland Waterways," *Geography,* 46, pt. 2 (April 1961), 165–167; and "Size as a Factor in Economic Growth: Some Reflections on the Case of Hong Kong," *Tijdschrift voor Economische en Sociale Geografie,* 56, no. 5 (September–October 1965), 186–192.

ELEGANT, ROBERT S., "China's Next Phase," *Foreign Affairs,* 46, no. 1 (October 1967), 137–150.

FAIRBANK, JOHN K., "The People's Middle Kingdom," *Foreign Affairs,* 44, no. 4 (July 1966), 574–586; and *The United States and China* (rev. ed.; American Foreign Policy Library. Cambridge, Mass.: Harvard University Press, 1958; New York: Viking Press, Compass Books, 1962).

FESSLER, LOREN, and the Editors of *Life, China* (Life World Library; New York: Time Inc., 1963).

FITZGERALD, C. P., *The Chinese View of Their Place in the World* (New York: Oxford University Press, 1964); also, *A Concise History of East Asia* (London: Heinemann, 1966); and "Tension on the Sino-Soviet Border," *Foreign Affairs,* 45, no. 4 (July 1967), 683–693.

FREEBERNE, MICHAEL, "Natural Calamities in China, 1949–61: An Examination of the Reports Originating from the Mainland," *Pacific Viewpoint,* 3, no. 2 (September 1962), 33–72;

and "Minority Unrest and Sino-Soviet Rivalry in Sinkiang, China's North-western Frontier Bastion, 1949–1965," in Charles A. Fisher, ed., *Essays in Political Geography* (London: Methuen and Co., 1968), pp. 177–209.

HARRINGTON, LYN, "Land beyond the Great Wall," *Geographical Magazine,* 40, no. 7 (November 1967), 561–569.

HARRINGTON, RICHARD, "Impressions of Mongolia," *Canadian Geographical Journal,* 74, no. 2 (February 1967), 64–75.

HEENAN, L. D. B., "The Petroleum Industry of Monsoon Asia," *Pacific Viewpoint,* 6, no. 1 (May 1965), 65–95.

HERMAN, THEODORE, "The Economic Development of China: Studies Reviewed in Tranquillity," *Geographical Review,* 50, no. 1 (January 1961), 114–120; and "Group Values toward the National Space: The Case of China," *Geographical Review,* 49, no. 2 (April 1959), 164–182.

HERRMANN, ALBERT, *An Historical Atlas of China,* general editor, Norton Ginsburg; prefatory essay by Paul Wheatley (Chicago: Aldine Publishing Company, 1966).

HO, SAMUEL PAO-SAN, "Agricultural Transformation under Colonialism. The Case of Taiwan," *Journal of Economic History,* 28, no. 3 (September 1968), 313–340.

HSIEH, CHIAO-MIN, *China: Ageless Land and Countless People* (Princeton, N.J.: D. Van Nostrand Co., Searchlight Books, 1967); also *Taiwan–ilha Formosa: A Geography in Perspective* (Washington, D.C.: Butterworth Inc., 1964); and "Taoyuan and Chianan: Two Irrigation Systems in Taiwan," in Richard S. Thoman and Donald J. Patton, eds., *Focus on Geographic Activity: A Collection of Original Studies* (New York: McGraw-Hill Book Company, 1964), pp. 27–30.

HU, CHANG-TU, in collaboration with others, *China: Its People, Its Society, Its Culture* (New Haven, Conn.: Human Relations Area Files Press, 1960).

HUGHES, R. H., "Hong Kong–Far Eastern Meeting Point," *Geographical Journal,* 129, pt. 4 (December 1963), 450–465.

HUNTER, HOLLAND, "Transport in Soviet and Chinese Development," *Economic Development and Cultural Change,* 14, no. 1 (October 1965), 71–84.

JACKSON, W. A. DOUGLAS, *Russo-Chinese Borderlands* (Princeton, N.J.: D. Van Nostrand Co., Searchlight Books, 1962).

JONES, F. C., *The Far East: A Concise History* (New York: Pergamon Press, 1966).

KARAN, P. P., "The Sino-Soviet Border Dispute," *Journal of Geography,* 63, no. 5 (May 1964), 216–222.

LAMB, ALASTAIR, *The China-India Border: The Origins of the Disputed Boundaries* (Chatham House Essays; New York: Oxford University Press, 1964).

LATOURETTE, KENNETH SCOTT, *The Chinese: Their History and Culture* (4th ed.; New York: The Macmillan Company, 1964).

LATTIMORE, OWEN, *From China, Looking Outward* (an inaugural lecture; Leeds, England: Leeds University Press, 1964); also, *Studies in Frontier History: Collected papers 1928–1958* (New York: Oxford University Press, 1962); and, with Eleanor Lattimore, *China, A Short History* (New York: W. W. Norton & Company, 1947).

LI, CHOH-MING, ed., *Industrial Development in Communist China* (New York: Frederick A. Praeger, 1964).

McCOLL, ROBERT W., "Development of Supra-Provincial Administrative Regions in Communist China, 1949–1960," *Pacific Viewpoint,* 4, no. 1 (March 1963), 53–64.

MURPHEY, RHOADS, *An Introduction to Geography* (2d ed.; Chicago: Rand McNally & Co., 1966), Chap. 26, "The Middle Kingdom: Splendid Isolation," and Chap. 27, "Changes on the Good Earth," pp. 420–459; also, "Man and Nature in China," *Modern Asian Studies,* 1, no. 4 (October 1967), 313–333.

MURPHY, GEORGE G. S., *Soviet Mongolia: A Study of the Oldest Political Satellite* (Berkeley: University of California Press, 1966).

NEEDHAM, JOSEPH, "The Past in China's Present: A Cultural and Social Background for Contemporary China," *Pacific Viewpoint,* 4, no. 2 (September 1963), 115–138.

NUTTONSON, M. Y., *The Physical Environment and Agriculture of Central and South China, Hong Kong and Taiwan* (Washington, D.C.: American Institute of Crop Ecology, 1963).

PETROV, VICTOR P., *China: Emerging World Power* (Princeton, N.J.: D. Van Nostrand Co., Searchlight Books, 1967).

REISCHAUER, EDWIN O., and JOHN K. FAIRBANK, *A History of East Asian Civilization:* Vol. 1, *East Asia: The Great Tradition* (1958); Vol. 2 (with Albert M. Craig), *East Asia: The Modern Transformation* (1965). (Boston, Mass.: Houghton Mifflin Company.)

RICHARDSON, S. D., *Forestry in Communist China* (Baltimore, Md.: Johns Hopkins Press for Resources for the Future, 1966).

ROSE, JOHN, "Sinjao: A Chinese Commune," *Geography,* **51**, pt. 4 (November 1966) 379–383.

SHABAD, THEODORE, *China's Changing Map: A Political and Economic Geography of the Chinese People's Republic* (New York: Frederick A. Praeger, 1956).

SPENCER, J. E., "The Prospects of Communist China Today," *Journal of Geography,* **66**, no. 7 (October 1967), 364–371; also, "Ancient China under Modern Communism: The Comparative Morphology of Pre-Communist and Communist Society," *California Geographer,* **3** (1962), 1–25; and "The Houses of the Chinese," *Geographical Review,* **37**, no. 2 (April 1947), 254–273.

STEWART, ROSEMARY, "Inside China's Factories and Communes," *Geographical Magazine,* **39**, no. 12 (April 1967), 967–976.

TREGEAR, T. R., *A Geography of China* (Chicago: Aldine Publishing Company, 1965).

TRUE, PHILIP A., "Nan-Ts'un: An Example of Changing Units of Rural Organization in Mainland China," in Richard S. Thoman and Donald J. Patton, eds., *Focus on Geographic Activity: A Collection of Original Studies* (New York: McGraw-Hill Book Company, 1964), pp. 19–26.

TUAN, Y. F., "A Preface to Chinese Cities," in R. P. Beckinsale and J. M. Houston, eds., *Urbanization and Its Problems* (New York: Barnes & Noble, 1968), pp. 218–253.

U.S. Army Area Handbook for Communist China (Department of the Army Pamphlet No. 550–60; Washington, D. C.: 1967).

U.S. BUREAU OF INTERNATIONAL COMMERCE, *Basic Data on the Economy of Hong Kong,* OBR 68–35 (May 1968). (Washington, D.C.: Government Printing Office, 1968.)

U.S. CONGRESS, 90th, 1st Session, *An Economic Profile of Mainland China* (Washington, D.C.: Government Printing Office, 1967). Studies prepared for the Joint Economic Committee, Congress of the United States.

VANDERMEER, CANUTE, "Changing Water Control in a Taiwanese Rice-Field Irrigation System," *Annals of the Association of American Geographers,* **58**, no. 4 (December 1968), 720–747.

VAUGHAN, T. D., and D. J. DWYER, "Some Aspects of Postwar Population Growth in Hong Kong," *Economic Geography,* **42**, no. 1 (January 1966), 37–51.

WATSON, FRANCIS, *The Frontiers of China: A Historical Guide* (New York: Frederick A. Praeger, 1966).

WIENS, HEROLD J., "Regional and Seasonal Water Supply in the Tarim Basin and Its Relation to Cultivated Land Potentials," *Annals of the Association of American Geographers,* **57**, no. 2 (June 1967), 350–366; and "The Historical and Geographical Role of Urumchi, Capital of Chinese Central Asia," *Annals of the Association of American Geographers,* **53**, no. 4 (December 1963), 441–464.

WU, YUAN-LI, *The Spatial Economy of Communist China: A Study on Industrial Location and Transportation* (1967); also, *The Economy of Communist China: An Introduction* (1965), and *The Steel Industry in Communist China* (1965). (New York: Frederick A. Praeger.)

ACKNOWLEDGMENT. The authors express their appreciation to Mr. Philip True for his assistance in revising Chapters 20 and 21. Most of the revision of these chapters for the Third Edition of this text was done by him. Final responsibility for the chapters, of course, rests with the authors.

Japan

and Korea

21   Until the middle of the last century the nation of Japan played little part in world affairs. In fact, from the beginning of the seventeenth century until the middle 1850s Japan lived in almost complete isolation, shut off by government decree from the outside world. In 1853, however, the visit of an American naval squadron under Commodore Matthew Perry led to a change of policy. The old era of isolation was ended, and Japan entered a new age of industrial and commercial development and territorial expansion. By the middle 1930s Japan had become a leading industrial and trading nation, acquired a sizable empire, and taken a prominent place among the military and naval powers of the world.

Had its leaders followed another course, Japan, the first Oriental nation to develop a modern economy along Occidental lines, might have led its Asian neighbors into an age of unprecedented progress and prosperity. But the influence of the military was increasingly strong as the twentieth century progressed, and Japan became committed to a program of territorial expansion through military conquest. Domestic economic crises in the late 1920s and early 1930s helped erode democratic institutions

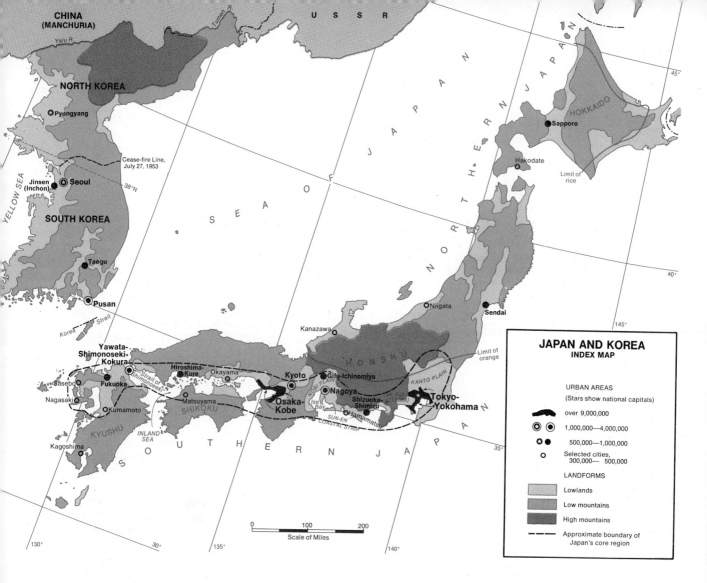

Index map of Japan and Korea. Limits of rice and orange after Edward A. Ackerman, *Japan's Natural Resources and Their Relation to Japan's Economic Future* (Chicago: University of Chicago Press, 1953), p. 23. City-size symbols are based on metropolitan area estimates.

and, along with actions of the Japanese Army that reduced Manchuria to a colony (1931), increased the strength of the militaristic and totalitarian elements in Japanese life. The role of the military was further enhanced after Japan became involved in a general war with China in 1937, and military considerations became more and more important in the country's political and economic planning and decisions. In 1938 Japan proclaimed its "New Order in East Asia," and in 1940–1941 this widened into the "Greater East Asia Co-Prosperity Sphere" –a euphemism for Japanese political and economic control over China and the rich tropical lands of Southeast Asia. Eventually, however, the Japanese leaders overreached themselves with an attack on the United States

at Pearl Harbor, and in 1945 Japan went down to complete military defeat at the hands of the United States and its Allies. Following the Japanese surrender the country was occupied by Allied forces, mainly American, and remained under occupation until a peace treaty went into effect in 1952.

Japan, like Germany, faced an enormous task of physical and economic reconstruction after the war. Most of the country's merchant shipping had been sunk, and American air raids had devastated or severely damaged many of the most important urban and industrial areas. But the Japanese recovery from the war was rapid. The cities were rebuilt, parliamentary government was reinstituted, and the economy expanded and grew with remarkable speed. Recovery was facilitated by American financial aid, but the primary credit goes to the skill and diligence of the Japanese themselves. In addition, the Japanese were favored by the general economic situation in the postwar world, with its strong tendencies toward freer trade, its expanding markets for industrial exports, and its increasingly efficient mechanisms for supplying the oil, ore, and other bulk commodities that Japan must import if her economy is to function.

Japan's eventful history has seen successive periods of racial and cultural immigration, nation building, withdrawal into seclusion, emergence as an industrial, commercial and imperial power, military defeat and occupation, and reemergence as a sovereign state, shorn of colonies but with greatly expanded industry, a thriving trade, and an overall position as one of the world's major economic powers. Some pertinent aspects of this history are summarized below.

HISTORICAL BACKGROUND

The racial and ethnic origins of the Japanese are somewhat obscure. They are thought to be descended from a number of different peoples, primarily Mongoloids, who reached Japan from other parts of eastern Asia at various times in the distant past. An earlier people, the Ainu, were driven into outlying areas where remnants still exist, principally in Hokkaido. Some of the Ainu may possibly have been assimilated by the invading peoples. The oldest surviving Japanese written records date from the eighth century A.D., but Japanese traditions extend back to the reign of Jimmu, the first emperor, whose accession is ascribed to the year 660 B.C., although modern scholars have placed the date some centuries later. The acceptance of these traditions has been an important factor in Japanese psychology, since they ascribe a divine origin to the imperial family (Jimmu was supposedly descended from the Sun Goddess) and thus call attention to Japan as "the Land of the Gods."

The early emperors gradually extended control over all of their island realm. A society developed based on subsistence agriculture and organized into warring clans. By about the twelfth century the emperors were being pushed into the background by powerful military leaders, or *shoguns,* who actually controlled the country while the emperors remained as figureheads. Meanwhile the provinces were ruled by nobles or *daimyo,* whose power rested on the military prowess of lesser nobles, the *samurai.* Thus evolved a governmental system resembling in some ways the European feudal system of medieval times.

Early Contacts with Europe

The adventurous and wide-ranging Europeans of the Age of Discovery reached Japan in the first part of the sixteenth century. In the 1540s came the Portuguese, and after them came representatives of other European nations. These early comers were mainly merchants and Roman Catholic missionaries. Both groups were well received. The merchants were allowed to set up trading establishments and open an active commerce, and the missionaries were allowed to preach freely and even received a certain amount of governmental favoritism. There were an estimated 300,000 Japanese Christians by the year 1600.

However, these early promising contacts were eventually nullified. Even while they were occurring, a series of strong military leaders were imposing internal unity and central authority on the disorderly feudal structure of Japanese society. By 1600 the Tokugawa family had acquired absolute power, which it retained until 1868. This period is generally known as the period of the Tokugawa Shogunate. In order to maintain power and stability the early Tokugawa shoguns desired to eliminate all disturbing social influences; the latter included the foreign traders and missionaries. In addition, the shoguns seem to have feared that the missionaries were the forerunners of an attempted conquest by Europeans, especially the Spanish, who held the Philippines. Consequently, the traders were driven out, and Christianity was almost completely eliminated in bloody persecutions during the early part of the seventeenth century. After 1641 a few Dutch traders were the only Occidentals allowed in Japan, and even they were segregated on the small island of Deshima in the harbor of Nagasaki. Japan settled under the Tokugawa into two centuries of isolation, peace, and stagnation.

Westernization and Expansion

When the foreigners again made a serious attempt to open Japan to trade two centuries later, the growing strength of the outside nations could not be thwarted by a Japan which had fallen far behind in the arts of both war and peace. This time the United States took the lead, and visits in 1853 and 1854 by American naval squadrons under Commodore Matthew Perry resulted in treaties opening Japan to trade with the United States. The major European powers were soon able to obtain similar privileges. Some of the great feudal authorities in southwestern Japan were strongly opposed to the new policy, but their opposition was quelled when coastal areas under their control were bombarded by American, British, French, and Dutch ships in 1863 and 1864.

These events so weakened the faltering power and prestige of the Tokugawa Shogunate that in 1868 the ruling shogun was overthrown by a rebellion. The revolutionary leaders restored the legitimate sovereignty of the emperor, who took the name Meiji or "Enlightened Rule." Thus the revolution of 1868 is generally termed the Meiji Restoration. Since that time Japan has generally been ruled, despite the forms of democracy, by relatively small groups of powerful men manipulating the machinery of government and the prestige of the emperor. The period since World War II has seen the reemergence of the prewar Japanese political parties and a return to parliamentary government, initially aided and guided by the occupation authorities. Since the signing of the Japanese Peace Treaty in 1951, the political system has functioned effectively, although a lack of consensus between the two principal political parties on fundamental issues is an omen of possible concern for the future. Without doubt, Japan is governed more democratically today (1968) than at any time in the past.

The men who came to power in 1868 and the able Emperor Meiji himself were true revolutionaries in that they aimed at a complete transformation of Japan's society and economy. They saw that if Japan were not to fall under the control of the Western nations, its demonstrated military impotence would have to be remedied. They saw also that this would require a reconstruction of the Japanese economy and of many aspects of the social order. These tasks were approached with energy and intelligence. Feudalism was abolished, though not without the necessity of suppressing a bloody revolt in 1877. Thus was cemented the power of a strong central government which could be used as a vehicle for remodeling the country. Such a government was an institution that no other Oriental people possessed as an instrument of modernization and resistance to foreign encroachment.

The new government set about the task of having its subjects learn and apply the knowledge and techniques which the Occidental countries had been accumulating during the centuries of Japan's isolation. Foreign scholars were brought to Japan, and Japanese students were sent abroad in large numbers. A constitution, modeled largely after that of imperial Germany, was promulgated, the forms of representative government were instituted, and the legal system was remodeled to bring it into greater conformance with Occidental practices. On the economic side the government used its financial power, largely derived from oppressive land taxes, to foster industry in a variety of ways. Railroads, telegraph lines, and a modern merchant marine were constructed, and banks and other financial institutions were developed. Light industry was stimulated as a means of providing exports, and basic heavy industries were gradually developed. Wherever private interests could not achieve the desired economic development unaided, the government provided subsidies to private companies, or else built plants and operated them until they could be acquired by private concerns.

So spectacular were the results achieved that in 40 years Japan had become the first Oriental nation in modern times to attain the status of a world power. The Japanese have often been spoken of in a derogatory fashion as mere imitators, but this seems unjust. It is obvious that rapid and excellent imitation was called for in the situation that confronted Japan, and the scope and success of that imitation represent a major accomplishment achieved in the face of great difficulties imposed by cultural inertia and the country's poverty in natural resources.

Japan's Prewar Empire

In the political sphere, after Japan abandoned the stay-at-home policy of the Tokugawa period, it gradually emerged as an imperial power. Between the early 1870s and World War II the country pursued, with only brief interruptions, a consistently expansionist policy, and by 1941 Japan controlled one of the world's most imposing empires (map, right). The major components of the Japanese Empire, in the order of their acquisition, were as follows:

1. The Kuril Islands. Acquired from Russia in exchange for Japanese abandonment of claims on Sakhalin, 1875.

2. The Bonin Islands. Annexed without opposition, 1876.

3. The Ryukyu Islands. These had had a semi-independent status under the overlordship of both Japan and China. They were occupied, despite local and Chinese protests, in 1879, and Japan was confirmed in ownership by China in 1895. The Ryukyus, together with the Bonins and the Kurils, were incorporated into Japan as part of the homeland. These islands, and others acquired later, provided sites for military and naval bases protecting Japan proper and serving as staging points for military expansion.

4. Taiwan (Formosa). Acquired from China in the Sino-Japanese War of 1894–1895 and developed to serve Japanese economic needs, particularly for rice and cane sugar.

5. Korea, the Liaotung peninsula, and southern Sakhalin. Acquired as a result of victory over Russia in

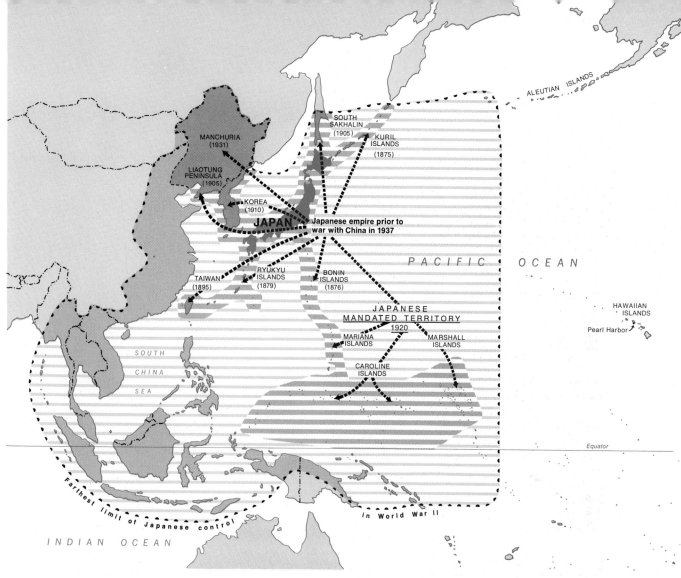

MANCHURIA
(1931)

LIAOTUNG
PENINSULA
(1905)

KOREA
(1910)

SOUTH
SAKHALIN
(1905)

KURIL
ISLANDS
(1875)

JAPAN

Japanese empire prior to
war with China in 1937

PACIFIC OCEAN

ALEUTIAN ISLANDS

TAIWAN
(1895)

RYUKYU
ISLANDS
(1879)

BONIN
ISLANDS
(1876)

SOUTH
CHINA
SEA

JAPANESE
MANDATED TERRITORY
1920

MARIANA
ISLANDS

MARSHALL
ISLANDS

CAROLINE
ISLANDS

HAWAIIAN
ISLANDS

Pearl Harbor

Equator

Farthest limit of Japanese control

in World War II

INDIAN OCEAN

Map showing overseas areas held by Japan prior to 1937 and the line of maximum Japanese advance in World War II.

the Russo-Japanese War, 1904–1905. With Russian influence and pressure checkmated, Korea became a Japanese protectorate, though it was not formally annexed to the Japanese Empire until 1910. It was exploited and developed for its considerable mineral and agricultural resources. China remained the nominal sovereign over the Liaotung peninsula. However, actual control over Port Arthur and nearby territory had passed to Russia in 1898, when she forced China to grant a lease on the area, and this leasehold was now transferred to Japan. With Russia no longer able to intercede, Japan was able to extort from China numerous economic and political privileges throughout southern Manchuria, placing that area definitely within a Japanese "sphere of interest," though leaving China still the legal sovereign.

6. The Caroline Islands, Marshall Islands, and Mariana Islands. These Pacific island groups were former German possessions, except for the island of Guam in the Marianas, an American possession since 1898. When World War I broke out in Europe, Japan joined the Allied side and seized the German holdings in the Orient and the Pacific north of the equator. The peace settlements after the war confirmed Japanese occupancy of the islands under a mandate of the League of Nations. While the other major powers were embroiled in Europe, Japan was able, in addition, to exert pressure successfully on China for increased economic privileges, especially in Manchuria, Inner Mongolia, and northern China proper. Although Japan had occupied former German-controlled territory in China on the Shantung peninsula, she was

induced by Chinese and international pressure to evacuate that area in 1922.

7. Manchuria. Conquered in the course of hostilities with China lasting from 1931 to 1933. Manchuria became the Japanese protectorate of "Manchukuo" and the Japanese intensified their development of its economy as an adjunct to that of Japan itself. In addition, further Japanese privileges were extorted in Inner Mongolia and northern China proper.

8. General war with China in 1937 saw Japan by 1938 in control of most of North China, the middle and lower Yangtze Valley, and pockets along the southeast China coast. The war with China which Japan precipitated in 1937 continued, however, until 1945 without additional significant Japanese gains. Instead, Japan found it difficult to control the territory already conquered due to steadily increasing guerrilla warfare by the Chinese.

9. French Indochina. Occupied without resistance during 1940 and 1941, after the conquest of France by Germany.

10. The empire was brought to its greatest extent by a series of rapid conquests between December 1941 and the summer of 1942, after the Japanese had entered World War II. These conquests included Guam and Wake islands, the Philippines, Hong Kong, Thailand, Burma, Malaya, Singapore, the East Indies, much of New Guinea, the Admiralty, Bismarck, Solomon, and Gilbert islands, and part of the Aleutian chain.

After 1942 Japan was generally in retreat, and 1945 found her shorn of the overseas territories acquired during nearly 70 years of successful imperialism. The motives of Japanese expansionism were mixed, including a tradition of national superiority and "manifest destiny," a desire for security and great-power recognition, the desire of military leaders to aggrandize themselves and gain control of the Japanese government, and a desire on the part of various elements in Japan to gain assured markets for Japanese goods and assured sources of materials. Though the methods of gaining them may have been illegitimate, the desire for materials and markets was solidly based on

A view of Sapporo, the main city of Japan's northern island, Hokkaido. A 300-foot-wide boulevard flanked by modern buildings dominates the center of the photo. Characteristic mountains rise in the distance. Mountains or hills are a prominent feature of most Japanese landscapes.

need. Expansion of industry and population on an inadequate base of domestic natural resources has made Japan vitally dependent on sales of industrial products outside the homeland. Such sales provide the principal funds with which the country purchases the imported foods, fuels, and materials that it requires for its large and growing population and its expanding industries. In its dependence on foreign markets and supplies Japan obviously resembles the United Kingdom. The trend toward militarism in prewar Japan was abetted by (1) the world depression that threatened the Japanese economy so perilously dependent on world trade, and (2) the growing political isolation of Japan caused by her continued expansion on the Asian mainland. These factors aided those who espoused the need for creation of an Asian realm controlled by Japan that, with the Western powers excluded, would insulate the Japanese economy from the cyclical trends of the world economy and, concurrently, would elevate the nation to the rank of a leading world power. Since World War II Japan, without benefit of colonies or spheres of influence, has become one of the world's leading industrial nations and has achieved a comparatively high level of living. But the general circumstances of the world economy in the postwar era have been far more favorable for Japanese industrial and commercial expansion than they were in the 1920s and 1930s.

THE JAPANESE HOMELAND

Japan is a nation of some 101 million people (1968 official estimate) occupying four main islands and numerous smaller islands with an aggregate of approximately 142,800 square miles. The Japanese homeland is briefly described in the following selection:[1]

The homeland of Japan . . . consists of four large volcanic islands—Honshu, Hokkaido, Shikoku, and Kyushu—and several hundred smaller islands . . . .

. . . Honshu, the largest island, has an area a little less than the combined areas of New York and Pennsylvania but with [two and one-half times] their population. Hokkaido, lying in about the same latitude as Maine [map, above], has approximately the same area and [over five] times the population. Shikoku, in the

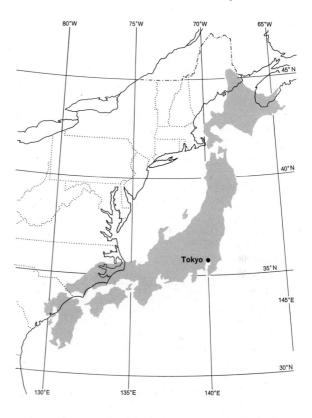

Japan compared with the eastern United States in latitude and area. (After a map by E. O. Reischauer.)

general latitude of South Carolina, has less than one-fourth the area but [nearly] double the population. Kyushu, between the same parallels as Georgia, has one-fourth the area and [over three] times the population. . . .

Climate and Basic Natural Resources

The climatic conditions in Japan . . . are not altogether comparable to those of points in corresponding latitudes in the eastern United States. On the whole, the winters are colder in Japan and the annual precipitation is heavier. This may be accounted for partly by the presence of a cold current which extends farther south in winter

[1] *Geographical Foundations of National Power* (Army Service Forces Manual M 103-2; Washington, D.C.: Government Printing Office, 1944), pp. 125–131. Six headings have been omitted and two headings slightly altered. Population figures and comparisons quoted in original source have been brought up to date. One editorial note has been added. The spelling of "Kwanto Plain" has been altered to read "Kanto Plain." Some changes in italicizing have been made, and supplementary material has been supplied in brackets in several places.

than in summer on both sides of the Japanese islands, and by their position in the path of the monsoon winds. Within the islands, differences in latitude produce marked climatic differences, and the great variation in altitude produces much local diversity irrespective of latitude. Along the southern and southwestern shores the climate is humid and subtropical; the winters are mild and the summers uncomfortably hot and oppressive. North of Tokyo the winters grow progressively colder. . . . Seasonal variations are particularly marked in the north and in the interior basins.

A winter scene in the mountains of western Honshu near the Sea of Japan. The buildings in the picture belong to a hot-spring and ski resort. (Japan Travel Bureau.)

Nature was not generous to the Japanese. In a country where [1 out of 4] people engage in agriculture, nearly five sixths of the land is too rugged for cultivation. The Japanese have been able partially to supply themselves with rice and other cereals, but only by the most intensive use of the arable soils. Forests cover [more than] half the land surface but do not supply all the kinds of wood needed. The many varieties of fish in the adjacent seas furnish a large part of the protein food consumed in Japan, as well as a valuable export.

. . . The hydroelectric energy generated from the mountain streams of Japan compensates only in part for the lack of petroleum and the poor quality of the coal. The raw materials for heavy industry are far from abundant. [Well over half of the country's requirements of nearly all the metals, including iron, must be imported.] The cotton textile [industry is] wholly dependent on foreign sources of raw cotton . . . .

Contrasts between North and South

The contrast between areas north and south of the 37th parallel is in part due to differences in climate and in part to the earlier settlement of the south. Southern Japan was settled early in the Christian era. There, particularly around the shores of the Inland Sea, the features considered typical of the Japanese landscape are found: a dense population living in lightly constructed dwellings, tiny farms, with fields of paddy rice on the bottomlands, small terraced fields on the hill slopes, and tea gardens and mulberry groves on the uplands. The settlement of the north began much later and proceeded gradually, in spite of the pressure of population and the scarcity of arable land in the south. The expansion of manufacturing after 1868 has been much more significant in the south, and the concentration of population in the lowlands there has been greatly augmented.

Highlands and Lowlands

About three quarters of the area of Japan . . . is composed of hill land or mountains of which the slopes are usually too steep and the soils too thin for normal cultivation. Scores of volcanic cones in various stages of activity provide some of the highest elevations. The mountain streams are short, swift, shallow, and generally unnavigable. They are chiefly important as sources of hydroelectric power and irrigation water for the populous lowlands.

The typical Japanese plain is an isolated area of alluvial soil deposited by rivers in mountain basins or by rivers and waves in coastal indentations. There is no continuous lowland belt along the shores of the islands. Because of shallowness and divided channels, the lowland streams are of little use for navigation, but their elevated beds make gravity irrigation simple. . . .

The marginal lowlands at the heads of bays or other openings upon the sea contain most of the arable land, most of the population, and the centers of trade, industry, and political and cultural life. Those of the north are less fully developed than those of the south.

Japan's Core Area

The core area of Japan . . . is an irregular zone about 800 miles long, extending westward from the Kanto Plain through Nagoya to Osaka and thence along both shores of the Inland Sea to [northern] Kyushu [map, p. 490].

The Kanto (Tokyo) Plain has an importance in the national life comparable in many ways to that of the English Lowland and the Paris Basin. . . . [Tokyo, the national capital, is the largest city in a metropolitan agglomeration which, including Japan's leading seaport, Yokohama, and various other satellite and surburban areas, numbered in excess of 16 million in 1968.] The Kanto Plain on which [the Tokyo agglomeration] stands is the largest lowland in Japan (2500 square miles). . . . Tokyo and Yokohama, both on tidewater, [together with their suburbs] comprise the [leading] industrial center of Japan and the commercial center not only for the plain itself, but for all of northern Japan.[2] The regions to the north and west are tapped by a network of railroads. . . .

South of the Kanto Plain the *Sun-en Coastal Strip* runs along the Pacific shore. This lowland, composed of rather small, isolated deltas separated by spurs of rocky hills, has been for several centuries a thoroughfare between Kyoto and Tokyo. It is traversed by the famous old Tokaido highway and the modern railroad. The mild winters and heavy precipitation in summer have favored the cultivation of mandarin oranges and tea. Half of the Japanese tea crop is grown there and is refined, blended, and packed in the castle town of Shizuoka [600,000].[3]

The *Nobi (Nagoya) Plain* which lies at the head of Ise Bay, 160 miles west of Tokyo, is second only to the Kanto Plain in area. Although third in importance as an industrial center, it [is very densely populated]. The harbor of Nagoya [2.4 million] is shallow and silted, and the city, unlike Tokyo and Osaka, lacks a connecting deep-water port.

Farther west, the *Kinki District* at the eastern end of the Inland Sea has the longest record of compact human habitation and contains the earliest capitals. The district is the [second] industrial area of the nation today and the site of three of the largest cities—Osaka, Kyoto, and Kobe.

Osaka [metropolitan area, with Kobe, 9.5 to 10 million] is 230 miles west of Tokyo on the bayhead delta of the broad, diked Yodo River and its numerous tributaries. Like Tokyo, the city has a network of canals and a shallow, silted harbor. Many of its modern Occidental buildings are supported on piles or metal drums. Its facilities for water transportation have contributed to the industrial pre-eminence of modern Osaka. So, too, have its excellent rail facilities, its level expanse of plain, and its large labor supply.

Kobe is on deep water, 16 miles down the bay from Osaka. Like Yokohama, its rival deep-water port, Kobe was brought into existence by the demands of the modern commercial and industrial era. It has become a relatively important manufacturing center, specializing in metal industry, especially shipbuilding. . . . The coastal strip on which it is built is too narrow to allow much industrialized settlement, however.

Inland Kyoto [1.7 million], the capital from 794 to 1869, has been little changed by the Industrial Revolution. Large factories are forbidden by law. The city retains much of its ancient splendor, making it a center . . . for millions of pilgrims and tourists. The handicrafts of the feudal period—the making of lacquer, porcelain, bamboo articles, cloisonné, bronze, and silk textiles—survive as cottage and workshop industries.

Along the deeply indented shores of the *Inland Sea* there are innumerable alluvial lowlands and river deltas. The sea is, in its way, as historic and beautiful as the Mediterranean. Settlements crowd the diminutive plains and adjacent hill slopes. The density of population and the scarcity of arable land have led to the artificial terracing of the hillsides, sometimes to elevations of several hundred feet. . . . The toil of generations of a single family is recorded in the stones or terraces inclosing a bare half acre of good loam. . . .

The sea varies from 20 to 40 miles in width. . . . Hundreds of small intercoastal and inter-island boats [carry] on a thriving local trade, and the entire length of 230 miles [is] traversed by trans-Pacific steamers and coastwise freighters plying between the coal fields of northern Kyushu and the factories of the eastern cities.

The *Northern Kyushu Coastal Strip* . . . extending . . . west from . . . the Strait of Shimonoseki [is a major center] of heavy industry in Japan. [This industrial region, based initially on local coal deposits, and relatively remote from Tokyo, Osaka, and Nagoya, has grown at a slower rate in recent times than the market-oriented industrial

---

[2] *Editorial note:* Yokohama developed originally as a deep-water port for Tokyo, which was handicapped by a shallow harbor. However, the harbor of Tokyo itself has now been improved to accommodate most deep-sea vessels. The two cities form a diversified industrial center. In parallel fashion, Kobe developed as the deep-water port for Osaka and has now coalesced with it to form a continuous urban and industrial area.

[3] Unless otherwise noted, city populations cited in this chapter are metropolitan-area approximations for 1968.

centers in central and southern Honshu. At least 3 million people live in a string of medium-sized industrial cities extending from Moji through Fukuoka (950,000) to Nagasaki (450,000). The greatest single agglomeration is formed by the iron and steel center of Yawata, largest in Japan, plus Kokura, Moji, and two other cities on Kyushu, and the city of Shimonoseki on the Honshu side of the narrow Strait of Shimonoseki. Most of the coal for the industries of northern Kyushu] is supplied by the Kyushu mines. Other raw materials—some coking coal . . ., iron ore, pig iron, steel scrap, petroleum, etc.—are brought in by sea. Coal from the Kyushu fields and steel for fabrication are sent out by the Inland Sea route to the Osaka and Nagoya industrial areas and even to Tokyo.

JAPAN'S POPULATION AND ECONOMY

Although a little smaller in area than the state of California, Japan is the world's seventh most populous country. A population aggregating about half that of the entire United States must be supported in a nation whose total acreage of cultivated land amounts to only about 24,000 square miles—approximately the area of the state of West Virginia. No other important country approaches the overall Japanese figure of approximately 4200 people to be supported per cultivated square mile. Furthermore, while the rate of population increase is relatively low (about 1.1 percent annually), this still results in the addition of over 1 million Japanese each year. In spite of great efforts to reclaim additional land during the past two decades, the amount of land actually cultivated has remained about the same. Gains made in reclaiming upland areas have been largely offset by losses, often consisting of better-quality ricelands sacrificed for new industry and urbanization. Thus under present conditions of agriculture there seems to be no real prospect of counterbalancing the rise in population simply through bringing additional land into cultivation. Japan imports about a fifth of its total food requirements as measured in calories, and long-range prospects suggest an increase in this figure.

A century and a half ago Japan, already a crowded land, had a population now estimated at around 30 million. However, increases in the population were kept in check by disease and starvation and by the almost universal practices of abortion and infanticide.

A rapid growth of population after 1868 was made possible by industrialization. New industries provided exports of manufactured goods with which to purchase imported food, resulting in, growing prosperity and a rising birth rate. Industrialization also undergirded the rising military power that enabled Japan to build an empire and thereby to gain assured markets and assured sources of food and industrial materials. Today the country no longer has an empire to exploit, but it must still continue to import industrial raw materials, foodstuffs, and fuels and to find ways to pay for them in the face of stiff competition from other industrial nations in export markets.

Under these conditions it is not surprising that the Japanese are emphasizing additional measures to help solve their food problem, including (1) maximization of agricultural output and (2) birth control. Food production has managed to increase at an annual average rate of 2 to 3 percent during the past 20 years. Rice, which is still the dominant crop though its share of total agricultural production is declining, has increased from an annual output of about 8 million metric tons just after World War II to 12 to 14 million metric tons since 1959. Japan is nearly self-sufficient (90 to 95 percent) in rice. This has been accomplished mainly through increased yields attributable to heavier applications of chemical fertilizers, improved varieties, better control of pests, and improvements in water control and irrigation; and secondarily through an increase in the acreage of paddy rice at the expense of less productive cereals. Meanwhile, Japan's birth rate of the 1930s has been halved, possibly portending a stabilization of the population within a few decades. Government encouragement of birth control, a tradition of population limitation, and the country's very rapid urbanization have been important factors contributing to the decline of the birth rate.

Agriculture and Fisheries

For Japan to supply even four-fifths of its current caloric needs from domestic production has required an intensive agriculture giving some of the world's highest yields, together with associated development of an extremely large fishing industry. Nearly nine-tenths of all cultivated land is devoted to food crops. Irrigated rice, which occupies about half of all cultivated land, is the basic crop nearly everywhere, including most parts of the northern island of Hokkaido. Wherever conditions of climate, water, and soil will permit—which includes the greater part of the irrigated acreage south of 38°N.—double cropping, or the planting of a second crop on the same field following the rice harvest, is practiced. In most areas rice is followed by winter wheat or winter barley, and in some cases by green-manure crops, vegetables, or white potatoes. Overall, about a third of the irrigated rice fields are sown to a second crop. Crops grown in unirrigated fields include a variety of grains, among which barley and wheat occupy the largest acreages. Various kinds of beans and peas, of which soybeans are the most important, stand next to the grains in acreage planted, and these are very closely followed by potatoes: white potatoes in northern Honshu and Hokkaido, and sweet potatoes in southern Honshu, Kyushu, and Shikoku. Most of the common vegetables are also grown. More than half of the unirrigated fields are double cropped—usually with fall-sown grains and vegetables. Intertillage—the growing of two or more crops simultaneously in alternate rows—is common. Most Japanese farms have at least a few fruit trees, and fruit production is rapidly increasing. Citrus fruits, mainly mandarin oranges, now lead, and are grown widely along the Pacific and the Inland Sea south of about 37°N. Apples, peaches, pears, grapes, and persimmons also are significant. Specialty crops grown in certain areas include mulberry trees—used for feeding silkworms—and such crops as tea, sugar cane, tobacco, peanuts, flax, hemp, pyrethrum, and peppermint. Sugar beets are raised in Hokkaido, and along the Inland Sea are grown the reeds used for making the mats which provide the floor covering in most homes throughout Japan. On the whole, the variety of Japanese agricultural production belies the customary picture of Japan as almost exclusively a rice-growing country. Nevertheless, rice is a basic component of every Japanese meal, and provides more than half of the total calories in the Japanese diet.

Nearly two-thirds of the calories in the diet of the Japanese people are supplied by starchy foods, including rice, other grains, and potatoes. Livestock products are of minor but growing importance. The number of farm animals, particularly dairy cattle, pigs, and chickens, has increased markedly in recent years. The caloric intake of the Japanese has been increasing and, more importantly, the diet has broadened in composition and is more nutritious. Fish are the principal source of protein and along with rice are an important element in most Japanese meals. A variety of valuable food fish are present in the waters surrounding Japan, and Japanese fishermen range widely throughout the world's major ocean fishing grounds. The annual fish catch of Japan ranked first in the world in both tonnage and value until 1962, when Peru assumed leadership in tonnage but not in value. Japan has pioneered in fishery technology, particularly the development of large fleet and processing facilities.

To the casual visitor, Japanese agriculture still presents a surface picture of tranquil tradition: small, carefully tended fields, little more than garden plots; night soil

A characteristic scene of rural Japan: transplanting rice seedlings by hand in a flooded paddy. (U.S. Department of State.)

carefully dipped from large wooden buckets and poured on growing crops; and women bending low to transplant rice seedlings in carefully aligned rows. But change has occurred and is accelerating. Rural labor, once overabundant, has been siphoned off by the higher wages available in urban employment. Part-time farming and supplemental nonfarm income have increased dramatically. Off-season employment is common, particularly where little winter cropping is done, but there also is a growing tendency for husbands and young adults to work in the towns and cities, returning on the weekends to help the womenfolk with the more arduous farm tasks. These changes, however, could not have occurred so rapidly or to such a degree without the great increase in mechanization. At present one farm family in three owns a small hand tractor, and less fortunate farmers can usually borrow from neighbors. At least four-fifths of the rice fields and more than half of the nonirrigated upland fields now are plowed with power cultivators; mechanization of threshing and hulling is universal; pumps distribute irrigation water; and power sprayers and dusters are increasingly evident. In addition, a three-wheel farm truck may be present, and in the farmhouse a radio and a television set are commonplace.

A significant trend in Japanese agriculture and a portent for the future is the growth in size of farm. Additional increases in size are recognized as essential by the Ministry of Agriculture, although changes in land-transfer regulations and in rural financing are required to further modernize agriculture. Another significant development of the last two decades has been the change in the composition of the Japanese diet, with less dependence on grains and very large increases in the consumption of milk, dairy products, eggs, meat, and fruit. This has caused many farmers to devote a portion of their land to growing profitable cash crops, such as fruit trees, and to raising livestock and poultry. Most of these changes on individual farms are minor in scale, since the average Japanese farm still is slightly under 3 acres in size—except in Hokkaido where the average rises to 13 acres. In sum, while Japanese agriculture shares many of the traditional characteristics of Asian agriculture, such as dominance of cereals and small size of farm, the trends of the past decade—mechanization, specialty crops, part-time farming, and so on—clearly point to a breakup of old patterns and a transition to new ones.

Importance of Forestry and Wood Products

The varied local environments of Japan support a wide range of valuable tree species, including broadleaf evergreens in southern Japan, a mixture of broadleaf deciduous and conifers in central and northern Honshu and southern Hokkaido, and conifers in northern Hokkaido and at higher elevations farther south. About three-fifths of the entire country is forested, and the Japanese, short of mineral resources, make use of wood in almost endless ways. Industrial wood uses—sawlogs, pulpwood, and mine timbers—account for about 75 percent of Japan's annual wood consumption. Use of wood for fuel has been dropping steadily in competition with other sources of energy. The significance of forestry and wood industries in Japan has been described by Reischauer:[4]

Next to men and the power resources of coal and water, wood is probably Japan's greatest natural resource. With over half the land devoted to forest, Japan is among the more heavily forested of the civilized countries of the world and unquestionably the most heavily forested of the populous lands. Her poverty in agricultural land has meant a larger proportion of terrain preserved from the farmer's hoe. The country is rugged, but little of it is too high for forest growth, and relatively warm temperatures throughout most of the forest lands plus ample rainfall everywhere mean excellent growing conditions.

But the ratio of men to forest land in narrow Japan, while not so disastrous as the ratio of men to agricultural land, is precarious. The per capita acreage of forest land is only about [one-sixth] that of the United States. With timber lands so limited and with so little else to treasure, the Japanese have been forced to tend their forests with a care undreamed of in a land like ours, which has been more bountifully provided by nature. Up until the outbreak of war with China, attempts were made to cut no more than the annual growth, and reforestation over the years had gradually expanded the forest area. In the three southern islands, artificially planted forests account for about one fifth of the total, making curious patterns of symmetrical rows of conifers, like some great corn field, in the mountain fastnesses of Japan. Careful cutting and planting practices have spread over the hills a patchwork pattern of different types and sizes of timber, resembling the patchwork patterns of agricultural land. The ratio of men to land in Japan has left man's imprint even on her wild mountains and rugged hills.

Japan is almost as dependent on a maximum yield from her forest lands as from her farms. While far richer in wood than are many of the lands of Asia and Europe, she depends upon her wood supply for greater and more vital services than does any other nation. The chief of these is for building material. . . . Despite tile, thatch, or tin roofs, mat floors, paper and glass sliding partitions, and walls made in part of mud plastered on bamboo lattice,

---

[4] Reprinted by permission of the publishers from Edwin O. Reischauer, *The United States and Japan.* Cambridge, Mass.: Harvard University Press, Copyright, 1950, 1957, 1965, by The President and Fellows of Harvard College. Pp. 74–77. Updated material has been supplied in brackets in two places.

wood forms the framework and bulk of the building materials in every house. The frequency of serious earthquakes throughout the land makes the utilization of stone or brick for houses somewhat hazardous. Lightly built wooden houses will sway safely with the motion of the tremor, where more rigid walls of stone or brick crack and fall. . . .

As an industrialized nation, Japan needs huge quantities of wood for railroad ties, telegraph poles, and mine timbers. The petty handicraft uses of wood and bamboo, which must be classed with it as a forest product, are numberless. Short of other materials, the Japanese use wood and bamboo for a thousand things for which we can afford to employ more costly materials. Japanese buckets may not be oaken but they are still made of wood. Simple wooden chopsticks take the place of our more complicated cutlery. Bamboo, an unbelievably versatile material, turns up in hundreds of surprising and ingenious uses. Japan also depends on wood, normally converted into charcoal, for [a substantial proportion of home heating and cooking requirements]. The Japanese, by holding their hands and wrists over charcoal braziers or by placing their feet in or close to specially designed heating devices, derive a considerable amount of warmth and cheer from a tiny spot of heat in an otherwise frigid house. . . .

And over and above these direct uses of forest products is the growing demand for wood pulp. The Japanese have had a famous paper industry for about a thousand years and have developed some ingenious uses for paper. Pasted on sliding partitions and screens, it is an important architectural material. Long ago the Japanese, who suffer a great deal from winter colds, developed the sanitary use of fine paper tissues in place of the handkerchief. As a nation of voracious readers today, they absorb huge quantities of paper for books, magazines, and newspapers. . . . As a great producer of paper as well as rayon, Japan's capacity for pulp consumption is almost unlimited.

Manufacturing Industries and Industrial Resources

Silk manufacturing based on domestic silk was the first Japanese industry to be adapted to large-scale mechanized production during the period of industrialization following 1868. Japanese silk found a ready market in Europe and the United States, and the revenues from silk exports helped greatly in providing the necessary capital for industrial development in general. Mechanized production of cotton textiles based on imported cotton was also introduced very early, and large quantities of cheap cotton goods were soon being marketed among the peasant populations of eastern and southern Asia. Textile production is still of major importance in the industrial structure of Japan, although Japanese industry

has shifted increasingly to the manufacture of metalwares in the twentieth century. The new emphasis has been achieved in spite of a striking poverty of mineral resources. Aside from ordinary grades of building stone, sand, and clay, the only major minerals present in sizable quantities are coal, copper, chromite, manganese, zinc, gold, silver, sulfur, gypsum, and magnesium. And some of these minerals, such as copper and zinc, must be imported in quantity as a supplement to domestic production. About nine-tenths of Japan's total iron-ore requirements are customarily met by imports. Power requirements of the country can be met in part from domestic coal or hydroelectricity, although both sources have been more or less fully exploited. Recently, increasing amounts of crude oil, of which Japan has almost none, have been imported, and petroleum has now become the leading source of energy. Japanese coal, found mainly in Kyushu and Hokkaido, is mostly lowgrade bituminous. It is difficult to mine because of thin and broken seams, and the industry is in the process of modernization and mechanization to make coal more competitive with other energy sources. Economically retrievable reserves are sufficient only for the next few decades at present production rates. They are supplemented by hydroelectricity produced at hundreds of small generating stations along the many short but torrential rivers of Japan. However, a rapid increase in thermal facilities in recent years has reduced the share of electric power generated by hydroelectric stations from 75 to about 40 percent. Costs of power, as well as dependence upon oil imports, has caused Japan to look towards atomic power; the first atomic generating station was completed in 1966 and several more are scheduled for the near future.

Besides a fair base of domestic power resources, adequate for Japanese needs until quite recently, Japan's industry traditionally benefited from a large reserve of rural labor. This situation also has sharply altered, and since 1961 Japan has suffered from labor shortages resulting from the effects of low birth rates and the continuing rapid expansion and manpower requirements of Japanese industry. Historically, the farmers of Japan also provided an early basis for industry in another respect, for much of the original government-sponsored industrialization was financed by revenues from land taxes. Such taxes were often oppressively high, and were an important factor in an increasing rate of farm tenancy and indebtedness after 1868.

The national pattern of small industrial establishments, many of them essentially artisan units, still persists, though there is a continuing trend towards larger factories employing several score to several hundred workers.

The patient, disciplined labor force available to Japanese industries is well illustrated by this photo of necklace manufacture. (Japan National Tourist Organization.)

The nature, significance, and future prospects of Japanese industrial development have been summarized as follows by Reischauer:[5]

Japan's industrial progress . . . has been no steady, inevitable growth, based solidly on obvious economic advantages, as in the case of America's rise to industrial leadership. It has been more like the erratic progress of a broken-field runner, fighting his way against great odds by a quick getaway, brilliant improvisations, and daring reversals of the field. Utilizing to the fullest her one advantage over the West of cheap manpower and her one advantage over Asia of greater industrial skills, she has threaded her precarious way toward industrialization between the far greater industrial potential of the West and the still cheaper labor of the rest of Asia. With her industry-swollen population, there can be no turning back. She cannot even stand still, for her less industrialized neighbors are always threatening to catch up with her and wipe out the technical advantages on which her people now live. . . .

Japanese industry, which has only men and power to support it, will inevitably be forced more and more into the fields in which materials count least and labor and skills count most. Japan must go increasingly into the fields in which her cheaper labor costs give her an advantage over the West while the greater skills of her workers give her the edge over the rest of Asia. This means less emphasis on textiles and a growing emphasis on industrial art products, precision instruments, fine machinery, shipbuilding and ship repairing, and the like, which require a high ratio of skilled labor to materials. But even with the most careful rationalization of her industrial potential, the future of Japanese industry is certainly precarious. Even with determined efforts to maintain technical leadership over the rest of Asia on one side, balanced on the other by a willingness to accept a somewhat lower standard of living than the industrialized peoples of the West, the Japanese have ahead of them, at best, a very uncertain future.

But, whatever may be the ultimate fate of Japanese industry, it has already laid a heavy imprint on Japan and the Japanese. It has made Japan into a land of huge cities and sprawling factories, of whirring machines and crowded commuters' trains.

Despite limitations and uncertainties, Japan's upward push in manufacturing after World War II enabled it to regain prewar output levels by the middle 1950s, and since then the country has risen to a rank commensurate with or exceeding that of West Germany or the United Kingdom with respect to many important industries, though still far below the industrial stature of the United States or the Soviet Union. Japan leads the world in the production of ships, cameras, and radios, ranks third in steel output and electric-power production, and stands very high in many other lines. The country's technically advanced economy is fast gaining a worldwide reputation for quality of workmanship as well as massive and diversified output of goods. It may be noted that the bulk of the production is marketed in Japan itself, where the purchasing power of the average citizen, already far above that of most Asians, is rising very rapidly and providing an expanding internal market

---

[5] Reprinted by permission of the publishers from Reischauer, *ibid.*, pp. 80–81.

for goods. But the manufactures marketed abroad, though only a minor part of the total output, are crucially important to Japan, as they pay for the indispensable raw materials, some foods, and certain types of manufactures that Japan imports. Japan's export trade is worldwide, but with a particularly heavy concentration of sales in the United States (which took in 1966 some 31 percent of Japanese exports by value and provided 28 percent of Japanese imports) and in countries around the eastern and southern margins of Asia.

Though manufacturing is spread through hundreds of cities and towns in Japan, an overwhelmingly large share (well over four-fifths) lies within the country's core region from Tokyo to Nagasaki. The concentration of manufacturing is especially great in the giant industrial nodes of Tokyo-Yokohama and Osaka-Kobe-Kyoto. Each of these areas is extremely diversified, producing iron and steel, machinery, vehicles, miscellaneous metal products, electrical and electronic equipment, ships, textiles, chemicals, refined petroleum products, processed foods, and many other lines of goods. Together the two districts, comprised of their central cities plus numerous suburbs and satellites, account for more than half of Japan's manufacturing by value, and the Nagoya district, directly between the two, accounts for more than a tenth of the national total. Tokyo, Nagoya, and Osaka are interconnected by the efficient New Tokaido Railroad, whose numerous daily trains cover the 320 miles from Tokyo to Osaka in slightly less than 4 hours (one of these trains is shown in photo, below).

The Tokaido Super Express leaving Tokyo Central Station for the run to Osaka. The prominent high-rise structure at the left beyond the station platforms is an office building. (Japan National Tourist Organization.)

KOREA

A discussion of Japan's former colony, Korea, is included in this chapter for convenience and also on the basis of certain instructive similarities between the two countries in location, physical characteristics, and resource position. Korea's history during the twentieth century, first as a colony of Japan and then as a divided land comprised of the separate states of North and South Korea, has tended to obscure its distinctive culture and contributions to the world. Although the Koreans have been strongly influenced by Chinese culture, and to a lesser extent by Japanese, they are racially and linguistically a separate people representing a distinct variant of East Asian civilization.

Liabilities of Korea's Geographical Position

Korea is a country with an unfortunate geographical position. This relatively small country is surrounded by larger and more powerful neighbors that have frequently been at odds with one another and with the Koreans. Korea adjoins China along a land frontier which follows the Yalu and Tumen rivers (map, p. 490). It faces Japan across the Korea Strait, which is only about 120 miles wide. In the extreme northeast Korea borders the Soviet Union for a short distance.

For many centuries the Korean peninsula has served as a bridge between Japan and the Asian mainland. From an early time both China and Japan, the latter more intermittently, have been interested in controlling this bridge, and since Korea has generally been weaker than either, its history has usually been that of a subject or vassal state. However, from the late seventh century to the twentieth, Korea was a unified state, sometimes invaded, ravaged, and forced to pay tribute, but never destroyed as a political entity. As noted above, Chinese influence has been the most pronounced and continuous throughout Korean history, with Japanese contacts limited mainly to occasional pirate raids and the establishment of a few Japanese enclaves in the southeastern port cities. The decline of Chinese power in the nineteenth century and the concurrent rise of modern Japan greatly increased the influence of the latter, and at the same time Russia began to make itself strongly felt in eastern Asia. Korea became an object of contention among these powers. Japan emerged victorious in the struggles that followed, ousting the Chinese from Korea, as well as from certain other areas, in the Sino-Japanese War of 1894–1895, and ousting Russia from Korea and southern Manchuria in the Russo-Japanese War of 1904–1905. From 1905 until 1945 Korea was firmly under Japanese control, being formally annexed to the Japanese Empire in 1910.

Japan lost Korea along with the rest of its empire when it surrendered to the Allies, ending World War II, in 1945. In accordance with prearranged agreements Russian forces occupied Korea north of the 38th parallel and American forces south of that line. Although it had been understood that Korea was to become a unified and independent country, no agreement could be reached between the occupying powers as to the establishment of a Korean government. Accordingly, separate governments were set up in north and south Korea under the aegis of the respective occupying powers: in the south, the Republic of South Korea, under the auspices of the United Nations; and, in the north, the Democratic People's Republic of Korea, which became a Communist satellite. The occupying powers withdrew the bulk of their forces in 1948 and 1949.

In 1950 North Korea attacked South Korea in force, and United Nations units, mainly American, entered the peninsula to repel the aggression. In the latter part of that year, when the North Koreans had been driven back from the southern part of the peninsula almost to the Manchurian border, Communist China entered the war and drove the United Nations forces south of the 38th parallel. Then the Chinese and North Koreans were in turn driven back slightly north of the parallel, where a stalemate developed while truce talks were under way from July 1951 until an armistice was arranged in the summer of 1953.

Few lands have ever been more devastated than Korea after several years of warfare covering the length and breadth of the peninsula. The tragedy was all the greater in that Korea is not a poor land by nature. During their period of control the Japanese developed transportation, agriculture, and industry, the latter from a sizable base of mineral and power resources. The industrial structure was to some extent integrated with that of southern Manchuria. The Korean people received few benefits from these developments, however, since the increased production was put mainly to Japanese uses. After 1945 the Korean economy was seriously disorganized and handicapped by the division between north and south, and after 1950 by the enormous physical destruction. The physical scars of the Korean conflict, however, have been erased, and the economies of both the North and South Korean governments rehabilitated and expanded with the help of considerable outside aid.

A Brief Geographical Sketch of Korea

With a total area of some 85,000 square miles and an estimated population (1968) of 43.7 million, Korea has three-fifths the area of Japan but only two-fifths as many people. Like Japan, Korea is predominantly mountainous or hilly, although the proportion of level lowland, found mainly along the west side of the peninsula facing the Yellow Sea, is a little greater than in Japan. On the east or Sea of Japan side, mountains rise steeply from a very narrow coastal plain. The highest elevations are found in the northeast, adjoining Manchuria.

The long axis of the Korean peninsula is oriented north-south, and thus Korea, like Japan, exhibits a climatic contrast between a continental north with severe winters and a relatively short growing season, and a subtropical south with milder winters and a longer frost-free season. However, because of its closer proximity to the continental mass of Asia, southern Korea has somewhat colder winters and a shorter growing season than southern Japan. Nevertheless, the climate of South Korean lowlands permits double cropping, with the main crops being irrigated rice, grown in the summer, and barley, planted in the fall on the drained rice fields or other dry fields. Korea's precipitation, as in China north of the Yangtze, is sharply concentrated in the June–September period, with consequent problems of water control and storage. The climate in most parts of North Korea is too severe for winter grains, and the growing of irrigated rice becomes increasingly less important toward the north—a fact related not only to the shorter growing season but also to the smaller acreages of level land and the lower precipitation of the north. However, in North Korea irrigated rice accounts for a greater total production than does any other single crop, and it accounts for at least half the total output of grains. Of the dry-field grains, corn is the most important, and it has been pushed by the North Korean government at the expense of wheat, barley, and millet—now comparatively minor in output. Vegetables and hardy fruits, particularly apples and pears, are grown extensively throughout the peninsula. Other important crops are cotton and sweet potatoes, grown mainly in South Korea, and white potatoes, restricted mainly to the north. Thus the general situation with regard to distribution of crops between north and south is somewhat similar to that of Japan, although the variety of agricultural production in South Korea is less than that of southern Japan.

Korea is less industrialized than Japan, and about two-thirds of its people gain a livelihood directly from the soil. The proportion of cultivated land is a little greater than in Japan, but nevertheless amounts to only a fifth of the total area of the peninsula. Farming methods are less advanced, less fertilizer is used, and the average yields are significantly lower than in Japan. With its more favorable climate for agriculture and larger proportion of level lowland, South Korea supports nearly two and a half times the population of North Korea on a considerably smaller total area. Estimated areas and populations in mid-1968 were: South Korea, 37,000 square miles, population 30.7 million; North Korea, 48,000 square miles, population 13 million.

Although South Korea is the more productive area agriculturally, North Korea is the more important industrially. Most of the iron and steel, chemicals, hydroelectric power, and forest manufactures of Korea are produced in the north. Industry in South Korea is confined largely to cotton textiles and processed foods, but other industries, particularly chemical fertilizer production and petroleum refining, have risen to considerable importance in recent years. The greater industrialization of North Korea is related to its larger endowment of industrial resources, including high-grade coal, much of it anthracite, substantial reserves of iron ore, most of the country's hydroelectric power potential, and the most valuable forests of Korea, principally coniferous species growing on the higher lands adjoining Manchuria. In contrast, the mineral output of South Korea, comprised primarily of anthracite coal, tungsten, and graphite, probably represents no more than a third the value of that in the North. However, some industrial progress is being made in South Korea, and a sizable iron and steel complex is under construction on the southeastern coast about 50 miles north of Pusan, with completion scheduled for the early 1970s.

The South Korean capital of Seoul (3.8 million) is the largest city of Korea. Other important cities of South Korea include the seaport of Pusan (1.5 million) on the Korea Strait, the inland textile manufacturing and agricultural market center of Taegu (850,000), and the port of Jinsen (Inchon; 500,000), located on the Yellow Sea 20 miles west of Seoul. Pyongyang (over 700,000) is the capital, largest city, and most important industrial center of North Korea.

North and south Korea were highly complementary in resources and production prior to partitioning, and this initially posed very difficult economic problems when the two were severed. In response to this situation both have attempted, with at least partial success, to develop more balanced economies. In North Korea a typical Communist-run economy is present, with the usual plans and targets, collectivization of agriculture, and emphasis on heavy industrial production–though some diversion of resources to light industry has taken place in recent years. Expansion of agricultural land in South Korea has about reached its limits, and to feed its fast-growing population the south is placing more emphasis on use of chemical fertilizers and increased irrigation and construction of additional reservoirs to guard against drought. In addition it is probable that stress will be laid on expanding the size of farm holdings as a means of increasing rural capital and encouraging the development of commercial agriculture. Increased production of traditional manufac-tures such as textiles and light consumer products is continuing, and there is emphasis on such newer industries as petrochemicals and chemical fertilizers. South Korea's low labor costs enable it to import bulky raw materials, such as logs, and process them into export products of higher value, such as plywood.

In spite of more than two decades of political and economic separation, however, the Koreans–north and south of the 38th parallel–are one people sharing a common culture and historical traditions, and future demands for reunification may be anticipated.

REFERENCES and READINGS

JAPAN

General, Historical, Political, and Cultural Geography

BORTON, HUGH, *Japan's Modern Century* (New York: The Ronald Press Company, 1955).

DULLES, FOSTER RHEA, *Yankees and Samurai: America's Role in the Emergence of Modern Japan, 1791–1900* (New York: Harper & Row, 1965). An informal history, emphasizing personalities.

EYRE, JOHN D., "Japanese-Soviet Territorial Issues in the Southern Kurile Islands," *Professional Geographer,* 20, no. 1 (January 1968), 11–15.

FISHER, CHARLES A., "The Expansion of Japan: A Study in Oriental Geopolitics: Part I, Continental and Maritime Components in Japanese Expansion; Part II, The Greater East Asia Co-Prosperity Sphere," *Geographical Journal,* 115, nos. 1–3 (January–March 1950), 1–19, and nos. 4–6 (April–June 1950), 179–193.

INTERNATIONAL GEOGRAPHICAL UNION, *Proceedings of IGU Regional Conference in Japan, 1957* (Tokyo: The Organizing Committee of IGU Regional Conference in Japan; The Science Council of Japan, 1959); contains numerous papers on Japanese geography. And, with Science Council of Japan, *Regional Geography of Japan,* 6 vols. (guidebook for field trips of the IGU Regional Conference in Japan, 1957; published in cooperation with the Society of Japanese Regional Geography and the Japanese Ministry of Education; Tokyo: Kokon Shoin, 1957).

"The Japanese Empire," *Fortune,* 14, no. 3 (September 1936); and "Japan and the Japanese," *Fortune,* 29, no. 4 (April 1944). Two complete issues, still valuable for perspective on Japan before and during World War II.

JAPANESE NATIONAL COMMISSION FOR UNESCO, comp., *Japan: Its Land, People, and Culture* (rev. ed.; Tokyo: Ministry of Finance, 1964). An encyclopedic reference work.

KIDDER, J. EDWARD, *Japan before Buddhism* (Ancient Peoples and Places; London: Thames and Hudson, 1959).

KUBLIN, HYMAN, "Japan," *Focus,* 14, no. 6 (February 1964), 6 pp.

MARAINI, FOSCO, *Meeting with Japan,* trans. from the Italian by Eric Mosbacher (New York: The Viking Press, 1959). Informal, personalized descriptions of Japanese life, with numerous photographs.

McCUNE, SHANNON, "The Ryukyu Islands," *Focus,* 15, no. 7 (March 1965), 6 pp.

OLSON, LAWRENCE, *Japan Today and Tomorrow* (Headline Series, No. 181; New York: Foreign Policy Association, 1967).

PELZER, KARL J., "Japanese Migration and Colonization," in Isaiah Bowman, ed., *Limits of Land Settlement* (New York: Council on Foreign Relations, 1937), pp. 155-195.

PITTS, FORREST R., "A Mirror to Japan," *Landscape,* 9, no. 3 (Spring 1960), 24-28.

REISCHAUER, EDWIN O., *The United States and Japan* (American Foreign Policy Library, 3d ed.; Cambridge, Mass.: Harvard University Press; New York: The Viking Press, Compass Books, 1965). Also, "Japan Is One of the Biggest Countries in the World," *New York Times Magazine,* October 16, 1966, pp. 34-35 ff.; and "Our Dialogue with Japan," *Foreign Affairs,* 45, no. 2 (January 1967), 215-228.

SANSOM, G. B., *The Western World and Japan: A Study in the Interaction of European and Asiatic Cultures* (New York: Alfred A. Knopf, 1950; fifth printing, 1965).

TAEUBER, IRENE B., *The Population of Japan* (Princeton, N.J.: Princeton University Press, 1958); and, "Japan's Population: Miracle, Model, or Case Study," *Foreign Affairs,* 40, no. 4 (July 1962), 595-604.

TAKAKURA, SHIN'ICHIRO, "Vanishing Ainu of North Japan," *Natural History,* 75, no. 8 (October 1966), 16-25.

Teikoku's Complete Atlas of Japan (Tokyo: Teikoku Shoin, 1964). Inexpensive atlas in English, with good general, economic, and urban maps.

TREWARTHA, GLENN T., *Japan: A Geography* (Madison and Milwaukee: University of Wisconsin Press, 1965). A standard geography of Japan: careful, thorough, and authoritative.

U.S. Army Area Handbook for Japan (Department of the Army, Pamphlet No. 550-30; Washington, D.C.: Government Printing Office, 1964).

YOSHIDA, SHIGERU, "Japan's Decisive Century," *Britannica Book of the Year 1967* (Chicago: Encyclopaedia Britannica, Inc., 1967), pp. 17-48.

Resources and Economy: General

ACKERMAN, EDWARD A., *Japan's Natural Resources and Their Relation to Japan's Economic Future* (Chicago: University of Chicago Press, 1953). A massive, authoritative survey, outdated in some respects, but still valuable.

ALLEN, G. C., *Japan's Economic Expansion* (New York: Oxford University Press, 1965).

DEMPSTER, PRUE, *Japan Advances: A Geographical Study* (London: Methuen and Co.; New York: Barnes and Noble, 1967). Emphasizes recent economic expansion.

HALL, ROBERT B., JR., *Japan: Industrial Power of Asia* (Princeton, N.J.: D. Van Nostrand Co., Searchlight Books, 1963).

HOLLERMAN, LEON, "Japan's Place in the Scale of Economic Development," *Economic Development and Cultural Change,* 12, no. 2 (January 1964), 139-157.

LOCKWOOD, WILLIAM W., *The Economic Development of Japan: Growth and Structural Change, 1868-1938* (Princeton, N.J.: Princeton University Press, 1954).

MIYAMOTO, MATAJI, and Others, "Economic Development in Preindustrial Japan, 1859-1894," *Journal of Economic History,* 25, no. 4 (December 1965), 541-564.

ORCHARD, JOHN E., "Industrialization in Japan, China Mainland, and India—Some World Implications," *Annals of the Association of American Geographers,* 50, no. 3 (September 1960), 193-215.

SMITH, THOMAS C., *The Agrarian Origins of Modern Japan* (Stanford, Calif.: Stanford University Press, 1959).

U.S. BUREAU OF INTERNATIONAL COMMERCE, *Basic Data on the Economy of Japan,* Overseas Business Reports, OBR 66-83 (December 1966). (Washington, D.C.: Government Printing Office, 1966.)

WAYS, MAX, "Why Japan's Growth Is Different," *Fortune,* 76, no. 6 (November 1967), 127-129 ff.

Rural Life, Agriculture, Forestry, and Fisheries

BEARDSLEY, RICHARD K., JOHN W. HALL, and ROBERT E. WARD, *Village Japan* (Chicago: University of Chicago Press, 1959). A detailed study of a village considered to be generally representative of rural Japan.

COMITINI, SALVATORE, "Marine Resources Exploitation and Management in the Economic Development of Japan," *Economic Development and Cultural Change,* 14, no. 4 (July 1966), 414–427.

DORE, R. P., *Land Reform in Japan* (New York: Oxford University Press, 1959).

EMBREE, JOHN F., *Suye Mura: A Japanese Village* (Chicago: University of Chicago Press, 1939). A classic study of traditional village life.

EYRE, JOHN D., "Japanese Inter-Prefectural Rice Movements," *Economic Geography,* 38, no. 1 (January 1962), 78–86; "Mountain Land Use in Northern Japan," *Geographical Review,* 52, no. 2 (April 1962), 236–252; "Sources of Tokyo's Fresh Food Supply," *Geographical Review,* 49, no. 4 (October 1959), 455–474; and "Water Controls in a Japanese Irrigation System," *Geographical Review,* 45, no. 2 (April 1955), 197–216.

"Fishing in Japan," Chap. 14 in Richard M. Highsmith, Jr., ed., *Case Studies in World Geography: Occupance and Economy Types* (Englewood Cliffs, N.J.: Prentice-Hall, 1961), pp. 95–101.

"Food Supply in Japan," *Foreign Agriculture,* 5, no. 39 (September 25, 1967), 3–17. A special issue.

FUKUTAKE, TADASHI, *Japanese Rural Society,* trans. by R. P. Dore (New York: Oxford University Press, 1967).

GLACKEN, CLARENCE J., *The Great Loochoo: A Study of Okinawan Village Life* (Berkeley: University of California Press, 1955).

HALL, ROBERT B., JR., "Hand Tractors in Japanese Paddy Fields," *Economic Geography,* 34, no. 4 (October 1958), 312–320.

HIYAMA, YOSHIO, "Mother-Ship Whaling Operations," Chap. 13 in Richard S. Thoman and Donald J. Patton, eds., *Focus on Geographic Activity: A Collection of Original Studies* (New York: McGraw-Hill Book Company, 1964), pp. 75–78.

HOUGH, RICHARD F., "Impact of the Decline in Raw Silk on the Suwa Basin of Japan," *Economic Geography,* 44, no. 2 (April 1968), 95–116; and "Impact of the Decline of Raw Silk on Two Major Cocoon-Producing Regions in Japan," *Annals of the Association of American Geographers,* 58, no. 2 (June 1968), 221–249.

JOHNSTON, BRUCE F., "Agriculture and Economic Development: The Relevance of the Japanese Experience," Stanford University, *Food Research Institute Studies,* 6, no. 3 (1966), 251–312; and "Agricultural Development and Economic Transformation: A Comparative Study of the Japanese Experience," *Food Research Institute Studies,* 3, no. 3 (November 1962), 223–276.

KAKIUCHI, GEORGE H., "Recent Developments and Trends in the Cultivation of Wet Rice in Japan," *Land Economics,* 41, no. 1 (February 1965), 66–73; "Early-Season Cultivation of Wet-Rice in Southwestern Japan," *Pacific Viewpoint,* 5, no. 1 (May 1964), 51–68; "Recent Developments in the Cultivation of Wet-Rice in Northeastern Japan," *Journal of Geography,* 63, no. 4 (April 1964), 155–161; "Toji: A Study of a Semiagricultural Community Exploiting Marine Resources," *Papers of the Michigan Academy of Science, Arts, and Letters,* 47 (1961), 507–515; and, with Setsutaro Murakami, "Satsuma Oranges in Ocho-mura: A Study of Specialized Cash Cropping in Southwestern Japan," *Geographical Review,* 51, no. 4 (October 1961), 500–518.

LADEJINSKY, WOLF, "Agrarian Revolution in Japan," *Foreign Affairs,* 38, no. 1 (October 1959), 95–109.

MANCHESTER, CURTIS A., "Igusa: A Critical Cash Crop in the Rural Economy of Okayama Prefecture," *Economic Geography,* 34, no. 1 (January 1958), 47–63.

MATSUO, TAKANE, *Rice and Rice Cultivation in Japan* (Tokyo: Institute of Asian Economic Affairs, 1961).

NISHI, MIDORI, "Regional Variations in Japanese Farmhouses," *Annals of the Association of American Geographers,* 57, no. 2 (June 1967), 239–266.

OKAZAKI, AYAAKIRA, *Forestry in Japan* (Corvallis: Oregon State University, School of Forestry, 1964).

THOMPSON, JOHN H., "Urban Agriculture in Southern Japan," *Economic Geography,* 33, no. 3 (July 1957), 224–237.

Manufacturing, Commerce, and Cities

DAVENPORT, JOHN, "Japan's Competitive Cutting Edge," *Fortune,* 78, no. 3 (September 1968), 90–95 ff.

EYRE, JOHN D., "Japan's Electric-Power Supply," *Geographical Review,* 55, no. 4 (October 1965), 546–562; also "Industrial Growth in the Suwa Basin, Japan," *Geographical Review,* 53, no. 4 (October 1963), 487–502; and "Tokyo Influences in the Manufacturing Geography of Saitama Prefecture," *Economic Geography,* 39, no. 4 (October 1963), 283–298.

FISHER, CHARLES A., "Japan's Great Cities: Nearing 'Standing Room Only,'" *Geographical Magazine,* 39, no. 7 (November 1966), 554–565.

GINSBURG, NORTON S., *Japanese Prewar Trade and Shipping in the Oriental Triangle,* University of Chicago, Department of Geography Research Paper No. 6 (Chicago: 1949).

HALL, PETER, *The World Cities* (London: Weidenfeld and Nicolson, World University Library, 1966), Chap. 8, "Tokyo," pp. 217–233.

NAKAYAMA, SADAYOSHI, "Japan's Phenomenal Shipbuilders," *United States Naval Institute Proceedings,* 92, no. 8 (August 1966), 27–39.

THOMPSON, JOHN H., "Manufacturing in the Kita Kyushu Industrial Zone of Japan," *Annals of the Association of American Geographers,* 49, no. 4 (December 1959), 420–442; and, with Michíhiro Miyazaki, "A Map of Japan's Manufacturing," *Geographical Review,* 49, no. 1 (January 1959), 1–17 (includes a very useful folded map).

KOREA

FISHER, CHARLES A., "The Role of Korea in the Far East," *Geographical Journal,* 120, pt. 3 (September 1954), 282–298.

GLICK, NORMAN D., *Basic Data on the Economy of the Republic of Korea,* U.S. Bureau of International Commerce, Overseas Business Reports, OBR 68–37 (June 1968). (Washington, D.C.: Government Printing Office, 1968).

McCUNE, SHANNON, *Korea's Heritage: A Regional and Social Geography* (Rutland, Vt.: Charles E. Tuttle Co., 1956); also, *Korea: Land of Broken Calm* (The Asia Library; Princeton, N.J.: D. Van Nostrand Co., 1966); and "Korea: Geographic Parallels, 1950–1960," *Journal of Geography,* 59, no. 5 (May 1960), 201–206.

TAYLOR, ALICE, "Korea," *Focus,* 12, no. 4 (December 1961), 6 pp.

U.S. Army Area Handbook for Korea, 2d ed. (Department of the Army Pamphlet No. 550–41; Washington, D.C.: Government Printing Office, 1964).

See also the reference lists for Chapters 1, 2, 3, and 17, especially the list of standard texts on Asia, p. 415.

PART VI

THE PACIFIC WORLD

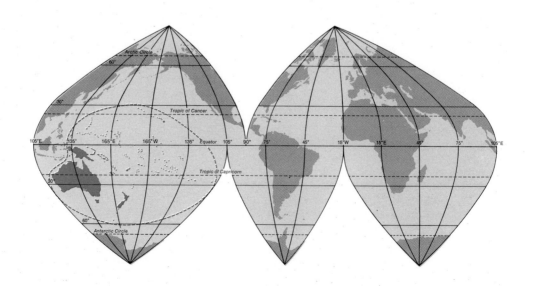

The Pacific
Islands

22   The Pacific world is mostly water. The Pacific itself, the largest of the oceans, is bigger than all the continents and islands of the world put together. But it has been a relative backwater of world commerce. Major shipping lanes reach mainly around the margins, and the products afforded by the multitudinous islands and the sparsely populated island continent of Australia do not bulk very large in world terms.

Prior to World War II the Western world had built up a legend about the Pacific and its islands as a kind of utopia. At an early time some of the islands did have an idyllic and appealing quality for visitors from Europe and America. However, on many islands the native peoples were reduced in numbers and their cultures disrupted by a long period of unrestrained exploitation on the part of traders, whaling crews, labor agents ("blackbirders"), and other opportunists before the outside governments claiming jurisdiction began to impose tardy measures of regulation. On many islands, also, whatever idyllic quality remained was shattered by the military actions of World War II.

As here conceived, the term "Pacific world" is generally restricted to Australia, New Zealand, and the islands of the mid-Pacific lying mostly between the Tropics. The Pacific islands bordering the mainland of the Orient, the Soviet Union, and

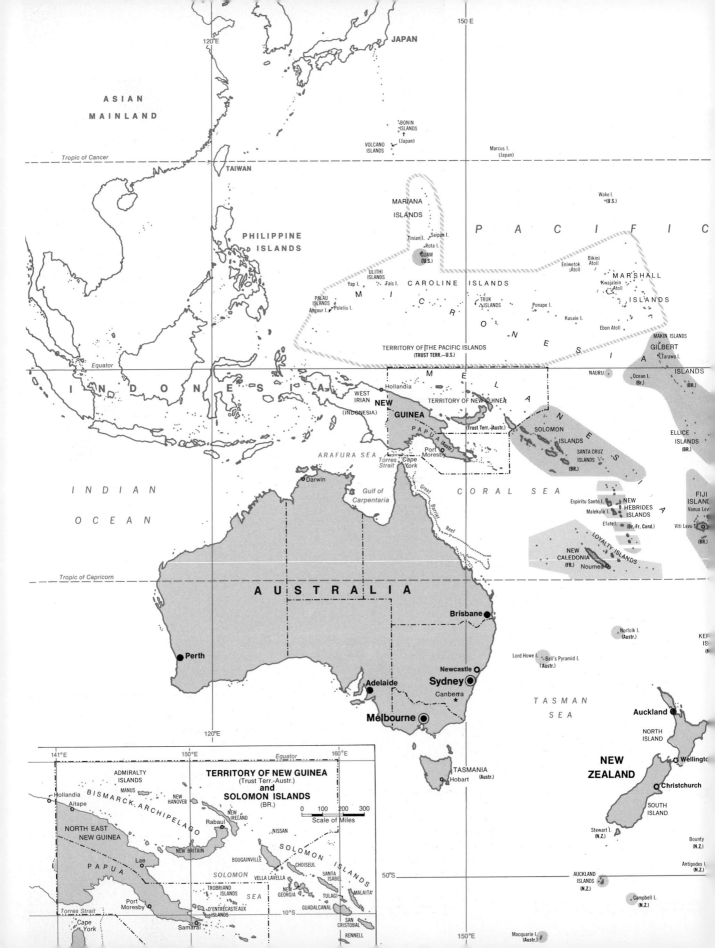

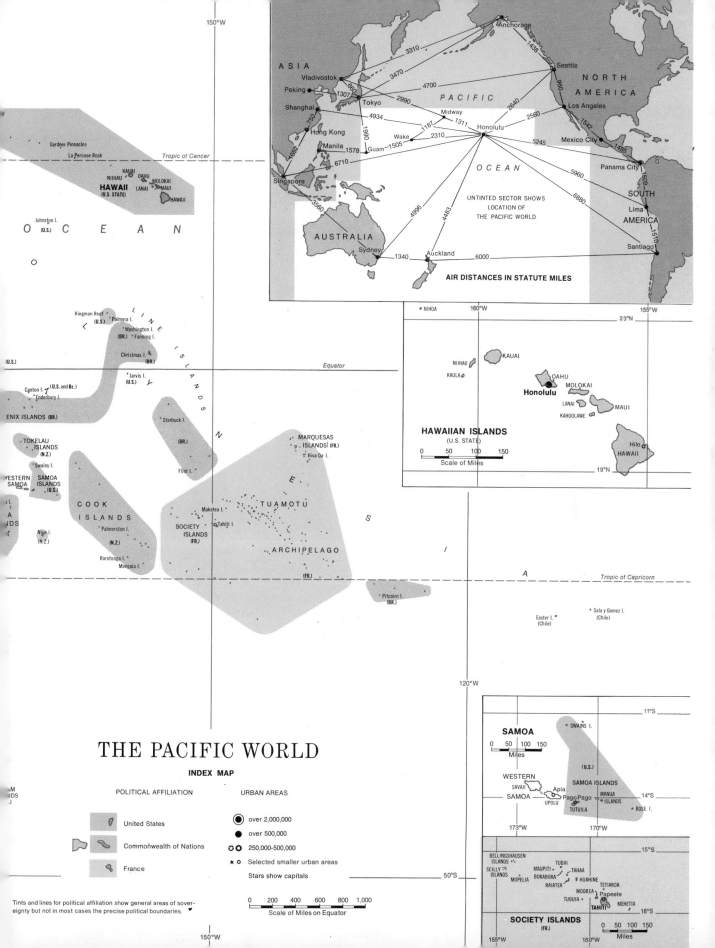

the Americas are excluded on the basis of their close ties with the adjoining continents, while large areas of the eastern and northern Pacific which contain few islands and are seldom visited are largely discounted. Because Australia and New Zealand are sufficiently different from the tropical island realms to the north, they could be considered a separate world region, but are included here in the Pacific world on the basis of their strong political and economic interest in the tropical islands, the ethnic affiliations of their original inhabitants with the peoples of those islands, and their insular character.

MELANESIA, MICRONESIA, POLYNESIA

The Pacific islands are commonly divided, mainly on an ethnic basis, into three principal realms: Melanesia, Micronesia, and Polynesia (map, pp. 514–515). The islands of *Melanesia* (Greek: "black islands"), bordering Australia, are relatively large. New Guinea, the largest, is about 1500 miles long and 400 miles across at the broadest point. In general, these islands are hot, damp, mountainous, and overgrown with dense vegetation (photo, p.

TABLE 17 SUMMARY OF MAIN PACIFIC ISLAND GROUPS

ISLAND REALMS	MAIN ISLAND GROUPS	WELL-KNOWN INDIVIDUAL ISLANDS OR ATOLLS	CONTROLLING POWERS
Melanesia		New Guinea	Australia, Indonesia
	Bismarck Archipelago	New Britain	Australia
		New Ireland	
	Solomon Islands	Guadalcanal	Britain, Australia
		Bougainville	
	New Hebrides Islands	Espiritu Santo	British-French condominium
	Fiji Islands	Viti Levu	Britain
		Vanua Levu	
	Loyalty Islands		France
		New Caledonia	France
Micronesia		Nauru	Nauru (see p. 524)
		Ocean	Britain
	Caroline Islands	Kusaie	United States
		Ponape	
		(Palau, Truk, and Yap are noteworthy island groups)	
	Mariana Islands	Guam	United States
		Saipan	
	Marshall Islands	Bikini	United States
		Eniwetok	
	Gilbert Islands	Tarawa	Britain
Polynesia	Hawaiian Islands	Oahu	United States
		Hawaii	(U.S. state)
		Maui	
	Ellice Islands		Britain
	Tonga Islands		Britain
	Samoa Islands		United States, Western Samoa (see p. 524)
	Marquesas Islands		France
	Tuamotu Archipelago		France
	Society Islands	Tahiti	France
	Cook Islands		New Zealand (controls foreign relations only; see p. 524)

519). *Micronesia* (Greek: "tiny islands") includes hundreds of small islands scattered about in the central and western Pacific north of the equator. Prior to World War II, Micronesia was held by Japan under a mandate from the League of Nations, except for the Gilbert Islands, which were a British possession, and Guam, a United States possession. Since the war the former Japanese islands have become the Pacific Islands Trust Territory, administered by the United States. *Polynesia* (Greek: "many islands") occupies a greater expanse of ocean than does either Melanesia or Micronesia. It is shaped like a rough triangle, with the corners at New Zealand, the Hawaiian Islands, and remote Easter Island.

Each of the three island realms contains a number of distinct island groups. The major island groups, together with some of the more noteworthy individual islands in the groups and the nations exercising political control, are indicated in Table 17.

THE NATIVE PEOPLES

It is commonly accepted among anthropologists that the original inhabitants reached the Pacific world from the mainland of Asia, the first migrants arriving in Melanesia and Australia thousands of years ago.[1] Four successive racial stocks are distinguished among the migrants into the islands: (1) Negritos, a short, dark, and frizzy-haired people; (2) Ainoids, a generally Caucasoid people, although excessively hairy, with lighter skins than the Negritos, and wavy hair; (3) Veddoids, dark-skinned and wavy-haired; (4) Indonesians,[2] a people of mixed Caucasoid and Mongoloid stock with light brown skin, black wavy hair, and medium to short stature. The Indonesian peoples, possibly intermixing with earlier immigrants, settled Micronesia and probably Polynesia, as well as mixing with other peoples in the more complex amalgam of Melanesia.

Over a long period the Micronesians and Polynesians became relatively distinct from each other and relatively homogeneous throughout their separate spheres. The people of each area came to speak dialects of a single basic language and to share many features of a common culture. In Melanesia, on the other hand, complex and varied mixing of successive migrant groups has led to an extremely confused racial and cultural pattern. Racially the Negroid element is generally dominant, though often diluted by other elements. A variety of languages and dialects are found, and anthropologists have distinguished literally hundreds of fairly distinct cultures. Some of the tribes, especially in the remote interiors of the larger islands, are extremely primitive.

TYPES OF ISLANDS

The almost countless islands scattered across the Pacific vary widely in size, type, and utility. Three main categories of islands are now recognized according to origin: coral islands, formed of the skeletons and living bodies of small marine organisms; volcanic islands, formed by submarine eruptions; and continental islands. These major categories, grouped into further subclasses, have been described as follows by an American anthropologist, Douglas L. Oliver, in an outstanding book on the Pacific islands:[3]

From man's point of view, physical setting consists of several elements: the form of the land and the nature of the soil, the quality of the climate, the types and dispositions of plants, and the presence of animal life. In Oceania these components are variously combined to produce seven major types of islands, differentiated according to the limits they imposed upon the islanders and their primitive technologies:

1. *Treeless Atolls or Coral Islands* (typical examples are Canton, Johnston, and Howland islands). Grass and herbs provide survival food for castaways and small parties of fishermen, but the lack of good soil and drinking water renders these land specks unsuitable for permanent native settlements. Sea birds and fish abound and, if drinking water were present, human life could be sustained indefinitely on these islands; but for the most part the vegetation-

---

[1] There is some evidence that at least a portion of the early inhabitants of Polynesia may have migrated there from the Americas. A spectacular attempt to lend credence to this theory was the voyage of the raft *Kon-Tiki*, navigated by Thor Heyerdahl and five companions from Peru to the Tuamotu Islands in 1947. An account of this journey was published by Thor Heyerdahl in *Kon-Tiki: Across the Pacific by Raft* (Chicago: Rand McNally & Co., 1950).

[2] This anthropological term should not be confused with the same term now used in a political sense to refer to citizens of the country of Indonesia.

[3] Reprinted by permission of the publishers from Douglas L. Oliver, *The Pacific Islands.* Cambridge, Mass.: Harvard University Press, Copyright 1951 by The President and Fellows of Harvard College. Pp. 12–14. Italicized portions numbered and capitalized in editing.

loving Oceanic natives have avoided them. On the other hand, the very conditions which discouraged native settlers and made these islands bird refuges resulted in the deposit of valuable guano fertilizer and produced ready-made runways for twentieth-century aircraft.

2. *Dry-forest Atolls or Coral Islands* (typical are most of the Marshall and Ellice islands, and many of the Tuamotus and northern Cooks). Strand flora, together with some arable soil and fresh water, make these islands habitable for limited numbers of people; rich marine resources supplement the scanty vegetable and fruit diets. However, overpopulation is a constant threat, and crowded native communities have sought solutions in mass emigration and in infanticide. Yet despite the paucity of resources on these islands, some communities have developed intricate and aesthetically pleasing arts and crafts, utilizing every material at their disposal. They have, moreover, appeared to compensate for the limitations imposed upon their economies by evolving exceedingly complex institutional relations.

3. *Luxuriant, Moist Atolls or Coral Islands* (typical examples are the Gilbert and Tokelau islands, Swains Island, Fanning, Ebon, Ulithi, and Nissan). These are the lush and beautiful "tropical isles" of romantic novel and motion picture. They have sufficient soil to support cultivation of taro and bananas, in addition to the coconut palms and breadfruit grown also on the dry islands.

4. *Raised-coral Islands* (typical examples are Makatéa, Niue, Ocean, Nauru, Angaur, Peleliu, and Fais). These islands were formed out of successive elevations of old coral reef and are composed of limestone covered by thin layers of soil which support low and dense stands of dryland flora (medium-sized trees, small trees, shrubs, vines, ferns, and so forth). Rough pitted limestone surfaces and tangled growth make these islands difficult to cross. Fresh water in the form of springs and streams is usually lacking, and cave pools or rain-catchment basins provide drinking water. Coconuts and other food plants grow fairly well in the scattered pockets of soil, but frequent droughts are a menace to native agriculture. Since many of these islands contain rich phosphate deposits, they have assumed an economic value far surpassing that of any other kind of island of comparable size.

5. *Unweathered Volcanic Islands* (typical examples are the northern Marianas and Niuafo'ou). Because weathering of these islands has not progressed far, there is considerably less soil on them, and hence less vegetation, than on the "older" weathered volcanic islands. Most forest growth is limited to the valleys and the ridges are covered with grasses and smaller plants. Native populations can and do thrive in moderate numbers on these islands; usually the little soil present is rich and supports luxuriant growths of coconuts and other food plants.

6. *Weathered Volcanic Islands* (typical examples are the Hawaiian, Society, and Samoan islands). These islands, some of them rising thousands of feet above sea level, contain many kinds of environment: wide strands, brackish and fresh water swamps, gentle and steep slopes, extensive lava fields, and so forth. They are subjected to great variation in rainfall. Plant life is extremely varied. Nearly every kind of growing thing or growing condition needed by man to develop complex technologies is to be found. Minerals alone are lacking.

Certain "mixed" islands, including Guam, Saipan, Rota, and Mangaia, possess characteristics of both weathered volcanic and raised-coral settings.

7. *"Continental" Islands* (New Guinea, Viti Levu, New Caledonia, Espiritu Santo, Guadalcanal, Bougainville,

A characteristic scene of the Pacific island world. The low coral islands in the view, located near New Guinea in the southwest Pacific, comprise an *atoll*. The islands and coral reefs surround a lagoon, visible beyond the large island in the center foreground. Coral reefs bordering the atoll are outlined by the white water. Countless numbers of atolls exist in the central and western Pacific. (Australian News and Information Bureau.)

Some of the wilder aspects of Melanesia are portrayed in this view from Papua, the southeastern quarter of the large island of New Guinea. Note the luxuriant, tangled growth of the tropical rain forest rising above the long thatched hut occupied by members of an entire village. (Australian News and Information Bureau.)

New Britain, New Ireland, and most other large islands in the Melanesian archipelagoes). These islands are formed of continental rocks complicated by volcanic intrusions and possess even wider varieties in environment and plant life than the volcanic islands. Their richer, mineral-bearing soils support nearly every kind of vegetation. Their high mountains, dense forests, and broad swamps have encouraged the existence of numerous isolated native communities, favoring the development of cultural and even racial diversity.

"High" and "Low" Islands

A somewhat simpler distinction is commonly made between the "high" and "low" islands of the Pacific.

Some of the characteristics of these two types of islands are elaborated in the following selection by an American geographer, the late Otis W. Freeman:[4]

In the Pacific one of the most distinctive things . . . is the difference between "high" and "low" islands. The "high" islands, mostly of volcanic origin, abound in food as compared with the "low" islands of coral origin. The mountainous islands are large in proportion to the land area of coral atolls, have a variety of trees, plenty of water, fertile soil, and can support considerable populations. The low islands have few plants except the coconut palm, little or no drinking water, scanty soil, and each coral isle can support few inhabitants. . . .

The major seaports of the mid-Pacific are located on the larger islands. No city of real size and importance

[4] Otis W. Freeman, "The Pacific Island World," *Journal of Geography,* 44 (1945), 22–28. Used by permission of the author and the *Journal of Geography.* City populations added in editing. Some supplementary and updated material supplied in brackets.

Fishing is a basic means of livelihood throughout the Pacific Islands. This view shows fishermen in the Trobriand Islands, east of New Guinea, arriving at a beach where part of their catch will be bartered for yams (center foreground) raised by inland villagers. Note the characteristic outrigger canoes. (Australian News and Information Bureau.)

is built on a coral atoll, although such localities may serve as . . . air and sea bases. Only Honolulu [575,000] and [four] cities in New Zealand exceed 100,000 in population. Hilo [26,000] on Hawaii, Noumea [40,000] in New Caledonia, Suva [55,000] on Fiji, Pepeete [30,000] on Tahiti, and Apia [25,000] on Samoa are smaller cities of importance as seaports and commercial centers. [Population figure for Honolulu is a 1968 estimate for the metropolitan area (see p. 714); figures for the other cities are estimates for the city proper.] The high islands in the trade winds are generally healthful and a considerable tourist trade exists in Hawaii and to a less extent in Fiji and Tahiti. Health conditions in the wet tropical islands of the Solomons and New Hebrides are not so favorable, . . . many . . . diseases being prevalent. . . .

On the volcanic islands large sugar plantations [were] developed by the Americans on Hawaii, the British at Fiji,

and the Japanese on [prewar] Saipan in the Marianas. . . . In none of these islands were the natives numerous enough or willing to work on the plantations so that outside labor had to be imported. In Hawaii Chinese and Portuguese were first used but today Japanese and Filipinos do the work. In Fiji, Indians, mostly of the Hindu faith, were brought in and now [outnumber] the native Fijians. At Saipan the Japanese brought in their own people. . . .

On the high islands the natives originally lived in villages rather than on individual farms, and the custom is still generally followed. Each valley commonly supported a village which by preference was located on the beach or as close to the sea as the topography permitted. . . .

As mentioned, a variety of food could be produced on the high islands. . . . Pigs furnished most of the meat to supplement the starchy foods, fruit, and fish. The seedless breadfruit, plantain, taro, yam, and sweet potato were the chief native sources of starch. These have now been supplemented by arrowroot, cassava, and improved varieties of bananas. Sugar cane, coffee, [cacao], papaya, mango, lime, orange, lemon, shaddock, and ginger are among the useful plants introduced by Europeans. Cattle and goats are found on the larger islands where grasslands exist. . . .

Coral Atolls

The "low" islands are made of coral and usually have an irregular ring shape around a lagoon. Such an island is called an atoll [(photo, p. 518)]. Generally the coral ring is broken into many pieces, separated by channels leading into the lagoon, but the whole circular group is commonly considered one island, although quite often individual names are given to the larger islets. . . . The coral atolls are wholly within the tropics as the reef-forming organisms can only survive in warm waters.

Two leading theories have been advanced to account for the coral atolls. That of Charles Darwin has three stages: (1) the coral builds a fringing reef around a volcanic island, (2) the island slowly sinks and the coral reef is built upward and forms a barrier reef separated by a lagoon from the shore, (3) the volcanic island has sunk out of sight and a lagoon occupies the former land area whose outline is reflected in the roughly oval form of the atoll. . . . Another theory is that of Sir James Murray who would have the coral forming a reef on the outer limits of a shallow submarine platform. Likely most such platforms would result from erosion of a volcanic island, so would have the approximately oval shape of a volcanic peak.

No matter how they may have been formed, the geographer finds that the atolls have similar characteristics wherever they occur, although in size they vary. Many are only a few miles in diameter, while a few huge affairs up to 50 to 100 miles across occur. . . . Some coral islands lack the central lagoon, which has either been filled to become land or the island itself has been elevated enough for the feature to disappear. . . .

. . . The greatest concentration of atolls is in the Tuamotu Archipelago (also called the Paumotu, Low, and Dangerous Archipelago). The atolls of the Marshalls are arranged in two parallel chains, an eastern and a western. The Gilbert, Ellice, and Tonga islands are atolls, as are the Union (Tokelau), Cook, and Phoenix groups. In addition there are many isolated low isles of coral. . . . Frequently the atolls follow an arcuate course and rise at intervals from a submarine platform or uplift, likely a zone of volcanic activity. . . . While the barrier reefs of the atolls protect ships in the lagoons from currents and waves, they are themselves a great danger to navigation, especially as charts of Polynesia and Micronesia are often incomplete and inaccurate. . . . Occasionally some atoll is completely overwhelmed by a hurricane and the giant waves associated with it that may wash entirely over the low land and destroy most of the palms and inhabitants. . . .

The soil of the low islands is broken from the coral by the waves and winds. It is thin and poor, and few plants except coconuts can grow on it or withstand the infiltration of the salt sea into the ground water supply. On the more barren islands, coconuts, the rather poor fruit of the pandanus, and fish are the only available foods. Sometimes the natives by much labor construct a garden of artificially fertile soil. Within a stone wall or pit . . . all available plant waste and other refuse is collected to supply needed humus to the sterile coral sand. . . .

The most significant animal life on the low islands consists of birds. . . . Birds have sometimes been slaughtered for their feathers, and both eggs and birds are eaten, although they have a fishy flavor. However, the most valuable product of the birds is guano and, indirectly, phosphate rock. On the low islands in the trade winds, rain is too scanty to wash away the bird droppings and other refuse which accumulate and change to the brownish guano that is a valuable fertilizer because it is high in nitrogen and phosphorus. On some islands, notably Nauru and Ocean, just south of the equator, the phosphorus from the guano has interacted chemically with the coral limestone to form phosphate rock. This is quarried and exported to the temperate zone [mainly to Australia and New Zealand] to use on depleted soils. . . .

RELATIONS WITH THE WEST

Europeans began to visit the Pacific islands early in the Age of Discovery. Spanish and Portuguese voyagers were followed by Dutch, English, French, and, later, American and German. Many famous names are connected with Pacific exploration, including those of Magellan, Tasman, Bougainville, La Perouse, and Cook. By the end of the eighteenth century virtually all of the important islands were known.

For a long period the European governments exercised only nominal control over the islands, and the native peoples were subjected to the unrestrained abuses of whaling crews, sandalwood traders, indentured labor contractors ("blackbirders"), and other adventurers. On island after island European penetration presented the dismal spectacle of decimation of the islanders and disruption of their cultures. The intruders introduced new diseases, alcohol, opium, forced labor, and firearms which greatly increased the slaughter in tribal wars. Although the introduction of better tools and medicines, as well as the work of missionaries and the belated attempts of the imperial governments to give the islanders a "new deal" must be entered on the other side, the balance sheet for four and a half centuries of Western influence in the Pacific islands does not reflect much credit on the outsiders.

In recent times Western personnel and capital have been attracted to the islands by mining and plantation agriculture. Although in general the islands are notably deficient in mineral resources, there are some exceptions. The phosphate resources have already been described. In addition, gold has been found in the Solomons, the Fijis, and, especially, New Guinea The French island of New Caledonia is particularly outstanding in mineral reserves and exploitation, with deposits of nickel, chrome, cobalt, and iron ore. Nickel accounts for most of the value of mineral production at present and is New Caledonia's main export.

A "COCONUT CIVILIZATION"

Most tropical plantation crops have been attempted in the islands at one time or another, with varying degrees of success. However, two crops—sugar and coconuts—have gained much greater importance than any others. Of the two, sugar leads in value of production, but coconuts are by far the more widespread and fundamentally important in the lives of the people. The growing of coconuts and the trade in copra provide a common denominator for a large part of the island world, which has been said with considerable justice to have a "coconut civilization." The latter is described in the following selection by Douglas L. Oliver:[5]

South Seas sugar statistics are far more impressive in tonnage and in value than those of the South Seas coconut, but sugar's influence is limited to Hawaii, Fiji, and prewar Saipan-Tinian, and only whites and immigrant Asiatics benefit directly from it. The influence of the coconut, however, stretches from Truk to Tonga and from Hiva Oa to Hollandia, and directly or indirectly affects the life of nearly every islander in this vast area. . . .

Coconuts require year-round warm temperature, a well-drained soil, and plenty of moisture and sunlight; they grow best of all in low altitudes near the coast. The palm grows out of the mature, fallen nut, and requires from eight to ten years to reach the bearing stage. After that it lives for nearly 80 years and bears nuts at the rate of about 50 a year for 60 or 70 years.

The mature nut consists of a hollow kernel of oily white meat, one half inch thick, encased in a hard woody shell. Around this is a fibrous husk one half to two inches thick. The cavity of the unripe nut is filled with a thin "milk," a nutritious and refreshing beverage with a tangy taste. As the nut matures, this "milk" is absorbed into the coconut meat, which when dried and removed from the shell becomes the copra of commerce.

Throughout Oceania the coconut leads a double life. In one way it is a source of food, shelter, and income which helps support native life nearly everywhere. In another way, however, it has been the instrument by which white men have done most to change native life.

To islanders the milk of the unripe coconut is a prized beverage, the only one, in fact, on many islands lacking potable water or rain catchments. The meat is scraped from the shell and eaten either by itself or mixed into puddings of taro and yams and sago. Or, the oil is squeezed from the meat and used as a food, an unguent, or a cosmetic. The hollowed shell becomes a flask, a clean-scraped shell a cup or a spoon or a material for carved ornaments. Cord is manufactured from the fibrous husk; furniture, utensils, and building timbers from the tough trunk. Leaves are used to thatch huts or weave baskets; and even the pith of the palm is eaten when for some drastic reason the palm is felled. On islands nearer to Asia's influence, the sugar-rich sap is drawn off by tapping the flower bud and is allowed to ferment into inebriating toddy. To deprive an islander of his coconut palms is to take away much of the basis of his living. Even the white man's world would be a much poorer place without them.

From the copra of commerce oil is expressed to produce margarine, cooking and salad oils, fine soap, and cosmetics. Copra cake, the residue after most oil has been expressed, is an important stock food rich in protein. Coir mats and rugs are manufactured from the fibrous husks, and coconut-shell charcoal is the best vapor absorbent known, especially valuable for absorbing industrial odors, for recovering gasoline and benzol from the air, and for use in gas masks. In the manufacture of soap, coconut oil plays a unique role; it hardens well, and its highly soluble acids possess exceptional lathering qualities, even in salt water. It has consequently become a favorite for toilet and saltwater soaps and for soap chips. And, in addition to all these, consider how empty life would be without coconut pie!

But most of these good things are end products, processed in factories thousands of miles from the islands. With the exception of a few tons of coconuts used in the local manufacture of desiccated coconut and some locally used cooking oils, most palm products leaving the islands do so in the form of copra.[6]

---

[5] Reprinted by permission of the publishers from Oliver, *op. cit.,* pp. 135–136. One footnote added in editing.

[6] Some copra producers, especially the Fiji Islands and the Australian Trust Territory of New Guinea, now crush a portion of their copra in local mills, for export in the form of coconut oil.

Copra is merely coconut meat removed from the shell and dried sufficiently for shipping. To obtain it, mature nuts may be picked from the palm or they may be collected after they have fallen. Then the whole coconut is split in half and the meat either immediately cut out of the shell or allowed to dry a bit before removal. Excess moisture must then be removed from the meat by some method of drying, and the resulting copra bagged in hundred-pound sacks for overseas shipment.

Organization of the Copra Industry

In 1966 the Pacific islands, exclusive of the Philippines and Indonesia, accounted for some 12 percent of total world exports by weight of copra and coconut oil (oil basis), as compared with 65 percent for the Philippines, 9 percent for Indonesia, and 6 percent for Ceylon. Despite the relatively minor role of Oceania on a world basis, however, copra production is of the utmost local importance on most of the islands, frequently being the leading industry in employment, value of product, and general effect on the lives of the islanders. The organization of the industry varies a great deal from one island group to another, but in general three principal types of production can be distinguished: by native islanders, by small planters, and by large corporations.

1. *Native production* is of some importance in the copra industry of every island group and is the dominant form of production in many groups. Islanders gather the nuts, prepare the copra, and sell it to middlemen for marketing. The middleman is often the local storekeeper and is frequently either a Chinese or a white man. He in turn disposes of the copra to larger firms for overseas shipment. Increasing amounts of copra, however, are being marketed through cooperative marketing societies fostered by the governing powers. In many areas a government marketing board controls all export copra sales, whether from native producers or plantations.

2. *Small-planter production* involves the "independent" operation of a single plantation by one of the many white men who have come into the islands for this purpose. Such a plantation will ordinarily comprise plantings amounting to several hundred acres of trees, and will employ up to a hundred workers living in barracks. The workers are usually natives but are sometimes imported Asians. Often the plantation is quite isolated. The small-planter type of enterprise is declining due to high production costs and hazards, coupled with widely fluctuating world prices which the planter can do nothing to control and against which he generally has little by way of a financial reserve. Planters complain of a chronic shortage of labor and its lack of efficiency, which they maintain is partly due to government regulation of working conditions, including wages. In addition, their position is made more precarious by a plenitude of diseases and insect pests capable of seriously damaging the groves and sometimes not even susceptible of control by organizations having much greater resources than those of the small planter.

3. *Production by large corporations* has been of increasing importance. British concerns are generally the most important, but those of a number of other nations are also in the field. Outstanding names are Lever Brothers, Burns Philip Company, Ltd., and W. R. Carpenter Company, Ltd. Such concerns own many separate plantations and usually operate such associated facilities as shipping lines, stores, and warehouses. They may, as in the case of Lever Brothers, have worldwide interests. Their financial resources apparently allow them to operate successfully in an industry becoming more and more difficult for the small planter.

Exports Other Than Copra

Although the export trade of most Pacific islands and island groups continues to be dominated by copra, a few have other industries that exceed the copra industry in importance, or at least make sizable contributions to export trade. Phosphate is the sole export of importance from Nauru and Ocean islands, nickel is overwhelmingly dominant in New Caledonia, and cane sugar is by far the largest export from Fiji and Hawaii, though Fiji has good-sized exports of copra and coconut oil, and gold as well, and Hawaii is the world's greatest exporter of pineapples. Rarotonga, the main island in the Cook Islands, grows citrus fruits, tomatoes, and pineapples, and exports them to New Zealand, either fresh or as canned juice. These fruit exports far overshadow copra in the total trade of the Cook group. Copra and coconut oil still lead the export list from Australia's Trust Territory of New Guinea (North East New Guinea, the Bismarck Archipelago, and the two northernmost islands in the Solomons) (map, pp. 514–515), and copra also leads in Western Samoa and the New Hebrides, but there are substantial exports of coffee, cacao, and gold from the New Guinea trusteeship, bananas and cacao from Western Samoa, and manganese and frozen fish from the New Hebrides. In Australia's external territory of Papua, occupying the southeastern part of the island of New Guinea, rubber is the leading export in most years, with copra being a close second, and sometimes first.

PROGRESS TOWARD SELF-GOVERNMENT

Today the various Pacific island groups exhibit many degrees of self-government. Most of them are gaining an increasing amount of freedom in the management of their internal affairs. In Hawaii, formerly a Territory of the United States, this progress culminated in statehood in 1959. Western Samoa, formerly a Trust Territory administered by New Zealand, was granted independence in 1962. The small, phosphate-rich island of Nauru, formerly a joint trusteeship of Australia, New Zealand, and Great Britain (with Australia the actual administrator), became independent in 1968. The Cook Islands have gained full self-government internally, though New Zealand continues to administer their foreign relations. Great Britain has granted a large degree of self-government to the Fiji Islands, though progress toward full independence is hampered by tensions between the Indian element (which has a slight majority) and the native Fijians. At the time of writing (1968) it seemed likely that quite a number of island areas in the Pacific would gain full independence within a few years, or at least would have a very large degree of autonomy while continuing to maintain a formal political tie with the present governing power. Opinions among islanders concerning the desirability of full independence vary widely; a great many feel that it would be unwise to cut all political ties with the overseas power which now provides economic, technical, and military support, and which often serves as a haven for migrants who are unable to make satisfactory living arrangements in the islands. Many island groups have been witnessing a persistent drift of population from the less-developed outer islands to more-developed central ones, or from the island group as a whole to the overseas governing country. The movement of Cook Islanders from the outer islands to Rarotonga and from the islands to New Zealand is a good example. What the future holds politically for the island world, made up of such a multitude of small, economically weak, and militarily powerless units, is a subject of much debate, both in the islands themselves and in the outside world. The experiences of the tiny new independent nations of Western Samoa and Nauru will be watched with close attention by all who are concerned with Pacific island affairs.

REFERENCES and READINGS

Pacific Islands as a Whole

GENERAL

BARRAU, JACQUES, *Subsistence Agriculture in Polynesia and Micronesia* (Bernice P. Bishop Museum, Bulletin 223; Honolulu: 1961); also, "Plant Introduction in the Tropical Pacific: Its Role in Economic Development," *Pacific Viewpoint,* 1, no. 1 (March 1960), 1–9.

BELSHAW, H., "Some Pacific Island Problems," *Pacific Viewpoint,* 1, no. 2 (September 1960), 125–142.

CHEPLO, NANDOR J., *Basic Data on the Economy of the British Pacific Islands.* U.S. Bureau of International Commerce, Overseas Business Reports, OBR 63–153 (December 1963). (Washington, D.C.: Government Printing Office, 1963.)

CUMBERLAND, KENNETH B., *Southwest Pacific: A Geography of Australia, New Zealand and Their Pacific Island Neighbors* (rev. ed.; New York: Frederick A. Praeger, 1968). A standard work, well illustrated with maps and photos.

FOSBERG, FRANCIS RAYMOND, ed., *Man's Place in the Island Ecosystem: A Symposium* (Tenth Pacific Science Congress, 1961; Honolulu: Bishop Museum Press, 1963).

FREEMAN, OTIS W., ed., *Geography of the Pacific* (New York: John Wiley & Sons, 1951). A standard reference work, out of date in some respects but still valuable.

GREAT BRITAIN, COLONIAL OFFICE. Reports on individual British colonial territories.

GREAT BRITAIN, NAVAL INTELLIGENCE DIVISION, *Pacific Islands,* Geographical Handbook Series: Vol. 1, *General Survey* (1945); Vol. 2, *Eastern Pacific* (1943); Vol. 3, *Western*

Pacific (Tonga to the Solomon Islands) (1944); Vol. 4, *Western Pacific (New Guinea and Islands Northwards)* (1945). An encyclopedic compilation of geographical and other information, produced during World War II. Still a very useful source. Many maps, diagrams, photos, and tables.

HARRIS, N. V., *The Tropical Pacific* (London: University of London Press, 1966). A regional geography.

KENNEDY, T. F., *A Descriptive Atlas of the Pacific Islands: Australia, Polynesia, Melanesia, Micronesia, Philippines* (Wellington, N.Z.: A. H. and A. W. Reed, 1966). Maps in black, white, and gray, with supporting text.

OLIVER, DOUGLAS L., *The Pacific Islands* (Cambridge, Mass.: Harvard University Press, 1951; reprinted, New York: Doubleday & Co., Anchor Books, 1962). A scholarly, well-written introduction to the island world.

Pacific Islands Year Book (Sydney, N.S.W.: Pacific Publications), annual.

Pacific Viewpoint (semiannual). A geographical journal of high quality.

TAYLOR, C. R. H., *A Pacific Bibliography: Printed Matter Relating to the Native Peoples of Polynesia, Melanesia, and Micronesia* (2d ed.; New York: Oxford University Press, 1965).

See also the reference lists for Chapter 1–3.

HISTORICAL

BEAGLEHOLE, J. C., *The Exploration of the Pacific* (3d ed.; Stanford, Calif.: Stanford University Press, 1966).

FRIIS, HERMAN R., ed., *The Pacific Basin: A History of Its Geographical Exploration* (American Geographical Society, Special Publication no. 38; New York: American Geographical Society, 1967). Incorporates a great deal of general geographical information about the Pacific islands. See especially Chap. 1, "The Pacific Basin: An Introduction," by William L. Thomas, Jr., pp. 1–17; Chap. 4, "Geographical Knowledge of the Pacific Peoples," by Gordon R. Lewthwaite, pp. 57–86; and Chap. 15, "The Intellectual Assumptions and Consequences of Geographical Exploration in the Pacific," by Wilcomb E. Washburn, pp. 321–334.

GRATTAN, C. HARTLEY, *The Southwest Pacific to 1900: A Modern History–Australia; New Zealand; The Islands; Antarctica;* and *The Southwest Pacific since 1900: A Modern History* (Ann Arbor: University of Michigan Press, 1963); also, *The United States and the Southwest Pacific* (Cambridge, Mass.: Harvard University Press, American Foreign Policy Library, 1961). These well-written histories, richly documented, focus primarily on Australia, but include much information on the islands.

MOOREHEAD, ALAN, *The Fatal Impact: An Account of the Invasion of the South Pacific, 1767–1840* (New York: Harper & Row, 1966). A book for the general reader by an informed and skillful historical journalist. Emphasizes early European contacts with Tahiti and Australia.

PRICE, A. GRENFELL, *The Western Invasions of the Pacific and Its Continents: A Study of Moving Frontiers and Changing Landscapes, 1513–1958* (New York: Oxford University Press, 1963).

CORAL REEFS AND ATOLLS

DARWIN, CHARLES, *The Structure and Distribution of Coral Reefs* (Berkeley: University of California Press, 1962; first published in 1889).

NEWELL, NORMAN D., "Questions of the Coral Reefs," *Natural History,* **68,** no. 3 (March 1959), 118–129; and "Biology of the Corals," *Natural History,* **68,** no. 4 (April 1959), 226–235. An article in two parts, with fine illustrations.

STODDART, D. R., "Catastrophic Human Interference with Coral Island Ecosystems," *Geography,* **53,** pt. 1 (January 1968), 25–40.

WIENS, HEROLD J., *Atoll Environment and Ecology* (New Haven, Conn.: Yale University Press, 1962). A major work, with extensive bibliography.

Melanesia

BARRAU, JACQUES, *Subsistence Agriculture in Melanesia* (Bernice P. Bishop Museum, Bulletin 219; Honolulu: 1958).

New Guinea

BROOKFIELD, H. C., "The Highland Peoples of New Guinea: A Study of Distribution and Localization," *Geographical Journal,* 127, pt. 4 (December 1961), 436–448; also, "Local Study and Comparative Method: An Example from Central New Guinea," *Annals of the Association of American Geographers,* 52, no. 3 (September 1962), 242–254; "The Chimbu: A Highland People in New Guinea," in S. R. Eyre and G. R. J. Jones, eds., *Geography as Human Ecology: Methodology by Example* (New York: St. Martin's Press, 1966), pp. 174–198; and, with Paula Brown, *Struggle for Land: Agriculture and Group Territories among the Chimbu of the New Guinea Highlands* (New York: Oxford University Press, 1963).

FISK, ERNEST KELVIN, ed., *New Guinea on the Threshold: Aspects of Social, Political and Economic Development* (London: Longmans, Green and Co.; Canberra: Australian National University Press, 1966).

INTERNATIONAL BANK FOR RECONSTRUCTION AND DEVELOPMENT, *The Economic Development of the Territory of Papua and New Guinea* (Baltimore, Md.: Johns Hopkins Press, 1965).

MATTHIESSEN, PETER, *Under the Mountain Wall: A Chronicle of Two Seasons in the Stone Age* [among the Kurelu people of New Guinea] (New York: The Viking Press, 1962).

MEAD, MARGARET, "The Rights of Primitive Peoples: Papua–New Guinea: A Crucial Instance," *Foreign Affairs,* 45, no. 2 (January 1967), 304–318.

PRICE, A. GRENFELL, *The Challenge of New Guinea: Australian Aid to Papuan Progress* (Sydney, N.S.W.: Angus and Robertson, 1965).

SPATE, O. H. K., "Problems of Development in New Guinea," *Geographical Journal,* 122, pt. 4 (December 1956), 430–440.

For references on Indonesian New Guinea, see Chapter 19.

Fiji Islands

COULTER, JOHN WESLEY, *The Drama of Fiji: A Contemporary History* (Rutland, Vt.: Charles E. Tuttle Co., 1967).

MAYER, ADRIAN C., *Peasants in the Pacific: A Study of Fiji Indian Rural Society* (Berkeley and Los Angeles: University of California Press, 1961).

SPATE, O. H. K., *The Fijian People: Economic Problems and Prospects* (Legislative Council of Fiji, Council Paper no. 15 of 1959; Suva: 1959).

WARD, R. GERARD, *Land Use and Population in Fiji: A Geographical Study* (Great Britain, Department of Technical Cooperation, Overseas Research Publication No. 9; London: H.M.S.O., 1965); also, "Cash Cropping and the Fijian Village," *Geographical Journal,* 130, pt. 4 (December 1964), 484–506; and "The Population of Fiji," *Geographical Review,* 44, no. 3 (July 1959), 322–341.

WATTERS, R. F., "Problems of Development in Fiji," *Pacific Viewpoint,* 2, no. 2 (September 1961), 155–176; and "Sugar Production and Culture Change in Fiji: A Comparison between Peasant and Plantation Agriculture," *Pacific Viewpoint,* 4, no. 1 (March 1963), 25–52.

WHITELAW, J. S., "Gold Production and the Mining Community in Fiji," *New Zealand Geographer,* 23, no. 1 (April 1967), 1–15.

Other Islands of Melanesia

EPSTEIN, T. S., "European Contact and Tolai Economic Development: A Schema of Economic Growth" [in New Britain], *Economic Development and Cultural Change,* 11, no. 3, pt. 1 (April 1963), 289–307.

LEANEY, CAROLINE, and D. A. M. LEA, "Some Recent Developments in New Hebridean Trade," *Australian Geographer,* 10, no. 4 (September 1967), 286–297.

PARSONSON, G. S., "Artificial Islands in Melanesia: The Role of Malaria in the Settlement of the Southwest Pacific," *New Zealand Geographer,* 22, no. 1 (April 1966), 1–21.

TEDDER, J. L. O., "The Solomon Islands: An Emerging Cash Economy," *Australian Geographical Studies,* 4, no. 1 (April 1966), 49–59.

WILSON, J. S. G., *Economic Survey of the New Hebrides* (Great Britain, Ministry of Overseas Development, Overseas Research Publication No. 15; London: H.M.S.O., 1966).

Micronesia

COULTER, JOHN WESLEY, *The Pacific Dependencies of the United States* (New York: The Macmillan Company, 1957).

HOPKINS, ROBERT S., "Micronesia," *Focus,* 13, no. 10 (June 1963), 6 pp.

WIENS, HEROLD J., *Pacific Island Bastions of the United States* (Princeton, N.J.: D. Van Nostrand Co., Searchlight Books, 1962).

CAROLINE ISLANDS

ALKIRE, WILLIAM H., *Lamotrek Atoll and Inter-Island Socioeconomic Ties* (Illinois Studies in Anthropology, no. 5; Urbana: University of Illinois Press, 1965).

BASCOM, WILLIAM R., *Ponape: A Pacific Economy in Transition* (Berkeley and Los Angeles: University of California Press, 1965). A detailed inventory (by an anthropologist) of population, culture, and livelihood on one of the main islands in the eastern Carolines.

BATES, MARSTON, and DONALD P. ABBOTT, *Coral Island: Portrait of an Atoll* [Ifaluk] (New York: Charles Scribner's Sons, 1958).

MURPHY, RAYMOND E., "' High' and 'Low' Islands in the Eastern Carolines," *Geographical Review,* 39, no. 3 (July 1949), 425–439; also, "Landownership on a Micronesian Atoll" [Mokil in the eastern Carolines], *Geographical Review,* 38, no. 4 (October 1948), 598–614; and "The Economic Geography of a Micronesian Atoll" [Mokil], *Annals of the Association of American Geographers,* 40, no. 1 (March 1950), 58–83.

OTHER ISLANDS OF MICRONESIA

BEARDSLEY, CHARLES, *Guam, Past and Present* (Rutland, Vt.: Charles E. Tuttle Co., 1964).

COUPER, A. D., "The Gilbert and Ellice Island Colony: Implications of Regional Trading Anomalies," *Pacific Viewpoint,* 8, no. 1 (May 1967), 68–86.

HUGHES, HELEN, "The Political Economy of Nauru," *Economic Record,* 40, no. 92 (December 1964), 508–534.

MAUDE, H. E., and EDWIN DORAN, JR., "The Precedence of Tarawa Atoll," *Annals of the Association of American Geographers,* 56, no. 2 (June 1966), 269–289.

MAUDE, H. E., and IDA LEESON, "The Coconut Oil Trade of the Gilbert Islands," *Journal of the Polynesian Society,* 74, no. 4 (December 1965), 396–437.

Polynesia

CUMBERLAND, KENNETH B., "The Future of Polynesia," *Journal of the Polynesian Society,* 71, no. 4 (December 1962), 386–396.

DENING, G. M., "The Geographical Knowledge of the Polynesians and the Nature of Inter-Island Contact," *Journal of the Polynesian Society,* 71, no. 4 (December 1962), 102–131.

SHARP, ANDREW, *Ancient Voyagers in Polynesia* (Berkeley: University of California Press, 1964; first published as *Ancient Voyagers in the Pacific,* 1956).

SUGGS, ROBERT C., *The Island Civilizations of Polynesia* (New York: The New American Library, Mentor Books, 1960).

FRENCH POLYNESIA

CUNNINGHAM, GLENN, "Food for Tahiti," *Economic Geography,* 37, no. 4 (October 1961), 347–352; also, "Manufacturing in Tahiti," *California Geographer,* 7 (1966), 19–27.

LEWTHWAITE, GORDON R., "Man and Land in Early Tahiti: Polynesian Agriculture through European Eyes," *Pacific Viewpoint,* 5, no. 1 (May 1964), 11–34; also, "Man and the Sea in Early Tahiti," *Pacific Viewpoint,* 7, no. 1 (May 1966), 28–53.

SCHMITT, ROBERT C., "Urbanization in French Polynesia," *Land Economics,* 38, no. 1 (February 1962), 71–75.

OTHER ISLANDS OF POLYNESIA

BASSETT, I. G., "The Cook Islands at Independence," *Perspective* [New Zealand Geographical Society], no. 2 (October 1965), 6 pp.

CROCOMBE, R. G., "Development and Regression in New Zealand's Island Territories" [Cook, Niue, and Tokelau islands], *Pacific Viewpoint,* 3, no. 2 (September 1962), 17–32.

FOX, JAMES W., and KENNETH B. CUMBERLAND, eds., *Western Samoa: Land, Life and Agriculture in Tropical Polynesia* (Christchurch, N.Z.: Whitcombe and Tombs, 1962). A major study, with numerous maps and photos.

JOHNSTON, W. B., "The Cook Islands: Land Use in an Island Group of the South-west Pacific," *Journal of Tropical Geography,* 13 (December 1959), 38–57.

KENNEDY, T. F., "Land, Food, and Population in the Kingdom of Tonga," *Economic Geography,* 37, no. 1 (January 1961), 61–71.

PIRIE, PETER, and WARD BARRETT, "Western Samoa: Population, Production, and Wealth," *Pacific Viewpoint,* 3, no. 1 (March 1962), 63–96.

For references on Hawaii, see Chapter 30.

23

Australia and New Zealand

The basic kinship of Australia and New Zealand is widely recognized and is indicated in common usage of the term "Anzac countries" to refer to the two collectively. This kinship is derived from similarities in population, cultural heritage, political problems and orientations, type of economy, and location.

Both Australia and New Zealand are products of British colonization and are strongly British in ancestry and culture.[1] The Australians and New Zealanders speak

---

[1] Both Australia and New Zealand contain surviving minorities of prewhite native inhabitants. The primitive Negroid inhabitants of Australia were slaughtered or driven into outlying areas by the whites. Today they numbered about 40,000 of unmixed blood and about 100,000 if mixed bloods are included, in contrast to an estimated 300,000 at the beginning of white settlement in 1788. Efforts are being made to improve the depressed condition of this remnant. Its members, found mainly in the tropical north, live partly on the fringes of white society and partly independently in the "bush." In New Zealand the native inhabitants were the Maori, a Polynesian people. Not so primitive as the Australian aborigines, and more warlike, their hold on the land was broken only by a bloody war between 1860 and 1870. At present the Maori, who seemed destined for extinction about 1900, are increasing rapidly in numbers, in great contrast to the Australian aborigines. A Maori population of about 40,000 at the turn of the century grew to over 170,000 by 1962 and is expected to number about 250,000—one-eighth of New Zealand's population—by 1972. While still something of a depressed group in New Zealand, their economic and social situation is much better than that of the Australian natives, and they are being increasingly integrated, both socially and biologically, with the rest of New Zealand's population.

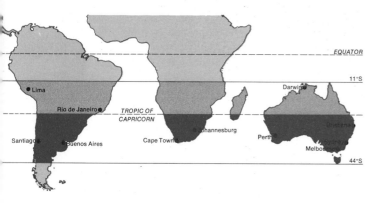

Australia compared in latitude and area with southern Africa and southern South America.

of the neighboring Asian countries to the north, whose teeming and land-hungry populations stand in striking contrast to the comparative emptiness and material abundance of Australia and New Zealand.

Australia and New Zealand are also tied closely to Britain and the United States by trade. Basic to the economies of both countries are exports of a few primary products (wool being the leader, with wheat a poor second, in the case of Australia, and wool, meat, and dairy products being the leaders in New Zealand), coupled with extensive imports of manufactured goods. This vital trade is carried on primarily with the United Kingdom and the United States and secondarily with other industrialized nations, especially Japan (Table 18).

English, live under parliamentary forms of government, claim the British sovereign as their own, attend schools patterned after those of Britain, and attend the Anglican Church in greater numbers than any other. Both countries, like Britain, are socialistic "welfare states," with comprehensive social security standards and benefits, although in the development of this aspect of their societies Australia and New Zealand actually preceded Britain and were world pioneers along with the Scandinavian countries. Both Australia and New Zealand have exceptionally high standards of living, ranking generally in a class with Canada and the leading countries of Europe in this respect, though somewhat below the United States.

Although they are fully independent nations, loyalty to Britain has been an outstanding historical characteristic of the Anzac countries. This loyalty is still expressed in their membership in the Commonwealth of Nations and was plainly evident at the outbreak of both world wars, which Australia and New Zealand entered immediately in support of Britain. However, since World War II, when a threatened Japanese assault on their homelands was frustrated mainly by American forces, the two countries have also sought closer relations with the United States. This has tended to weaken their British ties somewhat. Close international ties with strong and friendly outside powers, especially naval powers, are extremely important to the Anzac countries due to their small populations and remote insular locations, far distant from other Western and white nations. Australia's 12 million people plus New Zealand's 2.8 million (1968 estimates for both countries) amount to a smaller total than the population of the New York metropolitan area. In view of this fact, it is understandable that the two countries look with some apprehension at the awakening masses

AUSTRALIA'S PHYSICAL AND CLIMATIC CHARACTERISTICS

Australia is not only a country but an island continent with an area, including the offshore island of Tasmania, of nearly 3 million square miles. It is exceeded in area by only five countries: the Soviet Union, Canada, China, the United States, and Brazil. However, most of the continent is very sparsely populated, and in terms of total population Australia is a relatively small country. The sparseness of Australia's population and its concentration into a comparatively small part of the total land area are closely related to the continent's physical characteristics, among which aridity (accentuated by high temperatures and rapid evaporation) and low average elevation are outstanding. On the basis of climate and relief, Australia may be divided into four major natural regions: (1) the humid eastern highlands, (2) the tropical savannas of northern Australia, (3) the "mediterranean" lands of southwestern and southern Australia, and (4) the dry interior (map, p. 532).

The Humid Eastern Highlands

Australia's only major highlands extend along the east coast from Cape York to southern Tasmania in a belt 100 to 250 miles wide. Although complex in form and often rugged, these highlands seldom reach elevations of 3000 feet. Their highest summit and the highest point in Australia is Mount Kosciusko, which attains only 7316 feet. The highlands and the narrow and fragmented coastal plains at their base constitute the only part of Australia which does not experience a

TABLE 18 PRINCIPAL TRADING PARTNERS OF AUSTRALIA AND NEW ZEALAND, 1966

	AUSTRALIA		NEW ZEALAND	
Percent of exports[a] to	Japan	18	United Kingdom	42
	United Kingdom	15	United States	16
	United States	13	Japan	8
	New Zealand	6	France	5
	All others	48	Australia	5
			All others	24
Percent of imports[a] from	United States	25	United Kingdom	37
	United Kingdom	25	Australia	19
	Japan	9	United States	11
	West Germany	6	Japan	6
	All others	35	All others	27

SOURCE: *Encyclopedia Britannica Book of the Year*, 1968.
[a] Percentages are based on total value, 1966.

considerable period of drought each year. However, although onshore winds from the Pacific bring appreciable rain each month, the strong relief reduces the amount of agricultural land in this most favored of Australia's climatic areas. South of Sydney and at higher elevations to the north the climate is commonly classified as marine west coast (humid marine), despite the location. North of Sydney higher summer temperatures change the classification to humid subtropical, while still farther north, beyond approximately the parallel of 20° S., hotter temperatures and greater seasonality of rain cause essentially subhumid conditions.

Tropical Savannas of Northern Australia

Northern Australia, from near Broome on the Indian Ocean to the coast of the Coral Sea, receives heavy rainfall during a portion of the (Southern Hemisphere) summer season, but experiences almost complete drought during the winter 6 months, or more, of the year. This highly seasonal distribution of rainfall is essentially the result of monsoonal winds which blow onshore during the summer and offshore during the winter. The seasonality of the rainfall, combined with the tropical heat of the area, has produced a savanna vegetation of coarse grasses with scattered trees and patches of woodland. The effect of the long season of drought in reducing agricultural possibilities is compounded by the poverty of the soils and by a lack of highlands sufficient to nourish large perennial streams for irrigation. The alluvial and volcanic soils which support large populations in some

tropical areas are almost completely absent in northern Australia.

"Mediterranean" Lands of Southwestern and Southern Australia

The southwestern corner of Australia and the lands around Spencer Gulf have a mediterranean or dry-summer subtropical type of climate with subtropical temperatures, winter rain, and summer drought. In winter the Southern Hemisphere belt of the westerly winds shifts far enough north to affect these districts, while in summer this belt lies offshore to the south and the land is dry. Crops introduced from the mediterranean lands of Europe generally do well in these parts of Australia, but the agricultural possibilities of the Australian areas are limited by the lack of high highlands to catch moisture and supply irrigation water to the lowlands.

The Dry Interior

The huge interior of Australia is desert, surrounded by a broad fringe of semiarid grassland (steppe) which is transitional to the more humid areas around the edges of the continent. Altogether, the interior desert and steppe cover more than half of the continent, and extend to the coast in the northwest and along the Great Australian Bight in the south. This tremendous area of arid or semiarid land is too far south to get much rain from the summer monsoon, too far north to benefit from rainfall brought by the westerlies in winter, and is

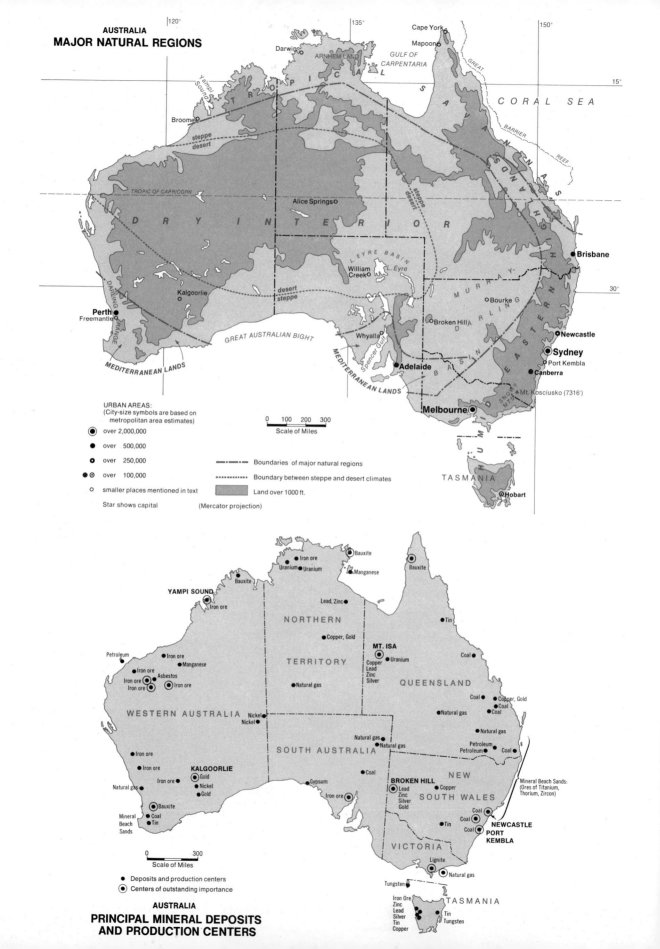

AUSTRALIA
MAJOR NATURAL REGIONS

120° 135° 150°

15°

Cape York
Mapoon
Darwin
ARNHEM LAND
GULF OF
CARPENTARIA
CORAL SEA
Yampi Sound
Broome
steppe
desert
TROPIC OF CAPRICORN
GREAT BARRIER REEF
Alice Springs
EYRE BASIN
L. Eyre
William Creek
Brisbane
Kalgoorlie
Perth
Freemantle
DRY INTERIOR
desert steppe
Bourke
Broken Hill
MURRAY
DARLING
Newcastle
Sydney
Port Kembla
Canberra
Whyalla
Spencer Gulf
GREAT AUSTRALIAN BIGHT
Adelaide
MEDITERRANEAN LANDS
MEDITERRANEAN LANDS
DARLING RANGE
TROPICAL SAVANNAS
EASTERN HIGHLANDS
Mt. Kosciusko (7316')
SNOWY MTS.
Melbourne

URBAN AREAS:
(City-size symbols are based on
metropolitan area estimates)
⊙ over 2,000,000
● over 500,000
◉ over 250,000
●⊙ over 100,000
○ smaller places mentioned in text

Star shows capital

0 100 200 300
Scale of Miles

—·—·— Boundaries of major natural regions
············ Boundary between steppe and desert climates
▨ Land over 1000 ft.

(Mercator projection)

TASMANIA
Hobart

AUSTRALIA
**PRINCIPAL MINERAL DEPOSITS
AND PRODUCTION CENTERS**

Bauxite
Iron ore
Uranium Uranium
Manganese
Bauxite
YAMPI SOUND
Bauxite
Iron ore
Lead, Zinc
NORTHERN
Tin
Petroleum
Iron ore
Manganese
Iron ore Asbestos
Iron ore Iron ore
Iron ore
Copper, Gold
TERRITORY
MT. ISA
Copper
Lead
Zinc
Silver
Uranium
Coal
QUEENSLAND
Natural gas
Coal Copper, Gold
Coal
Coal
WESTERN AUSTRALIA
Nickel
Nickel
Natural gas
Natural gas
Natural gas
Natural gas
Petroleum
Petroleum
Coal
Iron ore
Iron ore
Natural gas
KALGOORLIE
Gold
Nickel
Gold
SOUTH AUSTRALIA
Gypsum
Coal
Iron ore
BROKEN HILL
Lead
Zinc
Silver
Gold
Copper
NEW
SOUTH WALES
Mineral Beach Sands:
(Ores of Titanium,
Thorium, Zircon)
Bauxite
Coal
Tin
Mineral
Beach
Sands
Coal
Coal
Coal
NEWCASTLE
PORT
KEMBLA
Tin
VICTORIA
Lignite
Natural gas
Tungsten
Iron Ore
Zinc
Lead
Silver
Tin
Copper
TASMANIA
Tin
Tungsten

0 300
Scale of Miles

● Deposits and production centers
◉ Centers of outstanding importance

TABLE 19 CLIMATIC DATA FOR SELECTED AUSTRALIAN STATIONS

NATURAL REGION CLIMATE TYPE, AND STATION	LATITUDE TO NEAREST WHOLE DEGREE	ELEVATION ABOVE SEA LEVEL (FEET)	AVERAGE TEMPERATURE (DEGREES F TO NEAREST WHOLE DEGREE)			PRECIPITATION	
			ANNUAL	JANUARY (OR WARMEST MONTH)	JULY (COOLEST MONTH)	ANNUAL AVERAGE (TO NEAREST INCH)	MONTHS AVERAGING LESS THAN ONE INCH OF PRECIPITATION
Humid eastern highlands							
Humid subtropical							
Brisbane	27°S.	137′	69°	77°	58°	48″	0
Marine west coast							
Melbourne	38°S.	115′	58°	67° (Feb.)	49°	26″	0
Tropical savannas							
Darwin	12°S.	97′	82°	86° (Nov.)	77°	59″	5 (May–Sept.)
Mapoon (Cape York Peninsula)	12°S.	20′	82°	85°	78°	62″	6 (May–Oct.)
Mediterranean areas							
Adelaide	35°S.	140′	63°	74° (Feb.)	52°	21″	2 (Jan., Feb.)
Perth	32°S.	197′	64°	74° (Jan., Feb.)	55°	35″	5 (Nov.–Mar.)
Dry interior							
Bourke	30°S.	361′	69°	84°	52°	13″	4 (July–Oct.)
William Creek	29°S.	247′	69°	83° (Feb.)	53°	5″	12

shielded from Pacific winds by the eastern highlands. Here again the lack of highlands is unfortunate from a climatic point of view. Approximately the western half of Australia is occupied by a vast plateau of ancient igneous, metamorphic, and hardened sedimentary rocks, but its general elevation is only 1000 to 1600 feet and its few isolated mountain ranges are too low to materially influence the climate or supply perennial streams for irrigation. To the east, between the plateau and the eastern highlands, the land is still lower in the great central lowland which stretches across the continent between the Gulf of Carpentaria and the Great Australian Bight. The effect of elevation on the climate is shown by the fact that the lowest part of this lowland and of Australia, the Lake Eyre Basin, is the driest part of the continent. Another part of the lowland, however, the Murray-Darling Basin, contains Australia's only major river system and has the most extensive development of irrigation works on the continent.[2]

Temperature and precipitation data for the four major natural regions of Australia are presented in Table 19.

The Small Proportion of Arable Land

One very important result of the widespread seasonal or total aridity, and of the occupation of most of the only truly humid area by highlands, is that Australia offers very little good agricultural land relative to its total area. Although some estimates place the proportion of potentially cultivable land as high as 15 percent of the total, the Australian government now clas-

[2] The irrigated areas scattered along these rivers are in process of expansion. Water for the expansion is being supplied by Australia's Snowy Mountains scheme. This project is located in the Snowy Mountains part of the eastern highlands just south of the federal capital of Canberra. Streams flowing in their natural courses eastward to the Pacific are diverted into the headwaters of the westward-flowing Murray River and one of its tributaries, the Murrumbidgee, thus providing more water downstream on the dry side of the mountain divide. The engineering works of the Snowy Mountains scheme were begun in 1949. Though still under construction the project is now largely complete. It includes numerous dams and reservoirs, nine hydroelectric power stations which will make a major contribution to the economy of southeastern Australia, and many miles of tunnels and aqueducts.

sifies only about 2 percent of the land as cropland, with another 2.4 percent cultivated pasture. Only 4 percent is classed as forest and woodland, while almost 60 percent is classed as natural grazing land and about one-third as complete wasteland. These figures give Australia an arable area only one-twelfth that of the conterminous United States, which is comparable to Australia in size. Australia actually has less arable land than a number of countries with a small fraction of its total area, such as France, Spain, and Poland.

The distribution of population in Australia generally follows that of arable land. Thus most of the country's small population is found in the humid eastern highlands and coastal plains, especially in the cooler south; in the areas of mediterranean climate; and in the more humid grasslands adjoining the southern part of the eastern highlands and the mediterranean areas.

AUSTRALIA'S ECONOMY AND POPULATION

Australia is one of a few countries in the world which are highly industrialized and at the same time important exporters of raw materials and food. There is an obvious similarity to the United States and Canada in this respect. But Australia, despite employment of a larger proportion of its population in industry than is true of the United States, is predominantly an importer of manufactured products rather than an exporter like the United States. In the latter respect Australia is more like Canada, although Australia is more industrialized in proportion to the size of its population than is Canada. There are various other points of similarity between these two countries. Both are large countries with relatively small

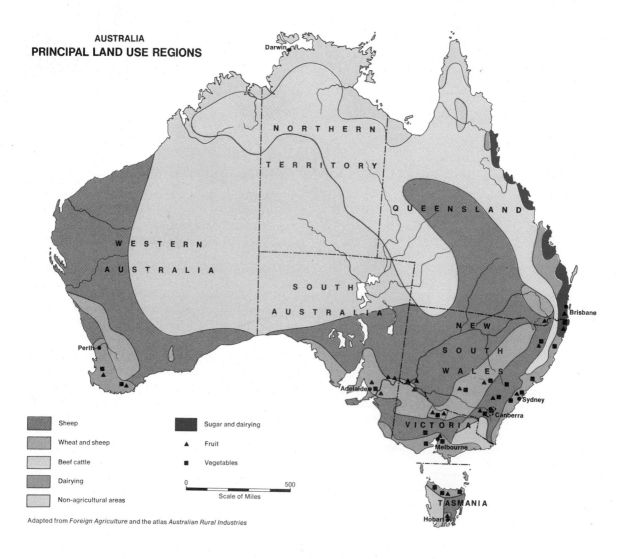

AUSTRALIA
PRINCIPAL LAND USE REGIONS

Legend:
- Sheep
- Wheat and sheep
- Beef cattle
- Dairying
- Non-agricultural areas
- Sugar and dairying
- ▲ Fruit
- ■ Vegetables

0   500
Scale of Miles

Adapted from *Foreign Agriculture* and the atlas *Australian Rural Industries*

Vast areas of grazing land are found in the central lowlands of Australia. This photo of sheep grazing in northwestern New South Wales offers a classic view of middle latitude steppe. (Australian News and Information Bureau.)

populations. In both, the population is mainly distributed around the periphery, a fact related to the large amounts of poor and undeveloped land which both possess in the interior. Both countries have a disproportionately high rank in world trade, both have close connections with Britain and with the United States, and both have high standards of living. One important difference, however, lies in the fact that Australia is able to produce a greater variety of foodstuffs than is Canada because of greater environmental variety. Canada, for example, has no tropical or subtropical climates.

Agriculture and Ranching

The sectors of the Australian economy which regularly produce a net surplus for export are grazing, agriculture, and mining. Such a result seems a logical outgrowth of a situation where 12 million people have at their command the natural resources of an entire continent, albeit a poor continent from the standpoint of agriculture. The existence of large and dependable markets for Australian primary products in Japan, Europe, the United States, and elsewhere—coupled with cheap

and efficient transportation by sea to those markets, and the possession by the Australian people of the necessary technical skills and equipment to exploit the continent on an extensive basis, utilizing a minimum of labor and a maximum of land and equipment—makes the emphasis on surplus production and export of primary commodities seem even more logical. Australian agricultural and pastoral output per man involved is notably higher than that of the United States.

Sheep and Cattle Ranching

Sheep ranching has always been the most important of Australia's rural industries (photo, above). When the first settlement was planted at Sydney in 1788, the intention of the British government was merely to establish a penal colony, with the expectation that enough agriculture would be carried on to make the colony self-sufficient in food and perhaps provide a small surplus of some products for export. While self-sufficiency proved difficult to obtain in the early days due to the poverty of the leached soils around Sydney, it was soon discovered that sheep did well. At the time the market for wool was rapidly expanding with mechanization of the woolen textile industry in Britain. Thus, Australia found its main export staple in the early years of the nineteenth century. Sheep graziers rapidly penetrated the interior of the continent, and by 1850 Australia was already the largest supplier of wool on the world market, a position it has never lost.

Sheep raising has been tried practically everywhere in Australia that there seemed to be any hope of success. Much of the interior has been found too dry, parts of the eastern mountain belt too rugged and wet, and most of the north too hot and wet and with too coarse a forage in summer. Thus the sheep industry has become localized. It is concentrated mainly in a crescentic belt of territory which follows roughly the gentle western slope of the eastern highlands, from the northern border of New South Wales to the western border of Victoria. New South Wales has over 40 percent of the Commonwealth's sheep, and Victoria between 15 and 20 percent. Beyond the main belt of concentration the sheep industry spreads on poorer pastures north almost to the Gulf of Carpentaria and west into South Australia. A minor area of production rims the west coast from about latitude 16°S. to beyond Perth, with the greatest concentration relatively near that city. The sheep ranches or "stations" in Australia vary in size from a few hundred to many thousands of acres. They are generally larger on the poorer pasture lands, that is, outside the main belt of production.

The main product of the Australian sheep stations is wool. The importance of wool in the country's economy is indicated by the fact that it ordinarily constitutes over one-quarter of the total value of all Australian exports. Its importance to the world is indicated by the fact that Australian wool normally amounts to over a quarter of the total world production and over half of the total production of high-quality Merino wool. In addition, Australia has been able, since the development of refrigerated transport, to export considerable quantities of mutton and lamb.

Lands in Australia which have not been found suitable for sheep or for a more intensive form of agriculture are generally devoted to cattle ranching. Such lands include particularly those of the north, with hot,

humid summers and coarse forage. The main belt of cattle ranching extends east-west across the northern part of the country (map, p. 534). The greatest concentration is in Queensland, which contains almost half of the cattle, other than dairy animals, in Australia. On many of the remote ranches, especially in the Northern Territory, the labor force is composed of remnants of the black aboriginal population of Australia.

The beef cattle industry is not as important to Australia as the sheep industry. It is difficult to market first-class beef from a remote tropical area. Variability in weather, poor pasturage in the dry season, tropical insect pests and diseases, and truck hauls of hundreds of miles to market—often over poor roads—are among the factors that retard the industry. Nevertheless, Australia supplies its own needs for beef, and exports appreciable quantities of meat, as well as hides for tanning.

Wheat Farming

After wool, wheat is Australia's most important export, generally supplying around 10 percent of all exports by value. As in the case of wool, wheat surpluses are largely a reflection of Australia's aridity and low density of population. In contrast to wool, however, wheat production did not begin as the result of a search for export staples. Wheat was the key crop in early attempts to make the settlement at Sydney self-sufficient. The soils along much of the coastal margin of Australia are leached and poor, though capable of much improvement through application of fertilizer. In the early years the colony almost starved as a result of wheat failures, and the problem of grain supply was not really solved until the development of the colony, now the state, of South Australia. There, fertile land and a favorable climate were found near the port settlement of Adelaide, so that wheat could be grown and shipped by sea to other coastal points.

In the decade of the 1860s, Australia passed from a deficit to a surplus position in grain production. Thereafter, wheat farming expanded rapidly. Among the factors making this expansion possible were (1) the development of the mechanical reaper and other types of machinery

These photos show different facets of Australia's beef cattle industry. The top photo is a view of mixed breeds of cattle being watered at the Alice Springs trucking yards in the Northern Territory. Brought by truck from immense ranches that may be several hundred miles away, these livestock will be transported south from Alice Springs by truck or rail to stockyards in South Australia's seaport and capital city, Adelaide. The bottom view shows stockmen taking a "mob" of Aberdeen Angus steers to water in humid, tree-dotted pasture land of New South Wales, on the Hunter River, inland from Newcastle. (Australian News and Information Bureau.)

suited to extensive wheat farming, (2) the building of railroads inland from the ports, (3) scientific work in plant breeding and soil fertilization, (4) the enactment of legislation under which some of the large grazing estates were divided into smaller farms, and (5) an expanding European market. The main belt of wheat production spread from the eastern coastal districts of South Australia into Victoria and New South Wales, generally following the semiarid and subhumid lands which lie inland from the crests of the eastern highlands. Thus the main wheat belt has come to occupy nearly the same position as the main belt of sheep production (maps, pp. 84, 86, and 534). In fact, the two types of production are very often combined in this area. A second, less important, wheat belt has developed in the southwestern corner of the continent inland from Perth. The combined exports from these two areas have normally been sufficient to give Australia high rank among the world's wheat exporters, after the United States, Canada, and sometimes Argentina. It should be emphasized that Australian wheat production is an extensive and highly mechanized form of agriculture, characterized by a small labor force, large acreages, and low yields per acre, although very high yields per man.

Dairy Farming

Though not as important as grazing or wheat farming for the export trade, dairying is now the leading type of agriculture in Australia in terms of employment. It has developed mainly in the humid coastal plains of Queensland, New South Wales, and Victoria (map, p. 534). Originally oriented entirely toward the local market, it has been mainly a response to the striking urbanization of the country's coastal districts and the necessity of a relatively high return per acre to defray the expense of clearing heavily forested land. Dairying has now become well established as the dominant type of farming in most of these coastal areas and regularly produces an export surplus beyond the requirements of Australia itself.

Fruit Growing

When Australia was settled, it was found not to have any native fruits of commercial value. Consequently, the fruit-growing industry which now exists is based entirely on plants transplanted to the Australian environment. The range of climates allows production of both tropical and mid-latitude fruits, the most important items being pineapples from Queensland, bananas from New South Wales and Queensland, citrus fruits mainly from New South Wales, grapes for wine and table use produced under irrigation in the Murray Basin and near Adelaide, and apples and pears produced especially in Tasmania and Victoria. The home market is abundantly supplied with fruit, and there is normally a substantial export.

Sugar Production

Australia is also an export sugar producer, though at a cost. Sugar production was begun on the coastal plain of Queensland in the middle of the nineteenth century to supply the Australian market. Laborers were imported from Melanesia to work in the cane fields on what often amounted practically to a slave basis. When the Commonwealth of Australia was formed in 1901 by union of the six states (New South Wales, Victoria, Queensland, South Australia, Western Australia, Tasmania), Queensland was required to repatriate the imported "Kanaka" laborers in the interest of a "white Australia." However, since it was felt that production costs would be considerably raised by the use of exclusively white labor, the Commonwealth gave Queensland in return a high protective tariff on sugar. Operating behind this protective wall, the sugar industry has prospered in a tropical area where it was once argued that no white man could survive for very long at manual labor in the fields. Most of the original large estates have been divided into family farms worked by individual white farmers. This system of agriculture is very unusual for a commercial cane-sugar area. The Queensland sugar area is also unusual in the fact that it is probably the most thoroughly tropical area yet settled by white men doing hard manual work without benefit of native labor.

Urbanization and Industrialization

While grazing and agriculture produce the great bulk of Australia's exports, most Australians are city dwellers, and the high degree of urbanization is one of the country's outstanding characteristics. In 1968 the estimated metropolitan-area populations of the five largest cities were as follows: Sydney, 2.6 million; Melbourne, 2.3 million; Brisbane, 800,000; Adelaide, 800,000; and Perth (including its port, Freemantle), 580,000. Thus around 40 percent, or two out of every five Australians, were in the two largest metropolitan areas, and around 60 percent in one of the five largest. When smaller cities and towns are added the Australian population may be counted as around 85 percent urban.

The central business district and wharves of Sydney, Australia's largest city and port. The famous Harbor Bridge connects the northern suburbs with the inner city. The bridge carries road and rail traffic across the broad inlet leading to the inner harbor at the right of the view. The modernistic new Sydney Opera House, still under construction when the photo was taken, is on Bennelong Point to the left of the Harbor Bridge. (Australian News and Information Bureau.)

It will be noted that all of the five largest cities are seaports and that each is the capital of one of the five mainland states of the Commonwealth. One of the reasons for the striking degree of urban development and its concentration in the port cities is the heavily commercial nature of the Australian economy. A large proportion of the production from the country's rural areas is destined for export by sea. In addition, much of Australia's internal trade moves from port to port by coastal steamer. Growth of the respective capital-city ports has been further stimulated by the fact that each state originally built its own individual rail system focusing on its particular port. Since the different states used different gauges in building their rail lines, necessitating expensive break of bulk at or near state frontiers, each

major port tended to effectively dominate the business of its own state. (The rail system is now being changed by substitution of a standard gauge throughout the Commonwealth.)

Among other things, the high degree of urbanization in Australia is indicative of a high degree of industrialization. A relatively small labor force is needed for the extensive forms of agriculture and grazing which are characteristic of Australia. As a means of supporting an increased population, the Australian government has encouraged industrial development. The latter has been viewed as highly desirable, not only as a means of giving Australia a population more in keeping with the size and resources of the continent, but also as a means of providing more adequate armaments for defense and of securing greater stability through a more diversified and self-contained economy.

Australia's Mineral Wealth

The rise of industry was facilitated by the formation of the Commonwealth, which gave a unified internal market and tariff protection for developing industries. It also has been facilitated by mineral wealth. As in the case of agriculture, Australia's small population is able to draw upon the resources of a continent; and in mineral resources the continent is relatively rich. Consequently, besides supplying most of its own needs for minerals Australia has long been an important supplier to the rest of the world. In recent years large new discoveries have been expanding this long-time role in international trade. (See map of minerals p. 532.)

Ample supplies of coal have been fundamental to Australian industrialization. Every Australian state mines some coal, but the major reserves are near the coast in New South Wales and Queensland. The main developed field outcrops on three sides of Sydney. Most of Australia's steel capacity is located near the coal mines at Newcastle (340,000) to the north of Sydney and at Port Kembla to the south. Ore is brought from distant deposits along or near other sections of the Australian coast. With materials either at hand or assembled by ocean freighter, production costs are among the lowest in the world, with the result that Australia is a net exporter of steel. In addition, an increasing export of coal, mainly to Japan, is coming from the fields near Sydney.

Abundant iron ore located near the sea and thus conveniently accessible to cheap ocean transportation is another mineral resource of prime importance. Most of the ore for the steel plants near Sydney comes from deposits slightly inland from the port of Whyalla in South Australia. Whyalla itself is a growing center of iron and steel production, making use of coal brought by ore freighters on the return trip from the east coast. Major new iron ore discoveries and developments in recent years

Coal is a major source of energy in Australia. The coal mine in the view is on the coast of New South Wales about 60 miles south of Sydney. (Australian News and Information Bureau.)

bid fair to make Australia one of the world's leading suppliers. As of the late 1960s ore was being shipped from two islands in Yampi Sound, on the remote northern coast of Western Australia, and other large deposits, newly discovered, were being equipped for production in two other areas of the same state and in northwestern Tasmania. Much of this greatly expanded production is designed for export to Japan under a long-term trade agreement with that country.

Major recent discoveries of bauxite also have been made; the north Queensland coast, the Darling Range inland from Perth, and coastal locations in Arnhem Land have been the areas of discovery. Production has commenced in all of them, with the result that Australia is an increasingly important exporter of bauxite. In addition, new aluminum plants are supplying the Australian market and providing increasing exports of the finished metal.

Among a host of other minerals produced in Australia those of unusual importance include lead, zinc, copper, titanium ores, and gold. Two famous interior mining settlements, Broken Hill in New South Wales and Mount Isa in Queensland, provide much of the production. Mount Isa produces copper, silver, lead, and zinc, while the complex ores at Broken Hill yield lead, zinc, silver, and gold as the primary products and other metals in smaller amounts. Mainly from the output of the two mining centers Australia normally accounts for 10 to 15 percent of the world's lead production and 5 to 10 percent of its production of zinc and silver. Broken Hill, located in a very dry environment, is one of the better known mining cities in the world. The Broken Hill Proprietary Company, which developed the mines, later branched into steel production and has become the largest Australian corporation. Titanium ores (rutile and ilmenite) are mined in coastal Queensland and Western Australia, with Australia being an exporter of considerable consequence. Gold has played a large role in Australian history and is still of some importance. Strikes in the 1850s in Victoria brought one of the world's major gold rushes and stimulated a threefold increase in the continent's population in 10 years. The present Australian output amounts to only 2 to 3 percent of the world total, with over half of it coming from the mines of Kalgoorlie in Western Australia.

The most serious deficiency marring the picture of Australia's mineral abundance is petroleum, the great bulk of which must be imported. By the late 1960s extensive exploration had uncovered only two small fields justifying commercial development. They lie relatively close together around 200 miles inland from Brisbane, with which they are connected by pipeline. At the time of writing (late 1968) exploration was being continued in many localities. A number of wells had brought up oil, but no large commercial fields had been discovered.

Relative Inefficiency of Australian Industries

Despite the extent of Australian industrialization, and despite considerable advantages with regard to raw materials, Australia is still an importer of most kinds of manufactured goods. This appears to be partly a reflection of a high standard of living bolstered by the earning power of grazing and agriculture. But it is also a reflection of the relative inefficiency of much of Australia's industry in terms of average production per worker. Available data indicate that the average Australian industrial worker produces much less than his American counterpart. This difference in productivity may be partly a matter of different standards—Australia's emphasis on leisure and leisureliness has often impressed American observers—but in all probability it is largely due to the fact that Australia's domestic market is so small and so dispersed that many Australian industrial establishments cannot produce and distribute on a large enough scale to minimize unit costs and use labor most efficiently. In the middle 1960s the average Australian industrial establishment had about 20 employees, as against over 50 in the average American establishment.

Nevertheless, Australia's policy has been to encourage industrial development, and it has done so with notable success. Political and defense considerations are involved in this policy. Australia has felt the need since World War II to build up its population rapidly by encouraging selective immigration, and industrialization has provided jobs for the expanding population. Growing industries have also bolstered the country's defensive capacity. It can be argued that expansion of the population, and hence the domestic market, will allow Australian industry to achieve greater efficiency as total output increases and thus to give the country a more diversified and soundly based economy. These things do appear to be happening. Between 1945 and 1965 approximately 2.3 million immigrants entered Australia, almost half of them under programs of the Australian government to assist financially the types of immigrants it desires. Most immigrants have been British, but many other nationalities—the Italians, Dutch, Greeks, Germans, Yugoslavs, Poles, Hungarians, and others—have been represented. The newcomers found jobs primarily in manufacturing industries, which increased greatly in output and efficiency during the period, and in mining. Both immigration and industrial growth are continuing.

Index map of New Zealand.

NEW ZEALAND

New Zealand, over 1000 miles southeast of Australia, consists of two large islands, North Island and South Island, separated by Cook Strait (map, above), and a number of smaller islands. North Island is smaller than South Island but contains the majority of the population, which numbered about 2.8 million in 1968. Since the total area is approximately 103,500 square miles, the population density averages about 27 per square mile. New Zealand is thus a sparsely populated country, although not so much so as Australia.

Physical and Climatic Characteristics

Rugged terrain is found throughout much of New Zealand. The topography of South Island is dominated by the Southern Alps, often cited as one of the world's most spectacular mountain ranges. These mountains present large areas above 5000 feet and many glaciated summits above 10,000 feet. The mountains of North Island are less imposing and extensive, but many peaks exceed 5000 feet.

The highlands of New Zealand lie in the Southern Hemisphere belt of westerly winds and receive abundant precipitation. Lowlands generally receive over 30 inches of precipitation, fairly evenly distributed throughout the year, and highlands often receive over 130 inches. Precipitation drops to less than 20 inches in small areas of rain shadow east of the Southern Alps. Temperatures are those of middle latitudes, moderated by the pervasive maritime influence. The result is a wet temperate climate commonly classified as marine west coast, with warm month temperatures generally averaging 60°–70°F and cool month temperatures 40°–50°F. Highland temperatures are more severe, and a few glaciers are found on both islands.

Importance of Pastoral Industries

Rugged terrain and excessive precipitation in the Southern Alps and the mountainous core of North Island have resulted in almost a total absence of population from a third of New Zealand and have rendered about a fifth of the country completely unproductive, except for the notable attractions offered by the mountains to a budding tourist industry. Well-populated areas are restricted to fringing lowlands around the periphery of North Island and along the drier east and south coasts of South Island.

The climate of New Zealand lowlands is ideal for growing grass and raising livestock, and almost half of the country's total area is maintained in pastures and meadows which support a major sheep and cattle industry. The outstanding importance and productivity of the pastoral industries in New Zealand is indicated in the following facts:

1. These industries completely dominate New Zealand's exports. In 1965, for example, wool, meat, and dairy products accounted for 29, 28, and 27 percent, respectively, of the total value of exports.

2. New Zealand ordinarily stands near the top among the world's countries in per capita trade.

3. New Zealand generally ranks first in the world in the export of mutton, lamb, and butter; second in cheese, preserved milk, and wool; and third or fourth in beef exports.

Dairy cattle on a well-kept pasture in New Zealand. The volcanic cone in the distance is Mount Egmont on North Island. (New Zealand Embassy.)

4. The earning power of its pastoral exports has been a major factor in New Zealand's exceptionally high standard of living, since they pay for the varied imports which the country requires. Not only manufactured goods are imported, but also considerable quantities of food. Only about 3 percent of New Zealand is ordinarily in cultivation, and most of that is devoted to the growing of supplemental animal feeds. Large imports of bread grains are necessary.

The Canterbury Plains on the drier eastern side of South Island contain the largest acreages of cultivated land and are the main wheat-growing section.

Industrial and Urban Development

In New Zealand as in Australia, however, basic dependence on the pastoral industries is accompanied by a high degree of urbanization and active attempts to develop manufacturing. The two countries are quite similar with respect to conditions and purposes fostering urban and industrial development. New Zealand's resources for manufacturing do not equal those of Australia, but coal,

iron, and a number of other minerals are present in at least modest quantities, and there are very considerable potentials for hydroelectric power development. Considerable possibilities for forest production also exist, although the magnificent natural forests of New Zealand are now seriously depleted and imports of lumber are required. To an even greater degree than Australia, New Zealand has the problem of a small internal market militating against mass production at low cost. Most of the present manufacturing industries are high-cost producers protected by tariffs.

A somewhat lower degree and a much smaller scale of urbanization than in Australia is indicated by 1968 population estimates for the metropolitan areas of New Zealand's four main cities: Auckland, 580,000; Wellington, the capital (with its major suburb, Hutt), 300,000; Christchurch, 260,000; Dunedin, 110,000. These figures assign about 31 percent of New Zealand's population to the two largest urban centers, and about 45 percent to all urban centers over 100,000. This is a smaller proportion in each case than in Australia, and the largest New Zealand metropolitan area is only about the same size as the fifth largest in Australia. In fact, Sydney has almost as many people as all of New Zealand.

REFERENCES and READINGS

Australia

Atlas of Australian Resources (Sydney: Angus and Robertson, 1952--).

Australian Geographer (two issues a year).

Australian Geographical Studies (two issues a year).

BEALE, JOHN V., "Broken Hill–A Living Legend," *Mining Engineering,* 16 (October 1964), 70–75.

BEARD, J. S., "Some Vegetation Types of Tropical Australia in Relation to Those of Africa and America," *Journal of Ecology,* 55, no. 2 (July 1967), 271–290.

BERNEY, MAURICE, ed., *Australia* (Bern, Switzerland: Kümmerly and Frey, 1965; distributed in the United States by Rand McNally & Co., Chicago, Ill.). Fine color photos; informative text for the general reader.

BIRD, JAMES, "The Foundation of Australian Seaport Capitals," *Economic Geography,* 41, no. 4 (October 1965), 283–299.

BIRKET-SMITH, KAJ, *Primitive Man and His Ways: Patterns of Life in Some Native Societies* (London: Odhams Press, 1960), "Australian Aborigines: Food Gatherers and Primitive Hunters; Isolation and Uniformity," pp. 15–58.

BJORKLUND, ELAINE M., *Focus on Adelaide: Functional Organization of the Adelaide Region, Australia* (University of Chicago, Department of Geography Research Paper No. 41; Chicago: 1955).

BROWN, D. A., and D. S. SIMONETT, "Integration and Locational Change in the Australian Electricity Industry: 1951–1965," *Economic Geography,* 43, no. 4 (October 1967), 283–302.

DAVIDSON, BRUCE ROBINSON, *The Northern Myth: A Study of the Physical and Economic Limits to Agricultural and Pastoral Development in Tropical Australia* (New York: Cambridge University Press, 1965).

DOWNER, ALEXANDER, "The Influence of Immigration on Australia's National Character," *Journal of the Royal Society of Arts,* 115, no. 5127 (February 1967), 190–202.

DRISCOLL, E. M., "Recent Developments in the Queensland Sugar Industry," *Geography,* 51, pt. 2 (April 1966), 147–149.

DUNCAN, CRAIG, "The Aluminum Industry in Australia," *Geographical Review,* 51, no. 1 (January 1961), 21–46.

GARZOLI, K. V., "Irrigated Agriculture in Australia," *World Crops,* 18, no. 4 (December 1966), 36–42; and "The Trend of Irrigated Agriculture in Australia," *World Crops,* 19, no. 1 (March 1967), 28–33.

HEATHCOTE, R. L., *Back of Bourke: A Study of Land Appraisal and Settlement in Semi-arid Australia* (New York: Cambridge University Press, 1965).

HEFFORD, R. K., and DEREK L. SMITH, "The Leigh Creek Coalfield, South Australia: Resource Development in a Difficult Environment," *Annals of the Association of American Geographers,* 57, no. 3 (September 1967), 503–518.

HORNE, DONALD, "Australia Looks Around," *Foreign Affairs,* 44, no. 3 (April 1966), 446–457.

KERR, ALEX, *Australia's North-West* (Nedlands: University of Western Australia Press, 1967).

LINGE, G. J. R., "Governments and the Location of Secondary Industry in Australia," *Economic Geography,* 43, no. 1 (January 1967), 43–63.

LOGAN, M. I., "Capital City Manufacturing in Australia," *Economic Geography,* 42, no. 2 (April 1966), 139–151; and "Locational Behavior of Manufacturing Firms in Urban Areas" [in Australia], *Annals of the Association of American Geographers,* 56, no. 3 (September 1966), 451–466.

LOGAN, W. S., "The Changing Landscape Significance of the Victoria-South Australia Boundary," *Annals of the Association of American Geographers,* 58, no. 1 (March 1968), 128–154.

MARSHALL, ANN, "Beef Roads: Revolution by Transportation," *Landscape,* 15, no. 3 (Spring 1966), 9–13.

MATHER, D. B., "Transport and Region in Australia," in J. W. House, ed., *Northern Geographical Essays in Honour of G. H. J. Daysh* (Newcastle upon Tyne, England: Oriel Press, 1966), pp. 310–330.

McLEOD, ALAN LINDSEY, *The Pattern of Australian Culture* (Ithaca, N.Y.: Cornell University Press, 1963).

MEINIG, DONALD W., *On the Margins of the Good Earth: The South Australian Wheat Frontier, 1869–1884* (Chicago: Rand McNally & Co. for the Association of American Geographers, 1962).

MULVANEY, D. J., "The Prehistory of the Australian Aborigine," *Scientific American,* 214, no. 3 (March 1966), 84–93.

PERRY, T. M., *Australia's First Frontier: The Spread of Settlement in New South Wales 1788–1829* (New York: Cambridge University Press, 1965).

ROBINSON, K. W., "The Political Influence in Australian Geography," *Pacific Viewpoint,* 3, no. 2 (September 1962), 73–86; and "Sixty Years of Federation in Australia," *Geographical Review,* 51, no. 1 (January 1961), 1–20.

ROSE, A. JAMES, *Dilemmas Down Under: Australia and the Southwest Pacific* (Princeton, N.J.: D. Van Nostrand Company, Searchlight Books, 1966); and "Dissent from Down Under: Metropolitan Primacy as the Normal State," *Pacific Viewpoint,* 7, no. 1 (May 1966), 1–27.

RUTHERFORD, J., M. I. LOGAN, and G. J. MISSEN, *New Viewpoints in Economic Geography* (Sydney: Martindale Press, 1966), chapters as follows: M. I. Logan, Chap. 18, "Manufacturing in Australia: A Survey of the Areal Distribution of Manufacturing Plants," pp. 353–368; Chap. 17, "Manufacturing Location within Cities: A Summary of Major Locational Factors with Special Reference to Sydney, Australia," pp. 332–352; and Chap. 22, "Metropolitan Sydney: A Case Study of the Internal Structure of the Large Western City," pp. 433–459. J. Rutherford, Chap. 10, "Commercial Dairying in Australia: An Outline of Structural and Locational Attributes with Special Reference to New South Wales," pp. 190–220; and Chap. 9, "Woolgrowing in the Western Division of New South Wales: A Study of Reactions to Climatic Uncertainty and of Government Measures to Conserve Resources," pp. 169–189.

SCOTT, PETER, "The Population Structure of Australian Cities," *Geographical Journal,* 131, pt. 4 (December 1965), 463–481; and "Population and Land Use in Australia: An Analysis of Metropolitan Dominance," *Tijdschrift voor Economische en Sociale Geografie,* 59, no. 5 (September–October 1968), 237–244.

SMITH, ROBERT H. T., "Transport Competition in Australian Border Areas: The Example of Southern New South Wales," *Economic Geography,* 39, no. 1 (January 1963), 1–13.

SOLOMON, R. J., "Locational Emphasis of the Australian Work Force," *Economic Geography,* 38, no. 2 (April 1962), 138–161.

VAN ZYL, F. D. WALLACE, "Water Resource Planning in Australia," *Tijdschrift voor Economische en Sociale Geografie,* 58, no. 6 (November–December 1967), 306–315. Focuses particularly on the Snowy Mountains scheme.

WILSON, M. G. A., "The Coal Traffic of Eastern Australia," *Economic Geography,* 43, no. 2 (April 1967), 128–142; also, "Changing Patterns of Pit Location on the New South Wales Coalfields," *Annals of the Association of American Geographers,* 58, no. 1 (March 1968), 78–90; and "Natural Gas Development in Australia," *Geography,* 51, pt. 4 (November 1966), 375–379.

New Zealand

FARRELL, BRYAN H., *Power in New Zealand: A Geography of Energy Resources* (Wellington: A. H. and A. W. Reed, 1962).

FIELDING, G. J., "The Role of Government in New Zealand Wheat Growing," *Annals of the Association of American Geographers,* 55, no. 1 (March 1955), 87–97.

FRANKLIN, S. H., "Immaturity and Affluence: The Social and Economic Problems of New Zealand's Colonial Status," *Geography,* 52, pt. 1 (January 1967), 1–11.

HEENAN, L. D. B., "The Changing South Island Maori Population," *New Zealand Geographer,* 22, no. 2 (October 1966), 125–165.

KING, LESLIE J., "Urbanization in an Agriculturally Dependent Society: Some Implications in New Zealand," *Tijdschrift voor Economische en Sociale Geografie,* 56, no. 1 (January–February 1965), 12–21.

LEWTHWAITE, GORDON R., "Wisconsin and the Waikato: A Comparison of Dairy Farming in the United States and New Zealand," *Annals of the Association of American Geographers,* 54, no. 1 (March 1964), 59–87.

LINGE, G. J. R., "The Concentration and Dispersion of Manufacturing in New Zealand," *Economic Geography,* 36, no. 4 (October 1960), 326–343; and, with R. M. Frazer, *Atlas of New Zealand Geography* (Wellington: A. H. and A. W. Reed, 1966).

MAUNDER, W. J., "Climatic Variations and Agricultural Production in New Zealand," *New Zealand Geographer,* 22, no. 1 (April 1966), 55–69.

METGE, JOAN, *The Maoris of New Zealand* (Societies of the World, ed. by John Middleton; London: Routledge and Kegan Paul, 1967).

New Zealand Geographer (two issues a year).

POOL, D. I., "The Rural-Urban Migration of Maoris: A Demographic Analysis," *Pacific Viewpoint,* 7, no. 1 (May 1966), 88–96.

RIMMER, PETER J., "The Changing Status of New Zealand Seaports, 1853–1960," *Annals of the Association of American Geographers,* 57, no. 1 (March 1967), 88–100.

RUTHERFORD, J., "Technical Change and Grassland Farming in New Zealand: A National Assessment of Some Features of Western Economic Growth," in J. Rutherford, M. I. Logan, and G. J. Missen, *New Viewpoints in Economic Geography* (Sydney: Martindale Press, 1966), pp. 137–168.

STOKES, EVELYN, "Kauri and White Pine: A Comparison of New Zealand and American Lumbering," *Annals of the Association of American Geographers,* 56, no. 3 (September 1966), 440–450; and "Timber, Pulp and Paper: Production and Trade, 1950–1964," *New Zealand Geographer,* 22, no. 1 (April 1966), 70–79.

WATTERS, R. F., ed., *Land and Society in New Zealand: Essays in Historical Geography* (Wellington, N.Z.: A. H. and A. W. Reed, 1965).

See also the References and Readings for Chapter 22.

PART VII

AFRICA

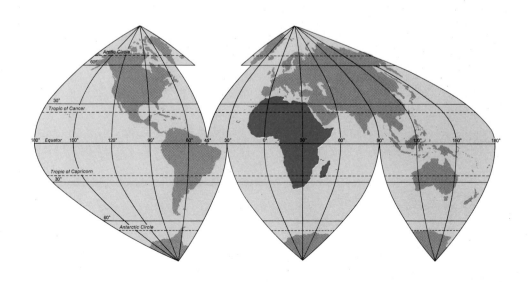

Introduction
to Africa

24

Recent decades have seen a worldwide surge of interest in the developing continent of Africa. Although many facets of African life have drawn attention, the world's interest has focused primarily on the continent's mineral wealth, its racial and tribal problems, its drive for political independence, and its increasing modernization.

Africa has long been famous for its gold and diamonds, and it still leads the world in the production of these minerals. But today it is the *diversity* of Africa's mineral wealth that is attracting more and more attention. The continent is a major producer of copper, cobalt, chromium, manganese, antimony, platinum, phosphate, and uranium; is a significant producer of petroleum, iron ore, tin, vanadium, lead, zinc, bauxite, and asbestos; has large natural gas reserves, exploited up to now in only a minor way; and has in its southern reaches impressive reserves and a substantial production of coal. A variety of other minerals are produced; for example the rare metal columbium (used in making alloys that will withstand extreme heat). Africa is the world's leading source. Large unexploited reserves of a great many valuable minerals are known to exist, and new discoveries are frequent.

AFRICA

POLITICAL UNITS AND URBAN AREAS

January, 1969

✪ ★ ◉	1,000,000 or over
◐ ●	500,000–1,000,000
◑ ○	250,000–500,000
★ ○	Selected smaller places

Stars show capitals

City-size symbols are based on metropolitan-area estimates

AZORES IS. (PORT.)

SPAIN
PORTUGAL
ITALY
GREECE

MEDITERRANEAN SEA

Strait of Gibraltar
Ceuta (Sp.)
Tangier
Melilla (Sp.)
Rabat
Fez

MADEIRA IS. (PORT.)

Algiers
Bougie Bône
Oran
Constantine
Tunis
TUNISIA
Sfax
Tripoli

Casablanca
Marrakesh
MOROCCO
Ghardaia

SANTA CRUZ de Tenerife
CANARY IS. (SP.)
Las Palmas

Benghazi
Beida

Alexandria
Port Said
Cairo
Suez
ISRAEL

UAR (EGYPT)

ALGERIA

LIBYA

SAUDI

ARABIA

45°E

SPANISH SAHARA (SP.)

Tropic of Cancer

Fort Gouraud

In Salah

Aswan

Port Etienne

MAURITANIA

Nouakchott

St. Louis
Dakar
SENEGAL
Bathurst
Kaolack
THE GAMBIA
Bissau
PORT. GUINEA (PORT.)
Conakry
Pepel
Freetown
SIERRA LEONE
Monrovia
Buchanan

Timbuktu
Gao

MALI
Kayes
Bamako

NIGER
Niamey

UPPER VOLTA
Ouagadougou

GUINEA

IVORY COAST
Kumasi
GHANA
Abidjan
Akosombo
TOGO
DAHOMEY
Lomé
Accra
Tema
Sekondi-
Takoradi

Wadi Halfa (submerged by High Dam reservoir)

CHAD

Kano
Ft.-Lamy

NIGERIA
Kaduna
Ogbomosho
Ibadan
Lagos
Enugu
Port Harcourt

Khartoum

SUDAN

Asmara
ERITREA
Aduwa
Massawa
Assab

YEMEN

FRENCH TERRITORY OF AFARS AND ISSAS (FR.)
GULF OF ADEN
Djibouti
Berbera

Addis Ababa

ETHIOPIA

SOMALI REPUBLIC

CAPE VERDE IS. (PORT.)

25°W

CAMEROON
Douala
Yaoundé
Edéa

Santa Isabel
FERNANDO PO
EQUATORIAL GUINEA
RIO MUNI
PRINCIPE (PORT.)
SÃO TOMÉ (PORT.)
ANNOBON

CENTRAL AFRICAN REPUBLIC
Bangui

REP. OF THE CONGO
Libreville
Port Gentil
GABON
Brazzaville
Pointe Noire
CABINDA (PORT.)

DEMOCRATIC

REPUBLIC OF

THE

CONGO

Kisangani (Stanleyville)
Port Francqui

Matadi
Kinshasa (Léopoldville)
KASAI REGION
Luluabourg

Juba

Kilembe
UGANDA
Kampala
Entebbe
RWANDA
Kigali
BURUNDI
Bukavu
Bujumbura
KIVU REGION
Kigoma

Mogadiscio

Equator

SEYCHELLES IS. (BR.)

KENYA
Nairobi

Arusha
Moshi
Shinyanga
Tanga
Tabora
PEMBA I.
Zanzibar
ZANZIBAR I.

Mombasa

15°N

ATLANTIC

OCEAN

0   50   100
Miles

LANDFORMS
(broadly generalized)

Nile R.

KATANGA REGION
Jadotville
Lubumbashi (Elisabethville)

Luanda

ANGOLA (PORT.)

Lobito

Moçamedes

Nova Lisboa

Cassinga

TANZANIA

Mpanda
Dar es Salaam

Mtwara

COMORO IS. (FR.)

INDIAN

OCEAN

Niger R.

Equator

Congo R.

Ndola
Broken Hill
ZAMBIA
Lusaka

MALAWI
Zomba
Blantyre

Nacala

Livingstone
Wankie
Maun
Bulawayo
RHODESIA
Salisbury
Que Que
Umtali

HIGH VELD
LOW VELD

MOZAMBIQUE (PORT.)

Beira

Tamatave
MALAGASY REPUBLIC (MADAGASCAR)
Tananarive

Port Louis
MAURITIUS
RÉUNION (FR.)

Zambezi R.

Orange R.

SOUTH WEST

AFRICA

Walvis Bay

Windhoek

BOTSWANA

Gaberones

Pretoria
Mbabane
SWAZILAND
Lourenço Marques

Johannesburg
Vereeniging
Kimberley

LESOTHO
Maseru

Durban

Tropic of Capricorn

Plains

Plateaus: Low elevation

Medium elevation

High elevation

Hills and low mountains

Mountains

0   500   1,000
Miles

Port Nolloth

Bloemfontein

REPUBLIC OF

SOUTH AFRICA

Cape Town

East London

Port Elizabeth

0   500   1,000
Scale of Miles

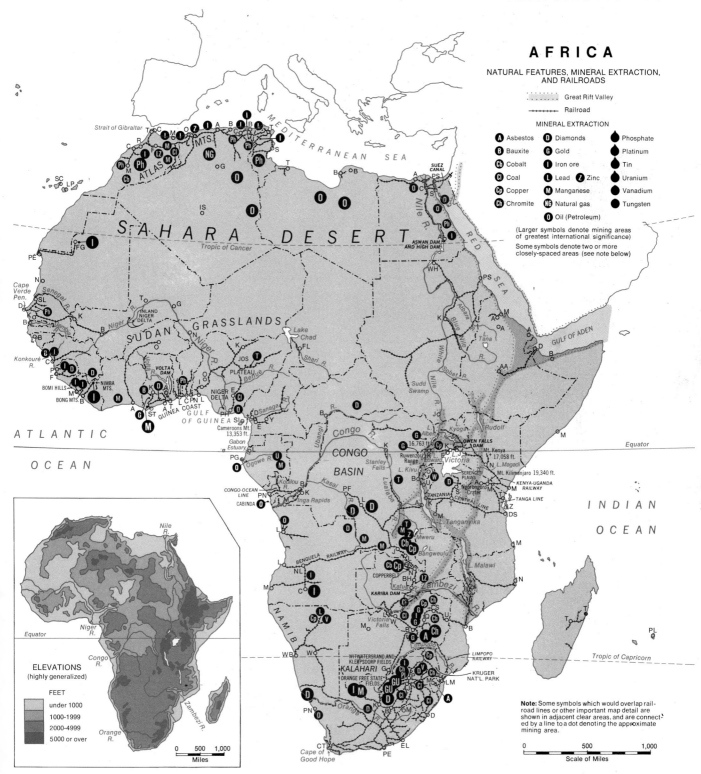

AFRICA

NATURAL FEATURES, MINERAL EXTRACTION, AND RAILROADS

Great Rift Valley

Railroad

MINERAL EXTRACTION

Ⓐ Asbestos	Ⓓ Diamonds	⬤ Phosphate
Ⓑ Bauxite	Ⓖ Gold	⬤ Platinum
Ⓒb Cobalt	Ⓘ Iron ore	⬤ Tin
Ⓒl Coal	Ⓛ Lead	Ⓩ Zinc	⬤ Uranium
Ⓒp Copper	Ⓜ Manganese	⬤ Vanadium
Ⓒh Chromite	Ⓝg Natural gas	⬤ Tungsten
	Ⓞ Oil (Petroleum)	

(Larger symbols denote mining areas of greatest international significance)

Some symbols denote two or more closely-spaced areas (see note below)

Note: Some symbols which would overlap railroad lines or other important map detail are shown in adjacent clear areas, and are connected by a line to a dot denoting the approximate mining area.

ELEVATIONS
(highly generalized)

FEET

under 1000
1000-1999
2000-4999
5000 or over

0   500   1,000
Miles

0   500   1,000
Scale of Miles

Many of Africa's largest and most modern cities are seaports. This view shows the downtown section of Durban, South Africa, the largest African seaport on the Indian Ocean. (South African Tourist Corporation.)

But Africa's minerals, important as they are, attract less attention than do its racial, tribal, and colonial problems. Tensions between Europeans and Africans[1] are the most familiar issues, but many others exist. Prior to World War II, most of Africa was controlled by Europeans. Only a handful of independent nations existed; the remaining units were possessions of European colonial powers. In these possessions Europeans were a privileged class, both socially and economically. But since the war, a drive for independence by African peoples has remade the political map of the continent and is bringing great social and economic changes. From 1951 through 1968 more than three dozen independent nations emerged from a colonial status (map, p. 65). Thus in most of Africa, European political control has ended and African governments have taken

---

[1] The term "Europeans" (persons of European extraction), as commonly used in reference to residents of Africa, includes both temporary and permanent residents. Some of these are newcomers, while others have family roots in Africa that go back for many generations. In its most common usage, the term "Africans" refers to the Negroid peoples who comprise the majority of the continent's inhabitants. It is not customarily used in reference to Arabs and Berbers, who make up most of the population of North Africa, nor to the Indian and Pakistani minority in eastern Africa. Due to racial and cultural mixing, neither "Europeans" nor "Africans" can be defined very precisely.

charge. Meanwhile a far-reaching modernization of Africa's economic, social, and cultural life is in progress. It is centered in the rapidly growing cities, but its influence is spreading to every corner of the continent. As horizons broaden, however, many traditional patterns persist, and the result is an intricate mixture of old and new influences in African life.

AREA, POPULATION, AND MAJOR AREAL DIVISIONS

Africa is the largest in land area of the major world regions discussed in this book. Its 11.6 million square miles (including Madagascar) make it more than three times times as large as the United States. But much of it is sparsely populated (though in relation to resources and technology many areas are actually overpopulated). The continent's rapidly increasing population was estimated in 1968 at 336 million, giving an average density of about 29 per square mile, as compared with an estimated population of 201 million or 56 per square mile for the United States. Seven main areas of comparatively dense population exist in Africa (map, p. 42). They are (1) the coastal belt bordering the Gulf of Guinea in West Africa, from southern Nigeria (including the secessionist area of Biafra) westward through the southern parts of Dahomey, Togo, and Ghana (for location of political units, see map, p. 550); (2) the savanna lands in the northern third of Nigeria; (3) the Nile Valley and Delta in Egypt; (4) the coastal fringes and some of the Atlas ranges of northwestern Africa in Algeria, Morocco, and Tunisia; (5) the highlands of Ethiopia; (6) the highland region surrounding Lake Victoria in Kenya, Uganda, Rwanda, Burundi, and Tanzania; and (7) the eastern coast and parts of the high interior plateau (High Veld) of the Republic of South Africa. Each of the foregoing regions is a key area in the political and economic geography of Africa. Certain areas in Africa containing smaller concentrations of people have been notable foci of mining and/or transportation and European influence. Prominent among these are (1) the Katanga mining region of the Congo and the adjoining Copperbelt of Zambia (map, p. 551); (2) the Salisbury–Bulawayo mineralized axis in the highlands of Rhodesia; and (3) the ocean and rail gateway to the Congo Basin formed by the port of Matadi and the capital city of Kinshasa (formerly Léopoldville) in the former Belgian Congo, and the port of Pointe Noire and capital city of Brazzaville in the former French Congo. Area and population data for the individual political units of Africa are given in Table 20.

Africa falls broadly into the two major divisions of North Africa, the predominantly Caucasoid Arab-Berber realm of the continent; and Africa south of the Sahara, which is dominantly Negro in population. North Africa is mostly comprised of desert or steppe with a fringe of mediterranean climate in northwestern coastal sections, while Africa south of the Sahara is mainly a region of tropical and subtropical grassland or deciduous forest, but with sizable areas of tropical rain forest and of dry lands, including some desert. Since North Africa has been considered in some detail in Chapters 15 and 16, major attention here will be focused on sub-Saharan Africa.

HISTORICAL BACKGROUND

By European or American standards, most parts of Africa were still very poorly developed at the end of the nineteenth century. This was least true of the northern and southern extremities. Parts of northern Africa have been civilized for thousands of years, and by 1900 large sections of South Africa had been occupied by European settlers. However, at the dawn of the twentieth century the vast spaces of interior tropical Africa were still primitive and little known. The isolation of much of Africa from contact with more advanced areas until comparatively recent times was due to a combination of factors. Among these were (1) the presence of the great Sahara Desert barrier in North Africa, making difficult the penetration of tropical Africa by land from the north, and even hindering early sailing expeditions due to the lack of water and fresh provisions along the shore; (2) the inhospitable character of the African shoreline, which is exceptionally regular, thus lacking protected harbors, sometimes is lined by submerged rocks and reefs, often has heavy surf, and is bordered in many places by extensive sandbars and/or thick tropical forest (including considerable stretches of mangrove swamp) or desert; (3) the character of African rivers, which are often shallow and full of sandbars near their mouths, generally fluctuate a great deal in water level from season to season, and are blocked inland at varying distances by falls and rapids, thus rendering difficult penetration of the interior by water; (4) the hot, humid climate and the many diseases encountered by outsiders in the coastal areas; (5) the hostility of many African tribes, due in considerable part to slave trading by Europeans, Americans, and Arabs; (6) the opposition of slave traders to penetration

TABLE 20 AFRICA: AREA AND POPULATION DATA

POLITICAL UNIT	AREA (THOUSAND SQUARE MILES)	POPULATION (MILLIONS: 1968 ESTIMATES)	DENSITY (PER SQUARE MILE)
Arab League States			
United Arab Republic (Egypt)	386.0	31.8	82
Republic of the Sudan	967.2	14.8	15
Kingdom of Libya	679.2	1.8	3
Republic of Tunisia	63.4	4.7	74
Democratic and Popular Republic of			
Algeria	919.4	12.9	14
Kingdom of Morocco	172.4	14.7	85
Totals	3187.5	80.7	25
Spanish and Portuguese Territories in			
Northwestern Africa			
Spanish Sahara (Spanish African province)	102.7	0.05	0.5
Spanish North Africa[a]	0.012	0.16	13333
Canary Islands (integral part of Spain)	2.8	1.1	392
Totals, Spanish-controlled Territories	105.6	1.3	12
Madeira Islands (integral part of Portugal)	0.31	0.27	880
Eastern Horn			
Empire of Ethiopia	471.7	23.8	50
Somali Republic	246.1	2.8	11
French Territory of Afars and Issas			
(French overseas territory)	8.5	0.10	12
Totals	726.3	26.7	37
West Africa			
Islamic Republic of Mauritania	397.9	1.1	3
Republic of Senegal	75.7	3.8	50
Republic of Mali	463.9	4.8	10
Republic of Niger	459.1	3.6	8
Republic of Upper Volta	105.9	5.2	49
Republic of Ivory Coast	124.5	4.1	33
Republic of Dahomey	43.5	2.5	57
Republic of Togo	21.9	1.8	81
Republic of Guinea	94.9	3.8	40
Totals, Former French-controlled Territories	1787.2	30.6	17
Republic of Nigeria[b]	356.7	62.0	174
Republic of Ghana	92.1	8.4	91
Sierra Leone (independent parliamentary state)	27.7	2.5	89
The Gambia (independent parliamentary state)	4.4	0.35	80
Totals, Former British-controlled Territories	480.8	73.1	152
Republic of Liberia	43.0	1.1	26
Portuguese Guinea (Portuguese overseas province)	13.9	0.53	38
Cape Verde Islands (Portuguese overseas province)	1.6	0.20	131
Totals, Portuguese-controlled Territories	15.5	0.73	47
Equatorial Africa			
Federal Republic of Cameroon	183.5	5.6	30
Gabon Republic	103.3	0.48	5
Republic of the Congo (Brazzaville)	132.0	0.88	7
Central African Republic	240.5	1.5	6
Republic of Chad	495.6	3.5	7
Totals, Former French-controlled Territories	1155.0	11.9	10

POLITICAL UNIT	AREA (THOUSAND SQUARE MILES)	POPULATION (MILLIONS: 1968 ESTIMATES)	DENSITY (PER SQUARE MILE)
Democratic Republic of the Congo (Kinshasa)	905.3	16.7	18
Republic of Rwanda	10.2	3.4	335
Republic of Burundi	10.7	3.4	317
Totals, Former Belgian-controlled Territories	926.2	23.4	25
Republic of Equatorial Guinea[c]	10.8	0.31	28
São Tomé and Principe (Portuguese overseas province)	0.37	0.10	269
South Central Africa			
Republic of Zambia	290.5	4.1	14
Republic of Malawi	46.1	4.2	91
Rhodesia	150.3	4.7	31
Totals, Former Federation of Rhodesia and Nyasaland	486.9	13.0	27
Angola (Portuguese overseas province)	481.2	5.5	11
Mozambique (Portuguese overseas province)	302.3	7.2	24
Totals, Portuguese-controlled Territories	783.5	12.7	16
East Africa			
Republic of Kenya	224.9	10.2	45
United Republic of Tanzania[d]	362.7	12.4	34
Uganda (Federal parliamentary state)	91.1	8.1	89
Totals	678.7	30.7	45
South and South West Africa and Former High Commission Territories			
Republic of South Africa	471.3	19.2	41
South West Africa	318.2	0.61	2
Totals, South and South West Africa	789.5	19.8	25
Republic of Botswana	219.9	0.61	3
Kingdom of Lesotho	11.7	0.91	78
Kingdom of Swaziland	6.7	0.39	59
Totals, Former British High Commission Territories (former British protectorates)	238.3	1.9	8
Indian Ocean Islands			
Malagasy Republic (Madagascar)	226.6	6.4	28
Comoro Islands (French overseas Territory)	0.84	0.21	245
Réunion (French overseas department)	0.97	0.41	426
Mauritius (independent state)	0.72	0.82	1135
Totals	229.1	7.9	34
Grand Totals, Africa[e]	11,644.5	336.4	29

[a] Comprised of five small enclaves along the Mediterranean coast of Morocco; the most important are the cities of Melilla (80,000) and Ceuta (75,000).

[b] Figures include Biafra, the southeastern part of Nigeria which declared itself independent in 1967. At the time of writing (early 1969) much of Biafra had been overrun by troops of the central Nigerian government and its future status was uncertain.

[c] Comprised of Rio Muni on the mainland, the island of Fernando Po, and smaller islands.

[d] Comprised of former Tanganyika on the mainland and the islands of Zanzibar and Pemba.

[e] Apparent discrepancies in totals and averages in this table are due to rounding of figures.

of the continent by other interests; and (7) the lack of readily apparent sources of wealth in the interior which would justify the risks of penetration.

Although Westerners did not penetrate the deep interior to any marked degree until the last half of the nineteenth century, a fringe of trading posts and of way stations to service passing ships began to be established around the Atlantic and Indian Ocean coasts of the continent early in the colonial period. The articles of trade in greatest demand by Europeans were slaves and ivory, though gold and spices such as pepper also were important. Slave trading, primarily by Arabs, was already extensively developed at the beginning of the European colonial age, but it was greatly intensified when the establishment of plantation agriculture in the Americas created a vast new market for slaves. One of the motives which led several European governments to claim territory in the interior of tropical Africa during the last quarter of the nineteenth century was a desire to stop this traffic at its source. Penetration of the interior was pioneered by a series of important journeys of exploration which began in the 1850s. These expeditions were undertaken by missionaries, traders, explorers motivated by scientific curiosity and often sponsored by geographical or other scientific associations, adventurers of various sorts, and in some cases government officials. Once the main outlines of inner African geography were revealed, a scramble for colonial territory in the interior took place among the European powers, beginning about 1881. By 1900 Africa had become predominantly a region of European dependencies—a status it retained for more than half a century.

Improved health services are gradually being brought to tropical Africa. The government nurse below is instructing a group of expectant mothers in Ghana regarding the proper care of the newborn child. (British Information Services.)

AFRICAN PROBLEMS

Recent decades have witnessed not only a transition from European colonialism to independence in most countries of Africa but also notable changes in the general conditions of life. Improved standards of health and literacy have resulted from the work of government agencies (photo, left) and, in sub-Saharan Africa, of Christian missions. Better transportation facilities, particularly a wide extension of roads and airways, have made possible the marketing of farm products, including perishable items, from formerly inaccessible areas. Stores and markets in both urban and rural areas stock an increasing variety of manufactured goods from overseas, and from African sources as well. Modern factories have been established in many urban centers, and improved agricultural techniques have been introduced in many areas. Today it is becoming very difficult to find peoples in Africa who are still essentially outside the range of modern influences.

The foregoing changes have affected some peoples and areas much more than others. Not all of the effects have been favorable: for example, one result of better transportation and increased opportunities for employment in mines and factories has been a movement of job seekers from villages to overcrowded slum areas in towns and cities, often in numbers too great for the local job market to absorb. Nor has the total impact of change been sufficient as yet to lift the continent (South Africa excepted) above its current status as one of the major underdeveloped regions of the world. Africa still shares many of the customary problems of such regions, including (1) a high incidence of illiteracy, poverty, hunger, and disease, (2) native customs, attitudes, and forms of social organization which tend to hinder the development of a modern economy, (3) inadequate overall facilities for transportation and communication, and (4) a lack of domestic capital which could be used to foster increased agricultural and industrial production and thus to raise the standard of living. Other problems arise from a heavy dependence on outside markets. Like most other underdeveloped regions, Africa's place in the commercial world still is basically that of a producer of foods and raw materials for sale outside the region. In the world markets where African goods are sold, there is a considerable fluctuation in demand and in prices over a period of years. Whenever the market is glutted and prices are low, unemployment and severe distress may result in African producing areas where the inhabitants have come to rely on cash returns from commercial production. Finally, there are problems of racial and tribal friction, which are among the most serious of all African problems at the present time.

THE DIFFICULT AFRICAN ENVIRONMENT

Many of the problems of Africa are due to the presence of a natural environment which is difficult to manage, particularly by peoples equipped with a rudimentary technology. This environment varies considerably from place to place, but its broad outlines are relatively simple.

Surface Configuration

Most of the continent is comprised of a vast plateau, or more precisely, a series of plateaus, at varying elevations. The plateau surfaces, predominantly level to rolling, are hollowed by shallow basins, often occupied by prominent river systems such as the Congo, Zambezi, and Orange. The principal areas of lowland plain are found in a narrow band around the coasts, averaging 20 to 100 miles in width, though in some places much narrower than this, and in other places considerably wider. Inland from the coast, abrupt or ragged escarpments, often hilly or mountainous in character, mark the transition to the plateau, which customarily lies at an elevation of 1000 feet or higher. In southern and eastern Africa the general elevation rises to 2000 or 3000 feet, with considerable areas lying at 5000 feet or higher (inset map, p. 551). The principal highlands above 5000 feet in elevation are found discontinuously within a broad zone extending from South Africa and South West Africa and Angola to Ethiopia. It is in this part of Africa that the largest number of Europeans reside. A great many live at altitudes of 4000 feet or higher, as indicated by the elevations of various cities containing substantial European populations: for example, in South Africa, Johannesburg (5700 feet); in Rhodesia, Salisbury (4800); and in Kenya, Nairobi (5500).

In the highland belt of southern and eastern Africa are located the highest peaks and the principal lakes of the continent. The principal mountainous areas are found in (1) Ethiopia, (2) the general region of the East African lakes, and (3) the eastern and southern parts of South Africa. The loftiest summits lie within a radius of 250 miles from Lake Victoria (map, p. 551); they include mounts Kilimanjaro (19,340 feet) and Kenya (17,058), which are volcanic cones, and Mount Ruwenzori (16,763), thought to be a block mountain produced by faulting.

Lake Victoria (26,828 square miles), the largest lake in Africa, is surpassed in size among inland waters of the world only by the Caspian Sea and Lake Superior. Other large lakes found in East Africa include lakes Tanganyika, Malawi (former Nyasa), Kyoga, Albert, and several others (map, p. 551). Although these lakes, particularly Lake Victoria, are used considerably for local water transportation and to some extent as connecting links in long-distance transportation routes, their total traffic is only a tiny fraction of that carried by the Great Lakes of North America.

Aside from the discontinuous mountain zone in eastern and southern Africa, the most extensive area of high mountains is the Atlas Mountains of northwestern Africa, described briefly in Chapters 15 and 16.

One of the most spectacular features of Africa's physical geography is the Great Rift Valley, a relatively broad, steep-walled trough, often with mountainous edges, which extends from Mozambique northward to the Red Sea and the valley of the Jordan River in southwestern Asia (map, p. 551). The Rift Valley has several branches and much of it is occupied by lakes, rivers, seas, and gulfs. It contains most of the larger lakes of Africa, although Lake Victoria, located in a depression between two of its principal arms, is an exception. The valley floor, generally 20 to 60 miles in width, is far below sea level in some places such as the bottom of Lake Tanganyika, the second deepest lake in the world, but in other places is 5000 feet or more above the sea.

African Rivers

The physical structure of Africa has significantly influenced the character of African rivers. The main rivers rise in interior uplands and descend by stages to the sea. At various points they descend abruptly, particularly at plateau escarpments, so that their courses are interrupted by rapids and waterfalls. These often block navigation a short distance inland from the sea. In addition, the navigation of many African rivers is hindered or prohibited by low water at certain seasons and by shallow and shifting delta channels. The maze of interlaced channels in some deltas made it difficult for early explorers to find the main stream. Among the important rivers which have built deltas are the Nile, Niger, Zambezi, Limpopo, and Orange. The Congo River, in contrast, has scoured a deep estuary 6 to 10 miles wide, which can be navigated by ocean vessels for a distance of about 85 miles to the seaport of Matadi in the Democratic Republic of the Congo (former Belgian Congo). The Congo is used more for transportation than is any other African river. A few

miles above Matadi navigation is blocked by rapids, and goods must be transshipped by rail a distance of 230 miles to Kinshasa. Above that city the Congo and its major tributaries are navigable for long distances by good-sized river craft, although the continuity of transportation is broken in places by falls and rapids which are bypassed by rail (map, p. 551) or road. The character of water transportation on the Congo system is generally representative of Africa's inland waterways, which are discontinuous and interconnected by rail or highway in a manner not duplicated in any other continent.

The frequent falls and rapids of African rivers have a more positive side. They represent a great potential reservoir of hydroelectric energy. Large power stations exist on the Nile River at the Aswan High Dam and the older Aswan Dam in Egypt and the Owen Falls Dam in Uganda (map, p. 368), on the Zambezi River at the Kariba Dam (p. 551), and at a large new dam on the Volta River in Ghana. Many other stations of varying size are scattered over the continent. But only a tiny fraction of the total hydroelectric potential has been developed thus far. Many of the best hydropower sites are remote from large markets for power. The largest single undeveloped source of hydroelectricity in Africa, and in the world, is the 220-mile stretch of rapids on the Congo River between Kinshasa and Matadi. According to some estimates, its hydroelectric potential is greater than that of the entire North American continent.

Climate and Vegetation

Africa is bisected by the equator. The distance from the equator to the northernmost point of the continent is approximately 2580 miles, and to the southernmost point, 2400 miles. Thus most of Africa lies within the low latitudes and is characterized by tropical climates. One of the most striking characteristics of Africa's climatic pattern is its symmetry or regularity. This is due mainly to the position of the continent athwart the equator, coupled with the generally level character of the surface and the comparative absence of extensive chains of high mountains.

Types of Climate

The broad pattern of climates in Africa may be outlined as follows. Areas of *tropical rain forest climate* in central and western Africa near the equator gradually merge into *tropical savanna climate* on the north, south, and east (endpaper climatic map). The savanna areas, in turn, trend into areas of *steppe* and *desert* on the north

The village of Bari near Juba in the zone of tropical savanna climate of the Sudan republic. The village, in its thorn-bush enclosure, is on the bank of a good-sized river; but the photo was taken at the height of the dry season, and the stream bed reveals no trace of water. (Photograph from R. U. Light, *Focus on Africa*, American Geographical Society.)

and southwest. Along the northwestern and southwestern fringes of the continent are found relatively small but important areas of *mediterranean climate*, while eastern coastal sections and adjoining interior areas of South Africa have a climate classified as *humid subtropical*. High elevations moderate the temperatures of extensive interior areas that lie within the realm of tropical savanna climate in the east and south of the continent.

TROPICAL RAIN FOREST CLIMATE. In Africa rainfall is heaviest, generally speaking, in areas near the equator in central and western Africa. In these areas abundant moisture and continuously high temperatures produce tropical rain forest as the characteristic type of climate and associated vegetation. The main area of rain forest occurs in a broad band along the equator from the Gulf of Guinea eastward to the highlands of East Africa. It includes most of southern Nigeria, plus Equatorial Guinea (Rio Muni and Fernando Po), most of Gabon, southern Cameroon, and large areas in the north-

ern part of the Congo Basin. A much smaller area is found farther to the west, centering in Liberia. This second area has a 3 to 4 months' dry season, but has enough total rainfall to support rain forest vegetation. Still a third area, remote from the main body, occurs along the eastern, windward side of Madagascar. The tropical rain forests of Africa have been highly publicized, but they occupy only about a tenth of the continent.

TROPICAL SAVANNA CLIMATE. Areas of tropical savanna climate are far more extensive than areas of tropical rain forest climate in Africa. The most characteristic feature of this type of climate is the alternation of well-marked wet and dry seasons. The natural vegetation is most often comprised of dry forest or scrub, intermixed with expanses of tall, coarse, tropical grass, although one of the best known of the endless variations in savanna-type vegetation is "park savanna," comprised of grass and scattered flat-topped trees. Near the

TABLE 21 CLIMATIC DATA FOR SELECTED AFRICAN STATIONS

STATION	POLITICAL UNIT	LATITUDE TO NEAREST WHOLE DEGREE	ELEVATION ABOVE SEA LEVEL (FEET)	TYPE OF CLIMATE	AVERAGE TEMPERATURE (DEGREES F TO NEAREST WHOLE DEGREE)			PRECIPITATION	
					ANNUAL	COOLEST MONTH	WARMEST MONTH	ANNUAL AVERAGE TO NEAREST INCH	NUMBER OF MONTHS AVERAGING LESS THAN 1 INCH OF RAIN
Casablanca	Morocco	33°N.	190'	Mediterranean	64°	54°	73°	17"	5
Ghardaïa	Algeria	32°N.	1725'	Desert	70°	51°	93°	2.7"	12
Djibouti	French Territory of Afars and Issas	12°N.	23'	Desert	85°	77°	94°	5"	11
Dakar	Senegal	14°N.	79'	Tropical steppe	76°	69°	82°	23"	8
Kano	Nigeria	12°N.	1561'	Tropical savanna	79°	71°	87°	34"	7
Accra	Ghana	5°N.	213'	Tropical savanna	80°	76°	82°	31"	3
Lagos	Nigeria	6°N.	125'	Tropical rain forest	79°	76°	82°	64"	0
Douala	Cameroon	4°N.	43'	Tropical rain forest	80°	76°	81°	162"	0
Nairobi	Kenya	1°S.	5897'	Tropical savanna (upland)	64°	59°	66°	36"	1
Lubumbashi	Congo (Kinshasa)	11°S.	4185'	Tropical savanna (upland)	68°	62°	74°	48"	5
Salisbury	Rhodesia	18°S.	4828'	Tropical savanna (upland)	65°	56°	70°	34"	5
Beira	Mozambique	20°S.	26'	Tropical savanna	76°	69°	82°	56"	0
Maun	Botswana	20°S.	3100'	Tropical steppe (upland)	71°	60°	80°	19"	6
Port Nolloth	South Africa	29°S.	23'	Desert	57°	53°	60°	2.3"	12
Cape Town	South Africa	34°S.	39'	Mediterranean	63°	55°	71°	26"	5
Durban	South Africa	30°S.	26'	Humid subtropical	68°	61°	75°	41"	0
Johannesburg	South Africa	26°S.	5556'	Humid subtropical (High Veld)	60°	50°	66°	30"	5
Tamatave	Malagasy Republic (Madagascar)	18°S.	20'	Tropical rain forest	75°	69°	79°	139"	0

desert margins of the savanna climate, rainfall is lighter and there is characteristically a belt of shorter grass or shrubs which is often referred to as tropical steppe.

A broad belt of tropical steppe and savanna bordering the Sahara Desert on the south is known as the Sudan. Only the eastern parts of this belt lie in the Republic of the Sudan. The rest of the belt stretches westward, through many different countries, to the Atlantic.

DESERT CLIMATE. Deserts border the tropical savannas and steppes on the north and southwest. The great Sahara Desert, the world's largest, extends across the north of Africa from the Atlantic Ocean to the Red Sea. It is bordered on the north by a narrow belt of middle latitude steppe climate. In South Africa and South West Africa, a coastal desert, the Namib, borders the Atlantic. The "Kalahari Desert," which lies inland from the Namib, has a considerable growth of low grasses and shrubs, and it is better described as steppe or semidesert than as true desert.

MEDITERRANEAN AND HUMID SUBTROPICAL CLIMATES. The coastal areas bordering the Mediterranean Sea in northwestern Africa and the southwestern tip of the continent around Cape Town have a mediterranean or dry-summer subtropical climate characterized by rainy winters and almost complete drought during the midsummer months. The southeastern coast of the continent in Natal Province of South Africa has rain at all seasons and is classed as humid subtropical. The high interior grasslands of South Africa, or High Veld, also have a subtropical climate, but one which resembles the tropical savanna climate in having a well-marked dry season during the period of low sun.

Climatic data for selected African stations are given in Table 21.

Water Resources and Problems

The total amount of precipitation received by the African continent is very large. Unfortunately, however, it is poorly distributed. Some parts of Africa receive too much rain, whereas other areas have scarcely any rain. Even in the rainier parts of the continent large areas have a dry season of considerable length, and wide fluctuations occur from year to year in the total amount of precipitation. In places considerable areas are inundated by waters which gather in swamps or marshes such as the Sudd Swamp along the Nile River in the Sudan.

One of the major needs of Africa is a better control over the available water, involving irrigation projects in some areas, drainage projects elsewhere, conversion of marshes and swamps to rice fields or pastures, and the development of a system of dams along both the major streams and small tributaries to control floods, regularize the flow of water between seasons, and provide hydroelectric power.

In the typical African village household water is carried laboriously by hand from a nearby stream or lake or a shallow, polluted well—a task ordinarily performed by women. Especially in the seasonally rainy areas, education is needed in the construction and use of small dams to provide water storage throughout the year, coupled with development of simple pipelines or flumes to carry water by gravity to village households or the fields. "The African can be taught to dig and line wells, and such a primitive device as an endless porous rope of fibers absorbing water and passing through two wooden rollers to squeeze out the supply is within the power of any African to make and maintain. Unfortunately, over much of Africa's underlying complex of ancient rocks the water table behaves irregularly, and a well is both difficult to dig and uncertain of its supply. There is much to be said, in hilly or rolling country, for horizontal 'wells' into the hillsides from which water would flow by gravity as an artificial spring. It should be noted that artesian conditions are rare in Africa."[2]

African Soils

The scientific study of African soils is still in an early stage. This represents one of the critical frontiers of human endeavor in Africa at the present time, inasmuch as further significant advances in the utilization of African lands for agriculture must be founded in large part on a better knowledge of the soils. Based on present knowledge, the following general observations can be made.

Among the most productive soils of the continent are some alluvial soils found on river plains. Soils along the Nile River in Egypt, for example, are especially noted for their fertility, and have supported agriculture for thousands of years. Other especially fertile soils are found in scattered areas (particularly in parts of the East African highlands) where certain types of volcanic parent material occur. A third group of better than average soils

[2] L. Dudley Stamp, *Africa: A Study in Tropical Development* (2d ed.; New York: John Wiley & Sons, Inc., 1953, 1964), p. 85. Used by permission.

is the grassland soils found in some areas of tropical steppe or tropical highland and in the mid-latitude grasslands of the High Veld in South Africa. These soils, however, do not appear to be entirely comparable to the chernozems, prairie soils, and chestnut soils of North America. They are more difficult to cultivate and seem to lose their fertility more quickly under continuous cropping.

Soils of the deserts and regions of mediterranean climate are generally thin and immature; over broad areas of desert, true soils are absent. In the mediterranean areas fertile soils are found in some valleys where transported materials from adjoining slopes have accumulated.

In the tropical rain forests and savannas reddish tropical soils are generally dominant. There is a common idea that these soils are exceptionally fertile when cleared of their natural vegetation and used for agriculture. This idea is apparently based on the luxuriant plant growth which they often support in their natural state. However, the truth appears to be that they are rather poorly supplied as a rule with plant nutrients and tend to lose their fertility quickly when used for crops. They are usually low in humus and deficient in lime. These disadvantages of tropical soils are at least partly due to the rapid chemical action induced by the abundant heat and moisture of tropical climates. Such conditions promote rapid leaching of nutrient materials by percolating waters, and cause organic matter exposed to the air to combine with oxygen rather quickly and thus to be lost into the atmosphere as carbon dioxide rather than being converted to humus and plant food by soil bacteria.

Diseases and Parasites

A high incidence of diseases and parasites affecting people, domestic animals, and cultivated crops has been one of the main hindrances to African development. Certain parts of tropical Africa, particularly the rainy lowlands of West Africa, long had a reputation as a "white man's graveyard." Among native Africans diseases and parasites take a heavy toll of strength and energy, even when they are not fatal. Many of the major diseases are carried by insects. Those carried by mosquitoes include malaria, yellow fever, and dengue or "breakbone fever." Sleeping sickness is carried by the tsetse fly, which also transmits nagana, a destructive disease affecting cattle and horses. Plague is spread by fleas, and relapsing fever by ticks and lice. Large numbers of Africans are afflicted by diseases and parasites of the digestive tract, including dysentery, typhoid and paratyphoid fever, bilharziasis, and hookworm and other types of intestinal worms. Such afflictions may be traced in most instances to the use of contaminated water or to other unsanitary conditions. Other diseases which are common in Africa include tuberculosis, filariasis, various nutritional deficiency diseases, pneumonia, yaws, leprosy, influenza, trachoma, the venereal diseases, and many fungoid diseases of the skin.

Despite the miserable conditions which still prevail in some areas, much progress has been made in recent decades toward the conquest of disease in Africa. At the present time enough is known about the control of tropical diseases and parasites to greatly reduce their incidence if means were available for the technical knowledge to be fully applied. But in the vast, poverty-stricken African continent, the need for medical assistance far outruns the money and personnel that are available for such assistance at the present time. Even where medical facilities are present, doctors and medical technicians are greatly handicapped by the ignorance, superstition, and fears of their African clientele. Thus education is a vital sector in the fight against ill-health and disease.

Native Animal Life

Africa's wild animals have been so well publicized that the continent is often thought of as a sort of giant zoo. However, it is easy to overstate the abundance and significance of African animal life at the present time. While reptiles, monkeys, and birds are still numerous, the total numbers of most of the larger animals—the elephant, lion, giraffe, zebra, hippopotamus, rhinoceros, and so on—have greatly declined. This decline has been partly the result of unrestricted hunting in past times, but has also been due to encroachment by an increasing human population and their domestic livestock on the habitats of wild species. Whatever the cause, the numbers of some species have now diminished to the point that they are often protected by law against excessive hunting. Such laws are difficult to enforce, however, and poaching on a large scale is reported to be making serious inroads on animal populations. In some parts of Africa attempts are being made to conserve the remaining wildlife in large game reserves or national parks such as the Kruger National Park in the extreme northeast of South Africa or the Serengeti National Park in Tanzania.

The tropical grasslands and open forests of Africa have been the principal habitat of the larger herbivorous animals, such as the elephant, buffalo, antelope, zebra, and giraffe, and also of carnivorous and scavenging animals, such as the lion, leopard, and hyena. The tropical rain forests have been much more deficient in the larger species of animals than is commonly realized. In the rain forest the most abundant species have been birds, monkeys, and snakes, together with the hippopotamus, the crocodile, and a great variety of fish in the innumerable streams.

Mineral Resources and Production

Africa has become increasingly important in recent times as a producer of minerals. Most of the production is exported, principally to Europe and the United States. In world terms, the continent is particularly important as a producer of (1) precious metals and precious stones, (2) ferroalloys, (3) copper, (4) phosphate, and (5) uranium. Very recently it has become a sizable producer of petroleum, and it is becoming more and more important as a producer of high-grade iron ore. Numerous other minerals needed by industrialized nations are produced and exported in quantity. For about a dozen African countries, minerals are the most important category of exports.

Table 22 shows African production as a percentage of world production for a number of important minerals and indicates the principal African producing countries for each mineral. It is apparent from this table and from the map on page 551 that four general areas in Africa are of primary importance in mineral production at the present time. These are (1) South Africa and South West Africa, (2) the Congo-Zambian-Rhodesian region, (3)

TABLE 22 PRODUCTION DATA FOR NINETEEN IMPORTANT AFRICAN MINERALS

MINERAL	PERCENTAGE OF WORLD OUTPUT MINED IN AFRICA, 1966	LEADING AFRICAN PRODUCING COUNTRIES, 1966 (FIGURES REPRESENT PERCENTAGE OF TOTAL AFRICAN PRODUCTION)
Diamonds		
Gem	93.0	South Africa, 33.3; South West Africa, 21.5; Angola, 12.9; Sierra Leone, 8.6; Tanzania, 6.5; Liberia, 4.3; Congo, 4.3; Ghana, 4.3; Central African Republic, 3.2
Industrial	88.0	Congo, 67.1; South Africa, 13.6; Ghana, 9.1; Sierra Leone, 3.4
Gold	69.0	South Africa, 94.9; Ghana, 2.3; Rhodesia, 1.9
Platinum	26.6	South Africa, 99.9
Manganese ore	24.2	South Africa, 38.4; Gabon, 28.9; Ghana, 13.2; Morocco, 8.3
Chromium	31.8	South Africa, 67.3; Rhodesia, 31.8
Cobalt	73.5	Congo, 76.3; Morocco, 13.5; Zambia, 10.2
Vanadium	32.9	South Africa, 55.9; South West Africa, 44.1
Copper	21.3	Zambia, 54.9; Congo, 27.7; South Africa, 10.8
Tin	9.2	Nigeria, 48.9; Congo, 26.1; South Africa, 8.7
Lead	7.3	South West Africa, 41.1; Morocco, 37.0; Tunisia, 8.2; Zambia, 8.2
Zinc	6.7	Congo, 38.8; Zambia, 20.9; Morocco, 19.4; Algeria, 8.9; South West Africa, 8.9
Asbestos	14.7	South Africa, 55.8; Rhodesia, 36.1; Swaziland, 7.5
Phosphate rock	18.8	Morocco, 56.4; Tunisia, 19.2; Togo, 6.9; Senegal, 6.9; South Africa, 6.4
Iron ore	6.6	Liberia, 40.9; Mauritania, 16.7; South Africa, 16.7; Sierra Leone, 6.1; Algeria, 4.5; Swaziland, 4.5
Coal (all grades)	1.8	South Africa, 94.4; Rhodesia, 5.6
Uranium oxide	20.1	South Africa, 83.1; Gabon, 15.4; Malagasy Republic, 1.5
Petroleum (crude)	8.5	Libya, 54.1; Algeria, 24.7; Nigeria, 15.3; Egypt, 4.7
Antimony	21.3	South Africa, 89.2; Morocco, 10.3
Bauxite	5.6	Guinea, 73.2; Ghana, 14.3; Sierra Leone, 12.5

SOURCE: U.S. Bureau of Mines, "World Mineral Production, 1966" (an annual mimeographed release). Percentages are based on volume; except for manganese ore, iron ore, and bauxite, figures for metallic minerals are based on metallic content. An annual table of data on world mineral production may also be found in the *Encyclopedia Britannica Book of the Year.* "Congo" in the above table refers to the former Belgian Congo.

northwestern Africa and Libya, and (4) West Africa, especially the areas near the Atlantic Ocean.

In *South Africa* gold is the leading mineral in value of production. Uranium secured as a by-product of gold mining makes South Africa the largest producer of this strategic mineral in Africa. The principal gold mining districts are found on the interior plateau in the southern Transvaal and the northwestern Orange Free State; the most famous and important district is the Witwatersrand or Rand, centering on Johannesburg (maps, pp. 551 and 608). South Africa is the world's leading gold-producing country and is a sizable producer of diamonds, manganese, chromium, vanadium, platinum, copper, asbestos, and several other important minerals. It has the continent's largest deposits of coal (in fact, they are Africa's only sizable deposits of good coal except for those in the neighboring country of Rhodesia) and also has large deposits of iron ore. Coal and iron have provided the basis for the largest iron and steel industry in Africa. South West Africa, held as a dependency of South Africa, mines diamonds, vanadium, lead, and small quantities of a number of other minerals. The remarkable variety of minerals in South and South West Africa is associated with a highly diversified group of rock formations, ranging in age from very recent to extremely ancient rocks.

The *Congo-Zambian-Rhodesian region* is also an extremely rich mineralized area. In few parts of the world do sizable deposits of so many important minerals occur in such close proximity. In the Congo (Kinshasa), mineral production is mainly centered in the Katanga region of the southeast, bordering Zambia. The Congo leads all countries in the production of industrial diamonds (from the Kasai region, west of the Katanga) and the important ferroalloy cobalt, and it ranked sixth among the world's countries in 1966 as a producer of copper. However, copper is by far its most valuable mineral, and it outranks any other export product by a very wide margin. The Congo also produces zinc, manganese, tin, gold, and various other minerals. Often a particular ore body will yield one main mineral and several others as by-products—a very common situation in the world's mining districts. Zambia exceeds the Congo in copper production, ranking fourth among the world's countries in 1966. The value of copper produced in Zambia is several times that of all other minerals combined. The principal mining district (Copperbelt) borders the Katanga mining region of the Congo (map, p. 551). Zambia, like the Congo, also produces cobalt. This ferroalloy is in demand for the manufacture of special steels, but is produced in only a few of the world's countries.

Rhodesia, separated from Zambia by the Zambezi River, has a more varied mineral production than that of Zambia; the leading minerals are gold, asbestos, coal, chromium, and copper. The total value, however, is considerably less than the value of Zambia's massive copper output. Aside from coal, found in the west at Wankie, Rhodesia's minerals come mainly from a belt in the central part of the country along or near the railroad connecting the country's two main cities, Bulawayo in the south and Salisbury, the capital, in the north. At Que Que, midway between the two, a small iron and steel industry utilizes local iron ore and limestone, and secures coking coal by rail from the large coal deposits at Wankie, about 300 rail miles to the west. It may be noted that the mining industries in Zambia and Rhodesia depend on a common source of electric power. The generating station at the immense Kariba Dam provides electricity for mines and other power users in both countries (see p. 599). In addition, there is thermal-electric production in both countries (comparatively minor in Zambia) fueled by coal (see pp. 598 and 599).

In *northwestern Africa and Libya* the leading minerals in value of production and exports are oil, almost all from Libya and Algeria, and phosphate from Morocco, Tunisia, and, to a lesser degree, Algeria. The region's phosphate deposits are among the largest in the world, and petroleum and natural gas reserves are also of major size. Phosphate exports amounted to nearly a fifth of the world total in 1966, and petroleum exports to nearly 7 percent. Natural gas is produced in Algeria for domestic consumption, and some liquefied gas is exported by tanker from both Algeria and Libya. Many other minerals are produced and exported in varying quantities; of special note are the high-grade iron ore of Algeria and Morocco, the lead of Morocco and Tunisia, and the manganese, cobalt, and zinc of Morocco (Table 22).

In *West Africa,* where mineral production comes almost entirely from countries that touch the Atlantic, the principal items currently mined are iron ore from Liberia, Mauritania, and Sierra Leone, petroleum and tin from Nigeria, manganese and gold from Ghana, bauxite, mainly from Guinea, and diamonds, mainly from Sierra Leone and Ghana. Perhaps the most critical items for the outside world at present are high-grade iron ore and manganese, both exported in large quantities to overseas nations that have iron and steel industries.

Some mineral production in Africa comes from countries other than those discussed here (for example, diamonds from Tanzania and Angola, iron ore from Angola, manganese from Gabon), but the total value is very minor when compared with the production and exports of the four main regions.

Most mining in Africa is done by large corporations financed primarily by investors in Europe or America. Mining has attracted far more investment capital to Africa than any other economic activity. Not only has money been invested in mines directly, but a great many of the transportation lines, port facilities, electric power stations, urban housing and business areas, manufacturing plants, and other elements in the continent's infrastructure have been developed primarily to serve the needs of the mining industry. Practically everywhere in Africa mining is carried on by modern large-scale methods. Managerial and supervisory personnel and skilled workers are commonly of European extraction, though the proportion of non-Europeans in these posts is increasing. Unskilled or semiskilled labor is done by native Africans (photo, p. 611) or Arabs. Great numbers of workers in the mines are temporary migrants from rural areas, often hundreds of miles away. They are recruited for specified periods, ranging from several months to several years, and normally are given free transportation to and from the mines and free housing in mine "compounds." When their contracts expire, they return home. The use of migrant labor, not only in mines but also on farms and plantations, is very widespread in Africa. The largest numbers of migrants work in South Africa; they come primarily from the nearby countries of Mozambique, Lesotho, Swaziland, Botswana, and Malawi. Married men who work in mines generally are not accompanied by their families, though there are many exceptions to this, and there is an increasing tendency in some areas to attract a more stable labor force by providing family housing. Thus the system of migrant labor tends to disrupt family life. On the other hand, it provides many Africans with better financing than they otherwise would have. The wages paid are not high by the standards of America or Europe, but they represent much larger amounts of money than it would be possible to earn in the home village. In addition, the mining companies generally provide free health care and an adequate diet (free or subsidized), and workers are apt to be in better physical condition on returning home than they were when they came to the mines. It may be noted that the recruitment of migrant workers has had important cultural effects. Many millions of Africans have worked for mining companies at one time or another, have come into contact with Western ideas, and have carried these with them on returning to their villages. Positive or negative aspects aside, it is a fact that the employers of migrant labor have been a powerful force in spreading Western influences throughout Africa. In addition, they have brought Africans from different localities into contact with each other and thus in some measure have leavened the parochialism of village life by making villagers more conscious of a wider African civilization. Needless to say, the radio —often bought with funds earned at a mine—also has had powerful effects in these directions.

AFRICAN PEOPLES AND THEIR AGRICULTURAL ECONOMY

Among the peoples who inhabit the African continent, excepting some 4 to 5 million Europeans and somewhat less than a million Asians (mainly Indians and Pakistanis), those south of the Sahara exhibit Negroid characteristics in varying degree, while those north of the desert and in Egypt are basically Caucasoid and speak languages, principally dialects of Arabic or Berber, which are quite distinct from the languages of Negro Africa. In the east-west grassland belt known as the Sudan, in the Sahara, and in the East African highlands and coastal areas, there has been considerable mixing of Negroid and non-Negroid peoples over a long period. Some peoples in these latter areas speak Hamitic or Semitic languages, while others speak more purely African tongues. Even the African languages, however, often reveal Hamitic or Semitic influences.

Most of the peoples of Negro Africa belong to one of two broad groupings: (1) the Bantu, most of whom live south of the equator, or (2) the Guinea Coast and Sudanese peoples. The *Bantu peoples* speak languages belonging originally to a common language stock, though tribal variations are often different enough to be mutually unintelligible. The *Guinea Coast and Sudanese peoples,* in contrast, speak a great variety of languages; in fact, the areas of forest and grassland they inhabit have often been termed a linguistic "shatter belt," comparable to parts of Melanesia in the confused pattern of languages.

Three minor and relatively primitive population elements, the Pygmies, Bushmen, and Hottentots, deserve mention. The *Pygmies,* averaging around 4½ feet in height, are found principally in the tropical rain forests of the Congo Basin. Traditionally, most Pygmies have lived by hunting (using poisoned arrows) and by gathering roots, nuts, fruits, and other food in the forest. Today some of them raise crops, but the food supply even of these is still augmented by gathering and hunting. The *Bushmen,* found primarily in the Kalahari region of southwestern Africa, are also below average in height,

This celebrated picture, published previously in a collection of outstanding photographs called *The Family of Man* (New York: Simon and Schuster for the Museum of Modern Art, 1955), shows a Bushman hunter in the desolate Kalahari semidesert of Botswana preparing to dispatch a cornered gemsbok. Africans who live primarily by hunting form a tiny minority of the continent's population, and their numbers, like those of the animals that support them, are dwindling. (Life Photo by N. R. Farbman.)

although a little taller than the Pygmies. They are nomadic hunters, using poisoned arrows and spears (photo, above), and constructing crude temporary shelters as they move about in search of game. They are expert trackers and stalkers of wild animals, and are also skilled in finding water. The latter is of vital importance in the semidesert country where they live. Like the Pygmies they are gatherers as well as hunters; their diet includes such items as roots, honey, gum from acacia trees, ant eggs, ostrich eggs, lizards, and locusts.

The *Hottentots* have nearly been eliminated as a separate people. The remnants live mainly on tribal reservations in South West Africa. They are primarily herdsmen, raising sheep, goats, and long-horned cattle. They live in semipermanent villages of oval huts built about a central open space. The village is surrounded by a thorn fence, and the animals are driven within the enclosure at night for protection. This type of village, known as a *kraal,* is often found among pastoral peoples of sub-Saharan Africa. Both the Hottentots and the Bushmen formerly ranged widely over southern Africa, but most of the lands they utilized were taken over by Bantu tribes pushing in from the north or by Europeans moving inland from the Cape of Good Hope.

African Agriculture

The peoples of the Guinea Coast live principally by tilling the soil and growing tree crops such as the cacao tree and the oil palm. Some tribes among the Sudanese peoples are also cultivators, some are herdsmen and many are both. The same is true of the Bantu. Some of the chief characteristics of the native agricultural and pastoral economy of Africa have been described as follows by the late L. D. Stamp:[3]

Along the Mediterranean borders we find an agriculture established from very early times, with cattle, the plow, cereals, and Mediterranean fruits—especially barley, wheat, olives, figs, and the grape. Southward toward the desert margins nomadic pastoralism based on sheep replaces cultivation; in the deserts cultivation is limited to oases, and human life depends on date palms and camels.

South of the deserts in the semiarid open lands of the Sudan are again pastoralists depending for their livelihood on cattle and sheep. Southward, with increasing rainfall and a modest rainy season, there is again some cultivation, precarious, and dependent mainly on sorghum and millet. Cultivation is by the hoe; use of the plow remains unknown.

The area of intermediate rainfall of the savanna between the steppes of the desert margins which are too dry and the forests of the equatorial margins which are too wet is favorable to cultivation. Large villages are the rule; there have been in the past empires of considerable size. Dependence for food is on millet and corn (maize); latterly on peanuts on the uplands and rice along the rivers. Tobacco is grown, sometimes cotton; kola nuts are gathered. Often cattle are bred, sometimes sheep and horses, and among the cultivators or mixed farming communities are tribes almost exclusively pastoral.

Unfortunately, where the rainfall becomes more reliable and where mixed farming should reach its maximum intensity, we encounter the "fly belts" where the tsetse fly virtually eliminates domestic animals. Here the basis of life becomes the yam, manioc, banana, and palm oil . . ., with cocoa as a "cash crop" now developed to a great extent in Ghana and elsewhere.

As the closer forest is reached, villages are smaller, population sparser and patches of cultivation (shifting cultivation) more scattered.

On the plateaus of East Africa, and formerly in South Africa, the balance is somewhat different. The settled cultivators tend to be less numerous than and subordinate to the aggressive, often warlike, pastoral tribes. It is here that wealth has come to be measured in head of cattle, independently of quality; indeed the cattle have little more practical value than the gold bars we civilized people bury in our vaults. Madagascar differs somewhat from the continental mainland in that cattle breeding is there associated with rice cultivation.

. . . . .

Shifting Cultivation

The system of agriculture commonly though somewhat misleadingly known as "shifting cultivation" is practiced through a very large part of tropical Africa. As a system it is better described as "land rotation" or "bush fallowing." In those areas inhabited by sedentary farming peoples, each village or settlement has proper to it a tract of surrounding land. The tract is probably only loosely defined except where settlements are close together and population dense. In a given year the villagers working together as a community clear a part of this village land, cutting and burning the woodland or scrub and then planting the crops appropriate to the climate and soil of the area. In due course the crops are harvested communally and the land used for a second and perhaps a third year. It is then abandoned, and a fresh tract of the village land is cleared. The abandoned land quickly becomes covered with a second-growth woodland or scrub. In due course the clearings reach a full cycle, and if a given tract has been allowed to lie fallow for about fifteen years it may be regarded as fully rested. Bush clearing is largely man's work. There is also surrounding the village itself, often as a series of enclosed gardens or "compounds" attached to individual huts, the "women's land," cultivated regularly to afford a supply of vegetables for the pot. Often the kokoyams, peppers, beans, melons, bananas, etc., are scarcely grown at all in the open farmland. The women's land is enriched by house sweepings, ashes, refuse, and manure afforded by chickens, goats, and human beings.

The system has often been condemned as wasteful of natural forest, of land, and of labor. But it has many good points. The natural forest is probably second growth of little value anyway. The land cleared in small patches protected by surrounding woodland escapes the evils of soil erosion, and its nutrient status temporarily enhanced by the ashes of the burnt bush is maintained by the fallowing, the soil not being exposed to the atmosphere long enough for serious oxidation. Expenditure of labor is minimized by burning, and no attempt is made to remove large stumps. The cultivation is by hand—by hoeing—so the stumps do

---

[3] Stamp, *op. cit.,* pp. 141–144. Used by permission.

not constitute the obstacles they would be if the plow were used. We may accordingly agree with Lord Hailey when he says that shifting cultivation is "less a device of barbarism than a concession to the character of a soil which needs long periods for recovery and regeneration."

Cash Crops in Low-Latitude Africa

Perhaps two-thirds or more of the people in Africa depend directly on agriculture or pastoralism for a livelihood. In many parts of the continent the proportion is much higher. In most areas a large share of the crops and livestock is grown for the use of the cultivators and their families or for a purely local sale. In other words, there is a strong subsistence component in agriculture. This is particularly true of low-latitude areas, though it applies also to various areas outside the tropics.

In low-latitude Africa the proportion of crops grown for export to overseas destinations or for sale in African urban centers has risen significantly in recent times. This has been partly a result of taxation and other governmental pressures, but has also been due to the desire of Africans for cash with which to purchase manufactured goods (hardware, utensils, bicycles, radios, clothing, cameras, phonographs) or food. In many areas farmers have become very dependent on the sale of cash crops.

Most export crops in low-latitude Africa are grown partly on small African farms and partly on plantations or estates,[4] but in the majority of cases the production comes primarily from small farms. Large plantations and estates have never become established to the same degree as in Latin America or Southeast Asia, and many of them have been forced out of business by political or economic pressures during the period of transition from European colonialism to independence. Some plantations have been nationalized by governments of new independent states, some have been obliged to sell their land to Africans, and others, for example in the Congo, have been unable to survive the civil warfare and other turmoil accompanying independence. Many of these units,

such as the well-known Firestone rubber plantations in Liberia, continue to operate, and for some countries they provide a very important source of tax revenue and foreign exchange. But throughout tropical Africa as a whole there is a definite trend for export production to come increasingly from small farms. The export crops of greatest value in low-latitude Africa as a whole are coffee, cacao, cotton, peanuts, and oil palm products. Crops of lesser overall significance, but very important in some regions, include sisal, tobacco, tea, rubber, bananas, cloves, vanilla, cane sugar, and cashew nuts. Within Africa itself, as transportation improves, cities grow, and the standard of living slowly rises, there are increasing possibilities for the sale of basic food crops such as corn, millets, manioc, rice, and fruit.[5]

Farming in Areas outside the Tropics

In the parts of Africa that lie outside the tropics, much of the farming by Africans or Arabs is on a subsistence or near-subsistence basis. But these sections of the continent also have a considerable development of commercial agriculture. It is carried on primarily by European growers of livestock, grain, fruit, and sugar cane in South Africa, by Egyptian cotton growers in the Nile Delta and Valley, and Arab or European fruit growers (citrus, grapes, olives) in Algeria, Morocco, and Tunisia. Most European farms in Africa, whether in low latitudes or middle latitudes, are worked by African or Arab labor.

Importance of Livestock in Sub-Saharan Africa

Many peoples in sub-Saharan Africa are pastoral. However, most of those who live by tilling the soil also keep some animals, even if only goats and poultry. The significance of livestock is portrayed in the following selection (pp. 569–570):[6]

---

[4] These terms, which are more or less interchangeable, cover enterprises that vary widely in type of ownership, scale of operations, organization of production, and types of products. Some plantations and estates are corporate owned, some are government owned, and many are owned by individuals. Most of these units were established by Europeans, though there are significant exceptions such as the clove plantations developed by Arabs in the islands of Zanzibar and Pemba (now a part of Tanzania). They are commercial enterprises, generally producing one crop, or at most a very few crops, for export, and employing large numbers of African workers.

[5] It is an interesting fact that many of the most important food and export crops of Africa are not native to the continent. For example, corn (maize), manioc, peanuts, cacao, tobacco, and sweet potatoes were introduced from the New World. The complex story of plant introductions and dispersals in Africa (and other world regions as well) is still being pieced together by scientists and historians.

[6] H. L. Shantz, "Agricultural Regions of Africa. Part I—Basic Factors," *Economic Geography*, 16, no. 2 (April 1940), 126–128. Used by permission of *Economic Geography*.

Masai herdsmen with their animals in the crater of an enormous extinct volcano (Ngorongoro Crater) in Tanzania. Wild animals may be seen grazing on the floor of the crater in the distance. Africa's increasing cattle population is steadily constricting the habitat of wild species. (Edgar Monsanto Queeny.)

Many [African peoples] rely largely on cattle. This is especially true of the . . . Basuto, [Botswana], Hereros, Watusi, Masai, Gallas, Nilotic Negroes, and [Fulani], some of whom have become so dependent on cattle that if the latter were destroyed their whole economic and social order would be disrupted and even their religion affected.

It is difficult to overestimate the importance of domestic animals in the lives of [Africans]. While wives represent to some extent real property or estates, cattle, sheep, and goats represent currency. In many tribes wives are secured only by the payment to the woman's family of a certain number of animals, say 30 goats or sheep or

10 cattle. . . . This is even the practice in the essentially crop-producing tribes of the Bantu peoples.[7] Among the tribes such as the Watusi, Masai, and others which are dependent on cattle for their daily food, the whole social, economic, and religious pattern of their society is dependent on cattle. . . . These are extreme cases, but even in a crop-raising tribe such as the Kikuyu sheep or goats constitute the basis of the most important social and economic events in the life of the individual. . . .

Cattle, sheep, and goats occupy the greater part of the African continent. They are absent or scarce in the extreme deserts . . . and the heavy tropical forest or coarse

---

[7] In African tribal society wives are not "bought," but it is a universal custom for the bridegroom to make a large gift to the bride's family in advance of the marriage. Payment of the "bride price" (or "bride wealth," as it is more properly called) has customarily been made in cattle or other livestock, but may also be made in cash or merchandise. The gift signifies that the groom's intentions are serious and that he has financial prospects for supporting the bride; in addition it compensates the bride's father for the loss of her labor in the household and the fields. Should the marriage break up on account of the wife's misdeeds, the husband is entitled to the return of the bride wealth.

high grass savannas of the Congo and the Guinea Coast. With these exceptions, they are distributed over the remainder of the continent. . . .

The great grasslands between the tropical rain forests and the deserts and the mountain grasslands of central and southern Africa were occupied by cattle-raising peoples. They dominate the adjacent agricultural tribes and often hold within their organization a dependent group of soil tillers. The Masai of East Africa depend entirely on cattle which they milk and bleed for their daily food. So close is this relationship that they cannot migrate without their herds. Moreover, they regard these cattle as almost a divine gift and husband them as carefully as they would their own children. The same can be said of the Banyoro and the Watusi. Each animal has a name, the herds are often carefully sorted as to color, and the greeting of the nobles of the tribe asks first about the welfare of the cattle, then about the welfare of the wives and children.

The great grasslands support the herds throughout the year in most places, and there is no set migration. In the drier portions migrations are often forced by lack of grass for the herds. . . .

Sheep and goat herdsmen have ranged nearer the deserts and been probably more nomadic than the cattlemen. . . . The . . . chicken [has] been a boon to the less favored [tribes]. Everywhere, even in the dense forests of tropical Africa, [chickens] constitute a part of the diet of the agricultural [tribes]. . . . One seldom sees [an African] moving without the accompanying coop with chickens inside or tied to the outside. . . .

One of the major needs of sub-Saharan Africa is the development of strains of grasses suited to African conditions which will be more nutritious for livestock than the native grasses now present. The latter are often coarse, tough, and not very nutritious, so that the carrying capacity of the native grasslands is relatively low. Considerable work with grasses has been done by government experiment stations in Africa, with results that seem promising. It is conceivable that Africa may eventually become a major exporter of livestock products, provided better grasses can be widely introduced, the menace of the tsetse fly brought under control, and the tribesmen induced to take an interest in improving the quality of their herds, rather than thinking of them in terms of mere numbers. At present the largest exports of livestock products are from European farms and ranches in South Africa.

REFERENCES and READINGS

Standard Geography Textbooks

CARLSON, LUCILE, *Africa's Lands and Nations* (New York: McGraw-Hill Book Company, 1967).

CHURCH, R. J. HARRISON, and Others, *Africa and the Islands* (2d ed.; New York: John Wiley & Sons, 1967).

de BLIJ, HARM J., *A Geography of Subsaharan Africa* (Chicago: Rand McNally & Co., 1964).

GROVE, A. T., *Africa South of the Sahara* (New York: Oxford University Press, 1967).

HANCE, WILLIAM A., *The Geography of Modern Africa* (New York: Columbia University Press, 1964). A detailed and authoritative work, stressing economic development.

MOUNTJOY, ALAN B., and CLIFFORD EMBLETON, *Africa: A New Geographical Survey*, with a contribution by W. B. Morgan (rev. ed.; New York: Frederick A. Praeger, 1967).

STAMP, L. DUDLEY, *Africa: A Study in Tropical Development* (2d ed.; New York: John Wiley & Sons, 1964).

Other General References

"Africa in Motion," *Annals of the American Academy of Political and Social Science,* 354 (July 1964). Entire issue; articles by various authors.

Africa Digest (six issues a year).

Africa Report (nine issues a year).

African Affairs (quarterly).

CARRINGTON, C. E., "Frontiers in Africa," *International Affairs,* **36,** no. 4 (October 1960), 424–439. An article in a special issue on Africa.

CARTER, GWENDOLEN M., ed., *National Unity and Regionalism in Eight African States: Nigeria, Niger, the Congo, Gabon, Central African Republic, Chad, Uganda, Ethiopia* (Ithaca, N.Y.: Cornell University Press, 1966).

DALE, EDMUND H., "Some Geographical Aspects of African Land-locked States," *Annals of the Association of American Geographers,* **58,** no. 3 (September 1968), 485–505.

GOULD, PETER R., ed., *Africa: Continent of Change* (Belmont, Calif.: Wadsworth Publishing Co., 1961).

GUNTHER, JOHN, *Inside Africa* (New York: Harper & Row, 1954). A book by a perceptive journalist; highly readable and still very useful.

HAILEY, LORD WILLIAM MALCOLM, *An African Survey: A Study of Problems Arising in Africa South of the Sahara* (rev. ed.; New York: Oxford University Press, 1957).

HAINES, CHARLES GROVE, ed., *Africa Today* (Baltimore, Md.: Johns Hopkins Press, 1955).

HODDER, B. W., and D. R. HARRIS, eds., *Africa in Transition: Geographical Essays* (London: Methuen and Co., 1967). Arranged by regions.

HODGSON, ROBERT D., and ELVYN A. STONEMAN, *The Changing Map of Africa* (rev. ed.; Princeton, N.J.: D. Van Nostrand Co., Searchlight Books, 1968).

KIMBLE, GEORGE H. T., *Tropical Africa* (Vol. 1, *Land and Livelihood;* Vol. 2, *Society and Polity*). (New York: Twentieth Century Fund, 1960; abridged ed., Garden City, N.Y.: Doubleday & Co., Anchor Books, 1962). Readable and enlightening volumes by a geographer.

LIGHT, RICHARD UPJOHN, *Focus on Africa* (New York: American Geographical Society, 1941). A book of photographs, mainly from the air; old, but still valuable.

LYSTAD, ROBERT A., ed., *The African World: A Survey of Social Research* (New York: Frederick A. Praeger, 1965). Eighteen contributions by authors in various fields, including Chap. 9, "Geography," by Benjamin E. Thomas, pp. 245–270.

McEWAN, PETER J. M., and ROBERT B. SUTCLIFFE, eds., *The Study of Africa* (London: Methuen and Co., 1965). 39 contributions by various authors.

MINER, HORACE, ed., *The City in Modern Africa* (New York: Frederick A. Praeger, 1967).

Oxford Regional Economic Atlas of Africa (New York: Oxford University Press, 1965).

RIVKIN, ARNOLD, *The New States of Africa* (New York: Foreign Policy Association, Headline Series, 1967).

SOMMER, JOHN W., *Bibliography of African Geography, 1940–1964* (Geography Publications at Dartmouth, No. 3; Hanover, N.H.: Department of Geography, Dartmouth College, 1965).

STILLMAN, CALVIN W., ed., *Africa in the Modern World* (Chicago, University of Chicago Press, 1955).

U.S. DEPARTMENT OF STATE, THE GEOGRAPHER, *Africa: Pattern of Sovereignty* (Geographic Bulletin No. 6; Washington, D.C.: revised June 1968).

U.S. DEPARTMENT OF THE ARMY, *Africa: Problems and Prospects, A Bibliographic Survey.* (Department of the Army Pamphlet No. 550–5; Washington, D.C.: 1967).

Peoples and Cultures: Historical Background

BOHANNON, PAUL, *Africa and Africans* (Garden City, N.Y.: Natural History Press, 1964); also, "Place for All Things: African Markets Fulfill Many Functions," *Natural History,* **73,** no. 8 (October 1964), 54–61; and, with George D. Dalton, eds., *Markets in Africa* (Evanston, Ill.: Northwestern University Press, 1962).

CUNNISON, IAN, *Baggara Arabs: Power and the Lineage in a Sudanese Nomad Tribe* (Oxford: Clarendon Press, 1966).

DAVIDSON, BASIL, *Black Mother: The Years of the African Slave Trade* (Boston: Atlantic-Little, Brown and Company, 1961); and, with the Editors of Time-Life Books, *African Kingdoms* (Great Ages of Man; New York: Time Inc., 1966).

GANN, LEWIS H., and PETER DUIGNAN, *White Settlers in Tropical Africa* (Baltimore, Md.: Penguin Books, 1962).

GIBBS, JAMES L., ed., *Peoples of Africa* (New York: Holt, Rinehart and Winston, 1965). Analyses (by anthropologists) of fifteen cultures of Africa south of the Sahara.

JOHNSON, HILDEGARD BINDER, "The Location of Christian Missions in Africa," *Geographical Review,* 57, no. 2 (April 1967), 168–202, with large folded map.

Journal of African History (quarterly).

Journal of Modern African Studies (quarterly).

Karte der Religionen und Missionen der Erde, 1:23,000,000 (4th ed.; Bern, Switzerland: Kümmerly and Frey, 1965). A multicolored world map of religions and missions.

LIENHARDT, GODFREY, "The Native Cultures of Africa," Chap. 12 in John Bowle, ed., *The Concise Encyclopedia of World History* (New York: Hawthorn Books, 1958), pp. 279–294.

MANNIX, DANIEL P., with MALCOLM COWLEY, *Black Cargoes: A History of the Atlantic Slave Trade* (New York: Viking Press, 1963).

MOOREHEAD, ALAN, *The White Nile* (1961), *The Blue Nile* (1962), and *No Room in the Ark* (1960) (New York: Harper & Row). Informative and entertaining works of history and travel by a journalist; well written and very worth while.

MURDOCK, GEORGE PETER, *Africa: Its Peoples and Their Culture History* (New York: McGraw-Hill Book Company, 1959).

OLIVER, ROLAND, and JOHN D. FAGE, *A Short History of Africa* (Baltimore, Md.: Penguin Books, 1962). An excellent, remarkably condensed history.

OLIVER, ROLAND, and ANTHONY ATMORE, *Africa Since 1800* (Cambridge: At the University Press, 1967).

RODRIGUES, JOSÉ HONÓRIO, "The Influence of Africa on Brazil and of Brazil on Africa," *Journal of African History,* 3, no. 1 (1962), 49–67.

SIMMONS, JACK, "The Transformation of Africa," a section of Chap. 18 in John Bowle, ed., *The Concise Encyclopedia of World History* (New York: Hawthorn Books, 1958), pp. 418–450.

THOMAS, ELIZABETH MARSHALL, *The Harmless People* (New York: Alfred A. Knopf, 1959). A description of the Bushmen in the Kalahari region.

TOUVAL, SAADIA, "Treaties, Borders, and the Partition of Africa," *Journal of African History,* 2, no. 2 (1966), 279–293; and "Africa's Frontiers: Reactions to a Colonial Legacy," *International Affairs,* 42, no. 4 (October 1966), 641–654.

TURNBULL, COLIN M., *The Peoples of Africa* (Cleveland, O.: The World Publishing Company, 1962); also, *The Forest People* (New York: Simon and Schuster, 1961); and "The Lesson of the Pygmies," *Scientific American,* 208, no. 1 (January 1963), 28–37.

Population, Resources, and Economic Development

BARBOUR, K. M., *Population in Africa: A Geographer's Approach* (Ibadan, Nigeria: Ibadan University Press, 1963).

CLARKE, V. de V., "The Threat of Bilharzia," *Geographical Magazine,* 38, no. 11 (March 1966), 865–872.

de KUN, NICOLAS, *The Mineral Resources of Africa* (New York: Elsevier Publishing Company, 1965).

EWING, A. F., *Industry in Africa* (New York: Oxford University Press, 1968).

HAMMING, EDWARD, "On Laterites and Latosols," *Professional Geographer,* 20, no. 4 (July 1968), 238–241.

HANCE, WILLIAM A., *African Economic Development* (rev. ed.; New York: Frederick A. Praeger for the Council on Foreign Relations, 1967); also, "The Race between Population and Resources" [in Africa], *Africa Report,* 13, no. 1 (January 1968), 6–12; and, with Vincent Kotschar and Richard J. Peterec, "Source Areas of Export Production in Tropical Africa," *Geographical Review,* 51, no. 4 (October 1961), 487–499, with large multicolored folded map.

HERSKOVITS, MELVILLE, and MITCHELL HARWITZ, eds., *Economic Transition in Africa* (Evanston, Ill.: Northwestern University Press, 1964).

HODDER, B. W., *Economic Development in the Tropics* (London: Methuen and Co., 1968).

KAMARCK, ANDREW M., *The Economics of African Development* (New York: Frederick A. Praeger, 1967).

McNEIL, MARY, "Lateritic Soils," *Scientific American,* 211, no. 5 (November 1964), 96–102.

MOSS, R. P., ed., *The Soil Resources of Tropical Africa* (Cambridge: At the University Press, 1968).

PROTHERO, R. MANSELL, "Continuity and Change in African Population Mobility," in Robert W. Steel and R. Mansell Prothero, eds., *Geographers and the Tropics: Liverpool Essays* (London: Longmans, Green and Co., 1964), pp. 189–213.

RILEY, BERNARD W., "Some Economic Problems of Africa: A Geographer's Point of View," *Journal of Geography,* 64, no. 2 (February 1965), 64–72.

ROBINSON, E. A. G., ed., *Economic Development for Africa South of the Sahara: Proceedings of a Conference Held by the International Economic Association* (New York: St. Martin's Press, 1964). 26 contributions by various authors.

TREWARTHA, GLENN T., and WILBUR ZELINSKY, "Population Patterns in Tropical Africa," *Annals of the Association of American Geographers,* 44, no. 2 (June 1954), 135–162.

UNESCO, *A Review of the Natural Resources of the African Continent* (Paris: UNESCO, 1963).

UNITED NATIONS, ECONOMIC COMMISSION FOR AFRICA, *Economic Bulletin for Africa* and other publications.

YATES, BARBARA A., "Railroads and Waterways of Africa," *Journal of Geography,* 60, no. 3 (March 1961), 120–135.

Agriculture and Animal Husbandry

"Agriculture in Africa," *Foreign Agriculture,* 4, no. 45 (November 7, 1966), 1–18. A special issue.

ALAN, WILLIAM, *The African Husbandman* (New York: Barnes & Noble, 1965).

BATCHELDER, ROBERT B., and HOWARD F. HIRT, *Fire in Tropical Forests and Grasslands* (U.S. Army Natick Laboratories, Earth Sciences Division, ES-23; Natick, Mass.: 1966).

BIEBUYK, DANIEL, ed., *African Agrarian Systems* (New York: Oxford University Press, 1963).

DARLING, F. FRASER, "Wildlife Husbandry in Africa," *Scientific American,* 203, no. 5 (November 1960), 123–134.

DE SCHLIPPE, PIERRE, *Shifting Cultivation in Africa: The Zande System of Agriculture* (New York: Humanities Press, 1956).

DESHLER, WALTER, "Cattle in Africa: Distribution, Types, and Problems," *Geographical Review,* 53, no. 1 (January 1963), 52–58, with large folded map.

DYSON-HUDSON, RADA, "Men, Women and Work in a Pastoral Society: A Man Is Known by the Cattle He Keeps," *Natural History,* 69, no. 10 (December 1960), 42–57.

HICKLING, CHARLES F., "The Cultivation of Tilapia," *Scientific American,* 208, no. 5 (May 1963), 143–152.

JONES, WILLIAM O., *Manioc in Africa* (Stanford, Calif.: Stanford University Press, 1959).

MILLIKAN, MAX, and DAVID HAPGOOD, *No Easy Harvest: The Dilemma of Agriculture in Underdeveloped Countries* (Boston: Little, Brown and Company, 1967).

MIRACLE, MARVIN P., *Maize in Tropical Africa* (Madison: University of Wisconsin Press, 1966).

MURDOCK, GEORGE PETER, "Staple Subsistence Crops of Africa," *Geographical Review,* 50, no. 4 (October 1960), 523–540.

PHILLIPS, JOHN, *Agriculture and Ecology in Africa: A Study of Actual and Potential Development South of the Sahara* (New York: Frederick A. Praeger, 1960).

UNITED NATIONS, ECONOMIC COMMISSION FOR AFRICA, *African Agricultural Development: Reflections on the Major Lines of Advance and the Barriers to Progress* (New York: United Nations, 1966).

Regions of Tropical Africa

25

As we have noted in the preceding chapter, a large majority of the political units in Africa were dependencies or quasi-dependencies of European nations prior to 1960. At the outbreak of World War II in 1939, only three units—South Africa, Egypt, and Liberia—were independent countries.[1] The remainder of the continent was divided among colonial possessions of six European nations: the United Kingdom, France, Belgium, Italy, Portugal, and Spain. But since the war, and particularly since 1955, the drive for independence has changed Africa from a continent of colonies to an area comprised of dozens of independent states, along with a handful of colonial remnants. Most of these units are located in the low-latitude parts of the continent. A few of the low-latitude units—the Republic of the Sudan and the countries of Africa's "Eastern Horn"—have been discussed in Part IV, The Middle East. The remainder fall conveniently into a few well-recognized regional groups: West Africa,

---

[1] Ethiopia was an Italian dependency from 1936, when it was overrun by an Italian invasion, until 1941. It was liberated by British forces in 1941 during World War II, and in 1942 resumed its original status as an independent state.

Equatorial Africa, East Africa, South Central Africa, and the Indian Ocean Islands. These regional divisions of Tropical Africa will be considered in the present chapter in the order named.

WEST AFRICA

West Africa—here defined as extending from the Cape Verde Islands, Senegal, and Mauritania eastward to Nigeria and Niger, and southward to the Guinea Coast (maps, pp. 550, 551)—was formerly a French or British colonial realm except for independent Liberia and the small Portuguese "overseas provinces" of Portuguese Guinea and the Cape Verde Islands. The latter islands, well out in the Atlantic off Senegal, are, like the Cape Verde peninsula on the mainland, of volcanic origin. Some of them rise to considerable heights, while others are low-lying. Influenced for most of the year by dry northeast trade winds, they suffer from moisture deficiencies, though favored areas have enough water for precarious cultivation of a variety of tropical crops. The islands of the Cape Verde group, discovered by Portugal in 1460 and occupied by the Portuguese since then, have a population that racially is West African—the descendants of slaves—or mulatto, along with a sprinkling of whites. Agriculture, largely subsistence in character, and fishing are the main occupations. In former times the islands were of considerable importance as a coaling point for ocean shipping on Atlantic routes. Today the Cape Verdes are characterized by heavy population pressure, much poverty, and a considerable emigration of islanders seeking work in other Portuguese territories or elsewhere. All in all, the islands of the Cape Verde group have a considerable number of physical, climatic, historical, cultural, and economic affinities with the West African mainland and are conveniently discussed in the context of West Africa. Coverage of them in this chapter, however, will be restricted to the brief summary here, and they will not be considered further in generalizations about West Africa as a whole.

Political and Ethnic Patterns and Problems

Of the present countries on the mainland, four—The Gambia, Sierra Leone, Ghana, and Nigeria—were British dependencies (for dates of independence, see map, p. 65), and nine—Mauritania, Mali (formerly French Sudan), Niger, Senegal, Guinea, Ivory Coast,

Upper Volta, Dahomey, and Togo—were under French control. Great Britain and France followed rather different policies in administering their African dependencies. It was the policy of the United Kingdom to foster the political and cultural development of its African peoples along lines chosen in a considerable measure by themselves, and to move the African units step by step toward responsible self-government and independence. Since independence, all of the United Kingdom's former possessions in West Africa have retained membership in the Commonwealth of Nations. Prior to the surge for independence which developed in Africa after World War II, France's policy was to draw her African peoples more and more within the orbit of French culture and thus to train them for citizenship in a Greater France. But this policy was gradually modified as insistent pressures for autonomy developed among these peoples. As previously noted in the section on the Middle East, two French possessions, the protectorates of Morocco and Tunisia in North Africa, gained independence in 1956. French Cameroons and French Togoland, held by France as trust territories under the United Nations, became the independent republics of Cameroun and Togo in 1960. The remaining African areas, except Algeria, held referendums in 1958 on the question of membership in a new political structure known as the French Community. Three options were available to each territory: (1) no change in current status, (2) immediate independence outside the Community, with loss of French financial support, and (3) membership in the Community as an autonomous republic handling its own internal affairs, but entrusting foreign relations, defense, and certain other functions to France. Only one territory opted for immediate independence. This was French Guinea in the federation of colonial territories called French West Africa. It became the independent Republic of Guinea in 1958. The remaining units of French West Africa, plus all units in French Equatorial Africa and Madagascar (now the Malagasy Republic), voted to have the status of autonomous republics. But this status was short-lived. In 1960 all of the republics petitioned France for complete independence, and by the end of the year they all had become independent states. Today (1969) the French Community appears to have little or no significance as a formal organization. Some of the independent African units withdrew from it, while others—Senegal, the Malagasy Republic, and the four republics of former French Equatorial Africa—did not, but in any event the organization, as such, has been deemphasized. However, some spirit of community pervades the French-speaking African states, and officials from some or all of these states gather periodically to discuss subjects of mutual interest and to coordinate governmental and other activities.

It should be noted that in West Africa—as in various other parts of Africa formerly controlled by Britain or France—the English and French languages continue in widespread use among educated people, are taught in the schools, and are, in fact, official languages in which the day-to-day transactions of national governments are largely carried on. It may be that in the course of time many of these countries will move away from English or French in favor of indigenous tongues (a few are already attempting to do so), but for the present, and with no connotation of continued colonialism, one may properly speak of "West Africa of the English expression" ("Anglophone") and "West Africa of the French expression" ("Francophone"). European languages are a highly useful means of communication in West Africa, a region where literally hundreds of indigenous languages and dialects, often unrelated, are spoken. It should be noted that economic relationships between most West African countries and the metropolitan power that formerly controlled them continue to be close, and in the case of the former French territories there are close defense relationships as well. France maintains small military forces in West Africa, as in some other parts of its former African empire, and is bound by treaty to render military assistance to these countries if called upon to do so. Both Britain and France were able to provide their West African dependencies with a nucleus of trained and experienced government servants drawn from the African population. Such men were in most instances able to take effective control of the many independent states that emerged after 1955. On the whole the transition to independence was handled with remarkable smoothness, particularly in the French possessions.

As so often is the case in Africa, however, the governments of the region must deal with serious problems growing out of the ethnic complexity of national populations. West Africa is inhabited by innumerable tribes who speak an extraordinary variety of languages and dialects, hold different religious beliefs, and often are characterized by a long history of suspicion and hostility toward one another. The majority of people within the region are animists, with endlessly varied tribal customs and rituals. In the drier lands of the north the Moslem faith has had a great impact, and it has been spreading southward over a long period. To add further to the religious complexity, many West Africans, particularly in areas near the coast, are converts of Protestant or Roman Catholic missions. Political affairs in the countries of the region generally are controlled by a Westernized African elite that often includes many graduates of European or American universities; but illiterate villagers living more or less in the ancestral way are far more typical of the population. Only small numbers of Europeans live in these countries, a circumstance initially due to a variety of factors: the inhospitable climate, the numerous diseases, the opposition of slave-trading interests to other forms of European enterprise, the opposition of well-organized African kingdoms to encroachment by white settlers, and, later, policies by European colonial powers that forbade or circumscribed the acquisition of land by Europeans. In the age of African independence West Africa was largely spared the difficulties that afflicted Kenya or Rhodesia, in which white settler classes held large blocks of farmland. Practically all the Europeans in West Africa have been of a governmental, professional, managerial, commercial, or technical class; and the overwhelming majority of them have long been found in large cities such as Lagos or Dakar. Thus the presence of an entrenched European settler class in rural areas has not been a circumstance contributing to the internal political troubles of West African countries since independence. However, European governments contributed initially to some of the present troubles by drawing boundary lines around colonial units in the absence of adequate ethnological knowledge concerning the populations thus enclosed. The typical West African country, like most countries elsewhere in Tropical Africa, is a jumble of tribes that have in most instances only a rudimentary sense of national purpose. Under such conditions progress toward national unity and a modern political system is bound to be slow.

One country that has had unusually severe difficulties in the foregoing respects is Nigeria. The country's three major population elements—the Moslem Fulani and Hausa of the north, the Yoruba of the southwest, and the Ibo of the southeast—have frequently been at odds with each other. One group, the Ibo, heavily localized in the area east of the Niger River and south of the Benue, actually attempted secession from Nigeria in 1967 by establishing a separate country called Biafra. This action was resisted by the Nigerian central government, and a bloody civil war resulted. By early 1969 central government forces had occupied important parts of Biafra, and food supplies had sunk so low in the latter area, one of Africa's most densely populated regions, that widespread starvation was feared. The Biafrans were receiving some relief supplies from outside agencies, and both sides were receiving military equipment from the outside. A few African governments had recognized Biafra, but most of the world's governments had not. No end to the struggle was in sight.

As a final point concerning the cultural and political geography of West Africa, it should be noted that this was one of the parts of Tropical Africa which was most advanced culturally and politically prior to the Age of Discovery, and that it is one of the areas which has been longest and most continuously subject to European and Islamic influences. A series of strong African kingdoms and empires developed here, both in the forest belt of the south and the grasslands of the north. Some of these states produced notable works of art, for example the bronze sculptures of the kingdom of Benin in what is now southern Nigeria. Empires in the grasslands, such as ancient Mali or ancient Ghana, were trading intermediaries between the Guinea Coast and Islamic North Africa in the days before this trade was largely preempted by Europeans approaching the Guinea Coast from the sea. In the nineteenth century, Freetown in Sierra Leone (which was founded, like Monrovia in Liberia, as a settlement for freed slaves), became a center for mission work and education along the Guinea Coast at a time when such efforts were absent or minimal in other sections of Tropical Africa. A much larger proportion of West Africans have received training in overseas universities than have peoples of other Tropical African regions. All in all, this was the section of Tropical Africa that was best prepared to meet the challenges of political independence in the modern age.

Area, Population, and Physical Environment

Africa is a large continent, and West Africa alone is larger than one might realize from looking at the ordinary map unless he pays careful attention to scale. The fifteen mainland political units in the region aggregate over 2.3 million square miles, or about three-quarters the area of the conterminous United States. The larger countries, by area, are Mali, Niger, Mauritania, and Nigeria (Table 20). Their areas range from 357,000 to 464,000 square miles, and the four countries combined have about one-seventh or one-eighth the area of the conterminous United States. Distances are correspondingly great. When using a map showing all of Africa, one may easily fail to realize that it is 600 miles from the coast of Nigeria to the country's northern border, or that the corresponding distance in the smaller country of Ghana is 400 miles, to say

nothing of the fact that the distance along the savanna belt between Kano in northern Nigeria and Dakar in Senegal is 1700 miles. The isolating effects of such distances in West Africa are compounded by inadequate transportation facilities.

The region is less impressive with respect to population totals. In 1968 the fifteen mainland units had an estimated total population of just over 105 million—a little more than half that of the United States. Ten of the fifteen had under 4 million people, and only two of them exceeded 6 million. The latter units—Ghana, with 8.4 million people, and Nigeria—are the two that are most characterized by intensive, commercialized agriculture tied into world markets. But the most striking single aspect of West Africa's population distribution is that over half of the region's total population is found within one country, Nigeria (map, p. 42). With 62 million people (1968), Nigeria is on an entirely different scale in population from the other West African countries. Not only is it one of the largest West African countries in area; it is also by far the most densely populated (174 per square mile), is one of the most commercialized in its economy, and contains more than twice as many cities of 100,000 population or more as the other fourteen units combined. The unusual concentration of people in Nigeria apparently existed even before European contact, with its attendant commercialization and urbanization; it may have been related to the existence of powerful kingdoms in the country which were able to impose a degree of internal order and security. But essentially the reasons for the great concentration of West Africa's people in Nigeria are incompletely understood and something of a mystery.

In most parts of West Africa population density is low. Few sections average more than 25 persons per square mile, and in the north the average density is less than 2 per square mile within a broad band of desert and semidesert stretching across sizable parts of Niger, Mali, and Mauritania. Most of Nigeria and Ghana, and areas near the coast in a number of other countries, are somewhat more densely populated than the average, and in a few places high densities occur. The principal areas in the latter category are the southern parts of Ghana, Togo, Dahomey, and Nigeria and an east-west belt of territory around Kano in northern Nigeria. In these areas the average density is over 100 per square mile and in some places rises very much higher.

Environmental contrasts within West Africa are extreme. Climatically the region ranges from parts of the Sahara Desert in the north through belts of tropical steppe, dry savanna, and wetter savanna, to some areas of tropical rain forest along the southern and south-western coasts (front endpaper climatic map and back endpaper vegetation map). Most West African countries have their most populous areas in the zone of tropical savanna climate, though such areas may differ considerably with respect to the timing and intensity of the yearly wet season or seasons, as well as in the relative proportions of grass and deciduous woodland in the vegetation cover. A few countries deviate significantly from the general pattern. Mauritania, for example, lies far enough north that its more populated southern section is only wet enough to be classed as steppe; while Liberia, on the other hand, lies in a climate that has a dry season but receives enough moisture to be classed as tropical rain forest. The large country of Nigeria has two very populous sections: a northern one in the savanna and a southern one in rain forest. These two sections are the largest concentrations of population in West Africa (map, p. 42). In many parts of West Africa the original pattern of vegetation has been greatly altered as a consequence of human use of the land for agriculture and grazing. In Sierra Leone, for example, tropical rain forest is thought to have been the original vegetation over the greater part of the country, but most of the original cover has been destroyed by cultivators. The present vegetation of Sierra Leone is comprised largely of secondary forests, often of poor quality, interspersed with grasslands.

Throughout West Africa temperatures are tropical. Along the southern coast, in the "coolest" month of the year temperatures average in the high 70s, and in the warmest month in the low 80s. Farther north and to the interior, in the "coolest" month the average tends to be in the low 70s, and in the hottest month it is often in the high 80s. Steaming heat is the rule in wetter areas and periods; blazing heat in drier areas and periods. Most uplands are not high enough to bring about significant reductions in temperatures.

The region exhibits much topographic as well as climatic variety, especially when considered in detail (maps, pp. 550–551). In the dry north are vast reaches of low plateau country, much of it with the appearance of plains. In places the monotonous plains landscape is broken by rougher terrain, including some areas of sand dunes, some sections dissected into hills, and a few areas of mountains. To the south lies an east-west belt of uplands, from interior Senegal to eastern Nigeria. It reaches almost to the coast in the south and extends some 400 to 600 miles northward into the interior. The detail of this area is very complex, including hills, low mountains, and plateaus in various stages of dissection. Along the immediate coast lies a narrow belt of coastal

plain, widening in some sections such as the Niger delta, and tending to extend inland somewhat along the lower courses of major rivers. Parts of this belt are mangrove swamp, and its straight reaches of coast and barrier sandbars provide only a handful of good natural harbors. Often the immediate coast is thinly populated, the coast of Nigeria being a good example.

Agriculture and Mining

The West African countries exhibit the customary economic pattern of underdeveloped nations. They depend heavily on subsistence or near-subsistence agriculture—or, in the drier north, livestock grazing—along with a few export specialties. In the wetter south, well exemplified by southern Nigeria, subsistence agriculture relies mainly on root crops such as manioc and yams, on corn, and on the oil palm. In some areas modern methods are being used to introduce or expand wet rice production; in Sierra Leone rice is now the most important subsistence crop. In the drier, seasonally rainy grasslands of the north (as, for example, in northern Nigeria or northern Ghana), nomadic and seminomadic herding of cattle and goats is important, along with the growing of subsistence grain crops of millet, sorghum, and corn. In a number of areas such as the flat "inland delta" of the Niger River in Mali (once an entrance to an extensive lake), irrigated rice has been introduced. Many livestock from the north are marketed in the cities farther south; the animals may be driven to market or transported at least part of the distance by rail or truck. The major export specialties of West African agriculture are oil-palm products and cacao in the wetter, forested south, and peanuts in the drier north. Southern Nigeria's massive exports of palm kernels and palm oil make that country the world's largest exporter of oil-palm products. The main belt of oil palms lies in the denser rain forest of the southeast (photo, right), parts of which receive up to 100 inches of rain per year. Located within the boundaries of the secessionist state of Biafra east of the Niger River, the most highly developed areas of commercial palm production have population densities that reach 400 to 500 per square mile in some localities. The Ibo peoples are the main population element here; they are largely village dwellers, in contrast to the Yoruba peoples in the cacao belt of southwestern Nigeria, a sizable proportion of whom live in cities. Besides Nigeria, the other coastal countries as far west as Portuguese Guinea—notably Dahomey and Sierra Leone—also export products of the oil palm, though on a

A market in Nigeria. Open-air markets are a universal institution in tropical Africa. Oil palms rise behind the market stalls. (British Information Services.)

much smaller scale than does Nigeria. The principal producer and exporter of cacao by a wide margin, in West Africa and in the world, is Ghana (photo, p. 580), which has become West Africa's "richest" country (1965 per capita income, $245). The cacao beans are roasted and then ground to a powder in overseas processing centers to produce cocoa. This product, grown mainly in the part of southern Ghana known as Ashanti, normally provides well over half of Ghana's exports by value. Nigeria, Dahomey, and the Ivory Coast are also cacao exporters; in fact, Nigeria ranks second in the world and the Ivory Coast third or fourth. Nigeria's main cacao belt is found in the drier, more open forests of the southwest. Besides oil-palm products Nigeria is

also the world's greatest exporter of peanuts. The latter are the great commercial product of the dry savannas of northern Nigeria, and are the main export from three other West African countries—Senegal, The Gambia, and Niger—that lie at least partly in the dry-savanna belt. This is the world's foremost peanut-exporting area, with Nigeria, Senegal, and Niger ranking first, second, and third, respectively, among world exporters of the product. The three countries accounted for an estimated 63 percent of world exports of peanuts and peanut oil in 1967, and other West African countries exported an additional 6 or 7 percent of the world total. Peanuts supply well over nine-tenths of all exports from The Gambia and more than three-quarters of Senegal's ex-

Preparing cacao for export on a family farm in Ghana. The pods are opened with machetes preparatory to removing the cacao beans, which are embedded in a mass of pulp. After being dried in the sun, the beans will be packed for shipment to overseas processing centers. (British Information Services.)

ports. From the eastern part of West Africa's peanut belt the product moves generally southward to ports on the Gulf of Guinea, especially Lagos and Port Harcourt (over 180,000) in Nigeria, while from the western part it moves mostly through Senegal's ports of Dakar and Kaolack. A variety of other agricultural exports not yet mentioned are of some importance in West Africa. They include coffee and bananas from the south (most of the coffee comes from the Ivory Coast and bananas especially from Guinea), as well as rubber from Firestone plantations in Liberia, and cotton and livestock products (mainly hides and skins) from the north. In addition, the forests of the south are supplying increasing quantities of timber for export. One peculiarity of West African commercial agriculture is the almost complete dominance of African smallholder production. With a few exceptions, notably the Firestone rubber plantations in Liberia, the foreign-owned and managed plantation never achieved a notable foothold in the area, although foreign traders and companies have stimulated commercialization of indigenous agriculture.

The other main export industry of West Africa, and an increasingly important one, is mining. In broad outline four main areas may be distinguished, as follows (map, p. 551).

1. In the desert of northwestern *Mauritania,* near Fort Gouraud, the existence of very large deposits of high-grade iron ore has long been known. Difficulties of access were such, however, that mining began only in 1960. The international company, headed by French interests, which developed the deposits had to build port facilities (at Port Étienne), a town in the mining area, and a railroad 419 miles long connecting the two. The rail line passes through country that is mostly waterless, and its construction involved a 5900-foot tunnel. Production from these mines is of great importance to a country as poor as Mauritania (1965 per capita income, $106).

2. At the southwestern extremity of the western protrusion of Africa, three countries—*Liberia, Sierra Leone,* and *Guinea*—are mineral producers of some importance. Each of the three exports iron ore and diamonds (Guinea's exports of both are small), and Guinea exports

alumina and bauxite as well. The most valuable product from this group of countries is iron ore, much of it high-grade. In Liberia the production comes from four mining areas: the Bomi Hills and Mano River areas northwest of Liberia's capital and largest city, Monrovia (over 85,000), and the newer Bong Mountains and Mt. Nimba districts located farther east. Ore from the first three districts moves by rail to Monrovia for shipment, while that from Mt. Nimba (the largest producer) is shipped a considerably greater distance by rail to the new ore port of Buchanan (map, p. 551). Sierra Leone's iron mines, located in the northwestern interior of the country, are served by a 57-mile line to the ore port of Pepel on the shore of the Sierra Leone Estuary opposite Freetown (over 150,000), Sierra Leone's capital, largest city, and main port. The estuary provides the finest natural harbor in West Africa; at Monrovia and Buchanan in Liberia, by contrast, it has been necessary to construct artificial harbors by dredging and building breakwaters. Alumina and bauxite from Guinea are exported through that country's capital, main city, and port, Conakry. The principal mines and the plant that processes bauxite into alumina lie some distance north of the city and are connected to it by rail. These materials comprise Guinea's largest export by value.

3. *Ghana* has been a significant producer of mineral wealth for centuries. As a colony it was known as the Gold Coast. The country now exports gold, manganese, diamonds, and bauxite. The new Volta River Project, a large hydroelectric dam and lake on the Volta River about 60 miles north of the capital city of Accra, may allow it to make a significant advance in exploiting its minerals by producing aluminum from its extensive bauxite deposits. Ghana's principal mining districts are located in the southwest, within the general area from which most of the country's cacao exports come.

4. *Nigeria* has for some time produced tin from the 4000-foot Jos Plateau in the central part of the country. Ores and concentrates are shipped by rail to Port Harcourt, at the eastern edge of the Niger Delta, and to Nigeria's main port, Lagos, farther to the west. From these ports they are exported to Great Britain for smelting. In the 1960s Port Harcourt became the headquarters for a new and more significant mineral industry as oil strikes were made in the Niger Delta area and production and exports rose rapidly. The Royal Dutch-Shell and British Petroleum companies, as well as American companies, are involved in developing Nigeria's new oil industry. One facet of this development has been the completion of a refinery to supply domestic needs for gasoline and kerosene—the latter a product much used for lighting in Africa. Nigeria has the only coal deposits of any consequence in West Africa. The coal is low grade, and the production, centered at Enugu (over 140,000) to the north of Port Harcourt, is relatively small.

Urbanization

The scale and degree of urbanization in West Africa are low. Five of the fifteen mainland units—Mauritania, Niger, The Gambia, Portuguese Guinea, and probably Liberia—have no urban area which reaches 100,000 in population. In five countries—Upper Volta, Dahomey, Togo, Sierra Leone, and Mali—the largest city is in the 100,000 to 200,000 class. Only five countries have urban areas of approximately 200,000 or more; and of these, three—Senegal, Ivory Coast, and Guinea—have only one such area, while a fourth country, Ghana, has two. In the remaining country, Nigeria, there are eight cities of over 200,000, and possibly a few more, but even this is a very modest urbanization for a country with more than 60 million people.

Sizable cities in the world are mostly associated with trade, centralized administration, and large-scale manufacturing. As West Africa has little manufacturing of more than minor dimensions, the region's principal cities are in its major areas of commercial agriculture, and many of them are seaports and/or national capitals. Nigeria has by far the most impressive urban development. Of the nine largest cities in West Africa, four are clustered in southwestern Nigeria in or near the densely populated belt of commercial cacao production. They include the two largest metropolitan areas in West Africa—Lagos (over 750,000), the capital of Nigeria and a major seaport; and the inland metropolis of Ibadan (over 700,000, and possibly 1 million), a city of traders and craftsmen which also contains a surprisingly large number of farmers. Ibadan, in fact, has the characteristics of both a city and a huge village—a circumstance also true of other sizable urban places in southwestern Nigeria. Farmers living in these cities may commute 10 to 20 miles or more to their fields outside the city. In addition to Lagos and Ibadan, two other cities in the southwest Nigerian cluster—Ogbomosho and Mushin—have well over 300,000 people. By way of contrast, it should be noted that Nigeria's main area of commercial oil-palm production, in the southeast, is remarkably lacking in large cities. Southern Ghana, another agricultural area with a relatively high degree of commercialization, contains two of the nine largest West African cities. Accra (over 600,000) is the national capital; it has been an important seaport despite the lack of deep-water harbor facilities and the consequent necessity of transferring cargo between ship and shore by means of surf boats or lighters. Recently a modern deep-water port has been developed at Tema, some 17 miles east of Accra. Kumasi (over 300,000), located farther inland, is the principal urban center of Ghana's main cacao-producing region. A smaller but rapidly growing urban center is Sekondi–Takoradi (perhaps 185,000), the locale of Ghana's first modern deep-water port at Takoradi. About 150 miles farther west is another of West Africa's nine largest cities—Abidjan (perhaps 275,000), the Ivory Coast's capital, largest city, and main seaport. Situated, like Lagos, Nigeria, on a lagoon, it has deep-water harbor facilities made accessible to ships by an artificial channel through a coastal sandbar. Of the nine largest cities, the remaining two—Dakar and Kano—are associated with the commercial peanut production of the dry savanna belt. Dakar (over 400,000), in Senegal, is the leading seaport in all of West Africa, as well as the capital of Senegal, and was formerly the capital of the immense group of eight colonial territories known as French West Africa. It is the commercial outlet for much of the peanut-producing belt, is a considerable industrial center by African standards, and contains the largest European community in West Africa. Kano (over 300,000) in northern Nigeria is the most important commercial and administrative center of Nigeria's peanut belt and has long been a major West African focus of Islamic culture.

Transportation Facilities and Problems

A common difficulty of underdeveloped areas such as West Africa is inadequate transportation facilities. In this region a notable problem has been the scarcity of good natural harbors—a circumstance due largely to the frequency of offshore sandbars and the tendency of river mouths to be choked with silt. Often it has been necessary to transfer goods by lighter between ship and shore, and large expenses are incurred by these poor countries as they attempt to provide, artificially, improved and modern port conditions and facilities. Inland from the coast lies the formidable problem of adequate connection between the ports and the interior. Rapids, together with annual seasons of low water, limit the utility of rivers for transport, although a number of the latter carry some traffic. Most notable are the Niger and its major tributary the Benue, the Senegal River, and the Gambia River. In fact the Gambia River is the main transport artery for The Gambia—a country consisting of a 20-mile-wide strip of savanna grassland and woodland extending inland for 300 miles along either side of

the river. On the whole, however, West Africa's rivers probably are more of an obstacle than a resource with respect to transportation in view of the bridging problem along land routes. Railways are much more important than rivers in the present transport structure, but are few and widely spaced. Instead of being a network, the rail pattern shows basically a series of lines that extend inland from various ports but that usually do not connect with other rail lines (map, p. 551). Two of these isolated fingers of railway are the main freight routes between landlocked capital cities and the sea: the line connecting Bamako (170,000), Mali, with Dakar; and the line from Ouagadougou (100,000), Upper Volta, to Abidjan. Other lines extend inland from Conakry (200,000), Guinea; Freetown and Pepel, Sierra Leone; Monrovia and Buchanan, Liberia; the various ports of Ghana; Lomé (130,000), Togo; Cotonou (120,000), Dahomey; and Lagos and Port Harcourt in Nigeria; there is also the iron-ore line from Port Étienne to Fort Gouraud in Mauritania. No rail line penetrates Niger, whose capital and largest city, Niamey, on the Niger River, has only an estimated 60,000 people in the city proper. The road network of the region is denser than the rail network and is growing rapidly, but most roads are not yet paved or reliable in all weather. Most of the region's area and many of its people are still quite remote from any motorable road. Nevertheless, truck traffic is becoming steadily more important relative to rail traffic. Not surprisingly, a very large part of the limited long-distance passenger travel within the region takes place by air. But either air freight must become much more economic than it now is, or road and rail facilities must be very greatly expanded, if much of West Africa is to become an effective part of the modern commercial world.

EQUATORIAL AFRICA

East and southeast of West Africa lies a varied group of countries to which the term "Equatorial Africa" often is applied. As used here, the name embraces former French Equatorial Africa, Cameroon, former Belgian Africa, the former Spanish possession of Equatorial Guinea, and the small Portuguese-held islands of São Tomé and Principe (map, p. 550). The equator crosses the heart of the area, and the climate is more truly "equatorial" than it is in African countries farther east (Uganda, Kenya, and Tanzania) that also lie on or near the equator but are, on the average, much higher in elevation. The latter are discussed separately in the section on East Africa. Three political units—Chad, Rwanda, and

Burundi—diverge from the basic pattern of the area, but for convenience are included with Equatorial Africa. Chad, the northernmost unit in the area, reaches the margins of the Sahara and is much drier and more dependent on livestock raising than are the other countries; and it has a larger proportion of Moslems in its population. Rwanda and Burundi, former Belgian dependencies located in the margins of East Africa, are much too high in elevation, temperate in climate, and densely populated to be typical of Equatorial Africa.

Only scattered parts of Equatorial Africa, primarily along the eastern and western margins, can be described as truly mountainous. On the other hand, only a narrow band along the immediate coast is comprised of plains below 500 feet in elevation. As in West Africa the coastline is fairly straight and has few natural harbors; shallow lagoons behind coastal sandbars are common; and lagoons and delta channels often are fringed by mangrove swamps. Most of Equatorial Africa is comprised of plateaus that lie, in general, at elevations of 1000 to 3000 feet but that in places rise considerably higher. Plateau surfaces usually are undulating, rolling, or hilly. They are capped in many places by remnants of older surfaces and are cut by innumerable stream valleys, often deeply incised. A conspicuous exception to the generally uneven terrain is a broad area of flat land in the inner Congo Basin, once the bed of an immense lake. The most prominent mountain ranges of Equatorial Africa lie along the margins of the Great Rift Valley in Rwanda, Burundi, and the eastern part of Congo-Kinshasa, and in the west of Cameroon along or near the Nigerian border. In both instances vulcanism has played an important role in mountain building. Strips of rough, hilly country or low mountains often are found along the dissected edges of plateaus, particularly in the west where the plateaus drop away to the coastal plain. The greater part of Equatorial Africa lies within the drainage basin of the Congo River. Most of the remainder drains either to the Ogowe River or to Lake Chad. The heart of the region has a tropical rain forest type of climate, grading into tropical savanna to the north and south (front endpaper climatic map). Most of Chad has a tropical steppe or desert climate; only in the extreme south does it reach into the zone of tropical savanna. Some mountains in the east of Congo-Kinshasa rise high enough to have permanent snow, and the climate of this area, as well as that of adjoining Rwanda and Burundi, is classed as undifferentiated highland.

Equatorial Africa is inhabited by hundreds of different tribes speaking a multiplicity of languages and dialects. Some of these peoples are among the most primitive in Africa. The majority of the region's inhabitants are classed as Bantu, though Sudanese peoples predominate in the drier areas of grassland to the north of the rain forest. The Watusi of Rwanda and Burundi are thought to have affinities with Hamitic peoples of the upper Nile and the East African highlands. Further complexity is added to the pattern of peoples by small bands of Pygmies in the deep rain forest of the Congo Basin.

Within Equatorial Africa as a whole, population densities are low, and populated areas are sporadically distributed. Population clusters tend to be isolated from each other by rough terrain, thick forests, swamps, or tsetse-infested areas. In general the savanna lands and highlands are more densely populated than the rain forests. Most countries have an overall density well below that of Africa as a whole; for example, the average density is about 5 per square mile in Gabon, 6 in the Central African Republic, 7 in Congo–Brazzaville, and 18 in Congo-Kinshasa, as compared with an overall figure of 29 for Africa as a whole (Table 20). An overwhelming majority of the population in Equatorial Africa is rural, and a very large majority is illiterate. Some 3.3 million children were reported to be enrolled in primary schools in 1965–1966, and about 110,000 in secondary schools, within the region as a whole.

Former French Equatorial Africa and Cameroon

Prior to independence, most of Equatorial Africa was under French or Belgian control. The units held by France included the United Nations trust territory of French Cameroons and the "overseas territories" which comprised French Equatorial Africa—Chad, Ubangi-Shari (now the Central African Republic), Middle Congo (now the Republic of the Congo [Brazzaville]), and Gabon. The four last-named units gained internal self-government in 1958 as Associated States of the French Community and then attained full independence in 1960. Today they remain more closely linked to France in economic, political, and military matters than most of France's former possessions. As a group they are less developed economically than France's former dependencies in West Africa, and their economies are very dependent on French subsidies of various kinds. Gabon, with its valuable exports of minerals and timber, is the most viable country economically. The Federal Republic of Cameroon is comprised of former French Cameroons (now East Cameroon) and the southern part of former British Cameroons (now West Cameroon). East Cameroon gained independence in 1960 as the Republic of Cameroun, and West Cameroon voted to join it in 1961. The two parts of the

Federal Republic have experienced some difficulty in functioning as a single country, due partly to disparities in size and economic strength, but mainly as the result of their previous development under very different colonial systems. East Cameroon accounts for about nine-tenths of the federation's area, four-fifths of its population, and nearly nine-tenths of its exports. Both French and English are recognized as official languages in Cameroon.

The former French units form a contiguous series bounded by Nigeria and Niger on the west and Congo-Kinshasa and the Republic of the Sudan on the east (map, p. 550). They lie mostly north of the equator, but the southernmost unit, Congo–Brazzaville, reaches to about 5°S. As in the former French territories of West Africa, several climatic and vegetational zones are encompassed:

1. *Tropical rain forest,* occupying most of Gabon, plus the northern part of the Republic of the Congo, the southwestern part of the Central African Republic, and the southern part of Cameroon.

2. *Tropical savanna,* occupying the southern part of the Republic of the Congo, most of the Central African Republic, the northern part of Cameroon, and southern Chad.

3. *Tropical steppe* in central Chad.

4. *Desert* in northernmost Chad.

In most parts of these units the population is relatively sparse and is supported by agriculture or pastoralism, generally on a subsistence or near-subsistence basis. Production of crops for export within the area as a whole is comparatively small, being most significant in southern Cameroon. In the latter area coffee and cacao are the most important export crops, although they are supplemented by a much smaller production of bananas, oil-palm products, natural rubber, and tea. These exports come partly from African small farms and partly from plantations: among the latter are plantations in West Cameroon that were developed during the period of German control before World War I and now are operated by a government development corporation. They are found in an area of volcanic soil at the foot of Mt. Cameroon (13,353 feet), a volcano which is still intermittently active. The mountain is notable for its extraordinarily heavy rainfall; a station on the lower slopes near the sea averages 392 inches of precipitation a year, and the average on higher slopes is probably well over 400 inches. A chain of extinct volcanoes stretches northward from Mt. Cameroon, and the line of mountains is continued southward in the islands of Fernando Po, São Tomé, Principe, and Annobon. Most agricultural exports from the former French units come from areas of

rain forest. Small amounts of cotton are exported from savanna lands in Chad, the Central African Republic, and Cameroon; and there are limited exports of livestock products from the drier savannas and tropical steppes. Millions of animals, primarily cattle, are raised in the grasslands by Moslem herdsmen such as the Fulani, but they are kept primarily for wealth and prestige, ceremonial uses, or local food supply. The main subsistence crops in Equatorial Africa are similar to those of West Africa: millets, sorghums, corn, and peanuts in the grasslands; manioc, yams, corn, rice, bananas, plantains, and oil palms in the rain forests.

All of the former French units except Chad have exports of timber. Such exports are comparatively minor in Cameroon and even more so in the Central African Republic, but they are of major significance in the economies of Gabon and Congo–Brazzaville. Up to the present minerals have been comparatively unimportant except in Gabon, though scattered discoveries of valuable minerals have been reported from several countries in recent years. In Gabon, large deposits of high-grade manganese ore and high-grade iron ore in interior locations, along with comparatively small deposits of petroleum on the coast and uranium deposits near the manganese of the interior, are the principal minerals discovered thus far (1968). Manganese led Gabon's mineral exports by a wide margin in 1965, followed by petroleum and uranium. The manganese ore, mined in the deep interior southeast in the upper basin of the Ogowe River (map, p. 551), is transported across a range of low mountains by a 47-mile cableway to a railhead in Congo-Brazzaville from which it moves to the Congo seaport of Pointe Noire. Here the ore is processed in a concentrating plant prior to export. Gabon's main reserve of iron ore, located in the remote northeast over 400 miles from the sea, awaits completion of a railroad through difficult terrain to connect the ore field with a shipping point on the Gabon Estuary. Development of the deposits is being undertaken by French and American steel interests.

Nearly all the overseas trade of former French Equatorial Africa and Cameroon is handled by four seaports: Pointe Noire (90,000) in the Congo, Port Gentil (40,000) and Libreville (60,000) in Gabon, and Douala (210,000) in Cameroon. Pointe Noire, the most important port, not only is the main sea outlet for the Republic of the Congo but handles a large share of the overseas trade of the Central African Republic and Chad as well. In addition, as noted above, it ships the manganese exports of Gabon. The 320-mile Congo-Ocean Railway connects Pointe Noire with Brazzaville (150,000) on the Congo River. Brazzaville, located directly across the river from the capital city of Kinshasa in the former Belgian

Congo, was the administrative center of French Equatorial Africa and is the capital and largest city of the present Republic of the Congo. The Congo River and its large tributary the Ubangi provide a 740-mile navigable water connection from the Brazzaville railhead to the capital city of Bangui (150,000) at the head of navigation in the Central African Republic. The latter country, a landlocked state with poor highway connections to the sea, depends on the waterway and the Congo-Ocean rail line for transportation of most of its small foreign trade. In Cameroon two comparatively short railroads connect the main agricultural exporting areas with the port of Douala. The longer of the two lines reaches Yaoundé (city proper, 110,000), Cameroon's capital, located in the heart of the country's main cacao-exporting district. An extension of this line toward northern Cameroon is under construction. In Gabon, Port Gentil, located on a distributary of the Ogowe River, is primarily a timber port. It handles large quantities of logs floated down the Ogowe. Libreville, located on the broad Gabon Estuary north of Port Gentil, is primarily a general-cargo port, but it also ships timber, as do Pointe Noire in the Congo and Douala in Cameroon. At Libreville the timber traffic is handled by a separate port some 10 miles upstream. Many kinds of hardwood enter the export trade from the former French units, but the most valuable wood shipped from the area as a whole is okoumé, a soft wood prized for plywood and paneling. Okoumé logs are light enough to be rafted down rivers such as the Ogowe. The bulk of the timber from these countries is shipped in the form of logs, but some is processed in sawmills or plywood mills before shipment.

Aside from the handful of lines mentioned above, the five countries considered here are devoid of railways, and most areas are poorly served by roads. Isolation and poor transport facilities are major limiting factors for economic development, particularly in Chad and the Central African Republic. The airplane is doing a great deal to break down isolation, but it can scarcely take the place of surface transport for bulk goods.

Manufacturing is conspicuously absent from the five countries, aside from such establishments as cotton gins, oil presses, sawmills, and a limited number of factories that manufacture simple consumer items, process sawlogs into plywood and veneer, process imported alumina to make aluminum, or refine oil. Africa's first aluminum plant was established in 1957 at Edéa on the Sanaga River in Cameroon. Electric power is supplied by a hydroelectric station at a dam on the river near by, and alumina for processing is brought from Guinea or from France. Bauxite deposits in Equatorial Africa itself may eventually be tapped by this plant and others, but lack of transportation precludes this possibility at present. It is anticipated that future industrial development in the five countries will make extensive use of hydroelectricity for power. The most ambitious project envisioned thus far is a large dam on the Kouilou River to the north of Pointe Noire in Congo-Brazzaville. The present status of this project is uncertain (1968). In Gabon a refinery has been opened at Port Gentil to process oil from small fields in the coastal zone.

Equatorial Guinea, São Tomé, and Principe

Until 1968, Spain's modest African empire included small areas in Equatorial Africa: Rio Muni on the mainland between Gabon and Cameroon, and Fernando Po and smaller islands in the Gulf of Guinea. These units were granted independence in October 1968 as the Republic of Equatorial Guinea. Rio Muni, a densely forested area, exports mahogany and okoumé logs, together with small amounts of coffee and other tree crops. It is less developed economically than Fernando Po, a volcanic island rising to a central peak some 9350 feet high. The tropic heat, abundant rainfall, high humidity, and fertile soils of the island provide fine conditions for cacao, the main export crop. Coffee and small amounts of other tree crops also are exported. Most of the production comes from plantations owned by Europeans, though African small farms play a minor role. Local supplies of labor are inadequate to operate the plantations, and thousands of migrant laborers from West African countries, primarily Nigeria, are employed. The island's plantations and small farms occupy lower slopes near the coast. At higher elevations, generally above 4000 feet, some land has been cleared for pastures on which dairy cattle are raised. The Portuguese-held islands of São Tomé and Principe, southwest of Fernando Po, are smaller than the latter island but bear some similarity to it in environment and economy. As in Fernando Po, cacao and coffee plantations provide the principal exports. A large part of the labor force is recruited from Angola, Mozambique, and the Cape Verde Islands.

The Democratic Republic of the Congo (Kinshasa)

On June 30, 1960, Belgium's vast Congo colony in Equatorial Africa became the independent Republic of the Congo (later the name was changed to Democratic Republic of the Congo). The colony was ill-prepared for this step, as events swiftly proved. But pressure for independence had built up rapidly in 1959, and Belgium had yielded to the demands of Congolese leaders. The new country was quickly faced with major crises that its government was unable to surmount. Among these were (1) a large-scale mutiny within the Belgian-trained and -officered Congolese army, (2) regional separatist movements and rebellions, the most serious being a declaration of independence by the mineral-rich Katanga Province in the southeast, and (3) outbreaks of tribal warfare, especially in Kasai Province, the country's main diamond-mining area. With the Congo apparently disintegrating into chaos amid scenes of violence and confusion, Belgian troops entered to protect Belgium's nationals, large numbers of whom were evacuated to Belgium. Shortly afterward a United Nations emergency force, composed mainly of troops from African nations, was dispatched to the Congo to restore order and assist the central Congolese government in repairing the country's damaged political and economic fabric. After sharp fighting between United Nations and Katangese forces, the secession of Katanga was ended in 1963. Withdrawal of United Nations military units in 1964 was followed by new rebellions and political crises, centering particularly in the eastern regions around Kisangani (city proper, 150,000), formerly Stanleyville, and Bukavu (city proper, 35,000 in 1959). Government forces eventually contained the rebels with the aid of white mercenaries, who themselves subsequently became disaffected and for a while exerted independent control over sizable areas in the east. This episode ended with the withdrawal of the mercenaries from the Congo under guarantees of safe passage—guarantees arrived at during prolonged negotiations involving outside countries and agencies as well as the Congo central government. Today (early 1969) rebel activity is confined to isolated pockets in remote areas, and the authority of the central government appears not to be seriously challenged in any of the main populated areas of the country. Restoration and development of the economy are proceeding, with financial and technical assistance from Belgium, the United Nations, the United States, and other quarters. Many thousands of Belgians who left the country during the worst crises have returned. Grave problems of political, economic, and social organization and development remain to be solved, but if internal peace and security can be permanently maintained and outside assistance continued for a considerable period, the Congo, richer in natural resources than most African countries, would seem to have good prospects for becoming a truly viable state.

Nearly all of the Congo lies within the drainage basin of the Congo River. This immense, shallow downwarp in the surface of the African plateau has a general elevation of 1000 to 2000 feet, but is rimmed, particularly on the east and south, by land that lies considerably higher (maps, pp. 550–551). Great thicknesses of sediments brought by streams from the bordering uplands have accumulated in the flat inner basin, most of which lies within a triangle with Kisangani at the eastern apex, Bangui in the Central African Republic at the northern apex, and the junction of the Congo and Kasai rivers at the southwestern apex.

Variety of the Congo

The Congo exhibits more variety in relief, climate, and vegetation than is commonly realized. Most of the northern half, including nearly all of the inner Congo Basin, is an area of tropical rain forest. The southern part of the country has a tropical savanna climate and a vegetation comprised partly of woodland and partly of tall tropical grasses. Strips of trees ("gallery forests") are found along the streams. To the north of the rain forest and mostly outside of the Congo state lies a second belt of tropical savanna. The areas of tropical savanna climate in or near the Congo have a dry season of 2 to 6 months. In general the rainy season comes at the time of high sun; because the Congo drainage basin is bisected by the equator, the northern savanna is experiencing the high-sun rainy season at the same time that the southern savanna is having the low-sun dry season, and vice versa. Since tributaries of the Congo River drain considerable areas both north and south of the equator, the main river tends to be much more constant in flow than such rivers as the Niger or the Zambezi, which are often very low at the height of the dry season in their respective hemispheres. In the extreme east of the Congo along the margins of the Great Rift Valley, and in neighboring parts of Rwanda, Burundi, and Uganda, there are highlands, frequently of volcanic origin, that rise above 5000 feet over sizable areas. The highest summits are found in the Ruwenzori Range of the Congo and Uganda, a nonvolcanic massif rising to nearly 17,000 feet and capped by permanent snow and ice. In this eastern region of the Congo there is considerable vertical zonation of climate and vegetation according to altitude. Here in the western arm of the Great Rift Valley are found lakes Tanganyika, Kivu, Edward, and Albert. In the vicinity of Lake Kivu there was a considerable development of resort facilities during the period of Belgian rule, including some hotels, as well as bungalows used by Europeans part-time or throughout the year. Among the attractions of this eastern area are the highland climate, cooler and more pleasant than the climate in most other parts of the Congo, and some of the finest natural scenery in Africa. Besides the spectacular Ruwenzori between lakes Edward and Albert, the mountains include a group of active volcanoes in the area between lakes Kivu and Edward. Along or near Lake Edward the Belgians established the scenic wildlife sanctuary of the Albert National Park. During the early and middle 1960s life in the eastern Congo was severely disrupted and there was much killing, looting, and disorder as a consequence of forays and clashes of rebel bands, central government troops, and white mercenaries. The southeastern part of the Congo, in the Katanga region, is an upland ranging in elevation from about 3000 to 6000 feet. Temperatures here are considerably moderated by the altitude, though not as much as in the eastern lakes region.

Historical Background

Belgian penetration of the Congo began in the last quarter of the nineteenth century, following exploratory journeys by an American, Henry M. Stanley. During this early period the colony was virtually a personal possession of King Leopold II, whose agents exploited it ruthlessly for wild rubber, ivory, and other tropical products gathered by Africans. In the 1890s a railroad was built connecting the port of Matadi with Léopoldville (now Kinshasa), thus avoiding a 220-mile stretch of rapids on the Congo River and providing access to the interior of the colony from the sea. In 1908 the Congo was formally annexed by Belgium.

Commercial Economy of the Congo

From the beginning of administration by the Belgian government, the economic life of the colony was dominated by large corporations licensed by the government to carry on business in certain lines. Often these companies were given concessions covering certain geographical areas and lines of business, so that in many sectors of the Congo economy the result was monopoly control rather than competition. The government kept a close watch, however, over the activities of these corporate groups, and in many cases economic development was carried on jointly by the government and the corporations. The large corporations owned and operated plantations, factories, mines, concentration plants and smelters, stores, hotels, railroads, river boats, and port facilities. They operated banks, served as representatives of European insurance companies, and dealt in real estate. They collected, processed, and marketed the oil-palm products, cotton, coffee, cacao, and other commodities produced on African farms. These companies paid taxes to the Congo government which were used exclusively to meet the expenses of the government, including investment in projects designed to further the economic and social development of the colony. None of the tax returns went to Belgium. However, the companies were mostly owned in Belgium, and dividends from company earnings went principally to Belgian shareholders. Many shares were owned by the Belgian government; some were owned in America or Britain.

The Congo government was headed by a governor-general appointed by the government of Belgium and directly responsible to it. Subordinate officials, likewise, were appointed. Even the African chiefs were appointees. Neither whites nor Africans were allowed to vote, though both groups had recourse to a system of courts of justice. The Belgian colonial administration took the position that the first requirement for the colony was a sound economic base and an acceptable standard of living for both whites and Africans, with political development to come later. Attempts were made to draw the Africans more fully into the general economic life of the colony than was the case in most African colonial possessions. The Africans were not confined to unskilled labor, but through education and on-the-job training were gradually brought into many occupations requiring considerable skill. They learned to operate trains, river boats, steam shovels, bulldozers, and electric furnaces, to work as carpenters, masons, telegraphers, and typists, and to serve as postal clerks, nurses, elementary school teachers, and pastors of African Christian churches. However, almost none received training as doctors, lawyers, or engineers. In their school system the Belgians emphasized elementary education, and a larger proportion of the children attended elementary schools (operated in large measure by Roman Catholic or Protestant missions) than was true in most African colonies. But Belgian policies barred all but a tiny handful from receiving a college education. The lack of well-educated African leaders trained in the techniques of government proved disastrous for the Congo when independence came. Today elementary and secondary education is expanding as rapidly as the none-too-plentiful supply of teachers will allow, and a beginning has been made in higher education.

During the period of Belgian rule, employers of African labor were required to conform to minimum wage scales set by the government. Workers received a share of their pay in the form of rations, clothing, housing, and services, such as free medical care and education. Actual cash wages were low. However, the Belgians hoped to gradually stimulate among the Africans a desire for material possessions and thus to bring them more fully within a money economy. It was hoped, in other words, that the ideal of leisure could be replaced to some degree by the ideal of consumption, so that Africans would come of their own volition to do regular and steady work in order to accumulate money with which to buy goods. In this way a large internal market was to be developed and production of a variety of goods stimulated. Thus the colony would come in time to have a more balanced economy and be less dependent on exports of a few primary products. But Belgium's political control was terminated before these purposes could be achieved.

The Congo, like many other units in Tropical Africa, has experienced a modest development of manufacturing plants, mainly producing the simpler types of goods. Typical commodities are cheap cotton textiles, shoes, bricks, cement, wood products, various types of processed foods, cigarettes, some chemicals, and simple metal products. Up to the present, however, the Congo has been mainly important in the economic world as

an exporter of minerals and tropical agricultural products. A large proportion of the minerals are smelted or concentrated in the Congo and then are further refined at plants in Belgium or other overseas industrial countries. The principal exports of the Congo are shown in Table 23. It will be noted that about a dozen primary commodities or commodity groups constituted most of the country's exports by value in 1965. Among the mineral exports, copper, cobalt, zinc, and manganese come principally from the Katanga region, tin from the Katanga and Kivu regions, diamonds from the Kasai, and gold from the Kivu and the northeast (map, p. 551). Copper, by far the country's leading export, is secured from both open-pit and underground mines. Most of the electric power needed for mining and smelting the ore is supplied by hydroelectric stations on the upper Congo (Lualaba) River or its tributaries. Ownership of the Katanga mining industry was taken over by the Congo government in 1967, and negotiations for a financial settlement with the Belgian-based mining company (Union Minière du Haut Katanga) have been in progress. At least for the present, outside interests will continue to operate the mines under government supervision. Diamonds, primarily industrial,

TABLE 23 CONGO (KINSHASA): PRINCIPAL EXPORTS, 1965

	PERCENT OF TOTAL EXPORTS BY VALUE
MINERALS	
Copper	52.4
Diamonds (excluding industrial)	7.1
Zinc (unwrought metal and alloys; ores and concentrates)	6.5
Tin (unwrought metal and alloys; ores and concentrates)	5.6
Cobalt	5.5
Manganese ores and concentrates	1.3
Cadmium	0.62
AGRICULTURAL AND FOREST PRODUCTS	
Palm oil and palm kernel oil; oilseed cake and meal	8.0
Coffee	5.2
Natural rubber	2.8
Veneer sheets and plywood	0.55
Tea	0.54
Cocoa beans	0.35
ALL OTHER EXPORTS	3.5

SOURCE: *United Nations Yearbook of International Trade Statistics,* 1965. Percentages in this table are based on an export total which excludes the values of gold and industrial diamonds.

but including some gem stones, are secured by open-pit mining of alluvial materials in the Kasai region to the east and west of Luluabourg (city proper, 145,000). Tin, formerly extracted in large measure from widely scattered alluvial deposits, now comes primarily from mining of lodes. Most nonmineral exports of the Congo, including palm products, rubber, some coffee, and wood, come from the areas of tropical rain forest. Prior to independence the country had fairly sizable exports of cotton from African small farms in the savanna lands and the drier fringes of the rain forests, but such exports have greatly declined since 1960. Coffee is grown in many different parts of the Congo; the variety known as *arabica,* grown in the eastern highlands, brings a particularly high price. A rather large share of the country's agricultural exports continue to be produced on corporate-owned plantations.

Matadi (city proper, 60,000 in 1959) is the Congo's only seaport of consequence. It handles well over half of the country's overseas trade. Most of the remaining trade, including a considerable share of the Katanga's mineral shipments, moves through the port of Lobito in Angola (p. 596). Kinshasa (750,000), formerly Léopoldville, located on the Congo River at the lower end of a thousand-mile stretch of navigable water, is the country's capital, largest city, and principal manufacturing center. The second largest city is Lubumbashi (city proper, 235,000), formerly Elisabethville, the main city of the Katanga mining region. One of the Congo's major arteries of transportation connects these two cities. It is comprised of a railway from Lubumbashi to Port Francqui on the Kasai River (map, p. 551) and a navigable water connection from Port Francqui to Kinshasa via the Kasai River and the Congo River. An eastward extension of the Lubumbashi–Port Francqui railway provides connections with navigable stretches of the upper Congo (Lualaba) River and with Albertville (city proper, 30,000 in 1958) on Lake Tanganyika. Lake transport, in turn, links Albertville with Kigoma, the railhead for Tanzania's Central Line to the Indian Ocean port of Dar es Salaam.

Kinshasa, Port Francqui, and Kisangani are the Congo's leading river ports. Kisangani, 1080 miles upriver from Kinshasa, lies at the foot of an unnavigable stretch of the Congo known as Stanley Falls. A short rail line from Kisangani bypasses the Falls to a navigable portion of the upper river.

The Inga Project deserves mention. This is a scheme to harness part of the immense hydroelectric power potential of the lower Congo at the Inga Rapids between Kinshasa and Matadi. The project is still in the planning stage (1968). Large-scale assistance from technically advanced countries will undoubtedly be required if actual construction is to be undertaken in the foreseeable future.

Rwanda and Burundi

In addition to its Congo colony, Belgium formerly administered Ruanda-Urundi, a densely populated area located in highlands between the Congo, Tanzania, and Uganda. Once a part of German East Africa, Ruanda–Urundi was given to Belgium as a League of Nations mandated territory after World War I. Following World War II it became a United Nations trust territory under Belgian administration. At independence in 1962 the territory divided into two countries: the Republic of Rwanda in the north and the Kingdom of Burundi in the south. Then, in 1966, the Burundi monarchy was overthrown by a military coup and that country, too, was proclaimed a republic. The largest population element in each of the two countries is the Bahutu, Bantu-speaking farmers and herders who formerly lived in subjection to the tall Watusi, a pastoral people of possible Hamitic antecedents. Fighting between the two groups following independence resulted in much turmoil and slaughter, especially in Rwanda, and many refugees (mainly Watusi) fled to neighboring countries.

With more than 300 persons per square mile (Table 20), Rwanda and Burundi have extraordinarily high population densities for Tropical African countries. The population of the two countries combined was nearly 7 million according to 1968 estimates, as compared with less than 17 million in the immensely larger Congo republic which they adjoin. Agriculture and cattle-keeping on a subsistence basis are the main components in the economies of the two republics. A serious degree of soil erosion has resulted from the tilling of steep slopes without proper protective measures, and the problem has been intensified by the grazing of excessively large numbers of cattle. Exports from the two countries are relatively small and consist mainly of coffee, with a fair amount of tin ore from Rwanda. Most parts of these small, mountainous, deforested countries are above 5000 feet in elevation, though elevations descend to somewhat lower levels in limited areas. One such area is the extreme west, where the boundary between the Congo and the two republics runs along the floor of the Great Rift Valley's western arm. Part of Lake Kivu lies in Rwanda and part of Lake Tanganyika in Burundi. The largest town in the two countries, Bujumbura (city proper, 80,000), formerly Usumbura, the capital of Burundi, lies at the northern end of Lake Tanganyika and has connections by lake shipping to railheads at Albertville in the Congo and Kigoma in Tanzania.

EAST AFRICA

In this chapter the term "East Africa" is restricted to three former British dependencies—Kenya, Tanzania, and Uganda—that lie between the Congo Basin and the Indian Ocean. Two of the countries—Kenya and Tanzania—front on the Indian Ocean, but Uganda is landlocked. Ethiopia, the Somali Republic, and the French Territory of Afars and Issas, occupying the "Eastern Horn" of Africa to the north and east of Kenya, have been discussed with the Middle East. The Portuguese possession of Mozambique, stretching along the eastern side of Tropical Africa south of Tanzania, is grouped with Rhodesia, Malawi, Zambia, and Portugal's large west-coast dependency, Angola, under the heading of "South Central Africa." In many respects the small, densely populated highland countries of Rwanda and Burundi fit well with East Africa, but for convenience have been included in this text with the Congo (Kinshasa) in a discussion of former Belgian Africa. The Malagasy Republic, occupying the large island of Madagascar southeast of Tanzania, is discussed with smaller islands in the Indian Ocean at the end of the chapter.

Prior to independence from Britain, the units comprising East Africa varied in political status. Uganda was a protectorate administered with African interests paramount, Kenya was a colony and protectorate in which a white settler class had great political influence, Tanganyika (former German East Africa) was a United Nations trust territory administered by Britain, and the islands of Zanzibar and Pemba, now a part of Tanzania, were a British-protected Arab sultanate. Tanganyika became independent in 1961, Uganda in 1962, Kenya in 1963, and Zanzibar and Pemba in 1963. In 1964 a political union of Tanganyika and Zanzibar was formed, for which the name "United Republic of Tanzania" subsequently was chosen. Kenya and Uganda also are republics.

Physiographic Characteristics

Although a coastal lowland fringed by coral reefs and, in many places, mangrove swamps, occupies the eastern margins of Kenya and mainland Tanzania (Tanganyika), the terrain of the three East African units is comprised mainly of plateaus. These lie generally at elevations of 3000 to 6000 feet, but rise higher in some places, and in much of eastern and northern Kenya descend to relatively low plains ranging from below 500 feet to 2500 feet in elevation. Individual plateau surfaces in East Africa exhibit great variety, not only in elevation, but in topography, climate, and vegetation. Often the land opens out broadly into great expanses of undulating or rolling country; for East Africa as a whole this is the most widespread single type of landscape. In many places remnant hills or mountains formed of unusually resistant rock rise from the plateau surface. Volcanic cones are prominent features in some areas. All three countries include sections of the East African rift valleys ("Great Rift Valley"). The Western Rift separates Uganda, Tanzania, Rwanda, and Burundi from Congo–Kinshasa, and portions of the valley in the south lie between Tanzania and Zambia or Tanzania and Malawi. International frontiers follow the floor of the Western Rift for long distances and divide lakes Albert, Edward, Kivu, Tanganyika, and Malawi among different sovereignties. The more discontinuous Eastern Rift crosses the heart of Tanzania and Kenya, being better developed and more conspicuous in the Kenyan section than in the Tanzanian. Lake Victoria, lying in a shallow downwarp between the two major Rifts, is shared by the three East African countries. Kenya has a very small portion of the northeastern sector, and the remainder of the lake is divided about equally between Uganda and Tanzania. Lake Victoria's fisheries make an important contribution to East Africa's food supply. None of the three countries is predominantly mountainous, but all have mountains in some areas, most commonly in close proximity to the rift valleys. Often these more rugged areas are rift-valley escarpments, or plateau escarpments elsewhere, that have been carved by erosion into mountains or hills. It may be noted that a sizable proportion of the rougher lands in Tropical Africa as a whole fall in this category. Formation of the rift valleys was accompanied in many places by vulcanism, often on a massive scale. Most of this activity occurred within 100 miles of the rifts, and often along the immediate edges of, or in, the rifts. The two highest mountains in East Africa, Mount Kilimanjaro (19,340 feet) and Mount Kenya (17,058 feet), are extinct volcanoes. Both are majestic peaks crowned by permanent snowfields and visible for great distances across the surrounding plains. The third highest summit in the region, however, is found in a nonvolcanic block mountain, the Ruwenzori Range (16,763 feet), which rises from the floor of the Western Rift along the Uganda–Congo border. In some places, most notably in southwestern Kenya, lava flows have been carved by erosion into a mass of hills or mountains. East Africa's most productive agricultural districts, including many that contribute greatly to the export trade, tend to be localized in volcanic areas. Soils formed from volcanic material are the most fertile large group of soils in the region; they are present in all three countries but are most extensive in Kenya.

Climate and Vegetation

In most of East Africa the climate is classified broadly as tropical savanna—a term that embraces a decided range of temperature and moisture conditions. Temperatures are, of course, much influenced by altitude; and some agricultural districts, even on the equator, are so tempered by elevation that mid-latitude crops like wheat, apples, or strawberries do well. There are great variations from place to place in the amount, effectiveness, and dependability of precipitation and in the length and time of occurrence of the dry season (or seasons). Over much of East Africa the available moisture is too scanty for a truly productive nonirrigated agriculture. There is an unfortunate tendency for long droughts to occur, even in what would normally be the rainy season. Relatively little of the present crop acreage is irrigated, but a number of large projects are under way or planned, especially in Tanzania. A correspondence of adequate moisture and good soil is found only in scattered areas, and such lands support the bulk of East Africa's population. Often they attain remarkably high densities, in contrast to far larger expanses of dry, infertile, or tsetse-infested land that have very sparse populations.

East Africa's vegetation is largely comprised of various types of deciduous woodland interspersed with grasslands. Forest growth of commercial value is scanty; softwoods, in particular, are in short supply, and some planting of such trees under government auspices is in progress. The most luxuriant forests in the region are found on rainy mountain slopes. Park savanna, comprised of grasses and scattered flat-topped trees, with belts of woodland along watercourses, stretches over broad areas and is the habitat for some of Africa's largest remaining communities of big game. Numerous parks and wildlife reserves have been established in the three countries, and these are visited by increasing numbers of tourists. Among the most famous areas of this sort are the Serengeti Plains and Ngorongoro Crater of Tanzania, and the Nairobi National Park, located just outside East Africa's largest city and providing the most convenient access to large animals in their habitat that is to be found anywhere in Africa.

Over large sections of East Africa the tsetse fly is a menace; such areas are largely devoid of cattle and generally have a sparse human population as well. Of the three countries Tanzania has the largest area and proportion of tsetse-infested land. Tsetse-control measures include the clearing of brush that harbors the fly; the slaughter or fencing-off of wild animals that serve as intermediate hosts to the organisms (trypanosomes) transmitted by the fly that cause sleeping sickness in humans and nagana in cattle; spraying; and the development and application of preventives and medicines. Such measures are applied in all three countries with some success, but the tsetse problem is still very far from being solved.

The driest large area in East Africa is northern and northeastern Kenya. Here the land, with its sparse vegetation of low grasses and shrubs, wears a desertlike aspect except in the brief season of the rains. Parts of this area are frequented by Somali herdsmen, and an active boundary controversy exists between Kenya and the neighboring Somali Republic.

Population and Agriculture

In general, East Africa is not a thickly populated area; indeed, most of it can fairly be described as sparsely, or at best moderately, populated. The three countries aggregate some 679,000 square miles—substantially larger than France, Spain, Italy, the United Kingdom, and the Low Countries combined—but their total population in 1968 was only an estimated 30.7 million (Table 20), or somewhat less than that of Spain. The average density in 1968 was an estimated 45 persons per square mile, but this average has relatively little meaning as the population tends to be heavily concentrated in a few areas of dense settlement. Such areas often support hundreds of people per square mile. The two largest and most important areas with a greater-than-average density of population are found in a belt along the northern end and eastern shores of Lake Victoria and in south central Kenya around and north of Nairobi.

Aside from comparatively small minorities of Asians, Europeans, and Arabs, the population of East Africa is comprised of a large number of African tribes, Tanzania alone having around 120 tribal groups. Among the better known tribes are the Kikuyu and Luo tribes of Kenya, the Baganda of Uganda, and the Sukuma tribe of Tanzania, each numbering more than a million people; the pastoral Masai tribe of Kenya and Tanzania; and the prosperous, coffee-growing Wachagga (Chagga)

In this view from Uganda, cotton grown on small African farms is being loaded on a truck which will take it to a local gin for processing. The cotton will then move to a Uganda cotton mill or to an overseas market. The cotton truck in the view is part of a spreading net of modern transportation facilities that is gradually breaking down the isolation and self-sufficiency of underdeveloped lands and is bringing them, for better or worse, within the orbit of a worldwide exchange economy. (Department of Information, Uganda, by courtesy of *Foreign Agriculture*.)

tribe on the slopes of Mount Kilimanjaro in Tanzania. Traditional rivalries and animosities among different tribes are hindrances to the creation of effective national units; for example, Uganda has encountered great difficulties due to suspicion, fear, and jealousy of the large and politically powerful Baganda tribe on the part of other tribal units. Most East African tribes speak Bantu languages, but Hamitic and Nilotic languages are present in some areas such as northern Uganda. Swahili, a Bantu language drawing heavily on Arabic for vocabulary, is a widespread *lingua franca,* as is English. In all three countries the population is divided among pagan, Moslem, and Christian elements. The Moslem religion reflects in part the long history of penetration of the region by Arab traders, particularly slave traders, prior to the age of European control

An overwhelming majority of East Africa's people are rural and agricultural. Most of them are cultivators, but a minority are pastoral. Practically the full range of subsistence and export crops typical of tropical Africa is to be found in the region, although there are endless variations from one area to another in the specific crops that are grown. For the area as a whole the main subsistence crops are maize, millets and sorghums, sweet potatoes, plantains, beans, and manioc. Manioc is a particularly valuable plant in areas prone to crop failure, since the roots can be left in the ground for several years as a reserve food supply. Cultivators may depend exclusively on crops for subsistence, or they may keep animals; the latter tend to be both poor in quality and unscientifically husbanded. The East African governments are endeavoring to induce farmers to improve the quality of their livestock (partly through crossbreeding of local cattle with imported strains), to sell excess stock where necessary to prevent overgrazing, to make more use of animals as draft power for breaking ground and cultivating crops, now done mainly with hoes, and to apply more animal manure to cropland. Progress in these directions tends to be slow, as it requires a considerable change in attitudes, values, and habits of long standing.

For East Africa as a whole coffee and cotton (map, p. 326, and photo, left) are the leading export crops. In Uganda, where almost all the agriculture is carried on by African small farms, the economy is built around these two crops. In 1966, excluding trade with Kenya and Tanzania, coffee represented 53 percent of the total value of exports from Uganda, and cotton 23 percent. Coffee is raised primarily in southern Uganda, on lands bordering Lake Victoria that have good soils and receive 40 inches or more of rainfall a year. Cotton cultivation is more widespread. In Tanzania cotton and coffee, again raised very largely by African small farmers, are leading exports, but they dominate the export trade to a smaller extent than in Uganda. In 1966, excluding trade with Uganda and Kenya, cotton accounted for 21 percent of mainland Tanzania's exports, and coffee for 18 percent; other prominent exports included sisal (14 percent) and diamonds (11 percent). Tanzania's cotton comes mainly from the northwestern area bordering Lake Victoria; the coffee is produced in a series of districts, often in volcanic highlands, that are scattered widely across the northern part of the country. Coffee was by far the largest single export by value from Kenya in 1966, representing 30 percent of all exports; other agricultural exports included tea, 14 percent; sisal, 5 percent; meat and meat products, 5 percent; and a wide variety of other products representing less than 5 percent each. Export production in Kenya is centered heavily in the south central and southwestern highlands; formerly most of it came from European estates, but production today is dominated increasingly by Africans. All three governments in East Africa are currently endeavoring to diversify their agricultural exports in order to reduce the dangerously large dependence on two or three export crops.

Prior to the new era of political independence in East Africa, the region contained around 300,000 Asians (Indians, Pakistanis, and Goans), 100,000 Europeans, and 60,000 Arabs (the figures are estimates for 1958). Most Asians were engaged in commercial, industrial, or clerical occupations, though some owned estates producing export crops; they dominated retail trade and owned many small factories and some large ones. Many were clerical workers in government offices. Some were wealthy, and practically all were more prosperous than most Africans. Arabs were primarily a mercantile class. Europeans were a professional, managerial, and administrative class; in the "White Highlands" of southwestern Kenya about 4000 European families operated estates that were worked by African labor and produced export crops and excellent beef or dairy cattle. Much smaller numbers of European estates were found in Tanzania; most of them were engaged in growing sisal, which was Tanganyika's main export until very recently. Asians, Europeans, and Arabs were most numerous in Kenya and least so in Uganda. Since independence many Europeans have left and much farmland formerly in European estates is now occupied by Africans. Most of the latter are small farmers, but some operate sizable units. Some large agricultural enterprises are operated by the government, particularly in Tanzania. In addition to the European exodus (primarily from Kenya) there has also been a considerable outflow of Asians. However, the majority of the non-African population still remains (1968). It is hard to predict what their status ultimately will be when their skills and services are less vital to the three new nations than at present.

Transportation and Major Cities

The most productive parts of East Africa are bound together by railways that form a connected system leading inland from the seaports of Mombasa in Kenya and Dar es Salaam and Tanga in Tanzania. Both the railways and ports are operated by the East African Railways and Harbors Administration, a component of the East African Common Services Organization. The latter body provides transportation, telecommunications, mail, tax collection, and various other services for the three countries, which have been cooperating with each other closely in economic matters. Mombasa (200,000), the most important seaport in East Africa and the second industrial center (after Nairobi) handles practically all of Kenya's overseas trade and almost all of Uganda's as well. In addition, some Tanzanian trade flows through the port. The city, situated on an island connected with the mainland by causeways, has two harbors: a picturesque old harbor still frequented by some Arab sailing vessels (*dhows*) and other small ships, and a deep-water harbor, Kilindini, which has modern facilities for handling large ocean vessels. From Mombasa the main line of the Kenya–Uganda Railway leads inland to Kenya's capital, Nairobi (325,000), the largest city and most important industrial center in East Africa. It is also the region's busiest crossroads of international air traffic and the main outfitting and departure point for safaris into East Africa's big game country. Tourism is Kenya's second largest earner of foreign exchange (after agriculture). Nairobi was founded early in the twentieth century as a construction camp on the Kenya–Uganda Railway, which continues westward to Kampala (150,000), the capital and largest city of Uganda. Kampala is located in southern Uganda near Lake Victoria. About 20 miles by road to the southwest is Entebbe (11,000), formerly the center of British administration in Uganda. Near Entebbe is Uganda's international airport; like the airports at Nairobi, Mombasa, and Dar es Salaam it is able to accommodate large jetliners. An extension of the Kenya–Uganda Railway from Kampala to the foot of the Ruwenzori Range gives access to a copper mine at Kilembe, 10 miles from the railhead. Copper has been Uganda's only mineral export of any consequence. It is feared, however, that the Kilembe deposits will be exhausted in a comparatively short time. Dar es Salaam (200,000) is the capital, largest city, main seaport, and main industrial center of Tanzania. Its new deep-water port facilities are able to handle good-sized ships. From the city the Central Line of Tanzania's rail system extends westward 780 miles to the railhead and lake port of Kigoma on Lake Tanganyika. From Tabora, once a major Arab slave-trading center, a branch line leads north to Lake Victoria and gives access to Tanzania's main cotton-growing district. About midway between Tabora and the lake, the branch line passes through Shinyanga, the site of the Williamson Diamond Mine which supplies most of the value of Tanzania's mineral output. The mine, discovered and formerly owned by a prospector, Dr. J. A. Williamson, is now owned in equal shares by the Tanzania government and an international corporation, De Beers Consolidated Mines. The diamonds are secured from a kimberlite "pipe"; actually most of the production has come from surface gravels overlying the pipe. The mining area covers about 250 acres. Dozens of other "pipes" have been discovered in Tanzania, but this is the only one that has produced many diamonds up to the present. Well to the north of Dar es Salaam near the Kenya border, the port of Tanga (40,000) provides the principal outlet for Tanzania's sisal exports. Sisal once supplied more than half of the country's exports by value, but the market price has declined in recent years and the crop has now been surpassed by cotton and coffee as earners of foreign exchange. A rail-road, the Tanga Line, leads inland through Tanzania's main area of sisal plantations to Moshi, the main urban center of an important coffee-growing region at the base of Mount Kilimanjaro, and, farther west, to Arusha, a tourist and coffee center. The Tanga Line has connections both with Tanzania's Central Line and with the Kenya–Uganda Railway. The interconnected rail system of the three East African countries makes possible efficient use of rolling stock by enabling the transfer of maximum numbers of freight cars to points where they are needed during seasonal periods of peak demand.

Minerals and Industry

In comparison with many other African countries, the East African nations are extremely lacking in known mineral reserves and in mineral production. The diamonds of Tanganyika, the copper of Uganda, and sodium carbonate extracted from Lake Magadi in southern Kenya are the minerals chiefly worthy of note, though various others are produced in small quantities. Manufacturing in all three countries is in an early stage of development, being confined primarily to agricultural processing, some textile milling, and the manufacture of miscellaneous consumer items. Both Mombasa and Dar es Salaam have refineries that process imported crude oil. The largest textile mill in East Africa is a plant at Jinja (30,000), Uganda, that operates with hydroelectricity from a station at the nearby Owen Falls Dam on the Nile. Practically all of Uganda's electricity comes from the Owen Falls installation, and additional amounts of power are exported to Kenya. All three of the East African governments are anxious to increase the industrial output of their countries as well as improve the performance of agriculture. Considerable amounts of economic and technical aid from both Western and Communist sources have been received, and the three governments are trying to create a political and economic "climate" that will be attractive to private investors from abroad. All are determined, however, to guide their own destinies and to favor those types of investment that are deemed most essential to national development. In all three countries, and particularly Tanzania, a considerable number of manufacturing plants and other economic enterprises are operated by the government.

Zanzibar and Pemba

The islands of Zanzibar and Pemba, formerly a British-protected sultanate, and now a component of the

United Republic of Tanzania, lie north of Dar es Salaam and some 25 to 40 miles off the Tanganyika coast. Zanzibar city (60,000) once was the major political center, slave-trading center, and entrepôt of a considerable Arab empire in East Africa. It still has something of an entrepôt function today, though it lacks modern port facilities. Tourism is of some importance. However, the islands rely mainly for support on millions of clove trees (primarily on Pemba) which provide the bulk of the world supply of cloves. Arabs formerly controlled the politics and economy of the islands, but the majority of the population is African, and Africans have been in control since 1964 when a revolution overthrew the ruling sultan and considerable numbers of Arabs were killed. Communist Chinese influence has been strong in Zanzibar during recent years.

SOUTH CENTRAL AFRICA

In the southern part of Tropical Africa are five countries –Rhodesia, Zambia, Malawi, Mozambique, and Angola– that share the basin of the Zambezi River and long have had important functional relationships with each other. The first three are former British dependencies (Rhodesia was still regarded by the British government in 1968 as being in a dependent status, though it had unilaterally declared its independence in 1965), while the latter two are dependencies ("overseas provinces") of Portugal. The five units are grouped in this section as the countries of South Central Africa.

Angola and Mozambique[2]

Portugal has played an active role as an imperial power in South Central Africa since a very early time in the age of European colonial expansion. The Portuguese were the earliest of the European colonial powers to build an African empire, and Portugal is the only power whose African possessions have survived intact in the new age of African independence. Indeed, aside from a handful of minor dependencies held by Spain or France, Portugal in 1969 was the only European country with any African colonial possessions left. The epochal voyage of Vasco da Gama to India in 1497–1499 by the Cape route was the culmination of several decades of Portuguese

exploration along the western coasts of Africa. During the sixteenth century Portugal controlled an extensive series of strong points and trading stations along both the Atlantic and Indian Ocean coasts of the continent. Later she was outdistanced by stronger powers, but managed to retain footholds along both coasts. Today her principal African possessions, Angola, or Portuguese West Africa, and Mozambique, or Portuguese East Africa, are the most significant colonial units still remaining in the world. In both units Portugal has been harassed by guerrilla warfare for several years, with large military forces required to meet the threat, and there have been insistent demands from newly independent African nations (supported by many states elsewhere) that the Portuguese be ousted from these territories by economic sanctions or military force. As this edition went to press, however, no voluntary surrender of Portuguese authority or effective action against Portugal by the international community of nations seemed in early prospect. Until recently the Portuguese did not do very much to develop their African territories, but their record since World War II has been better, and both Angola and Mozambique now compare very favorably in overall development with many African countries. Portugal has been assisting immigration into these units from the homeland. Probably 250,000 Europeans, mainly Portuguese, now live in Angola, and perhaps 150,000 in Mozambique.

Like many other units in tropical Africa, *Mozambique* consists of a coastal lowland rising to interior uplands and plateaus. Most interior areas are relatively low in elevation, although one peak in the northeast reaches 8200 feet. The country has a tropical savanna type of climate, with a natural vegetation comprised primarily of grasslands with scattered trees; deciduous woodlands occur in places, but dense forests are infrequent. Marshy and mangrove vegetation is conspicuous along the coast. Mozambique is divided into two fairly equal parts by the lower Zambezi River. There are two main port cities: Lourenço Marques (200,000), the capital, in the extreme south, and Beira (50,000), located about 125 miles south of the Zambezi mouth. A third deep-water port, Nacala, located on an excellent harbor in northern Mozambique, is in an earlier stage of development. The two main ports have rail connections with adjoining landlocked countries and handle much international transit trade as well as overseas trade for Mozambique itself. Lourenço Marques has long

[2] For late information on Angola and Mozambique, this section has drawn heavily on William A. Hance, "Three Economies," *Africa Report,* Special Issue on Portuguese Africa, **12,** no. 8 (November 1967), 23–30.

been an important port for the Transvaal, one of the four provinces of South Africa, and is also the main sea outlet for Swaziland. It lies closer than any South African port to the mining, industrial, urban, and agricultural districts which make the Transvaal southern Africa's most important generator of freight traffic. Under an arrangement with South Africa that has endured for half a century, Lourenço Marques is guaranteed almost half of the overseas freight traffic moving to and from a major segment of the Transvaal, and Mozambique in return permits South Africa to recruit specified numbers of laborers for work in gold mines. In addition to its rail connections with South Africa and Swaziland, the port has an important connection with Rhodesia via the Limpopo Railway (map, p. 551), completed in 1956. Beira is another important outlet for Rhodesia and also handles most of the overseas trade of Malawi. It has rail connections to both countries. British capital initially played a major role in financing the port facilities at Lourenço Marques and Beira, as well as the rail lines which connect these ports with the interior. Much traffic for these ports formerly was generated by Zambia, particularly the Copperbelt (p. 599), but this has declined considerably in the last years. In order to avoid dependence on the rail link across Rhodesia, Zambia is routing some traffic through Angola and is trying to develop stronger transportation links by road, air, and pipeline, and hopefully by rail, with Dar es Salaam in Tanzania.[3] An extensive program of highway improvement, dictated partly by military pressures, has been in progress for some time in Mozambique.

As in other African countries, the European population of Mozambique is mainly urban. However, there are several settlement schemes for Portuguese immigrant farmers, and there is a limited development of European plantation-type agriculture. Much of the country's African population lives essentially on a subsistence basis in agricultural villages, although the south has fairly progressive African farming. Exploited mineral resources have long been scanty in Mozambique, aside from some coal. Recently natural gas has been found in quantity in the southern coastal lowland by a subsidiary of Gulf Oil, which plans to export it by pipeline to the Transvaal; and Japanese interests have secured iron-ore mining concessions. Up to the present, however, Mozambique's exports have consisted largely of a few products of tropical agriculture. Cotton, formerly the largest export

by value, is produced for the most part on African small farms; cashews, which have led in most years since 1964, are collected by Africans from natural groves and also are grown on some European plantations. Sugar, tea, copra, and sisal are largely plantation crops. Mozambique derives considerable revenue from fees and transportation charges paid by international transit traffic using the country's rail lines and seaports. Good revenue is obtained from foreign tourists, especially visitors to Lourenço Marques and Beira. Mozambique's economy also benefits from the earnings of large numbers of migrant workers in South Africa and Rhodesia.

Like Mozambique, Portugal's west-coast possession of *Angola* links neighboring political units with the sea. Its seaport of Lobito (60,000), located on one of Africa's finest natural harbors, is connected by a rail line, the Benguela Railway, with the Katanga mining region of the Congo and with the adjacent Copperbelt of Zambia. The Katanga supplies a considerably larger traffic than does the Copperbelt. The future of the railway seemed somewhat uncertain as this edition went to press. It has suffered a good deal of harassment and damage due to guerrilla warfare, and recent Congolese policy has favored the shipment of Katanga copper via the Congo's own seaport of Matadi (p. 589). However, Lobito still handles sizable amounts of Congo transit trade. Zambia, which formerly depended very largely on the ports of Mozambique, has increased its traffic through Lobito in order to lessen shipments across Rhodesia, but it is developing alternative links across neighboring Tanzania. Thus it is apparent that political and strategic considerations are playing a large role in determining the pattern of freight movement in South Central Africa. The seaport of Luanda (over 350,000), Angola's capital and largest city, is primarily a port for Angola itself. In the far south the rapidly developing port of Moçâmedes is the largest of several ports that serve Angola's sea fisheries, the third largest in Africa (after those of South Africa and Southwest Africa). The fishing fleet operates in the plankton-rich waters of the cool Benguela Current. Moçâmedes has gained additional significance as the sea outlet for the railway from Angola's new iron mines at Cassinga, about 400 miles to the east. The ore, much of it high-grade, is present in very large quantities and is mined for export, primarily to Japan and West Germany.

Physically, Angola is comprised of a narrow coastal lowland and expanses of plateau in the interior,

---

[3] An oil products pipeline from the refinery at Dar es Salaam to Ndola in Zambia's Copperbelt became operational in 1968. The possibility of a railroad to link Zambia with Tanzania's Central Line was being considered in 1968 and preliminary surveys had been made.

including a fairly extensive area in the south central part of the country that averages over 5000 feet in elevation. The climate is tropical savanna, with a fringe of tropical steppe in the west. At the southwest the steppe grades into coastal desert. The vegetation associated with the tropical savanna climate consists of savanna grasses with scattered woody growth in some areas and open woodland or scrub in others. Most Africans in Angola are still supported by subsistence agriculture and cattle raising. European farms and plantations and a number of African farmers grow coffee as the leading agricultural export; in 1966 coffee comprised 48 percent of the country's total exports by value. Other agricultural exports are sisal, maize, cotton, sugar, oil-palm products, dried manioc, and bananas. Diamonds were Angola's leading mineral export up to 1968. The diamond-mining area is a southward extension of the Kasai diamond fields in the Congo (map, p. 551). Oil was discovered in the 1950s near Luanda, but the production, though sizable by Tropical African standards, has been minor on a world scale. However, in the 1960s very large deposits of oil were found off the coast of Cabinda, a small exclave of Angola north of the Congo River estuary. The production not only promises to give Portugal considerably greater financial resources, but is potentially important to South Africa and Rhodesia as a means of rendering them less vulnerable to economic sanctions. Sanctions on oil were applied against Rhodesia by Britain, with endorsement by the United Nations, following Rhodesia's unilateral declaration of independence in 1965, and attempts have been made repeatedly in the United Nations to invoke sanctions against South Africa. Such efforts, motivated by the white-supremacist policies of the South African and Rhodesian governments, tend to strengthen relationships between Portugal, itself the target of much criticism, and the two African countries.

Rhodesia, Zambia, and Malawi

Prior to the post-World War II movement for African independence, Rhodesia, Zambia, and Malawi (then known as Southern Rhodesia, Northern Rhodesia, and Nyasaland) were British dependencies, Southern Rhodesia being a self-governing colony while Northern Rhodesia and Nyasaland were protectorates. In 1953 the three units became linked politically in the Federation of Rhodesia and Nyasaland, commonly known as the Central African Federation. The Federation had its own parliament and prime minister, but did not achieve full independence. Serious strains within the Federation were created by the racial policies and attitudes of Southern Rhodesia's white-controlled government, and these helped bring about dissolution of the Federation in 1963. Northern Rhodesia then achieved full independence in 1964 as the Republic of Zambia. Nyasaland became independent in the same year, and subsequently declared itself the Republic of Malawi. Both of the latter countries have remained members of the Commonwealth of Nations. The government of Southern Rhodesia, the main area of European settlement in the Federation, was unable to come to an agreement with Britain concerning the future political status of the colony's African majority. In 1965 the government issued a Unilateral Declaration of Independence and now terms the country "Rhodesia." Britain had not recognized Rhodesian independence at the time of writing (early 1969), nor had any other country. The British government had placed an embargo on trade with Rhodesia, but the Rhodesian economy, though damaged by the loss of British and other Commonwealth trade, was managing to operate. Unofficial assistance of various kinds was being given the Rhodesians by South Africa and Mozambique. Rhodesia has become estranged from its former partners in the Federation, particularly Zambia, but its economy is so linked to theirs that a certain amount of economic intercourse is, at least for the present, unavoidable. Strained political relations between Rhodesia and Zambia have been exacerbated by guerrilla warfare waged against the Rhodesians by forces entering from Zambia.

Rhodesia, Zambia, and Malawi occupy highlands with a tropical savanna type of climate. Over nine-tenths of the annual rainfall comes in the six months from November to April. The natural vegetation is comprised primarily of open woodland, although tall-grass savanna predominates on the High Veld of Rhodesia. The High Veld, lying generally at 4000 feet or higher, forms a broad divide between the drainage basins of the Zambezi and Limpopo rivers. Comprising a band of territory about 50 miles broad by 400 miles long, oriented southwest-northeast across the center of Rhodesia, the High Veld is the main area of European settlement in the three countries. Most of Rhodesia's 225,000 whites live on it. European farms, which comprise nearly half of the country's land, are especially clustered in this central strip of highlands, although the very largest holdings, including some huge cattle ranches, tend to be found in other areas. Other parts of Rhodesia are, in general, drier than the High Veld, and productive farming generally requires irrigation. The Rhodesian government has constructed a number of dams on rivers draining from the High Veld to the Limpopo, in order to facilitate the growing of irrigated sugar cane and other crops in the "Low Veld" of the southeast near the Mozambique border.

African mothers at work on a European-owned tobacco farm in Rhodesia. Suckers that grow from the tobacco stalks are being removed in order to promote the maximum development of leaf. The cultivation and harvesting of tobacco requires a great deal of painstaking hand labor. *(Foreign Agriculture.)*

European farmers on the High Veld are engaged primarily in growing maize, beef and dairy cattle, and Rhodesia's main agricultural export, tobacco. The tobacco growers were badly hurt by the British embargo, as Britain had been by far the main market. Tobacco is also the largest agricultural export from Zambia and Malawi, although in Zambia it is a very minor item compared to the massive exports of copper. In all three countries maize occupies the largest acreage of any crop and is the most important subsistence crop for the African population.

The main axis of economic development in Rhodesia is found along the railway which connects the largest city and capital, Salisbury (330,000), in the northeast, with the second city, Bulawayo (230,000), in the southwest. Along or near the axis are found the principal areas of European farming (especially around Salisbury), and also a succession of important mines producing gold, asbestos, chromium, copper, and other minerals. Most of the minerals produced along the Salisbury-Bulawayo axis are exported, mainly through Beira or Lourenço Marques. An exception is iron ore, which is mined at Que Que, about midway between Salisbury and Bulawayo, and utilized at

that place to produce steel in South Central Africa's only iron and steel plant. Coke comes to the plant by rail from the Wankie coal field in western Rhodesia. This field is the most important coal-producing area on the African continent outside of South Africa, though its output is relatively minor compared to that of the latter country. Rhodesia's manufacturing industries are far more important, both in value of output and in diversity of production, than those of any other country in South Central Africa. Factories are especially clustered in Salisbury and Bulawayo, but a number of important plants are located in other places. An oil refinery at Umtali in eastern Rhodesia had to be closed when the British embargo stopped the flow of crude oil by pipeline from Beira.

Zambia, separated from Rhodesia by the middle course of the Zambezi River, has a much smaller European population (perhaps 75,000 in the middle 1960s) and is far less developed economically than its neighbor to the south. Much of Zambia is a tsetse-infested wilderness with a sparse population. Swampy areas exist in several parts of the country, particularly around Lake Bangweulu in the north and in the upper Zambezi River basin in the west. As in Rhodesia the African population lives mainly on a subsistence basis, but in Zambia African agriculture and cattle keeping are even more traditional, uncommercialized, and unproductive than in Rhodesia. European farms worked by African labor play a far less important role in Zambia than in Rhodesia. They are confined almost entirely to a narrow strip of land along the railway that connects Rhodesia with the Zambian Copperbelt.

Zambia's economy depends heavily on copper exported from several large mines developed by British and American capital. The reserves of copper ore are the largest yet discovered in Africa. Production is mainly from underground mines, though there is some open-pit mining. An urban area has developed at each mine, and the mining area, known as the Copperbelt, consists of a series of separate population nodes strung close to the frontier with the adjacent Katanga copper mining region of the Congo. The mine at the largest urban place, Ndola (90,000), is closed, but the city is the site of the principal copper refinery. Railroads and surfaced roads interconnect the different mining nuclei, which are separated from each other by open country. The Copperbelt, which has the majority of Zambia's manufacturing plants as well as its copper mines, contains around 350,-000 people, including a high proportion of the country's European population. Copper, most of which is shipped in refined form, supplied 93 percent of Zambia's exports by value in 1966. Other mineral production in the country includes cobalt, a by-product of copper mining; lead and zinc mined at Broken Hill, located on the railway about midway between the Copperbelt and Zambia's capital and largest city, Lusaka (140,000), and, very recently, coal production from newly opened deposits in extreme southern Zambia. The coal beds, located some 80 miles northeast of Rhodesia's coal field at Wankie, are of lower quality than Wankie coal and are rather difficult of access, but are being developed to relieve Zambia of dependence on Rhodesian coal. The coal is used mainly as a fuel for processing copper and other metal-bearing ores.

One of the most publicized aspects of Rhodesia and Zambia during recent times has been the construction of the huge Kariba Dam on the Zambezi River. Primarily a power dam, the largest in Africa except Egypt's High Dam at Aswan, the Kariba installation supplies large quantities of hydroelectricity to both Zambia (primarily the Copperbelt) and the cities, mines, and manufacturing plants of Rhodesia. The site lies directly between the Copperbelt and Salisbury (map, p. 551), and

is roughly equidistant from them. The dam, completed in 1958, is a joint enterprise of Zambia and Rhodesia. The generators are on the Rhodesian side, but Zambia plans to install additional generators on its side. In addition, Zambia is contemplating the construction of its own large power dam across the Kafue River, a tributary of the Zambezi. At present (1968) about two-thirds of Zambia's electricity is supplied by the Kariba station, but in Rhodesia a number of large thermal-electric stations fueled by Wankie coal produce as much power as comes from Kariba.

Malawi, the third of the former partners in the Federation, is a long, narrow, densely populated country which lies along the western and southern margins of Lake Malawi (formerly Lake Nyasa). Its population is almost entirely African, the European population numbering perhaps 9000 and the Asian population perhaps 11,000. Tobacco and tea comprised two-thirds of Malawi's exports by value in 1965. Tobacco is produced mainly on African small farms, while tea, along with some tobacco, comes mainly from a small number of European plantations in the southern highlands. Malawi has very little mineral wealth, and its manufacturing industries are exceptionally meager. The country has long supplied many migrant laborers to mines in Rhodesia and South Africa, and their earnings are an important adjunct to the internal economy. A recent estimate credited Malawi with an average per capita income of $38 a year, as compared with $206 for Rhodesia and $174 for Zambia.[4] Among the independent countries of Africa, only Upper Volta ($35) ranked lower than Malawi.

INDIAN OCEAN ISLANDS

A number of islands or island groups in the Indian Ocean—the Malagasy Republic (Madagascar), the Comoro Islands, Réunion, Mauritius, and the Seychelles—are briefly described in this section. They exhibit African, Asian, Arab, and European ethnic and cultural influences in varying proportions. The Malagasy Republic is a former dependency of France. The Comoro Islands are a French "overseas territory" and Réunion a French "overseas department." Mauritius, formerly a British colony, became an independent state within the Commonwealth of Nations in 1968. The Seychelles are a British colony.

Madagascar (French, "La Grande Ile") is one of the largest islands in the world. It is nearly 1000 miles long and about 350 miles wide. It lies off the southeast

---

[4] See David Simpson, "The Dimensions of World Poverty," *Scientific American,* 219, no. 5 (November 1968), 27–35.

coast of Africa and has geological formations similar to those of the African mainland, though it has developed a distinctive flora and fauna of its own. In 1960 the island received full independence as the *Malagasy Republic.*

Some of the early inhabitants of Madagascar appear to have migrated from the East Indies, bringing with them the cultivation of irrigated rice and various other culture traits that may still be found in both areas. There seems also to have been a considerable influx of African peoples from the mainland. Small numbers of Frenchmen, Indians, Chinese, Arabs, Greeks, and immigrants from Réunion and Mauritius are resident on the island.

The east coast of Madagascar rises steeply from the Indian Ocean to heights of over 6000 feet. Since the island lies in the path of trade winds blowing across the Indian Ocean, the east side receives the heaviest rain and has a natural vegetation of tropical rain forest. The remainder of the island has a vegetation consisting primarily of savanna grasses with scattered woody growth. An eastern coastal strip, low, flat, and sandy, is backed by hills and then by escarpments of the central highlands. Paddy rice is the principal food crop of the coastal zone, and coffee, vanilla, sugar, cloves, some rice, and pepper are grown for export. The most densely populated part of the republic is the central highlands. Here the economy is principally a combination of rice growing in valleys and cattle raising on higher lands. Tananarive (350,000), the capital and largest city, is located in the central highlands. It is connected by rail with Tamatave (50,000), the main seaport, located on the east coast. The Malagasy Republic has experienced a much smaller degree of economic development than such tropical islands as Java, Taiwan, or Ceylon.

The *Comoro Islands,* located about midway between northern Madagascar and northern Mozambique, are of volcanic origin. They exhibit a complex mixture of African, Arab, Malayan, and European influences. Most of the exports—vanilla, cinnamon, cloves, and a variety of other tropical crops—come from plantations. *Réunion,* east of Madagascar, is a volcanic tropical island with a population mainly of French and African origin. It was uninhabited prior to the coming of the French in the seventeenth century. Elevations on the island reach 10,000 feet. The soils are fertile, and tropical crops in great variety are grown. Cane sugar, however, mostly from plantations, accounts for around four-fifths of all exports by value, with rum contributing an additional 4 or 5 percent. Sugar is also the mainstay of the larger island of *Mauritius,* which lies near by. In fact, unrefined cane sugar comprises over 95 percent of all exports by value from Mauritius, and molasses an additional 2 percent. Sugar cane, grown mainly on large plantations, is reported to occupy 85 percent of all cultivated land. Mauritius has a larger proportion of lowland than Réunion. Its population is comprised mainly of Indians, but also includes descendants of eighteenth-century French planters, and in addition, some Negroes, Chinese, and mixed bloods. The isolated, coral-rimmed *Seychelles Islands,* far out in the Indian Ocean to the east of Kenya, have a small population of French, Mauritian, and African descent. Copra from plantations is the main export. Many food requirements have to be imported.

REFERENCES and READINGS

West Africa

ABBOTT, ELAINE R., *Basic Data on the Economy of the Republic of Ivory Coast,* U.S. Bureau of International Commerce, Overseas Business Reports, OBR 67–45 (July 1967). (Washington, D.C.: Government Printing Office, 1967.)

BOATENG, E. A., *A Geography of Ghana* (2d ed.; Cambridge: At the University Press, 1966).

BRIERLY, TIM, "Mauritania," *Geographical Magazine,* 37, no. 10 (February 1965), 754–765.

CHURCH, R. J. HARRISON, *West Africa: A Study of the Environment and of Man's Use of It* (6th ed.; New York: John Wiley & Sons, 1968); also, *Environment and Policies in West Africa* (Princeton, N.J.: D. Van Nostrand Co., Searchlight Books, 1963); *Some Geographical Aspects of West African Development: An Inaugural Lecture* (London: G. Bell and Sons, 1966); "Urban Problems and Economic Development in West Africa," *Journal of Modern African Studies,* 5, no. 4 (December 1967), 511–520; "Gambia and Senegal: Senegambia?" *Geographical Magazine,* 39, no. 5 (September 1966), 339–350; "Guinea," *Focus,* 17, no. 7 (March

1967), 6 pp.; "The Niger Republic," *Focus*, 16, no. 1 (September 1965), 6 pp.; "Senegal," *Focus*, 15, no. 1 (September 1964), 6 pp.; "The Islamic Republic of Mauritania," *Focus*, 12, no. 3 (November 1961), 6 pp.; "Port Étienne: A Mauritanian Pioneer Town," *Geographical Journal*, 128, pt. 4 (December 1962), 498–504; and "Some Problems of Regional Economic Development in West Africa," *Economic Geography*, 45, no. 1 (January 1969), 53–62.

CLARKE, JOHN I., ed., *Sierra Leone in Maps* (London: University of London Press, 1966).

COPPOCK, J. T., "Agricultural Developments in Nigeria," *Journal of Tropical Geography*, 23 (December 1966), 1–18.

DALTON, GEORGE, "History, Politics, and Economic Development in Liberia," *Journal of Economic History*, 25, no. 4 (December 1965), 569–591.

DICKSON, K. B., "Trade Patterns in Ghana at the Beginning of the Eighteenth Century," *Geographical Review*, 56, no. 3 (July 1966), 417–431; "Evolution of Seaports in Ghana: 1800–1928," *Annals of the Association of American Geographers*, 55, no. 1 (March 1965), 98–111; and "Background to the Problem of Economic Development in Northern Ghana," *Annals of the Association of American Geographers*, 58, no. 4 (December 1968), 686–696.

FAGE, J. D., *An Introduction to the History of West Africa* (3d ed.; Cambridge: At the University Press, 1962).

FLOYD, BARRY, "The Federal Republic of Nigeria," *Focus*, 15, no. 2 (October 1964), 6 pp.; and, with Monica Adinde, "Farm Settlements in Eastern Nigeria: A Geographical Appraisal," *Economic Geography*, 43, no. 3 (July 1967), 189–230.

GLEAVE, M. B., and H. P. WHITE, "The West African Middle Belt: Environmental Fact or Geographer's Fiction?" *Geographical Review*, 59, no. 1 (January 1969), 123–139.

GOULD, PETER R., "Man Against His Environment: A Game Theoretic Framework," *Annals of the Association of American Geographers*, 53, no. 3 (September 1963), 290–297; and, with Edward J. Taaffe and Richard L. Morrill, "Transport Expansion in Underdeveloped Countries: A Comparative Analysis," *Geographical Review*, 53, no. 4 (October 1963), 503–529. The first article centers on Ghana and the second on Ghana and Nigeria.

GREEN, R. H., and S. H. HYMER, "Cocoa in the Gold Coast: A Study in the Relations between African Farmers and Agricultural Experts," *Journal of Economic History*, 26, no. 3 (September 1966), 299–319.

HALLETT, ROBIN, *People and Progress in West Africa* (New York: Pergamon Press, 1966).

HANCE, WILLIAM A., *African Economic Development* (rev. ed.; New York: Frederick A. Praeger, 1967), "The Volta River Project: A Study in Industrial Development," pp. 87–114.

HILLING, DAVID, "Tema, The Geography of a New Port," *Geography*, 51, pt. 2 (April 1966), 111–125; "The Volta River Project," *Geographical Magazine*, 37, no. 11 (March 1965), 830–841; and "Politics and Transportation: The Problems of West Africa's Land-locked States," in Charles A. Fisher, ed., *Essays in Political Geography* (London: Methuen and Co., 1968), pp. 253–269.

HILTON, T. E., "Akosombo Dam and the Volta River Project," *Geography*, 51, pt. 3 (July 1966), 251–254; also, "Ivory Coast," *Focus*, 16, no. 2 (October 1965), 6 pp.; "Mali," *Focus*, 18, no. 1 (September 1967), 6 pp.; and "The Coastal Fisheries of Ghana," *Oriental Geographer*, 8, no. 1 (January 1964), 61–78.

HIRSCH, ANNE-ROSE, "The Development and Organisation of Commerce in the Ivory Coast and Senegal," *Pacific Viewpoint*, 6, no. 2 (September 1965), 167–178.

HODDER, B. W., "Tin Mining on the Jos Plateau of Nigeria," *Economic Geography*, 35, no. 2 (April 1959), 109–122; "The Ewe Problem: A Reassessment," in Charles A. Fisher, ed., *Essays in Political Geography* (London: Methuen and Co., 1968), pp. 271–283; and, with C. W. Newbury, "Some Geographical Changes along the Slave Coast of West Africa," *Tijdschrift voor Economische en Sociale Geografie*, 52, no. 3 (March 1961), 77–84.

HOPKINS, BRIAN, *Forest and Savanna: An Introduction to Tropical Plant Ecology with Special Reference to West Africa* (London: Heinemann and Co., 1965).

JENNINGS, J. H., and S. O. ODUAH, *A Geography of the Eastern Provinces of Nigeria* (Cambridge: At the University Press, 1966).

JORDAN, H. D., "Rice in the Economy of Sierra Leone," *World Crops*, 17, no. 4 (December 1965), 68–74.

KUPER, HILDA, ed., *Urbanization and Migration in West Africa* (Berkeley and Los Angeles: University of California Press, 1965). Contributions by various authors.

LAWSON, G. W., *Plant Life in West Africa* (London: Oxford University Press, 1966).

LLOYD, P. C., A. L. MABOGUNJE, and B. AWE, eds., *The City of Ibadan* (Cambridge: At the University Press, 1967).

McDONELL, GAVAN, "The Dynamics of Geographic Change: The Case of Kano," *Annals of the Association of American Geographers,* 54, no. 3 (September 1964), 355–371.

McGOUGH, BONNIE, *Basic Data on the Economy of Ghana,* U.S. Bureau of International Commerce, Overseas Business Reports, OBR 68-51 (July 1968). (Washington, D.C.: Government Printing Office, 1968.)

MACHLIN, STEPHANIE, and ASHIDA, JAMES H., *Basic Data on the Economy of Liberia,* U.S. Bureau of International Commerce, Overseas Business Reports, OBR 68-27 (March 1968). (Washington, D.C.: Government Printing Office, 1968.)

MARTELLI, GEORGE, "Progress in Portuguese Guinea," *Geographical Magazine,* 40, no. 2 (June 1967), 128–137.

MAY, JACQUES M., *The Ecology of Malnutrition in the French Speaking Countries of West Africa and Madagascar: Senegal, Guinea, Ivory Coast, Togo, Dahomey, Cameroon, Niger, Mali, Upper Volta, and Madagascar* (New York: Hafner Publishing Company, 1968).

MELAMID, ALEXANDER, "The Geography of the Nigerian Petroleum Industry," *Economic Geography,* 44, no. 1 (January 1968), 37–56.

MORGAN, W. B., "The Influence of European Contacts on the Landscape of Southern Nigeria," *Geographical Journal,* 125, pt. 1 (March 1959), 48–64.

Nigerian Geographical Journal (semiannual). Innumerable articles on the geography of Nigeria.

OBOLI, H. O. N., *An Outline Geography of West Africa* (5th ed.; London: George G. Harrap and Co., 1967).

OJO, G. J. AFOLABI, *Yoruba Culture: A Geographical Analysis* (London: University of London Press, 1966).

PETEREC, RICHARD J., *Dakar and West African Economic Development* (New York: Columbia University Press, 1967).

PORTER, PHILIP W., "Liberia," *Focus,* 12, no. 1 (September 1961), 6 pp.

PRESCOTT, J. R. V., "Nigeria's Regional Boundary Problems," *Geographical Review,* 49, no. 4 (October 1959), 485–505.

RAYMOND, W. D., "The Palm Oil Industry," *Tropical Science,* 3, no. 2 (1961), 69–89.

STEMBRIDGE, JASPER, *Nigeria: A Descriptive Geography* (3d ed.; Ibadan, Nigeria: Oxford University Press, 1966).

SWINDELL, KENNETH, "Iron Ore Mining in West Africa: Some Recent Developments in Guinea, Sierra Leone, and Liberia, *Economic Geography,* 43, no. 4 (October 1967), 333–346; also, "Iron-Ore Mining in Liberia," *Geography,* 50, pt. 1 (January 1965), 75–78; and "Diamond Mining in Sierra Leone," *Tijdschrift voor Economische en Sociale Geografie,* 57, no. 3 (May–June 1966), 96–104.

TRIMINGHAM, J. SPENCER, *Islam in West Africa* (Oxford: Clarendon Press, 1959).

UDO, R. K., "Sixty Years of Plantation Agriculture in Southern Nigeria: 1902–1962," *Economic Geography,* 41, no. 4 (October 1965), 356–368; also, "Disintegration of Nucleated Settlement in Eastern Nigeria," *Geographical Review,* 55, no. 1 (January 1965), 53–67; and, with B. Ogundana, "Factors Influencing the Fortunes of Ports in the Niger Delta," *Scottish Geographical Magazine,* 82, no. 3 (December 1966), 169–183.

U.S. Army Handbook for Liberia (Department of the Army Pamphlet No. 550-38; Washington, D.C.: 1964).

WHITE, H. P., "Dahomey: The Geographical Basis of an African State," *Tijdschrift voor Economische en Sociale Geografie,* 57, no. 2 (March–April 1966), 61–67; also, "The Movement of Export Crops in Nigeria," *Tijdschrift voor Economische en Sociale Geografie,* 54, no. 11 (November 1963), 248–253; and "The Ports of West Africa: Some Geographical Considerations," *Tijdschrift voor Economische en Sociale Geografie,* 50, no. 1 (January 1959), 1–8.

WILLS, J. BRIAN, ed. *Agriculture and Land Use in Ghana* (London: Oxford University Press, 1962).

WINDER, R. BAYLY, "The Lebanese in West Africa," *Comparative Studies in Society and History,* 4, no. 3 (April 1962), 296–333.

WOOD, DONALD S., "The Pattern of Settlement and Development in Liberia," *Journal of Geography,* 62, no. 9 (December 1963), 406–413.

ZARTMAN, I. WILLIAM, "The Politics of Boundaries in North and West Africa," *Journal of Modern African Studies,* **3,** no. 2 (August 1965), 155–173.

Equatorial Africa

BEDERMAN, S. H., "Plantation Agriculture in Victoria Division, West Cameroon," *Geography,* **51,** pt. 4 (November 1966), 349–360; and "The Tole Tea Estate in West Cameroon," *Tijdschrift voor Economische en Sociale Geografie,* **58,** no. 6 (November–December 1967), 316–323.

CARROLL, DOUGLAS F., *Basic Data on the Economy of the Democratic Republic of the Congo (Kinshasa),* U.S. Bureau of International Commerce, Overseas Business Reports, OBR 68–6 (April 1968). (Washington, D.C.: Government Printing Office, 1968.)

CLARKE, JOHN I., "Cameroon," *Focus,* **16,** no. 8 (April 1966), 6 pp.

HANCE, WILLIAM A., and IRENE S. van DONGEN, "Matadi, Focus of Belgian African Transport," *Annals of the Association of American Geographers,* **48,** no. 1 (January 1958), 41–72.

HILLING, D., "The Changing Economy of Gabon: Developments in a New African Republic," *Geography,* **48,** pt. 2 (April 1963), 155–165.

LE VINE, VICTOR T., "The Central African Republic: Insular Problems of an Inland State," *Africa Report,* **10,** no. 10 (November 1965), 17–23.

MAY, JACQUES M., *The Ecology of Malnutrition in Middle Africa (Ghana, Nigeria, Republic of the Congo, Rwanda, Burundi, and the Former French Equatorial Africa)* (New York: Hafner Publishing Co., 1965).

MERRIAM, ALAN P., *Congo: Background of Conflict* (Evanston, Ill.: Northwestern University Press, 1961).

MIRACLE, MARVIN P., *Agriculture in the Congo Basin: Tradition and Change in African Rural Economies* (Madison: University of Wisconsin Press, 1967).

PATTEN, GEORGE P., "Republic of the Congo (Brazzaville)," *Focus,* **13,** no. 2 (October 1962), 6 pp.; and "Gabon," *Focus,* **12,** no. 2 (October 1961), 6 pp.

"Why Belgium Quit the Congo," *Fortune,* **62,** no. 5 (November 1960), 128–131 ff.

East Africa

BAILEY, P. J. M., "The Changing Economy of the Chagga Cultivators of Marangu, Kilimanjaro," *Geography,* **53,** pt. 2 (April 1968), 163–169.

BAKER, S. J. K., "The East African Environment," Chap. 1 in Roland Oliver and Gervase Mathew, *History of East Africa,* Vol. 1 (Oxford: Clarendon Press, 1963), pp. 1–22.

BROOKE, CLARKE, "Types of Food Shortages in Tanzania," *Geographical Review,* **57,** no. 3 (July 1967), 333–357.

CAREY JONES, N. S., "The Decolonization of the White Highlands of Kenya," *Geographical Journal,* **131,** pt. 2 (June 1965), 186–201.

COLWELL, T. C., and E. J. ALEXANDER, *Basic Data on the Economy of Uganda,* U.S. Bureau of International Commerce, Overseas Business Reports, OBR 66–19 (April 1966). (Washington, D.C.: Government Printing Office, 1966.)

de BLIJ, HARM J., *Dar es Salaam: A Study in Urban Geography* (Evanston, Ill.: Northwestern University Press, 1963).

DESHLER, WALTER, "Livestock Trypanosomiasis and Human Settlement in Northeastern Uganda," *Geographical Review,* **50,** no. 4 (October 1960), 541–554; and "The Dodos: A Cattle-Keeping Tribe in East Africa," Chap. 2 in Richard S. Thoman and Donald J. Patton, eds., *Focus on Geographic Activity: A Collection of Original Studies* (New York: McGraw-Hill Book Company, 1964), pp. 7–11.

DUMONT, RENÉ, "Julius Nyerere and Tanzanian Socialism," *Pacific Viewpoint,* **9,** no. 1 (May 1968), 1–11.

East African Geographical Review (annual).

GILDEA, RAY, JR., and ALICE TAYLOR, "Rwanda and Burundi," *Focus,* **13,** no. 6 (February 1963), 6 pp.

GRENFELL, JULIAN, "Kenya's New Breed of Tea Planter," *The Fund and Bank Review* (Washington, D.C.: International Monetary Fund and International Bank for Reconstruction and Development), 4, no. 2 (June 1967), 81–91.

GRIMEZEK, BERNARD, "The Last Great Herds of Africa: An Aerial Census of Animals in the Serengeti National Park," *Natural History,* 70, no. 1 (January 1961), 8–21.

HANCE, WILLIAM A., and IRENE S. van DONGEN, "Dar es Salaam: The Port and Its Tributary Area," *Annals of the Association of American Geographers,* 48, no. 4 (December 1958), 419–435.

HOLLINGSWORTH, L. W., *The Asians of East Africa* (London: Macmillan and Co., 1960).

HOYLE, B. S., *The Seaports of East Africa* (Nairobi, Kenya: East African Publishing House, 1967); also, "Early Port Development in East Africa: An Illustration of the Concept of Changing Port Hierarchies," *Tijdschrift voor Economische en Sociale Geografie,* 58, no. 2 (March–April 1967), 94–102; "The Expansion of Electricity Production and Distribution in Uganda," *Geography,* 52, pt. 2 (April 1967), 196–199; and "The Economic Expansion of Jinja, Uganda," *Geographical Review,* 53, no. 3 (July 1963), 377–388.

HUXLEY, ELSPETH, *With Forks and Hope: An African Notebook* (New York: William Morrow & Co., 1964); and, with Margery Perham, *Race and Politics in Kenya* (new and rev. ed.; London: Faber and Faber, 1956).

INTERNATIONAL BANK FOR RECONSTRUCTION AND DEVELOPMENT, *The Economic Development of Tanganyika* (1961), *The Economic Development of Uganda* (1962) and *The Economic Development of Kenya* (1963). (Baltimore, Md.: Johns Hopkins Press.)

KENWORTHY, JOAN M., "Rainfall and Water Resources of East Africa," in Robert W. Steel and R. Mansell Prothero, eds., *Geographers and the Tropics: Liverpool Essays* (London: Longmans, Green and Co., 1964), pp. 111–137.

LANGLANDS, B. W., "Nationalism, Regionalism and Federalism: The Geographical Basis of Some Conflicting Political Concepts in East Africa," in Charles A. Fisher, ed., *Essays in Political Geography* (London: Methuen and Co., 1968), pp. 285–309.

LARIMORE, ANN E., *The Alien Town: Patterns of Settlement in Busoga, Uganda; An Essay in Cultural Geography* (University of Chicago, Department of Geography Research Paper No. 55; Chicago: 1958); and "Nyanza Textile Industries Ltd.: A Modern Producer of Cotton Fabrics in an Underdeveloped Economy," Chap. 23 in Richard S. Thoman and Donald J. Patton, eds., *Focus on Geographic Activity: A Collection of Original Studies* (New York: McGraw-Hill Book Company, 1964), pp. 146–153.

McMASTER, DAVID N., "Kenya," *Focus,* 16, no. 6 (February 1966), 6 pp.; also, "Speculations on the Coming of the Banana to Uganda," *Journal of Tropical Geography,* 16 (1962), 57–69; and "Change of Regional Balance in the Bukoba District of Tanganyika," *Geographical Review,* 50, no. 1 (January 1960), 73–88.

McNAMARA, F. T., *Basic Data on the Economy of Tanzania,* U.S. Bureau of International Commerce, Overseas Business Reports, OBR 66-4 (February 1966). (Washington, D.C.: Government Printing Office, 1966.)

MORGAN, W. T. W., "The 'White Highlands' of Kenya," *Geographical Journal,* 129, pt. 2 (June 1963), 140–155; and *Nairobi: City and Region* (London: Oxford University Press, 1967).

O'CONNOR, A. M., *An Economic Geography of East Africa* (New York: Frederick A. Praeger, 1966); and "New Railway Construction and the Pattern of Economic Development in East Africa," Institute of British Geographers, *Transactions No. 36* (1965), pp. 21–30.

OJANY, FRANCIS F., "The Physique of Kenya: A Contribution in Landscape Analysis," *Annals of the Association of American Geographers,* 56, no. 2 (June 1966), 183–196.

PORTER, PHILIP W., "East Africa: Population Distribution–As of August, 1962" (Map Supplement No. 6), *Annals of the Association of American Geographers,* 56, no. 1 (March 1966). A large dot map.

SCHNEIDER, HAROLD K., "The Subsistence Role of Cattle among the Pakot and in East Africa," *American Anthropologist,* 59, no. 2 (April 1957), 278–300.

SHAFFER, N. MANFRED, "Land Resettlement in Kenya," *Yearbook of the Association of Pacific Coast Geographers,* 29 (1967), 121–139.

SIMKO, ROBERT A., "The United Republic of Tanzania," *Focus,* 18, no. 6 (February 1968), 6 pp.

SOJA, EDWARD W., *The Geography of Modernization in Kenya* (Syracuse, N.Y.: Syracuse University Press, 1968).

TAYLOR, D. R. F., "New Tea-Growing Areas in Kenya," *Geography,* 50, pt. 4 (November 1965), 373–375.

TRIMINGHAM, J. SPENCER, *Islam in East Africa* (Oxford: Clarendon Press, 1964).

TURNER, BRENDA J., and P. RANDALL BAKER, "Tsetse Control and Livestock Development: A Case Study from Uganda," *Geography,* 53, pt. 3 (July 1968), 249–259.

U.S. Army Handbook for Kenya (Department of the Army Pamphlet No. 550-56; Washington, D.C.: 1967).

van DONGEN, IRENE S., *The British East African Transport Complex* (University of Chicago, Department of Geography Research Paper No. 38; Chicago: 1954); and "Mombasa in the Land and Sea Exchanges of East Africa," *Erdkunde,* 17, no. 1-2 (1963), 16–38.

WHITTLESEY, DERWENT, "Kenya, the Land and Mau Mau," *Foreign Affairs,* 32, no. 1 (October 1953), 80–90.

South Central Africa; Portuguese Africa

BOTTING, DOUGLAS, "Triumph of the Benguela Railway," *Geographical Magazine,* 40, no. 4 (August 1967), 255–269.

CHILCOTE, RONALD H., *Portuguese Africa* (The Modern Nations in Historical Perspective; Englewood Cliffs, N.J.: Prentice-Hall, 1967).

COLE, MONICA M., "The Kariba Project," *Geography,* 45, pts. 1-2 (January–April 1960), 98–105; and "The Rhodesian Economy in Transition and the Role of Kariba," *Geography,* 47, pt. 1 (January 1962), 15–40.

DUFFY, JAMES, *Portuguese Africa* (Cambridge, Mass.: Harvard University Press, 1959); and "Portugal in Africa," *Foreign Affairs,* 39, no. 3 (April 1961), 481–493.

DURAND, LOYAL, Jr., COLIN A. BAKER, and PHILIP P. DURAND, "Malawi," *Focus,* 19, no. 3 (November 1968), 1–9; and "Development Schemes" (based in part on material by Colin A. Baker), 10–11.

FLOYD, BARRY N., "Land Apportionment in Southern Rhodesia," *Geographical Review,* 52, no. 4 (October 1962), 566–582.

GRIFFITHS, IEUAN L., "Zambian Coal: An Example of Strategic Resource Development," *Geographical Review,* 58, no. 4 (October 1968), 538–551; and "Zambia's New Coalfields," *Geography,* 53, pt. 4 (November 1968), 415–418.

HAMILTON, P., "The Changing Pattern of African Land Use in Rhodesia," in J. B. Whittow and P. D. Wood, eds., *Essays in Geography for Austin Miller* (Reading, England: Reading University, 1965), pp. 247–271.

HANCE, WILLIAM A., and IRENE S. van DONGEN, "Beira, Mozambique Gateway to Central Africa," *Annals of the Association of American Geographers,* 47, no. 4 (December 1957), 307–335; and "The Port of Lobito and the Benguela Railway," *Annals of the Association of American Geographers,* 46, no. 4 (October 1956), 460–487.

KAY, GEORGE, "The Towns of Zambia," in Robert W. Steel and Richard Lawton, eds., *Liverpool Essays in Geography: A Jubilee Collection* (London: Longmans, Green and Co., 1967); and *A Social Geography of Zambia: A Survey of Population Patterns in a Developing Country* (London: University of London Press, 1967).

KAY, HUGH, "The Portuguese Way in Africa," *Fortune,* 69, no. 1 (January 1964), 112–115 ff.

PELLETIER, R. A., *Mineral Resources of South-Central Africa* (London: Oxford University Press, 1964).

PERHAM, MARGERY, "The Rhodesian Crisis: The Background," *International Affairs,* 42, no. 1 (January 1966), 1–13.

PIKE, J. G., and G. T. RIMMINGTON, *Malawi: A Geographical Study* (London: Oxford University Press, 1965).

POLLOCK, N. C., "Industrial Development in Malawi," *Geography,* 52, pt. 3 (July 1967), 316–319; and "Irrigation in the Rhodesian Lowveld," *Geographical Journal,* 134, pt. 1 (March 1968), 70–77.

PRESCOTT, J. R. V., "Population Distribution in Southern Rhodesia," *Geographical Review,* 52, no. 4 (October 1962), 559–565; and "Overpopulation and Overstocking in the Native Areas of Matabeleleland," *Geographical Journal,* 127, pt. 2 (June 1961), 212–225.

RODER, WOLF, *The Sabi Valley Irrigation Projects* (University of Chicago, Department of Geography Research Paper No. 99; Chicago: 1965); and "The Division of Land Resources in Southern Rhodesia," *Annals of the Association of American Geographers,* **54,** no. 1 (March 1964), 41–58.

van DONGEN, IRENE S., "Coffee Trade, Coffee Regions, and Coffee Ports in Angola," *Economic Geography,* 37, no. 4 (October 1961), 320–346; and "Sea Fisheries and Fish Ports in Angola," *Boletim,* Sociedade de Geografia de Lisboa, 80, nos. 1–6 (January–June 1962), 30 pp. In English, with Portuguese summary.

WHITTINGTON, G., "Some Effects on Zambia of Rhodesian Independence," *Tijdschrift voor Economische en Sociale Geografie,* 58, no. 2 (March–April 1967), 103–106.

WHITTLESEY, DERWENT, "Southern Rhodesia–An African Compage," *Annals of the Association of American Geographers,* 46, no. 1 (March 1956), 1–97.

Indian Ocean Islands

BENEDICT, BURTON, *Mauritius: The Problems of a Plural Society* (London: Pall Mall Press, 1965).

BOURDÉ, ANDRE, "The Comoro Islands: Problems of a Microcosm," *Journal of Modern African Studies,* 3, no. 1 (May 1965), 91–102.

BROOKFIELD, H. C., "Problems of Monoculture and Diversification in a Sugar Island: Mauritius," *Economic Geography,* 35, no. 1 (January 1959), 25–40.

HANCE, WILLIAM A., "Transportation in Madagascar," *Geographical Review,* 48, no. 1 (January 1958), 45–68.

See also the lists of references and readings for Chapters 1–3 and 24, especially the relevant chapters in geographies of Africa listed on p. 570.

26

South Africa

The far south of the African continent is occupied by a state that has become one of the most controversial units of the political world. Organized in 1910 as the Union of South Africa, it was for five decades a self-governing constitutional monarchy under the British crown. But in 1960 its white population voted to make it a republic. The Republic of South Africa was officially proclaimed May 31, 1961, and on the same day it withdrew from the Commonwealth of Nations (more precisely, it decided not to apply for readmission upon assuming republican status). The country's official policy of "separate development of the races" (commonly known as *apartheid*) has drawn much criticism from other nations, some economic boycotts, and censure by the United Nations. The decision to break away from Great Britain politically was in part a reaction against liberal racial policies of Britain. It was also occasioned by hostility to South Africa among African and Asian members of the Commonwealth.

SOUTH WEST AFRICA (label)
Windhoek
Walvis Bay
Lüderitz
Port Nolloth

BOTSWANA

Gaberones
Mafeking
Thabazimbi
T R A N S V A A L
Pretoria • Belfast
Witbank
Johannesburg ◉ Germiston
Springs
Vanderbijl
Park Vereeniging
Klerksdorp Sasolburg
Sishen
Odendaalsrus
Welkom O R A N G E
F R E E
Postmasburg Kimberley
Bloemfontein
S T A T E Maseru
LESOTHO
RHODESIA
MOZAMBIQUE
Lourenço Marques
Mbabane
SWAZILAND
Newcastle
N A T A L
Pietermaritzburg
Durban

C A P E   O F   G O O D   H O P E   P R O V I N C E
TRANSKEI
East
London

A T L A N T I C
OCEAN

I N D I A N   O C E A N

Cape Town Paarl
Cape of Good Hope
Port Elizabeth

Scale of Miles
(for both maps)
0 50 100 150 200

URBAN AREAS: City-size symbols are
based on metropolitan area estimates.
Stars show national capitals.
Province capitals are underlined.

◉ over 1,000,000
★● 500,000–1,000,000
○ 250,000–500,000
★○ Selected smaller places

African Reserves and Transkei
Selected railroads
Witwatersrand gold mining district

SOUTH AFRICA INDEX MAPS

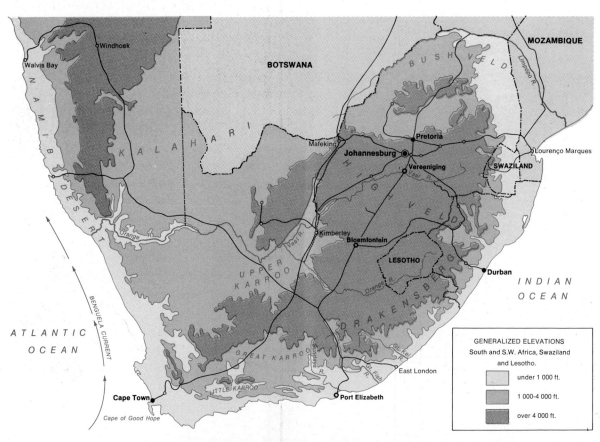

Windhoek
Walvis Bay

BOTSWANA

MOZAMBIQUE

N A M I B   D E S E R T

K A L A H A R I

B U S H   V E L D

Limpopo R.

Mafeking
Pretoria
Johannesburg ◉
Vereeniging
Vaal R.

Lourenço Marques
SWAZILAND

H I G H

V E L D

Kimberley
Bloemfontein
LESOTHO
Durban

Orange R.
Vaal R.

U P P E R
K A R R O O

D R A K E N S B E R G

B E N G U E L A   C U R R E N T

A T L A N T I C
OCEAN

I N D I A N
OCEAN

G R E A T   K A R R O O
Gt. Kei R.
Gt. Fish R.
LITTLE KARROO
East London

Cape Town
Cape of Good Hope
Port Elizabeth

GENERALIZED ELEVATIONS
South and S.W. Africa, Swaziland
and Lesotho.

under 1 000 ft.
1 000–4 000 ft.
over 4 000 ft.

DISTINCTIVE QUALITIES OF SOUTH AFRICA

South Africa contains well over two-thirds of all the Europeans in Africa. A technically progressive, prosperous, and Westernized country, it represents the main center of European settlement and modern industry in an underdeveloped continent that has been dominated by Europeans until very recently but is mainly populated by non-Europeans. Lying almost entirely in middle latitudes, South Africa has mean annual temperatures ranging generally between 60° and 70°F except in the higher mountains. Thus it is on the average a cooler land than Tropical Africa; indeed most places in South Africa have a period of frost in winter. From the standpoint of temperature, as in many other respects, the country has proved well suited to European settlement, although it should not be forgotten that the temperatures have also been favorable for the Africans who form a very large majority in the population.

South Africa's Europeans, who have had complete control in political and economic affairs for a long time, have developed the country according to their own conceptions. The Republic is today a land where the visitor from western Europe or Anglo-America will find most of the institutions and facilities to which he is accustomed. Johannesburg and Cape Town, the largest urban places, are imposing cities; downtown Johannesburg has many towering office buildings that might have been

The central business district of Johannesburg. Piles of waste material from gold mines are seen at the top of the view. (Scope, South African Information Service.)

transplanted from a large American city. Hundreds of smaller places manifest the European or American influence in the arrangement of their business districts and residential sections.

In all the major types of economic activity, South Africa is in a class by itself among African countries. In the middle 1960s it accounted for approximately one-fifth of the continent's international trade by value and more than two-fifths of its manufactural output. It generated over half of Africa's electric power, and had about half of the continent's telephones, over two-fifths of its registered automobiles, and its best network of paved highways. It had about 30 percent of the railway mileage in all of Africa and handled over half of the continent's tonnage of rail freight. It accounted for about two-fifths of the total value of mineral production and was easily the leading country in total value of agricultural products. The scale and diversity of its resources and production, the extremely rapid growth of its economy in recent years, the massive amounts of foreign capital (mainly British and American) invested in mining and manufacturing, and the possession of Africa's strongest military forces are important assets to South Africa's government as it carries forward its racial policies in the face of criticism and reprisals or threats from the outside world.

Comparisons with the United States

Visitors to South Africa often say that they are reminded of the United States. However, such impressions are apt to be superficial, and comparisons between the two countries should not be pushed too far. For one thing, they are not in the same class with respect to size. The United States has roughly seven and one-half times the area (four and one-half times if South West Africa is counted as part of the Republic; see p. 619), and ten times the population of South Africa. The difference in population is even more striking when Europeans alone are considered. The United States has about fifty times as many people of European descent as does South Africa. Instead of being overwhelmingly the majority group, as in the United States, the Europeans in South Africa represent only 19 percent of the total population, being outnumbered more than three to one by native Africans. In addition, there are two other sizable racial groups, the Coloureds, of mixed origin, and the Asians,

TABLE 24 POPULATION ELEMENTS OF SOUTH AFRICA

RACIAL GROUP	NUMBER (1968 ESTIMATES— FOUR PROVINCES)	PERCENT OF TOTAL POPULATION
European	3,650,000	19.0
African	13,075,000	68.1
Coloured	1,900,000	9.9
Asian	575,000	3.0
Totals	19,200,000	100.0

who are mostly Indians. Numbers and proportions of the four major population groups are given in Table 24.

The Africans

It is the existence of the large African majority in South Africa which provides perhaps the most essential point of difference between that country and the United States. There is, of course, a sizable minority of African descent in the United States, but only in the South and some of the larger metropolitan centers outside of the South does this group form a large enough proportion of the population to be of major political, social, and economic significance. In South Africa, by contrast, the African is omnipresent.

Africans are to be met with everywhere in South Africa. There is no farm or factory or town where they cannot be found. Where building and road-making are in progress, where European-owned shops and offices are run, where ships are being loaded and unloaded, there the African will be. The traveller by train will see Africans at every station and siding; the traveller by road will, from time to time, pass an African who is walking to a town or city hundreds of miles away. And when the traveller stays at a hotel for the night he will probably find Africans waiting on him at table and waking him with an early-morning cup of tea.[1]

Sizable numbers of Africans in the Republic live permanently or seasonally on European-owned farms. For the country as a whole they provide the main supply of farm labor, though much of the labor on farms in the Cape of Good Hope Province is provided by the Coloured group. The dependence of South African farmers on non-

---

[1] Leo Marquard, *The Peoples and Policies of South Africa* (3d ed.; London, Cape Town, New York: Oxford University Press, 1960, 1962), p. 33. Used by permission of the publisher.

Workers in a South African diamond mine. Holes are being drilled for dynamite to blast the "blue ground" that yields diamonds. (N. W. Ayer & Son, Inc.)

white laborers is very heavy, although it is gradually diminishing as agriculture becomes more mechanized. Several million Africans live in towns and cities, where they are mainly employed as unskilled or semiskilled laborers in mines and factories, as house servants, or in other low-paid occupations. The remaining Africans live on reserves. The Europeans in the Republic have placed legal restrictions on the African population and, to a slightly smaller extent, on the Coloured and Asian populations, which have no parallel in the United States. Africans are barred by law from voting, aside from participation in selecting officials of their own race within areas specifically reserved for African residence. They may not legally reside outside of the African reserves unless they are employed by Europeans, and even then they are required to live in designated places. They may not own land outside the reserves or join trade unions or political parties in "white" South Africa. They are generally precluded by law or administrative ruling from the better paying jobs. They may not intermarry or have sexual relations with Europeans. They are required to use separate public facilities and to attend segregated schools. Even their freedom of movement is strictly controlled by a system of passes. As a final constraint every African (and Coloured, Asian, and European as well) is required to have on his person

at all times an identification card on which his race is stated. Penalties for infringement of the all-encompassing racial laws include fines, jail terms, and whipping. The South African government actively discourages Africans from aspiring to citizenship in South Africa proper; its official policy is to strengthen the ties of Africans with their tribes (even those numerous Africans who have become detribalized) and to develop semiautonomous states ("Bantustans") in which political aspirations of Africans will find expression. The central intent of these laws and policies is to maintain the separate racial and cultural identity of the Europeans, and to guarantee permanent European supremacy in the 87 percent of South Africa that surrounds the African reserves and is regarded as a white preserve.

Economic Comparisons

Further points of difference between South Africa and the United States are revealed when the resources and economies of the two countries are compared. The United States has far more diversified and abundant resources (though South Africa's mineral resources are remarkable for a country of its size), and the American economy is more fully developed. Probably the United

States comes closer than any other country to a completely balanced economy with a full development of all the principal types of economic activity. South Africa, in contrast, was long extremely dependent on a single economic activity, mining (particularly gold mining), though in recent years it has lessened its dependence on the mining industry very markedly. This has been accomplished primarily by a steadily increasing development of diversified types of manufacturing. Today manufacturing employs far more persons than does mining and accounts for a much greater value of production. In addition, the country's agriculture has been made more productive by the introduction of improved livestock and crop varieties (such as hybrid corn), increased irrigation, fertilization, and mechanization, better control of erosion, and other measures. Despite these changes, however, minerals continue to dominate the country's export trade and thus to provide foreign exchange with which to purchase the many kinds of manufactured goods and some raw materials and foods that South Africa still must import. In addition, the mining industry provides revenues (taxes and royalties) which are used by the government to subsidize agriculture, manufacturing, and services.

In trade emphases the United States and South Africa are opposites. The United States is predominantly an importer of raw materials and foodstuffs and an exporter of manufactured goods. In contrast, over nine-tenths of South African imports are manufactured goods, and exports consist mainly of primary products of mining, agriculture, grazing, and fishing, along with some manufactures. The leading exports in 1966 were gold, diamonds, wool, nonferrous metals, and fruit.

SOUTH AFRICAN REGIONS

Most countries in the world are characterized by a certain degree of physical, cultural, and economic regionalism. In South Africa this characteristic is very pronounced. To gain much of an insight into the geography of the country, it must be considered not only as a whole, but in terms of regional units. The latter may be delimited in various ways according to the purpose in view. For an introductory overview, as in this book, the four provinces and South West Africa will serve reasonably well as major units for study. Although the provinces are distinct from each other in various respects, they have a tendency to fall into three groups: (1) the Transvaal and Orange Free State, (2) Natal, and (3) the Cape of Good Hope Province and South West Africa.

Such a grouping has advantages from the standpoint of simplicity and eliminating unnecessary repetition, and will be adhered to in the following discussion.

Transvaal and Orange Free State

The interior of South Africa is a plateau lying at a general elevation of 3000 to 6000 feet above the sea. It represents the southernmost extension of the interior plateaus which occupy most of the African continent. In South Africa the plateau surface is highest at the east and tapers off gradually toward the west. Two river systems, the Orange (together with its large tributary the Vaal) and the Limpopo, drain most of the plateau area. At the extreme east of the plateau is the *High Veld,* originally an expanse of grassland somewhat similar to the original prairies of midwestern United States. The general elevation of the High Veld is 4500 to 6000 feet. It occupies most of the Orange Free State and the southern third of the Transvaal and extends into Lesotho (formerly Basutoland). The northern two-thirds of the Transvaal is primarily an area of woodlands and savanna grasses—the "Bush Veld."

Gold and Diamond Mining

South Africa's gold-mining industry is almost entirely confined to the High Veld. The great bulk of the gold mined thus far has come from the Witwatersrand (Rand), an area in the southern Transvaal about 115 miles in an east-west direction by 5 to 30 miles from north to south. Johannesburg, the largest city in Africa south of the Sahara, is located in the central part of the Rand. To the east and west are a number of smaller mining towns and cities, the largest of which are Germiston and Springs. Johannesburg, its suburbs and satellites, and nearby Pretoria, comprise a metropolitan region of well over 2½ million people–by far the most important mining and industrial complex in Africa, though second in population to the Cairo metropolitan area.

The Witwatersrand has been the principal area of gold mining since the original discovery of gold in 1886, though production is now shifting increasingly to other South African fields (see below). Immense amounts of capital, primarily British, are invested in the gold mines, which are operated exclusively by large corporations. In a few places mining operations have now reached depths of 9000 to over 11,000 feet—among the deepest mines in the world. The African laborers employed in the mines are recruited from other African countries—notably Lesotho, Mozambique, Botswana, Swaziland, and Malawi —or from reserves in South Africa.

It seems certain that the Witwatersrand mines have now passed their peak of production, but sufficient gold reserves probably remain to support some large-scale operations for at least two decades. In addition, South Africa will be able to rely increasingly on the important Klerksdorp gold field in the southwestern Transvaal and an even larger field discovered in 1946 around Odendaalsrus and Welkom in the Orange Free State. The Orange Free State deposits may in time rival those of the Rand in importance. Besides the value of the gold itself, recovery of uranium from mine tailings and as a by-product of current gold mining has made South Africa the largest uranium producer in Africa and one of the three or four largest in the world.

Since 1867 South Africa has been an important producer of diamonds. Formerly the major production was from diamond-bearing kimberlite "pipes" (circular rock formations of volcanic origin) which were mined at Kimberley, Pretoria, and a number of other places. Since 1927, however, production from alluvial deposits in the Transvaal, Cape Province, and South West Africa has given increasing competition. The most important area for this type of mining is found on closely guarded beaches along the coast of South West Africa (the diamonds are mixed with beach sands and gravels). Some of the mining is done offshore by dredging under shallow waters. Kimberley (90,000) in the Cape Province is the principal administrative center for the diamond industry, although actual production of diamonds in the vicinity of the city is now somewhat smaller than it was at the peak of production. South and South West Africa are the leading producers of gem diamonds in the world. In production of industrial diamonds, however, they are less important, the unchallenged world leader being the Congo (Kinshasa).

Coal and Steel

The mining industry in South Africa requires large quantities of electricity, most of which is supplied by thermal-electric plants. These are powered by bituminous coal, which is present in vast quantities and can be mined cheaply since much of it lies near the surface. The coal deposits are immensely important to South Africa, as the country has no known petroleum or natural gas (though active prospecting for these fuels is in

The first atomic reactor on the African continent, located in dry hill country near Pretoria, South Africa. The reactor symbolizes the industrial and scientific advancement of South Africa as compared with other African countries. (Scope, South African Information Service.)

progress) and has only a modest hydroelectric potential. Most of the important coal deposits are in the southern Transvaal and northwestern Natal, with some in the northern Orange Free State; they include various grades of bituminous and some anthracite coal. Much of the coal is of a rather poor grade and some of it is difficult to mine because it lies underneath water-bearing rocks (which also create difficulties for some gold mines). But the deposits have been adequate for South African needs and long have supported a considerable export trade.

Coking coal secured from fields around Witbank (Transvaal) and Newcastle (Natal) has facilitated development of Africa's leading iron and steel industry (the Transvaal coal cannot be used alone, but must be blended with true coking coal from Natal to make acceptable coke). The industry is mainly concentrated within a radius of about 40 miles from Johannesburg. Blast furnaces and associated steelworks are located at Pretoria (530,000) and at Vanderbijl Park, near Vereeniging (250,000 with Vanderbijl Park), and additional steelworks are found at Vereeniging and Johannesburg. Iron ore comes by rail from large deposits around Thabazimbi (Transvaal) and Postmasburg (Cape Province), and the Postmasburg area also supplies manganese.

Various districts in the Transvaal produce chromium, platinum, vanadium, copper, antimony, phosphate, and other minerals. The Transvaal and adjoining sections of the other three South African provinces comprise a remarkably rich and varied mining region in which the known reserves of a great many minerals are still very large. Only a handful of the world's largest nations possess a greater storehouse of mineral wealth.

Manufacturing

The government of South Africa, which is anxious to foster a greater self-sufficiency in manufactured goods, as well as to increase export revenues from this source, has given assistance in various ways to the iron and steel industry (a large part of which is government-owned) and to other manufacturing industries as well. Manufacturing now accounts for a greater value of national product than do mining, agriculture, and grazing combined. Foreign investments in manufacturing are important, but South Africa itself contributes a large and increasing share of the capital. The country's diversified industries include iron and steel, machinery (including much mining machinery), miscellaneous metal manufactures, automobiles (assembled from foreign and South African components, though progress is being made toward the manufacture of an all-South African car), tractors, textiles, chemicals, processed foods, and

others. The rapidly growing chemical industry initially was developed to supply explosives for the mines; today it also manufactures a wide range of industrial chemicals, fertilizers, paints, and numerous other products. At Sasolburg, just south of Vereeniging, gasoline and petrochemicals are manufactured from coal. South Africa has no petroleum or natural gas fields, though a search for these minerals is being pressed vigorously, with government support, by the international oil industry. Production at Sasolburg supplies only a minor share of the country's needs for petroleum products, the remainder coming primarily from Middle Eastern sources. Crude oil and its products are stockpiled in large quantities as a means of withstanding possible boycotts deriving from South Africa's racial policies. The country's main oil-refining complex, with associated petrochemical industries, is on the Indian Ocean coast a short distance south of the seaport of Durban. In addition to crude oil, South Africa imports roughly the same value of refined products.

Well over two-thirds of the manufacturing industries in South Africa are concentrated in the southern Transvaal, with Johannesburg as the main center. Outside of the Transvaal, manufacturing is mainly localized in or near the seaports of Durban, Cape Town, Port Elizabeth, and East London. Kimberley and Bloemfontein also are industrial centers of some note.

The Agriculture of the High Veld

The High Veld is not only the country's principal focus of mining and manufacturing but also the leading area of crop and livestock production. The main field crop in acreage and value of production is maize. Around 85 or 90 percent of the nation's maize acreage is located in the Orange Free State and Transvaal, mainly on the High Veld. Much of the production is fed to livestock, although in good years large amounts are exported to Japan or west European countries. Maize is also a major item of diet for the Republic's African population. The principal farm animals of the High Veld are beef cattle, dairy cattle, and sheep. This is the only part of South Africa where commercial crop and livestock farming ("mixed farming") has been developed on an extensive scale. Individual farms are large, averaging between 500 and 1000 acres. Farmsteads are widely spaced and customarily include a number of planted shade trees around the house. Windmills are extensively used to pump water from wells for stock and household use. The High Veld is rather short of surface water, but underground supplies are relatively abundant.

Some parts of the Transvaal, mostly outside of the High Veld, have a more specialized agriculture based on tobacco, wheat, alfalfa, fruit, vegetables, or cotton. These districts, which often depend on irrigation water from the rivers to supplement the summer rains, are generally at a lower elevation than the High Veld and have a longer frost-free season. The northern half of the Transvaal, in fact, is essentially an area without frost. Under these conditions extensive production of citrus fruits has been possible, and the Transvaal now has around half of the total citrus acreage in South Africa.

The Moisture Problem

Although temperatures nearly everywhere in South Africa are favorable for agriculture, much of the country is too dry for nonirrigated farming. Most of the western half, and virtually all of South West Africa, receives less than 15 inches of precipitation annually. Due to the high rate of evaporation, this amount is too small for most types of nonirrigated cropping. The High Veld gets about 20 to 30 inches of precipitation on the average. But the total amount varies considerably from year to year, and there are frequent periods of drought during what would normally be the rainy season. Fortunately for agriculture, the High Veld receives most of its precipitation during the summer half year, when it is most needed for crop production.

Precipitation in South Africa is mainly derived from the Indian Ocean. Moisture-bearing air masses blowing off the sea beat against the mountainous eastern es-

carpment of the plateau, are forced upward and cooled, and produce an annual rainfall of 40 inches or more in most of Natal, northeastern Cape Province, and Lesotho. But moisture decreases sharply in the western part of South Africa. Western Cape Province and South West Africa are mainly semidesert or desert.

Climatic data for some representative South African stations are given in Table 25.

The Orange River Scheme

As a means of conserving the country's none-too-plentiful water supplies and to provide hydroelectricity, increasing numbers of dams are being constructed on South Africa's rivers. The most important component in this program from a long-range point of view is a scheme to control the Orange River by a series of large dams (one of which already exists) on the middle and upper reaches of the main river and dams of various sizes on tributaries. Water from the reservoirs will be used for urban and industrial needs, but a major share will be utilized to irrigate fields and orchards, not only in the drainage basin of the Orange River itself, but also in districts to the south. Canals and tunnels will transfer water to rivers that empty into the Indian Ocean in the Cape of Good Hope Province. These in turn will supply water for expanded irrigation within their drainage basins. The entire scheme will require several decades to complete.

TABLE 25 CLIMATIC DATA FOR SELECTED STATIONS IN SOUTH AFRICA AND SOUTH WEST AFRICA

STATION	CLIMATIC AREA	ELEVA-TION (FEET)	AVERAGE TEMPERATURE		AVERAGE PRECIPITATION		
			JAN. °F	JULY °F	ANNUAL (INCHES)	PERCENT OCT.-MAR.	PERCENT DEC.-FEB.
Cape Town (Cape Province)	Mediterranean region	39'	71°	55°	25.7″	19.3%	6.9%
Port Nolloth (Cape Province)	Namib desert	23'	60°	53°	2.3″	30.4%	13.0%
Windhoek (South West Africa)	Upland steppe	5669'	73°	55°	14.6″	88.6%	53.2%
Kimberley (Cape Province)	High veld	3927'	76°	50°	16.4″	76.3%	41.2%
Pretoria (Transvaal)	High veld	4491'	70°	51°	29.4″	85.0%	48.5%
Belfast (Transvaal)	High veld	6134'	62°	44°	32.2″	84.3%	46.8%
Pietermaritzburg (Natal)	Humid subtropical upland	2243'	71°	56°	36.5″	79.5%	43.3%
Durban (Natal)	Humid subtropical coast	26'	74°	61°	41.1″	69.0%	35.9%

Natal

Natal occupies a region of uneven terrain, often hilly in character, between the Indian Ocean and the Drakensberg, a bold, mountainous escarpment marking the edge of the interior plateau. The highest elevations in South Africa are found in the Drakensberg. Several peaks exceed 10,000 feet, and some in neighboring Lesotho exceed 11,000.

The leading commercial crop of Natal is sugar cane, most of which is grown in a narrow coastal belt lying mainly north of Durban, but also extending for some distance to the south. Nearly all of South Africa's cane sugar comes from Natal. The production is sufficient for the country's needs, and also supplies exports.

The Indians

Today, sugar cane is grown mainly on large European farms worked by African labor. Formerly, however, the majority of workers in the cane fields were indentured laborers brought from India. The latter began coming to Natal in the 1860s. The majority chose to remain in South Africa when their terms of indenture ended, and the total number of Indians has steadily increased. In addition to indentured workers, some Indians came as free immigrants. No new immigration has been permitted since 1913, and internal migration of Indians from one province to another is severely restricted by law. About nine-tenths of the total Indian population is found in Natal, mainly in Durban or its vicinity. Today most Indians are employed in commercial, industrial, or service occupations; some are market gardeners, and a few are still employed in the sugar industry. A great many have small stores and shops. The presence of over 500,000 Indians is an important distinguishing characteristic of Natal. It may be noted that the Indians of South Africa, like the Europeans, Africans, and Coloureds, do not form a unified cultural group. They are divided among Hindu, Moslem, and Christian elements (nearly three-fourths are Hindu), and they speak a variety of Indian languages (Hindi, Tamil, etc.), although English and Afrikaans are gaining ground, especially among younger Indians, as the languages customarily spoken.

Outside of Natal the main concentrations of Indians are found on the Witwatersrand and in Cape Town. In these places they are employed almost exclusively as tradesmen. Indians do not have equal political and economic rights with Europeans, although they rank somewhat higher than the Africans in these respects. Their average level of living is higher than that of Africans, but well below that of whites. A good many Indians, however, are well off, and some are wealthy. Relatively few have elected to emigrate to India or Pakistan, although the South African government provides free passage for this purpose. It may be noted that most of this population element came originally from parts of the Indian subcontinent that lie today within the Republic of India.

The Economy of Natal

Aside from sugar cane, the agriculture of Natal is built around maize and cattle, raised on a commercial basis on European farms and largely on a subsistence basis on African reserves. There is some production of citrus fruits and vegetables, though in neither product does Natal compare with the Transvaal or the Cape Province. Except for the coal fields of the northwest, mineral wealth is largely absent.

Durban (750,000), in southern Natal, is the province's largest city and main industrial center and port (photo, p. 552). It handles a larger tonnage of freight than any other South African port, being an important outlet for the High Veld as well as Natal itself. It is growing in importance as a diversified manufacturing center, second only to Johannesburg within South Africa, and has in addition a thriving resort business centering in an impressive line of beachfront hotels. The population of Durban proper is about equally divided among Europeans, Indians, and Africans.

Cape Province

The Cape of Good Hope Province, like Natal, is distinct in various ways from the rest of South Africa. It is by far the largest in area of the four provinces, being half again as large as the other three provinces combined. It is also the most deficient in rainfall. Except for a mountainous, humid fringe along the southern and southeastern coasts, nearly all of the Cape Province is semidesert or desert. Parts of the province adjoining Natal share the humid subtropical climate of the latter area. The vicinity of the Cape of Good Hope has a mediterranean type of climate, being the only African area south of the Sahara where this climatic type occurs. Average temperatures at Cape Town are similar to those of Los Angeles, in the mediterranean climatic zone of southern California. The annual rainfall at Cape Town is greater, however, and the tendency to summer drought is less pronounced.

Cape Town (850,000) is the principal city of the Cape Province and the second largest city of South Africa. Of the country's four main ports, it is the only one which fronts on the Atlantic. The others, including Durban in Natal and Port Elizabeth (350,000) and East London (150,000) in Cape Province, are Indian Ocean ports. After the closure of the Suez Canal as a result of the Arab-Israeli war in the summer of 1967, South Africa's seaports, and particularly Cape Town, had a considerable increase in ship traffic. Re-routing of Suez traffic around Africa enabled these ports to enlarge their roles in re-fueling, provisioning, and maintenance of ships.

Cape Town ranks third in South Africa as a manufacturing city, after Johannesburg and Durban. Its industries are predominantly of lighter types such as textiles, clothing, food processing, and secondary metal manufacturing. As in the case of southern California, the climate and scenery of Cape Town and its vicinity have been an important attraction for tourists and retired folk, and have stimulated resort development.

The Cape Coloured

The Cape Province, like Natal, has a distinctive racial group, the Cape Coloured. Around nine-tenths of the total Coloured population of South Africa is found in the province, primarily in or near Cape Town. This element had its origin in the early days of white settlement as a product of miscegenation among Europeans (Dutch East India Company servants, settlers, or sailors) and non-Europeans, including slaves. Non-European racial stocks represented include Hottentot, Malagasy, Bantu, West African, and various south Asian peoples. The term "Coloured" is reserved for these mixed-blood people, the Bantu being referred to as "Natives" or "Africans." The Coloureds vary in appearance from persons with pronounced African features to others who are physically indistinguishable from Europeans. About nine-tenths speak Afrikaans as a customary language and most of the remainder speak English; like great numbers of other South Africans, many are bilingual. Culturally they are much closer to the Europeans than to Africans or Hottentots, and many Coloureds with light skins have successfully "passed" into the European community. One intent of the national Population Registration Act, under which every South African must be officially registered as to his race and carry an identification card on which his race is shown, was to prevent further "passing" by Coloureds. Included among the Coloureds are around 100,000 Moslem Cape Malays, and the few Bushmen and Hottentots left in South Africa are counted with the Coloureds also.

Most of the Coloured inhabitants of South Africa work as domestic servants, factory workers, and farm laborers, or perform other types of unskilled or semiskilled labor. Many have found employment in the fishing industry. There is a small but growing professional and white-collar class. The Coloureds have always had a higher

social standing and greater political and economic rights than the Africans, although ranking considerably below the Europeans in these respects, and far below them in level of living. However, in recent decades legislation has been enacted to divest them of some of their rights and privileges and thus to reduce their status to a level more similar to that of Africans. It may be noted that within the Coloured community there is a pronounced social stratification, generally on the basis of lightness or darkness of skin. Here we note again an outstanding characteristic of South Africa: that its population not only is divided among four major racial groups, but within each of these groups there is further cultural or social splintering. It is a classic instance of a "plural society."

Distinctive Agriculture of Cape Province

Within the area of mediterranean climate, agriculture is adjusted to the characteristic regime of winter rain and summer drought. The early settlers established vineyards, for which the climate proved well adapted. Today,

A pleasant agricultural landscape in a valley of the Cape region. Regularly spaced rows of grapevines in the foreground give evidence of the mediterranean climate of this area. Paarl, the main town in the valley, was founded in 1688. (Scope, South African Information Service.)

grape growing and wine making are a form of production helping to distinguish the Cape Province from the rest of the country. The province contains nearly all of the vineyards in South Africa (photo, p. 617).

Although grapes are the most important fruit, the Cape Province also produces a variety of others, including citrus fruits (mostly oranges), apples, plums, pears, peaches, pineapples, and others. These are grown on lowlands near the coast or in valleys and basins among the Cape Ranges. The various fruit-growing areas lie at different elevations and exhibit great diversity as to topography, moisture, temperature, sunshine, and winds. Each area specializes in fruits adapted to local conditions. Thus some valleys specialize in grapes for light, dry wines, others in grapes for heavy sweet wines, still others in grapes for raisins or for table use; and many areas specialize in deciduous tree fruits or in citrus fruits. Most vineyards and orchards are irrigated. Much of the production is marketed within South Africa in the form of fresh fruits, canned or dried fruits, wines, brandies, and jams or jellies; but there is also a sizable export trade. Since seasons in South Africa are the reverse of those in the Northern Hemisphere, fresh fruits can be marketed advantageously in western Europe when that area is having its off season for a particular fruit. Great Britain is by far the largest market for South Africa's fruit exports.

The principal field crop of the Cape Province is winter wheat, some of which is irrigated. Most of it is grown on small plains that lie between the main Cape Ranges and the sea within 100 miles of Cape Town. The production is dependable, and Cape Province regularly harvests at least half of South Africa's wheat acreage. The proportion may rise to two-thirds when a drought restricts production in the neighboring Orange Free State.

The Inland Districts

Most of the population of Cape Province resides in a band of relatively continuous settlement within 100 miles of the southern and eastern coast, and extending from the vicinity of Cape Town to the border of Natal. This area, comprised mainly of low mountains, valleys, and small plains, receives most of the province's rainfall– winter rain in the west, but gradually changing to rainfall at all seasons in the east. Inland, the surface of the province rises by stages to the escarpment of the interior plateau, which is lower and less spectacular here than in the Drakensberg. Two well-known inland valleys located beyond the main belt of settlement are the Little Karroo and Great Karroo, respectively. Both are semidesert areas, with a vegetation chiefly composed of scattered gray shrubs. This type of vegetation is known as "karroo"

(a Hottentot word meaning dry), and extends onto the western part of the interior plateau or Upper Karroo. Most of interior Cape Province is so dry that field agriculture is confined to scattered irrigated districts. However, sheep are able to forage on the sparse vegetation, and sheep ranching has become a very successful and major pursuit. The province has about two-thirds of the sheep in South Africa. Many sheep are raised on farms in the coastal areas as well as on large ranches in the interior districts. Sheep are predominantly of the Merino breed, known for their fine grade of wool. South Africa is second only to Australia in production of Merino wool. In some parts of the province considerable numbers of goats are raised for mohair. The Cape Province has about a third of the country's cattle, mostly raised in the more humid eastern areas.

Aridity in the province increases toward the west, and northwestern Cape Province is desert. Most of the Atlantic shore of the province is barren and sparsely settled. Aridity here is thought to be due primarily to high atmospheric pressures that block the entry of rain-bearing air masses from the Indian Ocean. It also results in some measure from the aridifying effects of the Benguela Current. Stable air masses moving inland from the cool offshore waters lower the temperatures of the coastal areas and produce much cloud and fog, but little rain.

Along this arid coast are found the most important fisheries in Africa. The circulation in depth over the continental shelf and continental slope results in an upwelling of cool waters bringing up nutrients from the ocean floor and providing a fertile habitat for plankton, the basic food of fish. These waters harbor many kinds of fish, with pilchards (sardines) being the most abundant and caught in the largest quantities. Some fish are marketed fresh, canned, or frozen, while others, primarily pilchards and a mackerel-like fish, the "massbanker", are processed into fish meal and fish oil. In addition, this area is the locale for the rock lobster (crawfish) industry. South African rock lobster tails, canned or frozen, are exported in large quantities and have become a prized delicacy in the United States and western Europe, particularly Great Britain and France. Some lobsters are transported live by air to markets in France. Cape Town is the largest center of South Africa's fisheries, but numerous fishing harbors and processing plants are scattered along the coast of South and South West Africa as far north as Walvis Bay. Actually, this fishing region extends still farther north, being shared by important fisheries along the shore of southern Angola. South Africa has some Indian Ocean fisheries, but they are very minor compared to those along the west coast. A rather large proportion of the country's fish catch is exported. The government of South Africa, through a national Fisheries Development Corporation, has given much assistance to the fishing industry.

South West Africa

The coastal desert of the western Cape Province extends northward into South West Africa. This former German colony was overrun by South African forces during World War I, and following the war it was mandated to South Africa by the League of Nations. The terms of the mandate called for administration of the territory as an integral part of South Africa. Since World War II the United Nations has repeatedly called upon South Africa to place South West Africa under the trusteeship system, but the South African government has refused to comply. In 1966 the General Assembly passed a resolution to end the mandate and subsequently appointed a governing council to administer South West Africa. South Africa declined to cooperate with the council, and continues to integrate the territory into its own political and economic system.

The Namib Desert extends the full length of South West Africa in the coastal areas. Inland is a broad belt of semiarid country, merging at the east with the Kalahari semidesert of Botswana. Furnishing a certain amount of sparse forage for stock, it is mainly peopled by Bantu herders or by remnants of the Hottentot population. Small and dwindling Bushman remnants also survive there. Europeans in South West Africa, perhaps 80,000 in number, are mostly found on an interior plateau lying at a general elevation of 5000 feet or higher. About a fourth live in the capital, Windhoek (40,000). Export production in South West Africa is mainly confined to minerals (diamonds, copper, lead, vanadium, and a miscellany of others); livestock products, including a sizable export of Karakul lambskins; and fish, primarily pilchards and rock lobsters.

SOUTH AFRICA'S ECONOMY AND THE RACIAL PROBLEM

For unskilled and semiskilled labor with which to operate their mines, factories, and farms, the Europeans of South Africa rely mainly on low-paid African workers. The white man's economy, in its present form, could not operate without them, and the Africans, in turn, are extremely dependent on the white man's payrolls. Thus the races are interlocked economically despite the intense political and social segregation imposed by innumerable laws and customs. Even in the economic realm, however, segregation is in force, as Africans, and to a somewhat smaller degree, other nonwhites, are excluded from the more desirable types of employment. Due to certain shortages of skilled labor in the rapidly evolving South African economy, some nonwhites are gradually moving into jobs of a higher order, but it will take a long time to fundamentally alter the present situation.

The British-Afrikaner Division

There is every evidence that the great majority of white South Africans are in favor of continuing white supremacy, at least in political affairs. Or to put the matter another way, the whites are not prepared to yield political power to the extent that they might be outvoted by any combination of the other racial groups. In matters considered less fundamental, however, there is a division within the white population between the British South Africans and the Afrikaners, or Boers (Dutch: "farmers"). The latter speak Afrikaans, a derivative of Dutch, as a preferred language. They outnumber the British approximately three to two.

Early Boer Settlement at the Cape

The Afrikaners are the descendants of Dutch, French Huguenot, and German settlers who began coming to South Africa over three centuries ago. The earliest permanent settlement, at Cape Town, was established by the Dutch East India Company in 1652 as a way station to provide water, fresh vegetables, meat, and repairs for company vessels plying the Cape of Good Hope route to the Orient. Although it had not been the intention of the company to annex large areas of land, agricultural settlement slowly expanded in valleys near Cape Town, and a pastoral frontier society developed in the back country. Most of the original Hottentot and Bushman inhabitants were killed, or driven out, or, in the case of the Hottentots, decimated by smallpox; the survivors were put to work as servants or slaves. However, their numbers were not adequate for the labor requirements of the colony, and from the very beginning slaves were brought in from West Africa; later arrivals came from Madagascar, East Africa, Malaya, India, and Ceylon.

The Boer colonists were Calvinists and based many of their social and legal arrangements on the teachings of the Old Testament. They held a firm belief in the

God-given inequality of different races, and could see little or nothing that was reprehensible in the institution of slavery. Many developed a conviction that the Afrikaner nation had a divinely instituted civilizing mission in Africa; the concept of Africans as special wards of the Afrikaners endures to the present time. Particularly in the frontier districts, the early Boers became hardy individualists with an active dislike for central authority. During the Napoleonic Wars the Cape Colony was acquired by Great Britain. Occupation by the British on a permanent basis began in 1806, although a temporary occupation had taken place in 1795. Friction developed almost immediately between many Boers and the British authorities, who imposed tighter administrative and legal controls than the Boers were accustomed to, took an unfavorable view of the stern discipline, including use of the whip, applied by many Boer masters to their slaves and servants, and often took the side of the Africans when there were frontier clashes between Boer ranchers and Bantu tribesmen. English was made the official language, despite the fact that most colonists did not speak it. Meanwhile an increasing shortage of range land developed as the population grew, and this added to Boer dissatisfaction. Then in 1833 slavery was officially abolished throughout the British Empire. The Boers were compensated for the loss of their slaves, but compensation was given in promissory notes redeemable only in London. Much of this paper was bought at a large discount by speculators, and the entire transaction further embittered relations between the Boers and the British governing officials.

The Great Trek

Boer discontent resulted in the Great Trek—a series of northward migrations through which groups of Boers, primarily from the eastern part of the Cape Colony, sought to find new grazing lands and establish new political units beyond the reach of the authorities at Cape Town. After some preliminary exploring expeditions in earlier years, the main trek by horse and ox-wagon began in 1836. It resulted in the founding of Natal, the Orange Free State, and the Transvaal as Boer republics. Natal was annexed by Britain in 1845, but Boer sovereignty in the Transvaal and the Orange Free State was officially recognized in 1852 and 1854 respectively. The northward migration of the *voortrekkers* (pioneers) is a dramatic and colorful story, comparable to the settling of the American Far West. It is a central event in the history of the Afrikaner nation and a major focus of Afrikaner national pride and cohesion. An estimated 12,000 Boers—men, women, and children—were involved in the movement during the decade following 1835. As a result of the Great Trek, roughly a fourth of the Boer population was withdrawn from the Cape Colony. At the time the interior plateau was comparatively empty, due to depredations by the formidable and disciplined Zulu tribe. The Boers clashed with remnants of tribes that had fled from the Zulus, and in Natal they encountered the main Zulu force. Though a good many Europeans were killed, the Bantu spearmen, fighting on foot, eventually proved no match for mounted Europeans equipped with guns, and the Boers were able to take firm possession of the High Veld, though the majority withdrew from Natal. On the grasslands they grazed their animals and reestablished their traditional way of living.

The Anglo-Boer War

Meanwhile, British settlers were coming to South Africa in increasing numbers. Port Elizabeth and East London were founded as British towns, and British influence also became dominant in Cape Town and Durban. It is conceivable that the British colonies and the Boer republics in South Africa might have developed peaceably side by side had not diamonds been discovered in the Orange Free State in 1867 and gold in the Transvaal in 1886. These discoveries set off a rush of prospectors and other fortune hunters and entrepreneurs from outside, and mining camps sprang up at Kimberley, Johannesburg, and other places. The Boers possessed little or no capital with which to work the deposits, but British capital soon poured in. The conservative Boer leaders did not welcome the influx of outsiders ("uitlanders"), the majority of whom were British. Ill feeling led to the Anglo-Boer War in 1899. It was an unequal contest since the British were able to draw support from the homeland and the Empire, whereas the Boers had to depend entirely on their own resources. After some early successes the Boer forces were decisively defeated, and the war ended in 1902. The Union of South Africa was established in 1910, following a display of constructive statesmanship by both sides. Dutch was recognized as an official language on a par with English (later this was altered to specify Afrikaans rather than Dutch), and Pretoria, in the Transvaal, was made the administrative capital as a concession to Boer sentiment. However, the national parliament meets at Cape Town and the supreme court sits at Bloemfontein (300,000).

Since 1910, and particularly since the Nationalist (now the National) Party came to power in 1948, the Afrikaners have dominated the political life of South Africa by virtue of their greater numbers and cohesion. All of the country's prime ministers have been Afrikaners.

Before 1948 the United Party, a coalition of Afrikaner and British elements, generally controlled the government, but since the triumph of the Nationalist Party the direction of affairs has been more purely in the hands of Afrikaners. Under this regime the policy of Separate Development for different races has been enunciated in innumerable laws and decrees, culminating in the scheme to establish a series of semiautonomous tribal states or "Bantustans."

The Bantustan Scheme

The South African government plans to form territorial units, perhaps eight in number, which will be reserved for Africans and will have elected African governments, though foreign relations and certain other functions will be controlled by the central government. These states are to be organized essentially on a tribal basis by amalgamating some 260 separate tracts of land contained in African reserves. Some additional land may be purchased for inclusion in the new states. The first Bantustan, known as the Transkei, commenced to function in 1963. It is the homeland of the large Xhosa tribe and is located within the boundaries of the eastern Cape Province between the Natal border and East London. In accordance with the present distribution of African reserves, the Bantustans will be arranged in two lines, oriented northeast-southwest, along the outer margins of "white" South Africa (map, p. 608). One line, including the Transkei, will lie in well-watered, hilly country between the Indian Ocean and the Drakensberg in eastern Cape Province and Natal; the other in drier plateau country in the northern and western Transvaal and northern Cape Province. The long-range plans for these units postulate that they will become permanent homelands for the Republic's Africans, including a great many who are to be transferred from their present places of residence in "white" South Africa. It is assumed that large numbers of Africans will continue to provide labor for the Republic's European-run enterprises, but these workers will be considered temporary visitors, to be permitted in "white" South Africa only in such numbers as needed by the white man's economy. During their stay in the white man's country they will live, as now, in segregated areas. The "Bantustan" scheme, branded by many (including many white South Africans) as unworkable, immoral, or even ludicrous, depends for its success on finding ways for the Bantu homelands to support much larger numbers of people than at present, and it is hoped they can be supported with a rising level of living. Unfortunately, the present African reserves and the Transkei have little in

the way of mineral resources, manufacturing, or even large towns, and their agriculture, built around livestock and maize, and operating largely on a subsistence basis, has long been substandard. To make these new units economically viable, it will be necessary to remove a large share of the population from agricultural occupations, to provide employment for them and for newcomers from "white" South Africa in industrial and service occupations, and to build up the productivity of agriculture by the removal of surplus livestock (many areas held in common tribal ownership are seriously overgrazed by cattle prized for numbers rather than quality), plus an energetic program of soil conservation, the introduction of improved varieties of livestock and crops, a greater use of chemical fertilizers and irrigation, and, in general, the development of a modern, diversified agriculture. In the units bordering the Indian Ocean (though not those in the Transvaal) agricultural progress will be facilitated by climatic and soil conditions which are among the most favorable in South Africa. These measures obviously will require extremely large outlays of capital, particularly in the development of new industries and the necessary infrastructure to support them. Up to the present South Africa's government has been chary of permitting white investors to develop industries within the reserves themselves, but instead has been trying to foster "border" industries located in "white" South Africa but near enough to the reserves to permit workers who live there to travel back and forth to work each day. As yet these efforts have had only the most modest success. The ultimate outcome of the Bantustan scheme is a question that occasions no end of debate and speculation, both in South Africa and the outside world. It may be noted that separate "homelands" on the Bantu model are not envisaged for the Coloured and Indian populations. They will continue to lead segregated lives and will not participate in national politics, but will have some voice in the management of their affairs through national councils, one for Indians and another for Coloureds, to which they will elect representatives.

The South African problem of race relations is undoubtedly one of the most difficult in the world. It would be a mistake to suppose that thinking Europeans in South Africa do not realize the deeper implications of their situation, or that they are unconcerned with the welfare of the other racial groups. Doubtless a substantial majority of whites would be found to favor a gradual advancement of the other groups in such matters as health, education, economic status, and general welfare, and, indeed, much progress is being made in these fields. However, the whites seem determined not to grant nonwhites the voting privilege on terms of full equality. They are resolved to preserve South Africa as a white man's country, and rightly or wrongly, they assume that the only way to secure this end is for them to keep a permanent superiority at the polls. They regard South Africa as a home, not only for themselves, but for their children and children's children, and they are afraid of being overwhelmed by numbers. Thus South Africa must continue to struggle with a problem which is full of explosive possibilities. By and large it seems a problem which must by resolved by the South Africans themselves, but since the South African situation is only one facet of a larger world problem of race relations, other nations and peoples will continue to view events in South Africa with a deep interest and concern.

BOTSWANA, LESOTHO, AND SWAZILAND

Three units bordering South Africa—Botswana (formerly Bechuanaland), Lesotho (formerly Basutoland), and Swaziland—were protectorates ("High Commision Territories") of Britain until recently but are now independent. Botswana and Lesotho gained independence in 1966 and Swaziland in 1968. Lesotho, a mountainous country containing the highest summits in the Drakensberg escarpment, is surrounded by South Africa, and Swaziland is nearly so, though it has a short frontier with Mozambique. Botswana, most of which is semidesert or, in the north, tsetse-infested swamp, is bordered on three sides by South and South West Africa and on the fourth side by Rhodesia. Of the three, Swaziland has the best prospects for supporting its population from its own resources. Mountains in the west along the Drakensberg escarpment receive heavy rainfall, and rivers that rise there and flow to the Indian Ocean are used for irrigation projects in the lower and drier east. Mountain slopes have been planted in conifers that form the basis for wood pulp and lumber industries. But Swaziland's export trade is based primarily on minerals, notably asbestos and iron ore. Most of the larger enterprises have been developed by Europeans, using Swazi labor, but racial relationships have been better than in many African countries and the economic benefits to Swaziland have been considerable. Though some of the population still lives outside of the money economy, increasingly the Swazi farmer is improving his methods and is marketing surpluses. Lesotho has no mineral production of any importance, and mineral output in Botswana is very modest. The African populations of these two countries raise livestock, largely on a subsistence basis; some Europeans own cattle ranches in eastern Botswana. Large numbers of African workers find employment in South Africa, primarily in mining. Both Lesotho and Botswana, and Swaziland to a somewhat smaller degree, are very dependent on South Africa for employment opportunities as well as trade and many kinds of services. They are likely to remain economic satellites of their larger neighbor. Since independence, South Africa has established amicable political relations with all three.

REFERENCES and READINGS

General

COLE, MONICA, *South Africa* (2d ed.; London: Methuen and Co., 1966). The most comprehensive and generally useful reference on South Africa's geography.

de BLIJ, HARM J., *Africa South* (Evanston, Ill.: Northwestern University Press, 1962).

KING, LESTER C., *South African Scenery: A Textbook of Geomorphology* (3d ed.; Edinburgh and London: Oliver and Boyd, 1963).

NIDDRIE, DAVID L., *South Africa: Nation or Nations?* (Princeton, N.J.: D. Van Nostrand Co., Searchlight Books, 1968); and "South Africa," *Focus,* **17,** no. 10 (June 1967), 6 pp.

South African Geographical Journal (annual).

TALBOT, A. M., and W. J. TALBOT, *Atlas of the Union of South Africa* (Pretoria: The Government Printer, 1960). A large reference atlas.

WELLINGTON, JOHN H., *Southern Africa: A Geographical Study,* 2 vols. Vol. 1, *Physical Geography* (1955); Vol. 2, *Economic and Human Geography* (1960). (Cambridge: At the University Press.) A basic geographical reference.

Historical Background

BROOKFIELD, H. C., "Trek and Laager: The Europeans in South Africa," in John Andrews, ed., *Frontiers and Men: A Volume in Memory of Griffith Taylor* (Melbourne, Australia: F. W. Cheshire, 1966), pp. 66–89.

COPE, JOHN, *South Africa* (2d ed.; New York: Frederick A. Praeger, 1967). A general survey, primarily historical and political.

GLUCKMAN, MAX, "The Rise of a Zulu Empire," *Scientific American,* 202, no. 4 (April 1960), 157–168.

HEATON, HERBERT, "Other Wests Than Ours," *Tasks of Economic History,* Supplement 6 (1946) to *Journal of Economic History,* pp. 50–62.

PLUMB, J. H., *Men and Centuries* (Boston: Houghton Mifflin Company, 1963), "Cecil Rhodes," pp. 203–213.

POLLOCK, N. C., and SWANZIE AGNEW, *An Historical Geography of South Africa* (London: Longmans, Green and Co., 1963).

WALKER, ERIC A., *The Great Trek* (4th ed.; London: Adam and Charles Black, 1960); and *A History of Southern Africa* (3d ed., new impression with corrections; London: Longmans, Green and Co., 1962).

Economic and Urban Geography

AGNEW, SWANZIE, "South African Farming and the Pioneer Legacy," in R. Miller and J. Wreford Watson, eds., *Geographical Essays in Memory of Alan G. Ogilvie* (London: Thomas Nelson and Sons, 1959), pp. 221–246.

BURLS, JOHN, "Diamonds," *Geographical Magazine,* 38, no. 2 (June 1965), 141–154.

COETZEE, J. A., "The Transvaal Competitive Area and the Distribution of Its Commercial Seaborne Imports via the Ports of South Africa and Lourenço Marques," *Tijdschrift voor Economische en Sociale Geografie,* 54, no. 8–9 (August–September 1963), 186–192.

COLE, MONICA M., "The Witwatersrand Conurbation: A Watershed Mining and Industrial Region," Institute of British Geographers, Publication No. 23, *Transactions and Papers, 1957,* pp. 249–265.

DAVENPORT, JOHN, "The Only Real Industrial Complex South of Milan," *Fortune,* 74, no. 7 (December 1966), 180–185 ff. A discussion of South Africa's economy and racial situation. Very concise and informative.

FAIR, T. J. D., "Vital Waters of the Vaal," *Geographical Magazine,* 40, no. 9 (January 1968), 776–784.

HARDINGER, CHRISTINE C., *Basic Data on the Economy of the Republic of South Africa,* U.S. Bureau of International Commerce, Overseas Business Reports, OBR 66–71 (October 1966). (Washington, D.C.: Government Printing Office, 1966.)

"Harry Oppenheimer's Industrial Africa," *Fortune,* 61, no. 5 (May 1960), 152–165.

HOUGHTON, D. HOBART, *The South African Economy* (New York: Oxford University Press, 1964).

LOMAS, P. K., and M. B. GLEAVE, "Recent Changes in the Distribution of Production in the South African Gold Mining Industry," *Geography,* 53, pt. 3 (July 1968), 322–326.

NUTTONSON, M. Y., *The Physical Environment and Agriculture of the Union of South Africa, with Special Reference to Its Winter-Rainfall Regions Containing Areas Climatically and Latitudinally Analogous to Israel* (Washington, D.C.: American Institute of Crop Ecology, 1961).

POLLOCK, N. C., "The Development of Urbanization in Southern Africa," in R. P. Beckinsale and J. M. Houston, eds., *Urbanization and Its Problems* (New York: Barnes & Noble, 1968).

RICHARDS, C. S., "Problems of Economic Development of the Republic of South Africa," in E. A. G. Robinson, ed., *Economic Development for Africa South of the Sahara* (New York: St. Martin's Press, 1964), pp. 246–278.

SCOTT, PETER, "The Witwatersrand Gold Field," *Geographical Review,* 41, no. 4 (October 1951), 561–589; also, "The Orange Free State Goldfield," *Geography,* 39, pt. 1 (January 1954), 13–20; "The Iron and Steel Industry of South Africa," *Geography,* 36, pt. 3 (July 1951),

137–149; "The Development of the Northern Natal Coalfields," *South African Geographical Journal,* 33 (1951), 53–68; "Cape Town: A Multiracial City," *Geographical Journal,* 121, pt. 2 (June 1955) 149–157; and "Some Functional Aspects of Cape Town," *Economic Geography,* 30, no. 4 (October 1954), 347–363.

SHAFFER, N. MANFRED, *The Competitive Position of the Port of Durban* (Northwestern University Studies in Geography, No. 8; Evanston, Ill.: Department of Geography, Northwestern University, 1965).

TALBOT, WILLIAM J., "Land Utilization in the Arid Regions of Southern Africa, Part 1, South Africa," in L. Dudley Stamp, ed., *A History of Land Use in Arid Regions* (Arid Zone Research, No. 17; Paris: UNESCO, 1961), pp. 299–331.

WILLIAMS, OWEN, "Sugar Growing and Processing in the Union of South Africa," *Economic Geography,* 35, no. 4 (October 1959), 356–366; and "The Premier: A Diamond Mine in South Africa," Chap. 16 in Richard S. Thoman and Donald J. Patton, eds., *Focus on Geographic Activity: A Collection of Original Studies* (New York: McGraw-Hill Book Company, 1964), pp. 93–97.

The Racial Situation and *Apartheid*

BROOKFIELD, H. C., "Some Geographical Implications of the *Apartheid* and Partnership Policies in Southern Africa," Institute of British Geographers, Publication No. 23, *Transactions and Papers, 1957,* pp. 225–247; and, with M. A. Tatham, "The Distribution of Racial Groups in Durban: The Background of Apartheid in a South African City," *Geographical Review,* 47, no. 1 (January 1957), 44–65.

BUCHANAN, KEITH, and N. HURWITZ, "The 'Coloured' Community in the Union of South Africa," *Geographical Review,* 40, no. 3 (July 1950), 397–414; and "The Asiatic Immigrant Community in the Union of South Africa," *Geographical Review,* 39, no. 3 (July 1949), 440–449.

CALPIN, G. H., ed., *The South African Way of Life: Values and Ideals of a Multi-racial Society* (New York: Columbia University Press, 1953).

CARTER, GWENDOLEN M., and Others, *South Africa's Transkei: The Politics of Domestic Colonialism* (Evanston, Ill.: Northwestern University Press, 1967).

DE KIEWIET, C. W., "Loneliness in the Beloved Country," *Foreign Affairs,* 42, no. 2 (April 1964), 413–427.

DUNCAN, PATRICK, "Toward a World Policy for South Africa," *Foreign Affairs,* 42, no. 1 (October 1963), 38–48.

HAMMOND-TOOKE, DAVID, "Chieftainship in Transkeian Political Development," *Journal of Modern African Studies,* 2, no. 4 (December 1964), 513–529.

HANCE, WILLIAM A., ed., with LEO KUPER, VERNON McKAY, and EDWIN S. MUNGER, *Southern Africa and the United States* (New York: Columbia University Press, 1968). Penetrating discussions of the racial situation and its relationship to American policies.

HELLMAN, ELLEN, and LEAH ABRAHAMS, *Handbook of Race Relations in South Africa* (London: Oxford University Press, 1949).

HILL, CHRISTOPHER R., *Bantustans: The Fragmentation of South Africa* (New York: Oxford University Press, 1964).

KAHN, E. J., Jr., *The Separated People* (New York: W. W. Norton & Company, 1968).

KUPER, LEO, HILSTAN WATTS, and RONALD DAVIES, *Durban: A Study in Racial Ecology* (New York: Columbia University Press, 1958).

LAWRIE, G. G., "South Africa's World Position," *Journal of Modern African Studies,* 2, no. 1 (March 1964), 41–54.

MARQUARD, LEO, *The Peoples and Policies of South Africa* (3d ed.; New York: Oxford University Press, 1962).

MUNGER, EDWIN S., *Afrikaner and African Nationalism: South African Parallels and Parameters* (New York: Oxford University Press, 1967); *African Field Reports* (Cape Town: Struik, 1961); and "South Africa: Are There Silver Linings?" *Foreign Affairs,* 47, no. 2 (January 1969), 375–386.

NEL, A., "Geographical Aspects of Apartheid in South Africa," *Tijdschrift voor Economische en Sociale Geografie,* 53, no. 10 (October 1962), 197–209.

NIEWENHUYSEN, JOHN, "Economic Development in the African Reserves of South Africa," *Land Economics,* 42, no. 2 (May 1966), 195–202.

NIXON, CHARLES R., "The Conflict of Nationalisms in South Africa," *World Politics,* 11, no. 1 (October 1958), 44–67.

PATON, ALAN, *Cry, the Beloved Country* (1948) and *Too Late the Phalarope* (1953). (New York: Charles Scribner's Sons.) Famous novels by a South African liberal.

SABBAGH, M. ERNEST, "Some Geographical Characteristics of a Plural Society: Apartheid in South Africa," *Geographical Review,* 58, no. 1 (January 1968), 1–28. A first-rate article.

SNELLEN, I. TH. M., "Apartheid: Checks and Changes," *International Affairs,* 43, no. 2 (April 1967), 293–306.

"South Africa and the World," *Foreign Affairs,* 43, no. 1 (October 1964): Charles A. W. Manning, "In Defense of Apartheid," 135–149, and Philip Mason, "Some Maxims and Axioms," 150–164.

South West Africa

D'AMATO, ANTHONY A., "The Bantustan Proposals for South-West Africa," *Journal of Modern African Studies,* 4, no. 2 (October 1966), 177–192.

FIRST, RUTH, *South West Africa* (Baltimore, Md.: Penguin Books, 1963).

LOGAN, RICHARD F., "South West Africa," *Focus,* 11, no. 3 (November 1960), 6 pp.; "Land Utilization in the Arid Regions of Southern Africa, Part 2, South West Africa," in L. Dudley Stamp, ed., *A History of Land Use in Arid Regions* (Paris: UNESCO, 1961), pp. 331–338; and *The Central Namib Desert, South West Africa* (National Research Council, Publication 758; Washington, D.C.: 1960).

WELLINGTON, JOHN H., *South West Africa and Its Human Issues* (Oxford: Clarendon Press, 1967).

Botswana, Lesotho, and Swaziland

BARKER, DUDLEY, *Swaziland* (1965); AUSTIN COATES, *Basutoland* (1966); B. A. YOUNG, *Bechuanaland* (1966). Volumes in the Corona Library under the sponsorship of the British Colonial Office (London: H.M.S.O.).

MUNGER, EDWIN S., *Bechuanaland: Pan-African Outpost or Bantu Homeland?* (New York: Oxford University Press, 1965).

SNEESBY, G. W., "Economic Development in Swaziland," *Geography,* 53, pt. 2 (April 1968), 186–189.

STEVENS, RICHARD P., *Lesotho, Botswana, and Swaziland: The Former High Commission Territories in Southern Africa* (New York: Frederick A. Praeger, 1967).

WHITTINGTON, G., "The Swaziland Railway," *Tijdschrift voor Economische en Sociale Geografie,* 57, no. 2 (March–April 1966), 68–73.

YOUNG, BRUCE S., "High Commission Territories of Southern Africa," *Focus,* 14, no. 4 (December 1963), 6 pp.

See also the list of references and readings for Chapter 24, particularly the relevant chapters in Hance, *The Geography of Modern Africa,* and Gunther, *Inside Africa.*

PART VIII

LATIN AMERICA

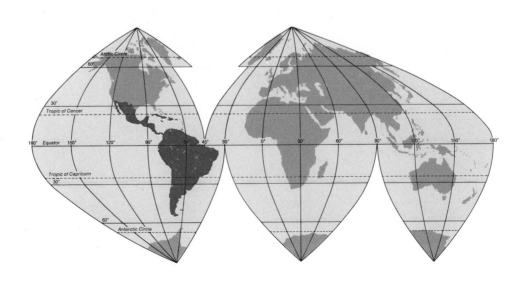

Introduction

to

Latin America

27   The land portion of the Western Hemisphere to the south and southeast of the United States has come to be known as Latin America (map, p. 631). Both the name of this region and its ways of life reflect the importance of culture traits inherited from the Latin-European nations of Spain, Portugal, and France. Spanish and Portuguese, for example, are overwhelmingly the major languages in Latin America as a whole. Spanish is the official language in 18 of the 24 independent states, the exceptions being Brazil (Portuguese), Haiti (French), and Barbados, Jamaica, Trinidad and Tobago, and Guyana (English). In several countries more than one language is in use; indeed some countries have a multiplicity of native Indian languages and dialects in addition to the prevailing Spanish or Portuguese. Roman Catholicism, also a heritage from Mediterranean Europe, is the main religious faith of the region and is the official religion in some nations. Nor are all the ties with Latin Europe indirect: newcomers from Portugal, Spain, and Italy are very prominent in Latin American immigrant lists.

        It should not be assumed that the transplanting of European cultures to Latin America has been achieved without essential modification or that a uniform culture prevails today throughout the region. Many Latin American culture traits are a heritage from native Indians, imported Africans, or other non-European groups, and the cultural impress of these groups has varied unequally from one part of Latin America to another. Nevertheless, many features of a common Latin culture are discernible in most parts of the region, and the civilization of Latin America as a whole stands in recognizable contrast to that of Anglo-America, which has primary roots in Great Britain and in Germany, the Scandinavian countries, and other countries of northwestern continental Europe.

PHYSICAL DIMENSIONS

With a land area of slightly more than 7.9 million square miles, Latin America is outranked in size by Africa, the Soviet Union, Anglo-America (including Greenland), and the Orient. However, its maximum latitudinal extent of more than 85 degrees or nearly 5900 statute miles is greater than that of any other world region, and its maximum east-west measurement, amounting to more than 82 degrees of longitude, is by no means unimpressive. Yet Latin America is not so large as these figures

A religious procession in a rural section of northern Venezuela some distance west of Valencia. A number of important characteristics of Latin America are exemplified in this view: the Roman Catholic faith, the tropical climate, underdevelopment, poverty, and the mixture of Spanish and Indian blood that characterizes a large share of Latin America's people. (Standard Oil Company, N.J.)

LATIN AMERICA

INTRODUCTORY LOCATION MAP

URBAN AREAS

over 4,000,000

1,000,000–3,000,000

500,000–1,000,000

★ ○ Selected smaller urban areas

Stars show capitals

City-size symbols are based on
metropolitan area estimates

UNITED STATES

GULF OF MEXICO

Rio Grande

MEXICO

Monterrey

Guadalajara

Mexico City

Veracruz

BAHAMA
ISLANDS
(BR.)

Havana

CUBA

Tropic of Cancer

DOMINICAN
REPUBLIC

HAITI

JAMAICA

San Juan

Santo
Domingo

PUERTO
RICO
(U.S.)

35°W

CARIBBEAN SEA

BRITISH
HONDURAS
(BR.)

GUATEMALA

Guatemala
City

HONDURAS

EL SALVADOR

NICARAGUA

COSTA RICA

PANAMA

Barranquilla

Maracaibo

Caracas

VENEZUELA

BARBADOS

TRINIDAD
& TOBAGO

ATLANTIC

OCEAN

PACIFIC

Cauca R.

Magdalena R.

Medellín

Cali

Bogotá

COLOMBIA

Guaviare R.

Ciudad
Bolívar

Orinoco R.

GUYANA

SURINAM
(NETH.)

FRENCH
GUIANA
(FR.)

OCEAN

GALAPAGOS
ISLANDS
(Ecu.)

Equator

Quito

ECUADOR

Guayaquil

Negro R.

Amazon

River

Manaus

Belém

Equator

PERU

Callao

Lima

Madeira R.

BRAZIL

Fortaleza

Recife

Salvador
(Bahia)

Lake Titicaca

La Paz

BOLIVIA

Sucre

Brasília

Belo Horizonte

São Paulo

Rio de Janeiro

Santos

Paraguay R.

110°W

85°W

70°N

Arctic Circle

60°N

NORTH
AMERICA

Pittsburgh

EUROPE

50°N

to New York (3500 miles)

to English Channel (3600 miles)

40°N

AFRICA

Tropic of Cancer

30°N

20°N

to Freetown
(1550 miles)

10°N

Equator

Natal

SOUTH
AMERICA

Longitude 80°W

Tropic of Capricorn

Tropic of Capricorn

PARAGUAY

Asunción

Paraná R.

Córdoba

Rosario

URUGUAY

Pôrto Alegre

CHILE

ARGENTINA

Valparaíso

Santiago

Buenos
Aires

Montevideo

Rio de la Plata

Paraná R.

Uruguay R.

ATLANTIC

OCEAN

RELATIVE POSITION

(Mercator projection)

Distances shown are in nautical miles

100°W

60°W

40°W

20°W

0°

20°E

40°E

60°S

50°S

40°S

30°S

20°S

10°S

Strait of

Magellan

FALKLAND ISLANDS
(BR.)

Cape Horn

85°W

60°W

35°W

0   200   400   600   800 1,000

Scale of Miles

might suggest, for its two main parts are offset from each other. The northern part of the region, known as Caribbean or Middle America, trends sharply northwest from the north-south oriented continent of South America. The latter continent is thrust much farther into the Atlantic Ocean than is the Caribbean realm or its northern neighbor, Anglo-America. In fact, the meridian of 80°W., which intersects the west coast of South America in Ecuador and Peru, passes through Pittsburgh, Pennsylvania (insert map, p. 631). Africa lies less than 200 statute miles east of Brazil.

DIVERSITY OF LATIN AMERICAN PEOPLES AND CULTURES

Statistics on physical size and shape, however, do not answer the basic questions of significance. What is Latin America? How many people live there and what are they like? What are their political, economic, and social viewpoints? Are they the masters of their own destiny? What is the character of the land they live upon? In short, how does Latin America fit into the world pattern of major regions?

Variations in Population Density in Latin America

The total population of Latin America was estimated at 268 million in mid-1968, as compared with a total of 201 million people for the fifty states of the United States. But as Latin America embraces well over twice the area of the United States, its overall population density is correspondingly less, amounting to 34 per square mile in 1968, as compared with an average overall density of some 56 per square mile for the United States. However, the Latin American population is more unevenly distributed than that of the United States. Some sections of Latin America exhibit rural population densities which are much heavier than any to be found within areas of comparable size in the United States. On the other hand, a much larger proportion of Latin America is sparsely inhabited. However, the population of Latin America, especially Caribbean America, is growing very rapidly, whereas the rate of population growth in the United States has declined somewhat in recent years.

Political Diversity

The 268 million people of Latin America are organized into political units that vary tremendously in size

and political status. Most of the population is contained in the twenty-four independent states. Aside from a few that have gained independence since World War II, these countries broke away from European control in the nineteenth century. In most of them, democratic processes have developed very slowly, and dictatorships of various kinds are still common. The republics vary in size from gigantic Brazil, which occupies more than a third of the total Latin American land area and boasts about a third of its total population, down to tiny El Salvador, which possesses only 0.1 percent of the land area, though containing a disproportionately high 1.2 percent of the aggregate population. The remainder of Latin America's population is found in units that are affiliated politically with the United Kingdom, France, the Netherlands, or the United States. All of these except the United Kingdom's Falkland Islands colony—a very marginal part of Latin America—are located in the Caribbean region or adjoining parts of northern South America. Most of them are even smaller than El Salvador in area, population, or both. The different units vary in their relationships to the overseas nations with which they are linked. Most of them, however, have much latitude in handling their internal affairs. Area and population data for the individual political units of Latin America are given in Table 26.

Economic Diversity

The people of Latin America are supported primarily by various forms of subsistence and commercial agriculture, supplemented in some areas by mining, manufacturing, and trade. A considerable proportion of the region's population lives on a hand-to-mouth basis bordering on mere subsistence, within a self-contained economic orbit that has little contact with foreign commerce and exchange. However, the economy of Latin America also reflects—with marked variations from area to area—the presence of the foreign investor. Large amounts of foreign capital have been invested in oil wells and refineries, factories, mines, transportation lines, commercial plantations, and other sources of profit. Most of the invested capital is from the United States, which has placed between 20 and 25 percent of its private long-term foreign investment in Latin America. United States investment is especially high in Venezuela, Brazil, Mexico, and Chile. Investment from West Germany is increasingly important. Britain and France also have investments in Latin America, but these are small and represent declining percentages of the total foreign investment in the region. Factories and other enterprises financed by Latin American private or public capital are still a third ingredient in the economic structure. Often such capital is borrowed from sources outside the country concerned.

Native Indians constitute a major population element in several Latin American countries. These Indian coffee pickers in Guatemala are engaged in separating the ripe red coffee berries from the immature green ones. (Foreign Agricultural Service, U.S. Department of Agriculture.)

Ethnic Diversity

The inhabitants of this region do not evince much ethnic uniformity. In only three nations—Argentina, Uruguay, and Costa Rica—have European strains been preserved on a large scale with little admixture by Indians or Negroes. (Scattered districts in other political units are predominantly European.)

Native Indians comprise more than half of the total population in the highland nations of Guatemala, Bolivia, Ecuador, and Peru. They are also a major population element in southern Mexico, El Salvador, and Paraguay. In outlying areas, especially in the basin of the Amazon River and in Panama, scattered lowland Indian tribes live more or less apart from the world about them.

The Negro is found in greatest numbers on the islands and along the hot, wet coastal lowlands of Caribbean America and areas near by. Except in Puerto Rico, Cuba, and the Dominican Republic, nearly all Caribbean islanders are Negroes. On the mainland, the east coast of Brazil between Cape Sâo Roque and Salvador and the northern hinterland of Rio de Janeiro have sizable Negro populations, as do small districts in French Guiana, Surinam (Netherlands Guiana), Guyana (former British Guiana), Venezuela, and Panama. A major concentration of Negroes is found along the lower Magdalena River valley and on the Caribbean coast of Colombia.

Lesser districts in Latin America support still other distinct ethnic groups. About one-half of Guyana's population, for example, is made up of people whose ances-

TABLE 26 LATIN AMERICA: AREA, POLITICAL STATUS, AND POPULATION DATA

POLITICAL UNIT[a]	POLITICAL STATUS (EARLY 1969)	AREA (THOUSAND SQUARE MILES)	POPULATION (MILLIONS: 1968 ESTIMATES)	DENSITY (PER SQUARE MILE: TO NEAREST WHOLE NUMBER)
CARIBBEAN AMERICA[b]		2005.7	118.0	59
Mexico	Independent republic	761.6	47.2	62
Guatemala	Independent republic	42.0	4.8	115
El Salvador	Independent republic	8.3	3.2	388
Honduras	Independent republic	43.3	2.5	58
Nicaragua	Independent republic	50.2	1.8	36
Costa Rica	Independent republic	19.7	1.7	84
Panama[c]	Independent republic	28.8	1.4	48
Totals: Central American republics		192.3	15.4	80
Colombia	Independent republic	439.5	19.8	45
Venezuela	Independent republic	352.1	9.7	28
Guyana	Independent parliamentary state within the Commonwealth of Nations	83.0	0.72	9
Panama Canal Zone	United States leased territory	0.55	0.060	109
British Honduras	British colony (internally self-governing)	8.9	0.117	13
Surinam (Netherlands Guiana)	Self-governing unit of Netherlands kingdom	55.1	0.366	7
French Guiana	Overseas department of France	35.1	0.037	1
Totals: Mainland affiliates of overseas nations		99.7	0.58	6
Barbados	Independent parliamentary state within the Commonwealth of Nations	0.17	0.252	1482
Cuba	Independent republic	44.2	8.2	185
Dominican Republic	Independent republic	18.8	4.0	214
Haiti	Independent republic	10.7	4.9	457
Jamaica	Independent parliamentary state within the Commonwealth of Nations	4.4	1.9	426
Trinidad and Tobago	Independent parliamentary state within the Commonwealth of Nations	2.0	1.0	515
Totals: Independent island countries		80.3	20.3	252

POLITICAL UNIT[a]	POLITICAL STATUS (EARLY 1969)	AREA (THOUSAND SQUARE MILES)	POPULATION (MILLIONS: 1968 ESTIMATES)	DENSITY (PER SQUARE MILE: TO NEAREST WHOLE NUMBER)
Puerto Rico	Self-governing commonwealth affiliated with the United States	3.4	2.8	809
U.S. Virgin Islands	United States dependency	0.133	0.046	346
Bahama Islands	British colony (internally self-governing)	4.4	0.156	35
British Virgin Islands	British colony	0.067	0.008	119
Cayman Islands	British colony	0.100	0.009	91
Montserrat	British colony	0.038	0.014	370
Turks and Caicos Islands	British colony	0.166	0.006	39
West Indies Associated States[d]	Self-governing states in association with the United Kingdom	1.120	0.487	434
Guadeloupe[e]	Overseas department of France	0.687	0.328	477
Martinique	Overseas department of France	0.426	0.336	789
Netherlands Antilles[f]	Autonomous unit of Netherlands kingdom	0.371	0.213	574
Totals: Island affiliates of overseas nations		10.9	4.4	399
NATIVE INDIAN COUNTRIES OF SOUTH AMERICA		*1183.2*	*24.5*	*21*
Bolivia	Independent republic	424.2	3.9	9
Peru	Independent republic	496.2	12.8	26
Ecuador[g]	Independent republic	105.7	5.7	54
Paraguay	Independent republic	157.0	2.2	14
BRAZIL	Independent republic	3287.2	89.8	27
COUNTRIES OF THE SOUTHERN MID-LATITUDES		*1431.3*	*35.4*	*25*
Argentina	Independent republic	1072.7	23.4	22
Chile	Independent republic	286.4	9.2	32
Uruguay	Independent republic	72.2	2.8	39
Grand totals		7907.4	267.8	34

[a]The Falkland Islands, a British colony in the South Atlantic located somewhat more than 400 miles to the east of Argentina and claimed by that country, are not included in the table. Area of the Falklands (excluding Antarctic dependencies): 4618 sq. mi.; estimated total population, about 2700 in 1968.

[b]Apparent discrepancies in totals and averages in this table are due to rounding of figures.

[c]Figures do not include the Canal Zone.

[d]Consists of Antigua (171 sq. mi.; est. pop. 60,960; density 356/sq. mi.), Dominica (290 sq. mi.; est. pop. 69,360; density 239/sq. mi.), Grenada (133 sq. mi.; est. pop. 98,164; density 738/sq. mi.), St. Christopher-Nevis-Anguilla (138 sq. mi.; est. pop. 61,793; density 448/sq. mi.), St. Lucia (238 sq. mi.; est. pop. 104,442; density 439/sq. mi., and St. Vincent (150 sq. mi.; est. pop. 91,800; density 612/sq. mi.).

[e]Includes two main and five lesser islands.

[f]Includes Curaçao, Aruba, Bonaire, and three smaller island units.

[g]Figures include the Galápagos Islands, a dependency of Ecuador.

tors immigrated from India. A small area in southern Brazil is predominantly Japanese. Still another is German.

In most of Latin America, however, mixed bloods predominate. Most of the region exhibits a primary mixture of Spanish and native Indian stocks, resulting in a heterogeneous group known as *mestizos*. Mixed bloods of Negro-native Indian ancestry are usually termed *zambos*. European-Negro mixtures, or *mulattoes,* are fewer in number and are found chiefly in the Caribbean units that are or have been affiliated with overseas nations.

LATIN AMERICA AND OTHER MAJOR WORLD REGIONS COMPARED

From the foregoing description one can infer that Latin America is an unusual amalgam of ideas, pursuits, and men. Yet this region is very much a part of the world geographic pattern, possessing characteristics which, however altered, tend to recur in other parts of the world. In fact, a comparison of Latin America with each of the other world regions considered in the present volume reveals many interesting and significant similarities and differences.

Latin America and the Soviet Union

In overall political and economic organization Latin America is unlike the Soviet Union. It is not, in other words, a unitary block of land controlled by a single dictatorial government and undergoing rapid and sometimes ruthless economic development according to a preconceived plan. Yet certain instructive comparisons can be made between recent economic developments in Latin America and the USSR. For example, in Mexico a government-enforced program of land reform has been introduced during the past six decades. Its primary objective has been the breaking up of large estates and redistribution of the property among the many landless farmers.

By subdividing large holdings Mexico is making an appeal to man's inherent territorial instincts and is trusting that the new responsibility will, with government aid and advice, result eventually in higher yields per unit of area and a higher standard of living per person. In the Soviet Union, where the former private holdings have been amalgamated into large collective farms or state farms, the stated objectives—higher agricultural production and a higher level of living—are the same as in Mexico. The methods, however, are quite different. The Soviet leaders have placed their dependence in a collectivized system of agriculture controlled from above and fitted into an overall economic plan for the entire nation. Private ownership of farmland has been abolished (though each family is allotted a garden plot to till as it sees fit), and farming has been made to resemble more and more a factory type of enterprise. Land reform in Mexico has also proceeded according to a plan—but a plan which depends for success on giving each peon access to land—*his* land. The cornerstone of the Mexican land reform program has been the expropriation of large, semifeudal private estates (*haciendas*), and their redistribution or restoration to the landless. Many estates were enlarged in the mid-nineteenth century by taking lands of Indian groups (*comunidades indigenas*); these have been restored, where possible, in the past half century. However, the largest amount of the ex-*hacienda* land was divided into agrarian communities known as *ejidos*.[1] Peasant farmers, as *ejido* members, are given rights of access and use of land and water on designated plots, provided the plots do not remain unused for more than 2 consecutive years. The parcels cannot be sold or mortgaged, though family rights of inheritance are enjoyed. Woodland and pasture are held by the *ejido*. In the past half century, more than 2.5 million heads of families (*ejidatarios*) have thus gained title to their land, which by the 1960s included somewhat less than half of Mexico's cropland. Laws of the Revolution guaranteed that under the process of *ejido* formation, a part of each estate could remain in private hands if the owner was cooperative. Such small to medium-sized individual properties comprised a little more than half of the cropland of the Republic in the 1960s.

Latin America and Europe

In political organization Latin America bears a certain resemblance to Europe. Each of these world regions contains a large number of individual political units which vary widely in population, area, mode of governance, and other respects. However, unlike some nations in Europe, no Latin American country has been able to boast a worldwide political and economic system with lifelines reaching over the globe toward raw materials and markets in outlying nations and political de-

---

[1] For material on the *ejido* incorporated in this paragraph, the authors are indebted to Dr. Herbert M. Eder, Department of Geography, University of California, Berkeley.

pendencies. Several Latin American republics have small island possessions,[2] but these are trivial when compared, for example, with even the remaining overseas dependencies of Great Britain, not to speak of the many former colonies that remain associated with Britain in the Commonwealth of Nations. Far from being generators of economic power, the independent countries of Latin America, as well as the politically dependent or semidependent units, exhibit a high degree of economic dependence on the United States and Europe. Most of the trade of Latin America, for example, is with the latter two areas.

Latin America and Anglo-America

Latin America bears a certain resemblance to Anglo-America in the nature and arrangement of its major topographic features. The major land mass within each of these world regions exhibits a gross physical pattern of high rugged mountains with associated plateaus and basins at the west, lower and generally older highlands and uplands at the east, and broad plains in the center. In each case more than half of the central region of plains lies within the drainage area of a single river system—the Amazon system in Latin America and the Mississippi-Missouri system in Anglo-America.

In political and economic respects, however, these New World regions are very different. The two massive, stable political units of Anglo-America—the United States and Canada—stand in marked contrast to the fragmented political order and notorious governmental instability of Latin America. The tremendous economic productivity and general prosperity of the Anglo-American nations place them in a different class from the underdeveloped and often poverty-stricken political units which comprise most of Latin America.

Latin America and the Orient

Latin America, like the Orient, is in part a region of heavy population pressure, primitive or intensive subsistence agriculture, poverty, and hunger.

Population Density

At first glance the region does not seem overpopulated, for its average population density of 34 per square mile (1968 estimate) is several times less than the average density in Europe or the Orient. Yet the island of Barbados has reported 1482 persons, chiefly rural, per square mile—a density comparable to that of the more heavily populated rural areas of the Orient. Also in the Caribbean realm are other, not so extreme, examples: Puerto Rico, a self-governing commonwealth voluntarily associated with the United States, has 809 persons per square mile, and Martinique, an overseas department of France, has 789 per square mile. Several other Latin American countries, especially island countries in the Caribbean, have population densities exceeding the overall average in the Orient. There are, of course, political units in Latin America which are very sparsely populated. French Guiana, for example, has an average density of a little over 1 person per square mile. Between these extremes, the larger countries occupy a position resembling that of the region as a whole: Argentina has approximately 22, Brazil 27, Chile 32, and Mexico 62 inhabitants per square mile.

Population Distribution

In its general pattern of population distribution Latin America differs somewhat from the Orient and from all other world regions. The Latin American pattern is one of heavy density in and around coastal cities (if local climatic handicaps, insects, diseases, and other liabilities are not too serious), a marked pressure in and near high mountain urban centers—in many cases, the capital of a country—and corridors of moderate to heavy concentration along the outlet routes from the mountains to the sea. There are, of course, variations and exceptions; yet the pattern is sufficiently consistent that nearly every political unit consists of a well-defined population core (or cores) with an outlying sparsely populated hinterland. In other words the total area of each country differs considerably from the total occupied area, with exceptions in the case of Uruguay, El Salvador, and most of the island units in the West Indies. In Brazil over three-fourths of the total land area is essentially without inhabitants other than aboriginal tribes. Most Brazilians live along or very near the eastern seaboard south of Recife. Around seven-tenths of the Argentine population is clustered in Buenos Aires or the adjacent humid pampa—an area containing slightly more than one-fifth of Argentina's total land. About one-half of Mexico's population is found in several mountain basins and valleys clustering around Mexico City—a district rep-

---

[2] The principal island possessions of the Latin American republics include the Galápagos Islands (Ecuador), and Easter Island and Sala y Gómez (Chile), all in the Pacific Ocean. In recent years Argentina has contested British claims to the Falkland Islands, in the Atlantic. Argentina and Chile have claims to portions of Antarctica.

The rural poverty which afflicts much of Latin America is well exemplified in this photograph, taken in Venezuela. The farm family in the picture is engaged in planting a field of upland rice amid the stumps in newly cleared land. With a sharpened stick the farmer punches holes into which his wife drops the rice seeds. (Standard Oil Company, N.J.)

and sources differ appreciably—this amount is about $700 per year in Argentina, $501 in Chile, $271 in Brazil, $470 in Mexico, and $378 in Peru. The same source cites a figure for the United States of $3648.[3] But income comparisons between the United States and Latin American countries must be made with great caution, because the measurement and comparison of national incomes are subject to wide margins of error for at least two reasons. (1) National income includes the value of produce consumed by farmers in their own households, and this is impossible to calculate accurately. (2) In calculating the figures cited above, Latin American currencies have been converted into United States dollars by the international exchange rates of the respective Latin American countries. It is well known that a United States dollar, thus converted into a Latin American currency, will often buy more in Latin America than it buys in the United States. Usually it will buy very much more of those goods and services which are consumed by the rank and file of Latin Americans. Thus the figures above do not indicate that Americans are, on the average, five times as rich as Argentinians and nearly ten times as rich as Peruvians; but we can say with assurance that compared to Americans the people of Argentina are poor, and those of Peru very poor.

resenting less than one-seventh of the country's entire area.

Widespread Poverty

Like most residents of the Orient, the great majority of Latin Americans are poor (photo, above). An inadequate, yet noteworthy, indicator of this poverty is the per capita gross national product (roughly, per capita income) of selected countries. According to one source—

Latin America and the Pacific World

Like the Pacific World, Latin America is comprised of a series of islands and a continental area supporting, all in all, a comparatively sparse population. The comparison can be carried still further. Like Australia, the largest land unit of the Pacific World, South America is a continent in which the most densely populated areas are situated on or near coasts, the interior being sparsely peopled. However, the settlement possibilities of

[3] The figures cited are 1966 estimates from Agency for International Development, Office of Program Coordination, Statistics and Reports Division, *Gross National Product: Growth Rates and Trend Data by Region and Country* (RC–W–138), March 31, 1967. For alternative estimates of per capita national incomes, see Population Reference Bureau. *World Population Data Sheet–1968* (Washington, D.C.), or David Simpson, "The Dimensions of World Poverty," *Scientific American*, 219, no. 5 (November 1968), 130–131.

the two continents differ. Australia possesses neither an Amazon River providing easy access to the interior nor a natural environment over most of its area which, by present standards, can be occupied successfully by more than small numbers of people. In short, Australia's unpopulated expanses are chiefly desert. In contrast, over 80 percent of Brazil, a country somewhat larger than Australia, may eventually be occupied, although at present only about 25 percent is used effectively—even for such extensive practices as the pasturing of livestock.

Ethnically, Latin America and the Pacific World offer some interesting contrasts. The islands of Latin America are occupied almost wholly by descendants of immigrants from Europe or Africa, the native Indians having been driven off or exterminated long ago. The islands of the Pacific World, however, except New Zealand, the Hawaiian Islands, and the Fiji Islands, are settled principally by native peoples who have lived for centuries in their present habitats. The mainland of Latin America is populated by an amalgam of native Indians, Africans, Europeans, and other ethnic groups; but the mainland of the Pacific World—Australia—is occupied almost entirely by descendants of European, mainly British, immigrants, the aborigines having been killed or driven to inaccessible places by the early European settlers.

There are also political and economic similarities between these two world regions. Both are governed, with few exceptions, by Europeans or descendants of Europeans. In each case most of the inhabitants are citizens of independent countries and yet must depend economically upon foreign commerce—primarily with Anglo-America and Europe—to take away surplus raw materials (chiefly from the farm or mine) and to supply certain manufactured goods. Each is an outlying part of the commercial world, the most important cores of which are found in the industrialized areas lying on either side of the North Atlantic Ocean.

Latin America and the Middle East

Like the Middle East, Latin America is a region that derives much of its regional unity from the nigh-omnipresence of a single religion—the Moslem faith in the Middle East and Roman Catholicism in Latin America. Each faith is dominant in most parts of its region. In Latin America, Roman Catholicism is the accepted religion of the upper classes, and is the official state religion in some countries. In somewhat altered form it filters down to the majority of the people, who tend to observe not only Catholic rites but also those learned from their native Indian or African forebears. This is particularly true in the communities of the native Indian.

The relationships between church and state in Latin America vary from country to country. The basic issues appear to be (1) the control of educational facilities, (2) the control of the marriage ceremony, and (3) continued exercise by the church of certain rights and property titles held during the colonial period—rights and titles whose validity has been challenged, with varying degrees of vigor, by the various national governments since independence.

COMPARISONS BETWEEN LATIN AMERICA AND AFRICA

Thus Latin America contains features or combinations of features which have counterparts in the Soviet Union, Europe, Anglo-America, the Orient, the Pacific World, and the Middle East. But perhaps the greatest number of similarities, especially of a locational, climatic, and economic nature, exist between Latin America and Africa.

Similarities in Shape and Location

Perhaps the most apparent similarity between Latin America and Africa is that of general shape. The major land mass of each assumes the form of a triangle with the apex pointing toward the South Pole. Moreover, in each case that land mass is connected to an even larger continent to the north—that is, to North America and Eurasia, respectively—by an isthmus which man has found it advantageous to canalize. There are also other similarities of location: north of the major land mass in each region is a sea—the Caribbean and the Mediterranean—which tends to separate it from, and concomitantly to provide shipping lanes to, its poleward neighbor. And, in either case, that northern continental neighbor is the home of a busy industrial society which has established strong political and/or economic ties with the region to the south. Thus Latin America tends to lie predominantly within the overseas economic orbit of the United States, and Africa within that of northwestern Europe.

The economic ties between Latin America and Africa and the industrialized regions to the north are associated with the fact that both Latin America and Africa are located predominantly in the lower latitudes, and hence are able to produce tropical crops which are in demand in Anglo-America and Europe but cannot be produced in quantity in the latter regions because of climatic handicaps.

Similarities and Differences in Landforms

Both Latin America and Africa are characterized by pronounced differences in elevation from one part of the region to another. However, contrasts in elevation are much greater in Latin America than in Africa. Nowhere in the latter continent is there a prominent lowlying plain series like that of the Orinoco-Amazon-Paraná-Paraguay—the river plains which dominate the interior of South America and separate the older, lower highlands of the east from the rugged Andes of the west. With high mountains, too, Latin America is the more generously endowed. This is evidenced by the nearly continuous Andes, Sierra Madre, and associated mountains which extend from northern Mexico to Tierra del Fuego (maps, pp. 654 and 664, and back endpaper landform map). Between the latitudes of 30°N. and 40°S. these mountains reach prevailing heights of at least 5000 feet above the sea. Within the Andes proper the highest crests exceed 9000 feet for over 3500 miles from northern Colombia and Venezuela to central Chile and Argentina. The latter's Mount Aconcagua—with an altitude of 22,834 feet that marks the highest point of the Western Hemisphere—is in the southern portion of this more rugged Andean belt. In Central America, the West Indies, and Mexico, mountains reaching higher than 9000 feet are found in northern Panama, Costa Rica, Guatemala, the Dominican Republic, and Mexico—in the last of these, especially near the capital city. Thus Latin America has very extensive areas of high mountains. But in Africa, mountains of comparable elevation are limited to a few rather small and erratically distributed ranges or isolated peaks.

Similarities in Types of Climate

In general, the types of climate and associated vegetation in Latin America as recognized in climatic classifications duplicate those of Africa. The differences most readily apparent are the much larger proportion of desert climate in Africa and the greater prevalence of middle latitude and high highland climates in Latin America. (See maps, pp. 654 and 664, and endpaper climatic map.)

Humid Tropical, Subtropical, and Marine Climates of Latin America

The *tropical rain forest climate* in the massive center of Latin America tends to be—like its African counterpart—located at or near the equator. However, the Latin American area of this climate type is somewhat larger than the African area, reaches farther poleward in the Northern Hemisphere, and its coastal segments are mainly found on the east coast, rather than on the west coast as in Africa. On either side of this tropical rainy climate —as in Africa—is found the *tropical savanna climate,* which in Latin America extends erratically to the vicinity of the Tropic of Cancer in the Northern and the Tropic of Capricorn in the Southern Hemisphere. Still farther poleward in the eastern portion of South America is found a sizable area of *humid subtropical climate.* Its Northern Hemisphere counterpart lies north of the Mexican border in southeastern United States. These three climate types— tropical rain forest (photo, p. 25), tropical savanna (photo, p. 27), and humid subtropical—characterize most of Latin America. Of the others, perhaps the most idyllic is a small strip of *mediterranean* or *dry-summer subtropical climate* in central Chile. To the south in Chile is a strip of *marine west coast (humid marine) climate* bordered by bleak, rainy, windswept, glaciated, essentially uninhabited mountain country.

Dry Climates of Latin America

The Latin American climates discussed above are those of more or less orderly, repetitious arrangement. One may expect to find generally similar climates in generally similar positions on all major land masses of the world. However, there are also in Latin America climates which are due at least partially to the presence of high landforms. These include some of the region's dry climates, and, of course, its mountain highland climates.

The dry climates, especially those of the arid and semiarid sections of Mexico (and southwestern United States), are to be associated partially with the global pattern of orderly climatic arrangement, and partially with local mountains and uplands. A glance at a climate map will reveal dry lands in generally similar positions in Africa (both north and south of the equator) and in Australia. However, the high mountain ranges on either side of the northern Mexican plateau, meeting in the vicinity of Mexico City, cannot be entirely overlooked in a consideration of causes for this particular area of dry land. The aridity of the dry lands of Argentina is largely the result of their location in the rain shadow of the Andes. They have counterparts in other sections of the world only where high ranges of mountains happen to block the path of prevailing winds and hence to cause a deficiency of rainfall on the lee side. In the west-coast tropics and subtropics of South America, the Atacama and associated deserts cannot be explained so simply; for shifting winds, cold offshore currents, and other complexities—as well as the Andes Mountains—are important to climatic conditions there. However, the mountains serve to restrict this area of desert to the coastal strip.

High Upland and Highland Climates of Latin America

Although the lowlands and low uplands of Latin America are extensive and support sizable populations in some countries, the highlands and high uplands are also very important to man's use of this region. They are significant not only for the mineral resources they contain but also—chiefly because of their cooler temperatures—as habitats for settlement, particularly in the otherwise persistently hot lands. To immigrants or descendants of immigrants from the middle latitudes such conditions have been especially attractive, for they resemble somewhat the climates from which the settlers came. However, the annual range of temperatures is much lower than in the middle latitudes. For example, Quito, Ecuador—an extreme case—records an average temperature of 55 or 56 degrees F every month of the year.

These upland and highland climates change markedly with increased elevation, and hence are subject to classification into vertical zones. Although each of the four basic weather and climatic elements—temperature, pressure, humidity, and winds—varies with altitude, temperature is the primary criterion for zonal classification. At least three major zones are commonly recognized in the higher lands of Latin America: the *tierra caliente* (hot country), the *tierra templada* (cool country), and the *tierra fría* (cold country).

Rising from the previously discussed lowland climates, the *tierra caliente* zone of hot, wet conditions reaches to approximately 3000 feet above sea level at or near the equator, and to slightly lower elevations in parts of Mexico and other areas near the margins of the tropics. It is, in effect, an upward projection of these lowland climates, from which it has never been satisfactorily isolated by geographers and climatologists. Thus it is the environment of nature's rain forest or tropical savanna and man's rice, sugar cane, and cacao. It tends to be the zone of the truly tropical plantation—whether owned locally or abroad—and of the Negro, zambo, and mulatto worker. In the islands of Caribbean America, it is the zone of major urban units, containing all of the cities of 100,000 or over.

The *tierra caliente* merges almost imperceptibly into the *tierra templada*. Although sugar cane, cacao, bananas, oranges, and other lowland products reach their respective uppermost limits at some point in this higher level, the *tierra templada* is most notably the zone of the coffee tree. Indeed, some scholars use the criterion of effective coffee culture as the dividing line from the *tierra caliente.* In the *tierra templada* coffee can be grown with relative ease; at lower altitudes the crop encounters difficulties occasioned by excessive heat and/or moisture. The upper limit of this zone—approximately 6000 feet above sea level—tends also to be the upper limit of European-induced plantation agriculture in Latin America. In its distribution the *tierra templada* flanks the rugged western mountain cordilleras and, in addition, is the uppermost climate in the lower uplands and highlands to the east. The broadleaf evergreen trees of its moister, hotter sections tend to resemble those of the *tierra caliente*. This is especially true of the eastern flanks of the high mountains. In its poleward margins, however, broadleaf trees are replaced to some degree by coniferous evergreens. In such places as the highlands of Brazil or Venezuela where there is less moisture, scrub forest or savanna grasses appear—the latter generally requiring the more water.

In brief, the *tierra templada* is a prominent zone of European-induced settlement and of commercial agriculture. Urban as well as rural settlement is very much in evidence: of Latin America's metropolitan areas of 100,000 or over about a fourth are in or very near the *tierra templada*. Four of the metropolises exceeding 1,000,000—São Paulo, Caracas, Medellín, and Guadalajara—are in this zone, while another—Mexico City—lies just above it. Others, like Rio de Janeiro, which are situated at lower elevations, have close ties with predominantly residential or resort towns in these cooler temperatures.

The *tierra fría,* or cold country, may be distinguished from the other zones by two criteria. First and perhaps most important in an agricultural region like Latin America, it is a zone where frost occurs. As one might expect, frosts are only occasional in the zone's lower reaches at approximately 6000 feet above the sea, but are much more frequent at higher elevations. The second criterion refers to type of economy: in contrast to the Europeanized *tierra templada* the *tierra fría* tends to be the habitat of a subsistence, native Indian economy —most extensive in Peru and Bolivia but also present in Ecuador, Colombia, Guatemala, El Salvador, and Mexico. The upper limit of the *tierra fría* is generally placed at about 10,000 feet above sea level for locations near the equator, and at lower elevations toward either pole. This line is usually drawn on the basis of two criteria: (1) the upper limit of agriculture, as represented by such hardy crops as potatoes, barley, or the locally important cereal quinoa; and (2) the upper limit of natural tree growth. Above are the alpine meadows, sometimes called *paramos;* still higher there may or may not exist barren rocks and snow or ice.

The *tierra fria* tends to be a last retreat and the major home of the native Indian and is characterized by small permanent settlements and by what Europeans or Americans might consider rather primitive ways of life. It is chiefly rural, containing only about one-tenth of the metropolitan areas in Latin America that number 100,000 or over. However, nature has placed here certain valuable minerals like tin and copper which have attracted modern types of large-scale mining enterprise into the *tierra fria* of Bolivia and Peru as well as some other Latin American countries.

Climatic data for representative Latin American stations are given in Table 27.

Latin American Minerals

Latin America is a large-scale producer of a somewhat shorter list of important minerals than is Africa. However, Latin America's mineral output is very significant to the outside nations that purchase most of it, despite the rather small number of key minerals that bulk large when viewed in a world perspective. The region's bauxite, for example, is critically important to the industrial economies of the United States and Canada. Currently, Latin America's known mineral resources of greatest consequence include iron ore, petroleum, bauxite, tin, copper, nitrate, sulfur, and silver. Supplies of good coal, especially coking coal, are almost entirely lacking.

TABLE 27 CLIMATIC DATA FOR SELECTED LATIN AMERICAN STATIONS

STATION	COUNTRY	LATITUDE TO NEAREST WHOLE DEGREE	ELEVATION ABOVE SEA LEVEL (FEET)	TYPE OF CLIMATE	AVERAGE TEMPERATURE (DEGREES F TO NEAREST WHOLE DEGREE)			PRECIPITATION	
					ANNUAL	COOLEST MONTH	WARMEST MONTH	ANNUAL AVERAGE TO NEAREST INCH	NUMBER OF MONTHS WITH LESS THAN 1 INCH OF RAIN
Monterrey	Mexico	26°N.	1752'	Steppe	72°	59°	83°	26"	5
Bridgetown	Barbados	13°N.	181'	Tropical rain forest	79°	77°	80°	54"	0
Caracas	Venezuela	11°N.	3418'	Tierra templada	69°	66°	71°	33"	3
Ciudad Bolivar	Venezuela	8°N.	164'	Tropical savanna	82°	80°	83°	38"	3
Manaus	Brazil	3°S.	144'	Tropical rain forest	80°	78°	82°	82"	0
Cuiaba'	Brazil	16°S.	541'	Tropical savanna	78°	73°	81°	54"	3
São Paulo	Brazil	24°S.	2608'	Tierra templada	65°	58°	70°	51"	0
Rosario	Argentina	34°S.	89'	Humid subtropical	62°	50°	75°	39"	0
Mendoza	Argentina	33°S.	2713'	Desert	61°	45°	76°	8"	11
Santiago	Chile	33°S.	1706'	Mediterranean	58°	46°	69°	13"	7
Valdivia	Chile	39°S.	43'	Marine west coast	54°	46°	63°	96"	0
Lima	Peru	12°S.	449'	Desert	65°	59°	72°	1"	12
La Paz	Bolivia	17°S.	13458'	Tierra fria	50°	46°	53°	22"	5
Quito	Ecuador	0°	9243'	Tierra fria	55°	55°	56°	49"	2

Petroleum is the leading Latin American mineral in value of production. The photo shows oil being extracted from beneath the shallow waters of Lake Maracaibo in western Venezuela. (Standard Oil Company, N.J.)

Deposits of high-grade *iron ore* in the eastern highlands of Brazil and Venezuela are the largest known in the Western Hemisphere and are among the largest in the world. Lesser and yet noteworthy producing deposits occur in northern Mexico, north-central Chile, and southern Peru. Brazil is the largest Latin American producer of iron ore, followed by Venezuela, Chile, and Peru.

Petroleum is significant in the Caribbean Sea–Gulf of Mexico area, particularly in northern Venezuela (the leading producing region by far), several parts of Colombia, the central and southern Gulf coast of Mexico, and Trinidad. Other fields exist (1) along the Peruvian-Ecuadorian coast, (2) along the Atlantic margins of Patagonia in Argentina, (3) along the eastern piedmont of the Andes in northern Argentina and southern Bolivia, (4) on the east coast of Brazil near Salvador, and (5) in the Chilean part of Tierra del Fuego. The large sedimentary basins east of the Andes, shared by Brazil and several other South American countries, may contain sizable petroleum deposits not currently known.

Most of the region's production of *bauxite*—the major source material for aluminum—comes from Jamaica, Surinam (Netherlands Guiana), or Guyana. The deposits in these countries are located relatively near the sea.

Most of the known reserves of *tin* in Latin America are in the Andes of Bolivia, and that country is the region's only significant producer.

Low-grade but comparatively abundant *copper* deposits occur in the Atacama Desert of northern Chile and the arid and semiarid sections of Mexico. Additional reserves of copper are found in the Andes, especially in the mountainous sections of Peru and Chile. Chile is overwhelmingly the largest copper producer in Latin America, and was the world's third largest producer in 1966 (after the United States and the Soviet Union).

Chile contains the only sizable reserves of natural *nitrate* that have ever been exploited commercially on a grand scale. Like the lower grades of copper, these are in the Atacama Desert.

Since 1955, Mexico has become a major source of native *sulfur*. Only the United States produces more. The main fields, like those of the United States, border the Gulf of Mexico. They lie in the vicinity of Veracruz.

The *silver* of Mexico, Peru, and Bolivia is principally found in mountains or rough plateau country. Mexico and Peru are by far the largest Latin American producers of silver, with Bolivia third, and Mexico competes each year with the United States for first place in the world. Mexico's deposits are mainly in the dry northern and north-central sections of the country.

Similarities in Economic Activities

Since both Latin America and Africa are outlying regions in the world's economy, their basic industries serve principally to supply the wants of their own inhabitants and to furnish certain raw materials and foods to more highly industrialized areas. In both regions agriculture is the leading livelihood industry.

Latin American Agriculture

In Latin America as a whole, agriculture employs approximately 50 percent of the total labor force. The percentage of the labor force that is employed in agriculture is declining rather rapidly in some countries but scarcely at all in others. There is an appreciable range in the proportionate current importance of agriculture among the various countries. For example, Argentina and Uruguay have less than one-fifth of their respective labor forces in agriculture, whereas over four-fifths is so engaged in Haiti. In other countries the figure is between these two extremes.

The systems of agriculture vary. In some tropical lowland areas, particularly in Caribbean America, single-crop commercial plantations owned by companies or syndicates dominate the scene. The capital to establish these enterprises has come principally from Anglo-America or Europe. Most plantations grow sugar cane or bananas. Sugar plantations are especially prominent in the Caribbean islands, while banana plantations are found principally in Central America. A considerable amount of

Beef cattle raised in the Latin American tropics are not generally of prime quality as compared with American or European cattle, but are adapted to the tropical climate, relatively resistant to disease, and able to subsist on coarse forage. The cattle pen in this view is in interior ranching country in Brazil. (Brazilian Government Trade Bureau.)

subsistence farming by individual families is carried on in plantation areas, often on plantation-owned land.

But despite the importance of commercial plantations in some areas, the bulk of Latin America's farm production is accounted for by individually owned and community-owned agricultural holdings belonging to citizens of the respective countries in which these properties are located. Communal holdings, reflecting Indian traditions or twentieth-century revolutionary plans of agrarian reform, are especially well represented in Mexico (see pp. 655–657). Discontinuous landholdings of indigenous groups form the dominant patterns on the land in southern Mexico and northern Central America, inhabited by descendants of the Aztecs, Zapotecs, Mayas, and their neighbors. A comparable situation is found in the central Andes of Peru with descendants of the Incas. Most of the communal Indian lands of southern Mexico are devoted to subsistence agriculture, as are the adjacent *ejidos*.[4]

Most individually owned holdings are of Latin European origin. They may be large *haciendas* (Spanish) or *fazendas* (Portuguese) such as are found in some form (although declining in number) in nearly all Latin American nations, or they may be small holdings like those of Costa Rica. Agricultural production on the large estates, whether called *haciendas, fazendas,* or some other name, involves the employment of large numbers of landless, often illiterate, workers. In some areas these workers are traditionally bound to the estate, and in others they are free to move at will. In actual fact, most of them remain on the estate. The estate owner is well educated—often in Europe. He is usually a citizen of the country in which his property is located. He may have a residence in a city in addition to that on the estate. He may also be engaged in business, or one of the professions, leaving the direction of the estate to a carefully chosen manager. Although one estate may produce several commodities commercially, there is a tendency toward specialization in only one or two. All in all, such estates produce a sizable proportion of the coffee, sugar, cotton, henequen, livestock products, and other Latin American agricultural commodities that enter world markets.

Types of production on small individual holdings vary with owner and place. The more active small farmers, like those of Costa Rica, tend to grow coffee, and other commercial specialties as well as subsistence crops of corn, beans, sorghums, and vegetables. Other,

more inefficiently managed holdings tend to concentrate on subsistence crops only. Various land-reform programs in Latin America are resulting in an increased number of small private holdings and collectives and a decreased number of large estates.

Not only does agriculture employ about half of Latin America's people; it is also the leading source of exports in most of the region's countries. Of the independent states, all but Venezuela, Boliva, Chile, Trinidad and Tobago, Jamaica, Guyana, and Peru export a larger value of agricultural commodities than of products from other livelihood industries. Coffee, sugar, cotton, meat, cacao, wheat, bananas, wool, and corn are among the major agricultural products leaving the region.

Land redistribution is very much at the forefront of attention in Latin America. A major effort toward this goal in Mexico has been mentioned briefly, and will be examined in more detail in the next chapter. Ten other Latin American countries—Chile, Colombia, Costa Rica, the Dominican Republic, Guatemala, Honduras, Nicaragua, Panama, Paraguay, and Peru—have passed legislation or issued decrees regarding land redistribution.[5] Others are attempting regional efforts which do not necessarily apply to entire national territories. Although success in implementation has varied sharply from country to country, this movement is gaining momentum. Some attention is being given to the breakup of existing properties, and other efforts are being exerted to bring more vacant land (whether owned by the public, by private individuals, or by the church) into cultivation, usually by small farmers.

Mining as a Source of Livelihood

The significance of mining to Latin America is not easily appraised. In terms of labor force it is comparatively unimportant, for the highly mechanized mining industry seldom employs more than 3 percent of the total labor force of a Latin American country. Since mining ventures are largely financed and usually managed by outside interests in order to get raw materials for the industrial economies of Anglo-America and Europe, few of the extracted products reach the Latin American people. However, revenues from mining—whether in the form of income, property, export tariff, or other taxes—are very important to some Latin American governments.

---

[4] This paragraph incorporates material supplied by Professor Herbert M. Eder, Department of Geography, University of California, Berkeley.

[5] See especially, Organization of American States, *Economic Survey of Latin America* (Baltimore, Md.: Johns Hopkins Press for OAS, 1964), pp. 217–227.

Some of the area's mining activities are of considerable importance on a world basis. For example, in 1966 over 10 percent of the world's petroleum output by volume and nearly 3 percent of its iron ore originated in Venezuela. About 17 percent of all silver, 16 percent of all sulfur, 6 percent of all lead, and 5 percent of all zinc was extracted in Mexico. About 23 percent of all bauxite came from Jamaica, 12 percent from Surinam, and 7 percent from Guyana. Approximately 12 percent of all copper and essentially all natural nitrate came from Chile. Over 12 percent of all tin was mined in Bolivia, and over 3 percent of all iron ore in Brazil. Some 13 percent of all silver and 6 percent of all zinc came from Peru. In addition, individual Latin American countries supplied large amounts of lesser known minerals such as bismuth, fluorspar, and strontium.

The Increasing Importance of Manufacturing

In both Latin America and Africa, manufacturing industries are making an increasingly important contribution to national economies. In fact, the industrial revolution which we so often place in the latter part of the eighteenth century is only now beginning to reach effectively to these outlying world regions. In a sense, a frontier—or a series of frontiers—of industrialization is migrating over them, just as the now legendary frontier of settlement in the United States once shifted westward. A leading authority on Latin America, Professor Preston E. James, has characterized this change as follows:[6]

The history of the Occidental world during the last few centuries has been involved with the impact of the new *industrial society* and the older *pre-industrial society*. This is what we call the industrial revolution. Beginning in Western Europe the new way of living, coupled with enormously increased productivity in all forms of economic activity through the use of controlled inanimate power, has gradually transformed whole sections of Europe and America. In some instances the transformation has taken place by gradual evolution; in not a few instances it has been accompanied by violence and warfare, both civil and international. The rapid increase in the need for raw materials of all kinds has produced the present intense rivalry for the control of the productive regions, especially of the sources of power. The English-speaking peoples, who were the first to adopt the new way of living, were able to gain control of about 75 percent of the developed power resources of the world; and the challenge to this control lies behind the present international turmoil. In Latin America, the impact of the industrial society with the traditional pre-industrial society is now going on. Where the industrial way of living has become established, a new and still more profound line of cleavage has been formed across all the previous diversities of Latin-American society.

The fundamental characteristics of the industrial society should be reviewed briefly. The use of controlled inanimate power changes the emphasis from production by cheap labor to production by machines—or, in terms of economics, capital investment assumes a position of preponderant importance, and the owners of capital rather than the owners of land assume places of highest prestige and political power. Production is enormously increased, not only total production, but also per capita production. This leads to specialization and exchange, and hence to interpendence over wide areas. Trade is transformed from a small-scale exchange of luxury goods or specialties to a large-scale exchange of staples, and as a result communities are no longer supported by the products of the territories immediately surrounding them, but from a wide variety of producing areas, most of them beyond the control of the community which absorbs the products. With life organized on such a pattern society reaches a much higher standard of material comfort than any previous society has been able to reach; but this standard can be maintained only if a nation accepts the fact of wide geographical interdependence, turns away from provincial isolation, and cooperates with other nations in the maintenance of a stable financial structure of money and credits.

The industrial society brings profound changes in the details of human life. Prestige, we repeat, is to be gained through the ownership of capital which brings power, rather than through the ownership of land which brings security. Life becomes more speculative, less certain, but with rewards for the successful which are in a material way far beyond anything the world has offered before. There comes a notable change in the time concepts. With the increased tempo of life the vague concepts of pre-industrial society, such as *por la mañana, por la tarde,* must be given up for more precise concepts, such as 9:45 A.M. or 3:10 P.M. Behavior of all sorts becomes more standardized. The picturesqueness of provincial costumes disappears under a uniform cover of blue denim overalls; people from Patagonia to Labrador watch the antics of Mickey Mouse; local differences in manners and customs are modified by the impact of the new patterns of life. In the big cosmopolitan centers of Latin America life follows the same routine as in North American or European cities—this uniformity is apparent in styles of architecture, styles of dress, forms of work and recreation—

---

[6] Preston E. James, *Latin America* (3d ed.; New York: The Odyssey Press, 1959), pp. 52–57. Used by permission of the author and the publisher. This book is highly recommended as a standard, very readable, and comprehensive work on the geography of Latin America.

in short, the whole aspect of life is changed from its variegated pre-industrial base to a uniformity repeated in all the Occidental urban centers.

These changes affect the distribution of people. As long as coal remains the chief source of power, manufacturing industry is carried on at the lowest cost in large concentrated units. People gather together in great cities—cities greater than any that the world ever knew before, cities of more than a million inhabitants. Although the use of electric power may have the effect of spreading manufacturing industry over a wider area, thus transforming the life in smaller towns and villages, the large concentrations of city people still perform more efficiently the urban functions of commerce and administration. These cities are still dependent on the productivity of the land for their support, but the land base has been greatly extended; and as a result the means of transportation which tie the cities together must be greatly elaborated.

The urban-industrial way of living has come to Latin America from outside, not by slow evolution from the earlier pre-industrial base. In parts of Europe and in Anglo-America where the cities and the urban life developed out of the rural background there is a certain normal relationship between the size of the city and the productivity of its rural hinterland. . . . In Latin America one finds cities which have become industrial and commercial centers with an industrial way of living, but which bear little relationship in size or in function to the rural districts back of them. The contrast between the cities and the rural districts is enormous: the average tourist who journeys by boat or airplane from one city to another scarcely catches a glimpse of the Latin America which is traditional, and which is still dominant in terms of area and numbers of people.

Urban-industrial growth has appeared at various places in Latin America. The largest developments of this cosmopolitan life center in Buenos Aires, Rio de Janeiro, São Paulo, Santiago, Lima, Mexico City, Caracas, and Habana [Havana]—all cities of a million or more people, all thoroughly modern metropolises, with a way of living entirely familiar to metropolitan dwellers throughout the Occidental world. Modern industrial development has appeared also in many smaller cities throughout Argentina, Chile, Brazil, Colombia, Venezuela, Mexico, and Cuba. Yet the industrial productivity of all Latin America is still very small compared with that of the United States or of Western Europe.

The Latin American countries which have the largest output of factory-made goods are Brazil, Argentina, and Mexico. Also prominent in this regard are Venezuela, Colombia, Chile, Uruguay, Cuba, and Puerto Rico. Although heavier types of industry have developed in scattered localities, factories in Latin America are engaged chiefly in the production of processed foods, textiles, and fabricated metal products. Most manufactures are sold in home markets.

Increasing Economic Cooperation

Modern economic growth usually requires large domestic markets, varied natural resources, and a dynamic system of production and exchange. Most countries in Latin America and Africa (or indeed in most other economically poor sections of the world) are seriously lacking in these basic ingredients. However, success in Europe with economic integration has stimulated similar movements, especially in Latin America. The Central American Common Market (CACM), established in 1960, involves Costa Rica, El Salvador, Guatemala, Honduras, and Nicaragua. Panama is considering membership. As implied in the title, this group aspires to a high measure of integration. Trade among the five members is growing rapidly. A Central American Bank of Economic Integration has been created at Tegucigalpa, Honduras, to provide capital financing for all members. A permanent secretariat is located at Guatemala City, and permanent councils are to be established on political policies, economic measures, education, defense, and other matters.

The Latin American Free Trade Association (LAFTA) involves Argentina, Bolivia, Brazil, Chile, Colombia, Ecuador, Mexico, Paraguay, Peru, Uruguay, and Venezuela. Established in 1961, this group aspires, at least for the time being, to cooperation rather than integration. The members have agreed to a progressive reduction of interparticipant trade barriers, reaching zero by 1973. Escape clauses have been provided for specified commodities.

To date, the Central American Common Market has moved more rapidly toward its goals than the Latin American Free Trade Association. Reasons for this include the small size, close proximity, urgent need, and relative cultural homogeneity of the CACM countries as contrasted with generally opposite characteristics of the countries of LAFTA.

Africa's record regarding economic integration, especially since the successful drive for independence in so many countries, has been varied. Some proposed organizations have remained in the discussion stage, whereas others have commenced to be implemented. Among the latter is the East Africa Community formed by Kenya, Tanzania, and Uganda in 1967. In the same year the first steps toward an Economic Community of West Africa were taken by 12 nations that signed articles of association and were subsequently joined by two other countries. Still another effort is the Maghrib Regional Economic Organization, formed in 1964, involving Algeria, Morocco, Tunisia, and Libya.

Although not restricted to Latin America and Africa, the trends toward economic integration in these continents are indicative of a worldwide search for meaningful solutions to underdevelopment. A major obstacle to be overcome involves the traditional agricultural economies of member nations. Two countries producing palm oil or coffee, for example, will scarcely find an exchange to the advantage of either. However, if one or the other will attempt to specialize in another set of products, trade may be possible. A second hurdle, related to the first, is the competitive history of these prevailingly agricultural economies. A spirit of competition can be expected to give way only slowly to one of cooperation.

Similarities in Overall Patterns of Settlement and Transport

Both Latin America and Africa tend to have a "rim" pattern of settlement. In other words, the main concentrations of people tend to be found around or relatively near the coasts. The principal exceptions are certain interior highlands which support dense, or moderately dense, populations. In both regions, large areas of sparsely occupied land, peopled by erratically distributed aboriginal tribes, exist in the interior. Large cities, with few exceptions, lie on or near the sea or at high elevations. These cities largely have preceded intensive factory development—in contrast to many large cities in Europe whose growth was stimulated initially by factory-type industrialization.

Land transportation is poorly developed in most parts of Latin America and Africa. Rail lines reach inland from seaports in many places, but in only a few sizable areas has a reasonably dense web of railroads developed (maps, pp. 654 and 665). Highways, also, are in an immature stage of development. Few roads are hard-surfaced, and many are passable only with difficulty. This is particularly true in the rainier climates of the low latitudes. In both regions, the airplane is becoming increasingly important as a carrier—not only of specialty items but also of meat and other staples.

Some Important Differences between Latin America and Africa

Similarities between Latin America and Africa should not be overstressed. Many of the apparent similarities are only superficial. The two regions exhibit profound differences in racial composition, in historical relationships, and in cultural heritage. The independent countries of Latin America have had much more experience in self-government than most independent African countries. Racial segregation and strife are less of a problem in Latin America than in Africa, due to the fact that many Latin American ethnic groups have tended to mingle through intermarriage.

There are also important environmental differences. Latin America has a larger area of mountain lands and of low-lying river plains and smaller areas of plateau and upland than does Africa. In addition, Latin America projects farther into the Southern Hemisphere and thus gains access to a larger area of humid middle-latitude climates. Africa has no effective counterpart of the rich pampas of Argentina, Uruguay, and southern Brazil, nor of the marine west coast climatic area of southern Chile.

SUMMARY OF LATIN AMERICAN CHARACTERISTICS

Latin America is a highly diversified region. Particularly in its landforms, climates, political arrangements, and population patterns, it is one of the most diverse of the eight world regions.

It is a region which does not use its natural and human resources as efficiently as do more highly industrialized regions elsewhere. Approximately 50 percent of its labor force depends upon some form of agriculture for a livelihood, yet the region produces only about one-tenth of the world's farm output. Many of its farmers have no knowledge of modern agricultural techniques, and most Latin American governments lack adequate capital to promote a revision of farming practices.

It is a region in which population increase is pronounced—particularly in the poorer, less developed sections such as most islands in Caribbean America.

It is also a region which depends markedly upon the industrial societies of Anglo-America and Europe with respect to exports, imports, and investment of capital. This dependence is currently being decreased, especially in the economically stronger countries, by the development of local manufacturing plants—of which many are sponsored by the respective national governments. In these countries a long-established landed aristocracy now sees its traditional rights, privileges, and power threatened by a rising merchant and factory class.

Finally, it is a region with potentialities for further development, which tend to take one of five primary forms: (1) the establishment of more factories in the region; (2) the breaking up of large landed estates into smaller parcels which can be worked by owners or quasi-owners; (3) the opening up of new agricultural and grazing lands, particularly in the interior of South America; (4) the intensified utilization of mineral resources, for domestic consumption wherever possible; and (5) the encouragement of cooperation among heretofore competitive, largely agricultural, economies. Whether Latin America's potentialities can be realized at a rate concomitant with the rapid increase in population numbers remains to be seen.

REFERENCES and READINGS

ALEXANDER, ROBERT J., "Nature and Progress of Agrarian Reform in Latin America," *Journal of Economic History,* **23**, no. 4 (December 1963), 559–573; and "Agrarian Reform in Latin America," *Foreign Affairs,* **41**, no. 1 (October 1962), 191–207.

AMERICAN ASSEMBLY, *Population Dilemma in Latin America,* ed. by J. Mayone Stycos and Jorge Arias (Washington, D.C.: Potomac Books, 1966).

ARCINIEGAS, GERMÁN, *Latin America: A Cultural History,* trans. from the Spanish by Joan MacLean (New York: Alfred A. Knopf, 1967).

An Atlas of Latin American Affairs. Text by Ronald M. Schneider; maps by Robert C. Kingsbury (New York: Frederick A. Praeger, 1965).

AUGELLI, JOHN, "Latin America and the Alliance for Progress: A Geographer's Appraisal," *Journal of Geography,* **66**, no. 8 (November 1967), 417–424; and "The Controversial Image of Latin America: A Geographer's View," *Journal of Geography,* **62**, no. 3 (March 1963), 103–112.

AVILA, FERNANDO BASTOS DE, *Immigration in Latin America* (Washington, D.C.: Pan American Union, 1964).

BAILEY, HELEN MILLER, and ABRAHAM P. NASATIR, *Latin America: The Development of Its Civilization* (Englewood Cliffs, N.J.: Prentice-Hall, 1960).

BARRACLOUGH, SOLON L., and ARTHUR L. DOMIKE, "Agrarian Structure in Seven Latin American Countries," *Land Economics,* **42**, no. 4 (November 1966), 391–424.

BATES, MARSTON, and the Editors of *Life, The Land and Wildlife of South America* (Life Nature Library; New York: Time Inc., 1964).

BEYER, GLENN H., ed., *The Urban Explosion in Latin America: A Continent in Process of Modernization* (Ithaca, N.Y.: Cornell University Press, 1967).

BRANCO, RAUL, "Land Reform: The Answer to Latin American Agricultural Development?" *Journal of Inter-American Studies,* **9**, no. 2 (April 1967), 225–235.

BROWN, ROBERT T., *Transport and the Economic Integration of Latin America* (Washington, D.C.: The Brookings Institution, 1966).

BURCK, GILBERT, Latin America: Bureaucracy or the Market?" *Fortune,* **65**, no. 2 (February 1962), 85–87 ff.

BUTLAND, GILBERT J., *Latin America: A Regional Geography* (2d ed.; New York: John Wiley & Sons, 1966); also, "Southern Affinities and Contrasts: A Comparative Review of the Cultural Landscapes of South America and Australia," *Revista Geográfica* (Rio de Janeiro), No. 63 (December 1965), 117–130; and "Frontiers of Settlement in South America," *Revista Geográfica,* No. 65 (December 1966), 93–108.

CANADA, DEPARTMENT OF MINES AND TECHNICAL SURVEYS, GEOGRAPHICAL BRANCH, *Colonization and Settlement in the Americas: A Selected Bibliography,* comp. by S. C. Wiley (Bibliographical Series, No. 25; Ottawa: 1960).

COLE, J. P., *Latin America: An Economic and Social Geography* (Washington, D.C.: Butterworth Inc., 1965).

Cornell Latin American Year Conference on the Potentials of the Hot-Humid Tropics in Latin American Rural Development, November 29–December 3, 1965 [Papers]. (Ithaca, N.Y.: New York State College of Agriculture of the State University at Cornell University, Office of International Agricultural Development, 1965.) Numerous papers by experts in many fields.

CRIST, RAYMOND E., "Tropical Subsistence Agriculture in Latin America: Some Neglected Aspects and Implications," Smithsonian Institution, *Annual Reports, 1963* (Publication 4530; Washington, D.C.: 1964), pp. 503–519; also, "The Latin American Way of Life, I: A Culturally Diverse Continent—Balkanized; II: Cultural Factors and Attitudes Making for Unity," *American Journal of Economics and Sociology,* **27**, no. 1 (January 1968), 67–76, no. 2 (April 1968), 171–183; cont.; and "Some Aspects of Human Geography in Latin American Literature," *American Journal of Economics and Sociology,* **21**, no. 4 (October 1962), 407–412.

DORN, GEORGETTE M., comp. *Latin America: An Annotated Bibliography of Paperback Books* (Hispanic Foundation, Bibliographical Series, No. 11; Washington, D.C.: Library of Congress, 1967).

FIGUERES, JOSE, "The Problems of Democracy in Latin America," *Journal of International Affairs,* **9**, no. 1 (1955), 11–23.

FOSTER, GEORGE M., *Culture and Conquest: America's Spanish Heritage* (Wenner-Gren Foundation for Anthropological Research, Viking Fund, Publications in Anthropology, No. 27; New York: 1960).

GLADE, WILLIAM, "Social Backwardness, Social Reform, and Productivity in Latin America," *Inter-American Economic Affairs,* **15**, no. 3 (Winter 1961), 3–32.

GONZALEZ, ALFONSO, "Some Effects of Population Growth on Latin America's Economy," *Journal of Inter-American Studies,* **9**, no. 1 (January 1967), 22–42.

GORDON, WENDELL C., *The Political Economy of Latin America* (New York: Columbia University Press, 1965).

GUERNSEY, LEE, and ALICE TAYLOR, "The Alliance for Progress," *Focus,* **14**, no. 8 (April 1964), 6 pp.

GUNTHER, JOHN, *Inside South America* (New York: Harper & Row, 1967). A book by a noted journalist, written for a mass audience. Mainly about personalities, but with much general information about the various countries.

GUZZARDI, WALTER, Jr., "The Crucial Middle Class" [in Latin America], *Fortune,* **65**, no. 2 (February 1962), 98–100 ff.

HAAR, CHARLES M., "Latin America's Troubled Cities," *Foreign Affairs,* **41**, no. 3 (April 1963), 536–549.

HANSON, EARL PARKER, *South from the Spanish Main* (New York: Delacorte Press, 1967).

HARING, C. H., *The Spanish Empire in America* (New York: Harcourt, Brace & World, 1947, 1963).

HAUSER, PHILIP M., ed., *Urbanization in Latin America* [proceedings of a United Nations seminar at Santiago, Chile, 1959] (New York: Columbia University Press, International Documents Service, 1961).

HERRING, HUBERT, *A History of Latin America* (2d ed.; New York: Alfred A. Knopf, 1961). A well-written, comprehensive history providing excellent background for geographical study of the region.

HOLMES, JERROLD, "The Pan American Highway," *Journal of Geography,* **62**, no. 4 (April 1963), 145–152.

HOY, DON R., and JAMES S. FISHER, "Latin America and the European Common Market," *Geographical Review,* **56**, no. 1 (January 1966), 90–97.

JAMES, PRESTON E., *Latin America* (3d ed.; New York: Odyssey Press, 1959); a standard text on the region. Also, *Introduction to Latin America* (New York: Odyssey Press, 1964); "Recent Developments in Latin America," *Journal of Geography,* **65**, no. 6 (September 1966), 260–265; and three chapters on Latin America in W. Gordon East and A. E. Moodie, eds., *The Changing World: Studies in Political Geography* (New York: Harcourt, Brace & World, 1956), pp. 881–941.

JAVITS, JACOB K., "Urgencies in Latin America: Last Chance for a Common Market," *Foreign Affairs,* **45**, no. 3 (April 1967), 449–462.

JOHNSON, JOHN J., ed., *Continuity and Change in Latin America* (Stanford, Calif.: Stanford University Press, 1964).

"Latin America Tomorrow," *Annals of the American Academy of Political and Social Science,* **360** (July 1965), 1–138. An entire issue, with articles by various authors.

LEONARD, JONATHAN NORTON, *Ancient America* (Great Ages of Man; New York: Time Inc., 1967).

LIBRARY OF CONGRESS, HISPANIC FOUNDATION, *Handbook of Latin American Studies* (annual; Gainesville: University of Florida Press). Primarily an annotated list of recent publications on Latin America.

LOUKOTKA, ČESTMÍR, comp., "Ethno-Linguistic Distribution of South American Indians" (Map Supplement No. 8), *Annals of the Association of American Geographers,* 57, no. 2 (June 1967), 437–438, with large folded multicolored map.

MASELLI, G. S., "Immigration as an Essential Element for the Development of Latin America," *International Migration,* 5, no. 2 (1967), 108–126.

MIRÓ, CARMEN A., "The Population of Latin America," *Demography,* 1, no. 1 (1964), 15–41.

MORSE, RICHARD M., "Some Characteristics of Latin American Urban History," *American Historical Review,* 67, no. 2 (January 1962), 317–338; and "Latin American Cities: Aspects of Function and Structure," *Comparative Studies in Society and History,* 4, no. 4 (July 1962), 473–493, reprinted in John Friedmann and William Alonso, eds., *Regional Development and Planning: A Reader* (Cambridge, Mass.: M.I.T. Press, 1964), pp. 361–381.

NYSTROM, J. WARREN, and NATHAN A. HAVERSTOCK, *The Alliance for Progress: Key to Latin America's Development* (Princeton, N.J.: D. Van Nostrand Co., Searchlight Books, 1966).

PAASSEN, CHR. van, "The General Situation of Latin and Especially Tropical Spanish America at Mid-Century: A Preliminary Introduction to a Macro-Geographical Image," *Tijschrift voor Economische en Sociale Geografie,* 56, no. 5 (September–October 1965), 161–170.

PAN AMERICAN UNION, *Plantation Systems of the New World* (Social Science Monographs, No. 7; Washington, D.C.: 1959).

PARSONS, JAMES J., and WILLIAM M. DENEVAN, "Pre-Columbian Ridged Fields," *Scientific American,* 217, no. 1 (July 1967), 93–100.

PIERSON, W. W., ed., "Pathology of Democracy in Latin America: A Symposium," *American Political Science Review,* 44, no. 1 (March 1950), 100–149.

PIKE, F. B., *The Conflict between Church and State in Latin America* (New York: Alfred A. Knopf, 1964).

PLATT, D. C. M., "British Agricultural Colonization in Latin America, I and II, *Inter-American Economic Affairs,* 18, no. 3 (Winter 1964), 3–38; 19, no. 1 (Summer 1965), 23–42.

PLATT, ROBERT S., *Latin America: Countrysides and United Regions* (New York: McGraw-Hill Book Company, 1942). Individual field studies of farms, ranches, mines, etc., set in a frame of large generalizations; dated in some respects, but filled with valuable insights, and indispensable for serious students of Latin American geography.

POHL, IRMGARD, and JOSEF ZEPP, *Latin America: A Geographical Commentary,* ed. by Kempton E. Webb (London: John Murray, 1966).

ROBINSON, HARRY, *Latin America: A Geographical Survey* (rev. ed.; New York: Frederick A. Praeger, 1967).

SABLE, MARTIN H., *A Guide to Latin American Studies* (Los Angeles: Latin American Center, University of California, 1967).

SAUER, CARL O., "Man in the Ecology of Tropical America," Pacific Science Congress, 9th, Bangkok, Thailand, 1957, *Proceedings,* 20 (1958), 104–110.

SCHURZ, WILLIAM LYTLE, *Latin America: A Descriptive Survey* (rev. ed.; New York: E. P. Dutton & Co., 1963); and *This New World: The Civilization of Latin America* (New York: E. P. Dutton & Co., 1954).

SMITH, T. LYNN, ed., *Agrarian Reform in Latin America* (New York: Alfred A. Knopf, 1965); and *The Process of Rural Development in Latin America* (Gainesville: University of Florida Press, 1967).

The South American Handbook (annual; London: Trade and Travel Publications; distributed in the United States by Rand McNally & Co., Chicago, Ill.). A yearbook and guide to the countries and resources of Latin America.

Statistical Abstract of Latin America (annual; Los Angeles: Latin American Center, University of California).

STEWARD, JULIAN H., and LOUIS C. FARON, *Native Peoples of South America* (New York: McGraw-Hill Book Company, 1959).

TURK, KENNETH L., and LOY V. CROWDER, eds., *Rural Development in Tropical Latin America* (Ithaca, N.Y.: New York State College of Agriculture of the State University at Cornell University, 1967).

UNITED NATIONS, ECONOMIC COMMISSION FOR LATIN AMERICA, "Hydro-electric Resources in Latin America: Their Measurement and Utilization," *Economic Bulletin for Latin America,* 7, no. 1 (February 1962), 73–113; also, *Economic Survey of Latin America* (annual).

U.S. DEPARTMENT OF AGRICULTURE, FOREIGN AGRICULTURAL SERVICE, *Agricultural Geography of Latin America* (Miscellaneous Publication No. 743; Washington, D.C.: Government Printing Office, 1958). An agricultural atlas.

U.S. DEPARTMENT OF STATE, THE GEOGRAPHER, *States and Regions of Latin America* (Geographic Bulletin, No. 7; Washington, D.C.: 1967).

URQUIDI, V. L., *The Challenge of Development in Latin America,* trans. from the Spanish by Marjory M. Urquidi (New York: Frederick A. Praeger, 1964).

WAGLEY, CHARLES, *The Latin American Tradition: Essays on the Unity and the Diversity of Latin American Culture* (New York: Columbia University Press, 1968); and, ed., *Social Science Research on Latin America* (New York: Columbia University Press, 1964). See especially James J. Parsons, "The Contribution of Geography to Latin American Studies," pp. 33–85.

WHITE, C. LANGDON, "Whither South America: Population and Natural Resources," *Journal of Geography,* 60, no. 3 (March 1961), 103–112.

WISH, JOHN R., *Economic Development in Latin America: An Annotated Bibliography* (New York: Frederick A. Praeger, 1965).

WOLF, CHARLES, "The Political Effects of Economic Programs: Some Indications from Latin America," *Economic Development and Cultural Change,* 14, no. 1 (October 1965), 1–20.

WYTHE, GEORGE, "The United States and the Latin-American Economy," *Journal of Inter-American Studies,* 3, no. 4 (October 1961), 451–467.

Latin American

Regions

28

One of many possible ways of subdividing Latin America for purposes of study results in four regions: a major northern realm known as Caribbean or Middle America; a second region comprised of the predominantly native Indian countries of Ecuador, Peru, Bolivia, and Paraguay; a third consisting of giant Brazil; and, fourth, mid-latitude South America—cored by Argentina, with the smaller country of Chile flanking its western border and still smaller Uruguay situated to the northeast. These regional subdivisions will be discussed in the order named. Area and population totals for each regional group of countries, as well as for individual countries, were given in Table 26, Chapter 27.

CARIBBEAN AMERICA

The northernmost of the four major Latin American realms is Caribbean or Middle America. Broadly conceived, it includes (1) the large mainland nations of Mexico,

NATURAL REGIONS, TRANSPORTATION ROUTES, AND MAJOR CITIES
OF
CARIBBEAN AMERICA

MAJOR CITIES

- 1,000,000 or over
- 500,000–1,000,000
- 250,000–500,000
- 100,000–250,000
- Places under 100,000 mentioned in text

City-size symbols are based on metropolitan area estimates

NATURAL REGIONS

MOUNTAIN AREAS
- Tierra Fría
- Tierra Templada

HUMID TROPICAL LOWLANDS
- Rain forest
- Savanna

DRY LANDS
- Semiarid
- Desert

TRINIDAD AND TOBAGO

0   20
Miles

PANAMA CANAL

CANAL ZONE

0   10 Miles

SELECTED TRANSPORTATION ROUTES
OF CARIBBEAN AMERICA

0   200   400   600   800
Scale of Miles

RAILWAYS

Area more than 20 miles from a railway

PAN AMERICAN HIGHWAY SYSTEM

——— in operation

- - - - - under construction or proposed

See separate map of Transportation Routes of South America

LESSER ANTILLES AND PUERTO RICO

WEST INDIES ASSOCIATED STATES

0   50   100   150   200
Scale of Miles

The island of Anguilla in the Leeward Islands, shown above as part of the West Indies Associated States (see page 663) has been a component of the British colony of St. Christopher-Nevis-Anguilla, but it attempted to secede in 1967. The resulting dispute remained unresolved in early 1969.

Colombia, and Venezuela, plus the smaller nation of Guyana (former British Guiana); (2) the small Central American republics of Guatemala, El Salvador, Honduras, Nicaragua, Costa Rica, and Panama; (3) the mainland units of British Honduras, Surinam, French Guiana, and the Panama Canal Zone, all affiliated politically with overseas nations and dependent in varying degree on those nations; and (4) the numerous islands, large and small, in the Caribbean Sea or near it. Of the insular political units, only Cuba (which encompasses the largest island), Haiti and the Dominican Republic (which share the second largest), and Jamaica, Barbados, and Trinidad-Tobago are independent nations.

Variations in Size and Population Density among the Component Political Units

Caribbean America is made up of 16 independent countries and 20 political units (6 of which comprise the West Indies Associated States) that are affiliated with foreign nations (map, p. 654, and Table 26). These units vary greatly in size, shape, and intensity of human settlement and use. The largest unit, Mexico, has an area of 761,600 square miles and a population of 47,200,000 (unless otherwise noted, population figures in this chapter are 1968 estimates). The smallest independent country, Barbados, has only 166 square miles, but has a population of 252,000. The population density of Mexico—approximately 62 people per square mile—is slightly greater than that of all Caribbean America (59 per square mile). The density in Barbados, on the other hand, amounts to 1482 per square mile. Great variations in population density also are found in the units affiliated with overseas nations. Surinam, the largest of these in area, has some 55,144 square miles, a population estimated at 366,000, and a density of only 7 persons per square mile. The tiny colony of the British Virgin Islands, by contrast, includes a total of only 67 square miles and an estimated population of a little over 8000, but has an average population density of about 119 persons per square mile. In appraising these statistics, the student should remember that the people of Caribbean America are predominantly rural. By way of comparison, the average population density of highly industrialized United States was 56 per square mile according to 1968 estimates.

In the remainder of this section, the numerous political units that comprise Caribbean America are grouped for discussion as follows: (1) Mexico, (2) the Central American republics, (3) Colombia, Venezuela, and Guyana, (4) the mainland units affiliated with overseas nations, (5) the independent island states, and (6) the island units affiliated with overseas nations.

Mexico

The following are striking characteristics of Mexico: (1) nearly half of its people live in, around, or between Mexico City (6.8 million)[1] and Guadalajara (1.2 million); (2) almost a third of the total population is native Indian, and most of the rest is *mestizo;* (3) the proportion of the labor force that is employed in agriculture (51 percent in 1960) has been declining for many years and is smaller than in most Latin American political units; and (4) about half of all tilled land grows corn, chiefly for human consumption.

Agriculture

Mexicans are a people who prize agricultural land. The following three major types of agricultural holdings are represented.

1. Properties owned by entire communities, whether allotted to (but not owned by) individual heads of families or, in rare cases, worked communally, are called *ejidos.* Unlike the other holdings, the *ejido* system can be traced to native Indian ways of life. The *ejidos* are most numerous in areas of dense population near Mexico City, but also are found in irrigated areas farther north, as well as in the Yucatán peninsula and in the south.

2. Small private holdings have various names: *solar* (if containing fewer than 2½ acres); *granja* (if involving from 2½ to 12½ acres); and *rancho* (if covering from 12½ to 2500 acres). These are far more numerous than the *ejidos,* amounting to over 1 million holdings as compared to less than 20,000 *ejidos.* However, acre for acre they contain far less cultivated land than do the *ejidos,* though the total amount of cultivated land in these properties within the nation as a whole is somewhat greater than in the *ejidos.* A high proportion of the privately owned land in Mexico is in forest, pasture, or other noncrop use.

---

[1] Unless otherwise noted, city populations in this chapter are metropolitan area estimates for 1968. In most cases the figures are not very accurate, but they do provide approximations that allow rough comparisons among the cities of the region as to relative orders of magnitude. Most figures cited in the chapter are believed to be on the conservative side.

A field worker walks along the edge of a canal through a cotton field which is being irrigated by means of the siphon pipe method in the Yaquí Valley on Mexico's northwestern coast. The field is part of a large irrigation project developed with the assistance of a loan from the World Bank. The project, supplied with water from the Yaquí River, is located about midway up the coast of the Gulf of California. (International Bank for Reconstruction and Development [World Bank].)

3. Individual *haciendas,* the feudalistic, large-scale private holdings, contain over 2500 acres and in some cases over 100,000 acres. They are scattered throughout the country, but are found in largest numbers in the semiarid northeast. Before the 1910 revolution *haciendas* contained nearly all of Mexico's exploited land and as late as 1940 encompassed possibly 60 percent. Their share of the total exploited land is still appreciable, but they contain only a very small portion of all cultivated land. The dominance of the *hacienda* in Mexican life is over.

Shortly before World War I, Mexico instituted a state-enforced agricultural reform program which is still continuing. Its major objective has been the breaking up of *haciendas* into small units. The estates usually have been left with no fewer than 250 acres of land, including the central buildings. Most farm families that have been granted access to land through this movement are members of *ejidos.* The *ejido*[2] is a twentieth-century institution with origins in the pre-Hispanic communal lands of Indian Mexico and in the medieval Iberian commons outside towns. With the cry of "Land and Liberty" the *ejido* became an element of the Mexican landscape (codified in 1915, and as Article 27 of the 1917 Constitution).

Compensation was paid to estate owners where title was established, but redistribution of land expropriated without compensation was the important mechanism of reform. Over 150 million acres were given to 2.5 million *ejidatarios* (heads of families)—some 15 million people in toto—from 1916 to 1966. The progress of agrarian reform is dependent upon the attitude of the President. Under Cardenas (1934–1940) nearly 50 million acres were distributed—twice as much as was given in the 18 years before and after his regime. The best lands were titled in the 1930s; those distributed since, including the 25 percent by López Mateos (1958–1963), have been of lower quality.

Two styles of land tenure and labor operate in the *ejido:* (1) communal lands divided among and husbanded by individuals, owning the fruits of their labor (see page 636), and (2) communal lands held and worked as a collective unit, with the members doing specialized jobs and sharing profits. The collectives, concentrated in the arid north and tropical south, amount to only 5 percent of all *ejidos.* Devoted to commercial agriculture, they have consumed large amounts of capital investment and support by the Federal government—for which they have been severely criticized. Noncommunal *ejidos,* too, are under attack by economists for being less productive than private properties. Attempts to turn *ejidos* into efficient producing units are hindered by (1) the low level of technology, (2) peasant unwillingness to accept innovation owing to limited horizons and poor education, (3) soil erosion and declining fertility, (4) faulty organization of credit, transportation, and marketing, and (5) the small size of plots—though from 1950 to 1960 the average increased from 20 to 22 acres. Considering the fact that Mexico has one of the highest rates of annual population increase in the world (3.5 percent), success in overcoming the foregoing obstacles seems unlikely. Certainly a massive nationwide educational, technological, and financial effort would be required. Perhaps

[2] The authors are grateful to Dr. Herbert M. Eder, Department of Geography, University of California, Berkeley, for preparing the material on the *ejido* for this and the following paragraph.

fulfillment of the promise "land for the landless," providing incentive and security for many rural Mexicans, should be the main criterion for evaluating the *ejido.* If so, land reform in Mexico is a qualified success.

In addition to corn, the primary subsistence crops include beans, wheat, rice, and barley. The 5 crops occupy over three-fourths of Mexico's cropland–most of it in the *tierra templada* and lower *tierra fria* climates of the heavily populated central section. Yields per acre are low, although they have improved somewhat in recent years. Corn, for example, averages less than 20 bushels per acre as compared with an average of 75 to 80 bushels per acre for the United States. Mexico is one of the larger Latin American producers of tobacco, citrus fruits, vegetables, cane sugar, oilseeds, and meat (primarily beef). Most of this production is utilized within the country, although exports of a few commodities are fairly sizable.

Cotton, grown principally in the dry north under irrigation (photo, left), is of increasing importance to both home and foreign markets. It is Mexico's leading export. The country's second agricultural export is coffee, grown in scattered districts throughout the southern half of Mexico, but primarily to the east of Mexico City. Mexico is Latin America's leading cotton exporter and vies with Guatemala for fourth place in coffee exports (after Brazil, Colombia, and El Salvador).

Manufacturing

Measured by labor force, manufacturing is far less significant than agriculture, accounting for only about 14 percent of the nation's total employment. Many of the workers are in handicraft industries. However, the total output of manufactured goods has been growing rapidly, with emphasis both on consumer goods and on production goods such as iron and steel, chemicals and chemical fertilizers, machinery, and cement. Manufacturing now exceeds agriculture in value of product. The principal modern manufacturing plants are found in two general areas: (1) the central region, in which labor-oriented and market-oriented industries (for example, cotton, woolen, and rayon textile plants; tobacco-processing plants) predominate, and (2) the dry lands of the north, in which the chief industries are oriented to raw materials (for example, concentrating and smelting mills for production of iron and steel, lead, zinc, copper, silver, and other metals). Except for iron and steel production, these latter

industries are often more a branch of mining than of manufacturing, as many of them simply enrich the ores sufficiently that they can be shipped elsewhere for further manufacture.

Of major importance to Mexico is the northeastern iron and steel industry centered at Monclova (city proper, 50,000) and at the regional metropolis of Monterrey (950,000). Most of the iron ore and coal are brought from fields that lie within a radius of 300 miles from Monterrey. This northeastern region–which also includes a plant manufacturing steel from scrap iron at Piedras Negras (city proper, 50,000) on the Rio Grande–produces the great bulk of Mexico's iron and steel output. Some steel is manufactured in small plants located in the Mexico City area. The national production is insufficient for the country's needs, and steel in various forms is one of Mexico's larger imports.

Mining

Less than 2 percent of Mexico's labor force is engaged in mining. This figure does not convey accurately the importance of the mining industry, for many of the mines are highly mechanized and thus employ comparatively few workers. Mined products comprise an important part of Mexico's exports. The leading commodities, of which most are exported in semifinished form, are sulfur (Frasch process),[3] fluorite, barites, zinc (concentrates and refined metal), lead (refined), copper (refined metal and concentrates), petroleum (largely refined), and silver (largely refined). Mexico's silver production, though of comparatively minor importance to the country's total economy, is sufficient to rank Mexico with the United States as one of the two leading producers in the world. The mining industry of Mexico is financed largely by investment from foreign countries. However, since 1961 only companies with majority Mexican ownership have been permitted new mining concessions.

Minor Livelihood Industries

Among minor livelihood industries are commercial fishing (along both coasts), forest industries, and tourism. Mexico began to develop fishing actively during and after World War II. Shrimp, anchovies, tuna, and a variety of mixed and unassorted types of fish are taken. The leading fishing grounds are on the west coast, especially in the

---

[3] In the Frasch process, underground deposits of practically pure sulfur, often deeply buried, are melted and forced through pipes to the surface, where the molten sulfur flows into vats and solidifies.

TABLE 28 DOMINANT EXPORT COMMODITIES OF THE INDEPENDENT COUNTRIES OF CARIBBEAN AMERICA

COUNTRY	COMMODITY AND PERCENT OF TOTAL EXPORTS BY VALUE, 1965
Mexico[a]	Cotton, 19; coffee, 7; maize, 7; sugar and molasses, 6
Guatemala	Coffee, 49; cotton, 18
El Salvador	Coffee, 51; cotton, 20
Honduras	Bananas, 42; coffee, 18
Nicaragua	Cotton, 44; coffee, 18
Costa Rica	Coffee, 42; bananas, 25
Panama	Bananas, 51; petroleum products, 30
Colombia	Coffee, 64; petroleum, 18
Venezuela	Petroleum and products, 93; iron ore, 5
Guyana	Bauxite and concentrates, 41; sugar and molasses, 28; rice, 14
Cuba	Sugar and molasses, 86; tobacco, 5
Haiti	Coffee, 62; sisal, 18
Dominican Republic	Sugar and molasses, 49; coffee, 16; bauxite and concentrates, 10
Jamaica	Bauxite and alumina, 47; sugar and molasses, 22; bananas, 8
Barbados	Sugar and molasses, 80; rum, 6
Trinidad and Tobago	Petroleum and products, 85; sugar, 6

[a] The combined total of lead, zinc, copper, and silver exports (ores, concentrates, and metal) accounted for 10 percent of all exports for 1965.

SOURCE: *United Nations Yearbook of International Trade Statistics, 1965.*

Gulf of California and the Pacific Ocean flanking the peninsula of Lower California. Other grounds are east of Veracruz (city proper, 190,000) along the Caribbean shore and on the southern shores of the Pacific near the Gulf of Tehuantepec. Commercial forest exploitation occurs in the mountainous north and the low-latitude south. A wide variety of tree types are cut, ranging from pine to mahogany.

The general economic status of Mexico is not unlike that of most other Latin American political units, particularly those of the Caribbean area. In most parts of Caribbean America the bulk of the people support themselves by subsistence or semisubsistence farming. A few commercial export crops or mined products are grown or extracted in scattered localities, mainly with the aid of foreign capital. In most countries one or two primary commodities dominate the export lists, as shown in Table 28.

The Central American Republics

The six small republics of Central America have economies and resources that are far less diversified than those of Mexico. Mineral wealth is very lacking, manufacturing is poorly developed, and exports are restricted in rather large measure to three agricultural commodi-

ties: coffee, bananas, and cotton (Table 28). Coffee is grown on both small and large holdings, most of which are owned by citizens of the country involved. Banana plantations, found principally along the hot, moist, coastal lowlands, are in large measure under the control of the United Fruit Company, an American concern. Because careful management and precision timing are needed to market the ripening bananas (which are picked while still green), the company's own employees manage both the plantations and a fleet of specially built ships plying between Caribbean and Pacific ports and wholesale markets, chiefly in coastal cities of Anglo-America. The main plantation districts are found in the Caribbean lowlands of Honduras and the Pacific lowlands of Panama, Costa Rica, and Guatemala. There is a widespread subsistence component in Central American agriculture, a characteristic that is most evident in the native Indian settlements of northern Guatemala and adjacent areas and least perceptible in Costa Rica, where a society of small farmers—relatively unmixed descendants of Spaniards—works most of the land.

These countries have a land problem even more serious than that of Mexico. Their average acreage of cultivated land per farm worker is less than half that of their northern neighbor—and, for comparison, less than one-twentieth that of the United States. Moreover, they are the masters of their own destiny to a much more limited degree than Mexico. Catering to tastes in the world market, they manage as best they can by attracting foreign capital when possible and selling such tropically grown products of farm and forest as will be purchased by peoples with standards of living higher than their own.

In every republic of Central America, the largest city is the national capital. The two largest are Guatemala City (650,000), and San Salvador (450,000), the capital of El Salvador.

Colombia, Venezuela, and Guyana

Colombia and Venezuela

The high landforms which dominate Mexico and Central America extend southward, with interruptions, through Colombia. An outlier reaches eastward in Venezuela to the offshore island of Trinidad. Population nuclei of these South American nations are distributed in the same erratic manner as are those of the northern mainland republics, with a preference for highlands and uplands very much in evidence.

In the two countries the basic geographic ingredients do not differ so much in nature as in distribution. Both have mountains, but Colombia's areas of high mountains are much the larger; both have populations that are predominantly *mestizo*, but that of Colombia is over twice as large and the ethnic divisions within the population are more pronounced; both contain a wide range of climates, but Colombia has more of the temperate or cool upland and highland types; in both countries agriculture is the leading source of employment, but a considerably larger proportion of Colombia's labor force is employed in agriculture than is the case in Venezuela. Colombia has more cultivated land, but also has a larger population, so that ratios between man and cultivated land are about the same for the two countries. Both countries have significant mineral resources, but Venezuela has most of the known reserves and output of the two outstanding minerals—petroleum and iron ore.

Of Colombia's exports, about 59 percent by value in 1967 was represented by coffee and 13 percent by petroleum. The coffee is grown principally along the mountainous middle sections of the Magdalena and Cauca river valleys. It is shipped northward by rail, truck, river, air, or a combination of these, to the Caribbean ports of Barranquilla (550,000) and Cartagena (250,000), or westward by truck to Colombia's main Pacific seaport, Buenaventura (city proper, 110,000) (map, p. 664).

Petroleum comes from a series of widely separated fields to the southwest of Venezuela's fields in or around Lake Maracaibo. Most petroleum leaves the country via two small Caribbean ports to the south of Cartagena.

The majority of Colombia's people, like their counterparts in Mexico and Central America, live in the highlands. Of the country's four cities with metropolitan populations of over 500,000, three are in the *tierra templada* or the *tierra fría* of the mountains. These are Bogotá (1.8 million), Medellín (1.1 million), and Cali (650,000). Colombia has twelve other cities or metropolitan areas with more than 100,000 people; of these eight are in the highlands. Each highland city serves as a market center for an agricultural basin or series of basins. One of these cities, Medellín, is the country's leading textile manufacturing center and one of the main centers of the textile industry in all of Latin America.

In broad outline, Venezuela is comprised of (1) a large, sparsely populated or unoccupied hinterland of tropical savanna climate in the basin of the Orinoco River, (2) an irregularly settled east-west trending mountain cordillera, largely *tierra templada,* in the northern and northwestern parts of the country, and (3) a discontinuous coastal lowland, which centers around petroleum-rich Lake Maracaibo in the northwest and around the Orinoco River delta in the northeast.

Except for the mining of iron ore by United States corporations at Cerro Bolívar and El Pao in the Guiana Highlands—operations supplying tidewater iron and steel plants in the United States—the savanna country is essentially the domain of the livestock ranch. Ranching, however, is largely confined to the half of the country that lies north and west of the Orinoco. Areas to the south and east of the river, largely in the Guiana Highlands, are mostly unoccupied except for small numbers of primitive and little-known Indians. In contrast, the *tierra templada* of northern Venezuela contains four of the country's seven cities with metropolitan populations of 100,000 or over—Caracas, the capital (2.1 million), Barquisimeto (280,000), Maracay (170,000), and San Cristóbal (120,000). (The remaining cities are Maracaibo [550,000] and Cabimas [115,000] on the northern coast and Valencia [210,000] in an intermontane basin.) It also contains the most important farming districts. Coffee from the *tierra templada* is Venezuela's principal export crop. However, corn, grown in all of the country's agricultural areas, occupies much more acreage than any other crop. It is primarily a subsistence crop.

Petroleum, amounting to over 10 percent of the estimated world output in 1966, constituted 92 percent of Venezuela's exports by value in that year. It is extracted in Venezuela by American and British firms, but refined, for the most part, on the islands of Aruba and Curaçao in the Netherlands Antilles or at overseas refineries. Only about one-third is refined in Venezuela itself. Most of the country's oil is procured in the northwestern fields centering on Lake Maracaibo (photo, p. 643), but petroleum extraction is also important in a rather large area between the Orinoco River and the northern mountains. By far the greater part of Venezuela's petroleum exports move to the United States or Europe.

In recent years Venezuela has experienced a rapid development of manufacturing, as Colombia has at a somewhat slower rate. Consumer-type industries are greatly in the majority in both countries, though Venezuela has an iron and steel plant (at Puerto Ordaz) on the Orinoco River in the port area which serves the iron mines of the Guiana Highlands. This plant is planned to become the center of an industrial complex. An aluminum reduction plant and a hydroelectric network in the area are now being completed and expanded. Venezuela's industries also include an impressive cluster of chemical plants on the coastal plain about 100 miles west of Caracas.

Guyana[4]

Guyana can be divided roughly into three regions: a coastal strip about 30 miles wide where the principal agricultural activity is found; a strip approximately 100 miles wide of undulating land, mainly forested and containing the mineral wealth of the country; and, farther inland, mountain ranges and savannas. Nearly 90 percent of the country is forested, although about only one-fifth of these forests are considered accessible on a commercial basis at the present time. In the coastal region there are still large tracts of unimproved land which would be suitable for agriculture or cattle raising.

The principal crops are sugar cane and rice, with most of the former and about one-third of the latter being exported.

Bauxite is the leading industrial product, accounting for about 7 percent of the total world production in 1966. Some of the bauxite is processed into alumina before export, and the rest is shipped abroad (mainly to the United States and Canada) in crude form. Small amounts of manganese and uncut diamonds are also produced and exported.

Guyana, formerly a British colony, gained independence in 1966 as a parliamentary state within the Commonwealth of Nations. Georgetown (150,000) is the capital and main city.

Mainland Units Affiliated with Overseas Nations

French Guiana, Surinam or Netherlands Guiana, British Honduras, and the Panama Canal Zone are the mainland units of Caribbean America that are affiliated with overseas nations. The British unit is a colony; French Guiana is an overseas department of France; and Surinam is officially an integral part of the Kingdom of the Netherlands. The Panama Canal Zone, a United States government reservation, is essentially a district 10 miles wide trending for some 40 miles northwest-southeast in conformance with the Panama Canal. Over one-half of its population of around 48,000 (including armed forces) lives in urban centers at either end of, or along, the canal. The Canal Zone enjoys substantial economic support from the United States.

The two Guiana units and British Honduras have much in common. They have been among the farthest outposts of European colonialism in Latin America. Their physical environment is chiefly that of tropical rainy, disease-ridden lowlands. On the whole they have received comparatively little attention from the European colonial powers which have controlled them. An important exception has been the development of large-scale bauxite mining in Surinam. In 1966 it supplied 12 percent of the total world production of this aluminum-bearing ore. Employment opportunities are relatively few, however, as most of the work is done with machines.

Only a tiny fraction of the total land area of the two Guianas has been cleared for cultivation, though agriculture supports many of the people. Plantations, clustered along the coast, grow irrigated rice and some sugar cane. The majority of these are in Surinam. (A substantial number of sugar cane plantations also are found in Guyana, as mentioned earlier.) Some farming of a subsistence nature is carried on, with rice and yams as the principal crops.

Both Surinam and French Guiana can be considered as largely undeveloped, with "pockets" of advanced technology. These "pockets" mainly involve coastal plantations or mines. Some 80 percent of the exports

[4] Although Guyana, like Surinam, French Guiana, the Bahama Islands, and El Salvador, does not touch the Caribbean Sea, it lies near by and like the other units listed it can be discussed without strain in the general context of Caribbean America.

from Surinam involve bauxite. Fishing and forestry, largely under native technology, also have considerable importance. Shrimps accounted for about two-thirds of the exports from French Guiana in 1966, and timber for 15 percent. There is some evidence of new activity. Hydroelectric energy is being developed, especially to refine bauxite now mined in Surinam and untapped reserves of bauxite in French Guiana.

The economy of British Honduras bears some resemblance to that of the Guianas. Until recently, timber was a leading export. Now sugar (28 percent in 1965), citrus fruit and products (22 percent), and timber (12 percent) are the main export items.

It may be noted that the new independent state of Guyana, previously discussed, shares many of the general characteristics of the colonial units described here.

The Islands

The islands of Caribbean America are an extremely varied group. A wide range of physical types and sizes is represented, racial and cultural variations are great, and there is notable variety from island to island in political arrangements and in economic mainstays.

Cuba, the largest island, is comprised primarily of lowlands with low to moderate relief, the principal exception being a mountainous area that occupies the southern end of the island. The islands that rank next in size—Hispaniola (Haiti), Jamaica, and Puerto Rico—are very mountainous or hilly. The island of Trinidad, a detached fragment of the South American continent, has a low mountain range in the north (a continuation of the Andes) and level to hilly areas elsewhere. The small islands of Aruba, Curaçao, and Bonaire, affiliated with the Netherlands, also lie near the mainland. They have low to moderate elevations and slopes. Most of the remaining islands of Caribbean America, all comparatively small, fall into two broad physical types. (1) Low, flat, limestone islands rimmed by coral reefs include the Bahamas, northeast of Cuba (see note 4, p. 660), and a few others. (2) Most islands in the Virgin Island, Leeward, and Windward groups, stretching along the eastern margin of the Caribbean from Puerto Rico toward Trinidad (map, p. 654), are of volcanic origin. Such islands consist of one or more volcanic cones, most of which are extinct or inactive, with limited amounts of cultivable land on lower slopes and small plains. Some cones are thousands of feet high, while others have been greatly worn down by erosive processes. Intermixed with the volcanic islands are a small number of low limestone islands. Barbados, located east of the Windward Islands, is a limestone island with moderately elevated, rolling surfaces.

All of the islands have warm temperatures throughout the year, though extremely hot weather is uncommon. The difference between the average temperatures of the warmest and coolest months ranges from about 4°F in Barbados to about 10°F in northern Cuba. Precipitation, however, varies notably from island to island, and from one section of mountainous islands to another. Windward slopes of mountains often receive very heavy precipitation, amounting in some instances to more than 200 inches annually, whereas leeward slopes and very low islands may have rainfall so scanty as to create semiarid conditions. In some areas a moisture deficiency is created or abetted by porous limestone bedrock into which precipitation rapidly disappears. The natural vegetation of the islands varies from luxuriant forests in areas of abundant moisture to a sparse, scrubby woodland in very dry areas. Most areas experience two rainy seasons and two dry seasons a year. Hurricanes, approaching from the east and then curving northward along rather well-defined tracks, are a scourge of the northern islands in the late summer and autumn.

The majority—in most islands a very large majority—of the inhabitants are Negroes or mulattoes. People of relatively unmixed European descent comprise most of the remainder, aside from a considerable Asian element (mainly Hindu or Moslem Indians) in Trinidad. The cultural heritage of the different islands varies greatly, and includes Spanish, British, French, Dutch, American, Danish, African, and (in Trinidad) Hindu and Moslem influences in practically endless combinations. Political arrangements also are diversified. Only six units—the republics of Cuba, Haiti, and the Dominican Republic, and the independent parliamentary states of Jamaica, Barbados, and Trinidad and Tobago—are sovereign nations; the remaining units exhibit varying degrees of dependence on overseas nations and are tied to those nations under a variety of governmental forms (Table 26 in Chapter 27). For the islands as a whole and for most individual islands, agriculture is by far the leading source of livelihood. A large share of the production is for subsistence or for sale in local markets. Corn, manioc, yams, rice, and bananas are among the prominent subsistence crops. But single-crop plantation agriculture is the economic activity for which these islands are best known, and it is still the commercial mainstay of most islands. Generally a particular island will specialize heavily in one crop—sugar cane being favored in the majority of cases, as it has been for centuries. Other commercial crops that are prominent on some islands include coffee, sea-island cotton, bananas (particularly in Jamaica), tobacco (particularly in Cuba), cacao, spices, limes and other citrus fruits, and coconuts. Though most of the value of commercial production for the islands as a whole has up to the present been accounted for by large plantations or estates worked by tenant farmers or hired laborers, some crops—such as the coffee of Haiti—are produced primarily by small land-owning farmers. On most islands mineral production is absent or unimportant, the most conspicuous exceptions being Jamaica (bauxite), Trinidad (petroleum), and Cuba (metals). Manufacturing has made very large gains in Puerto Rico in recent years and significant gains in Cuba, but in most of the other islands industrial progress has been slow or absent. Tourism is of growing significance in the island area as a whole, though American tourism ceased in Cuba following the revolution which brought the government of Fidel Castro to power in 1959. Most tourists in the islands are from the United States.

The Island Republics

The largest political units in the Caribbean islands are the island republics of Cuba, the Dominican Republic, and Haiti. In area they fall into a general class with the independent countries of Central America. Population densities, however, are generally much higher than on the mainland, except in the case of densely-populated El Salvador.

Cuba's has been essentially a sugar economy. Prior to the Castro revolution in 1959, approximately one-fifth of the world's sugar was produced here each year, mostly for export. The immediate aftermath of that revolution was marked by a period of experimentation in many aspects of Cuba's economy, but particularly in its mainstay, agriculture. By 1963, the output of sugar was less than one-half of the 1959 figure. Land that had formerly grown sugar was converted to the growing of varied subsistence and specialty crops, including rice, tomatoes, coffee, tobacco, potatoes, and still others. Such a drastic conversion is not made easily, however, and crop yields in many instances were not satisfactory. Meanwhile, the world market price of sugar rose rapidly and Cuban planners decided to reinvest in sugar. Although sugar prices subsequently fell, Cuba has continued to emphasize sugar, and also has continued growing in lesser amounts the diversified crops mentioned above. Today the island supplies about one-fourth of the world's export sugar, sending it mainly to the Soviet Union, Communist China, and other Communist countries. Cane sugar and products accounted for 88 percent of all Cuban exports in 1964.

Meanwhile, major changes have occurred in the ownership of Cuban land. The First Agrarian Reform, initiated in 1960, resulted in ownership of 40 percent of all agricultural land by the state. The Second Agrarian Reform, begun in 1963, has emphasized heavily the trend toward state ownership, and more than three-fourths of all harvested land is now publicly owned. Interestingly, there are very few collective farms on the Soviet model in Cuba. The change of ownership has been almost entirely from private to state-owned land.

Tobacco and metalliferous ores (primarily nickel and copper ores, with some manganese, chromite, and cobalt ore) account for most of Cuba's exports other than sugar and molasses. All mineral resources were nationalized in 1960. Mining is largely confined to two mountainous areas, one in the far south of the island and the other in the extreme northwest. Nickel comes primarily from the first area and most copper from the second. The island has experienced a considerable development of manufacturing industries in recent years. Prominent among these are industries that process food and tobacco and that manufacture cotton and rayon textiles, cement, chemical fertilizers, and automobile tires. Havana (Habana; 1.8 million) is Cuba's main industrial center, as well as its political capital, largest city, and main seaport.

Haiti and the Dominican Republic share uneasily the island of Hispaniola or Haiti. Although the diversity of natural features on this mountainous island is not inconsequential, that of culture and society is even more pronounced. Haiti is a Negro republic in which the rapidly growing population has already attained an average density that is exceeded among the independent Latin American countries only by Barbados and by Trinidad and Tobago. Its present-day agricultural economy, largely of a subsistence character, reflects little influence from the French who once controlled the country. The Dominican Republic, though containing some Negroes and a large majority with some Negro blood, is proud of its Spanish cultural heritage. Although the Dominican Republic has nearly twice the area of Haiti, its small farm-plantation economy supports a population which is somewhat smaller than that of its western neighbor.

The boundary line between these two countries is one of long-standing instability, and local border incidents have been rather frequent.

The largest cities of Hispaniola are the two national capitals, Santo Domingo (550,000) in the Dominican Republic and Port-au-Prince (350,000), in Haiti.

Jamaica, Trinidad-Tobago, and Barbados

In the 1950s an attempt was made to form a self-governing federation of numerous West Indian islands affiliated politically with the United Kingdom. The federation, known officially as The West Indies, commenced to function in 1958 and was expected to achieve full independence within the Commonwealth of Nations. But its career proved to be short and it collapsed in the early 1960s when its three largest components—Jamaica, Trinidad and Tobago, and Barbados—withdrew. They were granted independence as separate parliamentary states—Jamaica and Trinidad-Tobago in 1962 and Barbados in 1966. Most of the remaining island units in the original federation continue to have a loose political relationship as the West Indies Associated States. Neither the association as a whole nor any of its component units seem destined for early independence, though most of the separate British colonies that comprise the association enjoy a large measure of internal self-government and could probably obtain independence soon if they desired. The economies of these small units, dependent on sugar and a few other export crops, are so fragile as to cast doubt on the desirability of full independence either as single units or as a federation.

The three units that have gained independence have better prospects, though even these islands exhibit disparities, with Jamaica being the strongest economically, Barbados the least strong, and Trinidad-Tobago intermediate. The differences reflect very markedly the presence or absence of important mineral resources, and attractions and facilities for the tourist trade. Minerals and tourism are the most important sources of overseas revenue for Jamaica and Trinidad-Tobago, whereas Barbados has no mineral exports, only a modest development of tourism, and is almost entirely dependent on exports of cane sugar and molasses.

Of the three island units, Jamaica has the largest total and per capita exports of minerals and agricultural products and the greatest development of tourism. Since 1952, when production began, Jamaica has become the world's largest producer and exporter of bauxite. Huge reserves of this aluminum-bearing ore have resulted from the weathering of limestones in central Jamaica. Localized in numerous solution hollows, the deposits are covered by a relatively thin overburden of earth and thus are accessible for large-scale mining with power shovels. The United States and Canada are the principal markets for the ore. The greater part is shipped in crude form, but sizable amounts are processed into alumina before shipment. Royalty and tax payments to the Jamaican government by the huge American and Canadian aluminum corporations that mine, ship, and process the bauxite have made possible a wide range of measures to advance the island's economy and raise the standard of living.

The island of Trinidad has oil deposits that provide a major source of revenue, though the wells are far less productive than those of neighboring Venezuela, and the known reserves are only a tiny fraction of Venezuela's huge deposits. Trinidad, however, increases the revenues from its oil industry by refining imported oil (mainly from Venezuela or Colombia) as well as much of its own domestic crude. Most of the island's exports consist of crude oil or refined products from the oil industry. A well-known Trinidad export associated with oil is asphalt. Much of this is a by-product of oil refining, but some of it is mined from Trinidad's famous Pitch Lake—the world's main source of natural asphalt.

Jamaica, with its distinctive cultural and historical attractions, its pleasant climate, scenic landscapes of mountain, beach, and sea, and extensive development of resorts, has long been an important focus of the West Indian tourist trade. This aspect of its economy has benefited from the decline of tourism in Cuba during the Castro era. Trinidad's tourist industry, though on a smaller scale, is growing rapidly. It has been assisted by government subsidies to resort hotels, and it benefits from large numbers of stopover tourists who pass through the island's busy airport. Barbados has less natural or cultural variety and fewer tourist facilities than Jamaica or Trinidad, but its tourist trade is gradually increasing nonetheless.

Agricultural exports from the three countries are sold primarily in the sheltered market provided by Great Britain. Preferential tariff rates and assured quotas in the British market are vitally important to Barbados, with its overwhelming dependence on cane sugar, but they are also important to Jamaica and, to a lesser degree, Trinidad and Tobago. Sugar cane is the leading export crop in all three countries. In Jamaica, bananas are also of consequence in the export trade, as are cacao and coconuts in Trinidad-Tobago. Estates, more efficient and productive than the farms of peasant smallholders, completely dominate the agriculture of Barbados, and they produce well over half of the farm exports from Jamaica and Trinidad-Tobago. Small-farm output is not static, however, though in rougher areas it often is handicapped by severe soil erosion. In Jamaica during recent years there have been notable increases in diversified food crops produced on small farms for the domestic market.

The only cities of much size in the three countries are the capitals: Kingston (450,000) in Jamaica, Port of Spain (300,000) in Trinidad, and Bridgetown (120,000) in Barbados.

Islands Affiliated with Overseas Nations

The statement that Caribbean America is not the master of its own destiny is most applicable to the islands

NATURAL REGIONS AND
MAJOR CITIES OF
SOUTH AMERICA

NATURAL REGIONS

MOUNTAIN AREAS
Tierra Fría
Tierra Templada

HUMID TROPICAL LOWLANDS
Rain forest
Savanna

HUMID SUBTROPICAL LOWLANDS
Prairie
Forest

MID-LATITUDE WEST COAST
MOUNTAINS AND VALLEYS
Mediterranean subtropical
Marine west coast

DRY LANDS
Semiarid
Desert

MAJOR CITIES

◉ 1,000,000 or over
● 500,000–1,000,000
◎ 250,000–500,000
○ 100,000–250,000

City-size symbols are based on
metropolitan-area estimates

PACIFIC OCEAN

ATLANTIC OCEAN

ATLANTIC OCEAN

PANAMA
San José
COSTA RICA
Panama City

Barranquilla
Santa Marta
Cartagena
Maracaibo
L. de Maracaibo
Montería
Cúcuta
Bucaramanga
Medellín
Manizales
Buenaventura
Cali
Armenia
Bogotá
Pasto
Quito
ECUADOR
Guayaquil

Caracas
Valencia
Maracay
Barquisimeto
VENEZUELA
Orinoco River
ORINOCO DELTA
Port of Spain
TRINIDAD AND TOBAGO

COLOMBIA

Georgetown
GUYANA
Paramaribo
SURINAM (NETH.)
FRENCH GUIANA (FR.)

Equator

Amazon River
Manaus
Belém

PERU
Lima
Callao

BRAZIL

São Luís
Fortaleza
Teresina
Cape São Roque
Natal
João Pessoa
Campina Grande
Recife
Maceió

Arequipa
La Paz
Lake Titicaca
Cochabamba

BOLIVIA
Goiânia
Brasília
São Francisco R.

Aracaju
Salvador

Antofagasta
Tropic of Capricorn
Salta
PARAGUAY
Paraguay River
Asunción

Belo Horizonte
Vitória
Ribeirão Prêto
Paraná R.
São Paulo
Campinas
Sorocaba
Volta Redonda
Juiz de Fora
Petrópolis
Rio de Janeiro
See detailed map, below
Santos
Curitiba

Tucumán

CHILE

San Juan
Córdoba
Santa Fe
Paraná
Rosario
Mendoza
Uruguay R.
Pôrto Alegre

Valparaíso
Santiago

ARGENTINA
URUGUAY
Buenos Aires
La Plata
See detailed map, below
Montevideo
Río de la Plata

Concepción

Bahía Blanca
Mar del Plata

FALKLAND ISLANDS
(U.K.)

GREATER RIO DE JANEIRO

Nova Iguaçu
Duque de Caxias
São João de Meriti
CITY OF RIO DE JANEIRO
Niterói
Guanabara Bay

0   10   20 Miles

GREATER BUENOS AIRES

San Isidro
Vicente López
San Martín
Morón
San Justo
Lanús
Lomas de Zamora
BUENOS AIRES CITY
Avellaneda
Quilmes
La Plata
Río de la Plata

0   10   20 Miles

0   200   400   600   800
Scale of Miles

10°N   80°W   60°W   40°W
Equator
10°S
20°S
30°S
40°S
50°S

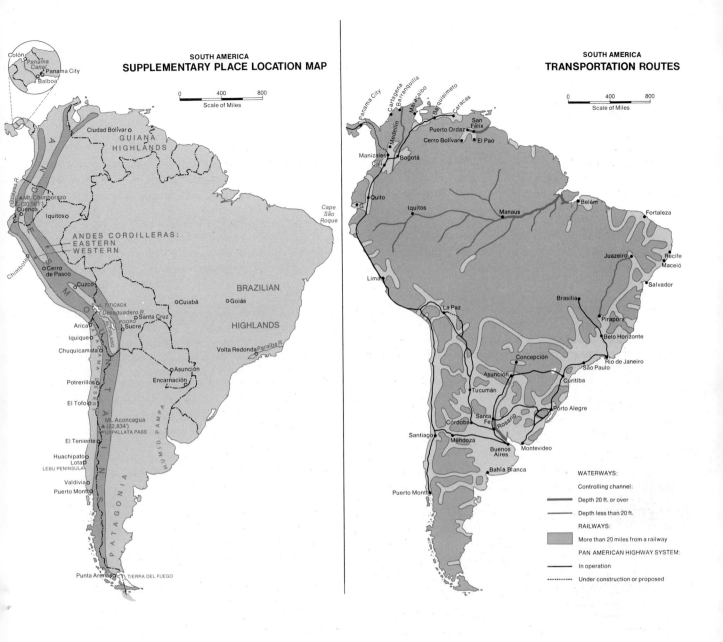

SOUTH AMERICA
SUPPLEMENTARY PLACE LOCATION MAP

0   400   800
Scale of Miles

Colón
Panama Canal
Panama City
Balboa

Ciudad Bolívar ○

GUIANA HIGHLANDS

Guayas R.
▲Mt. Chimborazo (20,561')
Cuenca ○
Iquitoso ○

ANDES CORDILLERAS:
EASTERN
WESTERN

BRAZILIAN
HIGHLANDS

Cape São Roque

Chimbote ○
Cerro de Pasco ○
Cuzco ○

L. TITICACA
Desaguadero R.
L. POOPÓ
Santa Cruz
Arica ○
Sucre
Iquique ○
Chuquicamata ○

○ Cuiabá
○ Goiás

Potrerillos ○

El Tofo ○

Volta Redonda   Paraíba R.

Asunción ○
Encarnación ○

HUMID PAMPA

Mt. Aconcagua (22,834')
USPALLATA PASS
El Teniente ○
Huachipato ○
Lota ○
LEBU PENINSULA
Valdivia ○
Puerto Montt ○

PATAGONIA

Punta Arenas ○   TIERRA DEL FUEGO

SOUTH AMERICA
TRANSPORTATION ROUTES

0   400   800
Scale of Miles

Panama City
Cartagena
Barranquilla
Maracaibo
Barquisimeto
Caracas
San Félix
Medellín
Puerto Ordaz
Cerro Bolívar ● El Pao
Manizales ○
Cali ○
Bogotá ○

Quito ○
G ○
Iquitos ○
Manaus
Belém
Fortaleza

Lima ○

Juazeiro
Recife
Maceió
Salvador

La Paz ○
Brasília
Pirapora
Belo Horizonte

Concepción
Asunción ○
Tucumán
Rio de Janeiro
São Paulo
Curitiba

Córdoba ○
Santa Fe
Rosário
Pôrto Alegre
Santiago ○
Mendoza ○
Buenos Aires
Montevideo

Bahía Blanca ○

Puerto Montt ○

WATERWAYS:
Controlling channel:
Depth 20 ft. or over
Depth less than 20 ft.
RAILWAYS:
More than 20 miles from a railway
PAN AMERICAN HIGHWAY SYSTEM:
In operation
Under construction or proposed

665

The map of natural regions and major cities (left) shows all of the cities in South America that had estimated metropolitan populations of 100,000 or over in 1968. Note the clearly defined arrangement of cities around the rim of the continent and the enormous expanses in the interior with no city over 100,000. The supplementary place location map (above) shows additional cities and other features that are mentioned in the text but in most cases are not shown on the main map.

that are affiliated politically with overseas nations. In these units, severe population pressures and poverty are almost omnipresent. One unit in which conditions are better is Puerto Rico. A determined effort on the part of Puerto Ricans to raise their standard of living has brought great changes to the island in recent decades. Hydroelectric power resources have been harnessed, a land classification program has provided a basis for sound agricultural planning, and a thriving development of manufacturing industries and tourism, financed in large measure by capital attracted from mainland United States, is under way. But in the islands generally, movements for political, economic, and social improvement tend to be sluggish, and in some cases are essentially absent. Most islands continue to suffer from the traditional ills of poverty, illiteracy, disease, excessive dependence on single-crop plantation agriculture, absentee ownership of land, and, particularly on some of the smaller islands, an undesirable measure of control by plantation or estate owners over the lives of the inhabitants.

In the islands affiliated with overseas nations, the only cities with metropolitan populations of more than 100,000 are San Juan (750,000) and Ponce (150,000) in Puerto Rico.

Transportation and Urbanization in Caribbean America

Caribbean America is not noteworthy for a highly developed system of transportation facilities or for a conspicuous degree of urbanization. However, connections with the outside world are afforded by the many important shipping lanes which converge on the region, attracted in no small measure by the great interocean passageway through Panama. A fair number of large cities exist, including one, Mexico City, with a metropolitan population of nearly 7 million, and five others—Caracas, Bogotá, Havana, Guadalajara, and Medellín—with metropolitan populations of over 1 million. Seven cities—Monterrey, San Juan, Cali, Guatemala City, Santo Domingo, Maracaibo, and Barranquilla—have between 500,000 and 1 million people in their metropolitan areas. Of the many political units in Caribbean America only three—Mexico, Colombia, and Venezuela—have more than one city of half a million or over.

The Panama Canal

The Panama Canal is one of the two major man-made waterways which have caused a gross reorientation of the world's ocean shipping lanes. Like its counterpart, the Suez Canal (not functioning in early 1969; see p. 372), its economic importance has lain chiefly in the intensity of ocean traffic it has attracted by providing a passage through a narrow land bridge connecting two continents. In the late 1960s it carried over 80 million long tons of cargo per year. It is a major focus of shipping lanes, being particularly important in the funneling of vessels from Atlantic ports of Anglo-America, Latin America and Europe into the Pacific, where Japan and western South America are the main destinations. Atlantic-bound cargo is much less.

Again like the Suez, the Panama Canal has had military as well as economic importance. Although the military implications vary with current events, they are never absent.

The Panama Canal and the Panama Canal Zone are currently being reassessed as to political, economic, and military implications. Political considerations directly involve the United States and Panama. Treaties agreed upon in 1967 provide for relinquishment of absolute authority by the United States over the Panama Canal Zone, and the creation of a binational governing authority made up of five members from the United States and four from Panama. Panama's share of tolls also would be increased. Meanwhile, the United States is seriously contemplating the building of a second canal, either in Panama or another Caribbean country. Some thought is being given to increasing the canal's present depth of 42 feet (actually, the minimum depth is now 37 feet at low tide in the harbor of Balboa) to at least 60 feet—thus permitting the largest naval and merchant ships, present and prospective future, to negotiate the channel. The possibility of converting the present vulnerable lock system into a sea-level waterway is also under consideration.

Unlike the other transportation facilities of Caribbean America, the Panama Canal tends to be superimposed upon, rather than meshed into, the local economies. The traffic of most Caribbean countries would not be seriously altered if the canal were not used.

Local Transportation

The remaining transportation facilities are, in the main, designed to serve Caribbean America. Mexico and Cuba have rather loosely woven rail networks with gauges that are generally the same as the standard gauge in the United States and Canada. Thus it is possible for a railway freight car to move from Canada to Mexico, or (by car ferry) to Cuba.

Highway systems tend to parallel the railways. In Honduras, Panama, Haiti, the Dominican Republic, and most small islands, highways take precedence in importance over railroads. The parts of the Pan American Highway System in Caribbean America are gradually reaching completion, although a few segments tend to be dry weather roads only, and a major break between Panama and Colombia still remains.

Ocean shipping and commercial aircraft are of primary importance as media of transportation within Caribbean America. Nearly every city of 100,000 or over has access to one or the other, or both.

Urbanization

Rapid urbanization is almost universal among the countries of Caribbean America, but the degree of urbanization reached thus far varies considerably from country to country. If attention is focused on larger cities—those over 100,000 in population—most of the larger countries have about one-fourth to one-fifth of their populations resident in such centers (United States: 62 percent in 1960). In each of the four most urbanized countries—Dominican Republic, Surinam, Cuba, Venezuela—about a third of the population resides in cities over 100,000. At the other end of the scale, some countries have no cities of this size, and Haiti and Honduras have less than 10 percent resident in the single large city each possesses.

NATIVE INDIAN COUNTRIES OF SOUTH AMERICA

Ecuador, Peru, Bolivia, and Paraguay may be conveniently grouped as the native Indian countries of South America. In each of them native Indians are the largest ethnic element. The proportion in Peru—46 percent—is the smallest; in Bolivia it is 57 percent, in Ecuador 58 percent, and in Paraguay 64 percent.[5] Except for Guatemala in Central America no other Latin American country contains a comparable percentage of native Indians.

These countries are further characterized by their low per capita gross national product ranging from $145 per year in Bolivia to $367 in Peru. They rank low on the Latin American scale in this respect (see p. 638).

Still another distinctive feature of these four countries is the very small contribution they make to Latin America's total exports. Peru, the most important exporting nation of the group, accounts for slightly more than 4 percent of all Latin American exports by value.

Furthermore, the native Indian countries have an unusually high ratio of rural to urban population. Cities with metropolitan populations of 100,000 or over include only Guayaquil and Quito in Ecuador, Lima-Callao and Arequipa in Peru, La Paz and Cochabamba in Bolivia, and Asunción in Paraguay.

In each of these countries except Paraguay, a major segment of the total population resides in mountain valleys of the Andes. In Bolivia most of the people live at high altitudes; in Ecuador and Peru the population is distributed among high, intermediate, and low elevations; and in Paraguay it is found at low elevations. Obviously the range from *tierra fría* through the *tierra templada* to the *tierra caliente* is well illustrated here.

Bolivia

Of Bolivia's 3.9 million people, most are residents of the *tierra fría* zone of the Andes, which constitute the western third of the country (maps, pp. 664 and 665). It is here that the Andean system reaches its widest extent—over 400 miles, if one includes the small portion to the west in Chile. Most of this highland country in Bolivia is above the level of the *tierra templada*. It is essentially comprised of two high mountain cordilleras, separated by a high intermontane plateau. The Western Cordillera, reaching elevations of 20,000 feet above the sea, is the most inhospitable of these areas to human settlement and the most formidable barrier to traffic routes. The intermontane plateau, the Altiplano, contains Lake Poopó, a salt-water lake at an elevation of about 12,120 feet; Lake Titicaca, a fresh-water lake which Bolivia shares with its neighbor Peru at a height of about 12,500 feet; and the Desaguadero River connecting the two lakes. The Eastern Cordillera, although attaining elevations of more than 20,000 feet, is more thoroughly dissected by streams than is its western counterpart. In addition, it contains some very rich vein deposits of tin and other metals. These metals were sought and to some extent exploited by the Spanish, and they still form Bolivia's principal exports.

Settlement in the Tierra Fría

Most of Bolivia's people live in either the Altiplano or the Eastern Cordillera. As has been stated, the major-

---

[5] Donald D. Brand, "The Present Indian Population of Latin America," in *Some Educational and Anthropological Aspects of Latin America* (Austin: University of Texas Press, 1948), p. 51. Other sources state that Paraguay's population is overwhelmingly *mestizo*. But all sources agree that the predominant racial strain in that country is Guarani Indian.

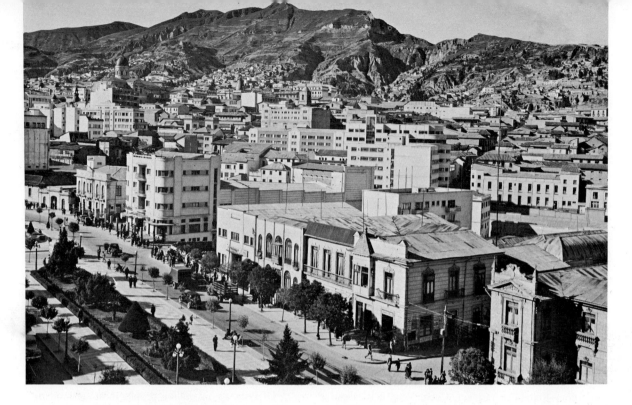

The downtown business and administrative district of Bolivia's largest city and *de facto* capital, La Paz. The city lies in a valley beneath the rim of the central Andean plateau or Altiplano, marked by the irregular skyline in the background. In the foreground is a spacious avenue—the only one in La Paz. Such a boulevard is frequently found in the central part of the larger Latin American cities. (Foreign Agricultural Service, U.S. Department of Agriculture.)

ity of these are native Indians whose ancestors, once the subjects of the Inca Empire, have inhabited these highlands for centuries. National population density averages 9 per square mile, and exceeds 25 per square mile only in spots, as around the shores of Lake Titicaca, which is surrounded by a closely spaced fringe of native Indian settlements. Although the lake with its attendant cluster of settlements is bisected by the political boundary line between Bolivia and Peru, the local inhabitants form a unitary community and customarily pay little heed to the boundary. Where left alone, they continue to live as their ancestors have done—grazing their sheep, llamas, and burros in the grassy area of the Altiplano and on the *paramos* above, and growing their meager crops of potatoes, quínoa, and barley. Throughout the year the average monthly temperatures vary comparatively little—ranging generally between 40°F and 50°F. Rainfall varies greatly from valley to valley, with a rough average of perhaps 30 to 50 inches annually. It is a dreary life in a dreary land—a life which the native Indians ease somewhat by chewing leaves of the coca plant. This is a cultivated plant that contains the drug cocaine.

Tin is the most important product taken from Bolivian mines. In 1966 the country was responsible for slightly less than 13 percent of the total world mine production of this commodity. The leading mines are in the Eastern Cordillera and, to a lesser extent, in the Altiplano. In addition, silver, tungsten, antimony, lead, zinc, and bismuth are extracted in the same general areas—often from the same mines as the tin. Until recently the mines have been privately owned—in some measure from abroad and in some measure by Bolivian citizens. However, control over most of them has lately been assumed by the Bolivian government. The native Indians provide the principal labor force of the mines. They are more effective than lowland dwellers, who become short of breath because there is less oxygen at these heights.

Bolivia has a small but promising production of oil and natural gas. Small fields near and south from Santa Cruz in the eastern lowlands supply national needs and provide minor exports.

La Paz

Several urban settlements at varying altitudes are situated in valleys of streams which drain eastward through the Eastern Cordillera. The most noteworthy of these is the city of La Paz (400,000). Lying in a gorge immediately beneath the rim of the Altiplano, this unusually placed city is the *de facto* capital of Bolivia, although Sucre (city proper, 60,000), 350 miles to the southeast, has long been officially designated as the capital, and the Bolivian Supreme Court actually sits there. La Paz is a true highland capital of a highland country. In addition, it is a redistribution point for agricultural products from the lower mountain valleys to the northeast and from the Altiplano, and for necessities brought by rail from either the Atlantic or the Pacific coast. Its manufacturing industries process food and make textiles, clothing, and other items needed by Bolivians.

Settlements of the Lower Tierra Fría and Upper Tierra Templada

At lower elevations, especially in the south and east in the *tierra templada,* are found a number of settlements where European-induced forms of agriculture prevail. The most heavily populated of these is centered around Cochabamba, a city of approximately 100,000 inhabitants. Food products from the small commercial agricultural district surrounding the city move chiefly to the mines in the highlands. In all of the four native Indian countries this district is the only sizable area within which the average population density exceeds 250 per square mile.

The eastern and northeastern wilds of Bolivia, a lowland countryside dominated by tropical savanna climate and ranging in vegetation from high forest in the north to scrub forest on the Paraguay border, is as yet very thinly inhabited. Some agricultural colonization by highland Indians is taking place under government sponsorship.

Peru

North and west of Lake Titicaca, the Altiplano comes to a rather abrupt end. The two cordilleras which have been distinct in Bolivia merge into a series of echelonlike ranges and groups of mountains. This series continues through Peru, where the Andean system becomes narrower until, at the Ecuadorian border, it is only about 150 miles wide.

The Tierra Fría

The native Indians living in the high mountain valleys of Peru follow ways of life similar to those of highland Bolivia. Indeed, the highland Indians of both countries are descended from subjects of the ancient Incas, who once controlled a considerable empire with its capital at Cuzco (city proper, 90,000). As in Bolivia, the *tierra fría* supports primarily a subsistence agricultural economy, especially to the northwest and to the southeast.

In the center of the Peruvian Andes is a rich mining district in the vicinity of Cerro de Pasco (city proper, 25,000), producing copper, silver, gold, lead, zinc, bismuth, vanadium, and coal. Some commercial farming is carried on in valleys not far from the mines. The wheat, corn, and other products grown here are shipped by rail to the mines or—farther downslope—to Lima.

The Arid Coast

The Andes drop sharply on the west to a narrow desert coast. Water from the numerous streams draining from the mountains into the Pacific provides irrigation for approximately 40 intermittently distributed oases. Some of these are the sites of commercial plantations specializing in cotton, others of similar plantations growing sugar cane. In still others the farming is predominantly of a subsistence character. Rice grown for local consumption is an important crop in most oases. No continuous rail route parallels the coast, and each oasis has its own seaport. Quality rather than quantity is emphasized in the production of both cotton and sugar. The cotton is chiefly a long staple variety similar to that of the Nile Valley in Egypt. Peru supplies approximately 1 percent of the total annual world production of both commodities.

Lima, the capital (2 million), is also on the arid coastal lowland. It was built by the Spanish and with its seaport Callao became the focus of routes leading not only from nearby mines of the Andes but also from more distant points. It is today an important manufacturing and university metropolis. Its manufacturing is mainly devoted to food, apparel, and other necessities for Peruvian consumption.

A series of small petroleum fields, situated at the northern edge of this arid coast, extends across the border into Ecuador.

Peru has recently become a major fishing nation. Since the early 1960s the country has been the world leader in tonnage of fish caught, accounting for somewhat less than 20 percent of the world total. (The *value* of the catch is exceeded by that of Japan, however.) Most of the catch is comprised of anchovies, a small fish sustained in enormous numbers along the Peruvian coast by abundant plankton in the cool waters of the northward–flowing Humboldt Current. Chimbote, north of Lima, is the leader among some 18 Peruvian ports that participate in the fisheries. Practically all of the anchovies landed are processed into fishmeal, which is exported in quantity to the United States and other developed countries, mostly for use in poultry feeds. Fish products were Peru's leading export in 1967, accounting for 27.0 percent of all exports by value (copper, 26.1 percent, ranked second).

An important problem for Peru's fisheries is to maintain a level of catch that will preserve the basic stock of fish unimpaired. There are indications that the catch in recent years—coupled with consumption of anchovies by sea birds and predatory fish—has exceeded the ability of the anchovy population to reproduce itself.

The Eastern Tierra Caliente

The eastern part of Peru lies beyond the Andes within the drainage basin of the Amazon. Here is a broad lowland area of rain forest and sluggish river, of untamed Indian and venturesome trader. Iquitos (city proper, 60,000), near the upper limit of effective steamboat navigation on the Amazon, is the major focus of commerce and transportation. From the city move wild animals and skins, crude rubber, certain gums used in the manufacture of chewing gum, and other extracted forest and animal products. Some cotton is also shipped. In turn, the city acts as a supply center for a large, sparsely settled hinterland. Its traders sell to the natives and a few immigrant settlers such items as drugs, groceries, hardware, and cloth. Besides being a port for ocean vessels (although 2300 miles by river from the Atlantic), Iquitos is connected to Lima by two-hour air service. A small oil refinery has been established to spur the development of the city and its region.

Ecuador

Ecuador, like Peru, has a lowland strip on its Pacific side, a high-mountain central segment, and a section of the Amazon River basin. The Pacific lowlands are more hilly than their Peruvian counterparts. In climate, the coastal desert of the south gives way to tropical savanna and, in the extreme north, to tropical rain forest.

The Andes Mountains, having narrowed increasingly from central Bolivia through Peru, are generally only 100 to 150 miles wide in Ecuador. They broaden again to the north in Colombia. Their areal arrangement in Ecuador becomes once again that of two rather distinct cordilleras, separated by a structural depression—or,

more strictly, by a series of mountain valleys which together constitute such a depression. Volcanic peaks, very much present in Bolivia but relatively few in Peru, reappear conspicuously in either cordillera. The highest, Mount Chimborazo in the Western Cordillera, reaches 20,561 feet. The Amazonian lowland of Ecuador to the east of the Andes is not unlike that of Peru and southeastern Colombia in general character, but it is smaller in extent.

Settlement Pattern and Economy of the Highlands

A majority of the people in Ecuador live in the Andean *tierra fría* at elevations higher than 7000 feet. They are clustered mainly in mountain valleys between the two cordilleras. Valleys of the far north, like their counterparts across the border in Colombia, are mainly occupied by highland Indians, as are those of the south. Quito, the capital and second largest city (elevation about 9250 feet; population 400,000), lies in a high basin in the north central part of the country. In the surrounding countryside the subsistence economy of the native Indian gives place to commercial farming dominated by *mestizos* and whites, who also form a majority of the population in Quito itself. Barley, corn, quínoa, potatoes, and livestock are grown, chiefly for sale in Quito. The manufacturing industries of the city, like those in so many isolated Latin American urban centers, concentrate on the processing of locally grown agricultural produce for local consumption. Another valley characterized by commercial forms of agriculture and a predominantly *mestizo* or white population surrounds Ecuador's third city, Cuenca (city proper, 75,000), approximately 300 miles south of Quito. The country's only noteworthy railroad connects Quito and Cuenca with the important seaport of Guayaquil on the Pacific.

The highland economy of Ecuador differs sharply from that of Peru and Bolivia in the comparative unimportance of minerals. Many different minerals are present, but few in deposits sufficiently large or rich to justify the costs of mining.

The Pacific Lowlands

On the Pacific coast lives the other substantial portion of Ecuador's population. Here is found the port of Guayaquil (650,000), the largest city in Ecuador. A low alluvial area inundated regularly by the Guayas River surrounds the port. Here, under a tropical savanna climate, is grown rice, primarily for domestic use but to some degree for export. Bananas, coffee, and cacao, respectively comprising 59, 15, and 11 percent of the country's exports by value (1967), also are grown on the Guayas River lowland or its margins. Most of the land of the drier section along the western margin of the lowland is devoted to the raising of livestock, chiefly for domestic consumption. Petroleum recovery supplies domestic needs, and sends nearly a third of each year's output into the world market.

Paraguay

Paraguay is lowland. Only along its western margin and in a few scattered hills to the east does the elevation exceed 1000 feet. It is sparsely populated. The three-fifths of the country situated to the west of the Paraguay River contains fewer than 2½ persons per square mile. Except for a corridor of settlement along the railway from the capital, Asunción (350,000), to Encarnación (city proper, 25,000), the density to the east of the river does not generally exceed 25 persons per square mile. Effective Paraguay is, therefore, along the railway.

In perhaps no other country of South America has a not unfavorable physical environment been used so ineffectually as in Paraguay. Many of the soils, though not of the highest caliber, are reasonably fertile. The climate of the west, it is true, is a rather inhospitable, dry tropical savanna. To the east, however, where agricultural potentialities are much higher, are humid subtropical conditions where the annual precipitation, rather evenly distributed, ranges generally between 40 and 80 inches each year. Yet the country has lagged behind in the development of commercial forms of agriculture.

The Paraguayans themselves, like their Bolivian neighbors, ascribe much of their low economic status to an interior position. Of the eleven independent countries in South America, only Paraguay and Bolivia are without a coastline. In 1865 Paraguay decided to remedy this situation by marching to the Atlantic Ocean. This proved a foolhardy undertaking, resulting in war with Brazil, Argentina, and Uruguay. In 5 years of war a Paraguayan population numbering over 500,000 was reduced to approximately 300,000, of which only 22,000 were males. The country is today strongly dependent on the seaports of Argentina for contact with world markets. Most of Paraguay's exports are consigned to Argentine ports for overseas shipment. The country's main port, Asunción, is located on the Paraguay River more than 900 miles from the sea.

Present-day Paraguayans live in a semisubsistence economy. Their small farms grow corn, cassava, beans, vegetables, some fruit—including citrus fruit—and cotton.

The much larger land area that is not tilled is devoted chiefly to the raising of cattle. Much of the country's land is held by large estates. Meat was Paraguay's largest export in 1966, with timber a strong second. The two products accounted for half of all exports by value.

Paraguay sends two interesting forest products to the regional and, to a lesser degree, the world market. One of these is extract (tannin) from the quebracho tree which thrives along the Paraná and Paraguay rivers. The extract is used in the tanning of leather. The second product is comprised of leaves from a South American holly plant. These leaves are used to make *yerba maté,* a form of "tea" which is an important item of diet not only in Paraguay but also in other—especially nearby—sections of South America.

BRAZIL

Brazil, although a single nation, forms a major subdivision of Latin America comparable in extent and importance to regions containing several different countries. Both the area and the population of Brazil are impressive, especially in comparison with other Latin American nations. Its area, exceeding that of conterminous United States, amounts to nearly half of all South America. Its population, while less than half as large as that of the United States, is nearly double that of Mexico, the second Latin American country in population. Brazil contains about a third of all the people in Latin America.

Brazil compared in area with conterminous United States.

Some Major Characteristics of Brazil

Brazil occupies a central position in Latin America. Situated on the equator, the country has an extensive coastline that fringes the Atlantic Ocean in both the Northern and Southern hemispheres. Its westernmost margin is not far from the Pacific. Smaller Latin American neighbors, including all of the other South American political units except Ecuador and Chile, flank its northern, western, and southern boundaries.

Much of its large area is undulating lowland or low upland. Unlike most other nations of comparable size, it is lacking in truly high mountains; at no place does Brazil reach far enough west to touch the Andes. Still, it has mountains and associated rough country in the Brazilian Highlands which flank most of the southeast coast; and, to the north across the Amazon River, in a small segment of the Guiana Highlands which reach over the border from Venezuela.

It encompasses nearly the whole of the drainage basin of the Amazon and its tributaries, one of the world's greatest river systems.

It includes the major portion of Latin America's tropical climates, vegetation, and soils. Except for the *tierra templada* of the highlands and the humid subtropical climate of the far south, Brazil is essentially a tropical nation.

It is the only Latin American nation in which Portuguese is the official language.

Its people are predominantly rural dwellers, yet it contains Rio de Janeiro and São Paulo, two of Latin America's four metropolises that exceed 4 million in population. In addition, it possesses some 25 metropolitan cities with populations ranging from 100,000 to slightly over a million.

It is a nation in which manufacturing industries and service occupations are growing rapidly in size and importance.

It contains the largest expanses of uninhabited or sparsely settled and yet potentially productive land in Latin America. According to some estimates, nearly four-fifths of Brazil could be made productive and support at least a moderately dense population. At present, however, only about a fourth of the country is effectively occupied.

Major Land Divisions

The land surface of Brazil falls generally into three major divisions: (1) the Atlantic coastal lowland strip; (2) the eastern highlands and uplands, sloping generally toward the continental interior; and (3) the lowlands of the Amazon, Paraná-Paraguay, and Uruguay rivers.

The Atlantic Coastal Strip

Although recognizable along the northeastern shorelines of Brazil, the Atlantic coastal strip is best developed on the country's southeastern margin between the cities of Recife (1.2 million) and Pôrto Alegre (900,000). It is narrow, varying in width from almost nothing to about 100 miles. From Salvador (700,000) almost to Pôrto Alegre its western side is flanked by a steep series of terraces—or, in a few places, by a single terrace—customarily referred to as "the great escarpment" (photo, p. 19). The coastal strip is predominantly a lowland of uneven surface, of tropical rainy or savanna climate and associated natural vegetation (much of which has been removed by man), and of rather infertile tropical soils. This coastal lowland is the most densely populated and intensively utilized of the three major surface divisions of the country.

The trend of the great escarpment is in conformance with that of the seacoast. Attaining elevations of 2500 feet—and in isolated mountains up to 8000 feet—above sea level, this escarpment is the "height of land" in Brazil. It tends to separate headwaters of the numerous short rivers draining to the Atlantic Ocean on the east from those of longer arteries extending to the Amazon and Paraná-Paraguay systems on the west. The single major exception is the large São Francisco River, which reaches the Atlantic Ocean between Recife and Salvador.

The Brazilian Highlands

The great escarpment forms the eastern edge of the Brazilian Highlands, which slope gently toward the interior. The irregular terrain has a tropical savanna climate, but with much variation in total rainfall from place to place and also in types of associated vegetation. This area is sparsely populated at present, except for parts of its seaward margin. It contains much of Brazil's potentially productive land.

The Interior Lowlands

In general the surface of the interior lowlands is uneven, although the flood plains proper are quite flat. To the north, the great Amazon system lies under tropical rainy climate, with tributary waterways extending into the adjacent tropical savanna—particularly that of the Southern Hemisphere. During the rainy summer months (December-February) these tributaries bring floods to the main river. Such floods are large enough to destroy sedimentary islands deposited the preceding year, together with the luxuriant jungle vegetation which establishes itself annually on these islands. Meanwhile, islands are again created, upon which the jungle will grow anew. Above the high-water mark are the tall broadleaf evergreen trees of the tropical rain forest proper.

To the south, the Paraguay, Paraná, and Uruguay rivers also flow through uneven terrain. The climate varies from tropical savanna along the middle Paraguay to humid subtropical along the middle Paraná and upper Uruguay. Settlement of the river lowlands ranges from almost empty areas of rain forest in the Amazon River basin to moderately peopled subtropical grasslands in the south.

Population Distribution

Brazil is predominantly an area of sparse, unevenly distributed, rural population. The majority of the inhabitants are concentrated within two major clusters along the southeast coast. One of these lies around and between the cities of Recife and Salvador. The other extends from Rio de Janeiro (4 million) and its hinterland to São Paulo (5.4 million) and its hinterland. Smaller clusters are found around the other major cities and at other, less prominent places, of which most are near the sea. In no sizable area does the population exceed 250 persons per square mile.

Brazilian Agriculture

The people of Brazil look primarily to agriculture and grazing for a livelihood. These engage about one-half of all employed workers, excluding a very large number who are domestic servants. The principal farming areas are found along, or relatively near, the southeast coast. They form a series of discontinuous districts separated by lightly occupied areas where grazing and shifting cultivation are the primary activities. There is considerable variety in the crops that are emphasized from one district to the next. The northernmost important district is a cotton-growing area that extends inland from Natal (170,000). Short-staple cotton is grown here largely as an adjunct of local cattle ranching, the pressed seeds (oilcake) being in great demand for cattle feed. Farther south, sugar cane is grown in a series of coastal districts extending discontinuously to the vicinity of Salvador. The largest district is near Recife. Immediately inland from it are districts producing coffee, cotton, and agave (the plant from which sisal fiber is secured). A small area to the west of Salvador specializes in tobacco. Well

to the south of Salvador, the country's main cacao district occupies an elongated strip parallel to the coast but some distance inland. Brazilian cacao accounts for 10 to 15 percent of the world's commercial output, and the country was the fourth largest cacao exporter (after three African countries) in 1966.

The climate of the foregoing districts is tropical rain forest around Salvador and in the cacao district to the south, and tropical savanna in districts north of Salvador. Commercial production tends to be dominated by *fazendas,* which are large holdings not greatly unlike the *haciendas* discussed elsewhere. Workers on *fazendas* may be tenants or hired employees, but are usually the former. Some are permanently attached to the land (more by custom than by law), while others tend to move about. Although definitely subordinate to *fazendas* in extent and importance, farms of a subsistence or semisubsistence nature are also present.

In the far north of this coastal strip the majority of the landowners are of relatively unmixed European descent. Around Salvador, however, Negroes and mulattoes constitute approximately two-thirds of the total population and a sizable proportion of the property owners.

A coastal area of tropical rain forest climate around Rio de Janeiro grows bananas, oranges, and truck crops, while irrigated rice is grown in an elongated strip along the Paraíba River valley to the west of the city. A large share of the food produced in the immediate hinterland of Rio de Janeiro is consumed by the 4 million people of that city's metropolitan area.

In the *tierra templada* and tropical savanna hinterland of São Paulo one-third to one-half of the world's annual coffee crop is produced. Unlike Brazil's agricultural districts discussed thus far, this area reaches far over the crest of the great eastern escarpment of the Brazilian Highlands. Coffee is grown here on units of varying size, from small farms to large *fazendas.* The trees are planted on slopes and ridgetops for good drainage and to minimize the chances of damage from occasional frosts. They commence bearing in 4 to 6 years, reach good productivity about the eighth year, and commence to decline some 15 to 20 years later. Some trees will produce beans for over half a century. Cotton is the second most important crop of the coffee region, and the bulk of Brazil's cotton is produced here (map, p. 326). The country vies with Mexico for top rank among Latin American cotton producers.

Coffee was by far the largest export from Brazil in 1965 (44 percent of all exports by value), as it has been for many years. Iron ore and cotton ranked second, each with 6 percent, followed by cane sugar with 4 percent, of the country's total export value.

In contrast to areas near Rio de Janeiro and Salvador, the coffee area, most of which lies in São Paulo state, is chiefly a district of Europeans and *mestizos* (in Brazil called *mamelucos*). There are few Negroes or mixed Negroid types. The proportion of small to medium-sized farm units in the area has been increasing.

As one proceeds toward Uruguay into humid subtropical climate, he finds himself first in an area of coniferous evergreen forest—a general counterpart to the softwoods of southeastern United States. Beyond this is a discontinuous belt of semideciduous woods which quickly gives way to prairie grasses. Colonies of Slavs (mostly Poles), Germans, Italians, and Japanese have developed in the forested areas—flourishing, after a slow start, in a diversified agricultural economy based on that of their homelands. We may note that Japan is a major source of current Brazilian immigration, thus strengthening this strain in Brazil's racially diverse population.

To the south, long-established cattle ranching dominates the rolling prairie lands. Important on the western side of all the Brazilian Highlands, this form of livelihood is perhaps the most picturesquely represented here where the Brazilian *gaucho*—in most cases, a *mameluco*—rules supreme.

Manufacturing and Mining

Brazil is increasingly significant in manufacturing and mining, especially the former. Under 15 percent of the labor force is as yet employed in manufacturing, but rapid industrial growth has been going on for some years. Brazil's industrial structure is the largest in Latin America, representing about one-quarter of the region's manufacturing industry.

As in other Latin American countries, manufactured products are made primarily for home consumption. In addition to iron and steel made at the Volta Redonda plant (largest in Latin America) on the Paraíba River and at other mills in the Rio–São Paulo region, there are important industries engaged in producing cotton textiles, foodstuffs, leather and leather products, cement, and a substantial variety of other commodities.

The booming city of São Paulo leads in this industrial trend, accounting for nearly one-third of Brazil's manufacturing, or one-twelfth of the manufacturing industry in all of Latin America. There is a secondary concentration of manufacturing in Rio de Janeiro, and a scattering of industries among the other cities of the country, especially those exceeding 100,000 in population.

Brazil's manufacturing industries lay heavy emphasis upon fiber and food necessities, especially for domestic consumption. However, the production of metals, fabricated metal goods, electrical machinery, and other commodities associated with modern aspects of technology is now appearing, especially in São Paulo, Rio de Janeiro, and other large cities.

Brazil lacks supplies of good coal (although some deposits of low to medium grade exist in the southeast), but has substantial reserves of waterpower (photo, right). The country has a large number of hydroelectric generating plants, the bulk of which are concentrated in the Brazilian Highlands. Two major recent power developments are on the Paraná River well to the interior from the core area around São Paulo and Rio de Janeiro.

Both the foreign investor and the Brazilian government—the latter acting often with borrowed funds—have been active in developing Brazil's manufacturing industries. Most foreign capital has come from the United States and West Germany. Especially in recent years, such government agencies as the Export-Import Bank of the United States, and the United States Agency for International Development (of which the well-known Alliance for Progress is a subordinate organization), and the International Bank for Reconstruction and Development of the United Nations have made funds available to the Brazilian government for development purposes.

Some Brazilian mines antedate European settlement. Yet except for precious metals and stones, the mineral deposits of the country have not been exploited to the same degree as many deposits in other parts of Latin America. The most important concentration of minerals lies in the Brazilian Highlands, in the general district of Belo Horizonte (800,000). This area contains one of the world's largest reserves of high-quality iron ore, now mined increasingly, both for export and for Brazilian mills. Manganese is also in good supply. Other Brazilian minerals include copper, bauxite, lead, zinc, nickel, chromium, quartz crystals, and diamonds.

Transportation and Urbanization

In a country where most of the inhabitants live on or near the seacoast and where many live within the framework of an essentially subsistence economy, inland transportation routes are less important than might be the case under another population pattern. Nevertheless, coffee, cotton, minerals, livestock, meat, and other commodities must get to market, and consumer goods and other necessities must be shipped back. Brazil's rather

The large waterpower potential of Brazil is mirrored in these rapids where the Rio São Francisco plunges from the interior plateau of northeastern Brazil to the coastal lowlands. This site was selected for Brazil's large Paulo Afonso hydroelectric power dam. (Brazilian Government Trade Bureau.)

poorly developed railways assume much of the responsibility for such shipments. The networks, which converge upon Rio de Janeiro and São Paulo, have been forged into a single system that is financed and administered by the national government. The pattern of paved highways conforms generally to that of the railways, but is less extensively developed. Building of gravel-surfaced or unsurfaced roads into Brazil's frontier districts is being vigorously pushed. The Amazon River, its tributaries, and the São Francisco River (the latter not navigable over the great escarpment) provide in-

land water transportation which varies in depth and therefore in utility with the season, for both river systems extend into the zone of seasonally rainy climate. The airplane is increasingly important as a means of quick access from the interior to the coast, where connections exist with the world's major air lanes.

Although a large proportion of Brazil's people live essentially outside of a commercial economy, there is a close coincidence between the general distribution of the country's population and that of its major cities. Brazil has 27 cities or metropolitan areas with populations of

100,000 or over. In areal distribution they form a sinuous linear pattern beginning at Manaus (170,000) in the central Amazon Basin, extending downstream to Belém (420,000), and thence southward along or near the coast. Each city tends to stand alone, acting as a manufacturing center and/or trade focus for a distinct hinterland.

The two metropolises of Rio de Janeiro and São Paulo, although only about 250 miles apart, differ rather markedly. Rio de Janeiro was the nation's capital until April 1960, as attested by its many beautiful government buildings. The city is a major tourist center, offering easy access to both the sea and a mountain countryside. It is particularly attractive during the drier winter months. It is also a major shipping, manufacturing, and trade center. Its port, which can accommodate ships drawing

as much as 40 feet of water, accounts for nearly one-third of the nation's total imports by value and ranks second to Santos in exports. Its railroads reach toward São Paulo and Santos on the southwest, past Belo Horizonte and the mining district of the Brazilian Highlands on the northwest, and toward Salvador on the northeast. Therefore, its hinterland is sizable in area and population, and it is diversified in economic activity.

São Paulo and its port, Santos (420,000), depend chiefly upon an economic base of commerce and manufacturing. Much of the commerce involves coffee, shipped through São Paulo to ocean ships at Santos. The hinterland of these cities includes not only the major Brazilian coffee district, but also some of the lumber district to the southwest and the northern portion of the livestock

One of the most strikingly modern urban scenes in Latin America. The photo shows the downtown section of Belo Horizonte, the capital of Minas Gerais state and the main urban center of Brazil's principal mining district. (Brazilian Government Trade Bureau.)

Brazil's colonial past with its strong Iberian influence is strongly reflected in this photo of an older section in the port of Salvador. The contrast with the dramatic modern architecture of downtown Belo Horizonte (opposite page) is evident. (Brazilian Government Trade Bureau.)

ranching area beyond. To the northeast, it merges with the trading area of Rio de Janeiro in the vicinity of the Volta Redonda steel mill. Through the port of Santos move nearly one-third of Brazil's exports by value and over one-half of its imports.

The manufacturing of São Paulo is comparatively new and growing rapidly, as, indeed, is the city itself. A major portion of the Volta Redonda iron and steel products come here for further fabrication into textile and electrical machinery, finished steel and wire, and railway rolling stock. Automobiles (final assembly only) are also produced. Textiles, including cotton, wool, rayon, jute, and silk, are important, as are chemicals, leather products, and processed foods.

Brazil's New Capital—Brasília

For many years, Brazilians gave consideration to the possibility of shifting their national capital to an interior location. In April 1960 this idea was consummated when Brasília, more than 500 miles inland, was officially proclaimed the capital. An entirely new, carefully planned, and spectacular creation in the midst of sparsely settled frontier country, Brasília has been constructed at enormous cost as a strictly governmental city. Construction was authorized by the Brazilian Congress in 1957, following forceful recommendations by a new president (Juscelino Kubitschek de Oliveira, inaugurated in 1956). By 1968 the metropolitan population of Brasília was reported to be around the 350,000 mark. The site, enclosed within a Federal District, lies in a gently sloping part of the Brazilian Highlands on the border between the states of Minas Gerais and Goiás. The tropical savanna climate, here associated with a vegetation of savanna grasses interspersed with patches of scrubby deciduous woods, is tempered by an elevation of about 4000 feet above the sea.

Brasília symbolizes the aspiration of Brazilians to occupy the empty interior lands of their enormous country. It is hoped that the government, in its new location, will be able to give effective direction to Brazil's westward movement and redevelopment of the northeast, and will be able to view the nation's overall problems in a broader and more balanced perspective than was possible

in the old seaboard location at Rio. It is further hoped that the shift will encourage settlement away from the coast.

COUNTRIES OF THE SOUTHERN MIDDLE LATITUDES

Argentina, Chile, and Uruguay are the Latin American countries of the southern middle latitudes. Their populations live mainly on subtropical lowlands, although sparsely settled portions of Argentina and Chile extend poleward into cooler climates.

Agriculture is an important element in the economies of these countries, although the proportion of the labor force that is employed in agriculture is considerably lower in Argentina and Uruguay than in any of the other Latin American republics (Chile or Venezuela would rank next to the two leaders in this respect). Unlike most other Latin American units, the three countries compete in world markets with the powerful agricultural economy of the United States, as well as with the agricultural economies of Canada, Australia, South Africa, New Zealand, and other middle-latitude nations. Argentina, Chile, and Uruguay lie largely outside the tropics (Uruguay entirely so), and thus in general they must grow crops and types of livestock that duplicate those of other subtropical and middle-latitude areas.

With respect to total exports, Argentina ranks very high among the countries of Latin America (third, in 1965, after Venezuela and Brazil), while Chile and Uruguay occupy a middle position. All three rank higher in value of exports per capita than in value of total exports. The exports of Argentina and Uruguay consist mainly of agricultural and pastoral products (Uruguay's exports almost entirely so), while Chile's exports consist predominantly of minerals. Total exports of all three nations have declined, both absolutely and relatively, since the end of World War II.

The three countries are among the few units in Latin America with economies that may be classified as technically or economically advanced. Among the Latin American republics, they are exceeded in per capita gross national product only by oil-rich Venezuela and (for Chile only) Panama, which derives the bulk of its revenues from the Panama Canal.

These countries have a further distinctive characteristic in their ethnic composition. Over 90 percent of the aggregate population in the three is either European or *mestizo*. In all of Latin America only Nicaragua, Costa Rica, Colombia, Venezuela, and Mexico approach this very high ratio. In Argentina and Uruguay approximately 90 percent of the people are of European descent. Chile, with 10 percent of its population native Indian and over 65 percent *mestizo,* is more diversified ethnically. In all three countries Negroes are few.

Argentina

Of the three countries, Argentina is by far the largest. But, as in the case of so many Latin American countries, effective Argentina is far smaller than total Argentina. Over two-thirds of the people live in the immediate hinterland of Buenos Aires (7.9 million)—an area encompassing only slightly more than one-fifth of the entire nation.

The Agriculture of the Humid Pampa

The populous core of Argentina is essentially the humid pampa, a crudely circular district bordering on the Atlantic Ocean and extending outward from Buenos Aires as far as 350 miles. It is characterized by a humid subtropical climate that merges into drier conditions on the south and west. Precipitation ranges from 20 to 40 inches annually. Toward the west and south there is a tendency for a decrease in total amount, for a summer maximum, and for a great degree of unreliability in rainfall. The most highly prized farmland is found southwest of the Paraná River between Buenos Aires and Rosario (700,000). Temperatures are moderate; monthly averages vary within a general range of 45° to 75°F. The black, waxy, and highly fertile prairie and chernozem soils have developed over a nearly flat residual accumulation of windblown and stream-deposited materials—a mantle that is, in places, over 1000 feet deep.

The humid pampa has been dominated historically by huge estates called *estancias*. Although the trend has been toward smaller units for some time, the large unit is still characteristic. In 1960 approximately 4 percent of the farms of the pampa held 52 percent of the land. These holdings were all over 1000 hectares (2471 acres), and they averaged over 3000 acres (4½ square miles). Some were very much larger. The other 48 percent of the land was held by the remaining 96 percent of the region's farmers. Even these smaller farms tend to be good-sized units, averaging between 100 and 150 acres. Owner operation of farms is the common pattern, but about 27 percent of the operators were tenants and sharecroppers in 1960.

Extending outward from Buenos Aires are a series of districts, each characterized by a rather distinct type of rural land use. Near the city itself is a fruit and vegetable district supplying urban residents with garden produce. To the northwest, the fertile lands between Buenos Aires and Rosario are devoted chiefly to corn, alfalfa, and livestock. Most of Argentina's corn is grown here (map, p. 84). Although its annual harvest of the grain is small compared to that of the United States, Argentina is normally the world's second exporter of corn, though its exports are far less than those of the United States.

Farther to the north, west, and south, a combination of wheat and livestock becomes more important. These same commodities also are predominant in a coastal district somewhat removed from Buenos Aires, between La Plata (450,000) and Mar del Plata (250,000). Argentina provides about one-fourth of all beef and 5 to 10 percent of all wheat entering world markets.

Between Buenos Aires and Mar del Plata are a series of swamps that have not as yet been drained sufficiently to be tilled. These soggy lands are devoted principally to the grazing of sheep, beef cattle, and dairy cattle.

Livestock, wheat, alfalfa, corn, and truck garden products are thus the principal agricultural commodities of the humid pampa. Except for the truck garden products (grown primarily for the Buenos Aires market) and the alfalfa (grown chiefly for livestock feed and pasture), these commodities are produced mainly for export. Products of lesser significance include barley and flax.

Agriculture in the Outlying Areas

The outlying countryside of Argentina beyond the humid pampa is sparsely settled. The existing population is generally clustered into districts of agricultural specialization—districts that are superimposed upon an extensive livestock grazing economy. Thus, in the far northeast of the area between the Paraguay-Paraná and Uruguay rivers, *yerba maté*—the tea of Argentina and nearby lands—is grown. In the north, along the Paraguay-Paraná River, the quebracho tree is cut—as it is in Paraguay—for its tannin. Beef cattle comparable to those of the savannas of Venezuela are also raised here. In the semiarid west and southward into even drier Patagonia, grazing is the characteristic land use, with the cattle of the west giving way to the sheep of Patagonia. Irrigated valleys support an intensive form of agriculture built around specialized crops. Tucumán (320,000) is the center of an irrigated sugar cane district, and Mendoza (320,000) has developed a reputation for its fine vineyards. Other, lesser oases are scattered along the eastern piedmont of the Andes.

Manufacturing and Mining

More workers are employed in manufacturing in Argentina than in any other Latin American country with the exception of Brazil. Nearly 2 million wage earners—approximately one in four of the entire Argentine labor force—are so engaged.

Manufacturing has been stimulated not only by local investment but also by large-scale investment of foreign capital. Early foreign investment was largely British and centered particularly on rail transport and other facilities to bring Argentina into the world trading system. Many of the facilities have now been nationalized for some years. Since World War II Argentine governments have been stressing the development of manufacturing, and much foreign capital has been attracted into this sphere. The leading sources in recent years have been the United States and continental Western Europe.

Mechanical energy is obtained largely from domestic or imported petroleum products and natural gas. Argentina's petroleum fields, which produced a little under 1 percent of the world's oil in 1966, are situated along the Patagonian seacoast, in the far northwest, and in the west near Mendoza. Most of these areas also produce natural gas. A small amount of energy comes from domestic hydroelectric plants, notably those clustered near Córdoba (700,000) and Mendoza—sites appreciably removed from the concentrated market for power at Buenos Aires. Most electricity is secured from thermal-electric plants that burn oil or natural gas. Electrical production of all kinds is being encouraged by the government to expand.

The older industries of Argentina are engaged primarily in the production of foodstuffs, textiles, boots, shoes, and other "soft" consumer goods. Many of the more recently established factories produce such hard durables as cement, iron and steel, machinery, and automobiles, as well as chemicals and electronic equipment. Argentine industry can now produce almost anything required by the domestic market, though its efficiency often leaves much to be desired. Improved overall efficiency, further expansion of the newer product lines, and growth of the steel industry to stem rising imports are major industrial objectives. Most of Argentina's manufacturing is concentrated within Greater Buenos Aires, an urban agglomeration containing 10 of the country's 21 cities of 100,000 or over.

Argentina is poor in known minerals. Besides modest deposits of oil and gas the country possesses some copper, lead, silver, tin, manganese, iron, and low-grade lignite. None of the country's mineral deposits is of much importance on a world basis, but the oil and gas are quite important to the economy of Argentina itself.

Transportation and Urbanization

Freight transport within Argentina is handled by water, road, rail, and pipeline in that order of importance. River and coastal movements handle just over one-third of total tonnage, with the Rio de la Plata and Paraná, Uruguay, and Paraguay rivers being the main arteries. Trucking has been increasing rapidly at the expense of the railways, despite the fact that about 60 percent of the highway net is comprised of dirt roads and 20 to 25 percent is gravel-surfaced. The rail net looks impressive on a map, being the most extensive in Latin America and connected with the systems of all neighboring countries. But it was built with varied gauges, making interchanges difficult, and has been inefficiently operated and poorly maintained in recent years. Between 1950 and 1965 its share of Argentine freight traffic decreased from 55 percent to 20 percent. Air transport is well developed, but is primarily a passenger service.

Argentina is very highly urbanized. Just over 10 percent of its people live on the farms that produce its exports. Around 60 percent of the total population lives in metropolitan areas of 100,000 or over. About one-third lives in the cluster of urban units comprising Greater Buenos Aires. It is in these and other sizable cities that most of Argentina's major industries are concentrated. Of these industries, many are either managed by, or provide employment for, Italians, Spaniards, and other recent European immigrants or descendants of comparatively recent immigrants. Although the main tide of immigration has waned, its effects are still felt in Argentina—particularly in the coastal cities.

Chile

Across the Andes from Argentina lies Chile, which has the distinction of possessing the greatest latitudinal extent of all Latin American countries except Brazil. In width, however, it seldom exceeds 225 miles. To the north, more than one-third of the country is in the exceptionally dry Atacama Desert. To the south, an even larger portion is comprised essentially of windswept, fiorded, inhospitable mountain country. In the center, primarily within an area of mediterranean climate, is effective Chile, where most of the people live. The summer deficiency and winter maximum of precipitation and the moderate temperatures resemble those previously described for the borderlands of the Mediterranean Sea. This district is bordered on the south by a habitable section of marine west coast climate, with conditions not unlike those in Portland, Oregon, or Vancouver, British Columbia.

The System of Agriculture

The populous middle section of Chile has traditionally been a land of the *hacienda*. A small, well-educated, wealthy upper class has held most of the land, which has been worked by a large, poverty-stricken tenant population. In recent decades a breakup of large properties has increased the number and proportion of small landowners, but many large units continue to function, and they hold the bulk of the farmland.

The main agricultural products of middle Chile are wheat, livestock, vegetables, grapes, and other products characteristic of mediterranean climates. The major portion of these commodities do not reach the export market. In fact, despite its relatively small population and its agricultural economy, Chile must import considerable amounts of food.

Importance of Mining for the Export Trade

Exports from Chile consist mainly of minerals. Of these, copper is by far the most important, accounting for two-thirds of all the country's exports by value in 1966. The primary deposits are in the Atacama Desert, as are those of two other mineral exports, iron ore and nitrate. Thus the major exports of the country come from one of its most sparsely populated sections. The copper ore of the Atacama, located in the Andes on the eastern edge of the desert, is poor in quality, averaging approximately 2 percent metal. Yet Chile contains over one-third of the world's known reserves of this important metal. Besides the desert production centered chiefly at Chuquicamata and near Potrerillos (map, p. 665), there is a third major production center at El Teniente, in the Andes south and east of Santiago in middle Chile.

Chile's iron ore, providing over a tenth of all exports in 1966, is mined near El Tofo and elsewhere in a narrow north-south zone, about 350 miles long, in the southern Atacama. Controlled by the Bethlehem Steel Company, the ore moves through the Panama Canal to one of the world's largest iron and steel plants at Sparrows Point near Baltimore, Maryland. Ore from this district is also carried southward to a small iron and steel plant in the Chilean city of Huachipato, a suburb of Concepción (360,000). Here it meets Chilean coal. This is a poor grade of coal that is mined still farther south at Lota, a coastal location on the Lebu Peninsula, 50 miles southwest of Concepción.

Nitrate currently accounts for about a twentieth of Chile's exports. The Atacama Desert contains the only extensive deposit of commercially recoverable nitrate in the world. It has been exploited for over a century. But Chilean nitrate faces an uncertain future, due to competition with synthetic nitrate—competition which first made itself felt effectively soon after World War I, and was abetted by an export tax levied by the Chilean government which amounted to nearly one-third of the total sale price.

Chilean Manufacturing

Chile has a larger proportion of its labor force employed in manufacturing than any other Latin American republic except Argentina and Uruguay, and it has the fourth largest industrial employment in the region after Brazil, Argentina, and Mexico. Recent progress has centered on engineering industries and oil refining, the latter concomitant with the growing exploitation of oil fields in Tierra del Fuego which now supply most of the country's demand for petroleum. As in Argentina, a wide range of manufacturing industries supply goods for the home market, but these suffer from the limitations of that market and are often small and inefficient. Most industry is in and near Santiago (2.5 million) and Valparaíso (500,000).

Transportation and Urbanization

With its 2600 miles of coastline and its narrow width, Chile is oriented toward the ocean. Ports along the Atacama Desert, however, are hampered by the fact that no good harbors exist in the northern third of the country. However, several ports have been equipped with special facilities for handling ore shipments. Valparaíso in Middle Chile is the seaport for Santiago, the capital. Lesser ports stud the coast as far south as Puerto Montt (45,000).

A railway line trends along the central valley of the country from a point north of Iquique (55,000) to Puerto Montt. Its usefulness is somewhat limited, however, by the fact that it is narrow gauge (3.28 ft) in the northern desert, and broad gauge (5.5 ft) from Santiago southward. Thus goods shipped the length of the network by rail must be transferred from one freight train to another. From Santiago and Valparaíso a railway climbs over high Uspallata Pass in the Andes to reach the Argentine piedmont city of Mendoza. It is not a heavily traveled route. From the Atacama two small northern lines also reach into the Andes, in Bolivia.

Of the roads, the most noteworthy is the Pan-American Highway, which provides Santiago and Valparaíso with access to coastal Peru on the north and, over Uspallata Pass, to Mendoza on the east. The remaining roads tend to form a pattern not unlike that of the rail net.

Airways from Santiago lead not only to other major world cities, but also to Arica (50,000) in Chile's far north, Punta Arenas (55,000) in the far south, and points in between.

About one-third of Chile's people live in the two major metropolitan areas of Santiago and Valparaíso. Two-thirds of Chile's population is urban, one of the highest figures in Latin America.

Uruguay

Uruguay, like Argentina, is an urbanized country dependent on agricultural exports from the farms and ranches of the humid pampa. By now less than one-fifth of the population is classed as agricultural. It is also like Argentina—and Brazil—in that more of its land is pastured than is cultivated. The agricultural district is mainly a small crescent around Uruguay's capital, Montevideo (1.5 million). Here, under a humid subtropical climate that encompasses the entire country, are grown wheat, corn, oats, flax, sunflowers, and some truck crops. The pastoral section includes nearly all of the remainder of the country. From it comes wool amounting to one-half of the country's exports, but only an insignificant part of the wool exports of the world. Meat, hides, leather and other animal products account for most of the balance of Uruguay's exports.

Over one-half of Uruguay's people live in the Montevideo metropolitan area. The rest are distributed across the entire national territory, with the density decreasing away from the capital and being lightest in the north. A rail network focusing on Montevideo connects with the rail systems of Brazil and Argentina. Steamship services, both passenger and freight, between Montevideo and Buenos Aires add to the capital's commerce. Manufacturing in Montevideo largely involves the processing of meat and other agricultural commodities from the hinterland, the making of textiles and beverages, and the refining of imported petroleum.

REFERENCES and READINGS

Caribbean America: General

AUGELLI, JOHN P., "The Rimland-Mainland Concept of Culture Areas in Middle America," *Annals of the Association of American Geographers,* **52,** no. 2 (June 1962), 119–129.

LOWENTHAL, DAVID, "The Range and Variation of Caribbean Societies," *Annals of the New York Academy of Sciences,* **83** (1960), 786–795, reprinted in Philip L. Wagner and Marvin W. Mikesell, eds., *Readings in Cultural Geography* (Chicago: University of Chicago Press, 1962), pp. 186–194; and "Caribbean Views of Caribbean Land," *Canadian Geographer,* **5,** no. 2 (Summer 1961), 1–9.

MITCHELL, HAROLD, *Caribbean Patterns: A Political and Economic Study of the Contemporary Caribbean* (London: W. and R. Chambers, 1967).

SAUER, CARL ORTWIN, *The Early Spanish Main* (Berkeley and Los Angeles: University of California Press, 1966).

WEST, ROBERT C., and JOHN P. AUGELLI, *Middle America: Its Lands and Peoples* (Englewood Cliffs, N.J.: Prentice-Hall, 1966). A standard text.

WILGUS, A. CURTIS, ed., *The Caribbean: Its Hemispheric Role* (1967) and *The Caribbean: Natural Resources* (1959). (Gainesville: University of Florida Press.)

Mexico

BRAND, DONALD D., *Mexico: Land of Sunshine and Shadow* (Princeton, N.J.: D. Van Nostrand Co., Searchlight Books, 1966); and "Mexico," *Focus,* **16,** no. 10 (June 1966), 6 pp.

BURLEY, T. M., "Mexico D.F. Since Diaz," *Tijdschrift voor Economische en Sociale Geografie,* **55,** no. 1 (January 1964), 19–23.

FLORES, PAUL S., *Basic Data on the Economy of Mexico,* U.S. Bureau of International Commerce, Overseas Business Reports. OBR 67–50 (August 1967). (Washington, D.C.: Government Printing Office, 1967.)

FOX, DAVID J., "Man-Water Relationships in Metropolitan Mexico," *Geographical Review,* **55,** no. 4 (October 1965), 523–545.

GILDEA, RAY Y., "Geographic Aspects of Industrial Growth in the Monterrey Region of Mexico," *Journal of Geography,* **59,** no. 1 (January 1960), 34–41.

HENDERSON, DAVID A., "Arid Lands under Agrarian Reform in Northwest Mexico," *Economic Geography,* **41,** no. 4 (October 1965), 300–312.

MILLON, RENÉ, CLARA HALL, and MAY DÍAZ, "Conflict in the Modern Teotihuacan Irrigation System," *Comparative Studies in Society and History,* **4,** no. 4 (July 1962), 494–524.

WAGLEY, CHARLES, and MARVIN HARRIS, *Minorities in the New World: Six Case Studies* (New York: Columbia University Press, 1958), "The Indians in Mexico," pp. 48–86.

Central America

BREWER, THOMAS K., *Basic Data on the Economy of Guatemala,* U.S. Bureau of International Commerce, Overseas Business Reports, OBR 66–31 (May 1966). (Washington, D.C.: Government Printing Office, 1966.)

CROSBIE, A. J., and P. A. FURLEY, "The New Belize: Prospects for British Honduras," *Scottish Geographical Magazine,* **83,** no. 1 (April 1967), 53–63.

FOX, DAVID J., "Prospects for the Panama Canal," *Tijdschrift voor Economische en Sociale Geografie,* **55,** no. 4 (April 1964), 86–101.

GOULD, PETER R., and JACK P. SPARKS, "The Geographical Context of Human Diets in Southern Guatemala," *Geographical Review,* **59,** no. 1 (January 1969), 58–82.

HORST, OSCAR H., "The Specter of Death in a Guatemalan Highland Community," *Geographical Review,* **57,** no. 2 (April 1967), 151–167.

JACOBSEN, CARL R., *Basic Data on the Economy of Costa Rica* and *Basic Data on the Economy of the Republic of Panama,* U.S. Bureau of International Commerce, Overseas Business Reports, OBR 67–12 (March 1967) and OBR 68–41 (June 1968). (Washington, D.C.: Government Printing Office, 1967, 1968.)

PEARSON, ROSS, "Zones of Agricultural Development in Guatemala: An Experiment in Land Reform," *Journal of Geography,* **62,** no. 1 (January 1963), 11–22.

RAYNOLDS, DAVID R., *Rapid Development in Small Economies: The Example of El Salvador* (New York: Frederick A. Praeger, 1967).

West Indies

AUGELLI, JOHN P., and HARRY W. TAYLOR, "Race and Population Patterns in Trinidad," *Annals of the Association of American Geographers,* 50, no. 2 (June 1960), 123–138.

BALCHIN, W. G. V., and A. COLEMAN, "Puerto Rico," *Geography,* 50, pt. 3 (July 1965), 274–286.

DALE, EDMUND H., "The State-Idea: Missing Prop of the West Indies Federation," *Scottish Geographical Magazine,* 78, no.3 (December 1962), 166–176.

FLOYD, BARRY, "Jamaica" and "Outlook for Agriculture" [in Jamaica], *Focus,* 19, no. 2 (October 1968), 1–8 and 8–11.

GILDEA, RAY Y., "Haiti," *Focus,* 17, no. 9 (May 1967), 6 pp.

GREEN, BARBARA, *Basic Data on the Economy of Jamaica,* U.S. Bureau of International Commerce, Overseas Business Reports, OBR 67–38 (July 1967). (Washington, D.C.: Government Printing Office, 1967.)

HENSHALL, JANET D., "The Demographic Factor in the Structure of Agriculture in Barbados," Institute of British Geographers, *Transactions No. 38* (June 1966), pp. 183–195.

JONES, DAVID, "Cuba's First Decade," *Geographical Magazine,* 40, no. 17 (September 1968), 1422–1433.

LOWENTHAL, DAVID, ed., *The West Indies Federation: Perspectives on a New Nation* (New York: Columbia University Press, 1961); also, "The West Indies Chooses a Capital," *Geographical Review,* 48, no. 3 (July 1958), 336–364; and "Race and Color in the West Indies," *Daedalus: Journal of the American Academy of Arts and Sciences,* 96, no. 2 (Spring 1967), 580–626.

NIDDRIE, D. L., "Eighteenth-Century Settlement in the British Caribbean," Institute of British Geographers, *Transactions No. 40* (December 1966), pp. 67–80.

PEARCY, G. ETZEL, *The West Indian Scene* (Princeton, N.J.: D. Van Nostrand Co., Searchlight Books, 1965).

SMITHERS, DAVID, "Barbados, the Singular Island," *Geographical Magazine,* 40, no. 5 (September 1967), 407–422.

WELCH, BARBARA, "Population Density and Emigration in Dominica," *Geographical Journal,* 134, pt. 2 (June 1968), 227–235.

YOUNG, B. S., "Jamaica's Bauxite and Alumina Industries," *Annals of the Association of American Geographers,* 55, no. 3 (September 1965), 449–464.

Colombia, Venezuela, and the Guianas

BLYDENSTEIN, JOHN, "Tropical Savanna Vegetation of the Llanos of Colombia," *Ecology,* 48, no. 1 (Winter 1967), 1–15.

FRIEDMANN, JOHN, *Regional Development Policy: A Case Study of Venezuela* (Cambridge, Mass.: M.I.T. Press, 1966).

GALBRAITH, W. O., *Colombia: A General Survey* (2d ed.; London: Oxford University Press, 1966).

LINDOW, HERBERT A., *Basic Data on the Economy of Colombia,* U.S. Bureau of International Commerce, Overseas Business Reports, OBR 66–45 (July 1966). (Washington, D.C.: Government Printing Office, 1966.)

LOWENTHAL, DAVID, "Population Contrasts in the Guianas," *Geographical Review,* 50, no. 1 (January 1960), 41–58.

MILLER, ELBERT E., "The Guayana Region, Venezuela: A Study in Industrial and Urban Development," *Yearbook of the Association of Pacific Coast Geographers,* 27 (1965), 77–88.

PROCTOR, N., "Economic Developments in Guayana, Venezuela," *Geography,* 52, pt. 2 (April 1968), 183–186.

RODWIN, LLOYD, "Ciudad Guayana: A New City," *Scientific American,* 213, no. 3 (September 1965), 122–132. (A planned city on the Orinoco in Venezuela's iron-mining area.)

STOKES, CHARLES J., "The Freight Transport System of Colombia, 1959," *Economic Geography,* 43, no. 1 (January 1967), 71–90.

VARGO, FRANKLIN J., *Basic Data on the Economy of Surinam,* U.S. Bureau of International

Commerce, Overseas Business Reports, OBR 67-51 (August 1967). (Washington, D.C.: Government Printing Office, 1967.)

VINING, JAMES W., "Guyana" and "Hydrologic Problems" [of Guyana], *Focus,* 19, no. 4 (December 1968), 1-8 and 8-11.

WATTERS, R. F., "Economic Backwardness in the Venezuelan Andes: A Study of the Traditional Sector of the Dual Economy," *Pacific Viewpoint,* 8, no. 1 (May 1967), 17-67.

Bolivia, Ecuador, Peru, and Paraguay

CRIST, RAYMOND E., "Politics and Geography: Some Aspects of Centrifugal and Centripetal Forces Operative in Andean America," *American Journal of Economics and Sociology,* 25, no. 4 (October 1966), 349-358; also, "Bolivia," rev. ed., *Focus,* 16, no. 7 (March 1966), 6 pp; "Paraguay," *Focus,* 18, no. 4 (December 1967), 6 pp.; with Alice Taylor, "Ecuador," *Focus,* 18, no. 2 (October 1967), 6 pp; and, with Alice Taylor, "Peru," *Focus,* 17, no. 8 (April 1967), 6 pp.

EIDT, ROBERT C., "Pioneer Settlement in Eastern Peru," *Annals of the Association of American Geographers,* 52, no. 3 (September 1962), 255-278.

FIFER, J. VALERIE, "Bolivia's Boundary with Brazil: A Century of Evolution," *Geographical Journal,* 132, pt. 3 (September 1966), 360-372; and "Bolivia's Pioneer Fringe," *Geographical Review,* 57, no. 1 (January 1967), 1-23.

KENNETT, AUDREY, "Peru's Anchovy See-Saw," *Geographical Magazine,* 39, no. 8 (December 1966), 621-631.

MANGIN, WILLIAM, "Squatter Settlements" [of Peruvian cities], *Scientific American,* 217, no. 4 (October 1967), 21-29.

MINNIES, MALACHY T., *Basic Data on the Economy of Ecuador,* U.S. Bureau of International Commerce, Overseas Business Reports, OBR 68-52 (July 1968.) (Washington, D.C.: Government Printing Office, 1968.)

PENDLE, GEORGE, *Paraguay: A Riverside Nation* (3d ed.; London: Oxford University Press, 1967).

PRESTON, DAVID A., "Changes in the Economic Geography of Banana Production in Ecuador," Institute of British Geographers, *Transactions No. 37* (December 1965), pp. 77-90.

SMITH, C. T., "Aspects of Agriculture and Settlement in Peru," *Geographical Journal,* 126, pt. 4 (December 1960), 397-412.

WEBSTER, WILFRID, ed., *Paraguay: 1852 and 1968* (Occasional Publication No. 2; contributions by Edward A Hopkins, Raymond E. Crist, and William P. Snow). (New York: American Geographical Society, 1968.)

Brazil

BOTTING, DOUGLAS, "Manaus," *Geographical Magazine,* 39, no. 6 (October 1966), 434-447.

DICKENSON, J. P., "The Iron and Steel Industry in Minas Gerais, Brazil, 1695-1965," in Robert W. Steel and Richard Lawton, eds., *Liverpool Essays in Geography: A Jubilee Collection* (London: Longmans, Green and Co., 1967), pp. 407-422.

DYER, DONALD R., "Growth of Brazil's Population," *Journal of Geography,* 65, no. 9 (December 1966), 417-428.

GALLOWAY, J. H., "The Sugar Industry of Pernambuco During the Nineteenth Century," *Annals of the Association of American Geographers,* 58, no. 2 (June 1968), 285-303.

HESSLER, EVELYN R., *Basic Data on the Economy of Brazil,* U.S. Bureau of International Commerce, Overseas Business Reports, OBR 68-68 (August 1968). (Washington, D.C.: Government Printing Office, 1968.)

JAMES, PRESTON E., and SPIRIDIÃO FAISOL, "The Problem of Brazil's Capital City," *Geographical Review,* 46, no. 3 (July 1956), 301-317.

JONES, CLARENCE F., "Brazil," *Focus,* 15, no. 3 (November 1964), 6 pp.

MOMSEN, RICHARD P., JR., *Brazil: A Giant Stirs* (Princeton, N.J.: D. Van Nostrand Co., Searchlight Books, 1968).

OBERG, KALERVO, "The Marginal Peasant in Rural Brazil," *American Anthropologist,* 67, no. 6, pt. 1 (December 1965), 1417–1427.

PLATT, ROBERT S., "Brazilian Capitals and Frontiers: Part I" *Journal of Geography,* 53, no. 9 (December 1954), 369–375; and Part II, 54, no. 1 (January 1955), 5–17.

SCHURZ, WILLIAM LYTLE, *Brazil, The Infinite Country* (New York: E. P. Dutton & Co., 1961).

SMITH, T. LYNN, *Brazil: People and Institutions* (rev. ed.; Baton Rouge: Louisiana State University Press, 1963).

STERNBERG, HILGARD O'REILLY, "Brazil: Complex Giant," *Foreign Affairs,* 43, no. 2 (January 1965), 297–311.

WAGLEY, CHARLES, and MARVIN HARRIS, *Minorities in the New World: Six Case Studies* (New York: Columbia University Press, 1958), "The Indians in Brazil," pp. 20–47.

Argentina, Chile, and Uruguay

KENWORTHY, ELDON, "Argentina: The Politics of Late Industrialization," *Foreign Affairs,* 45, no. 3 (April 1967), 463–476.

McGANN, THOMAS F., *Argentina. The Divided Land* (Princeton, N.J.: D. Van Nostrand Co., Searchlight Books, 1966).

REDDING, DAVID C., "The Economic Decline of Uruguay," *Inter-American Economic Affairs,* 20, no. 4 (Spring 1967), 55–72.

THIESENHUSEN, WILLIAM C., "Agrarian Reform and Economic Development in Chile: Some Cases of Colonization," *Land Economics,* 42, no. 3 (August 1966), 282–292.

VON STROEBEL, JAMES MICHAEL, *Basic Data on the Economy of Argentina,* U.S. Bureau of International Commerce, Overseas Business Reports, OBR 66–78 (December 1966). (Washington, D.C.: Government Printing Office, 1966.)

WHITE, C. LANGDON, and RONALD H. CHILCOTE, "Chile's New Iron and Steel Industry," *Economic Geography,* 37, no. 3 (July 1961), 258–266.

WINSBERG, MORTON D., "Buenos Aires," *Focus,* 18, no. 7 (March 1968), 6 pp.

WRIGHT, A. C. S., "The Earthquake of May 1960 in South-Central Chile," *Pacific Viewpoint,* 2, no. 2 (September 1961), 137–154.

See also the lists of references and readings for Chapters 1–3 and 27.

PART IX

ANGLO-AMERICA

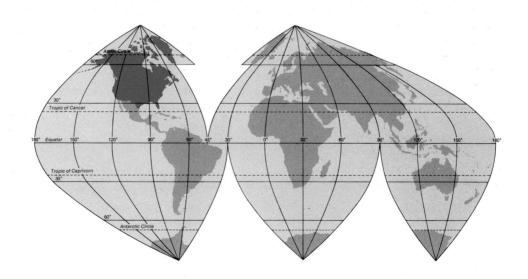

Introduction to Anglo-America

29

The colonizing activities of Great Britain have left a major imprint on the United States and Canada. Hence these two countries are commonly distinguished, under the term "Anglo-America," from the rest of the Americas, which have received primarily the Latin imprint of Spain and Portugal. For some purposes the large Arctic island of Greenland is conveniently included in the concept of Anglo-America. This procedure is justified by Greenland's proximity to Canada (map, p. 690) and its strategic importance to both of the Anglo-American countries, but not by the island's culture (a blend of native Eskimo and imported Danish elements) nor by its present political status as a self-governing part of the Kingdom of Denmark. The same general considerations apply to the small French island possession of St. Pierre and Miquelon, south of Newfoundland, except that the latter is French in culture and is administered as a dependency of France.

ANGLO-AMERICA

INTRODUCTORY LOCATION MAP

URBAN AREAS

◑ ● over 1,000,000

● 500,000—1,000,000

✶ ○ Selected smaller urban units

Stars show national capitals

*Albany-Schenectady-Troy

On this map, and subsequent Anglo-America maps, city-size symbols for Canadian cities are based on 1968 metropolitan area estimates. Symbols for United States cities are based on 1968 estimates for Standard Metropolitan Statistical Areas (see footnote, p. 714). A number of SMSAs contiguous to the Los Angeles, New York, or Chicago SMSAs are not shown separately on this map.

TIES WITH BRITAIN

Effective British settlement of the present United States and of Newfoundland began in the early seventeenth century. Effective settlement of the Canadian mainland by Britain occurred in the later eighteenth century, after Canada had been wrested from France in the French and Indian Wars. Continued immigration from the homeland and a high rate of natural increase in the newly colonized areas led to a rapid growth in the number of British settlers. Consequently, a basically British form of culture became firmly established and has remained dominant in Anglo-America to the present day, despite the entry of large numbers of settlers from continental Europe and smaller numbers from Africa, Latin America, the Orient, and the Middle East. With a few exceptions these non-British elements in the Anglo-American population have now been largely assimilated into the general culture of the United States and of Canada.

The most obvious and probably most important evidence of the cultural tie with Britain is the dominance of the English language in Anglo-America. Other evidences of the British heritage are found in the political and legal institutions of the Anglo-American countries, as well as in a multitude of everyday customs and practices among the people. Present-day cultural and political affiliations with Great Britain are closer and more openly acknowledged in Canada than in the United States, despite the large French-Canadian element in Canada's population. Canada retains membership in the Commonwealth of Nations and gives formal allegiance to the British crown. The basic cultural bonds between the United States and Britain, on the other hand, were weakened by a long period of political antagonism beginning in colonial times and continuing for many years after the American Revolution. Indeed, this antagonism is still evident at times among some segments of the American and British populations, although the two peoples as a whole have been drawn closer together by the great world crises and military conflicts of the twentieth century.

ANGLO-AMERICAN WEALTH AND POWER

In the twentieth century Anglo-America has become the main center of wealth and power in the world. Consequently, this region exerts an immense influence in world affairs. The United States, although hard pressed by the competition of the Soviet Union, is still in all probability the world's most powerful nation.

Canada, with a population of only 20.7 million (1968 estimate) is a less powerful nation in its own right but is an exceedingly valuable ally of the United States and wields an influence disproportionately greater than might be expected of a country with such a small number of inhabitants.

The material wealth of the Anglo-American countries finds an expression in their exceptionally high average material standards of living. Recent estimates place the per capita income of the United States second only to Kuwait's, and that of Canada behind only those two countries, Sweden, Switzerland, and Iceland.

Although the success of the Anglo-American nations in achieving such high levels of wealth, power, and influence is a matter too complicated to be fully explained, it is clear that these two countries possess a number of rather specific assets on which they have been able to capitalize. Some of the more important assets may be stated as follows:

1. The Anglo-American countries are large in area.
2. They possess comparatively effective internal unity.
3. They are outstandingly rich in natural resources.
4. Their combined population is impressively large (the United States alone being the fourth most populous nation of the world), yet neither country is overpopulated.
5. They possess highly mechanized economies with high overall production per capita.
6. They occupy a strong defensive position, due to relative isolation from other major centers of population and military power.
7. Their relations with each other are generally friendly and cooperative.

Each of the foregoing statements will be discussed at some length in the remainder of the present chapter. These points, when elaborated, not only help to explain the success of the United States and Canada in reaching their present eminence in world affairs, but also throw much light on the general character of Anglo-America as a world region.

LARGE AREA OF THE ANGLO-AMERICAN COUNTRIES

Possession of a large area assures a country of neither wealth nor power. But it does afford at least the possibility of finding and developing a wider variety of resources and, other things being equal, of supporting a larger population than might be expected in a small country. In the twentieth century two of the world's largest countries, the United States and the Soviet Union, have surpassed the older and smaller states of western Europe as world powers. Both the United States and the Soviet Union have resources which are much superior to those of the west European nations, but until the twentieth century neither country had yet found the means to occupy and organize its national territory effectively and thus to take adequate advantage of its resources.

That there is no direct relationship between area on the one hand and wealth and power on the other is amply attested by the comparative areas of the Anglo-American countries themselves. Canada is the larger country of the two, but is much the lesser in total wealth and political, economic, and military power. The United States—the world's leading nation in total wealth and productive capacity, and probably its most influential nation politically—measures 3,615,000 square miles, while the area of Canada is 3,852,000 square miles. Canada's territorial extent is second in the world only to that of the Soviet Union, while the United States ranks behind these two countries and China (see n. 1, p. 462). Even the United States, however, has nearly twice the area of Europe outside the Soviet Union. France, the largest European country in area, is only about one-seventeenth the size of the United States.

INTERNAL UNITY OF THE UNITED STATES AND CANADA

The welding of their large national territories into effectively functioning units represents a major accomplishment by the Anglo-American countries and a major source of their wealth and power. They are not weakened by chronic political separatism in any of their parts to a degree such as that which exists in many of the world's countries. Their economic welfare and, indirectly, their political unity are promoted by constant and large-scale interchange of goods between sections far removed from one another.

Political separatism in the world's countries is often based on ethnic differences. The degree of cultural unity that exists in Anglo-America is strikingly in contrast to the large number of different ethnic groups

found in such areas as Europe and the Orient. The aboriginal population of American Indians was overwhelmed by the tide of European settlement and has now been reduced to a politically insignificant minority concentrated mainly in the western and northern parts of Anglo-America. The Indian population is gradually being absorbed into the main stream of American and Canadian life, though much of it still lives as a depressed group under conditions difficult to justify. Non-British immigrants have been gradually absorbed also, though not without making valuable contributions to the Anglo-American society they have entered.

Each of the Anglo-American countries has one major ethnic minority, and in each case this minority has been, and still remains, the focus of serious problems of national unity. Almost one-third of Canada's people belong to the French-Canadian group, which is differentiated from the rest of the Canadian population not only by language but by religion, being overwhelmingly Roman Catholic while the rest of Canada's population is predominantly Protestant. The French-Canadians are concentrated, for the most part, in the lowlands along the St. Lawrence River in the province of Quebec, although they have also spread to some extent into adjoining areas. The ancestors of this group, some 60,000 to 70,000 strong, were left in British hands when France was expelled from Canada in 1763. They did not join the English-speaking colonists of the Atlantic seaboard in the American Revolution and by their abstinence laid the basis for the division of Anglo-America into two separate countries. Since the eighteenth century the French-Canadians have increased rapidly in numbers and have clung tenaciously to their distinctive language and culture. At times major controversies arise between this group and other Canadians, and separatist sentiments are occasionally voiced by elements of the French minority. The Canadian government has achieved and preserved national unity through the recognition of French as a second official language and the provision of legal protection for French-Canadian institutions, but the degree, details, and effectiveness of this bicultural policy are matters of frequent and sometimes acrimonious review. Although not assimilated in a cultural sense, French Canada has been incorporated into the nation to the extent that it has provided a number of outstanding Canadian leaders, including prime ministers.

The principal minority group in the United States is the American Negroes, numbering just over one-tenth of the total population. Unlike the French-Canadians they do not form a coherent minority which is numerically dominant in any major political division or section of the country. Neither have black Americans as a group been an active force toward sectionalism in the United States. Nevertheless, a conflict of attitudes toward Negro slavery was among the important factors which nearly split the United States into two nations in the nineteenth century, and differing attitudes toward the status and aspirations of black people are still a major source of friction in American politics. During the 1960s the Negro protest against discrimination and depressed status within the American nation, along with the white reaction, began a period of recurrent racial violence. This centered in the big American cities where an increasingly large proportion of the country's black population is concentrated.

Even without strong ethnic contrasts and antagonisms, however, regional conflicts of interest are bound to occur in countries of such large extent and varied physical and economic conditions. In this connection it is a striking and significant aspect of Anglo-America that both of its countries employ a federal structure of government. A multitude of powers and functions are assigned to the fifty states of the United States and the ten provinces of Canada in an attempt to give latitude for governmental expression of regional differences and the solution of regional problems which might otherwise work toward the disruption of national unity. Of course, in both countries such federal structures themselves involve frequent disputes concerning divisions of power between the central governments and the state or provincial governments.

Another notable and important political characteristic of Anglo-America lies in the fact that this region represents an outstanding stronghold of democratic representative government. Governmental responsiveness to majority opinion, combined with safeguards for the minority, has tended to prevent revolutionary pressures from building up. The effectiveness of the political system is strengthened by the fortunate economic circumstances of the region, which lead most of its inhabitants to congratulate themselves on being Americans or Canadians and to feel that their institutions have provided opportunities and results beneficial to themselves.

The result is that Anglo-America has been characterized thus far by great stability of government. The American Civil War is the only large-scale violent civil conflict that has ever occurred in either nation. In neither country has the national government ever been overthrown by force. Thus the governments of the United States and Canada have not been obliged to devote large energies merely to preserving the state's existence or its territorial integrity against internal stresses.

The governments of both countries have assiduously fostered national unity through the provision of adequate means of internal transportation. In both, the development of transportation networks has had to be accomplished not only over long distances but primarily against the "grain" of the land, since effective transport links between east and west have been most imperative, while most of the mountains and valleys in Anglo-America have a north-south trend, thus opposing themselves as a series of obstacles to cross-country transport. Both the United States and Canada were tied together as effectively functioning units by heavily subsidized transcontinental railroads built during the latter half of the nineteenth century. These fostered national unity not only directly, through the connections they afforded between sections, but indirectly through their promotion of settlement and of general economic development. Later, automobile transportation on publicly constructed highway networks and, still later, the development of air transportation further solidified internal unity.

RESOURCE WEALTH OF ANGLO-AMERICA

The natural resources which the Anglo-American countries have within their boundaries are outstanding in both variety and abundance. In part, this fact is simply a reflection of the large territorial extent of each nation. Countries of such magnitude are almost bound to contain a sufficient diversity of physical conditions to produce resources that are out of the ordinary in size and utility. Nevertheless, it is probable that no area of the earth's surface of comparable extent contains so much natural wealth as Anglo-America. The abundance of resources also reflects the inventiveness and skill of the Anglo-American peoples in the exploitation of nature, in the "making" of resources. It should not be forgotten, however, that the value of Anglo-American resources is partly due to scientific and technological advances in other lands. The usefulness of coal, for example, was discovered in England long before the superior coal resources of Anglo-America were more than vaguely known.

Certain important aspects of the world position of the Anglo-American countries with regard to resources and production are summarized in Table 29.

TABLE 29 ANGLO-AMERICA'S SHARE IN THE WORLD'S KNOWN WEALTH AND PRODUCTION: SELECTED ITEMS

ITEM	APPROXIMATE PERCENT OF WORLD TOTAL [a]		
	UNITED STATES	CANADA	ANGLO-AMERICA
Area	7	7	16
Population	6	0.6	7
Agricultural resources and production			
Arable land	17	4	21
Permanent meadows and pastures	11	1	12
Production of			
Wheat	14	7	21
Corn (grain)	49	under 1	50
Soybeans	73	under 1	74
Cotton	22	–	22
Tobacco	20	2	22
Oranges	24	–	24
Milk	16	2	18
Number of			
Cattle	10	1	11
Hogs	10	1	11
Forest resources and production			
Area of forests and woodlands	7	11	18
All wood cut	16	6	22
Sawn-wood production	23	6	29
Wood-pulp production	39	16	55
Minerals, mining, and smelting			
Estimated coal reserves (bituminous and better grades)	32	1	33
Coal production (bituminous and better grades)	17	under 1	18
Estimated petroleum reserves	9	1	10
Crude-oil production	26	3	29
Crude-oil refinery capacity	42	4	46
Natural-gas production	65	6	71
Potential iron-ore reserves	36	9	45
Iron ore mined	15	6	21
Pig iron produced	25	2	27
Steel produced	26	2	28
Bauxite reserves	2	–	2
Bauxite mined	5	–	5
Aluminum produced	37	11	48
Copper reserves (copper content)	15	4	19
Copper mined (copper content)	24	9	33
Copper smelted	24	7	31
Lead reserves (lead content)	12	18	30
Lead mined (lead content)	10	10	20
Lead smelted	15	6	21
Zinc reserves (zinc content)	14	22	36
Zinc mined (zinc content)	12	20	32
Zinc smelted	23	9	32
Nickel mined (nickel content)	3	54	57
Gold mined	4	7	11
Phosphate rock mined	40	–	40
Potash produced (K_2O equivalent)	21	11	32
Native sulfur produced	65	–	65
Asbestos mined	4	41	45
Waterpower			
Undeveloped waterpower	6	8	14
Capacity of existing hydroelectric plants	21	10	31
Manufacturing other than metals			
Raw cotton consumed industrially (yarn)	23	1	24
Raw wool consumed industrially (yarn)	12	1	13
Sulfuric acid produced	34	3	37
Motor-vehicle production (all units)	46	4	50
Steel consumed industrially	27	2	29

[a] The notation (–) indicates none or no appreciable amount.

Agricultural and Forest Resources

Large areas of Anglo-America are suited by topography and climate for cultivation. The United States, although the fourth largest country in total area, is second only to the USSR in total area of arable land and is easily first in the world in arable land of high quality. A much smaller proportion of Canada is arable, but even in that nation the total arable acreage is still greater than that of any country other than the Soviet Union, the United States, India, or China. With only 20 million people Canada is the world's second richest country in arable land per capita (almost 5 arable acres per person). The United States, despite its much larger population (201 million: 1968 estimate), ranks behind only Australia, Canada, and the Soviet Union in the amount of arable land it possesses per capita, which is over 2½ acres. In contrast, no country in Europe has even 2 acres of arable land to support each member of its population. Furthermore, the United States ranks third, after the USSR and Australia, in total area of meadow and pasture lands, while in Canada such lands, although much less extensive than in the United States, are sufficient to allow a perennially important export of meat and live cattle in addition to a high domestic consumption of livestock products.

Anglo-America is estimated to contain about 18 percent of the world's forested area. Canada and the United States rank second and fourth among the nations of the world in this regard, with the Soviet Union ranking first and Brazil third. The intensively exploited forests of the United States can no longer supply the needs of the country for wood, although American forest resources are still extremely large and varied. In fact, the United States now ranks first in the world as an importer of forest products, with Canada, the largest exporter of such commodities, being the main source of supply. Canadian forest resources are larger in extent than those of the United States, but are not so varied in type. For example, Canada has no forests comparable to the redwoods of California or the pine forests of southeastern United States, and is far behind the United States in both acreage and variety of deciduous hardwoods.

Mineral Resources

Not only is Anglo-America rich in agricultural and forest resources, but estimates of mineral reserves give a picture of a region possessing outstanding mineral wealth, probably greater than that found in any part of the earth of comparable size.

Power resources are especially plentiful. The United States is estimated to possess about one-third of the world's known coal of bituminous grade or better, ranking ahead of any other country in total reserves. Canada, however, is relatively poor in coal, although not totally devoid of it. Various ways of defining and estimating petroleum resources lead to estimates which differ substantially, but most estimates credit the United States with about one-tenth of the world's petroleum reserves. It normally produces around one-fourth of the world's total output of petroleum, although even this immense production is not sufficient to supply the nation's needs, and substantial imports are required. Canada is a growing producer of petroleum and has noteworthy reserves, though in neither reserves nor production is it comparable to the United States (Table 29). The United States currently produces an estimated two-thirds of the world's natural gas, and has very large reserves. Estimates of the world's total hydroelectric potential do not give Anglo-America an outstandingly high proportion. Canada is credited with 8 percent of the world total and the United States with 6 percent. But they lead the world in *developed* water power, with the United States having about one-fifth and Canada about one-tenth of the world total, and at present neither country has developed more than a small fraction of its potential.

With respect to metals the general picture is also one of outstanding wealth and great variety. The United States originally had huge deposits of high-grade iron ore, but the enormous scale of exploitation by the iron and steel industry has seriously depleted these reserves, and dependence on imported ore has increased considerably in recent years. However, the country still has an estimated one-third of the world's "potential" iron ore reserves—reserves only approximately estimated as to extent, and often composed of lower grade ores than those usually mined in the past—and new methods of treating

TABLE 29 SOURCE

SOURCE: A considerable number of standard statistical sources, such as those listed on pages 12–13 were used in the preparation of this table. While the data are believed to convey as accurate an impression as is possible with a single set of figures, the student must be warned that they can give a false impression of stability, reliability, and detail. Actually, many of the percentages are subject in varying degree to (1) errors in estimation and compilation, (2) fluctuations over a period of time, and (3) misinterpretations due to differences in quality or ease of exploitation of particular resources from one area to another. The percentages show *volume* rather than *value* of production and reserves. In general, percentages given for production were computed on the basis of 2-year or 3-year averages for the middle 1960s.

ore now are making more and more of these "potential" reserves economically usable. Canada also has substantial deposits of iron ore. Canadian reserves are smaller in total quantity than those of the United States, as far as is known, but are larger per capita and are at a much earlier stage of exploitation. A selection of other important metals shows the following estimated percentages of world reserves in Anglo-America: copper, United States 15 percent, Canada 4 percent; lead, United States 12 percent, Canada 18 percent; zinc, United States 14 percent, Canada 22 percent. The region is a major producer of these metals and also of some others such as magnesium, nickel, uranium, molybdenum, titanium, and silver. Anglo-America is also outstandingly rich in a number of important nonmetallic minerals other than power minerals: the list includes phosphate, potash, sulfur, industrial salt, gypsum, and asbestos. Anglo-American reserves and production of the latter minerals are concentrated mainly in the United States, except for asbestos, of which Canada is the world's greatest producer.

Such is the wealth of the region that it is easier to summarize the gaps in the resource structure than to describe the resources themselves. Agriculturally, the lack of any large producing area for certain tropical specialty crops is felt, necessitating the import of such commodities as coffee and tea, bananas, rubber, and tropical vegetable oils. The major mineral deficiencies are in chromite, manganese, tin, diamonds, graphite, and high-grade bauxite. However, the industrial development of the United States has progressed to the point that large imports are normally required even of minerals with which the United States is itself well endowed and of which it is a major producer. Canada is an important source of supply for some of these imported minerals.

Such estimates as those given in the foregoing paragraphs and in Table 29 serve adequately to outline the general picture of Anglo-American advantages with respect to resources. However, these estimates should be understood as only approximations. They are subject to a certain amount of error and change and tend to underemphasize the factor of quality. The contrasts and rankings they show are the result not only of differing natural conditions, but also of differences in technical knowledge and ability, in general economic conditions, and in cultural emphases among the countries concerned. Estimates of mineral reserves, in particular, must be used with caution, since the figures cited by various authorities often differ widely, and the known quantities and distributions of minerals are constantly changing as a consequence of new ore discoveries and new processes for making use of ores previously considered to be substandard.

ADVANTAGES WITH RESPECT TO POPULATION

In mid-1968 the United States had an estimated population of 201.3 million, while Canada had an estimated 20.7 million inhabitants. Thus Anglo-America contained about 222 million people. Three of the world's countries—China, India, and the Soviet Union—surpass all of Anglo-America in population. Nevertheless, the United States ranks fourth among all countries, with a population total nearly as large as the combined totals of Europe's four most populous countries—West Germany, the United Kingdom, Italy, and France. Most of Anglo-America's people live in the southeastern quarter of the region, from the St. Lawrence Lowlands and Great Lakes south, and this part of Anglo-America is one of the world's four extensive areas of dense population, along with the continent of Europe (including European Russia), China, and the Indian subcontinent.

From the beginning of settlement Anglo-America has been a region of rapidly growing population, and this growth continues at the present time. By 1800, approximately two centuries after the first settlements were firmly established at Jamestown and at Quebec, there were over 5 million people in the United States and several hundred thousand more in Canada. Population growth since that time is summarized in Table 30. Continuing rapid—though slowing—growth is evident in an increase of one-fifth in the United States between the 1950 and 1960 censuses, and 12 percent in the 8 years following 1960.

The population of this region long increased at a rate more rapid than that of the world as a whole. Between 1900 and 1950 the increase in world population is estimated at around 50 percent, while that in Anglo-America was over 100 percent. Thus one element in the mounting world importance and power of Anglo-America, and particularly of the United States, was its possession of an increasing share of the world's population. The older world powers of western Europe, especially, have been surpassed. In 1870, at the time of the Franco-Prussian War, the United States had almost exactly the same population as each of the two belligerents, Germany and France, but in 1968 it had over 1½ times as many people as France and the two Germanies combined. Recently, however, the growth rate of the Anglo-American population has fallen below that of the world as a whole, though not of its more industrialized parts.

Despite rapid growth in the past the Anglo-American share of the world's population is still not an excessively large one. The United States and Canada have

TABLE 30 POPULATION GROWTH IN THE UNITED STATES AND CANADA, 1850–1968

YEAR	U.S. POPULATION (MILLIONS)	CANADIAN POPULATION (MILLIONS)	TOTAL (MILLIONS)	TOTAL INCREMENT	
				MILLIONS	PERCENT
1850	23.3	2.4	25.7		
1900[a]	76.1	5.4	81.5	55.8	217
1950[a]	151.1	14.0	165.1	83.6	103
1960[a]	179.3	18.2	197.5	32.4	20
1968	201.3[b]	20.7[b]	222.0	24.5	12

[a] Actual dates for Canada are 1901, 1951, and 1961. The decennial census is taken one year later in Canada than in the United States. The table uses census data except for 1968.

[b] Mid-year estimate.

together about 6 percent of the estimated world total. This compares with their possession of about 14 percent of the world's land area. The average population density of the two countries considered as a whole is only 30 per square mile, as compared with about 67 per square mile for the world as a whole excluding Antarctica. However, the figure for average density obscures the contrast between the United States, with 56 persons per square mile, and Canada, with only 5 per square mile. The very low figure for Canada is due to its tremendous expanse of sparsely settled northern lands, the effectively occupied part of the country in the south being much more densely populated, although far smaller in areal extent.

It is evident that judged by the world as a whole even the United States is far from overpopulated, considering its exceptional resources and enormous production, and Canada, if anything, is underpopulated. Thus the Anglo-American countries have a combined population sufficiently large to exploit the resources of the region effectively but not so large as to press heavily on those resources.

ECONOMIC PRODUCTIVITY OF ANGLO-AMERICA

On the whole, economic production per person involved is higher in Anglo-America than in any other part of the world. This regional characteristic is reflected in, and to some extent can be measured by, the high per capita incomes of the United States and Canada. It is attested by the results of comparative studies of industries in the United States and in overseas countries, as well as by the efforts of other nations to learn American production methods. High production per worker is a basic factor leading not only to a high income level and high standard of living in Anglo-America, but also to military power, since it means that there is a large surplus of productive capacity available beyond that needed to supply the necessities of the people. This surplus capacity can be used to produce military equipment.

Anglo-American production methods are not easy to imitate in the rest of the world because they are so closely related to other outstanding characteristics of the region and to its distinctive historical development. The use of machines and mechanical energy on a scale greater than anywhere else in the world is basic to Anglo-American productivity. But this superior mechanization has been achieved in a region which has had an abundance of resources and a shortage of labor throughout most of its history; which has possessed resources sufficiently large and rich both to attract foreign capital and, through large-scale and often wasteful exploitation, to make possible large domestic accumulations of capital; and which has been superlatively endowed with the necessary fuels to drive machinery. Furthermore, it has been achieved in a region where a free and fluid society has encouraged full development of the abilities of the people; where free and untrammeled economic progress has generally been encouraged, fostered, and subsidized by government; where the energies of the people did not, through much of their history, have to be too much diverted into war; and where large segments of the population have subscribed with an almost religious fervor to the ideals of hard work and economic success since early times. Finally, in the United States, a large and unified internal market allowed great economic organizations to

Advanced scientific technology in Anglo-America comes to one peak of intensity along Route 128, the "Highway of Science," in Waltham, Massachusetts, a suburb of Boston. For example, most of the buildings to the immediate right of the cloverleaf are devoted to electronics research or the manufacture of electronic equipment. Many different industrial and research firms are represented in the view. The diamond-shaped plot (with center circle) near the overpass is a heliport. Accessibility to the Massachusetts Institute of Technology and Harvard University, with their outstanding scientists and research facilities, is an important locational consideration for the companies along the highway. (New England Survey Service, Inc.)

specialize in the mechanized mass production of a few articles, thus lowering the unit costs of production. The foregoing combination of circumstances has been essentially unmatched in any other part of the world, much of which now seeks mechanization under comparatively severe handicaps.

DEFENSIVE ADVANTAGES OF RELATIVE ISOLATION

Anglo-America is essentially a huge island, bordered on the east by the Atlantic Ocean, on the north by the Arctic Ocean and on the west by the Pacific. The only land connection with another world region is along the boundary between the United States and Mexico, which mostly runs through sparsely populated territory and is far removed from the core area of either country. No good land route yet connects the United States and Canada with South America, and Anglo-America's contacts with that continent, as well as with other world regions, are made by sea and air.

The closest land approaches of Anglo-America to Eurasia are at the northeastern and northwestern corners of the region. A series of islands bridges the North Atlantic between Canada and Europe, and includes Greenland, Iceland, the Faeroes, the Shetlands, and Svalbard. Only the Bering Strait separates Asia from Alaska. The first historic contact between Europe and North America, that of the Vikings from the ninth to the eleventh centuries, was made via the island stepping stones of the North Atlantic. Contacts with Asia occurred earlier, since the North American Indians are thought to have entered the continent in prehistoric times from Asia via the Bering Strait area. In the eighteenth century Russian fur traders entered Alaska, which passed from Russian to American control by purchase only in 1867.

But harsh climatic conditions in the latitudes of these land approaches have prevented them from serving as major avenues of movement between Anglo-America and Eurasia during historic times. The North Atlantic approach is dangerous to navigate and leads, on the American side, only to the wasteland of tundra and coniferous forest which occupies northern Canada. Bering Strait is backed on both sides by similar stretches of wasteland. Between the two approaches Eurasia and North America are separated by the ice-jammed Arctic Ocean, again backed by tundra and taiga on both sides. Consequently, the main connections between the two great land masses of the Northern Hemisphere have lain across the oceans farther south, although the routes generally followed do swing north to approximate great circle courses and some air services now cross the Arctic. From the Atlantic seaboard ports of Anglo-America south of the taiga the sea distance to Britain is about 3000 miles (only about 2000 between the outer extremities of Newfoundland and Ireland), and from Puget Sound to Japan is about 4000. If the Pacific is crossed via Hawaii to Japan, the distance becomes well over 5000 miles. These have been the real distances separating the states of Anglo-America from the world's other centers of political and military power—distances not overlong for peaceful commerce but constituting a major obstacle to military operations. Such distances (added to inherent strength) have so reduced the danger of a successful military attack on Anglo-America that the region was able in the past to proceed with its domestic development while paying relatively little attention to considerations of defense and the entanglements of world politics.

However, three new factors have arisen in relatively recent times to lessen the value of Anglo-America's protected position. In the first place, the destruction of the balance of power in Eurasia has raised the possibility that while no individual Eurasian power could successfully overcome Anglo-America, a Eurasia united wholly or in large part under one dominion might well aspire to do so. A united Eurasia would change the position of Anglo-America from that of an isolated region protected by distance to that of an area semisurrounded by its possible antagonist. Second, the development of air power, including guided missiles, has seriously compromised the protective function of the oceans and the northern wastes. Finally, the development of nuclear armaments has raised the possibility of devastating, or even decisive, effect by attack from the air.

The relatively isolated position of Anglo-America has allowed the region to develop in relative peace and security, and remains a valuable asset and source of strength, since it is still harder to attack effectively over long than over short distances. But the protective value of this position has declined sharply in recent years, and the people of the region can no longer afford to view the rest of the world with the degree of detachment that was common in the past.

COOPERATIVE RELATIONS OF THE ANGLO-AMERICAN STATES

For many years after the American Revolution the political division of Anglo-America between the independent United States to the south and a group of British colonial possessions to the north was accompanied by serious friction between the peoples and governments on either side of the boundary. A heritage of antagonism was present due to the failure of the northern colonies to join the Revolution, their use as British bases during that war, the large element of their population which was composed of Tory stock driven from American homes during the Revolution, and uncertainty and rivalry as to ultimate control of the central and western reaches of the continent. The War of 1812 was largely an American effort to conquer Canada. Even after the failure of this effort a series of border disputes occurred, and American ambitions to possess this remaining British territory in North America were openly expressed; suggestions and threats of annexation were made in official quarters throughout the nineteenth and into the twentieth century.

In fact, Canada as a unified nation is in good part a result of American pressure. After the American Civil War the military power of the United States took on a threatening aspect in Canadian and British eyes. Suggestions were made in some American quarters that Canadian territory would be a just recompense to the United States for British hostility toward the Union during the war. The British North America Act, passed by the British Parliament in 1867, brought an independent Canada into existence. Britain sought, and as the event has proved, successfully, to establish in Canada an independent nation capable of achieving transcontinental unity, of maintaining its independence and integrity, and of relieving Britain of some of the burden of defense in North America.

Hostility between the United States and its northern neighbor did not immediately cease with the establishment of an independent and unified Canada; a certain amount of friction is evident from time to time even today. Nevertheless, relations between the two Anglo-American nations have improved gradually to the point where these countries are often cited as an outstanding example of international amity and cooperation, and the frontier between them has ceased to be a source of mutual insecurity and weakness. This frontier, stretching completely unfortified across a continent, has become more a symbol of friendship than of enmity.

The bases of Canadian-American friendship are cultural similarities, the material wealth of both nations, and the mutual need for and advantages of cooperation. Despite certain vexatious tariff restrictions, a very large volume of trade moves across the frontier and strengthens both countries economically and militarily. Each country is the other's principal foreign market and source of supply, although Canada is much more dependent on the United States in this respect than is the United States on Canada. In 1966, for example, Canada supplied 24 percent of all United States imports by value and took 22 percent of all United States exports; while the United States supplied 72 percent of Canada's imports and took 61 percent of its exports. In addition, the United States supplies large quantities of capital which have been an important factor in the rapid economic development of Canada during recent decades, but are also a frequent source of disquiet in Canada due to the amount of control over the Canadian economy they involve. Major items moving from the United States to Canada include automobiles and parts, and other machinery; steel; chemicals; and coal. Those moving from Canada to the United States include paper, pulp, and lumber; nonferrous ores and metals, including ferroalloys; and iron ore.

In defense arrangements the two nations work together so closely as to form almost a single unit. Important aspects of their relationship include (1) a structure of joint command, (2) cooperative military maneuvers in far northern areas, (3) cooperative installations to warn against air attack, and (4) cooperation in the production and supply of military equipment. These arrangements mean essentially that Canada is protected by its more powerful neighbor, which would certainly regard any attack on Canada as an attack on itself. But at the same time, of course, Canada's resources and strategic assets greatly strengthen that neighbor. Meanwhile, as an independent nation Canada continues to guard its sovereignty and cultural identity, to differ with American foreign policies at times, and to maintain its ties with the United Kingdom and the other states of the Commonwealth of Nations.

REFERENCES and READINGS

Anglo-America: General

ATWOOD, WALLACE W., *The Physiographic Provinces of North America* (Boston: Ginn and Company, 1940).

BREBNER, JOHN BARTLET, *North Atlantic Triangle: The Interplay of Canada, the United States, and Great Britain* (New Haven: Yale University Press, 1945).

BRUBAKER, STERLING, *Trends in the World Aluminum Industry* (Baltimore, Md.: Johns Hopkins Press for Resources for the Future, 1967).

CHAPMAN, JOHN D., and JOHN C. SHERMAN, advisory eds., *Oxford Regional Economic Atlas: The United States and Canada* (Oxford: Clarendon Press, 1967).

DARLING, F. FRASER, and JOHN P. MILTON, eds., *Future Environments of North America* (Garden City, N.Y.: Natural History Press, 1966).

GRIFFIN, PAUL F., DONALD L. CHATHAM, and ROBERT N. YOUNG, *Anglo-America: A Systematic and Regional Geography* (2d ed.; Palo Alto, Calif.: Fearon Publishers, 1968).

LOGAN, M. I., "The North American Iron and Steel Industries: Three Case Studies of the Location of Complex Manufacturing Industries," in J. Rutherford, M. I. Logan, and G. J. Missen, *New Viewpoints in Economic Geography* (Sydney, Australia: Martindale Press, 1966), pp. 314–331.

MEAD, W. R., and E. H. BROWN, *The United States and Canada: A Regional Geography* (London: Hutchinson Educational, 1962).

MONKHOUSE, F. J., and A. V. HARDY, *North American Landscape* (Cambridge: At the University Press, 1965).

OLMSTEAD, CLARENCE W., "The Application of a Concept to the Understanding of a Region: People, Time, Space and Ideas in the Economic Core Region of Anglo-America," *Journal of Geography,* **59,** no. 2 (February 1960), 53–61.

PARKER, W. H., *Anglo-America: Canada and the United States* (London: University of London Press, 1962).

PATERSON, J. H., *North America: A Regional Geography* (3d ed.; New York: Oxford University Press, 1965).

SHELFORD, V. E., *The Ecology of North America* (Urbana: University of Illinois Press, 1963).

STARKEY, OTIS P., and J. LEWIS ROBINSON: *The Anglo-American Realm: A Geographical Analysis of the Economies of the United States and Canada* (New York: McGraw-Hill Book Company, 1969).

WARNER, DONALD F., *The Idea of Continental Union* (Lexington: University of Kentucky Press, 1960). A history of the idea of a union between the United States and Canada.

WATSON, J. WREFORD, *North America, Its Countries and Regions* (rev. ed.; New York: Frederick A. Praeger, 1967).

WHITE, C. LANGDON, EDWIN J. FOSCUE, and TOM L. McKNIGHT, *Regional Geography of Anglo-America* (3d ed.; Englewood Cliffs, N.J.: Prentice-Hall, 1964).

WOHLSTETTER, ALBERT, "Illusions of Distance," *Foreign Affairs,* **46,** no. 2 (January 1968), 242–255.

Canada: General

Atlas of Canada (Ottawa: Department of Mines and Technical Surveys, Geographical Branch, 1957).

BESLEY, LOWELL, "Canada's Pulp and Paper Industry," *Journal of Forestry,* **65,** no. 10 (October 1967), 699–705.

BREBNER, JOHN BARTLET, *Canada, A Modern History* (Ann Arbor: University of Michigan Press, 1960).

CAMPBELL, H. C., ed., *How to Find Out about Canada* (New York: Pergamon Press, 1967). Numerous bibliographical references.

CAMU, PIERRE, E. P. WEEKS, and Z. W. SAMETZ, *Economic Geography of Canada, with an Introduction to a 68-Region System* (New York: St. Martin's Press, 1964).

CANADA, DEPARTMENT OF MINES AND TECHNICAL SURVEYS, GEOGRAPHICAL BRANCH, *Selected Bibliography of Canadian Geography* (Ottawa: annual).

CANADA, DOMINION BUREAU OF STATISTICS, CANADA YEAR BOOK, HANDBOOK AND LIBRARY DIVISION, *Canada Year Book* (Ottawa: annual); and *Canada One Hundred: 1867–1967* (Ottawa: 1967).

Canadian Geographer (quarterly).

Canadian Geographical Journal (monthly).

CLARK, ANDREW H., "Geographical Diversity and the Personality of Canada," in Murray McCaskill, ed., *Land and Livelihood–Geographical Essays in Honour of George Jobberns* (Christchurch, N.Z.: New Zealand Geographical Society, 1962), pp. 23–47.

CORBETT, EDWARD M., *Quebec Confronts Canada* (Baltimore: Johns Hopkins Press, 1968).

GAJDA, ROMAN T., *The Canadian Ecumene: Inhabited and Uninhabited Areas* (Ottawa: Canada, Department of Mines and Technical Surveys, 1960).

GENTILCORE, R. LOUIS, ed., *Canada's Changing Geography* (Scarborough, Ont.: Prentice-Hall of Canada, 1967). A book of readings.

HAIG-BROWN, RODERICK, "The Land's Wealth," Chap. 13 in J. M. S. Careless and R. Craig-Brown, eds., *The Canadians, 1867–1967* (New York: St. Martin's Press, 1967), pp. 419–448. A concise summation of Canada's major natural resources, their use and conservation.

HARRIS, RICHARD COLEBROOK, *The Seigneurial System in Early Canada: A Geographical Study* (Madison: University of Wisconsin Press, 1966).

KERR, D. G. G., ed., *A Historical Atlas of Canada* (Toronto: Thomas Nelson and Sons [Canada], 1960).

KERR, DONALD, "Some Aspects of the Geography of Finance in Canada," *Canadian Geographer,* 9, no. 4 (1965), 175–192; and "The Geography of the Canadian Iron and Steel Industry," *Economic Geography,* 35, no. 2 (April 1959), 151–163.

KRUEGER, RALPH R., and Others, eds., *Regional and Resource Planning in Canada* (Toronto: Holt, Rinehart and Winston of Canada, 1963).

LLOYD, TREVOR, "A Water Resource Policy for Canada," *Canadian Geographical Journal,* 73, no. 1 (July 1966), 2–17.

MacKIRDY, K. A., "Conflict of Loyalties: The Problem of Assimilating the Far Wests to the Canadian and Australian Federations," *Canadian Historical Review,* 32, no. 4 (December 1951), 337–355.

McARTHUR, NEIL, and MARTIN E. GERLAND, "The Spread and Migration of French Canadians," *Tijdschrift voor Economische en Sociale Geografie,* 52, no. 6 (June 1961), 141–147.

MATTHEWS, ROY A., "A New Atlantic Role for Canada," *Foreign Affairs,* 47, no. 2 (January 1969), 334–347.

MEGILL, W. J., ed., *Patterns of Canada* (Toronto: Ryerson Press, 1967).

MOORE, BRIAN, and the Editors of *Life, Canada* (Life World Library; New York: Time Inc., 1963).

MORTON, W. L., *The Canadian Identity* (Madison: University of Wisconsin Press, 1961); and "The Geographical Circumstances of Confederation," *Canadian Geographical Journal,* 70, no. 3 (March 1965), 74–87.

NICHOLSON, NORMAN L., *Canada in the American Community* (Princeton, N.J.: D. Van Nostrand Co., Searchlight Books, 1963).

PUTNAM, DONALD F., and DONALD P. KERR, *A Regional Geography of Canada* (rev. ed.; Toronto: J. M. Dent and Sons, 1965).

PUTNAM, DONALD F., ed., BENOIT BROUILLETTE, DONALD P. KERR, and J. LEWIS ROBINSON, *Canadian Regions: A Geography of Canada* (Toronto: J. M. Dent and Sons, 1952).

RYAN, CLAUDE, "The French-Canadian Dilemma," *Foreign Affairs,* 43, no. 3 (April 1965), 462–474.

SIEKMAN, PHILIP, "The Revolt of French Canada," *Fortune,* 71, no. 2 (February 1965), 156–162.

U.S. Bureau of International Commerce, *Basic Data on the Economy of Canada,* Overseas Business Reports, OBR 67–98 (December 1967). (Washington, D.C.: Government Printing Office, 1967.)

URQUHART, M. C., and K. A. H. BUCKLEY, eds., *Historical Statistics of Canada* (Cambridge: At the University Press, 1965).

WAGLEY, CHARLES, and MARVIN HARRIS, *Minorities in the New World: Six Case Studies* (New York: Columbia University Press, 1958), "The French Canadians," pp. 169–202.

WARKENTIN, JOHN, ed., *Canada: A Geographical Interpretation* (Toronto: Methuen Publications, 1968). Essays by 23 geographers, stressing geographical change 1867–1967.

WATSON, J. WREFORD, "Canada and Its Regions," *Scottish Geographical Magazine,* 78, no. 3 (December 1962), 137–149; and "Canadian Regionalism in Life and Letters," *Geographical Journal,* 131, pt. 1 (March 1965), 21–33.

WILSON, GEORGE W., SCOTT GORDON, and STANISLAW JUDEK, *Canada: An Appraisal of Its Needs and Resources* (New York: Twentieth Century Fund, 1965).

WOLFE, ROY I., "Transportation and Politics: The Example of Canada," in W. A. Douglas Jackson, *Politics and Geographic Relationships* (Englewood Cliffs, N.J.: Prentice-Hall, 1964), pp. 221–240; reprinted from *Annals,* A.A.G., 52, no. 2 (June 1962), 176–190.

WOOD, W. DONALD, and RICHARD S. THOMAN, eds., *Areas of Economic Stress in Canada* (Proceedings of a Conference, Queen's University, at Kingston, Ontario; Kingston: Queen's University Industrial Relations Centre, 1965).

United States: General

PHYSICAL, HISTORICAL, POPULATION, AND CULTURAL GEOGRAPHY

"American Indians and American Life," *Annals of the American Academy of Political and Social Science,* vol. 311 (May 1957). An entire issue, with contributions by various authors.

BEALE, CALVIN L., "Rural Depopulation in the United States: Some Demographic Consequences of Agricultural Adjustments," *Demography,* 1, no. 1 (1964), 264–272.

BROWN, RALPH H., *Historical Geography of the United States* (New York: Harcourt, Brace, & World, 1948).

CALEF, WESLEY C., and HOWARD J. NELSON, "Distribution of Negro Population in the United States," *Geographical Review,* 46, no. 1 (January 1956), 82–97.

CHRISTENSEN, D. E., "The Auto in America's Landscape and Way of Life," *Geography,* 51, pt. 4 (November 1966), 339–348.

CLAWSON, MARION, "Factors and Forces Affecting the Optimum Future Rural Settlement Pattern in the United States," *Economic Geography,* 42, no. 4 (October 1966), 283–293.

COAN, OTIS W., and RICHARD G. LILLARD, *America in Fiction: An Annotated List of Novels That Interpret Aspects of Life in the United States, Canada, and Mexico* (5th ed.; Palo Alto, Calif.: Pacific Books, 1967).

COHEN, SAUL B., ed., *Problems and Trends in American Geography* (New York: Basic Books, 1967). A methodological and substantive survey, with contributions by 19 American geographers.

Fortune (14 issues a year). Innumerable articles on the economy and society of the United States.

HANDLIN, OSCAR, and others, eds., *Harvard Guide to American History* (Cambridge, Mass.: Harvard University Press, 1954). A bibliographical guide.

HART, JOHN FRASER, "The Changing Distribution of the American Negro," *Annals of the Association of American Geographers,* 50, no. 3 (September 1960), 242–266.

JACKSON, J. B., "The New American Countryside: An Engineered Environment," *Landscape,* 16, no. 1 (Autumn 1966), 16–20.

JENSEN, MERRILL, *Regionalism in America* (Madison: University of Wisconsin Press, 1951).

KNIFFEN, FRED, and HENRY GLASSIE, "Building in Wood in the Eastern United States: A Time-Place Perspective," *Geographical Review,* 56, no. 1 (January 1966), 40–66.

KOELSCH, WILLIAM A., ed., *Lectures on the Historical Geography of the United States as Given in 1933 by Harlan H. Barrows* (University of Chicago, Department of Geography Research Paper No. 77; Chicago: 1962).

LA FARGE, OLIVER, "The Enduring Indian," *Scientific American,* 202, no. 2 (February 1960), 37–45.

Landscape (quarterly).

LERNER, MAX, *America as a Civilization: Life and Thought in the United States Today* (New York: Simon and Schuster, 1957).

LOWENTHAL, DAVID, "The American Scene," *Geographical Review,* 58, no. 1 (January 1968), 61–88. Deals with American attitudes toward the American landscape.

McMANIS, DOUGLAS R., *Historical Geography of the United States: A Bibliography–Excluding Alaska and Hawaii* (Ypsilanti, Mich.: Division of Field Services, Eastern Michigan University, 1965).

MORRILL, RICHARD L., "The Negro Ghetto: Problems and Alternatives," *Geographical Review,* 55, no. 3 (July 1965), 339–361.

National Atlas of the United States. Available as separate map sheets from the U.S. Geological Survey, Washington, D.C.

NATIONAL RESEARCH COUNCIL, COMMITTEE TO SELECT TOPOGRAPHIC QUADRANGLES ILLUSTRATING CULTURAL GEOGRAPHY, *Rural Settlement Patterns in the United States as Illustrated on One Hundred Topographic Quadrangle Maps* (Washington, D.C.: 1956).

O'DONOVAN, PATRICK, and Others, *The United States* (Life World Library; New York: Time Inc., 1965).

PATTON, DONALD J., *The United States and World Resources* (Princeton, N.J.: D. Van Nostrand Co., Searchlight Books, 1968).

ROSE, ARNOLD M., ed., "The Negro Protest," *Annals of the American Academy of Political and Social Science,* vol. 357 (January 1965). Entire issue.

ROSE, HAROLD M., "The All-Negro Town: Its Evolution and Function," *Geographical Review,* 55, no. 3 (July 1965), 362–381.

SHIMER, JOHN A., *This Sculptured Earth: The Landscape of America* (New York: Columbia University Press, 1959).

THORNBURY, WILLIAM D., *Regional Geomorphology of the United States* (New York: John Wiley & Sons, 1965).

Time-Life Library of America (New York: Time Inc.). Individual volumes on different sections of the United States. Good photos and bibliographies.

U.S. BUREAU OF THE CENSUS, *County and City Data Book, 1967: A Statistical Abstract Supplement; Statistical Abstract of the United States* (annual); *Historical Statistics of the United States, Colonial Times to 1957: Continuation to 1962 and Revisions* (1965); *Pocket Data Book, USA, 1967;* and Geography Division, maps, 1:5,000,000, in color, on various topics.

WAGLEY, CHARLES, and MARVIN HARRIS, *Minorities in the New World: Six Case Studies* (New York: Columbia University Press, 1958), "The Negro in the United States," pp. 119–159.

WHITTLESEY, DERWENT, Chap. 9, "The United States: The Origin of a Federal State," and Chap. 10, "The United States: Expansion and Consolidation," in W. Gordon East and A. E. Moodie, eds., *The Changing World* (New York: Harcourt, Brace & World, 1956), pp. 239–260 and 261–284.

ZELINSKY, WILBUR, "Changes in the Geographic Patterns of Rural Population in the United States 1790–1960," *Geographical Review,* 52, no. 4 (October 1962), 492–524; and "An Approach to the Religious Geography of the United States: Patterns of Church Membership in 1952," *Annals of the Association of American Geographers,* 51, no. 2 (June 1961), 139–193.

NATURAL RESOURCES; ECONOMIC AND URBAN GEOGRAPHY

ALEXANDERSSON, GUNNAR, "Changes in the Location Pattern of the Anglo-American Steel Industry: 1948–1959," *Economic Geography,* 37, no. 2 (April 1961), 95–114; and *The Industrial Structure of American Cities* (Lincoln: University of Nebraska Press, 1956).

AUSTIN, MORRIS E., *Land Resource Regions and Major Land Resource Areas of the United States (Exclusive of Alaska and Hawaii).* (U.S. Department of Agriculture, Soil Conservation Service, Agriculture Handbook No. 296; Washington, D.C.: Government Printing Office, 1965.)

BARLOON, MARVIN J., "The Interrelationship of the Changing Structure of American Transportation and Changes in Industrial Location," *Land Economics,* 41, no. 2 (May 1965), 169–179.

BERRY, BRIAN J. L., and THOMAS D. HANKINS, *A Bibliographic Guide to the Economic Regions of the United States* (University of Chicago, Department of Geography Research Paper No. 87; Chicago: 1963).

BOGUE, DONALD J., AND CALVIN L. BEALE, *Economic Areas of the United States* (New York: The Free Press, 1961).

BORCHERT, JOHN R., "American Metropolitan Evolution," *Geographical Review,* 57, no. 3 (July 1967), 301–332.

BURCK, GILBERT, "The American Genius for Productivity," *Fortune,* 52, no. 1 (July 1955), 86–87 ff.

BURTON, IAN, and ROBERT W. KATES, *Readings in Resource Management and Conservation* (Chicago: University of Chicago Press, 1965).

CARTER, RICHARD E., "A Comparative Analysis of United States Ports and Their Traffic Characteristics," *Economic Geography,* 38, no. 2 (April 1962), 162–175.

GOODWIN, WILLIAM, "The Management Center in the United States," *Geographical Review,* 55, no. 1 (January 1965), 1–16.

HARRIS, CHAUNCY D., "Agricultural Production in the United States: The Past Fifty Years and the Next," *Geographical Review,* 47, no. 2 (April 1957), 175–193; and "The Market as a Factor in the Localization of Industry in the United States," *Annals of the Association of American Geographers,* 41, no. 4 (December 1954), 315–348.

HIGBEE, EDWARD, *American Agriculture: Geography, Resources, Conservation* (New York: John Wiley & Sons, Inc., 1958); and *The Squeeze: Cities without Space* (New York: William Morrow & Co., 1960).

HOGERTON, JOHN F., "The Arrival of Nuclear Power," *Scientific American,* 218, no. 2 (February 1968), 21–31.

LANDSBERG, HANS H., and SAM H. SCHURR, *Energy in the United States: Sources, Uses, and Policy Issues* (Baltimore, Md.: Johns Hopkins Press for Resources for the Future, 1968).

McDONALD, JOHN, "Steel Is Rebuilding for a New Era," *Fortune,* 74, no. 5 (October 1966), 130–137ff.

MARSCHNER, F. J., *Land Use and Its Patterns in the United States* (U.S. Department of Agriculture, Handbook No. 153; Washington, D.C.: Government Printing Office, 1959).

MOUZON, OLIN T., *Resources and Industries of the United States* (New York: Appleton-Century-Crofts, 1966).

MURPHY, RAYMOND E., *The American City: An Urban Geography* (New York: McGraw-Hill Book Company, 1966).

NORTON-TAYLOR, DUNCAN, "What the United States Can Do about World Hunger," *Fortune,* 73, no. 6 (June 1966), 111–114 ff.

PASSONNEAU, JOSEPH R., and RICHARD SAUL WURMAN, *Urban Atlas: 20 American Cities* (Cambridge, Mass.: M.I.T. Press, 1966).

PATTON, DONALD J., "General Cargo Hinterlands of New York, Philadelphia, Baltimore and New Orleans," *Annals of the Association of American Geographers,* 48, no. 4 (December 1958), 436–455.

PRED, ALLAN, "The Concentration of High-Value-Added Manufacturing," *Economic Geography,* 41, no. 2 (April 1965), 108–132; also, "Industrialization, Initial Advantage, and American Metropolitan Growth," *Geographical Review,* 55, no. 2 (April 1965), 158–185; "The Intra-metropolitan Location of American Manufacturing," *Annals of the Association of American Geographers,* 54, no. 2 (June 1964), 165–180; and "Toward a Typology of Manufacturing Flows," *Geographical Review,* 54, no. 1 (January 1964), 65–84.

STONE, JOSEPH K., "Oxygen in Steelmaking," *Scientific American,* 218, no. 4 (April 1968), 24–31.

TAAFFE, EDWARD J., "The Urban Hierarchy: An Air Passenger Definition," *Economic Geography,* 38, no. 1 (January 1962), 1–14; and "Air Transportation and United States Urban Distribution," *Geographical Review,* 46, no. 2 (April 1956), 219–238.

ULLMAN, EDWARD L., *American Commodity Flow, A Geographical Analysis of Rail and Water Traffic Based on Principles of Spatial Interchange* (Seattle: University of Washington Press, 1957); and "Amenities as a Factor in Regional Growth," *Geographical Review,* 44, no. 2 (April 1954), 119–132.

VAN DERSAL, WILLIAM R., *The American Land: Its History and Uses* (New York: Oxford University Press, 1943).

WALLACE, WILLIAM H., "Freight Traffic Functions of Anglo-American Railroads," *Annals of the Association of American Geographers,* 53, no. 3 (September 1963), 312–331; and "Railroad Traffic Densities and Patterns," *Annals of the Association of American Geographers,* 48, no. 4 (December 1958), 352–374.

WINSTON, OLIVER C., "An Urbanization Pattern for the United States: Some Considerations for the Decentralization of Excellence," *Land Economics,* 43, no. 1 (February 1967), 1–9.

WOYTINSKY, EMMA S., *Profile of the U.S. Economy: A Survey of Growth and Change* (New York: Frederick A. Praeger, 1967).

ZIVNUSKA, JOHN A., *U.S. Timber Resources in a World Economy* (Washington, D.C.: Resources for the Future, 1967; distributed by the Johns Hopkins Press).

See also the lists of references and readings for Chapters 1–3 and 30.

Anglo-American
Regions

30   Among the most notable characteristics of Anglo-America as a world region are a certain broad uniformity of culture and a simplicity of political organization which contrast sharply with the cultural and political complexity of such areas as Europe or the Orient. Nevertheless, even Anglo-America is sufficiently different from place to place that a multiplicity of cultural and political regions and subregions can be distinguished. The regional complexity becomes much greater when physical and economic factors are considered. Since a regional treatment of some type seems essential in presenting the geography of such a vast and varied segment of the earth's surface, the problem arises as to what scheme, or schemes, of regional division should be chosen.

One possible solution to this problem is provided by certain grand divisions which are commonly recognized by scholars in various fields and the public generally: in the United States the Northeast (including its distinctive subrealm of New England), the South, the Middle West, and the West; in Canada the Maritime Provinces, Newfoundland, Quebec, Ontario, the Prairie Provinces, British Columbia, and the vast,

Major landform divisions of Anglo-America. Only certain portions of Canada's Arctic islands share the rock formations of the Canadian Shield. However, the overall physical conditions of the islands are sufficiently similar to conditions in the Shield that it is convenient for the purposes of this text to regard the entire island group as an extension of the Shield. (Physiographic boundaries on the map after Fenneman, Lobeck, and others.)

empty Northern Lands; with Alaska and Greenland representing essentially continuations of the Canadian Northern Lands in physical respects, though politically separate from the latter. But these grand divisions, defined essentially on a cultural and political basis, are cut across, subdivided, and given great internal variety by a series of major landform divisions and climatic regions, each of which tends to have certain characteristic economic activities associated with it. For example, one landform division, the Rocky Mountains, extends not only completely across Canada and nearly across conterminous United States but also into Alaska. Yet through-

out the Rockies there are certain broad similarities in the aspect of the land and a recurring pattern of economic development, whether the particular segment of the mountains be located in the American West, the Canadian West, or Alaska (although in northern Canada and Alaska a severe climate greatly restricts the range of economic opportunities). A somewhat comparable, though economically more complex, situation prevails with respect to the great Interior Plains of Anglo-America, which occupy portions of the American Middle West, South, and West, and extend into the Prairie Provinces of Canada and parts of Ontario and Quebec.

On the whole, the physiographic and associated climatic and economic complexity of Anglo-America appears sufficiently greater than the cultural and political complexity that a basic scheme of regions utilizing landform boundaries seems a suitable means of avoiding repetition and generally giving sharpness to an introductory discussion. Thus the regional treatment in the present chapter is organized under eleven major landform divisions: the Canadian Shield, the Gulf-Atlantic Coastal Plain, the Piedmont, the Appalachian Highlands, the Interior Highlands, the Interior Plains, the Rocky Mountains, the Intermountain Basins and Plateaus, the Pacific Mountains and Valleys, the Arctic Coastal Plains, and Greenland (map, p. 707). The discussion of these divisions is prefaced by a brief survey of the major climatic regions of Anglo-America, whose boundaries only rarely coincide with those of the landform divisions, and which give much internal variety to the latter. Data with respect to regional economic and urban development are included in the discussion of the respective landform divisions, and the entire treatment is viewed, insofar as possible, against a general background of the grand cultural-political divisions, such as the South and Middle West, which have been outlined earlier.

CLIMATIC TYPES AND REGIONS

The climatic pattern of Anglo-America exhibits both diversity and largeness of scale. Every major nontropical type of climate is found within the region, usually over broad expanses (endpaper climatic map). The United States includes a greater number of major climatic types within its boundaries than does any other country of the world. In addition to the country's middle and high latitude climates, the state of Hawaii and extreme southern Florida give it small areas of tropical climates. Even Canada is more diversified in climate than is commonly assumed. The variety of economic opportunities and possibilities afforded by the wide range of climates in Anglo-America is one of the basic factors underlying the economic strength of this world region.

Polar and Subarctic Climates

Northern Anglo-America is handicapped by cold. A belt of *tundra climate,* with its associated vegetation of mosses, lichens, sedges, hardy grasses, and low bushes, rims the coast from Alaska to Labrador and extends into the Arctic islands of Canada and the coastal sections of Greenland. In interior Greenland an *ice-cap climate* prevails, and the land is covered by an enormous continental glacier. In the ice-cap climate every month has an average temperature below 32°F. In the tundra climate from 1 to 4 months average above freezing, although no month averages above 50°F. South of the tundra a vast expanse of *subarctic climate* spreads across most of Alaska and occupies about half of Canada. Like its counterpart in Eurasia, this belt extends completely across the continent between the Atlantic and Pacific oceans. The subarctic region of Anglo-America, with its long, cold winters and short, mild summers, has a natural vegetation of coniferous snow forest resembling the Russian taiga. Population is extremely sparse in the zones of tundra and subarctic climate, and practically nonexistent in the ice-cap climate. Small groups of people are supported by trapping, hunting, fishing, mining, lumbering, and military activities.

Humid Continental and Humid Subtropical Climates

Approximately the eastern half of Anglo-America south of the subarctic zone is an area of humid climate which accounts for most of Anglo-America's agricultural production and within which most of the people of this world region are found. Temperature variations associated with a spread in latitude of more than 25 degrees provide the basis for a division of this humid eastern area into three climatic regions, each taking the form of an east-west belt. At the north is the region of *humid continental climate with short summers,* in the center is the *humid continental climate with long summers,* and to the south is the *humid subtropical climate.* The pattern of natural vegetation within these climatic regions is complex, each being associated with areas of coniferous evergreen softwoods, broadleaf deciduous hardwoods, mixed hardwoods and softwoods, and prairie grasses. Variations in temperature and growing season have affected agriculture in such a way that each of the three regions has come to be particularly identified with one or two characteristic crops and/or types of production: spring wheat and dairy farming in the humid continental short-summer climate; a combination of corn, soybeans, small grains, and livestock in the humid continental long-summer climate; and cotton in the humid subtropical climate.

Marine West Coast and Mediterranean Climates

Due to the barrier effect of high mountains near the sea, ocean waters offshore which are warm in winter and cool in summer relative to the land, and winds prevailingly from the west throughout the year, a narrow coastal strip from southern Alaska to northern California has a distinctive *marine west coast climate,* which is strikingly similar to that of the corresponding climatic region in northwestern Europe. Most of the land along the coast is too rugged for much agriculture, and the mountains prevent the penetration of marine conditions for any great distance inland. Nevertheless, a certain economic resemblance to northwestern Europe is shown in a considerable development of dairying which has taken place in western Oregon and Washington and southwestern British Columbia. However, dairy farming in this section is much less widely practiced than in its European counterpart. The mild, moist conditions of the Anglo-American region of marine west coast climate have produced a magnificent growth of giant conifers, including the famed redwoods of northern California and Douglas fir of Oregon and Washington. A great development of lumbering based on these forests is perhaps the activity for which this section of Anglo-America is best known. In central and southern California, south of the area of marine climate, the rain comes in winter and temperatures are subtropical, giving the United States a small but very important area of *mediterranean* or *dry-summer subtropical climate.* Irrigated production of fruit and many other crops constitutes the principal form of agriculture in this climatic region.

Dry Climates and Highland Climates

In the immense region between the western littoral of the United States and the landward margins of the humid East, the dominant climatic characteristic is lack of moisture. The only large area of true *desert climate* occurs along the southern edge, in Arizona and adjoining parts of California, Nevada, Utah, New Mexico, and Texas. The rest is generally classified as having a semiarid *steppe climate,* although scattered areas of desert occur west of the Rocky Mountains as far north as southeastern Washington. The region of steppe climate extends northward into Canada both east and west of the Rockies. The steppe and desert areas of Anglo-America are inhabited by a sparse population with an economy based principally on extensive grain farming, irrigated agriculture, mining, and the grazing of cattle and sheep on vast areas of range land. The prevailing natural vegetation of short grass, bunch grass, shrubs, and stunted trees supplies forage varying greatly in utility from place to place. High, rugged mountains rising in the midst of these dry lands have *highland climates* varying with latitude, altitude, and exposure to sun and to moisture-bearing winds. Mountain slopes having sufficient precipitation are forested up to the timber line with conifers, principally pine and fir.

Climatic data for selected Anglo-American stations are given in Table 31.

THE CANADIAN SHIELD

The landform division called the Canadian Shield,[1] or sometimes the Laurentian Shield or Laurentian Upland or Plateau, occupies an immense area in northeastern Anglo-America. It extends from the Arctic Ocean to the line of the Great Lakes and the St. Lawrence Lowlands, and from the Atlantic to a line on the west traversing Great Bear Lake, Great Slave Lake, Lake Athabaska, and Lake Winnipeg (maps, pp. 707 and 711). The Shield covers more than half of Canada, and extends into the United States to the west and south of Lake Superior; this American section is known as the Superior Upland. The Adirondack Mountains in New York State are geologically an extension of the Shield, but for most purposes they are more conveniently discussed with the Appalachian Highlands. In general, the topography of the Shield is that of a rolling plain or an endless succession of low, rounded hills. Its surface was repeatedly scoured by the continental glaciers which spread from both sides of Hudson Bay in the Great Ice Age. Thus the ancient rocks composing the Shield are generally exposed at the surface or covered by a very thin layer of poor soil.

---

[1] "Shield" is a term applied to a number of areas in the world that are underlain by extremely ancient igneous and metamorphic types of rock. Such areas have remained above the level of the sea for most or all of their known geologic history. Hence they are devoid of sedimentary formations, or nearly so. Besides the Canadian Shield, prominent examples include the Fennoscandian Shield in northern Europe and the Angara Shield in Siberia.

TABLE 30 CLIMATIC DATA FOR SELECTED ANGLO-AMERICAN STATIONS

STATION	REGION	ELE-VATION ABOVE SEA LEVEL (FEET)	TYPE OF CLIMATE	AVERAGE TEMPERATURE (DEGREES F. TO NEAREST WHOLE DEGREE)			AVERAGE ANNUAL PRECIPITATION TO NEAREST INCH
				AN-NUAL	JANU-ARY	JULY	
Barrow, Alaska	Arctic Coastal Plains	22'	Tundra	10°	– 18° (Feb.)	39°	4"
Fairbanks, Alaska	Yukon River Basin	436'	Subarctic	26°	– 11°	60°	11"
Kapuskasing, Ontario	Canadian Shield	752'	Subarctic	34°	0°	63°	33"
Regina, Saskatchewan	Interior Plains (Spring Wheat Belt)	1884'	Humid continental (short summer)	36°	2°	67°	16"
Montreal, Quebec	St. Lawrence Lowlands	187'	Humid continental (short summer)	45°	16°	71°	41"
Portland, Maine	Northern New England	61'	Humid continental (short summer)	45°	22°	68°	43"
Des Moines, Iowa	Interior Plains (Corn Belt)	948'	Humid continental (long summer)	49°	20°	76°	30"
Concordia, Kansas	Interior Plains (Winter Wheat Belt)	1375'	Humid continental (long summer)	55°	28°	80°	25"
Lexington, Kentucky	Southeastern Interior Plains	979'	Humid subtropical	56°	35°	77°	45"
Norfolk, Virginia	Atlantic Coastal Plain	26'	Humid subtropical	60°	41°	79°	45"
Atlanta, Georgia	Southern Piedmont	975'	Humid subtropical	62°	45°	79°	47"
Miami, Florida	Atlantic Coastal Plain	8'	Humid subtropical	75°	67°	82°	60"
Houston, Texas	Gulf Coastal Plain	41'	Humid subtropical	70°	55°	84°	46"
Dallas, Texas	Gulf Coastal Plain	487'	Tropical savanna	66°	46°	85°	35"
Laredo, Texas	Lower Rio Grande Valley	500'	Steppe	74°	58°	88°	19"
Denver, Colorado	Great Plains	5292'	Steppe	50°	30°	73°	15"
Miles City, Montana	Great Plains	2629'	Steppe	46°	17°	75°	12"
Calgary, Alberta	Great Plains	3540'	Steppe	39°	16°	62°	18"
Spokane, Washington	Columbia Plateau	2357'	Steppe	48°	25°	71°	17"
Tucson, Arizona	Basin and Range Country	2558'	Desert	68°	50°	86°	11"
Los Angeles, California	Southern California Coast	312'	Mediterranean	62°	54°	69°	13"
Sacramento, California	California Central Valley	25'	Mediterranean	60°	45°	75°	16"
Seattle, Washington	Puget Sound Lowland	14'	Marine west coast	52°	38°	67°	36"
Juneau, Alaska	Southern Alaska Panhandle	15'	Marine west coast	40°	25°	55°	55"
Alamosa, Colorado	Rocky Mountains	7536'	Highland	42°	17°	65°	6"

Glaciation disrupted the preexisting drainage system, resulting in a plethora of lakes, swamps, and wandering streams with many rapids and waterfalls. Today this tangled wilderness area often seems to consist more of water than of land.

Much of the Shield has a rugged quality, and a few sections are mountainous. Most elevations are between 1000 and 2000 feet. The edge of the Shield north of the St. Lawrence Lowlands in Quebec rises rather abruptly to about 2000 feet and is called the Laurentide Mountains. The Torngat Mountains of northern Labrador

exceed 4000 feet and are reported to be twice that high in their little-known extension in Baffin Island.

Economic and Urban Development

Most of the Shield has either a tundra or a subarctic climate. Economic development is largely concentrated along the southern fringe, which lies in the southernmost reaches of the subarctic climate zone or in areas of humid continental short-summer climate. This more developed section of the Shield is relatively close to densely populated areas of Canada and the United States farther south. The immense, cold, northern reaches are inhabited by small numbers of Eskimos, Indians, and whites who live mainly by trapping, fishing, mining, and military employment. Most of the permanent settlements are tiny, widely scattered trading-post villages, mining settlements, or military installations. Small, isolated commercial fishing villages dot the rocky coast of Labrador, where the eastern margins of the Shield reach the Atlantic. As the search for minerals is pressed, mining settlements are appearing in increasing numbers. The most important ones in the northern Shield thus far are on the western side; they include Yellowknife (photo, bottom right), a gold-mining town on Great Slave Lake; Discovery, another gold-mining settlement north of Great Slave Lake; and Uranium City and Beaverlodge Lake on Lake Athabaska. Yellowknife, the largest of these places, has a population of only about 4000. The northern Shield is an area of critical importance to both of the Anglo-American nations from the standpoint of aerial warfare, and is the scene of many joint operations and installations of the Canadian and American armed forces.

Importance of the Forest Industries, Hydroelectricity, and Tourism in the Southern Shield

The southern fringe of the Shield makes invaluable contributions to the Anglo-American economy, although it is still a thinly populated area. These contributions stem principally from its forests, its minerals, and its rivers. The Shield is forested, mainly with conifers, almost everywhere south of the tundra. But the best and most accessible stands are along the southern margins, and here the forest industries reach their maximum

development, especially in Quebec and Ontario. Pulp and paper, destined mainly for United States markets, are the most valuable products (map, p. 722), but large quantities of lumber are also produced. Most of the mills are scattered in small riverside communities near their sources of timber. Quebec and Ontario are responsible for almost 60 percent of Canada's total forest production, and most of this comes from the southern Shield.

A multitude of hydroelectric plants are found along the rivers of the southern Shield. Canada generates 80 percent of its electricity from hydropower, and produces more hydroelectricity than does any country except the United States. About half of the country's installed hydroelectric capacity is located along or near the edge of the Shield in Quebec and Ontario. Shield power, now strongly supplemented by large installations on the St. Lawrence River, has been a key element in the development of manufacturing in the populous lowlands of Quebec and Ontario to the south.

Another industry which has been developing rapidly throughout the southern Shield is the tourist and resort business. This area offers both summer recreation and winter sports. Tourism has been fostered by public recreational developments on both the national and provincial level. Income from American visitors is an important factor helping Canada to achieve a balance in trade with the United States.

The Highly Developed Mining Industry

The ancient and highly metamorphosed rocks of the Shield are rich in metallic ores. Iron, nickel, copper, zinc, gold, and uranium are the principal metals extracted, though a number of others are also produced in considerable quantities. Mining activity is found in scattered parts of the southern Shield from Labrador to Saskatchewan, but two mining districts are outstanding—the Sudbury district in Ontario north of Lake Huron and the Mesabi district in Minnesota.

Two views of mining enterprise in the Canadian Wilderness: the new iron-mining town of Gagnon in eastern Quebec (above) and the older gold-mining town of Yellowknife in the Northwest Territories (below). Gagnon is much farther south than Yellowknife, and coniferous forest of commercial utility surrounds the town. Yellowknife, in contrast, is in the border zone where the northern coniferous forest (boreal forest) of Canada merges gradually into Arctic tundra. In the view of Gagnon, the water occupying hollows in the surface of the Canadian Shield beside the town typical of the innumerable lakes and reservoirs that are a characteristic aspect of the shield. High-grade iron ore from open-pit mines near the town is shipped southward by rail to Port Cartier on the St. Lawrence River. The watercourse in the photo of Yellowknife is an arm of Great Slave Lake. The view looks eastward across the Canadian Shield. (*Gagnon photo:* Quebec Cartier Mining Company. *Yellowknife photo:* Canadian Government Travel Bureau.)

Sudbury (110,000)[2] is the Shield's largest mining settlement. It produces about one-third of the world's nickel and large quantities of other metals, especially copper and platinum. The Superior Upland has long supplied most of the iron ore for the United States steel industry. Within the Upland the Mesabi "Range"[3] of Minnesota is overwhelmingly the main producer, although mines also are found in the Michigan and Wisconsin portions. About 60,000 to 70,000 people live in mining villages, towns, and small cities near the open-pit Mesabi mines that have made this section of Minnesota one of the world's greatest iron-mining districts. The ore is shipped east on the Great Lakes, and the port which dispatches most of it, the twin-city district of Duluth, Minnesota, and Superior, Wisconsin (260,000), is the largest urban settlement on the Canadian Shield.

The best Mesabi ores are now approaching exhaustion. New processes for concentrating its abundant low-grade ores will keep the area a major producer, but more and more of the iron ore needs of the United States are having to be met by imports. Consequently, iron resources in the Canadian sections of the Shield are increasing steadily in significance. Two important Canadian mines are operating in southwestern Ontario, one near Michipicoten and the other at Steep Rock Lake, but the great new iron ore source is the one first opened in 1954 and continuing to develop in the interior of the Shield in eastern Quebec and adjacent Labrador. Ore from new mining towns in this area (photo, p. 713), which extends north to the railhead of Schefferville, is brought by railways built for the purpose to shipping ports on the estuary of the St. Lawrence River (map, p. 711). Sept Îles (Seven Islands; about 20,000) is the largest of the ports. In recent years shipments from the Quebec-Labrador field have risen to over half of Canada's total iron ore output.

Canadian metal mining is tied closely to the United States market, in that most of the production is exported, and primarily to the United States. Growth of the American economy since World War II has led to a tremendous expansion of mining, with many new settlements as well as expansion of output at older settlements on the Shield. However, this situation can be troublesome to Canadian mining centers, as illustrated by the case of Elliot Lake. During the 1950s this settlement, a few miles north of Lake Huron and 80 miles west of Sudbury, developed from a wilderness into the world's leading uranium-mining area. However, new discoveries in the United States contributed to a glutted uranium market, and Elliot Lake's mining activities declined in the 1960s. The new town became a distressed area and required government assistance. It remains the main center of Canadian uranium output, but with lower levels of employment and output than previously.

Urban Development in the Shield

Aside from mining towns, the principal urban places of the Shield are ports and manufacturing centers. In the east a cluster of small manufacturing towns and cities has developed along the Saguenay River, a north bank tributary of the St. Lawrence. The tremendous power resources of the Saguenay have been harnessed to provide electricity for large-scale aluminum production based on ores brought in by ship. The urban centers of this district aggregate about 150,000 people. Farther west, along the "Soo" Locks connecting Lakes Superior and Huron, a pair of small Canadian and American cities with the identical name, Sault Ste. Marie, have grown to a combined population of about 95,000. The Canadian city, the larger of the two, is an iron and steel manufacturing center. On the northwestern shore of Lake Superior another twin-city district, Fort William and Port Arthur, Ontario (combined population about 100,000), is the main Great Lakes shipping point for wheat from the Prairie Provinces moving toward the east and Europe.

---

[2] As in other chapters of this book, city populations are 1968 estimates for metropolitan areas unless otherwise noted (see n. 2, Chap. 5). In the case of United States cities, 1968 estimates for standard metropolitan statistical areas are cited. A "standard metropolitan statistical area (SMSA)," as defined by the Census Bureau, is a county, or group of contiguous counties, containing at least one city ("central city") with 50,000 inhabitants or more. (In New England, standard metropolitan statistical areas are based on civil divisions other than counties.) Besides the county or counties that actually contain the central city or cities, other contiguous counties are included if they meet specified criteria as to metropolitan character. Several SMSAs include counties in more than one state. In most cases population figures for standard metropolitan statistical areas give a useful approximation of the number of people whose daily activities are closely tied to the needs and services of the central city. But of course they often include many more people than actually live in the central city or cities and their immediate suburbs. Estimates in this text for metropolitan areas in Canada or other foreign countries are not directly comparable to estimates for SMSAs. Different countries use various bases for defining metropolitan areas; often they do not include as large a proportionate extent of land outside the continuous built-up area of the city as in the case of SMSAs.

[3] The term "range" refers to the ore-bearing formation, which is long and narrow (roughly 100 miles long by 5 to 10 miles wide).

Retarded Agriculture of the Shield

Agriculture is little developed on the Shield, being hindered by short growing seasons, and soils that are thin, low in fertility, and inadequately drained. The southern fringe has been penetrated by some agricultural settlement, especially in Quebec, where the rapid natural increase of the French-Canadian population has pressed hard on the limited areas of cultivable land in the lowlands. A few outlying pioneer farming communities have been established farther north, generally in districts where the soil is formed on filled and exposed lake beds. Such ventures have not been notably successful, on the whole. Although a few dairy farming districts seem established and moderately prosperous, many of these pioneer areas have the aspect of rural slums. Often the settlers farm part time and eke out the remainder of their livelihood by work in the mining or logging camps near which their farms are usually located.

THE GULF-ATLANTIC COASTAL PLAIN

An extensive coastal plain occupies the seaward margin of Anglo-America from New Jersey to the Rio Grande River (maps, pp. 707, 718, 725, and 726). North of New Jersey it appears in two disconnected sections, Long Island and Cape Cod, and south of the Rio Grande it continues into Mexico. On the landward side it is bordered by the higher ground of the Appalachian Highlands, the Piedmont, and the Interior Highlands, and merges in certain sections with the Interior Plains. The Coastal Plain is narrow in the northeast, but widens to the south and west. Beyond the Appalachians it extends northward along the line of the Mississippi River to the mouth of the Ohio before its margin again swings off southwestward toward the Rio Grande. The landward margin of the plain may be traced through the states of Massachusetts, New York, New Jersey, Delaware, Maryland, Virginia, North Carolina, South Carolina, Georgia, Alabama, Tennessee, Kentucky, Missouri, Arkansas, Oklahoma, and Texas. Three states—Florida, Mississippi, and Louisiana—lie entirely within the Coastal Plain. This extensive portion of Anglo-America includes a substantial part of both the South and the Northeast, and reaches the margins of the Middle West and the West.

Physiography and Climate

The Gulf-Atlantic Coastal Plain is generally low in elevation. Few places exceed 500 feet above sea level, and most lie considerably lower. In general, the surface is relatively level, although there are a few belts of rolling hills. Most of the soils are sandy and of low to medium fertility. Especially fertile soils are associated, for the most part, with limited areas of river alluvium or of chalky limestone bedrock. The plain is crossed by numerous sluggish rivers, and contains many swamps and marshes, especially along its seaward edge.

The climate of most of the Coastal Plain is commonly classified as humid subtropical. However, this rather broad category embraces some significant differences, as, for example, between Virginia and Florida (where tropical savanna climate is reached in the extreme south). North of Chesapeake Bay the climate is humid continental with a considerable marine influence, and near the Rio Grande it is classed as steppe. Pine (photo, p. 26) or mixed oak-pine forest is the prevailing natural vegetation, except for stands of swamp hardwoods in poorly drained areas and prairie grasslands in certain areas underlain by limestone and along the coasts of Texas and southwestern Louisiana. The coastal prairies of Texas grade into a steppe vegetation in the vicinity of the Rio Grande.

The Varied, Specialized Agriculture of the Coastal Plain

Long, hot summers, mild winters, and abundant, well-distributed annual precipitation, amounting to 40 inches or more in all areas except the extreme west, compensate to some extent for the poverty of Coastal Plain soils, and this area has long been important in American agriculture. Much of the plain is in forest, however, and the general pattern is one of islands of intensive agricultural development scattered through a sea of woods. These intensively farmed areas vary greatly in size, productivity, and type of agricultural development. Most of them are rather highly specialized on one or a few products, but the variety of these specialties for the Coastal Plain as a whole is very great. Thus the overall pattern of agricultural activity is most intricate. Only the major aspects of this pattern are discussed below.

Truck Farming

The growing of vegetables and small fruits is important in many Coastal Plain communities, represent-

MAJOR USES OF LAND
IN THE CONTERMINOUS
UNITED STATES

Desert, mostly ungrazed
Nonforest grazing land
Crop land and pasture land
Crop land, pasture and woodland
Forest land
Swamp land and marsh land
Irrigated land

✚ Metropolitan cities, over 7,000,000

● Metropolitan cities, 500,000 or over

○ Selected smaller urban areas

----- Boundary of major landform division

0 100 200 300 400
Scale of Miles

SOURCE: Bureau of Agricultural Economics, U.S. Department of Agriculture.

ing often the principal source of income. Most truck-farming communities are near the coast. They dot almost the whole of the coastal margin from Cape Cod, with its famous cranberries, to the lower Rio Grande Valley, with its large irrigated production of early vegetables. Three general areas of truck growing are outstanding. (1) The *northeastern area* includes Long Island in New York, the southern part of New Jersey, the Delmarva Peninsula to the east of Chesapeake Bay, and extends with lesser intensity to the vicinity of Norfolk, Virginia. Nearness to the great urban markets of the industrialized Northeast is the principal advantage of truck farms in this section. (2) The

Florida Peninsula, with its warm winters and consequent advantages for early harvesting and marketing, produces over 10 percent of all United States vegetables by value. In truck growing Florida is second as a state only to California. (3) The *lower Rio Grande Valley* of Texas has climatic advantages for truck farming similar to those of Florida except that scanty rainfall makes irrigation a necessity.

Poultry Farming

The northeastern part of the Coastal Plain, from Long Island southward through Virginia, is the most

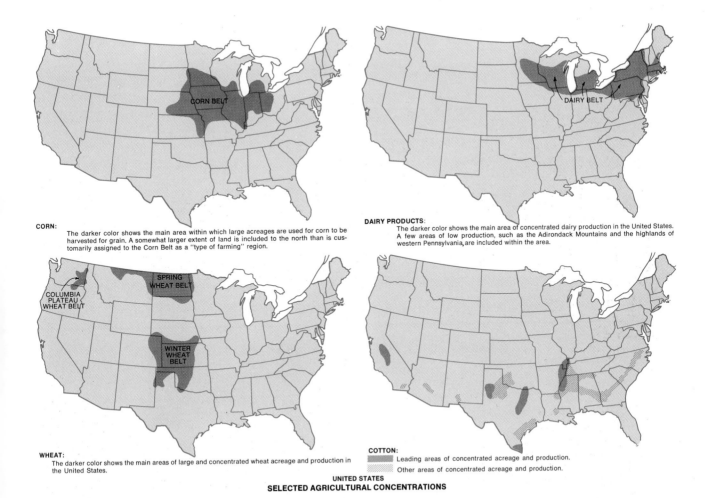

CORN:
The darker color shows the main area within which large acreages are used for corn to be harvested for grain. A somewhat larger extent of land is included to the north than is customarily assigned to the Corn Belt as a "type of farming" region.

DAIRY PRODUCTS:
The darker color shows the main area of concentrated dairy production in the United States. A few areas of low production, such as the Adirondack Mountains and the highlands of western Pennsylvania, are included within the area.

WHEAT:
The darker color shows the main areas of large and concentrated wheat acreage and production in the United States.

COTTON:
▰ Leading areas of concentrated acreage and production.
░ Other areas of concentrated acreage and production.

UNITED STATES
SELECTED AGRICULTURAL CONCENTRATIONS

intensively developed area of poultry production in the United States. Poultry farming has risen rapidly in recent years to displace truck growing as the main type of agriculture in many communities. Over half of the total farm income of Delaware, for instance, generally comes from the sale of poultry products. As with truck farming, the major stimulus is the demand of nearby urban markets. However, in recent decades chilling techniques and refrigerated transport have also allowed the development of poultry producing areas in sections of the South which are more remote from the major markets.

Peanuts

Over four-fifths of the peanut crop of the United States is grown on the Coastal Plain. Peanuts are the mainstay of agriculture in two widely separated sections— one on the Virginia–North Carolina border and the other in southern Georgia and adjoining parts of Alabama and Florida (map, p. 718). Farmers in these areas grow peanuts mainly as a cash crop, but also use them to some extent as forage for hogs. Peanuts are one of several major sources of vegetable oils used for making lard and butter substitutes, soap, and kindred products.

Tobacco

Tobacco was established in early colonial times as the first great agricultural staple of the Coastal Plain. Today the plain accounts for over two-fifths of American tobacco production by volume. Practically all of this is raised in a discontinuous belt extending from the Maryland counties between the Potomac River and Chesapeake Bay to northernmost Florida. Soils on the Coastal Plain of Virginia, where American tobacco culture originated, were so depleted and eroded by a one-crop system of agriculture that this area produces almost no tobacco today. The major areas of present-

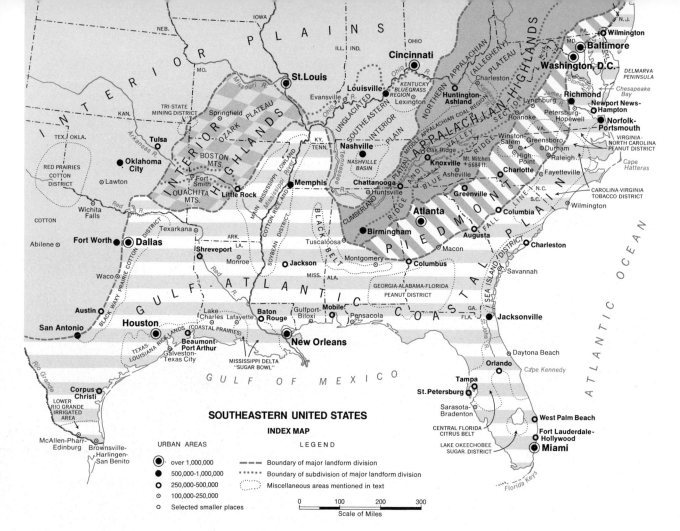

SOUTHEASTERN UNITED STATES

INDEX MAP

URBAN AREAS

- ◉ over 1,000,000
- ● 500,000-1,000,000
- ○ 250,000-500,000
- ⊙ 100,000-250,000
- ○ Selected smaller places

LEGEND

- — — — Boundary of major landform division
- ⋯⋯⋯ Boundary of subdivision of major landform division
- ⸳⸳⸳⸳ Miscellaneous areas mentioned in text

0   100   200   300
Scale of Miles

day production on the Coastal Plain are in North and South Carolina (map, above).

Cotton

The greater part of the American Cotton Belt is located on the Gulf–Atlantic Coastal Plain (maps, pp. 326 and 717). The only important cotton-producing areas outside the plain are found in the southern Piedmont, the Tennessee Valley, the northwestern Texas and adjoining Oklahoma sections of the Interior Plains, and irrigated districts scattered from Texas to California. Within the Coastal Plain cotton is widely grown from southern Virginia to Texas, except that it is generally unimportant near the immediate coast. However, most of the annual harvest comes from limited areas of concentrated production within the main belt. Throughout the history of the American South the center of cotton production has tended to migrate toward the west. To a certain extent this migration has been merely a phase of the general westward expansion of American settlement. However,

it has also been a response to progressive exhaustion of the soil and to the extensive ravages of the boll weevil, a remarkably destructive insect pest which thrives best in the more humid eastern sections of the Cotton Belt.

AREAS OF CONCENTRATED COTTON PRO-DUCTION. The first great center of cotton production in the United States was the offshore islands and adjoining mainland sections of Georgia and South Carolina. This "Sea Island District" (map, above) had lost its primacy by the time of the Civil War, but continued to be an important cotton producer until devastated by the boll weevil in the twentieth century. Very little cotton is now grown along this coast. By 1860 the Black Belt of Alabama and Mississippi and the flood plain of the Mississippi River had become the principal centers of cotton production, although the crop had already spread farther westward into Texas. The Black Belt, so called for the color of its soils, is a crescent-shaped area extending east and west through central Alabama and then curving northward into northeastern Mississippi (map, above). The virgin soils of this belt, developed on limestone bedrock,[4] were exceptionally fertile, but were worn out by excessive cropping in cotton, which, like tobacco, is notoriously hard on soils. Exhaustion of the soil plus the ravages of the boll weevil ruined cotton production in the Black Belt almost as completely as it was ruined in the Sea Island District. In recent decades the Black Belt has become an important Southern area in the production of beef cattle—a phase of agriculture which is rapidly developing in the Southern states.

The only area of concentrated cotton production now found on the Coastal Plain east of the Mississippi Lowland is a long, curving arc of land which follows the inner margin of the plain from southern Virginia into Alabama. The most important center of production on the plain, however, lies in the alluvial lowland along the Mississippi River and the lower courses of its tributaries (maps, pp. 718 and 725). The Mississippi Lowland, some 600 miles long from the mouth of the Ohio to the Gulf of Mexico and often over 50 miles in width between the bordering bluffs, is a remarkable feature in the physical geography of Anglo-America, comparable in many ways to the great flood plain and delta areas which support dense populations in the Orient. On its deep and fertile soils, shared by six states, about a quarter of the American cotton crop and nearly 10 percent of the world's crop by volume is produced. Mississippi and Arkansas, whose principal cotton-growing areas are found in the Mississippi Lowland, are the third- and fourth-ranking states in cotton production, with Texas first and California second. Almost one-third of the American cotton crop generally comes from Texas; this state contains three outstanding areas of concentrated cotton production, two of which are on the Coastal Plain. One of the two Coastal Plain districts extends along the Gulf Coast from the Rio Grande Valley, where the crop is irrigated, to the vicinity of Houston. The other Coastal Plain cotton district in Texas is the Black Waxy Prairie, an area which resembles physically the Black Belt of Alabama and Mississippi, but which, unlike the latter, is still a major cotton producer. It extends as a belt some 50 miles wide from near San Antonio to the northern border of the state. The third district is located outside the Coastal Plain in northwestern Texas and adjoining parts of Oklahoma. Production in

This mechanical cotton picker in operation near the eastern edge of the Mississippi Lowland between Memphis and Vicksburg symbolizes the high degree of mechanization which characterizes the agriculture of the United States as a whole. It also symbolizes one aspect of the agricultural revolution that has overtaken large sections of the old Cotton Belt. (Standard Oil Company, N.J.)

---

[4] There is a common idea that the soils of the Alabama–Mississippi Black Belt developed under a relatively continuous cover of tall grass prairie. Recent research, however, has indicated that the Black Belt, like the southeastern part of the United States in general, was originally a forested area with pockets of grassland. See Erhard Rostlund, "The Myth of a Natural Prairie Belt in Alabama: An Interpretation of Historical Records," *Annals of the Association of American Geographers,* 47, no. 4 (December 1957), 392–411.

this Great Plains district is mostly irrigated and exceeds that in either of the two Coastal Plain districts of the state. In recent years it has accounted for over half of Texas' total cotton crop.

EFFECTS OF COTTON GROWING ON THE REGIONAL LIFE OF THE SOUTH. More than any other economic activity, cotton production has given distinctive regional qualities to Southern life. The large Negro population of the South, in many ways the most important element differentiating this section culturally and politically from the rest of the nation, is descended from African slaves who became mainly concentrated on cotton plantations in the period between the invention of the cotton gin (1793) and the close of the Civil War. Today the heaviest densities of rural Negro population are commonly found in present or former areas of concentrated cotton production. The prevalence of cotton culture has had a primary bearing on the relatively low income levels and standards of living, the long-retarded development of other forms of economic activity, and the general economic difficulties which have been outstanding characteristics of the South, at least up to recent years. The attraction of cotton as a cash crop, its adaptability to systems of farm tenancy, especially sharecropping, and its heavy demand for hand labor, coupled with the poverty-stricken condition of the post-Civil War South and the lack of alternative sources of employment, created a situation which allowed a relatively small class of large landowners, cotton ginners and dealers, storekeepers, fertilizer merchants, and bankers to accumulate profits and even become wealthy, while at the same time the much larger group of actual cultivators, unable to acquire sufficient capital to improve their situation, were living fairly close to the margin of subsistence. Continuous cropping in cotton wore out the soil, and dependence on overseas markets resulted in increasing difficulties under the stress of a growing foreign competition. The depths of distress in the cotton South were reached in the Great Depression of the early 1930s, when the price of cotton sank to unprecedented lows and large numbers of cultivators and their families were forced to seek government relief in order to live. In the twentieth century, particularly since 1930, cotton culture in the South has experienced a drastic decline in the face of these mounting difficulties and under the stimulus of a concentrated attack on Southern problems. Great acreages have been removed from cotton cultivation, and communities which were formerly dependent on cotton have groped, with varying degrees of success, toward new types of agriculture and other means of support. Now the mechanical cotton picker is rapidly being adopted (photo, p. 719) and is releasing labor from cotton (and sometimes stranding it in unemployment) in a Southern economy that was long nearly static but today is characterized by vigorous growth and change.

Citrus Fruit

Most of the output of citrus fruit in the United States is concentrated in a few localities on the Gulf-Atlantic Coastal Plain. The danger of frost in areas farther north restricts citrus growing on the plain to the extreme southern sections of the zone of humid subtropical climate. There is some production scattered along most of the Gulf Coast, but the southernmost areas, Florida and the lower Rio Grande Valley, contain the major producing districts. Florida, in which fruit growing is the most important type of agriculture, normally accounts for about 55 percent of the total United States output of citrus. Most of the Florida production comes from a belt in the center of the peninsula east of Tampa Bay (map, p. 718). This part of Florida has many lakes and low hills which give a certain amount of added protection against frost damage. Citrus production is the principal element in the irrigated agriculture of the lower Rio Grande Valley. It is supplemented by truck farming, as in Florida, and by cotton. The only other major area of citrus growing in Anglo-America is found in southern California.

Sugar

No part of the conterminous United States is sufficiently tropical for really efficient production of cane sugar. However, two districts specializing in this product have developed behind tariff protection. The larger and more important of the two is the so-called "Sugar Bowl" in the southern Mississippi Delta of Louisiana; the other is near Lake Okeechobee in Florida (map, p. 718). Their combined production amounts to only a small percentage of American sugar consumption, considerably less than the percentage supplied by the beet-sugar industry, which is localized in other parts of the country. Most sugar used in conterminous United States is cane sugar from tropical sources, especially the Philippines, Hawaii, and Puerto Rico.

Rice

The United States supplies its own needs for rice and even exports a certain amount to the Orient. Rice consumption per capita in the United States is relatively small, and American rice is produced with extreme efficiency by machine methods. About half of the American crop comes from the prairies along the coasts of Louisiana and Texas and another fourth from Arkansas and Mississippi portions of the Mississippi Lowland (map, p. 718). The remainder comes almost entirely from California. The total American production is quite small when measured on a world scale, but rice growing is an important phase of the economy in the areas where production is centered.

Other Crops

In addition to the major specialties of Coastal Plain agriculture described above, many other crops are grown, sometimes as basic specialties in small areas, sometimes in combination (often in rotation) with the major crops. The most widespread and important of these secondary crops are corn and soybeans. Among the others are a certain amount of wheat, sorghums toward the drier west, tree crops such as pecans and tung, and a variety of grasses and legumes grown as hay, forage, and soil-building crops.

The Growing Importance of Beef Production

A growing emphasis is being placed on the production of beef cattle in many parts of the Coastal Plain, and the proportion of farmland devoted to feed crops, hay, and improved pasture has increased in recent decades. However, cattle production has become dominant in Coastal Plain agriculture only in certain areas that tend to be especially unproductive agriculturally. Prominent examples are parts of northern Florida and southern Mississippi, and much of eastern Texas and adjoining sections of Louisiana and Arkansas. In these areas some counties that derived less than a fourth of their farm income from cattle before World War II now gain over three-quarters from this source.

Forestry, Tourism, and Fishing

Three other important economic activities are widely scattered over the Coastal Plain or along its shores. These are the forest industries, the tourist and resort industry, and commercial fishing. Most of the original natural forest of the Coastal Plain has been cut, but very large areas are in second growth timber of varying size and quality. One of the great natural assets of the southern Coastal Plain, as well as of other areas in the South, is the rapid growth of trees in the subtropical climate. The forests of southeastern United States furnish about one-third of the country's

total production of lumber and 60 percent of its wood pulp (map, p. 722), which is used chiefly, and generally in the same integrated mills, to make kraft paper and containers. At least half of the South's lumber and pulp production comes from the Coastal Plain. Southern pine is the outstanding Southern source both of lumber and of pulpwood, and in addition supplies more than half of the world output of naval stores. But the production of hardwoods from river bottom lands is also important. Memphis, Tennessee, on the Mississippi, is a major commercial center for hardwood lumber. Thus over wide areas wood industries vie with agriculture as the main basis of economic life. Only in areas of natural prairie or steppe and in the sections north of Virginia, where trees grow more slowly, are forest industries absent or unimportant on the Coastal Plain.

Resort establishments and fishing villages dot the shore from Cape Cod to Texas. Miami (1.1 million), St. Petersburg (with Tampa, 910,000), Fort Lauderdale–Hollywood (490,000), Orlando (380,000), and West Palm Beach (300,000), all in Florida, are highly developed resort and retirement centers and are the largest cities of the United States in which the resort function plays so great a role. Atlantic City, New Jersey (190,000), is the most important resort city on the Atlantic coast outside of Florida. Many of the larger cities along the coast besides those named count the resort business as an important element in their economies, while many smaller settlements are primarily dependent on it. Commercial fishing is sometimes carried on in the same communities as the resort trade, sometimes in separate villages and towns. About half of the commercial fish catch of the United States by value is ordinarily landed along this coast. Major areas of concentration are found in Chesapeake Bay and off the Texas coast, where oysters and shrimp, respectively, are the main specialties.

Mineral Resources

Most of the Coastal Plain is poor in mineral resources. A conspicuous exception is phosphate. Several parts of the plain have phosphate deposits, but over four-fifths of the American production by volume, and approximately a third of the total world production, comes from a narrow belt inland and north from Tampa in western Florida. Much of the production is marketed in the South, which uses more chemical fertilizer per acre than most regions of Anglo-America. A small area in central Arkansas near the Interior Highlands produces almost all the bauxite mined in the United States. This

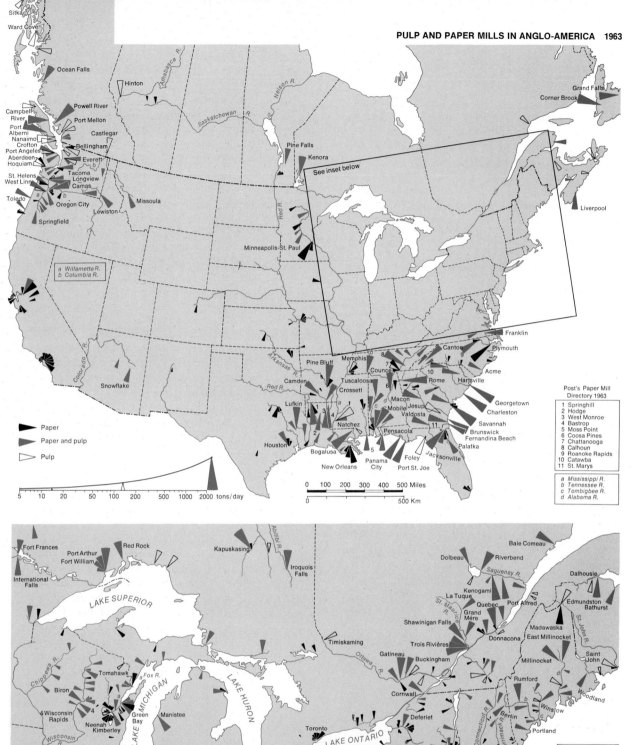

Sitka
Ward Cove
Ocean Falls
Hinton
Campbell River
Powell River
Port Mellon
Port Alberni
Nanaimo
Crofton
Castlegar
Bellingham
Port Angeles
Aberdeen
Hoquiam
Everett
Tacoma
St. Helens
West Linn
Longview
Camas
Toledo
Oregon City
Springfield
Lewiston
Missoula
Grand Falls
Corner Brook
Liverpool
Pine Falls
Kenora
See inset below
Minneapolis-St. Paul

a Willamette R.
b Columbia R.

Snowflake

Franklin
Canton
Plymouth
Memphis
Acme
Pine Bluff
Counce
Rome
Hartsville
Camden
Tuscaloosa
Crossett
Macon
Georgetown
Charleston
Lufkin
Mobile
Jesup
Valdosta
Natchez
Pensacola
Savannah
Brunswick
Fernandina Beach
Palatka
Houston
Bogalusa
Panama City
Port St. Joe
Foley
Jacksonville
New Orleans

Paper
Paper and pulp
Pulp

0 100 200 300 400 500 Miles
0   500 Km

5 10 20 50 100 200 500 1000 2000 tons/day

Post's Paper Mill Directory 1963
1 Springhill
2 Hodge
3 West Monroe
4 Bastrop
5 Moss Point
6 Coosa Pines
7 Chattanooga
8 Calhoun
9 Roanoke Rapids
10 Catawba
11 St. Marys

a Mississippi R.
b Tennessee R.
c Tombigbee R.
d Alabama R.

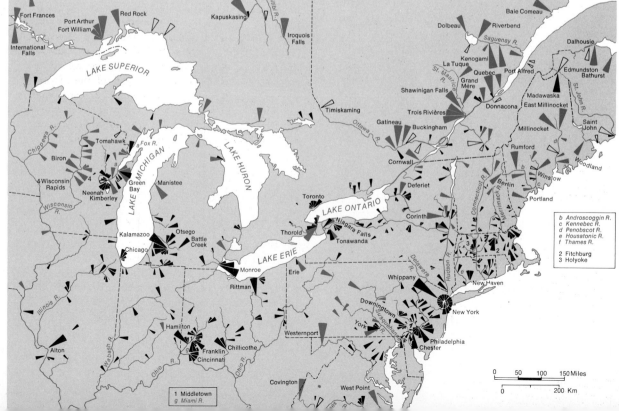

Fort Frances
Red Rock
Kapuskasing
Baie Comeau
Port Arthur
Fort William
Iroquois Falls
Dolbeau
Riverbend
International Falls
Dalhousie
Kenogami
La Tuque
Edmundston
Bathurst
Quebec
Port Alfred
LAKE SUPERIOR
Grand Mère
Madawaska
East Millinocket
Shawinigan Falls
Donnacona
Saint John
Tomahawk
Trois Rivières
Timiskaming
Gatineau
Millinocket
Biron
Green Bay
Manistee
Buckingham
Rumford
Wisconsin Rapids
Neenah
Kimberley
Cornwall
Deferiet
Berlin
Winslow
Portland
Kalamazoo
Otsego
Battle Creek
Toronto
Corinth
Chicago
Thorold
Niagara Falls
Tonawanda
Monroe
Erie
New Haven
Rittman
Whippany
Downingtown
Hamilton
Westport
New York
Alton
Franklin
Cincinnati
Chillicothe
York
Philadelphia
Chester
Covington
West Point

b Androscoggin R.
c Kennebec R.
d Penobscot R.
e Housatonic R.
f Thames R.

2 Fitchburg
3 Holyoke

1 Middletown
g Miami R.

0 50 100 150 Miles
0   200 Km

production, however, represents only a small percentage of the total American consumption of bauxite ore, most of which must be imported.

Superior Resources and Industrial Development of the Texas-Louisiana Coast

The most valuable mineral resources of the Coastal Plain are found in Texas and Louisiana. Areas of the Coastal Plain in these states, and to a lesser extent in Mississippi and Arkansas, produce an important share of the total American output of petroleum and natural gas. Major oil fields are widely scattered along the immediate coast and also inland in eastern Texas and in Louisiana. Increasing amounts of oil are being produced offshore from beneath shallow waters overlying the continental shelf. Texas, which has many fields outside the Coastal Plain, accounted for about 35 percent of the total United States petroleum production by volume and Louisiana for another 21 percent during the middle 1960s. The Coastal Plain contribution to American oil production can hardly be stated as less than about one-third, which is about 8 percent of the petroleum production of the world. To their wealth in oil, coastal Louisiana and Texas add large deposits of sulfur and salt. Around two-thirds of the total world production of natural sulfur comes from this area.

The abundant resources of oil and natural gas, which are important chemical raw materials as well as fuels, plus sulfur, salt, inland timber, agricultural wealth, and access to ocean and coastwise water transportation go far toward explaining the rapid development of industry along this coast in recent years. During this period most of the Coastal Plain has experienced industrial development on a much smaller scale, and it has been confined primarily to the processing of local agricultural products or timber. The major industries that have developed in coastal Texas and southern Louisiana include oil refining, chemical and synthetic rubber production, aluminum manufacture from imported

bauxite, and the smelting of other imported ores such as zinc and tin. However, a scattering of other types of factories exist, including even a small iron and steel plant at Houston and another one near Texarkana (100,000).

Urban Development

The Coastal Plain is not, on the whole, an outstanding area with respect to the level of industrial and urban development, despite rapid urban and industrial growth in recent decades. According to 1968 estimates, the United States included twenty-eight metropolitan areas of over 1 million population (counting New York and northeastern New Jersey as one). Of these only four—Houston, Dallas, Miami, and New Orleans (in order of population)—are Coastal Plain cities, and the largest of these ranked thirteenth among the twenty-eight. With respect to manufacturing, the larger cities rank even lower. In the middle 1960s only two Coastal Plain cities were among the top twenty-eight metropolitan areas in the United States in manufacturing employment; and these two—Houston and Dallas—ranked twenty-first and twenty-third, respectively.

Most of the principal cities are on or near the coast and are either ports or resort and retirement centers, or both. The Florida resort-retirement cities have already been mentioned. Among the major urban centers which are also major ports are Houston, New Orleans, and the cluster of cities around Hampton Roads at the mouth of the James River. The growth of Houston (1.8 million) to a major city has occurred almost entirely in the twentieth century. Completion of a ship canal to the city in the early part of the century has allowed spectacular growth in ocean shipping. The city is also the main commercial and supply center for the Gulf Coast oil fields and a considerable center for the manufacture of oil field equipment and petro-

Pulp and paper mills in Anglo-America are concentrated primarily in three areas providing abundant supplies of pulpwood from coniferous softwood trees: the Gulf-Atlantic Coastal Plain from Virginia to eastern Texas; the Pacific Mountains and Valleys of Oregon, Washington, and British Columbia; and the southeastern margins of the Canadian Shield and adjacent St. Lawrence Valley in Quebec Province. The numerous small paper mills in the northeastern quarter of the United States are devoted in large part to the manufacture of specialized, relatively expensive types of paper. The map was redrafted from Gunnar Alexandersson and Bengt Geijer, "The Paper and Pulp Industry in Anglo-America and the Nordic Countries: a Comparison of Mill Size and Degree of Integration," *Skandinaviska Banken Quarterly Review*, 1966:4, p. 111, reprinted in Gunnar Alexandersson, *Geography of Manufacturing* (Englewood Cliffs, N.J.: Prentice-Hall, Inc., 1967), p. 107. Reproduced by permission of Gunnar Alexandersson, *Skandinaviska Banken Quarterly Review*, and Prentice-Hall, Inc.

chemicals, and for oil refining. New Orleans (1.1 million) became a major port for the Mississippi River basin in the early 1800s, and continues in this role, as well as sharing in the Gulf Coast oil business. Hampton Roads, the magnificent harbor at the mouth of the James River in southeastern Virginia, lies between the metropolitan areas of Norfolk–Portsmouth (680,000), south of the harbor, and Newport News–Hampton (290,000), north of it. Collectively these cities comprise the greatest coal exporting port in the world, are one of the principal shipbuilding centers in the United States, and are the main Atlantic base of the United States Navy. As a commercial seaport Hampton Roads ranks fourth in the United States in total tonnage handled, after New York, Philadelphia, and Los Angeles. Two urban clusters of the Coastal Plain are major ports although not so outstanding in population. These are Beaumont–Port Arthur (320,000) in eastern Texas and Baton Rouge (270,000) in Louisiana. Beaumont and Port Arthur together comprise the fifth ranking seaport in the United States in tonnage handled. Their traffic consists mainly of outbound oil shipments, especially for east coast American markets. Baton Rouge, likewise, is mainly an oil port and refining center, although it also receives bauxite for a large alumina plant. Many other cities along both the Atlantic and Gulf margins of the Coastal Plain are seaports, but on a lesser scale. These include all of the major metropolitan areas on the coasts of central and south Florida, plus such cities as Charleston (300,000), Savannah (195,000), Jacksonville (520,000), Mobile (410,000), Galveston (165,000), and Corpus Christi (290,000).

Six of the nineteen largest metropolitan centers of the Coastal Plain are not on the coast. These are mostly cities which have grown as commercial centers for particularly productive agricultural areas, and have developed oil, industrial, or resort interests. Thus Dallas (1.4 million) and San Antonio (850,000) are associated with the Black Waxy Prairie area of Texas. Both of these cities share in the state's oil industry, and San Antonio is a commercial and processing center for extensive grazing and agricultural areas to the west. Dallas is not only a major commercial metropolis but also a diversified manufacturing center in such lines as aircraft, automobiles, machinery, and clothing. Two of the larger inland metropolitan areas, Memphis (770,000) and Little Rock (330,000) are business centers for the Mississippi Alluvial Plain, on whose edges they are located; while Orlando, Florida (385,000), is a combined commercial center for the Florida citrus industry and resort-retirement center. Shreveport, Louisiana (290,000), though not in a particularly good

agricultural area, is in the midst of an important cluster of oil and natural gas fields in northwestern Louisiana and adjacent Arkansas and Texas.

The Coastal Plain, with only four metropolitan areas of more than a million people, does not bulk very large in the total picture of urbanization in the United States. But its cities are outstanding in one significant respect: practically all of them are among the most rapidly growing of the country's urban areas. This is a reflection of the sudden dynamic growth of the Southern economy since the end of the Great Depression of the 1930s and the beginning of World War II— a growth through which the South is narrowing the gap in economic development and income which has long existed between it and several other sections of the country.

THE PIEDMONT

The Piedmont lies between the Coastal Plain and the Appalachian Highlands (maps, p. 707, 718, 725, and 726). Its elevation, varying generally between 400 and 1500 feet, is distinctly higher than that of the Coastal Plain, but lower than that of the Appalachians. Most of its surface is rolling to hilly. The northern end of the Piedmont is found in the vicinity of New York City, from which it extends generally southwestward, widening toward the south, until its southern end is reached in east central Alabama. Parts of ten states—New York, New Jersey, Delaware, Pennsylvania, Maryland, Virginia, the Carolinas, Georgia, and Alabama—lie within the Piedmont.

Climate, Vegetation, and Soils

The climate of the Piedmont is similar to that of adjacent sections of the Coastal Plain, being humid continental in the north and humid subtropical in the center and south. This landform division was originally an area of mixed forest, with hardwoods, especially oak, predominating. Once largely cleared for agriculture, large acreages of the Piedmont are now reverting to woodland, and there is a considerable production of hardwood lumber from some sections. Piedmont soils vary greatly in fertility, depending largely on the character of the underlying rock and the kind of treatment received since the beginning of settlement. Certain limestone-derived soils of southeastern Pennsylvania, in the vicinity of Lancaster, are among the most

Major physiographic features of eastern United States and southeastern Canada. (The base map is a portion of A. K. Lobeck's Physiographic Diagram of the United States, copyright, The Geographical Press, a division of C. S. Hammond & Company, Maplewood, N.J.)

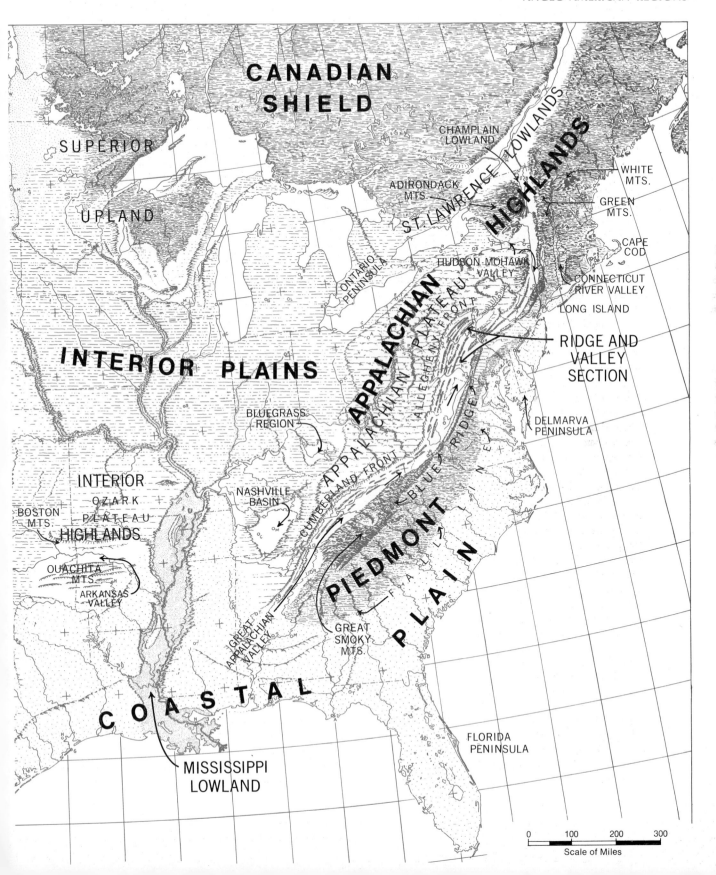

CANADIAN SHIELD

SUPERIOR

UPLAND

CHAMPLAIN LOWLAND

ST. LAWRENCE LOWLANDS

HIGHLANDS

WHITE MTS.

ADIRONDACK MTS.

GREEN MTS.

CAPE COD

ONTARIO PENINSULA

HUDSON-MOHAWK VALLEY

CONNECTICUT RIVER VALLEY

LONG ISLAND

INTERIOR PLAINS

APPALACHIAN PLATEAU

ALLEGHENY FRONT

RIDGE AND VALLEY SECTION

BLUEGRASS REGION

APPALACHIAN

DELMARVA PENINSULA

INTERIOR

OZARK

PLATEAU

NASHVILLE BASIN

CUMBERLAND FRONT

BLUE RIDGE

BOSTON MTS.

HIGHLANDS

OUACHITA MTS.

ARKANSAS VALLEY

PIEDMONT

PLAIN

GREAT APPALACHIAN VALLEY

GREAT SMOKY MTS.

FALL LINE

COASTAL

PLAIN

FLORIDA PENINSULA

MISSISSIPPI LOWLAND

0   100   200   300
Scale of Miles

NORTHEASTERN UNITED STATES and SOUTHEASTERN CANADA

INDEX MAP

Boundary of major landform division
Boundary of subdivision of major landform division

Scale of Miles

URBAN AREAS

over 4,000,000
1,000,000–4,000,000
500,000–1,000,000
250,000–500,000
100,000–250,000
Smaller places mentioned in text

City-size symbols are based on 1968 estimates for metropolitan areas (Standard Metropolitan Statistical Areas [SMSAs].)

productive in the United States. These naturally fertile soils are farmed largely by the "Pennsylvania Dutch" (of German extraction), long known as some of America's most skillful and industrious farmers. Lancaster County, the heart of the Pennsylvania Dutch country, is renowned as an outstanding agricultural area. But most Piedmont soils are derived from metamorphic or igneous rocks, were never exceptionally fertile, and have been seriously depleted and eroded

(especially in the central and southern Piedmont) as a result of agriculture stressing the production of clean-tilled row crops—such as tobacco, cotton, and corn—in a hilly area subject to violent downpours of rain.

Agriculture

The principal agricultural contributions of the Piedmont to the American economy are dairy products, tobacco, livestock, and poultry. Dairy farming is the principal type of agriculture as far south as Virginia, that is, in the section closest to the largest urban markets. Hay and feed grains, especially corn, barley, and wheat, are the principal crops grown on Piedmont dairy farms. Dairying is supplemented or supplanted in various areas by the production of tobacco, beef cattle, potatoes, apples, poultry, or some combination of these. The Piedmont of southern Virginia and North Carolina is a major tobacco farming area. North Carolina, with its Piedmont and Coastal Plain production areas, is the leading American state in tobacco growing, accounting for about 37 percent of the national production by value. The Virginia Piedmont adds another 7 percent. All in all, over 60 percent of this important American crop comes from Piedmont or Coastal Plain areas between northern Florida and Maryland. From North Carolina to Alabama the Piedmont was until recently a part of the Cotton Belt, but has now been converted principally to livestock and poultry production, while much of it has returned to woodland. Corn is generally the leading secondary crop in the tobacco districts, and the main crop in the livestock and poultry districts.

Urban and Industrial Development in the Northern Piedmont

The northern Piedmont is a very highly urbanized and industrialized area. New York City and its environs, which lie at the extreme northern end, are discussed in the section on the Appalachian Highlands. South of New York, five important cities, including three of the country's greatest metropolitan districts, are closely spaced in a northeast-southwest line of about 150 miles. They include Trenton, New Jersey (300,000), Philadelphia, Pennsylvania (4.8 million), Wilmington, Delaware (480,000), Baltimore, Maryland (2.0 million), and Washington, D.C. (2.7 million). Washington is almost entirely a governmental city. The others are industrial and commercial centers.

It will be noted that each of these cities lies at or near the outer edge of the Piedmont. The zone of contact between this landform division and the Coastal Plain is known as the Fall Line, a name deriving from the falls and rapids which mark the course of rivers as they descend from the ancient hard rocks of the Piedmont uplands to the lower, sedimentary Coastal Plain. These falls and rapids mark the head of navigation on many rivers and have long supplied waterpower. Consequently, many urban areas originated and developed into cities at sites where a river crosses the

Fall Line, and a line of cities may now be traced along the latter. Besides the five already mentioned, the largest Fall Line cities are Richmond, Virginia (515,000); Columbia, South Carolina (310,000); and Augusta (290,000) and Columbus (270,000), Georgia. Smaller Fall Line cities, all over 200,000 in metropolitan population, include Raleigh and Fayetteville in North Carolina, Macon in Georgia, and Montgomery in Alabama.

However, the only place where really large cities have developed along the Fall Line is in the north. Here the Delaware River, Delaware Bay, and Chesapeake Bay, combined with feasible passages through the Appalachians to the west, have provided a setting for the development of two of the country's major seaports. Philadelphia, located on the lower Delaware River, and Baltimore, located near the head of Chesapeake Bay, are, respectively, the second- and eighth-ranking United States ports as measured by tonnage of goods handled. The cities of the northern Piedmont have developed extremely important and varied industries. Their combined production comprises about 5 percent of the United States total of manufactures. Philadelphia, including its New Jersey satellite, Camden, is especially important in the manufacture of woolen textiles, apparel, magazines, ships, chemicals, and locomotives. In addition, one of the country's great concentrations of oil refineries is located along the Delaware River in this vicinity. In recent years several new iron and steel plants, exemplifying a trend toward tidewater location in the steel industry, have been built along the river. Baltimore has important steel, apparel, aircraft, and shipbuilding industries. Wilmington is primarily a chemical manufacturing city, the administrative and research center of the Du Pont Company. Trenton's major products are metal goods, pottery, and rubber goods. In addition, there are automobile assembly plants in all of these centers except Trenton, and such other manufactures as sugar refining, canning, and copper refining, as well as a very large output of machinery and miscellaneous metal goods.

Urban and Industrial Development in the Central and Southern Piedmont

South of Washington, D.C., the Piedmont has few large cities. Atlanta, Georgia (1.3 million), a commercial center located in an area where railroads from the east skirt the southern end of the high Appalachians, is the largest. The metropolitan area centering on the three cities of Greensboro, Winston-Salem, and High

Large industrial plants surrounded by open country are a characteristic feature of recent industrialization in the American South. The photo shows the plant of Owens-Corning Fiberglas Corporation at Anderson, South Carolina. Located in the Piedmont to the southwest of Greenville and Spartanburg, the plant has been in production since 1951. It produces Fiberglas continuous filament textile yarns and employs over 2000 people. An irregular patchwork of fields, woods, and winding roads—the typical rural landscape of the Southern Piedmont—stretches into the distance. Note the even skyline which gives evidence of the former flatness of the ancient Piedmont surface. Although the geomorphic history of the Piedmont is complex, this physical region is in essence an extremely old erosional plain which has been uplifted and then moderately dissected by renewed downcutting of streams. (Owens-Corning Fiberglas Corporation.)

Point in North Carolina is next in size, aggregating 600,000 people. Richmond, Virginia (520,000) is third, followed by Charlotte, North Carolina (380,000). Nine smaller SMSAs are scattered through the Piedmont and along the Fall Line.

Nevertheless, the central and southern Piedmont is an area of considerable importance with regard to manufacturing. Major stimuli to industrial development have been the waterpower of the Piedmont and the adjacent Appalachians, the surplus of cheap labor offered by overcrowded, eroded, and often poverty-stricken rural areas, and the availability of certain raw materials from both the Piedmont and the adjoining regions, notably tobacco, timber (both hardwood and softwood), and mostly in the past, cotton. Much of the manufacturing is dispersed throughout a large number of small cities and mill towns and villages.

The largest industry is the manufacture of cotton textiles. In the twentieth century the central and southern Piedmont has displaced New England as the leading area of cotton milling in the United States. South Carolina and North Carolina, now the leading textile-milling states, have over half of the installed capacity of the American cotton-textile industry, and the Piedmont between Virginia and Alabama over 80 percent.

However, in certain sections other industries are quite important. Rayon and nylon plants in its Piedmont and Appalachian sections make Virginia the leading American state in the production of synthetic fibers. Another major American manufacturing industry, cigarette production, is almost entirely concentrated in a number of Piedmont cities and towns located in the Virginia-North Carolina tobacco-growing district. These cities include Winston-Salem, North Carolina, Richmond, Virginia, and others. This same Piedmont section in North Carolina and southern Virginia has developed an important furniture-manufacturing industry, based partly on local resources of mixed woods, which now totals over 15 percent of the entire United States output by value. Other manufactures which have gained a foothold in various parts of the central and southern Piedmont are pulp and paper, chemicals, boots and shoes, and apparel.

THE APPALACHIAN HIGHLANDS

The Appalachian Highlands extend from Newfoundland to northeastern Alabama. On the east they face the Atlantic and the Piedmont, on the west the Interior Plains (maps, pp. 707, 718, 725, and 726). The eastern boundary of the Highlands is the Atlantic shore as far south as the vicinity of New York City and then may be traced across the states of New York, New Jersey, Pennsylvania, Maryland, Virginia, North Carolina, South Carolina, and Georgia into Alabama. The western boundary extends from Alabama through the states of Tennessee, Kentucky, Ohio, Pennsylvania, and New York and through the Canadian province of Quebec to the south shore of the St. Lawrence estuary.

The mountains which dominate the topography of this landform division are not very high. The loftiest summits are below 7000 feet, and most of the Appalachian area lies below 3000 feet. But the Highlands form a very extensive and complex system in which mountain ranges, ridges, isolated peaks, and rugged dissected plateau areas are interspersed with narrow valleys, lowland pockets, and rolling uplands. Climatic conditions range from subarctic in Newfoundland to humid subtropical in the south, and some sections are high enough to experience abnormal coolness and precipitation. The natural vegetation is forest (map, p. 716): coniferous and mixed coniferous and deciduous as far south as Pennsylvania, predominantly deciduous (though with some mixture of coniferous stands) south from Pennsylvania.

The Appalachian Highlands exhibit some very great internal contrasts in population density and economic development, often within short distances, but as a whole they have a surprisingly large population for a highland area and are of enormous importance in the Anglo-American economy. Many parts of the Highlands have a considerable rural population, farming such level lands as are available, and sometimes extending cultivation up steep slopes better left in forest. A great variety of agricultural products and types of farming are found in different sections, although most of the land is wooded. But the principal significance of the Appalachian region lies in nonagricultural resources and activities: tourism, wood, coal, waterpower, metals, and manufacturing, especially the last.

Both the diversity and the economic importance of this landform division are brought out in the descriptions of the major subdivisions of the Highlands which follow. These subdivisions, as considered herein, are (1) Newfoundland, (2) the Maritime Provinces and Southeastern Quebec, (3) Northern New England, (4) Southern New England, (5) the New York Metropolitan Area (only partly within the Appalachians, but presented with this landform division for convenience), (6) the Hudson-Mohawk Valley and the Adirondacks, (7) the Northern Appalachian Plateau, (8) the Appalachian Ridges and Valleys of Pennsylvania, and (9) the Southern Appalachians.

Newfoundland

Newfoundland became Canada's tenth province in 1949. As a province, it includes the dependent territory of Labrador on the mainland (map, p. 711), but in common usage the name Newfoundland is restricted to the large island which lies opposite the mouth of the St. Lawrence. The island, which has an area of some 43,000 square miles, is rugged and rocky, although the highest elevation is only about 2600 feet. The climate is subarctic, with strong marine modifications, and the vegetation a spruce-fir forest, except for the higher areas, where considerable expanses of tundra occur. Although climate and soil have thus far conspired to make any extensive development of agriculture next to impossible, Newfoundland was the earliest center of permanent British settlement in Anglo-America. Today the island's 508,000 people (1968 estimate) represents the largest population concentration in the Anglo-American sub-arctic.

At the beginning of European settlement in the sixteenth century, fishing became established as the basic economic activity, and it is still of major importance today. Elevated portions of the sea bottom known as "banks" are found off the coast of Anglo-America from near Cape Cod to the Grand Bank, largest of all, which lies just off the southeastern corner of Newfoundland. The elevated, soft sea bottom and the mixing of waters from the cold Labrador Current and the warm Gulf Stream which occurs over the banks provides an excellent habitat for schools of cod and other economically valuable fish, and this part of the Atlantic has become one of the world's greatest centers of commercial fishing. Not only Newfoundland, but the Maritime Provinces, Maine, and Massachusetts send fishing fleets to the banks. However, in none of these other areas is fishing an economic mainstay to the same extent as in Newfoundland. Inshore fisheries supplement the catch from the banks in all of the areas mentioned.

In the 1950s, the pulp and paper and mining industries of the island surpassed fishing in value of output, though not in employment. However, in 1966 mining received a severe setback with the closure of the last of several iron mines that had operated for many years on Bell Island in Conception Bay. The Bell Island mines were by far the largest mining enterprise in Newfoundland proper. Some copper, lead, and zinc continue to be produced in various places, and exploration and development of iron deposits in Labrador continue. The largest center of paper milling is Corner Brook (27,000) on Newfoundland's west coast.

Most of the island's people live in hundreds of small, isolated fishing villages scattered along its lengthy coastline. St. John's (105,000), the capital and principal seaport, is the only city larger than Corner Brook. It is located on the southeast coast.

The Maritime Provinces and Southeastern Quebec

Nova Scotia, New Brunswick, and Prince Edward Island are Canada's Maritime Provinces (map, p. 711). Not only do these provinces border the Atlantic, but their traditional outlook is overseas rather than toward the rest of Canada. They are somewhat isolated physically by the mountains of northern New Brunswick and adjacent Quebec, politically by the northward projection of Maine, and socially by French Canada, which lies between the Maritimes and other dominantly British areas beyond Quebec. The Quebec French have been penetrating the Maritimes, and now make up over one-third of the population of New Brunswick.

Prince Edward Island, the smallest Canadian province, has only about 2200 square miles and 110,000 people (1968 estimate). It is a lowland area which is almost completely in farms, and has little other economic activity except for some fishing. Dairying and potato growing are the main forms of agriculture. Nova Scotia and New Brunswick resemble Newfoundland in that their land is generally rugged, although the highest summit, found in northwestern New Brunswick, is only about 2700 feet. Over four-fifths of the land in the combined provinces is still forested, and lumbering and pulp and paper milling are of great importance in their economies. However, in Nova Scotia, the value of manufactured wood products is outranked by the value of steel produced at the seaport of Sydney. Coal, amounting to over one-third of the total Canadian output, is mined here, and iron ore is brought by sea. The steel industry of Sydney is a small one when viewed against the background of Anglo-American production as a whole, but, nevertheless, the city ranks as one of Canada's three main centers of steel production, along with Hamilton, Ontario, and Sault Ste. Marie. Agriculture in New Brunswick and Nova Scotia is hindered by a cold climate and poor soils, as well as by the generally rugged character of the terrain. Many farms are part-time subsistence enterprises, although there are a limited number of commercial farming areas specializing in dairy products, potatoes, or fruit. The Annapolis–Cornwallis Valley, a long narrow lowland just inland from the Bay of Fundy in southwestern Nova Scotia, is one of Anglo-America's main apple-growing districts.

The economic development of the Maritime Provinces has been hindered by relative remoteness from major markets—a remoteness partly physical, but partly economic owing to tariffs imposed by the United States. The population of the three provinces, totaling only 1.5 million (1968 estimates), has been growing more slowly for some time than that of the rest of Canada. The scale of their economic development is indicated to some extent by the respective sizes of their major cities, which include Halifax (200,000), a Nova Scotian port with a magnificent harbor but remote from interior Canadian centers of production; Sydney (city proper 32,000; not recognized in Canada as a metropolitan area, but lies near many smaller places that aggregate, with Sydney, about 125,000); and St. John (103,000), the largest city and main port of New Brunswick.

A strip of Quebec south of the St. Lawrence River lies within the Appalachians. It is generally similar to New Brunswick and Nova Scotia in physical character and economic development. A noteworthy element in its economy is the production of about 40 percent of the world's asbestos, and practically all the asbestos of Anglo-America, from mines in the vicinity of the town of Thetford Mines (22,000).

Northern New England

Although New England is one of the most commonly recognized regional units in Anglo-America, there are actually sharp contrasts—physically, economically, and socially—between its three northern states and its three southern states. The three states of northern New England—Maine, New Hampshire, and Vermont—contain more rugged and elevated land and more land considerably removed from the sea. Their economies are less intensively developed, they are less urbanized, and are far less densely populated than the states of southern New England—Massachusetts, Connecticut, and Rhode Island. In northern New England the native "Yankee" stock is a more distinctive element in the population than in southern New England, which has become one of the greatest "melting pots" of the United States.

Northern New England contains two major subranges of the Appalachians, each running in a general north-south direction (maps, pp. 725 and 726). The Green Mountains occupy most of Vermont, although the state also contains a part of the Champlain Lowland to the west. Though generally somewhat subdued and rounded in shape, these mountains have some peaks reaching over 4000 feet. Most of New Hampshire is occupied by the White Mountains, which extend on the north into interior Maine. Northern New Hampshire, where Mount Washington reaches 6288 feet, contains

Many aspects of rural New England are reflected in this view: the grazing dairy animals; the rocks protruding at intervals through the thin soil; the large, neat frame structures of the village; the stone fences; the prominent spire of the church; the low, rounded uplands bearing a mixed second growth of coniferous and deciduous forest. The village in the picture is Tamworth, New Hampshire, in the eastern foothills of the White Mountains. (A. Devaney, Inc.)

the highest land in northeastern United States. The Green and White mountains are separated by the upper valley of the Connecticut River. Northernmost Maine consists of a rolling upland which extends into adjacent New Brunswick. The coastal area of Maine and New Hampshire is a hilly, stony lowland which extends south along the Atlantic to the vicinity of New York City.

Maine and New Hampshire are strongly industrial states from the standpoint of employment, income, and value of production, although their industries are only a small northward extension of the really intensive manufacturing development of southern New England. Most of the factories and main population clusters are found in small towns and cities in the coastal sections of New Hampshire and southern Maine. Wood industries, primarily pulp and paper milling, rank first in value of production, followed by the manufacture of boots and shoes, and by textile production, largely of woolens.

Relatively little of the rugged and rocky land which makes up Maine and New Hampshire is farmed, and many of the farms that do exist are part-time subsistence operations. The main types of commercial farming are dairying and poultry production, except in the Aroostook Valley of far northern Maine, which is one of the most important areas of concentrated potato production in Anglo-America.

In contrast to its industrialized neighbors, Vermont is the only state in all New England which is to a large degree agricultural in its economy, although even in Vermont a variety of small industrial enterprises in small towns and cities give employment to more people for the state as a whole than does agriculture. The state lies within the milkshed of New York City, and dairy farming is relatively more important in its economy than in that of any other American state. Over 40 percent of the land in the state is in farms, a great contrast to the situation in Maine and New Hampshire, where the proportion is nearer one-tenth. Throughout northern New England, with its mountains, seacoast, and large wilderness expanses, a noted and highly developed tourist and resort industry—comprising both winter ski resorts and many summer facilities—is part of the economy.

The estimated total population of northern New England in 1968 was only 2.1 million. Despite the basic dependence on manufacturing in the region only two metropolitan areas, those of Portland, Maine (145,000), and Manchester, New Hampshire (115,000), have over 100,000 people.

Southern New England

Very little of the land in the three states of southern New England is actually mountainous. Most of the area is hilly and rocky, however. The southern fringe of the White Mountains extends into Massachusetts, while the Green Mountains extend across Massachusetts into Connecticut. The southern section of the Green Mountains, generally called the Berkshire Hills, contains the highest elevations in southern New England, somewhat over 3000 feet.

Southern New England is an area of great industrial development, dense population, and a high degree of urbanization. Within a total area of about 14,500 square miles are contained some 9.3 million people (1968 estimate), or nearly 5 percent of the population of the United States. The overall density of population is around 640 per square mile. The metropolis of southern New England is Boston, which includes 2.6 million people in its metropolitan area.

The United States phase of the industrial revolution began in the late eighteenth and early nineteenth centuries with the rise of the mechanized cotton-textile industry in southern New England, and the area has been an important manufacturing region ever since. Before machine production commenced, it was already the leading center of commerce and handicraft industries in the country. Rapid industrial development was facilitated by the accumulated capital of the area, its maritime location, its wood, its waterpower, and the surplus labor provided by an agriculture beginning to fail under competition from cheap and fertile lands beyond the Appalachians. Now the original forests of the area are largely gone, the available waterpower is inadequate for present needs, and much of the plant equipment in the older industries is out of date. Both management and labor in the older industries are often accused of being unduly conservative, and the region is in competition with newer industrial areas often better placed with regard to raw materials and markets. Similar difficulties are felt in northern New England. In recent decades New England has often been regarded as something of an industrial problem area in the American economy, as its industry and population have not grown so rapidly as those of other major sections of the United States.

Nevertheless, southern New England is still an important industrial area. It retains only a small remnant of the once-dominant cotton-textile industry—now fled to the Piedmont—but still has approximately 40 percent of the woolen textile industry of the United States.

The main cities to which textile manufacturing is still of some importance are Providence, Rhode Island (870,000); New Bedford (150,000) and Fall River (145,000) in southeastern Massachusetts; and Lowell (185,000) and Lawrence (215,000 with neighboring Haverhill) on the Merrimack River in northeastern Massachusetts. Another relatively old industry that is still important is the shoe industry. Southern New England accounts for almost a third of the national output of boots and shoes, mostly from the Boston metropolitan area and Brockton, Massachusetts (185,000).

But these older industries are overshadowed in the larger cities (except Providence) and in southern New England as a whole by a varied assortment of more rapidly growing metal-goods and electronics industries. The region produces 7 to 8 percent of the United States output of machinery and fabricated metal goods, offsetting to some extent its handicaps with regard to natural resources and geographical position by specializing in products which derive a large part of their value from the expenditure of skilled labor. Office and store equipment, firearms, precision instruments, and watches are good examples, though they are exceeded in value of output by various types of industrial machinery and electrical equipment. The principal metropolitan centers in which these metal goods are produced include (besides Boston and Providence) Worcester (340,000) and Springfield–Holyoke (520,000) in Massachusetts; and Hartford (635,000), Bridgeport (370,000), New Haven (360,000), Waterbury (200,000), and New Britain (140,000), all in the tiny state of Connecticut. Stamford, Connecticut (220,000), is often considered a part of the New York metropolitan area. Altogether there are seventeen metropolitan areas (SMSAs) of over 100,000, all basically industrial in nature, and many smaller cities in southern New England—an area which measures only about 100 by 150 miles.

Naturally in such a region agriculture is greatly overshadowed by other activities and is strongly influenced by the close proximity of large cities. Except for the Connecticut River Valley, which is a specialty area producing truck crops and shade-grown tobacco, the main forms of agriculture are (1) subsistence farming on a part-time basis by people with city employment, (2) dairy farming, and (3) poultry farming. Agricultural development is so intensive that a very high value of products per unit of land is achieved. But total production falls far short of meeting the area's needs, and much food must be brought from other parts of the country.

A section of the Port of New York. The view looks toward the East River from the Upper Bay. Governors Island in the foreground is a United States military reservation. The skyscrapers of the Wall Street district appear at the tip of Lower Manhattan to the left of the view. The Manhattan waterfront is lined with narrow "finger piers" which jut into the water at right angles to the shore and thus maximize the berthing space available to ships. Most of the piers used by passenger liners are on the Hudson River side of Manhattan, not visible in this view. The three bridges across the East River in the photo connect Manhattan with Brooklyn: in order from bottom to top they include the Brooklyn Bridge, Manhattan Bridge, and Williamsburg Bridge. The Brooklyn waterfront between the Brooklyn Bridge and the lower right corner of the photo is lined with piers of the Brooklyn–Port Authority Marine Terminal. Like the piers on the New Jersey side of the Hudson, they are devoted almost entirely to handling cargo vessels. (Port of New York Authority.)

The New York Metropolitan Area

Anglo-America's greatest city is not enclosed within the Appalachians, but is located where the Appalachians, the Piedmont, and the Coastal Plain converge (map, p. 726). New York City proper centers on Manhattan Island (map, p. 734). This island is narrowly separated by the Harlem River from the southern tip of the mainland in New York State. On the east Manhattan is separated from Long Island by the East River; on the west from New Jersey by the lower Hudson River, here often called the North River. Manhattan is one of five boroughs composing the city proper. Of the remaining boroughs, the Bronx lies on the mainland to the north, Brooklyn and Queens are on the western end of Long Island, and Richmond is on Staten Island. The city proper contains some 8 million people (1968 estimate), but the metropolitan area contains a large number of suburban and satellite cities and towns lying to the north, east, and, especially, to the west in New Jersey. The population included within

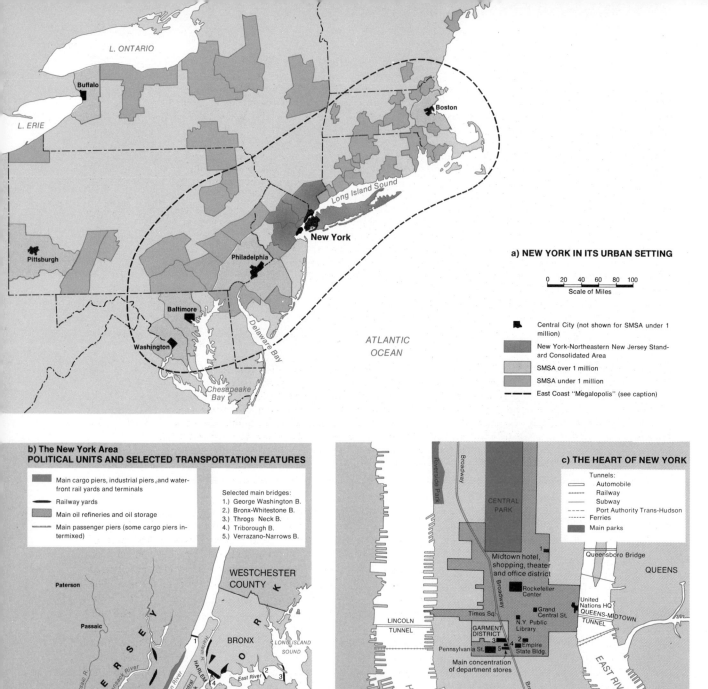

a) NEW YORK IN ITS URBAN SETTING

L. ONTARIO

L. ERIE

Buffalo

Boston

Pittsburgh

Philadelphia

New York

Baltimore

Washington

Long Island Sound

ATLANTIC OCEAN

Delaware Bay

Chesapeake Bay

0 20 40 60 80 100
Scale of Miles

▪ Central City (not shown for SMSA under 1 million)

New York-Northeastern New Jersey Standard Consolidated Area

SMSA over 1 million

SMSA under 1 million

– – – East Coast "Megalopolis" (see caption)

b) The New York Area
POLITICAL UNITS AND SELECTED TRANSPORTATION FEATURES

Main cargo piers, industrial piers and waterfront rail yards and terminals

Railway yards

Main oil refineries and oil storage

Main passenger piers (some cargo piers intermixed)

Selected main bridges:
1.) George Washington B.
2.) Bronx-Whitestone B.
3.) Throgs Neck B.
4.) Triborough B.
5.) Verrazano-Narrows B.

Paterson

Passaic

WESTCHESTER COUNTY

N E W   Y O R K

BRONX

LONG ISLAND SOUND

N E W   J E R S E Y

Passaic R.

Hackensack River

Hudson River

Harlem River

Central Park

East River

Manhattan

La Guardia Airport

Newark

Jersey City

NEWARK AIRPORT

Bayonne

NEWARK BAY

UPPER NEW YORK BAY

QUEENS

LONG ISLAND

Elizabeth

Kill Van Kull

JOHN F. KENNEDY INTERNATIONAL AIRPORT

BROOKLYN

JAMAICA BAY

RICHMOND

Arthur Kill

STATEN ISLAND

LOWER NEW YORK BAY

Perth Amboy

ATLANTIC OCEAN

0 2 4 6
Scale of Miles

The five boroughs of New York City are shown in light blue.

Department stores:
1.) Bloomingdale's
2.) Altman's
3.) Macy's
4.) Korvette's (34th St.)
5.) Gimbel's

c) THE HEART OF NEW YORK

Tunnels:
Automobile
Railway
Subway
Port Authority Trans-Hudson
Ferries

Main parks

Riverside Park

Broadway

CENTRAL PARK

Midtown hotel, shopping, theater and office district

Rockefeller Center

Times Sq.

N.Y. Public Library

Grand Central St.

GARMENT DISTRICT

Pennsylvania St.

Empire State Bldg.

Main concentration of department stores

LINCOLN TUNNEL

Queensboro Bridge

QUEENS

United Nations HQ

QUEENS-MIDTOWN TUNNEL

EAST RIVER

NEW JERSEY

HUDSON RIVER

N E W   Y O R K

M A N H A T T A N

Broadway

HOLLAND TUNNEL

East River Park

Williamsburg Bridge

Lower Manhattan financial and office district

Wall St.

Manhattan Bridge

Brooklyn Bridge

BROOKLYN

Battery Park

BROOKLYN-BATTERY TUNNEL

Staten Island Ferry

0 ½ 1
Scale of Miles

New York stands at the center of a strip of highly urbanized land extending along the northeastern seaboard from just north of Boston, Massachusetts, to Washington, D.C. This urban strip was christened "Megalopolis" by a French geographer, Jean Gottmann, who described and analyzed it in his book *Megalopolis: The Urbanized Northeastern Seaboard of the United States* (New York: Twentieth Century Fund, 1961). The dashed line surrounding "Megalopolis" is not intended to indicate a precise boundary, but shows broadly the area included by Gottmann in the "Megalopolis" concept. The five boroughs of New York City proper (Manhattan, Brooklyn, Queens, the Bronx, and Richmond) form the inner nucleus and focus of activity for the New York–Northeastern New Jersey Standard Consolidated Area—a grouping of several Standard Metropolitan Statistical Areas in the two states. The impressive cluster of waterfront facilities, rail yards, and airports on the map of political units and transportation features points up sharply the enormous importance of New York as a transportation center. Note the line of oil refineries and storage areas for crude oil and petroleum products along the water channels separating Staten Island from the New Jersey mainland. The large oil-refining industry of the New York area, located a great distance from the fields that supply its raw material, is a classic instance of a *market-oriented* industry. Transportation facilities in the metropolitan area consume most of the products of the area's refineries, plus large quantities of refined products brought by sea from other refining centers. In the map "The Heart of New York," the shading for the "Midtown Hotel, Shopping, Theater, and Office District" covers the area in which most of Manhattan's well-known hotels, restaurants, department stores, specialty shops, theaters, concert halls, and museums are found. The Midtown area also includes numerous office buildings, most notably the cluster of skyscrapers in Rockefeller Center. At the southern end of the area near the main department stores is found New York's famous Garment District with its huge work force and numerous workrooms and showrooms crowded into a remarkably small segment of Manhattan. The Lower Manhattan Financial and Office District includes the imposing cluster of office skyscrapers in the Wall Street area and, farther north, City Hall and a large assemblage of other government buildings occupied by city, county, state, and federal offices. Note the numerous bridges, tunnels, and ferries that tie Manhattan Island to the rest of metropolitan New York. On a normal weekday, some 3 million commuters or other persons enter the parts of Manhattan shown on the map, and there is a return outflow of about the same size.

the New York metropolitan area is about 16.2 million,[5] or about eight percent of the population of the entire United States.

New York owes its development in good part to the fact that it stands at the seaward end of the only continuous east-west lowland route through the Appalachians. This route follows the Hudson River Valley north to Albany, then runs west along the valley of the Mohawk River to the plain bordering Lake Ontario. Thus it affords an easy connection with the Great Lakes and the productive Interior Plains. The enormous and protected harbor of New York, fashioned by an intricate interpenetration of land and water, plus a central location between the agricultural South and the commercial and manufacturing region of New England, had already allowed the city to become the leading seaport of the United States before the trans-Appalachian route became of major importance. However, it was the completion of the Erie Canal along the Mohawk Valley in 1825, connecting Buffalo on Lake Erie with Albany on the Hudson and thus giving an all-water route from the Great Lakes to the port of New York, that provided the real impetus for the rise of New York to a position of clear and unchallenged commercial leadership among American cities. The New York Central Railroad (now part of the Penn Central) was built along the Erie Canal route, and this and other railroads through the Appalachians, as well as trucks on arterial highways such as the New York State Thruway (which follows the Hudson-Mohawk corridor), today far surpass the water route in the transportation of goods to and from New York. Today the Erie Canal, deepened and otherwise modified, forms a part of the New York State Barge Canal.

New York is still basically a commercial city. It generally handles one-quarter to one-third of all United States foreign trade by value, although the proportion is less by bulk because the port specializes in relatively light and valuable freight. But from its commercial leadership has flowed preeminence in such fields as

[5] This estimate is for the New York–Northeastern New Jersey Standard Consolidated Area, which includes several contiguous SMSAs. Inclusion of the Norwalk and Stamford SMSAs in Connecticut would increase the figure to 16.5 million.

finance, business administration, advertising, and entertainment. And its enormous local market and population, its transportation facilities, and its wealth have made it the greatest manufacturing center in the United States, with over 10 percent of the country's total industrial employment. Practically every manufacturing industry is represented to some extent in the metropolitan area, but the leading industry is the manufacture of clothing. Approximately one-third of the national output of clothing comes from Manhattan's crowded and famous garment district (map, p. 734). It produces a considerably larger percentage of all women's clothing, but a smaller percentage of men's clothing. The scale of the industry is indicated by the fact that there are over twice as many people employed in the Manhattan garment industry as there are in automobile factories in Detroit. The dominant position of Manhattan in the nation's clothing industry reflects a number of factors, but especially the prestige of New York as a style center. Other industries in which New York leads the nation are printing and publishing, chemical production, and the manufacture of electrical machinery and equipment. A major share of the factories within the metropolitan area are located in the New Jersey section, where space is more abundant and cheaper and where the main rail-freight terminals are located (map, p. 734).

The Hudson-Mohawk Valleys and the Adirondacks

Traffic along the Hudson and Mohawk valleys has fostered the growth of a string of industrial towns and cities, mostly small and extremely varied in types of production. The only large metropolitan area is the tri-city area of Albany-Schenectady-Troy (705,000), which occupies an especially strategic position with respect to natural transportation routes, being at the junction of the Hudson Valley south to New York City, the Champlain Lowland north to the St. Lawrence and Montreal, the Mohawk Valley west toward the Great Lakes, and routes through mountain passes to the east that are followed by important rail lines and an arterial throughway to New England. Administration is a major function of these cities, Albany being the state capital and Schenectady the headquarters of the General Electric Company. Their main manufactures are machinery, electrical equipment, and clothing. The next largest metropolitan area along the valley route is Utica-Rome (350,000) at the western end of the Mohawk Valley. The principal manufactures here include machinery, brassware, textiles, and paper.

The circular mass of the Adirondack Mountains is geologically an upraised outlier of the Canadian Shield. The hard-rock formations which tie it to the main body of the Shield emerge in the St. Lawrence River as the picturesque "Thousand Islands" and are responsible for the rapids which blocked navigation on the St. Lawrence above Montreal until they were bypassed by small canals in the nineteenth century. Very recently the possibilities for navigation of the St. Lawrence have been greatly enlarged by construction of the St. Lawrence Seaway, which permits good-sized oceangoing vessels to enter the Great Lakes. The highest peak of the Adirondacks exceeds 5000 feet in elevation. These mountains form an island of rugged, forested land and sparse population, surrounded on all sides by more populous lowlands. The Adirondacks are important as a recreational area, however, and there is some production of pulp and paper around their margins, as well as a small production of iron ore.

Subdivisions of the Central and Southern Appalachians

South of the Hudson-Mohawk line the Appalachians fall into three major subdivisions. From east to west these are the Blue Ridge, the Ridge and Valley Section, and the Appalachian Plateau (maps, pp. 718, 725, and 726). Each of the three trends northeast-southwest, and they roughly parallel each other.

The *Blue Ridge,* long and narrow, can be traced from near the mouth of the Hudson to Alabama. Often it rises abruptly from the Piedmont to the east, and throughout much of its length it is but a single ridge. It has been given various local names. In some places, especially in the north, the Blue Ridge is discontinuous, and lowland gaps provide important east-west routes for transportation. One of these gaps to the east of Harrisburg, Pennsylvania, was utilized in constructing both the main line of the Pennsylvania Railroad (today a part of the Penn Central System) and by the Pennsylvania Turnpike. The two arteries connect Philadelphia and the eastern seaboard with the Interior Plains via Pittsburgh, Pennsylvania. The Blue Ridge widens into several ranges and becomes higher and more continuous toward the south. Although the northern sections of the Blue Ridge often fail to reach 1000 feet in elevation, the section in western North Carolina known as the Great Smoky Mountains has a number of peaks over 6000 feet. Mount Mitchell in the Great Smokies, at 6684 feet, is the highest peak in the entire Appalachian system and in eastern United States.

Characteristic folded topography of the Appalachian Ridge and Valley Section seen from above. At the upper left is a narrow, knife-edged mountain ridge, oriented northeast-southwest and flanked by parallel valleys. At the lower right one side of a second ridge, similar to the first one, is visible. Dozens of these folded ridges and valleys lie parallel to each other between the Blue Ridge on the east and the Appalachian Plateau on the west. At the lower right of the above view, taken near the resort town of White Sulphur Springs in southeastern West Virginia, farms occupy the narrow strip of alluvium that has been laid down by the stream which winds along the valley floor. The farms extend into the lower slopes of the adjoining ridges, but the upper slopes are occupied by Federal government land of the Monongahela National Forest. The black line outlines a farm, 182 acres in size. (U.S. Department of Agriculture photo; farm outline added by Jesse H. Wheeler, Jr.)

West of the Blue Ridge, the *Ridge and Valley Section* of the Appalachians extends from southern New York to Alabama. Long, narrow, roughly parallel ridges trend generally north and south and are surrounded and separated by narrow valleys (photo, p. 737). The ridges are often 3000 to 4000 feet high, the valley floors 1000 to 2000 feet lower. On the east, immediately adjacent to the Blue Ridge, a valley somewhat wider than the rest extends almost continuously throughout the length of the Ridge and Valley Section. It is known in general as the Great Appalachian Valley, although it has many local names such as Shenandoah Valley in Virginia and Lehigh Valley in northeastern Pennsylvania. The Great Valley is a historic north-south passageway, and its productive limestone soils have provided the basis for an agricultural development superior to that of most parts of the Appalachians.

The *Appalachian Plateau* lies west of the Ridge and Valley Section. It covers most of southern New York State and extends southward into Alabama, narrowing toward the south. Although geologically a plateau (more precisely a series of plateaus), it is so deeply and thoroughly dissected in most places that it is actually an area of tangled hills and low mountains separated by narrow, twisting stream valleys. Most summit elevations are between 2000 and 3500 feet above sea level. The northern part is often called the Allegheny Plateau and the southern part the Cumberland Plateau. Notable escarpments, the Allegheny Front in the north and the Cumberland Front in the south, mark the eastern edge of the plateau area, while the frayed western edge merges into the Interior Plains.

The Northern Appalachian Plateau (Allegheny Plateau)

The Northern Appalachian Plateau occupies parts of the states of New York, Ohio, Pennsylvania, West Virginia, Maryland, Virginia, and Kentucky. It has important forest and waterpower resources and some oil and natural gas. But its greatest importance lies in the fact that it contains what is probably the world's best coal field (map, right), considering the quantity and quality of coal available and the extraordinary ease of working it. Among the states sharing the Northern Appalachian Plateau, only New York has no noteworthy coal resources or production. About two-thirds of the United States output of bituminous coal, or some 12 percent of the world's total coal production, comes from the Northern Appalachian Plateau. The greatest production within the field comes from West Virginia,

followed by Pennsylvania, Kentucky, and Ohio ranking in that order.

This enormous power resource fuels industry and feeds electric generating plants on the Atlantic Coast and in the Middle West, and its use has made the Northern Appalachian Plateau itself one of the leading areas of heavy industry in the world. Pittsburgh, Pennsylvania (2.4 million), strategically located where the navigable Allegheny and Monongahela rivers, flowing from the coal fields, unite to form the navigable Ohio River leading to the Midwest and South, is the area's outstanding steel-producing and urban center. Long the greatest center of primary iron and steel production in the world, the Pittsburgh metropolitan area has now been slightly surpassed by the Chicago area and by West Germany's Ruhr. But the plants strung along the river valleys that radiate from the city still represent about 16 percent of all United States steel capacity, and approximately 5 percent of world capacity. The Northern Appalachian Plateau includes one other iron and steel center of major importance. This is Youngstown, Ohio (530,000 with nearby Warren), located in a valley which leads from Lake Erie to Pittsburgh. Smaller centers include the Steubenville, Ohio–Weirton, West Virginia, area (170,000), the Huntington, West Virginia–Ashland, Kentucky, area (260,000), the Canton, Ohio, area (365,000), and the Johnstown, Pennsylvania, area (265,000). In the aggregate, the steel industry of the Northern Appalachian Plateau represents about one-third of all United States steel capacity. The iron and steel plants of the area draw their ore mainly from the Superior Upland, and probably could not have reached their present great development without the fortunate interposition of the Great Lakes to allow cheap transportation of ore most of the way to their various locations in or near the Appalachian coal fields.

Other important industries in the Northern Appalachian Plateau manufacture secondary metal products, rubber, chemicals, glass, and clay products. Akron, Ohio (660,000), the original center of the rubber industry, is still the leading center of rubber manufacturing, with nearly a tenth of the national output by value, while nearby Canton, Ohio (365,000), specializes in rolling and finishing steel and machinery manufacture. A district extending along the valley of the Kanawha River in the vicinity of Charleston, West Virginia (240,000), and westward to Huntington, specializes in chemicals, glass, and synthetic fabrics. Wheeling, West Virginia (180,000), another Ohio River city, carries on secondary processing of steel and manufactures clay products and glass. The only sizable

▲ Anthracite and semianthracite

  Low-volatile bituminous coal

  Medium and high-volatile bituminous coal

  Sub-bituminous coal

  Lignite

Coal fields of the United States. (After a map by W. S. and E. S. Woytinsky in *World Population and Production: Trends and Outlook*, New York: Twentieth Century Fund, 1953.)

industrial center of the Northern Appalachian Plateau that is clearly outside of the coal-producing area is Binghamton, New York (300,000), which is atypical in its industrial structure as well as in location, being primarily an electrical machinery and shoe-manufacturing center.

Agriculture is overshadowed by industry and mining in the Northern Appalachian Plateau, although much of the land is in farms and there is a fair-sized farm population scattered along many of the valleys. Level land for crops is scarce, and farming is for the most part not a very rewarding or productive enterprise. West Virginia is fairly typical. Here two thirds of all farms are part-time operations, and a sizable part of the state's farm products are used at home. Livestock are the principal source of cash income to farmers. In the northern part of the region, especially in New York, the topography is somewhat more subdued, adequate

markets are more easily reached, and a considerable dairy industry has developed. As a state New York is second only to Wisconsin in dairy products, and its Appalachian Plateau section makes a considerable contribution to the state's total dairy production. The Finger Lakes district at the northern edge of the Plateau is a well-developed fruit- and truck-farming area.

The Appalachian Ridges and Valleys of Pennsylvania

A small district in the Ridge and Valley Section of northeastern Pennsylvania produces almost the entire United States output of anthracite coal. Two main metropolitan areas, Wilkes-Barre–Hazelton (345,000) and Scranton (225,000) have developed on the anthracite

In an age of mechanized farming, remnants of traditional agriculture, often with a strong subsistence component, still persist in many Appalachian valleys. Frequently such farms are worked by owners or tenants whose main employment is in a manufacturing plant or some other nonagricultural activity. The men in the above photo are husking corn which has been cut and shocked by hand. The view was taken in eastern West Virginia near the Greenbrier River, a tributary of the Kanawha. The valley floor in the view lies at the western edge of the Ridge and Valley Section. The unpainted barn in the photo is a characteristic feature of this area. Lower slopes of the valley wall in the distance have been cleared for use as pasture. Crops are grown in the valley primarily for livestock feed, and most cash income from farming is derived from sales of cattle, calves, sheep, and lambs. (Standard Oil Company, N.J.)

field. With the twentieth-century decline in the demand for anthracite these have become major problem cities, suffering chronic unemployment and a considerable decrease in population. Some firms, especially apparel manufacturers, have come in to take advantage of the surplus labor supply created by the decline of the anthracite industry.

The early development of the anthracite fields, plus small local deposits of iron ore, led to the rise of a sizable iron and steel industry in eastern Pennsylvania which far antedates the recent construction of tidewater plants in the Philadelphia area. Three urban areas in the Ridge and Valley Section of the state are still important steel producers: Allentown–Bethlehem–Easton (510,000),

Harrisburg (380,000), and a much smaller producer, Reading (290,000). These cities are all located in the Great Valley on rivers whose valleys provide superior passageways into the rugged country to the west and through the Blue Ridge to the east. Their steel industries have now remained relatively static for some time, and other industries, especially textiles and clothing, have developed which at present tend to overshadow steel. The Allentown–Bethlehem–Easton area has also had an extraordinary development of the cement industry, which has been favored by its access to coal, the limestone of the Great Valley, and the very large and easily reached markets near by. The area has the largest concentration of cement output to be found within a comparable extent of land anywhere in the United States, or probably in the world.

The valleys of this section of eastern Pennsylvania are almost completely occupied by farms, primarily dairy farms, though livestock and subsistence farms exist in valleys more remote from markets.

The Southern Appalachians

Most parts of the Appalachians south of the Potomac River and Northern Appalachian Plateau are still in the process of emerging from relative isolation and a poor agricultural economy. Except for the Great Valley, agriculture is largely on a subsistence or semi-subsistence basis (photo, left), although commercial livestock production is increasing. The Great Valley has long been a superior agricultural section. It is now devoted principally to livestock and dairy farming or to general commercial farming. The Shenandoah Valley section is in addition one of the principal apple-producing areas of the United States. Alabama sections of the Appalachians are far enough south to have a considerable cotton production.

Some coal production exists in Virginia, Tennessee, and Alabama, but a more important initial stimulus to increased industrialization in recent decades was waterpower, produced largely by the Tennessee Valley Authority (TVA) as a major aspect of its program to control the Tennessee River and its tributaries and to develop the resources of the Tennessee River Basin.[6] The main industrial growth has been in the fields of chemicals, aluminum, rayon and other textiles, pulp and paper, and aerospace industries. Only five cities in the Southern Appalachians reach 150,000 in metropolitan population. Birmingham, Alabama (750,000), at the extreme southern end of the Appalachians, is by far the largest and is a special case. Here large deposits of coal and iron ore in close proximity present one of the world's most favorable natural situations for the development of an iron and steel industry, and the city is the main center of that industry in the southern part of the United States. The remaining cities include Knoxville (400,000 with adjacent Oak Ridge), the main commercial center of the Tennessee portion of the Great Valley and an important center of federal government activity connected with the Tennessee Valley Authority and the production of atomic energy; Chattanooga, Tennessee (300,000), located where the Tennessee River cuts a natural passageway through the southern end of the Appalachian Plateau; Huntsville, Alabama (245,000), a very rapidly growing center of government rocket production, and Roanoke, Virginia (185,000), a rail center and industrial city located at a gap through the Blue Ridge opposite the port of Norfolk.

THE INTERIOR HIGHLANDS

The Interior Highlands, often called the Ozark or Ozark–Ouachita Highlands, occupy most of southern Missouri, the northwestern half of Arkansas, and adjoining parts of eastern Oklahoma (maps, pp. 707, 725, and 742). They constitute an island of hill country and low mountains in the midst of a sea of plains: the Interior Plains on the north and west and the Coastal Plain on the east (Mississippi Lowland) and south. The Interior Highlands are divided into two major segments by the east-west valley of the Arkansas River. The northern segment consists of the Boston Mountains, overlooking the Arkansas Valley, plus the more extensive Ozark Plateau of northern Arkansas, southern Missouri, and northeastern Oklahoma. South of the Arkansas Valley are the Ouachita Mountains, constituting the other principal segment of the Highlands.

The Interior Highlands display much internal variety in form and relief. North of the Arkansas Valley most of the area consists of dissected plateau surfaces (photo, p. 47) bearing a certain physical resemblance to the Appalachian Plateau. However, the plateau section of the Interior Highlands is lower in elevation and relief and lacks the coal deposits of the Appalachian area. The Ouachita Mountains consist of roughly parallel ridges and valleys resembling the Ridge and Valley Section of the Appalachians, except that in the Ouachitas the trend of the topography is east and west. The peak elevations in the Interior Highlands lie generally between 1000 and 2000 feet. No crest in the entire area reaches 3000 feet, although the highest summits of the Ouachitas approach that elevation. There are notable differences in local relief within the Highlands. A number of fairly

---

[6] Although a series of great dams and associated hydroelectric stations along the Tennessee and various tributaries are the best-known aspect of TVA, the demand for current has far outstripped the existing hydroelectric capacity, and TVA has moved heavily into the production of electricity at coal-powered thermal stations. In addition, a huge nuclear power station was under construction in 1968.

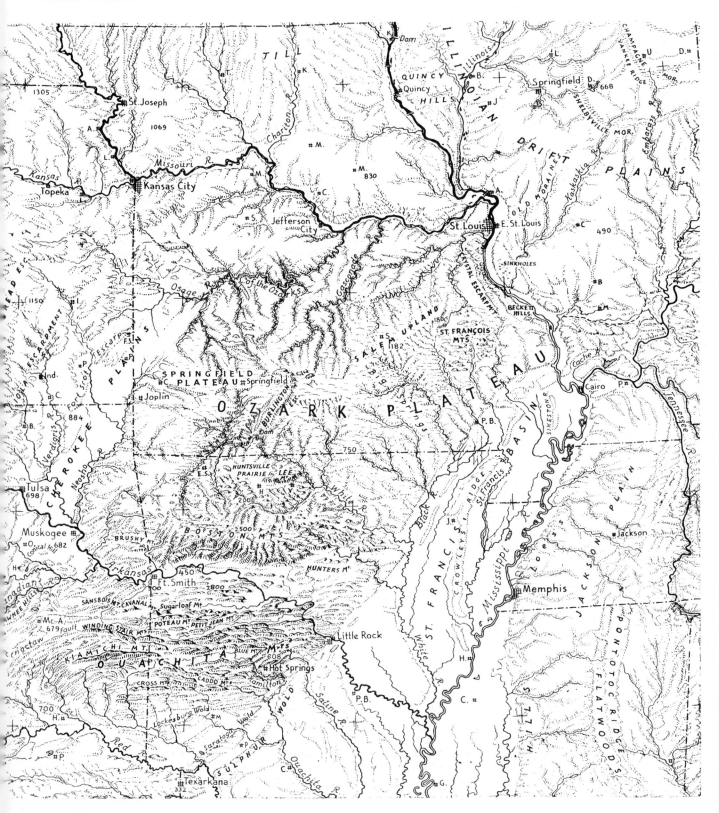

A distinctive method of portraying topography is shown in this representation (by the late Erwin Raisz) of the Interior Highlands and adjacent areas. Compare the method used here with that used by A. K. Lobeck in the map of eastern United States earlier in this chapter, and with the method used to portray the topography of western United States later in the chapter. The above map is a section from Erwin Raisz's map "Landforms of the United States," prepared originally to accompany Wallace W. Atwood, *The Physiographic Provinces of North America* (Boston, Mass.: Ginn and Company, 1940). Copyright by Erwin Raisz; reproduced by permission of Ginn and Company. On the original map the area in northern Missouri marked "till" was shown as "dissected loess-covered till prairies."

level upland surfaces occur, but other areas are quite rugged. Most areas fall between these topographic extremes.

Farming, sawmilling, and the production of a variety of wood products are the economic mainstays in most parts of this landform division. Most of the agriculture is of a semisubsistence or part-time character, and in most areas the number of farms and the total population are decreasing rapidly. On the better lands, especially where adequate transportation is available, an important commercial livestock industry has developed, emphasizing dairying in some areas and the raising of beef cattle in others. Commercial poultry farming on a sizable scale is a relatively recent development in some districts. Cotton is the major crop of the Arkansas River Valley.

The Interior Highlands were originally in forest, and most parts are still forested (map, p. 716), generally with a rather poor grade of second-growth timber. Deciduous hardwoods predominate north of the Arkansas Valley and coniferous softwoods in the Ouachita Mountains.

Many deposits of minerals exist in the Interior Highlands, but often are not of sufficient value to justify exploitation at present. One area, however, is an important producer of metals. In the northeastern part of the Highlands, south and southwest of St. Louis, a small district supplies over 40 percent of the annual United States production of lead. It is the leading lead-mining district in the country and, in addition, produces small quantities of other minerals, notably iron ore. In northeastern Oklahoma and adjoining parts of Kansas and Missouri is the Tri-State mining district, for many years the leading area of zinc production in the United States, but now declining as a mining area. A number of other mineral resources are exploited on a small scale in various parts of the Highlands.

There are no major industrial centers, but many small towns have one or more small factories producing wood products, apparel, shoes, or food products. Springfield, Missouri (140,000), located in a well-developed dairy-farming area, is the largest city of the Highlands

and the most important focus of highways and rail lines.

Recent years have witnessed a marked expansion in the tourist and resort industry. A considerable number of private resort developments cater to populations of the surrounding lowlands. In addition, state and federal government agencies have been active in the development of public recreational facilities. The impoundment of lakes behind federally built dams has been an important factor in increased tourist trade, and the electricity created in generating stations at these dams appears to offer a basis for increased industrial development in the Highlands.

THE INTERIOR PLAINS

The interior of Anglo-America, between the Rocky Mountains on the west and the Appalachian Highlands and Canadian Shield on the east, is essentially composed of a vast expanse of plains (map, p. 744). These plains lie mostly within the drainage basins of four important rivers: the Mississippi, the Mackenzie, the Saskatchewan, and the St. Lawrence. Most parts of the Interior Plains are many hundreds of miles from the sea; their location, in other words, is essentially continental rather than maritime. However, the Interior Plains reach the sea in the northwest and northeast of Anglo-America, respectively, along two narrow corridors—the Mackenzie River Lowlands and the St. Lawrence Lowlands. In the south the Interior Plains extend into Mexico, merge with the Gulf-Atlantic Coastal Plain, or abut against the Interior Highlands and Appalachian Highlands. The western edge of the plains can be traced through the Northwest Territories, British Columbia, Alberta, Montana, Wyoming, Colorado, New Mexico, and Texas; their northeastern edge through the Northwest Territories, Alberta, Saskatchewan, Manitoba, Minnesota, Wisconsin, Michigan, Ontario, and Quebec; and their southeastern edge through Quebec, New York, Pennsylvania, Ohio, Kentucky, Tennessee, Illinois, Alabama, Missouri, Oklahoma,

The names of certain cities shown only by symbol may be found on the map of Northeastern United States and Southeastern Canada, p. 726. City-size symbols are based on 1968 estimates for metropolitan areas.

and Texas. The Interior Plains include the whole of six states (the Dakotas, Nebraska, Kansas, Iowa, and Indiana), practically all of Illinois, and parts of 23 other states, provinces, and territories. Approximately the western third of the Interior Plains, from Mexico north to the southern part of Saskatchewan and Alberta, is known as the Great Plains.

Physiography, Climate, Vegetation, and Soils

The Interior Plains are seldom flat, although considerable areas are very nearly so. Most of the land is gently rolling, but some areas are hilly and occasional areas rather rugged. Most parts of the Interior Plains north of the Ohio and Missouri rivers were covered with a mantle of glacial debris during the Great Ice Age and the topography was somewhat smoothed thereby. Thus the more hilly sections tend to be outside the area of glaciation, although some unglaciated portions are also fairly level. The principal unglaciated hilly areas are in the southeast from southern Indiana across Kentucky into Tennessee, and in an "island" of unglaciated terrain in southwestern Wisconsin. Other areas of hilly terrain exist (1) in eroded strips along major stream courses, (2) in areas of terminal moraine deposition, (3) along cuesta escarpments separating different plain levels, (4) in fantastically eroded badlands of the western sections such as the famous ones in western South Dakota, and (5) in small, isolated mountain areas near the western border, of which the Black Hills of Sourth Dakota are the most notable example. These various exceptions, however, do not essentially damage but only refine the overall concept of a huge plains area with enormous expanses of land level enough for cultivation.

An interior area of this size extending from the margins of the low latitudes to the margins of the high latitudes naturally has a great range of climatic conditions, natural vegetation, and soils. The Mackenzie Delta is in the tundra zone, and the subarctic climatic zone with its coniferous forest reaches southward into central Alberta, Saskatchewan, and Manitoba. Southern Alberta and Saskatchewan and areas west of approximately the 100th meridian in the Great Plains section of the United States are semiarid and have a short grass, or steppe, vegetation. South of the general latitude of the Ohio River and east of the belt of steppe, considerable sections of the Interior Plains fall within the zone of humid subtropical climate. But the most important and productive agricultural areas of this landform division lie within the humid continental climatic zones: humid continental with short summers in a belt from Alberta to the Atlantic which includes southern Canada and large sections of the northern tier of states in the United States; and humid continental with long summers in a belt reaching from Nebraska and Kansas to the Appalachians. Large areas within the humid continental zones were originally in forest, with deciduous hardwoods predominating in most places, but with a belt of mixed hardwoods and conifers in the north. However, tall grass, or prairie, vegetation occupied a roughly triangular area with its apex in Illinois and widening westward toward the steppe. The deep, black soils originally developed under a vegetation of tall grass are outstandingly fertile and some soils developed under the deciduous forest are above average in natural fertility,

though not equal to the grassland soils. Such soils, combined with adequate rainfall and wide expanses of land sufficiently level for cultivation, have provided a natural setting for the development of one of the world's most important and impressive agricultural areas. All things considered, the Interior Plains represent the agricultural heartland of Anglo-America.

Agricultural Regions

Relatively smooth topography, good soils, and the existence of a number of major climatic regions, each occupying a broad area, have permitted the development of several large and rather distinct agricultural regions. These display considerable internal uniformity rather than the intricate and broken pattern of agriculture characteristic of many of Anglo-America's landform regions. The major agricultural regions of the Interior Plains, as considered herein, include (1) the Corn Belt, (2) the Dairy Belt, (3) the Spring and Winter Wheat Belts, (4) the Unglaciated Southeastern Interior Plain, (5) the Texas-Oklahoma Red Prairies, and (6) the Great Plains Region of Ranching and Irrigated Agriculture.

The Corn Belt

The famous Corn Belt of the United States lies wholly within the Interior Plains (map, p. 717). It extends from western Ohio to eastern South Dakota, Nebraska, and Kansas, and includes most of Indiana, all of Iowa, and parts of Missouri, Minnesota, Michigan, and Wisconsin. From Minnesota and Iowa eastward, its northern margins interlock with the southern edges of Anglo-America's main belt of dairy farming. The Corn Belt lies essentially within the humid continental long-summer climatic zone, and except for its Indiana, Ohio, and Michigan sections is generally within the region of prairie or chernozem soils. Physically its limits are set (1) on the east by the more rugged land and poorer soils of the Appalachian Plateau, (2) on the south by the eroded northern margin of the Ohio River Valley and by the Ozarks, (3) on the west by semiaridity (approximately the line of 20 inches average annual rainfall), and (4) on the north by summers too short and cool to ensure a dependable crop of mature corn.

The Corn Belt system of agriculture emphasizes the production of hogs and cattle for cash sale. The principal feedstuff is corn, which has a high feed value and grows exceedingly well in this region of fertile soil, long hot summers, and ample moisture. Most of the livestock are bred within the area, but many Western range cattle are brought into the Corn Belt for fattening before their final trip to the slaughterhouses. Soybeans, oats, winter wheat, hay, and pasture grasses and legumes are grown in a variety of rotations with corn. These latter crops contribute to livestock production, or they may, if the market favors it, be sold as cash crops, especially the wheat and soybeans. In addition, corn is often marketed as a cash crop if the market is more favorable than for livestock. This practice is the usual one in some parts of the Corn Belt that are especially level and near major markets, as in large sections of Illinois where cash grain farming of corn has largely replaced livestock production.

Iowa and Illinois may be considered the heart of the Corn Belt. In total farm production, by value, Iowa ranks second among the states and Illinois fourth. Together these states account for about 13 to 14 percent of the country's total agricultural production by value. They produce about 40 percent of the nation's corn and hogs, and the Corn Belt as a whole accounts for around three-quarters of the national output of these items. Although quite important within the Corn Belt, often more so than hogs (especially toward the west), the beef cattle industry of the United States is not as highly centered in the area as is the production of corn and hogs. Nevertheless, Iowa ranks second only to the much larger state of Texas in the production of beef cattle.

The Dairy Belt

Dairy farming tends to be the main form of agriculture in those portions of the Interior Plains that lie in Wisconsin and adjacent parts of Minnesota, in most Michigan and New York sections of the Plains, in Illinois and Indiana sections near Chicago (map, p. 717), and in sizable parts of the Ontario and Quebec sections. Outside the Plains, dairying is also important, though generally less intensive, in adjoining sections of the Appalachians and the southern margins of the Canadian Shield. Thus dairy farming is particularly an enterprise of the humid continental short-summer climatic zone in Anglo-America. The southern boundary of the

A landscape along the southern margin of the Dairy Belt section of the Interior Plains. This view was taken in southeastern Minnesota near the Wisconsin border. The tilled field in the foreground bordered by pasture contains, in order from the bottom of the photo, contour strips of hay, wheat (in shocks), and corn. (United States Information Service, Department of State.)

Dairy Belt within the Interior Plains occurs where a combination of warmer summers, better soils, and more level topography favor heavier concentration on intensive corn production; its northern boundary, lying generally outside the Plains, is essentially the northern climatic limit of agriculture itself; and its western limit is set by semiaridity and distance from adequate markets.

The principal items of sale from dairy farms in this agricultural region are fluid milk, where city markets are sufficiently near, or butter and cheese in sections more distant from large cities. However, calves, nonproductive milk cows, hogs, poultry, and cash crops are also marketed. Feed for animals on dairy farms is supplied mainly by crops of hay, oats, and corn (mostly for silage), by permanent and rotation pastures, and by creamery wastes and surplus milk. About one-fourth of United States dairy products by value originate in the Interior Plains sections of the Dairy Belt, Wisconsin alone accounting for about 12 or 13 percent. The Interior Plains of eastern Canada—comprised of lowlands along the St. Lawrence River in Quebec and the Ontario Peninsula, between the edge of the Shield and the Great Lakes, in Ontario—are Canada's most productive agricultural area. Much of this area may be classed as part of the Dairy Belt. But in the Ontario Peninsula, the southernmost section of Canada, feed grain and livestock production somewhat like that of the Corn Belt overshadows dairying. In addition, small and scattered specialty areas producing such items as fruits and vegetables, tobacco, potatoes, and sugar beets are relatively prominent in these Canadian dairy and livestock regions. Such specialty areas are also found in the United States portions of the Dairy Belt, though their relative importance is less than in the Canadian portions. The main ones are fruit and truck districts that lie close to the Great Lakes and benefit from the moderating climatic influence of the lakes.

The Spring and Winter Wheat Belts

Most of Anglo-America's large production of wheat originates in two zones of specialized wheat growing within the Interior Plains (map, p. 717). One of these zones grows spring wheat, the other winter wheat. The Spring Wheat Belt occupies the western edge of Minnesota, nearly all of North Dakota, part of northeastern South Dakota, much of northern Montana, and most of the plains area of the Canadian Prairie Provinces of Manitoba, Saskatchewan, and Alberta. It thus occupies the drier margins of the humid continental short-summer climatic zone and the wetter

margins of the steppe. In general its limits are set by an increasingly short growing season on the north, sufficient moisture for more intensive forms of agriculture on the east, summers too warm for spring wheat on the south, and increasing aridity on the west. The Winter Wheat Belt centers in central and western Kansas and also extends into adjoining parts of Nebraska, Colorado, and Oklahoma and into the Texas Panhandle. This belt occupies principally the drier margins of the humid continental long-summer climatic zone in the United States section of the Interior Plains and the wetter margins of the steppe. In general its limits are set by winters too cold for winter wheat on the north, adequate moisture for more intensive forms of agriculture on the east, and insufficient moisture for wheat on the west and south. The actual limits of wheat growing undergo marked changes from one period of years to another due to severe climatic and market fluctuations.

Statistics vary considerably from year to year, but in general the Winter Wheat Belt supplies close to 40 percent of the United States wheat crop, with Kansas (the leading wheat state) accounting for almost a fifth of the national output. The Spring Wheat Belt supplies approximately a quarter of the American crop, with North Dakota (the second-ranking wheat state) accounting for about half of this. Practically all of the Canadian wheat crop is grown in the Spring Wheat Belt.

Commercial wheat farming in the Spring and Winter Wheat belts is a risky and uncertain type of enterprise. These areas are climatically marginal, and relatively slight deviations from normal conditions can result in disastrous crop failures or, alternatively, bumper crops. Most of the Canadian wheat crop and a considerable share of the American are destined for export markets, in which the demand for Anglo-American wheat fluctuates considerably. The result of these factors can be a great variation in farm incomes from year to year, sometimes leading to the adoption of radical political programs by state and provincial governments in the wheat belts. North Dakota and the three Prairie Provinces of Canada are particularly known for such programs.

The high-risk, "boom or bust" character of wheat belt agriculture has often been manifest in the past, but has been somewhat tempered by increased diversification in recent decades. Over most of the Winter Wheat Belt cattle raising is now such an important part of agriculture that sales of livestock are at least as important to agricultural income as are sales of the wheat for which the region is named. Large acreages are used for pasture, and, while wheat is still the dominant crop, increasing acreages are used for hay and drought-resistant sorghums for feed. In the Spring Wheat Belt there is somewhat less emphasis on livestock but more diversification of crops, with hay, barley, oats, and flax all of some importance. Although an unusual emphasis on wheat and a huge production of wheat continue to be characteristic of these areas, and while wheat is still a major foundation of their economies, wheat is no longer so completely dominant, especially in the Winter Wheat Belt, as the use of the term "wheat belt" implies.

The Unglaciated Southeastern Interior Plain

The Unglaciated Southeastern Interior Plain lies generally to the south of the Corn Belt. Its northern edge is found in dissected lands near the Ohio River in southern Indiana and Illinois, and it extends southward to include sizable areas in Kentucky and Tennessee and between the Appalachians and the Mississippi River embayment of the Gulf-Atlantic Coastal Plain, plus a small part of northern Alabama (map, p. 744). The topography of this southeastern section of the Interior Plains has not been smoothed by glaciation, and it is essentially a hilly area. The climate is mostly humid subtropical, albeit a cool phase of this climatic type.

The Unglaciated Southeastern Interior Plain is distinguished from other large agricultural regions of the Interior Plains by the fact that no well-established, highly developed, and specialized system of farming has evolved in the area as a whole. Rather, each part of the region tends to have a relatively varied and balanced mixture of different types of farms. In general the main types of farms in the mixture are tobacco farms and livestock farms, though in some sections dairy and poultry farms are important. Corn is an extremely widespread and important crop throughout the region. Compared to the glaciated areas to the north, this is an agricultural region of inferior surface and soils, lower productivity, and poorer people. Included within the general region, however, are two small but outstanding agricultural districts, both developed in basinlike areas where soils of more than average fertility have been derived from underlying limestone formations. These are the Bluegrass Region of north central Kentucky, famous as the home of fine horses, and the Nashville Basin of central Tennessee.

The Texas-Oklahoma Red Prairies

South of the Winter Wheat Belt in Texas and Oklahoma a significant area of cotton cultivation has developed on reddish prairie soils (map, p. 718). Cotton can be grown without irrigation here, and extensive mechanization helps to compensate for low average yields; however, in recent years a considerable amount of irrigated cotton acreage has been developed. Until the rise of large-scale cotton culture in the twentieth century, this was principally a ranching district. Today some areas exhibit a mixture of cotton cultivation, ranching, and some wheat and sorghum production; nonetheless, cotton is the dominant agricultural interest for the region as a whole.

The Great Plains Region of Ranching and Irrigated Agriculture

A north-south belt of country which is too dry for wheat except in abnormally wet years extends along the western edge of the Interior Plains from Alberta and Saskatchewan to the Mexican border. This area of steppe climate in the western portion of the Great Plains is culturally and economically a zone of transition between the farming country of the humid and subhumid East and the range livestock country of the West. Sheep and cattle ranches occupy most of the land, although goat ranching is important in the extreme south. The population is generally sparse, but islands of greater density are found in scattered irrigated areas. The largest of these in terms of population are in Colorado, especially along the South Platte River, and in the High Plains of West Texas around Lubbock. The agricultural mainstays of most Great Plains irrigated areas are hay, grown as supplemental feed for range animals, and sugar beets, which supply both a cash crop and fodder for livestock. The West Texas area, however, has recently become a major cotton region due to the rapid expansion of irrigation from deep wells (maps, pp. 716, 717, and 755).

Urban and Industrial Development

The Interior Plains are not only the leading agricultural area of Anglo-America, but contain many of the largest cities and much of the manufacturing industry of this world region. Most of the urban and industrial development is found in the eastern half of the Plains. Of 64 metropolitan areas (SMSAs) in the Interior Plains numbering 150,000 inhabitants or more, 47 lie either on the Mississippi or east of it. Nineteen of the 25 metropolitan areas numbering 500,000 or more are on the Mississippi or east of it, as are all but 2 of the 13 that number more than a million inhabitants.

Most of the major cities in the eastern part of the Interior Plains lie within the American Manufacturing Belt, a zone of concentrated industrial activity extending eastward from the Mississippi River to the northeastern Atlantic seaboard. The western limit of the belt lies in the vicinity of a line connecting St. Louis and Milwaukee; the northern limit along or near the line Milwaukee–Toronto–Montreal–Portland, Maine; and the southern limit along the general line St. Louis–Cincinnati–Charleston, West Virginia–Baltimore. The American Manufacturing Belt thus includes large portions of the American Middle West and Northeast, as well as several districts in southeastern Canada. It represents one of the two largest concentrations of relatively continuous industrial development in the world, the other being the great manufacturing region of western Europe. In terms of major landform divisions, the American Manufacturing Belt lies partly in the Appalachian Highlands, partly in the northern Piedmont and adjacent Coastal Plain, and partly in the Interior Plains.

In general, the individual cities of the Interior Plains have been favored in their development by one or more of the following factors: (1) access to the agricultural wealth of the plains; (2) access to coal, iron, and steel; (3) access to oil; (4) advantageous location with respect to transportation. Most of these cities began as commercial centers serving an agricultural hinterland. Many of them had particular advantages with respect to water transportation, which was of exceptional importance during the time they were first developing. Later the coming of railroads provided a further stimulus to growth, although many cities have continued to be important centers of water transportation as well as railroad centers. The manufacturing industries which have been developed in these cities of the Interior Plains fall largely under two main categories: (1) industries engaged in processing the agricultural products of the plains, and (2) industries manufacturing machinery, vehicles, and other secondary metal products from iron and steel produced in or near the Interior Plains. Position between the Appalachian coal fields and the iron-ore resources of the Canadian Shield and its extension in the Superior Upland, with the Great Lakes–St. Lawrence waterway providing cheap and efficient water transportation for both iron ore and coal, gives the manufacturing centers of the eastern Interior Plains a superior access to the basic raw materials needed by the iron and steel industry.

For purposes of description and analysis, the major cities and industries of the Interior Plains may be conveniently divided into the following four groups according to location: (1) cities and industries of the Ontario and Quebec lowlands, (2) United States cities and industries on or near the Great Lakes, (3) cities and industries of the southeastern section of the Interior Plains, and (4) cities and industries of the Interior Plains west of the Great Lakes.

Cities and Industries of the Ontario and Quebec Lowlands

The core region of Canada is found in a group of lowlands bordering the Great Lakes or the St. Lawrence River from the vicinity of Detroit to Quebec city. The entire lowland area is often termed loosely the "St. Lawrence Lowlands," although the Ontario Peninsula between Lakes Huron, Erie, and Ontario is frequently recognized as a separate section (maps, pp. 725, 726, and 744). Well over half of Canada's manufacturing is done in the Ontario and Quebec lowlands, and most of the large Canadian cities are found here. The industries of this Canadian section of the Interior Plains are extremely diversified, although generally smaller in scale than those found in the United States sections. The major industries of the Quebec section are clothing, textiles, pulp and paper, and meat packing; those of the Ontario section are automobiles, electrical machinery, and iron and steel. But almost the full range of industrial products expected of the economy of a "developed" industrial country are turned out to some extent. Among the factors that have favorably influenced the development of these industries are (1) the availability of waterpower and minerals from the Canadian Shield, (2) the availability of coal from the Northern Appalachian Plateau, (3) access to good transportation facilities, especially water transportation, (4) the agricultural wealth of the surrounding lowlands, and (5) markets provided by the relatively large population of these lowlands.

Most of the main cities in the Ontario and Quebec lowlands are located either on the Great Lakes or on the St. Lawrence River. The St. Lawrence is navigable for large ships as far as Montreal, although it is blocked by ice for a considerable period in winter. Beyond Montreal, canals bypassing rapids on the river gave small ships access to Lake Ontario before completion of the St. Lawrence Seaway in 1959. The new Seaway allows relatively large oceangoing ships to enter the Great Lakes. The obstacle of Niagara Falls, between Lakes Ontario and Erie, is circumvented on the Canadian side by the Welland Ship Canal.

The largest of Canada's cities, aggregating more than one-tenth of the country's population in its metropolitan area, is Montreal (2.5 million). It lies at the convergence of several major passageways: the St. Lawrence River and Valley leading to the Atlantic and the Great Lakes; the Champlain Lowland leading south toward New York City; and the Ottawa River Valley along which Canada's transcontinental railroads find a natural passage west and north into the Shield. Montreal is Canada's chief seaport and its largest manufacturing center. Its industries are highly diversified, with the manufacture of textiles and clothing the most important. Quebec (420,000), on the St. Lawrence below Montreal, was the original center of French settlement and administration in Canada, chosen because of its fortress position at the head of the estuary where the river became narrow enough to be controlled by the cannon of the time. Its advantages for commerce, however, have proved far inferior to those of Montreal, though it, too, is an active ocean port.

Montreal's great rival among Canadian cities is Toronto (2.2 million), which has the finest natural harbor on Lake Ontario. Toronto has become the financial center of Canada and the commercial center of Ontario, as well as Canada's second city in population and manufacturing. Together Montreal and Toronto account for over a third of Canada's total employment in manufacturing. Toronto's manufacturing industries, like those of Montreal, are highly diversified.

Other Ontario cities include Ottawa (505,000), Canada's capital, which is located on the Ottawa River in the section where the river forms the boundary between the country's two largest provinces; Hamilton (455,000), a Lake Ontario port which accounts for over half of Canada's steel output; and Windsor (215,000), located opposite Detroit and thus well situated for its role as a major Canadian center of automobile production. London (210,000) and Kitchener-Waterloo (195,000) are inland commercial and industrial centers in one of the most productive agricultural areas of lowland Ontario. In addition to the cities mentioned, a great deal of manufacturing is carried on in smaller industrial centers of the Ontario and Quebec lowlands.

United States Cities and Industries on or near the Great Lakes

Some of the largest cities in the United States, including the largest ones in the Interior Plains, are found along or near the southern shore of the Great Lakes. Their locations generally reflect past or present advantages deriving from such factors as natural harbors on the lakes, portages between rivers, river valley passageways leading from the lakes to the interior, or other factors relating to water transportation.

Ready access to coal and steel has been an advantage of the greatest importance to the manufacturing industries of the lake cities. Most of the coal comes from the Appalachian Plateau, as does much of the steel. However, a number of the lake ports are themselves major steel producers. The leading ones in this respect are Chicago (where the industry is largely centered in the Indiana suburbs, especially Gary), Detroit, Buffalo, and Cleveland. These four cities and their suburbs alone, not including smaller installations in other lake ports, contain about a third of the United States total steel capacity. The Chicago metropolitan area is the world's greatest center of steel production, with about 18 percent of all United States steel-manufacturing capacity. The steel mills of the lake ports draw most of their ore from the Superior Upland via the lakes, and their coal mainly from the Appalachians, but Chicago benefits also from relative proximity to the Eastern Interior coal field of central and southern Illinois and adjoining parts of Indiana and Kentucky. This field produces about one-fifth of the United States output of bituminous coal. Much of the coal can be used for coking when mixed with higher-grade Appalachian coal. Most of the lake cities are highly industrialized, and the group as a whole is of great importance in the iron and steel, machinery, vehicle, and other metal industries. In addition, two cities in the group, Chicago and Buffalo, are major world centers of food processing, and another city, Rochester, is the world's greatest center of the photographic equipment industry.

Chicago's metropolitan area (7.4 million including Gary and other Indiana satellites and suburbs) is the second largest in the United States and in Anglo-America. Chicago is by far the largest city of the Interior Plains. It has developed on a harbor formed by the mouth of the Chicago River near the southern end of Lake Michigan. Here the Great Lakes make their farthest penetration into the agricultural heartland of the United States. An old portage, first canalized a century ago, connects Lake Michigan drainage and Mississippi drainage in the vicinity of the city. Today a modern commercial waterway, following the old canal route, connects Chicago with the Mississippi and the Gulf of Mexico via the Illinois River, and a series of artificial harbors have been developed to handle most of the Lake Michigan traffic. The city has become the world's greatest railway center, a result partly of the diversion of east-west tracks around the southern end of Lake Michigan, but more largely due to Chicago's original importance as a lake port. The relation of Chicago to the agriculture of the Corn Belt and the wheat belts is seen in its function as a major grain market and in its important meat-packing and farm machinery industries. However, most of the city's industries are engaged in producing a diverse array of metal products, including many different types of machinery.

Milwaukee (1.4 million), north of Chicago, is the second most important Lake Michigan port and industrial center. It has a better natural harbor than Chicago, but in other respects its location is less advantageous. Nevertheless it is a major industrial center, known for its large production of heavy machinery, beer (probably the result of its large population of German origin), automobile bodies, and a miscellany of other goods.

The second largest city of the Interior Plains, and second largest on the Great Lakes, is Detroit (4.1 million, excluding the Canadian suburb of Windsor). Its position on the Detroit River linking Lakes Huron and Erie has long made it a city of considerable importance. However, its meteoric rise in the twentieth century into the top group of American population and industrial centers was connected mainly with the equally meteoric rise of the American automobile industry. Despite much decentralization of the industry in recent years, Detroit still has no significant rival as a center of automobile manufacturing. In the economies of many smaller cities in the eastern Midwest the manufacture of motor vehicles and/or parts is either dominant or of extreme and unusual importance. Most of these cities are on the periphery of, or within, a rough circle with Detroit near its eastern edge. The periphery of the circle is marked by Flint (480,000), Saginaw (215,000), and Muskegon (150,000), Michigan; South Bend (280,000) and Ft. Wayne (260,-000), Indiana; and Lima (170,000) and Loraine-Elyria (265,000), Ohio. Within the perimeter are other centers of the auto industry, notably Jackson (135,000) and Lansing (345,000), Michigan, and Toledo (670,000), Ohio, but also many smaller places. In only a few of the larger cities of this area is the auto industry distinctly smaller in importance, and even in these the industry is by no means entirely absent. These places include the Michigan cities of Grand Rapids (505,000), famous as a furniture manufacturing center but mainly making metal goods and machinery today; Kalamazoo (190,000), primarily a paper-milling center; and Ann Arbor (210,000), the site of the University of Michigan as well as an industrial center.

Two other major Great Lakes ports of the United States are Cleveland, Ohio (2 million), and Buffalo, New York (1.3 million). Between them lies the much smaller port city of Erie (255,000), the only Pennsylvania port on the lakes. Cleveland has a good harbor and lies in the path of rail lines skirting the Appalachian Plateau along the level coastal plain of Lake Erie, and others fol-

lowing relatively easy valley routes into the interior of the Plateau. The city is important in both the steel and automobile industries, and is noteworthy for the production of machinery and assorted metalwares. Several ports along the Ohio shore of Lake Erie handle large tonnages of iron ore (inbound) or coal (outbound). Cleveland and the small cities of Conneaut and Ashtabula (map, p. 726) are the main ore ports, while Toledo is by far the leading coal port. Buffalo is preeminently the port where transportation along the Hudson-Mohawk Valley and the Lake Ontario Plain intersects that of the Great Lakes. Until the building of the Welland Canal it was the eastern end of navigation on the upper Great Lakes, and it was the western terminus of the Erie Canal. Buffalo has important steel and automobile industries, and is the world's greatest center of flour milling. Hydroelectricity from Niagara Falls has provided a basis for a considerable development of electrochemical and electrometallurgical manufacturing. East of Buffalo, along the Erie Canal route, are two other important industrial centers, each connected by canal with a small port on Lake Ontario. Rochester (835,000) grew originally where the lowland route passed a waterfall on the Genessee River. It was primarily a flour-milling center in the early days, but it has now become the world's leading center for the production of photographic equipment. Its other manufactures are dominantly of a type requiring highly skilled labor. Syracuse (630,000) is located where the lowland route along the Mohawk Valley widens into the Lake Ontario Plain. Large salt deposits in the vicinity of the city fostered an early growth of chemical manufacturing, but the chemical industry is now greatly outweighed in importance by a variety of secondary metal products.

Cities and Industries of the Southeastern Section of the Interior Plains

During the early days of settlement in the Middle West, people and goods moved westward along three great routes issuing from the Appalachians into the Interior Plains. One of these routes was the Great Lakes, one was the Ohio River, and the other was the first land highway in the Middle West, the National Road, which pursued its westward course between the other two. Important cities are now found at intervals along each of these routes. Those along the Great Lakes have already been described. The principal Ohio River cities are Cincinnati, Ohio (1.4 million), located on the great northern bend of the river where tributaries enter from both sides; Louisville, Kentucky (810,000), located at a stretch of falls and rapids that long interrupted traffic until bypassed by a canal; and Evansville, Indiana (230,000), located not far from the mouth of the Ohio's major north bank tributary, the Wabash, which once carried a considerable amount of commerce. The main cities along the old National Road are Columbus, Ohio (890,000), located where the road issued from the Appalachian Plateau and intersected the Scioto River leading toward the Ohio; Dayton, Ohio (825,000), approximately where the road intersected the Miami River leading southward to the Ohio at Cincinnati; and Indianapolis, Indiana (1 million), located at the intersection of the road with an Ohio tributary thought, erroneously, to be navigable at the time the city was founded. Despite the error with regard to the availability of water transport, Indianapolis became the capital of Indiana, for which its central location favored it, and also became a major rail center. The only large city in the southeastern part of the Interior Plains that is not on either of these old transport routes is Nashville, Tennessee (535,000), which lies considerably farther south and is the commercial center of the fertile Nashville Basin, besides being the state capital and a minor industrial center with emphasis on processing agricultural products.

With the exception of Nashville, each of the cities described in the preceding paragraph has a diversified industrial structure in which the manufacture of machinery and other secondary metal products is important. In addition, some of them are known for other specialties such as soap in Cincinnati, whiskey and tobacco products in Louisville, and drugs and meat products in Indianapolis. Cincinnati is the greatest single center of the machine-tool industry in the United States, Columbus is noted for its mining machinery, and Dayton for light manufactures, such as calculators, cash registers, and various automobile components, which require a high degree of skill to make.

Cities and Industries West of the Great Lakes

In the Interior Plains west of the Great Lakes, cities are more widely scattered and smaller than farther east, and, in general, this becomes increasingly so the farther west one progresses. The cities located in the western parts of the plains tend to be basically commercial centers serving agricultural hinterlands, or cities closely related to the oil industry. Such manufacturing as they possess is often closely related to the needs and products of their agricultural and/or oil-producing surroundings. In the eastern part of this area the manufacture of machinery, especially agricultural machinery, continues to be of considerable importance, but farther west grain milling and meat packing, and in some places

oil-field industries, become relatively more and more important.

The largest cities are on the Mississippi River. St. Louis (2.3 million), located near the junction of the Missouri, Mississippi, and Illinois rivers, ranks first in size. It has long been a famous and important river port, and has also become a major railroad center. It can draw on the coal of the Eastern Interior Field for power, and has developed an extremely diversified list of manufactures, including aircraft and space-flight equipment, automobiles, and many different types of machinery, as well as clothing, chemical products, shoes, meat, beer, steel, grain products, and alumina. It is also one of the country's principal markets for grain and livestock. The twin cities of Minneapolis and St. Paul (1.7 million), comprising the second largest urban center on the Mississippi, have developed at a point where rapids mark the head of navigation on the river and in addition furnish hydroelectric power. The Minneapolis–St. Paul urban district is primarily a commercial center, not only for adjacent areas but for the Spring Wheat Belt and areas beyond. For years the Twin Cities led the world in flour milling, and are still of great importance in that industry. They are also a major center of the meat-packing industry and have a growing machinery industry.

A number of smaller cities in Illinois, Iowa, and Wisconsin are major centers for the manufacture of farm equipment. The largest of these are Peoria, Illinois (350,000), the headquarters of a large tractor company; the Davenport, Iowa–Rock Island–Moline, Illinois, metropolitan area (350,000), which is almost entirely specialized in the manufacture of agricultural machinery; and Rockford, Illinois (275,000). However, in the same general area Madison, Wisconsin (275,000), is primarily a governmental, educational, and food-processing center.

Toward the western edge of the Corn Belt three sizable cities have developed, primarily as market centers for adjoining sections of the Corn Belt, the Winter Wheat Belt, and ranching areas to the west. These are Kansas City, Missouri and Kansas (1.2 million), Omaha, Nebraska (530,000), and Wichita, Kansas (400,000). Each is an important market and processing center for grain and livestock. In general other industries are little developed in the three cities, though Kansas City has automobile assembly plants and Wichita a large aircraft industry. Des Moines, Iowa (285,000), also in the western Corn Belt, is a commercial, political, and educational center with little significant industry.

The two largest cities of Oklahoma have a character somewhat different from that of the other cities of the western Interior Plains. Oklahoma City (610,000) and Tulsa (450,000) owe their growth largely to the oil industry. The Mid-Continent oil field extends in this part of the Interior Plains from Texas across Oklahoma and into southern Kansas. In addition to serving the oil industry as commercial, administrative, and refining centers, both cities are agricultural market centers, and Tulsa is the site of an aircraft plant. Farther south another prominent oil, aircraft, and agricultural market center, Fort Worth (650,000), is so close to the Coastal Plain–Interior Plains boundary that it might legitimately be classed with either area. In the Great Plains of northwestern Texas, Lubbock (190,000) is the commercial center for a major irrigated cotton district, while Amarillo (170,000), farther north, serves the Texas part of the Winter Wheat Belt. Both lie near oil and natural gas fields.

In the western Great Plains the thinning of urban development and manufacturing toward the west in the Interior Plains reaches its climax. Only one United States city in this area exceeds 200,000 in metropolitan population. This is Denver, Colorado (1.1 million), at the foot of the Rocky Mountains and economically almost as much a city of the mountains as of the plains. It is the commercial center for mining districts scattered through the Rockies to the west and for the irrigated areas and ranches of the plains to the east. The agricultural production of the Great Plains flows mainly toward the east and has not been intensive enough to stimulate the growth of large commercial centers within the Great Plains proper. Nor have the mineral resources of this area led to any large development of manufacturing. These mineral resources include oil and natural gas, mostly but not entirely in the south from Kansas to eastern New Mexico; enormous potash beds in New Mexico which furnish most of the United States supply; and considerable deposits of coal and lignite in a number of places. Most of the coal, however, is relatively low in quality.

The metropolis of the Canadian Prairie Provinces is Winnipeg, Manitoba (515,000), which is similar in function to such cities as Minneapolis–St. Paul, Omaha, and Kansas City. At Winnipeg the extensive rail network of the prairies focuses on two major lines which cross the Shield to the east (map, p. 711). Consequently, this city has become the principal market and distributing center for the three provinces. Toward the west, the Canadian part of the Great Plains has resources of coal and oil. In recent years oil production has increased rapidly in Alberta, and the two main cities of that province have experienced rapid growth. These cities are Edmonton (405,000) and Calgary (335,000). In their combination of oil and agricultural interests they somewhat resemble Tulsa and Oklahoma City.

THE ROCKY MOUNTAINS

The Rocky Mountains, broadly defined, may be regarded as extending from northwestern Alaska to northern New Mexico (map, p. 707). Within the mountain area are many distinct ranges. Some of the latter are known locally as the Rocky Mountains while others are not, but the whole may be viewed as a single enormous system. On the north the Rockies are bordered by the Arctic Coastal Plains, on the east by the Mackenzie River Lowlands and Great Plains sections of the Interior Plains, and on the west and south by various sections of the Intermountain Basins and Plateaus. In Canada these mountains occupy the border zone between Yukon Territory and the Northwest Territories, and they extend south through eastern British Columbia and southwestern Alberta. In the United States they occupy northeastern Washington, northern and eastern Idaho, western Montana, western Wyoming, northeastern Utah, central Colorado, and north central New Mexico (maps, pp. 754 and 755).

The Rockies are lofty, rugged mountains with many snowcapped peaks. The highest elevations are found in Colorado, which contains a number of summits above 14,000 feet. Mount Elbert, the highest peak in the entire system, has an elevation of 14,431 feet. Throughout the Rockies, peak elevations of 10,000 feet or higher are very frequent, and the valleys and passes are generally above 5000 feet. Elevations decline toward the extreme north, where the Brooks Range of Alaska rises to general summit levels of 5000 to 6000 feet. The mountains generally form series of linear ranges trending roughly in conformance with the system as a whole. Valleys of varying size and shape are enclosed between and within the ranges. Only in the United States section north of Yellowstone Park does the linear outline break down. Here the mountains rise in jumbled masses.

Easy passes through the Rockies are almost nonexistent, and the system has been a major barrier to transportation. Seven east-west trunk railroads cross the mountains: the Canadian National, Canadian Pacific, Great Northern, Northern Pacific, Milwaukee, Union Pacific, and Denver and Rio Grande Western (map, p. 755). The easiest route, followed by the Union Pacific, lies through southern Wyoming, where the Wyoming Basin almost breaks the mountain system in two, and thence through low passes in the Wasatch Range to the Great Salt Lake district. Elsewhere, three routes in Canada (the Canadian Pacific has two separate lines crossing through separate passes; one of these is not shown on the map), three through Montana, and two through Colorado cross Rockies via passes varying from 3700 to 10,000 feet.

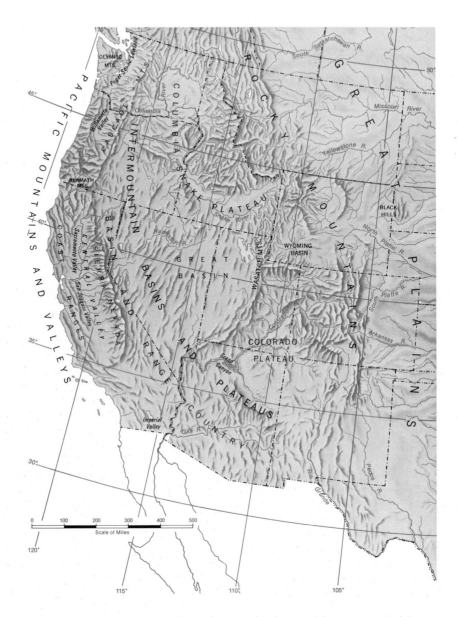

Major physiographic features of western United States.

Key to railroads: CN, Canadian National; CP, Canadian Pacific; DRW, Denver and Rio Grande Western; GN, Great Northern; MW, Milwaukee; NP, Northern Pacific; RI, Rock Island; SF, Santa Fe; SP, Southern Pacific; SPS, Spokane, Portland and Seattle; TP, Texas and Pacific; UP, Union Pacific; WP, Western Pacific. Urban areas are grouped by population according to 1968 estimates for metropolitan areas (Standard Metropolitan Statistical Areas [SMSAs] in the case of United States Cities).

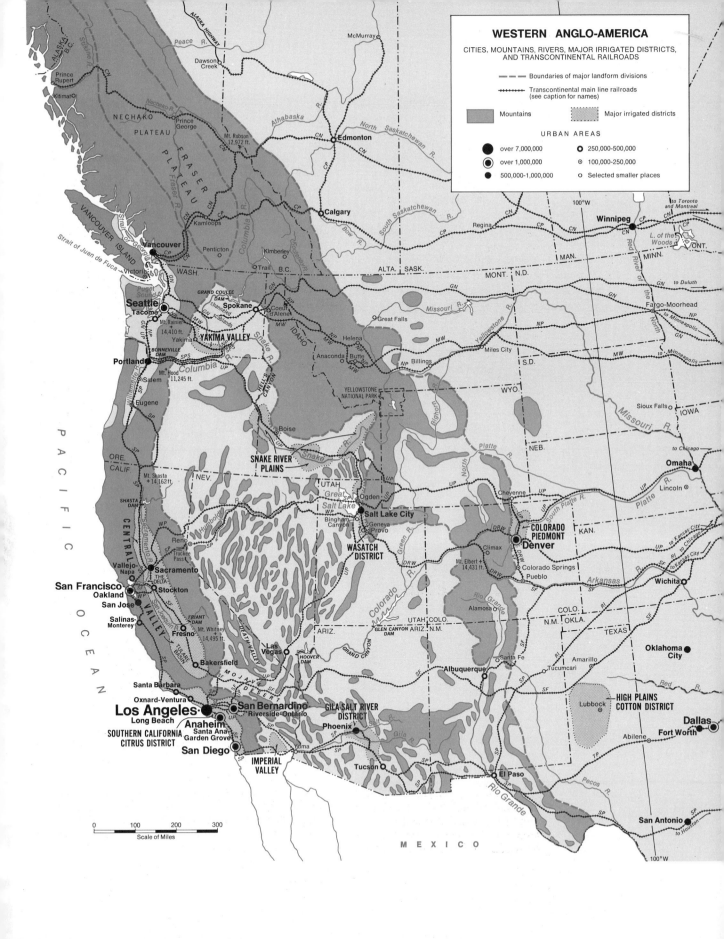

PACIFIC OCEAN

MEXICO

0   100   200   300
Scale of Miles

100°W

Prince Rupert
Kitimat
ALASKA HWY. B.C.
Peace R.
McMurray
Dawson Creek
Stikine R.
NECHAKO PLATEAU
Nechako R.
Prince George
Mt. Robson 12,972 ft.
Athabaska R.
North Saskatchewan R.
Edmonton
CN
FRASER PLATEAU
Fraser R.
Kamloops
Columbia R.
Calgary
Bow R.
South Saskatchewan R.
Regina
CP
Winnipeg
L. of the Woods
ONT.
to Toronto and Montreal
VANCOUVER ISLAND
Strait of Georgia
Penticton
Kimberley
Trail B.C.
ALTA. SASK.
MAN.
Red River of the North
to Duluth
Vancouver
Victoria
Strait of Juan de Fuca
WASH.
Columbia R.
MONT. N.D.
Fargo-Moorhead
MINN.
Puget Sound
GRAND COULEE DAM channeled scablands
Spokane
Coeur d'Alene
Great Falls
Missouri R.
Yellowstone R.
NP
Minneapolis
to Minneapolis
Seattle
Tacoma
Mt. Rainier 14,410 ft.
Yakima
YAKIMA VALLEY
IDAHO
Helena
Anaconda Butte
Miles City
S.D.
Sioux Falls
IOWA
Portland
BONNEVILLE DAM
Columbia R.
Snake R.
HELLS CANYON
Billings
Bighorn R.
Salem
Mt. Hood 11,245 ft.
Willamette R.
WYO.
Missouri R.
Eugene
YELLOWSTONE NATIONAL PARK
ORE. CALIF.
Boise
SNAKE RIVER PLAINS
Snake R.
Platte R.
NEB.
Omaha
to Chicago
Mt. Shasta 14,162 ft.
NEV.
Cheyenne
Lincoln
SHASTA DAM
Humboldt R.
UTAH
Great Salt Lake
Ogden
UP
South Platte R.
COLORADO PIEDMONT
Denver
KAN.
to Kansas City
Reno
Truckee Pass
Salt Lake City
Bingham Canyon
Geneva Provo
WASATCH DISTRICT
Green R.
Climax
Mt. Elbert 14,431 ft.
Colorado Springs
Pueblo
Arkansas R.
Wichita
Vallejo Napa
Sacramento
THE DELTA
Stockton
CENTRAL VALLEY
San Francisco
Oakland
San Jose
Salinas Monterey
FRIANT DAM
Fresno
Mt. Whitney 14,495 ft.
DEATH VALLEY
TULARE BASIN
ARIZ.
Colorado R.
GLEN CANYON DAM
UTAH COLO.
ARIZ. N.M.
Rio Grande
Alamosa
N.M. OKLA.
COLO.
TEXAS
Oklahoma City
Bakersfield
Las Vegas
HOOVER DAM
GRAND CANYON
Santa Fe
Albuquerque
Tucumcari
Amarillo
RI
to Kansas City
Santa Barbara
Oxnard-Ventura
MOJAVE DESERT
Lubbock
HIGH PLAINS COTTON DISTRICT
Los Angeles
Long Beach
San Bernardino
Riverside Ontario
GILA SALT RIVER DISTRICT
Phoenix
Gila R.
Abilene
Dallas
Fort Worth
Anaheim
Santa Ana Garden Grove
SOUTHERN CALIFORNIA CITRUS DISTRICT
San Diego
IMPERIAL VALLEY
Yuma
Gila R.
Tucson
El Paso
Pecos R.
Rio Grande
Red R.
San Antonio
to Houston

The Important Mining Industry

Mining has always been of great economic importance in the Rocky Mountains from the days in which it attracted the first permanent settlers into a region which is still sparsely settled. Due to the small development of other activities, scattered mining towns from New Mexico to British Columbia tend to have a significance in the region that is disproportionate to their size. They produce a variety of minerals, often from complex ores yielding several end products. The major minerals in terms of value are molybdenum, copper, silver, lead, zinc, and, in the Wyoming Basin, oil, uranium, and natural gas. But many others are produced. Among the many small and scattered mining communities, a few are of outstanding importance; they include Butte, Montana, where the famous copper mines also yield silver, lead, and zinc; Coeur d'Alene, Idaho, producing silver, lead, zinc, and gold; Kimberley, British Columbia, which mines most of Canada's lead and much of its zinc; and Climax, Colorado, the site of the world's largest molybdenum mine. Some of the ores are refined and smelted in or near the mining districts. Anaconda, Montana, near Butte, and Trail, British Columbia, are among the world's great centers of nonferrous metallurgy. Many "ghost towns" in the Rockies tell of mineral deposits once important but now exhausted. Conversely, the region frequently sees the opening of new mines. Recent notable developments have exploited oil, natural gas, and uranium in the Wyoming Basin and phosphate in Wyoming, Utah, Idaho, and Montana. Attention is increasingly drawn to the possibilities for utilizing large reserves of oil shale in the central and northern U.S. Rockies and adjoining parts of the Intermountain Basins and Plateaus.

Other Economic Activities

Ranchers graze sheep and cattle in many parts of the Rockies (photo, right), and many valleys contain small agricultural communities. Agriculture is generally dependent on irrigation, however, and is limited by short growing seasons. Fodder crops are grown to supplement natural forage for grazing animals; and specialties such as tree fruits, potatoes, and vegetables have developed in some areas. The Rockies are forested, largely with varieties of pine, fir, and other coniferous softwoods, up to the tree line, and a certain amount of lumbering is found in various sections. However, large-scale lumbering has developed only in northern Idaho and adjoining areas. A very important and growing activity is the tourist and resort trade. Most of the land in the Rockies, in both the United States and Canada, is owned by the respective national governments. The latter exert control over the exploitation of mineral, timber, and water resources within their properties in the interests of conservation and sustained economic development of their countries. They have contributed to the growth of tourism by setting aside areas containing spectacular scenery

The enormous open-pit copper mine of the Kennecott Copper Corporation at Bingham Canyon, Utah; in the background, a range in the Intermountain Basins and Plateaus. (Courtesy of Utah Copper Division, Kennecott Copper Corporation.)

A roundup of range cattle in the Montana Rockies south of Butte. This semiarid landscape in the upper valley of the Ruby River is characteristic in many ways of vast areas of range land in the Rocky Mountains and Intermountain West. Note the small trees lining the watercourse in the foreground. (Charles W. Herbert, Western Ways Features.)

or other natural or historical attractions as national parks or monuments. Tourist facilities have been developed in many of these recreational areas. In addition, a large number of privately owned resorts are found in the Rockies. They are often in close proximity to areas that have been developed as public recreational facilities.

On the whole, the economy of the Rocky Mountains is handicapped by the relative isolation of much of the area. This isolation is partly a consequence of remoteness from important centers of population, and is partly due to the difficulty of constructing adequate transportation lines in such rugged terrain. The mountain area is sparsely populated, and urban development is very limited. The largest city, Butte, has a population of only about 27,000 in the city proper. The commercial capitals of the various sections of the Rockies generally lie just outside the mountains, in such cities as Denver, Colorado (1.1 million), Salt Lake City, Utah (550,000), Spokane, Washington (270,000), and Calgary, Alberta (335,000).

THE INTERMOUNTAIN BASINS AND PLATEAUS

The Rocky Mountain System and the mountains near the western shore of Anglo-America are separated from each other, from Alaska to Mexico, by the Intermountain Basins and Plateaus (map, p. 707). This landform division occupies an immense part of western Anglo-America. In general, while lower than the bordering mountains, the Intermountain Basins and Plateaus lie at comparatively high elevations above the sea. Few sections lie below 3000 feet, and very few below 2000. There is a great deal of variety in elevations, landforms, and local relief from place to place. Most of the land is composed of plateau surfaces in various stages of dissection so that rolling uplands and rugged hilly and mountainous sections are included, in addition to the large areas which are comparatively level or actually flat. Over much of the area the river valleys are deeply incised, forming

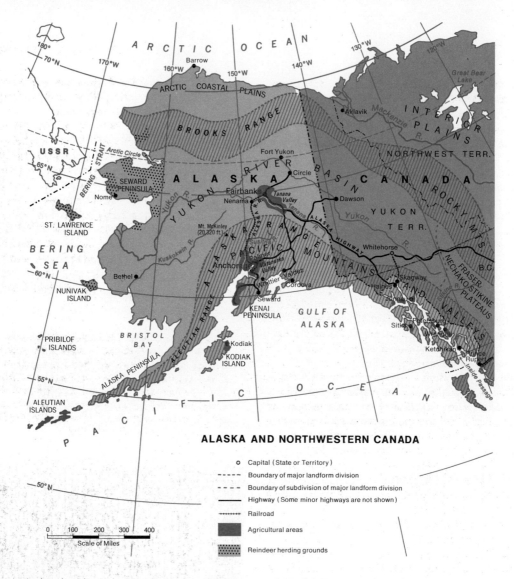

ALASKA AND NORTHWESTERN CANADA

○	Capital (State or Territory)
----	Boundary of major landform division
- - -	Boundary of subdivision of major landform division
——	Highway (Some minor highways are not shown)
+++++	Railroad
▓	Agricultural areas
░	Reindeer herding grounds

0 100 200 300 400
Scale of Miles

Agricultural and reindeer-herding areas of Alaska are adapted from a map in Karl E. Francis, "Outpost Agriculture: The Case of Alaska," *Geographical Review*, Vol. 57 (1967), p. 498. Copyrighted by the American Geographical Society of New York; by permission of the author and the *Review*.

spectacular canyons and gorges. The most famous of these, the Grand Canyon of the Colorado River, is an enormous gash cut in the surface of the Colorado Plateau in northern Arizona. In many places isolated mountain ranges project far above the general surface level.

Climatically, the Intermountain Basins and Plateaus are distinguished by low rainfall due to the position of this landform division between two shielding mountain systems. Most of the Intermountain area has a semiarid (steppe) climate, although there are a number of sizable desert areas. The heaviest precipitation occurs on the higher surfaces. In the more northerly sections the effects of low precipitation are offset to some degree by the less-ened evaporation attendant on lower temperatures. Temperature conditions range from subarctic in Alaska and northern Canada to subtropical along the Mexican border. Locally, there are great temperature as well as rainfall contrasts resulting from differences in elevation.

Major Subdivisions

The Intermountain area may be divided into a number of major subsections, as follows (maps, pp. 754, 755, and 758): (1) the Basin and Range Country, (2) the Colorado Plateau, (3) the Columbia–Snake Plateau,

(4) the Fraser–Nechako–Stikine Plateaus, and (5) the Yukon River Basin.

The *Basin and Range Country* extends from southern Oregon and southern Idaho to the Mexican border. It includes parts of Oregon, Idaho, California, Utah, Arizona, New Mexico, and Texas, as well as practically the entire state of Nevada. The section lying between the Wasatch Mountains of Utah and the Sierra Nevada of California is often referred to as the Great Basin, although the latter includes many smaller basins. The basins of the Basin and Range Country, many of which have no external drainage, are often separated from each other by blocklike mountain ridges rising high above the general level. Semiarid climatic conditions of northern sections and most eastern sections produce such vegetation forms as short grass, bunch grass, and sagebrush. In the southwestern part, true desert with extremely sparse vegetation prevails over a wide area in California, Nevada, and Arizona, and extends eastward into southern New Mexico and southwestern Texas.

The *Colorado Plateau* occupies parts of Colorado, Utah, Arizona, and New Mexico. Rolling uplands lie at varying levels, often separated by steep escarpments. Mountain areas rise above the general surface in various places. Rivers, principally the Colorado and its tributaries, flow in deep canyons. The latter have been a great hindrance to transportation and have kept many sections extremely isolated. The climate is generally semiarid, although some higher sections have sufficient precipitation to produce a forest growth.

The *Columbia–Snake Plateau* is found in eastern Oregon and Washington and southern and western Idaho. It is characterized by extensive areas of level land, the result of massive lava flows in the past which buried the previous topography. The soils formed from these volcanic materials are exceptionally fertile. Isolated mountains occur, and the streams often flow in canyons. The Columbia River, the master stream of the area, has cut many alternative channels in the past, which now form rugged "scablands" in parts of Washington. The Columbia's giant tributary, the Snake, flows along the border between Oregon and Idaho in a canyon rivaling the Grand Canyon of the Colorado. The climate is semiarid.

The *Fraser–Nechako–Stikine Plateaus* of British Columbia form a considerably more narrow and constricted section of the Intermountain region than do their counterparts in the United States. The Fraser Plateau in the south is a deeply dissected and rugged area, but in the north the Intermountain section of British Columbia presents large areas of rolling upland, interrupted by several mountain ranges. The semiaridity of the

area is moderated by lower temperatures toward the north, where the grasslands and parklike forests of the south trend into the subarctic or taiga forest.

The *Yukon River Basin,* which occupies most of Canada's Yukon Territory and the greater part of Alaska (map, left), has a varied topography of rolling uplands, hill country, low mountains, intrenched streams, and comparatively small areas of flat alluvial plains. As the elevations decline toward the sea in the Alaska section of the Basin, marshy areas become more prominent. The extreme subarctic climate of the Yukon Basin sets it apart from other areas in the Intermountain Basins and Plateaus. Cold rather than aridity is the dominant climatic factor, and the subarctic taiga forest is the prevailing vegetation type (photo, below).

A field of potatoes in the Yukon River Basin near Fairbanks, Alaska. Potatoes are one of the few crops that can be grown successfully in the subarctic climate zone. Vast areas of the Alaskan and Canadian subarctic are covered by a straggling growth of small trees, mainly conifers, such as cover the low uplands in the background. (United States Information Service, Department of State.)

Population, Economy, and Urban Development

Population is very sparse in the Intermountain Basins and Plateaus, being almost nonexistent over wide stretches, especially in the north. The major economic activities are grazing, irrigated agriculture, and mining. Most of the land as far north as the subarctic area of British Columbia is grazed, but the pasturage is generally poor. Ranches must perforce be large, and the ranching population, in consequence, is small and scattered. Here and there definite clusters of population are associated with irrigated areas or, to a lesser extent, with mining districts. Among the many scattered irrigated districts or oases, the following are outstanding in size and importance (map, p. 755): (1) the Imperial Valley of California, watered by the lower Colorado River, (2) the Gila–Salt River district around Phoenix, Arizona, (3) the Wasatch district in Utah at the western foot of the Wasatch Range, (4) the Snake River Plains in Idaho, and (5) a series of irrigated valleys along the eastern slopes of the Cascade Mountains in Washington and Oregon, the most important of which is the Yakima Valley with its massive production of apples, pears, peaches, and other fruits. There is much variety in types of production from one irrigated district to another. Different oases specialize in one or a combination of such crops as alfalfa, cotton, citrus fruits, early vegetables, sugar beets, potatoes, and apples.

Mining activity characterizes almost the whole extent of the Intermountain area in greater or lesser degree, being least important in the Columbia–Snake Plateau and the plateaus of British Columbia. The leading mineral product by value for the area as a whole is copper by a wide margin, followed by natural gas, uranium, and oil. A great variety of other minerals are produced in varying quantities. Arizona mines over half and Utah just under a fifth of the United States output of copper. The main producing centers include mines in southeastern Arizona and a great open-pit operation at Bingham Canyon, Utah (photo, p. 756). Production of natural gas, oil, and uranium in the Colorado Plateau sections of Utah, Colorado, and New Mexico has developed rapidly in recent years and is breaking down some of the isolation that has characterized large parts of the Plateau. In addition to copper, oil, and uranium, Utah mines both coal and iron ore, and has, south of Salt Lake City at Geneva, the only iron and steel plant in the Intermountain region. In the Yukon Basin section of the Intermountain area, gold mining attracted the first settlement and has long been a mainstay of the economy, but is now over-shadowed by silver output in the part of the Basin that lies in the Yukon Territory.

Nonirrigated cropping is important in only one district, the Columbia Plateau, which is able to produce large quantities of wheat thanks to soils that are exceptionally fertile and water retentive. The tourist and resort industry is also limited mainly to one section –the southern deserts. Visitors are attracted to the latter for reasons of health as well as recreation. The intrenched rivers of the Intermountain area, flowing from adjacent mountain regions, produce a tremendous hydroelectric potential, much of which has been developed. The Columbia River and its tributaries have a greater hydroelectric potential than any other river system in Anglo-America has. The largest and most famous power installations in the Intermountain area are Grand Coulee Dam on the Columbia River and Hoover Dam (formerly Boulder Dam) on the Colorado. Power from these dams and others such as Glen Canyon Dam on the Colorado is largely transmitted to points outside of the Intermountain Basins and Plateaus and has given rise to relatively little manufacturing within the area itself.

The low population densities of the region are indicated by the small size and small number of appreciable urban centers scattered over such a vast area. Only ten cities have metropolitan (SMSA) populations of 100,000 or more:

Phoenix, Arizona	895,000
Salt Lake City, Utah	545,000
El Paso, Texas	355,000
Tucson, Arizona	335,000
Albuquerque, New Mexico	315,000
Spokane, Washington	270,000
Las Vegas, Nevada	255,000
Reno, Nevada	125,000
Provo, Utah	120,000
Boise, Idaho	100,000

Except for Provo, which is an industrial community with the Geneva steel plant in its metropolitan area, these cities are primarily trade and service centers rather than manufacturing centers. They tend to be located in or near the region's scattered areas of high agricultural production: Phoenix in the Salt River area of Arizona; Salt Lake City in the Salt Lake oasis at the foot of the Wasatch; Albuquerque and El Paso in proximity to areas irrigated from the Rio Grande; and Boise in the Snake River Plains. Reno is associated with a smaller irrigated area; while Spokane is on the northern edge of the Columbia Plateau wheat belt and also serves mining and lumbering areas in the adjacent Rockies. There is also some tendency for these cities to occupy notably strategic sites with respect to long-distance transport routes: Salt Lake City and Spokane near the western end of passes through the Rocky Mountains; El Paso where the Rio Grande cuts through mountains; Albuquerque where the Santa Fe Railroad rounds the southern end of the Rockies; and Reno where the major pass over the Sierra Nevada reaches the floor of the Basin and Range Country. Three cities— Las Vegas, Reno, and Tucson—are somewhat different from the others in their major functions. Las Vegas and Reno are well known as service centers for the whole of the United States in the fields of gambling, entertainment, and divorce, while Tucson is a resort, retirement, and educational center. There are no large cities in the Canadian and Alaskan sections of the Intermountain region. In Yukon Territory the largest settlement is Whitehorse with 5000 people, while in Intermountain Alaska, Fairbanks with 14,000 people is the metropolis.

THE PACIFIC MOUNTAINS AND VALLEYS

The Pacific shore of Anglo-America is bordered by a series of mountain ranges extending from Mexico to the Aleutian Islands. In the United States several large lowlands are included within this mountainous region.

Physical Description of the Major Subdivisions

For purposes of introductory physical description, the Pacific Mountains and Valleys may be conveniently discussed in terms of the following major subdivisions (maps, pp. 754, 755, and 758): (1) the Coast Ranges of California, Oregon, and Washington, (2) the Sierra Nevada, (3) the Central Valley of California, (4) the Klamath Mountains, (5) the Cascade Mountains, (6) the Willamette–Puget Sound Lowland, (7) the Coastal Ranges of British Columbia and southeastern Alaska, and (8) the Alaska Range and other mountains of southern Alaska north of the southeastern "Panhandle."

The coasts of California, Oregon, and Washington are fronted by the *Coast Ranges.* The section in northwestern Washington known as the Olympic Mountains has peaks reaching approximately 8000 feet, and in the south some sections east of Los Angeles exceed 10,000 feet. However, the peak elevations more commonly lie at 3000 to 5000 feet in California and at only 1000 to

3000 feet farther north. Along most of the western coast of the United States there is no coastal plain, or almost none, and even the lower parts of the Coast Ranges are often quite rugged. However, a few valleys are available for agriculture, especially in California; and from Los Angeles to the Mexican border the ranges lie a few miles inland and a lower, hilly district containing much of California's population fronts the sea.

The *Sierra Nevada* forms the inland edge of the Pacific Mountains and Valleys in central California, merging with the Coast Ranges north of Los Angeles and with the Klamath and Cascade ranges in northern California. The Sierra Nevada is an immense upraised, tilted, broken, and eroded block presenting a long and comparatively gentle slope to the west and a precipitous face eastward toward the Basin and Range Country. It is very high and rugged, and constitutes a major barrier both climatically and with regard to transportation. Mount Whitney (14,495 feet) in the southern Sierra Nevada is the highest peak in the conterminous United States.

The *Central Valley,* a level-floored alluvial trough some 500 miles long by 50 miles wide, occupies the center of California between the Coast Ranges and the Sierra Nevada. It is completely surrounded by mountains except where San Francisco Bay breaks the continuity of the Coast Ranges and brings the Central Valley into contact with the Pacific.

The *Klamath Mountains* form a link between the Coast Ranges and the Sierra Nevada and Cascade ranges, and separate the northern end of the Central Valley of California from the southern end of the Willamette Valley of Oregon. The Klamath Mountains are physiographically a dissected plateau. The valleys are deeply incised, giving an extremely rugged aspect to the terrain. Summit levels are frequently at 6000 to 7000 feet or higher.

The *Cascade Mountains* extend northward from the Sierra Nevada and the Klamath Mountains across Oregon and Washington and into British Columbia. Much of the area of the Cascades lies between 5000 and 9000 feet in elevation, and the mountains are surmounted by a series of volcanic cones reaching much higher elevations. Mt. Rainier (14,410 feet) and Mount Hood (11,245 feet) are probably the most famous of these. The Cascades are broken into two sections at the Oregon-Washington boundary by a spectacular gorge through which the Columbia River flows westward toward the Pacific.

The *Willamette–Puget Sound Lowland* lies between the Coast Ranges and the Cascades. Its Oregon section, from the Klamath Mountains in the south to the lower Columbia River in the north, is the valley of the Willamette River, while the northern section is commonly known as the Puget Sound Lowland. This latter section extends all the way across western Washington and north into British Columbia to include a small area along the lower course of the Fraser River.

In British Columbia and the southeastern "Panhandle" of Alaska a northward extension of the Cascades lies along the coast and is known as the *Coastal Ranges*. These Ranges are generally higher and more rugged than the Coast Ranges in the United States. Many peaks reach 9000 or 10,000 feet. The mountains rise abruptly from the sea and are penetrated by fjords, resembling in this respect the coasts of Norway and southern Chile. West of the Coastal Ranges a valley analogous in position to the Central Valley of California and the Willamette–Puget Sound Lowland has subsided below sea level and now forms the famous "Inside Passage" to Alaska. The mountains in British Columbia and southeastern Alaska which correspond in position to the Coast Ranges of the United States are partly submerged, and form a string of rugged islands along the outer edge of the Inside Passage.

Most of southern Alaska is mountainous. Just north of the Panhandle, spectacular glaciers descend the mountains to the sea. The highest mountains, however, are found farther to the north and west in the *Alaska Range,* where Mount McKinley reaches 20,320 feet, the highest elevation in Anglo-America. In a more subdued form the mountains continue from the Alaska Range into the Alaska Peninsula and the Aleutian Islands. A fair-sized area of lower land is found south of the Alaska Range in the Susitna River Valley and along the western side of the Kenai Peninsula.

Distribution of Population

From southern California to southern British Columbia the Pacific Mountains and Valleys contain a rather large population, especially considering that most of the included land is mountainous. This Pacific littoral forms an island of relatively dense population separated from the populous eastern areas of Anglo-America by the thinly peopled Intermountain, Rocky Mountain, and Great Plains regions. In 1968 about 25 million people lived in the parts of the Pacific Mountains and Valleys that lie within conterminous United States, as compared with a total of about 8 million in all the interior states eastward to the western boundaries of the Dakotas, Nebraska, Kansas, and Texas. California is by far the largest center of population

in the western third of Anglo-America. With 19.7 million people (1968 estimate), it ranks first in population among the states. In Canada around three-quarters of the 1.9 million people of British Columbia live in small lowland areas along or near the lower Fraser River and on the southern part of Vancouver Island. North of this southwestern corner of British Columbia there is little population in the Pacific Mountains and Valleys, but even so, the population is larger than in the adjacent regions of the interior. Two-thirds or more of Alaska's 280,000 people (1968 estimate) live along or near the Pacific Coast, the most concentrated area of settlement being in the Anchorage (city proper, 49,000) district.

The Highly Developed Agriculture of California

The marked concentration of people along the Pacific littoral is partly due to the agricultural advantages of this area as compared with interior sections of the Anglo-American West. These advantages are much more pronounced in California than in any other section of the Pacific Mountains and Valleys. The lowlands of California are areas of mediterranean or dry-summer subtropical climate, except for especially sheltered spots in the southern part of the Central Valley, which grade into steppe and desert. The winter rains and mild winter temperatures permit nonirrigated grain farming to be carried on in the winter and spring, and there is enough moisture to provide fairly good pasturage for grazing. During the dry summers, nearby mountain pastures often provide valuable supplementary grazing for livestock. But California's position as the leading agricultural state in the United States derives mainly from the most extensive development of irrigation in the United States. This development has been made possible by the close proximity of high mountain ranges (especially the Sierra Nevada but also parts of the Coast Ranges) which receive very heavy precipitation. Rain and snow falling in these mountains provide large amounts of water for irrigating the lowlands of the Central Valley (photo, right) and southern California, and some valleys in the Coast Ranges.

California's agriculture was long known largely as a supplier of fruits and vegetables to the rest of the United States. But farm production in the state has become increasingly oriented toward supplying the state's own urban markets, which have expanded enormously in recent decades. Thus the single most valuable crop today is hay, and barley also is of major importance. These and other feedstuffs supply a livestock production

A view of Friant Dam in the western foothills of the Sierra Nevada. The irrigation canal from the reservoir supplies water to farmland in the San Joaquin Valley. The installations in the photo are part of a coordinated net of facilities through which water derived from heavy precipitation in the Sierra Nevada is stored and then channeled to users in California's Central Valley. (Bureau of Reclamation, Department of the Interior.)

which now ranks fifth among the states (after Texas, Nebraska, Iowa, and Illinois) and a dairy industry which ranks third (after Wisconsin and New York). However, fruit and vegetable production, largely for "export" to the rest of the country, is still a very important element in the state's economy. Among an astonishing variety of such products grapes and peaches are the most valuable. The state accounts for over 40 percent of United States fruit production and over one-third of the country's sales of vegetables. The famous citrus crops, however, have come to be a very poor second to those of Florida, as many groves have been destroyed in the urban expansion of southern California. However, the state still accounts for about a quarter of the oranges produced in the United States. In addition to its livestock, dairy, fruit, and vegetable output California also produces over 10 percent of the American cotton crop and about a fifth of the country's rice crop. Most of the cotton is grown in the southern part of the Central Valley, known as the San Joaquin Valley, while the rice comes from the northern part, known as the Sacramento Valley.

Agriculture in the Area of Marine West Coast Climate

North of California the lowlands of the Pacific area have a marine west coast climate characterized by mild, wet conditions similar to those of the corresponding climatic region in northwestern Europe. In the Willamette–Puget Sound Lowland, agricultural development has emphasized principally dairy farming and the production of tree and bush fruits, with some truck gardening. But agriculture has been handicapped, especially in the Puget Sound Lowland, by the difficulty and expense of clearing the land of immense stumps in cutover forest areas, and by soils which are often poorly drained and heavily leached. Nor does this section have a climate permitting it to produce specialties of the type which have allowed California to largely overcome its problem of great distance from major consuming centers in eastern United States. Consequently, agricultural development here is on a considerably smaller scale than in California, and there is much farming on a part-time subsistence basis. North of the Fraser lowlands and Vancouver Island, there is very little land sufficiently level for agriculture, and the largest agricultural area, found in lowlands north of Anchorage (Matanuska Valley) (map, p. 758), contains only a few hundred farms.

Nonagricultural Resources

The principal nonagricultural resources of the Pacific Mountains and Valleys are waterpower, oil, timber, and fish. Hydroelectricity is a major source of power for both domestic and industrial uses, and helps to overcome the region's shortage of coal. Electricity is supplied by many installations within the Pacific area, of which Bonneville Dam in the Columbia gorge and Shasta Dam on the Sacramento River in northern California are probably the most famous examples, though now surpassed in size and power output by other dams. In addition, large amounts of power are transmitted from installations in the Intermountain Basins and Plateaus such as Hoover Dam and Grand Coulee Dam. The location of a large part of the United States aluminum industry in Washington and Oregon, and the presence of a very large plant at Kitimat on the coast of British Columbia, are closely related to the superior waterpower resources of these areas.

Oil is another major source of power along the Pacific Coast, but its production is confined to California. Southern California and the southern end of the Central Valley supply about 11 percent of the national total, giving California third rank, after Texas and Louisiana, as an oil-producing state. Los Angeles and San Francisco and their environs are major centers of oil refining.

The softwood forests which occupy the mountainous portions of the Pacific Coast region from the Sierra Nevada and San Francisco Bay northward are the most valuable in Anglo-America from the standpoint of size of trees and total reserves of saw timber. These large trees can be cut into long, high-quality boards, and represent the greatest reserve of saw timber in the United States. The famous Douglas fir of Oregon and Washington is the outstanding species, although the immense redwoods of northern California and various other species are important. Oregon, California, and Washington, in that order, are the first-, second-, and third-ranking American states in sawmilling, accounting together for about 45 percent of the total United States production, while British Columbia accounts for 70 percent of Canada's output in this phase of the forest industries. Pulp and paper milling is relatively less important here than in most other forested areas in Anglo-America, but the Pacific Northwest does have a large concentration of plants and ranks second to the South as a region of wood pulp production (map, p. 722). In Alaska the forests are, on the whole, less impressive. Nevertheless, they offer a major resource which has been very little exploited as yet, although there is some pulp production in the Panhandle section.

The fishing industry is important along the Pacific Coast from San Diego to the Aleutian Islands. In terms of value of catch, Alaska and California are the two leading fishing states, with just under 15 percent of the United States total catch landed in each. The other Pacific Coast states are of much less importance in the industry, but British Columbia also has important fisheries, being in fact the leading Canadian province in total value of fishery production, and approaching Alaska and California in the scale of the industry. Salmon and herring are the major species landed by the Pacific fisheries north of California, while California's fisheries rely primarily on tuna and sardines. The fishing industry is relatively more important as a phase of the total economy in northern British Columbia and Alaska than in areas farther south. Little other economic development has occurred in these more northerly sections, and fishing affords the principal support for most of the small and scattered coastal communities. The Anchorage area, dependent largely on United States military installations, is an exception to this generalization.

Finally, the scenery of the Pacific area and the climate of southern California must be considered as major resources since they afford the principal basis for a large and growing tourist trade.

Urban and Industrial Development

The Pacific border from Vancouver southward has an impressive number of large cities, most of which are growing rapidly. Almost all of this urban development is concentrated into four major groups of cities: a dense cluster in southern California; a second cluster around San Francisco Bay; a line of cities in the Central Valley; and a line of seaports in the Pacific Northwest.

Southern California contains six metropolitan areas (SMSAs) within a narrow crescent of territory whose maximum length is approximately 180 miles (Santa Barbara to San Diego) and whose maximum width is approximately 60 miles (San Bernardino to the Pacific). Los Angeles–Long Beach (7.1 million), the greatest of these, has developed from a Spanish agricultural town founded on the Los Angeles River in the northern and widest part of the hilly coastal lowland of southern California. The city has constructed a large artificial harbor and has become a great seaport. It grew originally by servicing the productive agricultural territory which surrounded it. Later impetus was given by discoveries of oil underneath the metropolitan area. In addition the area has attracted the movie industry and aircraft, automobile, apparel, and other industries. The outstanding feature in its present economy is the presence of approximately one-third of the United States aircraft industry, with the heavy dependence on government contracts which this implies. The adjacent metropolitan area of Anaheim–Santa Ana–Garden Grove (1.3 million) is largely suburban to Los Angeles, shares its character, and might well be combined with it in a single population total. San Diego (1.3 million) has grown as the main Pacific Coast naval base of the United States. The city is located on a tremendous natural harbor whose hinterland is almost nonexistent due to the difficulty of crossing the Coast Ranges to the eastward. San Diego is deeply involved in the aircraft and aerospace industries. San Bernardino (1.1 million) is generally considered to form the core of another metropolitan area, which includes important secondary cities to the west and south. San Bernardino grew at the Cajon Pass, through which most transportation from the interior crosses the Coast Ranges to enter the coastal lowland. The metropolitan area has developed a varied industrial structure which includes, just west of San Bernardino, the only large iron and steel plant of the Pacific Coast area (photo, below). Two smaller metropolitan areas, Oxnard–Ventura (335,000) and Santa Barbara (250,000), have developed on the coast just to the

The iron and steel plant of the Kaiser Steel Corporation at Fontana, California, near the eastern edge of the Los Angeles metropolitan region. The view looks eastward across a gently sloping piedmont plain to the rugged San Bernardino Mountains in the Coast Ranges. Fontana City is in the center of the view, and San Bernardino city is in the distance. (Kaiser Steel Corporation.)

northwest of Los Angeles. The former is a center of the oil business and of agricultural processing; the latter a resort and university center and a town much favored for residence by the very wealthy. Altogether the six metropolitan areas of southern California aggregated over 11 million people by 1968 and were steadily coalescing toward one huge urban and semiurban sprawl.

The second dense urban cluster of the Pacific border has grown around San Francisco Bay. It includes the metropolitan areas of San Francisco–Oakland (3.1 million), west and east of the Bay; San Jose (985,000) in the Santa Clara valley south of the Bay; and Vallejo–Napa (250,000) north of the Bay. These aggregated well over 4 million people by 1968. To them may be added the Salinas–Monterey area (250,000) on the shores of Monterey Bay not far to the south. This cluster of cities surrounding San Francisco Bay has developed on a magnificent natural harbor which is opposite the important Truckee Pass over the high Sierra Nevada and which offers the only break in the mountain wall surrounding the agriculturally wealthy Central Valley. In addition the Coast Range valleys themselves are areas of unusual agricultural productivity, now being encroached on by urban development. The cluster grew as the product of a many-sided economic development. During and after the gold rush to the western Sierra Nevada in the nineteenth century the Bay area experienced rapid growth as a port and center of wealth. It became the main financial center of the western United States, and it developed important agricultural servicing and processing industries. Further development was stimulated by oil from the southern Central Valley. In the course of time the area evolved a highly diversified industrial structure in which food-processing and electrical equipment industries are especially notable.

The Central Valley is frequently thought of as subdivided into four sections, with each section containing a sizable metropolitan center as its economic capital. All these centers are subordinate to either Los Angeles or San Francisco. In the extreme south the Tulare Basin, an arid region of cotton and oil production, centers economically on Bakersfield (340,000). The San Joaquin section of the Central Valley, notably varied in agricultural production but especially famed for grapes, is served by Fresno (425,000). Stockton (285,000), is a service center for the northern end of the San Joaquin Valley and the tangle of watercourses and reclaimed lands to the north known as the Delta. The city is located in the border zone between the two areas. But the largest of the Valley cities is Sacramento (795,000), which is not only the economic capital of the whole northern half of the Central Valley (Sacramento Valley), but is also the political capital of the most populous American state and is on the main transport routes between San Francisco and the interior via the Truckee Pass.

Five ports of the Pacific Northwest form a final group of major cities in the region. Portland, Oregon (920,000), is in the Willamette Valley. Located well up the Columbia River, the city is favored in access to the interior by the Columbia gorge through the Cascades. In Washington the two largest cities are ports with harbors on Puget Sound. Seattle (1.2 million) is not only a considerable port but also a major center of the aircraft industry, being the headquarters and principal production center of the Boeing Company. The industrial structure of Tacoma (355,000) is built around wood industries and food processing. A bit farther to the north, Vancouver, British Columbia (905,000), is Canada's second busiest port (after Montreal) and a major center of wood industries. It has an excellent natural harbor near the mouth of the Fraser River. The Fraser Valley affords a passageway through the mountains to the east, allowing Vancouver to tap some of the trade of the Prairie Provinces as well as that of its own immediate hinterland. The largest inland city of the Pacific Mountains and Valleys north of California is Eugene, Oregon (200,000), an especially large center of wood industries, but smaller in population than any of the five port cities named in the preceding paragraph.

THE ARCTIC COASTAL PLAINS

Fairly extensive coastal plains fronting on Arctic waters are found in two distinct sections of Anglo-America, one north of the Brooks Range (map, p. 758) and the other along the southern shores of Hudson Bay (map, p. 711). The Alaskan coastal plain is an area of tundra, while that along Hudson Bay is largely subarctic in climate and forested, though grading into tundra in some sections. These areas are extremely thinly populated and play almost no part in the economic life of Anglo-America, although there are a few fur-trading posts in the Hudson Bay section. Subsistence hunting and fishing are the principal means of support of the Eskimo population of the Alaska section. However, a major oil discovery was made in northern Alaska in 1968, giving promise that this area may soon assume some economic importance.

GREENLAND

Greenland is the world's largest island, with an area of approximately 840,000 square miles. While not culturally or politically a part of Anglo-America, it may be regarded as a marginal part of the latter region from the standpoint of proximity and strategic geography. Roughly 85 percent of the surface is covered by an ice cap and most of the island's 40,000 permanent inhabitants live in rugged strips of tundra along the southwestern and southeastern coasts. These people are primarily Eskimo but with a strong admixture of Scandinavian blood. Fishing (carried on both commercially and on a subsistence basis), hunting, trapping, and sheep herding are their chief means of livelihood. The population includes about 3000 Danes, mainly traders and government officials. Greenland is known to have considerable mineral wealth, but little of this has yet been exploited except for cryolite—a material useful in the glass, chemical, and aluminum industries—and a small production of lead and zinc. Greenland has practically a world monopoly of natural cryolite.

Politically, Greenland is an integral part of Denmark. The latter, a member of the North Atlantic Treaty Organization, granted permission for the establishment of a large NATO air base at Thule on Greenland's far northwestern coast. The United States has played a leading role in the development of this base.

HAWAII

Hawaii, admitted to the American Union as the fiftieth state in 1959, has been mentioned briefly in Chapter 22 on the Pacific Islands. A volcanic, mountainous tropical area populated originally by Polynesians, the new state is comprised of eight main islands (inset map, pp. 514–515), one of which is uninhabited. The islands lie in the path of trade winds which bring heavy rainfall to windward slopes throughout the year. Leeward and low-lying areas, however, may receive as little as 10 to 20 inches of rainfall annually.

The Hawaiian Islands lie just south of the Tropic of Cancer and slightly more than 2000 miles from California. The island of Hawaii is by far the largest of the group and contains the only active volcanic craters. The state's population, however, is heavily concentrated in the island of Oahu, where the capital and only large city, Honolulu (595,000), is located. The area of the state, 6424 square miles, is greater than the combined areas of Connecticut and Rhode Island, and its population of 755,000 (1968 estimate) is larger than that of ten of the mainland states. Caucasian (39 percent) and Japanese (28 percent) elements comprise the principal groups in the population, but are augmented by much smaller groups of native Hawaiian, Filipino, Chinese, Korean, Puerto Rican, or Samoan ancestry. Much racial and ethnic mixing has taken place. Most of the population is thoroughly Americanized.

Hawaii's economy leans heavily on commercial agriculture, defense, and tourism. The main agricultural products and exports—cane sugar and pineapples—are grown on large estates owned and administered by Caucasian family and corporate interests but actually worked by non-Caucasian labor. Both products are marketed primarily in mainland United States, though some exports go to foreign countries. Most sugar cane is grown in irrigated fields, while pineapples are largely an unirrigated crop. Military installations, including the famous naval base at Pearl Harbor, are a vital source of income for the state, as is the rapidly expanding tourist industry. In 1967 it was estimated that the temporary population in Hawaii at a given time was one-tenth as large as the permanent population. Urban development, tourism, and business administration are overwhelmingly concentrated in Honolulu, and most of the main defense installations, including the Pearl Harbor base, are near the city.

REFERENCES and READINGS

Climate and Vegetation

KIMBLE, GEORGE H. T., *Our American Weather* (New York: McGraw-Hill Book Company, 1955; Bloomington: Indiana University Press, Midland Books [paperbound], 1961).

KÜCHLER, A. W., *Potential Natural Vegetation of the Conterminous United States* (Special Publication No. 36; New York: American Geographical Society, 1964). A large multicolored map with accompanying text.

U.S. DEPARTMENT OF AGRICULTURE, *Climate and Man: Yearbook of Agriculture, 1941* (Washington, D.C.: 1941), especially pt. 2, "Climate and Agricultural Settlement," pp. 157–236.

VISHER, STEPHEN SARGENT, *Climatic Atlas of the United States* (Cambridge, Mass.: Harvard University Press, 1954).

Canadian Shield, Canadian North, and Greenland

BLADEN, V. W., ed., *Canadian Population and Northern Colonization* (Toronto: University of Toronto Press, 1962). 12 contributions by various authors.

BROWN, L. CARSON, "Ontario's Mineral Heritage," *Canadian Geographical Journal,* 76, no. 3 (March 1968), 80–101.

de NEVERS, NOEL, "Tar Sands and Oil Shales," *Scientific American,* 214, no. 2 (February 1966), 21–29.

FINE, M. M., "The Beneficiation of Iron Ores" [in North America], *Scientific American,* 218, no. 1 (January 1968), 28–35.

HARE, FREDERICK KENNETH, "New Light from Labrador-Ungava," *Annals of the Association of American Geographers,* 54, no. 4 (December 1964), 459–476.

KENT, JON, "Agriculture in the Clay Belt of Northern Ontario," *Canadian Geographer,* 10, no. 2 (1966), 117–126.

LLOYD, TREVOR, "Iron Ore Production in Quebec-Labrador," Chap. 15 in Richard S. Thoman and Donald J. Patton, eds., *Focus on Geographic Activity: A Collection of Original Studies* (New York: McGraw-Hill Book Company, 1964), pp. 85–92.

MACDONALD, R. ST. J., ed., *The Arctic Frontier* (Toronto: University of Toronto Press; London: Oxford University Press, 1966). 11 contributions by various authors.

McDERMOTT, GEORGE L., "Frontiers of Settlement in the Great Clay Belt, Ontario and Quebec," *Annals of the Association of American Geographers,* 51, no. 3 (September 1961), 261–273.

PUNCH, KATHERINE, "Sault Ste. Marie," *Canadian Geographical Journal,* 73, no. 6 (December 1966), 198–207.

REA, K. J., *The Political Economy of the Canadian North* (Toronto, Ont.: University of Toronto Press, 1968).

ROBINSON, IRA M., *New Industrial Towns on Canada's Resource Frontier* (University of Chicago, Department of Geography Research Paper No. 73; Chicago: 1962).

UNDERHILL, FRANK H., ed., *The Canadian Northwest: Its Potentialities* (Toronto: University of Toronto Press 1959). Contributions by various authors.

WINTERS, ROBERT H., "The Hamilton Falls Project," *Canadian Geographical Journal,* 70, no. 5 (May 1965), 146–153.

U.S. Northeast and Atlantic Canada

ALEXANDER, LEWIS M., *The Northeastern United States* (Princeton, N.J.: D. Van Nostrand Co., Searchlight Books, 1967).

BIRCH, J. W., "On the Stability of Farming Systems, with Particular Reference to the Connecticut Valley," in J. B. Whittow and P. D. Wood, eds., *Essays in Geography for Austin Miller* (Reading, England: Reading University, 1965), pp. 225–246.

CAREY, GEORGE W., LENORE MACOMBER, and MICHAEL GREENBERG, "Educational and Demographic Factors in the Urban Geography of Washington, D.C.," *Geographical Review,* 58, no. 4 (October 1968), 515–537.

CHINITZ, BENJAMIN, "New York: A Metropolitan Region," *Scientific American,* 213, no. 3 (September 1965), 134–148.

CLARK, ANDREW H., "Acadia and the Acadians: The Creation of a Geographical Entity," in John Andrews, ed., *Frontiers and Men: A Volume in Memory of Griffith Taylor (1880–1963)* (Melbourne, Australia: F. W. Cheshire, 1966), pp. 90–119; and *Three Centuries and the Island: A Historical Geography of Settlement and Agriculture in Prince Edward Island, Canada* (Toronto: University of Toronto Press, 1959).

DEASY, GEORGE F., and PHYLLIS R. GRIESS, "Effects of a Declining Mining Economy on the Pennsylvania Anthracite Region," *Annals of the Association of American Geographers,* **55,** no. 2 (June 1965), 239–259; also, "Local and Regional Differences in Long-Term Bituminous Coal Production Prospects in Eastern United States," *Annals of the Association of American Geographers,* **57,** no. 3 (September 1967), 519–533.

DURAND, LOYAL, JR., "The Historical and Economic Geography of Dairying in the North Country of New York State," *Geographical Review,* **57,** no. 1 (January 1967), 24–47; and "The Major Milksheds of the Northeastern Quarter of the United States," *Economic Geography,* **40,** no. 1 (January 1964), 9–33.

ESTALL, R. C., *New England: A Study in Industrial Adjustment* (London: G. Bell and Sons, 1966); also, "Changing Industrial Patterns of New England," *Geography,* **46,** pt. 2 (April 1961), 120–138; and "The Electronic Products Industry of New England," *Economic Geography,* **39,** no. 3 (July 1963), 189–216.

GOTTMANN, JEAN, *Megalopolis: The Urbanized Northeastern Seaboard of the United States* (New York: Twentieth Century Fund, 1961); also, "Megalopolis or the Urbanization of the Northeastern Seaboard," *Economic Geography,* **47,** no. 3 (July 1957), 189–200; and "Why the Skyscraper?" *Geographical Review,* **56,** no. 2 (April 1966), 190–212.

HALL, MAX, ed., *New York Metropolitan Region Study,* 10 vols. by various authors (Cambridge, Mass.: Harvard University Press, 1959–1960); reviewed, together with Jean Gottmann, *Megalopolis* (cited above), in Howard J. Nelson, "Megalopolis and New York Metropolitan Region: New Studies of the Urbanized Eastern Seaboard," *Annals of the Association of American Geographers,* **52,** no. 3 (September 1962), 307–317.

HALL, PETER, *The World Cities* (London: Weidenfeld and Nicolson, World University Library, 1966), Chap. 7, "New York," pp. 182–216.

HIGBEE, EDWARD C., "The Three Earths of New England," *Geographical Review,* **42,** no. 3 (July 1952), 425–438.

KENYON, JAMES B., *Industrial Localization and Metropolitan Growth: The Paterson-Passaic District* (University of Chicago, Department of Geography Research Paper No. 67; Chicago: 1960); and "The Industrial Structure of the New York Garment Center," Chap. 25 in Richard S. Thoman and Donald J. Patton, eds., *Focus on Geographic Activity: A Collection of Original Studies* (New York: McGraw-Hill Book Company, 1964), pp. 159–166.

KLIMM, LESTER E., "The Empty Areas of the Northeastern United States," *Geographical Review,* **44,** no. 3 (July 1954), 325–345.

LEMON, JAMES T., "The Agricultural Practices of National Groups in Eighteenth-Century Southeastern Pennsylvania," *Geographical Review,* **56,** no. 4 (October 1966), 467–496.

THOMPSON, JOHN H., ed., *Geography of New York State* (Syracuse, N.Y.: Syracuse University Press, 1966).

WALLACE, WILLIAM H., "Merrimack Valley Manufacturing: Past and Present," *Economic Geography,* **37,** no. 4 (October 1961), 283–308.

U.S. South, Appalachia, and Interior Highlands

Atlas of Florida by Erwin Raisz and associates with text by John R. Dunkle (Gainesville: University of Florida Press, 1964).

BEARDWOOD, ROGER, "The Southern Roots of Urban Crisis," *Fortune,* **78,** no. 2 (August 1968), 80–87 ff.

BOWMAN, MARY JEAN, and W. WARREN HAYNES, *Resources and People in East Kentucky: Problems and Potentials of a Lagging Economy* (Baltimore, Md.: Johns Hopkins Press for Resources for the Future, 1963).

CASH, W. J., *The Mind of the South* (New York: Alfred A. Knopf, 1941).

CAUDILL, HARRY M., *Night Comes to the Cumberlands: A Biography of a Depressed Area* (Boston: Little, Brown and Company, 1962).

CLARK, THOMAS D., *The Emerging South* (2d ed.; New York: Oxford University Press, 1968); and, with Albert D. Kirwan, *The South Since Appomattox: A Century of Regional Change* (New York: Oxford University Press, 1967).

de LAUBENFELS, D. J., "Where Sherman Passed By," *Geographical Review,* 47, no. 3 (July 1957), 381–395.

DURAND, LOYAL, " 'Mountain Moonshining' in East Tennessee," *Geographical Review,* 46, no. 2 (April 1956), 168–181.

ESTALL, R. C., "Appalachian State: West Virginia as a Case Study in the Appalachian Regional Development Problem," *Geography,* 53, pt. 1 (January 1968), 1–24.

FORD, ROBERT N., *A Resource Use Analysis and Evaluation of the Everglades Agricultural Area* (University of Chicago, Department of Geography Research Paper No. 42; Chicago: 1956).

FORD, THOMAS R., ed., *The Southern Appalachian Region: A Survey* (Lexington: University of Kentucky Press, 1962).

GOTTMANN, JEAN, *Virginia at Mid-Century* (New York: Holt, Rinehart and Winston, 1955).

HAMILTON, C. HORACE, "The Negro Leaves the South," *Demography,* 1, no. 1 (1964), 273–295.

HART, JOHN FRASER, *The Southeastern United States* (Princeton, N.J.: D. Van Nostrand Co., Searchlight Books, 1967); and "Loss and Abandonment of Cleared Farm Land in the Eastern United States," *Annals of the Association of American Geographers,* 58, no. 3 (September 1968), 417–440.

JORDAN, TERRY G., "The Imprint of the Upper and Lower South on Mid-Nineteenth-Century Texas," *Annals of the Association of American Geographers,* 57, no. 4 (December 1967), 667–690; and "Population Origins in Texas, 1850," *Geographical Review,* 59, no. 1 (January 1969), 83–103.

KEY, V. O., JR., *Southern Politics in State and Nation* (New York: Alfred A. Knopf, 1949).

KNIFFEN, FRED B., *Louisiana: Its Land and People* (Baton Rouge: Louisiana State University Press, 1968).

LAMB, ROBERT BYRON, *The Mule in Southern Agriculture* (University of California Publications in Geography, Vol. 15; Berkeley and Los Angeles: University of California Press, 1963).

LONSDALE, RICHARD E., *Atlas of North Carolina* (Chapel Hill: University of North Carolina Press, 1967).

McLAUGHLIN, GLENN E., and STEFAN ROBOCK, *Why Industry Moves South: A Study of Factors Influencing Recent Location of Manufacturing Plants in the South* (Washington, D.C.: National Planning Association, Committee of the South, 1949).

MADDOX, JAMES G., *The Advancing South: Manpower Prospects and Problems* (New York: Twentieth Century Fund, 1967).

MERRENS, HARRY ROY, *Colonial North Carolina in the Eighteenth Century: A Study in Historical Geography* (Chapel Hill: University of North Carolina Press, 1964).

MILLER, E. JOAN WILSON, "The Ozark Culture Region as Revealed by Traditional Materials," *Annals of the Association of American Geographers,* 58, no. 1 (March 1968), 51–77.

O'HANLON, THOMAS, "The Great Sulfur Rush," *Fortune,* 77, no. 3 (March 1968), 107–111 ff.

PARKINS, A. E., "The Antebellum South: A Geographer's Interpretation," *Annals of the Association of American Geographers,* 21, no. 1 (March 1931), 1–33.

PARSONS, JAMES J., "Recent Industrial Development in the Gulf South," *Geographical Review,* 40, no. 1 (January 1950), 67–83.

PRUNTY, MERLE, JR., "Land Occupance in the Southeast: Landmarks and Forecast," *Geographical Review,* 42, no. 3 (July 1952), 439–461; also "Recent Quantitative Changes in the Cotton Regions of the Southeastern States," *Economic Geography,* 27, no. 3 (July 1951), 189–208; "The Renaissance of the Southern Plantation," *Geographical Review,* 45, no. 4 (October 1955), 459–491; and "The Woodland Plantation as a Contemporary Occupance Type in the South," *Geographical Review,* 53, no. 1 (January 1963), 1–21.

ROSE, WARREN, "Catalyst of an Economy: The Houston Ship Channel," *Land Economics,* 43, no. 1 (February 1967), 32–43.

SAUER, CARL O., *The Geography of the Ozark Highland of Missouri* (Chicago: University of Chicago Press for the Geographic Society of Chicago, 1920).

SCHROEDER, WALTER A., *The Eastern Ozarks: A Geographic Interpretation of the Rolla 1:250,000 Topographic Map* (Normal, Ill.: National Council for Geographic Education, 1967).

"The South's Big Bet on Technology," *Fortune,* 45, no. 3 (March 1952), 92–95 ff.

"The Splendid Anachronism of Huntsville," *Fortune,* 65, no. 6 (June 1962), 150–155 ff.

THOMAN, RICHARD S., *The Changing Occupance Pattern of the Tri-State Area, Missouri, Kansas, and Oklahoma* (University of Chicago, Department of Geography, Research Paper No. 31; Chicago: 1953).

VAN ROYEN, WILLIAM, and S. MORYADAS, "The Economic Geographic Basis of Appalachia's Problems," *Tijdschrift voor Economische en Sociale Geografie,* 57, no. 5 (September–October 1966), 185–193.

ZELINSKY, WILBUR, "The Changing South," *Focus,* 2, no. 2 (October 15, 1951), 6 pp.; and "Where the South Begins: The Northern Limit of the Cis-Appalachian South in Terms of Settlement Landscape," *Social Forces,* 30, no. 2 (December 1951), 172–178.

U.S. Middle West and Great Plains; St. Lawrence Lowlands and Prairie Provinces of Canada

AKIN, WALLACE E., *The North Central United States* (Princeton, N.J.: D. Van Nostrand Co., Searchlight Books, 1968).

BOGUE, ALLAN G., *From Prairie to Corn Belt: Farming on the Illinois and Iowa Prairies in the Nineteenth Century* (Chicago: University of Chicago Press, 1963).

BORCHERT, JOHN R., "The Twin Cities Urbanized Area: Past, Present, Future," *Geographical Review,* 51, no. 1 (January 1961), 47–70.

BOWEN, WILLIAM, and EDGAR SHOOK, "Why They Call Those Soybeans Golden," *Fortune,* 74, no. 3 (August 1966), 127–129 ff.

COLBY, CHARLES C., *Pilot Study of Southern Illinois* (Carbondale, Ill.: Southern Illinois University Press, 1956).

GARLAND, JOHN H., ed., *The North American Midwest: A Regional Geography* (New York: John Wiley & Sons, 1955). Contributions by 15 authors.

HELBURN, NICHOLAS, "The Ronald Holtz Farm: A Commercial Crop Operation in the Northern Great Plains," Chap. 9 in Richard S. Thoman and Donald J. Patton, eds., *Focus on Geographic Activity: A Collection of Original Studies* (New York: McGraw-Hill Book Company, 1964), pp. 49–54.

HEWES, LESLIE, "Causes of Wheat Failure in the Dry Farming Region, Central Great Plains, 1939–1957," *Economic Geography,* 41, no. 4 (October 1965), 313–330.

HIGHSMITH, RICHARD M., JR., ed., *Case Studies in World Geography: Occupance and Economy Types* (Englewood Cliffs, N.J.: Prentice-Hall, 1961): Chap. 7, "A Beef Cattle Farm in the Corn Belt," pp. 47–54; Chap. 24, "Iron and Steel—the Indiana Harbor Plant of the Inland Steel Company," pp. 173–180; and Chap. 47, "The Soybean, a Versatile Product of American Farms," pp. 196–200.

ILLINOIS, DIVISION OF INDUSTRIAL PLANNING AND DEVELOPMENT, *Atlas of Illinois Resources* (Springfield: 1958–).

JOHNSON, HILDEGARD BINDER, "The Location of German Immigrants in the Middle West," *Annals of the Association of American Geographers,* 41, no. 1 (March 1951), 1–41; and "Rational and Ecological Aspects of the Quarter Section: An Example from Minnesota," *Geographical Review,* 47, no. 3 (July 1957), 330–348.

JORDAN, TERRY G., "Between the Forest and the Prairie," *Agricultural History,* 38, no. 4 (October 1964), 205–216.

KOLLMORGEN, WALTER M., and DAVID S. SIMONETT, "Grazing Operations in the Flint Hills-Bluestem Pastures of Chase County, Kansas," *Annals of the Association of American Geographers,* 55, no. 2 (June 1965), 260–290.

LEWTHWAITE, GORDON R., "Wisconsin and the Waikato: A Comparison of Dairy Farming in the United States and New Zealand," *Annals of the Association of American Geographers,* 54, no. 1 (March 1964), 59–87.

MALIN, JAMES C., *The Grassland of North America: Prolegomena to Its History* (rev. ed.; Lawrence, Kan.: The Author, 1956).

MAYER, HAROLD M., "Politics and Land Use: The Indiana Shoreline of Lake Michigan," *Annals of the Association of American Geographers,* 54, no. 4 (December 1964), 508–523; also, *The Port of Chicago and the St. Lawrence Seaway* (University of Chicago, Department of Geography Research Paper No. 49; Chicago: 1957); and "Prospects and Problems of the Port of Chicago," *Economic Geography,* 31, no. 2 (April 1955), 95–125.

McMANIS, DOUGLAS R., *The Initial Evaluation and Utilization of the Illinois Prairies, 1815–1840* (University of Chicago, Department of Geography Research Paper No. 94; Chicago: 1964).

MILLER, E. JOAN WILSON, "Hybrid Corn and More Food for More People," *Journal of Geography,* 63, no. 9 (December 1964), 413–417.

PHILBRICK, ALLEN K., "The Nodal Water Region of North America," *Canadian Geographer,* 8, no. 4 (1964), 182–187.

SAARINEN, THOMAS FREDERICK, *Perception of the Drought Hazard on the Great Plains* (University of Chicago, Department of Geography Research Paper No. 106; Chicago: 1966).

SHARER, CYRUS J., "A New Peninsula in the New World," *Tijdschrift voor Economische en Sociale Geografie,* 57, no. 2 (March–April 1966), 73–77.

SPENCER, J. E., and RONALD J. HORVATH, "How Does an Agricultural Region Originate?" *Annals of the Association of American Geographers,* 53, no. 1 (March 1963), 74–92.

VANDERHILL, BURKE G., "The Decline of Land Settlement in Manitoba and Saskatchewan," *Economic Geography,* 38, no. 3 (July 1962), 270–277; also, "Post-War Agricultural Settlement in Manitoba," *Economic Geography,* 35, no. 3 (July 1959), 259–268; "Trends in the Peace River Country," *Canadian Geographer,* 7, no. 1 (1963), 33–41; and, with David E. Christensen, "The Settlement of New Iceland," *Annals of the Association of American Geographers,* 53, no. 3 (September 1963), 350–363.

WEAVER, JOHN C., "Changing Patterns of Cropland Use in the Middle West," *Economic Geography,* 30, no. 1 (January 1954), 1–47; also, "Crop-Combination Regions in the Middle West," *Geographical Review,* 44, no. 2 (April 1954), 175–200; and, with Leverett P. Hoag and Barbara L. Fenton, "Livestock Units and Combination Regions in the Middle West," *Economic Geography,* 32, no. 3 (July 1956), 237–259.

WEIR, THOMAS R., *Economic Atlas of Manitoba* (Winnipeg: Manitoba Department of Industry and Commerce, 1960).

WERTENBAKER, THOMAS J., "The Molding of the Middle West," *American Historical Review,* 53, no. 2 (January 1948), 223–234.

U.S. West, Canadian West, Alaska, and Hawaii

"Alaska in 1967," *Inter-Nord,* 10 (March 1968), 133–154. A series of articles.

ARRINGTON, LEONARD J., "Science, Government, and Enterprise in Economic Development: The Western Beet Sugar Industry," *Agricultural History,* 41, no. 1 (January 1967), 1–18.

ASCHMANN, HOMER, "The Head of the Colorado Delta," in S. R. Eyre and G. R. J. Jones, eds., *Geography as Human Ecology* (New York: St. Martin's Press, 1966), pp. 231–263.

ATHERTON, LEWIS, *The Cattle Kings* (Bloomington: Indiana University Press, 1961).

ATWOOD, WALLACE W., *The Rocky Mountains* (New York: Vanguard Press, 1945).

BAILEY, HARRY P., *The Climate of Southern California* (Berkeley and Los Angeles: University of California Press, 1966).

BOWDEN, LEONARD W., *Diffusion of the Decision to Irrigate: Simulation of the Spread of a New Resource Management Practice in the Colorado Northern High Plains* (University of Chicago, Department of Geography Research Paper No. 97; Chicago: 1965).

CALEF, WESLEY, *Land Associations and Occupance Problems in the Uinta Country* (Chicago: University of Chicago, Department of Geography, 1948); and *Private Grazing and Public Lands: Studies of the Local Management of the Taylor Grazing Act* (Chicago: University of Chicago Press, 1960).

CANTOR, L. M., "Irrigated Agriculture in the Columbia Basin Project," *Geography,* 52, pt. 1 (January 1967), 67–72.

COOKE, R. U., and I. G. SIMMONS, "Some Recent Changes in California," *Tijdschrift voor Economische en Sociale Geografie,* 57, no. 6 (November–December 1966), 232–242.

COOLEY, RICHARD A., *Alaska: A Challenge in Conservation* (Madison: University of Wisconsin Press, 1966).

CURTI, G. PHILIP, "The Colorado River: Its Utilization by Southern California," *Geography,* 42, pt. 4 (November 1957), 230–238.

DE VOTO, BERNARD, "The Anxious West," *Harper's Magazine,* 193, no. 1159 (December 1946), 381–391; also, "The West Against Itself," *Harper's Magazine,* 194, no. 1160 (January 1947), 1–13; and "Sacred Cows and Public Lands," *Harper's Magazine,* 197, no. 1178 (July 1948), 44–55 (also in the same issue, "The Easy Chair," 108–112).

FIELDING, GORDON J., "Dairying in Cities Designed to Keep People Out," *Professional Geographer,* 14, no. 1 (January 1962), 12–17; and "The Los Angeles Milkshed: A Study of the Political Factor in Agriculture," *Geographical Review,* 54, no. 1 (January 1964), 1–12.

FRANCIS, KARL E., "Outpost Agriculture: The Case of Alaska," *Geographical Review,* 57, no. 4 (October 1967), 496–505.

GREGOR, HOWARD F., "Push to the Desert: The Pressure of Agriculture on California's Arid Lands Illustrates the Law of Diminishing Returns," *Science,* 129, no. 3359 (May 15, 1959), 1329–1339; also, "Spatial Disharmonies in California Population Growth," *Geographical Review,* 53, no. 1 (January 1963), 100–122; and "Urbanization of Southern California Agriculture," *Tijdschrift voor Economische en Sociale Geografie,* 54, no. 12 (December 1963), 273–278.

HARRIS, CHAUNCY D., *Salt Lake City, a Regional Capital* (Chicago: University of Chicago, Department of Geography, 1940).

HERRERA, PHILIP, "Megalopolis Comes to the Northwest," *Fortune,* 76, no. 7 (December 1967), 118–123 ff.

HIGHSMITH, RICHARD M., JR., ed., *Case Studies in World Geography: Occupance and Economy Types* (Englewood Cliffs, N.J.: Prentice-Hall, 1961): Chap. 12, "Irrigation Farming in the Yakima Valley," pp. 78–86; Chap. 25, "The Kitimat Aluminum Development," pp. 181–187; Chap. 28, "Vancouver, British Columbia," pp. 203–212; and, with J. M. Leverenz, *Atlas of the Pacific Northwest: Resources and Development* (3d ed.; Corvallis: Oregon State University Press, 1962).

HODGE, CARLE, and PETER C. DUISBERG, ed., *Aridity and Man: The Challenge of the Arid Lands in the United States* (American Association for the Advancement of Science Publication No. 74; Washington, D.C.: 1963). 24 contributions by various authors.

HOLLON, W. EUGENE, *The Great American Desert, Then and Now* (New York: Oxford University Press, 1966).

HOLMES, CHARLES H., "Factors Affecting Development of the Steel Industry in Intermountain America," *Journal of Geography,* 58, no. 1 (January 1959), 20–31.

HUDSON, JAMES, *Irrigation Water Use in the Utah Valley, Utah* (University of Chicago, Department of Geography Research Paper No. 79; Chicago: 1962).

JAEGER, EDMOND C., *The California Deserts* (4th ed.; 1965), and *The North American Deserts* (1957). (Stanford, Calif.: Stanford University Press.)

KNIGHT, OLIVER, "Correcting Nature's Error: The Colorado–Big Thompson Project," *Agricultural History,* 30, no. 4 (October 1956), 157–169.

LANTIS, DAVID W., "California," *Focus,* 8, no. 2 (October 1957), 6 pp.; and, with Rodney Steiner and Arthur E. Karinen, *California: Land of Contrast* (Belmont, Calif.: Wadsworth Publishing Company, 1963).

LARGE, DAVID C., "Cotton in the San Joaquin Valley: A Study of Government in Agriculture," *Geographical Review,* 47, no. 3 (June 1957), 365–380.

LEIGHLY, JOHN, "John Muir's Image of the West," *Annals of the Association of American Geographers,* 48, no. 4 (December 1958), 309–318.

LOEFFLER, M. JOHN, "Beet-Sugar Production on the Colorado Piedmont," *Annals of the Association of American Geographers,* 53, no. 3 (September 1963), 364–390; and "The Population Syndromes on the Colorado Piedmont," *Annals of the Association of American Geographers,* 55, no. 1 (March 1965), 26–66.

LUTEN, DANIEL B., "The Use and Misuse of a River" [the Colorado], *The American West,* 4, no. 2 (May 1967), 47–53 ff.

MEINIG, D. W., "The Mormon Culture Region: Strategies and Patterns in the Geography of the American West, 1847–1964," *Annals of the Association of American Geographers,* **55,** no. 2 (June 1965), 191–220.

MILLER, ELBERT E., "Economic and Social Changes in the Columbia Basin, Washington," *Land Economics,* **41,** no. 4 (November 1965), 335–346.

NELSON, HOWARD J., "The Spread of an Artificial Landscape over Southern California," *Annals of the Association of American Geographers,* **49,** no. 3, pt. 2, Supplement (September 1959), 80–100.

PEATTIE, RODERICK, ed., *The Sierra Nevada: The Range of Light* (New York: Vanguard Press, 1947).

POMEROY, EARL, *The Pacific Slope: A History of California, Oregon, Washington, Idaho, Utah, and Nevada* (New York: Alfred A. Knopf, 1965).

PRESTON, RICHARD E., "Urban Development in Southern California between 1940 and 1965," *Tijdschrift voor Economische en Sociale Geografie,* **58,** no. 5 (September–October 1967), 237–254.

QUINN, FRANK, "Water Transfers: Must the American West Be Won Again?" *Geographical Review,* **58,** no. 1 (January 1968), 108–132.

RAND, CHRISTOPHER, *Los Angeles, the Ultimate City* (New York: Oxford University Press, 1967).

RIDD, MERRILL K., *Landforms of Utah—in Proportional Relief,* Map Supplement No. 3, *Annals of the Association of American Geographers,* **53,** No. 4 (December 1963).

ROGERS, GEORGE W., *Alaska in Transition: The Southeast Region* (1960), and *The Future of Alaska: Economic Consequences of Statehood* (1962). (Baltimore, Md.: Johns Hopkins Press.)

SMITH, DAVID A., "Interaction within a Fragmented State: The Example of Hawaii," *Economic Geography,* **39,** no. 3 (July 1963), 234–244.

SMITH, DUANE A., *Rocky Mountain Mining Camps: The Urban Frontier* (Bloomington: Indiana University Press, 1967).

SMITH, RICHARD AUSTIN, "Los Angeles, Prototype of Supercity," *Fortune,* **71,** no. 3 (March 1965), 98–101 ff.; and "Alaska: The Last Frontier," *Fortune,* **52,** no. 3 (September 1955), 104–113 ff.

STEINER, RODNEY, "Reserved Lands and the Supply of Space for the Southern California Metropolis," *Geographical Review,* **56,** no. 3 (July 1966), 344–362.

STONE, KIRK H., "Alaskan Problems and Potentials," *Journal of Geography,* **50,** no. 5 (May 1951), 177–189; and "Populating Alaska: The United States Phase," *Geographical Review,* **42,** no. 2 (July 1952), 384–404.

THOMAS, FRANK H., *The Denver and Rio Grande Western Railway: A Geographic Analysis* (Northwestern University Studies in Geography, No. 4; Evanston, Ill.: Northwestern University Press, 1960); and "Evolution of Railroad Route-to-Region Relationships: A Case Study of the Denver and Rio Grande Western Railroad," *Journal of Geography,* **62,** no. 9 (December 1963), 389–397.

THOMAS, WILLIAM L., JR., ed., "Man, Time and Space in Southern California," *Annals of the Association of American Geographers,* **49,** no. 3, pt. 2, Supplement (September 1959), 120 pp. Contributions by various authors.

THROWER, NORMAN J. W., *California Population: Distribution in 1960* (1:1,000,000), Map Supplement No. 7, *Annals of the Association of American Geographers,* **56,** no. 2 (June 1966).

VANCE, JAMES E., JR., "The Oregon Trail and Union Pacific Railroad: A Contrast in Purpose," *Annals of the Association of American Geographers,* **51,** no. 4 (December 1961), 357–379.

WEBB, WALTER PRESCOTT, *The Great Plains* (Boston: Ginn and Company, 1931; Waltham, Mass.: Blaisdell Publishing Co., 1959).

ZIERER, CLIFFORD M., ed., *California and the Southwest* (New York: John Wiley & Sons, 1956). A geography, with 32 contributors.

775

NOTE: All page references to maps are in *italics* following general entry for topic.